International Handbook of Research in History,
Philosophy and Science Teaching

Michael R. Matthews
Editor

International Handbook of Research in History, Philosophy and Science Teaching

Volume II

 Springer

Editor
Michael R. Matthews
School of Education
University of New South Wales
Sydney, NSW, Australia

ISBN 978-94-007-7653-1   ISBN 978-94-007-7654-8 (eBook)
DOI 10.1007/978-94-007-7654-8
Springer Dordrecht Heidelberg New York London

Library of Congress Control Number: 2013958394

© Springer Science+Business Media Dordrecht 2014
This work is subject to copyright. All rights are reserved by the Publisher, whether the whole or part of the material is concerned, specifically the rights of translation, reprinting, reuse of illustrations, recitation, broadcasting, reproduction on microfilms or in any other physical way, and transmission or information storage and retrieval, electronic adaptation, computer software, or by similar or dissimilar methodology now known or hereafter developed. Exempted from this legal reservation are brief excerpts in connection with reviews or scholarly analysis or material supplied specifically for the purpose of being entered and executed on a computer system, for exclusive use by the purchaser of the work. Duplication of this publication or parts thereof is permitted only under the provisions of the Copyright Law of the Publisher's location, in its current version, and permission for use must always be obtained from Springer. Permissions for use may be obtained through RightsLink at the Copyright Clearance Center. Violations are liable to prosecution under the respective Copyright Law.
The use of general descriptive names, registered names, trademarks, service marks, etc. in this publication does not imply, even in the absence of a specific statement, that such names are exempt from the relevant protective laws and regulations and therefore free for general use.
While the advice and information in this book are believed to be true and accurate at the date of publication, neither the authors nor the editors nor the publisher can accept any legal responsibility for any errors or omissions that may be made. The publisher makes no warranty, express or implied, with respect to the material contained herein.

Printed on acid-free paper

Springer is part of Springer Science+Business Media (www.springer.com)

Contents

Volume 1

1 Introduction: The History, Purpose and Content of the Springer *International Handbook of Research in History, Philosophy and Science Teaching..................................... 1
Michael R. Matthews

Part I Pedagogical Studies: Physics

2 Pendulum Motion: A Case Study in How History and Philosophy Can Contribute to Science Education..................... 19
Michael R. Matthews

3 Using History to Teach Mechanics.. 57
Colin Gauld

4 Teaching Optics: A Historico-Philosophical Perspective.................. 97
Igal Galili

5 Teaching and Learning Electricity: The Relations Between Macroscopic Level Observations and Microscopic Level Theories....................... 129
Jenaro Guisasola

6 The Role of History and Philosophy in Research on Teaching and Learning of Relativity.. 157
Olivia Levrini

7 Meeting the Challenge: Quantum Physics in Introductory Physics Courses.. 183
Ileana M. Greca and Olival Freire

v

vi Contents

8 Teaching Energy Informed by the History and Epistemology of the Concept with Implications for Teacher Education 211
Manuel Bächtold and Muriel Guedj

9 Teaching About Thermal Phenomena and Thermodynamics: The Contribution of the History and Philosophy of Science 245
Ugo Besson

Part II Pedagogical Studies: Chemistry

10 Philosophy of Chemistry in Chemical Education: Recent Trends and Future Directions 287
Sibel Erduran and Ebru Z. Mugaloglu

11 The Place of the History of Chemistry in the Teaching and Learning of Chemistry 317
Kevin C. de Berg

12 Historical Teaching of Atomic and Molecular Structure 343
José Antonio Chamizo and Andoni Garritz

Part III Pedagogical Studies: Biology

13 History and Philosophy of Science and the Teaching of Evolution: Students' Conceptions and Explanations 377
Kostas Kampourakis and Ross H. Nehm

14 History and Philosophy of Science and the Teaching of Macroevolution 401
Ross H. Nehm and Kostas Kampourakis

15 Twenty-First-Century Genetics and Genomics: Contributions of HPS-Informed Research and Pedagogy 423
Niklas M. Gericke and Mike U. Smith

16 The Contribution of History and Philosophy to the Problem of Hybrid Views About Genes in Genetics Teaching 469
Charbel N. El-Hani, Ana Maria R. de Almeida,
Gilberto C. Bomfim, Leyla M. Joaquim,
João Carlos M. Magalhães, Lia M.N. Meyer,
Maiana A. Pitombo, and Vanessa C. dos Santos

Part IV Pedagogical Studies: Ecology

17 Contextualising the Teaching and Learning of Ecology: Historical and Philosophical Considerations 523
Ageliki Lefkaditou, Konstantinos Korfiatis,
and Tasos Hovardas

Contents   vii

Part V   Pedagogical Studies: Earth Sciences

18   **Teaching Controversies in Earth Science: The Role of History and Philosophy of Science** 553
Glenn Dolphin and Jeff Dodick

Part VI   Pedagogical Studies: Astronomy

19   **Perspectives of History and Philosophy on Teaching Astronomy** 603
Horacio Tignanelli and Yann Benétreau-Dupin

Part VII   Pedagogical Studies: Cosmology

20   **The Science of the Universe: Cosmology and Science Education** 643
Helge Kragh

Part VIII   Pedagogical Studies: Mathematics

21   **History of Mathematics in Mathematics Education** 669
Michael N. Fried

22   **Philosophy and the Secondary School Mathematics Classroom** 705
Stuart Rowlands

23   **A Role for Quasi-Empiricism in Mathematics Education** 731
Eduard Glas

24   **History of Mathematics in Mathematics Teacher Education** 755
Kathleen M. Clark

25   **The Role of Mathematics in Liberal Arts Education** 793
Judith V. Grabiner

Volume 2

26   **The Role of History and Philosophy in University Mathematics Education** 837
Tinne Hoff Kjeldsen and Jessica Carter

27   **On the Use of Primary Sources in the Teaching and Learning of Mathematics** 873
Uffe Thomas Jankvist

Part IX Theoretical Studies: Features of Science and Education

28 Nature of Science in the Science Curriculum: Origin, Development, Implications and Shifting Emphases 911
Derek Hodson

29 The Development, Use, and Interpretation of Nature of Science Assessments 971
Norman G. Lederman, Stephen A. Bartos, and Judith S. Lederman

30 New Directions for Nature of Science Research 999
Gürol Irzik and Robert Nola

31 Appraising Constructivism in Science Education 1023
Peter Slezak

32 Postmodernism and Science Education: An Appraisal 1057
Jim Mackenzie, Ron Good, and James Robert Brown

33 Philosophical Dimensions of Social and Ethical Issues in School Science Education: Values in Science and in Science Classrooms 1087
Ana C. Couló

34 Social Studies of Science and Science Teaching 1119
Gábor Kutrovátz and Gábor Áron Zemplén

35 Generative Modelling in Physics and in Physics Education: From Aspects of Research Practices to Suggestions for Education 1143
Ismo T. Koponen and Suvi Tala

36 Models in Science and in Learning Science: Focusing Scientific Practice on Sense-making 1171
Cynthia Passmore, Julia Svoboda Gouvea, and Ronald Giere

37 Laws and Explanations in Biology and Chemistry: Philosophical Perspectives and Educational Implications 1203
Zoubeida R. Dagher and Sibel Erduran

38 Thought Experiments in Science and in Science Education 1235
Mervi A. Asikainen and Pekka E. Hirvonen

Part X Theoretical Studies: Teaching, Learning and Understanding Science

39 Philosophy of Education and Science Education: A Vital but Underdeveloped Relationship 1259
Roland M. Schulz

Contents

40 Conceptions of Scientific Literacy: Identifying and Evaluating Their Programmatic Elements 1317
Stephen P. Norris, Linda M. Phillips, and David P. Burns

41 Conceptual Change: Analogies Great and Small and the Quest for Coherence 1345
Brian Dunst and Alex Levine

42 Inquiry Teaching and Learning: Philosophical Considerations 1363
Gregory J. Kelly

43 Research on Student Learning in Science: A Wittgensteinian Perspective 1381
Wendy Sherman Heckler

44 Science Textbooks: The Role of History and Philosophy of Science 1411
Mansoor Niaz

45 Revisiting School Scientific Argumentation from the Perspective of the History and Philosophy of Science 1443
Agustín Adúriz-Bravo

46 Historical-Investigative Approaches in Science Teaching 1473
Peter Heering and Dietmar Höttecke

47 Science Teaching with Historically Based Stories: Theoretical and Practical Perspectives 1503
Stephen Klassen and Cathrine Froese Klassen

48 Philosophical Inquiry and Critical Thinking in Primary and Secondary Science Education 1531
Tim Sprod

49 Informal and Non-formal Education: History of Science in Museums 1565
Anastasia Filippoupoliti and Dimitris Koliopoulos

Part XI Theoretical Studies: Science, Culture and Society

50 Science, Worldviews and Education 1585
Michael R. Matthews

51 What Significance Does Christianity Have for Science Education? 1637
Michael J. Reiss

x   Contents

Volume 3

**52 Rejecting Materialism: Responses to Modern Science
in the Muslim Middle East** 1663
Taner Edis and Saouma BouJaoude

**53 Indian Experiences with Science: Considerations
for History, Philosophy, and Science Education** 1691
Sundar Sarukkai

**54 Historical Interactions Between Judaism and Science
and Their Influence on Science Teaching and Learning** 1721
Jeff Dodick and Raphael B. Shuchat

**55 Challenges of Multiculturalism in Science Education:
Indigenisation, Internationalisation, and *Transkulturalität*** 1759
Kai Horsthemke and Larry D. Yore

**56 Science, Religion, and Naturalism: Metaphysical
and Methodological Incompatibilities** 1793
Martin Mahner

Part XII Theoretical Studies: Science Education Research

**57 Methodological Issues in Science Education Research:
A Perspective from the Philosophy of Science** 1839
Keith S. Taber

**58 History, Philosophy, and Sociology of Science
and Science-Technology-Society Traditions
in Science Education: Continuities and Discontinuities** 1895
Veli-Matti Vesterinen, María-Antonia Manassero-Mas,
and Ángel Vázquez-Alonso

**59 Cultural Studies in Science Education:
Philosophical Considerations** 1927
Christine L. McCarthy

60 Science Education in the Historical Study of the Sciences 1965
Kathryn M. Olesko

Part XIII Regional Studies

**61 Nature of Science in the Science Curriculum
and in Teacher Education Programs in the United States** 1993
William F. McComas

Contents

62 The History and Philosophy of Science in Science Curricula and Teacher Education in Canada 2025
Don Metz

63 History and Philosophy of Science and the Teaching of Science in England 2045
John L. Taylor and Andrew Hunt

64 Incorporation of HPS/NOS Content in School and Teacher Education Programmes in Europe 2083
Liborio Dibattista and Francesca Morgese

65 History in Bosnia and Herzegovina Physics Textbooks for Primary School: Historical Accuracy and Cognitive Adequacy 2119
Josip Slisko and Zalkida Hadzibegovic

66 One Country, Two Systems: Nature of Science Education in Mainland China and Hong Kong 2149
Siu Ling Wong, Zhi Hong Wan, and Ka Lok Cheng

67 Trends in HPS/NOS Research in Korean Science Education 2177
Jinwoong Song and Yong Jae Joung

68 History and Philosophy of Science in Japanese Education: A Historical Overview 2217
Yuko Murakami and Manabu Sumida

69 The History and Philosophy of Science and Their Relationship to the Teaching of Sciences in Mexico 2247
Ana Barahona, José Antonio Chamizo, Andoni Garritz, and Josip Slisko

70 History and Philosophy of Science in Science Education, in Brazil 2271
Roberto de Andrade Martins, Cibelle Celestino Silva, and Maria Elice Brzezinski Prestes

71 Science Teaching and Research in Argentina: The Contribution of History and Philosophy of Science 2301
Irene Arriassecq and Alcira Rivarosa

Part XIV Biographical Studies

72 Ernst Mach: A Genetic Introduction to His Educational Theory and Pedagogy 2329
Hayo Siemsen

73 Frederick W. Westaway and Science Education: An Endless Quest 2359
William H. Brock and Edgar W. Jenkins

74 E. J. Holmyard and the Historical Approach to Science Teaching 2383
Edgar W. Jenkins

75 John Dewey and Science Education 2409
James Scott Johnston

76 Joseph J. Schwab: His Work and His Legacy 2433
George E. DeBoer

Name Index 2459

Subject Index 2505

Chapter 26
The Role of History and Philosophy in University Mathematics Education

Tinne Hoff Kjeldsen and Jessica Carter

26.1 Introduction

In this chapter we discuss the roles of the history and philosophy of mathematics in the learning of mathematics at university level. University mathematics is organised differently in different universities and countries. In some universities mathematics is separated into different programmes: masters in pure mathematics, in applied or industrial mathematics, in financial mathematics, in teacher training education, etc. In this paper we consider mathematics programmes that lead to a graduate degree in mathematics, i.e. mathematics programmes where pure mathematics plays an essential role.[1] In the context of the present handbook, the following three questions immediately come to mind: (1) Why do we need a chapter that focuses especially on mathematics? (2) Why do we need a chapter that focuses especially on university level mathematics? (3) Why combine history and philosophy?

The first question has also been addressed by Michael N. Fried in the Chap. 21. Fried pointed out that there are differences between the sciences and mathematics that justify the inclusion of this question in the present handbook of separate chapters focusing on history and philosophy of mathematics in mathematics education. Here we will mention the picture of mathematics as the epitome of timeless truths and mathematical objects as ideal, timeless entities – named by some as an

---

[1] For the roles of the history and philosophy of mathematics in liberal arts education, we refer to the previous chapter by Judith Grabiner.

T.H. Kjeldsen (✉)
Department of Science, Systems and Models,
IMFUFA, Roskilde University, Roskilde, Denmark
e-mail: thk@ruc.dk

J. Carter
Department of Mathematics and Computer Science,
University of Southern Denmark, Odense, Denmark
e-mail: jessica@imada.sdu.dk

M.R. Matthews (ed.), *International Handbook of Research in History, Philosophy and Science Teaching*, DOI 10.1007/978-94-007-7654-8_26,
© Springer Science+Business Media Dordrecht 2014

absolutist philosophy of mathematics. One consequence of this picture, which is conveyed to students by traditional mathematics educations at all levels (François and Van Bendegem 2007), is that it portrays mathematics as a cumulative science of a seemingly static and infallible character of knowledge (Otte 2007, p. 243).

An essential difference between the history and philosophy of mathematics in primary and secondary mathematics education and mathematics at university level is that at university level these subjects often have their own courses within the mathematics programme: with their own learning goals, curriculum, disciplinary standards and agendas which are not restricted by a mathematics curriculum that has to be taught in the same courses as well. Hence, the history and philosophy of mathematics can play very different roles in mathematics education at the university level than at the primary and secondary level. These differences justify a separate chapter within the present handbook that focuses on the university level.

Finally, during the past few decades, research in the history and philosophy of mathematics has witnessed a trend towards a focus on mathematical practices from which historical and philosophical investigations and analyses have taken a point of departure. On the one hand this has strengthened the relationship between the professional academic disciplines of history and philosophy of mathematics, with historical investigations serving as cases for philosophical studies and vice versa philosophical ideas serving as inspiration and tools for historical analyses.[2] On the other hand, studying the history and philosophy of mathematics from the practices of mathematics brings these subjects close to mathematical research activities, to processes of knowledge production in mathematics and hence to mathematics education at university level. It therefore makes sense to look at the history and philosophy of mathematics in a common perspective in relation to the roles they (can) play in mathematics education at university level.

In the following we first briefly introduce a historiographic framework of a multiple perspective approach to the history of mathematics from its practices together with some reflections about uses of history, and we introduce the direction of research in the philosophy of mathematics that is denoted 'Philosophy of Mathematical Practice'. We then link the history and philosophy of mathematical practices to recent ideas in mathematics education in order to identify different roles history and philosophy can play in mathematics education at university level. This is followed by presentations, analyses and discussions of different examples of the inclusion of history and philosophy in university programmes in mathematics. These presentations are divided into courses in history and philosophy, since this is the main way they are organised at the universities. We shall see, however, that the history courses address philosophical questions and that the philosophy courses employ historical material. The chapter is rounded off with comments on how mathematics educations at university level can benefit from history and philosophy of mathematics.

---

[2] Our joint paper (Kjeldsen and Carter 2012) serves as an example of this mutual beneficial relationship.

26 The Role of History and Philosophy in University Mathematics Education 839

26.2 History and Philosophy of Mathematics from Its Practices

The history and philosophy of mathematics are two independent professional disciplines that have followed their own trajectories. As mentioned in the introduction, during the past few decades the research in the history and philosophy of mathematics has witnessed a trend towards a focus on mathematical practices. In the following we will introduce a historiographical framework of a multiple perspective approach to history of mathematics from its practices and introduce one direction of research in philosophy of mathematics that is denoted 'Philosophy of Mathematical Practice'. These are methodological issues in the professional disciplines of history and philosophy of mathematics. In mathematics education at university level, approaches to history can be found that are not necessarily aimed at purely historiographical goals, but are directed towards the teaching and learning of mathematics. In order to analyse such approaches, we also introduce parts of a framework for uses of the past.

26.2.1 History of Mathematics from Its Practices and Uses of the Past

Research into people's uses of history has shown that people use history in many different contexts, for different purposes and with different approaches; see, e.g. Ashton and Kean (2009) and Jensen (2010). In this context, the Danish historian Eric Bernard Jensen conceive of history as an umbrella term for related forms of knowledge and practises people uses in their lives, and he defines history accordingly by saying that we are dealing with history 'when a person or a group of people is interested in something from the past and uses their knowledge about it for some purpose' (Jensen 2010, p. 39). Historians (or users of history) might have different perspectives on history depending on their aims. This is also the case for history of mathematics which is, and has been, studied and used in different contexts with different goals.[3] In his book *What is History*, Jensen (2010) provides a framework in which different uses of the past can be characterised. The framework can be used to analyse and characterise implementations of history in mathematics education. Here we will only introduce his distinction between a *pragmatic* and a *scholarly* approach to history.[4]

A historian who studies history from a utility perspective is said to have a *pragmatic* approach to history. It is an approach in which history is conceived of as 'the master of life', i.e. we can learn from history. In a pragmatic approach to history,

---

[3] See, e.g. Lützen and Purkert (1989) where the different historiographical views of Cantor and Zeuthen are discussed.

[4] In Kjeldsen (2012) Jensen's terminology is outlined and used as lens through which we can identify and distinguish between different conceptions and uses of history of mathematics.

the historian will try to make history relevant in a present-day context. In Jensen's terminology, historians who focus on understanding and interpreting the past on its own terms, regardless of the present situation, is said to have a *scholarly*[5] approach to history. According to Jensen the scholarly approach has been the dominant one in academic, professional history since the mid-nineteenth century.

In the following we will present a multiple perspective approach to history of mathematics from its practices. The multiple perspective approach to history is inspired by Jensen's (2003) thinking about historiography. His underlying premise is that people produce history and are shaped by history. In order to understand social-historical processes and gain insights into the past, history is studied from perspective(s) of the historical actors. The historian pays attention to the historical actors' motivations, their projects, their intentions as well as unintended consequences of their actions. The perspective of the actors is taken into account which means that perspectives such as the actors' placement in space and time, in a certain society and/or in a particular intellectual context are considered as part of historical investigations.

If we think of mathematics as a historical and cultural product of knowledge that is produced by human intellectual activities, then a multiple perspective approach to history can be adapted to historiography of mathematics. Studying the history of mathematics then also involves searching for explanations for historical processes of change, such as, but not limited to, changes in our perception of mathematics as such, in its status and function in society, in our understanding of mathematical notions or objects and in our idea of what counts as legitimate arguments for mathematical statements.

Studying processes in the development and shaping of mathematics from mathematical practices means asking why mathematicians introduced specific concepts and definitions; why they studied the problems they did, in the way they did; and what were the driving forces behind their mathematical investigations; see, e.g. Epple (1999, Chap. 1) and Epple (2004, p. 133). One approach is to study concrete episodes of mathematical research activities focusing on the 'workplace' of the involved mathematicians in order to uncover and understand the dynamics of the knowledge production. The methodological framework of epistemic objects, techniques and configurations (Epple 2004), originally developed by the philosopher of science Hans-Jörg Rheinberger (1997), has been used recently in the historiography of mathematics; see Epple (2004, 2011) and Kjeldsen (2009). In short, epistemic objects refer to the mathematical objects about which new knowledge is searched for. Epistemic techniques refer to the methods and mathematical techniques that are used to investigate the mathematical objects in question; and epistemic configuration of mathematical research refers to the total of intellectual resources present in a specific episode. These terms are not intrinsically given, but are bound to concrete episodes of mathematical research. They are to be understood functionally, they change during the course of mathematical research and they might shift place. They are excellent tools for analysing the production of knowledge and the understanding of mathematical

---

[5] This is our translation of the Danish word 'lærd'.

entities in historical texts, because they are constructed to differentiate between how problem-generating and answer-generating elements of specific research episodes functioned and interacted. In Kjeldsen (2011a) it is suggested how this framework can be used in connection with student project work in history of mathematics for the learning of and about mathematics in university mathematics education.

In bridging the history of mathematics with mathematics education, we find a multiple perspective approach to the history of mathematics, from its practices, that is particularly relevant to mathematics education at university level. This is due to its striving for understanding the dynamics of mathematical knowledge production, the status and functions of mathematics in society and in concrete episodes from the past. This will be discussed below in the examples from our own approach of specific implementations of history in mathematics programmes at university level.

26.2.2 *Philosophy of Mathematical Practice*

Similar to the history of mathematics, the philosophy of mathematics poses different kinds of questions and offers ways of dealing with these. New questions may arise because of changes in the practice of mathematics, and changes in perspectives are often the outcome of perceived limitations of previous methods. At present there is a growing interest in what is denoted 'Philosophy of Mathematical Practice'. There are many motivations for this shift in interest; some are indicated below. Since this perspective is our focus, we will describe it in more detail.

It is no easy task to define philosophy. One way to describe it is that it poses fundamental questions concerning the world and our place in it. In addition philosophy seeks to answer these questions and, equally important, find arguments for the given answers. Philosophy – unlike mathematics – is a discipline where there seems to be very little agreement about answers. This does by no means entail that 'anything goes'. There are certain standards measured, for example, by the coherence of one's proposal, soundness of arguments as well as quality and sensibility of assumptions. Ideally, philosophy should advance our knowledge by critically examining our ideas, assumptions and arguments. Mathematics also gives rise to philosophical queries. Traditionally, philosophers have mainly asked questions within ontology and epistemology: questions such as 'What kind of entities are mathematical objects?' and 'Do they exist independently of the activities of human beings?' and epistemological questions like 'how do we obtain knowledge in mathematics?' Philosophers at all times have been fascinated with the apparent necessity of mathematical truths and the fact that mathematics is applicable to the real world while its subject matter seems remote from anything real. Philosophers such as Plato, Aristotle, Descartes, Kant and Mill all found that something should be said about mathematics.[6]

---

[6] Their views on mathematics are very different. For a presentation of their positions, see Shapiro (2000). For a nontraditional description of Plato's philosophy, in line with the perspective of mathematical practice, see McLarty (2005).

During the nineteenth century, mathematics changed drastically. From being conceived as somehow describing the real world, mathematics changed into an autonomous body of its own ideas. Gray (2008) characterises modern mathematics as being remote from the real world, having a strong 'emphasis on formal aspects of the work and maintaining a complicated – indeed anxious- rather than a naive relationship with the day-to-day world' (p. 1). For one thing, these changes led *mathematicians* to pose fundamental questions regarding the nature of mathematics and – perhaps more importantly – concerning how to obtain a secure foundation.[7] These questions led to the three foundational schools, Logicism, Intuitionism and Formalism; see, e.g. Benacerraf and Putnam (1983), Mancosu (1998), and van Heijenoort (1967). Outcomes of these programmes were the development of logic and proof theory[8] as well as a (one-sided) focus on questions pertaining to the justification of mathematics. Philosophy of mathematical practice can be seen as a reaction to the philosophers' one-sided stress on formal mathematics. Among the first to enter this lane was Lakatos (1976) who found that answers could be found by studying the practice of mathematics, writing (explicitly referring to Kant's famous line from Critique of Pure Reason A51,B75) 'Philosophy of science without history of science is empty; history of science without philosophy of science is blind' (Lakatos 1970, p. 91). Lakatos held that philosophy should also deal with questions concerning discovery or more precisely he argued that the processes of discovery and justification are intertwined. This approach to the philosophy of mathematics has gradually increased in popularity. Pioneers are Kitcher (1984), Maddy (1990), Tymoczko (1985), and later Corfield (2003). In what could be denoted as 'Philosophy of mathematical practice' today,[9] a (rough) distinction can be made between three approaches. These outlooks can be termed social, historical and epistemological. Since our approach to the philosophy of mathematics lies mainly within the epistemological strand, we describe this in more detail below.

The strand that has a sociological focus takes as a starting point the view that mathematics is a human activity (Hersh 1979) and as such can be described by sociological tools (Heinz 2000). Others are closer to mathematics education; see

---

[7] Another interesting development in mathematics around the turn of the century was the move towards structuralism. What is studied in mathematics is not the objects as such – it is the relations between objects. This is most famously described by Hilbert saying 'one must be able to say "tables, chairs, beer-mugs" each time in place of "points, lines, planes"' (Blumenthal 1935, pp. 402–403), expounded mathematically in his Foundations of Geometry (Hilbert 1899). Traces of this conception about mathematics can still be found in today's philosophies of mathematics; see Benacerraf (1965), Hellman (1996), Resnik (1999), and Shapiro (1997). More recently philosophers have argued that category theory provides a sound basis of a 'top-down' structuralist view (Awodey 1996; Landry and Marquis 2005).

[8] Since the original foundational schools failed for a variety of reasons, other ways of obtaining a foundation were looked for. It was, for example, proved by Gentzen during 1930s that if sufficiently strong methods (induction over ε_0) are used, then it is possible to prove the consistency of arithmetic. A different approach is to find a weaker system than Primitive Recursive Arithmetic where completeness and consistency are provable.

[9] See, e.g. Ferreiros and Gray (2006), Mancosu (2008), Van Kerkhove and Van Bendegem (2007), and Van Kerkhove et al. (2010).

26 The Role of History and Philosophy in University Mathematics Education   843

Bloor (1994), Ernest (1998), and Restivo (1993). This outlook is also the basis of the fairly recently formed Phimsamp group.[10] It is already integrated with mathematics education, especially through the work of Ernest (see http://people.exeter.ac.uk/PErnest/ and references throughout this chapter).

The second strand has history of mathematics at its core. One such perspective notices that mathematics itself at different times has posed philosophical questions and seeks to bring out the historical circumstances for these questions. This was the case in the period during which the foundational schools were formed – and as noted above, the people asking the questions and providing answers were in fact mathematicians themselves. A nice example of this outlook is presented by J. Tappenden (2006). He shows that there are certain misconceptions regarding Frege's mathematical motivation for engaging in his project.[11] Another perspective deals with the philosophical conceptions of the mathematicians themselves and how these conceptions help form the development of mathematics.

The final strand asks traditional philosophical questions and seeks answers to these by considering mathematical practice. Mancosu (2008) presents a number of excellent papers within this strand. For us taking this approach means that both questions and means for answering these questions are taken from mathematical practice. We acknowledge that questions may arise within mathematical practice itself and that assumptions should to some extent agree with practice. This is in part a reaction to traditional philosophical approaches, where one starts with an assumption about mathematics, such as 'mathematical statements are necessarily true', and then argues that it follows that mathematical objects exist (by necessity). There are two objections to this procedure. First, mathematics itself is missing from the picture. Second, the assumption needs to be examined. It is not clear whether mathematical statements are necessary or, if so, in which sense they are necessary (Carter 2008). When taking this approach, the aim is to obtain a better understanding of the mathematics that we (as human beings) know and use. The focus of Carter is to understand better contemporary mathematics; but in principle any part of mathematics could be the object of study. Which practice – or case – to study depends (in part) on which question is posed. This approach also has as consequence that it may not make sense to talk about the 'right' picture. Instead one may talk of useful pictures, in terms of determining for a given picture what is achieved by adopting this picture. We are still concerned about the coherence of pictures of mathematics, and that sound arguments should be provided. Thus in addition to the triple of standards – sensible assumption, valid arguments and coherent theory – a fourth component is added, namely, value of theory to a practice. This outlook may not be so different from the social outlook. A major difference concerns the set of

---

[10] Philosophy of Mathematics: Sociological Aspects and Mathematical Practice.

[11] One misconception is that when Frege started worrying about the foundation of analysis, it had already been settled by the work of Weierstrass in Berlin. The fact of the matter, Tappenden argues, is that problems of *real* analysis were being solved, but Frege knew of the (revolutionary) work of Riemann from the 1850s integrating geometry and complex analysis, opening up whole new fields of study.

assumptions built in. On the social outlook, for example, it would be assumed that there exist a community of learners and teachers (Ernest 2009) and some mathematics, created by human beings, which need to be taught/learned. In addition it is a helpful assumption when teaching mathematics that communication is possible (Carter 2006). Our interest focuses on the nature of mathematical objects and development of mathematics in general and thus implicitly requires the assumption of the existence of mathematicians doing mathematics. On this perspective, the relevant practices to study are the different contexts in which mathematics is developed. It is also important to note that what we study is *how we humans acquire knowledge of mathematics*. By doing this, our intention is not to assume anything about the true ontology of mathematics.

Since we take as a starting point mathematical practice, the history of mathematics is an important ally in providing case studies. The dependency relation, however, is a two-way relation which was already pointed out by Lakatos. Even though philosophy and history 'feed on each other', our aims are different. The philosopher tries to establish whatever general can be stated by considering particular cases. But sometimes interesting things can also be said about the particular cases – see below for an example. In contrast, the historian seeks to bring out the particular in each practice. As argued in the section above, when doing this the historian needs certain (philosophical) tools or concepts. To conclude this section, we give an example of how this 'philosophy' works in practice. When addressing the question concerning the nature of mathematical objects, a way to obtain an answer is to look into the practice of introducing mathematical objects (Carter 2004). As a result of such studies,[12] Carter (2013) concludes that mathematical objects are often introduced with reference to, or even as representations of, already accepted objects. The importance of contexts both for introducing objects and reasoning with them is also pointed out. These are general categories that can again be tested against historical studies. The aim of Kjeldsen and Carter (2012) is to test these claims against a case study on the introduction of convex bodies in the work of Minkowski. We find that overall this case fits the given description. In addition we find that the cases display important differences, which also provides insight about the general development of mathematics. One such difference concerns the type of relation between the new object and its referent. The convex body is defined as a set having certain properties singled out as important when solving problems within number theory, whereas a Riemann surface is an actual representation of part of the defining expression of an Abelian function.

When teaching the philosophy of mathematics for university students, we find that it is particularly important that this teaching takes as its starting point the actual *practices* of doing mathematics. In their ordinary mathematics courses, students are exposed to one picture of mathematics. We believe they should be exposed to different pictures. We stress, though, that students should also be aware that different pictures have different assumptions and that these are, useful or not, merely *assumptions*.

---

[12] The introduction of Riemann surfaces and K-theory.

26.3 History and Philosophy in Mathematics Education: Mathematical Competence, Critical Mathematics Education, Interdisciplinarity and Thinking as Communicating

In this section we link history and philosophy of mathematics with conceptions of mathematics education and learning in order to identify different roles history and philosophy can play in mathematics education at university level.

The cultural argument for mathematics in education and a need for students to develop interdisciplinary competences both provide roles for the history and philosophy of mathematics in mathematics education. Mathematical knowledge is a historical and cultural product of human intellectual activity. Its development and thoughts are tied to arts, philosophy and science. By integrating history and philosophy in mathematics education, the cultural argument for mathematics in education is emphasised. Through interdisciplinary teaching, history and philosophy can play a role in mathematics teaching and learning in developing students' interdisciplinary competences by counteracting disciplinary narrow-mindedness (Beckmann 2009).

Both of these roles for the history and philosophy of mathematics in mathematics education are embedded in the competence-based view of mathematics education as developed by Mogens Niss (2004) in the Danish KOM-project.[13] In Niss' competence-based description of mathematics education, mathematics curricula on all levels are based on mathematical competencies instead of a catalogue of subjects, notions and results. In the KOM-project, eight main competencies were identified. They are divided into two groups: (1) a group that has to do with the ability to ask and answer questions within and with mathematics (thinking, problem tackling, modelling and reasoning competencies) and (2) a group that concerns abilities and familiarities with language and tools in mathematics (representing, symbol and formalism, communicating, aids and tools competencies). Besides developing students' mathematical competencies, mathematics education should also provide students with three second-order competencies, so-called overview and judgement, regarding mathematics as a discipline. The first concerns actual applications of mathematics in other areas, the second concerns the historical development of mathematics in culture and societies, and the third concerns the nature of mathematics as a discipline. The second one explicitly requires knowledge about history of mathematics, though not as an individual discipline (the goal is not to educate competent historians of mathematics), but to develop students' overview and judgement regarding the historical development of mathematics resting on concrete examples from the history of mathematics. The third one explicitly requires knowledge related to the philosophy of mathematics. In the examples of actual implementations and incorporations of history and/or philosophy of mathematics given in the next section, we will discuss the

---

[13] The project was called Competencies and Mathematical Learning. It was initiated by the Danish National Council for Science Education in 2000. For a shortened English version of the original report, see Niss and Højgaard (2011).

roles history and philosophy of mathematics (can) play in university mathematics programmes in the framework of mathematical competence.

The three types of overview and judgement concern the character of mathematics and its functions and roles in the world. This relates to issues that have been raised and researched in the field of critical mathematics education. In his paper 'Critical mathematics education for the future', Ole Skovsmose (2004, p. 10) points towards

> an important concern in mathematics education: Mathematics must be reflected on and criticised in its variety of forms of actions.

One of the aims in critical mathematics education is to develop in students the ability to critique the uses of mathematics. This relates to the first of the second-order competencies introduced above. In the next section we give an example of how history can function in mathematics education at university level with authentic cases that have the potential to develop students' ability to critique the uses of mathematics. Since critique is at the core of philosophy, it is also developed in the philosophy courses.

The last theoretical framework from mathematics education that we want to link to the role history of mathematics can play in mathematics education, is Anna Sfard's (2008) *Theory of thinking as communicating*. In Kjeldsen (2012) it is argued that, within this theory, history of mathematics can function at the core of what it means to learn mathematics. Sfard defines thinking as 'the individualized version of interpersonal communication' (Sfard 2008, p. xvii). Her theory is also referred to as the theory of commognition where the term commognition captures the combination of communication and cognition. Mathematical thinking is a human activity and Sfard treats mathematics as a type of discourse, where discourse 'refers to the totality of communicative activities, as practiced by a given community' (Sfard 2000, p. 160). Learning mathematics means to become a participant in mathematics discourse. Discursive patterns are the results of communicative processes that are regulated by rules. Sfard distinguishes between object-level rules and meta-level rules of mathematics discourse. Object-level rules concern the content of the discourse. Meta-level rules have the discourse itself as an object. They govern 'when to do what and *how* to do it' (Sfard 2008, pp. 201–202) – they are implicitly given. To develop proper meta-level rules is essential for becoming a participant in mathematics discourse. It is an important aspect of teaching and learning mathematics to create situations where meta-discursive rules are exhibited and made into explicit objects of reflection for students. These rules are contingent. They develop and change over time and as such they can be subject to historical investigations.

This is demonstrated in Kjeldsen and Blomhøj (2012) and Kjeldsen and Petersen (forthcoming) where it is shown how historical sources, investigated and interpreted within the mathematical practice of the historical actors, can function as 'interlocutors' that are following a set of meta-level rules within the mathematical community of their times. Through such historical investigations, students can become confronted with differences in metarules between rules that governed the mathematician(s) of the past episode they are studying, their own (maybe) and the rules of their textbooks and/or their teacher. In this way, meta-level rules can be revealed and turned into objects for students' reflections. This will be illustrated by one of the examples given in the next section of some students' investigations of a concrete mathematical episode from the past. From a philosophical view, we note that Sfard's position is just *one* possible view about mathematics in line with a social outlook on mathematical practice.

Within general mathematics education, philosophy of mathematics plays different types of roles. It is generally acknowledged that the teacher's conception of mathematics forms his or her teaching; see, e.g. Hersh (1979) and Lerman (1990). Chassapis (2007) therefore argues that it is relevant to train teachers in some kind of philosophy of mathematics and also shows one way to do it. Correspondingly it is also acknowledged that students' beliefs[14] on mathematics influence their learning. Prediger (2007) convincingly shows how addressing themes from philosophy of mathematics help pupils make sense of mathematical problems. She even argues that philosophical reflections *must* play a prominent role in the learning process.

With respect to beliefs about mathematics, two major camps are usually described. The discussion between these two sides is termed the 'science wars' (Ernest 2004). In one camp are the 'absolutist' views; these are often taken to include Platonism and Formalism. Ernest (1994) describes it as the 'Euclidean paradigm of mathematics as an objective, absolute, incorrigible, and rigidly hierarchical body of knowledge' (p. 1). In the other camp are the 'fallibilist' views. The often mentioned hero of this programme is Lakatos (1976). Other fallibilist views include Ernest (1998) and Bloor (1994), i.e. views that stress that mathematics is created by human beings and mathematical knowledge is as fallible as the rest of our knowledge. As a reason for the shift from absolutist views to fallibilist views is pointed to 'Gödel's theorems' which show that 'formal axiomatic systems can never be regarded as ultimate' (Ernest 1994, p.1). Another reason (which is more in line with the outlook of this paper) is the desire for a philosophy to pay attention to mathematical practice.

The roles philosophy play can be divided into three levels: the level of the individual, mathematics and society. On the mathematical level, one could claim that pupils/students should be able to reflect on the nature of mathematical objects and knowledge. As we have seen above, it is argued that such reflections are vital for both the learning and teaching of mathematics. On the level of society, as indicated above, educators have pointed to the social and political role that mathematics education (can) play. The 'Mathematics Education and Society' (MES) conferences were started in 1998 in Nottingham, UK, in order to focus on these roles. This perspective believes that there is much more to be said in mathematics education than a narrow picture of learning accomplished by an interrelation of the mind of a learner and (the value free and objective) mathematics to be learned. For one thing, it leaves out the class room and the teacher and the social relations between these which also affect learning. On a much broader scale are questions pertaining to the role of mathematics education in a particular society, such as who is included and what determines whether you are 'in' or 'out'? In some (most?) countries

---

[14]It is generally held that students' beliefs influence their learning, for example, that affective beliefs play a major role (Burton 2004). Here we are only interested in philosophical beliefs.

knowledge of mathematics is a 'gatekeeper' to the inclusion of the society (see Skovmose 2004). We will not delve more on these roles, but merely state that they are important and complex. Skovsmose (2004), for example, describes how mathematics education can empower or disempower, include or exclude and discriminate, advancing what he denotes critical mathematics education. Finally, at the level of the individual, it could be the case that awareness of philosophical matters concerning mathematics could help the individual learn and – as we will discuss later – even become better mathematicians.[15]

On the actual implementation of philosophy in the teaching of mathematics, not much is written. In secondary level mathematics, Flanders, François and Van Bendegem (2010) conclude there is little room for philosophy of mathematics. Concerning philosophy of mathematics for university education, the authors ask:

1. Is there room for philosophy of mathematics at university level? We answer YES!
2. If so, what kind of philosophical approach? Should one stress the fallibility of mathematical knowledge, should one stress the social nature of mathematics or should one stress the curious mechanisms that have led to such a strong consensus among mathematicians (François and Van Bendegem 2010)? We answer neither! All pictures could be presented. The point would be to introduce the questions to which these are answers as well as tools to deal with them, so that students may form their own conclusions.

In this chapter our task is to consider implementation and possible roles for history and philosophy for university mathematics students. It seems clear that these students must be considered among the included people of the society. They already know how to learn mathematics, and hopefully even like it. The role that philosophy should play in university education is thus clearly different. However, the roles introduced above are relevant in the following ways. We will argue below that both philosophy and history of mathematics will make mathematics students better as mathematicians, not necessarily because they learn more mathematics, but since they will be able to get a wider picture of their subject. We also find that mathematics students should *be made aware of* some of the social implications of mathematics, even though they themselves may not personally be affected by them. We return to a discussion of these roles when we have presented actual examples of implementing courses in history and philosophy of mathematics.

26.4 Examples of History in University Mathematics Programmes

In this section we will present and discuss some specific implementations of history in mathematics programmes at university level. We have chosen three examples that illustrate different ways and approaches of integrating history as well as different

---

[15] It has also been argued that knowledge of philosophy can turn you into a better person. Philosophy teaches rational thinking, and in particular, ethics deals with the good and bad.

roles history plays in these programmes. The third example comes from our own approach and will be treated in more depth. The examples will be presented, analysed and compared with respect to their aims, their learning objectives, their use of sources and the significance of history. The function of history in these implementations will be analysed within the conceptions of mathematics education and learning of mathematics that were introduced above. We will point out and explain situations where we find the approach of history and philosophy of mathematics from the perspective of mathematical practices particularly relevant for mathematics education at university level.

26.4.1 *Ex. 1: History of Mathematics 1: Copenhagen University*

History of Mathematics 1 is a course that is offered in the mathematics programme at Copenhagen University, Denmark. It has been developed and taught by Jesper Lützen, who is a historian of mathematics. The history course is placed at the bachelor's (undergraduate) level in the mathematics programme. The students who follow the course are mathematics students, who will finish with a university degree with a master's in mathematics or a master's in another subject and a bachelor's in mathematics. The study programme in mathematics is not divided into pure, applied or teacher education programmes, but History of Mathematics 1 (or a similar course in the history of mathematics) is required for students who (later on) decide to become high school teachers in mathematics.

The course is a general history of mathematics course and its main purpose is to teach a survey of the history of mathematics from ancient times to the present. The course book is Victor Katz's (2009) *A History of Mathematics: An Introduction* which is supplemented with a booklet (Lützen and Ramskov 1999) with selected sources and exercises comprised and developed for the course. The objectives of the course are formulated in terms of what the students should be capable of doing after following the course, namely, to[16]:

1. Communicate orally as well as in written form about the history of mathematics
2. Use the history of mathematics in connection with mathematics teaching and more generally reflect on the development of mathematics
3. Place a concrete piece of mathematics in its historical context
4. Find literature (primary as well as secondary) on the history of mathematics
5. Give a historical analysis of a mathematical text from the past
6. Independently formulate and analyse historical questions within a limited field
7. Use the history of mathematics as a background for reflections about the philosophical and social status of mathematics

These objectives are reached on one hand through broad lectures on various cultures and time periods following (more or less) the outline of Katz's book

---

[16] http://sis.ku.dk/kurser/viskursus.aspx?knr=121117&sprog=2&forrige=57876

and on the other hand through a small group-organised project work on either tangent and max-min methods or methods of quadrature and curvature. The aim of the project work

> is [subjectwise] to give the student insight into the early history of the differential and integral calculus. Methodologically, the aim is to give the students a chance to work together in a group on a subject from the history of mathematics, to interpret primary sources, assess the secondary literature, chose important aspects, formulate a written report, and constructively criticize the work of another group.[17]

All objectives are formulated with respect to history (of mathematics). The overall goal is to provide students with historical knowledge about the development of mathematics and develop their historical awareness. The connections to subject matter of mathematics come about through reading of secondary literature (Katz's book), through analyses of sources and through the project work. The content matter of mathematics is subordinate to the content matter of history and historiographical issues. The overall impression is that the cultural argument lies underneath this implementation of history in the mathematics programme at Copenhagen University. The description of the objectives of the course is rounded off with the following declaration of expected outcome:

> Moreover the course will show connections between different mathematical fields that may appear unconnected in the more specialized mathematics courses. It will help students to formulate and form opinions about meta-mathematical questions and will counteract the tendency to absolutism that can result from ordinary text books. The students will see that during history there have been many different approaches to mathematics and they will meet cases where there are still different views about mathematical and meta-mathematical questions. That will ripen the student's view of mathematics.[18]

Historiographically, the ambition is to have a scholarly approach to history where historical episodes are interpreted on their own terms with an emphasis on differences between now and then. We also see indications of arguments in line with Beckmann's (2009) argument for interdisciplinary teaching as a way to counteract disciplinary narrow-mindedness and history as a method for revealing connections between mathematical fields that appear autonomous and disconnected in mathematics study programmes in universities. There is also a focus on the changing of meta-level rules. Philosophical issues are addressed in the course in connection with the historical development of mathematics, and we have here a clear interaction between history and philosophy.

26.4.2 Ex 2: Teaching with Original Historical Sources in Mathematics: New Mexico State University

The second example we have chosen comes from the developmental work that has been going on at New Mexico State University, USA, from the late 1980s spearheaded by the Professors David Pengelley and Reinhard Laubenbacher.[19] Their idea

---

[17] http://sis.ku.dk/kurser/viskursus.aspx?knr=121117&sprog=2&forrige=57876

[18] http://sis.ku.dk/kurser/viskursus.aspx?knr=121117&sprog=2&forrige=57876

[19] http://sofia.nmsu.edu/~history/; http://www.cs.nmsu.edu/historical-projects

was to teach mathematics through primary historical sources. The group has developed and taught two undergraduate mathematics courses that are based on students' study of original sources from the history of mathematics. One of the courses is a lower division course in which students are introduced to 'great problems of mathematics'.[20] According to Laubenbacher and Pengelley (1992, p. 2), the course 'serves as an "Introduction to Mathematics" drawing good students to the subject [mathematics]'. The other course is called 'Great Theorems: The Art of Mathematics', and it functions as 'a capstone course for college juniors and seniors with substantial mathematics background' (Pengelley 2002, p. 1). A book for each course has been completed based on annotated original sources (Laubenbacher and Pengelley 1999; Knoebel et al. 2007). The courses (and the books) are centred around selected problems and theorems from different mathematical subjects. Each problem/theorem (chapter) comes with an extended introduction in which the authors present a chronicle of the problem/theorem often extending over several centuries.

Laubenbacher and Pengelley have presented the ideas behind their developmental work at conferences and in articles of which most can be found on their website 'Teaching with Original Sources in Mathematics'.[21] In contrast to what was the case at Copenhagen University, their aim is not to teach history of mathematics per se. Their aim is to teach mathematics through the use of mathematical sources from the past. Their work originated out of a critique of traditional undergraduate mathematics instruction in which they found a lack of motivation for abstract concepts and an approach in modern textbooks and typical instruction that 'deprives students of the sense that mathematics is a process … [and] … fail to illustrate the way mathematicians actually think about and work on problems' (Laubenbacher et al. 1994, p. 1). They wanted to remedy this by introducing a historical perspective in which the study of original sources is firmly integrated into 'all our courses, presenting these sources to motivate the modern theories they have spawned'. They provide two arguments for this:

> First, by reading original sources students are brought as close as possible to the experience of mathematical creation. … [Second], when students read original sources, they are initiated into the way mathematics is practiced. … Mathematicians at the cutting edge of their field don't read textbooks; they read research papers. (Laubenbacher et al. 1994, p. 2)

Hence, their argument for history in mathematics teaching at university level is pedagogical. Students should learn from the masters of the past.

In Jensen's terminology, we are dealing with the use of history that is guided by the idea that we can learn from history, i.e. a pragmatic use of history. The historical sources are subordinate to the mathematics. The selection and the reading of the sources are guided not by historical questions, but with respect to how central they are for the curriculum and their utility with respect to the learning of modern mathematics. Many of the exercises presented to the problems/theorems the courses are evolving around are mathematical questions aimed at understanding the modern theories.

---

[20] http://sofia.nmsu.edu/~history/

[21] http://sofia.nmsu.edu/~history/

In the preface to their second book, they describe some of the benefits they have observed of using the past in this way as an approach to teaching mathematics. They write:

> Although teaching and learning with primary historical sources requires a commitment of study, the investment yields the rewards of a deeper understanding of the subject, an appreciation of its details, and a glimpse into the direction research has taken. (Knoebel et al. 2007, p. v)

> Primary sources also inject students directly into the process of mathematical research. They become active participants at the cutting edge of their own knowledge, experiencing actual research through grappling with the writings of great thinkers of the past. This creative immersion into the challenges of the past helps students better understand the problems of today. (Knoebel et al 2007, p. vi)

The main point is to learn mathematics, and history is used in the sense that the reading of the sources from the past is used as a pedagogical teaching method to teach students mathematics in an inquiry, research-like way. It can be argued that this approach to teaching has the potential to train, evoke and develop many of the eight mathematical competencies from Niss' conception of mathematics education. Historiographic and philosophical issues do not seem to play any significant role. In this respect, the courses differ fundamentally from the history of mathematics course taught at Copenhagen University. The selected sources also play very different roles in the two settings. In the booklet completed for the course in Copenhagen, the selection of the sources have been guided by historical and philosophical issues regarding the development and understanding of mathematics in the corresponding time period. The students are guided in their reading of the sources through questions that also point out how the mathematics of the past differs from our modern understanding and how the rules of the game have changed over time. In the courses developed at New Mexico State University, the selection of the sources has been guided by pedagogical principles (Barnett et al. 2011, p. 188). The sources play the role of authentic pieces of mathematics at research level for their time. They provide a context in which students can gain experiences with mathematical research processes. The students are guided in their reading of the sources through mathematical questions, and often they are asked to connect the mathematics of the source with the way it is presented in modern textbooks. The sources function as motivation for our modern theories and concepts.

Pengelley and his group have by now integrated their pedagogical approach of learning from the masters into many of the regular mathematics courses in the curriculum at New Mexico State University. This is done mostly in the form of modular projects.[22] Their goal is to allow students to learn all their mathematics in regular courses from primary sources. According to Pengelley, the 'team has now taught at least 3 of the regular dept. courses entirely from the projects we have developed, no more textbook'.[23] Some of these projects have also been implemented at Colorado State University (Barnett 2012).

---

[22] See http://www.cs.nmsu.edu/historical-projects.

[23] Personal e-mail correspondence between David Pengelley and Tinne Hoff Kjeldsen on Monday the 25. of June, 2012.

26.4.3 Ex 3: Problem-Oriented Student-Directed Project Work: The RUC Model, Roskilde University

The last example of specific implementations of history (and philosophy) in mathematics education comes from the educational practice at Roskilde University in Denmark. Historical and philosophical perspectives on mathematics are implemented through problem-oriented, student-directed and group-organised project work. We have chosen to present three such projects. One of us (Tinne Hoff Kjeldsen) was supervising professor for two of the projects and was consulted as a supervisor by the group of students who completed the third project. The project works are exemplars of our own approach.

As will be explained below, the problem-oriented, student-directed and group-organised project work (the RUC model) as it is carried out at Roskilde University creates very complex learning situations for students. Hence, each project work has potential for multifaceted learning outcomes. However, in our presentation and discussion of the three projects, we have singled out in each of them one particular aspect of (possible) roles history and/or philosophy play in the mathematics education at Roskilde University.

All study programmes at Roskilde University are based on the four overarching pedagogical principles of problem orientation, student-directed project work, interdisciplinarity and exemplarity which constitute the RUC model (Salling Olesen and Højgaard Jensen 1999, pp. 16–17). In each semester the students participate in project work of their own choice. At the beginning of the semester, the students in a particular study programme form groups of 3–8 students in accordance with their interests.[24] They formulate a problem that they want to work on throughout the semester. A problem is eligible if it fulfils the requirements for the students' semester, e.g. in the mathematics programme each student participates in three projects, fulfilling three semester requirements: a 'modelling' requirement, a 'mathematics as a discipline' requirement and a 'profession'[25] requirement.[26] The justification for the project requirements is a mixture of the cultural argument, an argument for interdisciplinarity in its own right and as a vaccination against disciplinary narrow-mindedness, and arguments similar to those of the critical mathematics education: in the project work the students come to reflect upon and criticise mathematics in some of its forms of actions. In the project work, the students develop their three second-order competencies of overview and judgement from Niss' competence description of mathematics education presented above.

---

[24] Three to eight students is the common group size, but students are allowed to perform a project on their own.

[25] Under the 'profession' requirement, the students have a choice between a modelling project, a pure mathematics project, a history and/or philosophy of mathematics project or a project on aspects of mathematics education, according to in what kind of direction, they want their future profession to move.

[26] See also Niss' (2001) narrative on his 25 years of experiences with the RUC model.

The projects where history and/or philosophy of mathematics enters are the 'mathematics as a discipline' projects and sometimes the profession projects. Before the students enter into the mathematics programme, they have completed a four semester interdisciplinary science programme (Blomhøj and Kjeldsen 2009). In this programme, the project requirement for the third semester is a 'meta' requirement, meaning that the students should work with a problem through which they will gain experiences with science as a cultural and social phenomenon. Of the three projects we discuss below, the first two are 'mathematics as a discipline' projects from the master's programme whereas the third one is a 'meta' project performed in the 2-year interdisciplinary science programme (Kjeldsen and Blomhøj 2009). The projects are only constrained by these requirements. The problem a group of students chose to work on in a project should fulfil the requirements, and the project should meet the academic level to be expected of students who have reached the corresponding semester. There are no requirements on the content of the project work. The content is determined by the problem the students decide to work on. During the first 1–2 weeks of each semester, the students form groups based on their interests. Suggestions for problems will be raised, discussed and qualified in discussions between students and the professors who are going to be assigned as supervisors for the semester.

When the groups are formed, they will write an application to the board of study seeking approval of their problem and their project. They will also indicate which professor(s) they would like to have as a supervisor. The supervisor will follow the group throughout the entire semester. Normally, the group will meet with the supervisor once or twice a week for 1–2 h. The agenda of the meetings often comes from the students. They decide what they want to discuss, what they need help with and how they want to 'use' their supervisor. The supervisor makes sure that the academic standards are met and will let the students know if they are on a false track. In the RUC model, the project work can be thought of as student research projects. In each semester every student is part of a research team of fellow students, who perform a research project guided by the problem they chose to work on and by a supervising professor.

In the following we will present and discuss three specific student projects from the RUC model. As should be clear by now, the RUC project work creates a very complex studying and learning environment for the students. However, in the following we will focus, as mentioned above, on only one aspect of learning outcome for each project and leave the rest aside.

Project 1

Generalisations in the Theory of Integration: An Investigation of the Lebesgue Integral, the Radon Integral and the Perron Integral

This project was performed by two students. The students documented their work in a written report of 75 pages.[27] It originated out of a curiosity about different

---

[27] The students' project report can be downloaded at the following address: http://milne.ruc.dk/ ImfufaTckstcr/pdf/403.pdf.

types of integrals. In the students' first analysis course, there was a footnote in the textbook that pointed out that there exists functions that are not Riemann integrable and that there are other types of integrals that can handle more functions than the Riemann integral, e.g. the Lebesgue integral. The two students wanted to investigate what these other types of integration can do. They immediately found out that the Lebesgue integral is just one of many different integrals. There is also the Denjoy, the Perron, the Henstock, the Radon, the Stieltjes and the Burkill integral, to mention just a few. The students noticed that they were often presented in the literature as generalisations of either the Riemann or the Lebesgue integral. These observations generated a bunch of questions (we are quoting from the students' project report): 'What do these integrals do? Why have so many types of integrals been developed? Why is it always the Lebesgue integral we hear about? What is meant by generalization in this respect? In what sense are the various integrals generalizations of former definitions of integrals? Are the generalizations of the same character?' (Timmermann and Uhre 2001, p. 1).[28]

In the end the students' project work was guided by the following problem:

> What were Lebesgue, Perron and Radon motivated by in their pursuit of their generalizations of the integral?

> What are the character and scope of the generalizations by Lebesgue, Perron and Radon, and what are the differences between them? (Timmermann and Uhre 2001, p. 3)

The students performed a historical study to answer the first part of their problem formulation. In Jensen's terminology, they had a scholarly approach to history. They studied a concrete episode from the history of mathematics from the perspective of the historical actors' motivation to extend and generalise the concept of the integral. The students read a selected variety of sources – Journal articles and books by Denjoy, Henstock, Lebesgue, Perron and Radon – with focus on the work of the last three mathematicians. For example, with respect to Lebesgue, the students read his note *Sur une généralisation de l'intégrale définie* which was published in *Comptes Rendus de l'Académie des Sciences de Paris* in 1901 and his thesis *Intégrale, Longueur, Aire* from 1902. They interpreted his motivation for the generalisation of the integral concept, as they explained in their report, 'detached from the context in which it is part of today and detached from our knowledge of the later significance of the concept' (Timmermann and Uhre 2001, p. 4).

As in the courses at New Mexico, the students studied the masters by reading research literature from a past episode in the history of mathematics, but in contrast to the courses at New Mexico, they were not guided by mathematical questions, but by historical questions. However, these questions were answered with reference to analyses of the mathematical content, theorems, definitions, proofs and techniques of the sources. In this way, the students gained first-hand experiences with processes and initiations of research in pure mathematics. With regard to mathematical competencies, an analysis of the students' work shows that six of the eight competencies

---

[28] All quotes from student reports have been translated into English by us.

were invoked and trained during this project work, but the main purpose of their project was to develop their second-order competency of overview and judgement regarding the historical development of mathematics.

Project 2

Fourier and the Concept of a Function: The Transition from Euler's to Dirichlet's Concept of a Function

This project was designed and completed by four master students.[29] Their project work was guided by the following interest and curiosity:

> We wish to investigate the significance of Fourier for the development of the concept of a function. (Godiksen et al. 2003, p. 2)

The students analysed relevant sources from the works of Euler, Fourier and Dirichlet with respect to changes in as well as discussions about the concept of a function and the proper way to argue with functions. The relevance of the reading of sources from these three mathematicians was explained in the following way in the students' project report:

> The strength of focusing on these three mathematicians is, that it has given us the opportunity to study their original works (sometimes in translations) in depth, which have given us a more direct impression of their thoughts than secondary literature could have given us. (Godiksen et al. 2003, pp. 2–3)

We are again dealing with an ambition of employing a scholarly approach to history. The students wanted to interpret the past on its own terms. In their project report, they compared the works in the sources by Euler, Fourier and Dirichlet with each other and with our modern approach. They used the sources as 'interlocutors' emphasising the central ideas and the differences. With respect to Euler's concept of a function, the students' wrote in their report:

> The main elements of Euler's conception of a function could easily be explained very shortly, but that would not contribute to any deep understanding of the concept. In order to obtain this, one has to *look* at how Euler worked with functions. (Godiksen et al. 2003, p. 17; italic in the original)

In order to understand Euler's conception of a function, the students point out that it is necessary to study Euler's mathematical practice – how *he* worked on and used functions. The students gave the following interpretation of Euler's conception of a function:

> The definition of a function [Euler's definition] does not contain any specific information about its domain and image. This is because in Euler's theory, variable quantities are ascribed a property that render specifications of such sets superfluous. ...
>
> ... Euler conceived a variable as an arbitrary element, quite like our conception, but no constraints are allowed. The variable should be able to take all values ... (it is universal). (Godiksen et al. 2003, p. 18)

---

[29] The students' project report can be downloaded at the following address: http://milne.ruc.dk/ImfufaTekster/pdf/416.pdf.

26 The Role of History and Philosophy in University Mathematics Education   857

Euler's analysis is global in nature – variables were universal, they were not limited in scope, and hence, Euler's functions had the property of analytic continuation. The students identified two meta-level rules of Euler's mathematical discourse: the generality of the variable and the general validity of analysis – two rules that were revealed and made into objects of the students' reflection through this historical project work, as can be seen from the following discussion in the students' report, where they wrote:

> This property which [...] has been named the criteria of the *generality of the variable* clearly reflects the earlier mentioned paradigm of *the general validity of analysis*.
> [...]
> Even though the use of the methods of analysis often created weird results the methods were used frequently in Euler's concept of a function. The reason why there weren't that many contradictions and paradoxes was that almost all Euler functions, which consists of analytical expressions, *have* all the above mentioned properties [they were nice], except maybe in isolated points. ... Hence, there was no natural driving force that led to a clarification of the concepts of continuity, differentiability, and integration, since these properties so to speak were built into the concept of a function. (Godiksen et al. 2003, p. 22; italic in the original)

As pointed out by the students:

> Euler ... was of the opinion that the analysis had to be developed such that it was able to describe all situations that occur in nature. (Godiksen et al. 2003, p. 23)

And in their treatment of Fourier's work, they continued:

> Fourier expresses clearly that mathematics is a tool for describing nature and mathematics had to be governed by nature. (Godiksen et al. 2003, p. 53)

The idea that mathematics has to be governed by nature is a third meta-level rule of past mathematical discourse that became exposed for the students and became an explicit object of reflection for them.

The students wrote a report of 88 pages explaining, analysing and interpreting the mathematics and the ideas about mathematics in the sources in order to answer their problem formulation. Again, it can be argued that six of the eight main mathematical competencies were invoked and trained during the project work together with the second-order competencies of history and philosophy. Here we have focused only on how this project work in history of mathematics functioned as a learning and teaching situation for students to experience meta-level rules in mathematics discourse and gain experiences with how these change over time (for further details, see Kjeldsen and Blomhøj (2012)). Within Sfard's theory of thinking as communicating, this project work is an example of how history of mathematics can function at the core of what it means to learn mathematics, as explained above.

Project 3

Rashevsky's Pride and Prejudice

The two projects discussed above relate to the two second-order competencies of history and the nature of mathematics. The project presented here on Nicolas Rashevsky's model for cell division from the 1930s relates to the third second-order competency of gaining knowledge about and experiences with actual applications of mathematics in other areas. In this case, it is the application of mathematics as a practice in other sciences. The project was conducted by four students. They read a paper by the physicist Nicolas Rashevsky (1934), where he discussed what he called *physico-mathematical aspects of cellular multiplication and development*. He presented the paper at a Cold Spring Harbor symposium on quantitative biology (see Keller (2002), and Abraham (2004)). His talk was followed by a discussion where the biologists in the audience were very critical of Rashevsky's approach in explaining cell division. Rashevsky's talk and the discussion were published from the proceedings of the meeting.

The students were curious about the hostile attitude of the biologists. They formulated the following problem that guided their project work:

> Why was Rashevsky unable to get through to the biologists of his time with his ideas? Was it because the biologists could not accept Rashevsky's scientific method? If so, was this then caused by a fundamental difference in biologists and physicists conception of biology and were/are controversies about the scientific method then a manifestation of this difference? (Andersen et al. 2003, p. 2)

The students studied the status of biology and of mathematics and physics in biology in the 1930s to understand the scientific culture of biologists and their conception of the significance of physics and mathematics in biology at the time. In reading Rashevsky's paper, the students became aware that he held a reductionist view of science. The students studied philosophy of biology and physics in order to understand whether the differences in opinion between the biologists and Rashevsky could be explained by differences in philosophical standpoints about science. Rashevsky's strategy was to take a general phenomenon that occurs in all cells and investigate its mathematical consequences. He chose cell metabolism. If, as he wrote in the paper, the process of division is found among such consequences, then cell division can be explained logically and mathematically as a direct consequence of the forces arising from cell metabolism.

In order to understand the critique raised by the biologists, the students had to read and understand Rashevsky's paper. It can be argued that in this process all eight mathematical competencies were invoked and trained. Regarding the second-order competency of gaining knowledge about actual applications of mathematics in other areas, the students experienced that the validity of arguments depends on the scientific context. For further details, see Kjeldsen (2010) and Kjeldsen and Blomhøj (2009). The students' historical and philosophical investigations in the project work contributed to critical mathematics education in the sense of Skovsmose, since the students came to reflect upon and criticise a form of action of mathematics, the action of mathematics in the production of knowledge in other scientific areas. For further details, see Kjeldsen and Blomhøj (2013).

The upper level undergraduate courses at Copenhagen University and New Mexico State University and the RUC model of project work at Roskilde University all address students who educationally wise are similar. However, the arguments, aims and objectives of the three different approaches of integrating history into the mathematics study programmes are, as we have seen, quite different. The project work in a sense combines the two approaches from Copenhagen and New Mexico. The problem, the students work on in the RUC model, is a so-called metaproblem. It is a problem *about* mathematics, historical and sometimes also philosophical, but the students' domain of inquiry is the mathematics of the past. In order to answer their problem formulation, the students work as historians (and philosophers) with historical (philosophical) problems. To get answers, they dig into the sources, studying and interpreting the mathematics – though not from a modern perspective, as is the case in the New Mexico approach, but from the historical actors' perspective within their mathematical practice.

26.5 Examples of Philosophy in University Mathematics Programmes

26.5.1 Philosophy of Science in University Education in Denmark: Aims

In 2000 the Danish Government and the Association of Vice-chancellors of Danish Universities decided to reintroduce a philosophy course for all university students.[30] This new course is called 'Fagets videnskabsteori' which translates to 'philosophy of science of the subject of study'. The overall aim of this course is to qualify students' specialisation in their subject by allowing them to see it from a broader and more general perspective. Ten specific points were listed as requirements; see http://www. nbi.dk/~natphil/FVT/i_Alment.html. We mention in particular 3 and 4 that address the particular implementation of these courses, requiring (i) that the courses should be research based and (ii) that the curriculum should 'take interesting questions from the field of study and combine them with questions of a more general kind'.

The overall aim was to address the question of how knowledge is tackled in a knowledge society. A university graduate should not only know narrowly his/her own subject but should also obtain competencies within values and perspectives and be able to reflect. These aims are elaborated on by Professor Hans Fink, who was one of the researchers involved in the discussions prior to the Bill (Fink 2001). He states that the overall aim is to produce better students and to prepare them for the job market. He says:

> We thought that a course that allows the students to reflect on their subject's distinctive philosophical character, seen in a wider, general, philosophical, and historical context,[31] would be well-founded, if it could forestall the risk that the students' professional absorption

---

[30] Until 1971 it was compulsory for all university students in Denmark to take an introductory philosophy course, so-called filosofikum. This course was mainly handled by philosophers.

[31] In addition 'videnskabsteori' is mentioned. It is best translated as 'theory of science' and is usually thought of more broadly than philosophy of science, including social science, history of science and ethics.

leads to narrow-mindedness that makes them less fit to engage with the interdisciplinary connections that they will later have to deal with. A course like this will also counter the often criticised schoolification of the universities by encouraging the students to systematic reflection about what a university, science and research actually are. ... Finally, the course could fulfil a large need of society, if it could ensure that everyone with a university degree is given the opportunity to make clear the social and ethical responsibility of both science and the individual researcher.

It is thus clear that the aims of this course are broad. One major requirement, however, is that all these reflections take as a starting point the actual subject. The course – in mathematics – should therefore:

1. Include philosophical reflections relevant to the subject in question, i.e. mathematics.
2. Place the subject in a wider context, e.g. discuss mathematics' distinctiveness in relation to other subjects.
3. Discuss the role of mathematics, research and universities in society.
4. Discuss ethical questions relevant to the subject.

Overall the stress is on the students' abilities to *reflect* and *critically examine ideas*. These are philosophical competencies, so the core of the course is a philosophical one. This means in particular that the course should not dictate any one perspective as the right perspective, but should aim at giving the students tools to handle these competencies. In what follows we present examples of how this course is implemented.

26.5.2   Implementation of Philosophy of Science Course in Aarhus University and University of Southern Denmark

In what follows we give concrete examples of how this course is handled at the Universities of Aarhus (AU) and Southern Denmark (USD). The overall structure of both courses is the same. Both courses' credit is 5 ECTS. They include common lectures, presentations by students, discussions and group projects, and they address the above-mentioned points. Many of the included topics are the same. But there are differences in how these are presented. We present Aarhus first, since the Centre for Science Studies at Aarhus University has devoted much time to develop a 'philosophy' on how to handle courses on the philosophy of natural science. In a recent paper Kragh Sørensen writes:

We have adopted a teaching philosophy of using historical and contemporary case studies to anchor broader philosophical discussions in the particular subject discipline under consideration. Thus, the courses are tailored to the interests of the students of the particular programme whilst aiming for broader and important philosophical themes as well as addressing the specific mandated requirements to integrate philosophy, some introductory ethics, and some institutional history. These are multiple and diverse purposes which cannot be met except by compromise. (Kragh Sørensen 2012, p. 1)

26 The Role of History and Philosophy in University Mathematics Education

The course starts by addressing the question 'what is mathematics?' and includes topics like the application of mathematics, discussing the question concerning 'The Unreasonable Effectiveness of Mathematics in the Natural Sciences' (Wigner 1960) and issues concerning modelling, the role of proofs and the foundation of mathematics and the role of mathematics and university in society, including mathematics and gender and ethical issues. To give a better sense of how this philosophy works in practice, we present two examples in more detail. One concerns the role of proof in mathematics, the other, the (unreasonable effectiveness of the) applications of mathematics.

Example 1

When dealing with the role of proofs, the course takes as a starting point a general held belief, presumably also among the students. A proof is taken to be[32]:

- Axiomatic deductive, obtained from accepted assumptions
- Employing logical steps in a specified logical system
- (In principal) fully formalised (or possible to formalise)

For homework the students are asked to consider the following questions:

- What is the difference between foundation and practice?
- What is the difference between formal and rigorous proofs?
- What is the foundation, and how certain is it?
- Can proofs prove or can they (at most) convince?

The aim of the treatment of this topic is to 'enable discussions about proofs that go beyond the idea that proofs merely guarantee truth'. In addition, the teacher 'wishes to emphasise the role of proofs as tools of communication in a mathematical discourse' (personal communication).

The first case study to challenge the standard conception of proof is Perelman's proof of the Poincaré assumption. This case shows, among other things, that the mathematical community may disagree on whether a presented proof is actually a proof. It also illustrates different motives for pursuing a mathematical career – fame, glory and money – or the pleasure of obtaining insight. Later the course considers the role of proof in more detail. In order to discuss this theme, for example, Wiles' proof of Fermat's last theorem is presented. This case challenges the standard conception of proof. For one thing, the proof is too long and too complicated for the average mathematician to follow it through. This leads to a discussion about how one trusts a given proof and which criteria should hold for a 'good proof'. The discussion brings in many examples from current mathematical practice. The phenomena of experimental mathematics (and use of computers in mathematical proofs in general) are treated in order to discuss, for example, the notion that mathematical proofs and knowledge are a priori.

---

[32] The following is taken – and translated – from slides used in the course.

Example 2

Questions pertaining to the conception and role of a proof could be taken to be internal to mathematics. In contrast are questions concerning the relation between mathematics and reality. One question that has puzzled philosophers and scientists concerns 'The Unreasonable Effectiveness of Mathematics in the Natural Sciences' (Wigner 1960). It is clear that mathematics is applied to the real world. Since the modern development of mathematics, however, it has been a challenge to explain actually why and how mathematics can be applied. The answers – and the degree of mystery involved – depend on which view of mathematics is taken. Wigner presents a view of mathematics, where concepts are developed with no connections to considerations of the world but in order to develop beautiful and interesting theorems. On this view, it is clearly a mystery that mathematics can say anything about the physical world. Several solutions to this problem have been offered; see, for example, Grattan-Guinness (2008). The course in Aarhus introduces this question and indicates how proponents of different views would account for the applicability (Platonism, Formalism, Kant, Empiricism). In addition, some case studies are introduced in order to qualify the discussion. One example illustrates interrelation between development of mathematics and – in this case – chemistry. It tells part of the story on the development of crystallography. In this theory crystals are modelled as certain lattices with symmetry properties (translation and rotation). According to the main theorem of crystallography, crystals can only have certain types of symmetries. In 1984, however, a crystal was found contradicting this result. This led to the notion of a quasicrystal and an ongoing search for a mathematical theory describing these.

The course also stresses the point made by several mathematicians (Hilbert 1902; Poulsen 2001, and Toft 2001) that mathematics solves *problems*. Problems can be either internal mathematical problems or external problems.

An important part of this topic is 'to point out and discuss the difference between a mathematical model and the part of reality that it is supposed to be a model of. In this lies also the task to point out the choices (perhaps even theory-laden) made when formulating the model, and the relations one could (wrongly) think there is between a model and reality. This leads to the possibility of discussing predictions made from mathematics and mathematical explanations in natural science' (personal communication with Kragh Sørensen). Mathematical models are widely used in today's society and for many different purposes. In some uses, models are tools to solve problems. As such, they can be very simple or extremely complicated giving rise to a whole range of problems.[33] Mathematical models are also used in arguments. When used in this way, it is important to be aware of the fact that a model can at most tell something about the actual data or assumptions put into the model. If the model is complex, meaning that the resulting mathematical problem cannot be solved by exact methods, it may not even give certain results about these. Even so, mathematical models are

---

[33] The topic of the role of models in society is also dealt with at USD, and the following description is mainly based on the treatment there.

26 The Role of History and Philosophy in University Mathematics Education   863

often used in, for example, political debates, where the certainty of mathematics (faultily) is attributed to the model (see Kjeldsen 2011b). When teaching this topic, the students are shown traditional steps of a modelling process and are asked to think these through for various examples. In addition they find examples of models used in society, discussing which roles they play and – if possible – which assumptions have been made when formulating the model.

Overall, the philosophy of allowing examples to generate the philosophical discussion seems to work. In order to generate and qualify the discussion, however, certain philosophical theories and notions need to be explained as the course moves along. It is also the case that certain convictions the students (may) hold need to be explicitly stated before they are discussed as was the case in the first example, where the standard conception of proof is discussed. This is not surprising, since we are dealing with mathematics students, not philosophers. It takes training to think philosophically, i.e. to be able to pose the relevant questions or to pose questions at all and to learn to use tools to handle these questions.

26.5.3   *Example from University of Southern Denmark*

In contrast, the course in Odense takes as a starting point certain *philosophical questions* and tries to answer these based on actual examples of mathematics. When entering the mathematics programme at USD, a student will take courses within natural science during the first year – together with all students who wish to study natural science. The philosophy behind this is to strengthen the interdisciplinarity of subjects within natural science. This philosophy carries through to the course in the philosophy of science, where there is a common core and the lectures are held for all students. The course also consists of a subject-based part so that, for example, mathematics students have a number of classes that addresses issues particularly relevant for mathematics. In the common part, it is possible to address general themes such as the distinctiveness of natural science and mathematics, ethics and the role of research and university in society. We also treat 'standard' schools in philosophy of science, such as inductivism, Popper's critical rationalism and Kuhn's paradigms and revolutions. The point here is not so much to discuss these from a philosophical point of view, rather it is to see which questions they raise and how they propose to answer them. The main idea is to consider these questions and answers in the light of their subject. As a supplement, the students are presented a number of case studies, illustrating and challenging conceptions of science. These examples range from the overthrow of the Phlogiston theory to discussions concerning Creationism and Intelligent Design and the status of research in particle physics.

In Danish, the word for science is 'videnskab'. It includes 'viden' which means knowledge and infers that it is a practice that yields knowledge. Students are given the philosophical definition of knowledge as 'justified true belief' and are encouraged throughout the course to reflect on whether the methods used in their discipline do in fact yield knowledge in this sense. It is also stressed many times during the course that it poses questions and provides *possible* answers and their job is to reflect on these answers, finding their own.

Example 3

The final example concerns the question about the nature of mathematical objects. In the lecture, the students are introduced to the realism-antirealism debate in the philosophy of science. One point is that even in natural science – that supposedly should concern the real world – there are questions to ponder about. For a first example, take the phlogiston theory that during part of the eighteenth century was held to explain processes of combustion. This theory was overthrown by Lavoisier's theory stating that oxygen is the fundamental matter at play in such processes. The question is whether our theories today could suffer the same fate or our current scientific theories are better? If they answer yes to this last question, the challenge is to argue just *how* these theories are better and even explain what a 'better' theory means. Most scientists' intuitions will tell them that there is progress in the development of scientific theories. The challenge is to explain in which sense. A more recent example concerns the theory of dark matter and energy. So far we (the scientists) only have circumstantial evidence that these types of 'objects' exist. Scientists have never seen (it cannot be observed, since it does not radiate electromagnetic force) or isolated samples of dark matter. Even so, they are convinced it exists (see Sannino 2009). This raises the challenge to formulate rational criteria to determine whether something exists. Within philosophy of mathematics is the dispute on whether mathematical objects exist independently of human beings, in case one is a realist – or Platonist – or whether they do not, in which case one is an antirealist. The way this question is tackled is to present the students with:

- A number of different versions of realist and antirealist positions. Pointing out that there are many possibilities for being of each kind.[34]
- Equally important, the motivation for and arguments typically given in favour both of realism and antirealism. Realists would, for example, argue that mathematical propositions are true, and then it follows from a correspondence theory of truth that mathematical objects exist. An influential argument which dates back to Frege but is usually credited to Quine is the indispensability argument (Colyvan 2001; Shapiro 2000). It combines the fact that mathematics is indispensable to natural science with a confirmational holism[35] and an ontological commitment to entities of any accepted theory. Anti-realists in turn argue (with Benacerraf 1973) that the existence of abstract mathematical objects outside of space and time makes it a mystery of how we obtain knowledge in mathematics.

---

[34] Maddy's (1990) set theoretic realism and Shapiro's (2000) ante rem structuralism are examples of what is presented. In addition we discuss Field's (1980) fictionalism and the empiricism of Mill.

[35] A scientific theory is confirmed as a whole. If mathematics is part of a confirmed scientific theory, then the included mathematics is also confirmed.

26 The Role of History and Philosophy in University Mathematics Education

A popular strand of anti-realism is fictionalism, denying that mathematical statements are true (else realism follows) rather they are a certain kind of fictitious statements.

Based on these positions, arguments and motivations, the students are asked to determine for themselves which kind of position they find most convincing. When doing this they are, of course, asked to take into account their own experience with mathematics. Taking any one position also requires producing arguments for this position and in some cases countering the arguments of the other side. As an example, one could question the statement that mathematical sentences are true in the same sense as ordinary sentences or that mathematical theories are confirmed alongside scientific theories.

In addition we have responsibility to show students the many sides of mathematics that they do not encounter in their traditional lectures. They are most often taught from textbooks.[36] These are re-presentations of the mathematical theories, definitions, theorems and proofs as presented in research articles, presented in a form that intends making it accessible for the student. In this form, however, the original presentation and motivation is often lost. What else is missing is the, often long, process leading to the results in question.

26.5.4 Are the Aims Fulfilled?

The examples given mainly address aim number 1, i.e. introducing students to philosophical questions to their subject. We hope to have conveyed in the above presentation that we aim at the following:

1. Making the students aware of some of the philosophical questions underpinning their subject.
2. Providing them with tools to tackle these questions, by teaching them a certain terminology, positions and ways of thinking distinct to philosophy.
3. An emphasis on the fact that they are mathematicians, so that their expertise when dealing with such questions lies in the ability to critically examine mathematical assumptions and point to misconceptions about mathematics.
4. It is also our task to show our students a variety of examples enabling them to fulfil 3.

These aims are in part achieved. In general, however, we find that students are not particularly motivated to take the courses in philosophy of science. It is our impression, though, that most students appreciate the course more after finding out what it is actually about. As we were about to write this chapter, the second author of the present chapter asked students from her last class about their thoughts

---

[36] In Aarhus and Copenhagen, it is very common for teachers to write their own textbook material.

concerning the relevance of such a course. One student complains that he/she does not find it relevant now, but still concludes that it might later lead to some reflections. Among the positive responses, a student found the course relevant, because it teaches them to think differently. Answers like this are particularly welcome!

We recall the three types of second-order competencies. The first competency addresses applications of mathematics whereas the other two concerned the historical development of mathematics as well as its nature. It is clear that the courses in philosophy of mathematics described above intend to provide students with competencies in order to deal with both the problems concerning various applications of mathematics as well as fundamental questions regarding mathematics. More specifically, we find that one major role (on the mathematical level) these courses play is to present *a wider picture of mathematics* than is usually given in regular mathematics courses. Students are shown that mathematics is more than 'definition-theorem-proof' – a picture objected to by the Fields medallist Thurston (1994). The intention is to get them behind the scenes of doing mathematics. A related role is to enable and convince mathematics students to become ambassadors of mathematics. For one thing, they, if no one else, should be able to say something about what mathematics is, its distinctiveness and importance (mathematics is actually used everywhere in our society), and that mathematics need not be frightening.

Students ought to also be aware of some of the ethical problems concerning application of mathematics – or applications in general – and have some tools to handle these.

Finally, the course may even help them on a personal level in their mathematical studies. Philosophy as well as mathematics seeks *arguments*. Training to argue within philosophy strengthens ability to argue correctly in mathematical reasoning.

26.6 Conclusion

Comparing the identified roles of the courses in history and philosophy of mathematics, we find that they (to a large extent) coincide. On the mathematical level, one motivation for both history and philosophy is to provide students with a wider picture of mathematics, in particular to include meta- mathematical considerations 'preventing the absolutist tendency of mathematics textbooks'. On a social, or cultural, level is the motivation of interdisciplinarity counteracting disciplinary narrow-mindedness (Beckmann 2009; Fink 2001). In addition is a responsibility to show the political and social roles of mathematics, which can be done by both history and philosophy. On a personal level it is argued that including historical and philosophical perspectives make the students better mathematicians. We have even seen that both perspectives can be used to strengthen their abilities in mathematics. Philosophy of mathematics does this, for example, by teaching them to argue.

Although history and philosophy are quite different (history analyses and interprets sources and searches for explanations for historical processes of change – philosophy poses questions, critically examines ideas and provides arguments), they do interact in fundamental ways. In the history courses, most notably in the group projects, the motivations for doing the historical studies were in some cases philosophical. For example, the question concerning different types of generality, posed by the first group, is a philosophical question. Similarly, the aims of the philosophy courses could not be fulfilled without a handful of good (historical) cases to show to the students.

One remaining challenge is to convince students that courses like these are important for their mathematical training. It is also an open question whether the students actually benefit and use tools obtained from these courses in their professional careers.

References

Abraham, T.H. (2004). Nicolas Rashevsky's Mathematical Biophysics. *Journal of the History of Biology,* 37, 333–385.

Andersen, L. D., Jørgensen, D. R., Larsen, L. F. and Pedersen, M. L. (2003). *Rashevsky's pride and prejudice* (in Danish). Report, 3rd semester, Nat-Bas, Roskilde University.

Ashton, P. & Kean, H. (eds.) (2009). *People and their Pasts. Public History Today.* Houndsmills: Palgrave Macmillan.

Awodey, S. (1996). Structures in mathematics and logic: A categorical perspective. *Philos. Math.,* 4, 29–237.

Barnett, J. (2012). Bottled at the Source: The Design and Implementation of Classroom Projects for Learning Mathematics via Primary Historical Sources. Plenary address. *Proceedings from the HPM International Congress* in Korea, 2012. http://www.hpm2012.org/Proceeding/Plenary/PL3.pdf

Barnett, J., Lodder, J., Pengelley, D., Pivkina, I. and Ranjan, D. (2011). Designing student projects for teaching and learning discrete mathematics and computer science via primary historical sources. In V. Katz and C. Tzanakis (eds.). *Recent developments on introducing a historical dimension in mathematics education*, Washington, DC: Mathematical Association of America, 2011.

Beckmann, A. (2009). A Conceptual Framework for Cross-Curricular Teaching. *The Montana Mathematics Enthusiast*, 6, Supplement 1.

Benacerraf, P. (1973). Mathematical Truth. *The Journal of Philosophy,* 70, 661–679.

Benacerraf, P. (1965). What numbers could not be. *The Philosophical Review,* 74, 47–73.

Benacerraf, P & Putnam H. (1983). *Philosophy of Mathematics Selected Readings.* Cambridge University Press, (2. ed.), NY.

Blomhøj, M. & Kjeldsen, T.H. (2009). Project organised science studies at university level: exemplarity and interdisciplinarity. *ZDM – International Journal on Mathematics Education*, 41 (1–2), 2009, 183–198.

Bloor, D. (1994). What can the sociologist say about $2+2=4$? In Ernest (1994), pp. 21–32.

Blumenthal, O. (1935). Lebensgeschichte. In D. Hilbert. *Gesammelte Abhandlungen* vol. 3, (pp. 388–435), Berlin: Springer.

Burton, L (2004). Confidence is Everything - Perspectives of Teachers and Students on Learning Mathematics. *Journal of Mathematics Teacher Education,* 7, 357–381

Carter, J. (2004). Ontology and mathematical practice. *Philos. Math.,* 12, 244–267.

Carter, J. (2006). A less radical constructivism. *Nomad* 11, 5–28.

Carter, J. (2008). Categories for the working mathematician. Making the impossible possible. *Synthese,* 162, 1–13.

Carter, J. (2013). Mathematical Objects: Representations and context. *Synthese.* doi: 10.1007/s11229-012-0241-5.

Chassapis, D. (2007). Integrating the Philosophy of Mathematics in Teacher Training Courses. In: François & Van Bendegem (eds.) (2007), pp. 61–80.

Colyvan, M. (2001). *The indispensability of mathematics*. New York: Oxford University Press.

Corfield, D. (2003). *Towards a Philosophy of Real Mathematics*. Cambridge University Press.

Epple, M. (1999). *Die Entstehung der Knotentheorie: Kontexte und Konstruktionen einer modernen mathematischen Theorie*. Wiesbaden: Vieweg.

Epple, M. (2004). Knot Invariants in Vienna and Princeton during the 1920s: Epistemic Configurations of Mathematical Research. *Science in Context,* 17, 131–164.

Epple, M. (2011). Between Timelessness and Historiality. On the Dynamics of the Epistemic Objects of Mathematics. *Isis,* 102, 481–493.

Ernest, P. (ed.) (1994). *Mathematics, Education and Philosophy,* London: Falmer Press.

Ernest, P. (1998). *Social Constructivism as a Philosophy of Mathematics*, Albany, New York: SUNY Press.

Ernest, P. (2004). What is the Philosophy of Mathematics Education? *Philosophy of mathematics Education Journal* 18, http://people.exeter.ac.uk/PErnest/pome18/PhoM_%20for_ICME_04.htm Accessed 1 July 2012.

Ernest, P. (2009). What is first philosophy in mathematics education? In: M. Tzekaki et al. (eds.) *Proceedings of the 33rd Conference of the International Group for the PME*, Vol 1, 25–42.

Ferreirós, J. and Gray, J. (eds.) (2006). *The Architecture of Modern Mathematics*. Oxford: Oxford University Press.

Field, H. (1980). *Science Without Numbers,* Princeton: Princeton University Press.

Fink, H. (2001). Fra Filosofikum til Studium Generale. http://udd.uvm.dk/200103/udd03-10.htm Accessed 28 June 2012.

François, K. & Van Bendegem, JP. (eds.) (2007). *Philosophical Dimensions in Mathematics Education*. Springer Science and Business Media.

François, K. & Van Bendegem, JP. (2010). Philosophy of mathematics in the mathematics curriculum, questions and problems raised by a case study of secondary education in Flanders. Contribution to Mathematics Education and Society 6. http://www.ewi-psy.fu-berlin.de/en/v/mes6/documents/research_papers/Francois_VanBendegem_MES6.pdf?1286354771. Accessed 1 July 2012.

Godiksen, R.B., Jørgensen, C., Hanberg, T.M. & Toldbod, B. (2003). *Fourier and the concept of a function – the transition from Euler's to Dirichlet's concept of a function*. (In Danish). IMFUFA, text 416, Roskilde University.

Gray, J. (2008). *Plato's Ghost: The Modernist Transformation of Mathematics*. Princeton: Princeton University Press.

Grattan-Guinness, I. (2008). Solving Wigner's mystery: The Reasonable (Though Perhaps Limited) Effectiveness of Mathematics in the Natural Sciences. *The mathematical intelligencer,* 30, 7–17.

Heinz, B. (2000). *Die innenwelt der mathematik. Zur kultur und Praxis einer beweisenden disziplin*. Wien: Springer-Verlag.

Hellmann, G. (1996). Structuralism without structures. *Philos. Math.,* 4, 129–157.

Hersh, R, (1979). Some proposals for revising the philosophy of mathematics. *Advances in Mathematics,* 31, 31–50.

Hilbert, D. (1899). *Grundlagen der Geometrie*. B.G. Täubner.

Hilbert, D. (1902). Mathematical Problems, *Bulletin of the American Mathematical Society*, 8, no. 10 (1902), 437–479.

Jensen, B.E. (2003). *Historie – livsverden og fag*. Copenhagen: Gyldendal.

Jensen, B.E. (2010). *Hvad er historie* (in Danish). Copenhagen: Akademisk Forlag.

Katz, V.J. (2009). *A History of Mathematics. An Introduction*. Boston: Addison-Wesley.

Keller, E.F. (2002). *Making Sense of Life*. Cambridge Massachusetts: Harvard University Press.

Kitcher, P. (1984). *The Nature of Mathematical Knowledge*. Oxford: Oxford University Press.

Kjeldsen, T.H. (2009). Egg-forms and Measure Bodies: Different Mathematical Practices in the Early History of the Development of the Modern Theory of Convexity. *Science in Context,* 22(01), 85–113.

26 The Role of History and Philosophy in University Mathematics Education

Kjeldsen, T.H. (2010). History in mathematics education - why bother? Interdisciplinarity, mathematical competence and the learning of mathematics. In B. Sriraman, and V. Freiman (eds.): *Interdisciplinarity for the 21st Century: Proceedings of the 3rd International Symposium on Mathematics and its Connections to Arts and Sciences* (pp. 17–48). Information Age Publishing, incorporated.

Kjeldsen, T.H. (2011a). History in a competency based mathematics education: a means for the learning of differential equations. In Katz, V., & Tzanakis, C. (eds.). *Recent developments on introducing a historical dimension in mathematics education*, Washington, DC: Mathematical Association of America, 2011, 165–173.

Kjeldsen, T.H. (2011b). *Hvad er matematik?* Copenhagen: Akademisk forlag.

Kjeldsen, T.H. (2012). Uses of History for the Learning of and about Mathematics. Towards a theoretical framework for integrating history of mathematics in mathematics education. Plenary address. *Proceedings from the HPM International Congress* in Korea, 2012. http://www.hpm2012.org/Proceeding/Plenary/PL1.pdf

Kjeldsen, T.H. & Blomhøj, M. (2009). Integrating history and philosophy in mathematics education at university level through problem-oriented project work. *ZDM Mathematics Education, Zentralblatt für Didaktik der Mathematik*, 41(1–2), 2009, 87–103.

Kjeldsen, T.H. & Blomhøj, M. (2012). Beyond Motivation - History as a method for the learning of meta-discursive rules in mathematics. *Educational Studies in Mathematics*, 80 (3), 2012, 327–349.

Kjeldsen, T.H. & Blomhøj, M. (2013). Developing Students' Reflections about the Function and Status of Mathematical Modeling in Different Scientific Practices: History as a Provider of Cases. *Science & Education*, 22 (9), 2013, 2157–2171.

Kjeldsen, T.H. & Carter, J. (2012). The growth of mathematical knowledge - introduction of convex bodies. *Studies in History and Philosophy of Science Part A*, 43, 359–36.

Kjeldsen, T.H. & Petersen, P.H. (forthcoming). Bridging History of the Concept of a Function with Learning of Mathematics: Students' meta-discursive rules, concept formation and historical awareness. *Science & Education* doi: 10.1007/s11191-013-9641-2.

Knoebel, A., Laubenbacher, R., Lodder, J. & Pengelley, D. (2007). *Mathematical Masterpieces: Further Chronicles by the Explorers*. Springer.

Kragh Sørensen, H. (2012). Making philosophy of science relevant for science students. *RePoSS: Research Publications on Science Studies* 18. Aarhus: Centre for Science Studies, University of Aarhus, Jan. 2012. URL: http://www.ivs.au.dk/reposs

Lakatos, I. (1970). History of science and its rational reconstructions. PSA: Proceedings of the Biennial Meeting of the Philosophy of Science Association, Vol. 1970, 91–136.

Lakatos, I. (1976). *Proofs and Refutations*. Cambridge University Press.

Landry, E. & Marquis, J-P. (2005). Categories in context: Historical, Foundational and Philosophical. *Philos. Math.*, 13, 1–43.

Laubenbacher, R. & Pengelley, D. (1992). Great problems of mathematics: A course based on original sources. *American Mathematical Monthly*, 99, 313–317.

Laubenbacher, R., Pengelley, D. and Siddoway, M. (1994). Recovering motivation in mathematics: Teaching with original sources. *Undergraduate Mathematics Education Trends* 6, No. 4.

Laubenbacher, R., & Pengelley, D. (1999). *Mathematical Expeditions: Chronicles by the Explorers*. Springer.

Lerman (1990). Alternative Perspectives on the Nature of Mathematics and their Influence on the Teaching of Mathematics. *British Educational Research Journal* 16 (1), 53–61.

Lützen, J. & Purkert, W. (1989). Conflicting Tendencies in the Historiography of Mathematics: M. Cantor and H. G. Zeuthen. In D. E. Rowe, J. McCleary, and E. Knobloch (eds.): *The History of Modern Mathematics*. Vol. 3. Proceedings of the Symposium on the History of Modern Mathematics (pp. 1–42), Vassar College, Poughkeepsie, New York, June 20–24, 1989. 3 volumes. Academic Press.

Lützen, J. & Ramskov, K. (1999). *Kilder til matematikkens historie*. Department of Mathematics, Copenhagen University.

Maddy, P. (1990). *Realism in Mathematics*. Oxford: Oxford University Press.

Mancosu, P. (1998): *From Brouwer to Hilbert. The debate on the Foundations of mathematics in the 1920's*. New York: Oxford University Press.

Mancosu, P. (ed.) (2008). *The Philosophy of Mathematical Practice*. Oxford University Press.

McLarty, C. (2005). Mathematical Platonism' Versus Gathering the Dead: What Socrates teaches Glaucon. *Philosophia Mathematica* 13, 115–134.

Niss, M. (2001). University mathematics based on problem-oriented student projects: 25 years of experience with the Roskilde model. In D. Holton (Ed.), *The teaching and learning of mathematics at University level*: An ICMI study (pp. 153–165). Dordrecht: Kluwer.

Niss, M. (2004). The Danish "KOM" Project and possible consequences for teacher education. In Strässer, R., Brandell, G., Grevholm, B., & Helenius, O. (eds.). *Educating for the Future: Proceedings of an International Symposium on Mathematics Teacher Education: Preparation of Mathematics Teachers for the Future. Malmö University, Sweden, 5–7 May 2003* (pp. 179–190). Göteborg: The Royal Swedish Academy of Sciences.

Niss, M. & Højgaard, T. (2011). *Competencies and Mathematical Learning. Ideas and inspiration for the development of mathematics teaching and learning in Denmark*. IMFUFA-text 485. http://milne.ruc.dk/ImfufaTekster/

Otte, M. (2007). Mathematical history, philosophy and education. *Educational Studies of Mathematics*, 66, 243–255.

Poulsen, E. T. (2001). Matematikken og virkeligheden. In Niss (ed.) *Matematikken og Verden*. Fremads debatbøger — Videnskaben til debat (pp. 19–36). København: Fremad.

Pengelley, D. (2002). The Bridge between the continuous and the discrete via original sources. In O. Bekken et al. (eds.) *Study the Masters: The Abel-Fauvel Conference*. National Center for Mathematics Education, University of Gothenburg, Sweden.

Prediger, S. (2007). Philosophical reflections in mathematics classrooms. In K. François & JP. Van Bendegem (eds.) (2007), 43–59.

Restivo, S. (1993). The social life of mathematics. In Restivo et al. (eds.); *Math worlds. Philosophical and Social Studies of Mathematics and Mathematics Education*. Albany: State University of New York Press.

Rashevsky, N. (1934). Physico-mathematical aspects of cellular multiplication and development. *Cold Spring Harbor symposia on quantitative biology*, II (pp. 188–198). Long Island, New York: Cold Spring Harbor.

Resnik, M. (1999). *Mathematics as a Science of Patterns*. Oxford: Oxford University Press.

Rheinberger, H-J. (1997). *Towards a History of Epistemic Things: Synthesizing Proteins in the Test Tube*. Standford: Standford University Press.

Salling Olesen, H. & Højgaard Jensen, J. (1999). Can 'the university' be revived in 'late modernity'? In H. Salling Olesen and J. Højgaard Jensen (Eds.), *Project studies: A late modern university reform?* (pp. 9–24). Roskilde: Roskilde University Press.

Sannino (2009). *Universe's Bright and Dark Side*. http://cp3-origins.dk/content/uploads/2009/09/Bright-Dark-Article-060920091.pdf Accessed 1 July 2012.

Sfard, A. (2000). On reform movement and the limits of mathematical discourse. *Mathematical Thinking and Learning*, 2(3), 157–189.

Sfard, A. (2008). *Thinking as Communication*. Cambridge: Cambridge University Press.

Shapiro, S. (1997). *Philosophy of Mathematics. Structure and Ontology*. Oxford: Oxford University Press.

Shapiro, S. (2000). *Thinking about mathematics*. Oxford: Oxford University Press.

Skovmose, O. (2004). Critical mathematics education for the future. Regular lecture at ICME 10. http://www.icme10.dk/proceedings/pages/regular_pdf/RL_Ole_Skovsmose.pdf Accessed 1 July 2012.

Tappenden, J. (2006). The Riemannian Background to Frege's Philosophy. In Ferreirós & Gray (2008), 97–132.

Timmermann, S. & Uhre, E. (2001). *Generalizations in the theory of integration – an investigation of the Lebesgue integral, the Radon integral and the Perron integral*. (In Danish). IMFUFA, text 403, Roskilde University.

Thurston, W. (1994). On Proof and progress in mathematics. *Bulletin of the American Mathematical Society* 30, 161–177.

Toft, B. (2001). Matematik løser problemer. In: M. Niss (Ed.) *Matematikken og Verden*. Fremads debatbøger — Videnskaben til debat (pp. 158–179). København: Fremad.

Tymoczko, T. (ed.) (1985). *New Directions in the Philosophy of Mathematics*. Boston: Birkhäuser.

van Heijenoort, J. (1967): *From Frege to Gödel. A source book in mathematical logic 1879–1931*. Harvard University Press.

Van Kerkhove, B. & Van Bendegem, J.P. (2007). *Perspectives on mathematical practices. Bringing Together Philosophy of Mathematics, Sociology of Mathematics, and Mathematics Education*. Logic, Epistemology, and the Unity of Science Vol. 5, Springer-Verlag, Dordrecht.

Van Kerkhove, B., De Vuyst & Van Bendegem (2010). *Philosophical Perspectives on Mathematical Practices*. Texts in Philosophy 12, College Publications, London.

Wigner, E. (1960). The Unreasonable Effectiveness of Mathematics in the Natural Sciences. *Communications in Pure and Applied Mathematics* 13, 1–14.

Tinne Hoff Kjeldsen is Associate Professor at IMFUFA, Department of Science, Systems and Models at Roskilde University, Denmark. Kjeldsen has a master's degree in mathematics from Copenhagen University and a Ph.D. degree in history of mathematics from IMFUFA, Roskilde University, from 1999. Her historical research has helped to identify, define and clarify forces of internal mathematical and external sociological nature that have driven developments of pure and applied mathematics in the twentieth century. In particular, she has focused on the history of the moment problem, game theory, the development of the theory of linear and nonlinear programming in the post-war USA and the emergence and development of the theory of linear inequality systems. She is currently working on the history of the modern theory of convexity; see, e.g. 'Egg-forms and Measure Bodies: Different Mathematical Practices in the Early History of the Development of the Modern Theory of Convexity'. (*Science in Context*, 2009). Her research in didactics of mathematics focuses on roles of modelling and history for the teaching and learning of mathematics; see, e.g. 'Beyond Motivation - History as a method for the learning of meta-discursive rules in mathematics' (*Educational Studies in Mathematics*, 2012). She has been an invited lecturer at numerous international conferences in Europe, North and South America and Asia. Her research is published in books and major journals of history, philosophy and didactics of mathematics and science.

Jessica Carter is Associate Professor at the Department of Mathematics and Computer Science at University of Southern Denmark, Odense. Carter obtained her Ph.D. degree in 2002, financed by the Centre for Educational Development in University Science, based at the University of Southern Denmark. Before obtaining the position in Odense, she worked 2 years at the Danish University of Education in Copenhagen. Her main research interests are within the philosophy of mathematical practice. Her current work concerns the role of representations in mathematical reasoning. The recent paper, 'Diagrams and Proofs in Analysis' (International Studies in the Phil. Sci., 2010), shows roles that visual representations play in mathematics. Another interest is the topic of mathematical understanding. Carter is one of 9 founders of the Association for the Philosophy of Mathematical Practice (see http://institucional.us.es/apmp/) and is also the secretary of this association.

Chapter 27
On the Use of Primary Sources in the Teaching and Learning of Mathematics

Uffe Thomas Jankvist

27.1 Introduction

Why would one go to the extent of using a, possibly very old and maybe even somewhat inaccessible, primary original source[1] in the teaching and learning of mathematics, when so many contemporary textbooks are ready-made and pedagogically prepared for dealing with the same mathematical topics using modern-day language, coherent notation, etc.?

The above question is the overall one to be addressed in this chapter. The question is, of course, not a new one. It is a question which has already been addressed on several occasions and by various researchers of mathematics, the history of mathematics, and of course mathematics education, notably in the ICMI[2] Study on *History in Mathematics Education* (Fauvel and van Maanen 2000), where Jahnke, Arcavi, Barbin, Bekken, Furinghetti, El Idrissi, Silva da Silva, and Weeks, in the chapter on *The use of original sources in the mathematics classroom,* state:

> Among the various possible activities by which historical aspects might be integrated into the teaching of mathematics, the study of an original source is the most demanding and the most time consuming. In many cases a source requires a detailed and deep understanding of the time when it was written and of the general context of ideas; language becomes important in ways which are completely new compared with usual practices of mathematics teaching. Thus, reading a source is an especially ambitious enterprise, but [...] rewarding and substantially deepening the mathematical understanding. (Jahnke et al. 2000, p. 291)

---

[1] Throughout the chapter the phrases "primary original sources," "primary sources," and "original sources" are used interchangeably but always to mean the same.

[2] ICMI is *International Commission on Mathematical Instruction.*

U.T. Jankvist (✉)
Aarhus University, Campus Emdrup, Denmark
e-mail: utj@dpu.dk

M.R. Matthews (ed.), *International Handbook of Research in History, Philosophy and Science Teaching*, DOI 10.1007/978-94-007-7654-8_27,
© Springer Science+Business Media Dordrecht 2014

This chapter addresses some of the reasons usually given for resorting to primary original sources in the teaching and learning of mathematics as well as exemplifies some of the approaches to doing so. At the end of the chapter, the use of original sources as a potential way of dealing with certain problems and issues in mathematics education in general is discussed, i.e., problems not only related to the teaching and learning of mathematics itself. Some illustrative examples on the use of original sources, and (empirical) didactic findings in relation to such, will be given later in the chapter. But before we get to that point, a bit of background information seems in order.

27.2 The Background and Academic Forums of Using Primary Original Sources

As mentioned, the discussion of using primary original sources in the teaching and learning of mathematics is not a new one. Historically speaking, it was the French, whom with the Commission Inter-IREM[3] first dealt with the use of original sources in a more productive manner and developed a wide range of related activities (see, e.g., Jahnke et al. 2000).[4] In the English-speaking part of the world, the HIMED movement[5] in England, initiated by John Fauvel, made several contributions, and so did the IHMT[6] initiative in the USA, one result being the massive collective work *Historical Modules for the Teaching and Learning of Mathematics,* edited by Katz and Michalowicz (2004). Traditionally, the discussion of using original sources is one which is embedded in the larger discussion of generally using history of mathematics in mathematics education (cf. Chap. 21 by Michael Fried). For that reason, much of the literature on using original sources is to be found within the literature on using history.[7] Still to this date, the most comprehensive volume available on the topic is the aforementioned ICMI Study (Fauvel and van Maanen 2000), which provides an account of the literature available up until the year 2000 as well as extensive examples on the use of history and original sources. The purpose of this chapter is therefore not to repeat what has already been done and said in the ICMI

---

[3] IREM is *Institut de Recherche sur l'Ensignement des Mathématiques.* Inter-IREM here refers to the particular Inter-IREM Commission on history and epistemology of mathematics under the co-ordination and leadership of Evelyne Barbin.

[4] Of course, smaller-scale initiatives had been taken in other places. As an example, the pioneering work of Abraham Arcavi from Israel is mentioned, beginning with his Ph.D. (Arcavi 1985) and the subsequent extensive publications in several journals.

[5] HIMED is *History in Mathematics Education.*

[6] IHMT stands for *Institute in the History of Mathematics and its use in Teaching.*

[7] A few examples of collections published before year 2000 including discussions of the use of original sources are Calinger (1996), Katz (2000), Laubenbacher and Pengelley (1999), and Swetz et al. (1995).

27 On the Use of Primary Sources in the Teaching and Learning of Mathematics

Study and its chapter on original sources,[8] but instead to provide a slightly different perspective on the use of original sources and refer to some of the more recent developments on this topic.

As a brief overview of literature in general on the use of original sources after the year 2000, the following account is offered: In 2002, a conference was held in Kristiansand as a tribute to John Fauvel who passed away one year earlier and also as a celebration of the bicentennial year of the birth of Niels Henrik Abel – *the Abel-Fauvel Conference* – the proceedings of which was published in the book *Study the Masters,* edited by Bekken and Mosvold (2003). Besides offering various contributions related to the history of mathematics, this book also offers some papers on the use of original sources in the classroom.[9] The Abel-Fauvel Conference was a Nordic preconference to ICME[10] 10 and the HPM[11] satellite conference of 2004. In the proceedings of HPM2004, which were edited by Furinghetti et al. (2007), there are also some contributions of interest, not least the report from a panel discussion coordinated by Barbin on original sources in the classroom. The HPM2004 was a joint meeting with the ESU4,[12] since their quadrennial and triennial conferences coincided in 2004. The proceedings from ESU5 edited by Barbin et al. (2008) also offer various input on the debate of original sources and do so in relation to various educational levels and teacher training.[13] The same goes for the proceedings from ESU6, edited by Barbin et al. (2011), from which special attention should be given to the plenary talk by Glaubitz (2011), which empirically compares some of the approaches for using original sources to be discussed later in this chapter. In fact, this plenary is the first entry in an entire section devoted to the use of original sources in the classroom and their educational effects.[14] At ICME there is a Topic Study Group (TSG) on *The role of history of mathematics in mathematics education.* Another recent initiative in relation to history, and therefore also original

---

[8] It should be mentioned that Chap. 9 in the ICMI Study is not the only one which addresses elements of using original sources, so do also Chap. 5 (in particular Sect. 5.3 in this), Chap. 7 (e.g., Sect. 7.3 which addresses matters of the genetic approach), and Chap. 8 in which many of the outlined examples rely on original sources.

[9] For example, those of Furinghetti and Somaglia, Horng, Pengelley, and Siegmund-Schultze.

[10] ICME stands for *International Congress on Mathematical Education,* the quadrennial world congress of ICMI, which took place in Copenhagen in 2004, and since in Monterrey, Mexico, in 2008 and in Seoul in 2012.

[11] HPM stands for ICMI-affiliated *International Study Group on the Relations between the History and Pedagogy of Mathematics,* which has its quadrennial international meeting at the time of ICME but also has yearly local meetings as, for example, those of the *HPM Americas.* The HPM satellite conference was in Uppsala in 2004 and since then in Mexico City in 2008 and in Daejeon in 2012.

[12] ESU stands for *European Summer University on History and Epistemology in Mathematics Education,* which is an initiative of the French IREM. ESU5 was in Prague in 2007; ESU6 is Vienna in 2010; and ESU7 is planned for Barcelona in 2014 and will now be held every fourth year so that it no longer coincides with HPM satellite meetings.

[13] Some examples are the contributions by Bastos and Veloso, Fried and Bernard, Glaubitz, Guichard, Katz, Poulos, Thomaidis and Tzanakis, and Weeks.

[14] Other contributions related to the use of original sources also occur outside this section of course, e.g., those by Kjeldsen, Rosas and Pardo, and others.

sources, is the working group on *History in Mathematics Education* at CERME.[15] The volume recently published in MAA Notes on *Recent developments on introducing a historical dimension in mathematics education*, edited by Katz and Tzanakis (2011), includes papers originally presented at the ICME11 TSG on history, HPM2008, and CERME6.[16] *From Calculus to Computers: Using the Last 200 Years of Mathematics History in the Classroom*, edited by Shell-Gellasch and Jardine (2005),[17] is yet a book which was published in the MAA Notes after the ICMI Study from 2000. Another stand-alone publication which is clearly of relevance for the use of original sources in the teaching and learning of mathematics is the collection of projects relying on primary historical sources by Knoebel et al. (2007). In 2007 a special issue of *Educational Studies in Mathematics* entitled "The history of mathematics in mathematics education: theory and practice" was edited by Furinghetti, Radford, and Katz. This issue includes several articles relevant for the topic of using original sources.[18]

In relation to the use of original sources, the Oberwolfach meeting organized by Furinghetti et al. (2006) must also be mentioned. One of the outcomes of this meeting was the formulation of a set of questions to guide future research on the use of original sources:

1. What are the possible epistemological/theoretical basis and frameworks for research and development towards the integration of original sources into the teaching and learning of mathematics?
2. What are the characteristics of viable models for implementing the integration of original sources in the teaching and learning of mathematics?
3. What is the actual impact of these models on students' and teachers' learning and understanding of mathematics, and on teachers' teaching practices?
4. How can historical research and practice inspire, impact, support or supply explanatory frameworks and working tools for research on learning and teaching mathematics?
5. How can research and practice in mathematics education inspire, support and broaden the research in the history of mathematics in general, and on original sources in particular? (Furinghetti et al. 2006, p. 1287)

As the last thing, the recent conference in Paris in honor of Michèle Artigue – *Colloque Artigue 2012* – is mentioned, where a workshop on epistemology – *Atelier 5: epistémologie et didactique* – was organized by de Hosson, Chorlay, and

---

[15] CERME is *Congress of European Research in Mathematics Education*, which is held every second year in between HPMs and ESUs. The WG on history was at CERME6 in Lyon, 2009; CERME7 in Rzeszów, 2011; and CERME8 in Antalya, 2013.

[16] The most important chapters in this volume in relation to original sources are those by Barbin (Chap. 2), Kjeldsen (Chap. 15), and Pengelley and colleagues (Chaps. 1 and 17); but other chapters are clearly also relevant.

[17] Examples of contributions discussing the use of original sources in this collection are Atzema and White, D'Antonio, Pengelley, and Rogers.

[18] In particular the articles by Arcavi and Isoda (2007), Barbin (2007), Katz (2007), and Thomaidis and Tzanakis (2007)

Jankvist. The first session of this workshop was devoted to young researchers accounting for the epistemological (and historical) dimension in their doctoral theses.[19] Often, it is the case that French doctoral students begin their doctoral work with an epistemological/historical account of the mathematical concepts which their thesis addresses, and in this work the study of original sources play an essential part, for example, in identifying epistemological obstacles (e.g., Bachelard 1938; Brousseau 1997). To end this section where it began, namely, with the French Inter-IREM Commission on history and epistemology of mathematics, Evelyne Barbin, who was invited to speak in the second session of this workshop, pointed out that in the IREM context around 1994 epistemology also functioned as a "weapon" against the widespread view of mathematics as a language, since epistemology puts emphasis on mathematics as an activity.[20]

It is now time to introduce some general constructs developed in the setting of history in mathematics education – constructs, which will prove themselves useful in our further discussion of primary original sources.

27.3 A Distinction Between In-Issues and Meta-issues of Mathematics

In the context of using history – and thus also original sources – in mathematics education, we may distinguish between the inner issues of mathematics and the metaperspective issues (Jankvist 2009a). When history is used as a *tool* for teaching and learning of mathematics,[21] the primary focus is on the inner issues – or *in-issues* – of mathematics such as mathematical ideas, concepts, theories, methods, algorithms, and ways of argumentation and proof. When history is used more in terms of a *goal*,[22] the focus is on the metaperspective issues – or *meta-issues* – of mathematics as a scientific discipline, regarding it as a goal to show the students something about how mathematics has come into being, its historical development, human and cultural aspects of this development, its interplay

---

[19] The presenters were Thomas Barrier, Patricia Crépin, Mathias Front, Eric Laguerre, and Caroline Poisard. Furthermore, posters on the role of epistemology and history, also in relation to original sources, were presented.

[20] Currently, *Science & Education* is preparing a special issue on "History and Philosophy of Mathematics in Mathematics Education," guest edited by Victor J. Katz, Uffe Thomas Jankvist, Michael N. Fried, and Stuart Rowlands. This issue will include new articles discussing the use of primary original sources in mathematics education.

[21] The history-as-a-tool arguments may be subcategorized into being concerned with history as a motivational and/or affective tool, history as a cognitive tool (e.g., the idea of epistemological obstacles), and the role of history in what may be referred to as the evolutionary arguments (the recapitulation argument or historical parallelism) (Jankvist 2009a).

[22] Note that this is not the same as teaching the history of mathematics *per se*.

with society through applications, or issues of epistemology, ontology, etc.[23] For example, learning about the number sets (N, Z, Q, R, C), their interrelations, their cardinalities, etc., is considered to be a study of in-issues. On the other hand, learning about the historical development of the different kinds of numbers, the difficulties regarding the acceptance of the irrational numbers, the negative numbers, or the complex numbers concerns aspects of the meta-issues of mathematics.

Because the distinction between in-issues and meta-issues applies not only to the use of history in mathematics education but also to the use of philosophy and actual applications of mathematics (Jankvist 2013), it will be relevant for the use of original sources in any of these contexts as well. As further examples of various meta-issues to be addressed from the point of view of history, application, and philosophy of mathematics in mathematics education, we may consider:

> How does mathematics evolve over time? What forces and mechanisms can be present in the evolution? Do society and cultural circumstances play a part in this evolution? If so, how? And does mathematics then depend on culture and society, place and time? Is old mathematics also obsolete mathematics? (Niss 2001a, p. 10, my translation from Danish)

> Who, outside mathematics itself, actually uses it for anything? What for? Why? How? By what means? On what conditions? With what consequences? What is required to be able to use it? Etc. (Niss and Højgaard 2011, p. 74)

> What is characteristic of mathematical problem formulation, thought, and methods? What types of results are produced and what are they used for? What science-philosophical status does its concepts and results have? How is mathematics constructed? What is its connection to other disciplines? In what ways does it distinguish itself scientifically from other disciplines? Etc. (Niss and Højgaard 2011, pp. 75–76)

The distinction between in-issues and meta-issues bears some resemblance to that of Davis and Hersh (1981), who talk about the "inner issues" and the "outer issues" of mathematics, or that of Niss (2001b), who talks about "knowledge of mathematics from the inside" and "knowledge of mathematics from the outside."

Even though a use of history (or application or philosophy) as a tool is concerned with students' learning of mathematical in-issues, this is of course not to say that meta-issues cannot act as a means (tool), for example, by motivating students, to reach an end (goal) of learning specific in-issues. In a similar manner, when concerned with history (or application or philosophy) as a goal, it is clear that certain meta-issues cannot be understood without comprehension of some mathematical in-issues (see Jankvist 2011 for examples). The distinction between in-issues and meta-issues is merely a useful way to talk about the primary foci of including history (or application or philosophy) and thus also of resorting to primary original sources when doing so.

---

[23] For a further discussion of history as a tool and history as a goal, see also Tzanakis and Thomaidis (2012).

27.4 Various Reasons for Using Primary Original Sources

In this section we take a look at some of the reasons usually given for using original sources in the teaching and learning of mathematics, relying on the distinction between in-issues and meta-issues. To illustrate how this notion may be applied to uncover some of the underlying motives for resorting to original sources, let us have a look at the following list of reasons provided by Pengelley, one of the pioneers of using original sources at US undergraduate level:

> [1] … motivation and deep connection along time, [2] understanding essence, origin, and discovery, [3] mathematics as humanistic endeavor; [4] practice moving from verbal to modern mathematical descriptions; [5] reflections on present day status and paradigms; [6] participating in the process of doing mathematics through experiment, conjecture, proof, generalization, publication and discussion; [7] more profound technical comprehension from initial simplicity; [8] also dépaysement (disorientation, cognitive dissonance, multiple points of view); [9] and a question-based curriculum that knows where it came from and where it might be going. Questions before answers, not answers before questions that have not been asked. (Pengelley 2011, p. 3, numbering not in original)

Clearly, reason 9 refers to the commonly seen structure of modern mathematics textbooks, where definitions and theorems are usually presented first, and then, motivation and application follow afterwards. As the reader will know, the historical development of mathematical theories, concepts, etc. often is the exact opposite; one is motivated by a problem or an application, the solution of which leads to theorems, proofs, and in the end definitions.

Pengelley's other eight reasons for using primary original sources may be ordered according to their focus being either on the in-issues or the meta-issues of mathematics. Reason 3 about "mathematics as a humanistic endeavor" and the part of reason 2 which reads "understanding origin and discovery" are mainly concerned with the meta-issues. Reasons 4, 6, and 7 on "practice moving from verbal to modern mathematical descriptions" and "participating in the process of doing mathematics…" and "more profound technical comprehension…" concern in-issues in one way or another. Of course, these reasons or arguments are subject to interpretation. This is the case for the part of reason 2 which reads "understanding essence"; does this refer to in-issue mathematical essence of a given concept or topic or does it, for example, refer to the meta-issue applicational essence of the concept in an extra-mathematical context? And similarly, does reason 5 about "reflections on present day status and paradigms" refer to, for example, inner mathematical concerns of unsolved problems, unproven conjectures, etc. or to aspects of a more metaperspective nature regarding, for example, the science-philosophical status of a given concept or changes in the notion of mathematical proof? In order to find out, one would have to scrutinize the concrete use of an original source in a specific educational setting. Something similar goes for reason number 8 on *dépaysement*.

In fact, the idea of *dépaysement* is one which was originally suggested by Barbin (1997) and later translated and rephrased in Jahnke et al. (2000), as one of three general ideas for describing the special effects of using original sources:

- *Replacement* (in the original French wording, *vicariante*), which refers to the replacement of the usual with something different, for example, by allowing mathematics to be seen as more than just a corpus of knowledge and techniques.
- *Reorientation* (a translation of *dépaysement*), which challenges one's perception by making the familiar unfamiliar, eventually causing a reorientation of the reader's views and thus a deepening of the mathematical understanding – also sources remind students that mathematical constructs have come into being at one point in time (and space) and that this did not happen by itself.
- *Cultural understanding* (*culturel*), which allows us to place the development of mathematics in a scientific, technological, or societal context of a given time and place and in the history of ideas and society.

As we shall see in the given examples of the following sections, the idea of *dépaysement* is indeed a central one regarding the benefits of using original sources. As the reader may already have noticed, there is some variation in the interpretations of Jahnke et al. (2000) and Pengelley (2011). While Jahnke et al. talk about a *re*orientation and making the familiar unfamiliar, Pengelley refers to a *dis*orientation and mentions cognitive dissonance and multiple points of views. One reason for this variation may be that the idea of *dépaysement* involves a *process*. The first element is of course that of putting the student on unfamiliar ground, which is done by exposing him or her to an original source. This can indeed involve making the familiar unfamiliar. An example of this might be that of the notion of limits in real analysis: A student, who is used to the modern-day common notion, relying on Weierstrass' ε-δ-approach, may find that this becomes very unfamiliar, when seeing the earlier ideas of infinitesimals as discussed by, for example, Leibniz in his version of calculus. According to Furinghetti, Jahnke, and van Maanen:

> There are many experiences which show that students are motivated to reflect about the limit approach to calculus when they study Leibniz' way of dealing with infinitely small quantities. Also the teacher may gain insight by concentrating on the unfamiliar. It is often difficult enough to cope with unexpected solutions by students; however, studying sources enables to the teacher and students to keep an open mind. (Furinghetti et al. 2006, p. 1286)

Indeed, such a study may cause a cognitive dissonance, because the student is suddenly exposed to two different mathematical discourses. Sfard (2008) refers to such a situation as a *commognitive conflict*[24] – a situation where different discursants are acting according to different *meta-discursive rules* (not to be confused with the previously introduced meta-issues of mathematics) of mathematical discourse, which are historically given rules about proper communicative actions shaping the mathematical discourse, rules governing when to do what and how to do

---

[24] The word commognitive relates to *commognition*, which is a contraction of *communication* and *cognition*, and which means to stress the fact that communication and cognition are different intrapersonal and interpersonal manifestations of the same phenomenon (Sfard 2008).

27 On the Use of Primary Sources in the Teaching and Learning of Mathematics

it, and which are implicitly present in discursive actions of the mathematists.[25] Surely, such experience of cognitive dissonance, or an encounter with a commognitive conflict, may cause disorientation on the students' behalf, which should be seen as a good thing, since disorientation often is a precursor for reorientation. And of course, such reorientation may involve the development of multiple points of view on a mathematical topic, as, for example, that of the notion of limit in real analysis.

The example of the notion of limit above may of course hold in it the dimension of replacement also, since the student's idea is replaced by a different one.[26] As for cultural understanding, however, more would be needed in order to develop a student's perception of it in relation to the notion of limits.[27] For example, one might focus on the relevance of extra-mathematical phenomena for the development of limits, applicational uses, or the scientific importance of the notion. Thus, while both replacement and *dépaysement* seem to hold in them a natural focus on mathematical in-issues, cultural understanding appears to focus more specifically on the meta-issues. However, in a setting of using original sources, a true understanding of the involved meta-issues may often not be reached if the involved mathematical in-issues are not understood to some degree. For instance, if one wishes to make sense of why it was Weierstrass' ε-δ-approach that eventually became the standardized one in real analysis and not the rival one of infinitesimals (except in non-standard analysis), one will need to understand the two different notions. Another way of phrasing this is to say that the meta-issue context is rooted or *anchored* in the in-issue context (Jankvist 2011). Of course, the converse may also be true that the in-issues of an original source can be rooted in the meta-issue context surrounding it.

The various reasons discussed above do, of course, *not* make up a complete list of every possible reason one might think of in relation to the use of original sources in the teaching and learning of mathematics. Still, it illustrates that the reasons for doing so are not only multiple but also multifaceted. And the analysis of the discussed reasons illustrates that they may be viewed as focusing mainly on aspects of meta-issues or of in-issues. Regarding the in-issues, we shall later see that these may also be discussed in terms of the use of original sources contributing to the development of students' mathematical competencies. Furthermore, by relating to the three general ideas of Barbin (replacement, *dépaysement*, cultural

---

[25] A *mathematist* is a participant in mathematical discourse including students, teachers, and mathematicians (Sfard 2008).

[26] Furinghetti et al. (2006) give the example of Newton's letter to Leibniz of 1676 in which he describes how as a young man of 22 years; he arrives at the general binomial formula (a cornerstone in his fluxional calculus) as an example of possible replacement, since it is a unique document for a process of mathematical innovation progressing by bold generalizations and analogies.

[27] In relation to cultural understanding, Furinghetti et al. (2006, p. 1286) mention Heron's textbook (first century A.D.) on land surveying called *The Dioptra*. Reading parts of this, they say, "connects the topic of similarity to the context of ancient surveying techniques and shows the astonishingly high achievements of ancient engineers in this and other areas. Such sources may as well provoke students to engage in practical activities (simulations, measurements, theatre), which otherwise would not come to their mind or to the mind of their teacher."

understanding), we are able to conduct a more detailed analysis of the various reasons – or purposes – of using original sources, by placing a given use within one of the three general ideas, and then try to sort out how much focus must be (or was) put on in-issues and meta-issues, respectively.

But before we get to some illustrative examples of actual uses of original sources with students, we should look at some of the approaches to using original sources.

27.5 Various Approaches to Using Primary Original Sources

In this section, we shall have a look at some of the approaches usually discussed in the literature for using original sources in the teaching and learning of mathematics. Surely a different selection could have been made, but the one below offers both longtime well-known and more recently developed approaches, the latter of which also relates to later examples. It is important to mention that there may be overlaps between the approaches given and that some of them may resemble "philosophies" about the inclusion of original sources (or history), while others may be seen more as methodologies, methods, or models, e.g., for addressing practical issues associated with using original sources or for reaching specific goals by using such sources. Nevertheless, since all usually are referred to as *approaches* in the literature, the same is done here, although the reader shall be notified which approaches are more heavy on philosophy and which on methodology, based on the given descriptions.

Before we get to the descriptions, a certain notion from historiography must be introduced, namely, that of *Whig* history. This notion is due to the British historian Herbert Butterfield who defined it as a way of "measuring" the past in terms of the present (Butterfield 1931/1951). A well-known example hereof is that of the Bourbaki approach to the history of mathematics, in particular with Jean Dieudonné and André Weil. In a discussion related to why and how history of mathematics should be conducted, Weil, for instance, claimed that

> ... it is impossible for us to analyze properly the contents of Book V and VII of Euclid's without the concept of group and even that of groups with operators, since the ratios of magnitudes are treated as a multiplicative group operating on the additive group of the magnitudes themselves. (Weil 1978, p. 232)

The problem partly consists in that one comes to follow a path from the present back to the past, where the only things which are considered are those which lead to something deemed significant today, whereas "from the perspective of the past, and that is the *historical* perspective, it is a zigzag path of a wanderer who does not know where exactly he is going" (Fried 2001, p. 396, italics in original). Thus, if we, in the context of introducing a historical dimension in mathematics education, wish to be honest to history, i.e., adopt a genuine historical perspective, we may be much better off resorting to primary original sources, since these are open to the reader's own interpretation – an interpretation which in its outset is not Whig. Also, the notion of Whig history may be used as a sort of measure of various approaches for using original sources; in the sense that while several in-issues might very well be brought forth in a Whig approach to history, the reaching of meta-issues often requires in its outset a non-Whig approach.

27.5.1 The Genetic Approach

One of the oldest approaches to including history through a use of original sources is the so-called genetic approach, which was strongly advocated by the German mathematician Felix Klein (Schubring 2011), for example, in his *Elementarmathematik vom höheren Standpunkte aus*,[28] where he argued for a genetic method of teaching wherever possible (Klein 1908). A more theoretically reflected conception of the genetic approach was formulated by another German mathematician, Otto Toeplitz. According to Burn (1999, p. 8), "the question which Toeplitz was addressing was the question of how to remain rigorous in one's mathematical exposition and teaching structure while at the same time unpacking a deductive presentation far enough to let a learner meet the ideas in a developmental sequence and not just a logical sequence." Worth noting is that Toeplitz distinguished between a *direct* genetic method and an *indirect* genetic method:

> [A]ll these requisites [...] must at some time have been objects of a thrilling investigation, an agitating act, in fact at the time when they were created. If one were to go back to the roots of the concepts the dust of times [...] would fall from them and they would again appear to us as living creatures. And from then on there would have been offered a double road into practice: Either one could directly present the students with the discovery in all of its drama and in this way let the problems, the questions, and the facts rise in front of their eyes – and this I shall call the *direct genetic method* – or one could by oneself learn such an historical analysis, what the actual meaning and the real core in every concept is, and from there be able to draw conclusions for the teaching of this concept which as such is no longer related to history – the *indirect genetic method*. (Toeplitz 1927, pp. 92–93, my translation from German)

In relation to the use of original sources, it is of course the direct genetic approach which is of interest. Tzanakis (2000) describes the genetic method as one in which there is no uniquely specified way of presentation in the sense of an algorithm; rather, it is a general attitude towards the presentation of a scientific subject. In that way, the motivation behind introducing new concepts, theories, or key ideas of proofs is based on their genesis and evolution. Toeplitz points out that nothing is further from his thoughts than giving a lecture on history. Instead, he wants to "...pick from history only the basics of those things that have stood the test of time and make use of them" (Toeplitz 1927, p. 95). Thus, the genetic method is not one concerned with getting messages related to meta-issues across to the students, but instead it is using history as a tool – and therefore also original sources – to teach and have the students learn mathematical in-issues.

---

[28] Usually translated to: *Elementary Mathematics from an Advanced Standpoint.*

However, as pointed out by Glaubitz (2011), Toeplitz' approach implicitly assumes that there is a "great ascending line" in history to follow, which there often is not, in which case "important stages" will have to be identified instead. Although it may be possible to identify such important stages – or key points or crucial steps, which acted as a catalyst for further progress in the historical development[29] – as in any situation where an interpretation is made of what is historically important and what is not, Whiggishness may be lurking, since such interpretation is often made from a contemporary point of view. Of course, when using primary original sources in the teaching and learning, this ought to be less likely to happen.

27.5.2 The Hermeneutic Approach

Somewhat in contrast to the genetic approach is the so-called hermeneutic approach,[30] as proposed by Hans Niels Jahnke (also in Jahnke et al. 2000) – an approach which concerns itself with meta-issues as well as in-issues of mathematics. Glaubitz describes the difference between the hermeneutic and the genetic approach as follows:

> … the big difference to the genetic approach is, that the students are not expected to trace a history of thoughts that leads them from past roots to the standards of today. Essentially that is, because the students should be very familiar with a topic before they even touch a historical text that deals with it. In the hermeneutic approach, students are asked to examine a source in close detail and explore its various contexts of historical, religious, scientific etc. nature. In hermeneutics this is called: to move within hermeneutic circles – and these are circles of ever new understanding. (Glaubitz 2011, p. 357)

Jahnke relates the idea about hermeneutic circles to original sources in the following manner:

> The process of interpreting an original mathematical source may be described by a twofold circle [...] where in the primary circle a scientist (or a group of scientists) is acting and in a secondary circle the modern reader tries to understand what is going on. (Jahnke 2000, p. 298)

Following Schubring (2005), the key idea is to approach the intended meaning of the original source, this being in the primary circle, and for the modern-day reader to perform hermeneutical reconstructions of how concepts (and other mathematical in-issues) developed. It is worth noticing that being this alert of the shifting within circles makes the reader's interpretation much less prone to becoming Whig.

Due to the fact that within the hermeneutic approach, students need a prerequisite knowledge of the topic, Glaubitz explains that the approach is often applied

---

[29] See in particular Sect. 7.3 of Fauvel and van Maanen (2000) and Tzanakis and Thomaidis (2000).

[30] The hermeneutic approach is sometimes also referred to as the historico-hermeneutic approach.

only to subject matters with which the students are already familiar. He further describes the hermeneutic approach as essentially consisting of three steps:

> Usually the structure is as follows: First, the students have a quite conventional introduction to the topic. No history is involved until the second step, in which the students read a historical source. In this source, the same topic is covered, but in a way, that is historically distant, different in its representation, used in strange contexts and so forth. This is the step where the students' epistemic curiosity is – hopefully – aroused. In the third and final step students are required to explore the source in even greater detail, perform a horizon merger and reflect upon questions that occur to them. (Glaubitz 2011, p. 358)

The term "horizon merger" refers to a student's development of deeper awareness by wondering and reflecting about what she/he never thought about before (Gadamer 1990). Thus, in relation to the previously discussed reasons for including primary original sources, we notice the presence of the arguments of replacement and *dépaysement* in the hermeneutic approach.

27.5.3 Multiple-Perspective Approach

Recently, Kjeldsen has been arguing strongly for what she refers to as a *multiple-perspective approach* to history, including also the use of primary original sources in the teaching and learning of mathematics (e.g., Kjeldsen 2009a, b and 2011a, b, c; Kjeldsen and Blomhøj 2012; Jankvist and Kjeldsen 2011). Contrary to the hermeneutic approach, students do not necessarily need to possess prerequisite knowledge of the modern interpretation of the topic they are about to study through original sources. One similarity between the two, though, is that they both approach original sources from a somewhat micro-historical point of view – or *views* – since Kjeldsen by multiple perspectives is referring to the analysis and consideration of the practice of mathematical activities at a given time and place from several points of observation or contexts:

> The perspectives can be of different kinds and the mathematics can be considered from different angles, such as various sub-disciplines of mathematics, techniques of proofs, applications, "nature of mathematics" positions, other scientific disciplines, sociological institutions, personal networks, genders, religious beliefs and so on. This approach to history raises the question of which perspectives to choose, since, of course, not every perspective that one can think of is necessarily interesting (or accessible) regarding a particular historical analysis. A way to handle this difficulty when using history in mathematics education is to adopt a problem-oriented approach, that is, to have clearly formulated historical research questions and then to focus on perspectives and to choose historical episodes that explicitly address and relate to the issues one wants students to reflect upon. (Kjeldsen and Blomhøj 2012, p. 332)

Examples of such problem-oriented "research questions" to focus on while reading and studying primary original sources could be

> … why mathematicians asked the questions they did; why they treated the problems they dealt with in the way they did; what kinds of proofs or arguments they gave, and how these were perceived and received within the mathematical community(ies); why they introduced

certain mathematical objects, definitions, areas of research etc.; and how all of the above influenced further developments in mathematics as well as changes in perceptions of mathematics. (Jankvist and Kjeldsen 2011, p. 836)

One historical framework which is particularly interesting in relation to the multiple-perspective approach is that of *epistemic objects and epistemic techniques*, originally developed by Rheinberger (1997) for the study of experimental sciences but adapted by Epple (2004) to the historiography of mathematics. Epistemic objects refer to the mathematical object(s) under investigation by the mathematician(s) in a given time and space. Epistemic techniques refer to the methods and tools that were used to study the object(s), either well-established techniques or techniques which have to be developed in the process of studying the object(s). Together, the epistemic objects and techniques constitute what Epple calls the *epistemic configuration*. This is where the (groups of) mathematicians (pure or applied), within a given space and time, perform their work – their "intellectual working place" or "mathematical workshop." It is important to notice the dynamical process inherent in the notions of epistemic objects and techniques, meaning that what in one historical setting functioned as objects might be turned into techniques later or in another setting (e.g., Epple 2004; Kjeldsen 2009b). Besides making up a non-Whig approach to the historiography of mathematics, the multiple-perspective approach is also extremely suitable for focusing on both mathematical in-issues and meta-issues when using primary original sources in mathematics education, e.g., in Epple's framework, by considering the epistemic objects and techniques themselves and then considering (cultural or societal) aspects surrounding the epistemic configuration, respectively. (The notion of objects and techniques shall be illustrated in the section on examples.)

27.5.4 Comparative Readings of Original Sources and Other (Kinds of) Texts

Whereas the three approaches discussed above are more general approaches – or "philosophies" – to the use and studying of original sources in mathematics education, the following ones are more specific in nature and more methodological. The first one concerns comparisons of different kinds of texts, which Jahnke et al. (2000) discuss in terms of the benefits of comparing original sources with secondary sources, for example, books on the history of mathematics. In contrast to merely relying on such secondary sources, the study of primary original sources helps to:

(a) Clarify and extend what is found in secondary material,
(b) Uncover what is not usually found there,
(c) Discern general trends in the history of a topic (secondary sources are usually all-topic chronological accounts, and some topics are very briefly treated or omitted altogether), and
(d) Put in perspective some of the interpretations, value judgments or even misrepresentations found in the literature. (Jahnke et al. 2000, p. 293)

As may be derived from the points above, a *comparative reading* of primary original sources with secondary literature may be quite rewarding for students (and teachers as well).

Comparison of an original source with a (modern) textbook's presentation of the topic in the original source may also be a sensible task to undertake. In relation to the previously mentioned *dépaysement*, Barnett, et al. (2011) state that it may

> Engender cognitive dissonance (dépaysement) when comparing a historical source with a modern textbook approach, which to resolve requires an understanding of both the underlying concepts and use of present-day notation. (Barnett et al. 2011, p. 188)

A more compelling reason for resorting to primary original sources over textbooks – and some secondary literature, depending on its nature – is the one pointed to by Fried (2001), namely, that textbooks by definition are *closed to interpretation*, since they merely set out to present already accepted knowledge. Primary original sources, on the other hand, which convey their own order of inquiry, are open to the reader's own reflections in a very different sense, through the author's original language of discovery and invention.

A comparative reading of primary sources with other primary sources may enhance the effect of these in relation to the originally intended purposes of resorting to primary original sources. If, for example, the intention was one of the meta-issues regarding the development of mathematics (and science), a comparison of two original sources addressing the same problem, and possibly developing the same or similar theoretical constructs, could lead to a discussion of *multiple discoveries* (e.g., Merton 1973). One example of that is the well-known one of Newton's and Liebniz' simultaneous, but very different, developments of the infinitesimal calculus. Another example of comparing different original sources might be when comparing various translations of the same source to observe how cultural aspects have influenced such translations. An example of this may be found in Siu (2011), who discusses the translation of the first European text in mathematics (Euclid's *Elements*) into Chinese, by comparing the different ways of mathematical thinking, the different style of presentation, and the different views of mathematics in the East and in the West.

27.5.5 Guided Readings of Primary Historical Sources

The second of the methodological approaches is the so-called *guided readings* of original sources developed by David Pengelley, Janet Barnett, Jerry Lodder, and colleagues (see, e.g., Barnett et al. 2011; Knoebel et al. 2007; Laubenbacher and Pengelley 1999).[31] This method offers a sensible way of dealing with the occasional inaccessibility of primary original sources. The idea is to supply or "interrupt" the students' reading of an original source by explanatory comments and illustrative tasks along the way. The group surrounding Pengelley, Barnett, Lodder, and colleagues

---

[31] See also http://www.cs.nmsu.edu/historical-projects/papers.html (Retrieved on February 1, 2012).

has developed numerous teaching modules (or projects), ready for implementation in class (undergraduate level or high school). And one of the beautiful things about their work is that they have an "open-source" policy, providing teachers and other researchers with free access to the LaTeX code of their modules, and invite them to adapt it for their own use and to share their experiences.[32] (Examples of sources which have been subject to such guided readings will be given in the following section.)

27.5.6 Essay Assignments Related to Readings of Original Sources

While the approach of guided readings originally was developed to make the mathematical in-issues of the original source more accessible and understandable to the students, the third methodological approach to be discussed here, that of *essay assignments*, was developed from the point of view of bringing out meta-issues. In a study I carried out in 2007 (and which is reported in Jankvist 2009b, 2010, 2011), focus was on creating a setting where upper secondary school students could come to discuss and reflect upon historical, philosophical, and/or applicational meta-issues of the mathematics they had studied through the use of excerpts from original sources. This was realized by having groups of students write essays (a couple of pages long) answering a series of questions on meta-issues as part of their mathematics class. For example, in a historical teaching module on the development of public-key cryptography (Diffie, Helman, and Merkle; Rivest, Shamir, and Adleman) and the underlying number of theoretical constructs (e.g., stemming to Sun Zi; Fermat, and Euler), students were asked to discuss these researchers' motivations for engaging into their studies of topics from the point of view of *inner driving forces* in the discipline of mathematics and *outer driving forces*, i.e., influence on the development of mathematics from the outside due to societal needs (Jankvist 2011). Thus, this study also adopted the multiple-perspective approach to history of mathematics (for further discussion in relation to this particular historical case, see Jankvist and Kjeldsen 2011). Also, within the setting of the multiple-perspective approach, I have used a design combining the idea of guided readings of primary original sources and the use of essay assignments (Jankvist 2013), which will be illustrated in the following section.

27.6 Illustrative Examples of Didactic Design and Findings from Using Primary Original Sources

To illustrate actual uses of original sources in the teaching and learning of mathematics, a small selection of more recent examples from the literature will be described and discussed. These examples have been chosen to illustrate some of the

---

[32] Links are http://www.math.nmsu.edu/hist_projects/ and http://www.cs.nmsu.edu/historical-projects/ (Retrieved on February 1, 2012).

different uses in terms of purposes and approaches (cf. previous sections), whereas not much emphasis has been put on choosing examples from many different national settings, since interested readers may already find a cohort of such examples displayed in the ICMI Study (Fauvel and van Maanen 2000).[33] Also, since the ICMI Study offers a long list of examples regarding uses of original sources to support understanding of various important mathematical concepts from the perspective of students as well as teachers, the following examples were not chosen with this in mind. Instead, focus has been on the illustration of didactic designs and didactic findings related to the use of original sources in the teaching and learning of mathematics. However, some effort has been made to illustrate the use of original sources at different educational levels: primary, secondary, and tertiary level. Furthermore, an important element in some of the chosen examples is the adherence to a theoretical framework from mathematics education research – an aspect to be discussed in the final section of this chapter.

27.6.1 An Interdisciplinary Reading of Bartolus of Saxoferrato in Dutch Grammar School

On several occasions, van Maanen has described his use of the history of mathematics through primary original sources in upper secondary classrooms (examples are van Maanen 1991, 1997). In a paper appearing in the collection edited by Swetz et al. (1995), van Maanen continues his descriptions based on personal experiences with three historical cases: seventeenth-century instruments for drawing conic sections, improper integrals, and one on the division of alluvial deposits in medieval times. The latter, which shall be discussed in more detail, is different from the other two in that it describes a project in three first-year classes of the Dutch grammar school, pupils about age 11, and because it was an interdisciplinary project with Latin.

In a medieval setting of a case of three landowners, who all had land on the bank of a river, fighting over an alluvial deposit bordering their land, the students were to investigate the problem by means of a method proposed by the Italian professor Bartolus of Saxoferrato in 1355 (the example with the landowners was the one used by Bartolus himself). The ideas of the project, as described by van Maanen, were to demonstrate the importance of mathematics in society, which relates to Barbin's idea of cultural understanding (meta-issues); to let pupils "invent" a number of constructions by ruler and compass, thus emphasizing the importance of understanding mathematical in-issues; to have them apply these "inventions" in order to solve the legal problem of the medieval example, which illustrates the applicational dimension of mathematics as a discipline; and to have them read excerpts from Bartolus' treatise in the original Latin language, illustrating "that it is impossible to interpret the sources of Western culture without knowledge of classical languages" (van Maanen 1995, p. 79), clearly, yet another meta-issue and as such, one which would

---

[33] Of course a factor in choosing these exact examples is my own familiarity with them, which enables me to perform a deeper analysis of the underlying motives for resorting to original sources.

be very difficult to illustrate without resorting to original sources. Van Maanen's evaluation of the implementations of the project is

> Making contact with Bartolus was only possible via deciphering and translating, but that was simply an extra attraction to most of the pupils. They learned to work with point-sets in plane geometry, and simultaneously their knowledge of general history increased. Last but not least, they were greatly stimulated to learn Latin. (van Maanen 1995, p. 80)

27.6.2 *The Early History of Error-Correcting Codes: Working with Excerpts in Upper Secondary School*

The next example is a teaching module for Danish upper secondary school (described in Jankvist 2009b; 2010). The module deals with the early history of error-correcting codes, and as part of it, the students were to read excerpts from original sources by Shannon, Hamming, and Golay. The historical background is that based on Shannon's display of Hamming's so-called (7,4) error-correcting code[34] in 1948, Golay was able to make the generalization of this into the entire family of Hamming codes in 1949 and discover a few additional codes himself. Due to the wish of the Bell Laboratories, wanting to patent Hamming's codes, his own publication was delayed until 1950 (which later caused a quarrel regarding who was in fact the originator of the codes). Having read excerpts from these sources (in translation), the students were to do an essay assignment, identifying and discussing epistemic objects and epistemic techniques in Hamming's development of the (7,4) code, the object being the code itself and the techniques the notion of metric and elements of linear algebra. One of the objectives of this essay assignment was for the students to make a distinction between the already available techniques, i.e., the well-established mathematics that Hamming was using and the techniques that he himself had to create in the process. One group of students who did rather well in making this distinction answered:

Hamming uses generalized concept of distance; elements of linear algebra; geometrical models; and unity n-dimensional squares.

A metric which is called a distance $D(x, y)$ is the definition of Hamming distance.

Since x and y must be different $(x \neq y)$, we can find how similar the two codes are, because we talk about a more generalized concept of distance.

Linear algebra is concerned with the study of addition and proportionality, and it is the concept of linearity that binds the concepts of addition and proportionality together.

We use geometrical figures to understand the n-dimensions, since these exceed what can be understood physically, i.e. 1st, 2nd, and 3rd dimension.

n-dimensional cube is the same as the metric space. (Group 5, translated from Danish. Quoted from Jankvist and Kjeldsen 2011, p. 853)

---

[34] In this binary code, all code words consist of seven bits, four of these being information bits, hence the name (7,4)-code.

Of course, not all groups were able to distinguish in an equally clear manner the already available techniques from those constructed by Hamming in the process and therefore might list Hamming's own constructs as being already available (for an example, see Jankvist and Kjeldsen 2011). A long discussion of the mathematical in-issues mentioned in the quote above shall not be given, but despite that, it should still be clear that without an understanding of these in-issues on the students' behalf, it is not possible for them to enter into the meta-issue micro-historical discussion of Hamming's error-correcting codes coming into being. This multiple-perspective approach to mathematics through its use of essay assignments thus illustrates mathematical understanding as a prerequisite condition for deeper cultural understanding. Or to phrase it differently, an understanding of the meta-issues is *anchored* in an understanding of the in-issues (Jankvist 2011).

27.6.3 *Project Work on Bernoulli's Solution of the Catenary at University Graduate Level*

In the setting of students' problem-oriented project work at Roskilde University (Denmark), Kjeldsen (2011a, b) has discussed the learning outcome of five graduate students enrolled in the mathematics program while doing a project based on readings of three original sources from the 1690s: the solutions to the brachistochrone problem[35] by Jakob Bernoulli and Johann Bernoulli, respectively, and Johann Bernoulli's solution to the catenary problem[36] (see also Jankvist and Kjeldsen 2011; Kjeldsen and Blomhøj 2009, 2012). More precisely, Kjeldsen has addressed the learning outcome from two different mathematics educational perspectives: the development of students' mathematical competencies (Niss and Højgaard 2011) and students' learning of meta-discursive rules in mathematics (Sfard 2008).

The framework of students' *mathematical competencies,* as described by Niss and Højgaard (2011), lists eight mathematical competencies: mathematical thinking competency, problem-solving competency, modeling competency, reasoning competency, representation competency, symbols and formalisms competency, communication competency, and aids and tools competency. A mathematical competency is defined as "having knowledge of, understanding, doing, using and having an opinion about mathematics and mathematical activity in a variety of contexts where mathematics plays or can play a role," or in other words a kind of "well-informed readiness to act appropriately in situations involving a certain type of mathematical

---

[35] The *brachistochrone problem* (from Greek *brachistos* meaning shortest and *chronos* meaning time) is to find the curve by which a particle under influence of gravity will travel the fastest from a given point A to a given point B. The curve is the so-called cycloid, and the solution to the problem was found not only by the Bernoullis but also by both Newton and Leibniz around the same time (another example of multiple discoveries).

[36] The *catenary problem* consists in describing the curve formed by a flexible chain hanging freely between two points.

challenge" (Niss and Højgaard 2011, p. 49). In relation to the students' reading of Johann Bernoulli's original text on the catenary problem, the students' development of four of these eight competencies took place in the following way:

> The students' *problem solving* skills were trained extensively in their work with Bernoulli's text on the catenary. As emphasized above, the students had to fill in all the 'gaps' in Bernoulli's presentation in the course of which the students derived several intermediate results themselves. [...] Parts of the students' *mathematical modeling* competency were developed through their struggle to understand Bernoulli's mathematization of the physical description of the catenary, and to justify his five assumptions from statics. [...] Their *mathematical reasoning* competency was developed, in particular through their struggle to clarify the apparent lack of rigor in Bernoulli's way of arguing compared with modern standards. Bernoulli's way of argumentation is not in accordance with what is considered 'proper' or rigorous mathematical reasoning today. [...] The students' ability to distinguish between and utilize different *representations* of e.g. mathematical objects was especially challenged in [...] their study of Bernoulli's construction of the solution to the differential equation of the catenary. Here they experienced a representation of a solution that is quite different from the analytical representation they usually see in modern textbooks. (Jankvist and Kjeldsen 2011, pp. 843–844)

As evident from the quote, the development of these competencies is tightly connected to the studying of the original source, some even a direct consequence thereof. (For a discussion of the students' development of the four remaining competencies due to this project work, see Kjeldsen 2011a.) From the point of view of using history as a tool for teaching and learning mathematical in-issues, students' competency development provide a different – and possibly more natural – means for evaluating the use of original sources than just assessing students' conceptual understanding.[37]

In relation to the previous discussion of *dépaysement*, the definitions of Sfard's (2008) notions of *meta-discursive rules* of mathematical discourse and *commognitive conflict* have already been given (please refer to these if needed). In the context of the same students' reading of Bernoulli's solution of the catenary, Kjeldsen and Blomhøj explain that the learning of meta-discursive rules

> ... is difficult, and as pointed out by Sfard, because of the contingency of meta-level rules, it is not likely that learners by themselves will begin a meta-level change. Such a change is most likely to happen when (or if) the learner experiences another (new) discourse that is governed by meta-rules other than those the learner so far has been regulated by. An experience like that constitutes what Sfard (2008, p. 256) calls a commognitive conflict, which she defines as "a situation in which different discursants are acting according to different metarules." (Kjeldsen and Blomhøj 2012, p. 330)

Such commognitive conflicts may of course also arise when students compare discourses of different discussants, for example, when comparing different original sources as mentioned earlier. However, the occurrence of commognitive conflicts is to a large extent dependent on and conditioned by a non-Whig approach to history, because

---

[37] Niss and Højgaard (2011) give three dimensions for assessing a person's mastery and development of a competency: degree of coverage, radius of action, and technical level.

> If one's reading and interpretation of historical sources are constrained by the way mathematics is perceived and conceptualised in the present, the historical text cannot play the role of an "interlocutor" that can be used to create commognitive conflicts [...] when students "communicate" with the text, since differences in the way of communicating in the past and in the present will have been "washed away" by the whig interpretation. (Kjeldsen and Blomhøj 2012, p. 331)

Thus, original sources, or alternatively secondary sources which are very close to the original, come to play an almost indispensible role in this setting of developing students' understanding of meta-discursive rules through a use of history of mathematics.

As an addendum, the role of the original source as an *interlocutor* is also discussed by Jahnke et al. (2000, p. 296), who say that "a source can be a trigger for establishing a dialogue with the ideas expressed" and it "then becomes an interlocutor to be interpreted, to be questioned, to be answered and to be argued with." Jahnke et al. state that the role of a source as an interlocutor in particular applies to issues such as the nature of mathematical objects and the essence of mathematical activity, that is to say matters related to the meta-issues of mathematics. But as seen from the example of Kjeldsen and Blomhøj (2012), original sources as interlocutors may play an equally important role in relation to mathematical in-issues. However, a word of caution should be provided, because until recently history of mathematics has primarily been carried out by professional mathematicians in the Whig tradition (cf. the quote by Weil above). Unfortunately, this also has consequences for translations of sources and consequently for their roles as interlocutors. Schubring (2008), for example, points out that contrary to their claim of being accurate, some translations of well-known historical classics exist that have distorted the original sources.

27.6.4 Projects on Boolean Algebra and Electric Circuit Design at Undergraduate Level

The next example is situated in the context of teaching US undergraduate students with projects based on guided readings of original sources (cf. earlier). Janet Barnett has developed several such projects (see Barnett et al. 2011), and in the following we shall look at two of them.

In the first one, she deals with the origin of Boolean algebra by guiding students through carefully chosen parts of Boole's 1854 treatise on the *"Laws of Thought"* (Barnett 2011a). Boole introduces a new algebra by explaining the use of "signs" to represent "classes" and by defining a system of symbols $(+,-,\times, 0, 1)$ to represent operations on these signs and finally deducing the basic laws that the operations must follow. While doing so, Boole continuously compares his new algebra to standard arithmetical algebra,[38] one of the main differences being that in Boole's algebra

---

[38] One might talk about a built-in replacement and *dépaysement* in Barbin's sense in Boole's own presentation of his algebra.

we have that $xx=x$ (or $x^2=x$), because if, for example, x stands for "good," then saying "good, good men" is the same as saying "good men." Knowing that this can only be so if x is either 0 or 1, Boole draws the consequence by conceiving an algebra where symbols x, y, z, etc. admit indifferently of the values 0 and 1, and of these values alone, and carries on to deduce a series of laws. Barnett's (2012, p. 344) ultimate goal of this project is to have the students "develop an understanding of the modern paradigm of elementary set theory as a specific example of a Boolean algebra," and for that reason she has students study original sources of both Venn and Peirce next, as part of the project.

In the other project of Barnett's which is mentioned here, she has students read part of Shannon's famous 1938 article on design of electric circuits by means of simplification through Boolean algebra (Barnett 2011b). Using a set of postulates from Boolean algebra ($0\cdot0=0$; $1+1=1$; $1+0=0+1=1$; $0\cdot1=1\cdot0=0$; $0+0=0$; and $1\cdot1=1$) and interpreting these in terms of circuits ($0\cdot0=0$ meaning that a closed circuit in parallel with a closed circuit is a closed circuit and $1+1=1$ meaning that an open circuit in series with an open circuit is an open circuit), Shannon is able to deduce a number of theorems which could be used to simplify electric circuits.

From studying these two projects by Barnett, it is clear that focus is on the in-issues as she guides the reader through the original texts and eventually leads him or her up to a, in the first project, modern account of the algebra of logic, set theory, and Boolean algebra of today and in the second project to relating Shannon's original representation of connections in series and in parallel to that of today as well as introducing truth tables in the guiding tasks. Regarding the first project, Barnett says that it

> … provides an opportunity for students to witness how the process of developing and refining a mathematical system plays out, the ways in which mathematicians make and explain their choices along the way, and how standards of rigor in these regards have changed over time. (Barnett 2012, p. 344)

However, the in-issue focus must also be seen as a natural consequence of the educational setting for which the projects are designed – undergraduate students following courses relying on such projects will eventually have to follow several traditionally taught courses relying on modern-day textbooks, for which they need to know modern-style notation and standards for proof. Nevertheless, the approach of guided reading still leaves the excerpts from the original sources completely untouched, potentially enabling students to make their own interpretations along the way.

27.6.5 *HAPh Modules and the Development of Upper Secondary Students' Overview and Judgment*

The final example concerns students' reading of some of the same original sources as those used by Barnett but in the setting of Danish upper secondary school. The motivation for this use of original sources was twofold: to design teaching modules focusing on meta-issues of history, applications, and philosophy of mathematics in unison (Jankvist 2013) and to measure the effect of such modules on the students' development of what Niss and Højgaard (2011) term "overview and judgment" (Jankvist 2012a). The notion of overview and judgment stands in contrast to the previously mentioned one of mathematical competency, in that these do not concern "readiness to act" but instead are "'active insights' into the nature and role of mathematics in the world" which "enable the person mastering them to have a set of views allowing him or her *overview and judgement of the relations between mathematics and in conditions and chances in nature, society and culture*" (Niss and Højgaard 2011, p. 73, italics in original). Niss and Højgaard deal with three types of overview and judgment, which have already been exemplified in the quotes given when introducing the notion of in-issues and meta-issues earlier:

- The historical evolution of mathematics, both internally and from a social point of view
- The actual application of mathematics within other subjects and practice areas
- The nature of mathematics as a subject

The design encompassing all three of these dimensions – referred to as *H*istory, *A*pplication, and *Ph*ilosophy (*HAPh*) – was based on an observation that the inclusion of these dimensions share an overlap in the so-called whys and hows, i.e., that similar reasons are provided for their inclusion and that similar approaches can be used (see Jankvist 2013). Further, it was argued that the use of original sources, one for each dimension, through a guided reading and a use of essay assignments was a suitable way of evoking the desired meta-issues.

In one such HAPh module, Boole's original text on the *Laws of Thought* made up the historical dimension and Shannon's 1938 article on electric circuit design the applicational dimension (cf. the previous illustrative example). The original source for the philosophical dimension was a paper by Hamming from 1980 in which he discusses *The unreasonable effectiveness of mathematics*[39] from the viewpoint of engineering (and computer science), asking why it may be that so comparatively simple mathematics suffices to predict so much, this question making up the "unreasonable" aspect. After discussing what mathematics is, Hamming provides some tentative explanations of why mathematics is in fact so effective (see Jankvist 2013). Based on their readings of these three original sources, students were to do essay assignments discussing if the idea of Boolean algebra and its later use in describing electric circuits may be seen as an example of the unreasonable effectiveness of mathematics and if so then why (for students answers and reflections, see Jankvist 2012b, 2013). In another HAPh module, the same class of students had previously worked with the beginning of graph theory (Euler), its later use in solving the shortest path problem (Dijkstra), and the role of mathematical problems (Hilbert).[40]

---

[39] This paper was a comment to a paper by the physicist Wigner from 1960, who addressed *The unreasonable effectiveness of mathematics in the natural sciences*.

[40] For a description of this HAPh module, see Jankvist (2012a, 2013). See also Barnett (2009) for a project on graph theory.

Now, in order to actually try and measure any effect these HAPh modules may have had on the development of a class of upper secondary students' overview and judgment, one must of course have ways of both *accessing* and *assessing* such. The means for accessing the students' overview and judgment were a combination of questionnaires, asking questions related to each of the three types of overview and judgment (see Jankvist 2012a) and follow-up interviews before, during, and after the implementation of the HAPh modules.[41] As for assessment, it is argued in Jankvist (2012a) that it makes sense to address the development of overview and judgment as a combination of students' actual *knowledge* and their *reflected images of mathematics as a (scientific) discipline*, where the latter may be measured in terms of:

- The growth in *consistency* between students' related beliefs/views
- The extent to which a student seeks to *justify* his or her beliefs and views
- The amount of provided *exemplifications* in support of the beliefs and views a student holds, i.e., the beliefs appear to be held more evidentially

The mentioning of beliefs here of course refers to the vast amount of literature on students' beliefs (e.g., Leder et al. 2002). While *beliefs* are usually taken to be something rather persistent, meaning that they are not likely to develop and change over a short-time period,[42] *views,* on the other hand, may be taken to be something less persistent but with the potential to develop into beliefs at a later point in time (Jankvist 2009b, forthcoming). Thus, development in, and changes of, students' beliefs *and* views, in unison the reflected images, provides an insight into their overview and judgment. However, it is important to stress again that overview and judgment does not equal beliefs and views, since it also consists of the actual *knowledge* which the students possess in relation to both mathematical in-issues and meta-issues.

One question or topic which has proven itself particularly helpful in uncovering students' overview and judgment is that of mathematics being discovered or invented (e.g., Hersh 1997). To illustrate this, an excerpt from an interview with the upper secondary student Larry from when he had been through both HAPh modules is offered;

Interviewer	And you say about Boolean algebra that you believe it to be invented? [...]
Larry	Yes, we define [it] ourselves. It's... We take a way of thinking and turn it into mathematics; this means this and that means that... Well, it might be that the way of thinking was already there, but we invent a way of writing it within mathematics. I'm not sure I can phrase it any better.
Interviewer	Okay. In question 26 you say that Boolean algebra, classes, etc. are human constructions, which of course is connected to what you just said. But do you believe mathematics in general to be something we invent, or are there also things in mathematics which we discover?

(continued)

---

[41] In some instances it was also possible to perform methodological triangulation with other data sources such as students' hand-ins, student group essay assignments, and video recordings as exemplified in Jankvist (2009b).

[42] Evidential (or evidence-based) beliefs are, however, more likely to change than non-evidentially held beliefs (Green 1971).

27 On the Use of Primary Sources in the Teaching and Learning of Mathematics

(continued)

Larry	Well… There can be connections in mathematics which we discover. For example the equation with Euler's number in the power of π times i minus or plus 1 equals 0 [$e^{i\pi} + 1 = 0$]. These are some interrelations which we have not made ourselves. It is a lot of independent things which we have found and which then fits together and reveals a beautiful connection. […] I think it is a good example of something which we just discover. As far as I know these [π, e, and i] were not that associated. But that they fit together in this way, it kind of shows… that there must be a… that no matter what, we did something right.
Interviewer	Yes?
Larry	So regarding invention or discovery in mathematics, I think… I think that some things are invented and some discovered. I will risk claiming that.
Interviewer	Alright. Can you give some examples?
Larry	Well… for example our way… in graph theory, to translate bridges into numbers and the way of writing it all up [the Königsberg bridge problem]. That is something we've made. While things as… what is a good example? Things as π is something we discovered. […]
Interviewer	Okay. Is it possible to say if one precedes the other? Does discovery precede invention or does invention precede discovery?
Larry	In most cases it must… well, not necessarily… With π, for instance, I guess that discovery was before invention, because… If we say that we invented, that we set a circle to 360°. But when we calculate π […] then we don't use the 360°, as far as I recall. […] It is different within different areas of mathematics, but with π I think we discovered that there was a connection first, and then we built on that. But it's quite related; when we choose something we quickly arrive at some further discoveries.

(Interview with Larry, November 16th, 2011, translated from Danish. Quoted from Jankvist 2012a, pp. 859–860)

The excerpt illustrates well how Larry is able to both justify and exemplify his beliefs/views regarding invention and discovery. In terms of *development* of overview and judgment, Larry had started off a year and a half earlier, before the HAPh modules, to believe mathematics to be mainly an invention. In between the modules, he shifted to believing in mainly discovery. But above we see him adopting a more reflected stance, arguing that mathematics can be both.[43] To illustrate discovery, he first mentions Euler's identity ($e^{i\pi} + 1 = 0$) as something that he finds unlikely to have been invented (Euler's identity was not part of the HAPh modules). Later he carries on to elaborate on the number π in relation to discovery, and we get an example of the interplay of beliefs/views and knowledge, when he refers to his actual knowledge of our convention of 360° in a circle being unrelated to the appearance of the number π. As examples of mathematics which is invented, he refers to the cases of the HAPh modules: first, Boole's introduction of his new algebra and later Euler's approach to dealing with the Königsberg bridge problem. Such exemplification helps Larry to hold his beliefs and views more evidentially,

---

[43] Although the dimension of consistency is not very present in the displayed interview excerpt above, a growth in consistency was present in general in the case of Larry (Jankvist 2012a).

which also assists him in justifying them as well as the development and changes in them.[44] Thus, in this respect the primary original sources play the role of *evidence* which, together with the knowledge students have acquired from studying them, comes to form the basis for developing their overview and judgment (For further discussion, see Jankvist 2012a).

27.7 Discussing the Past, Present, and Future of Using Primary Original Sources

Having now looked at the background, the various reasons and approaches for using primary original sources, and the selected illustrative examples of didactic findings, it is time to recapitulate and look at possible directions which the use of original sources may take in the future, i.e., other reasons than those already discussed. A handful of such different perspectives on the use of original sources shall be pointed to.

27.7.1 Expanding the Emphasis

As indicated, the main emphasis of the use of original sources has, until recently, been on the teaching and learning of mathematical in-issues, both in terms of students' understanding or sensemaking of mathematical concepts and notions as well as in terms of motivating these; it is clear that the experience of seeing how a certain mathematical construct was motivated historically can provide students with a deeper understanding of why this construct came to look the exact way it did.[45] What should be added is that original sources may offer *a truer mode of presentation* when compared to that of textbooks, which usually go as follows: definition(s), theorem(s), proof(s), and application(s). In reality, the historical development of a mathematical topic is close to being the reverse – as mentioned when discussing Pengelley's (2011) reason 9 earlier. With original sources' often built-in relation between theory and practice, definitions of concepts, theorems, etc. are motivated. They do not appear out of nowhere. This means that the study of original sources may have as an outcome that students become more willing to accept abstract mathematical constructs, since these come to appear as natural consequences of the mathematical investigations presented in the source.

With the ICMI Study of 2000, it became clear that *more empirical research was needed* on introducing a historical dimension into the teaching and learning of

---

[44] In the case of non-evidentially held beliefs, justification might very well take place without exemplification (see Jankvist 2009b, 2012a).

[45] Take, for example, the concepts of function, limit, continuity, and not least uniform convergence (see, e.g., Katz 1998 or the appendix on uniform convergence in Lakatos 1976).

mathematics, including the use of original sources to do so.[46] One recent example is the study by Glaubitz (2011), in which he compared different approaches to the teaching of quadratic equations and the quadratic formula and found that pupils who had been taught with original sources through the hermeneutic approach scored higher in standard tests than a control group who had been taught with a conventional approach. Also, as illustrated with the previously provided examples, some of the empirical work which eventually followed has contributed in broadening the scope of using original sources. With the studies of Kjeldsen, for instance, focus is not only on students' understanding of mathematical in-issues but also on their development of mathematical competencies as a result of studying and working with primary original sources (Kjeldsen 2011a; Jankvist and Kjeldsen 2011). Further, the examples illustrated how original sources (and excerpts of such) may play a role in teaching students something about the meta-issues of mathematics as a scientific discipline; how to do this in such a manner that the students' discussions of these meta-issues are anchored in their understanding of the related mathematical in-issues (Jankvist 2011); and how teaching modules involving readings of original sources related to history, application, and philosophy in unison may assist in developing students' overview and judgment (Jankvist 2012a).[47]

27.7.2   *The Role of Theoretical Constructs from Mathematics Education Research*

Another important aspect, which the chosen examples illustrate, is the relationship between empirical studies on resorting to original sources and *the use of theoretical frameworks and constructs from mathematics education research*. The discussion and "measuring" of original sources' effect on students' development of mathematical competencies and overview and judgment as defined by Niss and Højgaard (2011) are examples of this (including the discussion of students' beliefs in relation to overview and judgment). And another example is the use of Sfard's (2008) theory of commognition to illustrate the (almost indispensable) role of original sources as different discussants in students' learning of meta-discursive rules (Kjeldsen and Blomhøj 2012). But it is not only the case that the use of mathematics education research frameworks may inform the use of original sources and history of mathematics in general, in the teaching and learning of mathematics; it is also the case that the history of mathematics and the use of original sources may inform mathematics education frameworks. An example of this may be found in the study by Clark (2012), who, in a setting of teacher education, has used excerpts from an original

---

[46] Since then a handful of PhD theses to some degree addressing empirical aspects of using original sources (or excerpts of such) have appeared: Clark (2006), Glaubitz (2010), Gulik-Gulikers (2005), Jankvist (2009b), and Ta'ani (2011).

[47] For a much more detailed account of the empirical studies available in the field of history in mathematics education, see Jankvist (2012c).

source by al-Khwariszmi, presenting his methods for solving quadratic equations. Based on her findings, Clark argues that history of mathematics (including the study of original sources) is part of the "something else" in the framework of mathematical knowledge for teaching (MKT) (Ball et al. 2008), since it informs teachers in their instructional practice. Taking into consideration the obvious benefits of discussing the use of history and original sources in teacher training within the framework of teachers' development of mathematical knowledge for teaching, there is thus a dual relationship between the MKT framework and the use of history and original sources in teacher training, which deserves to be investigated further (Jankvist et al. 2012).

27.7.3 Interdisciplinarity and Original Sources

The example of van Maanen's activity of students' reading Bartolus of Saxoferrato also indicates an important role for primary original sources, namely, that of providing a natural environment for *interdisciplinary teaching and learning*. Over the past decades, interdisciplinarity in both research and education has become a frequently debated topic. However, it shall be my claim that from an educational perspective, interdisciplinarity often presents a didactic dilemma: On the one hand, students are told that interdisciplinary work is extremely important (for various different reasons); on the other hand, the students are often only shown somewhat artificial and situational constructed examples, which make the dimension of interdisciplinarity appear pasted on. But because original sources deal with reality – even if it is a historical one – any given interdisciplinary elements within them are likely to illustrate much better to students the importance of interdisciplinarity in research and society. The example by Kjeldsen and colleagues of students' reading of Bernoulli's work on the catenary is actually also an example of interdisciplinarity between mathematics and physics, since the students had to struggle with "Bernouli's mathematization of the physical description of the catenary" – in fact, the students' purpose of studying these sources was to investigate physics' influence on the development of differential equations (see Kjeldsen and Blomhøj 2012). In short, original sources bring authenticity to the dimension of interdisciplinarity in teaching and learning.

27.7.4 Recruitment, Transition, and Retention: A Triple Aspect of a Potential Role for Original Sources

Just as the role of original sources as a means for authentic interdisciplinarity is as relevant for mathematics as it is for the natural sciences (and engineering), so are the following three possible future roles, on which is speculated. The first of these is the role of original sources in relation to the *recruitment problem*. As we know, in the Western world, we have been experiencing problems with recruiting students for the mathematical sciences. Very often, upper secondary school students do not have an accurate idea of what mathematics is all about, when practiced as a scientific discipline at the tertiary levels, e.g., by pure and applied mathematicians at universities. Students' answers to the question of what professional mathematicians do typically range from having no clue at all to believing that they perform some kind of "clean-up job" consisting in finding "errors" in already existing formulas and proofs, more efficient ways of calculating already known quantities, etc. (Jankvist 2009b, forthcoming). Often such views have to do with the students' impression of mathematics as something a priori given, static and rigid – a belief not unrelated to textbooks' usual presentation of mathematical topics of course; only very few students seem to believe it possible that mathematicians can come up with actual *new* mathematics. Therefore, the students know neither what they accept to study nor what they reject to study, if they choose to engage with mathematics. Studying primary original sources can enlighten the students in this respect.[48]

The second role of original sources is related to the *transition problem* between educational levels. From the design and implementation of the HAPh modules as discussed above (Jankvist 2013) and the projects by Barnett using the same original sources, we can make the observation that choosing original sources presenting novel mathematical ideas can assist in making otherwise complicated mathematical topics more accessible to students. One of the reasons for this is that when stepping into uncharted territory, researchers may be more careful in explaining what they are doing and why, as well as why they are approaching a problem in a certain manner. Boole's text on the *Laws of Thought* is one such original source and Euler's text on the Königsberg bridge problem is certainly another. Thus, my point here is that the use of primary original sources may be seen as a way of dealing with the transition problem between, for example, upper secondary level and undergraduate university level. Because, on the one hand, when upper secondary students are presented with such material, they encounter actual research papers from mathematics and/or science, texts, in which the motivation for this piece of research and/or the application of it is sometimes discussed, and even though this happens in a historical setting, it illustrates the nature of research in academic communities in general. And, on the other hand, if such carefully chosen primary sources presenting novel ideas are used in introductory courses at undergraduate level, they may ease the students into the academic way of thinking and thereby make the transition to university level less "harsh."[49]

The third possible role of original sources which is mentioned has to do with the problem of retaining the students once they have already been recruited and

---

[48] In fact, student interview data from the HAPh module study support this claim; students expressed that due to their work with the two HAPh modules and hence their reading of the six original texts, they were able either to select or deselect a future engagement with mathematics and mathematical sciences on a more enlightened basis (Jankvist, preprint).

[49] This was also discussed in Jankvist's presentation "The use of original sources and its possible relation to the transition problem" at Colloque Artigue 2012 in Paris.

undergone transition, i.e., the *retention problem*.[50] The authenticity of the sources is also a key element in relation to this problem, as it was for the problem of interdisciplinarity. Because original sources offer a variety of real-life, that is, non-artificial, applications of various mathematical in-issues, such sources can provide *meaning* for the students in their learning. Although not taken from university level, the following quote from an interview with one of the students in the HAPh modules study illustrates this:

> For me, I personally think that I get much more interested, when I see it all, than if I'm only told that now we are studying vectors and we must learn how to dot these vectors and then we must be able to calculate a length, right. That's all very good, but what am I to use it for? Whereas, when you know about the background, the development up till today, that I think was exciting. Because when we began with the first text [Boole's text], it was kind of like, yeah, that's alright, he can figure out this thing here, and this equals that, I can follow that, and 'white sheep' and so... That was good for starters. Then more is built on top, and all of a sudden we see: Why, it's a [electric] circuit we are doing! You could begin to relate it to your own reality; that is, something you knew already. So, the thing about starting from scratch [...] and suddenly seeing it form a whole, what it was used for today – and be able to relate it to something, something you knew about – that I think was way cooler. (Interview with Nikita, November 3rd, 2011, translated from Danish. Quoted from Jankvist 2012b, pp. 139–140)

Recently, Pengelley (2012) and Barnett (2012) have designed entire courses in elementary number theory and abstract algebra, respectively, around the use of original sources only – courses thought possibly to play a role also in terms of retention. But nevertheless, what may be still be needed from an empirical point of view is more quantitative data on the effect and efficacy of using primary original sources, data which can underpin and support the many positive qualitative statements, as, for example, that of Nikita above, which are present in the relatively larger number of qualitative empirical studies available in the field.

27.7.5 *Additional and Concluding Remarks*

In the present chapter, not much effort has been taken to distinguish between using original sources at the various educational levels, except from choosing the illustrative examples to more or less cover the possible spectra. When actually choosing which sources to use in a given situation, the educational level is of course essential, because as Jahnke et al. (2000) point out:

> Incorporating primary sources is not good or bad in itself. We need to establish the aims, including the target population, the kind of source that might be suitable and the didactical methodology necessary to support its incorporation. (Jahnke et al. 2000, p. 293)

---

[50] This role of original sources was suggested by David Pengelley, who discussed it during the panel "Empirical research on history in mathematics education: current and future challenges for our field" at HPM2012 in Deajeon.

27 On the Use of Primary Sources in the Teaching and Learning of Mathematics 903

Also, not much emphasis has been put on the use of original sources in teacher education, except when referring to the study of Clark (2012). However, based on what has been said and discussed, it is possible to provide some comments to at least four of the five questions for future research on the use of original sources formulated by Furinghetti et al. (2006) (cf. Sect. 27.2). Regarding question 1, it seems clear to me that if a use of primary original sources is to gain any real impact in the field of mathematics education, then the evaluation of such uses must build and rely on theoretical constructs and frameworks from mathematics education research. Some examples of such potentially relevant frameworks have been put forth in this chapter (e.g., Ball et al. 2008; Niss and Højgaard 2011; Sfard 2008), but clearly it is possible to find many more that applies in one way or another. As for question 2, it appears that some kind of guidance of the students is a characteristic of many viable approaches for using original sources, whether this is in the form of guided readings, familiarizing the students with the mathematical in-issues before reading a source, as done in the hermeneutic approach, or structuring students' discussions of meta-issues through a use of essay assignments. In terms of question 4, we have seen that a historical research stance, such as the multiple-perspective approach as described by Kjeldsen, indeed can support the use of original sources in the teaching and learning of mathematics. As one of the working tools of this approach, we saw that the notion of epistemic objects and techniques can assist in anchoring students' discussions of meta-issues in the related in-issues (e.g., Jankvist 2009b). Also, the notion of Whig history helped in qualifying our discussion of the various approaches to using original sources. Although question 5 makes up an interesting question, I am not aware of any examples of mathematics education having supported or broadened research in the field of history of mathematics. What is more common is that people in mathematics education, who have taught using original sources, may eventually be drawn towards doing some research in the history of mathematics. Regarding question 3 on the actual impact of the various viable approaches to students' learning and understanding of mathematics, the account given above illustrates that this is not simply a matter of getting students to learn and understand mathematical in-issues or motivate them to do so. Even though this may be the overall end goal, there are other crops to be harvested from using primary original sources, for example, the development of students' mathematical competencies and the development of their overview and judgment. And as mentioned, the development of mathematical competencies may be a much more natural way of assessing the efficiency and efficacy of using history and original sources in mathematics education (Jankvist and Kjeldsen 2011). As for potential crops to be harvested in the future field of using primary original sources, the aforementioned educational problems of interdisciplinarity, recruitment, transition, and retention make up interesting and promising research areas. Already, these problems have the attention of educational researchers, curriculum designers, and policy makers, so if positive empirical results (quantitative and qualitative) could be produced in support of these roles of primary original sources, then surely this would assist in the use of original sources gaining impact in mathematics education in general – and science education too, since these four problems are not restricted to mathematics education alone.

References

Arcavi, A.: 1985. *History of mathematics as a component of mathematics teachers' background.* Ph.D. thesis. Rehovot: Weizmann Institute of Science.

Arcavi, A. & Isoda, M.: 2007. Learning to listen: From historical sources to classroom practice. *Educational Studies in Mathematics*, 66(2), 111–129. Special issue on the history of mathematics in mathematics education edited by L. Radford, F. Furinghetti & V. Katz.

Ball, D. L., Thames, M. H., & Phelps, G.: 2008. Content knowledge for teaching: What makes it special? *Journal of Teacher Education*, 59(5), 389–407.

Barbin, E.: 1997. Histoire des Mathématiques: Pourquoi? Comment? *Bulletin AMQ*, Montréal, 37(1), 20–25.

Barbin, E.: 2007. On the argument of simplicity in 'elements' and schoolbooks of geometry. *Educational Studies in Mathematics*, 66(2), 225–242. Special issue on the history of mathematics in mathematics education.

Barbin, E., Stehlíková, N. & Tzanakis, C. (Eds.): 2008. *History and Epistemology in Mathematics Education Proceedings of the 5th European Summer University.* Plzen: Vydavatelský servis.

Barbin, E. Kronfellner, M. & Tzanakis, C. (Eds.): *History and Epistemology in Mathematics Education Proceedings of the 6th European Summer University.* Vienna: Holzhausen Publishing Ltd.

Barnett, J. H.: 2009. Early Writings on Graph Theory: Euler Circuits and the Königsberg Bridge Problem. In B. Hopkins (Ed.): *Resources for Teaching Discrete Mathematics.* MAA Notes 74. Washington, DC: The Mathematical Association of America. 197–208.

Barnett, J. H.: 2011a. *Origins of Boolean algebra in logic of classes: George Boole, John Venn and C. S. Peirce.* http://www.cs.nmsu.edu/historical-projects/projects.php.

Barnett, J. H.: 2011b. *Applications of Boolean algebra: Claude Shannon and circuit design.* http://www.cs.nmsu.edu/historical-projects/projects.php.

Barnett, J. H.: 2012. Bottled at the source: the design and implementation of classroom projects for learning mathematics via primary historical sources. In *HPM2012: The HPM Satellite Meeting of ICME-12*, Proceeding Book 1, pp. 335–347.

Barnett, J. H., Lodder, J., Pengelley, D., Pivkina, I., & Ranjan, D.: 2011. Designing student projects for teaching and learning discrete mathematics and computer science via primary historical sources. In V. Katz & C. Tzanakis (Eds.): *Recent Developments on Introducing a Historical Dimension in Mathematics Education.* MAA Notes 78. Washington DC: The Mathematical Association of America. 187–200.

Bachelard, G. (1938). La formation de l'esprit scientifique. Paris: Vrin.

Bekken, O. & Mosvold, R. (Eds.): 2003. *Study the Masters: The Abel-Fauvel Conference.* Göteborg: NCM.

Brousseau, G. (1997). Theory of didactical situations in mathematics. Dordrecht: Kluwer Academic.

Burn, R. P.: 1999. Integration, a genetic approach. *NOMAD*, 7(1), 7–27.

Butterfield, H.: 1931/1951. *The Whig Interpretation of History.* New York: Charles Scribners Sons.

Calinger, R. (Ed.): 1996. *Vita mathematica: historical research and integration with teaching.* MAA Notes 40. Washington DC: The Mathematical Association of America.

Clark, K. M.: 2006. *Investigating teachers' experiences with the history of logarithms: a collection of five case studies.* Ph.D. thesis. College Park: University of Maryland.

Clark, K. M.: 2012. History of mathematics: illuminating understanding of school mathematics concepts for prospective mathematics teachers. *Educational Studies in Mathematics*, 81(1), 67–84.

27 On the Use of Primary Sources in the Teaching and Learning of Mathematics

Davis, P. J., & Hersh, R.: 1981. *The mathematical experience*. London: Penguin Books.

Epple, M. (2004). Knot invariants in Vienna and Princeton during the 1920s: Epistemic configurations of mathematical research. *Science in Context*, **17**(1/2), 131–164.

Fauvel, J., & van Maanen, J. (Eds.): 2000. *History in mathematics education—The ICMI study*. Dordrecht: Kluwer.

Fried, M. N.: 2001. Can history of mathematics and mathematics education coexist? *Science & Education*, 10(4), 391–408.

Furinghetti, F., Jahnke, H. N. & van Maanen, J.: 2006. Mini-workshop on studying original sources in mathematics education. *Oberwolfach Reports* 3(2), 1285–1318.

Furinghetti, F., Kaijser, S. & Tzanakis, C. (Eds.): 2007. *Proceedings HPM2004 & ESU4: ICME Satellite Meeting of the HPM Group & Fourth European Summer University*. Revised edition. Rethymnon: University of Crete.

Gadamer, H. G.: 1990. *Wahrheit und Methode. Gesammelte Werke* (6. Aufl., Bd. 1). Tübingen: Mohr.

Glaubitz, M. R.: 2010. *Mathematikgeschichte lesen und verstehen: Eine theoretische und empirische Vergleichsstudie*. Ph.D. thesis. Duisburg-Essen: Universität Duisburg-Essen.

Glaubitz, M. R.: 2011. The use of original sources in the classroom: empirical research findings. In E. Barbin, M. Kronfellner, & C. Tzanakis (Eds.): *History and Epistemology in Mathematics Education Proceedings of the 6th European Summer University*, pp. 351–362. Vienna: Holzhausen Publishing Ltd.

Green, T. F.: 1971, *The Activities of Teaching*. New York: McGraw Hill Book Company.

van Gulik-Gulikers, I.: 2005. Meetkunde opnieuw uitgevonden: Een studie naar de waarde en de toepassing van de geschiedenis van de meetkunde in het wiskundeonderwijs. Ph.D. thesis. Groningen: Rejksuniversiteit Groningen.

Hersh, R.: 1997. *What is mathematics really?* Oxford: Oxford University Press.

Jahnke, H. N., Arcavi, A., Barbin, E., Bekken, O., Furinghetti, F., El Idrissi, A. Silva da Silva, C. M. & Weeks, C.: 2000. The use of original sources in the mathematics classroom. In: J. Fauvel and J. van Maanen (eds.): *History in Mathematics Education*, The ICMI Study, pp. 291–328. Dordrecht: Kluwer Academic Publishers.

Jankvist, U. T.: 2009a. A categorization of the 'whys' and 'hows' of using history in mathematics education. *Educational Studies in Mathematics,* 71(3), 235–261.

Jankvist, U. T.: 2009b. Using history as a 'goal' in mathematics education. Ph.D. thesis, IMFUFA, Roskilde: Roskilde University. *Tekster fra IMFUFA, no. 464.*

Jankvist, U. T.: 2010. An empirical study of using history as a 'goal'. *Educational Studies in Mathematics,* 74(1), 53–74.

Jankvist, U. T.: 2011. Anchoring students' meta-perspective discussions of history in mathematics. *Journal of Research in Mathematics Education*, 42(4), 346–385.

Jankvist, U. T.: 2012a. History, application, and philosophy of mathematics in mathematics education: Accessing and assessing students' overview and judgment. Regular lecture at ICME-12 in Seoul, Korea. Available in pre-proceedings as RL4-10, pp. 845–864.

Jankvist, U. T.: 2012b. A historical teaching module on "the unreasonable effectiveness of mathematics" – case of Boolean algebra and Shannon circuits. In *HPM2012: The HPM Satellite Meeting of ICME-12*, Proceeding Book 1, pp. 131–143.

Jankvist, U. T.: 2012c. A first attempt to identify and classify empirical studies on 'History in Mathematics Education'. In Sriraman, B. (Ed.) *Crossroads in the History of Mathematics and Mathematics Education*. The Montana Mathematics Enthusiast Monographs 12. pp. 295–332. Charlotte: IAP.

Jankvist, U. T.: 2013. History, Applications, and Philosophy in mathematics education: HAPh – a use of primary sources. *Science & Education*, 22(3), 635–656.

Jankvist, U. T.: preprint. The use of original sources and its possible relation to the recruitment problem. Accepted paper for *CERME 8 – the Eight Congress of the European Society for Research in Mathematics Education*, held in Antalya, February, 2013.

Jankvist, U. T.: forthcoming. Changing students' images of mathematics as a discipline. (Under review).

Jankvist, U. T. & Kjeldsen, T. H.: 2011. New avenues for history in mathematics education – mathematical competencies and anchoring. *Science & Education*, 20(9), 831–862.

Jankvist, U. T., Mosvold, R., Jakobsen, A. & Fauskanger, J.: 2012. Mathematical knowledge for teaching in relation to history in mathematics education. Paper presented at Topic Study Group 20 at ICME12 in Seoul, Korea. Available in pre-proceedings as TSG20-4, pp. 4045–4052.

Katz, V.: 1998. A history of mathematics: An introduction. Second edition. Reading: Addison Wesley Educational Publishers, Inc.

Katz, V. (Ed.): 2000. *Using history to teach mathematics: an international perspective.* MAA notes 51. Washington DC: The Mathematical Association of America.

Katz, V. & Barton, B.: 2007. Stages in the history of algebra with implications for teaching. *Educational Studies in Mathematics*, 66(2), 185–201. Special issue on the history of mathematics in mathematics education edited by L. Radford, F. Furinghetti & V. Katz.

Katz, V. J., & Michalowicz, K. D. (Eds.): 2004. *Historical modules for the teaching and learning of mathematics.* Washington DC: The Mathematical Association of America.

Katz, V. & Tzanakis, C. (Eds.): 2011. *Recent Developments on Introducing a Historical Dimension in Mathematics Education.* MAA Notes 78. Washington DC: The Mathematical Association of America.

Kjeldsen, T. H. 2009a.: Abstraction and application: New contexts, new interpretations in twentieth-century mathematics. In E. Robson & J. Stedall (Eds.), *The Oxford handbook of the history of mathematics*, pp. 755–780. New York: Oxford University Press.

Kjeldsen, T. H.: 2009b. Egg-forms and measure-bodies: Different mathematical practices in the early history of the modern theory of convexity. *Science in Context*, 22(1), 85–113.

Kjeldsen, T. H.: 2011a. History in a competency based mathematics education: A means for the learning of differential equations. In V. Katz & C. Tzanakis (eds.), *Recent developments on introducing a historical dimension in mathematics education*, pp. 165–174. Washington, DC: The Mathematical, Association of America.

Kjeldsen, T. H.: 2011b. Uses of history in mathematics education: development of learning strategies and historical awareness. In Pytlak, M., Rowland, T. & Swoobda, E. (eds.): *CERME 7, Proceedings of the Seventh Congress of the European Society for Research in Mathematics Education*, pp. 1700–1709. Rzeszów: University of Rzeszów.

Kjeldsen, T. H.: 2011c. Does history have a significant role to play for the learning of mathematics? Multiple perspective approach to history, and the learning of meta-level rules of mathematical discourse. In E. Barbin, M. Kronfellner, & C. Tzanakis (eds.): *History and Epistemology in Mathematics Education Proceedings of the 6th European Summer University*, pp. 51–62. Vienna: Holzhausen Publishing Ltd.

Kjeldsen, T. H., & Blomhøj, M.: 2009. Integrating history and philosophy in mathematics education at university level through problem-oriented project work. *ZDM Mathematics Education, Zentralblatt für Didaktik der Mathematik*, 41, 87–104.

Kjeldsen, T. H. & Blomhøj, M.: 2012. Beyond motivation: history as a method for learning meta-discursive rules in mathematics. *Educational Studies in Mathematics*, 80(3), 327–349.

Klein, F.: 1908. *Elementarmathematik vom höheren Standpunkte aus.* First edition. Ausgearbeitet von Ernst Hellinger. Vol. 1: Arithmetik, Algebra, Analysis. Vorlesung, gehalten im Wintersemester 1907–08, Leipzig: Teubner, 1908.

Knoebel, A., Laubenbacher, R., Lodder, J. & Pengelley, D.: 2007. *Mathematical Masterpieces – Further Chronicles by the Explores.* New York: Springer.

Lakatos, I.: 1976. *Proofs and Refutations, the Logic of Mathematical Discovery.* Cambridge: Cambridge University Press.

Laubenbacher, R. & Pengelley, D.: 1999. *Mathematical Expeditions – Chronicles by the Explorers.* New York: Springer.

Leder, G. C., Pehkonen, E. & Törner, G. (Eds.): 2002. *Beliefs: A Hidden Variable in Mathematics Education?* Dordrecht: Kluwer Academic Publishers.

van Maanen, J.: 1995. Alluvial deposits, conic sections, and improper glasses, in F. Swetz, J. Fauvel, O. Bekken, B. Johansson & V. Katz (Eds), *Learn From The Masters*, pp. 73–92. The Mathematical Association of America.

27 On the Use of Primary Sources in the Teaching and Learning of Mathematics

van Maanen, J.: 1991. L'Hôpital's weight problem. *For the Learning of Mathematics*, 11(2), 44–47.

van Maanen, J.: 1997. New maths may profit from old methods. *For the Learning of Mathematics*, 17(2), 39–46.

Merton, R. K.: 1973. *The Sociology of Science – Theoretical and Empirical Investigations*. Chicago: The University of Chicago Press. Chapter 16. Previously published as "Singletons and Multiples in Scientific Discovery" in Proceedings of the American Philosophical Society 105, no. 5, 1961, pp. 470–486.

Niss, M.: 2001a. Indledning. In M. Niss (ed.), *Matematikken og verden*, Fremads debatbøger – Videnskab til debat. Copenhagen: Forfatterne og Forlaget A/S.

Niss, M.: 2001b. University mathematics based on problem-oriented student projects: 25 years of experience with the Roskilde model. In D. Holton (ed.), *The teaching and learning of mathematics at university level, an ICMI study*, pp. 153–165. Dordrecht: Kluwer Academic.

Niss, M., & Højgaard, T. (eds.): 2011. Competencies and mathematical learning – ideas and inspiration for the development of mathematics teaching and learning in Denmark. English Edition, October 2011. *Tekster fra IMFUFA*, no. 485. Roskilde: Roskilde University. (Published in Danish in 2002).

Pengelley, D.: 2011. Teaching with primary historical sources: should it go mainstream? Can it? In V. Katz & C. Tzanakis (eds.): *Recent Developments on Introducing a Historical Dimension in Mathematics Education*, pp. 1–8. MAA Notes 78. Washington: The Mathematical Association of America.

Pengelley, D.: 2012. Teaching number theory from Sophie Germain's manuscripts: a guided discovery pedagogy. In *HPM2012: The HPM Satellite Meeting of ICME-12*, Proceeding Book 1, pp. 103–113.

Rheinberger, H.-J.: (1997). *Toward a history of epistemic things: Synthesizing proteins in the test tube*. Stanford: Stanford University Press.

Schubring, G.: 2005. *Conflicts between Generalization, Rigor and Intuition. Number Concepts Underlying the Development of Analysis in 17th-19th Century France and Germany*. New York: Springer.

Schubring, G.: 2008. The debate on a 'geometric algebra' and methodological implications. Paper presented at the HPM 2008 satellite meeting of ICME 11, Mexico.

Schubring, G.: 2011. Conceptions for relating the evolution of mathematical concepts to mathematics learning – epistemology, history, and semiotics interacting. *Educational Studies in Mathematics*, **77**(1), 79–104.

Sfard, A.: 2008. *Thinking as communicating*. Cambridge: Cambridge University Press.

Shell-Gellasch, A. & Jardine, D.: 2005. *From calculus to computers: using the last 200 years of mathematics history in the classroom*. MAA Notes 68. Washington DC: The Mathematical Association of America.

Shannon, C. E.: 1938. *A symbolic analysis of relay and switching circuits*. Master's thesis. Cambridge: Massachusetts Institute of Technology.

Siu, M. K.: 2011. 1607, a year of (some) significance: Translation of the first European text in mathematics – Elements – into Chinese. In E. Barbin, M. Kronfellner, & C. Tzanakis (Eds.): *History and Epistemology in Mathematics Education Proceedings of the 6th European Summer University*, pp. 573–589. Vienna: Holzhausen Publishing Ltd.

Swetz, F., Fauvel, J., Bekken, O., Johansson, B., & Katz, V., (Eds.): 1995. *Learn from the masters*. Washington, DC: The Mathematical Association of America.

Ta'ani, O. H.: 2011. *An analysis of the contents and pedagogy of Al-Kashi's "Key to Arithmetic" (Miftah al-Hisab)*. Ph.D. thesis. Las Cruces: New Mexico State University.

Thomaidis, Y. & Tzanakis, C.: 2007. The notion of historical "parallelism" revisited: historical evolution and students' conception of the order relation on the number line. *Educational Studies in Mathematics*, 66(2). 165–183. Special issue on the history of mathematics in mathematics education edited by L. Radford, F. Furinghetti & V. Katz.

Toeplitz, O.: 1927. Das Problem der Universitätsvorlesungen über Infinitesimalrechnung und ihrer Abgrenzung gegenüber der Infinitesimalrechnung an den höheren Schulen. *Jahresbericht der Deutschen Mathematiker-Vereinigung*, XXXVI, 88–100.

Tzanakis, C.: 2000. 'Presenting the relation between mathematics and physics on the basis of their history: a genetic approach'. In: V. Katz (ed.): *Using History to Teach Mathematics – An International Perspective*, pp. 111–120. MAA Notes 51. Washington: The Mathematical Association of America.

Tzanakis, C. & Thomaidis, Y.: 2000. Integrating the close historical development of mathematics and physics in mathematics education: some methodological and epistemological remarks. *For the Learning of Mathematics*, 20(1), 44–55.

Tzanakis, C. & Thomaidis, Y.: 2012. Classifying the arguments and methodological schemes for integrating history in mathematics education. In Sriraman, B. (Ed.) *Crossroads in the History of Mathematics and Mathematics Education*. The Montana Mathematics Enthusiast Monographs 12, pp. 247–295. Charlotte: IAP.

Weil, A.: 1978. History of mathematics: Why and how. In O. Lehto (Ed.), *Proceedings of the international congress of mathematicians*, pp. 227–236. Hungary: Academia Scientiarum Fennica.

Uffe Thomas Jankvist is associate professor of mathematics education at Aarhus University, Dept. of Education, Emdrup, Denmark. He holds a M.Sc. in mathematics and computer science as well as a Ph.D. in mathematics education from Roskilde University and a postdoc from the University of Southern Denmark. His research interests include the use of history of mathematics, applications of mathematics, and philosophy of mathematics in mathematics education, both from a theoretical and an empirical point of view, including also students' beliefs about and images of mathematics as a (scientific) discipline. Uffe Thomas Jankvist has published in a variety of journals, e.g. *The Mathematical Intelligencer, Journal for Research in Mathematics Education, Educational Studies in Mathematics, Science & Education, For the Learning of Mathematics, Mathematical Thinking and Learning, Historia Mathematica, BSHM Bulletin*, and *ReLIME*. Besides being the Danish editor of *NOMAD – Nordic Studies in Mathematics Education* – he is currently involved in a new initiative to design and implement an educational program for in-service high school mathematics teachers to become 'math counselors', whose purpose shall be to provide special aid to high school students with learning difficulties in mathematics.

Part IX
Theoretical Studies: Features of Science and Education

Chapter 28
Nature of Science in the Science Curriculum: Origin, Development, Implications and Shifting Emphases

Derek Hodson

28.1 Introduction

Before proceeding to the substance of this chapter, it is important to clarify what I mean by *nature of science* (NOS) and note the ways in which I use the term differently from some others. A number of authors seek to restrict its use to the characteristics of scientific knowledge (i.e. to epistemological considerations) and to exclude consideration of the nature of scientific inquiry.[1] This might strike some as an odd decision, given that much of our scientific knowledge and, therefore, consideration of its status, validity and reliability is intimately bound up with the design, conduct and reporting of scientific investigations. Moreover, teaching activities focused on NOS often include empirical investigations and/or critical scrutiny of existing data. Thus, as Ryder (2009) points out, the conduct of scientific inquiry and epistemological considerations are related conceptually, procedurally and pedagogically. Lederman (2006) has acknowledged that 'the phrase 'nature of science' has caused the confusion and the phrase 'nature of scientific knowledge' might be more accurate. The conflation of NOS and scientific inquiry has plagued research on NOS from the beginning' (p. 2). In other words, it would be less confusing to readers if authors used the term 'nature of scientific knowledge (NOSK)' when referring to strictly and/or solely epistemological matters. In common with

---

[1] Abd-El-Khalick (2001, 2004, 2005), Abd-El-Khalick and Akerson (2004, 2009), Abd-El-Khalick et al. (1998, 2008), Bell (2004), Flick and Lederman (2004), Hanuscin et al. (2006), Khishfe and Abd-El-Khalick (2002), Khishfe and Lederman (2006, 2007), Lederman (2006, 2007), Lederman and Abd-El-Khalick (1998), Lederman et al. (2001, 2002).

D. Hodson (✉)
Ontario Institute for Studies in Education of the University of Toronto, Toronto, Canada

University of Auckland, Auckland, New Zealand
e-mail: d.hodson@auckland.ac.nz

M.R. Matthews (ed.), *International Handbook of Research in History, Philosophy and Science Teaching*, DOI 10.1007/978-94-007-7654-8_28, © Springer Science+Business Media Dordrecht 2014

several other recent publications,[2] the definition of NOS deployed in this chapter encompasses the characteristics of scientific inquiry; the role and status of the scientific knowledge it generates; the modelling that attends the construction of scientific theories; the social and intellectual circumstances of their development; how scientists work as a social group; the linguistic conventions for reporting, scrutinizing and validating knowledge claims; and the ways in which science impacts and is impacted by the social context in which it is located.

Given this much broader definition of NOS, it is quickly apparent that arguments for including NOS in the science curriculum have a long and chequered history. The long-standing tradition of concern for 'the public understanding of science' in the United Kingdom, encompassing much of what I refer to as NOS, dates back to the early years of the nineteenth century. As Jenkins (1990) notes, science was vigorously promoted through the activities of the numerous Mechanics' Institutes and Literary and Philosophical Societies and further supported by public lectures, scientific demonstrations and 'a remarkable variety of books, journals, tracts, pamphlets and magazines, many of which would be categorized today as 'teach yourself publications" (p. 43). Perhaps the earliest proposal for an NOS-oriented curriculum at the school level was Henry Armstrong's heuristic approach,[3] published in 1898, although it is important to note that Armstrong's interest in NOS was mainly pedagogical and motivational; the real purpose was to acquire and develop scientific knowledge. In contrast, John Dewey (1916) argued that familiarity with scientific method was substantially more important than acquisition of scientific knowledge, particularly for those who do not intend to study science at an advanced level. Similarly, Frederick Westaway (1929), an influential HM Inspector of Schools in the United Kingdom in the 1920s, made a strong case for a curriculum focus on NOS:

> Now that science enters so widely and so intimately into every department of life, especially in all questions relating to health and well-being, it is important that the community should have a general knowledge of its *scope and aims*. (p. 9, emphasis added)

Some years later, similar rhetoric formed the basis of Joseph Schwab's (1962) advocacy of a shift of emphasis for school science education in the United States away from the learning of scientific knowledge (the products of science) towards an understanding of the processes of scientific inquiry and the structure of scientific knowledge – a line of argument that eventually led to a string of innovative curriculum projects (PSSC, BSCS, CHEM Study, CBA, ECSP, etc.). NOS-oriented developments in the United Kingdom during the 1960s included the Nuffield Science Projects (with their emphasis on 'being a scientist for the day' and 'developing a proper attitude to theory') and the Schools Council Integrated Science Project (SCISP). However, as a direct consequence of their reliance on an impractical pedagogy

---

[2] Allchin (2011), Bartholomew et al. (2004), Clough (2006, 2011), Clough and Olson (2008), Elby and Hammer (2001), Hodson (2008, 2009, 2011), Kelly (2008), Matthews (2012), Osborne et al. (2003), Rudolph (2000), van Dijk (2011), and Wong and Hodson (2009, 2010).

[3] See Brock (1973), Jenkins (1979), Layton (1973) and van Praagh (1973).

of discovery learning and the naïve inductivist model of science underpinning it, these somewhat elitist courses failed to deliver on their rhetoric and promise. Those of us who were required to adopt the pedagogy of discovery learning during its heyday in the 1960s will vividly recall the frustrations of not being allowed to provide students with any guidance or suggest alternative lines of approach when investigating phenomena and events.[4] Subsequently attention shifted towards the so-called process approaches to science education, exemplified by *Warwick Process Science* (Screen 1986, 1988), *Science in Process* (ILEA 1987) and *Active Science* (Coles et al. 1988), which envisaged scientific inquiry as the application of a generalized, all-purpose algorithmic method. A similar shift occurred in Australia, with the publication of the *Australian Science Education Project* (ASEP 1974), and in the United States, with initiatives such as *Science-A Process Approach* (AAAS 1967) being developed on the basis of Robert Gagné's (1963) claim to have identified thirteen basic skills of scientific inquiry.

After a period of decline, interest in NOS underwent a remarkable revival in the decade and a half between 1977 and 1992, with the publication of a number of opinion pieces and commissioned reports,[5] the establishment of the International History, Philosophy and Science Teaching Group (1987) and the first of the now biennial IHPST conferences in Tallahassee in 1987 – developments that led, through the prodigious efforts of Michael Matthews, to the foundation in 1992 of *Science & Education*, the first journal devoted primarily to NOS issues in education. Of particular significance during this period was the incorporation of NOS as a key component in the National Curriculum for England and Wales, established in 1989 following the Education Reform Act of 1988. Another landmark was the publication of Matthews' book *Science Teaching: The Role of History and Philosophy of Science* (Matthews 1994).

Although there has been continuing controversy about what the NOS component of the curriculum should comprise and how it should be implemented (Donnelly 2001), the overall curricular importance of NOS understanding per se is no longer in dispute. Indeed, it has been subsumed within the wider discussion of scientific literacy,

---

[4] For example, early on in the original *Nuffield Physics* course, students are provided with a lever, a fulcrum and some weights (uniform square metal plates) and are invited to 'explore' and to 'find out what you can'. No particular problem is stated; no procedure is recommended. It is assumed that the Law of Moments will simply emerge from undirected, open-ended exploration. Nothing could be further from the truth. First, the system does not balance in the way the students expect because the pivot is below the centre of gravity. If the weights are suspended *below* the pivot, as in a set of scales, the beam will balance. However, there is little chance that children will discover this for themselves. Second, children tend to spread the weights irregularly along the entire length of the beam. The complexity of this arrangement obscures the simple relationship that is sought. Consequently, teachers begin to proffer advice on how to make the problem simpler and to issue instructions about the best way to proceed. Similar things happen whenever children are presented with this kind of open-ended situation. See Hodson (1996) for an extended discussion of these issues.

[5] See, for example, Cawthron and Rowell (1978), Hodson (1985, 1986, 1988a, b, 1990, 1991), Matthews (1991, 1992), Nadeau and Désautels (1984)) and Royal Society (1985).

a term that first appeared in the US educational literature about 50 years ago in papers by Paul Hurd (1958) and Richard McCurdy (1958) and in the Rockefeller Brothers Fund (1958) report *The Pursuit of Excellence*, and is now regarded as a key feature of most science curricula.

Despite the term scientific literacy being enthusiastically adopted by many science educators as a useful slogan or rallying call (see Roberts 1983, 2007), there was little in the way of precise or agreed meaning until Pella et al. (1966) suggested that it comprises an understanding of the basic concepts of science, the nature of science, the ethics that control scientists in their work, the interrelationships of science and society, the interrelationships of science and the humanities and the differences between science and technology. Almost a quarter century later, the authors of *Science for All Americans* (AAAS 1989) drew upon very similar categories to define a scientifically literate person as 'one who is aware that science, mathematics, and technology are interdependent human enterprises with strengths and limitations; understands key concepts and principles of science; is familiar with the natural world and recognizes both its diversity and unity; and uses scientific knowledge and scientific ways of thinking for individual and social purposes' (p. 4). It is significant that these perspectives are now an integral part of the US *National Science Education Standards* (National Research Council 1996) and a central plank of the framework for the *Programme for International Student Assessment* (PISA) studies (OECD 1999, 2006, www.pisa.oecd.org). Detailed review of the literature focused on defining notions of scientific literacy is outside the scope of this chapter,[6] save to note that elements of the history of science, philosophy of science and sociology of science that constitute a satisfactory understanding of the nature of science (NOS) have now become firmly established as a major component of scientific literacy and an important learning objective of science curricula in many countries.[7] Indeed, the promotion of NOS in official curriculum documents has become so prominent that Dagher and BouJaoude (2005) have stated: 'improving students' and teachers' understanding of the nature of science has shifted from a *desirable* goal to being a *central* one for achieving scientific literacy' (p. 378, emphasis added). It follows that all arguments for scientific literacy become arguments for NOS.

---

[6] Extensive discussion of the history and evolving definition of scientific literacy can be found in Bybee (1997a, b); Choi et al. (2011), De Boer (2001), Dillon (2009), Feinstein (2011), Gräber and Bolte (1997), Hodson (2008, 2011), Hurd (1998), Laugksch (2000), Lehrer and Schauble (2006), Lemke (2004), Linder et al. (2012), McEneaney (2003), Miller (2000), Norris and Phillips (2003), Norris et al. (2013), Oliver et al. (2001), Roberts (2007), Roth and Calabrese Barton (2004), and Ryder (2001). Teachers' understanding of scientific literacy is explored by Smith et al. (2012).

[7] For example, AAAS (1993), Council of Ministers of Education (1997), Department of Education (RSA) (2002), Goodrum et al. (2000), Millar and Osborne (1998), National Research Council (1996), Organization for Economic Cooperation and Development (1999, 2003), Osborne and Dillon (2008), and UNESCO (1993).

28.2 Arguments for NOS/Scientific Literacy in the School Science Curriculum

Reviewing what they describe as an extensive and diverse literature, Thomas and Durant (1987) identify three major categories of argument for promoting scientific literacy (and, therefore, aspects of NOS understanding): (i) benefits to science, (ii) benefits to individuals and (iii) benefits to society as a whole. Driver and colleagues (1996) contend that in addition to its intrinsic value, NOS understanding enhances learning of science content, generates interest in science and develops students' ability to make informed decisions on socioscientific issues based on careful consideration of evidence, while Erduran and colleagues (2007) argue that NOS knowledge (and the wider HPS understanding subsumed in the notion of scientific literacy) is of immense value to teachers, making them more reflective and more resourceful.

Benefits to science are seen largely in terms of increased numbers of recruits to science-based professions (including medicine and engineering), greater support for scientific, technological and medical research and more realistic public expectations of science. A related argument is that confidence and trust in scientists depend on citizens having some general understanding of what scientists do and how they do it – in particular, about what they choose to investigate, the methods they employ, how they validate their research findings and theoretical conclusions and where, how and to whom they disseminate their work.

Arguments that scientific literacy brings benefits to *individuals* come in a variety of forms. First, it is commonly argued that scientifically literate individuals will have access to a wide range of employment opportunities and are well positioned to respond positively and competently to the introduction of new technologies in the workplace. Second, it is widely assumed that those who are scientifically literate are better able to cope with the demands of everyday life in an increasingly technology-dominated society, better positioned to evaluate and respond appropriately to the scientific evidence and arguments (sometimes authentic and relevant, sometimes biased, distorted, fallacious or irrelevant) used by advertizing agencies and deployed by politicians and better equipped to make important decisions that affect their health, security and economic well-being.

Arguments that enhanced scientific literacy brings benefits to *society as a whole* include the familiar and increasingly pervasive economic argument and the claim that it promotes democracy and responsible citizenship. The first argument sees scientific literacy as a form of human capital that builds, sustains and develops the economic well-being of a nation. Put simply, continued economic development brought about by enhanced competitiveness in international markets (regarded as incontrovertibly a 'good thing') depends on science-based research and development, technological innovation and a steady supply of scientists, engineers and technicians. The case for scientific literacy as a means of enhancing democracy and responsible citizenship is just as strongly made as the economic argument, though by a very different assembly of stakeholders and interest groups. In the words of Chen and Novick (1984), enhanced scientific literacy (and its attendant components of NOS understanding) is a means 'to avert the situation where social values, individual involvement, responsibility, community participation and the very heart of democratic decision making will be dominated and practiced by a small elite' (p. 425).

This line of argument maintains that democracy is strengthened when *all* citizens are equipped to confront and evaluate socioscientific issues (SSI) knowledgeably and rationally, as well as emotionally, and are enabled to make informed decisions on matters of personal and public concern. Those who are scientifically illiterate are in many ways disempowered and excluded from active civic participation. For these reasons, Tate (2001) declares that access to high-quality science education, with its increasing emphasis on NOS, is a civil rights issue. Of course, as both Levinson (2010) and Tytler (2007) remind us, the notion of science education for citizenship raises a whole raft of questions about the kind of citizen and the kind of society we have in mind and about what constitutes *informed* and *responsible* citizenship – matters well outside the scope of this chapter.

A number of writers have claimed, somewhat extravagantly, that appreciation of the ethical standards and code of responsible behaviour that the scientific community seeks to impose on practitioners will lead to more ethical behaviour in the wider community – that is, the pursuit of scientific truth regardless of personal interests, ambitions and prejudice (part of the traditional image of the objective and dispassionate scientist) makes science a powerful carrier of moral values and ethical principles: 'Science is in many respects the systematic application of some highly regarded human values – integrity, diligence, fairness, curiosity, openness to new ideas, skepticism, and imagination' (AAAS 1989, p. 201). Shortland (1988) summarizes this rationale as follows: 'the internal norms or values of science are so far above those of everyday life that their transfer into a wider culture would signal a major advance in human civilization' (p. 310). The authors of *Science for All Americans* (AAAS 1989) present a similar argument: 'Science is in many respects the systematic application of some highly regarded human values – integrity, diligence, fairness, curiosity, openness to new ideas, skepticism, and imagination' (p. 201). Studying science, scientists and scientific practice will, they argue, help to instill these values in students. In other words, scientific literacy doesn't just result in more skilled and more knowledgeable people, it results in *wiser* people, that is, people well-equipped to make morally and ethically superior decisions. Whether contemporary scientific practice does impose and instill these values is discussed later in the chapter.

28.3 Establishing NOS Priorities

Once the lens of NOS became focused on the school science curriculum, it was quickly apparent that whatever confused and confusing views of science are held by students are compounded by conventional science education. There are particularly powerful messages about science embedded in all teaching and learning activities, especially laboratory activities. These messages too often convey distorted or over-simplified views of the nature of scientific investigations, especially with respect to the role of theory. These 'folk theories' of science, as Windschitl (2004) calls them, are also held by teachers (as a consequence of their own science education) and have substantial influence on their day-to-day curriculum decision-making, thus reinforcing similar messages embedded in school science textbooks and other curriculum materials.

As part of a major survey of Canadian science education conducted by the Science Council of Canada, Nadeau and Désautels (1984) identified what they called five mythical values stances suffusing science education:

- *Naïve realism* – science gives access to truth about the universe.
- *Blissful empiricism* – science is the meticulous, orderly and exhaustive gathering of data.
- *Credulous experimentation* – experiments can conclusively verify hypotheses.
- *Excessive rationalism* – science proceeds solely by logic and rational appraisal.
- *Blind idealism* – scientists are completely disinterested, objective beings.

The cumulative message is that science has an all-purpose, straightforward and reliable method of ascertaining the truth about the universe, with the certainty of scientific knowledge being located in objective observation, extensive data collection and experimental verification. Moreover, scientists are rational, logical, open-minded and intellectually honest people who are required, by their commitment to the scientific enterprise, to adopt a disinterested, value-free and analytical stance. In Cawthron and Rowell's (1978) words, the scientist is regarded by the science curriculum as 'a depersonalized and idealized seeker after truth, painstakingly pushing back the curtains which obscure objective reality, and abstracting order from the flux, an order which is directly revealable to him through a distinctive scientific method' (p. 32). While much has changed in the intervening years, many school science curricula and school textbooks continue to project these images.[8] For example, Loving (1997) laments that all too often

(a) science is taught totally ignoring what it took to get to the explanations we are learning – often with lectures, reading text, and memorizing for a test. In other words, it is taught free of history, free of philosophy, and in its final form. (b) Science is taught as having one method that all scientists follow step-by-step. (c) Science is taught as if explanations are the truth – with little equivocation. (d) Laboratory experiences are designed as recipes with one right answer. Finally, (e) scientists are portrayed as somehow free from human foibles, humor, or any interests other than their work. (p. 443)

At about the same time, Hodson (1998) identified ten common myths and falsehoods promoted, sometimes explicitly and sometimes implicitly, by the science curriculum: observation provides direct and reliable access to secure knowledge;

---

[8] Abd-El-Khalick (2001), Abd-El-Khalick et al. (2008), Clough (2006), Cross (1995), Knain (2001), Kosso (2009), Lakin and Wellington (1994), McComas (1998), van Eijck and Roth (2008) and Vesterinen et al. (2011).

science always starts with observation; science always proceeds by induction; science comprises discrete, generic processes; experiments are decisive; scientific inquiry is a simple algorithmic procedure; science is a value-free activity; science is an exclusively Western, post-Renaissance activity; the so-called scientific attitudes are essential to the effective practice of science; and all scientists possess these attitudes. A broadly similar list of falsehoods was generated by McComas (1998) from his critical reading of science textbooks: hypotheses become theories that in turn become laws; scientific laws and other such ideas are absolute; a hypothesis is an educated guess; a general and universal scientific method exists; evidence accumulated carefully will result in sure knowledge; science and its methods provide absolute proof; science is procedural more than creative; science and its methods can answer all questions; scientists are particularly objective; experiments are the principal route to scientific knowledge; scientific conclusions are reviewed for accuracy; acceptance of new scientific knowledge is straightforward; science models represent reality; science and technology are identical; and science is a solitary pursuit. In quite startling contrast, Siegel (1991) states that

> Contemporary research… has revealed a more accurate picture of the scientist as one who is driven by prior convictions and commitments; who is guided by group loyalties and sometimes petty personal squabbles; who is frequently quite unable to recognize evidence for what it is; and whose personal career motivations give the lie to the idea that the scientist yearns only or even mainly for the truth. (p. 45)

Two questions spring to mind. First, is this a more authentic portrayal of scientific practice? Second, is it an appropriate view for the school science curriculum? Sweeping away an old and (for some) discredited view is one thing; finding an acceptable set of alternatives is somewhat different. Finding a list appropriate for the school curriculum is even more difficult. Many science educators will share Israel Scheffler's alarm at some of the alternatives that have been advanced:

> The extreme alternative that threatens is the view that theory is not controlled by data, but that data are manufactured by theory; that rival hypotheses cannot be rationally evaluated, there being no neutral court of observational appeal nor any shared stock of meanings; that scientific change is a product not of evidential appraisal and logical judgment, but of intuition, persuasion and conversion; that reality does not constrain the thought of the scientist but is rather itself a projection of that thought. (Scheffler 1967, p. v)

Longbottom and Butler (1999) express similar concerns when they state that 'if we go along with those who deny that modern science provides a privileged view of the world… we fall into an abyss where skeptical postmodernists, who have lost faith in reason, dismiss all knowledge claims as equally arbitrary and assume the universe to be unreliable in its behavior and incapable of being understood' (p. 482). Stanley and Brickhouse (1995) regard such remarks as examples of what Bernstein (1983) called 'Cartesian anxiety': the fear that if we do not retain our belief in the traditional objective foundations of scientific method we have no rational basis for making any knowledge claims. In short, fear that belief in scientific *progress* will be replaced by scientific *change* consequent upon power struggles among competing groups, with 'victory' always going to the better resourced. Fear that scientific knowledge is no longer to be regarded as the product of a rigorous method or

28 Nature of Science in the Science Curriculum...

set of methods; instead, it is merely the way a particular influential group of scientists happens to think and can persuade, cajole or coerce others into accepting.

In building a school science curriculum, are we faced with a stark choice between the traditional and the postmodern? Are we required to choose between the image of a scientist as a cool, detached seeker-after-truth patiently collecting data from which conclusions will eventually be drawn, when all the evidence is in hand, and that of 'an agile opportunist who will switch research tactics, and perhaps even her entire agenda, as the situation requires' (Fuller 1992, p. 401). Which view is the more authentic? Equally important, what should we tell students? What is in their interests? Some years ago, Stephen Brush (1974) posed the question: 'should the history of science should be rated X?' The question is just as pertinent to the philosophy of science and the sociology of science. Should we expose students to the anarchistic epistemology of Paul Feyerabend? Should we lift the lid off the Pandora's Box that is the sociology of science? Would students be harmed by too early an exposure to these views? When we seek to question (and possibly reject) the certainties of the traditional view of science, are we left with no firm guidance, no standards and no shared meaning? Does recognition of the sociocultural baggage of science entail regarding science as just one cultural artefact among many others, with no particular claim on our allegiance? Is any kind of compromise possible between these extremes and among this diversity? Can we retain what is still good and useful about the old view of science (such as conceptual clarity and stringent testing) while embracing what is good and useful in the new (such as sensitivity to sociocultural dynamics and awareness of the possibility of error, bias, fraud and the misuse of science)? Can the curriculum achieve a balance that is acceptable to most stakeholders? In short, what particular items from all the argument and counter argument would constitute an educationally appropriate and teachable selection? Later discussion touches on the age appropriateness of a number of NOS items, while attention at this point focuses on whether there is any nature of science understanding that can be taken for granted and regarded as no longer in dispute. Is there any consensus among scholars about an acceptable alternative to the traditional view that will allay the fears expressed by Scheffler and others?

Responses to a 20-item Likert-type questionnaire on '15 tenets of NOS' led Alters (1997a, b) to conclude that *there is no consensus* – at least, not among the 210 philosophers of science he surveyed. In the words of Laudan and colleagues (1986),

> The fact of the matter is that we have no well-confirmed general picture of how science works, no theory of science worthy of general assent. We did once have a well developed and historically influential philosophical position, that of positivism or logical empiricism, which has by now been effectively refuted. We have a number of recent theories of science which, while stimulating much interest, have hardly been tested at all. And we have specific hypotheses about various cognitive aspects of science, which are widely discussed but wholly undecided. If any extant position does provide a viable understanding of how science operates, we are far from being able to identify which it is. (p. 142)

Interestingly, despite this categorical denial of any consensus, it seems that the authors of several important science curriculum reform documents (AAAS (1989, 1993) and NRC (1996), among others) seem to be in fairly substantial agreement on the elements of NOS that should be included in the school science curriculum (McComas and Olson 1998):

- Scientific knowledge is tentative.
- Science relies on empirical evidence.
- Observation is theory laden.
- There is no universal scientific method.
- Laws and theories serve different roles in science.
- Scientists require replicability and truthful reporting.
- Science is an attempt to explain natural phenomena.
- Scientists are creative.
- Science is part of social tradition.
- Science has played an important role in technology.
- Scientific ideas have been affected by their social and historical milieu.
- Changes in science occur gradually.
- Science has global implications.
- New knowledge must be reported clearly and openly.

In an effort to shed further light on this matter, Osborne and colleagues (2003) conducted a Delphi study to ascertain the extent of agreement among 23 participants drawn from the 'expert community' on what ideas about science should be taught in school science. The participants included five scientists, five persons categorized as historians, philosophers and/or sociologists of science, five science educators, four science teachers and four science communicators. Although there was some variation among individuals, there was broad agreement on nine major themes: scientific method and critical testing, scientific creativity, historical development of scientific knowledge, science and questioning, diversity of scientific thinking, analysis and interpretation of data, science and certainty, hypothesis and prediction, and cooperation and collaboration. A comparison of these themes with those distilled from the science education standards documents in McComas and Olson's (1998) study reveals many similarities. A broadly similar but shorter list that has gained considerable currency among science educators can be found in Lederman and colleagues (2002): scientific knowledge is tentative, empirically based, subjective (in the sense of being theory dependent and impacted by the scientists' experiences and values), socioculturally embedded and, in part, the product of human imagination and creativity.

28.4   Some Problems with the Consensus View

Useful as consensus can be in assisting curriculum planning and the design of assessment and evaluation schemes, a number of questions should be asked. For example, is the apparent consensus deliberately pitched at such a trivial level that nobody could possibly quibble with it? Most of the items in the list are not specific to science, either individually or collectively. All human knowledge is tentative;

all forms of knowledge building are creative. This is not to say that these characteristics are not applicable to science; but it is to say that they do not distinguish it from several other human activities. It is the sheer banality and unhelpfulness of some of the items that many teachers find frustrating. For example, statements such as 'science is an attempt to explain natural phenomena' and 'science has played an important role in technology' – items in the consensus list developed by McComas et al. (1998) – do not claim anything particularly insightful or helpful for students trying to understand what science is all about. Of course, some would argue that a list of relatively trivial items is better than no list at all. Perhaps it is, although items in the consensus list can sometimes be very puzzling or even irrelevant to an understanding of scientific practice and the capacity to function as a scientist. For example, several writers who advocate the consensus view also argue that students should understand the functions of and relationships between theories and laws and draw a distinction between observation and inference. Drawing a distinction between laws and theories is certainly not a high priority for practising scientists, as informants in the study conducted by Wong and Hodson (2009) pointed out very clearly. As far as students are concerned, one is led to wonder in what ways knowledge of a supposed difference between a law and a theory would help them to make decisions on where they stand in relation to controversial socioscientific issues.

The naïve proposition that there is a crucial distinction between observation and inference is singularly unhelpful to students trying to make sense of contemporary technology-supported investigative work. Superficially the distinction sounds fine and seems to accord with what we consider to be good practice in scientific inquiry: having respect for the evidence and not claiming more than the data can justify. However, closer examination in the light of the theory-laden nature of scientific observation suggests that the supposed demarcation is not always as clear as some would claim. When a new theory appears or when new scientific instruments are developed, our notion of what counts as an observation and what counts as an inference may change. As Feyerabend (1962) points out, observation statements are merely those statements about phenomena and events to which we can assent quickly, relatively reliably and without calculation or further inference because we all accept, without question, the theories on which they are based. Thus, where individuals draw the line between observation and inference reflects the sophistication of their scientific knowledge, their confidence in that knowledge and their experience and familiarity with the phenomena or events being studied. When theories are not in dispute, when they are well understood and taken for granted, the theoretical language *is* the observation language, and we use theoretical terms in making and reporting observations. Terms like *reflection* and *refraction*, *conduction* and *nonconduction*, and *melting, dissolving* and *subliming*, all of which are used regularly in school science as observation terms, carry a substantial inferential component rooted in theoretical understanding. The key point is that unless some theories are taken for granted (and deemed to be no longer in dispute) and unless theory-loaded terms are used for making observations, we can never make progress. We would forever be trying to retreat to the raw data, to some position that we could regard as theory-free.

Too literal an interpretation of statements about the tentative nature of science can be counterproductive, leading students to regard *all* science as no more than temporary (Harding and Hare 2000). Scientific knowledge is tentative because it is based, ultimately, on empirical evidence that may be incomplete and because it is collected and interpreted in terms of current theory – theory that may eventually be changed as a consequence of the very evidence that is collected. In all these endeavours, the creative imagination of individual scientists is impacted by all manner of personal experiences and values. Moreover, the collective wisdom of the scientific community that supports the practice, scrutinizes the procedures and evaluates the products is also subject to complex sociopolitical, economic and moral-ethical forces. In consequence, there can be no certainty about the knowledge produced. However, to admit that absolute truth is an impossible goal is not to admit that we are uncertain about everything. We *know* many things about the universe even though we recognize that many of our theoretical systems are still subject to revision, or even rejection.

Regarding the issue of tentativeness, there are several closely related issues to consider. First, very specific claims about phenomena and events may be regarded as 'true' (in a scientific sense) even though the theories that account for the events are regarded as tentative. Because the whole necessarily extends beyond the parts of which it is comprised, the whole may be seen as tentative while the parts (or some of them) are regarded as certain. Most theories are tentative when first developed, but are accepted as true when they have been elaborated, refined and successfully used and when they are consistent with other theories and strongly supported by evidence. Teachers make a grave mistake when they encourage students to regard all science as tentative. Indeed, if scientists did not accept some knowledge as well established, we would be unable to make progress.

We should also ask whether the consensus list includes consideration of the 'big issues' with which philosophers of science have traditionally grappled. Apparently not, according to Abd-El-Khalick and BouJaoude (1997), Abd-El-Khalick, Bell and Lederman (1998) and Lederman et al. (2002), who state that while philosophers and sociologists might disagree on some aspects of NOS, these disagreements are irrelevant to K-12 students and their teachers. Many other scholars would disagree. Some of these disputes focus on the most interesting features of science, for example, the status of scientific knowledge in terms of realism and instrumentalism, the extent to which science is socially constructed/determined and the nature of scientific rationality. Another major concern with the consensus view is that it promotes a static picture of science and fails to acknowledge important differences among the sciences. In reality, the practices and procedures of science change over time. As a particular science progresses and new theories and procedures are developed, the nature of scientific reasoning changes. Indeed, we should seriously question whether views in the philosophy of science that were arrived at some years ago can any longer reflect the nature of twenty-first-century science, especially in rapidly developing fields such as genetics and molecular biology, where there is now substantial research related to the generation of data and subsequent data mining (e.g. generation of genomic sequences of a number of living things) rather than the kind of hypothesis-driven inquiry promoted by the consensus view – developments that are, of course, driven by technological advances.

In a little known but very insightful and educationally significant article, Michael Clough (2007) urges teachers to shift emphasis away from teaching the 'tenets of NOS', because they are easily misinterpreted, oversimplified and become something to be memorized rather than understood and utilized, and towards asking important questions such as the following: In what sense is scientific knowledge tentative and in what sense is it durable? To what extent is scientific knowledge socially and culturally embedded? In what sense does it transcend society and culture? How are observations and inferences different? In what sense can they not be differentiated? A recent essay by Michael Matthews (2012) subjects the consensus view (specifically, the 'Lederman Seven', as he calls it) to rigorous critical scrutiny, concluding that the items need to be 'much more philosophically and historically refined and developed' (p. 12) if they are to be genuinely useful to teachers and their students. As a way forward, he advocates a shift of terminology and research focus from the 'essentialist and epistemologically focussed 'Nature of Science' (NOS) to a more relaxed, contextual and heterogeneous 'Features of Science' (FOS)' (p. 4). Such a change, he argues, would avoid many of the pitfalls and shortcomings of current research and scholarship in the field – in particular, the confused conflation of epistemological, sociological, psychological, ethical, commercial and philosophical aspects of science into a single list of items to be taught and assessed, the avoidance of debate about contentious issues in HPS, the neglect of historical perspective and the failure to account for significant differences in approach among the sciences. In response to this and other criticism, Lederman, Antinck and Bartos (2012) state 'We (my colleagues and fellow researchers) *are not* advocating a definitive or universal definition of the construct [of NOS]. We have never advocated that that our "list" is *the* only list/definition… What we prefer readers to focus on are the understandings we want students to have. The understandings need not be limited to those we have selected' (p. 2).

28.5 Diversity Among the Sciences

Many philosophers of science hold that there is no universal nature of science because the sciences themselves have no unity. The best that can be said is that there is a 'family resemblance among the sciences' (Wittgenstein 1953), with common interests and some areas of methodological and conceptual agreement – what Loving (1997) calls a 'loose configuration of critical processes and conceptual frameworks, including various methods, aims, and theories all designed to shed light on nature' (p. 437). The consensus view specifically disallows consideration of diversity among the sciences and chooses to disregard the substantial differences between the day-to-day activities of palaeontologists and epidemiologists, for example, or between scientists researching in high energy physics and those engaged in molecular biology. There are significant differences among the subdisciplines of science in terms of the kind of research questions asked, the methods and technologies employed to answer them, the kind of evidence sought, the extent to which they use experimentation, the ways in which data for theory building are collected, the standards by which investigations and conclusions are judged and the kinds of arguments deployed. Jenkins (2007) puts it succinctly when he says that 'the criteria for deciding what counts as evidence, and thus the nature of an explanation that relies upon that evidence, may also be different' (p. 225). There are substantial differences in the extent to which mathematics is deployed (Knorr-Cetina 1999), and there may even be differences, as Cartwright (1999) notes, in the values underpinning the enterprise. In other words, the specifics of scientific rationality change between subdisciplines, with each subdiscipline playing the game of science according to its own rules, a view discussed at some length in Hodson (2008, 2009).

Like Sandra Harding (1986), Ernst Mayr (1988, 1997, 2004) has criticized the standard or consensus NOS views promoted in many curriculum documents on grounds that they are nearly always derived from physics. Biology, he argues, is markedly different in many respects, not the least significant of which is that many biological ideas are not subject to the kind of falsificationist scrutiny advocated by Karl Popper (1959) and given such prominence in school science textbooks: 'It is particularly ill-suited for the testing of probabilistic theories, which include most theories in biology... And in fields such as evolutionary biology... it is often very difficult, if not impossible, to decisively falsify an individual theory' (Mayr 1997, p. 49).

The procedures of investigation in a particular subdiscipline of science are deeply grounded in the field's substantive aspects and the specific purposes of the inquiry. For example, while physicists may spend time designing critical experiments to test daring hypotheses, as Popper (1959) states, most chemists are intent on synthesizing new compounds:

> Chemists make molecules. They do other things, to be sure – they study the properties of these molecules; they analyze... they form theories as to why molecules are stable, why they have the shapes or colors that they do; they study mechanisms, trying to find out how molecules react. But at the heart of their science is the molecule that is made, either by a natural process or by a human being. (Hoffmann 1995, p. 95)

Moreover, as a particular science progresses and new theories and procedures are developed, the nature of scientific reasoning may change. Indeed, Mayr (1988, 2004) has distinguished two different fields even within biology: *functional* or mechanistic biology and *evolutionary* biology, distinguished by the type of causation addressed. Functional biology addresses questions of proximate causation; evolutionary biology addresses questions of ultimate causation:

> The functional biologist is vitally concerned with the operation and interaction of structural elements, from molecules up to organs and whole individuals. His ever-repeated question is 'How?'... The evolutionary biologist differs in his method and in the problems in which he is interested. His basic question is 'Why?' (Mayr 1988, p. 25)

In similar vein, Ault (1998) argues that the geosciences are fundamentally historical and interpretive, rather than experimental. The goal of geological inquiry, he argues, is interpretation of geologic phenomena based on observations, carefully warranted inferences and integration or reconciliation of independent lines of inquiry, often conducted in diverse locations. These interpretations result in a description of historical sequences of events, *sometimes* accompanied by a causal model.

Elby and Hammer (2001) argue that the widely adopted consensus list of NOS items is too general and too broad and that it is neither philosophically valid nor productive of good learning of science: 'a sophisticated epistemology does not consist of blanket generalizations that apply to all knowledge in all disciplines and contexts; it incorporates contextual dependencies and judgments' (p. 565). Essentially the same point is made by Clough (2006) when he says that 'while some characteristics [of NOS] are, to an acceptable degree uncontroversial… most are contextual, with important and complex exceptions' (p. 463). In short, the differences in approach are just too extensive and too significant to be properly accounted for by generic models of inquiry. Instead of trying to find and promote broad generalizations about the nature of science, scientific inquiry and scientific knowledge, a position recently given renewed emphasis by Abd-El-Khalick (2012), teachers should be building an understanding of NOS from examples of the daily practice of diverse groups of scientists engaged in diverse practices and should be creating opportunities for students to experience, explore and discuss the differences in knowledge and its generation across multiple contexts. It is for this reason that NOS-oriented research needs to study the work of scientists active at the frontier of knowledge generation (Schwartz and Lederman 2008; Wong and Hodson 2009, 2010). Student understanding of the complexity and diversity of scientific practice would be immeasurably helped by adoption of the notion of a 'family resemblance' among the sciences, as in Irzik and Nola's (2011) organization of the cognitive aspects of science into four categories: (i) *activities* (planning, conducting and making sense of scientific inquiries), (ii) *aims and values*, (iii) *methodologies and methodological rules*, and (iv) *products* (scientific knowledge) (see also Nola and Irzik 2013). These four categories of cognitive aspects could and perhaps should be extended to accommodate the noncognitive institutional and social norms which are operative within science and influence science (see below).

In brief, it is time to replace the consensus view of NOS, useful though it has been in promoting the establishment of NOS in the school science curriculum, with a philosophically more sophisticated and more authentic views of scientific practice, as advocated by Elby and Hammer (2001), Hodson (2008, 2009), Matthews (2012), Rudolph (2000) and Wong and Hodson (2013). Interestingly, children regard diversity of approach in scientific investigations as inevitable. They have no expectations of a particular method; it is the teachers who create the expectation of a single method through their continual reference to *the* scientific method (Hodson 1998) and, by extension, establish the belief that there are particular and necessary attributes (the so-called scientific attitudes) for engaging in it.

28.6 Some Recent NOS-Oriented Initiatives

The past decade has seen a remarkable growth in research and curriculum development in two important NOS-related areas: *scientific argumentation* and *modelling*. Both these aspects of NOS (as defined at the beginning of this chapter) warrant some attention here. My concerns relate to both students' knowledge of these processes as used by scientists and the development of their ability to use them appropriately and productively for themselves.

What is often unrecognized by science teachers, science textbooks and curricula, and by the wider public, is that *dispute* is one of the key driving forces of science. Real science is impregnated with claims, counter claims, argument and dispute. Arguments concerning the appropriateness of experimental design, the interpretation of evidence and the validity of knowledge claims are located at the core of scientific practice. Arguments are used to address problems, resolve issues and settle disputes. Moreover, our day-to-day decision-making with regard to socioscientific issues is based largely on the evaluation of information, arguments, conclusions, views, opinions and reports made available via newspapers, magazines, television, radio and the Internet. Citizens need to know the kinds of knowledge claims that scientists make and how they advance them. They need to understand the standards, norms and conventions of scientific argumentation in order to judge the rival merits of competing arguments and engage meaningfully in debate on SSI. In particular, they need a robust understanding of the form, structure and language of scientific arguments, the kind of evidence invoked, how it is organized and deployed and the ways in which theory is used and the work of other scientists cited to strengthen a case.

Neglect of scientific argumentation in the school science curriculum gives the impression that science is the unproblematic accumulation of data and theory. In consequence, students are often puzzled and may even be alarmed by reports of disagreements among scientists on matters of contemporary importance. They may be unable to address in a critical and confident way the claims and counter claims impregnating the SSI with which they are confronted in daily life. A number of science educators have recently turned their attention to these matters and to what had previously been a shamefully neglected area of research and curriculum development.[9] The research agenda focuses on the following questions: Why is argumentation important? What are the distinctive features of scientific argumentation? How can it be taught? What strategies are available? To what extent and in what

---

[9] For example, Arduriz Bravo (2013), Berland and Hammer (2012), Berland and Lee (2012), Berland and McNeill (2010), Berland and Reiser (2009, 2011), Böttcher and Meisertt (2011), Bricker and Bell (2008), Driver et al. (2000), Duschl (2008), Duschl and Osborne (2002), Erduran et al. (2004), Evagorou and Osborne (2013), Ford and Wargo (2012), Jiménez-Aleixandre and Erduran (2008), Khishfe (2012a), Kuhn (2010), Newton et al. (1999), Nielsen (2012a, b, 2013), Osborne (2001), Osborne and Patterson (2011), Osborne et al. (2004), Passmore and Svoboda (2012), Pluta et al. (2011), Sampson and Clark (2008, 2011), Sampson and Blanchard (2012), Sampson and Walker (2012), Sampson et al. (2011), Sandoval and Cam (2011), Sandoval and Millwood (2005, 2008), Simon et al. (2006), and Ryu and Sandoval (2012)

ways are the strategies successful? What problems arise and how can the difficulties be overcome? This research is discussed at length in Hodson (2009).

Another significant NOS-related growth area in recent years has been the focus on models and modelling. Because scientific literacy entails a robust understanding of a wide range of scientific ideas, principles, models and theories, students need to know something of their origin, scope and limitations; understand the role of models in the design, conduct, interpretation and reporting of scientific investigations; and recognize the ways in which a complex of cognitive problems and factors related to the prevailing sociocultural context influenced the development of key ideas over time. They also need to experience model building for themselves and to give and receive criticism in their own quest for better models. As Matthews (2012) comments, 'It is difficult to think of science without models' (p. 19).

The nature of mental models has long been an area of research in cognitive psychology, dating back to the seminal work of Johnson-Laird (1983) and Gentner and Stevens (1983), but in recent years, the topic of models and modelling has generated considerable interest among science educators.[10] This interest can be categorized into three principal areas of concern: the particular models and theories produced by scientists as explanatory systems, including the history of their development; the ways in which scientists utilize models as cognitive tools in their day-to-day problem solving, theory articulation and theory revision; and the role of models and modelling in science pedagogy.

The emergence of curricula oriented towards the consideration of socioscientific issues (SSI), in which NOS plays a key role, is discussed later in the chapter.

28.7 Assessing NOS Understanding

Given the perennial concern of education policy makers with assessment and accountability measures and the need for teachers to ascertain students' knowledge and understanding both prior to and following instruction, there has been a long-standing interest in researching students' NOS views. Also, given the commonsense understanding that teachers' views will inevitably and profoundly impact the kind of teaching and learning experiences they provide, interest has been high in ascertaining

---

[10] Bamberger and Davis (2013), Clement and Rea-Ramirez (2008), Coll (2006), Coll and Taylor (2005), Coll and Treagust (2002, 2003a, b), Coll et al. (2005), Davies and Gilbert (2003), Duschl and Grandy (2008), Erduran and Duschl (2004), Franco et al. (1999), Gilbert (2004), Gilbert and Boulter (1998, 2000), Gilbert et al. (1998a, b), Gobert and Pallant (2004), Gobert et al. (2011), Greca and Moreira (2000, 2002), Halloun (2004, 2007), Hansen et al. (2004), Hart (2008), Justi and Gilbert (2002a, b, c, 2003), Justi and van Driel (2005), Kawasaki et al. (2004), Khan (2007), Koponen (2007), Lehrer and Schauble (2005), Lopes and Costa (2007), Maia and Justi (2009), Manz (2012), Nelson and Davis (2012), Nersessian (2008), Oh and Oh (2011), Perkins and Grotzer (2005), Russ et al. (2008), Saari and Viiri (2003), Shen and Confrey (2007), special issue of *Science & Education* (2007, 16, issues 7–8), Svoboda et al. (2013), Taber (2003), Taylor et al. (2003), Treagust et al. (2002, 2004), and van Driel and Verloop (1999)

teachers' NOS views. Given suitable modification in terms of language and theoretical sophistication, the two tasks can utilize many of the same instruments.

Methods employed include questionnaires and surveys, interviews, small group discussions, writing tasks and classroom observations (particularly in the context of hands-on activities). Each has its strengths and weaknesses. Necessarily, researchers who use questionnaire methods must decide what counts as legitimate research data *before* the data collection process begins; those who use classroom observation (and, to a lesser extent, those who use interview methods) are able to make such decisions *during* or *after* data collection. They also have the luxury of embracing multiple perspectives and can readily update their interpretive frameworks to take account of changes in our understanding in history, philosophy and sociology of science.

More than 30 years ago, a review by Mayer and Richmond (1982) listed 32 NOS-oriented assessment instruments, among the best known of which are the *Test on Understanding Science* (TOUS) (Cooley and Klopfer 1961), the *Nature of Science Scale* (NOSS) (Kimball 1967), the *Nature of Science Test* (NOST) (Billeh and Hasan 1975) and the *Nature of Scientific Knowledge Scale* (NSKS) (Rubba 1976; Rubba and Anderson 1978), together with a modified version (M-NSKS) developed by Meichtry (1992). Instruments dealing with the processes of science, such as the *Science Process Inventory* (SPI) (Welch 1969a), the *Wisconsin Inventory of Science Processes* (WISP) (Welch 1969b) and the *Test of Integrated Process Skills* (TIPS) (Burns et al. 1985; Dillashaw and Okey 1980) could also be regarded as providing valuable information on some key aspects of NOS.

While questionnaires are the most commonly used research methods, largely because they are quick and easy to administer, they can be overly restrictive, incapable of accommodating subtle shades of meaning and susceptible to misinterpretation. Sometimes the complexity and subtlety of NOS issues makes it difficult to find appropriate language for framing questions. If it is difficult for the researcher to find the right words, how much more difficult is it for the respondent to capture the meaning they seek to convey? It cannot be assumed that the question and/or the answer will be understood in exactly the way it was intended, especially by younger students and those with poor language skills. Multiple-choice items and other objective instruments leave little or no scope for expressing doubt or subtle shades of difference in meaning and rarely afford respondents the opportunity to explain *why* they have made a particular response to a questionnaire item. It may even be that the same response from two respondents arises from quite different understanding and reasoning, while similar reasoning by two respondents results in different responses.

Further, many instruments are constructed in accordance with a particular philosophical position and are predicated on the assumption that all scientists think and behave in the same way. Hence, teacher and/or student responses that do not correspond to the model of science assumed in the test are judged to be 'incorrect', 'inadequate' or 'naïve'. Alters (1997a, b), Koulaidis and Ogborn (1995), Lucas (1975) and Lederman et al. (2002) provide extended discussions of this issue. It is also the case that many of the early instruments predated significant work in the philosophy and sociology of science, and so are of severely limited value in contemporary studies. Reviews by Lederman (1992, 2007), Lederman et al. (1998, 2000, 2013) describe several NOS instruments that take into account the work of more recent and even contemporary scholars in the philosophy and sociology of science, including *Conceptions of Scientific Theories Test* (COST) (Cotham and Smith 1981), *Views on Science-Technology-Society* (VOSTS) (Aikenhead et al. 1989), the *Nature of Science Survey* (Lederman and O'Malley 1990), the *Nature of Science Profile* (Nott and Wellington 1993) and the *Views of Nature of Science Questionnaire* (VNOS) (Lederman et al. 2002) and its several subsequent modifications (see Flick and Lederman 2004; Lederman 2004, 2007; Schwartz and Lederman 2008). A recent review by Deng and colleagues (2011) reports and critiques 105 research studies of students' NOS views, using a wide range of instruments, though lack of space precludes discussion here. Constraints on space also preclude discussion of the recent critical review by Guerro-Ramos (2012) of research approaches for ascertaining teachers' views of NOS and their relevance to classroom decision-making.

The designers of VOSTS attempted to circumvent some of the common questionnaire design problems identified by psychometricians by constructing a number of different 'position statements' (sometimes up to ten positions per item) derived from student writing and interviews, including 'I don't understand' and 'I don't know enough about this subject to make a choice' (Aikenhead et al. 1987; Aikenhead and Ryan 1992). It is the avoidance of the forced choice and the wide range of aspects covered (definitions, influence of society on science/technology, influence of science/technology on society, characteristics of scientists, social construction of scientific knowledge, social construction of technology, nature of scientific knowledge, and so on) that give the instrument its enormous research potential. Lederman and O'Malley (1990) utilized some of the design characteristics of VOSTS to develop the *Nature of Science Survey*, an instrument comprising just seven fairly open-ended items (e.g. 'Is there a difference between a scientific theory and a scientific law? Give an example to illustrate your answer'), to be used in conjunction with follow-up interviews to further explore and clarify students' responses.

At present, the most widely used and most extensively cited contemporary instrument for ascertaining students' NOS views is the *Views of Nature of Science Questionnaire* (VNOS). While it has provided much valuable information on both students' and teachers' NOS views, it suffers from all the drawbacks attending the so-called consensus view of NOS, as discussed earlier. The *Views on Science and Education* (VOSE) questionnaire, developed by Chen (2006) for use with preservice teachers, focuses on the same seven NOS elements as VNOS (tentativeness of scientific knowledge; nature of observation; scientific methods; hypotheses, laws and theories; imagination; validation of scientific knowledge; objectivity and subjectivity in science) but seeks to address some perceived weaknesses of VOSTS – principally, the overgeneralization and ambiguity of some items and its failure to fully ascertain the reasons underlying a respondent's choice of response. It also seeks to accommodate differences in student teachers' views about what science is likely to be in practice and what science ought to be and to distinguish between NOS views they hold and NOS views they seek to teach.

As Abd-El-Khalick and BouJaoude (1997) point out, VOSTS was conceived and written within a North American sociocultural context and, in consequence, may have limited validity in non-Western contexts. In response to such concerns, Tsai and Liu (2005) have developed a survey instrument that is more sensitive to sociocultural influences on science and students' views of science. It focuses on five characteristics of scientific knowledge and its development: (i) the role of social negotiations within the scientific community; (ii) the invented and creative nature of science; (iii) the theory-laden nature of scientific investigation; (iv) cultural influences on science; and (v) the changing and tentative nature of scientific knowledge. Rooted in similar concerns about the socioculturally determined dimensions of NOS understanding is the *Thinking about Science* instrument designed by Cobern and Loving (2002) as both a pedagogical tool (for preservice teacher education programmes) and a research tool for assessing views of science in relation to economics, the environment, religion, aesthetics, race and gender.

Before leaving this brief survey of questionnaire instruments, it is important to draw attention to the *Views of Scientific Inquiry* questionnaire (Schwartz et al. 2008), which speaks directly to the problems of NOS definition discussed at the beginning of this chapter and is designed to gather information on students' understanding of some key elements of NOS, including (i) scientific investigations are guided by questions and theoretical perspectives; (ii) there are multiple purposes for scientific inquiry and multiple methods for conducting them; (iii) there is an important distinction between data and evidence; (iv) the validation of scientific knowledge involves negotiation of meaning and achievement of consensus; and (v) scientific inquiry is embedded within multiple communities, each with its own standards, values and practices.

28.8 Alternatives to Questionnaires

Frustrated by the seemingly intractable problems of designing effective questionnaires, some researchers and teachers incline to the view that more useful information can be obtained, especially from younger students, by use of open-ended methods such as the Draw-a-Scientist Test (DAST) (Chambers 1983). In his initial study, Chambers used this test with 4,807 primary (elementary) school children in Australia, Canada and the United States. He identified seven common features in their drawings, in addition to the almost universal representation of the scientist as a man: laboratory overall; spectacles (glasses); facial hair; 'symbols of research' (specialized instruments and equipment); 'symbols of knowledge' (books, filing cabinets, etc.); technological products (rockets, medicines, machines); and captions such as 'Eureka' (with its attendant lighted bulb) and $E = mc^2$, and think bubbles saying 'I've got it' or 'A-ah! So that's how it is'.

In the years since Chambers' original work, students' drawings have changed very little,[11] with research indicating that the stereotype begins to emerge at about grade 2 and is well-established and held by the majority of students by grade 5.

---

[11] Barman (1997, 1999), Farland-Smith (2009a), Finson (2002), Fort and Varney (1989), Fralick et al. (2009), Fung (2002), Huber and Burton (1995), Jackson (1992), Losh et al. (2008), Mason et al. (1991), Matthews (1994a, 1996), Newton and Newton (1992, 1998), Rahm and Charbonneau (1997), Rosenthal (1993), She (1995, 1998), and Symington and Spurling (1990)

Not only are these images stable across genders, they seem to be relatively stable across cultural differences,[12] although Song and Kim (1999) suggest that Korean students produce 'slightly less stereotypical' drawings, especially with respect to gender and age, than students in the United States. Generally, they draw younger scientists than their Western student counterparts – drawings that probably reflect the reality of the Korean scientific community. In a study of 358 students in grades 1–7 in Southwest Louisiana, Sumrall (1995) found that African American students (especially girls) produced less stereotyped drawings than Euro-Americans with respect to both gender and race. Interestingly, the drawings of African American boys showed an equal division of scientists by race but an 84% bias in favour of male scientists. Many researchers have pointed out that girls are generally less stereotyped in their views about science and scientists than are boys. However, Tsai and Liu (2005) note that female Taiwanese students are less receptive than male students to the idea that scientific knowledge is created and tentative rather than discovered and certain. There are some encouraging indications that students, and especially male students in the age range 9–12, produce drawings with fewer stereotypical features following the implementation of gender-inclusive curriculum experiences (Huber and Burton 1995; Losh et al. 2008; Mason et al. 1991).

Of course, there is a strong possibility that researchers can be seriously misled by the drawings students produce. As Newton and Newton (1998) point out, 'their drawings reflect their stage of development and some attributes may have no particular significance for a child but may be given undue significance by an adult interpreting them' (p. 1138). Even though young children invariably draw scientists as bald men with smiling faces, regardless of the specific context in which the scientist is placed, it would be unwise to assume that children view scientists as especially likely to be bald and contented. As Claxton (1990) reminds us, children compartmentalize their knowledge and so may have at least three different versions of the scientist at their disposal: the everyday comic book version, the 'official' or approved version for use in school and their personal (and perhaps private) view. It is not always clear which version DAST is accessing or how seriously the drawer took the task. Simply asking students to 'draw a scientist' might send them a message that a 'typical scientist' exists (Boylan et al. 1992). There is also the possibility that students in upper secondary school or university use their drawings to make a sociopolitical point – for example, that there are too few women or members of ethnic minority groups engaged in science.

Scherz and Oren (2006) argue that asking students to draw the scientist's workplace can be helpful, while Rennie and Jarvis (1995) suggest that students should be encouraged to annotate their drawings in order to clarify meaning and intention. Further insight into students' views can be gained by talking to them about their drawings and the thinking behind them, asking them if they know anyone who uses science in their work (and what this entails), or presenting them with writing tasks based on scientific discovery. Miller (1992, 1993) advocates the

---

[12] Chambers (1983), Farland-Smith (2009b), Finson (2002), Fung (2002), Laubach et al. (2012), Parsons (1997), She (1995, 1998), and Walls (2012)

following approach: 'Please tell me, in your own words, what does it mean to study something scientifically?' When given the opportunity to discuss their drawings and stories with the teacher, even very young children will provide detailed explanations and rationales (Sharkawy 2006; Sumrall 1995; Tucker-Raymond and colleagues 2007). Interestingly, it is increasingly evident that young children's responses to open-ended writing tasks involving science, scientists and engineers are not stable and consistent: accounts and stories of science produced in science lessons are very different from those produced in language arts lessons (Hodson 1993). Students may even provide significantly different oral and written responses to nature of science questions (Roth and Roychoudhury 1994).

While less restrictive, instruments designed for more flexible and open-ended responses, such as the *Images of Science Probe* (Driver et al. 1996), concept mapping, small group discussion and situated-inquiry interviews (Ryder et al. 1999; Welzel and Roth 1998), sometimes pose major problems of interpretation for the researcher. So, too, do observation studies, unless supported by an interview-based follow-up capable of exploring the impact of context on student understanding. While interviews hold out the possibility of accessing underlying beliefs, their effectiveness can be severely compromised by the asymmetric power relationship between interviewer and interviewee, regardless of whether the interviewer is the teacher or an independent researcher. In an interview situation, some students may be shy or reluctant to talk; they may feel anxious or afraid; they may respond in ways that they perceive to be acceptable to the interviewer, or expected by them. Observation via audio or video recording of group-based tasks involving reading, writing and talking, practical work, role play, debating and drama constitute a less threatening situation for students, though even here there can be problems. Indeed, any classroom activity can be impacted by complex and sometimes unpredictable social factors. These complicating factors can mask or distort the NOS understanding we hope to infer from conversations and actions. In short, all approaches to ascertaining NOS views carry a risk that the characterization or description of science ascribed to the research subject is, in some measure, an artefact of the research method.

28.9 Problems Relating to Authenticity and Context

The context in which an interview question, questionnaire item or assessment task is set and, indeed, whether there is a specific context at all can have a major impact on an individual's response. Decontextualized questions (such as 'What is your view of a scientific theory?' or 'What is an experiment?') can seem infuriatingly vague to students and can be met with seeming incomprehension. Use of such questions can pose major problems of interpretation for the researcher. Conversely, context-embedded questions have domain-specific knowledge requirements that may sometimes preclude students from formulating a response that properly reflects their NOS views. Moreover, respondents may feel constrained by restriction of the question to one context and, in consequence, unable to communicate what they know about the many significant differences in the ways that scientists in different fields conduct investigations. Familiarity with the context, understanding of the underlying science concepts, interest in the situation and opportunity to utilize knowledge about other situations are all crucial to ensuring that we access students' authentic NOS understanding. Put simply, questions set in one context may trigger different responses from essentially the same questions set in a different context (Leach and colleagues 2000) – a finding that is especially significant in research that addresses NOS views in the context of scientific controversies (Smith and Wenk 2006) and socioscientific issues (Sadler and Zeidler 2004). It should also be noted that further important perspectives and issues relating to assessment are raised by recent curricular interest in scientific argumentation[13] and modelling,[14] though constraints on space preclude discussion here.

It would be surprising if students didn't have different views about the way science is conducted in school and the way science is conducted in specialist research establishments. Hogan (2000) refers to these different views as students' *proximal* knowledge of NOS (personal understanding and beliefs about their own science learning and the scientific knowledge they encounter and develop in science lessons) and *distal* knowledge of NOS (views they hold about the products, practices, codes of behaviour, standards and modes of communication of professional scientists). Sandoval (2005) draws a similar distinction between students' *practical* and *formal* epistemologies. Contextualized questions that ask students to reflect on their own laboratory experiences are likely to elicit the former, questions of a more general, de-contextualized nature ('What is science?' or 'How do scientists validate knowledge claims?') are likely to elicit the latter. The problem for the researcher is to gauge the extent to which these differences exist and how they are accessed by different research probes. The problem for the teacher is to ensure that students are aware of the crucial distinctions as well as the similarities between science in school and science in the world outside school. It may also be the case that students hold significantly different views of science as they perceive it to be and science as they believe it *should* be – a distinction that Rowell and Cawthron (1982) and Chen (2006) were able to accommodate in their research.

A further complication to ascertaining students' NOS views is the significant potential for mismatch between what individuals say about their NOS understanding and what they do in terms of acting on that understanding. Thus, the question arises: Should we seek to ascertain *espoused* views or views *implicit in actions?*

---

[13] Important literature sources include Duschl (2008), Erduran (2008), Erduran et al. (2004), Kelly and Takao (2002), Naylor et al. (2007), Osborne et al. (2004), Sampson and Clark (2006, 2008), Sandoval and Millwood (2005), Shwarz et al. (2003), Takao and Kelly (2003), and Zeidler et al. (2003).

[14] Suitable references include Acher et al. (2007), Chittleborough et al. (2005), Coll (2006), Coll and Treagust (2003a), Duschl et al. (2007), Hart (2008a), Henze et al. (2007a, b), Justi and Gilbert (2002a), Justi and van Driel (2005), Kawasaki et al. (2004), Lehrer and Schauble (2000), Lin and Chiu (2007), Maia and Justi (2009), Perkins and Grotzer (2005), Prins et al. (2008), Raghavan et al. (1998a, b), Saari and Viiri (2003), Schauble (2008), Smith et al. (2000), Taylor et al. (2003), Treagust et al. (2002, 2004), van Driel and Verloop (1999), and Webb (1994).

The former would probably be best served by questionnaires, writing tasks and interviews; the latter would require inferences to be drawn from observed behaviours and actions – for example, responding to scientific texts, searching the Internet and formulating reports of investigations. The crucial distinction between *teachers'* NOS views implicit in action and those supposedly revealed by pencil-and-paper tests is explored at length by Guerra-Ramos (2012). Of particular value for use with teachers and student teachers is Nott and Wellington's (1996, 1998, 2000) 'Critical Incidents' approach. In group settings, or in one-on-one interviews, teachers (or student teachers) are invited to respond to descriptions of classroom events, many related to hands-on work in the laboratory, by answering three questions: What would you do? What could you do? What should you do? Responses, and the discussion that ensues, may indicate something about the teachers' views of science and scientific inquiry and, more importantly perhaps, how this understanding is deployed in classroom decision-making. Similar approaches using video and multimedia materials have been used by Bencze and colleagues (2009a), Hewitt and colleagues (2003), Wong and colleagues (2006) and Yung and colleagues (2007).[15]

Even if we solve all these problems, we are still confronted with decisions about how to interpret and report the data. Should we adopt a *nomothetic* approach that focuses on the extent to which the students' or teachers' views match a prespecified 'ideal' or approved view? Attempts to distinguish 'adequate' NOS views from 'inadequate' views involve judgement about the rival merits of inductivism and falsificationism, Kuhnian views versus Popperian views, realism versus instrumentalism, and so on. None of these judgements is easy to make and may even be counterproductive to good NOS learning. Does it make more sense, then, to opt for an *ideographic* approach? Should we be satisfied to describe the views expressed by students and seek to understand them 'on their own terms'?

A major complicating factor is that students will not necessarily have coherent and consistent views across the range of issues embedded in the notion of NOS. Rather, their views may show the influence of several different and possibly mutually incompatible philosophical positions. As Abd-El-Khalick (2004) points out, what researchers see as inconsistencies in the NOS views of students at the undergraduate and graduate levels may be seen by the students as 'a collection of ideas that make sense within a set of varied and personalized images of science' (p. 418). Moreover, older students, with more sophisticated NOS understanding, will have recognized that inquiry methods vary between science disciplines and that the nature of knowledge statements varies substantially with content, context and purpose. Few research instruments are sensitive to such matters. By assigning total scores rather than generating a profile of views, the research conflates valuable data that could inform the design of curriculum interventions.

Rather than assigning individuals to one of several predetermined philosophical positions, it might make more sense to refer to their *Personal Framework of NOS*

---

[15] Other important studies of video-based teacher professional development programmes include Borko and colleagues (2008), Rosaen and colleagues (2008), Santagata and colleagues (2007) and Zhang and colleagues (2011).

Understanding and seek to highlight its interesting and significant features, an undertaking that could be facilitated by the use of repertory grids (as in the study by Shapiro 1996).[16] One such recent study by Ibrahim et al. (2009) seeks to consolidate data from a purpose-built questionnaire into NOS profiles. The questionnaire, *Views about Scientific Measurement* (VASM), which comprises six items addressing aspects of NOS and eight items dealing with scientific measurement, uses a common context (in earth sciences) and allows space for students to elaborate on their response or compose an alternative. The data, obtained from 179 science undergraduates, were found to cluster into four partially overlapping profiles, which the authors refer to as *modellers, experimenters, examiners* and *discoverers*. For *modellers*, theories are simple ways of explaining the often complex behaviour of nature; they are constructed by scientists and tested, validated and revised through experimentation. Creativity plays an important role in constructing hypotheses and theories and in experimentation. When there are discrepancies between theoretical and experimental results, both theory and the experimental data need to be scrutinized. *Experimenters* also believe that scientists should use experimental evidence to test hypotheses and theories but should do so in accordance with a strict scientific method. In situations of conflict, data have precedence over theories. *Examiners* regard the laws of nature as fixed and 'out there' waiting to be discovered through observation, rather than constructed by scientists. Experimental work is essential; it is not informed by theory. Scientists may use both the scientific method and their imagination, but experimental data always have precedence over theories. *Discoverers* also believe that the laws of nature are out there waiting to be discovered through observation. Only experiments using the scientific method can be used to generate laws and theories. If experimental data conflict with a previously established theory, then both the theory and the data need to be checked.[17] Profiling could solve many of the problems associated with the compilation and interpretation of data on NOS understanding among both students and teachers.

28.10 Some Current Emphases in NOS-Oriented Curricula

Despite the many caveats concerning the validity and reliability of research methods, it is incumbent on teachers, teacher educators and curriculum developers to pay attention to the rapidly growing number of studies indicating that both students and

---

[16] Repertory grids enable researchers to ascertain links between different facets of an individual's knowledge and understanding (and between understanding and actions) in quantitative form (Fransella and Bannister 1977). Using them over the lifetime of a research project enables a developmental record of students' (or teachers') views to be built up. Because repertory grids often produce surprising data and highlight inconsistencies in respondents' views, they provide a fruitful avenue for discussion and exploration of ideas. For these reasons, Pope and Denicolo (1993) urge researchers to use them as 'a procedure that facilitates a conversation' (p. 530).

[17] Interestingly, as a percentage of the total, the modeller profile was more common among students following a 4-year science foundation course than among physics majors.

teachers have inadequate, incomplete or confused NOS understanding.[18] Two points are worth making. First, the goal of improving NOS understanding is often prejudiced by stereotyped images of science and scientists consciously or unconsciously built into school science curricula[19] and perpetuated by science textbooks.[20] This should be a relatively easy problem to fix, and it is fair to say that the situation is not nearly so dire as it was a decade or so ago. Second, research has shown that, in general, an *explicit* approach is much more effective than an *implicit* approach in fostering more sophisticated conceptions of NOS.[21]

In an explicit approach, NOS understanding is regarded as curriculum content, to be approached carefully and systematically, just like any other lesson content. This does not entail a didactic or teacher-centred approach or the imposition of a particular view through exercise of teacher authority, but it does entail rejection of the belief that NOS understanding will just develop in students as a by-product of engaging in other learning activities. Most effective of all are approaches that have a substantial reflective component.[22] Adúriz-Bravo and Izquierdo-Aymerich (2009), Howe and Rudge (2005) and Rudge and Howe (2009) argue that an explicit reflective approach is particularly effective when historical case studies are used to engage students in the kinds of reasoning used by scientists originally struggling to make sense of phenomena and events and to construct satisfactory explanations, while Wong and colleagues (2008, 2009) have shown the value of embedding explicit teaching of NOS within a consideration of important socioscientific issues.

---

[18] Abd-El-Khalick and Lederman (2000a, b), Abell and Smith (1994), Aikenhead and Ryan (1992), Akerson and Buzzelli (2007), Akerson and Hanuscin (2007), Akerson et al. (2008), Barman (1997), Apostolou and Koulaidis (2010), Brickhouse et al. (2002), Carey and Smith (1993), Carey et al. (1989), Chambers (1983), Dagher et al. (2004), Dogan and Abd-El-Khalick (2008), Driver et al. (1996), Duveen et al. (1993), Finson (2002, 2003), Fung (2002), Griffiths and Barman (1995), Hodson (1993), Hofer (2000), Hogan and Maglienti (2001), Honda (1994), Irez (2006), Kang et al. (2005), Koren and Bar (2009), Larochelle and Desautels (1991), Leach et al. (1996, 1997), Lederman (1992, 1999), Liu and Lederman (2002, 2007), Liu and Tsai (2008), Lubben and Millar (1996), Lunn (2002), Mbajiorgu and Iloputaife (2001), Meichtry (1992), Meyling (1997), Moseley and Norris (1999), Moss et al. (2001), Palmer and Marra (2004), Parsons (1997), Paulsen and Wells (1998), Rampal (1992), Rubin et al. (2003), Ryan (1987), Ryan and Aikenhead (1992), Ryder et al. (1999), Sandoval and Morrison (2003), Schommer and Walker (1997), She (1995, 1998), Smith and Wenk (2006), Smith et al. (2000), Solomon et al. (1994), Solomon et al. (1996), Song and Kim (1999), Sumrall (1995), Tucker-Raymond et al. (2007), Tytler and Peterson (2004), Vázquez and Manassero (1999), Vázquez et al. (2006), and Windschitl (2004)

[19] Bell et al. (2003), Hodson (1998), and Milne (1998).

[20] Abd-El-Khalick (2001), Abd-El-Khalick et al. (2008), Knain (2001), Kosso (2009), McComas (1998), van Eijck and Roth (2008), and Vesterinen et al. (2011).

[21] Abd-El-Khalick (2001, 2005), Abd-El-Khalick and Lederman (2000a), Akerson and Abd-El-Khalick (2003, 2005), Akerson and Hanuscin (2007), Bell (2004), Bell et al. (2000, 2011), Faikhamta (2012), Hanuscin et al. (2006, 2011), Khishfe (2008), Khishfe and Abd-El-Khalick (2002), Lederman and Abd-El-Khalick (1998), Lin et al. (2012), Morrison et al. (2009), Posnanski (2010), Ryder (2002), Scharmann et al. (2005), Schwartz and Lederman (2002), and Schwartz et al. (2004).

[22] Akerson and Donnelly (2010), Akerson and Volrich (2006), Akerson et al. (2000, 2010), Heap (2006), and Lucas and Roth (1996).

Other notable research studies include the finding by Schwartz et al. (2004) that preservice teachers' NOS understanding was favourably enhanced when their course included a research component and journal-based assignments; the report by Morrison et al. (2009) that substantial gains in NOS understanding are achieved when explicit, reflective instruction in NOS is augmented by opportunities to interview practising scientists about their work and/or undertake some job sharing; and the study by Abd-El-Khalick and Akerson (2009) that notes major gains in the NOS understanding of preservice elementary teachers when explicit, reflective instruction is supported by use of metacognitive strategies (especially concept mapping), opportunities to research the development of their peers' NOS understanding and the chance to discuss case studies of elementary science classes oriented towards NOS teaching. A further raft of studies point to the key role played by teachers' NOS-oriented pedagogical content knowledge, curriculum awareness, confidence, self-efficacy and access to appropriate curriculum resources (Hanuscin et al. 2011; Lederman et al. 2012; Ryder and Leach 2008). My own views on how we can build and implement a curriculum to achieve enhanced levels of NOS understanding are discussed at length in Hodson (2009).

It is both notable and disappointing that the gains in NOS understanding consequent on exposure to explicit, reflective instruction are considerably less substantial in relation to the sociocultural dimensions of science than for other NOS elements.[23] The drive to equip students with an understanding of science in its social, cultural, economic and political contexts is, of course, the underpinning rationale of the so-called science-technology-society (STS) approach – more recently expanded to STSE (where E stands for environment). STS(E) has always been a purposefully ill-defined field that leaves ample scope for varying interpretations and approaches, and much has changed over the years in terms of its priorities and relative emphases.[24]

Aikenhead (2005, 2006) describes how the early emphasis on values and social responsibility was systematized by utilizing a theoretical framework deriving from sociology of science and encompassing two key aspects of NOS: (i) the social interactions of scientists *within* the scientific community and (ii) the interactions of science and scientists with social aspects, issues and institutions *external* to the community of scientists. In the terms used by Helen Longino (1990), this is a distinction between the *constitutive* values of science (the drive to meet criteria of truth, accuracy, precision, simplicity, predictive capability, breadth of scope and problem-solving capability) and the *contextual* values that impregnate the personal, social and cultural context in which science is organized, supported, financed and conducted. Allchin (1999) draws a similar distinction between the *epistemic* values of science and the *cultural* values that infuse scientific practice. Both emphases

---

[23] Akerson et al. (2000), Dass (2005), Lederman et al. (2001), Moss et al. (2001), Tairab (2001), and Zémplen (2009).

[24] Aikenhead (2003, 2005), Barrett and Pedretti (2006), Bennett et al. (2007), Cheek (1992), Fensham (1988), Gallagher (1971), Gaskell (2001), Hurd (1997), Kumar and Chubin (2000), Lee (2010), Nashon et al. (2008), Pedretti (2003), Pedretti and Nazir (2011), Solomon and Aikenhead (1994), and Yager (1996).

have remained strong, though much has changed with respect to the sociopolitical and economic contexts in which educators and scientists work, our understanding of key issues in the history, philosophy and sociology of science and our theoretical knowledge concerning concept acquisition and development.

Drawing on the metaphor deployed by Sauvé (2005) in her analysis of trends in environmental education, Pedretti and Nazir (2011) describe variations and shifts in the focus of STSE in terms of 'a vast ocean of ideas, principles, and practices that overlap and intermingle one into the other' (p. 603). The six currents identified are as follows: *application/design* (practical problem solving through designing new technology or adapting old technologies), *historical* (understanding the sociocultural embeddedness of science and technology), *logical reasoning* (using a range of perspectives, including many outside science, to understand scientific and technological developments), *value-centred* (addressing the multidimensionality of socioscientific issues, including moral-ethical concerns), *sociocultural* (recognizing and critiquing science and technology as social institutions) and *socio-ecojustice* (critiquing and addressing socioscientific issues through direct and indirect action). Five of these categories include elements of NOS, as defined above.

Concern with constitutive and contextual values, and the ways in which these values have shifted in recent years, has been the trigger for renewed interest in the changing nature of NOS – in particular, the key differences between contemporary practice at the cutting edge of scientific research and what might be called 'classical scientific research' (the focus for much of school science), especially with regard to methods, publication practices, sponsorship and funding. Forty years ago, sociologist Robert Merton (1973) identified four 'functional norms' or 'institutional imperatives' that govern the practice of science and the behaviour of individual scientists, whether or not they are aware of it. These norms are not explicitly taught; rather, newcomers are socialized into the conventions of scientific practice through the example set by more senior scientists. Merton argued that these norms constitute the most effective and efficient way of generating new scientific knowledge and provide a set of 'moral imperatives' that serves to ensure good and proper conduct:

- *Universalism* – science is universal (i.e. its validity is independent of the context in which it is generated or the context in which it is used) because evaluation of knowledge claims in science uses objective, rational and impersonal criteria rather than criteria based on personal, commercial or political interests and is independent of the reputation of the particular scientist or scientists involved.
- *Communality* – science is a cooperative endeavour and the knowledge it generates is publicly owned. Scientists are required to act in the common good, avoid secrecy and publish details of their investigations, methods, findings and conclusions so that all scientists may use and build upon the work of others.
- *Disinterestedness* – science is a search for truth simply for its own sake, free from political or economic motivation or strictures, and with no vested interest in the outcome.
- *Organized scepticism* – all scientific knowledge, together with the methods by which it is produced, is subject to rigorous scrutiny by the community of scientists in conformity with clearly established procedures and criteria.

In the traditional forms of basic or fundamental research, usually located in universities and/or government research institutes, the so-called pure scientists constitute their own audience: they determine the research goals, recognize competence, reward originality and achievement, legitimate their own conduct and discourage attempts at outside interference. In the contemporary world, universities are under increasing public pressure to deliver more obvious value for money and to undertake research that is likely to have practical utility or direct commercial value. There are increasingly loud calls for closer links between academia and industry. In this changed sociopolitical environment, scientists are now required to practice what Ziman (2000) calls *post-academic* science.[25] Because contemporary scientific research is often dependent on expensive technology and complex and wide-ranging infrastructure, it must meet the needs and serve the interests of those sponsors whose funds provide the resources. Research is often multidisciplinary and involves large groups of scientists, sometimes extending across a number of different institutions, working on problems that they have not posed, either individually or as a group. Within these teams, individual scientists may have little or no understanding of the overall thrust of the research, no knowledge of their collaborators at a personal level and no ownership of the scientific knowledge that results. A number of governments and universities have moved to privatize their research establishments, that is, sell institutes or laboratories engaged in potentially commercially lucrative research areas to industry and business interests or turn them into independent companies. In consequence, scientists have lost a substantial measure of autonomy. In many universities, the research agenda no longer includes so-called blue skies research (i.e. fundamental research), as emphasis shifts to *market-oriented research, outcome-driven research* and ever-shortening *delivery times.* Many scientists are employed on contracts that prevent them from disclosing all their results. Indeed, there is a marked trend towards patenting, privatization and commodification of knowledge. As Ziman (2000) comments, many scientists have been forced to trade the academic kudos of publication in refereed journals for the material benefit of a job or a share in whatever profit there might be from a patented invention.

Varma's (2000) study of the work of scientists in industry paints a vivid picture of disturbing changes in the way research is conducted: customization of research to achieve marketable outcomes, contract funding and strict budget constraints, flexible but strictly temporary teams of researchers assembled for specific projects and a shift in the criteria for research appraisal from the quality and significance of the science to cost-effectiveness. The vested interests of the military and commercial sponsors of research, particularly tobacco companies, the petroleum industry, the food processing industry, agribusiness institutions and pharmaceutical companies, can often be detected not just in research priorities but also in research design, especially in terms of what and how data are collected, manipulated and presented. More subtly, in what data are *not* collected, what findings are omitted from reports and whose

---

[25] While Ziman (2000) refers to contemporary scientific practice as *post-academic science,* Funtowicz and Ravetz (1993) call it *post-normal science,* and Gibbons and colleagues (1994) and Nowotny et al. (2003) use the term *mode 2 science.*

voices are silenced. Commercial interests may influence the way research findings are made public (e.g. press conferences rather than publication in academic journals) and the way in which the impact of adverse data is minimized, marginalized, hidden or ignored – issues explored at length in Hodson (2011).

In summary, science can no longer be regarded as the disinterested search for truth and the free and open exchange of information, as portrayed in many school textbook versions of science. Rather, it is a highly competitive enterprise in which scientists may be driven by self-interest and career building, desire for public recognition, financial inducements provided by business and commerce or the political imperatives of military interests. Some would argue that one of the most disturbing features of contemporary science is the effective privatization of knowledge. Science is increasingly conducted behind closed doors, in the sense that many procedures and findings remain secret or they are protected by patenting, thus removing them from critical scrutiny by the community of scientists. The scope of what can be patented has been progressively and systematically broadened, such that the very notion of public accessibility to the store of contemporary scientific knowledge is under threat (Mirowski and Sent 2008). It seems that the realities of contemporary science are in direct contradiction of three, if not all four, of the functional norms identified by Merton. Communality, disinterestedness and organized scepticism have been replaced by 'the entrepreneurial spirit and economic growth, such that scientific intellectual creativity seems to have become synonymous with commodity' (Carter 2008, p. 626). Our definitions of NOS and the teaching/learning activities we provide in school need to take account of these matters.

28.11 SSI-Oriented Teaching and Its Curriculum Implications

Interestingly, as consideration of the nature of science has become a much more prominent part of regular science curricula, even a central part in many educational jurisdictions, so emphasis in STSE education has shifted much more towards confrontation of socioscientific issues (SSI), what Pedretti and Nazir (2011) call the value-centred current in STSE. Zeidler and colleagues (2005) contrast this orientation with earlier forms of STS or STSE education in terms of its emphasis on developing habits of mind (specifically, developing scepticism, maintaining open-mindedness, acquiring the capacity for critical thinking, recognizing that there are multiple forms of inquiry, accepting ambiguity and searching for data-driven knowledge) and 'empowering students to consider how science-based issues reflect, in part, moral principles and elements of virtue that encompass their own lives, as well as the physical and social world around them' (p. 357). They argue that while STSE education emphasizes the impact of scientific and technological development on society, it does not focus explicitly on the moral-ethical issues embedded in decision-making: 'STS(E) education as currently practiced... only 'points out' ethical dilemmas or controversies, but does not necessarily exploit the inherent pedagogical power of discourse, reasoned argumentation, explicit NOS considerations, emotive, developmental, cultural or epistemological connections within the issues themselves… nor does it consider the moral or character development of students' (p. 359).

Bingle and Gaskell (1994) had earlier noted that STS education tends to emphasize what Bruno Latour (1987) calls 'ready-made science' (with all its attendant implicit messages about certainty) rather than 'science in the making' (with its emphasis on social construction). Simmons and Zeidler (2003) argue that it is the priority given to science in the making through consideration of *controversial* SSI that gives the SSI approach its special character and its unique power to focus on NOS understanding: 'Using controversial socioscientific issues as a foundation for individual consideration and group interaction provides an environment where students can and *will* develop their critical thinking and moral reasoning' (p. 83, emphasis added). In a further attempt at delineation, Zeidler and colleagues (2002) claim that the SSI approach has much broader scope, in that it 'subsumes all that STS has to offer, while also considering the ethical dimensions of science, the moral reasoning of the child, and the emotional development of the student' (p. 344).[26] Robust understanding of NOS is a clear prerequisite for addressing SSI critically and systematically; importantly, enhanced NOS understanding (both *distal* and *proximal*) is also a significant learning outcome of an SSI-oriented approach (Schalk 2012).

If students are to address SSI thoroughly and critically and deal with the NOS issues they raise, they will need the language skills to access knowledge from various sources and the ability to express their knowledge, views, opinions and values in a form appropriate to the audience being addressed. Thus, teachers need to focus students' attention very firmly on the language of science, scientific communication and scientific argumentation and on students' capacity to become critical readers of a wide variety of texts. Because meaning in science is also conveyed through symbols, graphs, diagrams, tables, charts, chemical formulae and equations, 3-D models, mathematical expressions, photographs, computer-generated images, body scans and so on, Lemke (1998) refers to the language of science as 'multimodal communication'. Any one scientific text might contain an array of such modes of communication, such that it may be more appropriate to refer to the *languages* of science:

> Science does not speak of the world in the language of words alone, and in many cases it simply cannot do so. The natural language of science is a synergistic integration of words, diagrams, pictures, graphs, maps, equations, tables, charts, and other forms of visual mathematical expression. (Lemke 1998, p. 3)

Because much of the information needed to address SSI is of the science-in-the-making kind, rather than a well-established science, and may even be located at or near the cutting edge of research, it is unlikely that students will be able to locate it

---

[26] See also Eastwood and colleagues (2012), Ekborg and colleagues (2012), Khishfe (2012b), Lee (2012), Lee and Grace (2012), Nielsen (2012b), Robottom (2012), Sadler (2009, 2011), Sadler and Donnelly (2006), Sadler and Zeidler (2005a, b), Sadler and colleagues (2004, 2006, 2007), Schalk (2012), Tytler (2012), Wu and Tsai (2007), Zeidler and Sadler (2008a, b), Zeidler and Schafer (1984), and Zeidler and colleagues (2003, 2005, 2009).

in traditional sources of information like textbooks and reference books. It will need to be accessed from academic journals, magazines, newspapers, TV and radio broadcasts, publications of special interest groups and the Internet, thus raising important issues of *media literacy*. Being media literate means being able to access, comprehend, analyse, evaluate, compare and contrast information from a variety of sources and utilize that information judiciously and appropriately to synthesize one's own detailed summary of the topic or issue under consideration. It means recognizing that the deployment of particular language, symbols, images and sound in a multimedia presentation can each play a part in determining a message's overall impact and will have a profound influence on its perceived value and credibility. It means being able to ascertain the writer's purpose and intent, determine any subtext and implicit meaning and detect bias and vested interest. It means being able to distinguish between good, reliable information and poor, unreliable information. It involves the ability to recognize what Burbules and Callister (2000) call *misinformation, malinformation, messed-up information* and *useless information*. Students who are media literate understand that those skilled in producing printed, graphic and spoken media use particular vocabulary, grammar, syntax, metaphor and referencing to capture our attention, trigger our emotions, persuade us of a point of view and, on occasions, bypass our critical faculties altogether.

Many SSI are highly controversial, sometimes because the scientific information required to formulate a judgement is incomplete, insufficient, inconclusive or extremely complex and difficult to interpret, sometimes because judgement involves consideration of factors rooted in social, political, economic, cultural, religious, environmental, aesthetic and/or moral-ethical concerns, beliefs, values and feelings. In other words, controversy may be *internal* or *external* to science. Teachers need to make a decision about how they will handle such issues. Should they try to avoid controversy altogether, take a neutral position, adopt the devil's advocate role, try to present a balanced view or advocate a particular position? These questions are discussed at length in Hodson (2011). Further, almost any discussion of a topical SSI is likely to raise questions not only about what we *can* or *could* do but also about what is the *right* decision and what we *ought* to do. Because many SSI have this moral-ethical dimension, teachers will also need to foster students' moral development and develop their capacity to make ethical judgments. Helpful discussion of these matters and strategies that teachers might employ can be found in Fullick and Ratcliffe (1996), Jones et al. (2007, 2010) and Reiss (1999, 2003, 2010).[27]

It is also likely that addressing SSI in class will generate strong feelings and emotions, with students' views and assumptions being strongly influenced by personal experiences and the experiences of friends and family and by socioculturally determined predispositions and worldviews. A student's sense of identity, comprising

---

[27] See also Beauchamp and Childress (2008), Clarkeburn (2002), Goldfarb and Pritchard (2000), Keefer (2003), Levinson and Reiss (2003), Sadler and Zeidler (2004), Sáez et al. (2008), and Saunders and Rennie (2013).

ethnicity, gender, social class, family and community relationships, economic status and personal experiences extending over many years, will necessarily impact on their values, priorities and preferences and influence the ways in which they engage in discussion and the conclusions they reach. Teachers introducing SSI into the curriculum need to be sensitive to these influences and will need to assist students in dealing with potentially stressful and disconcerting learning situations. It is here that notions of *emotional intelligence, emotional literacy* and *emotional competence* can be helpful.[28] Although these three terms are closely related, Matthews (2005) chooses to draw a distinction between the individualistic nature of emotional intelligence and the strongly social nature of emotional literacy. Thus, he argues emotional intelligence refers to an individual's ability to perceive, describe, appraise and express emotions, understand emotions and emotional knowledge, access and/or generate appropriate feelings when they facilitate thought or manage them productively when they might inhibit, while emotional literacy is the capacity to be receptive to a wide range of feelings, empathize with others and continuously monitor the emotional climate in which one is located. Emotional competence may be seen as an amalgam of the two. In general, the goal of emotional literacy is awareness and management of one's emotions in both joyful and stressful situations, the confidence and self-assurance to understand one's own emotions and the capacity to deal with them in a positive and intentional way. It is closely related to notions of self-awareness, self-image, self-esteem and sense of identity, and less directly with self-efficacy and agency.

28.12 Future Developments

In a chapter dealing with the origin, development, implications and shifting emphases of NOS-oriented curricula, it is perhaps appropriate to speculate on future developments or even to promote one's own ideas for further development. On this latter count, I count myself among those authors who argue that current conceptions of STSE or SSI-oriented science education do not go far enough, among those who advocate a much more radical, politicized form of SSI-oriented teaching and learning in which students not only address complex and often controversial SSI, and formulate their own position concerning them, but also prepare for, and engage in, sociopolitical actions that they believe will 'make a difference', asking critical questions about how research priorities in science are determined, who has access to science, how science could (and perhaps should) be conducted differently, how scientific and technological knowledge are deployed, whose voices are heard and whose reading

---

[28] Goleman (1985, 1996, 1998), Matthews et al. (2002), Matthews and colleagues (2004a, b), Saarni (1990, 1999), Salovey and Meyer (1990), Salovey and Shayter (1997), Steiner (1997), Sharp (2001) and Zeidner et al. (2009).

of a situation are considered.[29] It is a curriculum clearly rooted in notions of equity and social justice.

The likelihood of students becoming active citizens in later life is increased substantially by encouraging them to take action *now* (in school), providing opportunities for them to do so and giving examples of successful actions and interventions engaged in by others. Students need knowledge of actions that are likely to have positive impact and knowledge of how to engage in them. A key part of preparing for action involves identifying action possibilities, assessing their feasibility and appropriateness, ascertaining constraints and barriers, resolving any disagreements among those who will be involved, looking closely at the actions taken by others (and the extent to which they have been successful) and establishing priorities in terms of what actions are most urgently needed (and can be undertaken fairly quickly) and what actions are needed in the longer term. It is essential, too, that all actions taken by students are critically evaluated and committed to an action database for use by others. From a teaching perspective, it is important that care is taken to ensure both the appropriateness of a set of actions for the particular students involved and the communities in which the actions will be situated and the overall practicality of the project in terms of time and resources. It is also essential that students gain robust knowledge of the social, legal and political system(s) that prevails in the communities in which they live and develop a clear understanding of how decisions are made within local, regional and national government and within industry, commerce and the military. Without knowledge of where and with whom power of decision-making is located and awareness of the mechanisms by which decisions are reached, effective intervention is not possible. Thus, an issue-based and action-oriented curriculum requires a concurrent programme designed to achieve a measure of *political literacy*, including knowledge of how to engage in collective action with individuals who have different competencies, backgrounds and attitudes, but shares a common interest in a particular SSI. It also includes knowledge of likely sympathizers and potential allies and strategies for encouraging cooperative action and group interventions.

Desirable as this approach may be in meeting the needs of citizens in the early twenty-first century, converting such curriculum rhetoric into practical action in real classrooms is an extraordinarily tall order for teachers to undertake. It is a tall order for three reasons. First, because it radically changes the nature of the school curriculum and puts a whole raft of new demands on teachers. Second, because it challenges many of the assumptions on which schooling is traditionally based. Third, because it is predicated on a commitment to bringing about extensive and wide-ranging social change at local, regional, national and international levels. It will only occur when sufficient teachers, teacher educators, curriculum developers and curriculum policy makers are convinced of the importance, desirability and feasibility of

---

[29] See also Alsop (2009), Alsop and colleagues (2009), Bencze and Alsop (2009), Bencze and colleagues (2009b, 2012), Bencze and Sperling (2012), Calabrese Barton and Tan (2009, 2010), Chawla (2002a, b), Hart (2008b, c), Hodson (2003, 2011, 2014), Mueller (2009), Mueller et al. (2013), Roth (2009a, b, 2010), Roth and Désautels (2002, 2004), and Santos (2008).

addressing SSI in the science classroom and encouraging sociopolitical action, and when there is commitment to teach and confidence in doing so through awareness of appropriate pedagogical strategies, capacity to organize the required classroom environment and access to suitable resources. The real breakthrough will come when individual teachers are able to find and work with like-minded colleagues to form pressure groups that can begin to influence key decision-making bodies. However, such matters are well outside the scope of this chapter.

28.13 Final Thoughts

The primary purpose of this chapter has been to convey something of the extraordinary rise and widening scope of curriculum interest in NOS understanding. From very humble beginnings (e.g. 'Let's ensure that we teach about the methods that scientists use as well as paying attention to content'), curriculum interest in NOS has developed into a major influence on science education in many parts of the world. Changing views of what counts as NOS knowledge have led to further extensive developments, including concern with the characteristics of scientific inquiry, the role and status of the scientific knowledge it generates, modelling and the nature of models, how scientists work as a social group, the linguistic conventions for reporting and scrutinizing knowledge claims, the ways in which science impacts and is impacted by the social context in which it is located and the centrality of NOS in addressing the science underpinning SSI. More recently, it has been extended in such a way that some educators see NOS as a central plank in citizenship education. In my view, the next development in the extension of NOS-oriented education is the establishment of an issue-based and action-oriented curriculum capable of directing critical attention to (i) the way contemporary research and development in science and technology is conceived, practised and funded and (ii) the ways in which scientific knowledge is accessed and deployed in establishing policy and priorities with respect to SSI.

A key issue concerns the NOS sophistication we should pursue via the school curriculum. It is unrealistic as well as inappropriate to expect students to become highly skilled philosophers, historians and sociologists of science. Rather, we should select NOS items for the curriculum in relation to important educational goals: the need to motivate students and assist them in developing positive but critical attitudes towards science, the need to pay close attention to the cognitive goals and emotional demand of specific learning contexts, the creation of opportunities for students to experience *doing* science for themselves, the capacity to address complex socioscientific issues with critical understanding, concern for values issues and so on. The degree of sophistication of the NOS items we include should be appropriate to the stage of cognitive and emotional development of the students and compatible with other long- and short-term educational goals. There are numerous goals for science education (and education in general) that can, will and *should* impact on decisions about the NOS content of lessons. Our concern is not just good philosophy of science, good sociology of science or good history of science, not just authenticity and preparation for sociopolitical action, but the educational needs and interests of the students – *all* students. Selection of NOS items should consider the *changing* needs and interests of students at different stages of their science education, as well as take cognizance of the views of 'experts' (philosophers of science, historians of science, sociologists of science, scientists, science educators) and the need to promote the wider goals of (i) authentic representation of science and (ii) pursuit of critical scientific literacy.

It is considerations like these that prompted Michael Matthews (1998) to advocate the pursuit of 'modest goals' concerning HPS in the school science curriculum. In his words, 'there is no need to overwhelm students with cutting edge questions' (p. 169). Perhaps so, but agreement with the notion of modest goals still raises a question of what they should comprise. At the very least, we should include the following: consideration of the relationship between observation and theory; the role and status of scientific explanations (including the processes of theory building and modelling); the nature of scientific inquiry (including experiments, correlational studies, blind and double-blind trials, data mining and all the other notable variations among the subdisciplines of science); the history and development of major ideas in science; the sociocultural embeddedness of science and the interactions among science, technology, society and environment; the distinctive language of science; the ways in which scientific knowledge is validated through criticism, argument and peer review; moral-ethical issues surrounding science and technology; error, bias, vested interest, fraud and the misuse of science for sociopolitical ends; and the relationship between Western science and indigenous knowledge. A number of these elements are present in some science curricula, but more often than not, they are implicit, part of the hidden curriculum, embedded in language, textbook examples, laboratory activities and the like, and so dependent, ultimately, on teachers' nature of science views.

This is a demanding prescription and I readily acknowledge that telling students too early in their science education that scientific inquiry is context dependent and idiosyncratic could be puzzling, frustrating and even off-putting. This is a similar point to Brush's (1974) concern that teaching history of science can have an adverse effect on young students by undermining their confidence in science and scientists. One approach is to take our cue from secondary school chemistry curricula, where we often begin with some very simple representations, such as 'elements are either metals or non-metals' or 'bonding is either covalent or electrovalent'. We then proceed to qualify these assertions in all manner of ways: 'there are varying degrees of metallic/non-metallic character, depending on atomic size and electron configuration' and 'there is a range of intermediate bond types, including polarized covalent bonds and lattices involving highly distorted ions, as well as hydrogen bonding, van der Waal's forces, and so on'. Similarly, in the early years, we may find it useful to characterize scientific inquiry as a fairly standard set of steps. Within this simple representation, we can emphasize the importance of making careful observations (using whatever conceptual frameworks are available and appropriate to the students' current stage of understanding), taking accurate measurements, systematically controlling variables, and so on. As students become more experienced, they can be introduced to variations in approach that are necessary as contexts change – for example, the startlingly different approaches adopted by experimental particle physicists, synthetic organic chemists and evolutionary biologists.

Matthews (2012) makes the same point when he states that students have 'to crawl before they can walk, and walk before they can run. This is no more than commonsensical pedagogical practice' (p. 21). The shift from nature of science (NOS) to features of science (FOS), with its inbuilt recognition of diversity among the sciences and the significant changes in constitutive values from 'classical' scientific research to contemporary, post-Mertonian scientific practice, would be a major step in assisting teachers to pitch their teaching at a level appropriate to the students and to the issues being addressed.

References

Abd-El-Khalick, F. (2001). Embedding nature of science instruction in preservice elementary science courses: Abandoning scientism, but... *Journal of Science Teacher Education, 12*(3), 215–233.

Abd-El-Khalick, F. (2004). Over and over again: College students' views of nature of science. In L.B. Flick & N.G. Lederman (Eds.), *Scientific inquiry and nature of science: Implications for teaching, learning, and teacher education* (pp. 389–425). Dordrecht: Kluwer.

Abd-El-Khalick, F. (2005). Developing deeper understandings of nature of science: The impact of a philosophy of science course on preservice science teachers' views and instructional planning. *International Journal of Science Education, 27*(1), 15–42.

Abd-El-Khalick, F. (2012). Examining the sources for our understandings about science: Enduring conflations and critical issues in research on nature of science in science education. *International Journal of Science Education, 34*(3), 353–374.

Abd-El-Khalick, F. & Akerson, V.L. (2004). Learning as conceptual change. Factors mediating the development of preservice elementary teachers' views of the nature of science. *Science Education, 88*(5), 785–810.

Abd-El-Khalick, F. & Akerson, V. (2009). The influence of metacognitive training on preservice elementary teachers' conceptions of the nature of science. *International Journal of Science Education, 31*(16), 2161–2184.

Abd-El-Khalick, F. & BouJaoude, S. (1997). An exploratory study of the knowledge base for science teaching. *Journal of Research in Science Teaching, 34*, 673–699.

Abd-El-Khalick, F. & Lederman, N.G. (2000a). Improving science teachers' conceptions of the nature of science: A critical review of the literature. *International Journal of Science Education, 22*(7), 665–701.

Abd-El-Khalick, F. & Lederman, N.G. (2000b). The influence of history of science courses on students' views of the nature of science. *Journal of Research in Science Teaching, 37*(10), 1057–1095.

Abd-El-Khalick, F., Bell, R.L. & Lederman, N.G. (1998). The nature of science and instructional practice: Making the unnatural natural. *Science Education, 82*(4), 417–437.

Abd-El-Khalick, F., Waters, M. & Le, A-P. (2008). Representations of nature of science in high school chemistry textbooks over the past four decades. *Journal of Research in Science Teaching, 45*(7), 835–855.

Abell, S.K. & Smith, D.C. (1994). What is science? Preservice elementary teachers' conceptions of the nature of science. *International Journal of Science Education*, 16, 475–487.

Acher, A., Arcà, M. & Sanmarti, N. (2007). Modeling as a teaching learning process for understanding materials: A case study in primary education. *Science Education*, 91(3), 398–418.

Adúriz-Bravo, A. & Izquierdo-Aymeric, M. (2009). A research-informed instructional unit to teach the nature of science to pre-service science teachers. *Science & Education*, 18(9), 1177–1192.

Aikenhead, G.S. (2003). A rose by any other name. In R. Cross (Ed.), *A vision for science education: Responding to the work of Peter Fensham* (pp. 59–75). New York: RoutledgeFalmer.

Aikenhead, G.S. (2005). Research into STS science education. *Educación Quimica*, 16(3), 384–397.

Aikenhead, G.S. (2006). *Science education for everyday life: Evidence-based practice.* New York: Teachers College Press.

Aikenhead, G.S., Fleming, R.W. & Ryan, A.G. (1987). High school graduates' beliefs about science-technology-society. I. Methods and issues in monitoring student views. *Science Education*, 71(2), 145–161.

Aikenhead, G.S. & Ryan, A.G. (1992). The development of a new instrument: "Views on Science-Technology-Society (VOSTS). *Science Education*, 76, 477–491.

Aikenhead, G.S., Ryan, A.G. & Fleming, R.W. (1989). *Views on Science-Technology-Society.* Saskatoon: Department of Curriculum Studies (Faculty of Education), University of Saskatchewan.

Akerson, V.L. & Abd-El-Khalick, F. (2003). Teaching elements of nature of science: A yearlong case study of a fourth-grade teacher. *Journal of Research in Science Teaching*, 40(10), 1025–1049.

Akerson, V.L. & Abd-El-Khalick, F.S. (2005). How should I know what scientists do?—I am just a kid: Fourth-grade students' conceptions of nature of science. *Journal of Elementary Science Education*, 17(1), 1–11.

Akerson, V.L., Abd-El-Khalick, F. & Lederman, N.G. (2000). Influence of a reflective activity-based approach on elementary teachers' conceptions of nature of science. *Journal of Research in Science Teaching*, 37(4), 295–317.

Akerson, V.L. & Buzzelli, C.A. (2007). Relationships of preservice early childhood teachers' cultural values, ethical and cognitive developmental levels, and views of nature of science. *Journal of Elementary Science Education*, 19, 15–24.

Akerson, V.L., Buzzelli, C.A. & Donnelly, L.A. (2008). Early childhood teachers' views of nature of science: The influence of intellectual levels, cultural values, and explicit reflective teaching. *Journal of Research in Science Teaching*, 45(6), 748–770.

Akerson, V.L., Buzzelli, C.A. & Donnelly, L.A. (2010). On the nature of teaching nature of science: Preservice early childhood teachers' instruction in preschool and elementary settings. *Journal of Research in Science Teaching*, 47(2), 213–233.

Akerson, V. & Donnelly, L.A. (2010). Teaching nature of science to K-2 students: What understandings can they attain? *International Journal of Science Education*, 32(1), 97–124.

Akerson, V.L. & Hanuscin, D.L. (2007). Teaching nature of science through inquiry: Results of a 3-year professional development program. *Journal of Research in Science Teaching*, 44(5), 653–680.

Akerson, V.L. & Volrich, M.L. (2006). Teaching nature of science explicitly in a first-grade internship setting. *Journal of Research in Science Teaching*, 43(4), 377–394.

Allchin, D. (1999). Values in science: An educational perspective. *Science & Education*, 8, 1–12.

Allchin, D. (2011). Evaluating knowledge of the nature of (whole) science. *Science Education*, 95(3), 518–542.

Alsop, S. (2009). Not quite the revolution: Science and technology education in a world that changed. In M. DeVries & A. Jones (Eds.), *International handbook of research and development in technology education* (pp. 319–328). Rotterdam: Sense Publishers.

Alsop, S., Ibrahim, S. & Members of the Science and the City Team (2009). Feeling the weight of the world: Visual journeys in science and technology education. *Journal of Activist Science & Technology Education*, 1(1), 85–104.

Alters, B.J. (1997a). Whose nature of science? *Journal of Research in Science Teaching*, 34(1), 39–55.

Alters, B.J. (1997b). Nature of science: A diversity or uniformity of ideas? *Journal of Research in Science Teaching*, 34(10), 1105–1108.

American Association for the Advancement of Science (AAAS) (1967). *Science- A process approach*. Washington, DC: Ginn.

American Association for the Advancement of Science (AAAS) (1989). *Science for all Americans*. A Project 2061 report on literacy goals in science, mathematics, and technology. Washington, DC: AAAS.

American Association for the Advancement of Science (AAAS) (1993). *Benchmarks for scientific literacy*. Oxford: Oxford University Press.

Apostolou, A. & Koulidis, V. (2010). Epistemology and science education: A study of In M.R. Matthews (ed.*), Handbook of historical and philosophical research in science education* (pp.). Dordrecht: Springer.

Ault, C.R. (1998). Criteria of excellence for geological inquiry: The necess epistemological views of teachers. *Research in Science & Technological Education*, 28(2), 149–166.

Arduriz Bravo, A. (2013). Argumentation in science and science classrooms. ity of ambiguity. *Journal of Research in Science Teaching*, 35(2), 189–212.

Australian Science Education Project (1974). *A guide to ASEP*. Melbourne: Government Printed.

Bamberger, Y.M. & Davis, E.A. (2013). Middle school science students' scientific modelling performances across content areas and within a learning progression. *International Journal of Science Education*, 35(2), 213–238.

Barman, C.R. (1997). Students' views about scientists and science: results from a national study. *Science and Children*, 35(9), 18–24.

Barman, C.R. (1999). Students' views about scientists and school science: Engaging K-8 teachers in a national study. *Journal of Science Teacher Education*, 10(1), 43–54.

Barrett, S.E. & Pedretti, E. (2006). Contrasting orientations: STSE for social reconstruction or social reproduction? *School Science & Mathematics*, 106(5), 237–245.

Bartholomew, H., Osborne, J. & Ratcliffe, M. (2004). Teaching students "ideas-about-science": Five dimensions of effective practice. *Science Education*, 88(5), 655–682.

Beauchamp, T. & Childress, J. (2008). *Principles of biomedical ethics (*6th edition). New York: Oxford University Press.

Bell, P. (2004). Promoting students' argument construction and collaborative debate in the science classroom. In M.C. Linn, E.A. Davis & P. Bell (Eds.), *Internet environments for science education* (pp. 115–143). Mahwah, NJ: Lawrence Erlbaum.

Bell, R.L., Blair, L.M., Crawford, B.A., & Lederman, N.G. (2003). Just do it? Impact of a science apprenticeship program on high school students' understandings of the nature of science and scientific inquiry. *Journal of Research in Science Teaching*, 40(5), 487–509.

Bell, R.L., Lederman, N.G. & Abd-El-Khalick, F. (2000). Developing and acting upon one's conception of the nature of science: A follow-up study. *Journal of Research in Science Teaching*, 37(6), 563–581.

Bell, R.L., Matkins, J.J. & Gansneder, B.M. (2011). Impacts of contextual and explicit instruction on preservice elementary teachers' understandings of the nature of science. *Journal of Research in Science Teaching*, 48, 414–436.

Bencze, J.L. & Alsop, S. (2009). Anti-capitalist/pro-communitarian science & technology education. *Journal of Activist Science & Technology Education*, 1(1), 65–84.

Bencze, L., Hewitt, J. & Pedretti, E. (2009a). Personalizing and contextualizing multimedia case methods in university-based teacher education: An important modification for promoting technological design in school science. *Research in Science Education*, 39(1), 93–109.

Bencze, J.L., Alsop, S. & Bowen, G.M. (2009b). Student-teachers' inquiry-based actions to address socioscientific issues. *Journal of Activist Science & Technology Education*, 1(2), 78–112.

Bencze, J.L. & Sperling, E.R. (2012). Student teachers as advocates for student-led research-informed socioscientific activism. *Canadian Journal of Science, Mathematics and Technology Education*, 12(1), 62–85.

Bencze, l., Sperling, E. & Carter, L. (2012). Students' research-informed socioscientific activism: Re/visions for a sustainable future. *Research in Science Education*, 42, 129–148.

Bennett, J., Hogarth, S. & Lubben, F. (2007). Bringing science to life: A synthesis of the research evidence on the effects of context-based and STS approaches to science teaching. *Science Education*, 91(3), 347–370.

Berland, L.K. & Hammer, D. (2012). Framing for scientific argumentation. *Journal of Research in Science Teaching*, 49(1), 68–94.

Berland, L.K. & Lee, V.R. (2012). In pursuit of consensus: Disagreement and legitimization during small-group argumentation. *International Journal of Science Education*, 34(12), 1823–1856.

Berland, L.K. & McNeill, K.L. (2010). A learning progression for scientific argumentation; Understanding student work and designing supportive instructional contexts. *Science Education*, 94(5), 765–793.

Berland, L.K. & Reiser, B.J. (2009). Making sense of argumentation and explanation. *Science Education*, 93(1), 26–55.

Berland, L.K. & Reiser, B.J. (2011). Classroom communities' adaptations of the practice of scientific argumentation. *Science Education*, 95(2), 191–216.

Bernstein, R.J. (1983). *Beyond objectivism and relativism: Science, hermeneutics, and praxis.* Philadelphia, PA: University of Philadelphia Press.

Billeh, V.Y. & Hasan, O. (1975). Factors affecting teachers' gain in understanding the nature of science. *Journal of Research in Science Teaching*, 12(3), 209–219.

Bingle, W.H. & Gaskell, P.J. (1994). Scientific literacy for decision making and the social construction of scientific knowledge. *Science Education*, 78(2), 185–201.

Borko, H., Jacobs, J., Eiteljorg, E. & Pittman, M.E. (2008). Video as a tool for fostering productive discussions in mathematics professional Development. *Teaching and Teacher Education*, 24(2), 417–436.

Böttcher, F. & Meisertt, A. (2011). Argumentation in science education: A model-based framework. *Science & Education*, 20(2), 103–140.

Boylan, C.R., Hill, D.M., Wallace, A.R. & Wheeler, A.E. (1992). Beyond stereotypes. *Science Education*, 76(5), 465–476.

Bricker, L.A. & Bell, P. (2008). Conceptualizations of argumentation from science studies and the learning sciences and their implications for the practices of science education. *Science Education*, 92(3), 473–498.

Brickhouse, N.W., Dagher, Z.R., Letts, W.J. (IV) & Shipman, H.L. (2002). Evidence and warrants for belief in a college astronomy course. *Science & Education*, 11, 573–588.

Brock, W.H. (1973). *H.E. Armstrong and the teaching of science 1880–1930.* Cambridge: Cambridge University Press.

Brush, S. (1974). Should the history of science be rated X? *Science*, 183 (4130), 1164–1172.

Burbules, N. & Callister, T. (2000). *Watch IT: The risks and promises of information technology.* Boulder, CO: Westview Press.

Burns, J.C., Okey, J.R. & Wise, K.C. (1985). Development of an integrated process skills test: TIPS II. *Journal of Research in Science Teaching*, 22(2), 169–177.

Bybee, R.W. (1997a). Towards an understanding of scientific literacy. In W. Graber & C. Bolte (Eds), *Scientific Literacy: An International Symposium* (pp. 37–68). Kiel: IPN, University of Kiel.

Bybee, R. (1997b). *Achieving scientific literacy.* Portsmouth, NH: Heinemann.

Calabrese Barton, A. & Tan, E. (2009). The evolution of da heat: Making a case for scientific and technology literacy as robust participation. In A.T. Jones & M.J. deVries (Eds.), *International handbook of research and development in technology education* (pp. 329–346). Rotterdam: Sense Publishers.

Calabrese Barton, A. & Tan, E. (2010). 'It changed our lives': Activism, science, and greening the community. *Canadian Journal of Science, Mathematics and Technology Education*, 10(3), 207–222.

Carey, S., Evans, E., Honda, M., Jay, E. & Unger, C. (1989). "An experiment is when you try it and see if it works": A study of grade 7 students' understanding of the construction of scientific knowledge. *International Journal of Science Education*, 11, 514–529.

Carey, S. & Smith, C. (1993). On understanding the nature of scientific knowledge. *Educational Psychologist*, 28(3), 235–251.

Carter, L. (2008). Globalization and science education: The implications of science in the new economy. *Journal of Research in Science Teaching*, 45(5), 617–633.

Cartwright, N. (1999). *The dappled world: A study of the boundaries of sciences*. Cambridge: Cambridge University Press.

Cawthron, E.R. & Rowell, J.A. (1978). Epistemology and science education. *Studies in Science Education*, 5, 31–59.

Chambers, D.W. (1983). Stereotypic images of the scientist: The draw-a-scientist test. *Science Education*, 67(2), 255–265.

Chawla, L. (2002a). Spots of time: Manifold ways of being in nature in childhood. In P.H. Kahn & S.R. Kellert (Eds.), *Children and nature: Psychological, sociocultural, and evolutionary investigations* (pp. 119–225). Cambridge, MA: MIT Press.

Chawla, L. (Ed.) (2002b). *Growing up in an urbanising world*. Paris/London: UNESCO.

Cheek, D.W. (1992). *Thinking constructively about science, technology, and society education*. Albany, NY: State University of New York Press.

Chen, D. & Novick, R. (1984). Scientific and technological education in an information society. *Science Education*, 68(4), 421–426.

Chen, S. (2006). Development of an instrument to assess views on nature of science and attitudes toward teaching science. *Science Education*, 90(5), 803–819.

Choi, K., Lee, H., Shin, N., Kim, S-W. & Krajcik, J. (2011). Re-conceptualization of scientific literacy in South Korea for the twenty-first century. *Journal of Research in Science Teaching*, 48(6), 670–679.

Chittleborough, G.D., Treagust, D.F., Mamiala, T.L. & Mocerino, M. (2005). Students' perceptions of the role of models in the process of science and in the process of learning. *Research in Science & Technological Education*, 23(2), 195–212.

Clarkeburn, H. (2002). A test of ethical sensitivity in science. *Journal of Moral Education*, 31(4), 439–453.

Claxton, G. (1990). *Teaching to learn: A direction for education*. London: Cassell.

Clement, J.J. & Rea-Ramirez, M.A. (Eds.) (2008). *Model based learning and instruction in science*. Dordrecht: Springer.

Clough, M.P. (2006). Learners' responses to the demands of conceptual change: Considerations for effective nature of science instruction. *Science Education*, 15, 463–494.

Clough, M.P. (2007). Teaching the nature of science to secondary and post-secondary students: Questions rather than tenets, *The Pantaneto Forum*, Issue 25, January, http://www.pantaneto.co.uk/issue25/front25.htm, Republished (2008) in *California Journal of Science Education*, 8(2), 31–40.

Clough, M.P. (2011). The story behind the science: Bringing science and scientists to life in post-secondary science education. *Science & Education*, 20, 701–717.

Clough, M.P. & Olson, J.K. (2008). Teaching and assessing the nature of science: An introduction. *Science & Education*, 17(2 & 3), 143–145.

Cobern, W.W. & Loving, C.C. (2002). Investigation of preservice elementary teachers' thinking about science. *Journal of Research in Science Teaching*, 39(10), 1016–1031.

Coles, M., Gott, R. & Thornley, T. (1988). *Active science*. London: Collins.

Coll, R.K. (2006). The role of models, mental models and analogies in chemistry teaching. In P.J. Aubusson, A.G. Harrison & S.M. Ritchie (Eds.), *Metaphor and analogy in science education* (pp. 65–77). Dordrecht: Kluwer.

Coll, R.K. & Taylor, I. (2005). The role of models and analogies in science education: Implications from research. *International Journal of Science Education*, 27, 183–198.

Coll, R.K. & Treagust, D.F. (2002). Learners' mental models of covalent bonding. *Research in Science & Technological Education*, 20(2), 241–268.

Coll, R.K. & Treagust, D.F. (2003a). Learners' mental models of metallic bonding: A cross-age study. *Science Education*, 87(5), 685–707.

Coll, R.K. & Treagust, D.F. (2003b). Investigation of secondary school, undergraduate and graduate learners' mental models of ionic bonding. *Journal of Research in Science Teaching*, 40(5), 464–486.

Coll, R.K., France, B. & Taylor, I. (2005). The role of models/and analogies in science education: Implications from research. *International Journal of Science Education*, 27(2), 183–198.

Cooley, W.W. & Klopfer, L.E. (1961). *Manual for the test on understanding science*. Princeton, NJ: Educational Testing Service.

Cotham, J. & Smith, E. (1981). Development and validation of the conceptions of scientific theories test. *Journal of Research in Science Teaching*, 18, 387–396.

Cross, R. (1995). Conceptions of scientific literacy: reactionaries in ascendancy in the state of Victoria. *Research in Science Education*, 25(2), 151–162.

Council of Ministers of Education, Canada (1997). *Common framework of science learning outcomes*. Toronto: CMEC Secretariat.

Dagher, Z.R. & BouJaoude, S. (2005). Students' perceptions of the nature of evolutionary theory. *Science Education*, 89(3), 378–391.

Dagher, Z., Brickhouse, N., Shipman, H. & Letts, W. (2004). How some college students represent their understanding of scientific theories. *International Journal of Science Education*, 26(6), 735–755.

Dass, P.M. (2005). Understanding the nature of scientific enterprise (NOSE) through a discourse with its history: The influence of an undergraduate 'history of science' course. *International Journal of Science and Mathematics Education*, 391), 87–115.

Davies, T. & Gilbert, J. (2003). Modelling: Promoting creativity while forging links between science education and design and technology education. *Canadian Journal of Science, Mathematics and Technology Education*, 3(1), 67–82.

DeBoer, G. (2001). Scientific literacy: Another look at its historical and contemporary meanings and its relationship to science education reform. *Journal of Research in Science Teaching*, 37(6), 582–601.

Dillashaw, F.G. & Okey, J.R. (1980). Test of integrated process skills for secondary school science students. *Science Education*, 64(5), 601–608.

Deng, F., Chen, D-T., Tsai, C-C. & Chai, C.S. (2011). Students' views of the nature of science: A critical review of research. *Science Education*, 95(6), 961–999.

Department of Education (Republic of South Africa) (2002). *Revised national curriculum statement for grades R-9 (schools) – Natural sciences*. Pretoria: Department of Education [*Government Gazette*, Vol. 443, No. 23406).

Dewey, J. (1916). *Democracy and education*. New York: Macmillan.

Dillon, J. (2009). On scientific literacy and curriculum reform. *International Journal of Environmental & Science Education*, 4(3), 201–213.

Dogan, N. & Abd-El-Khalick, F. (2008). Turkish grade 10 students' and science teachers' conceptions of nature of science: A national study. *Journal of Research in Science Teaching*, 45(10), 1083–1112.

Donnelly, J. (2001). Contested terrain or unified project? 'The nature of science' in the National Curriculum for England and Wales. *International Journal of Science Education*, 23(2), 181–195.

Driver, R., Leach, J., Millar, R. & Scott, P. (1996). *Young people's images of science*. Buckingham: Open University Press.

Driver, R., Newton, P. & Osborne, J. (2000). Establishing the norms of scientific argumentation in classrooms. *Science Education*, 84(3), 287–312.

Duschl, R.A. (2008). Quality argumentation and epistemic criteria. In S. Erduran & M.P. Jiménez-Aleixandre (Eds.), *Argumentation in science education: Perspectives from classroom-based research* (pp. 159–178). Dordrecht: Kluwer.

Duschl, R. & Grandy, R. (Eds.) (2008). *Teaching scientific inquiry: Recommendations for research and implementation*. Rotterdam: Sense Publishers.

Duschl, R.A. & Osborne, J. (2002). Supporting and promoting argumentation discourse in science education. *Studies in Science Education*, 38, 39–72.

Duschl, R.A., Schweingruber, H.A. & Shouse, A.W. (Eds.) (2007). *Taking science to school: Learning and teaching science in grades K-8*. Washington, DC: The National Academies Press.

Duveen, J., Scott, L. & Solomon, J. (1993). Pupils' understanding of science: Description of experiments or "a passion to explain"? *School Science Review*, 75(271), 19–27.

Eastwood, J.L., Sadler, T.D., Zeidler, D.L., Lewis, A., Amiri, L. & Applebaum, S. (2012). Contextualizing nature of science instruction in socioscientific issues. *International Journal of Science Education*, 34(15), 2289–2315.

Ekborg, M., Ottander, C., Sifver, E. & Simon, S. (2012). Teachers' Experience of working with socioscientific issues: A large scale and in depth study. *Research in Science Education*. Available online.

Elby, A. & Hammer, D. (2001). On the substance of a sophisticated epistemology. *Science Education*, 85(5), 554–567.

Erduran, S. (2008). Methodological foundations in the study of argumentation in science classroom. In S. Erduran & M.P. Jiménez-Aleixandre (Eds.), *Argumentation in science education: Perspectives from classroom-based research* (pp. 47-69). Dordrecht: Kluwer.

Erduran, S. & Duschl, R.A. (2004). Interdisciplinary characterizations of models and the nature of chemical knowledge in the classroom. *Studies in Science Education*, 40, 105–138.

Erduran, S., Simon, S. & Osborne, J. (2004). TAPping into argumentation: developments in the application of Toulmin's argument pattern for studying science discourse. *Science Education*, 88(6), 915–933.

Erduran, S., Bravo, A.A. & Naaman, R.M. (2007). Developing epistemologically empowered teachers. Examining the role of philosophy of chemistry in teacher education. *Science & Education*, 16(9 & 10), 975–989.

Evagorou, M. & Osborne, J. (2013). Exploring young students' collaborative argumentation within a socioscientific issue. *Journal of Research in Science Teaching, 50(2), 189–208.*

Faikhamta, C. (2012). The development of in-service science teachers' understandings of and orientations to teaching the nature of science within a PCK-based NOS course. *Research in Science Education*. Available online.

Farland-Smith, D. (2009a). Exploring middle schools girls' science identities: Examining attitudes and perceptions of scientists when working 'side-by-side' with scientists. *School Science and Mathematics*, 109(7), 412–421.

Farland-Smith, D. (2009b). How does culture shape students' perceptions of scientists? Cross-national comparative study of American and Chinese elementary students. *Journal of Elementary Science Teacher Education*, 21(4), 23–41.

Feinstein, N. (2011). Salvaging science literacy. *Science Education*, 95(1), 168–185.

Fensham, P.J. (1988). Approaches to the teaching of STS in science education. *International Journal of Science Education*, 10(4), 346–356.

Feyerabend, P.K. (1962). Explanation, reduction and empiricism. *Minnesota Studies in the Philosophy of Science*, 3, 28–97.

Finson, K.D. (2002). Drawing a scientist: What we do and do not know after fifty years of drawing. *School Science & Mathematics*, 102(7), 335–345.

Finson, K.D. (2003). Applicability of the DAST-C to the images of scientists drawn by students of different racial groups. *Journal of Elementary Science Education*, 15(1), 15–27.

Flick, L.B. & Lederman, N.G. (Eds.) (2004). *Scientific inquiry and nature of science: Implications for teaching, learning, and teacher education*. Dordrecht: Kluwer.

Ford, M.J. & Wargo, B.M. (2012). Dialogic framing of scientific content for conceptual and epistemic understanding. *Science Education*, 96(3), 369–391.

Fort, D.C. & Varney, H.L. (1989). How students see scientists: Mostly male, mostly white, and mostly benevolent. *Science and Children*, 26, 8–13.

Fralick, B., Kearn, J., Thompson, S. & Lyons, J. (2009). How middle schoolers draw engineers and scientists. *Journal of Science Education and Technology*, 18, 60–73.

Franco, C., Barros, H.L., Colinvaux, D., Krapas, S., Queiroz, G. & Alves, F. (1999). From scientists' and inventors' minds to some scientific and technological products: Relationships between theories, models, mental models and conceptions. *International Journal of Science Education*, 21(3), 277–291.

Fransella, F. & Bannister, D. (1977). *A manual for repertory grid technique*. New York: Academic Press.

Fuller, S. (1992). Social epistemology and the research agenda of science studies. In A. Pickering (Ed.), *Science as practice and culture* (pp. 390–428). Chicago, IL: University of Chicago Press.

Fullick, P. & Ratcliffe, M. (Eds.) (1996). *Teaching ethical aspects of science*. Totton: Bassett Press.

Fung, Y.Y.H. (2002). A comparative study of primary and secondary school students' images of scientists. *Research in Science & Technological Education*, 20(2), 199–213.

Funtowicz, S.O. & Ravetz, J. (1993). Science for the post-normal age. *Futures*, 25, 739–755.

Gagné, R.M. (1963). The learning requirements for inquiry. *Journal of Research in Science Teaching*, 1(2), 144–153.

Gallagher, J.J. (1971). A broader base for science education. *Science Education*, 55, 329–338.

Gaskell, P.J. (2001). STS in a time of economic change: What's love got to do with it? *Canadian Journal of Science, Mathematics and Technology Education*, 1(4), 385–398.

Gentner, D. & Stevens, A. (1983). *Mental models*. Hillsdale, NJ: Lawrence Erlbaum.

Gibbons, M., Limoges, C., Nowotny, H., Schwarzman, S., Scott, P. & Trow, M. (1994). *The new production of knowledge: The dynamics of science and research in contemporary societies*. London: Sage.

Gilbert, J.K. (2004). Models and modelling: Routes to more authentic science education. *International Journal of Science and Mathematics Education*, 2(1), 115–130.

Gilbert, J.K. & Boulter, C.J. (1998). Learning science through models and modelling. In B.J. Fraser & K.G. Tobin (Eds.), *International handbook of science education* (pp. 53–66). Dordrecht: Kluwer.

Gilbert, J.K., Boulter, C. & Rutherford, M. (1998a). Models in explanations, Part 1: Horses for courses? *International Journal of Science Education*, 20(1), 83–97.

Gilbert, J.K., Boulter, C. & Rutherford, M. (1998b). Models in explanations, Part 2: Whose voice? Whose ears? *International Journal of Science Education*, 20(2), 187–203.

Gilbert, J. & Boulter, C. (2000). *Developing models in science education*. Dordrecht: Kluwer.

Gobert, J.D. & Pallant, A. (2004). Foster students' epistemologies of models via authentic model-based texts. *Journal of Science Education and Technology*, 13(1), 7–21.

Gobert, J.D., O'Dwyer, L., Horwitz, P., Buckley, B.C., Levy, S.T. & Wilensky, U. (2011). Examining the relationship between students' understanding of the nature of models and conceptual learning in biology, physics and chemistry. *International Journal of Science Education*, 33(5), 653–684.

Goldfarb, T. & Pritchard, M. (2000). *Ethics in the science classroom: An instructional guide for secondary school science teachers with model lessons for classroom use*. Available from: www.wmich.edu/ethics/ESC/index.html

Goleman, D. (1985). *Emotional intelligence: Why it can matter more than IQ*. New York: Bantam Books.

Goleman, D. (1996). *Emotional intelligence*. London: Bloomsbury Publishing.

Goleman, D. (1998). *Working with emotional intelligence*. New York: Bantam Books.

Goodrum, D., Hackling, M. & Rennie, L. (2000). *The status and quality of teaching and learning science in Australian schools*. Canberra: Department of Education, Training and Youth Affairs.

Gräber, W. & Bolte, C. (Eds.) (1997). *Scientific literacy: An international symposium*. Kiel: Institut fur die Padagogik der Naturwiseenschaften (IPN) an der Universitat Kiel.

Greca, I.M. & Moreira, M.A. (2000). Mental models, conceptual models and modelling. *International Journal of Science Education*, 22(1), 1–11.

Greca, I.M. & Moreira, M.A. (2002). Mental, physical and mathematical models in the teaching and learning of physics. *Science Education*, 86(1), 106–121.

Griffiths, A.K. & Barman, C.R. (1995). High school students' views about the nature of science: Results from three countries. *School Science & Mathematics*, 95(2), 248–255.

Guerra-Ramos, M.T. (2012). Teachers' ideas about the nature of science: A critical analysis of research approaches and their contribution to pedagogical practice. *Science & Education*, available online, DOI: 10.1007/s11191-011-9395-7.

Halloun, I. (2004). *Modeling theory in science education*. Dordrecht: Kluwer.

Halloun, I.A. (2007). Mediated modeling in science education. *Science & Education*, 16(7 & 8), 653–697.

Hansen, J.A., Barnett, M., MaKinster, J.G. & Keating, T. (2004). The impact of three-dimensional computational modelling on students' understanding of astronomy concepts: A qualitative analysis. *International Journal of Science Education*, 26, 1555–1575.

Hanuscin, D.L., Akerson, V.L. & Phillipson-Mower, T. (2006). Integrating nature of science instruction into a physical science content course for preservice elementary teachers: NOS views of teaching assistants. *Science Education*, 90(5), 912–935.

Hanuscin, D.L., Lee, M.H. & Akerson, V.L. (2011). Elementary teachers' pedagogical content knowledge for teaching nature of science. *Science Education*, 95(1), 145–167.

Harding, P. & Hare, W. (2000) Portraying science accurately in classrooms: Emphasizing open-mindedness rather than relativism. *Journal of Research in Science Teaching*, 37(3), 225–236.

Harding, S. (1986). *The science question in feminism*. Ithaca, NY: Cornell University Press.

Hart, C. (2008). Models in physics, models for physics learning, and why the distinction may matter in the case of electric circuits. *Research in Science Education*, 38(5), 529–544.

Hart, P. (2008a). What comes before participation? Searching for meaning in teachers' constructions of participatory learning in environmental education. In A. Reid, B.B. Jensen, J. Nikel & V. Simovsla (Eds.), *Participation and learning: Perspectives on education and the environment, health and sustainability* (pp. 197–211). New York: Springer.

Hart, P. (2008b). Elusive participation: Methodological challenges in researching teaching and participatory learning in environmental education. In A. Reid, B.B. Jensen, J. Nikel & V. Simovsla (Eds.), *Participation and learning: Perspectives on education and the environment, health and sustainability* (pp. 225–240). New York: Springer.

Heap, R. (2006). *Myth busting and tenet building: Primary and early childhood teachers' understanding of the nature of science*. Auckland: Unpublished MEd thesis, University of Auckland.

Henze, I., van Driel, J. & Verloop, N. (2007a). The change of science teachers' personal knowledge about teaching models and modelling in the context of science education reform. *International Journal of Science Education*, 29(15), 1819–1846.

Henze, I., van Driel, J.H. & Verloop, N. (2007b). Science teachers' knowledge about teaching models and modelling in the context of a new syllabus on public understanding of science. *Research in Science Education*, 37(2), 99–122.

Hewitt, J., Pedretti, E., Bencze, L., Vaillancourt, B.D. & Yoon, S. (2003). New applications for multimedia cases: Promoting reflective practice in preservice teacher education. *Journal of Technology and Teacher Education*, 11(4), 483–500.

Hodson, D. (1985). Philosophy of science, science and science education. *Studies in Science Education*, 12, 25–57.

Hodson, D. (1986). Rethinking the role and status of observation in science education. *Journal of Curriculum Studies,* 18(4), 381–396.

Hodson, D. (1988a). Toward a philosophically more valid science curriculum, *Science Education*, 72, 19–40.

Hodson, D. (1988b). Experiments in science and science teaching. *Educational Philosophy & Theory*, 20, 53–66.

Hodson, D. (1990). Making the implicit explicit: A curriculum planning model for enhancing children's understanding of science. In D.E. Herget (Ed.), *More history and philosophy of science in science teaching* (pp. 292-310). Tallahassee, FL: Florida State University Press.

Hodson, D. (1991). Philosophy of science and science education. In M. Matthews (Ed.) *History, philosophy and science teaching*. Toronto: OISE Press/Columbia University Press.

Hodson, D. (1993). Philosophic stance of secondary school science teachers, curriculum experiences, and children's understanding of science: Some preliminary findings. *Interchange*, 24(1 & 2), 41–52.

Hodson, D. (1996). Laboratory work as scientific method: Three decades of confusion and distortion. *Journal of Curriculum Studies*, 28(2), 115–135.

Hodson, D. (1998). Science fiction: The continuing misrepresentation of science in the school curriculum. *Curriculum Studies*, 6(2), 191–216.

Hodson, D. (2003). Time for action: Science education for an alternative future. *International Journal of Science Education*, 25(6), 645–670.

Hodson, D. (2008). *Towards scientific literacy: A teachers' guide to the history, philosophy and sociology of science*. Rotterdam/Taipei: Sense Publishers.

Hodson, D. (2009). *Teaching and learning about science: Language, theories, methods, history, traditions and values*. Rotterdam/Taipei: Sense Publishers.

Hodson, D. (2011). *Looking to the future: Building a curriculum for social activism*. Rotterdam/Taipei: Sense Publishers.

Hodson, D. (2014). Don't be nervous, don't be flustered, don't be scared. Be prepared. *Canadian Journal of Science, Mathematics and Technology Education*, in press.

Hofer, B.K. (2000). Dimensionality and disciplinary differences in personal epistemology. *Contemporary Educational Psychology*, 25, 378–405.

Hoffmann, R. (1995). *The same and not the same*. New York: Columbia University Press.

Hogan, K. (2000). Exploring a process view of students' knowledge about the nature of science. *Science Education*, 84(1), 51–70.

Hogan, K. & Maglienti, M. (2001). Comparing the epistemological underpinnings of students' and scientists' reasoning about conclusions. *Journal of Research in Science Teaching*, 38(6), 663–687.

Honda, M. (1994). *Linguistic inquiry in the science classroom: 'It is science, but it's not like a science problem in a book'*. Cambridge, MA: MIT working papers in linguistics. Cited by Smith et al. (2000).

Howe, E.M. & Rudge, D.W. (2005). Recapitulating the history of sickle-cell anemia research. *Science & Education*, 14, 423–441.

Huber, R.A. & Burton, G.M. (1995). What do students think scientists look like? *School Science & Mathematics*, 95(7), 371–376.

Hurd, P.D. (1958). Science literacy: Its meaning for American schools. *Educational Leadership*, 16(1), 13–16.

Hurd, P.D. (1997). *Inventing science education for the new millennium*. New York: Teachers College Press.

Hurd, P.D. (1998). Scientific literacy: New minds for a changing world. *Science Education*, 82(3), 407–416.

Ibrahim, B., Buffler, A. & Lubben, F. (2009). Profiles of freshman physics students' views on the nature of science. *Journal of Research in Science Teaching*, 46(3), 248–264.

Inner London Education Authority (ILEA) (1987). *Science in process*. London: Heinemann.

Irez, S. (2006). Are we prepared?: An assessment of preservice science teacher educators' beliefs about nature of science. *Science Education*, 90(6), 1113–1143.

Irzik, G. & Nola, R. (2011). A family resemblance approach to the nature of science for science education. *Science & Education*, 20(7–8), 591–607.

Jackson, T. (1992). Perceptions of scientists among elementary school children. *Australian Science Teachers Journal*, 38, 57–61.

Jenkins, E.W. (1979). *From Armstrong to Nuffield: Studies in twentieth century science education in England and Wales*. London: John Murray.

Jenkins, E. (1990). Scientific literacy and school science education. *School Science Review*, 71 (256), 43–51.

Jenkins, E. (2007). School science: A questionable construct? *Journal of Curriculum Studies*, 39(3), 265–282.

Jiménez-Aleixandre, M.P. & Erduran, S. (2008). Argumentation in science education; An overview. In S. Erduran & M.P. Jiménez-Aleixandre (Eds.), *Argumentation in science education: Perspectives from classroom-based research* (pp. 3–27). Dordrecht: Kluwer.

Johnson-Laird, P.N. (1983). *Mental models*. Cambridge: Cambridge University Press.

Jones, A., McKim, A., Reiss, M., Ryan, B., Buntting, C., Saunders, K. et al. (2007). *Research and development of classroom-based resources for bioethics education in New Zealand*. Hamilton: Wilf Malcolm Institute of Educational Research, University of Waikato.

Jones, A., McKim, A. & Reiss, M. (Eds.) (2010), *Ethics in the science and technology classroom: A new approach to teaching and learning*. Rotterdam: Sense Publishers.

Justi, R.S. & Gilbert, J.K. (2002a). Modelling, teachers' views on the nature of modelling and implications for the education of modellers. *International Journal of Science Education*, 24(4), 369–388.

Justi, R.S. & Gilbert, J.K. (2002b). Science teachers' knowledge about and attitudes towards the use of models in learning science. *International Journal of Science Education*, 24(12), 1273–1292.

Justi, R.S. & Gilbert, J.K. (2002c). Models and Modelling in Chemical Education. In J.K. Gilbert, O. de Jong, R. Justi, D.F. Treagust & J.H. van Driel (Eds.), *Chemical education: Towards research-based practice* (pp. 47–68). Dordrecht: Kluwer.

Justi, R.S. & Gilbert, J.K. (2003). Teachers' views on the nature of models. *International Journal of Science Education*, 25(11), 1369–1386.

Justi, R. & van Driel, J. (2005). The development of science teachers' knowledge on models and modelling: Promoting, characterizing, and understanding the process. *International Journal of Science Education*, 27(5), 549–573.

Kang, S., Scharmann, L.C. & Noh, T. (2005). Examining students' views on the nature of science: Results from Korean 6th, 8th, and 10th graders. *Science Education*, 89(2), 314–334.

Kawasaki, K., Herrenkohl, L.P. & Yeary, S.A. (2004). Theory building and modelling in a sinking and floating unit: A case study of third and fourth grade students' developing epistemologies of science. *International Journal of Science Education*, 26(11), 1299–1324.

Keefer, M. (2003). Moral reasoning and case-based approaches to ethical instruction in science. In D. Zeidler (Ed.), *The role of moral reasoning on socioscientific issues and discourse in science education* (pp. 241–260). Dordrecht: Kluwer.

Kelly, G. J. (2008). Inquiry, activity, and epistemic practice, In R. Duschl & R. Grandy (Eds.), *Teaching scientific inquiry: Recommendations for research and implementation* (pp. 99–117). Rotterdam: Sense Publishers.

Kelly, G.J. & Takao, A. (2002). Epistemic levels in argument: An analysis of university oceanography students' use of evidence in writing. *Science Education*, 86(3), 314–342.

Khan, S. (2007). Model-based inquiries in chemistry. *Science Education*, 91(6), 877–905.

Khishfe, R. (2008). The development of seventh graders' views of nature of science. *Journal of Research in Science Teaching*, 45(4), 470–496.

Khishfe, R. (2012a). Relationship between nature of science understanding and argumentation skills: A role for counterargument and contextual factors. *Journal of Research in Science Teaching*, 49(4), 489–514.

Khishfe, R. (2012b). Nature of science and decision-making. *International Journal of Science Education*, 34(1), 67–100.

Khishfe, R. & Abd-El-Khalick, F. (2002). Influences of explicit and reflective versus implicit inquiry-oriented instruction on sixth graders' views of nature of science. *Journal of Research in Science Teaching*, 39(7), 551–578.

Khishfe, R. & Lederman, N. (2006). Teaching nature of science within a controversial topic: Integrated versus non-integrated. *Journal of Research in Science Teaching*, 43(4), 395–418.

Khishfe, R. & Lederman, N. (2007). Relationship between instructional context and views of the nature of science. *International Journal of Science Education*, 29(80, 939–961.

Kimball, M.E. (1967). Understanding the nature of science: A comparison of scientists and teachers. *Journal of Research in Science Teaching*, 5(2), 110–120.

Knain, E. (2001). Ideologies in school science textbooks. *International Journal of Science Education*, 23(3), 319–329.

Knorr-Cetina, K. (1999). *Epistemic cultures: How the sciences make knowledge*. Cambridge, MA: Harvard University Press.

Koponen, I.T. (2007). Models and modelling in physics education: A critical re-analysis of philosophical underpinnings and suggestions for revisions. *Science & Education*, 16(7 & 8), 751–773.

Koren, P. & Bar, V. (2009). Pupils' image of 'the scientist' among two communities in Israel: A comparative study. *International Journal of Science Education*, 31(18), 2485–2509.

Kosso, P. (2009). The large-scale structure of scientific method. *Science & Education*, 18(1), 33–42.

Koulaidis, V. & Ogborn, J. (1995). Science teachers' philosophical assumptions: How well do we understand them? *International Journal of Science Education*, 17(3), 273–283.

Kuhn, D. (2010). Teaching and learning science as argument. *Science Education*, 94(5), 810–824.

Kumar, D.D. & Chubin, D.E. (Eds.) (2000). *Science, technology, and society: A sourcebook on research and practice*. New York: Kluwer/Plenum.

Lakin, S. & Wellington, J. (1994). Who will teach the 'nature of science'? Teachers' views of science and their implications for science education. *International Journal of Science Education*, 16(2), 175–190.

Larochelle, M. & Désautels, J. (1991) "Of course, it's just obvious": Adolescents' ideas of scientific knowledge. *International Journal of Science Education*, 13, 373–389.

Latour, B. (1987). *Science in action: How to follow scientists and engineers through society*. Cambridge, MA: Harvard University Press.

Laubach, T.A., Crofford, G.D. & Marek, E.A. (2012). Exploring Native American students' perceptions of scientists. *International Journal of Science Education*, 34(11), 1769–1794.

Laudan, L., Donovan, A., Laudan, R., Barker, P., Brown, H., Leplin, J., Thagard, P. & Wykstra, S. (1986). Scientific change: Philosophical models and historical research. *Synthese*, 69, 141–223.

Laugksch, R.C. (2000). Scientific literacy: A conceptual overview. *Science Education*, 84(1), 71–94.

Layton, D. (1973). *Science for the people: The origins of the school science curriculum in England*. London: George Allen & Unwin.

Leach, J., Driver, R., Millar, R. & Scott, P. (1996). Progression in learning about 'the nature of science': Issues of conceptualisation and methodology. In M. Hughes (Ed.), *Progress in learning* (pp. 109–139). Clevedon (UK): Multilingual Matters.

Leach, J., Driver, R., Millar, R. & Scott, P. (1997). A study of progression in learning about 'the nature of science': Issues of conceptualisation and methodology. *International Journal of Science Education*, 19, 147–166.

Leach, J., Millar, R., Ryder, J. & Séré, M-G. (2000). Epistemological understanding in science learning: The consistency of representations across contexts. *Learning and Instruction*, 10(6), 497–527.

Lederman, J., Lederman, N., Kim, B.S. & Ko, E.K. (2012). Teaching and learning of nature of science and scientific inquiry: Building capacity through systematic research-based professional development. In M.S. Khine (Ed.), *Advances in nature of science research* (pp. 235–238). Dordrecht: Springer.

Lederman, N.G. (1992). Students' and teachers' conceptions of the nature of science: A review of the research. *Journal of Research in Science Teaching*, 29(4), 331–359.

Lederman, N.G. (1999). Teachers' understanding of the nature of science and classroom practice: Factors that facilitate or impede the relationship. *Journal of Research in Science Teaching*, 36(8), 916–929.

Lederman, N.G. (2004). Syntax of nature of science within inquiry and science instruction. In L.B. Flick & N.G. Lederman (Eds.), *Scientific inquiry and nature of science: Implications for teaching, learning, and teacher education*. (pp. 301–317). Dordrecht: Kluwer.

Lederman, N. G. (2006). Research on nature of science: Reflections on the past, anticipations of the future. *Asia-Pacific Forum on Science Learning and Teaching*. Available on-line at: http://www.ied.edu.hk/apfslt/v7_issue1

Lederman, N.G. (2007). Nature of science: Past, present, and future. In S.K. Abell & N.G. Lederman (Eds.), *Handbook of research on science education* (pp. 831–879). Mahwah, NJ: Lawrence Erlbaum Associates.

Lederman, N.G. & Abd-El-Khalick, F. (1998). Avoiding de-natured science: Activities that promote understandings about the nature of science. In W.F. McComas (Ed.), *The nature of science in science education: Rationales and strategies* (pp. 83–126). Dordrecht: Kluwer.

Lederman, N.G., Abd-El-Khalick, F., Bell, R.L. & Schwartz, R. S. (2002). Views of nature of science questionnaire: Toward valid and meaningful assessment of learners' conceptions of nature of science. *Journal of Research in Science Teaching*, 39(6), 497–521.

Lederman, N.G. & O'Malley, M. (1990). Students' perceptions of tentativeness in science: Development, use and sources of change. *Science Education*, 74(2), 225–239.

Lederman, N.G., Schwartz, R.S., Abd-El-Khalick, F. & Bell, R.L. (2001). Pre-service teachers' understanding and teaching of nature of science: An intervention study. *Canadian Journal of Science, Mathematics and Technology Education*, 1(2), 135–160.

Lederman, N., Wade, P. & Bell, R.L. (1998). Assessing the nature of science: What is the nature of our assessments? *Science & Education*, 7(6), 595–615.

Lederman, N., Wade, P. & Bell, R.L. (2000). Assessing understanding of the nature of science: A historical perspective. In W.F. McComas (Ed.), *The nature of science in science education: Rationales and strategies* (pp. 331–350). Dordrecht: Kluwer Academic.

Lederman, N.G., Antink, A. & Bartos, S. (2012). Nature of science, scientific inquiry, and socioscientific issues arising from genetics: A pathway to developing a scientifically literate citizenry. *Science & Education*, available online, June, DOI: 10.1007/s11191-012-9503-3.

Lederman, N.G., Bartos, S.A. & Lederman, J. (2013) The development, use and interpretation of NOS assessments. In M.R. Matthews (ed.*,), Handbook of historical and philosophical research in science education* (pp.). Dordrecht: Springer.

Lee, Y.C. (2010). Science-technology-society or technology-society-science? Insights from an ancient technology. *International Journal of Science Education*, 32(14), 1927–1950.

Lee, Y.C. (2012). Socioscientific issues in health contexts: treading a rugged terrain. *International Journal of Science Education*, 34(3), 459–483.

Lee, Y.C. & Grace, M. (2012). Students' reasoning and decision making about a socioscientific issue: A cross-context comparison. *Science Education*, 96(5), 787–807.

Lehrer, R. & Schauble, L. (2000). Modeling in mathematics and science. In R. Glaser (Ed.), *Advances in instructional psychology: Vol. 5 Educational design and cognitive science* (pp. 101–159). Mahwah, NJ: Erlbaum.

Lehrer, R. & Schauble, L. (2005). Developing modeling and argument in the elementary grades. In T. Romberg & T.P. Carpenter (Eds.), *Understanding mathematics and science matters* (pp. 29–53). Mahwah, NJ: Erlbaum.

Lehrer, R. & Schauble, L. (2006). Scientific thinking and science literacy. In K.A. Renninger & I.E. Sigel (Eds.), *Handbook of child psychology: Vol. 4 Child psychology and practice* (pp. 153–196). Hoboken, NJ: John Wiley.

Lemke, J.L. (1998). *Teaching all the languages of science: Words symbols, images, and actions.* Available at http://academic.brooklyn.cuny.edu/education/jlemke/papers/barcelon.htm

Lemke, J.L. (2004). The literacies of science. In E.W. Saul (Ed.), *Crossing borders in literacy and science instruction: Perspectives on theory and practice* (pp. 33–47). Newark, DE: International Reading Association/National Science Teachers Association.

Levinson, R. (2010). Science education and democratic participation: An uneasy congruence? *Studies in Science Education*, 46(1), 69–119.

Levinson, R. & Reiss, M. (Eds.) (2003). *Key issues in bioethics*. London: Routledge.

Lin, J-W. & Chiu, M-H. (2007). Exploring the characteristics and diverse sources of students' mental models of acids and bases. *International Journal of Science Education*, 29(6), 771–803.

Lin, S-F., Lieu, S-C., Chen, S., Huang, M-T. & chang, W-H. (2012). Affording explicit-reflective science teaching by using an educative teachers' guide. *International Journal of Science Education*, 34(7), 999–1026.

Linder, C., Ostman, L. & Roberts, D.A. (Eds.) (2012). *Exploring the landscape of scientific literacy*. New York: Routledge.

Liu, S-Y. & Lederman, N.G. (2002). Taiwanese gifted students' views of nature of science. *School Science & Mathematics*, 102(3), 114–123.

Liu, S-Y. & Lederman, N.G. (2007). Exploring prospective teachers' worldviews and conceptions of nature of science. *International Journal of Science Education*, 19(10), 1281–1307.

Liu, S-Y. & Tsai, C-C. (2008). Differences in the scientific epistemological views of undergraduate students. *International Journal of Science Education*, 30(8), 1055–1073.

Longbottom, J.E. & Butler, P.H. (1999). Why teach science? Setting rational goals for science education. *Science Education*, 83(4), 473–492.

Longino, H.E. (1990). *Science as social knowledge: Values and objectivity in scientific inquiry*. Princeton, NJ: Princeton University Press.

Lopes, J.B. & Costa, N. (2007). The evaluation of modelling competencies: Difficulties and potentials for the learning of the sciences. *International Journal of Science Education*, 29(7), 811–851.

Losh, S.C., Wilke, R. & Pop, M. (2008). Some methodological issues with 'draw a scientist tests' among young children. *International Journal of Science Education*, 30(6), 773–792.

Loving, C.C. (1997). From the summit of truth to its slippery slopes: Science education's journey through positivist-postmodern territory. *American Educational Research Journal*, 34(3), 421–452.

Lubben, F. & Millar, R. (1996). Children's ideas about the reliability of experimental data. *International Journal of Science Education*, 18, 955–968.

Lucas, A. (1975). Hidden assumptions in measures of 'knowledge about science and scientists'. *Science Education*, 59(4), 481–485.

Lucas, B. & Roth, W-M. (1996). The nature of scientific knowledge and student learning: Two longitudinal case studies. *Research in Science Education*, 26(1), 103–127.

Lunn, S. (2002). 'What we think we can safely say…': primary teachers' views of the nature of science. *British Educational Research Journal*, 28(5), 649–672.

Maia, P.F. & Justi, R. (2009). Learning of chemical equilibrium through modelling-based teaching. *International Journal of Science Education*, 31(5), 603–630.

Manz, E. (2012). Understanding the codevelopment of modelling practice and ecological knowledge. *Science Education*, 96(6), 1071–1105.

Mason, C.L., Kahle, J.B. & Gardner, A.L. (1991). Draw-a-scientist test: Future implications. *School Science & Mathematics*, 91(5), 193–198.

Matthews, B. (1994a). What does a chemist look like? *Education in Chemistry*, 127–129.

Matthews, B. (1996). Drawing scientists. *Gender and Education*, 8(2), 231–243.

Matthews, B. (2005). Emotional development, science and co-education. In S. Alsop (Ed.), *Beyond Cartesian dualism: Encountering affect in the teaching and learning of science* (pp. 173–186). Dordrecht: Springer.

Matthews, B., Kilbey, T., Doneghan, C. & Harrison, S. (2002). Improving attitudes to science and citizenship through developing emotional literacy. *School Science Review*, 84(307), 103–114.

Matthews, G., Roberts, R.D. & Zeidner, M. (2004). Seven myths about emotional intelligence. *Psychological Inquiry*, 15(3), 179–196.

Matthews, G., Zeidner, M. & Roberts, R.D. (2004) *Emotional intelligence: Science and myth*. Cambridge, MA: MIT Press.

Matthews, M.R. (Ed.) (1991). *History, philosophy and science teaching*. Toronto: OISE Press/ Columbia University Press.

Matthews, M.R. (1992). History, philosophy and science teaching: The present rapprochement. *Science & Education*, 1(1), 11–48.

Matthews, M.R. (1994/2014). *Science teaching: The role of history and philosophy of science*. New York: Routledge.

Matthews, M.R. (1998). In defence of modest goals when teaching about the nature of science. *Journal of Research in Science Teaching*, 35(2), 161–174.

Matthews, M.R. (2012). Changing the focus: From nature of science (NOS) to features of science (FOS). In M.S. Khine (Ed.), *Advances in nature of science research: Concepts and methodologies* (pp. 3–26). Dordrecht: Springer.

Mayer, V.J. & Richmond, J.M. (1982). An overview of assessment instruments in science. *Science Education*, 66(1), 49–66.

Mayr, E. (1988). *Towards a new philosophy of biology: Observations of an evolutionist*. Cambridge, MA: The Belknap Press of Harvard University Press.

Mayr, E. (1997). *This is biology: The science of the living world*. Cambridge, MA: Harvard University Press.

Mayr, E. (2004). *What makes biology unique? Considerations on the autonomy of a scientific discipline*. Cambridge: Cambridge University Press.

Mbajiorgu, N.M. & Iloputaife, E.C. (2001). Combating stereotypes of the scientist among pre-service science teachers in Nigeria. *Research in Science & Technological Education*, 19(1), 55–67.

McComas, W.F. (1998). The principal elements of the nature of science: Dispelling the myths. In W.F. McComas (Ed.), *The nature of science in science education: Rationales and strategies* (pp. 41–52). Dordrecht: Kluwer.

McComas, W.F. & Olson, J.K. (1998). The nature of science in international education standards documents. In W.F. McComas (Ed), *The nature of science in science education: Rationales and strategies* (pp. 53–70). Dordrecht: Kluwer.

McComas, W.F., Clough, M.P. & Almazroa, H. (1998) The role and character of the nature of science in science education. In W.F. McComas (Ed), *The nature of science in science education: Rationales and strategies* (pp. 3–39). Dordrecht: Kluwer.

McCurdy, R.C. (1958). Towards a population literate in science. *The Science Teacher*, 25(7), 366–369+408.

McEneaney, E.H. (2003). The worldwide cachet of scientific literacy. *Comparative Education Review*, 47(2), 217–237.

Meichtry, Y.J. (1992). Influencing student understanding of the nature of science: data from a case of curriculum development. *Journal of Research in Science Teaching*, 29, 389–407.

Merton, R.K. (1973). *The sociology of science: Theoretical and empirical investigations*. Chicago, IL: University of Chicago Press.

Meyling, H. (1997). How to change students' conceptions of the epistemology of science. *Science & Education*, 6, 397–416.

Millar, R. & Osborne, J. (Eds.) (1998). *Beyond 2000: Science education for the future*. London: King's College London School of Education.

Miller, J.D. (1992). Toward a scientific understanding of the public understanding of science and technology. *Public Understanding of Science*, 1, 23–26.

Miller, J.D. (1993). Theory and measurement in the public understanding of science: A rejoinder to Bauer and Schoon. *Public Understanding of Science*, 2, 235–243.

Miller, J.D. (2000). The development of civic scientific literacy in the United States. In D.D. Kumar & D. Chubin (Eds.), *Science, technology, and society: A sourcebook on research and practice* (pp. 21–47). New York: Plenum Press.

Milne, C. (1998). Philosophically correct science stories? Examining the implications of heroic science stories for school science. *Journal of Research in Science Teaching*, 35(2), 175–187.

Mirowski, P. & Sent, E-M. (2008). The commercialization of science and the response of STS. In E.J. Hackett, O. Amsterdamska, M. Lynch & J. Wajcman (Eds.), *The handbook of science and technology studies* (pp. 635–689). Cambridge, MA: MIT Press.

Morrison, J.A., Raab, F. & Ingram, D. (2009). Factors influencing elementary and secondary teachers' views on the nature of science. *Journal of Research in Science Teaching*, 46(4), 384–403.

Moseley, C. & Norris, D. (1999). Preservice teachers' views of scientists. *Science and Children*, 37(1), 50–53

Moss, D.M., Abrams, E.D. & Robb, J. (2001). Examining student conceptions of the nature of science. *International Journal of Science Education*, 23(8), 771–790.

Mueller, M.P. (2009). Educational reflections on the 'ecological crisis': Ecojustice, environmentalism, and sustainability. *Science & Education*, 18(8), 1031–1056.

Mueller, M.P., Tippens, D.J. & Stewart, A.J. (Eds.) (2013), *Assessing schools for generation R (responsibility): A guide to legislation and school policy in science education.* Dordrecht: Springer.

Nadeau, R. & Désautels, J. (1984). *Epistemology and the teaching of science.* Ottawa: Science Council of Canada.

Nashon, S., Nielsen, W. & Petrina, S. (2008). Whatever happened to STS? Pre-service physics teachers and the history of quantum mechanics. *Science & Education*, 17(4), 387–401,

National Research Council (1996). *National science education standards.* Washington, DC: National Academy Press.

Naylor, S., Keogh, B. & Downing, B. (2007). Argumentation and primary science. *Research in Science Education*, 37(1), 17–39.

Nelson, M.M. & Davis, E.A. (2012). Preservice elementary teachers' evaluations of elementary students' scientific models: An aspect of pedagogical content knowledge for scientific modelling. *International Journal of Science Education*, 34(12), 1931–1959.

Nersessian, N.J. (2008). Model-based reasoning in scientific practice. In R.A. Duschl & R.E. Grandy (Eds.), *Teaching scientific inquiry: Recommendations for research and implementation* (pp. 57–79). Rotterdam/Taipei: Sense.

Newton, D.P. & Newton, L.D. (1992). Young children's perceptions of science and the scientist. *International Journal of Science Education*, 14, 331–348.

Newton, L.D. & Newton, D.P. (1998). Primary children's conceptions of science and the scientist: Is the impact of a National Curriculum breaking down the stereotype? *International Journal of Science Education*, 20(9), 1137–1149.

Newton, P., Driver, R. & Osborne, J. (1999). The place of argumentation in the pedagogy of school science. *International Journal of Science Education*, 21(5), 553–576.

Nielsen, J.A. (2012a). Arguing from nature: The role of 'nature' in students' argumentations on a socioscientific issue. *International Journal of Science Education,* 34(5), 723–744.

Nielsen, J.A. (2012b). Science in discussions: An analysis of the use of science content in socioscientific discussions. *Science Education*, 96(3), 428–456.

Nielsen, J.A. (2013). Dialectical features of students' argumentation: A critical review of argumentation studies in science Education. *Research in Science Education.* 43(1), 371–393.

Nola, R. & Irzik, G. (2013). A new paradigm for understanding the nature of science: A family resemblance approach. In M.R. Matthews (ed.*), Handbook of historical and philosophical research in science education* (pp.). Dordrecht: Springer.

Norris, S. & Phillips, L. (2003). How literacy in its fundamental sense is central to scientific literacy. *Science Education*, 87(2), 224–240.

Norris, S.P., Phillips, L.M. & Burns, D.P. (2013). Conceptions of scientific literacy: Identifying and evaluating their programmatic elements. In M.R. Matthews (ed.*), Handbook of historical and philosophical research in science education* (pp.). Dordrecht: Springer.

Nott, M. & Wellington, J. (1993). Your nature of science profile: An activity for science teachers. *School Science Review*, 75(270), 109–112.

Nott, M. & Wellington, J. (1996). Probing teachers' views of the nature of science: How should we do it and where should we be looking? In G. Welford, J. Osborne & P. Scott (Eds.), *Science education research in Europe* (pp. 283–294). London: Falmer Press.

Nott, M. & Wellington, J. (1998). Eliciting, interpreting and developing teachers' understandings of the nature of science. *Science & Education*, 7, 579–594.

Nott, M. & Wellington, J. (2000). A programme for developing understanding of the nature of science in teacher education. In W.F. McComas (Ed.), *The nature of science in science education: Rationales and strategies* (pp. 293–313). Dordrecht: Kluwer.

Nowotny, H., Scott, P. & Gibbons, M. (2003). 'Mode 2' revisited: The new production of knowledge. *Minerva*, 41(3), 179–194.

Oh, P.S. & Oh, S.J. (2011). What teachers of science need to know about models: An overview. *International Journal of Science Education*, 33(8), 1109–1130.

Oliver, J.S., Jackson, D.F., Chun, S., Kemp, A., Tippens, D.J., Leonard, R., Kang, N.H. & Rascoe, B. (2001). The concept of scientific literacy: A view of the current debate as an outgrowth of the past two centuries. *Electronic Journal of Literacy through Science*, 1(1). Available at; *ejlts.ucdavis.edu*

Organization for Economic Cooperation and Development (OECD) (1999). Scientific literacy. In OECD, *Measuring student knowledge and skills.* (pp. 59–75). Paris: OECD.

Organization for Economic Cooperation and Development (OECD) (2003). *Assessment framework – Mathematics, reading, science and problem solving knowledge and skills.* Paris: OECD.

Organization for Economic Cooperation and Development (OECD) (2006). *Assessing scientific, reading and mathematical literacy: A framework for PISA 2006.* Paris: OECD.

Osborne, J. (2001). Promoting argument in the science classroom: A rhetorical perspective. *Canadian Journal of Science, Mathematics and Technology Education*, 1(3), 271–290.

Osborne, J.F. & Patterson, A. (2011). Scientific argument and explanation: A necessary distinction? *Science Education*, 95(4), 627–638.

Osborne, J., Collins, S., Ratcliffe, M., Millar, R. & Duschl, R. (2003). What "ideas-about-science" should be taught in school science? A Delphi study of the expert community. *Journal of Research in Science Teaching*, 40(7), 692–720.

Osborne, J. & Dillon, J. (2008). *Science education in Europe: Critical reflections.* London: The Nuffield Foundation.

Osborne, J., Erduran, S. & Simon, S. (2004). Enhancing the quality of argumentation in school science. *Journal of Research in Science Teaching*, 41(10), 994–1010.

Palmer, B. & Marra, R.M. (2004). College student epistemological perspectives across knowledge domains: A proposed grounded theory. *Higher Education*, 47, 311–335.

Parsons, E.C. (1997). Black high school females' images of the scientist: Expression of culture. *Journal of Research in Science Teaching*, 34(7), 745–768.

Passmore, C.M. & Svoboda, J. (2012). Exploring opportunities for argumentation in modelling classrooms. *International Journal of Science Education*, 34(10), 1535–1554.

Paulsen, M.B. & Wells, C.T. (1998). Domain differences in the epistemological beliefs of college students. *Research in Higher Education*, 39(4), 365–384.

Pedretti, E. (2003). Teaching science, technology, society and environment (STSE) education: Preservice teachers' philosophical and pedagogical landscapes. In D.L. Zeidler (Ed.), *The role of moral reasoning on socioscientific issues and discourse in science education* (pp. 219–239). Dordrecht: Kluwer.

Pedretti, E. & Nazir, J. (2011). Currents in STSE education: Mapping a complex field, 40 years on. *Science Education*, 95(4), 601–626.

Pella, M.O., O'Hearn, G.T., & Gale, C.W. (1966). Referents to scientific literacy. *Journal of Research in Science Teaching*, 4, 199–208.

Perkins, D.N. & Grotzer, T.A. (2005). Dimensions of causal understanding: The role of complex causal models in students' understanding of science. *Studies in Science Education*, 41, 117–166.

Pluta, W.J., Chinn, C.A. & Duncan, R.G. (2011). Learners' epistemic criteria for good scientific models. *Journal of Research in Science Teaching*, 48(5), 486–511.

Pope, M. & Denicolo, P. (1993). The art and science of constructivist research in teacher thinking. *Teaching & Teacher Education*, 9(5 & 6), 529–544.

Popper, K.R. (1959). *The logic of scientific discovery.* London: Hutchinson.

Posanski, T.J. (2010). Developing understanding of the nature of science within a professional development program for inservice elementary teachers: Project nature of elementary science teaching. *Journal of Science Teacher Education*, 21, 589–621.

Prins, G.T., Bulte, A.M.W., van Driel, J.H. & Pilot, A. (2008). Selection of authentic modelling practices as contexts for chemistry education. *International Journal of Science Education*, 30(14), 1867–1890.

Raghavan, K., Sartoris, M.L. & Glaser, R. (1998a). Impact of the MARS curriculum: The mass unit. *Science Education*, 82, 53–91.

Raghavan, K., Sartoris, M.L. & Glaser, R. (1998b). Why does it go up? The impact of the MARS curriculum as revealed through changes in student explanations of a helium balloon. *Journal of Research in Science Teaching*, 35, 547–567.

Rahm, J. & Charbanneau, P. (1997). Probing stereotypes through students' drawings of scientists. *American Journal of Physics*, 65(8), 774–778.

Rampal, A. (1992). Image of science and scientist: A study of school teachers' views. *Science Education*, 76(4), 415–436.

Rennie, L.J. & Jarvis, T. (1995). Children's choice of drawings to communicate their ideas about technology. *Journal of Research in Science Teaching*, 37, 784–806.

Reiss, M. (1999). Teaching ethics in science. *Studies in Science Education*, 34, 115–140.

Reiss, M. (2010). Ethical thinking. In A. Jones, A. McKim & M. Reiss (Eds.), *Ethics in the science and technology classroom: A new approach to teaching and learning* (pp. 7–17). Rotterdam: Sense Publishers.

Reiss, M.J. (2003). Science education for social justice. In C. Vincent (Ed.), *Social Justice, Education and Identity* (pp. 153–165) London: RoutledgeFalmer.

Roberts, D.A. (1983). *Scientific literacy: Towards balance in setting goals for school science programs*. Ottawa: Science Council of Canada.

Roberts, D.A. (2007). Scientific literacy/science literacy. In S.K. Abell & N.G. Lederman (Eds.), *Handbook of research on science education* (pp. 729–780). Mahwah, NJ: Lawrence Erlbaum Associates.

Robottom, I. (2012). Socioscientific issues in education: Innovative practices and contending epistemologies. *Research in Science Education*,. Available online.

Rockefeller Brothers Fund (1958) *The pursuit of excellence: education and the future of America*. Garden City, NY: Doubleday.

Rosaen, C.L., Lundeberg, M., Cooper, M., Fritzen, A. & Terpstra, M. (2008). Noticing noticing: How does investigation of video records change how teachers reflect on their experiences? *Journal of Teacher Education*, 59(4), 347–360.

Rosenthal, D.B. (1993). Images of scientist: A comparison of biology and liberal studies majors. *School Science & Mathematics*, 93(4), 212–216.

Roth, W-M. (2009a). Activism or science/technology education as a by product of capacity building. *Journal of Activist Science & Technology Education*, 1(1), 16–31.

Roth, W-M. (2009b). On activism and teaching. *Journal of Activist Science & Technology Education*, 1(2), 33–47.

Roth, W-M. (2010). Activism: A category for theorizing learning. *Canadian Journal of Science, Mathematics and Technology Education*, 10(3), 278–291.

Roth, W-M. & Calabrese Barton, A. (2004). *Rethinking scientific literacy*. New York: RoutledgeFalmer.

Roth, W-M. & Désautels, J. (2002). *Science education as/for sociopolitical action*. New York: Peter Lang.

Roth, W-M. & Désautels, J. (2004). Educating for citizenship: Reappraising the role of science education. *Canadian Journal of Science, Mathematics and Technology Education*, 4(2), 149–168.

Roth, W-M. & Roychoudhury, A. (1994). Physics students' epistemologies and views about knowing and learning, *Journal of Research in Science Teaching*, 31, 5–30.

Rowell, J.A. & Cawthron, E.R. (1982). Images of sciences: An empirical study. *European Journal of Science Education*, 4(1), 79–94.

Royal Society, The (1985). *The public understanding of science*. London: Royal Society.

Rubba, P. (1976). *Nature of scientific knowledge scale*. Bloomington, IN: Indiana University School of Education.

Rubba, P.A. & Anderson, H.O. (1978). Development of an instrument to assess secondary school students' understanding of the nature of scientific knowledge. *Science Education*, 62(4), 449–458.

Rubin, E., Bar, V. & Cohen, A. (2003). The images of scientists and science among Hebrew- and Arabic-speaking pre-service teachers in Israel. *International Journal of Science Education,* 25(7), 821–846.

Rudge, D.W. & Howe, E.M. (2009). An explicit and reflective approach to the use of history to promote understanding of the nature of science. *Science & Education,* 18(5), 561–580.

Rudolph, J.L. (2000). Reconsidering the 'nature of science' as a curriculum component. *Journal of Curriculum Studies,* 32(3), 403–419.

Russ, R.S., Scherr, R.E., Hammer, D. & Mikeska, J. (2008). Recognizing mechanistic reasoning in student scientific inquiry: A framework for discourse analysis developed from philosophy of science. *Science Education,* 92(3), 499–525.

Ryan, A.G. (1987). High school graduates' beliefs about science-technology-society. IV. The characteristics of scientists. *Science Education,* 71, 489–510.

Ryan, A.G. & Aikenhead, G.S. (1992). Students' preconceptions about the epistemology of science. *Science Education,* 76, 559–580.

Ryder, J. (2001). Identifying science understanding for functional scientific literacy. *Studies in Science Education,* 36, 1–44.

Ryder, J. (2002). School science education for citizenship: Strategies for teaching about the epistemology of science. *Journal of Curriculum Studies,* 34(6), 637–658.

Ryder, J. (2009). Enhancing engagement with science/technology-related issues. In A.T. Jones & M.J. deVries (Eds.), *International handbook of research and development in technology education* (pp. 287–296). Rotterdam/Taipei: Sense Publishers.

Ryder, J., Leach, J. & Driver, R. (1999). Undergraduate science students' images of science. *Journal of Research in Science Teaching,* 36(2), 201–219.

Ryder, J. & Leach, J. (2008). Teaching about the epistemology of science in upper secondary schools: An analysis of teachers' classroom talk. *Science & Education,* 17(2–3), 289–315.

Ryu, S. & Sandoval, W.A. (2012). Improvements to elementary children's epistemic understanding from sustained argumentation. *Science Education,* 96(3), 488–526.

Saarni, L. (1990). Emotional competence: How emotions and relationships become integrated. In R.A. Thompson (Ed.), *Socioemotional development: Nebraska symposium on motivation (Vol. 36)* (pp. 115–182). Lincoln, NE: University of Nebraska Press.

Saarni, C. (1999). *The development of emotional competence.* New York: Guilford Press.

Saari H. & Viiri, J. (2003). A research-based teaching sequence for teaching the concept of modelling to seventh-grade students. *International Journal of Science Education,* 25(11), 1333–1352.

Sadler. T.D. (2009). Situated learning in science education: Socioscientific issues as contexts for practice. *Studies in Science Education,* 45(1), 1–42.

Sadler, T.D. & Donnelly, L.A. (2006). Socioscientific argumentation: The effects of content knowledge and morality. *International Journal of Science Education,* 28(12), 1463–1488.

Sadler, T.D. (2011). *Socioscientific issues in the classroom: Teaching, learning and research.* Dordrecht: Springer.

Sadler, T.D. & Zeidler, D.L. (2004). The morality of socioscientific issues: Construal and resolution of genetic engineering dilemmas. *Science Education,* 88(1), 4–27.

Sadler, T.D. & Zeidler, D.L. (2005). The significance of content knowledge for informal reasoning regarding socioscientific issues: Applying genetics knowledge to genetic engineering issues. *Science Education,* 89(1), 71–93.

Sadler, T.D. & Zeidler, D.L. (2005b). Patterns of informal reasoning in the context of socio-scientific decision making. *Journal of Research in Science Teaching,* 42(1), 112–138.

Sadler, T.D., Amirshokoohi, A., Kazempour, M. & Allspaw, K.M. (2006). Socioscience and ethics in science classrooms: teacher perspectives and strategies. *Journal of Research in Science Teaching,* 43(4), 353–376.

Sadler, T.D., Barab, S.A. & Scott, B. (2007). What do students gain by engaging in socioscientific inquiry? *Research in Science Education,* 37(4), 371–391.

Sadler, T.D., Zeidler, D.L. & Chambers, F.W. (2004). Students' conceptualizations of the nature of science in response to a socioscientific issue. *International Journal of Science Education*, 26(4), 387–409.

Sáez, M.J., Niño, A.G. & Carretero, A. (2008). Matching society's values: Students' views of biotechnology. *International Journal of Science Education*, 30, 112–138.

Salovey, P. & Meyer, M.V. (1990). Emotional intelligence. *Imagination, Cognition and Personality*, 9, 185–211.

Salovey, P. & Shaytor, D. (Eds.) (1997). *Emotional development and emotional intelligence: Educational implications*. New York: Basic Books.

Sampson, V. & Blanchard, M.R. (2012). Science teachers and scientific argumentation: trends in views and practice. *Journal of Research in Science Teaching*, 49(9), 1122–1148.

Sampson, V. & Clark, D. (2006). Assessment of argument in science education: A critical review of the literature. In *Proceedings of the 7th international conference of the learning sciences* (pp. 655–661). Bloomington, IN, June. International Society of the Learning Sciences (www.isls.org).

Sampson, V. & Clark, D.B. (2008). Assessment of the ways students generate arguments in science education: Current perspectives and recommendations for future directions. *Science Education*, 92(3), 447–472.

Sampson, V. & Clark, D.B. (2011). A comparison of the collaborative scientific argumentation practices of two high and two low performing groups. *Research in Science Education*, 41(1), 63–97.

Sampson, V., Grooms, J. & Walker, J.P. (2011). Argument-driven inquiry as a way to help students learn how to participate in scientific argument and craft written arguments: An exploratory study. *Science Education*, 95(2), 217–257.

Sampson, V., & Walker, J. (2012). Argument-driven inquiry as a way to help undergraduate students write to learn by learning to write in chemistry. *International Journal of Science Education, 34(9–10)*, 1443–1486.

Sandoval, W.A. (2005). Understanding students' practical epistemologies and their influence on learning through inquiry. *Science Education*, 89(4), 634–656.

Sandoval, W.A. & Cam, A. (2011). Elementary children's judgments of the epistemic status of sources of justification. *Science Education*, 95(3), 383–408.

Sandoval, W.A. & Millwood, K.A. (2005). The quality of students' use of evidence in written scientific explanations. *Cognition and Instruction*, 23(1), 23–55.

Sandoval, W.A. & Millwood, K.A. (2008). What can argumentation tell us about epistemology? In S. Erduran & M.P. Jiménez-Aleixandre (Eds.), *Argumentation in science education: Perspectives from classroom-based research* (pp. 71–90). Dordrecht: Kluwer.

Sandoval, W. & Morrison, K. (2003). High school students' ideas about theories and theory change after a biological inquiry unit. *Journal of Research in Science Teaching*, 40, 369–392.

Santagata, R., Zannoni, C. & Stigler, J. (2007). The role of lesson analysis in preservice teacher education: An empirical investigation of teacher learning from a virtual video-based field experience. *Journal of Mathematics Teacher Education, 10*, 123–140.

Santos, W.L.P. dos (2008). Scientific literacy: A Freirean perspective as a radical view of humanistic science education. *Science Education*, 93, 361–382.

Saunders, K. & Rennie, L.J. (2013). A pedagogical model for ethical inquiry into socioscientific issues in science. *Research in Science Education*. 43(1), 253–274.

Sauvé, L. (2005). Currents in environmental education: Mapping a complex and evolving pedagogical field. *Canadian Journal of Environmental Education*, 10, 11–37.

Schalk, K.A. (2012). A socioscientific curriculum facilitating the development of distal and proximal NOS conceptualizations. *International Journal of Science Education, 34*(1), 1–24.

Scharmann, L.C., Smith, M.U., James, M.C. & Jensen, M. (2005). Explicit reflective nature of science instruction: Evolution, intelligent design, and umbrellaology. *Journal of Science Teacher Education*, 16, 27–41.

Schauble, L. (2008). Three questions about development. In R.A. Duschl & R.E. Grandy (Eds.), *Teaching scientific inquiry: Recommendations for research and implementation* (pp. 50–56). Rotterdam/Taipei: Sense.

Scheffler, I. (1967) *Science and subjectivity*. Indianapolis, IN: Bobbs-Merrill.

Scherz, Z. & Oren, M. (2006). How to change students' images of science and technology. *Science Education*, 90(6), 965–985.

Schommer, M. & Walker, K. (1997). Epistemological beliefs and valuing school: Considerations for college admissions and retentions. *Research in Higher Education*, 38(2), 173–186.

Schwab, J.J. (1962). The teaching of science as enquiry. In J.J. Schwab & P.F. Brandwein (Eds.), *The teaching of science* (pp. 3–103). Cambridge, MA: Harvard University Press.

Schwartz, R.S. & Lederman, N.G. (2002). 'It's the nature of the beast': The influence of knowledge and intentions on learning and teaching the nature of science. *Journal of Research in Science Teaching*, 39(3), 205–236.

Schwartz, R. & Lederman, N. (2008). What scientists say: Scientists' views of nature of science and relation to science context. *International Journal of Science Education*, 30(6), 721–771.

Schwartz, R.S., Lederman, N.G., & Crawford, B.A. (2004). Developing views of nature of science in an authentic context: An explicit approach to bridging the gap between nature of science and scientific inquiry. *Science Education,* 88, 610–645.

Schwartz, R.S., Lederman, N.G. & Lederman, J.S. (2008). *An instrument to assess views of scientific inquiry: The VOSI questionnaire.* Paper presented at the annual meeting of the National Association for Research in Science Teaching, Baltimore, MD, April.

Screen, P. (1986). *Warwick process science*. Southampton: Ashford.

Screen, P, (1988). A case for a process approach: The Warwick experience. *Physics Education*, 23(3), 146–149.

Shwarz, B.B., Neuman, Y., Gil, J. & Ilya, M. (2003). Construction of collective and individual knowledge in argumentative activity. *Journal of the Learning Sciences*, 12(2), 219–256.

Shapiro, B.L. (1996). A case study of change in elementary student teacher thinking during an independent investigation in science: Learning about the 'face of science that does not yet know'. *Science Education*, 80, 535–560.

Sharkawy, A. (2006). *An inquiry into the use of stories about scientists from diverse sociocultural backgrounds in broadening grade one students' images of science and scientists.* Toronto: Unpublished PhD thesis, University of Toronto.

Sharp, P. (2001). *Nurturing emotional literacy*. London: David Fulton.

She, H-C. (1995). Elementary and middle school students' image of science and scientists related to current science textbooks in Taiwan. *Journal of Science Education and Technology*, 4(4), 283–294.

Shen, J. & Confrey, J. (2007). From conceptual change to transformative modelling: A case study of an elementary teacher in learning astronomy. *Science Education*, 91(6), 948–956.

Shortland, M. (1988) Advocating science: Literacy and public understanding. *Impact of Science on Society*, 38(4), 305–316.

Siegel, H. (1991) The rationality of science, critical thinking, and science education. In M.R. Matthews (Ed.), *History, philosophy and science teaching: Selected readings* (pp. 45–62). Toronto: OISE Press.

Simon, S., Erduran, S. & Osborne, J. (2006). Learning to teach argumentation: Research and development in the science classroom. *International Journal of Science Education*, 28(2–3), 235–260.

Simmons, M.L. & Zeidler, D. (2003). Beliefs in the nature of science and responses to socio-scientific issue. In D.L. Zeidler (Ed.), *The role of moral reasoning on socioscientific issues and discourse in science education* (pp. 81–94). Dordrecht: Kluwer.

Smith, C.L., Maclin, D., Houghton, C. & Hennessey, M.G. (2000). Sixth grade students' epistemologies of science: The impact of school science experiences on epistemological development. *Cognition and Instruction*, 18(3), 349–422.

Smith, C.L. & Wenk, L. (2006). Relations among three aspects of first-year college students' epistemologies of science. *Journal of Research in Science Teaching*, 43(8), 747–785.

Smith, K.V., Loughran, J., Barry, A. & Dimitrakopoulos, C. (2012). Developing scientific literacy in a primary school. *International Journal of Science Education*, 34(1), 127–152.

Solomon, J. & Aikenhead, G. (Eds.) (1994) *STS education: International perspective on reform*. New York: Teachers College Press.

Solomon, J., Duveen, J. & Scott, L. (1994). Pupils' images of scientific epistemology. *International Journal of Science Education*, 16, 361–373.

Solomon, J., Scott, L. & Duveen, J. (1996). Large-scale exploration of pupils' understanding of the nature of science. *Science Education*, 80(5), 493–508.

Song, J. & Kim, K.-S. (1999). How Korean students see scientists: The images of the scientist. *International Journal of Science Education*, 21(9), 957–977.

Stanley, W.B. & Brickhouse, N.W. (1995). Science education without foundations: A response to Loving. *Science Education*, 79(3), 349–354.

Steiner, C. (1997). *Achieving emotional literacy*. London: Bloomsbury.

Sumrall, W.J. (1995). Reasons for the perceived images of scientists by race and gender of students in grades 1–7. *School Science & Mathematics*, 95(2), 83–90.

Svoboda, J., Passmore, C. & Giere, R. (2013). Models in science and in learning science: Organizing and elevating scientific practice. In M.R. Matthews (ed.*), Handbook of historical and philosophical research in science education* (pp.). Dordrecht: Springer.

Symington, D. & Spurling, H. (1990). The 'draw a scientist test': Interpreting the data. *Research in Science & Technological Education*, 8(1), 75–77.

Taber, K.S. (2003). Mediating mental models of metals: Acknowledging the priority of the learner's prior learning. *Science Education*, 87(5), 732–756.

Tairab, H.H. (2001). How do pre-service and in-service science teachers view the nature of science and technology. *Research in Science & Technological Education*, 19(2), 235–250.

Takao, A.Y. & Kelly, G.J. (2003). Assessment of evidence in university students' scientific writing. *Science & Education*,12(4), 341–363.

Tate, W. (2001). Science education as a civil right: Urban schools and opportunity-to-learn considerations. *Journal of Research in Science Teaching*, 38(9), 1015–1028.

Taylor, I., Barker, M. & Jones, A. (2003). Promoting mental model building in astronomy education. *International Journal of Science Education*, 25(10), 1205–1225.

Thomas, G. & Durant, J. (1987). Why should we promote the public understanding of science? In M. Shortland (Ed.), *Scientific literacy papers* (pp. 1–14). Oxford: Oxford University Department for External Studies.

Treagust, D.F., Chittleborough, G. & Mamiala, T.L. (2002). Students' understanding of the role of scientific models in learning science. *International Journal of Science Education*, 24(4), 357–368.

Treagust, D.F., Chittleborough, G. & Mamiala, T.L. (2004). Students' understanding of the descriptive and predictive nature of teaching models in organic chemistry. *Research in Science Education*, 34(1), 1–20.

Tsai, C-C. & Liu, S-Y. (2005). Developing a multi-dimensional instrument for assessing students' epistemological views toward science. *International Journal of Science Education*, 27(13), 1621–1638.

Tucker-Raymond, E., Varelas, M., Pappas, C.C., Korzah, H. A. & Wentland, A. (2007). 'They probably aren't named Rachel': Young children's scientist identities as emergent multimodal narratives. *Cultural Studies of Science Education*, 1(3), 559–592.

Tytler, R. (2007). *Re-imaging science education: Engaging students in science for Australia's future*. Camberwell, Victoria: Australian Council for Educational Research.

Tytler, R. (2012). Socioscientific issues, sustainability and science education. *Research in Science Education*. Available online.

Tytler, R. & Peterson, S. (2004). From 'try it and see' to strategic exploration: Characterizing young children's scientific reasoning. *Journal of Research in Science Teaching*, 41(10, 94–118.

UNESCO (1993). *International forum on scientific and technological literacy for all*. Final Report. Paris: UNESCO.

van Driel, J.H. & Verloop, N. (1999). Teachers' knowledge of models and modelling in science. *International Journal of Science Education*, 21(11), 1141–1153.

van Dijk, E.M. (2011). Portraying real science in science communication. *Science Education*, 95(6), 1086–1100.

van Eijck, M. & Roth, W-M. (2008). Representations of scientists in Canadian high school and college textbooks. *Journal of Research in Science Teaching*, 45(9), 1059–1082.

Van Praagh, G. (1973). *H.E. Armstrong and science education*. London: John Murray.

Varma, R. (2000). Changing research cultures in U.S. industry. *Science, Technology & Human Values*, 25(4), 395–416.

Vázquez, A. & Manassero, M.A. (1999). New response and scoring models for the "Views on Science-Technology-Society" instrument (VOSTS). *International Journal of Science Education*, 21, 231–247.

Vázquez, A., Manaserro, M.A. & Acevedo, J.A. (2006). An analysis of complex multiple-choice science-technology-society items: Methodological development and preliminary results. *Science Education*, 90(4), 681–706.

Vesterinen, V-M., Aksela, M. & Lavonen, J. (2011). Quantitative analysis of representations of nature of science in Nordic upper secondary school textbooks using a framework of analysis based on philosophy of chemistry, *Science & Education*. Available online

Webb, M.E. (1994). Beginning computer-based modeling in primary schools. *Computers in Education*, 22(1), 129–144.

Welch, W.W. (1969a). *Science process inventory*. Minneapolis, MN: University of Minnesota.

Welch, W.W. (1969b). *Wisconsin inventory of science processes*. Madison, WI: University of Wisconsin Scientific Literacy Research Center.

Walls, L. (2012). Third grade African American students' views of the nature of science. *Journal of Research in Science Teaching*, 49(1), 1–37.

Welzel, M. & Roth, W-M. (1998). Do interviews really assess students' knowledge. *International Journal of Science Education*, 20, 25–44.

Westaway, F.W. (1929). *Science teaching*. London Blackie & Son.

Windschitl, M. (2004). Caught in the cycle of reproducing folk theories of 'inquiry': How pre-service teachers reproduce the discourse and practices of an atheoretical scientific method. *Journal of Research in Science Teaching*, 41(5), 481–512.

Wittgenstein, L. (1953). *Philosophical investigations*. (trans: G.E.M. Anscombe) Oxford: Blackwell.

Wong, S.L. & Hodson, D. (2009). From the horse's mouth: What scientists say about scientific investigation and scientific knowledge. *Science Education* 93(1), 109–130.

Wong, S.L. & Hodson, D. (2010). More from the horse's mouth: What scientists say about science as a social practice. *International Journal of Science Education* 32(11), 1431–1463.

Wong, S.L. & Hodson, D. (2013). From the horse's mouth: Why scientists' views are crucial to nature of science understanding. *International Journal of Science Education*, in press.

Wong, S.L., Hodson, D., Kwan, J. & Yung, B.H.W. (2008). Turning crisis into opportunity: Enhancing student-teachers' understanding of nature of science and scientific inquiry through a case study of the scientific research in severe acute respiratory syndrome. *International Journal of Science Education*, 30(11), 1417–1439.

Wong, S.L., Kwan, J., Hodson, D. & Yung, B.H.W. (2009). Turning crisis into opportunity: Nature of science and scientific inquiry as illustrated in the scientific research on severe acute respiratory syndrome. *Science & Education*, 18(1), 95–118.

Wong, S.L., Yung, B.H.W., Cheng, M.W., Lam, K.L. & Hodson, D. (2006). Setting the stage for developing pre-service science teachers' conceptions of good science teaching: The role of classroom videos. *International Journal of Science Education*, 28(1), 1–24.

Wu, Y-T. & Tsai, C-C. (2007). High school students' informal reasoning on a socioscientific issue: Qualitative and quantitative analyses. *International Journal of Science Education*, 29, 1163–1187.

Yager, R.E. (Ed.) (1996). *Science/technology/society as reform in science education*. Albany, NY: State University of New York Press.

Yung, B.H.W., Wong, A.S.L., Cheng, M.W., Hui, C.S. & Hodson, D. (2007). Benefits of progressive video reflection on pre-service teachers' conceptions of good science teaching. *Research in Science Education*, 37(3), 239–259.

Zeidler, D.L., Sadler, T.D., Simmons, M.L. & Howes, E.V. (2005). Beyond STS: A research-based framework for socioscientific issues education. *Science Education*, 89(3), 357–377.

Zeidler, D.L. & Sadler, T.D. (2008a). The role of moral reasoning in argumentation: Conscience, character and care. In S. Erduran & M.P. Jiménez-Aleixandre (Eds.), *Argumentation in science education: perspectives from classroom-based research* (pp. 201–216). Dordrecht: Springer.

Zeidler, D.L. & Sadler, T.D. (2008b). Social and ethical issues in science education: A prelude to action. *Science & Education*, 17(8 & 9), 799–803.

Zeidler, D.L. & Schafer, L.E. (1984). Identifying mediating factors of moral reasoning in science education. *Journal of Research in Science Teaching*, 21, 1–15.

Zeidler, D.L., Osborne, J., Erduran, S., Simon, S. & Monk, M. (2003). The role of argument during discourse about socioscientific issues. D.L. Zeidler (Ed.), *The role of moral reasoning on socio-scientific issues and discourse in science education* (pp. 97–116). Dordrecht: Kluwer.

Zeidler, D.L., Sadler, T.D., Applebaum, S. & Callahan, B.E. (2009). Advancing reflective judgment through socioscientific issues. *Journal of Research in Science Teaching*, 46(1), 74–101.

Zeidler, D.L., Sadler, T.D., Simmons, M.L. & Howes, E.V. (2005) Beyond STS: A research-based framework for socioscientific issues education. *Science Education*, 89(3), 357–377.

Zeidler, D.L., Walker, K.A., Ackett, W.A. & Simmons, M.L. (2002). Tangled up in views: Beliefs in the nature of science and responses to socioscientific dilemmas. *Science Education*, 86(3), 343–367.

Zeidner, H., Matthews, G. & Roberts, R.D. (2009). *What we know about emotional intelligence.* Cambridge, MA: MIT Press.

Zemplén, G.A. (2009). Putting sociology first: Reconsidering the role of the social in 'nature of science' education. *Science & Education*, 18, 525–559.

Zhang, M., Lundeberg, M., Koehler, M.J., Eberhardt, J. (2011). Understanding affordances and challenges of three types of video for teacher professional development. *Teaching and Teacher Education,* 27(20), 454–462.

Ziman, J. (2000). *Real science: What it is, and what it means.* Cambridge: Cambridge University Press.

Derek Hodson is Emeritus Professor of Science Education at the Ontario Institute for Studies in Education (OISE), Professor of Science Education at the University of Auckland (part-time) and Visiting Professor of Science Education at the University of Hong Kong. After completing a B.Sc. in chemistry and a Ph.D. in synthetic organic chemistry at the University of Manchester and a M.Ed. degree at the University College of North Wales (Bangor), he taught chemistry and physics for 12 years in secondary schools in England, Scotland and Wales and spent 8 years in preservice education at the University of Manchester, 6 years at the University of Auckland, 17 years at OISE (including Directorship of the Imperial Oil Centre for Studies in Science, Mathematics and Technology Education) and four six-month periods at the University of Hong Kong. He was Founding Editor and General Manager of the *Canadian Journal of Science, Mathematics and Technology Education* – a position he held from 2001 to 2008. His research interests include history, philosophy and sociology of science and science education, science curriculum history, multicultural and antiracist education and action research approaches to science teacher education. Recently published books include *Towards Scientific Literacy: A Teachers' Guide to the History, Philosophy and Sociology of Science* (2008); *Teaching and Learning about Science: Language, Theories, Methods, History, Traditions and Values* (2009); and *Looking to the Future: Building a Curriculum for Social Activism* (2011) – all published by Sense Publishers.

Chapter 29
The Development, Use, and Interpretation of Nature of Science Assessments

Norman G. Lederman, Stephen A. Bartos, and Judith S. Lederman

29.1 Introduction

In the end, assessment becomes critical when considering the various goals of science curricula and instruction. This is as true for nature of scientific knowledge (NOS), typically considered synonymous with nature of science, as it is for any science subject matter. Hence, it is critical to delineate both the rationale for teaching the construct and its meaning.

The construct "nature of scientific knowledge" has been and continues to be an advocated goal of science education, as reflected in numerous US reform documents (e.g., American Association for the Advancement of Science [AAAS] 1990, 1993; National Research Council [NRC] 1996; National Science Teachers Association [NSTA] 1982) as well as other reform documents globally. Although conceptions of NOS, as reflected in these documents, have changed as much as the scientific knowledge they characterize, in general, an understanding of NOS is defended as being a critical component of scientific literacy. In spite of the arguments presented by a handful of researchers (e.g., Allchin 2011, 2012; Wong and Hodson 2009, 2010) contending that the views of NOS presented in these documents, and undergirding much of current NOS research, are not representative of the real work of scientists, it is important to note that the aspects of NOS outlined in the sections that follow are derived from careful examination of the writings of scientists, historians of science, and philosophers of science. Regardless, it is important not to lose sight of the audience for the often cited aspects of NOS, K-12 students. Consequently,

---

N.G. Lederman (✉) • J.S. Lederman
Department of Mathematics and Science Education,
Illinois Institute of Technology, Chicago, USA
e-mail: ledermann@iit.edu

S.A. Bartos
Womack Educational Leadership Department, Middle Tennessee State University,
Murfreesboro, USA

M.R. Matthews (ed.), *International Handbook of Research in History, Philosophy and Science Teaching*, DOI 10.1007/978-94-007-7654-8_29,
© Springer Science+Business Media Dordrecht 2014

when one begins to consider what aspects of NOS should be included in school curricula, developmental appropriateness and relevance/importance to daily life must be addressed. Most importantly, there does exist a relatively clear consensus supporting a group of scientifically, developmentally, and educationally appropriate (K-12) aspects of NOS. But, before unpacking the construct, it would seem appropriate to first explicate what is meant by "scientific literacy."

The view of scientific literacy used here is informed by Roberts' (2007) two "visions" of literacy that have been exemplars within the science education community. The first, "science literacy," is related to an understanding of the traditional science content, namely, the specific knowledge, processes, and products of a discipline, as it focuses "inward at the canon of orthodox natural science" (Roberts 2007, p. 2). "Scientific literacy" includes the ability to apply this conceptual knowledge and understanding of the processes of science to help inform personal decision-making and participation in a scientifically and technology-driven culture and economy (AAAS 1993; NRC 1996). While formalized by Showalter (1974) and the National Science Teachers Association (1982), the various arguments regarding the importance of NOS in the development of scientific literacy can best be understood by examining Driver et al. (1996). Specifically, Driver and colleagues contended that understanding NOS is necessary (1) in helping individuals comprehend "everyday" science and the related technology and process of their lives (utilitarian), (2) to aid in making informed decision-making when confronted with socio-scientific issues (democratic), (3) in fostering an appreciation of the value that science adds to contemporary culture (cultural), (4) in helping cultivate an understanding of moral commitments of the scientific community and their value to society as a whole, and (5) in facilitating the learning of science subject matter.

While scientific literacy has arguably become a principal and overarching goal of science education worldwide (Roberts 2007), unfortunately, and in spite of the preponderance of NOS as an objective of science education for over 100 years (Central Association of Science and Mathematics Teachers 1907; Kimball 1967–68; Lederman 1992), it is largely intuition that underpins these five arguments in favor of developing learners' understandings of NOS. Little empirical support can be found in the science education literature to support the various rationales for developing understandings of NOS. This is due, in no small part, to the challenges of improving individual's conceptions of NOS, as "the longevity of this educational objective has been surpassed only by the longevity of students' inability to articulate the meaning of the phrase 'nature of science,' and to delineate the associated characteristics of science" (Lederman and Niess 1997, p. 1). The obstacles involved in seeing teachers' informed views of NOS translated into their classroom practice further complicate this process. Moreover, without a sufficient number of NOS-informed individuals, there is no way to know if or how NOS contributes to the development of a scientifically literate populace. Unfortunately, existing data concerning understandings of NOS still support Shamos (1984) who, when speaking to the necessity of developing students' understandings of what we now refer to as NOS, concluded that "in spite of taking science classes, few students come through this experience with more than a fleeting glimpse of science, and fewer still retain any lasting impression of the scientific world" (p. 333).

29.2 What Is Nature of Scientific Knowledge?

Fortunately, in spite of the dearth of research specifically relating NOS to the development of scientific literacy, there is over 60 years of research on NOS that has, in part, sought to assess teachers' and students' understandings of NOS and investigate the efficacy of various approaches to improving these conceptions (Lederman 2007). But, irrespective of this ever growing body of research, the continued support for NOS in the science education and scientific communities and explicit statements regarding NOS in various reform documents, there are still unproductive disagreements regarding the meaning of NOS. Prior to delineating the specific conception of NOS espoused here, a few issues must be clarified.

First, the myriad views of NOS reflected on the pages of refereed journals and conference proceedings, which almost invariably contradict the aforementioned reform documents, do not provide direct support for the contention that there is, therefore, no consensus about the meaning of NOS as some have contended (e.g., Alters 1997). Not only is there more consensus than disagreement about the definition and/or meaning of NOS (Smith et al. 1997; Smith and Scharmann 1999), these disagreements, while providing fodder for a lively argument among philosophers, historians, or science educators, are irrelevant to K-12 classroom practice (Lederman 1998, 2007). What is necessary when considering NOS, as is the case with typical science content, is its educational and developmental appropriateness, as well as its presentation in a way that is connected with students' lives, but at an acceptable level of generality, as reflected in the aforementioned authors as well as others (e.g., Elby and Hammer 2001; Rudolph 2003). Little disagreement exists among philosophers, historians, and science educators for the characteristics of NOS that fit these criteria.

Second, it is important to stress that a definitive description of NOS, contrary to the assumptions of our critics (Irzik and Nola 2011; Matthews 2012), is not presented here or elsewhere. It is recognized, and it should be obvious, that other researchers may include or delete various aspects of NOS resulting in equally valid representations of NOS that are educationally and developmentally appropriate for learners (Osborne et al. 2003; Smith and Scharmann 1999). Far too much time has recently been spent arguing about what aspects of NOS should and should not be included in various lists of desired outcomes and standards. The discussion should be more centered on the value of the knowledge that is being considered, not the construction of a definitive definition of NOS. The focus here is not to simply promote the definition of the construct provided by favored colleagues and fellow researchers[1] but to assist the reader in delineating NOS from both the process of science and the scientific knowledge that results. The conflation of the processes of science (scientific inquiry), with the characteristics of scientific knowledge that are inherently derived from these processes (nature of science scientific knowledge, NOS), is an avoidable, yet common, characteristic of research done on NOS.

---

[1] See, for instance, Abd-El-Khalick (2005), Akerson et al. (2000), Bell and Lederman (2003), Khishfe and Abd-El-Khalick (2002), Lederman and Neiss (1997), and Schwartz and Lederman (2002).

Lastly, it should be reiterated that a focus on learning outcomes that are developmentally appropriate have a preponderance of empirical support for inclusion in K-12 curricula and are arguably essential if students are to achieve the goal of scientific literacy should be in the forefront of discussions about NOS. Furthermore, at this level of appropriate generality, there are few disagreements about aspects of NOS, as evidenced by their congruence with numerous reform documents worldwide. Consider the issue of the existence of an objective reality versus that which is purely phenomenal. This debate may certainly be situated in a philosophy of science class but is misplaced, misaligned, and counterproductive to the goals and objectives of K-12 science curricula. The reader is reminded that the goal of the K-12 science teacher is not to create philosophers of science, but rather to develop informed citizens so decisions can be made concerning personal and societal issues that are scientifically based. This goal is sometimes overlooked by participants in NOS disputes.

29.3   The Nature of Scientific Knowledge (NOS)

In general, the phrase "nature of scientific knowledge" or NOS refers to the characteristics of scientific knowledge that are inherently derived from the manner in which it is produced (i.e., scientific inquiry). These general characterizations aside, philosophers and historians of science, scientists, and science educators do not, nor should they be expected to, share a common consensus on a specific definition of NOS. This, as previously mentioned, should not be cause for alarm, as over the last century conceptions of NOS have changed just as conceptions of science have done, with the definition of NOS changing as much as the knowledge it intends to characterize.[2]

It is of utmost importance that the 1980s saw the phrase "nature of scientific knowledge" shortened to "nature of science," a modification that may have introduced some unnecessary confusion. In the research literature, "nature of science" more aptly refers to "nature of scientific knowledge" and is consistent with the definition used in the current chapter. Lastly, to reiterate, the "list" of characteristics of scientific knowledge that will be explicated in what follows is educationally and developmentally appropriate and has a wealth of evidence in support of their inclusion in K-12 science instruction – they should not be construed as representing the definitive "NOS catechism," as some have decreed (Matthews 2012). Furthermore, while some researchers have maintained that these aspects of NOS do not present the "whole picture" of science as it is practiced by scientists (e.g., Allchin 2011, 2012; Wong and Hodson 2009, 2010), these aspects are, the reader is reminded, derived

---

[2] As evidenced by AAAS (1990, 1993), Center of Unified Science Education (1974), Central Association for Science and Mathematics Teachers (1907), Klopfer and Watson (1957), and NSTA (1982).

from the writings and recommendations of scientists and are not intended to help inform efforts to create a new population of bench scientists, but to aid K-12 classroom science teachers and science education researchers in the development of a scientifically literate populace. With these caveats, let us return to unpacking the construct of NOS.

First, learners should develop an understanding of the crucial distinction between observation and inference. In the K-12 science classroom, observations are presented as descriptive statements about natural phenomena that are "directly" accessible to the senses, or extensions of the senses, and for which observers can reach consensus with relative ease (e.g., an object, once released, falls to the floor). Inferences, by contrast, are statements that are not "directly" accessible to the senses and can only accessed and/or measured though related manifestations or effects (e.g., gravity). Beyond developing explanations, at a higher level, scientists can infer models and/or mechanisms that serve to explain observations of complex phenomena (e.g., weather modeling, evolution).

Second is the distinction between scientific theories and laws, a pair of categories of scientific knowledge that is closely related to the distinction between observation and inference. This point is critical as the majority of individuals hold a simplistic, hierarchical view of the relationship between theories and laws, whereby theories, once they have "accumulated" sufficient supportive evidence, become laws. It follows from this misconception that scientific laws have a higher status than scientific theories, when in fact scientific theories and laws are different types of knowledge. A theory is not formulated with the hope that someday it will acquire the status of "law," as theories are not developed or transformed into laws, nor is a law ever demoted to being "just a theory." Scientific laws are statements or descriptions of the relationships among observable phenomena. Boyle's law, which relates the pressure of a gas to its volume at a constant temperature, is a case in point. Scientific theories, by contrast (and in contrast to the common usage of the word "theory"), are inferred explanations for observable phenomena. The kinetic molecular theory, which explains Boyle's law, is one example. Moreover, theories are as legitimate a product of science as laws. Scientific theories, in their own right, serve important roles, such as guiding investigations and generating new research problems, in addition to explaining relatively huge sets of seemingly unrelated observations in more than one field of investigation. For example, the kinetic molecular theory serves to explain phenomena that relate to changes in the physical states of matter, others that relate to the rates of chemical reactions, and still other phenomena that relate to heat and its transfer, to mention just a few. While some philosophers and historians of science (e.g., Allchin 2012; Wong and Hodson 2009, 2010) may contend that these descriptions of laws and theories leave something to be desired, this level of generality has evidenced itself as appropriate and accessible to K-12 science students. Indeed, these same critics of including the distinction of theories and laws under the rubric of NOS base their positions on the idea that scientists do not enter discussions about such differences in knowledge claims. In spite of this, the audience of such NOS instruction cannot be ignored. The commonly held misconception that "evolution is just a theory" is case in point. Although the distinction between theories and laws may not be important to scientists, it is certainly important for the general public, teachers, and students.

Third, the development of scientific knowledge involves human imagination and creativity. Science, contrary to common belief, does not rely solely on observations of the natural world (i.e., empirically based), nor is it totally lifeless, rational, and orderly. In addition to devising creative investigatory methodologies and data reduction techniques, science involves the invention of explanations and the generation of ideas that involve considerable creativity by scientists. The "leap" from atomic spectral lines to Bohr's model of the atom with its elaborate orbits and energy levels is one example. This aspect of science, coupled with its inferential nature, entails that scientific concepts, such as atoms, black holes, and species, are functional theoretical models rather than faithful copies of reality.

Fourth, scientific knowledge, owing to scientists' theoretical commitments, beliefs, previous knowledge, training, experiences, and expectations, is unavoidably subjective. These background factors form a mind-set that affects the problems scientists investigate and how they conduct their investigations, what they observe (and do not observe), what they consider as evidence, and how they make sense of and interpret their observations. It is this (sometimes collective) individuality or mind-set that accounts for the role of subjectivity in the production of scientific knowledge. It is noteworthy that, contrary to common belief, science rarely starts with neutral observations (Chalmers 1982). Observations (and investigations) are motivated by, guided by, and acquire meaning in reference to questions or problems, which, in turn, are derived from within certain theoretical perspectives. Often, hypothesis or model testing serves as a guide to scientific investigations. For example, a researcher operating from a Darwinian framework might focus his/her efforts on the location of transitional species. By contrast, from a punctuated equilibrist perspective, transitional species would not be expected, nor would what a Darwinian considered a transitional species be considered as such (see Gould and Eldridge 1977).

Fifth, science as a human enterprise is practiced in the context of a larger culture, and its practitioners (scientists) are the product of that culture. Science, it follows, affects and is affected by the various elements and intellectual spheres of the culture in which it is embedded. These elements include, but are not limited to, social fabric, power structures, politics, socioeconomic factors, philosophy, and religion. Telling the story of the evolution of humans (*Homo sapiens*) over the course of the past seven million years is central to the biosocial sciences and serves to illustrate how social and cultural factors impact scientific knowledge. Scientists have formulated several elaborate and differing story lines about this evolution. Until recently, the dominant story was centered about "the man-hunter" and *his* crucial role in the evolution of humans to the form we now know (Lovejoy 1981). This scenario was consistent with the white-male culture that dominated scientific circles up to the 1960s and early 1970s. As the feminist movement grew stronger and women were able to claim recognition in the various scientific disciplines, the story about hominid evolution started to change. One story that is more consistent with a feminist approach is centered about "the female-gatherer" and *her* central role in the evolution of humans (Hrdy 1986). It is noteworthy that both story lines are consistent with the available evidence.

Sixth, it follows from the previous discussions that scientific knowledge is never absolute or certain. This knowledge, including "facts," theories, and laws, while durable, is tentative and subject to change. Scientific claims change as new evidence, made possible through advances in *theory* and technology, is brought to bear on existing theories or laws or as old evidence is reinterpreted in the light of new theoretical advances or shifts in the directions of established research programs. It should be emphasized that tentativeness in science does not only arise from the fact that scientific knowledge is inferential, creative, and socially and culturally embedded. There are also compelling logical arguments that lend credence to the notion of tentativeness in science. Indeed, contrary to common belief, scientific hypotheses, theories, and laws can *never* be absolutely "proven." This holds irrespective of the amount of empirical evidence gathered in the support of one of these ideas or the other (Popper 1963, 1988). For example, to be "proven," a certain scientific law should account for *every single instance* of the phenomenon it purports to describe *at all times*. It can logically be argued that one such future instance, of which we have no knowledge whatsoever, may behave in a manner contrary to what the law states. As such, the law can never acquire an absolutely "proven" status. This equally holds in the case of hypotheses and theories. This philosophical aside, while not intended for inclusion with younger learners, can help highlight both the tentative nature of certain scientific knowledge and the durability of other knowledge as a function of the weight of empirical evidence.

Before moving on to specifically address the development, use, and interpretation of various NOS assessments, it is important to note that science educators and science education researchers often conflate NOS with science processes, practices, or scientific inquiry (SI). Although these aspects of science overlap and interact in important ways, it is nonetheless important to distinguish between them. Scientific processes are activities related to collecting and analyzing data and drawing conclusions (AAAS 1990, 1993; NRC 1996). For example, observing and inferring are scientific processes. More complex than individual processes, scientific inquiry involves various science processes used in a cyclical manner. On the other hand, NOS refers to the epistemological underpinnings of the activities of science and the characteristics of the resulting knowledge. As such, realizing that observations are necessarily theory-laden and are constrained by our perceptual apparatus belongs within the realm of NOS. Distinguishing NOS from SI for the purpose of providing focus to this chapter should in no way be construed to mean that NOS is considered more important for students to learn about. Certainly, NOS and SI, although different, are intimately related and are both important for students to understand, though making a distinction between NOS and SI is not meant to imply that the two constructs are distinct. Furthermore, there is much evidence that NOS is best taught within a context of SI or activities that are reasonable facsimiles of inquiry. That is, inquiry experiences provide students with foundational experiences upon which to reflect about aspects of NOS.

The conflation of NOS and SI has plagued research on NOS from the beginning. Hence, the reader will note that many NOS assessments are actually more focused on SI than NOS. These studies are nevertheless reviewed, rather than excluded, since they have become an accepted part of the history of research on assessment of NOS. The definition used by these studies for NOS is just not consistent with current usage of the construct. Again, the aspects of NOS presented here are not meant to be exhaustive, as other listings certainly exist. However, what has been presented is directly consistent both with what current reform documents state students should know about NOS and also with the perspective taken by an overwhelming majority of the research literature.

Lastly, and as has been communicated previously, NOS can be a moving target, as it becomes clear to anyone who considers the works of Popper (1959), Kuhn (1962), Lakatos (1970), Feyerabend (1975), Laudan (1977), and Giere (1988) that perceptions of NOS are as tentative, if not more so, than scientific knowledge itself. NOS is, in effect, analogous to scientific knowledge. Some individuals, unfortunately, have dwelled too heavily on such differing perceptions (e.g., Alters 1997) without consideration of the overarching goal of research on improving conceptions of NOS. The recognition that our collective views of NOS have changed and will continue to change is not a justification for ceasing all NOS-related research until total agreement is reached or for avoiding recommendations or identifying what we think students should know. As educators, we have no difficulty including certain theories and laws within our science curricula even though we recognize that these may change in the near or distant future. What is important is that students understand the evidence for current beliefs about natural phenomena and are aware that evidence has similarly lead to our current beliefs about NOS. Just as with "traditional" subject matter, these perceptions may change as additional evidence is collected or the same evidence is viewed in a different way.

Regardless of the various "problems" associated with reaching consensus on specific aspects of NOS, and issues created by the tentativeness of the construct itself, NOS has been the object of systematic educational research for approximately 60 years. While there have been numerous reviews of research related to the teaching, learning, and assessment of nature of scientific knowledge (e.g., Abd-El-Khalick and Lederman 2000; Lederman 1992, 2007; Meichtry 1992), this review will focus on assessment of NOS. For practical reasons, the research reviewed is restricted to published reports and to those studies with a primary focus on NOS and to those assessments that have at least attempted to establish validity and/or reliability.

29.4 Assessing Conceptions of NOS

The development and assessment of students' and teachers' conceptions of nature of scientific knowledge has been a concern of science educators for nearly 60 years and arguably constitutes a line of research in its own right. Although there have been numerous criticisms of the validity of various assessment instruments over the years, students' and teachers' understandings have consistently been found lacking. This consistent finding, regardless of assessment approach, supports the notion that student and teacher understandings are not at the desired levels.

The history of assessment of NOS mirrors the changes that have occurred in both psychometrics and educational research design over the past few decades. The first formal assessments, beginning in the early 1960s, emphasized quantitative approaches, as was characteristic of the overwhelming majority of science education research investigations. Prior to the mid-1980s, with few exceptions, researchers were content to develop instruments that allowed for easily "graded" and quantified measures of individuals' understandings. In some cases, standardized scores were derived. Within the context of the development of various instruments, some open-ended questioning was involved in construction and validation of items. More recently, emphasis has been placed on providing an expanded view of an individual's knowledge regarding NOS. In short, in an attempt to gain more in-depth understandings of students' and teachers' thinking, educational researchers have resorted to the use of more open-ended probes and interviews. The same has been true with the more contemporary approaches to assessment related to NOS. Unfortunately, and in accordance with the pressures of high-stakes testing, momentum appears to be building for a return to more quantitative measures of NOS, which allows for large-scale administration and assessment of students' and teachers' understandings. In addition to this shift back to more "traditional" assessments, the *Next Generation Science Standards* (NRC 2011) in the USA are improvements over the original but, by a large degree, still ignore the long-standing empirical research on NOS, equating the "doing of science" with developing understandings about NOS. Consequently, the resulting assessment is likely not to be a valid measure of what has been known as NOS.

Although critical evaluations of assessment instruments have been provided elsewhere (Lederman 2007; Lederman et al. 1998), the purpose here is to summarize the various instruments and identify trends in the assessment of NOS. Table 29.1 presents a comprehensive list of the more formal instruments constructed and validated to assess various aspects of NOS. Most of the instruments address only certain aspects of NOS and often inappropriately confuse the issue by addressing areas other than NOS, including science process skills and attitudes toward science. Instruments considered to have poor validity have the following characteristics:

1. Most items concentrate on a student's ability and skill to engage in the process of science (e.g., to make a judgment and/or interpretation concerning data).
2. Emphasis is on the affective domain (the realm of values and feelings) rather than knowledge (i.e., over 50 % of items deal with attitude toward or appreciation of science and scientists).
3. Primary emphasis is placed upon "science as an institution" with little or no emphasis placed upon the epistemological characteristics of the development of scientific knowledge.

As mentioned before, the validity of many of these instruments is questionable because their primary focus is on areas beyond the scope of "nature of scientific knowledge." Those instruments with questionable validity (as measures of NOS) include the Science Attitude Questionnaire (Wilson 1954); Facts About Science

980   N.G. Lederman et al.

Table 29.1 Nature of science instruments

Date	Instrument	Author(s)
1954	Science Attitude Questionnaire	Wilson
1958	Facts About Science Test (FAST)	Stice
1959	Science Attitude Scale	Allen
1961	Test on Understanding Science (TOUS)	Cooley and Klopfer
1962	Processes of Science Test	BSCS
1966	Inventory of Science Attitudes, Interests, and Appreciations	Swan
1967	Science Process Inventory (SPI)	Welch
1967	Wisconsin Inventory of Science Processes (WISP)	Scientific Literacy Research Center
1968	Science Support Scale	Schwirian
1968	Nature of Science Scale (NOSS)	Kimball
1969	Test on the Social Aspects of Science (TSAS)	Korth
1970	Science Attitude Inventory (SAI)	Moore and Sutman
1974	Science Inventory (SI)	Hungerford and Walding
1975	Nature of Science Test (NOST)	Billeh and Hasan
1975	Views of Science Test (VOST)	Hillis
1976	Nature of Scientific Knowledge Scale (NSKS)	Rubba
1978	Test of Science-Related Attitudes (TOSRA)	Fraser
1980	Test of Enquiry Skills (TOES)	Fraser
1981	Conception of Scientific Theories Test (COST)	Cotham and Smith
1982	Language of Science (LOS)	Ogunniyi
1987	Views on Science-Technology-Society (VOSTS)	Aikenhead, Ryan, and Fleming
1990	Views of Nature of Science A (VNOS-A)	Lederman and O'Malley
1992	Modified Nature of Scientific Knowledge Scale (M-NSKS)	Meichtry
1995	Critical Incidents	Nott and Wellington
1998	Views of Nature of Science B (VNOS-B)	Abd-El-Khalick, Bell, and Lederman
2000	Views of Nature of Science C (VNOS-C)	Abd-El-Khalick, and Lederman
2002	Views of Nature of Science D (VNOS-D)	Lederman and Khishfe
2004	Views of Nature of Science E (VNOS-E)	Lederman and Ko
2006	Student Understanding of Science and Scientific Inquiry (SUSSI)	Liang et al.

Test (Stice 1958); Science Attitude Scale (Allen 1959); Processes of Science Test (BSCS 1962); Inventory of Science Attitudes, Interests, and Appreciations (Swan 1966); Science Support Scale (Schwirian 1968); Test on the Social Aspects of Science (Korth 1969); Science Attitude Inventory (Moore and Sutman 1970); Science Inventory (Hungerford and Walding 1974); Test of Science-Related Attitudes (Fraser 1978); the Test of Enquiry Skills (Fraser 1980); and the Language of Science (Ogunniyi 1982). Recently, Allchin (2012) proposed a prototype to assess what he called "whole science," but the instrument was never fully developed and it clearly conflated NOS with scientific inquiry. Hence, it is not discussed further here.

The remaining instruments have generally been considered to be valid and reliable measures of NOS by virtue of their focus on one or more ideas that have been traditionally considered under the label of "nature of scientific knowledge," as well as their reported validity and reliability data. These instruments have been used in numerous studies, and even the more traditional instruments (e.g., TOUS) continue to be used even though there is a significant movement away from such types of paper-and-pencil assessments. The validity of some of the assessment instruments listed and briefly described below has been justifiably criticized in the past few years. However, they are presented here as being the most valid (in terms of assessment focus) attempts to assess understandings of NOS using a written response format. What follows is a brief discussion of each instrument.

29.4.1 Test on Understanding Science (TOUS)

This instrument has been the most widely used assessment tool in NOS research (Cooley and Klopfer 1961). It is a four-alternative, 60-item multiple-choice test. In addition to an "overall" or "general" score, three subscales can be scored regarding understandings about (I) the scientific enterprise, (II) the scientist, and (III) the methods and aims of science. During the past few decades, the content of the TOUS has been criticized and has fallen into disfavor.[3]

29.4.2 Wisconsin Inventory of Science Processes (WISP)

The WISP consists of 93 statements that the respondent evaluates as "accurate," "inaccurate," or "not understood." However, in scoring the exam, "inaccurate" and "not understood" responses are combined to represent the opposite of "accurate." The WISP was developed and validated for high school students (Scientific Literacy Research Center 1967). Although this instrument has excellent validity and reliability data, a few concerns should be considered prior to its use. Of primary concern is its length. The 93-item test takes over an hour to administer, which precludes it from use in a single class period. In addition, this instrument does not possess discrete subscales, which, unfortunately, means that only unitary scores can be calculated.

29.4.3 Science Process Inventory (SPI)

This instrument is a 135-item forced-choice inventory (agree/disagree) purporting to assess an understanding of the methods and processes by which scientific knowledge evolves (Welch 1967; Welch and Pella 1967–1968). The content of the

---

[3] See Aikenhead (1973), Hukins (1963), Welch (1969), and Wheeler (1968), among others.

SPI is almost identical to that of WISP and TOUS subscale III. The validation of the SPI was achieved in the usual manner for such instruments: literature review, devising a model, employing the judgment of "experts," getting feedback from pilot studies, and testing the instrument's ability to distinguish among different groups of respondents. The length (135 items) is a concern as well as its forced-choice format. Students are unable to express "neutral" or uncertain answers. Finally, like the WISP, the SPI does not possess subscales.

29.4.4 Nature of Science Scale (NOSS)

This instrument was developed to determine whether science teachers have the same view of science as scientists (Kimball 1968). It consists of 29 items to which the respondent may "agree," "disagree," or register a "neutral" response. Kimball's model of NOS is based upon the literature of the nature and philosophy of science and is consistent with the views of Bronowski (1956) and Conant (1951). The specific content of the NOSS was validated by nine science educators who judged whether the items were related to the model. The development, validation, and reliability measures were carried out with college graduates. Thus, it lacks reliability and validity data with respect to high school populations. Another concern is that the instrument lacks subscales and is, therefore, subject to the same criticism as any other unitary measure of the nature of scientific knowledge.

29.4.5 Nature of Science Test (NOST)

This instrument consists of 60 multiple-choice items addressing the following components of NOS: assumptions of science (8 items), products of science (22 items), processes of science (25 items), and ethics of science (5 items) (Billeh and Hasan 1975). The test consists of two types of items. The first type measures the individual's knowledge of the assumptions and processes of science and the characteristics of scientific knowledge. The second type of question presents situations that require the individual to make judgments in view of his/her understanding of NOS. The major shortcoming of this instrument is not its content, but, rather, that no subscales exist. Again, only a global or unitary score can be calculated.

29.4.6 Views of Science Test (VOST)

This instrument was developed specifically to measure understanding of the tentativeness of science (Hillis 1975). It consists of 40 statements that are judged to imply that scientific knowledge is either tentative or absolute. Respondents express their agreement with either view using a five-option Likert scale response format.

29.4.7 Nature of Scientific Knowledge Scale (NSKS)

This instrument is a 48-item Likert scale response format consisting of five choices (strongly agree, agree, neutral, disagree, and strongly disagree) (Rubba 1976). The test is described as an objective measure of secondary students' understanding of NOS. The NSKS and its subscales are based upon the nine factors of NOS specified by Showalter (1974). Rubba (1976) listed these nine factors as tentative, public, replicable, probabilistic, humanistic, historic, unique, holistic, and empirical. He noted a certain amount of shared overlap between the factors and proceeded to collapse them into a six-factor or six-subscale model of the nature of scientific knowledge. These six factors are amoral, creative, developmental (tentative), parsimonious, testable, and unified. The instrument was developed, validated, and found to be reliable for high school level students. The five-option Likert scale response format affords maximum freedom of expression to the respondent. The *NSKS* has generally been viewed positively by the research community; however, there is reason for some concern about its face validity. Many pairs of items within specific subscales are identical, except that one item is worded negatively. This redundancy could encourage respondents to refer back to their answers on previous, similarly worded items. This cross-checking would result in inflated reliability estimates which could cause erroneous acceptance of the instrument's validity.

29.4.8 Conceptions of Scientific Theories Test (COST)

The structure of this instrument was dictated by the developers' concern that previously existing instruments were based on single (supposedly enlightened) interpretations of NOS (Cotham and Smith 1981). Thus, the COST supposedly provides for nonjudgmental acceptance of alternative conceptions of science. The instrument is an attitude inventory consisting of 40 Likert scale items (with four options) and four subscales, each corresponding to a particular aspect of scientific theories. These include (I) ontological implications of theories, (II) testing of theories, (III) generation of theories, and (IV) choice among competing theories. The COST provides a theoretical context for four-item sets by prefacing each set with a brief description of a scientific theory and some episodes drawn from its history. The items following each theory description refer to that description. The four theoretical contexts are (1) Bohr's theory of the atom, (2) Darwin's theory of evolution, (3) Oparin's theory of abiogenesis, and (4) the theory of plate tectonics. A fifth context contains items that refer to general characteristics of scientific theories and is, therefore, not prefaced by a description. Two concerns must be addressed prior to using COST as an instrument to assess high school students' understandings of NOS. The first of these is the cognitive level of the instrument. It was designed for teachers and validated with undergraduate college students. The four theory descriptions used to provide context for the items are presented at a level that may be above the capabilities of many high school students.

A second concern with the COST instrument rests with the authors' claim that it, as opposed to all extant instruments, is sensitive to alternative conceptions of science. However, the authors actually specify which subscale viewpoints are consistent with a tentative and revisionary conception of science. Thus, although they claim to place no value judgments upon the various conceptions of science, Cotham and Smith actually do just that by linking certain viewpoints to the "highly prized" tentative and revisionary conception of scientific knowledge.

29.4.9 Views on Science-Technology-Society (VOSTS)

The VOSTS was developed to assess students' understanding of nature of scientific knowledge, technology, and their interactions with society (Aikenhead et al. 1987). It consists of a "pool" of 114 multiple-choice items that address a number of science-technology-society (STS) issues. These issues include Science and Technology, Influence of Society on Science/Technology, Influence of Science/Technology on Society, Influence of School Science on Society, Characteristics of Scientists, Social Construction of Scientific Knowledge, Social Construction of Technology, and Nature of Scientific Knowledge. The VOSTS was developed and validated for grades 11 and 12 students. A fundamental assumption underlying the development of this instrument was that students and researchers do not necessarily perceive the meanings of a particular concept in the same way. Aikenhead and Ryan (1992) recognized the importance of providing students with alternative viewpoints based upon student "self-generated" responses to avoid the "constructed" responses offered by most of the previous nature of scientific knowledge assessment instruments. Unlike most other instruments, the VOSTS does not provide numerical scores; instead it provides a series of alternative "student position" statements. Extensive work was done on the careful validation of the instrument over a period of 6 years.

29.4.10 Views of Nature of Science, Form A (VNOS-A)

In an attempt to ameliorate some of the problems that each of the seven items focused on different aspects of tentativeness noted by Aikenhead et al. (1987) during the development of the VOSTS and those noted in the use of the NSKS (Rubba 1976) relative to the use of paper-and-pencil assessments, Lederman and O'Malley developed an open-ended survey consisting of seven items (Lederman and

O'Malley 1990). This instrument was designed to be used in conjunction with follow-up interviews and in science. Several problems were noted in the wording of some of the questions, resulting in responses that did not necessarily provide information on students' views of "tentativeness." The authors claimed that follow-up interviews alleviated this problem.

29.4.11 Modified Nature of Scientific Knowledge Scale (M-NSKS)

This instrument is a modified NSKS instrument with 32 statements from four of the NSKS subscales (Meichtry 1992). These subscales are (I) creative, (II) developmental, (III) testable, and (IV) unified. M-NSKS was developed, with reliability and validity reported, for use with 6th, 7th, and 8th graders.

29.4.12 Critical Incidents

The use of "critical incidents" to assess teachers' conceptions of NOS was a significant departure from the usual paper-and-pencil assessment (Nott and Wellington 1995). In particular, Nott and Wellington are of the opinion that teachers do not effectively convey what they know about nature of scientific knowledge in "direct response to abstract, context-free questions of the sort, 'What is science?'" (Nott and Wellington 1995). Instead, they created a series of "critical incidents" that are descriptions/scenarios of actual classroom events. Teachers are expected to respond to the incidents by answering the following three questions: (1) What would you do?, (2) What could you do?, and (3) What should you do? Although the use of critical incidents appears to be an excellent instructional tool to generate meaningful discussions in preservice and in-service courses, whether the teachers' responses are related to their views about NOS is still questionable. In short, the approach is based on the assumption that teachers' views of NOS automatically and necessarily influence classroom practice, an assumption that is simply not supported by the existing literature.

29.4.13 Views of Nature of Science B, C, D, E (VNOS-B, VNOS-C, VNOS-D, VNOS-E)

This series or buffet of instruments has stemmed from the same research group and was meant to be variations and improvements upon the original VNOS-A (Lederman and O'Malley 1990). In particular, each instrument contains open-ended questions that focus on various aspects of NOS with the differences being either the additional context-specific questions in forms B and C or the developmental appropriateness and language of VNOS-D. From a practical standpoint, VNOS-B and VNOS-C are too lengthy to be administered easily during a regular class period. Consequently, VNOS-D and VNOS-E were created with the aid of focus groups of secondary ($n = 10$) and elementary ($n = 10$) teachers and their students. The resulting instruments are easily administered in less than one hour and yield the same results as the longer VNOS-B and VNOS-C. VNOS-E is the most recently developed instrument and it has been designed for very young students (grades K-3). The items can also be used with students that cannot read or write (using a focus group format), and it represents the first measure of NOS designed for such a young audience.

29.4.14 Student Understanding of Science and Scientific Inquiry (SUSSI)

This instrument was developed, as was the case with the majority of standardized, quantitative approaches to assessing NOS, to overcome the time constraints and provide a potential tool for large-scale assessments (Liang et al. 2006). The SUSSI targets tentativeness, observations and inferences, subjectivity, creativity, social and culturally embeddedness, theories and laws, and scientific methods. Extensive evidence for the validity of the SUSSI is provided by its authors. The SUSSI is a combination of four Likert-scaled items followed by an open-response question similar in nature to the VNOS. As such, the SUSSI does not alleviate the time constraints associated with scoring the VNOS or similar instruments yet complicates the development of individual profiles of NOS understandings by introducing issues regarding interpretation of quantitative results and understandings of NOS. The SUSSI does not appear to be capable of providing meaningful inferences at the grain size that their developers intend. Although offering a means to utilize inferential statistics to assess instructional interventions, the guidelines for interpreting these quantitative data, and how the results of the two components of the questionnaire (i.e., Likert and free response) are "married," are not clearly explicated.

29.5 Development of Assessments for NOS

It should go without saying that any assessment instrument for NOS, or anything for that matter, should go through a systematic and extensive process for the establishment of both validity and reliability. Unfortunately, this has not been the case during the more than 50 years of assessment development. The establishment of reliability of any assessment is a fairly straightforward process; however, the establishment of validity is more complicated than most consider it to be. Simply gathering a group of "experts" together to chime in on whether the assessment items measure what the assessment developer intends is not the whole story. Pursuing construct validity in addition to the aforementioned content validity does not complete the picture either. The context and the target audience for the assessment are critical for any assessment. This has been an area in need of much attention. In specific, researchers would be better served by focusing on what is appropriate for K-12 students to know and be able to do, in contrast to arguing about why "lists" of outcomes or outcomes derived from others than scientists are anathema.[4]

29.5.1 *Why Can't We Agree to Disagree?*

Far too much discussion and journal pages have focused on the lack of consensus on a definition or characterization of NOS. In short, scholars would rather argue about the need to reach consensus before an assessment of NOS can be developed. Why is NOS held to a higher standard than other content in science? How many of the concepts and ideas in science have achieved absolute consensus before we attempt to teach them to students and assess what they have learned? As previously discussed, when one considers the developmental level of the target audience (K-12 students), the aspects of NOS stressed herein are at a level of generality that is not at all contentious. Nevertheless, if one is not willing to let go of the idea that the various aspects of NOS lack consensus and that assessment of NOS is, therefore, problematic, the "problem" is easily handled. One's performance on a NOS assessment can simply be used to construct a profile of what the student knows/believes about scientific knowledge. In terms of the aspects of NOS to be assessed, there is no reason to require that all assessments measure the exact same understandings. If the focus is just upon the assessment of understandings that are considered to be important for scientifically literate individuals to know, then there is no reason to require an agreed upon domain of NOS aspects. Different assessments may stress, to one degree or another, different aspects of NOS. This is no different than assessing students' understandings of the human heart. Different valid and reliable assessments stress and include different structures.

29.5.2 *What Is So Bad About Lists?*

Many researchers point out that lists are problematic (e.g., Allchin 2012; Matthews 2012), but lists serve an important function as they help provide a concise organization of the often complex ideas and concepts they include. Each item on a list is just a label or symbol for a much more in-depth and detailed elaboration. If "tree" is included in a list, it is simply a referent for all the structures and process that are involved in what is involved in being a "tree." There is the temptation to think that

---

[4] As evidenced in Allchin (2012), Duschl and Grandy (2012), Irzik and Nola (2011), and Wong and Hodson (2009, 2010), among others.

lists are defined as consisting of very short (1–2 word) entries. There are numerous science education reform documents that specify and delineate what students should know and be able to do (i.e., standards). These are also lists of learning outcomes, even though the standards can be as long as a paragraph. The only problem with a list is how it is often used. If students are asked to simply and mindlessly memorize the list, then there is a problem. But, the problem is with pedagogy and not with the list. Irzik and Nola (2011) claim to have produced a depiction of NOS that is much more informative and comprehensive than a list. However, it is no different than a list. Their outcomes are formatted as a matrix as opposed to a linear format, but it is still a list.

Other researchers, most notably Duschl and Grandy (2012), label these "Consensus-based Heuristic Principles" as out-of-date and too general, in contrast to their "scientific practices in domain-specific contexts." Their description of how these lists are used unfortunately not consistent with the way they are intended to guide classroom practice, and it is difficult to image a thoughtful teacher using the aspects of nature of science in the manner assumed by Duschl and Grandy. On the contrary, the researchers criticized by Duschl and Grandy (e.g., Abd-El-Khalick 2012; Lederman et al. 2002; Niaz 2009) strongly advocate NOS as an overarching instructional theme that permeates not simply a single activity but hopefully an entire school science curriculum.

29.5.3 Knowing Versus Doing

There has been a perennial problem with developing assessments of nature of science that is connected to the research literature. All too often assessments include students' performance or inquiry skills/procedures within instruments on NOS. In spite of over a half century of research on NOS, some science education researchers (Allchin 2011, 2012) continue to conceptualize NOS as a skill as opposed to knowledge and espouse the belief that engagement in the practices of science is sufficient for developing understandings of NOS. The view that NOS is a skill, thus conflating it with scientific inquiry, minimizes the importance of understanding both of these constructs and their related characteristics and further obfuscates their associated nuances and interrelationships. Moreover, this view is not consistent with the National Science Education Standards (NRC 1996) and the Benchmarks for Science Literacy (AAAS 1993), which both describe NOS as knowledge, or the NRC's Framework for K-12 Science Education (NRC 2011). While focusing on scientific inquiry, the Benchmarks stress that students should develop understandings about SI beyond the ability to do SI, as this understanding is sine qua non to being scientifically literate, as is the case for understandings of NOS. Unfortunately the Next Generation Science Standards (NGSS) derived from the new framework are not so clear regarding their "vision" for promoting understandings of NOS. Although aspects such as tentativeness, creativity, and subjectivity in science are included in the framework, no clear distinction is made in the NGSS regarding how NOS explicitly fits into the crosscutting themes.

NOS has been a central theme underling science reforms since the 1950s for a good reason: NOS understandings (irrespective of how these are defined at the time of reform) are central to scientific literacy because NOS is metacognitive knowledge about science. Almost every other meaningful theme underlying past reform documents, such as AAAS Benchmark, NRC Standards, and NSTA Framework, appear in the NGSS, but the same cannot be said for NOS. This exclusion is simply not justified, nor is it justifiable. Unfortunately, regarding assessments of NOS, we may indeed be heading forward…into the past.

The conflation described is inherently linked to the assumption that NOS is learned by having students DO science. That is, if students are involved in authentic scientific investigations, they will also come to an understanding about NOS. The empirical research has consistently shown this assumption to be false for the past three decades (Lederman 2007). Clearly, students' ability to DO science is an important educational outcome, but it is not the same as having students reflect on what they have done. In terms of developing assessments of NOS, there must be a more concerted effort to realize that NOS is a cognitive outcome, not a "performance" outcome.

Related to this last issue is that a small minority of individuals (e.g., Sandoval 2005) insist that students' and teachers' understandings of NOS are best assessed through observations of behavior during inquiry activities (i.e., knowledge in practice). The literature clearly documents the discrepancies that often exist between one's beliefs/knowledge and behavior. More concretely, if an individual believes that scientific knowledge is tentative (subject to change) and another individual believes the knowledge to be absolute/static, how would this be evident in their behavior during a laboratory activity? If a student recognizes that scientific knowledge is partly subjective, how would this student behave differently during a laboratory investigation than a student with differing beliefs? This assessment approach adds an unnecessary layer of inference to one's research design. In the end, we must not forget that NOS is a cognitive outcome, not a behavior as some continue to insist (Allchin 2012). Hence, understandings of NOS are not appropriately assessed through observation of behaviors.

29.6 Uses and Interpretations of NOS Assessments

29.6.1 How Should Assessments Be Used?

Over the history of research on NOS, assessments have been primarily used as summative as opposed to formative assessments. There are few studies that make a systematic attempt to use assessment results to guide the development and enactment of instructional strategies related to NOS. The literature is replete with studies indicating that teachers and students do not possess what are considered adequate conceptions of NOS. It is safe to say that the research community can accurately predict what kinds of understandings teachers and students have about

NOS prior to instruction. Hence, the field needs to move forward and focus more attention on specific strategies to improve conceptions and to assess progressions of understandings, over time, from less to more sophisticated understandings. Unfortunately, there remains a consistent perceived need by researchers to develop an assessment instrument that can be administered to the masses in a short period of time and scored just as easily. Within all of us, it appears, is an "inherent" need to make our lives easier. Interviews and open-ended assessments are time-consuming to conduct and score. However, a quick perusal of recent programs from the Annual Meeting of the National Association for Research in Science Teaching indicates that the desire to create an instrument that can be mass administered and scored in a short period of time or allows for efficient scoring of existing ones continues (e.g., Abd-El-Khalick et al. 2012). Again, what is driving this approach is the perceived need for a more efficient summative assessment. Overall, we must not forget the current needs of researchers and the uses of assessments. It does not appear that there is a warranted need or justification for more traditional paper-and-pencil assessments of NOS.

29.6.2 The Devil Is in the Details: Interpreting the Data We Collect

Much has been said in this chapter about the problems with "traditional" paper-and-pencil assessments of an individual's understanding of NOS. One solution has advocated more open-ended questions followed by interviewing of respondents. Naturally, this approach directly contradicts the desire by some researchers to have easily administered and scored assessments. Although not a new insight, Lederman and O'Malley's (1990) investigation clearly highlighted the problem of paper-and-pencil assessments. They documented discrepancies between their own interpretations of students' written responses and the interpretations that surfaced from actual interviews of the same students. This unexpected finding (i.e., the purpose of the interviews was to help validate the paper-and-pencil survey that was used) was quite timely, as it occurred when educational researchers were making a serious shift toward more qualitative, open-ended approaches to assess individuals' understanding of any concept. Although the VNOS-A was created to avoid some of the concerns about "traditional" assessments (as were the subsequent series of VNOS forms), the problem of researchers interpreting responses differently than intended by the respondent remains to this day. The problem exists at all age levels (K-adult), with increasing levels of uncertainty as the age of the respondent decreases. It is for this reason that researchers should not abandon the interviewing of individuals about their written responses. Consequently, a clear issue when it comes to assessment of such complex constructs as NOS is that we get the most valid data possible. Just using paper-and-pencil assessments increases the possibility of a misinterpretation of respondents' understandings. In summary, the issue of using interviews as part of one's assessment of NOS is relevant to the development, use, and interpretation of assessments.

There has been an ongoing debate about the scoring and representation of data on understandings of NOS. In the early history of assessment development (i.e., 1960–1980), there were strong concerns about the bias inherent in each instrument (Cotham and Smith 1981; Lederman 2007). In particular, the value given to particular responses was directly related to whether the respondent held a view consistent with one philosophical view or another. Given the ever changing conceptualization of the construct NOS (Lederman 1992), many were concerned that scoring responses as if they were correct or incorrect was inappropriate. The solution to this problem was fairly easy, as the "scores" could simply be used to construct a profile of what an individual believes as opposed to a measure of whether they had an informed view of NOS. This approach seems valid, but it ignores the reality that educational systems have goals that specify what we want students to know. Even though conceptions of NOS may change (as is true with any science knowledge), at any point in time, we have an understanding we want our students to develop. In short there are "correct" and "incorrect" answers. Some profiles are more acceptable than others.

Perhaps a more important problem with interpretation is whether it is more accurate to develop numerical scores to represent what an individual knows or whether to develop profiles and then categorize the profiles as informed, naïve, etc. Numerical values allows for easy statistical analysis, but as with any numerical value assigned to a complex construct, much information is lost if we simply have a number. Consequently, it seems wise to have rich descriptions of what individuals know and how this knowledge becomes more or less sophisticated as opposed to simply providing a numerical value.

29.7 What a Long and Tortured Journey This Has Been

Research on students' and teachers' understandings of NOS has been pursued in earnest since 1957 (Mead and Métraux 1957). Naturally, an assortment of assessments for NOS has been developed to support this long line of research. The first assessment was rather informal with the most formalized, even standardized, assessment (i.e., TOUS) appearing in 1961 (Cooley and Klopfer 1961). This review, and others, clearly has shown that assessments of NOS began as more traditional convergent paper-and-pencil tests and then slowly transitioned to the use of more open-ended questionnaires that provide respondents more freedom to express their views. Finally, the past two decades have been characterized by open-ended questionnaires followed by interviews. The interviews, it is believed, help to clarify respondents' written answers as well as avoid some of the problems associated with researchers' misinterpretations of respondents' written answers. Although this chapter is focused on assessment, the discussion of assessment is inextricably connected to the scholarship on NOS, its conceptualization, instructional approaches, etc. As such and as long as this assessment journey has been, we are at a crossroad that threatens to transform assessments into tools that are irrelevant to the question at hand, namely, measuring K-12 students' and teachers' understandings of NOS.

Only a summary of the most critical issues facing NOS assessments discussed in this chapter will be highlighted here. Because of the inextricable link between assessments and the body of research on NOS understandings, these same issues are relevant to the direction of future research as well.

29.7.1 NOS Is a Cognitive Outcome, Not a Behavior

NOS is often confused with, or combined with, scientific inquiry (SI). NOS refers to characteristics of scientific knowledge and SI refers to what scientists do to develop scientific knowledge. Both are clearly important, but assessments of knowledge are different than assessments of performance behaviors, and we know from volumes of research that it is quite difficult to infer knowledge from behavior. Some of the current efforts to assess NOS knowledge based on students' performance of laboratory activities and participation in argumentation about ideas are a step backwards, and they ignore the results of over 50 years of empirical research. NOS is a complex construct that does not lend itself to easily administered and scored assessments.

Although this has been recognized by the trends we have seen over the years in how NOS is assessed, there is a continued desire to develop assessments that are convergent, easily administered to large samples, and easily scored. Moving in this direction is another step backwards. Is the goal expediency or accurately assessing what students and teachers know?

29.7.2 Do Not Lose Sight of the Target Audience, K-12 Students and Teachers

There is a continuing debate about "whose NOS we are measuring." This discussion began with Alters (1997), and it continues with the recent writings of Wong and Hodson (2009, 2010). In the end, the argument always rests on the voice of scientists, and how what they think is important regarding NOS is not being heard or used. Actually, the international reform documents specifying outcomes regarding NOS have had strong input from the scientific community. More importantly, the audience for which these outcomes have been specified is the consumers of science, not scientists. We need to continually remind ourselves for whom the NOS outcomes have been written. What has been specified is not directed at scientists, historians, or philosophers. The knowledge specified is what is considered important for the attainment of scientific literacy by the general citizenry. To dissect the construct of NOS down to its very esoteric levels reveals a construct that is far too abstract for the general public. This really is no different than why we do not expect all high school graduates to understand the most in-depth aspects of the dark reactions of photosynthesis.

29 The Development, Use, and Interpretation of Nature of Science Assessments

There is little doubt that the arguments described in this chapter will continue. At times it appears that our goal in academia is more about the debate than the purpose we are trying to accomplish. It is not all productive to argue about what should be included under the rubric of SI and NOS. It makes little sense to argue about whether lists are good or bad. The focus of our attention should always be on what we consider important for students and teachers and the general public to know, not the label we put on the knowledge. And when we consider the knowledge to be known and assessed, let us not forget the audience, their emotional and cognitive developmental levels, and their needs as citizens.

References

Abd-El-Khalick, F. (2005). Developing deeper understandings of nature of science: The impact of a philosophy of science course on preservice teachers' views and instructional planning. *International Journal of Science Education, 27* (1), 15–42.

Abd-El-Khalick, F. (2012). Examining the sources for our understandings about science: Enduring conflations and critical issues in research on nature of science in science education. *International Journal of Science Education*, 34(3), 353–374.

Abd-El-Khalick, F., Belarmino, J., & Summers, R. (2012). *Development and validation of a rubric to score the views of nature of science (VNOS) questionnaire.* Paper presented at the Annual Conference of the National Association for Research in Science Teaching (NARST), Indianapolis, IN.

Abd-El-Khalick, F., & Lederman, N.G. (2000). The influence of history of science courses on students' views of nature of science. *Journal of Research in Science Teaching*, 37(10), 1057–1095.

Aikenhead, G. (1973). The measurement of high school students' knowledge about science and scientists. *Science Education*, 57(4), 539–549.

Aikenhead, G., & Ryan, A. (1992). The development of a new instrument: "Views on science-technology-society" (VOSTS). *Science Education, 76*, 477–491.

Aikenhead, G., Ryan, A.G., & Fleming, R.W. (1987). High school graduates beliefs about science-technology-society: Methods and issues in monitoring student views. *Science Education*, 71, 145–161.

Akerson, V.L., Abd-El-Khalick, F., & Lederman, N.G. (2000). Influence of a reflective activity-based approach on elementary teachers' conceptions of nature of science. *Journal of Research in Science Teaching, 37* (4), 295–317.

Allchin, D. (2011). Evaluating knowledge of the nature of (whole) science. *Science Education*, 95(3), 518–542.

Allchin, D. (2012). Toward clarity on Whole Science and KNOWS. *Science Education, 96*(4), 693–700.

Allen, H. Jr. (1959). *Attitudes of certain high school seniors toward science and scientific careers.* New York: Teachers College Press.

Alters, B.J. (1997). Whose nature of science? *Journal of Research in Science Teaching, 34*(1), 39–55.

American Association for the Advancement of Science [AAAS]. (1990). *Science for all Americans.* New York: Oxford University Press.

American Association for the Advancement of Science [AAAS]. (1993). *Benchmarks for science literacy.* New York: Oxford University Press.

Bell, R.L., Lederman, N.G. (2003). Understandings of the nature of science and decision making in science and technology based issues. *Science Education*, 87(3), 352–377.

Billeh, V. Y., & Hasan, O. E. (1975). Factors influencing teachers' gain in understanding the nature of science. *Journal of Research in Science Teaching*, 12(3), 209–219.

Biological Sciences Curriculum Study [BSCS]. (1962). *Processes of science test.* New York: The Psychological Corporation.

Bronowski, J. (1956). *Science and Human Values.* New York: Harper & Row.

Center of Unified Science Education. (1974). *The dimensions of scientific literacy.* Columbus, OH: The Ohio State University.

Central Association of Science and Mathematics Teachers (1907). A consideration of the principles that should determine the courses in biology in the secondary schools. *School Science and Mathematics, 7,* 241–247.

Chalmers, A. F. (1982). *What is this thing called science?* (2nd ed.). Queensland, Australia: University of Queensland Press.

Conant, J.B. (1951). On understanding science. New York: New American Library.

Cooley, W. W., & Klopfer, L. E. (1961). *Test on understanding science.* Princeton, NJ: Educational Testing Service.

Cotham, J., & Smith, E. (1981). Development and validation of the conceptions of scientific theories test. *Journal of Research in Science Teaching, 18*(5), 387–396.

Driver, R., Leach, J., Millar, R., & Scott, P. (1996). *Young peoples' images of science.* Buckingham, UK: Open University Press.

Duschl, R. A., & Grandy, R. (2012). Two Views About Explicitly Teaching Nature of Science. *Science & Education,* DOI 10.1007/s11191-012-9539-4

Elby, A., & Hammer, D. (2001). On the substance of a sophisticated epistemology. *Science Education, 85* (5), 554–567.

Feyerabend, D. (1975). *Against method.* London: Verso Publishing.

Fraser, B. J. (1978). Development of a test of science-related attitudes. *Science Education, 62,* 509–515.

Fraser, B. J. (1980). Development and validation of a test of enquiry skills, *Journal of Research in Science Teaching, 17,* 7–16.

Irzik, G., & Nola, R. (2011). A family resemblance approach to the nature of science for science education. *Science & Education, 20(7–8),* 591–607.

Giere, R. N. (1988). *Explaining science: A cognitive approach.* Chicago: The University of Chicago Press.

Gould, S. J., & Eldridge, N. (1977). Punctuated equilibria: The tempo and model of evolution reconsidered. *Paleobiology, 3,* 115–151

Hillis, S. R. (1975). The development of an instrument to determine student views of the tentativeness of science. In *Research and Curriculum Development in Science Education: Science Teacher Behavior and Student Affective and Cognitive Learning* (Vol. 3), Austin, TX: University of Texas Press.

Hrdy, S. B. (1986). Empathy, polyandry, and the myth of the coy female. In R. Bleier (Ed.), *Feminist approaches to science* (pp. 119–146). Perganon Publishers.

Hukins, A. (1963). *A factorial investigation of measures of achievement of objectives in science teaching.* Unpublished doctoral thesis, University of Alberta, Edmonton.

Hungerford, H. & Walding, H. (1974). *The modification of elementary methods students' concepts concerning science and scientists.* Paper presented at the Annual Meeting of the National Science Teachers Association.

Khishfe, R., & Abd-El-Khalick, F. (2002). Influence of explicit and reflective versus implicit inquiry-oriented instruction on sixth graders' views of nature of science. *Journal of Research in Science Teaching, 39*(7), 551–578.

Kimball, M. E. (1967–68). Understanding the nature of science: A comparison of scientists and science teachers. *Journal of Research in Science Teaching, 5,* 110–120.

Klopfer, L. E., & Watson, F. G. (1957). Historical materials and high school science teaching. *The Science Teacher, 24*(6), 264–293.

Korth, W. (1969). *Test every senior project: Understanding the social aspects of science.* Paper presented at the 42nd Annual Meeting of the National Association for Research in Science Teaching.

Kuhn, T. S. (1962). *The structure of scientific revolutions.* Chicago: The University of Chicago Press.

Lakatos, I. (1970). Falsification and the methodology of scientific research programs. In I. Lakatos & A. Musgrave (Eds.), *Criticism and the growth of knowledge.* Cambridge: Cambridge University Press.

Laudan, L. (1977). *Progress and its problems*. Berkeley: University of California Press.

Lederman, N.G. (1992). Students' and teachers' conceptions of the nature of science: A review of the research. *Journal of Research in Science Teaching, 29* (4), 331–359.

Lederman, N.G. (1998). The state of science education: Subject matter without context. *Electronic Journal of Science Education* [On-Line], 3(2), December. Available: http://unr.edu/homepage/jcannon/ejse/ejse.html

Lederman, N. G. (2007). Nature of science: Past, present, and future. In Abell, S.K. and N.G. Lederman (Eds.), *Handbook of research on science education*. Mahwah, N.J.: Lawrence Erlbaum Associates, Inc.

Lederman, N., Abd-el-Khalick, F., Bell, R.L., & Schwartz, R.S. (2002). Views of Nature of Science Questionnaire: Towards valid and meaningful assessment of learners' conceptions of the nature of science. *Journal of Research in Science Teaching, 39*, 497–521.

Lederman, N.G., & Niess, M.L. (1997). The nature of science: Naturally? *School Science and Mathematics, 97*(1), 1–2.

Lederman, N. G., & O'Malley, M. (1990). Students' perceptions of tentativeness in science: Development, use, and sources of change. *Science Education*, 74, 225–239.

Lederman, N. G., Wade, P. D., & Bell, R. L. (1998). Assessing understanding of the nature of science: A historical perspective. In W. McComas (Ed.), *The nature of science and science education: Rationales and strategies* (pp. 331–350). Dordrecht, The Netherlands: Kluwer Academic Publishers.

Liang, L.L., Chen, S., Chen, X., Kaya, O.N., Adams, A.D., Macklin, M., & Ebenezer, J. (2006). *Student understanding of science and scientific inquiry: revision and further validation of an assessment instrument*. Paper presented at the Annual Conference of the National Association for Research in Science Teaching (NARST), San Francisco, CA.

Lovejoy, C.O. (1981). The origin of man. *Science, 211*, 341–350.

Matthews, M.R. (2012). Changing the focus: From nature of science (NOS) to features of science (FOS). In M.S. Khine (Ed.), *Advances in nature of science research: Concepts and methodologies*. Dordrecht, Netherlands: Springer.

Mead, M., & Métraux, R. (1957). Image of the Scientist among High-School Students. *Science, 126*, 384–390.

Meichtry, Y. J. (1992). Influencing student understanding of the nature of science: Data from a case of curriculum development. *Journal of Research in Science Teaching*, 29, 389–407.

Moore, R., & Sutman, F. (1970). The development, field test and validation of an inventory of scientific attitudes. *Journal of Research in Science Teaching*, 7, 85–94.

National Research Council [NRC]. (1996). *National science education standards*. Washington, DC: National Academy Press.

National Research Council [NRC]. (2011). *A framework for K-12 science education: Practices, crosscutting concepts, and core ideas*. Washington, D.C.: National Academy Press

National Science Teachers Association [NSTA]. (1982). Science-technology-society: Science education for the 1980s. (An NSTA position statement). Washington, DC.

Niaz, M. (2009). Critical appraisal of physical science as a human enterprise: Dynamics of scientific progress. Milton Keynes: Springer.

Nott, M., & Wellington, J. (1995). Probing teachers' views of the nature of science: How should we do it and where should we be looking? *Proceedings of the Third International History, Philosophy, and Science Teaching Conference*, 864–872.

Ogunniyi, M. B. (1982). An analysis of prospective science teachers' understanding of the nature of science. *Journal of Research in Science Teaching*, 19(1), 25–32.

Osborne, J., Collins, S., Ratcliffe, M., Millar, R., & Duschl, R. (2003). What "ideas-about-science" should be taught in school science? A Delphi study of the expert community. *Journal of Research in Science Teaching, 40*(7), 692–720.

Popper, K. R. (1959). *The logic of scientific discovery*. New York: Harper & Row.

Popper, K. R. (1963). *Conjectures and refutations: The growth of scientific knowledge*. London: Routledge.

Popper, K. R. (1988). *The open universe: An argument for indeterminism*. London: Routledge.

Roberts, D.A. (2007). Scientific literacy/science literacy. In S.K. Abell & N.G. Lederman (Eds.), *Handbook of research on science education* (pp. 729–780). Mahwah, NJ: Lawrence Erlbaum.

Rubba, P. (1976). Nature of scientific knowledge scale. School of Education, Indiana University, Bloomington, Indiana.

Rudolph, J.L. (2003). Portraying epistemology: School science in historical context. *Science Education, 87* (1), 64–79.

Sandoval, W. A. (2005). Understanding students' practical epistemologies and their influence on learning through inquiry. *Science Education*, 89(5), 634–656.

Schwartz, R.S., & Lederman, N.G. (2002). "It's the nature of the beast": The influence of knowledge and intentions on learning and teaching nature of science. *Journal of Research in Science Teaching, 39* (3), 205–236.

Schwirian, P. M. (1968). On measuring attitudes toward science. Science Education, 52, 172–179.

Scientific Literacy Research Center. (1967). *Wisconsin inventory of science processes*. Madison, WI: The University of Wisconsin.

Shamos, M.H. (1984). Exposure to science vs. scientific literacy. *Journal of College Science Teaching, 13(5)*, 333–393.

Showalter, V. (1974). What is unified science education? Program objectives and scientific literacy, *Prism II, 2(3–4),* 1–6.

Stice, G. (1958). *Facts about science test*. Princeton, NJ: Educational Testing Service.

Smith, M.U., Lederman, N.G., Bell, R.L., McComas, W.F., & Clough, M.P. (1997). How great is the disagreement about the nature of science: A response to Alters. *Journal of Research in Science Teaching, 34*(10), 1101–1103.

Smith, M. U., & Scharmann, L. C. (1999). Defining versus describing the nature of science: A pragmatic analysis for classroom teachers and science educators. *Science Education, 83*, 493–509.

Swan, M.D. (1966). Science achievement as it relates to science curricula and programs at the sixth grade level in Montana public schools. *Journal of Research in Science Teaching, 4*, 102–123.

Welch, W. W. (1967). *Science process inventory*. Cambridge, MA: Harvard University Press.

Welch, W.W. (1969). Curriculum evaluation. *Review of Educational Research, 39(4)*, 429–443.

Welch, W. W., & Pella, M. O. (1967–68). The development of an instrument for inventorying knowledge of the processes of science. *Journal of Research in Science Teaching*, 5(1), 64.

Wheeler, S. (1968). *Critique and revision of an evaluation instrument to measure students' understanding of science and scientists*. Chicago, IL: University of Chicago.

Wilson, L. (1954). A study of opinions related to the nature of science and its purpose in society. *Science Education*, 38(2), 159–164.

Wong, S. L., & Hodson, D. (2009). From the horse's mouth: What scientists say about scientific investigation and scientific knowledge. *Science Education, 93*, 109–130.

Wong, S.L. & Hodson, D. (2010). More from the horse's mouth: What scientists say about science as a social practice. *International Journal of Science Education, 32(11)*, 1431–1463.

Norman G. Lederman is chair and professor of mathematics and science education at the Illinois Institute of Technology. Dr. Lederman received his Ph.D. in science education and he possesses M.S. degrees in both biology and secondary education. Prior to his 20+ years in science teacher education, Dr. Lederman was a high school teacher of biology and chemistry for 10 years. Dr. Lederman is internationally known for his research and scholarship on the development of students' and teachers' conceptions of nature of science and scientific inquiry. He has been author or editor of 10 books, written 15 book chapters, published over 200 articles in professional journals, and made over 500 presentations at professional conferences around the world. He is the coeditor of the recently published *Handbook of Research on Science Education* (Routledge 2014).

Dr. Lederman is a former president of the National Association for Research in Science Teaching (NARST) and the Association for the Education of Teachers in Science (AETS). He has also served as director of teacher education for the National

Science Teachers Association (NSTA). He has received the Illinois Outstanding Biology from the National Association of Biology Teachers and the Outstanding Mentor Award from AETS. Most recently, he has been named a fellow of the American Association for the Advancement of Science and the American Educational Research Association and has received the Distinguished Career Award from the National Association for Research in Science Teaching.

Stephen A. Bartos is an associate professor in science education at Middle Tennessee State University, Murfreesboro, Tennessee. He is a certified teacher in mathematics and physics, with over a decade of classroom experience, and has worked extensively with preservice mathematics and science teachers. His research interests include improving students' and teachers' conceptions of nature of science and scientific inquiry and how teachers translate and communicate their subject matter during instruction (pedagogical content knowledge). Among his recent presentations and publications is a paper about using socio-scientific issues arising in genetics (Lederman N.G., Antink, A., & Bartos, S., 2012. Nature of Science, Scientific Inquiry, and Socio-Scientific Issues Arising From Genetics: A Pathway to Developing a Scientifically Literate Citizenry. *Science & Education*, online first, DOI 10.1007/s11191-012-9503-3) and how teachers include NOS in their classroom practice (Lederman, J.S., Bartos, S.A, Lederman, N.G., Meyer, D.Z., Antink-Meyer, A., & Holliday G., 2012. Interaction of knowledge and pedagogical decisions in teaching NOS. Paper presented at the NARST Annual International Conference, Indianapolis, IN.)

Judith S. Lederman is an associate professor in science education and director of teacher education in the Department of Mathematics and Science Education at Illinois Institute of Technology, Chicago, Illinois. Dr. Lederman received her Ph.D. in science education from Curtin University in Perth, Australia, and an M.S. degree in natural sciences from Worchester Polytechnic Institute. Her experience with informal education includes her work as curator of education at the Museum of Natural History and Planetarium in Providence, RI. Prior to her work in science teacher education, Dr. Lederman taught secondary and middle level physics and biology, as well as bilingual elementary science.

Dr. Lederman is known nationally and internationally for her work on the teaching and learning of scientific inquiry and nature of science in both formal and informal settings. She has over 600 presentations/publications on scientific literacy. She has authored an elementary science education methods textbook as well as an elementary (K-5) science textbook series for National Geographic Press. In 2008, Dr. Lederman was awarded a Fulbright Fellowship to work with South African university faculty, museum educators, and science teachers to develop research and curriculum that connects informal education to K-12 science classrooms. She has served on the Board of Directors of the National Science Teachers Association (NSTA) and is past president of the Council for Elementary Science International (CESI). She is currently chair of the NSTA International Advisory Board and the NSTA Task Force on International Outreach and the NARST Outstanding Dissertation Award. Her honors include Rhode Island State Teacher of the Year and the Milken Foundation National Educators Award.

Chapter 30
New Directions for Nature of Science Research

Gürol Irzik and Robert Nola

30.1 Introduction

Calls for the inclusion of the nature of science (NOS for short) into science education have a long history. A number of distinguished scientists, philosophers and education theorists such as John Dewey, James Conant, Gerald Holton, Leo Klopfer, Joseph Schwab, James Robinson, James Rutherford, Michael Martin, Richard Duschl, Derek Hodson, Norman Lederman, Michael Matthews and Norman McComas throughout the twentieth century emphasised the importance of teaching science's conceptual structure and its epistemological aspects as part of science education (Matthews 1998a; McComas et al. 1998). Today, science education curriculum reform documents in many parts of the world underline that an important objective of science education is the learning of not only the content of science but its nature.[1] The rationale is that scientific literacy requires an understanding of the nature of science, which in turn facilitates students' learning of the content of science, helps them grasp what sort of a human enterprise science is, helps them appreciate its value in today's world and enhances their democratic citizenship, that is, their ability to make informed decisions, as future citizens, about a number of controversial issues such as global warming, how to dispose nuclear waste, genetically modified food and the teaching of

---

[1] See, for example, American Association for the Advancement of Science (1990, 1993), Council of Ministers of Education (1997), National Curriculum Council (1988), National Research Council (1996), Rocard et al. (2007), and McComas and Olson (1998).

G. Irzik (✉)
Sabancı University, Istanbul, Turkey
e-mail: irzik@sabanciuniv.edu

R. Nola
The University of Auckland, Auckland, New Zealand
e-mail: r.nola@auckland.ac.nz

M.R. Matthews (ed.), *International Handbook of Research in History, Philosophy and Science Teaching*, DOI 10.1007/978-94-007-7654-8_30,
© Springer Science+Business Media Dordrecht 2014

intelligent design in schools.[2] Allchin expressed this idea succinctly: 'Students should develop an understanding of how science works *with the goal of interpreting the reliability of scientific claims in personal and public decision making*' (Allchin 2011, p. 521; emphasis original).

There is a voluminous literature on what NOS is, how to teach it and what views of NOS students and teachers hold. The aim of this chapter is not to review this literature. The interested reader can refer to other chapters of this handbook and earlier useful surveys (Abd-El-Khalick and Lederman 2000; Deng 2011 and others; Lederman 2007). Teachers' and students' views of NOS are also beyond the scope of this chapter, in which we focus exclusively on what NOS is. In the next section we summarise the consensus NOS theorising in science education has produced. Making use of the existing consensus, we then provide, in Sect. 30.3, a structural description of all the major aspects of science in terms of eight categories. Applying the idea of family resemblance to these categories, we obtain what we call 'the family resemblance approach'. We articulate it in some detail in Sect. 30.5. We believe that the family resemblance approach provides a systematic and unifying account of NOS. We discuss this and other virtues of the family resemblance approach in Sect. 30.6. We end the chapter by making some suggestions about how to use this approach in the classroom.

We would like to emphasise that the present chapter does not deal with empirical matters such as what teachers and pupils might understand of NOS. Rather, our task is one within the theory of NOS: it is to provide a new way of thinking about what is meant by the 'nature of science'. Nevertheless, we do hope that theorists of science education and science teachers familiar with NOS discussions will find our approach not only theoretically illuminating but also pedagogically useful.

30.2 Consensus on NOS

NOS research in the last decade or so has revealed a significant degree of consensus amongst the members of the science education community regarding what NOS is and which aspects of it should be taught in schools at the precollege level. This consensus can be highlighted as follows.

Based on considerations of accessibility to students and usefulness for citizens, Lederman and his collaborators specified the following characteristics of NOS:

- Scientific knowledge is empirical (relies on observations and experiments).
- Is reliable but fallible/tentative (i.e. subject to change and thus never absolute or certain).
- Is partly the product of human imagination and creativity.

---

[2] This point is commonly made, for example, in Driver et al. (1996), McComas et al. (1998), Osborne 2007, and Rutherford and Ahlgren (1990).

- Is theory-laden and subjective (i.e. influenced by scientists' background beliefs, experiences and biases).
- Is socially and culturally embedded (i.e. influenced by social and cultural context).[3]

They also emphasised that students should be familiar with concepts fundamental to an understanding of NOS such as observation, inference, experiment, law and theory and be also aware of the distinctions between observing and inferring and between laws and theories and of the fact that there is no single scientific method that invariably produces infallible knowledge. Others added that science is theoretical and explanatory; scientific claims are testable and scientific tests are repeatable; science is self-correcting and aims at achieving values such as high explanatory and predictive power, fecundity (fruitfulness), parsimony (simplicity) and logical coherence (consistency) (Cobern and Loving 2001; Smith and Scharmann 1999; Zeidler and others 2002).

A number of researchers propose a similar list of characteristics by studying the international science education standards documents. These documents also indicate substantial consensus on two further matters: the ethical dimension of science (e.g. scientists make ethical decisions, must be open to new ideas, report their findings truthfully, clearly and openly) and the way in which science and technology interact with and influence one another (McComas et al. 1998; McComas and Olson 1998). Based on a Delphi study of an expert group consisting of scientists, science educators and science communicators, philosophers, historians and sociologists of science, Osborne and others (2003) found broad agreement on the following eight themes:

- Scientific method (including the idea that continual questioning and experimental testing of scientific claims is central to scientific research)
- Analysis and interpretation of data (the idea that data does not speak by itself, but can be interpreted in various ways)
- (Un)certainty of science (i.e. scientific knowledge is provisional)
- Hypothesis and prediction (the idea that formulating hypotheses and drawing predictions from them in order to test them is essential to science)
- Creativity in science (the idea that since scientific research requires much creativity, students should be encouraged to create models to explain phenomena)
- Diversity of scientific thinking (the idea that science employs different methods to solve the same problem)
- The historical development of scientific knowledge (i.e. scientific knowledge develops historically and is affected by societal demands and expectations)
- The role of cooperation and collaboration in the production of scientific knowledge (i.e. science is a collaborative and cooperative activity, as exemplified by teamwork and the mechanism of peer review).

---

[3] See Abd-El-Khalick (2004), Abd-El-Khalick and Lederman (2000), Bell (2004), Khishfe and Lederman (2006), Lederman (2004, 2007). Note that all of these characteristics pertain to scientific knowledge. For that reason, Lederman suggested replacing the phrase 'nature of science' with 'nature of scientific knowledge' in his recent writings (Lederman 2007).

Wong and Hodson (2009, 2010) came up with very similar themes (but with slightly different emphasis) on the basis of in-depth interviews with well-established scientists from different parts of the world who worked in different fields:

- Scientific method (different disciplines employ different methods of investigation)
- Creativity in science (creative imagination plays an important role in every stage of scientific inquiry from data collection to theory construction, and absolute objectivity in the sense of freeing oneself from biases completely is impossible)
- The importance of theory in scientific inquiry (scientific activity is highly theoretical)
- Theory dependence of observation (scientific data is theory laden and can be interpreted in various ways)
- Tentative nature of scientific knowledge (science does not yield certainty)
- The impact of cultural, social, political, economic, ethical and personal factors on science (such factors greatly influence the direction of scientific research and development and may cause biased results and misconduct) and the importance of cooperation, peer review and shared norms (such as intellectual honesty and open mindedness) in knowledge production

The overlap between the findings of these studies indicates a substantial consensus regarding NOS amongst education theorists. However, there has been some debate as to whether processes of inquiry (such as posing questions, collecting data, formulating hypotheses, designing experiments to test them) should be included in NOS. While Lederman (2007) suggested leaving them out, other science education theorists disagreed arguing that they constitute an inseparable part of NOS (Duschl and Osborne 2002; Grandy and Duschl 2007). Indeed, research summarised in the above two paragraphs do cite processes of inquiry as an important component of NOS.

Of course, much depends on how the various aspects and themes of NOS are spelled out. Osborne and his collaborators warn that various characteristics of NOS should not be taken as discrete entities, so they emphasise their interrelatedness (Osborne and others 2001, 2003, p. 711). In a similar vein, others note that blanket generalisations about NOS introduced out of context do not provide a sophisticated understanding of NOS (Elby and Hammer 2001; Matthews 2011); rather, the items within NOS ought to be elucidated in relation to one another in 'authentic contexts'. Accordingly, many science educators have called for 'an authentic view' of science, which aims to contextualise science and focuses on science-in-the-making by drawing either on science-technology-society (STS) studies or on the interviews with scientists themselves about their day-to-day activities; this underlines the heterogeneity of scientific practices across scientific disciplines through historical and contemporary case studies.[4]

---

[4] See Ford and Wargo (2007), McGinn and Roth (1999), Rudolph (2000), Samarapungavan et al. (2006), Wong and Hodson (2009, 2010), and Wong et al. (2009).

A number of science education theorists also urged that issues arising from science-technology-society interactions, the social norms of science and funding and fraud within science all be allotted more space in discussions of NOS; a focus on these is especially pertinent when educating citizens who will often face making hard decisions regarding socio-scientific problems in today's democracies. These topics have been raised earlier in some detail (Aikenhead 1985a, b; Kolsto 2001; Zeidler and others 2002) and are receiving increasing attention in recent years, in line with calls for an authentic view of science.[5]

30.3 NOS Categories: A Structural Description

The consensus on NOS highlighted above reveals that science is a multifaceted enterprise that involves (a) processes of inquiry, (b) scientific knowledge with special characteristics, (c) methods, aims and values and (d) social, historical and ethical aspects. Indeed, science is many things all at once: it is an investigative activity, a vocation, a culture and an enterprise with an economic dimension and accordingly has many features (cognitive, social, cultural, political, ethical and commercial) (Weinstein 2008; Matthews 2011). What is needed then is a systematic and unifying perspective that captures not just this or that aspect of science but the 'whole science' (Allchin 2011). This is no easy task, and there is certainly more than one way of carrying it out. Our suggestion is to begin with a broad distinction between *science as a cognitive-epistemic system of thought and practice* on the one hand and *science as a social-institutional system* on the other. This distinction is actually implicit in the aspects of NOS expressed (a) through (d) above: science as a cognitive-epistemic system incorporates (a), (b) and (c), while science as a social-institutional system captures (d). We hasten to add that we intend this as an analytical distinction to achieve conceptual clarity, not as a categorical separation that divides one from the other. In practice, the two constantly interact with each other in myriad ways, as we will see.

30.3.1 Science as a Cognitive-Epistemic System

We spell out science as a cognitive-epistemic system in terms of four categories obtained by slightly modifying (a)–(c): processes of inquiry, aims and values, methods and methodological rules and scientific knowledge. We explain these categories briefly below.[6]

---

[5] See Sadler (2011), Weinstein (2008), Wong and Hodson (2010), Zemplen (2009); see also the special issue of the journal *Science & Education* vol. 17, nos. 8–9, 2008.

[6] For a more detailed discussion of these, see Nola and Irzik (2005, Chaps. 2, 4, 6, 7, 8, 9, and 10).

30.3.1.1 Processes of Inquiry

This includes posing questions (problems), making observations, collecting and classifying data, designing experiments, formulating hypotheses, constructing theories and models and comparing alternative theories and models (Grandy and Duschl 2007).

30.3.1.2 Aims and Values

This will include items such as *prediction, explanation, consistency, simplicity* and *fruitfulness*; these are amongst the well-known aims of science recognised in the science education literature, as we saw in the previous section. With regard to prediction and explanation, we would like to make two points, which the science education literature tends to neglect. First, scientists value *novel* predictions more than other kinds of predictions because novel predictions of a theory give greater support to it than those that are not (Nola and Irzık 2005, pp. 245–247). (A prediction is novel if it is a prediction of a phenomenon that was unknown to the scientists at the time of the prediction.) Second, although there are different kinds of explanations and therefore different models of explanations, all scientific explanations are naturalistic in the sense that natural phenomena are explained in terms of other natural phenomena, without appealing to any supernatural or occult powers and entities (Lindberg 1992, Chap. 1; Pennock 2011).[7]

Other aims of science include the following: *viability* (von Glasersfeld 1989), *high confirmation* (Hempel 1965, Part I), *testability* and *truth* or at least *closeness to truth* (Popper 1963, 1975) and *empirical adequacy* (van Fraassen 1980). Aims of science are sometimes called (cognitive-epistemic) values since scientists value them highly in the sense that they desire their theories and models to realise them (Kuhn 1977). Values in science can also function as shared criteria for comparing theories and be expressed as methodological rules. For example, we can say that given two rival theories, other things being equal, the theory that has more explanatory power is better than the one that has less explanatory power. Expressed as a methodological rule, it becomes, given two rival theories, other things being equal, *choose*, or *prefer*, the theory that is more explanatory. Similar rules can be derived from other values. These enable scientists to compare rival theories about the same domain of phenomena rationally and objectively (Kuhn 1977).

30.3.1.3 Methods and Methodological Rules

Science does not achieve its various aims randomly, but employs a number of methods and methodological rules. This point emerges clearly in many studies on NOS. Historically, there have been proposals about scientific method from Aristotle,

---

[7] See Godfrey-Smith (2003) for a succinct summary of different models of explanations in science.

Bacon, Galileo, Newton to Whewell, Mill and Peirce, not to mention the many theories of method proposed in the twentieth century by philosophers, scientists and statisticians. For many of them, deductive, inductive and abductive reasoning form an important part of any kind of scientific method. Additional methods for testing hypotheses include a variety of inductive and statistical methods along with the hypothetico-deductive method (Nola and Sankey 2007; Nola and Irzık 2005, Chaps. 7, 8, and 9). The idea of scientific methodology also includes methodological rules; these have not received sufficient attention in the science education literature. Methodological rules are discussed at length by a number of philosophers of science such as Popper (1959) and Laudan (1996, Chap. 7). Here are some of them:

- Construct hypotheses/theories/models that are highly testable.
- Avoid making ad hoc revisions to theories.
- Other things being equal, choose the theory that is more explanatory.
- Reject inconsistent theories.
- Other things being equal, accept simple theories and reject more complex ones.
- Accept a theory only if it can explain all the successes of its predecessors.
- Use controlled experiments in testing casual hypotheses.
- In conducting experiments on human subjects, always use blinded procedures.

Two general points about scientific methods and methodological rules are in order. First, although they certainly capture something deep about the nature of methods employed in science, it should not be forgotten that they are highly idealised, rational constructions. As such, they do not faithfully mirror what scientists do in their day-to-day activities; nor can they always dictate to them what to do at every step of their inquiry. Nevertheless, they can often tell them when their moves are, or are not, rational and do explain (at least partially) the reliability of scientific knowledge. Second, we presented the above rules of method as if they are categorical imperatives. This needs to be qualified in two ways. The first is that some of the rules can, in certain circumstances, be abandoned. Spelling out the conditions in some antecedent clause in which the rules can be given up is not an easy matter to do; so such rules are best understand to be defeasible in unspecified circumstances. The second is that such categorical rules ought to be expressed as hypothetical imperatives which say: rule R ought to be followed if some aim or value V will be (reliably) achieved (see Laudan 1996, Chap. 7). Often reference to the value is omitted or the rule is expressed elliptically. For example, the rule about ad hocness has an implicit value or aim of high testability. So, more explicitly it would look like: 'If you aim for high testability, avoid making *ad hoc* revisions to theories'. When rules are understood in this way, then the link between the methodological rules of category 3 and the aims of category 2 becomes clearly visible.

30.3.1.4 Scientific Knowledge

When processes of inquiry achieve their aims using the aforementioned methods and methodological rules, these processes culminate in some 'product', viz.

scientific knowledge. Such knowledge 'end products' are embodied in laws, theories and models as well as collections of observational reports and experimental data. Scientific knowledge is the most widely discussed category of NOS, as we have seen in the previous section.

30.3.2 Science as a Social-Institutional System

Science as a social-institutional system is investigated less than science as a cognitive-epistemic system, and for that reason it is harder to categorise. We propose to study it in terms of the following categories: professional activities, the system of knowledge certification and dissemination, scientific ethos and finally social values. We discuss them in some detail below, taking into account the findings of the NOS research on this topic indicated in Sect. 30.2.

As decades of science-technology-society studies have shown, science not only is a cognitive system but is, at the same time, both a cooperative and a competitive community practice that has its own ethos (i.e. social and ethical norms) and its own system of knowledge certification and dissemination. It is a constantly evolving social enterprise with intricate relationships with technology and with the rest of the society, which both influences and is influenced by it. Scientists form a tight community and are engaged in a number of professional activities, interacting both with each other and the larger public. In short, science is a historical, dynamic, social institution embedded within the larger society. Categories of science as social-institutional system can be described as follows.

30.3.2.1 Professional Activities

Scientists do not just carry out scientific research. Qua being scientists, they also perform a variety of professional activities such as attending academic meetings, presenting their findings there, publishing them, reviewing manuscripts and grant proposals, writing research projects and seeking funds for them, doing consulting work for both public and private bodies and informing the public about matters of general interest. In this way, they perform various cognitive-epistemic and social functions such as certifying knowledge and serving certain social goals. Whether they are engaged in cognitive-epistemic or professional activities, they are expected to conform to a number of social and ethical norms. We discuss these below.

30.3.2.2 The Scientific Ethos

Part of the meaning of the claim that science is a social institution is that it has its own social (institutional) and ethical norms, which refer to certain attitudes scientists are expected to adopt and display in their interactions with their fellow scientists as well as in carrying out their scientific activities. We call them 'the scientific ethos' (or, equivalently, 'the ethos of science') for convenience, a phrase coined by the famous sociologist of science Robert Merton. However, as we will see below, the scientific ethos as we understand it is not confined to what is known as the 'Mertonian norms' in the literature. Merton was one of the first to study the institutional norms of science in the 1930s and formulated some of them as follows, based on his extensive interviews with scientists (Merton 1973, Chap. 13):

- *Universalism*: Science is universal in the sense that scientific claims are evaluated according to pre-established objective, rational criteria so that characteristics of scientists such as ethnic origin, nationality, religion, class and gender are irrelevant when it comes to evaluation.
- *Organised scepticism*: Scientists subject every claim to logical and empirical scrutiny on the basis of clearly specified procedures that involve scientific reasoning, testability and methodology and suspend judgement until all the relevant facts are in and bow to no authority except that of critical argumentation.
- *Disinterestedness*: Scientists should evaluate and report their findings independently of whether they serve their personal interests, ideologies and the like. The norm of disinterestedness has the function of preventing scientists from hiding or fudging the results of their inquiries even when they go against their personal biases, interests and favoured ideology.
- *Communalism* refers to the common ownership of scientific discovery or knowledge. The rationale is that science is a cooperative endeavour: new scientific knowledge always builds upon old knowledge and that scientific discoveries owe much to open and free discussion and exchange of ideas, information, techniques and even material (such as proteins).

Although Merton arrived at these norms through an empirical study, we should not lose sight of the fact that they can be taken as both descriptive and prescriptive *qua* being norms. In other words, they tell us how scientists ought to behave, not just how they do behave when they do science. Their normative nature and power is evident from the fact that scientists often face the sanctions of the scientific community when they violate them.[8]

In time, the scientific community has become increasingly self-conscious of the norms of conduct in science, as a result of which they have proliferated and been codified under the banner 'ethical codes of conduct'. There is now a whole subfield called the 'ethics of science' devoted to this topic. Amongst other things, these norms include the following (Resnik 2007, Chap. 2):

- Intellectual honesty (or integrity): Scientists should not fabricate, distort or suppress data and should not plagiarise. They should bow to no authority except that of evidence and critical argumentation.

---

[8] STS scholars are generally critical of Mertonian norms and claim that there is a counter-norm for every Mertonian norm, with the implication that Mertonian norms do not guide scientific practice and therefore are simply functionless. See, for example, Sismondo (2004, Chap. 3) and the literature cited therein. However, there are also excellent critiques of these critiques such as Radder (2010).

- Respect for research subjects: Scientists should treat human and animal subjects with respect and dignity. This involves getting the informed consent of human subjects and not inflicting unnecessary pain on animal subjects and the like.
- Respect the environment: Avoid causing harm to the environment.
- Freedom: Scientists should be free to pursue any research, subject to certain constraints (e.g. as implied by the previous two ethical principles).
- Openness: Scientists should be open to free and critical discussion and to share ideas, data, techniques and even materials (such as proteins). They should be willing to change their opinion when presented with good reasons.

Today many scientific institutions (universities, academies, funding organisations, etc.) have such ethical codes which they announce on their websites.

None of this is meant to suggest that there is no misconduct, fraud, data suppression or misrepresentation and the like, or fierce competition, especially for scarce resources such as funding, which sometimes results in secrecy (the opposite of openness) in science. Scientists are not saints. Nevertheless, when they violate the norms of science, they often face sanctions. Science has developed a social mechanism of certification and dissemination to eliminate or at least reduce misconduct and promote collaboration amongst scientists.

30.3.2.3 The Social Certification and Dissemination of Scientific Knowledge

When a scientist or a team of scientists completes their research, they are hardly finished with their work. Their findings need to be published; this requires a process of peer review. When published, they become public and are now open to the critical scrutiny of the entire community of relevant experts. Only when they prove their mettle during this entire ordeal are their findings accepted into the corpus of scientific knowledge and can, amongst other things, be taught at schools. This is in a nutshell the *social* system of certification and dissemination of scientific knowledge, which involves the collective and collaborative efforts of the scientific community (Kitcher 2011, Chap. 4). This system functions as an effective *social quality control* over and above the *epistemic control* mechanisms that include testing, evidential relations and methodological considerations described in Sect. 30.3.1. They jointly work to help reduce the possibility of error and misconduct.

30.3.2.4 Social Values of Science

Science embodies not only cognitive-epistemic values but also social ones. Some of the most important social values are freedom, respect for the environment and social utility broadly understood to refer to improving people's health and quality of life as well as to contributing to economic development. Without sufficient freedom of research, scientific development would be stifled. Respect for the environment involves both the negative duty of not damaging it and the positive duty of protecting it by saving biodiversity and reducing carbon emissions that cause climate change. As a species we are unlikely to survive if we do not respect the environment. Science that does not contribute to better lives for people would not enjoy their support; the social legitimation of science today depends crucially on its social utility. Social utility then serves as an important social goal of science.

This completes our description of the eight categories of science which can be tabulated as below.

Science							
Science as a cognitive-epistemic system				Science as a social system			
1	2	3	4	5	6	7	8
Processes of inquiry	Aims and values	Methods and methodological rules	Scientific knowledge	Professional activities	Scientific ethos	Social certification and dissemination of scientific knowledge	Social values

Although we believe that the categories that make up science as a cognitive-epistemic system are pretty exhaustive, we admit the possibility that other categories might perhaps be added or new categories might emerge as science develops. We do not think, however, that categories of science as a social system is exhaustive in any way. Nor do we claim that this is the only or the best way of describing science as a social system. Others may carve it out differently. Nevertheless, we do believe that it captures an important part of science as social practice. Similarly, we do not pretend to have listed all the items that fall under each of the eight categories above. In fact, we consider them open-ended; that is, the characteristics of science that fall under each category are not fixed and develop historically. Overall, we believe that the eight categories capture the structural features of NOS in a systematic and comprehensive way.

30.4 Clarifying the Meaning of 'Nature of Science' and the Idea of Family Resemblance

Although we suggested that the above eight categories characterise nature *of science*, we have not explored the meaning of term 'nature' that occurs in that phrase. What do we mean by '*nature* of science'? To our knowledge, this is a question that is hardly raised in the science education literature. Here we briefly mention three conceptions of what such a nature might be.

First, the *nature* of science could be taken to be the specification of a natural kind of thing which has an essence, where an essence is a set of properties which a thing *must* have and without which it is *not possible* for that thing exist and to be that *kind* of thing. Triangles have an essence in this sense, but it is very doubtful that science has an essence of this sort. We can agree with Rorty's negative answer to the title of his paper 'Is natural science a natural kind?' (Rorty 1991, pp. 46–62).

A second suggestion about 'nature' is to claim that it is a (small) set of necessary and sufficient properties that something should possess if it is to be deemed science. Here strong modal claims found in the essentialist approach mentioned above are downplayed or eschewed in favour of the mere possession of the set of features shared by all sciences and only by them. However, so far all attempts to define science in terms of necessary and sufficient conditions have failed. Some have restricted their approach to the nature of science by focusing narrowly on just the fourth category of science, viz. scientific knowledge, and then have attempted to define what is to count as science as what is verifiable (some positivists) or what is falsifiable (Popper) and so on.[9] This is not the approach we advocate here in characterising science.

A third approach might be simply to list a number of items falling under the concept of science without pretending to give a set of necessary and sufficient properties or to specify essence for science. Thus one common approach to the *nature* of science in science education lists some salient features of science as in Sect. 30.2. This is also the approach we have adopted by setting out the eight categories of science and listing the items that fall under each. However, there is a problem to be tackled: not all sciences share these features or items all at once. Indeed, a number of science education theorists have drawn attention to important differences amongst scientific disciplines (Samarapungavan et al. 2006; Wong and Hodson 2009). If some sciences lack some of the features others share, what justifies the label 'science' for them? Merely providing a list of preferred items is powerless to answer this question.

Luckily, there is a satisfactory answer within philosophy that invites one to have a quite different approach to what counts as a 'nature' in talk of 'NOS'. In fact it takes us well away from the three ways of understanding 'nature' listed above in using the important idea of family resemblance (Eflin and others 1999; Hacking 1996; Dupre 1993). In a nutshell, the nature of science consists of a set of family resemblances amongst the items that fall under the eight categories of science. In an earlier article, we articulated this approach in some detail for the purposes of science education (Irzik and Nola 2011). In this chapter, we develop it further.

The idea of family resemblance was developed by the philosopher Ludwig Wittgenstein in recognition of the fact that not all terms can be defined in terms of necessary and sufficient conditions or by specifying essences or natures (Wittgenstein 1958, Sects. 66–71). To see this, compare 'triangle' with 'game'. The former can be defined explicitly as a closed plane figure with three straight sides. This definition not only gives six characteristics that specify the necessary and sufficient conditions for being a triangle but also determines the 'essence' of being a triangle or the analytic meaning of the term 'triangle'. In this definition, those properties that are shared by all triangles and only by triangles are specified explicitly. By contrast,

---

[9] See some of the following who may be, in addition, critical of the idea of the demarcation of science from non-science but whose focus in so doing is just upon the fourth category, viz. what is to count as a scientific statement: (Alters 1997; Hacking 1996; Laudan et al. 1986; Stanley and Brickhouse 2001; Ziman 2000).

Wittgenstein argued, the term 'game' cannot be defined in this way. Any attempt to define the term 'game' must include games as different as ball games, stick games, card games, children's games that do not involve balls, sticks or cards (such as tag or hide-and-seek), solo games (hopscotch) and mind games. Unlike the term 'triangle', there is no fixed set of necessary and sufficient conditions which determine the meaning of 'game' and thus no set of properties that cover all games and at the same time admit nothing which is not a game.[10] Nevertheless, Wittgenstein argued, all games form 'a family resemblance', forming a complicated network of similarities, overlapping and criss-crossing. It is these similarities that justify the use of the term 'game' to all those diverse activities from baseball to hopscotch.

Consider a set of four characteristics {A, B, C, D}. Then one could imagine four individual items which share any three of these characteristics taken together such as (A&B&C) or (B&C&D) or (A&B&D) or (A&C&D); that is, the various family resemblances are represented as four disjuncts of conjunctions of any three properties chosen from the original set of characteristics. This example of a polythetic model of family resemblances can be generalised as follows. Take any set S of n characteristics; then any individual is a member of the family if and only if it has all of the n characteristics of S, or any (n-1) conjunction of characteristics of S, or any (n-2) conjunction of characteristics of S, or any (n-3) conjunction of characteristics of S and so on. How large n may be and how small (n-x) may be is something that can be left open as befits the idea of a family resemblance which does not wish to impose arbitrary limits and leaves this to a 'case by case' investigation. In what follows we will employ this polythetic version of family resemblance (in a slightly modified form) in developing our conception of science.

Consider the following limiting case. Suppose an example like that above but in which there is a fifth characteristic E which is common to all the disjunctions of conjunctions as in the following: (A&B&C&E) or (B&C&D&E) or (A&B&D&E) or (A&C&D&E). Would this be a violation of the kind of family resemblance definition that Wittgenstein intended? Not necessarily. We might say as an example of characteristic E in the case of games that games are at least activities (mental or physical). Nevertheless, being an activity is hardly definitional of games, nor does it specify a criterion of demarcation; there are many activities that are not games, such as working or catching a bus.

We will see in the case of science that there are characteristics common to all sciences, but are such that they cannot be definitional of it. They cannot be used for demarcating science from other human endeavours either. An example would be observing. We cannot think of a scientific discipline which does not involve making or relying on observations at some point. But then not everything that involves observing is a science (such as being observant when crossing a road in heavy traffic). Similarly, we cannot think of a science that does not involve making some

---

[10] John Searle has disputed this example, arguing that 'game' can be defined as follows: a series of attempts to overcome certain obstacles that have been created for the purpose of overcoming them (Searle 1995, 103). However this dispute is resolved, there might still be other cases where the family resemblance idea gets some traction, as we think it does in the case of the term 'science'.

kinds of inference at some point; if it did not, it would not get beyond naive data collecting. Nevertheless, as before, inferring, though common to the sciences, is not exclusive to them. Judges in a court or speculators on the stock market make inferences as well, but they are not doing science.

In the light of these points we can say that there are a few core characteristics that all sciences share (collecting data and making inferences, for instance). Nevertheless, even though they are generic, they are not sufficient either to define science or to demarcate it from other human endeavours. It is the other characteristics that accompany observing and inferring that make an important contribution to the family-forming characteristics that characterise scientific disciplines. It is this modified version of polythetic family resemblance that we will employ in what follows.

30.5   The Family Resemblance Approach to Science

There are many items called 'science', ranging from archaeology to zoology. (Here we will exclude the special case of mathematics from our discussion because of its non-empirical character.) So what do these many things called 'science' have in common? The idea of family resemblance will tell us that this is a wrong question to ask. What we need to do is to investigate the ways in which each of the sciences are similar or dissimilar, thereby building up from scratch polythetic sets of characteristics for each scientific discipline. The science categories we have introduced in Sect. 30.3 will come in handy for this task.

Begin with the items data collecting, making inferences and experimenting that fall under the category 'processes of inquiry'. Although all disciplines employ the first two and most (such as particle physics and chemistry) are experimental, there are a few disciplines that are not. Astronomy and earthquake science are cases in point since experiments are simply impossible in these fields. We cannot manipulate celestial objects; nor can we carry out experiments in earthquake science by manipulating earthquakes (though there are elaborate techniques for seismic detection which are not strictly experimental in the sense of experimentation as manipulation that we intend). Consider next the category 'aims and values' and the item prediction falling under it. Again, most sciences aim to make predictions, especially novel ones, but not all of them succeed. For example, astronomy is very good indeed in predicting planetary positions. In contrast, even though earthquake science does a good job of predicting the approximate locations of earthquakes, it fails badly with respect to predicting the time of their occurrence. Medicine can statistically predict the occurrence of many diseases under certain conditions without being able to tell who will develop them and when.

Let us now explore the similarities and differences amongst various scientific disciplines in terms of the items under the category 'methods and methodological rules'. Many sciences employ the hypothetico-deductive method, which can be roughly described as drawing out observable consequences of theories and then checking them against observational or experimental data. For example, particle physics and earthquake science use this method, but there does not appear to be any place for randomised double-blind experiments in these disciplines. In contrast, in evidence-based clinical medical science, the hypothetico-deductive method appears not to be of common use, while the methods of randomised double-blind experiments are the ubiquitous gold standard for testing. Similarly, some very important scientific research projects like sequencing the human genome do not involve much hypothesis testing, but rather are data-driven, inductive inquiries where most of the work is done by computer technologies.

Finally, consider the category of scientific knowledge and the items like laws, theories and models that fall under them. The idea of family resemblance applies here as well since not all sciences may have laws. For example, while there are clearly laws in physics, it is a contested issue as to whether there are laws in biology (Rosenberg 2008).

In the above we have mentioned a number of individual sciences and a number of characteristics. As can be seen for any chosen pair of these sciences, one will be similar to the other with respect to some of these characteristics and dissimilar to one another with respect to other characteristics. If we think of these characteristics as candidates for defining science, then no definition in terms of necessary and sufficient conditions would be forthcoming. If we take a family resemblance approach, however, things look very different and promising. To see this more concretely, let us represent data collection, inference making, experimentation, prediction, hypothetico-deductive testing and blinded randomised trials as D, I, E, P, H and T, respectively. Then we can summarise the situation for the disciplines we have considered as follows:

$$Astronomy = \{D, I, P, H\}; Particle\ physics = \{D, I, E, P, H\};$$
$$Earthquake\ science = \{D, I, P', H\}; Medicine = \{D, I, P'', E, T\},$$
$$where\ P'\ and\ P''\ indicate\ differences\ in\ predictive\ power\ as\ indicated.$$

Thus, none of the four disciplines has all the six characteristics, though they share a number of them in common. With respect to other characteristics, they partially overlap, like the members of closely related extended family. In short, taken altogether, they form a family resemblance.

Note that in order to convey the core idea that 'science' is a family resemblance concept, we have so far considered characteristics of science understood only as a cognitive-epistemic system. Does the idea of family resemblance apply to science as a social-institutional system as well? We believe that it does, at least to some degree. All scientific disciplines have a peer review system and a system of knowledge certification and dissemination. However, not all of them share exactly the same social values or the same elements of the scientific ethos. For example, the norm 'respect human and animal subjects' would not apply to disciplines such as physics and chemistry that do not deal with human and animal subjects, but 'avoid damaging the environment' certainly would. Similarly, although many sciences serve social utility, there are some fields (such as cosmology and parts of particle physics such as unified field theory) that are not obviously socially useful in any way; they are practised merely to satisfy our curiosity about the workings of nature. In short, the sciences form a polythetic family resemblance set with respect to their social and ethical dimensions as well.

30.6 Virtues of the Family Resemblance Approach

We believe that the family resemblance approach to science has several virtues, both theoretical and pedagogical. Perhaps the most important theoretical virtue of this approach is the systematic and comprehensive way it captures the major structural features of science and thereby accommodates, in a pedagogically useful way, almost all of the findings of NOS research in science education summarised in Sect. 30.2. As we shall illustrate in the next section, both the categories themselves and the items that fall under them do not dangle in the air as discrete entities; rather, they are tightly related to each other in a number of ways, forming an integrated whole. Thus, we can say that

> Science is a cognitive and social system whose investigative activities have a number of aims that it tries to achieve with the help of its methodologies, methodological rules, system of knowledge certification and dissemination in line with its institutional social-ethical norms, and when successful, ultimately produces knowledge and serves society.

This generic description is not meant as a definition of science, but rather as indicating how various aspects of science can be weaved together systematically as a unified enterprise.

By including science as a social institution as part of the family resemblance approach, the social embeddedness of science emphasised in the NOS literature in science education is captured in a novel way. A significant part of what it means to say that science is socially embedded is to say that noncognitive values are operative in science and influence science. No social institution, not even science, exists in a vacuum, so all kinds of social, cultural, historical, political and economic factors may influence it. Just to give an obvious example, funding strongly affects the choice of scientific problems and research agendas. Noncognitive factors of all sorts (gender biases, ideologies, economic considerations, etc.) may influence data description, hypotheses and even evidential relations in certain areas such as primatology and research on sex differences, as noted by feminist scientists and philosophers (Longino 1990). Sometimes these factors may cause scientists to deviate from the ethical norms of science (they may, e.g. fabricate or suppress data) and thus have a distorting effect on scientific conduct. However, not all social factors have a negative impact on science. Indeed, one of the most important functions of the ethos of science and mechanisms like peer review along with open and free critical discussion is precisely to minimise the negative effects on science. The ethos of science and the social system of scientific knowledge production contribute to the reliability of scientific knowledge as much as scientific methods and methodological rules do.

In practice, scientific inquiry is always guided by both cognitive-epistemic and social-institutional 'rules of the game', so to speak. This gives substance to our earlier claim that the distinction between science as a cognitive-epistemic system and science as a social institution is a conceptual one introduced for analytical purposes; but in practice the two are inseparable.

The historical, dynamic and changing nature of science can be accommodated naturally by the family resemblance approach through its open-ended categories that allow for the emergence of new characteristics of science within each category. For example, from a historical perspective we see that many scientific disciplines such as physics, chemistry, electricity and magnetism became mathematical only after the scientific revolution that occurred in the sixteenth and seventeenth centuries. Similarly, the hypothetico-deductive method was first clearly formulated and became established during the same period. New methodological rules like the one that tells the scientist to use blind procedures in conducting experiments on human subjects in life sciences came about only in the twentieth century. So did many ethical norms of science. The family resemblance approach therefore incorporates the dynamic, open-ended nature of science.

A unique virtue of the family resemblance approach is that it does justice to the differences amongst scientific disciplines and yet at the same time explains their unity by emphasising the similarities and partial overlaps amongst them. It is the existence of these 'family ties' that justify the label 'science' that we apply to various disciplines from archaeology to zoology. The unity of science is a unity-within-diversity. Earlier we pointed out that observing and inferring are common to all scientific disciplines even though they are not unique to the sciences. Another particularly important common feature of all scientific disciplines is the naturalism inherent in them—a feature that has not received sufficient attention in the NOS literature. We have touched upon this in discussing the notion of scientific explanation in Sect. 30.3.1 and are now in a position to articulate it more fully.

Science appeals to only natural entities, processes and events; its mode of explanation, aims and values, ethos, methods and methodological rules and the system of knowledge certification contain nothing that is supernatural or occult. Scientific naturalism is not an addendum to science invented by philosophers; rather, it is inherent to science. As Robert Pennock aptly puts it, it is a 'ground rule' of science so basic that it seldom gets mentioned explicitly (Pennock 2011, p. 184). One of the important science reform documents that does draw attention to this aspect of science is the National Science Teachers Association's statement on NOS: 'Science, by definition, is limited to naturalistic methods and explanations and, as such, is precluded from using supernatural elements in the production of scientific knowledge' (quoted from Pennock 2011, p. 197). Scientific naturalism pervades the whole of science from A to Z. As such, it describes a core aspect of science that contributes to its unity.

A final virtue of the family resemblance approach is that it is free of philosophical commitments such as realism, positivism, empiricism and constructivism. One can adopt any one of these, depending on how one wants to spell out each item that falls under each category of the family resemblance approach. For example, while realist educators may wish to emphasise truth as an aim of science with respect to both observable and unobservable entities, those who are sympathetic to constructivism may settle for viability, provided that they inform students of the existence of alternative views on this issue. Thus, they can add content to the family resemblance approach according to their philosophical orientation or else completely avoid discussing these philosophical issues due to the pressure of limited time, the level of the class and so on.

30.7 Teaching the Family Resemblance Approach: Some Suggestions

Teaching NOS from the perspective of family resemblance can begin by introducing the categories of science and then showing how they are related to one another. A natural place to start is processes of inquiry since all students are engaged in them to varying degrees. A host of interesting questions can be pursued in this context. Is observing a passive activity (raised to illustrate the point that data collection is often driven by scientific problems and theories)? How does observation differ from experimentation? What are the different ways in which a given set of data be interpreted? And so on. Next, the teacher can explore the connection between processes of inquiry, aims and hypotheses (or models and theories). This could be motivated very naturally since processes of inquiry are activities and virtually all activities have some aim or other. Some of the questions that can be asked are as follows. What is the point of doing an experiment? How are observational and experimental data related to hypotheses, theories and models? Does this theory explain that set of data? How would an experiment be set up to test some claim? These and similar questions enable the teacher to make several points: data provide evidence for or against hypotheses, theories and models; experiments are conducted to test them; testing can be done (as in the hypothetico-deductive method) by deducing test predictions from them. The aforementioned questions also provide excellent opportunities for the teacher to discuss key scientific notions like 'testing', 'experiment', 'theory', 'law' and 'model'.

Another fruitful question that prompts the exploration of the relationships amongst various science categories is to ask how science achieves its aims. This may lead to the idea of scientific method and methodological rule. In this context, at least three points can be made. First, science does not achieve its various aims haphazardly, but by employing a number of methods and methodological rules. With their help, science produces reliable (though fallible) knowledge. The hypothetico-deductive method, in particular, enables students to see this clearly. Scientific predictions do not always come out right, and when that is the case, it means that scientists have made a mistake somewhere and they must revise some of their claims. In this way, science can eliminate its errors and produce more reliable results.

Second, methods and methodological rules do not dictate to scientists what to do at every step of their inquiry. A discussion of this point may help students appreciate the fact that scientific methods and rules are not mechanical procedures that generate theories (or models) from data. Hence, theory construction always requires much imagination and creativity. To stimulate creativity, students may be invited to come up with different hypotheses that fit or explain the same data.

Third, despite the existence of methods, methodological rules and values functioning as criteria for evaluating rival theories, scientists may sometimes come to reach different conclusions on the basis of the same body of evidence. This may happen when no single theory embodies all the cognitive-epistemic values equally well and when different scientists place different emphasis on them when faced with a choice amongst rival theories. One scientist may give more weight to fruitfulness, say, and another may value simplicity more due to the priority given to aesthetic considerations (in which case there will be disagreement about which theory is the better one). A historical example that comes close to this scenario is the debate scientists had between Aristotelian-Ptolemaic geocentric system and the Copernican heliocentric system during the early stages of the scientific revolution. The teacher may discuss this case as example of *rational disagreement* amongst scientists, a disagreement which in no way implies that they are acting arbitrarily, though they might have subjective (personal) preferences in weighing values. Properly understood, then, being subjective does not mean acting arbitrarily, which is the whole point of Kuhn (1977). In this way, students can see how both personal (subjective) and intersubjective (objective) factors play a role in scientific theory choice.

Once the students grasp the categories 'processes of inquiry', 'aims and values' and 'methods and methodological rules', then the fourth category can be introduced in a straightforward way: scientific knowledge, especially in the form of theories and models, is the end product of successful scientific inquiry pursuing the aims of truth, testability, prediction and the like under the guidance of scientific methods and methodological rules. The teacher can then draw attention to and explain the characteristics of scientific knowledge which have emerged (such as its empirical, objective and subjective nature, its reliability or tentativeness, its dependence on creativity).

As for the teaching of science as a social-institutional system, we foreground two categories: the scientific ethos and the social certification of scientific knowledge. What must be especially emphasised with respect to these categories is their function in scientific knowledge production. Students must understand that ethical norms like intellectual honesty and openness and social mechanisms of peer review and free and critical discussion are as important as processes of inquiry such as experimenting or in using methods, like the hypothetico-deductive method of testing, in producing *reliable* knowledge. This point can be made forcefully by inviting students to think about what happens if scientists were to fabricate data or to accept an idea or a theory without sufficient critical discussion.

30.8 Conclusion

The main point of this chapter is to suggest a new way of understanding the term 'nature' as it gets employed in the phrase 'nature of science' (NOS). The word 'science' is a broad umbrella term which, in the context of science education, cannot be unproblematically captured by proposing accounts of 'nature' which are essentialist or by specifying a set of necessary and sufficient conditions for science. Nor can it be captured by drawing up some small list of features. The problem with a list is that it remains arbitrary as to why some features are included on the list and not others; and it remains unclear how, when given such a list, one is to go on to features not mentioned on the list. Our answer is to suggest the family resemblance or cluster account of a definition—an account developed within philosophy to overcome problems with essentialism, necessary and sufficient conditions and lists already mentioned. As such our enterprise is more philosophical and is not directed upon empirical matters such as the kinds of understanding teachers and pupils might have of NOS, or what level matters pertaining to NOS might be discussed in classrooms. Nevertheless, the family resemblance conception of 'nature' that we have proposed is not irrelevant to these empirical matters. What it does is 'free up' one's approach to them in what we hope is an illuminating way which a too rigid conception of 'nature' might obscure.

References

Abd-El-Khalick, F. (2004). 'Over and over and over again: College Students' Views of Nature of Science'. In: L. B. Flick & N. G. Lederman (eds.) *Scientific Inquiry and Nature of Science*. Dordrecht: Kluwer Academic Publishers, pp 389–426

Abd-El-Khalick, F., & Lederman, N. G. (2000). 'Improving science teachers' conceptions of the nature of science: A critical review of the literature'. *International Journal of Science Education* 22: 665–701.

Aikenhead, G. S. (1985a). 'Training teachers for STS education'. In: R. James (ed.) *Science, Technology and Society: Resources for Science Educators*. The Association for the Education of Teachers in Science 1985 Yearbook. Cookeville, Tennessee: Tennessee Technological University.

Aikenhead, G. S. (1985b). 'Science curricula and preparation for social responsibility'. In: R. Bybee (ed.) *Science, Technology, Society. The National Science Teachers Association 1985 Yearbook*, Washington, D.C.: NSTA.

Allchin, D. (2011). 'Evaluating Knowledge of the Nature of (Whole) Science'. *Science Education* 95: 518–542.

Alters, B. J. (1997). 'Whose nature of science?' *Journal of Research in Science Teaching* 34: 39–55.

American Association for the Advancement of Science (AAAS) (1990). *Science for all Americans*. New York: Oxford University Press.

American Association for the Advancement of Science (AAAS) (1993). *Benchmarks for science literacy*. New York: Oxford University Press.

Bell, R. (2004). 'Perusing Pandora's box: exploring the what, when, and how of nature of science'. In: L. B. Flick & N. G. Lederman (eds.) *Scientific Inquiry and Nature of Science*, Dordrecht: Kluwer Academic Publishers, pp 427–446.

Cobern, W. & Loving, C. (2001). 'Defining "Science" in a multicultural world: Implications for Science Education', *Science Education* 85: 50–67.

Council of Ministers of Education, Canada (1997). *Common framework of science learning outcomes*. Toronto, Canada: CMEC Secretariat.

Deng, F. (2011). 'Students' views of the nature of science: a critical review of research', *Science Education* 95: 961–999.

Driver, R., Leach, J., Miller, A. & Scott, P. (1996). *Young people's images of science*, Buckingham, England: Open University Press.

Dupre, J. (1993). *The Disorder of Things*. Cambridge, Mass.: Harvard University Press.

Duschl, R. A. & Osborne, J. (2002). 'Supporting and promoting argumentation discourse in science education'. *Studies in Science Education* 38: 39–72.

Eflin, J. T., Glennan, S. & Reisch, G. (1999). 'The Nature of Science: A Perspective from the Philosophy of Science'. *Journal of Research in Science Teaching* 36: 107–116.

Elby, A. & Hammer, D. (2001). 'On the substance of a sophisticated epistemology'. *Science Education* 85: 554–567.

Ford, M. J. & Wargo, B. M. (2007). Routines, roles, and responsibilities for aligning scientific and classroom practices'. *Science Education* 91: 133–157.

Godfrey-Smith, P. (2003). *Theory and Reality*. Chicago: The University of Chicago Press.

Grandy, R. E. & Duschl, R. A. (2007). 'Reconsidering the character and the role of inquiry in school science: analysis of a conference'. *Science & Education* 16: 141–166.

Hacking, I. (1996). 'The Disunities of the Sciences'. In: P. Galison & D. Stump (eds.) *The Disunity of Science*. Stanford: Stanford University Press, pp 37–74.

Hempel, C. G. (1965). *Aspects of Scientific Explanation and Other Essays in the Philosophy of Science*, New York: Free Press.

Irzik, G. & Nola, R. (2011). 'A Family Resemblance Approach to the Nature of Science for Science Education', *Science & Education* 20: 567–607.

Khishfe, R. & Lederman, N. G. (2006). 'Teaching Nature of Science within a Controversial Topic: Integrated versus Nonintegrated'. *Journal of Research in Science Teaching* 43: 395–418.

Kitcher, P. (2011). *Science in a Democratic Society*, New York: Prometheus Books.

Kolsto, S. D. (2001). 'Scientific literacy for citizenship: tools for dealing with the science dimension of socio-scientific issues'. *Science Education* 85: 291–310.

Kuhn, T. S. (1977). 'Objectivity, Value Judgment, and Theory Choice'. In: *The Essential Tension*. Chicago: University of Chicago Press, pp 320–339.

Laudan, L., Donovan, A., Laudan, R., Barker, P., Brown, H., Leplin, J., Thagard, P., & Wykstra, S. (1986). 'Scientific change: Philosophical models and historical research'. *Synthese* 69: 141–223.

Laudan, L. (1996). *Beyond Positivism and Relativism: Theory, Method and Evidence*. Boulder: Westview.

Lederman, N. G. (2007). 'Nature of science: Past, present, and future'. In: S. K. Abell & N. G. Lederman (eds.) *Handbook of research on science education*. Mahwah, NJ: Erlbaum, pp 831–879.

Lederman, N. G. (2004). 'Syntax of nature of science within inquiry and science instruction'. In: L. B. Flick & N. G. Lederman (eds.) *Scientific Inquiry and Nature of Science*. Dordrecht: Kluwer Academic Publishers, pp ix–xviii.

Lindberg, D. (1992). *The Beginnings of Western Science*. Chicago: The University of Chicago Press.

Longino, H. (1990). *Science as Social Knowledge*. Princeton: Princeton University Press.

Matthews, M. R. (1998a). 'The nature of science and science teaching'. In: B. Fraser & K. Tobin (eds) *International Handbook of Science Education*. Dordrecht: Springer, pp 981–999.

Matthews, M. R. (2011). 'Changing the focus: from nature of science (NOS) to features of science (FOS)'. In: M. S. Khine (ed.) *Advances in Nature of Science Research*, Dordrecht: Springer, pp 3–26.

McComas, W. F., Clough, M. P. & Almazroa, H. (1998). 'The role and character of the nature of science in science education'. In: W. F. McComas (ed.) *The Nature of Science in Science Education: Rationales and Strategies*. Hingham: Kluwer Academic Publishers, pp 3–40.

McComas, W. F. & Olson, J. K. (1998). 'The nature of science in international science education standards documents'. In: W. F. McComas (ed.) *The Nature of Science in Science Education: Rationales and Strategies*. Hingham: Kluwer, pp 41–52.

McGinn, M. K. & Roth, W. M. (1999). 'Preparing students for competent scientific practice: Implications of recent research in science and technology studies.' *Educational Researcher* 28: 14–24.

Merton, R. (1973). *The Sociology of Science: Theoretical and Empirical Investigations*, Chicago: Chicago University Press.

National Curriculum Council (1988). *Science in the National Curriculum*. York, UK: NCC.

National Research Council (1996) *National Science Education Standards*. Washington, DC: National Academic Press.

Nola, R. & Irzık, G. (2005). *Philosophy, Science, Education and Culture*, Dordrecht: Springer.

Nola, R. & Sankey, H. (2007). *Theories of Scientific Method*. Stocksfield: Acumen.

Osborne, J. (2007). Science education for the twenty-first century. *Eurasian Journal of Mathematics, Science and Technology Education*, 3: 173–184.

Osborne, J., Collins, S., Ratcliffe, M., Millar, R. & Duschl, R. (2003). 'What "Ideas-about-Science" Should Be Taught in School Science? A Delphi Study of the Expert Community'. *Journal of Research in Science Education* 40: 692–720.

Osborne, J., Ratcliffe, M., Collins, S., Millar, R. & Duschl, R. (2001). *What Should we teach about science? A Delphi Study*. London: King's College.

Pennock, R. T. (2011). 'Can't philosophers tell the difference between science and religion? Demarcation revisited'. *Synthese* 178: 177–206.

Popper, K, R. (1959). *The Logic of Scientific Discovery*. London: Hutchinson.

Popper, K. R. (1963). *Conjectures and Refutations*. London: Routledge and Kegan Paul.

Popper, K. R. (1975). *Objective Knowledge*. Oxford: Clarendon Press.

Radder, H. (2010). 'The commodification of academic research'. In: H. Radder (ed) *The Commodification of Academic Research*. Pittsburgh: University of Pittsburgh Press, pp 1–23.

Resnik, D. (2007). *The Price of Truth*. New York: Oxford.

Rocard, M. et al. (2007). *Science education now: a renewed pedagogy for the future of Europe*. EU report Rocard on science education_en.pdf. Retrieved 31 January 2011.

Rorty, R. (1991). *Objectivity, Relativism and Truth: Philosophical Papers Volume 1*, Cambridge: Cambridge University Press.

Rosenberg, A. (2008). 'Biology'. In: S. Psillos & M. Curd (eds) *The Routledge Companion to Philosophy of Science*. London: Routledge, pp 511–519.

Rudolph, J. L. (2000). 'Reconsidering the 'nature of science' as a curriculum component'. *Journal of Curriculum Studies* 32: 403–419.

Rutherford, J. F. & Ahlgren, A. (1990). *Science for all Americans*. New York: Oxford University Press.

Sadler, T. D. (2011). *Socio-scientific Issues in the Classroom*. Dordrecht: Springer.

Samarapungavan, A., Westby, E. L., & Bodner, G. M. (2006). 'Contextual Epistemic Development in Science: A Comparison of Chemistry Students and Research Chemists'. *Science Education* 90: 468–495.

Searle, J. (1995). *The Construction of Social Reality*. London: Allen Lane Penguin Press.

Sismondo, S. (2004). *An Introduction to Science and Technology Studies*. Oxford: Blackwell.

Smith, M. U. & Scharmann, L. C. (1999). 'Defining versus describing the nature of science: a pragmatic analysis for classroom teachers and science educators'. *Science Education* 83: 493–509.

Stanley, W. B. & Brickhouse, N. W. (2001). 'Teaching science: The multicultural question revisited', *Science Education* 85: 35–49.

van Fraassen, B. (1980). *The Scientific Image*. Oxford: Clarendon Press.

von Glasersfeld, E (1989). 'Cognition, Construction of Knowledge and Teaching'. *Synthese* 80: 121–40.

Weinstein, M. (2008). 'Finding science in the school body: Reflections on transgressing the boundaries of science education and the social studies of science'. *Science Education* 92: 389–403.

Wittgenstein, L. (1958). *Philosophical Investigations*. Oxford: Blackwell.

Wong, S. L. & Hodson, D. (2009). 'From horse's mouth: What scientists say about scientific investigation and scientific knowledge'. *Science Education* 93: 109–130.

Wong, S. L., Hodson, D., Kwan, J., & Yung, B. H. W. (2009). Turning crisis into opportunity: Nature of science and scientific inquiry as illustrated in the scientific research on severe acute respiratory syndrome. *Science & Education* 18: 95–118.

Wong, S. L. & Hodson, D. (2010). 'More from horse's mouth: What scientists about science as a social practice'. *International Journal of Science Education* 32: 1432–1463.

Zeidler, D. N., Walker, K. A. & Ackett, W. A. (2002). 'Tangled up in views: beliefs in the nature of science and responses to socio-scientific dilemmas'. *Science Education* 86: 343–367.

Zemplen, G. A. (2009). 'Putting sociology first–Reconsidering the role of the social in nature of science'. *Science & Education* 18: 525–560.

Ziman, J. (2000). *Real Science: What it is and What it Means*. Cambridge: Cambridge University Press.

Gürol Irzik is a professor of philosophy at Sabanci University, Turkey. He has published papers in philosophy of science, social aspects of science and science education. His books include the following: with co-author Robert Nola, *Philosophy, Science, Education and Culture* (Dordrecht, Springer, 2005) and, with co-editor Güven Güzeldere, *Turkish Studies in the History and Philosophy of Science* (Dordrecht, Springer, 2005). He edited a special issue of the journal *Science & Education* on the topic of commercialisation of academic science (vol. 22, no. 10, 2013).

Robert Nola is a professor of philosophy at The University of Auckland, New Zealand. He has published papers in philosophy of science, metaphysics, the sociology of science and science education. His recent books include the following: with co-author Gurol Irzik, *Philosophy, Science, Education and Culture* (Dordrecht, Springer, 2005); with co-author Howard Sankey, *Theories of Scientific Method* (Chesham, Acumen Press 2007); and with co-editor David Braddon-Mitchell, *Conceptual Analysis and Philosophical Naturalism* (Cambridge MA, MIT Press, 2009). His current work continues in philosophy of science with emphasis on scientific naturalism and the religion/science conflict.

Chapter 31
Appraising Constructivism in Science Education

Peter Slezak

31.1 Part I: Psychological or "Radical" Constructivism

Referring to von Glasersfeld's (1995a) radical theories, Cobb (1994a, p. 4) writes of the "fervor that is currently associated with constructivism" and Ernest (1995, p. xi) has described it as "the most important theoretical perspective to emerge," receiving "widespread international acceptance and approbation." A review of research in mathematics education noted "In the second half of the 1980s public statements urging the introduction of radical constructivist ideas in school mathematics programs also began to assume bandwagon proportions" (Ellerton and Clements 1991, p. 58). And Catherine Twomey Fosnot observed "Most recent reforms advocated by national professional groups are based on constructivism" (Fosnot 1996, p. x). For example, she cites the National Council of Teachers of Mathematics and the National Science Teachers Association. Fosnot's account has been supported by many other researchers:

> As any glance at contemporary educational literature demonstrates, the concept of "constructivism" carries with it enormous appeal. Contemporary literature also reveals that many current educational reform initiatives encourage teaching practices that many people refer to as constructivist (Null 2004, p. 80).

Denis Phillips (1997a, p. 152) has said of this kind of radical or psychological constructivism, "arguably it is the dominant theoretical position in science and mathematics education," and he remarks, "Across the broad fields of educational theory and research, constructivism has become something akin to a secular religion" (Phillips 1995, p. 5). Indeed, Tobin (1991, p. 1) explains the transformative effects of the doctrine which becomes "a referent for thoughts and actions" that "assume a higher value than other beliefs."

P. Slezak (✉)
University of New South Wales, Sydney, Australia
e-mail: p.slezak@unsw.edu.au

M.R. Matthews (ed.), *International Handbook of Research in History, Philosophy and Science Teaching*, DOI 10.1007/978-94-007-7654-8_31,
© Springer Science+Business Media Dordrecht 2014

Despite its significant influence among educationalists,[1] radical constructivism has generated severe controversy and polarization (see contributions to Matthews (1998) and Phillips (2000)). Indeed, von Glasersfeld himself has remarked "to introduce epistemological considerations into a discussion of education has always been dynamite" (quoted in Ernest (1995, p. xi)). Critics including Kelly (1997), Suchting (1992), Matthews (1998, 2000, 2012), Nola (1998), Olssen (1996), Small (2003), and Slezak (2010) have argued that the "radical constructivism" of von Glasersfeld has serious, if not fatal, philosophical problems, and further, it can have no benefit for practical pedagogy or teacher education. Indeed, there are well-conducted studies showing that constructivist-type teaching has a deleterious effect on student learning in science and mathematics (Kirschner et al. 2006; Mayer 2004).[2]

31.1.1 Recoiling into Metaphysical Fantasy

von Glasersfeld (1995a) sees the origins of radical constructivism in Kant, Berkeley, and Piaget, among others. However, the range of philosophical issues raised in the constructivist literature includes abstruse questions whose relevance to practical or theoretical problem in education has been questioned. Thus, among the topics discussed include Berkeleyan idealism, Cartesian dualism, Kantian constructivism, Popperian falsifiability, Kuhnian incommensurability, Quinean underdetermination, truth, relativism, instrumentalism, rationalism, and empiricism, inter alia. Gale (1995, p. xii) identifies "Cartesian epistemology" and the "mind–body split" as having educational relevance. He suggests that constructivist approaches "differed from the Cartesian model in viewing knowledge in a nondualistic manner so as to avoid the mind–body split of endogenic (mind-centered) and exogenic (reality-centred) knowledge" (Gale 1995, p. xiii).

Arguably, good educational theory and practice have flourished despite the persistent obduracy of these problems. Indeed, there is a sharp contrast between such esoteric philosophical matters and the practical recommendations taken to follow from them. Thus, drawing the morals of his constructivism, von Glasersfeld says "Rote learning does not lead to understanding" (Cardellini 2006, p. 182). Similarly, he suggests "after a while they [i.e. students] will become interested in why certain things work and others do not; and it is then that teachers can help to foster this interest that leads to understanding" (Cardellini 2006, p. 182). Critics of constructivism note that such insights are familiar to teachers who are ignorant of constructivism or any other philosophy, for that matter (Nola 1998, p. 33).

von Glasersfeld suggests that his conception of constructivism arose "out of a profound dissatisfaction with the theories of knowledge in the tradition of Western philosophy" and he has suggested that adopting his constructivism "could bring

---

[1] See Niaz et al. (2003).

[2] See also Gil-Pérez et al. (2002). Nevertheless, the practical question remains controversial with positive evidence also available.

31 Appraising Constructivism in Science Education

about some rather profound changes in the general practice of education" (Glasersfeld 1989, p. 135). For example, he recommends, "Give up the requirement that knowledge represents an independent world" (Glasersfeld 1995b, p. 6–7). This is, of course, Berkeley's notorious idealism. von Glasersfeld has addressed objections invited by his frequent allusions to Berkeley. He complains "superficial or emotionally distracted readers of the constructivist literature have frequently interpreted this stance as a denial of 'reality'" (von Glasersfeld quoted by Phillips 1995, p. 6).

However, von Glasersfeld himself encourages the attribution of idealism where he misleadingly claims "all the great physicists of the twentieth century ... did not consider [... their theories] descriptions of an observer-independent ontological reality" (quoted in Cardellini 2006, p. 181). Evidently endorsing von Glasersfeld's constructivism, John Shotter (1995, p. 41) says "We also take it for granted that it no longer makes sense to talk of our knowledge of an absolute reality – of our knowledge of a world independent of us – because for us there is no 'external world,' as it used to be called." von Glasersfeld (1989, p. 121) writes of a person's "cognitive isolation from reality" and says Berkeley's insight

> ... wipes out the major rational grounds for the belief that human knowledge could represent a reality that is independent of human experience (von Glasersfeld 1995a, p. 34).

However, Putnam (1994, p. 446) has warned that we must find a way "to do justice to our sense that knowledge claims are responsible to reality without recoiling into metaphysical fantasy."

von Glasersfeld's worry appears to address what Rorty has called "the philosophical urge," namely, the urge to say that assertions and actions must not only cohere with other assertions and actions but "correspond" to something apart from what people are saying and doing (Rorty 1979, p. 179). By contrast, in the spirit of Putnam's (1994) "second naiveté," Rorty says that a Quinean naturalism questions

> ... whether, once we understand ... when and why various beliefs have been adopted or discarded, there is something left called 'the relation of knowledge to reality' left over to be understood (Rorty 1979, p. 178).

In the same vein, Quine has written:

> ... it is meaningless, I suggest, to inquire into the absolute correctness of a conceptual scheme as a mirror of reality. Our standard for appraising basic changes of conceptual scheme must be, not a realistic standard of correspondence to reality, but a pragmatic standard (Quine 1961b, p. 79).

von Glasersfeld is evidently led into his idealist worries by failing to distinguish questions concerning the warrant for our beliefs from questions of metaphysics about the existence of a mind-independent world (von Glasersfeld 1989, p. 122).

However constructivism is to be understood, it remains that the relevance of these matters to education remains unclear at best. Ruhloff (2001, p. 64) concludes "There are no compelling reasons ... to draw a lesson or practical pedagogical instruction from the results of skeptical analysis." Indeed, in the same spirit, Kant himself wrote, "The only thing necessary is not theoretical learning, but the Bildung [education] of human beings, both in regard to their talents and their character."

31.1.2 Piaget's "Construction of Reality"?

Although the title of Piaget's (1999) book *The Construction of Reality in the Child* is suggestive of von Glasersfeld's doctrines, Piaget's own text leaves little doubt about the significant difference between these two. Piaget clearly acknowledges a knowable objective world beyond our sense-data. Despite possibly "flirting with idealism," Piaget (1972a, p. 57) says that his epistemological position is "very close to the spirit of Kantianism," both in its constructivism and in its sensitivity to the need to avoid Berkeleyan idealism. Thus, Margaret Boden writes:

> Piaget is aware that as a constructivist he must be careful to avoid idealism – or, to put it another way, that he must answer the sceptic's challenge that perhaps all our so-called 'knowledge' is mind-dependent illusion. He tries to buttress his commonsense realism by appealing to the biological basis of knowledge (Boden 1994, p. 79).

Piaget himself explains clearly that his views are not an idealistic overestimation of the part played by the subject:

> ... the organism is not independent of the environment but can only live, act, or think in interaction with it (Piaget 1971, p. 345).

Thus, while von Glasersfeld is at pains on every occasion to emphasize the unknowability of reality and the need to abandon notions of objectivity and truth, Piaget by contrast writes of "The Elaboration of the Universe" and he asks how the world is constructed by means of the instrument of the sensorimotor intelligence. In particular, Piaget speaks of the shift from an egocentric state to one "in which the self is placed ... in a stable world conceived as independent of personal activity" (Piaget 1999, p. 395). Piaget explains:

> The universe is built up into an aggregate of permanent objects connected by causal relations that are independent of the subject and are placed in objective space and time ...
> ... step by step with the coordination of his intellectual instruments ... [the child] discovers himself in placing himself as an active object among the other active objects in a universe external to himself (Piaget 1999, p. 397).

Elsewhere, Piaget writes:

> The theory of knowledge is therefore essentially a theory of adaptation of thought to reality ... (Piaget 1972b, p. 18).

Although invoking Piaget, these are ways of talking that von Glasersfeld has repudiated.

31.1.3 The Philosophical Urge

von Glasersfeld sees his version of constructivism as a departure from traditional conceptions. Meyer (2008) notes that von Glasersfeld and followers such as Gergen (1995) view constructivism "as a replacement for a whole field of philosophy." von

31 Appraising Constructivism in Science Education

Glasersfeld (1995a) explains that radical constructivism is "an unconventional approach to the problem of knowledge and knowing" that

> ... starts from the assumption that knowledge, no matter how it is defined, is in the heads of persons, and that the thinking subject has no alternative but to construct what he or she knows on the basis of his or her own experience (Glasersfeld 1995a, p. 1).

However, we might ask: Who has ever doubted that knowledge is in the head? von Glasersfeld appears to have fallen victim to a notorious problem in philosophy concerning the "veil of ideas" that is supposed to intervene between the mind and the world as the direct objects of perception and knowledge. Von Glasersfeld explains:

> One of Vico's basic ideas was that epistemic agents can know nothing but the cognitive structures they themselves have put together. ... God alone can know the real world ... In contrast, the human knower can know only what the human knower has constructed.
>
> For constructivists, therefore, the word knowledge refers to a commodity that is radically different from the objective representation of an observer-dependent world that the mainstream of the Western philosophical tradition has been looking for. Instead, knowledge refers to conceptual structures .... (von Glasersfeld 1989, p. 123).

It is precisely this idea that we know only our own ideas or "conceptual structures" directly rather than the world that is the source of the traditional puzzle since the seventeenth century. Putnam (1994) provides a succinct diagnosis of this "disastrous idea":

> ... our difficulty in seeing how our minds can be in genuine contact with the 'external' world is, in large part, the product of a disastrous idea that has haunted Western philosophy since the seventeenth century, the idea that perception involves an interface between the mind and the 'external' objects we perceive (Putnam 1994).

31.1.4 Epistemology or Pedagogy?

In an interview, von Glasersfeld was asked whether constructivism is to be understood as an epistemology or pedagogy. His answer is most revealing for von Glasersfeld simply restates the formula of Berkeley as if this serves as an answer to the question: "there is no way of checking knowledge against what it was supposed to represent. One can compare knowledge only with other knowledge" (Glasersfeld 1993, p. 24). Other interviewer questions sought to clarify the "differences between constructivism and idealism," but again, von Glasersfeld reiterates that "we can only know what our minds construct" and that "the "real" world remains unknowable" and that "I could be one of Leibniz' monads" (ibid., p. 28). We might reasonably wonder how this insight could help teachers in the classroom.

When pressed on the question concerning "the implications of constructivism for a theory of instruction," von Glasersfeld suggests that there are many. These include the following: "It is ... crucial for the teacher to get some idea of where [the students] are," that is, "what concepts they seem to have and how they relate them" (von Glasersfeld 1993, p. 33). This modest recommendation is far from the "rather profound changes" promised. Similarly, von Glasersfeld says:

> ... asking students how they arrived at their given answer is a good way of discovering something about their thinking (von Glasersfeld 1993, p. 33).

> Whatever a student says in answer to a question (or 'problem') is what makes sense to the student at that moment. It has to be taken seriously as such, regardless of how odd or 'wrong' it might seem to the teacher. To be told that it is wrong is most discouraging and inhibiting for the student (ibid., p. 33).

> If you want to foster students' motivation to delve further into questions that, at first, are of no particular interest (from the students' point of view), you will have to create situations where the students have an opportunity to experience the pleasure inherent in solving a problem (ibid., p. 33).

We may assume that such insights are what Tobin (1993, p. ix) has in mind when he refers to constructivism as "a paradigm for the practice of science education." Tobin has his own contributions to offer:

> A most significant role of the teacher, from a constructivist perspective, is to evaluate student learning. In a study of exemplary teachers, Tobin and Fraser found that these teachers routinely monitored students in three distinctive ways: they scanned the class for signs of imminent off task behavior, closely examined the nature of the engagement of students, and investigated the extent to which students understood what they were learning. If teachers are to mediate the learning process, it is imperative that they develop ways of assessing what students know and how they can represent what they know (Tobin and Tippins 1993, p. 12).

In brief, good teachers make sure students pay attention and understand the lesson. We may wonder how differently a teacher might do things if not operating "from a constructivist perspective."

31.1.5   From the Metaphysical to the Mundane

von Glasersfeld had promised that constructivism "could bring about some rather profound changes in the general practice of education." (von Glasersfeld 1989, p. 135). Elsewhere he has suggested that, "taken seriously," radical constructivism "is a profoundly shocking view" that requires that "some of the key concepts underlying educational practice have to be refashioned." However, among these "profoundly shocking" recommendations, he suggests the following:

> ... students will be more motivated to learn something, if they can see why it would be useful to know it (von Glasersfeld 1995a, p. 177).

> Teaching and training are two practices that differ in their methods and, as a consequence, have very different results. ... rote learning does not lead to 'enlightenment' (ibid., p. 178).

> ...in order to modify students' thinking, the teacher needs a model of how the student thinks (ibid., p. 186).

> Students should be driven by their own interest (ibid., p. 188).

31 Appraising Constructivism in Science Education

...talking about the situation is conducive to reflection (ibid., p. 188).

To engender reflective talk requires an attitude of openness and curiosity on the part of the teacher, a will to 'listen to the student' (ibid., p. 188).

These are all undoubtedly sound, indeed platitudinous, recommendations, though hardly deserving to be regarded as "profoundly shocking." In this regard, the writing of Driver and colleagues (1995) is instructive:

...learning science involves being initiated into scientific ways of knowing. scientific entities and ideas, which are constructed, validated, and communicated through the cultural institutions of science, are unlikely to be discovered by individuals through their own empirical inquiry; learning science thus involves being initiated into the ideas and practices of the scientific community and making these ideas and practices meaningful at an individual level. The role of the science educator is to mediate scientific knowledge for learners, to help them make personal sense of the ways in which knowledge claims are generated and validated, rather than to organize individual sense-making about the natural world (Driver et al. 1995, p. 6).

Following the model of C.W. Mills' (1959), Slezak (2010) argues that the passage may be reduced without remainder to the following brief claim: "Learning science involves learning science. Individuals cannot rediscover science by themselves. So, the role of teachers is to teach." Such illustrations have been offered by critics to indicate a tendency to recast truisms in pretentious jargon to create the illusion of deep theory. Tobin and Tippins (1993) provide another typical illustration:

Constructivism suggests that learning is a social process of making sense of experience in terms of what is already known. In that process learners create perturbations that arise from attempts to give meaning to particular experiences through the imaginative use of existing knowledge. The resolution of these perturbations leads to an equilibrium state whereby new knowledge has been constructed to cohere with a particular experience and prior knowledge (Tobin and Tippins 1993, p. 10).

Translation: Students sometimes learn new things. Tobin and Tippins conclude their article with the following remarks:

... it is our contention that constructivism is an intellectual tool that is useful in many educational contexts. ...We do not claim that use of constructivism as a referent is the only way to initiate changes of ... a comprehensive and significant scope, but from our experience we can assert that constructivism can assume a dialectical relationship with almost every other referent in a process that culminates in a coherent world view consisting of compatible referents for action (Tobin and Tippins 1993, p. 20).

Translation: Constructivism is consistent with some other theories.

Constructivist terms have ordinary synonyms which reveal the truisms they assert. Instead of merely saying "talking among teachers and students," we can refer to "the discursive practices that support the coconstruction of scientific knowledge by teachers and students" (Driver et al. 1994, p. 9). Instead of saying simply that "teachers explain new ideas," we can say the "teacher's role is characterized as that of mediating between students' personal meanings and culturally established mathematical meanings of wider society" (Cobb 1994b, p. 15). Rather than the truism "teachers and students exchange ideas," we can say that "speaking from the sociocultural perspective, [we] define negotiation as a process of mutual appropriation in which the teacher and students continually coopt or use each others' contribution" (Cobb 1994b, p. 14). Saying that "students learn different things at different times" may be recast as "Rather than successive equilibrations, ... learning may be better characterized by parallel constructions relating to specific contexts" (Cobb 1994a, p. 7).[3]

Such translations make an important point that goes beyond mere ridicule. They reveal what has been referred to as the "paradox of pedagogy." In a survey of the psychological literature, Slezak (2007) suggests that theories taken to be important for education are without bearing despite the ritual claims of relevance. As foregoing truisms suggest, teaching and learning are among the natural, intuitive mental skills that humans display through a tacit knowledge rather than explicit theory or doctrine. In the light of centuries of successful teaching, it seems clear that teachers and learners manage effectively without knowing modern theories of psychology, much less epistemology or metaphysics. Teacher and learner are perhaps best conceived on the analogy of speaker and hearer in a conversation. von Glasersfeld himself makes the more modest point:

> ... in summary, the best teachers have always known and used all this information, but they have known and used it more or less intuitively and often against the official theory of instruction. Constructivism does not claim to have made earth-shaking inventions in the area of education; it merely claims to provide a solid conceptual basis for some of the things that, until now, inspired teachers had to do without theoretical foundation (von Glasersfeld 1995b, p. 15).

It is noteworthy that the foremost advocate of von Glasersfeld's constructivism, Ken Tobin, has published an article titled "Constructivism in Science Education: Moving On" in which he writes:

> The critical mass of science educators are still making sense of their praxis in terms of constructivism, but in a short time we will be in another theoretical epoch (Tobin 2000, p. 250).

But nevertheless:

> As an axiom, however, constructivism is the ether for an expanding constellation of theories that illuminate my praxis in science education (p. 251).

Who knows what "is the ether" might mean, but hopefully "the critical mass" of those "illuminating their praxis" by the new "constellation of theories" might subject these new theories to more informed philosophical appraisal than has been evident in the embrace of von Glasersfeld's psychological constructivism.

31.2 Part II: Social Constructivism

Phillips (1995, p. 5) distinguishes the sociological form of constructivism from the psychological variety, and he observes, "It is the work of the social constructivists that had drawn the most dramatic attention in recent years; clearly they have touched

---

[3] For further examples, see Matthews (2000).

a raw nerve" (1997a, p. 154). Indeed, the sociological variety of constructivism has been described by Laudan (1990a, x) as "the most prominent and pernicious manifestation of anti-intellectualism in our time" and has been at the center of the "science wars." The book *Higher Superstition* (Gross and Levitt 1994) brought the polemics surrounding social constructivism to wide popular attention. Adding to the controversy, a scandal arose surrounding the "Sokal Hoax" (Sokal and Bricmont 1997)—the unwitting publication of an article by the mathematical physicist Alan Sokal, which was deliberately nonsense, written in the postmodern style.

Although not concerned primarily with educational matters, inevitably, social constructivism has dramatic implications for pedagogy (see Slezak 1994a, b, 2000, 2007). As Phillips notes, "There is a lot at stake. For it can be argued that if the more radical of the sociologists of scientific knowledge … are right, then the validity of the traditional philosophic/epistemological enterprise is effectively undermined, and so indeed is the pursuit of science itself" (1997b, p. 86). The doctrines of social constructivism take scientific theories to reflect the social milieu in which they emerge, and therefore, rather than being founded on logic, evidence, and reason, beliefs are taken to be the causal effects of the historically contingent, local context. Accordingly, if knowledge is intrinsically the product of "external" factors rather than "internal" considerations of evidence and reason, then it is an illusion to imagine that education might serve to instill a capacity for critical thought or rational belief. On these views, education becomes indoctrination, pedagogy is propaganda, and ideas are merely conventional conformity to social consensus. There could be no more fundamental challenge to education than the one posed by social constructivism, since it purports to overturn the very conception of knowledge in the Western tradition: The self-advertising grandiosely proclaims, "The foundations of modern thought are at stake here" (Pickering 1992).

Even among the more temperate critiques in the academic literature, the disputes have been unusually acrimonious. For example, Mario Bunge (1991) has described most of the work in the field as "a grotesque cartoon of scientific research." In a similar vein, the philosopher David Stove (1991) has written of these doctrines as a form of lunacy which is "so absurd, that it eludes the force of all argument," a "philosophical folly," and "a stupid and discreditable business" whose authors are "beneath philosophical notice and unlikely to benefit from it." In his scathing remarks, Stove describes such ideas as an illustration of the "fatal affliction" and "corruption of thought" in which people say things which are bizarre and which even they must know to be false.

31.2.1 Ideas or Ideology? Pedagogy or Propaganda?

Laudan's (1990) charge of anti-intellectualism points to the source of concern for educators. Where traditional views see scientific knowledge as involving insight, inspiration, creativity, and aesthetic judgment, sociologists see something more prosaic and utilitarian. Thus, Collins and Pinch (1992), writing specifically on science education in schools, suggest, "It is nice to know the content of science—it helps one to do a lot of things such as repair the car, wire a plug, build a model aeroplane." This conception leaves out the intellectual dimension and the creative role of the mind in providing an understanding of the world. Instead of conceiving science education as fostering understanding and critical thinking, Collins and Pinch (1992, p. 150) recommend that science education should attend to the social negotiation, "myths," and "tricks of frontier science" as "the important thing."

Above all, the relativism inherent in social constructivist theories makes it impossible for teachers to offer the usual intellectual grounds for distinguishing science from nonsense. Since the rational, cognitive virtues of theories are taken to be irrelevant to their status, one cannot complain that some views are false or implausible or otherwise lacking intellectual, explanatory merit. For example, one cannot teach that Soviet Lysenkoism or Hitler's racialism were perversions of scientific truth. Their success in winning consensus must count ipso facto as exemplary scientific achievement according to social constructivist doctrines. Pseudosciences such as parapsychology and astrology are merely unfashionable rather than scientifically wrong.

31.2.2   *What Is Social Constructivism?*

The sociology of science and its constructivist doctrines emerged with post-Positivist developments in the philosophy of science and as elaboration of themes found in Kuhn's (1962) influential book *The Structure of Scientific Revolutions.*[4] However, as Zammito (2004, p. 181) notes, Kuhn himself emphatically repudiated these ideas as "absurd," as "deconstruction gone mad," and he "willingly enlisted in the 'science wars' on the side of the scientists."

David Bloor's (1976) book *Knowledge and Social Imagery* launched the so-called Edinburgh Strong Program in the sociology of scientific knowledge (SSK). Bloor was heralding a radical enterprise intended to displace traditional philosophy and epistemology. The essential, astonishing stance was the rejection of "the very idea" of science as a distinctive enterprise, effacing any distinction or boundaries between science and other institutions as merely "rhetorical accomplishments." Typically, Woolgar rejects the traditional "core assumption," namely, "The persistent idea that science is something special and distinct from other forms of cultural and social activity" (Woolgar 1988). A further "assumption" to be rejected is the curiously persistent view "that the objects of the natural world are real, objective and enjoy an independent pre-existence" (Woolgar 1988, p. 26). In place of the traditional "misconceptions" about science and the independent preexistence of the world, social constructivism proposed an amalgam of idealism and relativism according to which scientific theories are merely "fictions," the product of social

---

[4] The most comprehensive account of post-Positivism and sociology of science is Zammito (2004).

forces, interests, and other contingent, historical aspects of the milieu in which they arise. That is, the very substantive ideas and content of scientific theories are not explanatory or descriptive of the world, but are socially negotiated by some community of discourse and constituted entirely by social consensus. Even scientific discovery is a matter of "interpretative practice," and "genius has no bearing on the pattern of discovery in science."

Sociologists had a ready explanation for the predictable incredulity of philosophers. Bloor's preface to the first edition of his book hints darkly that the inevitable resistance by philosophers to his doctrines will be due not to the implausibility of the ideas but to uncomfortable secrets that philosophers would wish to hide. Bloor asserts that his approach to science from a sociological point of view encounters resistance because "some nerve has been touched." He announces his intention to "despoil academic boundaries" which "contrive to keep some things well hidden" (Bloor 1976, p. ix). In retrospect, this is somewhat ironic in view of Bloor's own deceptive attempt to hide the commitments of his Strong Program (see Slezak 1994c). Despite making substantive changes in the second edition of his book designed to avoid fatal criticisms by Laudan (1981), in his preface, Bloor denies that he had made any alterations (Bloor 1991). Bloor devotes an entire chapter of his landmark book to a kind of psychoanalysis of his opponents by speculating about the "sources of resistance" to the Strong Program which he attributes to hidden, indeed primitive, motives involving the fear of sociology's desacralizing of science and its mysteries.

One might suggest alternative reasons for the resistance to his sociological doctrines, but Bloor sees only repressed impulses concerning the "sacred" and the "profane" leading to "a superstitious desire to avoid treating knowledge naturalistically" (Bloor 1976, p. 73). Bloor imagines that the "threatening" nature of any investigation into science itself has been the cause of a "positive disinclination to examine the nature of knowledge in a candid and scientific way" (Bloor 1976, p. 42). However, this disinclination to examine knowledge and the need to keep it mystified is difficult to reconcile with the fact that every philosopher since Plato has been centrally concerned with the problem of knowledge and its justification.

31.2.3 "Knowledge as Such": Contexts, Contents, and Causes

In his manifesto, Bloor (1976, p. 3) had declared that the central claims of the Strong Program were "beyond dispute," and Barry Barnes (1981, p. 481) begins an article asserting that in the short time since its advent "developments have occurred with breathtaking speed" and "the view that scientific culture is constructed like any other is now well elaborated and exemplified."

This level of self-congratulatory hyperbole prompted critics such as Gieryn (1982, p. 280) to comment upon such "defenses and reaffirmations" as "expressions of hubris" and "exaggerations passing as fact." Gieryn (1982, p. 293) has suggested that the radical findings of the new sociology of science "are 'new' only in a fictionalized reading of antecedent work." In particular, Robert Merton's (1957) chapter on "The Sociology of Knowledge" had specifically enunciated the very central doctrine of the Strong Program. Merton wrote:

> The "Copernican revolution" in this area of inquiry consisted in the hypothesis that not only error or illusion or unauthenticated belief but also the discovery of truth was socially (historically) conditioned...The sociology of knowledge came into being with the signal hypothesis that even truths were to be held socially accountable, were to be related to the historical society in which they emerged (Merton 1957, p. 456).[5]

Since the Logical Positivists have been a particular target of criticism by social constructivists, it is noteworthy that the Positivist Philipp Frank remarked that in judging the philosophy of science, "we must not ignore the extrascientific factors" since "Every satisfactory philosophy of science has to combine logic of science with sociology of science" (Frank 1949 quoted in Perla and Carifio 2009).

Though it had appeared earlier in different guises in Hegel, Marx, and Durkheim, the radical idea at the heart of the Strong Program was to go beyond those sociological studies which stopped short of considering the actual substantive content, the ideas, of scientific theories as an appropriate domain for sociological investigation. Previously, sociological studies paid attention only to such things as institutional politics, citation patterns, and other such peripheral social phenomena surrounding the production of science, but had not ventured to explain the cognitive contents of theories in sociological terms. Since this crucial point has been obscured, its importance for appreciating subsequent developments cannot be overstated. The opening sentence of Bloor's book asks, "Can the sociology of knowledge investigate and explain the very content and nature of scientific knowledge?"—that is, of "knowledge as such, as distinct from the circumstances of production" (Bloor 1976, p. 1).

The failure of previous sociological studies to touch on the contents of scientific belief was portrayed by Bloor as a loss of nerve and a failure to be consistent. Karl Mannheim, among the founders of the sociology of knowledge, is characterized as failing to make the logical extension of his approach from knowledge of society to the knowledge of nature as well. The epistemological pretensions of the Strong Program—its relativist challenge—derive from this thoroughgoing application of the sociological principle which seeks to explain the hitherto exempted knowledge claims. The ambitions of Bloor's program are explicit from the outset, for he complains that previous sociologists, in "a betrayal of their disciplinary standpoint," had failed to "expand and generalize" their claims to all knowledge: "the sociology of knowledge might well have pressed more strongly into the area currently occupied

---

[5] The work of Merton and others who had already formulated the ideas of the current sociology of science are largely ignored today, and so, there is some irony in Merton's remarks which acknowledge, "The antecedents of *Wissenssoziologie* only go to support Whitehead's observation that 'Everything of importance has been said before by somebody who did not discover it'" (1957, p. 456).

by philosophers, who have been allowed to take upon themselves the task of defining the nature of knowledge" (Bloor 1976, p. 1).

31.2.4 Causes and Case Studies

The extensive body of case studies repeatedly invoked by sociologists has been taken to establish the thesis that the contents of scientific theories and beliefs have social causes, in contradistinction to psychological ones. The causal claim concerns such things as "connections between the gross social structure of groups and the general form of the cosmologies to which they have subscribed" (Bloor 1976, p. 3). That is, the very cognitive content of the beliefs is claimed to be causally connected with immediate, local aspects of the social milieu. Of this general thesis, Bloor and Shapin (1979) were evidently unable to believe that anyone might question the causal claims of the Strong Program except on the assumption that they must be unfamiliar with the extensive literature of the case studies. However, in a precise parallel with Durkheim and Mauss (1903) to be noted presently, the claims of social determination of beliefs are all the more extraordinary in view of the failure of these case studies to support them. Critics have challenged precisely the bearing of these studies on the causal claims, and so, repeatedly citing the burgeoning literature is to entirely miss the point.

Of course, scientific discoveries have always necessarily arisen in some social milieu or other, but this is merely a truism holding equally for most human activity not thought to have been actually caused in this way by social factors. However, to the extent that social factors are indeed ubiquitous, establishing a causal connection requires more than merely characterizing in detail the social milieu which must have existed. These more stringent demands have not been met anywhere in the voluminous case studies in the SSK literature. Thus, although Shapin has acknowledged that "the task is the refinement and clarification of the ways in which scientific knowledge is to be referred to the various contextual factors and interests which produce it," and that "we need to ascertain the exact nature of the links between accounts of natural reality and the social order," nevertheless, his much-cited case study of phrenology offers only a variety of anthropological approaches leading at best to a postulation of "homologies" between society and theories which may serve as "expressive symbolism" or perhaps function to further social interests in their "context of use."

This falls far short of demonstrating the strong claims of social determination which abound in the rhetoric of programmatic statements and their "social epistemology." Thus, it is a truism to assert, as Shapin does, merely that "Culture [taken to include science] is developed and evaluated in particular historical situations" (Shapin 1979, p. 42). Shapin undertakes to refute the accusations of empirical sterility by a lengthy recounting of the "considerable empirical achievements" of the sociology of scientific knowledge (Shapin 1979, p. 65). But he is simply begging the question with his advice that "one can either debate the possibility of the sociology of scientific knowledge or one can do it" (Shapin 1982, p. 158).

31.2.5   When Is a Cigar Just a Cigar?

The claimed contingent, historical determination of scientific theories by local social context entails that the substantive content of theories would have been different had the milieu been different. We are inevitably led to ask: Would Isaac Newton have enunciated an inverse *cube* law of gravitation had the society been different? The model of such empirical studies was Forman's (1971) much-cited work which attributes the development of quantum physics to the prevailing milieu in Weimar Germany. However, in the same vein, we might inquire: Did Kurt Gödel's Incompleteness Theorem arise from some lacunae in the Viennese social order of 1930? This admittedly facetious example merely invokes the same suggestive metaphorical connections adduced by social constructivist case studies.

There is, at best, a kind of affinity claimed between the social context and the contents of the theory in question. Thus, Shapin cites "homologies between society and nature" and sees theories as "expressive symbolism" which can be exploited to serve social interests. Given the tenuous nature of such "homologies" between theories and the *zeitgeist*, the distinction between parody and serious claims is difficult to discern. Shapin's *recherché* homologies between theory content and social context recall the Freudian interpretation of dreams, which involved a similar decoding of an allegedly symbolic connection. Likewise, sociology pretends to disclose the hidden meaning underlying our scientific theories. We may have imagined that nineteenth-century theories of phrenology were about the brain, but they were really "expressing a social experience" and were about the "differentiation and specialization [in the social order] perceived by the bourgeois groups" (Shapin 1979, p. 57). Gödel's Incompleteness Theorem, too, undoubtedly expresses a collective longing for wholeness and fulfillment among the Viennese intelligentsia. However, in the spirit of Freud's famous remark, one is tempted to ask: When is a cigar just a cigar?

31.2.6   The Social Construction of Social Constructivism

It is instructive to look at an authoritative and sympathetic statement of social constructivism in a book whose coauthors include two of its founders—*Scientific Knowledge: A Sociological Analysis* by Barnes and colleagues (1996). These authors are uniquely well qualified to offer the book to anyone "seeking a text in the sociology of scientific knowledge." However, borrowing earlier words of one of its authors, this sociological enterprise appears to "contrive to keep some things well hidden" (Bloor 1976, p. ix). A study of the index is revealing. Georg Cantor, infinite cardinal numbers, and the continuum hypothesis get several entries whereas social constructivism and the Strong Program get none at all. In view of the Strong Program being proclaimed with great fanfare as the radical new approach revolutionizing the study of science and epistemology, its omission from the index is revealing. The Duhem–Quine thesis, mentioned *en passant* in an obscure footnote, gets no index entry either, though the book is, in fact, an extended essay on the alleged consequences of this philosophical doctrine. Other omissions from the index are equally curious. In view of the decisive, foundational status of the diametrical opposition between the rationalist "teleological" account and that of the Strong Program, it is striking that this issue, too, has disappeared without trace. This rewriting of history makes it impossible to understand both the social constructivist doctrines themselves and the scandal they have generated (see Slezak 1997).

31.2.7 Relativism

Despite characterizing their book as focused on "basic foundations," Barnes and colleagues (1996) explain that it "gives little prominence" to such issues as relativism, inter alia. Indeed, a prefatory mention of relativism is the only one in the book. However, relativism has been the central, distinctive theoretical doctrine of social constructivism and the source of most disputes. Neglecting to discuss it is somewhat like a text on evolution professing to concentrate on basic foundations and yet choosing to ignore natural selection. The authors' reticence about their own central, and previously explicitly embraced, doctrines is a telling feature of their work (see Barnes and Bloor 1982). Relativism is at the heart of social constructivism because the supposed absence from the constraints of independent "reality" is assumed to warrant appeal to a sociological account of theory acceptance. Relativism, then, is the spurious assumption that there can be nothing more to say about the goodness of our theories if one cannot meaningfully compare them to an independent, inaccessible reality.

However, the question of realism has been the subject of a vast philosophical literature, and both sides of these philosophical arguments accept the rational force of evidence and the usual considerations of explanatory virtue such as comprehensiveness, coherence, and simplicity as grounds for rational theory choice. Social constructivists mistakenly conclude that the inaccessibility of "things in themselves" behind the veil of our theories (whatever this might mean) precludes saying anything sensible about their cognitive virtues.

31.2.8 Theory Choice: Underdetermination of Theory by Evidence

One consideration, above all, has been widely taken to warrant the appeal to sociological factors in the explanation of scientific theory choice. This is an argument which attempts to exploit the underdetermination of theory by evidence—the

Quine–Duhem thesis that there can be no direct inference from observational data to any particular theory, since indefinitely many theories are equally compatible with the same empirical evidence (see Laudan 1990b, p. 6). Therefore, other considerations must be invoked to explain the preference of scientists for one theory over others which are equally consistent with the observational or experimental data. However, a non sequitur from this thesis has become one of the foundational tenets of the social constructivist enterprise. Thus, when distilled to its essence, the entire case underlying Bloor's (1976) manifesto is a spurious inference from underdetermination to social construction.

However, underdetermination is completely neutral among the various alternative resources which might be invoked to explain theory choice beyond conformity with the evidence. Clearly, it has to be shown independently why it might be social factors rather than some others (say, astrological or theological) which are the operative ones in determining theory choice. Boorse (1975) has pointed out that the underdetermination of theories by all possible observational evidence does not make them indistinguishable on other criteria such as simplicity, fecundity, coherence, comprehensiveness, and explanatory power. These are, of course, the kinds of rational considerations typically invoked by the rationalist or teleological account of the growth of scientific knowledge.

Part of the problem may have arisen from an excessively literal construal of theory choice which cannot be considered an actual selection among equivalent available alternatives. Historians, above all, should recognize that the problem in science is typically to find even a single theory which is consistent with the observations. Accordingly, what is termed "choice" is more appropriately described as the psychology of scientific invention or discovery—the subject of a burgeoning research literature (see Langley et al. 1987; Tweney et al. 1981; Gorman 1992; Giere 1992; Slezak 1989).

31.2.9 Consensus as Conventional

Social constructivism rests on this idea that alternative "choices" are equally "good," for theories are adopted by convention—a view that opens the way to sociological relativism. Barnes et al. (1996, p. 154) assert, "Conventions could always be otherwise" suggesting that knowledge might have been negotiated differently had the local interpretative milieu been different and, thereby, inviting the facetious question about Newton's inverse cube law. Indeed, undaunted, the authors embrace precisely such a paradoxical idea even in the case of arithmetical laws (Barnes et al. 1996, p. 184). According to their own account, given the underdetermination of theory by evidence, sociologists must be committed to the possibility of a consensus settling on a vast range of possible laws via the contingent "collective accomplishment" of "fact production" by "local cultural traditions." They suggest that the consensus on "$2+2=4$" is due merely to "pragmatic reasons connected with the organization of collective action" and the fact that "it is probably easier to organize" than a different convention such as "$2+2=5$."

31 Appraising Constructivism in Science Education

31.2.10 *Revisiting Durkheim and Mauss: Recoiling in Dismay*

These central ideas of social constructivism have a notable pedigree. Emile Durkheim and Marcel Mauss (1903) in their work *Primitive Classification* claimed that the cosmologies of groups such as the Zuñi reflected precise features of their social structure. In his paper "Revisiting Durkheim and Mauss," Bloor (1982, p. 267) invokes them in support of "one of the central propositions of the sociology of knowledge"—namely, their view that "the classification of things reproduces the classification of men." Bloor recommends that Durkheim and Mauss should be rehabilitated after having been consigned to the history books since their work is important for "showing not merely how society influences knowledge, but how it is constitutive of it" (Bloor 1982, p. 297).

It is understandable, of course, that Bloor should commend the virtues of Durkheim and Mauss, for they offer essentially the same metaphorical links between concepts and contexts which have been the stock-in-trade of the sociology of science. However, the Strong Program emulates *Primitive Classification* to the extent of exactly reproducing its severe shortcomings. Thus, a rather different picture emerges if one takes Bloor's invitation seriously to revisit Durkheim's work in the edition cited by him—including the introduction by Rodney Needham. Needham makes trenchant criticisms of Durkheim and Mauss, which are identical with those which have subsequently been leveled against Bloor's Strong Program.

Needham draws attention to Durkheim's claim which Bloor characterizes without demurral as a "bold unifying principle" but which Needham describes as an unwarranted, abrupt inference and logical error which flaws the entire work. On the alleged parallelism between primitive societies and their concepts, Needham writes:

> Now society is alleged to be the model on which classification is based, yet in society after society examined no formal correspondence can be shown to exist. Different forms of classification are found with identical types of social organization, and similar forms with different types of society. ... There is very little sign of the constant correspondence of symbolic classification with social order which the argument leads one to expect, and which indeed the argument is intended to explain (Durkheim and Mauss 1903, p. xvi).

Needham notes further that with respect to one of their claims, their "evidences on this point lend their argument no support whatever" and on another claim "nowhere in the course of their argument do the authors report the slightest empirical evidence, from any society of any form, which might justify their statement" (1903, p. xxii). Needham suggests "Durkheim and Mauss's entire venture to have been misconceived" (1903, p. xxvi). In view of the more recent airing of identical concerns, the following remarks are worth quoting in full:

> Yet all such particular objections of logic and method fade in significance before two criticisms which apply generally to the entire argument. One is that there is no logical necessity to postulate a causal connection between society and symbolic classification, and in the absence of factual indications to this effect there are no grounds for attempting to do so. ... If we allow ourselves to be guided by the facts themselves, i.e. by the correspondences, we have to conclude that there are no empirical grounds for a causal explanation. In no single case is there any compulsion to believe that society is the cause or even the model of the

classification; and it is only the strength of their preoccupation with cause that leads Durkheim and Mauss to cast their argument and present the facts as though this were the case (1903, pp. xxiv–xxv).

Although not mentioned by Bloor, these remarks take on special significance in light of the fact that identical claims of the Strong Program have been repeatedly asserted and repeatedly challenged. Needham draws attention to the extensive evidence which actually suggests a conclusion exactly the reverse of that which Durkheim and Mauss suppose. "That is, forms of classification and modes of symbolic thought display very many more similarities than do the societies in which they are found" (1903, p. xxvi). Needham's sober judgment is:

> We have to conclude that Durkheim and Mauss's argument is logically fallacious, and that it is methodologically unsound. There are grave reasons, indeed, to deny it any validity whatever (1903, p. xxix).

Bloor's enthusiasm for reviving the thesis of Durkheim and Mauss is difficult to reconcile with Needham's judgment, "It is difficult not to recoil in dismay" from their "unevidenced and unreasoned" explanations for the complexities of social and symbolic classification (1903, p. xxiii).

31.2.11 Impartiality

Robert Merton, like Karl Mannheim, argued that theories judged to be correct and founded on rational considerations are not in need of sociological explanation in the way that false and irrational theories are. In this sense, traditional conceptions relegated sociological accounts to the residue of false and irrational beliefs. Bloor's revival of the Durkheimian view was explicitly rescuing sociology from this ignominious role by asserting the appropriateness of sociological explanations for all of science regardless of evaluative judgments such as truth and falsity, rationality and irrationality, and success or failure. Our own cosmology and science in general, like those of the Zuñi, were to be shown to be in their entirety reflections of the social milieu.

Bloor's complaint is directed at asymmetrical approaches such as Lakatos's "rational reconstruction" of episodes in the history of science which sought to explain correct scientific theories as products of reasoned thought and, therefore, not requiring resort to sociological explanations. Bloor regards this approach as having the effect of rendering science "safe from the indignity of empirical explanation" altogether (Bloor 1976, p. 7), but for Lakatos, only sociology was to be excluded from accounts of successful science since reasons are a species of explanation themselves. Analogously, veridical perception does not need explanation in the same way as misperception or illusion. We do not ordinarily seek explanatory causes in the case of normal veridical perception, not because we assume that there is no scientific explanation, but because we assume it to be of a certain general sort. Thus, we seek the cause of perceptual failure (such as the influence of alcohol or disease). In the same way, we do not seek to explain why the train stays on the tracks but only why it fails to do so. Again, this asymmetry does not mean that we believe there is no cause or no explanation for the train staying on the tracks. However, this is the view which Bloor (1976, p. 7) imputes to rationalist philosophers such as Lakatos. In his *Knowledge and Social Imagery*, Bloor characterized the "autonomy of science" view he is opposing:

> One important set of objections to the sociology of knowledge derives from the conviction that some beliefs do not stand in need of any explanation, or do not stand in need of a causal explanation. This feeling is particularly strong when the beliefs in question are taken to be true, rational, scientific or objective (1976, p. 5).

Elsewhere, Bloor characterizes the "rationalistic" view that he opposes as "the claim that nothing makes people do things that are correct but something does make, or cause, them to go wrong" and that in the case of true beliefs, "causes do not need to be invoked" (1976, p. 6). Bloor intends to make an absolute distinction between the "teleological" or "rationalist" view which inclines its proponents to "reject causality" (1976, p. 10), on the one hand, and "the causal view," that is, the sociological approach of the Strong Program. On Bloor's own account, the viability of the Strong Program rests on the tenability of this diametrical opposition and, in particular, the falsity of the "teleological model." There could be no more crucial issue for the constructivist program.

Laudan (1981, p. 178) has characterized Bloor's acausal attribution to philosophers as an absurd view which cannot plausibly be attributed to any philosopher at all. In particular, the approach of Lakatos does not deny the existence of causes in cases of rationally held beliefs, but only assumes that reasons are themselves a species of cause (see Phillips 1997b, p. 100). However, Bloor (1981) responded to Laudan by denying these quite explicit earlier intentions. Bloor's discomfort was understandable, since the entire edifice of the Strong Program rests on this claimed opposition. Indeed, in the second edition of his classic book, in the crucial section on the "Autonomy of Knowledge" dealing with the problem of causation, we discover judicious changes to the original text whose rationale is clearly to avoid the criticisms made by Laudan (see Slezak 1991a).

These alterations are impossible to reconcile with Bloor's prefatorial assertion that "attacks by critics have not convinced me of the need to give ground on any matter of substance" and, therefore, "I have resisted the temptation to alter the original presentation of the case for the sociology of knowledge" apart from minor spelling and stylistic changes (Bloor 1991, p. ix). Despite their significance and implications for the entire sociological Strong Program, the exposé of these alterations (Slezak 1994c) has received little attention in the subsequent literature. Bloor had declared forthrightly, "There is no doubt that if the teleological model is true then the strong programme [sic] is false. The teleological and causal models, then, represent programmatic alternatives which quite exclude one another" (Bloor 1976, p. 9).

31.2.12 Social Constructivism as Born-Again Behaviorism

If the "rationalist, teleological, autonomy" view is not the acausal, anti-empirical straw man that Bloor ascribed to philosophers, then its merits need to be confronted seriously. This means acknowledging the full weight of considerations from cognitive science. This, in turn, means rejecting the hostility to internal, mental, or psychological accounts of rational belief which was a central plank of the Strong Program and other varieties of social constructivism.

The purported causal connection between ideas and social context is actually a version of stimulus-control theory akin to that of Skinnerian behaviorism, and not surprisingly, in his later work, Bloor (1983) explicitly endorses such notorious theories. In characterizing opposing rationalist or teleological views, and quoting Ludwig Wittgenstein, Bloor (1983, p. 6) refers to explanations which postulate mental states as infected by the "disease" of "psychologism." Bloor's frontal assault on the explanatory force of mental states is an intrinsic part of the defense of the alternative sociological approach to explaining science, but this bold stance left his program vulnerable to a case on the other side whose strength he had grievously underestimated. For example, anachronistically, Bloor's program depends on rejecting the reality of mental states such as images.

However, this position is 30 years and a major scientific revolution too late. Thus, Bloor (1983, p. 191) has dismissed Noam Chomsky's review of B. F. Skinner's *Verbal Behavior* with a passing footnote and a reference to it as the "fashionable" and "standard" criticism of behaviorism. But this reveals a failure to comprehend its significance. One might have expected some indication of the weaknesses of the review and why this merely "fashionable" criticism is to be ignored—particularly since Skinner himself never replied to it. In fact, the Chomsky review is generally regarded as having precipitated the downfall of the entire tradition of behaviorism in psychology.

Bloor's handwaving is rather more misleading than these comments suggest. Chomsky's ideas foreshadowed in this review became the foundations of the dramatic developments of the so-called Cognitive Revolution (see Gardner 1987). Bloor's failure to indicate the magnitude and import of these developments is comparable to defending Creationism today by dismissing the *Origin of Species* as merely "fashionable" and failing to let one's readers know anything of modern biology founded on Darwin's theory.[6]

31.2.13 Newton's **Principia** as Conditioned Response

Since behaviorism is a doctrine concerning psychology, it is at first sight surprising that it has been recruited to the cause of social constructivism. However, behaviorism serves Bloor as an ally, since it denies the explanatory role of internal mental

---

[6] For further discussion, see Papayannakos (2008).

states and is thereby in diametrical opposition to the rationalist or teleological point of view that the Strong Program is also battling. If scientific beliefs are to be construed as the causal effects of an external stimulus, they are precisely analogous to Skinnerian "respondents" or "operants" and, therefore, science is the result of conditioning. In short, the deep insight of social constructivism is that Isaac Newton's *Principia* is to be explained just like a rat's bar-pressing in response to food pellets.

Bloor's recent protest that his views are entirely consistent with cognitive science cannot be taken seriously and can be asserted at all only because Bloor (1991) now pretends that the sociological thesis at stake is merely whether or not there are social aspects to science. This is significantly different from the claim that knowledge is entirely socially constructed and constituted. This new weak and uncontroversial thesis is not the original doctrine he propounded, whose inconsistency with cognitive science was evident from the accompanying assault on rationalism and the postulation of mental states. The truism that there are social dimensions to science would not have generated the opposition and controversy evoked by the Strong Program. Significantly, Bloor's sociological colleagues have reacted differently: their vehement attacks on cognitive science and artificial intelligence have been both telling and more ingenuous. Their strenuous attempts to discredit the claims of cognitive science have given tacit acknowledgment to the threat these pose to the central sociological doctrines (Slezak 1989). Indeed, Collins (1990), among others, has been perfectly explicit on this point, seeing the claims of artificial intelligence as a crucial test case for the sociology of scientific knowledge (Slezak 1991b).

31.2.14 Revolt Against Reason

Social constructivism is essentially the doctrine characterized in an earlier generation by Karl Popper (1966) as the "revolt against reason"—a rejection of certain ideals of truth and rationality which, however difficult to explicate, are nonetheless central to the Western heritage since the Milesian Pre-Socratics. Popper saw the same tendencies in Hegel which he bitterly denounced as "this despicable perversion of everything that is decent." There can be little doubt about the close affinities between Hegel's doctrines and those of social constructivism: Popper (1966, p. 49) observes that for Hegel, "History is our judge. Since History and Providence have brought the existing powers into being, their might must be right." The parallel is seen in their essentially similar answers to Popper's fundamental question, "who is to judge what is, and what is not, objective truth?" He reports Hegel's reply, "The state has, in general…to make up its own mind concerning what is to be considered as objective truth," and adds, "With this reply, freedom of thought, and the claims of science to set its own standards, give way, finally, to their opposites" (1966, p. 43).

Hegel's doctrine expressed in terms of the "state" is essentially the idea that political success is ipso facto the criterion of truth. As we will see presently, this idea is resuscitated by Latour and Woolgar, Pinch and Collins, and the entire enterprise of contemporary social constructivism. This is a historical relativism according to which truth is merely political and dependent on the *zeitgeist* or spirit of the age. It is a view which Popper (1966, p. 308) charges with helping to destroy the tradition of respecting the truth, and his discussion of Hegel's "bombastic and mystifying cant" is striking in its aptness to recent sociology of science and is echoed by Gross and Levitt, Laudan and Stove, and among others. Popper warns against the "magic of high-sounding words" and the "power of jargon" to be found in doctrines which are

> ... full of logical mistakes and of tricks, presented with pretentious impressiveness. This undermined and eventually lowered the traditional standards of intellectual responsibility and honesty. It also contributed to the rise of totalitarian philosophizing and, even more serious, to the lack of any determined intellectual resistance to it (Popper 1966, p. 395).

31.2.15 Laboratory Life *Under the Microscope*

Perhaps the most obvious cause for such concern is another celebrated, foundational classic of social constructivism, *Laboratory Life* by Latour and Woolgar (1979). This work is self-consciously subversive, rejecting the rules of logic and rationality as a merely "coercive orthodoxy" (Woolgar 1988). It has the avowed goal of deflating the pretensions of science both in its knowledge claims and in its possession of a special method. Among its iconoclastic goals, the book professes to "penetrate the mystique" (Latour and Woolgar 1979, p. 18), dissolve the appearances, and reveal the hidden realities of science-in-the-making at the laboratory workbench. This study purports to give an exposé of the "internal workings of scientific activity" (Latour and Woolgar 1979, p. 17).

Discovering puzzling questions concerning science, Latour and Woolgar conclude that all of science is merely the "construction of fictions" (1979, p. 284). Latour explains the insights emerging from the new discipline:

> Now that field studies of laboratory practice are starting to pour in, we are beginning to have a better picture of what scientists do inside the walls of these strange places called "laboratories." ... The result, to summarize it in one sentence, was that nothing extraordinary and nothing "scientific" was happening inside the sacred walls of these temples (Latour 1983, p. 141).

> ... The moment sociologists walked into laboratories and started checking all these theories about the strength of science, they just disappeared. Nothing special, nothing extraordinary, in fact nothing of any cognitive quality was occurring there (Latour 1983, p. 160).

Needless to say, if warranted, the implications of such insights must be revolutionary, not least for science education. Indeed, the foregoing remarks have been approvingly quoted in a teachers' journal recommending a radical new vision of "the reality of the scientific process" (Gough 1993). Science education is presumably only socialization into power, persuasion, and propaganda. Rather than learning as a cognitive process involving reasoning, logic, and understanding, education involves merely the observance of arbitrary practices and political interest. Although Latour and Woolgar do not explicitly address the questions of most direct interest to educators as such, their characterization of science clearly suggests the appropriate role of the teacher:

> Each text, laboratory, author and discipline strives to establish a world in which its own interpretation is made more likely by virtue of the increasing number of people from whom it extracts compliance (Latour and Woolgar 1986, p. 285).

On this conception, the function of science teacher is extraction of compliance, more like camp commandant than teacher.

31.2.16 Constructing the World

As a façon de parler, the thesis of "constructing facts" permits a sensible reading according to which a theory or description is settled upon and in a certain sense perhaps even "socially negotiated." However, one can also choose to construe such truisms as something more paradoxical—namely, that objects and substances themselves did not have an independent existence and were socially constructed. In like manner, one might say that Copernicus "removed the earth from the center of the universe," but intending this literally would be an attempt at humor or evidence of derangement. Nevertheless, it is just this sort of claim for which the work of Latour and Woolgar has been acclaimed as a defining text in the genre of ethnomethodology of science.

31.2.17 Witchcraft, Oracles, and Magic Among the Academics

On the face of it, the authors' own description of their project in *Laboratory Life* reads like a parody. Upon entering the Salk Institute for a 2-year study, "Professor Latour's knowledge of science was non-existent; his mastery of English was very poor; and he was completely unaware of the existence of the social studies of science" (1986, p. 273). It is from this auspicious beginning that the "revolutionary" insights into science were to emerge.

Of course, these apparent liabilities are portrayed by Latour and Woolgar as a unique advantage, since "he was thus in the classic position of the ethnographer sent to a completely foreign environment" (1986, p. 273). However, the idea that the inability to understand one's human subjects is a positive methodological virtue is surely a bizarre conception. For Latour and Woolgar, however, it is intimately connected with their doctrine of "inscriptions." The meaninglessness of the "traces, spots, points," and other recordings being made by workers in the laboratory is a direct consequence of Latour's admitted scientific illiteracy. Predictably, all these symbols are indiscriminable to an observer who is completely ignorant, and they must, therefore, be placed in the category of unintelligible markings or "inscriptions." Avoiding the possibility of understanding their subjects' behavior is justified on the grounds that just as the anthropologist does not wish to accept the witch doctor's own explanations, so one should remain uncommitted to the scientists' rationalizations too. However, this attitude follows from the simple failure to appreciate the difference between *understanding* the native and *believing* him.

31.2.18 Persuasion by Literary Inscription and Achieving Objects by Modalities

It is from a point of view of ignorance and incomprehension that Latour comes to rely on a "simple grammatical technique" in order to discern the true significance of the papers accumulating in the laboratory in which he was doing the fieldwork. Undeniably, this method has great merit as an alternative to undertaking many years of undergraduate study and postgraduate work as preparation for his fieldwork. On this grammatical basis, then, Latour and Woolgar obtain their insight: "Activity in the laboratory had the effect of transforming statements from one type to another" (1986, p. 81). Specifically, the rationale of the laboratory activities was the linguistic exercise of transforming statements in various ways in order to enhance their "facticity."

Thus, we see how Latour and Woolgar arrive at their celebrated social constructivist conclusions. They maintain that "a laboratory is constantly performing operations on statements," (1986, p. 86) and it is through this process that "a fact has then been constituted" (1987, p. 87) by social negotiation and construction. In short, the laboratory must be understood "as the organization of persuasion through literary inscription" (1986, p. 88). These are the grounds on which we must understand their claims that substances studied in the laboratory "did not exist" prior to operations on statements (1986, p. 110, 121). "An object can be said to exist solely in terms of the difference between two inscriptions" (1986, p. 127).

31.2.19 Poison Oracles and Other Laboratory Experiments

From the meaninglessness of the "inscriptions" and his revelation that "the 'scientificity' of science has disappeared" (Latour 1983, p. 142), Latour is led inexorably to a "naive but nagging question"—namely, "if nothing scientific is happening in laboratories, why are there laboratories to begin with and why, strangely enough, is the society surrounding them paying for these places where nothing special is produced" (Latour 1983, p. 141)? This is undoubtedly a deep mystery if one systematically refuses to understand the meaningfulness of the "inscriptions" on these papers.

On the analogy of the "anthropologist's refusal to bow before the knowledge of a primitive sorcerer" (1986, p. 29), Latour and Woolgar refuse to accept the authority of our best science, saying, "We take the apparent superiority of the members of our

31 Appraising Constructivism in Science Education

laboratory in technical matters to be insignificant, in the sense that we do not regard prior cognition ... as a necessary prerequisite for understanding scientists' work" (1986, p. 29). The affectation that Latour was like Evans-Pritchard among the Azande is "anthropological strangeness" in a rather different sense of the term: No anthropologist was ever so strange. Given his method, predictably, Latour finds the activities in the laboratory completely incomprehensible. Undaunted, and unwilling to allow this to become a liability, it becomes, in fact, the deep insight of *Laboratory Life*. The behavior of the scientists not only appears meaningless, it is meaningless. In their conclusion, Latour and Woolgar reveal that "[a] laboratory is constantly performing operations on statements" (1986, p. 86) and the activities of the laboratory consist in manufacturing "traces, spots, and points" with their "inscription devices." The production of papers with meaningless marks is taken to be the main objective of the participants in essentially the same way that the production of manufactured goods is the goal of any industrial process.

31.2.20 *"Derridadaism": Readers as Writers of the Text*

Concern about the perversity of this work arises from the fact that in the new edition of their book, Latour and Woolgar (1986) tell us that laboratory studies such as their own should, after all, not be understood as providing a closer look at the actual production of science at the workbench, as everyone had thought. This view would be "both arrogant and misleading" (1986, p. 282) and would presume they had some "privileged access to the 'real truth' about science" which emerged from a more detailed observation of the technical practices. Instead, Latour and Woolgar explain that their work "recognizes itself as the construction of fictions about fiction constructions" (1986, p. 282). This is the textualism of Jacques Derrida combined with a much-vaunted "reflexivity." They continue, "all texts are stories. This applies as much to the facts of our scientists as to the fictions 'through which' we display their work." Their own work, then, just like all of science, has no determinate meaning since "[i]t is the reader who writes the text" (1986, p. 273).

Here, we see a notorious deconstructionist affectation which conveniently serves to protect Latour and Woolgar against any conceivable criticism. The contrast with the work of Bloor is interesting: Where Bloor professes to adhere to the usual principles of scientific inquiry, Latour and Woolgar engage in a game David Lehman (1991) has aptly called "Derridadaism." They manage to evade criticism only by adopting deconstructionist double-talk and affecting a posture of nihilistic indifference to the ultimate cogency of their own thesis. In keeping with the principle of reflexivity, they embrace the notion that their own text (like the science they describe) has no "real meaning," being "an illusory, or at least, infinitely renegotiable concept" (Latour and Woolgar 1986, p. 273).

Reflecting on the controversies surrounding their work, Latour and Woolgar observe that defenders and critics alike have been duped into engaging in this futile "spectacle" in which they have debated the presumed intentions of the authors.

This "spectacle" is, of course, just the traditional exercise of scholarly criticism. Latour and Woolgar now reveal that the "real" meaning of a text must be recognized as illusory and indeterminate. Questions of what the authors intended or what is reported to have happened "are now very much up to the reader."

This Rorschach inkblot view of their own work is undoubtedly correct in one sense, if only because *Laboratory Life* is in many respects completely incoherent and unintelligible. For example, some of the diagrams offered as explanatory schemas are impossible to decipher. Above all, it is sobering to consider how science teaching might be conducted in accordance with this model of scholarship. Perhaps an indication can be seen in the notorious constructivist claim of Sandra Harding (1986) that Isaac Newton's *Principia* is a "rape manual."

31.2.21 Balance of Forces

Though the implications of social constructivism are not drawn out by the authors, these are not difficult to discern. Thus, once Latour and Woolgar reject "the intrinsic existence of accurate and fictitious accounts per se," the only remaining criterion for judgment is judgment itself. They say "the degree of accuracy (or fiction) of an account depends on what is subsequently made of the story, not on the story itself" (1986, p. 284). There are no grounds for judging the merits of any claim besides the "modalizing and demodalizing of statements," a purely political question of persuasion, propaganda, and power.

Thus, they suggest that the very idea of "plausibility" of any work, including their own, is not a rational, intellectual, or cognitive question, but simply a matter of political redefinition of the field and other such transformations involving shift in the "balance of forces." In particular, the current implausibility of their own theory is only due to its relative political disadvantages rather than the lack of any intellectual merits. Apart from being a self-serving justification of any nonsense at all, one could hardly find a more open endorsement of the doctrine that "might is right." The very distinction between education and indoctrination becomes impossible to draw.

31.2.22 Education: Truth as Power

The bearing of these doctrines on educational questions is starkly brought out in Chomsky's remarks:

> It is the responsibility of intellectuals to speak the truth and to expose lies. This, at least, may seem enough of a truism to pass without comment. Not so, however. For the modern intellectual, it is not at all obvious (Chomsky 1969, p. 257).

Chomsky quotes Martin Heidegger in a pro-Hitler declaration, echoing social constructivist ideas that "truth is the revelation of that which makes a people certain, clear and strong in its action and knowledge." Chomsky remarks ironically that for

Heidegger, it seems that it is only this kind of "truth" that one has a responsibility to speak, that is, the "truth" which comes from power. In the same vein, we have seen Latour and Woolgar assert that the success of any theory is entirely a matter of not persuasion but politics and power extracting compliance. On this theory, a repressive totalitarian regime must count as a model of scientific success.

31.2.23 Mertonian Norms: The Ethos of Science

On such a theory, it is impossible to distinguish fairness from fraud in science since, after all, both are ways of constructing fiction. In the absence of the usual distinctions, the scientist who fraudulently manufactures his evidence cannot be meaningfully distinguished from the honest researcher whose data are also "constructed," albeit in different ways. The problem arises from the social constructivists' rejection of the famous Mertonian norms of universalism, communism, disinterestedness, and organized skepticism which constitute the "ethos of science" (Merton 1942). Merton described these as institutional imperatives, being "moral as well as technical prescriptions," that is, "that affectively toned complex of values and norms which is held to be binding" on the scientist. As Merton observes, these institutional values are transmitted by precept and example, presumably in the course of the scientist's education. It is difficult to see how someone committed to the social constructivist view can either teach or conduct science according to the usual rules in which truth, honesty, and other intellectual and ethical measures of worth are taken seriously.

31.2.24 Facticity and Maintaining One's Position

In articulating the same political view of scientific claims, social constructivist authors stop short of openly encouraging cheating and other forms of dishonesty in science, but there can be no mistake about what their theory entails. Thus, when examining a dispute concerning the claims of parapsychology or astrology, Pinch and Collins (1984) draw attention to symmetries in the attempts of opponents to maintain their commitments—in one case, to orthodox science and, in the other, to the paranormal. However, from the standpoint of scrupulous sociological "neutrality" or "impartiality" regarding the intellectual merits of the case on each side, there can be no way to discriminate the relative merits of either the arguments or the evidence itself.

In the case study offered by Pinch and Collins, both sides make questionable attempts to protect their favored theory against contrary evidence and, indeed, the scientists appear to have been less than completely forthright about some disconfirming evidence. Pinch and Collins wish to generalize from this to a thesis about science as a whole by construing it as a typical case, that is, as evidence of the way in which public scrutiny removes the mystique of science and exposes its socially constructed, negotiated character. Such exposé serves to "dissolve the facticity of the claims."

Pinch and Collins are unwilling to see such episodes as anything other than the way science always operates—not because all scientists are dishonest, but because the very distinction relies on being able to discriminate fact from fiction. When the scientists finally admit their error and revise their earlier stance in the light of falsifying evidence, they are ridiculed by Pinch and Collins for their grandiose, mythical pretensions and for appearing to adopt "a mantle of almost Olympian magnanimity" (Pinch and Collins, 536). The scientists are reproached for failing to "re-appraise their understanding of scientific method" and to learn about its "active" character, that is, about the way in which "facts, previously established by their presentation in the formal literature [sic], can be deconstructed" (1984, p. 538) by public scrutiny of the informal, behind-the-scene reality of science.

Remarkably, however, Pinch and Collins suggest that the right lesson about science was that "provided they had been prepared to endorse the canonical model in public while operating in a rather different way in private, they could have maintained their position" (1984, p. 539). In other words, if they had been even more dishonest, they would have been right—in the only sense of "right" possible, that is, they would have "maintained their position." The status or "facticity" of a claim is just a matter of how the claim is publicly presented, and the literature can either construct or "dissolve the facticity of the claims" (1984, p. 523). If we drop all this jargon, their point is simply that truth is what you can get away with.

31.2.25 Altering the Grounds of Consensus: Affirmative Action?

In practice, through the feigned suspension of judgment, social constructivism has led to explicit advocacy of discredited or disreputable pseudoscience. Pinch (1993) and Ashmore (1993) go so far as to defend the supposed "merits" of unorthodox and rejected theories on the grounds of equity. Not least, this policy is evidently taken to include the case of fraud since this "is to be seen as an attributed category, something made in a particular context which may become unmade later" (Pinch 1993, p. 368). Ashmore proposes a radical skepticism concerning the exposé of notorious cases of misguided science such as that of Blondlot's N-rays. Amid the usual jargon-laden pseudotechnicality, such an approach amounts to actually promoting the alleged scientific merits or deserts of such discredited cases. Thus, Pinch writes of "making plausible the rejected view" (1993, p. 371) and Ashmore is perfectly explicit: "To put it very starkly, I am looking for justice! … in a rhetorically self-conscious effort to alter the grounds of consensus" (Ashmore 1993, p. 71). Again, the educational implications for the curriculum should hardly need drawing out. The "impartiality" defended by radical social constructivism has come to mean something like affirmative action for bullshit.[7]

---

[7] This may be regarded as a technical philosophical term since Frankfurt's (2005) celebrated article. However, my use of the term does not fit Frankfurt's taxonomy.

31.3 Conclusion

The two main varieties of constructivism considered here have different, but dramatic, consequences for science education. The psychological or "radical" constructivism of von Glasersfeld has had a direct and wide influence on educational theory and practice and makes extravagant claims to overturn the entire tradition of Western epistemology and pedagogical theory. However, when examined critically, radical constructivism appears to offer only commonplaces and platitudes.

By contrast, there could be no more fundamental challenge to science education than the one posed by the sociology of scientific knowledge, since social constructivism purports to overthrow conceptions of rationality, truth, and evidence to be replaced by fashion, negotiation, and consensus. On this view, science is a social institution just like any other. According to social constructivism, there can be no difference between true and false, fact and fiction, fair and fraudulent. On this account, the greatest achievements of the creative human intellect are merely accidents of social context. Isaac Newton was just lucky to be in the right place at the right time.

References

Ashmore, M. (1993). The Theatre of the Blind: Starring a Promethean Prankster, a Phoney Phenomenon, a Prism, a Pocket and a Piece of Wood, *Social Studies of Science*, 23, 67–106.

Austin J. L. (1962) *Sense and Sensibilia*. Oxford: Oxford University Press.

Barnes, B. (1981). On the Hows and Whys of Cultural Change, *Social Studies of Science*, 11, 481–498.

Barnes, B. and Bloor, D. (1982). Relativism, Rationalism and the Sociology of Knowledge," in M. Hollis and S. Lukes, eds., *Rationality and Relativism*, Oxford: Basil Blackwell, 21–47.

Barnes B., Bloor D. & Henry J. (1996) *Scientific Knowledge: A Sociological Analysis*. Chicago: The University of Chicago Press.

Bloor, D. (1976) *Knowledge and Social Imagery*, London: Routledge & Kegan Paul.

Bloor, D. (1981). The Strengths of the Strong Programme, *Philosophy of Social Sciences*, 11, 199–213.

Bloor, D. (1982). Durkheim and Mauss Revisited: Classification and the Sociology of Knowledge, *Studies in History and Philosophy of Science*, 13, 4, 267–297

Bloor, D. (1983). *Wittgenstein: A Social Theory of Knowledge*, New York: Columbia University Press,

Bloor, D. (1991). *Knowledge and Social Imagery*, Second Edition, Chicago: University of Chicago Press.

Boden M. A. (1994) *Piaget*. London: Fontana.

Boorse, C. (1975). The Origins of the Indeterminacy Thesis, *Journal of Philosophy*, 72, 369–887.

Bunge, M. (1991). A Critical Examination of the New Sociology of Science, *Philosophy of the Social Sciences*, 21, 4, 524–60.

Cardellini L. (2006) The foundations of radical constructivism: an interview with Ernst von Glasersfeld. *Foundations of Chemistry*, 8: 177–187.

Chomsky, N. (1969). *American Power and the New Mandarins*, Harmondsworth: Penguin Books.

Cobb P. (1994a) Constructivism in mathematics and science education, *Educational Researcher*, 23, 7, 4.

Cobb P. (1994b) Where is the mind? Constructivist and sociocultural perspectives on mathematical development. *Educational Researcher*, 23, 7, 13–20.

Collins, H.M. (1990). *Artificial Experts: Social Knowledge and Intelligent Machines*, Cambridge, MA: MIT Press.

Collins, H.M. and Pinch, T. (1992) *The Golem: What Everyone Should Know About Science*, Cambridge: Cambridge University Press.

Driver R., Asoko H., Leach J., Mortimer E. & Scott P. (1994) Constructing scientific knowledge in the classroom. *Educational Researcher*, 23, 7, 5–12.

Durkheim, E. and Mauss, M. (1903) *Primitive Classification*, translated and edited with introduction by Rodney Needham (1963), Chicago: University of Chicago Press.

Ellerton, N. & Clements, M.A. (1991), *Mathematics in Language: A Review of Language Factors in Mathematics Learning*, Deakin University Press, Geelong, Victoria.

Ernest P. (1995) Preface by series editor. In: E. von Glasersfeld, *Radical Constructivism: A way of knowing and learning*. London: Falmer Press, xi–xii.

Forman, P. (1971). Weimar Culture, Causality and Quantum Theory 1918–1927, in R. McCormmach, ed., *Historical Studies in the Physical Sciences*, Philadelphia: University of Philadelphia Press, 1–115.

Fosnot, C.T. (1996), 'Constructivism: A Psychological Theory of Learning'. In C.T. Fosnot (ed.), *Constructivism: Theory, Perspectives, and Practice*, Teachers College Press, New York, pp. 8–33.

Frank, P. (1949). *Modern Science and its History*, New York: George Braziller.

Frankfurt H. G. (2005). *On Bullshit*. Princeton: Princeton University Press.

Fraassen B. van (1980) *The Scientific Image*. Oxford: Oxford University Press.

Gale J. E. (1995) Preface. in L.P. Steffe & J.E. Gale eds., *Constructivism in Education*. Hillsdale, N.J.: Lawrence Erlbaum, xi–xvii.

Gardner, H. (1987). *The Mind's New Science: A History of the Cognitive Revolution*, New York: Basic Books.

Gergen K. J. (1995) Social construction and educational process. In L.P. Steffe & J. Gale eds., *Constructivism in Education*. Hillsdale, N.J.: Lawrence Erlbaum, 17–40.

Giere, R. ed. (1992). Cognitive Models of Science, *Minnesota Studies in the Philosophy of Science, Volume XV,* Minneapolis: University of Minnesota Press.

Gil-Pérez, D. et al., (2002). Defending Constructivism in Science Education, *Science & Education*, 11, 6, 557–571

Gieryn, T.F. (1982). Relativist/Constructivist Programmes in the Sociology of Science: Redundance and Retreat," *Social Studies of Science*, 12, 279–97.

Glasersfeld E. von (1989) Cognition, construction of knowledge, and teaching. *Synthese*, 80: 121–140. available at http://www. vonglasersfeld.com/118

Glasersfeld E. von (1993) Questions and answers about radical constructivism. In K. Tobin ed., *The Practice of Constructivism in Science Education*. Hillsdale, N.J.: Lawrence-Erlbaum, 23–38. Originally published in 1992. Available at http://www.vonglasersfeld.com/151

Glasersfeld E. von (1995a) *Radical Constructivism: A way of knowing and learning*. London: Falmer Press.

Glasersfeld E. von (1995b) a constructivist approach to teaching. in: Steffe L. P. & Gale J. (eds.) Constructivism in education. Lawrence Erlbaum, Hillsdale NJ: 3–16.

Glasersfeld E. von (2001) The radical constructivist view of science. Special issue on "The impact of radical constructivism on science," Part 1, edited by A. Riegler. *Foundations of science*, 6, 1–3, 31–43. Available at http:// www.univie.ac.at/constructivism/pub/fos/ pdf/glasersfeld.pdf

Gorman, M.E. (1992). *Simulating Science: Heuristics, Mental Models and Technoscientific Thinking*, Bloomington: Indiana University Press.

Gough, N.W. (1993). Laboratories in Schools: Material Places, Mythic Spaces, *The Australian Science Teachers' Journal*, 39, 29–33.

Gross, P. and Levitt, N. (1994). *Higher Superstition: The Academic Left and its Quarrels with Science*, Baltimore: Johns Hopkins University Press.

Harding, S. (1986). *The Science Question in Feminism*, Ithaca: Cornell University Press.

Kelly, G.J. (1997), 'Research Traditions in Comparative Context: A Philosophical Challenge to Radical Constructivism', *Science Education* 81(3), 355–375.

Kirschner, P., Sweller, J. & Clark, R.E., (2006). Why Minimally Guided Learning Does Not Work: An Analysis of the Failure of Discovery Learning, Problem-Based Learning, Experiential Learning and Inquiry-Based Learning, *Educational Psychologist,* 41, 2, 75–96.

Kitcher P. (2001) Real realism: The Galilean strategy. *The Philosophical Review*, 110, 2, 151–197.

Kuhn, T.S. (1962). *The Structure of Scientific Revolutions*, Chicago: The University of Chicago Press.

Langley, P., Simon, H.A., Bradshaw G.L. and Zytkow J.M., (1987). *Scientific Discovery: Computational Explorations of the Creative Process*, Cambridge, MA: MIT Press.

Latour, B. (1983). Give Me a Laboratory and I Will Raise the World, in K. Knorr-Cetina and M. Mulkay, eds., *Science Observed: Perspectives on the Social Study of Science*, New York: Sage,

Latour, B. and Woolgar, S. (1979). *Laboratory Life: The Social Construction of Scientific Facts*, London: Sage.

Latour, B. and Woolgar, S. (1986) *Laboratory Life: The Construction of Scientific Facts, 2nd Edition*, Princeton: Princeton University Press.

Laudan, L. (1981). The Pseudo Science of Science, *Philosophy of the Social Sciences*, 11. Reprinted in J. R. Brown, ed., (1984) *Scientific Rationality: The Sociological Turn*, Dordrecht: Reidel.

Laudan L. (1990a) *Science & Relativism*. Chicago: Chicago University Press.

Laudan, L. (1990b). Demystifying Underdetermination, In C. Wade Savage, ed., *Scientific Theories, Minnesota Studies in the Philosophy of Science, Vol. XIV* Minneapolis: University of Minnesota Press.

Lehman, D. (1991). *Signs of the Times: Deconstruction and the Fall of Paul de Man*, New York: Simon & Schuster.

Mayer, R.E. (2004). Should there be a Three-Strikes Rule Against Pure Discovery Learning? The Case for Guided Methods of Instruction, *American Psychologist* 59, 1, 14–19.

Matthews M. R. ed. (1998) *Constructivism in science education: A philosophical examination.* Dordrecht: Kluwer,.

Matthews M. R. (2000). Appraising constructivism in science and mathematical education. In D.C. Phillips ed., *Constructivism in Education: Ninety-Ninth Yearbook of the National Society for the Study of Education, Part 1*. Chicago: University of Chicago Press, 161–192.

Matthews, M.R.: 2012, 'Philosophical and Pedagogical Problems with Constructivism in Science Education', *Tréma* 38, 41–56.

Merton, R.K. (1942). Science and Technology in a Democratic Order, *Journal of Legal and Political Sociology*, 1; reprinted as 'Science and Democratic Social Structure,' in his *Social Theory and Social Structure*. New York: Free Press, 1957.

Merton, R.K. (1957). The Sociology of Knowledge, in his *Social Theory and Social Structure*, New York: Free Press.

Meyer D. L. (2008) The poverty of constructivism. *Educational Philosophy and Theory*, 41, 3, 332–341.

Mills C. W. (1959) *The Sociological Imagination*. Oxford: Oxford University Press.

Munzel F. (2003) Kant on moral education, or "englightenment" and the liberal arts. *The Review of Metaphysics*, 57, 1, 43–73.

Niaz, M., Abd-el-Khalick, F., Benarroch, A., Cardellini, L., Laburú, E., Marín, N., Montes, L.A., Nola, R., Orlik, Y., Scharmann, L.C., Tsai, C.-C. & Tsaparlis, G.: (2003), Constructivism: Defense or a Continual Critical Appraisal – A Response to Gil-Pérez et. al., *Science & Education,* 12, 8, 787–797.

Nola R. (1998) Constructivism in science and science education: a philosophical critique. In M.R. Matthews ed. *Constructivism in Science Education*. Dordrecht: Kluwer, 31–60; originally published in *Science & Education* 6 (1997), 1-2, 55–83.

Nola, R. (2003), Naked Before Reality; Skinless Before the Absolute: A Critique of the Inaccessibility of Reality Argument in Constructivism, *Science & Education*, 12, 2, 131–166.

Null, J.W. (2004), 'Is Constructivism Traditional? Historical and Practical Perspectives on a Popular Advocacy', *The Educational Forum*, 68(2), 180–188.

Olssen, M. (1996), 'Radical Constructivism and Its Failings: Anti-realism and Individualism', *British Journal of Educational Studies* 44, 275–295.

Orwell G. (1946) Politics and the English language. In *The Penguin Essays of George Orwell.* Harmondsworth: Penguin Books, (1984), 348–360.

Papayannakos, D.P., (2008), Philosophical Skepticism not Relativism is *the* Problem with the Strong Programme in Science Studies and with Educational Constructivism, *Science & Education*, 17, 6, 573–611.

Perla, R.J. and Carifio, J. (2009) Toward a General and Unified View of Educational Research and Educational Evaluation, *Journal of Multi Disciplinary Evaluation*, 6, 11, 38–55

Phillips, D. C. (1995) The good, the bad, and the ugly: The many faces of constructivism. *Educational Researcher*, 24, 7, 5–12.

Phillips, D. C. (1997a) How, why, what, when, and where: Perspectives on constructivism in psychology and education. *Issues in Education*, 3, 2, 151–194.

Phillips, D.C. (1997b) Coming to Grips with Radical Social Constructivism, *Science & Education*, 6, 1–2, 85–104. Reprinted in M.R. Matthews ed. (1998) *Constructivism in science education: A philosophical examination.* Dordrecht: Kluwer.

Phillips D. C. ed. (2000) *Constructivism in Education: Ninety-Ninth Yearbook of the National Society for the Study of Education, Part* 1, Chicago: University of Chicago Press.

Piaget J. (1971) *Biology and Knowledge.* Edinburgh: Edinburgh University Press.

Piaget J. (1972a) *Insights and Illusions of Philosophy.* London: Routledge.

Piaget J. (1972b) *Psychology and Epistemology. Towards a Theory of Knowledge.* Harmondsworth: Penguin.

Piaget J. (1999) *The Construction of Reality in the Child.* London: Routledge.

Pickering, A. (1992) *Science as Practice and Culture*, Chicago: University of Chicago Press.

Pinch, T.J. (1993). Generations of SSK, *Social Studies of Science*, 23, 363–73.

Pinch T.J. and Collins, H.M. (1984). Private Science and Public Knowledge: The Committee for the Scientific Investigation of the Paranormal and its Use of the Literature, *Social Studies of Science*, 14, 521–46.

Popper, K.R. (1966). *The Open Society and Its Enemies, Volume 2, Hegel and Marx*, London: Routledge & Kegan Paul.

Putnam H. (1994) Sense, Nonsense, and the Senses: An inquiry into the powers of the human mind. The Dewey Lectures at Columbia University, *The Journal of Philosophy*, 91, 9, 445–517.

Quale, A., (2007). Radical Constructivism and the Sin of Relativism, *Science & Education*, 16, 3–5, 231–266.

Quine W. V. (1960) *Word and Object.* Cambridge, Mass.: MIT Press.

Quine W. V. (1961a) Two Dogmas of Empiricism. In W.V. Quine, *From a Logical Point of View.* Second Edition. Harper, New York: 20–46. Originally published in 1951.

Quine W. V. (1961b) Identity, Ostension, and Hypostasis. In W.V. Quine, *From a Logical Point of View.* Second Edition. New York: Harper, 65–79.

Rorty R. (1979) *Philosophy and the Mirror of Nature.* Princeton: Princeton University Press.

Ruhloff J. (2001) The Problematic Employment of Reason in Philosophy of Bildung and Education, F. Heyting, D. Lenzen & J. Ponsford White eds., *Methods in Philosophy of Education.* London: Routledge, 57–72.

Shapin, S. (1979). Homo Phrenologicus: Anthropological Perspectives on an Historical Problem, in B. Barnes and S. Shapin, eds., *Natural Order: Historical Studies of Scientific Culture*, London: Sage Publications.

Shapin, S. (1982). History of Science and Its Sociological Reconstructions, *History of Science,* 20, 157–211.

Shotter J. (1995) In Dialogue: Social Constructionism and Radical Constructivism. In L.P. Steffe & J. Gale eds. *Constructivism in Education.* Hillsdale, N.J.: Lawrence Erlbaum, 41–56.

Slezak, P. (1989). Scientific Discovery by Computer as Empirical Refutation of the Strong Programme, *Social Studies of Science,* 19, 563–600.

Slezak, P. (1991a). Bloor's Bluff: Behaviourism and the Strong Programme, *International Studies in the Philosophy of Science*, 5, 3, 241–256.

Slezak, P. (1991b). Review of Collins' Artificial Experts, *Social Studies of Science* 21, 175–201.

Slezak P. (1994a) The sociology of science and science education. Part 1. *Science & Education*, 3, 3, 265–294.

Slezak P. (1994b) Laboratory life under the microscope: The sociology of science & science education. Part 2. *Science & Education*, 3, 4, 329–355.

Slezak, P. (1994c). The Social Construction of Social Constructionism, *Inquiry*, 37, 139–57.

Slezak P. (1997) Review of Barnes, Bloor & Henry: scientific knowledge: a sociological analysis. *Metascience – New Series*, 11: 44–52.

Slezak P. (2000) Radical social constructivism. In D.D. Philips ed., *National Society for the Study of Education (NSSE) Yearbook*. Chicago: University of Chicago Press, 283–307.

Slezak P. (2007) is cognitive science relevant to teaching? *Journal of Cognitive Science*, 8, 171–205.

Slezak, P. (2010). Radical constructivism: epistemology, education and dynamite, *Constructivist Foundations*, 6, 1, 102–111.

Small, R. (2003), 'A Fallacy in Constructivist Epistemology', *Journal of Philosophy of Education* 37(3), 483–502.

Sokal A. & Bricmont J. (1997) *Intellectual Impostures*. London: Profile Books.

Steffe L. P. & Gale J. (eds.) (1995) Constructivism in education. Lawrence Erlbaum, New Jersey: Hillsdale.

Stove, D. (1991) *The Plato Cult and Other Philosophical Follies*, Oxford: Basil Blackwell.

Suchting, W.A.: 1992, 'Constructivism Deconstructed', *Science & Education*, 1, 3, 223–254.

Tobin, K.: 1991, Constructivist Perspectives on Research in Science Education, paper presented at the annual meeting of the *National Association for Research in Science Teaching*, Lake Geneva, Wisconsin.

Tobin K. (ed.) (1993) *The Practice of Constructivism in Science Education*. Hillsdale, NJ: Lawrence Erlbaum.

Tobin, K. (2000), 'Constructivism in Science Education: Moving On'. In D.C. Phillips (ed.) *Constructivism in Education'*, National Society for the Study of Education, Chicago, pp. 227–253.

Tobin K. & Tippins D. (1993) Constructivism as a referent for teaching and learning. In K. Tobin ed., *The Practice of Constructivism in Science Education*. Hillsdale, N.J.: Lawrence-Erlbaum, 3–21.

Tweney, R., Doherty, M.E. and Mynatt, C.R. eds., (1981). *On Scientific Thinking*, New York: Columbia University Press.

Woolgar, S. (1988). *Science: The Very Idea*, London: Tavistock Publications.

Zammito, J.H. (2004). *A Nice Derangement of Epistemes*, Chicago: University of Chicago Press.

Peter Slezak is Associate Professor in Philosophy at the University of New South Wales. His B.A. was from University of New South Wales, and his M.Phil. and Ph.D. in Philosophy from Columbia University, New York. Dr. Slezak is Series Editor for Elsevier's "Perspectives on Cognitive Science" and on the editorial board of several leading journals. Dr. Slezak's teaching areas include philosophy of science, philosophy of religion, Galileo, and philosophy of mind and cognitive science. His research publications are in the areas of and philosophy of mind and philosophy of language and cognitive science with special interests in problems of mental representation and the work of René Descartes. He has also published widely on science education and the sociology of science as a combatant in the "Science Wars."

Chapter 32
Postmodernism and Science Education: An Appraisal

Jim Mackenzie, Ron Good, and James Robert Brown

32.1 Introduction

Though not easy to define, postmodernism has elicited strong reaction, both laudatory and critical. Some see it as liberating us from the tyranny of science, while others see it as a new form of insanity. It has kinship with some feminist and social constructivist approaches to science, though the overlap is limited. This three-part article will attempt to outline and evaluate some of the main ideas.

Postmodernism in different forms has had, inevitably, an impact on education and specifically science education. A number of countries have explicitly stated that their national science curricula are based on postmodern understanding of science and human knowledge claims. Constructivism, which has been a most influential force in science and mathematics education, is one manifestation of the educational reach of postmodernism. A great deal of research and curriculum construction in multicultural education is predicated on postmodernist epistemological assumptions.

This chapter will first document and give a sense of the impact of postmodernism in science education curriculum and research, then delineate and give some evaluation of the rise of postmodernist positions in philosophy and philosophy of science, and finally return to appraise some of the chief postmodernist arguments and claims in science education. The separate contributions by Good, Brown and Mackenzie have been interwoven to some extent to minimise overlap in their discussion. It will

---

J. Mackenzie (✉)
Faculty of Education and Social Work, University of Sydney, Sydney, Australia
e-mail: dijimpaddy@gmail.com

R. Good
College of Education, Louisiana State University, Baton Rouge, USA

J.R. Brown
Philosophy Department, University of Toronto, Toronto, Canada

M.R. Matthews (ed.), *International Handbook of Research in History, Philosophy and Science Teaching*, DOI 10.1007/978-94-007-7654-8_32,
© Springer Science+Business Media Dordrecht 2014

be readily appreciated in a complex and controversial topic such as this that the authors will stress different features and not be in full agreement on every detail. Their respective contributions are indicated by initials in the text.

32.2 Part One [R.G.]

32.2.1 Postmodernism in Science Education

In November 1992, the US National Research Council published *National Science Education Standards: A Sampler* and on page A-2 was this statement: 'The National Science Education Standards are based on the postmodern [PM] view of the nature of science'.

For our purposes in this chapter, we will use this statement to mark the beginning of the recognition of PM as an underlying force in the US effort to reform science education. As soon as this statement became known to scientists and others familiar with the history and nature of PM, it was opposed and omitted from later versions of the *Standards*. Among the public statements of opposition, there are two that appeared as editorials in professional journals and each is summarised here. The first, entitled 'The Slippery Slopes of Postmodernism', appeared in the May 1993 issue of *The Journal of Research in Science Teaching* and the following two statements (p. 427) are representative of the overall editorial:

> To question the *objectivity* of observation or the *truth* of scientific knowledge, one does not need to travel to the wispy world of postmodernism. Logical positivism and postmodernism are at the extremes of a long continuum of positions taken by scholars of the nature of science. It is not necessary to carry along the unwanted baggage of either logical positivism or postmodernism to place oneself, as did the authors of *Science for All Americans*, in a more *scientifically* defensible position.
>
> Science education research, like the science education standards being developed by the National Research Council, should be well-grounded on defensible assumptions about the nature of science. The postmodernism of Feyerabend, Foucault, and their followers offers very little insight about the nature of biology and chemistry and physics and so on, that can help in the reform of science education.

Following this editorial, an editorial entitled 'Postmodernism' appeared in the July 9, 1993, issue of *Science*, the journal of the American Association for the Advancement of Science (AAAS). *Science* editor Richard Nicholson begins by questioning whether there is a growing anti-science attitude, providing examples of recent criticisms. He then goes on to ask:

> Are these [criticisms] just isolated events or is something more going on? Harvard's Gerald Holton recently addressed this question from the historian's perspective in a Sigma Xi speech. Holton says "the discussion about science and values has been shifting in remarkable ways" and in this he sees a trend. The trend even has a name: The Postmodern Movement. It is decidedly anti-science. Holton acknowledges that today this movement represents "a minority view." However, he goes on to warn, "but a view held in prominent circles, among persons who can indeed influence the direction of a cultural shift.

An early systematic contribution of postmodernist thinking to education was William Doll's book *A Post-Modern Perspective on Curriculum* (Doll 1993). As a curriculum theorist known for his postmodern perspective, Doll reaches into science and mathematics to the uncertainty theory of Heisenberg, the incompleteness proof of Godel, and especially the chaos theory of Prigogine to craft his vision of a postmodern world. In a separate chapter on 'Prigogine and Chaotic Order', Doll uses a variety of examples, from chaotic pendula to chaotic change in gypsy moth populations to entropy interpretations that supposedly show how evolution can achieve 'perfectibility and perfection of humankind' (p. 100) in his attempt to envision a new postmodern world. According to Doll the paradigm he sees emerging from the insights of chaos theory 'requires of us nothing less than a brand new start in the description of nature—a start which will affect our metaphysics as well as our physics, our cosmology as well as our logic' (pp. 90–91).

In his 1993 book *Science and Anti-Science*, Holton discusses the anti-science phenomenon in detail in the last chapter and some of that discussion is used to focus attention on what he calls the single most malignant part: 'the type of pseudo-scientific nonsense that manages to pass itself off as an *alternative science, and does so in the service of political ambition*' (p. 147). He goes on to describe the relatively poor level of scientific literacy in the USA and warns of the dangers to democracy of a poorly informed public, especially in the sciences:

> Today there exist a number of different groups which from their various perspectives oppose what they conceive of as the hegemony of science-as-done-today in our culture. These groups do not form a coherent movement, and indeed have little interest in one another; some focus on the epistemological claims of science, others on its effects via technology, others still long for a romanticized pre-modern version of science. But what they do have in common is that each, in its own way, advocates nothing less than the end of science as we know it. That is what makes these disparate assemblages operationally members of a loose consortium. (p. 153)

In science education, Holton's postmodernist *loose consortium* consists of radical constructivism, queerism, variants of multiculturalism, some versions of feminism and more recently 'cultural studies of science education' (CSSE). They all share a family resemblance of postmodern ontology, epistemology, psychology and social theory. The following four claims are representative of the convictions of this consortium and indicate what is being contested in the 'science education wars'.

Claim One:

> We have to learn how to de-privilege science in education and to free our children from the *regime of truth* that prevents them from learning to apply the current cornucopia of simultaneous but different forms of human knowledge with the aim to solve the problems they encounter today and tomorrow (Van Eijck and Roth 2007, p. 944).

This first statement was published in *Science Education,* a well-established, highly regarded professional journal.

Claim Two:

> In the field of science education, the current views of scientific language and scientific literacy are based on an epistemology that begins the [sic] presupposition of the identity of a thing with itself – both the phenomenon of representation (inscription) and the figure of

the scientist as rational thinker and actor are premised on this identity. However, recent philosophical scholarship generally and the French philosophers of difference particularly – including Gilles Deleuze, Jacques Derrida, Didier Franck, Jean-Luc Nancy, and Paul Ricœur – take a very different perspective on the question of language. This perspective emphasizes the opposite, that is, the non-self-identity of a thing or person with it or him/herself, which is a conception more compatible with our experiential reality. This perspective also allows us to better theorize the learning of science and scientific literacy in the indeterminate manner in which it is actually experienced and observed (Roth et al 2008, p. 153).

Claim Three:

> Even more hidden and therefore more difficult to recover is the epistemological ground that presupposes equality (e.g., of gender) and sameness (identity of, for example, A and A in the equation $A=A$) rather than recognizing the inherent plural singularity of each human being. ...
>
> If, on the other hand, we begin with the ontological assumption of difference that exists in and for itself, that is, with the recognition that $A \downarrow A$ (e.g., because different ink drops attached to different paper particles at a different moment in time), then all sameness and identity is the result of work that not only sets two things, concepts, or processes equal but also deletes the inherent and unavoidable differences that do in fact exist. This assumption is an insidious part of the phallogocentric epistemology undergirding science as the method of decomposing unitary systems into sets of variables, which never can be more than external, one-sided expressions of a superordinate unit (Roth and Tobin 2007, pp. 99–100).

The second and third statements were published in a book series *New Directions in Mathematics and Science Education.*

Inasmuch as the third claim can be understood, it lays out what is at stake if postmodernism becomes further established in the science education community. The very achievements of Galileo and Newton depend on studying wholes (falling bodies and planetary motion) by dissolving them into parts (horizontal and vertical motions on an inclined plane and the moon-earth-sun orbits as a 3-body that neglected the effects of other planets, not a multi-body problem as it actually was). Without abstraction and idealisation, science goes nowhere, likewise of course for social science. The latter advances and discovers some things by dividing populations into classes and ascertaining average class weights, health, longevity, etc. and then looking for causal factors in a controlled manner. That no one embodies the 'average weight' does not mean that the construct is mythical and of no use in promotion of public health. In most scientific investigations, the whole cannot be grasped in toto, only in parts. The apple remains red, juicy and attractive even if physicists studying its rate of fall ignore all of this and even if economists likewise abstract from these real features and study the apple's exchange value in a given economic context. In none of this is the reality of the whole apple denied.[1]

This should be a basic lesson learned in science classes, not the reverse lesson that the above authors want students to learn, namely, that such method is 'insidious' and 'phallogocentric'. A historically and philosophically literate science teacher can assist students to grasp just how science captures, and does not capture, the real, subjective, lived world—the 'life world' as it is called by phenomenologists.

---

[1] On this important topic, see at least Harré (1989), McMullin (1985) and Nowak (1980).

An HPS-illiterate teacher, or science educator, leaves students with the unhappy choice between disowning their own world as a fantasy and rejecting the world of science as a fantasy, with, sadly, many doing the latter. Aldous Huxley, at the end of World War Two, commented on this matter saying:

> The scientific picture of the world is inadequate, for the simple reason that science does not even profess to deal with experience as a whole, but only with certain aspects of it in certain contexts. All of this is quite clearly understood by the more philosophically minded men of science (Huxley 1947, p. 28).

Claim Four:

> This centripetal tendency of science—a hegemonic, homogenizing force—is well described in the 'pasteurization' of France: Louis Pasteur's science works only when stables in the countryside are made to resemble laboratories, where re-presentations of nature come to be the same irrespective of time and space (Latour 1984); these re-presentations constitute, in Bakhtinian terms, a particular form of chronotope (time-space) (van Eijck and Roth 2011, p. 825).

> Novelization therefore models a process of cultural change toward democracy. In fighting 'for the renovation of an antiquated … language,' novelization is in the interest of those who are located 'outside the centralizing and unifying influence of … the ideological norm established by the dominant literary language.' Novelization constitutes a process of continuous 'linguistic stratification and differentiation' (Bakhtin 1981, p. 67) toward heteroglossia, a process that others much later have referred to as the 'multiplication of meaning' that comes about as the same and different means of expression are produced and stratified (Lat. *Stratum*, spread, layer) one on top of the other (e.g., Lemke 1998) (ibid, pp. 831–2).

> We propose abandoning the dominant notion of science curricula as inculcating the canonical scientific discourse of yesteryear, since this notion comes with its maintenance of a unitary language and hence cultural centralization that does not allow for valuing and keeping cultural diversity in science education (ibid, 840).

This article—'Cultural Diversity in Science Education Through "Novelization": Against the Epicization of Science and Cultural Centralization'—was published in the *Journal of Research in Science Teaching* which is considered by many to be the top professional journal of research devoted to science education and is the official journal of the world's largest science education research organisation (NARST).

After repeatedly warning that traditional science and science education are 'epics' that must be changed through 'novelisation' to ensure the future of multiculturalism and democracy, van Eijck and Roth warn against becoming too successful because that will 'institute the epicisation of novelisation'. In their words:

> We conclude this text with a word of caution. If novelization were to be the name of a form of science education that we aim at and eventually achieve, a new canonical form, then we would have done nothing other than institute the epicization of novelization (p. 843).

It is not difficult to see why one might dismiss this 'epic' project, and more generally, the CSSE project represented in the foregoing four claims, with words like 'a stupid and discreditable business', as Stove (1991) did when summarising the SSK project; 'Fashionable nonsense', the phrase used by Sokal and Bricmont in the title of their 1998 book that describes the postmodern agenda in SSK; or simply as pieces of execrable and inexcusably bad writing designed not to communicate but to obfuscate.

The postmodern movement in science education created its academic home in 2006 with publication of the journal *Cultural Studies of Science Education* with Kenneth Tobin and Wolff-Michael Roth as founding editors. Their aspirations for the journal were expressed as:

> In many ways, this new journal departs from the trodden paths in our discipline. CSSE is unique in focusing on the publication of scholarly articles that employ social and cultural perspectives as foundations for research and other scholarly activities in science education and studies of science. The journal encourages empirical and non-empirical research that explores science and science education as forms of culture, enacted in a variety of fields that are formally and informally constituted. ....We anticipate that the forms of dissemination will make visible the non-linearity of doing research and the recursive nature of delineating problems, deploying theoretical frameworks, constructing data, and adopting dynamic approaches to methodology, design, analysis, interpretation, and writing (Tobin and Roth 2006, p. 1).

What 'dynamic approaches' to methodology, design, analysis and interpretation might mean is left unstated, but the passage itself exhibits what a postmodern 'dynamical' approach to writing might be and why it should be avoided. George Orwell long ago warned of what happens in societies and cultures when such obfuscatory writing goes unchecked and becomes normalised (Orwell 1945).

It is difficult to measure the impact of the many CSSE books and articles in professional journals on the actual science curriculum and instruction in our schools; but the professional recognition given to its proponents is significant (see Chap. 39). The wider cultural and philosophical background that gave rise to postmodern enthusiasm in science education will be outlined in the following sections.

32.3 Part Two [J.M.]

32.3.1 What Is Postmodernism?

The term 'postmodern' was imported into discussions of knowledge (from architecture) by Jean-François Lyotard in *The Post Modern Condition* (1979, tr. 1984) and signified, if nothing else, a suspicion of, and scepticism about, grand narratives or universal claims. In practice, this extended to an opposition to rationality, a rejection of the notion of objective truth and an enthusiastic endorsement of localism (see Brown 2001, p. 76). The opposition to rationality took the form of supposing that scientists' behaviour served only their own social interests and discounting any suggestion that it was related to scientific evidence. In taking this position, the investigator is claiming to know what the scientists' interests are and attributing to those interests efficacy as causes. This is exactly the kind of knowledge of the topic under investigation that they denied scientists could have.[2] The rejection of objective truth

---

[2] Harry Collins and Steven Yearley argued: 'Natural scientists, working at the bench, should be naïve realists—that is what will get the work done. Sociologists, historians, scientists away from the bench, and the rest of the general public should be social realists. Social realists must experience

recalled the grand tradition of scepticism and its traditional tropes, which goes back to Sextus Empiricus in around the year 200 CE.

> Those who claim for themselves to judge the truth are bound to possess a criterion of truth. This criterion, then, either is without a judge's approval or has been approved. But if it is without approval, whence comes it that it is truth worthy? For no matter of dispute is to be trusted without judging. And, if it has been approved, that which approves it, in turn, either has been approved or has not been approved, and so on *ad infinitum* (Sextus Empiricus, *Adv. Log.* I 340 = 1935, p. 179).

But Lyotard and others formulation was more than usually conceptually muddled. The localism was adopted as a measure to combat totalising theory which was blamed for the wars and violence of the nineteenth and twentieth centuries. 'We have paid a high price for the nostalgia of the whole and the one …', said Lyotard (1983, p. 46), characteristically exaggerating just how consequential the role of those like himself had been. How academic papers in the humanities and social sciences had made any significant contribution to promoting wars and violence was left unexplained.

32.3.2   *Thomas Kuhn and the Origins of Social Constructivism*

Science is a social construction. It is an account of the world that has been, and is still being, put together by people. Many would regard this as so obvious as to be hardly worth saying. Nevertheless, over the past thirty or forty years, it has become the slogan of some science educators. They clearly think that it is somehow an important thing to say, one which is contrary to accepted ideas. Keith Tobin proclaimed: 'In 1997 I took a bold step in pronouncing that learning involved cultural production' (2010, p. 23). (It is difficult to think of any pronouncement which would have been less bold.) From their slogan the science educators draw a number of substantive conclusions about science education, such as an emphasis on the social aspects of how children learn science, and advise teachers also to concentrate on these aspects of what they do in the classroom. We need to ask how an idea which would hitherto have been regarded as banal could have come to seem so controversial and significant.

The stage for the conventional landscape in philosophy of science in the twenty-first century was set by a conference on July 13, 1965, at the then Bedford College of the University of London organised by Imre Lakatos. This conference was intended to oppose two thinkers. One was Karl Popper, an Austrian educated philosopher of science, whose seminal book *The Logic of Scientific Revolutions*, though published in German in 1934, had only been available in English since 1959. Popper was more widely known in the English-speaking world for his works on political

---

the social world in a naïve way, as the day to day foundation of reality (as natural scientists naively experience the natural world). That is the way to understand the relationship between science and the rest of our cultural activities' (Collins and Yearley 1992, p. 308).

philosophy, *The Poverty of Historicism*, published in three parts in the journal *Economica* in 1944–1945 and in book form only in 1957, and *The Open Society and its Enemies*, first edition 1945. The other was Thomas Kuhn, a Harvard historian of science, whose book *The Structure of Scientific Revolutions* was published in 1962. In organising the debate, Lakatos may have hoped that the debate would cast Popper and Kuhn as the opposing poles, allowing his own position to come through the middle as a less extreme compromise, embodying the strengths of each side. Also present was Paul K. Feyerabend, a former student of Popper, a former colleague of Kuhn at the University of California Berkeley and commentator on drafts of Kuhn's book and a correspondent of Lakatos himself.

But what were the two sides? In conventional accounts of philosophy of science today, they are generally represented with Popper as standing for a narrowly conceived prescriptive view of science, modelled on physics, and as rejecting almost everything apart from physics as insufficiently scientific to merit the name and with Kuhn as a radical, open-minded thinker, firmly based on a sound understanding of the history of the sciences and accepting a wide range of disciplines. That, after all, is how things would surely develop, from narrow to broad, from rigid to flexible and from prescriptive to empirical, and that is how those who compose textbooks most easily organise things. The temptation was perhaps exacerbated by the ages of the principals and the climate of the times: Popper was on the verge of retirement and Kuhn was some 20 years younger, and the social and intellectual turmoil of the 60s was beginning.

Nevertheless, this conventional picture has things almost exactly the wrong way round. Popper had indeed studied physics and used it as an example, and he did draw careful lines of demarcation among theories between those he would accept as scientific and those which were not, framed in terms of falsification by observation. He did not, unlike the positivists of the Vienna Circle, reject what was not scientific as meaningless nonsense: mathematics, history and ethics were all important and reputable disciplines, even though they did not count as science by his criterion. His criterion did, however, exclude from among the sciences not only such soft targets as astrology but the psychoanalysis of Freud and the dialectical materialism of the followers of Marx (Marx's own version may have been scientific, but if so it had already been falsified). Popper rejected traditional ideas of science having to have secure foundations. He was very aware of the need for a social structure to enable scientists to compare, test and above all criticise ideas—science for Popper could not be a merely personal or subjective activity.

Kuhn had begun his academic career as a physicist, and one of the motives behind his work was a desire to explain, or at least to characterise, the difference separating the natural sciences like physics, where practitioners largely agree about what is good work and what is not, from history and other areas in the social sciences, where the very criteria for evaluation are contested. Kuhn proposed as the distinguishing mark of a science that a scientific community shared what he called a *paradigm*, by which he meant an exemplar that serves as a model for future research. To be more specific would be risky: Margaret Masterman (1970) enumerated more than twenty different meanings of the term 'paradigm' in Kuhn's 1962 book. In non-sciences such as history or the study of society, there was no paradigm, and therefore no agreement about what constituted acceptable research. In a science governed by a paradigm, research was merely the working out of puzzles within the universe of the paradigm.

Like the positivists of the Vienna Circle, Kuhn required that science have a solid foundation, but whereas they had sought a foundation in eternally true principles of logic, for Kuhn the foundation lay in the historically situated social practices of interpretation and understanding of a professional community. Since these were not eternal, they were subject to change. For Kuhn, this occurred when 'normal science' broke down, increasing numbers of puzzles resisted solution, anomalies multiplied and the community began to be eroded by feelings of anxiety and insecurity. Half-formed ideas about how to proceed would be produced and developed and compared with one another. For a Popperian, this was how a science should be at any time; for Kuhn, it was a suspension of science, a reversion to pre-scientific confusion and a crisis. Popper would hope that the crisis might be resolved, if at all, by members of the scientific community becoming rationally convinced by evidence and arguments that at least some of the competing views should be abandoned and perhaps that all but one should be. Kuhn (1962, p. 150) cited the reminiscences of Max Planck, who in his *Autobiography* (1950, pp. 33–4) had said that a new scientific theory succeeds not by convincing its opponents but by the opponents dying and a new generation familiar with the theory growing up. Though he immediately (p. 151) insisted that scientists were sometimes persuaded by arguments and that on occasion a scientific community would change its mind in advance of biological succession of the next generation, Kuhn's underlying account of the adoption of a new paradigm is by comparison to the phenomenon of religious conversion.

Initially Kuhn's work was faced with severe criticisms, ranging from positivists and Popperians to more straightforward historians of science, let alone those worried by its ambiguities and inconsistencies. And of course Kuhn has retracted and qualified many of his claims and has regretted writing 'the purple passages' (Kuhn 1991/2000; 1993). But over the last twenty years, a new generation has grown up in the relevant disciplines who take Kuhn unproblematically as the paradigm for the study of science, and who are quite unaware of the original, and often still unanswered, criticisms. The uncritical embrace of Kuhn is especially apparent in the science education community (Matthews 2004).

32.3.3   *The Possibility of Objective Truth*

A major rhetorical weapon of sceptical positions is to sow doubt about the concept of objective truth. Objectivity presupposes the ability to distinguish the significance a remark has from the perspective of the person to whom it is attributed from its significance from the perspective of the one doing the attributing. As Donald Davidson pointed out, this becomes especially clear when a thought is attributed to a non-verbal creature. 'The dog, we say, knows that its master is home. But does it know that Mr. Smith (who is his master), or that the president of the bank (who is that same master), is at home?' (1984, p. 163). Brandom elucidates, 'The dog knows *of* the president of the bank that he is home, he just does not know *that* the president is home. ... [O]ne wants to appeal to the belief that his master is home to explain why the dog is so happy, and to its being a belief *of* the president of the bank (whether the dog knows this or not) in order to explain why one result of the dog's happiness is that he slobbers on the president of the bank' (1994, p. 710, n. 95). The same claim is specified differently depending on whether we consider its inferential antecedents and consequences in the context of what is admitted by the attributer or by the one to whom the claim is attributed.

It is this notion of objective truth conditions that makes explicit the possibility of mistaken belief, and so of the difference between what is merely held to be true (believed) and what is correctly held to be true. But objectivity is undermined if the objective correctness of a claim is taken to be what is endorsed by a privileged point of view, such as that of 'we', or of the community as a whole. That privileging would leave no possibility for the chosen point of view to be itself mistaken. For objectivity to be possible, no point of view can be globally privileged. Objectivity consists in a perspectival form, rather than any possibility of a non-perspectival content. 'What is shared by all discursive perspectives is *that* there is a difference between what is objectively correct in the way of concept application and what is merely taken to be so, not *what* it is—the structure, not the content' (Brandom 1994, p. 600).

This structure is symmetrical. Person A distinguishes between what is to be treated as specifying the objective content and what A regards as specifying the attributee B's subjective attitude. B does so too but the other way round. This symmetry is what prevents any one perspective from being privileged over all others. 'Sorting out who should be counted as correct, whose claims and applications of concepts should be treated as authoritative, is a messy retail business of assessing the comparative authority of competing evidential and inferential claims' (p. 601). A lack of understanding of this perspectival structure leads students of scientists' behaviour to focus only on what is agreed between them and to neglect of what it is that the scientists are agreeing or failing to agree about: to fail to take account of the *objects* of their discussion, its *object-ivity*.

> From the perspective of our students we might ask, 'Why should there be so much investment in teaching and learning science?' Maybe then we will be able to address conceptual change more adequately and for clearer and more significant purposes (Reis 2010, p. 239).

Our investment in science education may be opaque from the perspective of a schoolchild, but it is hardly from that of anybody else. Scientific knowledge is spread through our economy, our health system and our agriculture. Without a solid core of scientifically educated technicians, we could not keep our populations safe from diseases, adequately fed and gainfully employed.

32.3.4 Western and Indigenous Science

Nowadays we routinely hear about 'Western science', whereby 'West' is meant Western Europe and those regions elsewhere predominantly settled by descendants of people from Western Europe. The notion of science as being a cultural product of these people rather than others may have had some validity in the nineteenth century, and indeed even into the 1940s.[3] Science as known today can be, and in Western countries usually is, traced back to ideas largely formulated in Western Europe in the seventeenth century (though see Needham et al. 1954–2004), but it has long ago outgrown that locale and many of the ideas of that time have been superseded or incorporated into subsequent developments. To refer to today's science as Western is to overemphasise its origins and mislead in somewhat the same way as one would by describing modern Christianity as a Middle Eastern religion or the potato as a Peruvian vegetable. The terminology 'Western Science' substantially misrepresents the social context in which science has been done over the last 60 years or so.

The need for this term is of course to draw a contrast: One talks about 'Western' science to contrast it with the sciences of indigenous cultures. In this vein, two advocates of indigenous science have recently written:

> In most countries of the world, a culturally specific (Western) form of science has masqueraded as universal, true and irrefutable. With the introduction of the first national Australian curriculum, Western science and its epistemological base have been challenged by formal expectations that Australian Aboriginal and Torres Strait Islander knowledges be included in formal school science programmes (Baynes and Austin 2012, p. 60).[4]

There is a familiar and pointless debate as to whether indigenous cultures can be said to have sciences. Manifestly every human community has some knowledge of the world in which it lives and how things in that world interact—which plants are poisonous, for example—though that knowledge may not be seen by that community as forming a unified system but as being parts of the lore of hunters, of healers and of midwives. So in that sense every society has its own science. Manifestly no pre-industrial society has had the sort of organised knowledge-sector on which economies like Japan, China, India, Brazil and Russia (as well as Europe and North America) rely in the twenty-first century. In that sense no early society had science. But no cultures, including indigenous ones, are static, and people raised in them often adopt ideas and practices from elsewhere. Science has of course adopted and adapted much knowledge from pre-industrial cultures, and continues to do so. Many people with an indigenous background have become scientists.

---

[3] The Nobel Prize for Physics was first awarded outside the 'West' so defined in 1930, to ChandrasekharaVenkata Raman of Calcutta University for his discovery that when light traverses a transparent material, some of the light that is deflected changes in wavelength, a phenomenon now called *Raman scattering*.

[4] Baynes and Austin's use of the word *irrefutable* suggests that their understanding of Karl Popper's contributions to philosophy of science is not very deep. Popper famously maintained that the very defining feature of scientific claims is their refutability (see following section). Such disregard if not ignorance of important philosophical and historical matters is characteristic of the PM 'loose consortium' in science education.

Surprisingly, the dogma that science is a cultural product of, and therefore confined to, Western societies is not opened to empirical test. When an attempt is made to do so, the inconvenient conclusion is often that Western and indigenous sciences agree to an extraordinary extent. Ernst Mayr wrote:

> Forty years ago, I lived all alone with a tribe of Papuans in the mountains of New Guinea. These superb woodsmen had 136 names for the 137 species of birds I distinguished (confusing only two nondescript species of warblers). That ... Stone Age man recognises the same entities of nature as Western university-trained scientists refutes rather decisively the claim that species are nothing but a product of the human imagination (Mayr 1963, p. 17, quoted Gould 1980, p. 173).

Subsequent investigations have provided further examples of indigenous taxonomies matching those of Western science.[5] One taxonomy which does diverge from the scientific is the taxonomy embodied in vernacular English. For example, the western class 'panda' is confused: cladistically, giant pandas are bears, but red pandas are a separate family more closely related to racoons (Flynn et al. 2005, p. 325a; O'Brien et al. 1985; O'Brien 1987). Charles Sibley and John Ahlquist showed that Australian birds evolved from a crow-like ancestor and that their similarity to various European birds whose names they were given (e.g., warblers and robins) is a matter of evolutionary convergence. Jack Pettigrew has argued (1986) that flying foxes are more closely related to primates than to the microbats. If compared to modern cladistics, the English language with its pandas, robins and bats might fare much less well than many indigenous taxonomies.

A very obvious fact about the sociology of different cultures is that some cultural products have wide appeal, and others remain confined to their original homes. The dramas of Shakespeare, for example, seem to work well in other languages and other dramatic traditions, notably those of speakers of German and Russian, whereas the dramas of Racine have had much less success outside their native French habitat. The music of societies from Africa, including those of Africans transported to the Western hemisphere, has wide appeal those of China and Japan seem not to. The students of the social aspects of science, however, have not investigated this aspect of science's influence.

32.4 Part Three [J.R.B.]

32.4.1 Postmodernism and Philosophy of Science

Let's begin with something like *the standard view of science*, which we can roughly express like this: *There is a way things are and scientists try to figure it out; they*

---

[5] See, for example, Berlin et al. (1974), Boster and D'Andrade (1989), Diamond (1966), Hunn (1976), Majnep and Bulmer (1977). Though contrast, for example, Björnsen Gurung (2003).

32 Postmodernism and Science Education: An Appraisal

have a variety of (fallible) techniques for doing so and thus far have been quite successful. If pressed for details, we might include the following[6]:

1. There is a world in which there are objects, processes and properties which are independent of us and our beliefs about them. Any statement we make about them is true or is false (or at least approximately so). Of course, we may never know which.
2. The *aim* of science is to give true descriptions of reality. Science can have other aims as well (usually associated with technology), but truth is the chief aim of pure science.
3. We have a variety of tools and techniques (observation, logic, statistical inference, etc.) for learning how things are. These methods have developed from earlier methods and very likely will themselves be developed further.
4. Such methods are fallible; they may lead us astray. Nevertheless, science has made remarkable progress so far. It is reasonable to continue to use these methods in the belief that they are the most reliable source of information about nature.
5. There are no alternatives to this. For instance, the Bible does not give us reliable information about human origins; astrology and precognition do not give us reliable information about our futures and so on.
6. The progress of science is tied to these principles. Social factors can and do influence science, but the main course of its development is based on the recognised methods of evaluation.

Postmoderns and social constructivists generally would consider these points delusional. And yet, this cluster of views is what we all more or less start out with and is what most working scientists believe (though the fifth point might be controversial for some). It is, in short, common sense realism. But, as we know only too well, common sense is sometimes wrong. It is seriously challenged by a number of people active in science studies. Even some who would reject any form of postmodernism will reject parts of the standard picture as sketched here. For instance, various antirealists (including instrumentalists, verificationists and pragmatists) would all reject the idea that science aims at truth that exists independently from us. An instrumentalist such as Duhem claims that a scientific theory aims at 'saving the phenomena', that is, getting all the observational claims right but is indifferent to the truth of the theory itself. We might not be able to tell whether the earth rotates or is stationary, while everything else goes around it. What matters, instrumentalists claim, is that we correctly predict the angle at which we see Mars at any specified time. When two theories make the same predictions, we choose to adopt one of them on the basis of convenience—truth (which is inaccessible) has nothing to do with it.

Karl Popper, famous for asserting falsifiability as the defining criterion of scientific theory, would also be critical of aspects of the standard picture but for a different reason. He thought the aim of science is indeed truth, but he didn't think we could have good inductive evidence for the truth of any theory. Instead, the method of science should be conjectures and refutations. We make a guess, then we try to

---

[6]This section draws heavily on Brown (2001).

find counterexamples. When we refute our theory, we then make a new conjecture and so the process goes.

Though Duhem and Popper challenge some aspects of the standard picture of science, they do not quarrel with those features that are most central, namely, the idea that reason and observation play a dominant role in theory evaluation. The postmodern challenge is really quite different. The very idea of scientific reason and objectivity is at issue. Consequently, when we talk about the standard picture of science, we will include Duhem and Popper and almost every other major philosopher of science as embracing that picture. Of course, they differ significantly in detail, but they all hold that reason and observation are at least in principle objective and play a dominant role in science. When we talk about the standard view of science, we mean to include most prominent philosophers of science as upholding some version of it. This would include Whewell, Mill, Mach, Poincaré, Pierce, Duhem, Russell, Carnap, Neurath, Popper, Quine, Lakatos, Putnam, van Fraassen and a great many others. Kuhn and Feyerabend might also be included but are somewhat problematic. Interestingly, they are often seen as postmoderns.

With this outline of the standard view of science in mind, we can better understand the challenge posed by postmoderns and other social constructivists. Let's start by asking: Who's involved? Why should we care? What are the main battle lines?

32.4.2 Antecedents to Postmodernism

In some ways the fight is quite old. The much-cited second-century CE views of Sextus Empiricus and his rejection of the possibility of objectivity have been mentioned above. Two and a half thousand years ago, Protagoras championed a kind of relativism when he said 'Man is the measure of all things'. Plato took up the challenge and fought for objective knowledge. The Enlightenment with its emphasis on progress through rationality was no sooner established in the eighteenth century, then early in the nineteenth it faced the Romantic rebellion which stressed feeling over intellect and emotion over rational inference. Much debate in this century has been stimulated by Karl Marx, though sometimes his writings pull in opposite directions. Marx sounds distinctly like a social constructivist when he famously declared: 'The mode of production of material life conditions the general process of social political and intellectual life. It is not the consciousness of men that determines their existence, but their social existence that determines their consciousness' (1859, 20f.). Yet Marx also thought that objective knowledge is possible; the constructive sentiment gives way to a sensible though subtle form of realism:

> With the change of economic foundation the entire immense superstructure is more or less rapidly transformed. In considering such transformations the distinction should always be made between the material transformations of the economic conditions or production which can be determined with the precision of natural science, and the legal, political, religious, aesthetic or philosophic — in short, ideological — forms in which men become conscious of this conflict and fight it out (*ibid.*, 21).

32.4.3 The Strong Programme in Sociology of Knowledge

Current social constructivism has plenty of antecedents, but it is also reasonable to think of it as mainly a product of the recent past (see Chap. 31). In the mid-1970s David Bloor (in Edinburgh) announced the *strong programme* in the sociology of knowledge. Why *strong*? It's in opposition to *weak* sociology of science, any account which focuses on institutions and various other social features of science but takes for granted that the *content* of science has nothing to do with sociology. By contrast, Bloor asserts that the very content of scientific theories is also to be understood in terms of social factors.

This is a point that must be stressed, since a great deal of sociology of science is quite compatible with the epistemology of scientific orthodoxy while at the same time is potentially embarrassing to the orthodox. So-called weak sociology, for example, can ask: Why are there so few women physicists?, Why do they feel they must sacrifice career or children, and can't (unlike their male colleagues) have both? However, weak sociology of science does not ask questions such as: Why do women believe that the trajectory of a cannon ball is a parabola? The answer to such a question is 'the evidence' and it has nothing to do with their sex, nor with any other sociological factor. Bloor's *strong programme* will have none of this hands-off attitude. He, too, will ask the background questions, of course. But as likely as not, he will relate those factors to the very content of the theory at hand. More on his views below.

There are many others who are like-minded. Shortly after Bloor started to make his mark in science studies, Bruno Latour (a French philosopher and anthropologist) adopted the role of an 'anthropologist in the lab'. With Steve Woolgar he wrote up his experiences of an exotic tribe—a team of California biochemists—explaining their behaviour in social, political and economic terms. Meanwhile in France, Michel Foucault was claiming that *knowledge = power*, not in the sense that by having knowledge one has power (a sense made famous by Bacon), but in the very different sense that having political power allows one to say what knowledge is and is not.

32.4.4 The Sociology of Scientific Knowledge (SSK)

David Bloor's now classic book, *Knowledge and Social Imagery* (Bloor 1976/1991), is perhaps the single most important and influential work in the current social constructivist literature. It contains the manifesto of the *Edinburgh School* known as 'the strong programme'. What (to repeat my earlier question) might the *weak* programme be? To elaborate on my former answer, before Bloor and his like-minded colleagues got to work, *traditional* sociology of scientific knowledge focussed on various issues surrounding science, such as institutions (Who funds them? Why did this one flourish and that one collapse?), scientists (What social class do they come from? Why are there so few women?), relations to governments and corporations

(What impact did the cold war have on science funding? How is the biotech industry influencing research?) and choices of research topics (Why did Galileo take an interest in projectile motion?). But traditional sociology of science would *not* try to account for the *content* of any scientific theory. This, according to Bloor, is what makes the traditional approach 'weak'.

Robert Merton and his school is Bloor's target. Merton's sociology of science does not challenge, but rather complements traditional history and philosophy. Merton, for example, would be happy to account for the growth of science in seventeenth-century England by linking it to Puritanism, as he did in his famous study (1970). But not for a moment would he think it appropriate to give a sociological explanation of why Newton's theory of universal gravitation was widely accepted. Merton formulated a rule of thumb that has come to be known as the *A-rationality Principle*: If a rational explanation for a scientific belief is available, that explanation should be accepted; we should only turn to non-rational, sociological or psychological explanations when rational accounts are unavailable.[7] This is part of the *weak* approach that Bloor explicitly opposes. He insists upon a uniform strategy in dealing with science, one that is utterly thoroughgoing and which penetrates into the very content of scientific theories—in short, he wants a *strong* programme.

Bloor's motivation is his naturalism and his attachment to science. The idea of naturalism is also popular among philosophers, especially philosophers of science, who hold a variety of versions. The general principle is this: The natural world is all there is; there are no special methods of investigating things except the fallible methods of empirical science; norms (whether they be the norms of morality or the norms of scientific method) must be explained away or reduced to the concepts and categories of ordinary science; they must be understood in terms of the natural world.

Naturalism has great appeal, and many would cheer him on, if Bloor said: we want to know about the atom? Study it scientifically! Want to know about disease? Study it scientifically! Want to know about religion? Study it scientifically! Want to know about human society? Study it scientifically! We seem to be tripping right along, and now that we're on a roll, why hesitate? It's hard to resist continuing in the same way: Want to know about science? Study it scientifically! That's what Bloor urges, and it's difficult to object. But what's involved in a scientific study of science itself? Bloor's answer is the four tenets of the strong programme. If you want to adhere to a scientific understanding of science, Bloor claims, then these are the main principles with which your account must comply (See Bloor 1991, p. 7).

Causality: A proper account of science would be causal, that is, concerned with the conditions that bring about belief or states of knowledge.

Impartiality: It would be impartial with respect to truth and falsity, rationality or irrationality and success or failure. Both sides of these dichotomies will require explanation.

---

[7] (Merton (1968, p. 516); the principle is also embraced and discussed at length by Laudan (1977, p. 202).

32 Postmodernism and Science Education: An Appraisal

Symmetry: It would be symmetrical in its style of explanation. The same types of cause would explain, say, true and false, [rational and irrational] beliefs.

Reflexivity: It would be reflexive. In principle its patterns of explanation would have to be applicable to sociology itself. Like the requirement of symmetry, this is a response to the need to seek for general explanations. It is an obvious requirement of principle; otherwise sociology would be a standing refutation of its own theories.

Two of these principles seem to be perfectly correct—impartiality and reflexivity. The other two either need serious qualification or are simply wrong. Bloor does not say so, but he seems to assume that reason and evidence are not the sort of things that could be a cause. If we take evidence to be a cause, then there is no objection to the first principle. Let's see what's right about impartiality and reflexivity, which we can do without having to reinterpret them.

Since we're in the business of explaining belief, we're interested in all beliefs, not just the true or rational ones, and not just the false or crazy ones. Optical illusions, for example, are an engaging curiosity and it's nice to have explanations for them. But ordinary veridical perception is also worthy of our intellectual interest. Bloor is not alone in saying this, but he does think the point is underappreciated. The reigning story of how I manage to correctly see a cup on the desk in front of me is a wonderful achievement of physics and physiology research. It involves photons coming from the cup and entering my eye, a signal is sent down the visual pathway into the cortex and so on. Events such as these play a role in explaining how I come to believe that there is a cup on the desk. Whether my perception is veridical or illusory, it needs explaining. The true and the false are in this respect on a par. This is Bloor's impartiality principle. And he's perfectly right to espouse it.

Would anyone think explaining both the true/rational and the false/irrational wasn't the proper thing to do? The impartiality principle hardly seems necessary, yet the A-rationality principle (mentioned above) might be thought to be in conflict. That principle called on giving sociological explanations for beliefs *only* when no rational explanation was available. Actually, there is no conflict between the two principles. Every belief requires explanation, but some will get one type of explanation (say, in terms of social factors), while others will get a different type account (say, in terms of evidence and reason). As we will soon see, this conflicts with Bloor's symmetry principle, but not with impartiality. All sides in this debate can cheerfully embrace the impartiality principle.

What about the principle of reflexivity? Bloor's principle is something that readers immediately pounce on. If all belief is merely the product of various social forces (so this argument goes), then the same can be said of the strong programme itself. There can't be any evidence in support of the strong programme, if Bloor is right, because he has argued that there is no such thing as genuine evidence. Bloor may well believe the strong programme but that (by his own lights) is because it serves his interests.

This sort of self-refutation problem plagues all sorts of views. The sceptic says no belief is justified; thus, the sceptic's own scepticism isn't justified, so we can ignore it. Marx says belief reflects class structure; thus, Marx's own theory merely reflects his social position, so we can ignore it. These kinds of quick rebuttals really won't do, though they are a favourite with beginning philosophy students. It might well be that a particular doctrine is basically right, but any formulation of it runs into problems. It might well be that none of our beliefs is in any way justified, even though *saying so* runs into paradox.

The reflexivity of the sociology of knowledge is a small problem, perhaps none at all. Yes, says Bloor, social factors cause all belief, and yes, social factors even cause the belief that social factors cause all belief. There is certainly no logical problem here. If there is any sort of difficulty, it stems from thinking that if we know a belief is caused by social factors, then our faith in that belief is undermined. So, if we know that belief in the strong programme itself is caused by social factors, then that belief is also undermined. Bloor simply denies this. He staunchly holds that we can simultaneously hold a belief *and* hold that the belief is caused by social factors. Perhaps this is implausible (at least for a wide range of cases); but even if Bloor hasn't answered the self-refuting objection, his reflexivity principle certainly defuses it. Tell him that social factors are making him accept the strong programme and he will smile pleasantly back at you.

The symmetry principle may be the most contentious. It demands the same type of explanation for rational and for irrational beliefs. We can explain your health or your illness in physiological terms. We can explain why a bridge is standing or why it collapsed in terms of its structural properties. These are instances of symmetrical explanations. So, in the same vein we should explain rational and irrational beliefs in the same way. This contradicts the A-rationality principle in that it would demand sociological explanations for all beliefs, not just the irrational ones. An opponent of Bloor could turn this around and demand an explanation in terms of reason for all beliefs, rational and irrational. How could this possibly work? By showing that the agent rationally believes that holding the irrational belief will promote her interests. The symmetry principle looks plausible initially, but on close inspection it crumbles.

SSK rests on two strands. One is the philosophical argument (or should that be anti-philosophical argument) presented by Bloor and others. The other strand is the support it gains from the perceived success of several case studies. These are historical examples where some episode is analysed in sociological terms of 'interest' rather than in terms of reason and evidence, the way a traditional intellectual historian would try to understand the same events. A famous study by Paul Forman well illustrates this.

32.4.5   *A Social Constructivist Case Study: Quantum Theory in the Weimar Republic*

How do we explain the rise of the quantum theory in the mid-1920s? Paul Forman, in his elaborately titled 'Weimar Culture, Causality and Quantum Theory, 1918–1927: Adaptation by German Physicists and Mathematicians to a Hostile Intellectual Environment', offered a sociological explanation: After the Great War, German scientists lost much of their prestige; Spengler had just published his wildly popular *Decline of the*

West and Spenglerism was everywhere. The spirit of the times was decidedly mystical and anti-mechanistic. The scientists of the Weimar Republic, says Forman, created non-causal, non-deterministic quantum mechanics to appeal to the German public's mystical and anti-mechanistic outlook and thereby to regain their high social standing.

By contrast, a more traditional, 'rational' explanation might look something like this: The old quantum theory of Bohr and Sommerfeld was not a coherent set of physical principles; the new theory of Heisenberg, Born, Schrödinger and others (1925–1927) accounted for a wide range of phenomena including the so-called anomalous Zeeman effect which had been the subject of much perplexity; consequently, scientists who worked in this field were won over by the explanatory successes of the new mechanics and completely accepted it for that reason.

Forman will have none of this. Where others see 'rational' factors, he sees social forces. One need only pay attention to the footnotes of sociological literature in the 1970s and later to see the great importance of Forman's work to the newly emerging style of science studies. To use a Kuhnian expression, it was a new paradigm. The general idea, manifest in Forman's account, is that scientists had social interests and their scientific beliefs are shaped by those interests, not by so-called rational factors. Let's examine Forman's case study in a bit more detail, so that we can clearly see the structure of his argument.

The scientists of the Weimar Republic were living in a hostile intellectual environment, according to Forman. World War I was over and Germany had lost. The public was seriously disillusioned with science and technology. The spirit of the times was mystical and antirational. Indeed there was considerable opposition to science which was seen as mechanical, rationalistic and linked to causality and determinism. Into this hostile intellectual climate came Oswald Spengler's *Decline of the West*, which claimed that physics expressed the 'Faustian' nature of current Western culture. According to Spengler, physics had run its course, exhausting all its possibilities. It stood condemned as a force in opposition to 'creativity', 'life' and 'destiny'. Salvation could only come if science returned to its 'spiritual' home.

Several leading Weimar physicists are cited by Forman stressing the importance of 'spiritual values' and acknowledging the 'mystery of things'. He concludes that the concessions were so numerous and extensive that they constituted a 'capitulation to Spenglerism' (1971, p. 55). And so the general 'crisis of culture' was embraced by the scientists themselves: 'The *possibility* of the crisis of the old quantum theory was dependent upon the physicists' own craving for crises, arising from participation in, and adaptation to, the Weimar intellectual milieu' (1971, p. 62).

Perhaps the most striking feature of quantum mechanics is the widely accepted belief that it abandons strict causality; quantum processes have various probabilities of occurring, but they are not invariably determined to do so. (This is one of the features that Einstein so disliked, claiming that God does not play dice.) Did this new theory which surrendered determinism result from the usual evidential considerations? Not at all, says Forman:

> Suddenly deprived by a change in public values of the approbation and prestige which they had enjoyed before and during Wold War I, the German physicists were impelled to alter their ideology and even the content of their science in order to recover a favorable public

image. In particular, many resolved that one way or another, they must rid themselves of the albatross of causality (1971, p. 109).

...the movement to dispense with causality expressed less a research program than a proposal to sacrifice physics, indeed the scientific enterprise, to the *Zeitgeist* (1971, p. 113).

Forman's celebrated study became a new model for many historians of science. According to this model, we understand events in the history of science, not in terms of the empirical evidence, not in terms of theoretical innovations, not in terms of conceptual breakthroughs but rather in terms of social factors. A group of scientists in Weimar Germany had a social goal—to regain lost prestige. That's why the old quantum theory was rejected and the new quantum mechanics of Heisenberg, Born and others was adopted.

It's difficult to say why social constructivism has flourished to the extent that it has. One of the reasons is the perceived success of historical case studies such as Forman's. But are they really successful? Lots of historians do think so and lots do not. It's not easy to make a decisive case one way or the other. Certainly, explanations by social factors tend to be more interesting than explanations via dry data and arid inductive inferences. One can read about the events, study the experimental data and laboriously work through the calculations that lead up to the revolution in quantum mechanics in, say, Max Jammer's history of the period (Jammer 1966). Of its kind it's a fine work, but it's also hard going. By contrast, Forman's account is a real page turner with its descriptions of the social atmosphere of post-war Germany, Weimar politics and so on. Social history is often more fun—but that, of course, doesn't mean that it's right.

32.4.6 Feminism and Science

Sandra Harding famously introduced a taxonomy of feminist critiques of science:

1. Feminist empiricism
2. Feminist standpoint theory
3. Feminist postmodernism

She saw feminist philosophers of science as falling into one of these categories. Though something of a simplification, the taxonomy has proved quite useful. The first of these views, *feminist empiricism*, holds that the standard methods of science are fine as they are. Sexist science is the result of not living up to the existing canons of good science. (They often add that the same can be said about racist science.) When looking at nineteenth-century accounts of hysteria or Nazi race science, one cannot help but think that these were appallingly bad researchers who violated every principle of good science.

Feminist standpoint theory comes from Hegel and Marx. A slave has a superior understanding to the slave owner, according to Hegel, because he must understand both his own situation and the owner's. Similarly, for Marx, the worker must understand the boss's view as well as his own. A standpoint is not a mere perspective or point of view. It is an accomplishment requiring a struggle to obtain. For that reason it is a superior understanding of how things are. It is not automatic that women will have a feminist standpoint (unlike a woman's perspective), but if they do achieve it, they will have a better understanding than their male counterparts for whom nothing is to be gained by acquiring an understanding of the position of women. Feminist standpoint theory is a challenge to the standard account of science, but it is important to stress that it remains wedded to the ideal of scientific objectivity. It is just that objectivity is more complex and difficult to obtain than previously thought.

Feminist postmodernism (the third of Harding's categories), as an approach to science, was inspired by a number of feminist postmoderns, such as Judith Butler, Julia Kristeva and Luce Irigaray. It is highly sceptical of general principles and objectivity and takes a very dim view of what is here called the standard view of science. The emphasis is on the local with scant regard for any inconsistencies among different 'local narratives'. It thus embraces a form of relativism. Different societies have their own stories, their own local narratives. An all-embracing or 'total narrative' is dismissed out of hand.

When she proposed the taxonomy, Harding allowed that all three were genuine feminist outlooks, but that she seemed to favour postmodernism. More recently she has pulled away into what she calls 'strong objectivity'. It is generally true that while many feminists have considerable sympathy for postmodernism, the vast majority of feminist philosophers of science do not. In terms of Harding's taxonomy, they adhere to some version of feminist empiricism or standpoint theory, though they may not use these labels.

We can illustrate the more objective feminist approaches with the example of Okruhlik (1994). She begins with two assumptions that are commonly, though not universally, accepted by philosophers of science. One is the distinction between discovery (having new ideas) and justification (putting them to the test). We can ignore the process of discovery, the argument often runs, because the justification process will filter out all the crazy and biased aspects that go into having ideas and only the evidentially supported will survive. Her second assumption is that the process of justification, theory evaluation, is comparative. That is, we do not evaluate a theory merely by testing it against nature. Instead, we test it and its rivals against nature and on that basis we can (objectively) rank order them. We cannot really say a theory is good or bad except with respect to a comparison group. When we say a tennis player is good, we mean she can beat most other tennis players. Imagine that only one tennis player existed. We would have no way of saying she is or isn't a good player; we need the comparison group.

The moral Okruhlik draws from this is rather straightforward. If there is some bias systematically built into the comparison group, it will not be filtered out in the process of justification. What is needed to improve the process of evaluation it to enlarge and diversify the set of rival candidate theories. There are nice illustrations of how this has happened.

There is an important and influential class of theories called 'man-the-hunter' that accounts for human evolution. The general claim is that our ancestors developed language and tool use through the practice of hunting. Male developed tools for hunting and the developed language in order to facilitate cooperative hunting. This even accounts for some aspects of our physiology: large incisors gave way in the evolutionary process to molars, which are better for digestion, because tools replaced the need for teeth that rip apart a prey's throat. There were variations on this general idea and the available evidence would tend to support some of these over others. This was science as usual, rational, objective and so on.

With a growing number of women in anthropology, there arose a different approach to this issue. 'Woman-the-gatherer' theories made a different claim about human evolution. The claim was that our female ancestors are chiefly responsible for our evolution. Tools that were thought to be for hunting were reinterpreted as for food preparation. Language was seen as arising out of sociability. And certain types of facts that had been ignored were investigated. For instance, among contemporary hunter-gatherer communities, it turns out that the female gatherers provide 75% of the family caloric intake.

It does not matter which type to theory is right—perhaps neither is. The philosophical moral is that the quality of the set of rival theories to be evaluated has improved greatly. Okruhlik would not say women researchers are free from bias; rather, they have different biases. But now in the process of comparative evaluation, there is some hope that these biases can be neutralised.

As an approach to understanding science, it is indeed a challenge to the standard view. However, it is more of a modification than a rejection. It requires looking at the scientific community and making sure it is appropriately diverse. This is not a brand of social constructivism, but it does take the society in which science is pursued to be of great importance. This is quite different from the standard account which is largely oblivious to society as long as it does not interfere. Feminist philosophers of science who insist on taking these sorts of social factors into account while still upholding scientific objectivity have probably improved the standard account of science considerably, especially as it applies to the social sciences. Their ranks include Anderson, Harding, Kourany, Longino, Nelson, Okruhlik, Wylie and many others.

There are common misconceptions about feminist critics of science. They are often portrayed as anti-objectivity, anti-science and so on. Of course, some are, but one needs to be, careful when passing judgement. Norette Koertge, a prominent philosopher of science and among the first to write on science education (Koertge 1969), maintains that science needs more unorthodox ideas and a greater plurality of approaches. This is a standard Popperian position which does not in itself constitute an argument for a new epistemology of science. She then sounds the alarm against certain feminists, warning that

> If it really could be shown that patriarchal thinking not only played a crucial role in the Scientific Revolution but is also necessary for carrying out scientific inquiry as we now know it, that would constitute the strongest argument for patriarchy that I can think of (Koertge 1981, p. 354).

And she goes on to say

> I continue to believe that science -- even white, upper-class, male-dominated science -- is one of the most important allies of oppressed people (Koertge 1981, p. 354).

She is quite right, but the problem is that many of those she attacks believe the same thing. Most feminist philosophers of science believe that science can be objective. They are trying to find ways to improve its objectivity, not to expose it as a fraud. Similar sentiments are echoed by many feminists, including Susan Haack (2003) and Cassandra Pinnick (2003, 2005, 2008). Pinnick believes that popular, or postmodernist, feminist philosophy of science is not only unsupportable, but it has done an immense disservice to science and the advancement of women in science. She is right to heavily criticise postmodern approaches, but sometimes she assimilates feminist philosophers who champion objectivity with postmoderns who do not. Pinnick writes

> Viewed by a philosopher of science, there is nothing short of a puzzle as to why, at this date, any group of science educators would invoke so patently flawed a philosophical position as 'epistemologies of feminism', in the hope that women in science will then benefit from a revamped theory of learning that is modelled on or guided by its flawed theoretical notions. It is time that science educators are told, bluntly, the conclusion which philosophers of science have reached after two decades or so of careful, and even hopeful, consideration of feminist standpoint theory. The conclusion, in brief, is that feminist standpoint theory is indefensible (Pinnick 2008, p. 1056).

Expressions such as 'feminist epistemology' cover a variety of views. Helen Longino would advocate doing science as a feminist, by which she means investigating nature with a concern of women's issues. There is no conflict with objectivity here. Thus, a feminist anthropologist might ask new questions about the female gatherers, such as how many calories did they contribute to family intake. They discovered that it is about 75%, which came as a shock to those who thought they had overwhelming evidence for understanding our evolutionary past in terms of man-the-hunter. There is not a hint of different facts of different forms of reasoning for men and women. We might acknowledge different biases that go into theory construction, but those biases can (and we hope will) be overwhelmed by evidence in the long run. Acknowledging this is to promote objectivity.

32.4.7 Postmodern Critics of Science

The foregoing has partially characterised postmodernism but more must be said. Postmodernism stands in opposition to the Enlightenment (which is taken to be the core of modernism). Of course, there is no simple characterisation of the Enlightenment any more than there is of postmodernism, but a rough and ready portrayal might go like this: Enlightenment is a general attitude fostered (on the heels of the Scientific Revolution) in the seventeenth and eighteenth centuries; it aims to replace superstition and authority by critical reason. Divine revelation and Holy Scripture give way to secular science; tradition gives way to progress. Enlightenment advance is of two sorts: scientific and moral. Our scientific beliefs are objectively better than before and are continuing to improve, and our moral and social behaviour is also improving and will continue to do so.

There is another aspect to Modernism which is often linked to the Enlightenment but seems to go well beyond. This is the doctrine that there is one true story of how things are. Jean-François Lyotard, one of the most prominent postmodern commentators, speaks of the 'incredulity about metanarratives' (1984). Science for him is just a game with arbitrary rules, and truth is nothing more than what a group of speakers say it is. While most Enlightenment figures that postmoderns attack would happily embrace the view that there is one true story (perhaps with qualifications), so would Aristotle and so would the Mediaevals. In attacking so-called grand narratives or metanarratives, postmoderns are attacking much more than Modernism. Roman Catholicism's fondness for tradition and authority may stand opposed to the Enlightenment, but it certainly disdains relativism and embraces the one-true-story outlook.

Just as critical reason is seen by postmoderns as a delusion, so are all attempts to generalise or universalise. In place of so-called 'totalising' accounts of nature, society and history, 'local' accounts are offered. *Localism* or *perspectivalism* is the view that only very limited accounts of nature, or society, (or whatever the subject of discourse is), are to be taken as legitimate; grand theories are invariably wrong or oppressive or both. (To repeat what was said above, standpoint theory is not the same as localism, since it claims that some perspectives are objectively better than others.)

Jacques Derrida, another leading postmodern figure, has pronounced that any attempt to say what postmodernism is (or what it is not) will invariably miss the point. Bruno Latour, a source of inspiration for some postmoderns, has declared that we have never been modern, much less postmodern. Such claims put the would-be expositor in a difficult position. Nevertheless, it seems reasonably fair to say that these three ideas are central to postmodernism: one is the *anti-rationality* stance, a second is the *rejection of objective truth* and the third is *localism*. There are other ingredients such as *anti-essentialism*, but they would seem to follow from the initial ideas. In any case, this short list is not meant to be exhaustive.

Postmodern accounts of science are not easily identifiable. Feminist and SSK approaches usually announce themselves as being feminist and sociological, respectively, but postmoderns, who are often playful with language, find telegraphing prose to be plodding. Nevertheless, we can probably say something by way of characterising it, realising that what we say might be contentious.

The principal characteristic of postmodernism is the rejection of modernism. Modernism, or its equivalent, the Enlightenment, holds that we can and do make progress. This progress is due to reason; tradition and authority are impediments that we can overcome. Needless to say, science and technology are central to this outlook. Postmodernism can now be easily characterised as the rejection of all that. This is true for postmodern science and philosophy, but something like it would be true for postmodern art, music, architecture and so on. Modernist aesthetic principles are similarly rejected. Whereas modernist architects would aim for some sort of unity or symmetry in a building, a postmodern work might be composed of very different styles and building materials. While postmodern science is far from postmodern art, there is still a common spirit underlying each.

One of the striking features of postmodernism—admirable or disconcerting, depending on one's outlook—is the cherry picking of parts of science. While there is widespread distain for science as a 'totalising narrative', particular achievements are celebrated. Heisenberg's uncertainty principle, chaos and catastrophe theory are warmly embraced. It's easy to see why, since these theories involve the unpredictable and the uncontrollable. If it seems strange to embrace some parts of science and not others, remember that unity and coherence are modernist values, cheerfully abandoned by postmodernists. Of course, it makes debate rather difficult, since there seems to be no common ground from which to start.

Obviously postmodernism and the social constructivism with which it is associated pose a challenge to standard views of science, the views most commonly embodied in science curriculum and in arguments for the compulsory study of science in schools. The role of social factors in science is increasingly acknowledged and is now admitted to some extent in all quarters. The stronger forms of constructivism and postmodernism, however, have not been accepted by the general academic community of those who do serious research into the nature of science. On the contrary, postmodern and constructivist views have often foundered and many of their early champions have significantly modified their views and now acknowledge that reason and evidence do after all play a significant or even a determining role in the development of science.

32.5 Part Four [R.G.]

32.5.1 *Postmodernism Exposed: The Sokal Hoax*

The attempt to subject science to postmodernist interrogation received a major setback when New York University physicist Alan Sokal submitted his parody paper titled 'Transgressing the Boundaries: Toward a Transformative Hermeneutics of Quantum Gravity' to the leading postmodern journal *Social Text*. Embarrassingly the manuscript which was full of gibberish and nonsense that anyone with decent high-school science and mathematics should have detected, passed review and was published (Sokal 1996). Seemingly the journal's readers knew no more about science than its reviewers. The gibberish was 'music to the ears' of postmodern critics of science, speeded its publication, and led to dancing in the Cultural Studies corridors, if not streets. The music can be heard in a few quotes from Sokal's original paper (reproduced in *Fashionable Nonsense*):

- Over the past two decades there has been extensive discussion among critical theorists with regard to the characteristics of modernist versus postmodernist culture; and in recent years these dialogues have begun to devote detailed attention to the specific problems posed by the natural sciences. In particular, Madsen and Madsen have recently given a very clear summary of the characteristics of modernist versus postmodernist science. They posit two criteria for a postmodern science: A simple criterion for science to qualify as postmodern is that it be free from any dependence on the concept of objective truth. By this criterion, for example, the complementarity interpretation of quantum physics due to Niels Bohr and the Copenhagen school is seen as postmodernist. (pp. 223–4)

- In Andrew Ross' words, we need a science that will be publicly answerable to progressive interests. From a feminist standpoint, Kelly Oliver makes a similar argument: …in order to be revolutionary, feminist theory cannot claim to describe what exists, or, 'natural facts.' Rather, feminist theories should be political tools, strategies for overcoming oppression in specific concrete situations. The goal, then, of feminist theory, should be to develop 'strategic' theories—not true theories, not false theories, but strategic theories. (p. 235)
- The teaching of science and mathematics must be purged of its authoritarian and elitist characteristics, and the content of these subjects enriched by incorporating the insights of the feminist, queer, multiculturalist, and ecological critiques. (p. 242)
- Finally, the content of any science is profoundly constrained by the language within which its discourses are formulated; and mainstream Western physical science has, since Galileo, been formulated in the language of mathematics. But whose mathematics? The question is a fundamental one, for, as Aronowitz (1988) has observed, neither logic nor mathematics escapes the 'contamination' of the social. And as feminist thinkers have repeatedly pointed out, in the present culture this contamination is overwhelmingly capitalist, patriarchal, and militaristic: mathematics is portrayed as a woman whose nature desires to be the conquered Other. Thus, a liberatory science cannot be complete without a profound revision of the canon of mathematics. (Aronowitz 1996, pp. 244–245)

Sokal used over 200 references in his parody paper as he repeatedly praised leading PM 'thinkers' for showing how to interpret quantum physics, relativity theory and even mathematics in ways that seemed to provide a sound basis for the various PM agendas. After publication Sokal revealed the hoax and lampooned many of the PM and SSK gurus, including Jacques Lacan, Bruno Latour, Stanley Aronowitz, Jacques Derrida, Sandra Harding and Steve Woolgar. Not surprisingly, *Social Text* did not publish Sokal's revelation and explanation (it did not meet the journal's 'intellectual standards').

Fashionable Nonsense (the work was also published with another title: *Intellectual Impostures*) tells the story behind the parody paper, and in the Epilogue explains why PM can be dangerous to our intellectual health. In doing this Sokal and Bricmont suggest seven lessons (pp. 185–189) that can be learned from the hoax:

1. It's a good idea to know what one is talking about. This is especially true of the natural sciences where technical, abstract ideas are involved for understanding at more than a superficial, popular level.
2. Not all that is obscure is necessarily profound. Much of the discourse of cultural studies is laden with obscure jargon.
3. Science is not a 'text'. Postmodernists often use terms like uncertainty, chaos, theory and nonlinearity in ways that mislead nonscientists. Pseudoscience uses technical, scientific terms to fool people into believing their products have a real scientific basis.

4. Don't ape the natural sciences. The social sciences study people and their institutions, while the natural sciences study nature and these domains often require different assumptions and research methods.
5. Be wary of argument from authority. The tendency to follow gurus like Lacan and Freud in the social sciences is much more prevalent than in the natural sciences. Nature is the final authority in the natural sciences, not sacred texts or respect for culture.
6. Specific scepticism should not be confused with radical scepticism. The relativism inherent in postmodernism allows followers to question the value of logic and evidence. Embracing radical scepticism can result in the absurd conclusion that astrology and astronomy are equally valid.
7. Ambiguity as subterfuge. Postmodernists are often ambiguous on purpose as this allows one to claim, I was misunderstood. Deliberate ambiguity in their writing is a common strategy among postmodern authors.

These lessons from the Science Wars and the exposure of the severe intellectual problems surrounding the sociology of science knowledge (SSK) programme should have meant that postmodernism would lose its appeal by the end of the twentieth century. However, that is not the case; postmodernism survives, and as documented in Part One of this chapter, even thrives in some science education circles under the guise of 'cultural studies' or 'radical constructivism'.

32.6 Part Five Conclusion [R.G., J.M. & J.R.B.]

The anti-science attitude fostered by postmodernism and relativism can lead to habits of mind that diminish concern for evidence, for logic, for clear writing and for finding out the truth of the matter. A good example of this is the widespread rejection of the findings of modern climate science. Despite overwhelming evidence that burning fossil fuels results in a warming of Earth's climate, with potentially devastating results for all living things, many people reject the scientific findings. When ideology trumps science, as in climate science debates where oil and coal companies resist scientific findings and related implications for action, we are left with no reasonable way to solve problems. The same pattern is repeated in campaigns against child and adult vaccination. When scientific knowledge is seen as a 'regime of truth' that endangers our freedom and democracy, as suggested by PM proponents, then political and religious ideologies can replace knowledge gained through scientific methods; it becomes much more difficult for people to recognise and reject pseudoscience. Science's most precious gift, the phrase used by Albert Einstein to describe the great value of modern science to society, is its ability to reduce the influence of cultural ideologies in judging the truth value of competing claims. Postmodernism offers little that can be used to improve the scientific literacy of our citizens, little that can be used to improve teacher education and a lot that can be used to diminish literacy and distract good teacher education.

References

Aronowitz, S. (1988) *Science as Power: Discourse and Ideology in Modern Society*, Minneapolis: University of Minnesota Press.

Aronowitz, S. (1996) "The Politics of the Science Wars", *Social Text*, 177–196.

Bakhtin, M. (1981). *The Dialogic Imagination*. Austin: University of Texas Press.

Baynes, R. & Austin, J. (2012), 'Indigenous Knowledge in the Australian National Curriculum for Science: From Conjecture to Classroom Practice'. In K. Bauer (ed.) *International Indigenous Development Research Conference 2012*, Auckland, pp. 60–66.

Berlin, B., Breedlove, D. E. & Raven, P. H. (1974) *Principles of Tzeltal Plant Classification: An introduction to the botanical ethnography of a Mayan speaking people of highland Chiapas* (New York; Academic Press).

BjörnsenGurung, A. 2003: "Insects – a mistake in God's creation? Tharu farmers' perception and knowledge of insects: A case study of Gobardiha Village Development Committee, Dang-Deukhuri, Nepal." *Agriculture and Human Values*, 20 (4), pp. 337–370.

Bloor, D. (1976/91) *Knowledge and Social Imagery*, (2nd ed.) Chicago: University of Chicago Press.

Boster, J., & D'Andrade, R. (1989) "Natural and Human Sources of Cross-Cultural Agreement in Ornithological Classification." *American Anthropologist*, 91 (1), pp. 132–142.

Brandom, R. B. (1994) *Making It Explicit: Reasoning. representing, and discursive commitment* (Cambridge, Massachusetts; Harvard University Press).

Brown, J. R. (2001) *Who Rules in Science? An Opinionated Guide to the Wars*, Cambridge, MA: Harvard University Press

Collins, H., & Yearley, S. (1992) "Epistemological chicken." in *Science as Practice and Culture*, ed. A. Pickering (Chicago; University of Chicago Press) pp. 301–326.

Davidson, D. (1984) *Inquiries into Truth and Interpretation* (New York; Oxford University Press).

Diamond, J. (1966) "Zoological classification system of a primitive people." *Science* 151, pp 1102–1104.

Doll, W. (1993). *A post-modern perspective on curriculum*. New York: Teacher's College Press.

Flynn, J. J., Finarelli, J. A., Zehr, S., Hsu, J., Nedbal, M. A. (2005). "Molecular phylogeny of the carnivora (mammalia): assessing the impact of increased sampling on resolving enigmatic relationships". *Systematic Biology* **54** (2): 317–337.

Forman, P. (1971) "Weimar Culture, Causality and Quantum Theory, 1918–1927: Adaptation by German Physicists and Mathematicians to a Hostile Intellectual Environment", *Historical Studies in the Physical Sciences*, vol. 3.

Gould, S. J. (1980) *The Panda's thumb* (Repr. Harmondsworth, Middlesex; Penguin, Pelican, 1983).

Haack, S.: 2003, 'Knowledge and Propaganda: Reflections of an Old Feminist'. In C. L. Pinnick, N. Koertge & R.F. Almeder (eds.) *Scrutinizing Feminist Epistemology: An Examination of Gender in Science*, Rutgers University Press, New Brunswick, pp. 7–19.

Harré, R. 1989, 'Idealization in Scientific Practice', *Poznań Studies in the Philosophy of the Sciences and the Humanities* 16, 183–191.

Holton, G. (1993). *Science and anti-science*. Cambridge, MA: Harvard University Press.

Hunn, E. (1976) "Toward a perceptual model of folk biological classification."*American Ethnologist*, 3 (3), pp. 508–524.

Huxley, A.: 1947, *Science, Liberty and Peace*, Chatto & Windus, London.

Koertge, N.: 1969, 'Towards an Integration of Content and Method in the Science Curriculum', *Curriculum Theory Network* **4**, 26–43. Reprinted in *Science & Education* 5(4), 391–402, (1996).

Koertge, N.: 1981, 'Methodology, Ideology and Feminist Critiques of Science'. In P. D. Asquith & R. N. Giere (eds.), *Proceedings of the Philosophy of Science Association 1980*, Edwards Bros, Ann Arbor, pp. 346–359.

Kuhn, T. S. (1962/70) *The Structure of Scientific Revolutions*, Chicago: University of Chicago Press.

Kuhn, T. S.: 1991/2000, 'The Trouble with Historical Philosophy of Science', The Robert and Maurine Rothschild Lecture, Department of History of Science, Harvard University.

In J. Conant & J. Haugeland (eds.) *The Road Since Structure: Thomas S. Kuhn*, University of Chicago Press, Chicago, pp. 105–120.

Kuhn, T. S.: 1993, 'Afterwords'. In P. Horwich (ed.) *World Changes: Thomas Kuhn and the Nature of Science*, MIT Press, Cambridge, MA, pp. 311–341.

Laudan, L. (1977) *Progress and Its Problems*, Berkeley: University of California Press.

Lyotard, J.-F. (1979) *La Condition postmoderne: Rapport sur le savoir* (Paris; Éditions de Minuit, 1979). Tr. by G. Bennington and B. Massumias as *The Postmodern Condition: A Report on Knowledge* (Minneapolis: University of Minnesota Press, 1984).

Lyotard, J.-F. (1983) "Answering the Question: What Is Postmodernism?" Repr. in *Postmodernism: A reader* ed. T. Docherty (London; Harvester Wheatsheaf), pp. 38–46.

Majnep, I. S. & Bulmer, R. (1977) *Birds of My Kalam Country: Månmonyad Kalamyakt* (London; Oxford University Press).

Marx, K. (1859) *A Contribution to the Critique of Political Economy*, (trans. from the German original, 1970), New York: International Publishers.

Masterman, M. (1970). "The nature of a paradigm." In *Criticism and the Growth of Knowledge*, ed. I. Lakatos & A. Musgrave (Cambridge; Cambridge University Press), pp. 59–89.

Matthews, M. R. (2004), 'Thomas Kuhn and Science Education: What Lessons can be Learnt?' *Science Education* 88(1), 90–118.

Mayr, E. (1963) *Animal species and evolution* (Cambridge, Mass.; Belknap Press of Harvard University Press).

McMullin, E.: 1985, 'Galilean Idealization', *Studies in the History and Philosophy of Science* 16, 347–373.

Merton, R. K. (1968) *Social Theory and Social Structure*, New York: Free Press.

Merton, R. K. (1970) *Science, Technology and Society in Seventeenth-Century England*, New York: Harper & Row (Originally published in 1938).

Needham, J. *et al.* (1954–2004) *Science and Civilization in China* (Cambridge; Cambridge University Press).

Nowak, L.: 1980, *The Structure of Idealization*, Reidel, Dordrecht.

O'Brien, S. J. (1987) "The ancestry of the giant panda." *Scientific American* 257(5), pp. 102–7.

O'Brien, S. J., Nash, W. G., Wildt, D. E., Bush, M. E. & Benveniste, R. E. (1985) "Molecular solution to the riddle of the giant panda's phylogeny." *Nature* 317 (6033) (12 September 1985), pp. 140–144.

Okruhlik, K. (1994) "Gender and the Biological Sciences", *Canadian Journal of Philosophy*, Supp. Vol. 20, 21–42 (Reprinted in Curd and Cover 1998).

Orwell, G. (1945), 'Politics and the English Language'. In his *Shooting an Elephant and Other Essays*, Harcourt, Brace & World, New York.

Pettigrew, J. D. (1986) "Flying primates? Megabats have the advanced pathway from eye to midbrain." *Science*: 231 (4743) (14 March 1986), pp. 1304–1306.

Pinnick, C. L.: 2003, 'Feminist Epistemology: Implications for the Philosophy of Science'. In C. L. Pinnick, N. Koertge & R. F. Almeder (eds.), *Scrutinizing Feminist Epistemology: An Examination of Gender in Science,* Rutgers University Press, pp. 20–30.

Pinnick, C. L.: 2005, 'The Failed Feminist Challenge to "Fundamental Epistemology"', *Science & Education* **14**(2), 103–116.

Pinnick, C. L.: 2008, 'Science Education for Women: Situated Cognition, Feminist Standpoint Theory, and the Status of Women in Science', *Science & Education* **17**(10), 1055–1063.

Planck, M. (1950) *Scientific Autobiography, and other papers* (London; Williams & Norgate).

Popper, K. R. (1945) *The Open Society and its Enemies* (London; Routledge & Kegan Paul).

Popper, K. R. (1957) *The Poverty of Historicism* (London; Routledge & Kegan Paul).

Popper, K. R. (1959 [1934]) *The Logic of Scientific Revolutions* (London; Hutchinson).

Reis, G. (2010) "Making science relevant: Conceptual change and the politics of science education." in *Re/Structuring Science Education: ReUniting sociological and psychological perspectives*, ed. W.-M. Roth (Dordrecht; Springer, 2010), pp. 233–241.

Roth, W.-M. & Tobin, K.: 2007, 'Introduction: Gendered Identities'. In W.-M. Roth & K. Tobin (eds) *Science, Learning, Identity. Sociocultural and Cultural-Historical Perspectives*, Sense Publishers, Rotterdam, pp. 99–102.

Roth, W.-M., Eijck, M. van, Reis, G. & Hsu, P.-L.: 2008, *Authentic Science Revisited: In Praise of Diversity, Heterogeneity, Hybridity*, Sense Publishers, Rotterdam.

SextusEmpiricus. (tr. 1935) *Against the Logicians*, tr. R.G. Bury (London; William Heinemann, Loeb Classical Library, 1935, repr. 1967).

Sokal, A. (1996) "Transgressing the boundaries: toward a transformative hermeneutics of quantum gravity." *Social Text* 46–7, pp. 217–252; repr. in Sokal & Bricmont, 1998, pp. 199–240.

Stove, D. (1991). *The Plato cult and other philosophical follies*. Oxford: Basil Blackwell.

Tobin, K. & Roth, W.-M. 2006, 'Editorial', *Cultural Studies of Science Education* vol. 1 no. 1, 1–5.

Tobin, K. (2010) "Tuning in to others' voices: Beyond the hegemony of mono-logical narratives." in *Re/Structuring Science Education: ReUniting sociological and psychological perspectives*, ed. W.-M.l Roth (Dordrecht; Springer, 2010), pp. 13–29.

Van Eijck, M. & Roth, W-M. (2007). Keeping the local local: Recalibrating the status of Science and traditional ecological knowledge (TEK) in education. *Science Education* 91: 926–47.

Van Eijck, M. & Roth, W-M. (2011). Cultural diversity in science education through "novelization": Against the epicization of science and cultural centralization. *Journal of Research in Science Teaching* 48 (7), pp. 824–47.

Jim Mackenzie is retired from the Faculty of Education and Social Work at the University of Sydney. He completed his Ph.D. in Philosophy at the University of New South Wales, had lectured for many years in Philosophy of Education and has published on the formal logic of dialogue and on various aspects of educational theory and practice.

Ron Good Professor Emeritus, Science Education, Louisiana State University (1987–2002) and in a similar position at Florida State University (1968–1986). He is a former editor of the *Journal of Research in Science Teaching* and a Fellow of the *American Association for the Advancement of Science* since 1995. Research interests include science misconceptions, nature of science and pseudoscience, evolution education and habits of mind in science and religion. His books include *How Children Learn Science* (1977) and *Scientific and Religious Habits of Mind* (2005) and he has contributed many articles, conference papers and symposia on science education. He began his career in 1962 teaching high-school physics and chemistry in schools near Pittsburgh, PA and currently enjoys the island life at his beach house on St. George Island in Florida.

James Robert Brown is Professor of Philosophy at the University of Toronto. His interests include foundational issues in mathematics and physics, thought experiments and the relations of science to society. The last of these includes religion and science, the effects of commercialization on medical research, and equity issues in science and in the larger community. His most recent book is: *Platonism, Naturalism, and Mathematical Knowledge*, Routledge. Earlier books include: *Who Rules? An Opinionated Guide to the Epistemology and Politics of the Science Wars*, Harvard, *Smoke and Mirrors: How Science Reflects Reality*, Routledge, new editions of *The Laboratory of the Mind: Thought Experiments in the Natural Sciences*, Routledge, *Philosophy of Mathematics: An Introduction to the World of Proofs and Pictures* Routledge. His forthcoming books are on philosophy of science, foundations of seismology, and a history of aether theories. He is a Fellow of the Royal Society of Canada and a member of *LeopoldinaNationaleAkademie der Wissenschaften, and AcadémieInternationale de Philosophie des Sciences*.

Chapter 33
Philosophical Dimensions of Social and Ethical Issues in School Science Education: Values in Science and in Science Classrooms

Ana C. Couló

33.1 Values in Science Classrooms?

To date, there is no unanimous consensus on the "fundamental" notions related to the nature of science that should be taught within a history and philosophy of science (HPS) frame. However, the role of values in science has usually been a key element of the list, both in curricular proposals and in science education research (Adúriz-Bravo 2005a; Clough 2007; McComas 2002). Science has a significant impact on the way we live our lives, either through its products and processes or through the impact of new ideas on the ways we think about ourselves and the world. Scientific and technological developments raise many controversial issues: cancer treatment isotopes, increased-yield crops and xenotransplantation which saves human lives, on one side, and nuclear warfare, pesticide-induced diseases, undesirable social impacts and suffering and death for animals, on the other.

The way we think about things is often shaped by ideas born and matured within scientific projects, while scientific questions are frequently situated and related to the philosophical debates faced by contemporary societies. Science prides itself on widening our knowledge of the world, but is the discovery of true propositions *always* a good thing? It can be argued that some scientific ideas, however well grounded, have harmed people or diminished their happiness by leading them to change or question their self-images, their aspirations and their self-conceptions. And though counter-intuitive, this assertion merits consideration (Forge 2008, pp. 149–151; Kitcher 2001, Chap. 12). In short, the many ways in which science affects us are impregnated with value issues, while the ethical and political responsibilities of scientific work and knowledge impact scientists and science as an institution.

A.C. Couló (✉)
Instituto de Filosofía – CEFIEC, Universidad de Buenos Aires,
Buenos Aires, Argentina
e-mail: anacoulo@yahoo.com.ar

M.R. Matthews (ed.), *International Handbook of Research in History,*
Philosophy and Science Teaching, DOI 10.1007/978-94-007-7654-8_33,
© Springer Science+Business Media Dordrecht 2014

1087

So what are the values manifested and expressed in scientists' behaviour and in scientific practices? How are they embodied in scientific institutions?[1] These questions should be contemplated in science education at all levels.

There are several compelling reasons for advocating teaching science and technology students the ethical and political questions they will have to address in their lives as professional researchers, designers and citizens. For example, values are implicit in the choice of research subjects and research methods. Philosophers, historians and sociologists have debated the scope and significance of values in science, including consideration of whether scientists are accountable for not anticipating the consequences of their enquiries. Kitcher rejects the "myth of purity" that assumes that "there is a straightforward distinction between pure and applied science, or between "basic research" and technology" (Kitcher 2001, p. 86). Forge (2008) argues that scientists can be held responsible for the foreseeable results of their research, whether the outcome is a technological object or a published paper detailing new information on a given phenomenon.[2] Rollin (2009) points out that the invasive use of animals in experiments presupposes moral choices. It could be argued (and it *has* been argued) that animals cannot or do not feel pain in the same way that humans do. Nonetheless, in psychological research, animals are used to model harmful or undesirable psychological states. Researchers are then confronted by a dilemma: if animals cannot feel fear, pain, addiction, etc. as humans do, then what would be the point of inducing that state in the animal? And if they can feel fear, pain, addiction, etc. as humans do, then why is it morally acceptable to induce those states in animals? Answers to this dilemma cannot escape an ethical dimension, i.e. that the knowledge gained from the experiment outweighs the discomfort, pain or death suffered by the animal. And what are the ethical and political dilemmas that should be faced in biomedical research on human beings? Serious issues about the different layers of vulnerability should be addressed by potential researchers (Luna 2009). Individual scientists choose to engage in certain kinds of research, while different societies and institutions (scientific or not) encourage some of them and discourage others (Forge 2008, pp. 179–183).

However, this chapter will focus on teaching the role of values in science at high school level. Several rationales have been given for bringing these issues into schools. For example, teaching and learning about the role of values in science in socio-scientific and controversial issues can play a role in humanising sciences and illustrating their ethical, cultural and political facets (Matthews 1994). It can help to foster an appreciation of the nature of science (Bell and Lederman 2003). It is dependent on and contributes to core abilities in reasoning, dialogue and argumentation (Simmoneaux 2008; Zeidler and Sadler 2008; Zohar 2008). Furthermore, it encourages a richer and more comprehensive construction of the

---

[1] I would like to thank John Forge for his helpful suggestions on this paragraph and successive references to this point.

[2] For an interesting example, see the discussion on the Manhattan Project and especially on Frédéric Joliot-Curie's refusal to join the moratorium in publishing results on neutron multiplication in an assembly of heavy water, in 1939 (Forge 2008, pp. 72–76).

social and political aspects of science, avoiding trivialised images (Adúriz-Bravo 2005b). In the last couple of decades,[3] curricula in several different countries have regarded understanding of values-related issues as an important goal. Moreover, the consideration of ethical, political and other value outlooks in science content and research creates a particularly fertile ground for interaction between teachers of different curriculum subjects, such as natural sciences and philosophy (which has long been a high school subject in its own right in many countries)[4]. Even where philosophy is not part of the curriculum (not even as an elective course),[5] humanities subjects and civic and citizenship education offer many possibilities for collaboration, provided teachers adopt a controversial issues perspective rather than a more descriptive one (Kolstø 2008; Ratcliffe and Grace 2003). Appropriate teaching and learning of values-related questions in the nature of science should include involvement in reasoning and rational debate about controversial issues. Then, students can become more ready, responsible and adept at participating as citizens in science and technology-related issues both inside and outside their local communities, and even globally. Also, environmental, medical, biotechnological and telecommunication issues usually stimulate interest, so teaching science in socio-scientific contexts enhances motivation to learn the relevant scientific content (Grace 2006).

For science education, it is relatively easy to find interesting and relevant material from a socio-scientific point of view (SSI),[6] on the role of noncognitive values in the funding of scientific research and on the technological consequences of scientific inquiry. It is much harder, though not impossible,[7] to find related works framed in a more closely philosophical perspective. But philosophical reflection on values-related dimensions of scientific knowledge and inquiry has been on the increase in contemporary philosophy of science since 1970.[8] Philosophical debates have often distinguished cognitive (or epistemic)[9] values from noncognitive (non-epistemic, such as moral, political, economic) ones. The significance of cognitive values has become more or less commonly accepted, although there are several different standpoints on which constitute the relevant cognitive values and which should have precedence (Lacey 1999, Chap. 3). The place of noncognitive values, on the other hand, is much more controversial. For instance, different viewpoints can be found pertaining to the difference between the external impact of such values

---

[3] AAAS (1993), Conseil de l'Education et de la Formation (1999), National Research Council (1996), and OECD (2001)

[4] Argentina, Brazil, (French-speaking) Canada, France, Germany, Italy, Mexico, Spain, Uruguay, etc. (UNESCO 2007).

[5] Usually, English-speaking countries such as Australia, (English-speaking) Canada, the USA and the UK.

[6] See, for instance, Sadler and Zeidler (2006), Zeidler and Sadler (2008), Zemplén (2009), also Kutrovátz and Zemplén (2014), Vesterinen, Manassero-Mas and Vázquez-Alonso (2014).

[7] For example, Davson-Galle (2002), Lacey (1999b, 2009), Machamer and Douglas (1999), Matthews (2009a).

[8] Douglas (2000), Dupré et al. (2007), Echeverría (1995), Kitcher (1993, 2001), Lacey (1999), Laudan (1984), Longino (1990, 2011)

[9] Throughout this paper we will take "cognitive" and "epistemic" as synonymous.

(on the funding or on the consequences of research) and the internal ones (the role of non-epistemic values in theory choice or theory validation, i.e. the presence of noncognitive values in science *content*).[10]

In sum, the thesis that science is not value-free has been steadily gaining acceptance. But this does not automatically mean abandoning every ideal of objectivity. Since the full meaning of the thesis that science is not value-free is the focus of heated philosophical debate, this chapter will aim first at presenting an overview of some issues pertaining to the role of cognitive values, and of the significance of noncognitive values on theory choice, validation or acceptability, from a philosophical point of view.[11] It will then address the question of why and how these philosophical issues and debates deserve to be engaged with in school science education.

33.2 Is Science Value-Free?

What are we speaking about when we discuss "values"? In the eighteenth and nineteenth centuries, a theory of economic value developed in the works of Smith, Ricardo and Marx, and even today, many people think of *economic* value when they hear the word "value". But the term has much wider scope: values (such as beauty, goodness, justice or sanctity) on one hand, and judgments of value on the other, have long been the focus of philosophical consideration.[12] Since the 1850s, debates conducted by philosophers and philosophical schools such as Nietzsche, Brentano, Dilthey, the utilitarians and the neo-Kantians at Baden have developed a theory of values that became a major concern in philosophy in its own right. The debates encompassed many issues related to the ontological nature of values; their scope; the existence of "intrinsic" values, as distinct from "extrinsic" ones; the notion of a polarity of values (for every positive value there is a related negative one); etc. (Ferrater Mora 1975; Frondizi 1982).

Although it isn't possible to address all these matters here, the question of the relationship of facts and values has led to a long-standing philosophical debate that has a special bearing on the central issue. It can be traced (at least) to Hume's discussion of the difference between matters of fact and matters of value: can *ought* be logically deduced from *is*? In some moral systems, reasoning goes from premises

---

[10] I take SSI to refer to issues based on scientific results or practices that have an actual or potential relevant impact on society (Ratcliffe and Grace 2003). They may be considered either from the (frequently descriptive and explanatory) social sciences point of view (sociology, economy, anthropology, some theories of psychology, etc.) or from a philosophical (usually normative) standpoint (ethics, political philosophy, philosophical anthropology, aesthetics, etc.).

[11] Given that the emphasis is on a general overview of the issue and on the way science education in schools may address it, some philosophical depth will be inevitably lost. Interested readers can find that depth in many of the books mentioned in the references list.

[12] Many Platonic dialogues contemplate the nature and scope of specific values such as justice (first book of *Republic*), beauty (*Greater Hippias*, *Phaedrus*) or piety (*Euthyphro*).

that are related by "is" to conclusions where the components are connected by "ought". This inference seems to be "inconceivable". There have been multiple interpretations of the relevant passages, but the most common one has been to assume that no moral judgment and, more generally, no judgment of value ("ought") can be logically derived from a judgment of fact ("is").

The logical positivists and the logical empiricists, in the Humean tradition, were concerned with emphasising the distinction between facts and values, both in general and in relation to science. Values-related statements were deemed to be neither factual nor analytical statements, and therefore they lacked truth value. Ayer (1952) and Stevenson (1960) thought that they expressed emotions: acceptance, support and approval or else refusal, denial, and rejection. In the Anglo-speaking philosophical community, the stance that values are subject-related and that they express subjective preferences was prevalent in the first half of the last century, though philosophers of ordinary language contested the image of value statements as mere expressions of emotion. But it had a significant impact on the scientific community where it has been a widespread belief that since ethical (and other values) judgments have no empirical content, and therefore cannot be tested and verified, neither values clarification nor ethical debate of any sort has any relevance in science (Rollin 2009, 2012 calls this stance "scientific ideology"). However, this conclusion missed some of the points that positivist philosophers of science tried to make.

The value-free ideal maintains that science should be axiologically *neutral*. Mainstream epistemology and philosophy of science, until the middle of the twentieth century, had stated that science is supposed to be objective and rational in a strong sense, implying that it should depict the world as it *is* and not concern itself with how it *ought to* be. Therefore, scientific knowledge is not value-laden, since it aims at an empirically grounded understanding of the world. In the 1950s several factors started to undermine this position. On one hand, Quine's holism contested the idea of a precise distinction between statements of fact and statements of reason. Hypotheses cannot be confirmed (or refuted) independently, and there are no algorithmic rules for theory choice. So considering the *actual practices* of scientists when they decide on theory choice becomes more relevant than it had previously seemed. Later, Nelson (2002) built on Quine's arguments for holism to argue for recognition of the role of non-epistemic values in scientific practice.

In an influential paper, Rudner (1953) discussed the idea that the "scientist qua scientist makes value judgments" insofar as she must decide, for instance, when the evidence is *strong enough* to accept a hypothesis or how to balance simplicity against generality. Further discussions brought to the fore the role of the epistemic values. The context of discovery/context of justification distinction restricted context of justification to epistemic values related to logical soundness and empirical evidence and excluded subjective, social or contextual particularities. On the other hand, non-epistemic values (moral, political, economical, etc.) were still assigned to contexts of discovery or application, while deemed inadmissible as criteria for theory validation or theory choice.

Douglas (2000) explains how Hempel, in his 1965 essay, *Science and Human Values*, supports the idea that science should be pursued in such a way that only epistemic values have relevance with regard to hypothesis confirmation. But when we come to hypothesis *acceptance*, a proviso should be made concerning those instances that have direct consequences on practical issues (ethical matters, social impact, safety risks and so on). Inductive risk, the possibility that a confirmed hypothesis might be (ultimately) false, or that a rejected hypothesis might be (ultimately) true, gives cause for concern regarding the outcomes of a wrong decision.[13] In those cases, non-epistemic values become very relevant, by way of consequences and risk assessment. But in other cases of "pure scientific research", with no practical applications, epistemic values would suffice for hypothesis acceptance. From Hempel's point of view, not only the physical sciences but the social sciences as well should be conducted in accord with these principles.

At the same time, questions about the responsibility of scientists as individuals and as members of institutions became a matter of common concern, especially after World War II. From a sociological point of view, Robert Merton suggested that the *ethos* of science could be described as being composed of a small set of "moral" norms that summed up the ideals into which scientists (actual or in the making) are socialised: communalism, universalism, disinterestedness, originality and organised scepticism (Ziman 2003).[14] These moral norms express the values that govern the way scientific activity is run. They constrain practices and provide a standard against which they can be measured. Confidence in Merton's *ethos* has usually underpinned confidence in science and scientists as objective, honest and free of bias. But it has also undergone discussion and criticism. What is the role these values actually play? Further discussion, even by Merton himself, suggests that this ideal is normative, not descriptive: scientists should behave in accord with it, even if their primary motivations go a different way. However, in 1974, sociologist Ian Mitroff published a study on the ambivalence of norms within scientific institutions. Mitroff's research was based on a case study conducted with 42 scientists who were part of the Apollo mission. For every Mertonian norm, he proposed a counter-norm.[15] Mitroff posited that ambivalence is not only a characteristic of science but that it seemed necessary both to the existence and to the rationality of science (Mitroff 1974). In sum, a wide range of questions about the value-ladenness of science may be addressed both from inside and outside the scientific community. Debates develop in a theoretical, highly technical philosophical context, and also in a practical, widely public one.

For instance, recent philosophical discussions of mind–body relationship and the theory of actions are wont to explain human action as intentional, that is, in terms of beliefs and desires. We perform the action A because we desire to achieve E, and we believe that A is conducive to attaining E. So we may say that desires are one of the

---

[13] Research on the safety of a new drug, for instance, may result in a false negative, with dangerous consequences for future users (see Douglas 2007, for an interesting example).

[14] There were also "technical" norms, pertaining to reliable empirical evidence and logical consistency.

[15] Emotional commitment, particularism, solitariness, interestedness and organised dogmatism (Mitroff 1974, p. 592).

causes of action. But individual desires do not stand alone: they are related to other desires (and beliefs) in a spreading network. Eventually, they depend on a person's basic beliefs and desires, that is, on the person's values. Discussing the fact–value dichotomy with regard to theory choice, Putnam (1990) states that terms such as "coherence" and "simplicity" are

> *action guiding* terms: to describe a theory as "coherent, simple, explanatory" is, in the right setting, to say that acceptance of the theory is *justified;* and to say that acceptance of a statement is (completely) justified is to say that one ought to accept the statement or theory (p. 139).

Therefore, values are a basis for action, particularly in terms of choice and decision. And they can become manifest in behaviours and expressed in practices both as personal and as social values (Lacey 1999).[16] If we try to explain how scientists choose between possible explanations of a phenomenon, we will eventually have to take values into consideration.

33.3 Science and Values

In the last half century, questioning of the idea of value-free science has raised many important issues in philosophy of science: realism, rationality, objectivity, demarcation, scientific change, scientific controversies and the role of gender, race or class (Doppelt 2008; Dupré et al. 2007; Machamer et al. 2000). It also has a significant role in discussions on applied science, technology, Big Science issues (such as trust and authority) and risk assessment. Even without rejecting empiricism and some conception of objectivity, science content and scientific activity (and not just its consequences) may be regarded as value-laden.[17]

With Kuhn's *Structure of Scientific Revolutions* (1962) a turning point was reached. Along with the renewed interest in the history and actual practice of science came the notion that values (whether those of society, scientific communities or individual scientists) have a relevant part to play. They are present not only in the choice of problems and in the technological or applied aspects of science but also in the evaluation of hypotheses, in theory choice and in conceptual change, which cannot be described simply in terms of logical inferences. A heated discussion ensued regarding whether the presence of noncognitive values resulted in less objectivity, leading to downright relativism.

---

[16] Lacey (1999) states that personal values may be *manifested* in behavior, *woven into* a life, *expressed* in a practice, *present* in consciousness, *articulated* in words and *embodied* in social institutions and in society (pp. 25–6). Social values are *manifested* in the programmes, laws and policies of a society; *expressed* in its practices; *articulated* in histories, traditions and institutions; *woven into* a society when they are manifested constantly and consistently; and can be *personalised* when persons act on behalf of a society where particular values are embodied (pp. 28–9).

[17] Kitcher (1993), Lacey (1999), Longino (1990), Machamer and Douglas (1999), and Wylie and Nelson (2007)

That science is value-laden is now accepted by most philosophers of science, although the scope of this assertion should be clarified. The Hempelian distinction between hypothesis confirmation and hypothesis acceptance is still part of the debate. The thesis that only cognitive values are necessary for scientific knowledge[18] clashes with the notion that other noncognitive values are constitutive of proper scientific practice. Cognitive values may be regarded as constitutive of theory choice both with regard to *significance* requirements (choice of problems, selection of hypotheses and theories) and in connection with *confirmation* requirements (assessing the relevance of the evidence supporting hypothesis or theories) (Carrier 2012). But does science exclusively aim at understanding the world, or is scientific knowledge inextricably entangled with the purpose of making objects or solving problems? Is there a multiplicity of possible goals interacting within scientific inquiry (such as maximising human happiness, or economic profit, or political success)? Even if the answer is weighted towards understanding, noncognitive values are still in evidence.

Dupré and colleagues (2007) suggest that arguments against the idea of a value-free science may be categorised into three main groups: "(1) arguments from denying the distinction between fact and value, (2) arguments from underdetermination, and (3) arguments from the social processes of science" (p. 14). The first set criticises the possibility of a clear demarcation between fact and values, either by offering counterexamples or by theoretical discussion against the independence of both terms. The second group alludes to underdetermination either of theory by data or of theory choice by epistemic values, drawing from the original Duhem-Quine thesis and from Kuhn's claims, respectively (Carrier 2012). The third set encompasses different studies aimed at showing how scientists interact among themselves and with society at large and how values, interests and commitments shape these interactions. As is usually the case with classifications, this one may help us organise the multiple discussions on the field, but it is not supposed to cover all possible standpoints on the value-ladenness of science (see discussion of Lacey's arguments below, for instance).

Dupré and colleagues also suggest four dimensions that appear in philosophical debates with regard to values: the *kind* of values involved, the *way* in which they are involved, *where* they are involved and what are the *effects* of their involvement. As for the kind of values, the distinction is usually made between cognitive values, directly related to truth and knowledge, and noncognitive values, such as moral or political ones, though this distinction is itself subject to debate. There is no agreement on which values should be included under the labels of "epistemic" or "cognitive" and what the standards for their application are or the relative importance of each one. For instance, in reworking Kuhn's 1977 list, McMullin (1982) proposes predictive accuracy, internal coherence, external consistency, unifying power,

---

[18] Space precludes discussion of this position here. For a survey of the relevant arguments, the reader is referred to Doppelt (2008), Haack (1993), Laudan (1984), and McMullin (1982, 2008). The very distinction between epistemic–non-epistemic values has been discussed at least since Rooney (1992).

fertility and simplicity. Doppelt enumerates: "epistemic values include properties of theories such as simplicity, unification, accuracy, novel in prediction, explanatory breadth, empirical adequacy, etc." (Doppelt 2008, p. 303). Lacey (1999) reviews a list of items suggested by a range of authors: empirical adequacy, explanatory and unifying power, power to encapsulate possibilities, internal consistency, connectivity or holism, inter-theory support, source of interpretive power, puzzle-solving power, simplicity and fertility.

How are values involved? A first question to address is whether the involvement is unavoidable (essential) or only possible. And, in the latter case, whether it is something to be avoided, i.e. whether value-laden science is bad science. Where are values involved in scientific inquiry and knowledge? The authors discern three broad areas of science: the fields under research, the hypotheses that are posed and the evidence that is taken to support one hypothesis over others and, finally, the use of these results to generate explanations. In the first case, a division is made between the natural and the social sciences, in the sense that it could be possible in principle to investigate in a value-free way in the first group but not in the second. Or values (both epistemic and non-epistemic) would be present in funding decisions: choosing which projects to fund implies evaluating both the soundness of the proposal and the priorities of the government or agency providing the funding. A deeper consideration goes into establishing the entities that populate the area to be researched. Also in explanations, the choosing of one factor over the (multiple) others as *the* cause of a phenomenon may be determined by value considerations (Longino 1990). In the second question, there is a particularly sensitive point: if non-epistemic values have a bearing on the selection and confirmation of hypotheses, then the term "value-laden science" acquires a much stronger meaning than in the previous cases. Lastly, what would be the effects of a value-laden science? The authors state that given the variety of issues and the multiple possible positions, this question will have many answers, related to the different ways in which values may be involved in science.

33.4 The Value-Ladenness of Science: Some Philosophical Perspectives

The following reviews a few examples of relevant philosophical approaches to the question of value-ladenness with regard to noncognitive values.[19]

---

[19] There are many interesting approaches to this problem in the recent literature in philosophy of science. Because it would have been impossible to address even a representative selection, three have been selected as a first approach to the range of views expressed. See Doppelt (2008); the papers in Dupré et al. (2007), Kitcher (2001), and Laudan (1984). Also, Douglas (2009) *Science, Policy and the Value-Free Ideal*, University of Pittsburgh Press, and Machamer and Wolters (2004) *Science, Values and Objectivity*, University of Pittsburgh Press.

In *Science as Social Knowledge* (1990), Longino sets out to address the relationship of science and values in the terms of her *contextual empiricism*. She calls it a "modest" empiricism related above all to epistemology and the notion that knowledge depends on experience, and much less to metaphysics (see later discussion). *Experience* constitutes the basis for knowledge claims, coupled with an emphasis on *context* in a twofold sense: with regard to background assumptions to reasoning; and to the cultural milieu in which scientific inquiry takes place. Data cannot be considered as evidence per se: whether some fact or state of affairs will be considered relevant evidence is determined not with reference to natural relations but in connection with background assumptions. These may convey, on one hand, "constitutive values" internal to science and expressing cognitive virtues; and on the other hand, "contextual values" expressing social or practical interests. Background assumptions introduce contextual values into proper scientific inquiry: contextual values "guide interpretations and suggest models within which the data can be ordered and organized" (1990, p. 219). Their presence is not the consequence of methodological limitation or error: it does not imply bad science. Nonetheless, methodology does have a role to play: not all values are admitted without restrictions. Judgment about data changes when the meaning of terms is adjusted, but if the meaning were the same, the same earlier judgments would be made. Also, observational judgments may change places regarding their significance within a theory when assumptions change. Then, central judgments may become peripheral, and previous seemingly unimportant judgments may become significant.

Hence, we can discern an empirical dimension of science, concerning evidence retrieved from observation and experiment, and a theoretical dimension of science. Both are linked by evidential reasoning: reasoning from and to data and hypothesis. Reasoning is understood by Longino as a practice. It is not mere decontextualised computation, but an interaction that takes place in a context. And hypothesis may also change when contextual assumptions change: there is interaction between background assumptions, general theoretical perspectives and experience.

Longino emphasises the differences between her outlook and those of positivism and realism: observation and reason on their own are not enough. They are supported by assumptions that express social and cultural values. But she also distinguishes her position from the relativism linked to holism: not all statements are context relative in the same way. The role of social and contextual values does not rule out objectivity. The scientist's desires for some kind of knowledge may configure the objects of her inquiry, but the existence of background assumptions that introduce social and contextual values becomes a basis for relativism only in an individualist conception of scientific inquiry. Scientific inquiry is the undertaking of a community, and not of individual researchers. It depends on the collaborative social interactions of "transformative interrogation". In this way, the impact of subjective preferences is minimised through criticism and interrogation by the scientific community. The more diverse and heterogeneous the community, the more diverse their assumptions will be. More of them will be made explicit, scrutinised and eventually modified. Discovering where inferences and experiences differ presupposes a minimum communicative context: shared standards for criticism, recognised public forums for its presentation, community responses and an equality of intellectual authority (Longino 1990, pp. 76–81). This also leads Longino to support a minimalist realism: "there is a world independent of our senses with which those senses interact to produce our sensations and the regularities of our experience"(Longino 1990, p. 222). Criticism and interrogation go a long way to minimise the impact of assumptions, but they cannot eliminate them: those assumptions shared by all the members of the community will not be made explicit; they will remain invisible and thus evade examination.

Furthermore, and directly relevant to science education aims, Longino states that the view that science is value-neutral may have the undesirable effect of disempowering non-scientists from understanding not only the technical, disciplinary content of inquiry that inform technologies but the contextual dimensions that shape inquiry. So they will have fewer possibilities of being adequately critical when dealing with the products of those technologies (Longino 1990, p. 225).

Lacey (1999, 2009) partially agrees with Longino's thesis, insofar as cognitive and noncognitive values may be clearly distinguished. Science should aim at empirically grounded and confirmed knowledge and understanding of phenomena. Inside these limits, it may be conducted within a plurality of worldviews and their associated value outlooks. "Science is value-free" is not meant in the sense that science and values don't touch. Science itself can be regarded as a value (as far as knowledge is a value); value judgments may be informed by scientific knowledge of the relationship between means and ends; scientists must display personal and moral values in their scientific practices (the "ethos of science"); and so on. However, these interplays should not touch the three component notions: impartiality, neutrality and autonomy. Neutrality means that scientific theories and practices do not imply or favour any value judgments or outlooks, cognitively or with regard to applications. Impartiality entails that criteria for the appraisal and acceptance of theories or the making of scientific judgments should not include noncognitive values. Autonomy asserts that scientific communities claim sole authority in the choice of problems, the evaluation of theories, the content of scientific education and in prerequisites for being admitted to the scientific community.

Lacey revises these three notions. Impartiality presupposes that cognitive and noncognitive values can be discriminated. Theories are accepted if and only if they display cognitive values to the highest degree, in agreement with relevant empirical data and other accepted theories. They should be subject to the most rigorous standards of evaluation. Neutrality presupposes, first, impartiality. Also it entails that there are accepted scientific theories that are significant for every viable value outlook and that no value outlook is noticeably favoured by accepted scientific theories. The notion of *value outlook* means that different kinds of values (moral, social, political, etc.) may be ordered coherently and rationally founded by a set of presuppositions about nature and human nature and about what is possible. Such presuppositions may be scientifically investigated to some extent. Inquiry may then support or oppose the presuppositions of a given value outlook. So, for a value outlook to be *viable*, it must be in accord with the results of accepted scientific knowledge. Scientific knowledge constrains the range of viable value outlooks, but still leaves a plurality of them open to explore. Lastly, autonomy is subordinated to impartiality and neutrality. Accepting a theory implies exclusively the play of cognitive values. But since the previous moment of adopting a strategy leaves an important role for social values to play, autonomy cannot be well embodied.

Modern science has been predominantly associated with a particular worldview: materialism. This has led to science being conducted under a certain strategy that constrains potentially admissible theories to those that can

> represent and explain phenomena [...] in terms that display their lawfulness, thus usually in terms of their being generated or generable from underlying structure and its components, process, interaction and the laws (characteristically expressed mathematically) that govern them (Lacey 2009, p. 843).

Consequently, empirical data are chosen and reported typically in terms of quantitative categories, related to measurement and instrumental and experimental operations. Both data and theoretical representation of phenomena are "stripped of all links with values and dissociated from any broader context of human practices and experience" (Lacey 2009, p. 843). Lacey calls this way of conducting science the "decontextualised approach" (DA), which is associated with the "modern value scheme of control" (Lacey 1999b). Materialism is widely associated with the values of technological progress (VTP) that favour control of natural objects. This way of conducting science means that science is not neutral, but it can still be impartial, as long as cognitive and social values have a role in scientific judgments at different logical moments (i.e. adopting a strategy or soundly accepting a theory). But carrying out science within the DA is not the only possibility of conducting science impartially. The exclusive association of scientific research and materialism must be challenged not only in philosophy of science but in sound science teaching as well. With regard to applied science and technological innovation, DA may be useful to explain its *efficacy*, but it is certainly not enough to establish its ethical or political *legitimacy*. There is an interest in conducting systematic empirical research under other strategies which do not uphold materialism. For example, Lacey discusses how genetically engineered transgenic crops, produced under DA and in accord with VTP, differ from organic or ecological agricultural alternatives. These alternative strategies privilege sustainable ecosystems, agency and community and, therefore, different socio-economic relations of production (Lacey 1999, 2005, 2009).

Neutrality (in the sense discussed above) is better served by a plurality of strategies that address the interests of different value outlooks, instead of reducing options to the materialist strategies which exclude value-laden terms from theories and methodologies. Nonetheless, admitting a plurality of options does not support any or every worldview and value outlook, which may mean abandoning impartiality. For instance, religious outlooks inconsistent with accepted scientific results, such as intelligent design, are incompatible with the scientific attitude.

Spanish philosopher Echeverría (1995, 2008) characterises scientific activity as a value-laden transformation of the material world and of human beings. He proposes a variation on Reichenbach's distinction between context of discovery and context of justification into a new categorisation of four contexts: innovation, evaluation, application and education. Contexts of innovation and evaluation loosely resemble the original proposal of Reichenbach. However, context of evaluation widens to comprise the different practices scientists perform when they evaluate products of scientific activity as they are achieved: not only hypotheses and theories but also data, measurements, experiments, proofs, papers and other publications. Context of application relates to science-related activities that aim at attaining changes in the world. This includes the manufacturing of artefacts but also the modification of images, languages and social relationships. Expert consultation in problem solving is also part of this context. Context of education consists of two reciprocal activities: teaching *and* learning, not only of conceptual and linguistic systems but also of representations, scientific images, notations, operating techniques, relevant problems and instrument handling procedures.

Echeverría advocates for axiological pluralism in science: scientists share a common collection of values, practices, habits, goals and, of course, knowledge. Scientific practice is axiologically meliorative: new actions improve on previous actions because they increase the degree of satisfaction of some value or decrease some disvalue. Different contexts imply different activities and agencies (educational, investigative, evaluative and application-oriented). Therefore, axiological rationality does not exclude conflict of values, but it requires specific procedures for its resolution.

Core values may differ in scientific practice in different contexts. For instance, science teaching and learning hold their own criteria and procedures for enunciation, justification, evaluation and application of scientific theories, which may occasionally diverge from those present in other settings of scientific activity. It does not entail mere transmission of information or even knowledge. In the context of education, the main value is *"communicability* of scientific content to every human being and therefore a requirement for publicity"[20] (Echeverría 1995, p. 125). Scientific content alludes to knowledge but also to skills and abilities. And also, and here we find an occasion of tension, on one side it should aim at the normalisation of those knowledge, skills and abilities, but on the other it should foster freethinking, criticism and creativity. Furthermore, within the context of education itself, criteria for evaluation of popular communication of science may differ sharply from those of investigative or academic education. Therefore, education becomes a pertinent frame for the philosophical consideration of the historical and conceptual development of scientific content and practices, and of the interplay and conflict of values within them.

33.5 Science, Values and Science Education: Two Debates

While the foregoing outlines some of the richness and relevance of the question of values in contemporary philosophy of science, it is important to consider how philosophical discussions about values and science relate to science education.

---

[20] "...la *comunicabilidad* de los contenidos científicos a cualquier ser humano; de este se deriva la exigencia de publicidad"

This section begins by synthesising some interesting debates in two special issues of *Science & Education* (1999, 8(1) and 2009, 18(6–7)) to illustrate some of the questions at stake. Both publications share the same structure: an author posits a thesis on a controversial philosophical and metascientific question, and a number of colleagues respond to that core article. Authors come from diverse professional origins: philosophers, scientists, educators and historians engage in multidisciplinary interchange. Finally, the first author makes a synthesis and response to the critics or to the alternative viewpoints stated.

Many of the topics addressed in the first special issue (*Science & Education*, 1999, 8(1)), under the title *Values in Science and Science Education: A Debate*, concern philosophical discussion about the relationship of science and values, as outlined above in the *Science and Values* paragraph. Discussion here highlights the relevance of these issues to science *education* and readers are advised to refer to the actual papers for deeper understanding of the philosophical points. The first round of debate starts with a paper by Lacey (1999b). He states that different standpoints on human values and human flourishing may affect the strategies under which science is conducted. So, science education should foster not only a sound understanding of science knowledge and practices but a critical self-consciousness about scientific activity and applications. This means understanding how cognitive and social values interact in scientific activity, what the limits and risks of that interaction in the making of theoretical judgments are and how noncognitive values may affect the achievements of scientific inquiry. This understanding should also encompass the knowledge and evaluation of different points of view. For example, this would mean challenging the exclusive association of scientific research with materialism within science education. Debates should include not only scientific practices and results under the materialist strategy but also the desirable strategies to further human well-being. This, however, would not mean accepting those worldviews or value outlooks that clash with the scientific attitude, such as creationism. Also, Lacey still upholds, as a main goal of scientific education, the teaching of scientific knowledge, the methods for discovering and the criteria for evaluating it obtained within DA. This may even constitute the core of first approaches and experiences with science learning (Lacey 1999, 2009).

As Machamer and Douglas (1999) and Davson-Galle (2002) emphasise, Lacey's position entails making the values that inform science knowledge and research explicit so that alternative value outlooks become visible and evaluation becomes possible. In a response to Lacey's paper, Cross (1999) indicates that these suggestions for science education belong within a rich tradition of like-minded proposals from different ideological perspectives. He points out that Lacey's call for sound and critical understanding of scientific activity and applications may be interpreted in two different ways. On one hand, it may be thought of as a somewhat insufficient proposal to discuss science from within a "traditional conception of scientific literacy". This option entails the difficulty of integrating these metadisciplinary aspects of science in the science curriculum, where they tend to be relegated to the periphery as mere illustration, and traditional content and values will still constitute core science teaching. Alternatively, he recommends a more "transformative" scheme to initiate a thorough revision of science education in schools in order to promote citizens' engagement and participation.[21] In his *Reply*, Lacey agrees with the desirability of teaching science in such a way that it enables an integration of the philosophical, historical, economic and social context of science. This would mean favouring a richer understanding of science for those who will go on to be active scientists, but also for those who will not but still need to be able to judge the significance of scientific research and knowledge with regard to cognitive and social values.

Though not directly referring to Lacey's paper, Allchin (1999) discusses the relationship of science and values with an emphasis on the educational point of view. He argues for science teachers to explicitly address the various ways in which values and science may intersect: epistemic values in scientific research; non-epistemic values in individual practitioners of science and methodological provisos against potential biases; and social and cultural values' challenges and novelties deriving from scientific results. Teachers should help students develop the relevant skills for a critical consideration of these roles of values in science. And he argues that this will be best achieved through reflexive analysis of students' own "modest" scientific modelling practices in classrooms and of historical cases. Reflexive analysis may also foster argumentative skills oriented to public rational justification of values (ethical or otherwise) avoiding common sense, naïve recourses to individual, personal "feelings" or values.

The second *Science & Education* issue we will consider, *Science, Worldviews and Education* (2009, 18(6–7)), has a much larger scope than the first one, since the idea of "worldview" encompasses not only values-related questions but also metaphysical, ontological and epistemological standpoints (see Chap. 50). The focus here is on some arguments linked to the relationship of values and science and its consequences for science education.

What is a worldview? If we accept that worldviews imply ontological, epistemological, ethical and sometimes religious commitments, is there a scientific worldview or is science worldview neutral? How does science engage with religious, philosophical, ideological or cultural worldviews? How do philosophical systems relate to science (and the putative scientific worldview)? What are the educational consequences of these questions? Should science education inform student worldviews, promoting worldview-related beliefs and ways of life or should science be learned only for instructional purposes? Should they promote students' *acting* upon those beliefs and ways of life? Some of these questions aim for descriptive, factual research; others require normative, regulative argument.

The lead essay by Hugh Gauch, Jr. (2009), delineates seven key features or "pillars" of science, derived from AAAS and the US NRC position papers: realism, the presupposition that the world is orderly and comprehensible, the role of evidence, use of logic, the limits of science, the universality of science and its ambition to contribute to a meaningful worldview. Reasoning from those "pillars", and from a

---

[21] See Kutrovátz and Zemplén (Chap. 34), and Vesterinen, Manassero-Mas and Vázquez-Alonso (Chap. 58), for a discussion on research in sociology of science and science education and of STS and HPS traditions in science education. Also, Aikenhead (2006), Hodson (2011), and Pedretti et al. (2008).

discussion of scientific method, Gauch argues against naturalism as a necessary commitment for science[22] that "the presuppositions and reasoning of science can and should be worldview independent, but empirical and public evidence from the sciences and humanities can support *conclusions* that are worldview distinctive" (p. 667, emphasis added). So, science does not imply distinctive worldview beliefs. Scientism is unacceptable. While science can say something about worldviews, it is not the exclusive provider of knowledge; philosophy, religion and art can offer it, too. This means that natural theology is not impossible in principle: empiric scientific evidence may address conclusions that derive from non-natural premises. So, from Gauch's standpoint, worldview-specific implications have a place in individual beliefs and in public debate, but not in institutional (including educational) requirements. In his paper, Fishman (2009) agrees that science does not presuppose naturalism as an a priori commitment, and therefore, science can evaluate supernatural theses or claims as far as evidence supports them. So, the rationale for not teaching intelligent design as an alternative to evolution in public schools (for instance) rests on the basis that evidence does not support it, regardless of whether it is labelled as "natural", "supernatural", "religious" or "paranormal".

An interesting historical illustration of the relationship between a theistic worldview and a demand for empirical evidence to support belief may be found in Matthews' article on the life and works of Joseph Priestley (Matthews 2009a). Irzik and Nola (2009) oppose the idea that science can be worldview independent: since a "pillar of science" answers a worldview question, and it also contradicts other possible answers (other worldviews), it cannot be assumed to be worldview independent. For example, they contend that Gauch presents methodological naturalism as a "mere stipulatory issue", while they claim it to be one of the "essential and distinctive features of science": abandoning it is tantamount to abandoning science (p. 741).

However relevant this discussion, worldviews issues should not be reduced to metaphysics questions such as the existence of God or the purpose of the universe. Irzik and Nola (2009) present a series of questions worldviews seek to provide answers for. Among them are questions such as "How should we live our lives?" "What is good and bad, right and wrong?" and "What is the best form of government?" that deal with ethical and political issues. The authors argue that it is possible to offer naturalistic answers even to these questions.

Do worldviews necessarily imply particular value outlooks? What are the consequences of adopting a scientific worldview for the understanding of moral behaviour? What are the consequences for the conduct of science in adopting value outlooks associated with non-scientific worldviews?

---

[22] For the sake of concision, we will refer to methodological naturalism as claiming that natural entities and means only can be called upon in scientific knowledge and practices and that natural sciences are a paradigm of epistemic research. This does not exclude by itself the existence of supernatural beings. Ontological naturalism states that only natural entities *exist* as a content of reality and no supernatural explanations whatsoever are acceptable. Irzik and Nola (2009, p. 733) point out that in some versions of naturalism, mentalistic and even mathematics items can be legitimately involved, since "natural" not necessarily implies "physical". Finally, materialism or physicalism affirms that only *material* (i.e. physical) entities exist.

33 Philosophical Dimensions of Social and Ethical Issues in School Science...

The relationship between religion, philosophy and science (or natural philosophy) has long been the object of examination. It fuelled heated philosophical debate in the Middle Ages and has been a leitmotif in studies of the Copernican Revolution. Also, evolution-related issues frequently involve arguing on this relationship. Recently, Stephen Jay Gould coined the acronym NOMA to refer to one of the stances in the debate: the idea that science and religion each has a legitimate *magisterium* or teaching authority (from *magister*, Latin for "teacher"), and these *magisteria* do not overlap. Science has nothing to say about the domain of ethics, while religion rightfully deals with questions of ultimate meaning and moral value. Furthermore, scientific and religious *magisteria* do not exclude other possible inquiries (for instance, philosophical ones).

On the other hand, it may be argued that deciding which theories belong in the scientific canon and whether they have worldview content is something that cannot be determined necessarily or a priori (Cordero 2009, pp. 757–8). In fact, the limits for scientific understanding of the world (Gauch's Pillar 5) change over time. The scope of naturalistic perspectives, such as Darwinist theory of evolution, has been widened to explain human mind and culture, including political and moral issues. Sociobiology and evolutionary psychology endorse the idea of a naturalistic approach to ethics and moral behaviour. For instance, Ruse and Wilson (2006) contend that offering materialistic explanations for the basis of human culture and human mind undermines the foundations of any a priori philosophical or religious (extra material) ethics. Evolutionary biology and cognitive psychology may be expected to explain feelings of right and wrong, and so offer a basis for morality; "a naturalistic ethic developed as an applied science" (Ruse and Wilson 2006, p. 558). But showing the links that go from factual premises to normative conclusions (from *is* to *ought*) is harder than it seems: "the connections between biological facts and questions about the status of morality are extremely complicated" (Kitcher 2006, p. 181)[23]. This cautions us against assuming that the relationship between biology and ethics in questions pertaining to, for instance, moral objectivity or moral progress, has been addressed in a way that does justice to its complexity. But it does not mean discarding research aimed at connecting biological knowledge and moral philosophy. Cordero (2009) argues that scientific content and scientific methodology can establish some constraints on which worldviews should be marginalised.

Conversely, a case can be made to show how religious worldviews imply an associated value outlook that includes a set of moral imperatives, some descriptive and normative framework of the mutual relationships of human beings and of human beings and nature. And Fishman argues (2009) that this poses a renewed challenge to science educators: how to maintain the desirable intellectual integrity without offending students in a way that impedes science education. Also, Reiss (2009) focuses on scientific and religious comprehensions of biodiversity to show how science and religion can be seen as distinct or related worldviews, depending on the aims of school science education. Arguments may be raised for and against

---

[23] See his outline of three meta-ethical questions for an example of how complicated this can become.

teaching about religion in science classes. Reiss reviews some of them, and contends that science teachers can be respectful of students' personal positions while at the same time fostering their engagement with science and helping them understand the strengths and limitations of science with regard to specific issues.

33.6 Science Education and Philosophy Education: Teaching Controversial Issues

Even with the previous brief summary, it can be seen that teaching about the role of values in science content and practice may present diverse challenges to science (and other) teachers. Certainly, it will seem controversial to those who still think science teaching should aim at the transmission of facts and the scientific method (or methods), and to those who think of science as value-free. They will probably argue that ethical, aesthetic, political or social issues are simply not relevant to science teaching. Even cognitive values may not be easily identified or taken into consideration by science teachers (Figueiredo Salvi and Batista 2008). Furthermore, those who are committed to taking nature of science into consideration in their classes may find it hard to decide how much time to devote to values-related issues. Teaching controversial issues with an acceptable degree of depth might take too much time, detracting from "regular" scientific issues demanded by the curriculum. This could be perceived to be a loss to "real" scientific content, and not even central to nature of science teaching. Also, value issues imply taking into consideration a multiplicity of variables: incomplete or insufficient information, uncertainty and risk, multiple and sometimes incompatible ethical and political philosophical, psychological and sociological frameworks (Ratcliffe and Grace 2003; McKim 2010). Science teachers may feel daunted by this complexity. Some of them may feel uncomfortable with their lack of information or expertise regarding such issues as ethical or political theories, or moral reasoning. And they may not have adequate resources and strategies to cope with the teaching of open-ended issues. Also, teachers may feel concerned about being suspected of bias or of having a hidden personal or political agenda. Some issues may be "too" controversial in particular institutions, for instance, those that collide with religious faith, which is the case with many bioethical questions such as abortion or stem cell research (Grace 2006). So why bring value issues into the science classroom?

Nowadays the idea that the aims of science education should include teaching some significant scientific knowledge along with some metascientific knowledge and understanding about the nature of science is broadly accepted. It has long been argued that the purposes of science teaching at school level must expand from the traditional initial training of the next generation of scientists. It should involve preparing students who will not go on to a scientific career for a more ready, responsible and adept participation as citizens in scientific and technologically related issues. In a well-known work, Driver and colleagues (1996) stated five rationales for teaching about the nature of science, two of which are relevant here: the democratic argument and the moral argument. On one side, students should leave school with enough understanding of scientific knowledge and practices that they may be able to appreciate and engage with the dilemmas posed by value-laden ethical, political and socio-scientific issues. On the other, they should develop a significant awareness of the norms that guide the activities of the scientific community, norms and values that are of general worth. Students should be able to explore their own (frequently tacit) knowledge, beliefs and values, and learn to assess science and technology knowledge and artefacts with regard to individual, social and global responsible action concerning that knowledge and those artefacts. Also, ethical and citizenship education is an educational aim that can be enhanced by including reflective consideration of the traditional Mertonian values, the so-called counter-values (Mitroff 1974), and other epistemic and non-epistemic values in scientific practice.

The teaching of the role and impact of contextual values, especially with regard to biotechnology and environmental challenges, has been fully addressed in the SSI literature,[24] and it is not possible to review it here. Discussion here focuses on some questions related to ethical issues and education that have been brought to the fore both in science and in philosophy education from a philosophical point of view. This does not imply teaching philosophy of science (or philosophy in any other sense) in the science classroom as a pure discipline (Adúriz-Bravo 2001; Matthews 1994/2014).

Although from a philosophical point of view it would be difficult to find any accord on the nature of science, from the standpoint of the wider community of science education (including philosophers but also historians of science, sociologists, natural scientists, science educators, science communicators, policymakers, science teachers), a corpus of themes or strands worth teaching on the nature of science can be found. And values-related issues regularly appear in this corpus.[25] Furthermore, noting that this description is intrinsically problematic (but not arbitrary) may encourage a richer and more comprehensive construction of the ethical, social and political aspects of science. In this way, "straw-man" perceptions of science as a deceiving harbinger of oppression, or as a faultless and providential supplier of truth and human progress may be avoided or at least discouraged (Adúriz-Bravo 2005b). Kitcher (2001, p. 199) describes these two images as (i) the version of the faithful, "which views inquiry as liberating, practically beneficial, and the greatest achievement of human civilization", and (ii) the image of the detractors that portrays science as "an expression of power, a secular religion with no claims to "truth," which systematically excludes the voices and the interests of the greater part of the species".

On the practical side, many curricula from different countries[26] regard understanding of values-related, socio-scientific and ethical issues as an important goal.

---

[24] Jones et al. (2010), Ratcliffe and Grace (2003), Zeidler and Sadler (2008), Sadler and Zeidler (2006), Zeidler and Keefer (2003), Zemplén (2009), also Kutrovátz and Zemplén (2014).

[25] Adúriz-Bravo (2005b), Osborne et al. (2003), and Lederman et al. (2002)

[26] Among others, AAAS (1993), McComas and Olson (2002), Conseil de l'Education et de la Formation (1999), National Research Council (1996), OECD (2001), Consejo Federal de Cultura y Educación (2006), and Secretaria de Educação Básica Brasília (2006).

Also, many of these issues occupy a relevant space in the media. They are present in social and political debates, and even day-to-day subjects of discussion, decision-making and social and political participation. The topics may range from global dimensions (e.g. nuclear plant safety, which was lately in the news after the Fukushima Daiichi incident), to regional or national dimensions (for instance, grandparent's DNA evidence in determining the filiation of the *desaparecidos* (Disappeareds') offspring in Argentina), to personal decisions (shall I vaccine my children against H1N1 influenza or MMR?). Science teachers may help students acquire the relevant scientific information that has to be considered in these controversies. They may provide conceptual and procedural knowledge that enables students to evaluate scientific content, distinguishing mere opinion from well-substantiated evidence. They may highlight the importance of sound scientific knowledge to appraise the premises in the arguments provided, therefore inspiring new interest in the relevant scientific content. Students would then become acquainted with a more accurate portrait of the nature of scientific activity and the work of scientists, and develop a more sophisticated awareness of the scope and limits of scientific knowledge that will help them to evaluate the reasons provided and the conclusions arrived at.

Conner (2010) proposes another rationale: the appreciation of cultural determination of the solutions to ethical problems. This would help "develop tolerance and an appreciation of other viewpoints". Some questions can be noted here. For instance, we find again the normative–descriptive choice of stance. What is and what should be the relationship between philosophical ethics and social sciences empirical research? What *empirical* psychological or social traits do philosophical ethics theories presuppose? Are they sound? What do anthropological or sociological differences between cultures tell us about the different values embraced and the distinctive ways of resolving conflicts? What are the *normative* principles about truth, duty, happiness, a good life and justice? What are the different perspectives on toleration and rights that may apply?[27] On one hand, philosophical ethics or theory of knowledge may be supported and influenced by empirical findings (though some philosophers will argue that this influence is inappropriate). On the other, empirical research may be shaped by (sometimes tacit) philosophical frameworks. This can be extended to the teaching and learning of ethics (in science classroom and elsewhere): should we present different ethical perspectives and help students make informed decisions about them? What would be the impact on actual behaviour? Should teachers aim at the internalisation of specific moral codes? Or should they present a plurality of moral systems and a diversity of ethical theories and help students develop the requisite skills and attitudes conducive to making autonomous, reflective choices? Ethics teaching in a vocational education setting, for instance, would strongly aim at an impact on students' beliefs and behaviour. Discussion on research ethics in most scientific schools, of the moral status of animals in Veterinary school, or reproductive ethics in Medical school (particularly in Obstetrics and

---

[27] For a first approach to the philosophical problems posed by the notion of tolerance, see Forst, R. (2012), Toleration, In E.N. Zalta (Ed.), *The Stanford Encyclopedia of Philosophy* (Summer 2012 Edition, forthcoming). http://plato.stanford.edu/archives/sum2012/entries/toleration/.

Gynaecology) (Gillam 2009; Rollin 2009), for instance, will aim at fostering behaviour and exceed the understanding of ethical theories and the training in ethical reasoning typical of Humanities education.

Reiss (2008) enumerates four possible aims for teaching ethics in the science classroom: heightening students' ethical sensitivity, increasing students' ethical knowledge, improving students' ethical judgment and making students better people. Increasing student knowledge does not necessarily imply moral transmission. Moral transmission aims at the internalisation of a set of principles, values and rules. It implies the choice of a particular value outlook and a specific conception of the good life; it intends for students to adopt that outlook and that conception, and act accordingly. Ethical inquiry approaches, instead, aim at helping students develop the requisite skills and attitudes conducive to making autonomous, critical and reflective evaluation and choices on specific values, actual moral codes and ethical dilemmas (Gregory 2009). Since there is no philosophical consensus regarding the relationship of values and science, a plurality of views can be addressed. However, this standpoint requires a certain engagement with a minimal set of moral principles and does not imply a relativistic stance. From a Habermasian point of view, for instance, it would presuppose the requisite conditions and rules to conduct a genuine dialogue, that is, rational argumentation as the only authority, open choice of problems, unrestricted participation of any interested party that can make a relevant contribution, no coercion exercised, equal possibilities for everyone of expressing themselves and everyone being internally free to be honest and not deceive others or oneself.[28]

Political literacy, as an indispensable stage in the development of a democratic society, presupposes an education that, among other things, fosters skills related to rational deliberation and argumentation in the elucidation and resolution of social and political conflict (Gutmann 1999). From this point of view, philosophy, civic and science teachers can be regarded as epistemic agents, able to evaluate beliefs (own or other's) and to rationally sustain or change them. They should also be prepared to educate their students to attain the same abilities and dispositions.

There are no shared criteria in philosophy (including ethics) to identify universal problems or methods (since Plato and the Sophists to the twentieth-century analytic–continental divide) and no reaching a satisfyingly wide and stable consensus (Rabossi 2008; Rescher 1985). Value-laden questions are common in the social sciences and the humanities, and explicitly explored and addressed in philosophy education. Philosophy and civics education entail teaching and learning to pose questions, to analyse potential answers and to make decisions and act responsibly upon them. Students who encounter philosophy for the first time frequently have difficulties in dealing, intellectually and emotionally, with the mix of lively debate; rigorous, sophisticated reasoning; and the impossibility of reaching a sole, commonly accepted answer, typical of the philosophical outlook.

---

[28] See also Ratcliffe and Grace (2003) (pp. 21–24 & pp. 29–32) for related questions in environmental education.

Since one or more philosophy courses (including ethics) have long been mandatory in the French, German, Italian, Spanish and most South and Central America curricula, a rich corpus of research and scholarship has been growing on philosophy education in the last few decades.[29] In particular, relevant work is taking place on topics such as how to teach and learn to ask philosophical questions and the art of posing relevant problems. Just as science education research has shown students (and even teachers) to display a tendency to "black or white" positions, from a naïvely realistic and dogmatic view of science to a (equally naïve) relativism or scepticism, a similar phenomenon has also been reported with regard to ethics teaching and learning (Allchin 1999; Paris 1994). Naïve dogmatic students display an unjustified belief in rules or principles learned from their family, friends, school, church or the media. They may be reinforced by teachers who adopt some normative ethical stance (say, Kantianism or rule utilitarianism) and present it as the best (or even the only) one. Crude relativistic students tend to regard everything as a matter of taste or opinion and lack any form of justification beyond personal likes or dislikes. Consideration of these issues is usually part of the training and practice of philosophy teachers. Research from the Philosophy for Children programme and from subsequent investigation in the field (UNESCO 2007; Kohan 2005) can offer interesting approaches to ethics education in primary and secondary education classrooms.[30] In science classes, tendencies to dogmatism could lead to a simple presentation of the "right" values in science (usually epistemic ones) coupled with a hagiographic (Adúriz-Bravo and Izquierdo-Aymerich 2005) view of scientists. On the other hand, presenting the class with a variety of ethical stances may lead them to an increasingly relativistic picture of ethics, wherein anything may be acceptable so long as the right ethical theory that supports a particular position may be found.

Value issues in science education share philosophical traits such as their intrinsic controversial nature, the lack of a single or even a more or less commonly accepted framework for analysis, the complexity of the discussions and the inexistence of a simple answer or commonly accepted conclusion for debates. This makes their teaching a particularly promising ground for the interaction and communication of different curriculum subjects that are usually kept apart. This interaction will be more fruitful on those occasions when science teachers adopt a controversial issue perspective rather than a descriptive one (Kolstø 2008). Different standpoints

---

[29] See *Diotime* (on-line magazine on the teaching of philosophy, in French) and *Paideia* (the magazine of the Spanish Association of Philosophy Teachers). In English, *Teaching Philosophy* devoted to the discussion of the teaching and learning of philosophy since 1975. Also the American Philosophical Association publishes an on-line *Newsletter on Teaching Philosophy*. UNESCO (2007) has put together a comprehensive study of the status of the teaching of philosophy in the world. Recently a new series of regional documents expanding on the data presented in the 2007 study have been published. http://unesdoc.unesco.org.

[30] See Kasachkoff (2005) for an example of how a class of ethics may proceed. Items in the APA *Newsletter on Teaching Philosophy* (free on-line access) outline other approaches to ethics teaching. For a perspective on moral education from a theory of care point of view, see Noddings and Slote (2003).

regarding specific moral and ethical vocabulary and problems, such as freedom, responsibility, diverse theories on moral good, duty and rights that are frequently attended to and discussed in philosophy classrooms, can be reviewed and engaged with in science classrooms.[31] For instance, Conner (2010) presents and comments on several approaches to bioethics education: values clarification and values analysis; individual or collaborative inquiry approaches; futures thinking models. Ratcliffe and Grace (2003) discuss several structured learning strategies for whole-class and small-group discussion.

A significant consideration of ethical issues also entails core skills in reasoning, dialogue and argumentation. Value issues in science education are both dependent on and can contribute to developing these abilities.[32] When debating values-related issues, students should have the opportunity to consider the different, and sometimes conflicting, reasons there might be for accepting or opposing a position. And they should be able to foster the abilities to reason soundly about them and evaluate others' reasoning. But argumentation on its own is not enough to ensure the development of good ethical thinking. Reiss (2010) argues for three complementary criteria: first, the arguments must be reasonable. This does not mean that they must exclusively answer to the formal (classical) deductive logic validity criteria. Rational revision and change of beliefs imply the capability of logically, pragmatically and rhetorically producing and evaluating arguments, that is, the knowledge and correct application of both formal (validity) and non formal (acceptable, relevant and sufficient) criteria (Govier 2010). Sound arguing also requires a good understanding of the meaning of the terms employed and of whether the premises are warranted.[33]

In social, ethical and political issues, the premises may originate in natural, social or philosophical theories, making communication across the disciplines not only possible but highly desirable. A working understanding of ethical frameworks that have been developed throughout the history of philosophy is a fundamental help in avoiding those naïve extremes sketched above or mere common sense exchanges. There is no unique, universally accepted set of moral values, and in most multicultural societies there is only a partially shared set of commonly accepted values. But there are moral traditions (religious or otherwise) and philosophical theories that can give a framework to moral reasoning. So, in the second place, moral argumentation should be placed within an established, explicit ethical structure. Virtue ethics (whether of an Aristotelian persuasion or not), consequentialism (for instance, utilitarianism), deontological theories (Kantianism) and other philosophical frameworks may provide teachers and students with sophisticated arguments and examples to support discussions on ethical issues in scientific activity and scientific outcomes. An additional question is whether moral reasoning can be independent from an ethical theory: Aristotle's way of reasoning may be thought of

---

[31] See Forge (1998, 2008) for a discussion of the issue of responsibility with regard to scientists and science practice.

[32] Zeidler (2003), Simmoneaux (2008), Zeidler and Sadler (2008), and Zohar (2008)

[33] Adúriz-Bravo and Revel Chion (2005), Adúriz-Bravo (2014), contributions to Erduran and Jiménez Aleixandre (2008), and Matthews (2009b)

as quite different from Kant's categorical imperative (Reiss 2010). Reiss outlines another issue worth exploring: the need to widen the moral community. He mentions two instances: interspecific ethics (Do animals have rights? Should those rights be taken into consideration, for instance, in experimentation with animals? (Rollin 2009; Lindahl 2010)); and intergenerational ethics (consequences for those far away in the geographical or in the historical sense, climate altered for generations, nuclear disposals). Lastly, conclusions must be arrived at as the outcome of genuine debate. Also, and this point would probably be contested by other philosophy educators, discussions should aim at consensus. Consensus should be understood as being coherent with one of the well-established traditions of ethical reasoning. At the same time, it cannot be permanent, but provisional, subject to discussion and change. Finally, Reiss emphasises that consensus should not be equated with a majority vote, since minority rights and interests and those of other interested parts without a voice (children, non-humans, the mentally infirm) ought to be taken into consideration.

Moral development of students is another relevant issue (Reiss 2010; Zeidler and Keefer 2003). Piaget (1948) presented his studies on the moral development of children in *The Moral Judgment of the Child*, based on how children viewed and responded to rules in the games they played. Later, Kohlberg developed Piaget's conclusions by conducting a series of interviews that included the discussion of several dilemmas (Kohlberg 1992; Kohlberg and Gilligan 1971). From the results of that research, he described three levels of development of moral judgment (preconventional, conventional and postconventional), each of them subdivided into two stages. The topmost stage presumed the notion of justice as a universal regulative principle, and the acceptance of universal, unconditioned ethical principles (akin to Kant's ethical theory). Therefore, his results should be considered as culture independent. Kohlberg's standpoint presupposed cognitive development, since changes in moral outlooks should be not only explicit but explained and justified by the individual. The passage from one stage to the other implied *learning*, in the sense of reconstructing and enriching the level below as a consequence of facing a dilemma that could not be solved satisfactorily. In the 1980s one of his disciples, Carol Gilligan (2003), criticised Kohlberg studies, indicating that all of the participants in them had been male. She proposed a new account that considered gender differences: a principle of justice on one side and a principle of caring on the other "voice" (as Gilligan called it) that constituted an acceptable basis for moral judgment. Caring implies a more contextualised point of view, different from the decontextualised, universal stance of Kant's deontological ethics. From a philosophical point of view, though, Kohlberg's project (and in this sense, Gilligan's too) should be reviewed. The idea of "development" implies a normative stance on morality, i.e. an ethical standpoint. It would be incorrect to assume simply that a Kantian conception of ethics could replace an Aristotelian or utilitarist outlook, or *vice versa*. Also, in each of the stages Kohlberg describes, morally responsible behaviour can coexist with a less responsible one, whether the underlying rationale corresponds to the Golden Rule or to a utilitarian analysis of consequences. Reiss (2010) suggests a set of indicators of progression in ethical reasoning that may be a useful tool for teachers in designing strategies that would help students move from a lower level of ethical thinking to a higher point. This set, though akin to Kohlberg stages, refers to different perspectives on normative ethics. Although it could be revisable from different philosophical and ethical points of view, it represents an interesting starting point for teachers to consider the sense of the progress they want to help their students achieve.

But it can be argued that in-depth teaching of ethics theories and moral reasoning cannot be the object of a natural science course: teachers will probably not have the time nor the specific knowledge or abilities to embark on it. A reasonable discussion of these theories would be impracticable, turning the science classroom into a course on normative ethics (Crosthwaite 2001). And other subjects in the curriculum, in the humanities or the social sciences, will probably find it more congenial and specific. Here again, the dilemma becomes less pressing in those educational systems that include a philosophy course in its own right. But, at the same time, natural, social sciences and philosophy teachers should have enough familiarity with their mutual frameworks to be able to orientate moral reasoning, and, where possible, open up fruitful collaboration between subjects or departments in schools. This does not mean ignoring that in most countries a discipline-based curriculum poses obstacles to cross-curricular cooperation. Although teacher education may not be a sufficient condition towards changing this situation, it may be thought of as a necessary one.

33.7 Conclusion

Science education does not need to encompass all aspects of education. Certainly, not every NOS issue may be fully addressed without running the risk of turning science classrooms into philosophy, history or sociology ones. Machamer and Douglas (1999, p. 53) put it nicely:

> We realize that no one can do this all the time, or else science would never get done. But to know how to do so, and that the possibility always exists for questioning the data, the reasons and the goals – this is what philosophy of science can teach, and what the history of science demonstrates is necessary.

On the other hand, philosophy teachers will address ethical or political problems from a different point of view. They would rightfully not feel constrained by the need to exclude perspectives that clash with the scientific worldview. So, there will probably be divergences with the aims and strategies of science education.

However, philosophy and science teachers, as all teachers, are educators in the broader sense of the term. As such, they need a wide knowledge of scientific, ethical and political issues, values and practices that would allow them to be critically aware and to avoid inadvertently conveying an unconsidered stance (a hidden curriculum), and here collaboration between science and philosophy education may have a fundamental role to play.

Also, this familiarity may enable teachers to respond to students' emergent or spontaneous questions in class and facilitate the interaction between colleagues from different disciplines in designing joint school projects which would benefit from a mutual acquaintance with each other's outlooks, values and languages. Active engagement in reflecting on standing practices and designing and implementing new ones seem to be particularly relevant to an adequate understanding of social, political and ethical issues in science knowledge and practices. Practising teachers may therefore find that these issues can be integrated in their own ongoing teaching strategies and, hopefully, act upon it.

Acknowledgments The author would like to acknowledge the support of Universidad de Buenos Aires and Programa UBACYT – *Programa para el Mejoramiento de la Enseñanza de la Filosofía*, UBACYT 01/W518. Also, she wants to extend her appreciation for helpful suggestions and comments to all *Science & Education* reviewers of this paper and to Cristina González of Universidad de Buenos Aires.

References

Adúriz-Bravo, A. (2014) Revisiting School Scientific Argumentation from the Perspective of the History and Philosophy of Science. In Matthews, M. R. (ed.) *International Handbook of Research in History, Philosophy and Science Teaching*. Dordrecht: Springer

Adúriz-Bravo, A. (2005a). *Una introducción a la naturaleza de la ciencia: la epistemología en la enseñanza de las ciencias naturales*. Buenos Aires: Fondo de Cultura Económica.

Adúriz-Bravo, A. (2005b). Methodology and politics: A proposal to teach the structuring ideas of the philosophy of science through the pendulum. In M.R. Matthews, C.F. Gauld & A. Stinner (eds.) (2005). *The pendulum: Scientific, historical, philosophical and educational perspectives* (pp. 277–291) Dordrecht: Springer.

Adúriz-Bravo, A. (2001). *Integración de la epistemología en la formación del profesorado de ciencias*. Tesis de doctorado. Departament de Didàctica de les Matemàtiques i de les Ciències Experimentals, Universitat Autònoma de Barcelona.

Adúriz-Bravo, A. & Izquierdo-Aymerich, M. (2005). A research-informed instructional unit to teach the nature of science to pre-service science teachers. *Science & Education* 18, 1177–1192.

Adúriz-Bravo, A. & Revel Chion, A. (2005). Sharing assessment criteria on school scientific argumentation with secondary science students. In R. Pintó & D. Couso (eds), *Proceedings of the fifth international ESERA conference on contributions of research to enhancing students' interest in learning science* (pp. 589–592). Barcelona: ESERA.

Aikenhead, G. S. (2006). *Science education for everyday life – Evidence-based practice*. New York: Teachers College Press.

Allchin, D. (1999). Values in science: An educational perspective. *Science & Education* 8, 1–12.

American Association for the Advancement of Science (AAAS) (1993) *Benchmarks for science literacy*. New York, Oxford University Press On line at http://www.project2061.org/publications/bsl/online/index.php. Last retrieved December 2012

Ayer, A. J. (1952). *Language, truth and logic*. New York: Dover (Trad. esp. (1971) *Lenguaje verdad y lógica*. Bs.As.: Eudeba).

Bell, R. L. & Lederman, N. G. (2003). Understandings of the nature of science and decision making on science and technology based issues. *Science Education* 87, 352–377.

Carrier, M. (2012). Values and objectivity in science: value-ladenness, pluralism and the epistemic attitude. *Science & Education*, doi 10.1007/s11191-012-9481-5.

33 Philosophical Dimensions of Social and Ethical Issues in School Science... 1113

Clough, M. (2007). Teaching the nature of science to secondary and post-secondary students: Questions rather than tenets, *The Pantaneto Forum* 25. http://www.pantaneto.co.uk/issue25/clough.htm. Last retrieved September 2011.

Conseil de l'Education et de la Formation (1999). *Education scientifique, education citoyenne. Réaliser une alphabétisation scientifique et technologique, composante essentielle de l'éducation à la citoyenneté démocratique.* Avis n°67, Conseil du Septembre 1999.

Consejo Federal de Cultura y Educación (2006). *Núcleos de aprendizaje prioritarios. 3er Ciclo EGB / Nivel Medio.* Ministerio de Educación, Ciencia y Tecnología. República Argentina http://portal.educacion.gov.ar/secundaria/files/2009/12/nap3natura.pdf

Conner, L. (2010). In the classroom: Approaches to Bioethics for senior students. In A. Jones, A. McKim & M. Reiss (eds.), *Ethics in the science and technology classroom.* (pp. 55–67). Rotterdam: Sense Publishers.

Cordero, A. (2009). Contemporary science and worldview-making. *Science & Education* 18, 747–764.

Cross, R. T. (1999). Scientific understanding: Lacey's "critical self-consciousness" seen as echoes of J.D. Bernal. *Science & Education* 8(1), 67–78.

Crosthwaite, J. (2001). Teaching ethics and technology – What is required? *Science & Education* 10, 97–105.

Davson-Galle, P. (2002). Science, values and objectivity. *Science & Education* 11, 191–202.

Doppelt, G. (2008). Values in science. In S. Psillos & M. Curd (eds.), *The Routledge companion to the philosophy of science* (pp. 302–313). Taylor & Francis e-Library.

Douglas, H. (2000). Inductive risk and values in science. *Philosophy of Science*, 67 (4), 559–579 Retrieved at http://www.jstor.org/stable/188707.

Douglas, H. (2007). Rejecting the ideal of value-free science. In J. Dupré, H. Kincaid & A. Wylie (eds.) *Value-Free Science? Ideals and Illusions* (pp. 120–139). New York: Oxford University Press.

Douglas, H. (2009) *Science, Policy and the Value-Free Ideal*, Pittsburgh: University of Pittsburgh Press.

Driver, R., Leach, J., Millar, R. & Scott, P. (1996). *Young people's images of science. Buckingham:* Open University Press.

Dupré, J.; Kincaid, H. & Wylie, A. (ed.) (2007) *Value-Free Science? Ideals and Illusions.* New York: Oxford University Press.

Echeverría, J. (1995). *Filosofía de la ciencia.* Madrid: Akal.

Echeverría, J. (2008). Propuestas para una filosofía de las prácticas científicas. In J.M. Esteban & S.F. Martínez Muñoz (eds.), *Normas y prácticas en la ciencia* (pp. 129–149). México: IIFs-UNAM.

Erduran, S. & Jiménez Aleixandre, P. (Eds.) (2008). *Argumentation in science education: Perspectives form classroom-based research.* Dordrecht: Springer.

Ferrater Mora, J. (1975). *Diccionario de filosofía.* Bs.As.: Sudamericana.

Figueiredo Salvi, R. & Batista, I.L. (2008). A análise dos valores na educação científica: contribuições para uma aproximação da filosofia da ciência com pressupostos da aprendizagem significativa. *Experiências em Ensino de Ciências* 3(1), 43–52.

Fishman, Y.I. (2009). Can science test supernatural worldviews? *Science & Education* 18, 813–837.

Forge, J. (1998). Responsibility and the scientist. In . Bridgstock, D. Burch, J. Forge, J. Laurent & I. Lowe (eds.), *Science, technology and society. An introduction* (pp. 40–55). Melbourne: Cambridge University Press.

Forge, J. (2008). *The responsible scientist.* Pittsburgh, PA.: University of Pittsburgh Press.

Forst, R. (2012), Toleration, In E.N. Zalta (Ed.), The Stanford Encyclopedia of Philosophy (Summer 2012 Edition, forthcoming). http://plato.stanford.edu/archives/sum2012/entries/toleration/

Frondizi, R. (1982). *¿Qué son los valores?* México: FCE.

Gauch Jr., H.G. (2009). Science, worldviews and education. *Science & Education* 18, 667–695.

Gillam, L. (2009). Teaching ethics in the health professions. In H. Kuhse & P. Singer (eds.), *A companion to bioethics* (pp. 584–593). Singapore: Wiley-Blackwell.

Gilligan, C. (2003, f.e.1982). *In a different voice.* Cambridge, MA: Harvard University Press. Trad. esp. (1985) *La moral y la teoría* México: FCE.

Govier, T. (2010) A practical study of argument. Belmont, CA: Wadsworth

Grace, M. (2006). Teaching citizenship through science: Socio-scientific issues as an important component of citizenship. *Prospero*, 12 (3), 42–53.

Gregory, M. (2009). Ethics education and the practice of wisdom. *Teaching Ethics* 9 (2):105–130. On line at www.uvu.edu/ethics/seac/Gregory%20%20Ethics%20Education%20and%20the%20Practice%20of%20Wisdom.pdf Last retrieved June 2012.

Gutmann, A. (1999). *Democratic education*. Princeton, NJ: Princeton University Press.

Haack, S. (1993). Epistemological reflections of an old feminist *Reason Papers* 18, 31–43.

Hodson D. (2011). *Looking to the future: Building a curriculum for social activism*. Rotterdam: Sense Publishers.

Irzik, G. & Nola, R. (2009). Worldviews and their relation to science. *Science & Education* 18, 729–745.

Jones, A., McKim, A. & Reiss, M. (eds.) (2010). *Ethics in the science and technology classroom*. Rotterdam: Sense Publishers.

Kasachkoff, T. (2005). How one might use "two lives" in the ethics classroom. *APA Newsletter on Teaching Philosophy*. 5 (2), 2–3 On line at http://www.apaonline.org/APAOnline/Publications/Newsletters/Past_Newsletters/Vol05/Vol._05_Fall_2005_Spring_2006_aspx Last retrieved June 2012.

Kitcher, P. (1993) *The advancement of science: Science without legend, objectivity without illusions*. Oxford : Oxford University Press (Trad. Esp. (2001) *El avance de la ciencia. Ciencia sin leyenda, objetividad sin ilusiones*. México DF: UNAM).

Kitcher, P. (2001). *Science, truth and democracy*. Oxford : Oxford University Press.

Kitcher, P. (2006). Biology and ethics. In D. Copp (ed.), *The Oxford handbook of ethical theory*. (pp. 163–185). Oxford : Oxford University Press.

Kohan, W. (2005). Brésil: regard critique sur la méthode Lipman (II). *Diotime*, 24. On line at http://www.educ-revues.fr/Diotime/affichagedocument.aspx?iddoc=32659&pos=0 Last retrieved May 2012

Kohlberg, L. & Gilligan, C. (1971). The adolescent as a philosopher: the discovery of the self in a postconventional world. *Dedalus*, 100 (4), 1051–1086.

Kohlberg, L. (1992) *Psicología del desarrollo moral*. Bilbao, Desclée de Brouwer. English version (1984) *Essays on moral development*. Vol. 2. *The psychology of moral development*. San Francisco: Harper & Row

Kolstø, S. D. (2008). Science education for democratic citizenship through the use of the history of science. *Science & Education*, 17 (8–9), 977–997.

Kuhn, T. (1962). *The structure of scientific revolutions*. Chicago, IL: University of Chicago Press. (Trad.esp. (1971) *La estructura de las revoluciones científicas*, México: F.C.E.).

Kuhn, T. (1977). *The essential tension*. Chicago, IL: University of Chicago Press. (Trad.esp. (1993) *La tensión esencial*, México: F.C.E.).

Kutrovátz, G. & Zemplén, G. (2014) Social Studies of Science and Science Teaching. In Matthews, M. R. (ed.) International Handbook of Research in History, Philosophy and Science Teaching. Dordrecht: Springer.

Lacey, H. (1999). *Is science value free? Values and scientific understanding*. London: Routledge.

Lacey, H. (1999b). Scientific understanding and the control of nature, *Science & Education* 8(1), 13–35.

Lacey, H. (2005). On the interplay of the cognitive and the social in scientific practices. *Philosophy of Science*, 72, 977–988 (Trad. Port. (2008) Aspectos cognitivos e sociais das práticas científicas. *Scientiæ Studia*, São Paulo 6 (1), 83–96.

Lacey, H. (2009). The interplay of scientific activity, worldviews and value outlooks. *Science & Education*, 18 (6–7) 839–860.

Laudan, L. (1984). *Science and values*. Berkeley, CA: University of California Press.

Lederman, N., Abd-El-Khalick, F., Bell, R. & Schwartz, R. (2002). Views of nature of science questionnaire: Toward valid and meaningful assessment of learners' conceptions of nature of science. *Journal of Research in Science Teaching,* 39 (6), 497–521.

Lindahl, M. G. (2010). Of pigs and men: Understanding student's reasoning about the use of pigs as donors for xenotransplantation. *Science & Education*, 19 (9) 867–894.

Longino, H. (1990). *Science as social knowledge: Values and objectivity in scientific inquiry*. Princeton, NJ: Princeton University Press.

Longino, H. (2011). The social dimensions of scientific knowledge. In E.N. Zalta (ed.), *The Stanford encyclopedia of philosophy (Spring 2011 edit.)*. http://plato.stanford.edu/archives/spr2011/entries/scientific-knowledge-social/>. Last retrieved March 2012.

Luna F. (2009). Elucidating the concept of vulnerability. *International Journal of Feminist Approaches to Bioethics*, 2 (1), 120–138.

Machamer, P. & Douglas, H. (1999). Cognitive and social values. *Science & Education*, 8 (1), 45–54.

Machamer and Wolters (2004) Science, Values and Objectivity. Pittsburgh: University of Pittsburgh Press.

Machamer, P., Pera, M. & Baltas, A. (eds) (2000). *Scientific controversies: Philosophical and historical perspectives*. New York: Oxford University Press.

Matthews, M.R. (1994/2014). *Science teaching. The role of history and philosophy of science*. New York: Routledge.

Matthews, M.R. (2009a). Science and worldviews in the classroom: Joseph Priestley and photosynthesis. *Science & Education*, 18 (6–7), 929–960.

Matthews, M.R. (2009b). Teaching the philosophical and worldviews components of science. *Science & Education*, 18 (6–7) 697–728.

McComas, W. F. (ed) (2002). *The nature of science in science education: Rationales and strategies*. Dordrecht: Kluwer (first edition 1998).

McComas, W.F. & Olson, J.K. (2002). The nature of science in international standards documents. In W.F. McComas (ed.), *The nature of science in science education: Rationales and strategies* (pp. 41–52). Dordrecht: Kluwer

McKim, A. (2010). Bioethics education. In Jones, A., McKim A. & Reiss, M. (eds.), *Ethics in the science and technology classroom* (pp. 19–36). Rotterdam: Sense Publishers.

McMullin, E. (1982). Values in science. *PSA: Proceedings of the Biennial Meeting of the Philosophy of Science Association*. Vol 2: Symposia and Invited Papers, pp. 3–28. URL http://www.jstor.org/stable/192030. Last accessed November 2011

McMullin, E. (2008). The virtues of a good theory. In S. Psillos & M. Curd (eds.), *The Routledge Companion to the Philosophy of Science* (pp. 498–508). Taylor & Francis e-Library.

Mitroff, I. (1974). Norms and counter-norms in a select group of the Apollo moon scientists: A case study of the ambivalence of scientists. *American Sociological Review*, 39, 579–595.

National Research Council. (1996). *National science education standards*. Washington, DC: National Academy Press.

Nelson, L. H. (2002). Feminist philosophy of science. In P. Machamer & M. Silberstein (eds.), *The Blackwell guide to the philosophy of science* (pp. 312–331). Malden, MA: Blackwell.

Noddings, N. & Slote, M. (2003). Changing notions of the moral and of moral education. In N. Blake, P. Smeyers, R. Smith & P. Standish, P. (eds.), *The Blackwell Guide to the Philosophy of Education* (pp. 341–355). Malden, MA: Blackwell.

OECD (2001). *Knowledge and skills for life. First results from PISA 2000*. Paris: Organisation for Economic Co-operation and Development.

Osborne, J., Ratcliffe, M., Millar, R., & Duschl, R. (2003). What "ideas-about-science" should be taught in school science? A Delphi study of the expert community. *Journal of Research in Science Teaching*, 40 (7), 692–720.

Paris, C. (1994). Réflexions sur l'enseignement de l'éthique. *Philosopher* 16, 63–71.

Piaget, J. (1948). *The moral judgment of the child*. Glencoe, IL: Free Press. (Trad. Esp. (1935) *El juicio moral en el niño*. Madrid: Francisco Beltrán Librería española y extranjera.

Pedretti, E., Bencze, L. , Hewitt, J., Romkey, L. & Jivraj, A. (2008). Promoting issues-based STSE perspectives in science teacher education: Problems of identity and ideology. *Science & Education*, 17, 941–960.

Putnam, H. (1990). *Realism with a human face*. Cambridge, MA: Harvard University Press.

Rabossi, E. (2008). *En el comienzo Dios creó el Canon. Biblia berolinensis.* Bs.As.: Gedisa.

Ratcliffe, M. & Grace, M. (2003). *Science education for citizenship. Teaching socio-scientific issues.* Maidenhead: Open University Press.

Reiss, M. (2008). The use of ethical frameworks by students following a new science course for 16–18 year-olds. *Science & Education* 17, 889–902.

Reiss, M. (2009). Imagining the world: The significance of religious worldviews for science education. *Science & Education* 18, 783–796.

Reiss, M. (2010). Ethical thinking. In A. Jones, A. McKim & M, Reiss (eds.), *Ethics in the science and technology classroom* (pp. 7–17). Rotterdam: Sense Publishers.

Rescher, N. (1985). *The strife of systems. An essay on the grounds and implications of philosophical diversity.* Pittsburgh, PA: University of Pittsburgh Press (trad. esp. (1995) *La lucha de los sistemas.* México: UNAM).

Rollin, B.E. (2009). The moral status of animals and their use as experimental subjects. In H. Kuhse & P. Singer (eds.), *A companion to bioethics* (pp. 495–509). Singapore: Wiley-Blackwell.

Rollin, B.E. (2012). The perfect storm - Genetic engineering, science and ethics. *Science & Education,* 1–9, doi:10.1007/s11191-012-9511-3. Last accessed November 2012

Rooney, Ph. (1992). On values in science: is the epistemic/non-epistemic distinction useful? *PSA: Proceedings of the Biennial Meeting of the Philosophy of Science Association. Volume One: Contributed Papers* (pp. 13–22) The University of Chicago Press. http://www.jstor.org/stable/192740. Last accessed December 2012

Rudner, R. (1953). The scientist *qua* scientist makes value judgments. *Philosophy of Science.* 20(1), 1–6.

Ruse, M. & Wilson, E.O. (2006, origin. pub. 1983). Moral philosophy as applied science. In E. Sober (ed.), *Conceptual issues in evolutionary theory*, (pp. 555–573). Cambridge, MA: MIT Press.

Sadler, T. & Zeidler, D. (2006). Patterns of informal reasoning in the context of socio-scientific decision making. *Journal of Research in Science Teaching,* 42(1), 112–138.

Secretaria de Educação Básica. Brasília Ministério da Educação (2006). *Orientações Curriculares para o Ensino Médio. Vol. 2: Ciências da Natureza, Matemática e suas Tecnologias.* URL: http://portal.mec.gov.br/seb/arquivos/pdf/book_volume_02_internet.pdf

Simmoneaux, L (2008). Argumentation in socio-scientific contexts. In S. Erduran & M.P. Jiménez-Aleixandre (eds.) *Argumentation in science education* (pp. 179–199). Dordrecht: Springer.

Stevenson, C. (1960). *Ethics and language.* New Haven, CT: Yale University Press. Trad. esp. de E. Rabossi (1971) *Ética y lenguaje* Bs.As.: Paidós.

UNESCO (2007). *La Philosophie, une École de la Liberté. Enseignement de la philosophie et apprentissage du philosopher : État des lieux et regards pour l'avenir.* Trans. in English as *Philosophy, a School of Freedom. Teaching Philosophy and learning to philosophize: Status and Prospects.* http://unesdoc.unesco.org/images/0015/001536/153601f.pdf (Fr) or http://unesdoc.unesco.org/images/0015/001541/154173e.pdf (Eng). Accessed April 2012.

Vesterinen, V.-M., Manassero-Mas, M.A. & Vázquez-Alonso, A. (2014) History, Philosophy and Sociology of Science and Science-Technology-Society Traditions in Science Education: Continuities and Discontinuities. In Matthews, M. R. (ed.) International Handbook of Research in History, Philosophy and Science Teaching. Dordrecht: Springer.

Wylie, A. & Nelson, L. (2007). Coming to terms with the values of science: Insights from feminist science studies scholarship. In J. Dupré, H. Kincaid & A. Wylie (eds.), *Value-free science? Ideals and illusions* (pp. 58–86). New York: Oxford University Press.

Zeidler, D. (ed.) (2003) The role of moral reasoning on socio-scientific issues and discourse in science education. Dordrecht: Kluwer.

Zeidler, D. & Keefer, M. (2003). The role of moral reasoning and the status of socio-scientific issues in science education. In D. Zeidler (ed.), *The role of moral reasoning on socio-scientific issues and discourse in science education* (pp. 7–38). Dordrecht: Kluwer.

Zeidler, D. & Sadler, T. (2008). Social and ethical issues in science education. *Science & Education* 17, 8–9.

Zemplén, G. (2009). Putting sociology first. Reconsidering the role of the social in "nature of science" education. *Science & Education* 18 (5), 525–559.

Ziman, J. (2003). *Real science. What it is and what it means.* Cambridge: Cambridge University Press. Virtual Publishing.

Zohar, A. (2008). Science teacher education and professional development in argumentation. In S. Erduran & M.P. Jiménez-Aleixandre (eds.), *Argumentation in science education.* (pp. 245–268). Dordrecht: Springer.

Ana C. Couló is Associate Professor of Didactics of Philosophy (i.e. Philosophy Education) in the Department of Philosophy at Universidad de Buenos Aires. She also teaches History and Philosophy of Science to in-service science teachers. Her undergraduate degree is in Philosophy, and she is an advanced student at the Ph.D. level at UBA. Her research interests include students' misconceptions and errors in logic, epistemology and philosophy of science and the professional development of philosophy, logic and science teachers. She has served as a high school teacher of Philosophy and has taught graduate courses in Logic and Philosophy of Science. She has published papers in educational journals such as *Paideia* and *Diotime, Revue Internationale de Didactique de la Philosophie*. Recent publications include "Systematic Errors as an Input for Teaching Logic", with Gladys Palau (2011), in Blackburn, van Ditmarsch, Manzano and Soler-Toscano (Eds.); *Tools for Teaching Logic*, Springer Verlag; and "Enseñar filosofía y aprender a filosofar: la experiencia del CPF". En Pérez y Fernández Moreno (comps) (2008) *Cuestiones filosóficas. Ensayos en honor de Eduardo Rabossi*, Catálogos, Bs.As. She has co-authored (with A. Cerletti) two compilations on philosophy teaching, *La enseñanza de la filosofía: teoría y experiencias* (2009) and *Enseñar Filosofía: enfoques y propuestas* (2013) OPFYL, Bs.As.

Chapter 34
Social Studies of Science and Science Teaching

Gábor Kutrovátz and Gábor Áron Zemplén

Sagredo: As educators of science, we face fundamental problems like what kind of science we ought to teach in schools, how much, in what manner, and for what purposes. In the past decades, these questions have received increasing scholarly attention, and the enormous complexity of the field of relevant issues and approaches has been broadly recognized (McComas 2000). "Nature of science" has been identified as a core problem: how the teaching *of* science is, or should be, related to, supported by, and reconciled with teaching *about* science. In other words, science education is now believed to serve several purposes, and providing students with scientific knowledge is only one of them. Another purpose is to convey a general understanding of science, the scientific method, the reliability of scientific knowledge, and how and in what form it is accessible and useful to nonscientists and what social, cultural, or educational roles science can and does have. These questions have been raised in various fields. Philosophers of science have been investigating the most general problems concerning scientific enterprise for at least a century now, and historians of science have their own, equally important and partly overlapping, tradition of studying the dynamics of science (Holton 2003; Machamer 1998). But what about sociology? Various views and approaches, together with their merits and drawbacks, limitations, and contexts, need to be studied to find out for ourselves what we, educators, can best learn from them. Here is my suggestion for today: let us have a look at some of the most important views and approaches within the social studies

---

G. Kutrovátz
Department of History and Philosophy of Science,
Eötvös Loránd University, Budapest, Hungary

G.Á. Zemplén (✉)
Department of Philosophy and History of Science, Budapest University
of Technology and Economics, Budapest, Hungary
e-mail: zemplen@filozofia.bme.hu

M.R. Matthews (ed.), *International Handbook of Research in History,
Philosophy and Science Teaching*, DOI 10.1007/978-94-007-7654-8_34,
© Springer Science+Business Media Dordrecht 2014

of science and submit them to the scrutiny of our discussion.[1] After all, we are all familiar with the pedagogical values of presenting ideas in dialogue form.

Simplicio: I see no reason why "philosophy" can't be replaced by "sociology" or "social studies" in a favorite quote of mine. According to Feynman, "philosophy of science is about as useful to scientists as ornithology is to birds" (Kitcher 1998, p. 32). Do you really think that "hard" sciences can be meaningfully analyzed with the toolkit of "soft" social sciences, having a far less secure methodological base than their subject? Should we teach "xxx studies" in the little time allotted to the "science of xxx"?

Salviati: "Most scientists tend to understand little more about science than fish *about* hydrodynamics," as the philosopher Imre Lakatos (1970, p. 148n) offered another concise judgment. That's a shame, as there are a number of things "about science" (Barnes 1985) that would-be scientists can find very important to learn in school. When they enter the scientific field, they are often unprepared to cope with what they encounter and what they did not learn from studying scientific theories. Moreover, not every student of science will become a scientist, and those who will not may find philosophical perspectives as useful as sociological ones. All in all, science seems to be similar to other aspects of human society.

Simplicio: Similar? Science is special exactly because it is dissimilar. In what relevant way would it be similar?

Sagredo: For example, power is concentrated in the hands of a small minority, just like in most areas of human activity. Sociological studies have revealed serious inequalities in science, an enterprise that seems, to many, to represent democratic values. A vast majority of all publications is written by a small minority of all scientists (Lotka 1926). And it is not only that some scientists try to publish more simply because they are more prolific, but the more credit one gets, the easier it is to publish. Moreover, publication forums also receive unbalanced attention: from the 30,000 scientific journals of the 1960s, approximately 170 were the most prestigious, and half of all the interest in libraries was focused on this tiny proportion of the entire body of journals (Price 1986, p. 67). Similar inequalities were identified in citation patterns: while, on average, every scientific paper is cited only once by later publications, in reality, most papers will never get cited later, which leaves us with a few influential and highly visible papers that others read, surrounded by a myriad of papers lost in collective ignorance (*ibid.* p. 73). Robert Merton, a founding father of the sociology of science, called this the Matthew effect (Merton 1968), based on a quote from the Bible: "For whosoever hath, to him shall be given, and he shall have more abundance: but whosoever hath not, from him shall be taken away even that he hath" (Matthew 13:12, King James Bible, Cambridge Ed.). Scientists learn to live with this distribution of power, but it may take them a long time to find

---

[1] The authors wish to thank the inspiring community of the EU 7th framework-funded HIPST project, acknowledge the influence of Prof.Art Stinner's didactic approach, and thank Zsófia Zvolenszky for commenting on the manuscript. Support from TÁMOP-4.2.2.B-10/1-2010-0009 and OTKA K84145, K109456.

out from bitter personal experience what they could have learnt already in school about the peculiar meritocracy of science.

Simplicio: I concur, every student should know a bit of sociology, probably about this much: if you look at science as an activity, you will see that it is an institutionalized subsystem or field of society, with its own rules to follow, values to respect, goals to pursue, norms to be governed by, etc. This activity is performed by a social group, that of scientists, which is structured both "vertically" in a hierarchical order, and through complex power relations, and "horizontally" according to disciplinary maps, national and geographical factors, to start with. Science relies on a number of different resources to comply with its primary function, i.e., the production of reliable knowledge, or empirically fruitful theories, theoretical and practical knowledge, or even legitimizing various democratic functions of the society, as Salviati may claim. Certain resources are "material" like sites of activity (institutes, universities, etc.), financial conditions (salaries, grants, scholarships), and other technical and instrumental resources. Others are "symbolic" like the structure of ranks, degrees, and honors or protocols of formal communication and publication and all kinds of norms and values as mentioned above. Sociology of science, as I see things, investigates precisely these things…

Sagredo:…and more. In the past few decades, all disciplines examining science have developed a growing sensitivity to insights coming from the social sciences. "Sociology of science," "sociology of (scientific) knowledge," "social studies of science and technology," and "science (and technology) studies" are promising labels referring to a complex array of research traditions investigating how the study of society and the study of science can inform and supplement one another.

Simplicio: Despite the field's popularity, it remains unclear to me from a science education perspective how relevant these approaches are and what lessons we can take home from consulting them. And here is my answer. If we introduce a distinction between the form of scientific activity and the content of scientific knowledge, I expect that sociologists have a lot to say about the first, but are wrong if they think they can teach us anything about the second. Sociology of science and sociology of knowledge are different enterprises. Proper sociologists of science restrict themselves to the study of the social dimensions of scientific activity, like the analysis of the Matthew effect, and this is what I find promising about sociology. On the other hand, sociology of knowledge addresses the problem on how social factors affect what and in what way a certain culture thinks about the world – their conceptual categories, systems of classification, and so on. I believe that this is where sociology is *not* able to examine science meaningfully. Even prominent figures of this field admitted that when it comes to scientific knowledge, these external factors cease to matter and the "sociology of knowledge" perspective loses all power (Mannheim 1936, p. 239, as interpreted by Bloor 1973).

Salviati: An outdated and narrow view, as I hope to show later. But supposing for a second that you are right, what are the "social dimensions of scientific activity" that sociologists can legitimately study?

Simplicio: Sociologists of science provided a number of statistical analyses of the institutional patterns of science. Investigations, similar to those of Merton, were carried out from the 1960s and 1970s by Harriet Zuckerman, Jonathan Cole, Stephen Cole, Joseph Ben-David, Derek J. de Solla Price, and many others. I would be very surprised if you told me that this quantitative trend, opening a perspective on a measurable science, is uninteresting or irrelevant to students who need to understand what science is.

Sagredo: True. For example, Merton's study of publication patterns included the analysis of rejection rates for the publication of manuscripts in various academic journals. By comparing these rejection rates in different academic fields he found that, perhaps contrary to our prior intuition, journals in the arts, humanities, and social sciences are more strict than journals in the natural sciences. While an average of 90 % of the manuscripts were rejected by the most important history journals considered by Merton in 1967 (closely followed by literature, philosophy, political science, and sociology journals), only 20–30 % of manuscripts were rejected in biological, physical, and geological journals (Zuckermann and Merton 1971). How interesting, as if candidates for publication found it far easier to comply with the publication standards of scientific fields than those within the humanities....

Salviati: This is a red herring... Surely, these isolated sets of data can be interpreted in many different ways (Hess 1997, pp. 65–66), and I am afraid it is too rash to jump to unjustified conclusions concerning the strictness of publication standards. I can readily offer other explanations for the phenomenon. Consider, for instance, a different statistical tool, introduced by another prominent sociologist of science, Derek de Solla Price (1986, pp. 166–179). The so-called Price Index tells us the proportion of those publications, cited by a scientific journal of field or discipline, that are not older than 5 years, relative to the number of all citations. What he found in the 1960s (and this hasn't changed significantly since his time) is that the proportion is 60–70 % for hard sciences like physics, while only 20 % for philosophy. What this means is that in hard sciences, the immediate knowledge base that researchers rely on is far more up-to-date and recent than in softer fields. In my opinion, this fact discourages amateurs (i.e., alleged contributors with insufficient professional background) from submitting their pet theories to science journals, seeing how unable they are to participate in the technical discussion. Self-made academics in the humanities, on the other hand, may believe that rules are less esoteric and therefore the boundaries of scholarly discourse seem less strict, and they try to put their voices in – just to find themselves rejected as well in the end....

Simplicio: I loathe to interrupt, dear Salviati, but do not forget the serious methodological problems inherent in social sciences. As opposed to the natural sciences, social sciences have not yet crystallized their sound methodological foundations, and high rejection rates of publication, as well as intensive use of older technical literature, stand witness to this methodological uncertainty. Around the beginning of the twentieth century, a number of authors argued for a fundamental methodological difference between the natural sciences on the one hand and humanities and social sciences on the other. Others believed that there is some kind of unity to all the sciences, and an overarching methodology unites all scientific enterprises. This *Methodenstreit* lasted at least three generations...

Sagredo: ... and participants were lost in an ocean of complicated problems. I wouldn't be happy to lose the focus of our discussion. Both of you made a point though. Simplicio is right when urging the importance of statistical, quantitative findings about science: they largely contribute to our complex understanding of the phenomenon. But Salviati was also right when he pointed out that statistics can be very misleading without sufficient care and precaution.

Salviati: It is a general problem that statistics leaves us in partial darkness about the significance of the data. For example, sticking to publication patterns, the numbers show that when scientists become referees of their peers' papers, the younger referees are usually more strict (and more likely to reject) than older colleagues. Similarly with respect to rank, researchers with higher ranks are more admissive of manuscripts than those of lower ranks (Zuckermann and Merton 1971). Now this is all fascinating, but what does it mean? Not even those publishing these statistics seem to know the precise reasons and underlying mechanisms. What do we gain if we teach such things to students who need to understand what science is?

Simplicio: As Salviati noticed, teaching nature of science (NOS) includes teaching the would-be scientists, the academic *Nachwuchs*, and if you think of performance assessment, the study of publication patterns has obvious advantages that we already utilize in our everyday scientific activity. If you want to assess scientific performance, the most obvious way is to look at individual researchers: how many papers and in what journals did they publish.

Sagredo: "Scientometrics," as they often call this field (and also one of its flagship journals), aims to measure scientific output by counting and ranking publications, calculating impact factors and other indicators. While it has its roots in quantitative sociology of science, by today it has become an independent field that we often find indispensable... even though indicators for institutions are more reliable than any measurement tool that compares individual researchers.... It is a tool that we frequently rely on when it comes to assessment, despite some fundamental problems like the gradual shift from descriptive measurement of scientific productivity to a normative prescription of what counts as scientific productivity. The numerous difficulties would boil our discussion down to nitty-gritty details, so let us declare that quantitative sociology of science and the study of scientific institutions have some clear lessons to teach....

Salviati: Like in the case of the Matthew effect, one of the things I tend to emphasize about science is that it is a human enterprise, with all the typical features and "imperfections" all human enterprises have in common. Power inequalities are part and parcel of any human activity, and science is no exception. It is a simple game of power with winners and losers.

Sagredo: Such a profane claim would do injustice to Merton who did more than studying statistics. Above all, he sought to identify what distinguishes science from other things that humans do. He found that it is a peculiar set of norms that govern scientific activity, amounting to a "scientific ethos" (Merton 1942). There are four elements to this. Universalism means that the acceptance or rejection of theories is an impersonal process, where the origin of the scientist making the claim (e.g., class, gender, ethnical identity) is irrelevant to the acceptability of the claim. Communalism (or communism) means that knowledge is the intellectual property of the community, and every member has equal right to access it, contribute to it, or criticize it. Disinterestedness means that the assessment of scientific claims should be kept independent of the local interests and biases of certain groups. Finally, organized skepticism means that all scientific claims must be submitted to the scrutiny of critical thinking, and no dogmas should be considered as beyond skepticism.

Salviati: I have to admit that I find these expectations entertaining, given the mentioned profound power inequalities in science. Can we really expect scientists to have no personal interests, ready to share their most cherished results with everyone, and eager to criticize even the most central beliefs of their disciplines? Do we get closer to what science is if we teach this utopistic vision? Or do we believe that scientific education should include some training in ethics to implant these noble norms in students?

Simplicio: Salviati, I am never sure whether you are ironic or simply ignorant. Merton's norms of scientific ethos have puzzled many authors, but it is clear from the original texts that Merton did not intend to present these as "personality traits" of actual or ideal scientists. On the contrary, Mertonian norms do not hold at the level of individuals, and they are certainly not meant as behavioral prescriptions. Scientists are humans; their local interests, biases, dogmas, and secrets of the trade motivate individual scientists in doing what they do and inform them how to do it. So Merton's norms are general and symbolic value orientations that, at the collective level, govern the formation of rules and behavioral patterns in such a way that the final result of the entire activity, i.e., scientific knowledge, will accord to these general expectations.

Salviati: And how does this happen? Through some "cunning of reason"– out of the sum of individual selfishness and social contingency, something emerges that can be seen as pure and disinterested knowledge serving the ends of humankind.

Sagredo: And indeed, something like that does happen. The conditions of the "miracle" are embedded in the institutional patterns of scientific activity. It is precisely by studying these patterns that we can tell how individual interests, biases, commitments, and contingencies can result, through an institutionalized collective cognitive process, in a beneficial advancement of global science and human knowledge. And this is one of the most important reasons why a scientific approach to the workings of science is essential today.

Salviati: Fair words. Still, I feel uneasy about overemphasizing the statistical study of institutions when addressing the question of what science is. First, science may be seen very differently from the perspectives of other metascientific approaches: while it is a system of organized action for sociologists, it is also a body of texts and instruments for historians, or a world of abstract propositions and their logical relations to philosophers, to mention just a few views. In order to grasp science, all these perspectives should be taken into account simultaneously. Science, for example, may be seen as an institutionalized social field in modern times, but before the nineteenth century, the general institutional structure was very different, and before the seventeenth century, there was nothing similar to what science seems to be from the perspective you described. Yet most histories of science will go back to far earlier times. Would you tell your students that Euclid didn't do mathematics, or Ptolemy astronomy, just because science – as the sociologists describe it – did not exist in premodern times?

Sagredo: You are provoking us again. To say that institutionalized social action is a fundamental aspect of what we mean by the term "science" is not to claim that it is all there is to science. But the point you've just made is important: science is not something static and unchanging. The development, or to use a more neutral term, the dynamics of science does not only consist in ever better theories or more precise and comprehensive knowledge, whatever these mean (see the chapters on philosophy). On the contrary, the real dynamics of science can be revealed only by addressing its social and cultural conditions. For example, when Merton sought to identify the norms of scientific ethos, his real question was, in what social circumstances do we find a wholesome scientific activity? He believed that the Nazi Germany or the Soviet Union did not provide ideal conditions for doing science, since some norms were not respected. The norm of universality is clearly violated when ethnic and cultural circumstances of the proponents influence the reception of ideas. Merton's fundamental question was to find the norms that can provide the most ideal circumstances for scientific development when implemented in institutional practice (Merton 1938a).

Simplicio: From early on, Merton's work was imbued with a historical perspective. In his doctoral thesis, he investigated how modern science was formed in seventeenth- and eighteenth-century England. His central claim was that Protestant ethics based on Puritan principles proved beneficial for the development of "natural philosophy" (Merton 1938b). The two basic tenets of this philosophy were doing things for the glory of God and for the utility of mankind. Merton quoted many passages from prominent scientists of the era to show how these principles lead to the intellectual pursuit we call science, being both rationalist (by seeking a theoretical understanding of the Creation, i.e., nature) and empiricist (by submitting this understanding to the special needs and purposes of humans). He also relied on statistical analyses to establish the connection between scientific performance and social institutions and found both that Protestants were overrepresented in scientific societies (like the Royal Society) and that novel scientific views and results found their way to Protestant school curricula faster than to Catholic ones.

Sagredo: And this is just part of the story. The "utility of mankind" aspect is clearly visible if you look at the practical problems that early scientists of the era sought to solve. As England's power was largely dependent on sailing the seas, it comes hardly as a surprise that scientists addressed problems relevant for sailing: they studied astronomy in order to improve navigation and determine the position of the ships; they studied the motion of the pendulum to build better clocks, as precise timekeeping appeared to be essential for the determination of longitude; and they studied hydrostatics to build ever larger and faster ships. Mining, another vital area behind industrial growth, was laden with problems that could be addressed by hydrostatics and aerostatics, like circulating water and the air in mines. And, of course, the power, efficiency, and accuracy of guns and cannons raised many questions related to projectile paths, the compression and explosion of gases, rigidity and constancy of metals, etc. – all popular research areas of the time.

Salviati: All this seems to indicate that the direction of scientific research is a function of society. Soviet historians, most prominently Boris Hessen (2009), held similar views already in 1931. They pinpointed the role of military and information technologies during the scientific revolution and claimed that all important results of the era were rooted in practical questions. But even if their findings were similar to that of Merton, their starting point was different. Hessen and his colleagues worked in a framework that Merton later found inadequate for the purposes of science.

Simplicio: They dogmatically followed the Marxist ideological assumption that it is modes of production in the material world that determine the "superstructure," i.e., the conditions of social, political, and intellectual processes. From their point of view, the interconnectedness of scientific research and social conditions is anything but surprising. But if you want to understand the real dynamics of science, you had better consult statistics. Did you know that the total population of scientists is growing more rapidly than the population of nonscientists? This tendency was recognized by Derek de Solla Price in 1963 (Price 1986), who claimed that science (number of scientists, journals, papers, institutes, etc.) had been growing exponentially since the end of the seventeenth century when modern science had been created. Most indicators of the size of science double in a period of approximately 15 years, and that means that, on average, science becomes 1,000 times more populous in a span of 150 years. Merton also realized that in the modern societies of his time, the age distribution of researchers is different from that of the entire population: in science younger age groups are overrepresented.

Sagredo: Surely this has profound consequences for our understanding of how science works. It is always in the present: taking any moment in time, about half the scientists of the history have worked in the past 15 years, so an old scientist having worked 45 years can say that he or she has been contemporary to more than 90 % of all the scientists who have ever lived so far. This may tell us why, in most research areas, the Price Index is so high: recent science is not only more developed, but is also far more detailed and populous, than past science. This inevitably leads to increasing specification, since a research area becomes too large to handle when the number of experts exceeds a practical limit, so new research areas split from others to make room for an ever-growing number of newcomers.

Salviati: And how long can this miracle last? If the ratio of scientists increases in modern societies, there must come a time when everybody becomes a scientist. True that human population is growing in a more or less exponential manner, but the doubling rate is closer to 30 years than 15, so scientists will have to fill up the future. Fine with me, but who will work then?

Sagredo: Price asked the same question. Surely, such a growth must cease before every single person becomes a scientist. Practically, science is becoming simply too expensive to be paid by the rest of society. By analyzing the growth curves, Price came to realize that the growth was about to end soon, in his lifetime. After that, science needs to transform radically in order to survive....

Salviati: So you are talking about what Nowotny et al. (2001) called Mode 2 science. They realized that the social context of science was changing in a way that favors multidisciplinary projects focused on specific real-world problems, while traditional long-term theoretical research confined within certain disciplines ("Mode 1 science") was losing its former ground. According to this view, contemporary scientists are problem-solvers in the context of technological, social, and everyday needs. Is that your solution?

Sagredo: Exactly. Or, to use another term, John Ziman (2000) coined "post-academic" science to describe how academic science, funded more or less independently by the state in order to conduct free research, is gradually replaced with a research culture that is supported by multinational industrial or corporate actors for dealing with well-defined and context-given problems. If the state can no longer quench the ever-growing appetite of science, researchers have to find other forms of support. This looks inevitable, but the consequences are as yet far from clear. It seems probable that the role of "basic research" (a free pursuit of knowledge) will be reduced radically, while "applied research" (seeking a solution to a specific given problem) will dominate science. Moreover, the primary social function of science is no longer providing knowledge of how the world is, but finding answers to "external" questions, given to researchers by whoever pays them. All this affects the institutional structure of science, since the primary sites of research are less and less the state-funded research institutes or the universities... Students who are considering a career in science would clearly find these trends essential in terms of their career choices.

Simplicio: And as the institutional structure is transforming, so are the norms themselves that guide behavior in science. Ziman (2000) argues that Merton's norms are becoming obsolete. Knowledge is not universal any more, since scientific answers are strongly influenced by the circumstances of production. Results are increasingly unavailable in a free and public way to other members of the community, as they are the property of the patentee. Science is obviously far from being disinterested, for research serves the immediate interests of the sponsor. Finally, there is no time for organized skepticism in a world of deadlines and short-term projects: a firm result is expected by the end of the given period of research. If Merton's norms describe an ideal state in which traditional science can proliferate, Ziman opens our eyes to a different and more mundane reality.

Salviati: I see no reason for despair. In my understanding, science is adapting to a changing social environment. The dynamics of science is analogous to biological evolution, to invoke another classic from the social studies of science (Thomas S. Kuhn 1970, pp. 171–173, 2000a, pp. 96–99). The evolution of a complex system of theories and paradigms is not a linear progress aimed at some desirable end like "truth" or "ever better theories"….

Simplicio: Are you sure that the science/evolution metaphor implies the lack of a final truth? You say that "paradigms" (a term that is used in radically different senses throughout Kuhn's main work; see Masterman 1970) are like biological species adapting to a changing environment. Yet biological evolution produced us, humans, and while other species are adapted to different specific environments, human culture is a universal solution to an unlimited range of environments. So, why rule out that science is a similarly successful, potentially universal epistemic tool…?

Salviati: … one that threatens us with a general epistemic and social collapse, just like human behavior may easily trigger a global ecological catastrophe?

Sagredo: Before science can look back on as long an evolutionary path as nature can boast with, let us suppose that the process is driven by the past failures, rather than drawn toward a goal. What Salviati's Kuhn implies to me is this: there is no need to look for an "essence" of science, like Merton's norms or methodological rules prescribed by philosophers. Science is an ever-changing enterprise, and turning our attention to its past development opens a perspective that is at least as informative as the quantitative study of its present form.

Salviati: And this approach is more profoundly historical than Merton's analysis of the seventeenth century or Price's statistics about growth.

Simplicio: Let me ask this: if we dispense with the idea that science has an "essence" or, to put it less loftily, it has some constant and unchanging core of commitments and orientations, then what are we left with to teach as "nature of science"? Can we avoid discussing Salviati's post-Kuhnian relativists at this point (see also the special issue Matthews 1997)?

Salviati: These "relativists" or, more precisely, constructivists have provided us with the most detailed case studies of science in action, how good (or bad) science is actually done. Didactic transposition of any one of these detailed case studies can give a glimpse of the whole of science, instead of focusing on declarative statements (Allchin 2003).

Sagredo: Overemphasizing one perspective can be as deceptive as sticking to another. Still, I find Kuhn's emphasis on differences insightful. If we want to understand what science is, we need to look at it in various times and cultures and to identify the contingencies that help us see better behind the contingent and ephemeral.

Simplicio: Everything seems to be contingent for Kuhn in my reading. Salviati was proud to drop in the term "paradigm," as if the popularity of a term would do justice to its adequacy. But in Kuhn's view, paradigms come and go, they replace one another, and their relation is described by him as "incommensurable" (Kuhn 1970). He argues that there is no rational stance from where we could tell that the new paradigm is better than the old, and thus overall progress disappears from science. Would you tell your students that Copernican theory of planetary motions is just as good as the Ptolemaic system, since there are no universal standards to compare them? As an educator, I strongly protest!

Sagredo: Kuhn is a tad more complex than that, as it is not theories he holds incommensurable but, as you say, paradigms. Paradigms are not theories and sets of propositions about the world, although they contain theories and propositions. They also involve more basic elements such as rules and methods to conduct inquiry, successful solutions that serve as models for solving future problems, and ontological commitments about the entities our explanations rely on that help us interpret phenomena. In short, paradigms are ways to see the natural world. Of course, we do not want to teach our students to see the world as the ancient Greeks did, and we can argue efficiently that the Copernican theory is better than the Ptolemaic, just as Newton's theory of motion is better than that of Aristotle. But it is important for us to bear in mind that these "theories" as we present them are interpreted in our paradigm, relying on our own questions and evaluation criteria, and our judgments are guided by our own interests and predispositions.

Salviati: Exactly. And I believe that this radically historical perspective may help students to see science from a healthy distance, by contextualizing those elements of modern science that they otherwise would see as evident. Take Newtonian physics as an example. In order to show its merits, we present to them the very same solutions that made the Newtonians so successful: we talk about swinging pendulums, colliding balls, the elongation of springs, or the motion of planets – and lo and behold, these are phenomena where Newton's theory provides simple and efficient explanations, so Newton was correct! But are pendulums, colliding balls, springs, and planets important to an average student? We implicitly suggest that the key to understanding the physical world is by addressing these phenomena, and many find this kind of physics irrelevant and boring. If we teach any history of science before the early modern period, we simply say that Aristotle's physics was obviously naïve and dull, but we do not tell students that the phenomena that the Aristotelians wanted to explain, and the criteria of what counts as successful explanation, were radically different then. My point is, if we told them about these differences, they would better understand what modern physics is good for (for a general appraisal, see Matthews 2000, 2004)…

Simplicio: …and we would have even less time left to teach them proper modern science. After all, Aristotle lived a long, long time ago, and I don't see why he would be of any interest to us if you are correct in claiming that his world was radically different. I don't believe in radical raptures, and I can't accept that every new paradigm starts everything completely anew. Otherwise, how could we learn from the past if there is nothing in common between him and us? The concept of incommensurability is highly questionable….

Sagredo: The concept has drawn similar reactions from many scholars, and Kuhn seems to have softened his views. The emergence of new paradigms is analogous to speciation in evolution, and this evolutionary metaphor became more prominent in his later writings. The gap between paradigms is radical, but "lexicons" of research communities can be translated, by learning the conceptual system guiding their world view. Communication between lexicons is possible (Kuhn 2000b).

Simplicio: And this is where I firmly protest. When Kuhn claims that general conceptual structures orient our scientific theories, he simply ignores that science is an enterprise that, by studying nature and nothing else, helps us get rid of specific social conditions and contingencies. Remember the distinction I made between the meaningful study of institutional structures and the misguided sociological analysis of the contents of scientific knowledge? While Kuhn was not a sociologist, his views prompted many tenets in what is called post-Kuhnian science studies, and I think these tenets were a change for the worse.

Sagredo: You probably think of the most explicit and most influential theoretical backing to post-Kuhnian science studies developed by the Edinburgh School including David Bloor, Barry Barnes, John Henry, and others. They launched what is called the Strong Program in the sociology of scientific knowledge (SSK), characterized by four planks: (1) causality, it is "concerned with the conditions which bring about belief, or states of knowledge"; (2) impartiality, "it would be impartial with respect to truth and falsity, rationality or irrationality, success or failure"; (3) symmetry, "the same types of cause would explain, say, true and false beliefs"; and (4) reflexivity, "its patterns of explanation would have to be applicable to sociology itself" (Bloor 1992, p. 7). The main project of the program was to construct the theoretical underpinning of a scientific understanding of science...

Salviati: ... and let us add that they seem to have failed in that, since most subsequent authors in the field of science studies disputed several central points in the theoretical core of SSK. Still, and more importantly, SSK motivated many basic trends in the social studies of science. Versions of this program were employed in studies of scientific laboratory work (Knorr-Cetina 1981; Latour and Woolgar 1979), analyses of scientific controversies (Collins and Pinch 1982), histories of older (Shapin and Schaffer 1985) and recent (Pickering 1984) science, theories of technology (Bijker et al. 1987), etc. They all acknowledge their debts to SSK.

Simplicio: Relativists! I find it hard to keep calm when I meet their views. The people you mentioned might be simply wrong but benevolent, but they encouraged a hoard of hostile intellectuals – think of the various postmodernist, post-structuralist, neo-Marxist, feminist, environmentalist, etc., criticisms of science. Remember the Science Wars! A great number of scientists and philosophers pointed out that (i) the SSK tenets and their intellectual offspring are simply wrong and (ii) these views give rise to a general social hostility toward science.

Salviati: But these charges are seriously mistaken, as the science studies authors attacked in the Science Wars clearly showed. But if you haven't read them, I am happy to accept your challenge right here and prove the validity of "relativist" approaches....

Sagredo: But that would lead us into philosophy and a dense discussion. Still, it is worth considering the main charge concerning why SSK is wrong, emphasized by Alan Sokal and Jean Bricmont (1998) in their sweeping criticism of relativist science studies. When studying scientific knowledge, sociologists tend to refer to all kinds of social and cultural factors when explaining why a certain community accepted specific beliefs, but they forget about nature and its causal responsibility for the beliefs we hold. Isn't it a strongly one-sided perspective? If we tell our students that scientific theories are caused by social factors, how can we encourage them to carry out empirical investigations and the study of nature itself?

Salviati: You've just committed a widespread mistake: you seem to read SSK inattentively. With regard to the first plank of SSK, Bloor and his colleagues stress that all beliefs have various causes, but not all of these causes are social (Bloor 1992, p. 7). Natural causes obviously play their role.

Simplicio: This sounds promising, yet does it not undermine the symmetry principle, since true beliefs surely have more natural causes and false beliefs must have more social causes? If sociologists examined all the causes prompting a belief, they would certainly realize that much (Slezak 1994a). So why don't they talk about natural causes?

Salviati: The reasons are manifold, and they vary from author to author. The simplest reason is that sociological research has no access to nature independently of natural science as the very object of its study, so "nature" appears only in beliefs about it. This is a purely methodological interpretation advocated, e.g., by Harry Collins, called "methodological relativism" (Collins 2001a, p. 184). David Bloor offers another methodological explanation when he claims that sociologists are interested only in the differences between rival theories, and since nature that researchers study is the same, it is the different social factors that result in different interpretations of the same thing (Bloor 1999, p. 93). But the same Bloor gives philosophical arguments too, similar to that of some constructivist authors such as Karin Knorr-Cetina (1993). According to this view, since nature's causal role (*an sich*) is entirely unspecified (and inaccessible independently of the beliefs we hold about it), everything we conceive is constituted by our prior beliefs and categories. Kuhn's epistemological perspective building on "lexicons" supports a similar tenet, no wonder that he saw himself as a Kantian with movable categories (Kuhn 2000a, p. 264). There are even more radical views like that of Bruno Latour who denies the distinction not only between nature and society but also between things and beliefs about them, at an ontological level (Latour 1993), or post-humanists like Andy Pickering who mix every possible cause in their narrative and call it the "mangle of practice" (Pickering 1995). These radical views come very handy when it comes to understanding and discussing atypical or Oriental science....

Simplicio: Please stop! This much has been more than enough to illustrate that there is no theoretical consensus within the social studies of science. Just think of the "epistemological chicken" debate between relativists, where the stake was who dares to come up with the more extreme epistemological position (Collins and

Yearly 1992), or the controversy between Bloor (1999) and Latour (1999) where, intimidated by the criticisms they received in the Science Wars, they reproached each other for adopting harmful philosophical views. This proves not only that relativists stand on highly problematic theoretical ground but also that teaching their theories to students would result in confusion and despair (Slezak 1994b and Chap. 31).

Salviati: Science studies are manifold, as the term "studies" indicates – and we haven't mentioned a number of approaches yet. But then, to return to an earlier problem we raised, do you really think that science is a single and unified entity? While the positivist illusion of unity was expressed by launching the ambitious *International Encyclopedia of Unified Science*, the planned 14 or 26 volumes (Morris 1960) never came out. The second and last volume ironically contained Kuhn's revolutionary book on revolutions (1970). Post-Kuhnian science studies authors argue for the disunity of science (Galison and Stump 1996), and constructivists like Knorr-Cetina (1999) view different scientific fields as disconnected epistemic cultures. And we can go further than simply acknowledging disunity: thanks to science studies, we now have conceptual tools to study boundary phenomena between and around scientific areas. Think of "boundary work" introduced by Merton's student Thomas Gieryn (1983, 1999) to describe the ideological and rhetorical efforts that draw and shift different boundaries. Or think of "boundary objects" (Star and Griesemer 1989), entities that connect various groups in the large networks of science, "boundary organizations" (Guston 1999), "boundary infrastructures" (Bowker and Star 1999), or the "trading zones" that emerge when groups with different enculturation do research together (Galison 1997)....

Simplicio: I can't think of so many things simultaneously, and I am repelled by this conceptual and methodological cacophony. But I can anticipate your point: if there is no such single entity as "science," but only networks of "sciences," then what are we left with to teach as "nature of science"? And here is my answer: everything we normally teach, perhaps supplemented with some insights from the sociology of scientific institutions. I simply cannot accept claims or conceptual tools from a field which is so heterogeneous that its representatives can't agree on how to start building a minimally common theoretical core.

Sagredo: Your favored sociology of scientific institutions was not free of fundamental divergences in theoretical and methodological questions – emerging fields usually aren't. I agree that introducing too many theoretical positions in science classes could prove less useful than harmful. Still, what unites these authors is not the theory but the general perspective and the resulting approach to science. As Salviati mentioned earlier, science can be seen as a set of human practices that are as contingent, political, interest-driven, opportunistic, and mundane as any other human enterprise.

Simplicio: And this is the second mistake of SSK I mentioned, and a very bitter one. Why do we do science? Because we believe that it is the most reliable cognitive enterprise we have, resulting in the most secure knowledge about the world. The question in this case is not how science is the same as other cultural fields, but how it is different, something more. If we teach our students anything "about science," the guiding question must be, what's so special about science (Bauer 2000)?

Sagredo: But then it is useful not to tell idealized half-truths about it. Those who become scientists will learn of the imperfections soon enough, but it may take them by surprise and make it difficult for them to adopt to the reality of actual practice, as opposed to the ideals they pursued when choosing this profession. And those who will not do science professionally may find, when confronted with science depicted in the media, that scientists are not the good guys in flawless white lab coats that popularizing efforts try to show, and they might get disappointed and disrespectful.

Salviati: Importantly, SSK-type relativists try to follow the most general scientific principles. Scientists explain phenomena by referring to causes, and sociologists of science do the same....

Simplicio: But they do not agree in what type of causes they use! Some of them explain the general with the general: for Marxists larger formations of knowledge (cognitive styles, conceptual schemes, paradigmatic theories, etc.) are explained with reference to broad social characteristics. Others explain particular knowledge claims with reference to general social settings. Here the particular cases to be explained can range from specific beliefs to small-scale theories, advocated by a particular scientist, and the general basis of explanation can comprise the norms within the wide community, or general interests of cognitive, economic, or political kind, or technological possibilities available at a wider sociohistorical level. In these cases the source of explanation expands far wider than the local topic and its immediate environment. Still others, and probably the largest bulk of recent work in science studies, are constructivists (Golinski 1998), and they explain specific instances from specific sources, i.e., the range of explanatory principles does not extend beyond the immediate environment of the local phenomena under study. What are we to do with this mess?

Salviati: Again, the point is not the details of their theories, but the general attitude. Causes are neutral, as opposed to reasons which always include some normative or evaluative element. Just as physicists don't evaluate natural processes but simply describe them, sociologists do not evaluate scientific beliefs. Descriptive neutrality requires that "true," "rational," "objective," "better," and other strongly normative concepts be withdrawn from scientific discourse (or, rather, better be made topics as connected to norms rather than resources). This lack of evaluative tone is often seen as evaluation itself, when it replaces a discourse which is usually evaluative.

Sagredo: A fine point. To cite a fitting analogy from Harry Collins (2001b, p. 157): sociologists want to explain why members of a certain community believe that wine can turn into blood, while members of another community don't. Because the "Yes" community maintains a positive evaluation of the norms that justify their belief, the analysis will be seen as downgrading their claim and its entire cognitive background together with it. At the same time, the "No" group will have the same impression regarding their own position, and for the very same reason, but the sociologist simply did not want to decide and hence take into account whether or not the wine can "indeed" turn into blood. Seeking a causal explanation as to why participants in the

Nazi regime behaved in the way they did, the account will be seen as making unwarranted excuses – this time because of the lack of the generally accepted negative evaluation attached to the matter. Now, we are accustomed to a discourse about science that is highly praising, and the lack of this positive tone seems degrading to many.

Simplicio: But this "neutral" discourse is useless when the educator's purpose is to build respect for science!

Sagredo: But it is only one of the purposes. Understanding how science works is another, and the heroic stories about Great Achievements of Great Minds do not do justice here.

Salviati: Let me offer another aspect here. Historian of science Steven Shapin (2001) added his voice to the Science Wars when, paraphrasing Sokal's famous hoax (Sokal 1996), he collected a list of statements expressing the credo associated with relativists and then revealed that all the statements were made by prominent scientists. One moral of the story is that while scientists agree in most scientific questions (i.e., their consensual knowledge base, excluding the very recent problems they dispute), they strongly disagree when it comes to metascientific commitments, since some share many relativist views (about the political nature of science, or the social contingency of our scientific concepts), while others stick to a picture of science as the inevitable voice of nature.

Sagredo: Statements, including these, "There is no such thing as the Scientific Method," "Scientists do not find order in nature, they put it there," "New knowledge is not science until it is made social," and others, are generally associated with constructivist and sociological approaches (Shapin 2001, pp. 99–100). A fine illustration of the point you cited from Lakatos early in our discussion.

Salviati: The claims that Simplicio would find degrading are made in specific contexts when "secrets of the trade" are revealed to attentive audiences, and this is the more relevant conclusion. In formal public settings where the esteem of science is at stake, the very same scientists pronounce more "polite" statements in better accordance with the usual expectations. Quite contrary to metascientific statements in mathematics (Davis and Hersh 1981, p. 321), many scientists are constructivist on weekdays and inevitabilist (Hacking 1999, p. 79) on Sundays. Just like in a family, problems are explicit only for friends, but everything is fine for the wider public. Trevor Pinch (1986), when interviewing researchers of the solar neutrino project, found that those who admitted possible weaknesses in their work did it in a form resembling secret initiation rituals. But if we rely on scientists' white lies in science classes, students are misled – and, as Sagredo pointed out, they get disappointed later when they face reality.

Simplicio: Alright, I concur that this enthusiastic but small circle of SSK offers their descriptions with fair intentions. But what about those numerous "xxx-ist" and "post-xxx-ist" people from the humanities I mentioned earlier, whose explicit goal is to criticize, desacralize, and "unmask" science? Am I supposed to tell our students that modern science is capitalist, masculinist, and politically oppressive and that it is an obsolete "grand narrative" that we should best dispense with?

Sagredo: It is virtually impossible to pretend that science is for the pure benefit of humankind, in an independent and culture-transcendent niche. Science has always been embedded in a cultural environment interwoven with technological, economical, social, political, and all kinds of aspects, and it is hard to deny that science has grown huge and powerful, fundamentally shaping modern technological and information societies. A large proportion of recent sociological case studies therefore focus on science policy, politics of science, innovations, risks, stakeholders, and all kinds of stuff that are clearly relevant to students, regardless whether they would become scientists or not.

Simplicio: I understand how these matters are relevant to would-be scientists, but I am not sure why they could be of great interest to laypeople.

Sagredo: Consider "the third wave" in science studies, advocated by Harry Collins and Robert Evans (2002) in a paper that has become one of the most cited works in the field. By blurring the boundary between experts and the public and replacing it with a "periodic table of expertises" (Collins and Evans 2007, p. 14) in a program called "Studies of Expertise and Experience" (SEE), the paper attempts to conceptualize decision making in mixed policy settings. The first wave tried to account for the success of science as taken for granted – and traditional sociology of science would represent this tenet. The second challenged the automatic respect and offered naturalized descriptions of science as a profoundly human endeavor – that would be SSK and its many subsequent versions.

Salviati: I like toying with the idea of different kinds and degrees of expertise, especially "meta-expertise" (Collins and Evans 2007, pp. 45–76) that refers to the skills by which we, laypeople, become able to assess expert claims….

Simplicio: You must be joking. If an expert, by definition, is far more knowledgeable than a layperson, how can the latter legitimately evaluate what the former states?

Salviati: By relying on social discrimination rather than technical knowledge. If two experts disagree, I can check their credentials, positions, ranks, track records, etc. Most people have no sufficient grasp, for instance, of the theory of evolution to judge whether Intelligent Design protagonists are probably right or probably wrong, but they can still assess the approximate proportion of experts supporting the confronting views, or the statuses of their publication forums within the scientific community, or their ranks and credentials, or the relevance of their research background, or the origin of funds supporting their institutes, and so on. And I claim that one can reach a pretty secure judgment here, without having to become an expert oneself. However, and this is what makes me uneasy here, SEE explicitly rejects the purely descriptive stance taken by what they call Wave 2 and advocates normative intentions to guide us in our meta-expert decisions.

Sagredo: And why does this bother you? Why not be normative in classrooms? Descriptive neutrality becomes impotent when your students ask you whether they should believe the theory of evolution or supporters of Intelligent Design, or when they ask your opinion about table dancing, or when they seek your advice in going to see what kind of doctor or spiritual healer or shaman…Would you leave them with a Feyerabendian riddle about "anything goes" for the lack of no rational decision standards?

Salviati: I may be an expert at science education, but I am no scientific expert in all questions. I want to prepare my students to be able to decide for themselves, rather than follow my authoritative voice.

Simplicio: And then they need to acquire criteria according to which they make their decisions, and I do not see how you can avoid being normative here. You can't introduce your students to every single approach that ever existed and say, I don't know which one is the best, see for yourselves! … I actually like the idea of social discrimination. If meta-expert assessments depend on judgments about ranks, publication forums, institutes, and such kind of things, then it is the first wave, sociology of institutions, that we need to teach – but only the essential basics, of course. The relativist second wave is useless here.

Sagredo: Far from that. The emphasis for most relativist authors is not on the theories we discussed, but on the detailed case studies they offer to reveal how science works. If you want to understand the social dimension of science, you need to see the contingencies inherent in laboratory research, communication and interpretation of results, choices of method and theory, settling controversies, etc. – and this is precisely what the so-called Wave 2 has taught us. The image of science offered by popular movies and books or, for that matter, the image offered by popularizing literature can hardly be matched with the actual practice of science. If SEE calls for the use of social intelligence and argues that modern people acquire a fair bit of this intelligence just by being immersed in society, then my answer is that this general discrimination becomes insufficient when facing science. Laypeople hardly know any details of how scientists work, so how could we expect them to be meta-experts in a world of scientific expertise?

Simplicio: Why not let them rely on expert testimony? Let them believe what expert say, for this is why we have experts: they bear the epistemic authority and responsibility.

Salviati: Your suggestion turns on the validity of the so-called deficit model for the public understanding of science. It sees laypeople as passive in receiving scientific knowledge, and it relies on a premise that is modeled on the traditional science classroom situation, namely, that laypeople (like students) are viewed as yet ignorant of science but capable of having their head "filled" with knowledge diffusing from centers of knowledge production.

Simplicio: Such a process is beneficial since it increases laypeople's scientific literacy (and their ability to solve related technical problems), and their degree of rationality (following the rules of scientific method), and finally, their trust in and respect for science. Can you offer something better?

Sagredo: Surely, if you consider the way laypeople and scientists meet in our present cultures. According to the far more plausible premise of the so-called contextual model (Gregory and Miller 2001), members of the public do not need scientific knowledge for solving their problems, nor do they have "empty memory slots" to receive scientific knowledge at all. Instead, the public's mind is filled with intellectual strategies to cope with problems they encounter during their lives, and some of these problems are related to science. So the active public turns to science, more precisely to scientific experts, with questions framed in the context of their everyday lives. After all, we live in a world where most tasks are best performed, and most questions best answered, by specific people with outstanding skill and knowledge base, called experts.

Simplicio: And what is the difference between saying that scientists teach us how the world is and saying that our questions are answered by scientific experts? Isn't it a fancy way to express old trivialities?

Salviati: Not at all. The questions the public is interested in can rarely be answered by "ready-made science" deposited in textbooks; they belong to "science in the making" (Latour 1987). Instead of asking how planets or pendulums precisely move, to which there exist answers that are consensual and yet mostly irrelevant to the public, they want to know, e.g., what materials or activities are healthy. These questions are (still) controversial in science, and nonexperts are faced with a plethora of different and partly contradicting expert opinions, from which they have to build their own system of beliefs.

Simplicio: I may object that scientific experts would always reach a consensus if they were given enough time, but you would probably object that decision making in politics and economics is usually faster than consensus making in science, if we take the post-academic scenario for granted…

Salviati: Or the post-normal scenario (Funtowicz and Ravetz 1993), according to which decisions in the science and technology of our age are achieved under the circumstances of high risk and uncertainty.

Sagredo: So many concepts, so many challenges and aspects and approaches… One thing we found is that the social studies of science has clearly developed, become a field of expertise that has increasing relevance for science educators. As our image of science and the institutions of science change, this is reflected in our educational system, in the mindset of students, teachers, policy makers, and other stakeholders. The lessons of this discipline – even when at times it appears that it is not too disciplined – can help inform both the teaching practice and the theory of science education. But I am afraid we need to conclude our discussion, as we all have our classes to teach. I hope that we will find our discussion inspiring.

Simplicio: No doubt. A major lesson for me is that sociologists have efficient tools to study the practices of science. To see their results is important not only to students planning to pursue a scientific profession but to the wider public also, for science is an essential vehicle of modern social progress.

Salviati: To which I may add the following: student understanding of science can fruitfully extend beyond popular and ideological representations, by utilizing the knowledge of the discipline that pursues the practice of scientists in a historical, cultural, technological, and economic context.

Sagredo: And all this implies that one-sentence retorts about science, sold as "nature of science," watery metaphors, fancy analogies, or simplified algorithms of methodology can hardly capture the knowledge needed to develop understanding of what science is. The knowledge that scientists know about nature and about science are both essential ingredients of science education.

References

Allchin, D. (2003). Should the Sociology of Science Be Rated X? *Science Education*, 88, 934–946.

Barnes, B. (1985). *About Science*. Oxford: Basil Blackwell.

Bauer, H. H. (2000). Antiscience in Current Science and Technology Studies. In U. Segerstråle (Ed.) *Beyond the Science Wars. The Missing Discourse about Science and Society* (pp. 41–61). Albany: State University of New York Press.

Bijker, W. E., Hughes, T. P. & Pinch, T. (1987). *The Social Construction of Technological Systems: New Directions in the Sociology and History of Technology*. Cambridge: MIT Press.

Bloor, D. (1973). Wittgenstein and Mannheim on the sociology of mathematics. *Studies in the History and Philosophy of Science*, 4, 173–191.

Bloor, D. (1992/1976). *Knowledge and Social Imagery*. London: Routledge and Kegan Paul.

Bloor, D. (1999). Anti-Latour. *Studies in History and Philosophy of Science*, 30(1), 81–112.

Bowker, G. C. & Star, S. L. (1999). *Sorting Things Out: Classification and Its Consequences*. Cambridge, MA: MIT Press.

Collins, H. (2001a). One More Round with Relativism. In J. A. Labinger & H. Collins (Eds.), *The One Culture? A Conversation about Science* (pp. 184–195). Chicago & London: University of Chicago Press.

Collins, H. (2001b). A Martian Sends a Postcard Home. In J. A. Labinger & H. Collins (Eds.), *The One Culture? A Conversation about Science* (pp. 156–166). Chicago & London: University of Chicago Press.

Collins, H. & Evan, R. (2002). The Third Wave of Science Studies: Studies of Expertise and Experience. *Social Studies of Science*, 32(2), 235–296.

Collins, H. & Evans, R.(2007). *Rethinking Expertise*. Chicago: The University of Chicago Press.

Collins, H. & Pinch, T. (1982). *Frames of Meaning: The Social Construction of Extraordinary Science*. New York: Routledge.

Collins, H. & Yearly, S. (1992). Epistemological Chicken. In A. Pickering (Ed.), *Science as Practice and Culture* (pp. 301–326). Chicago: Chicago University Press.

Davis, P. J. & Hersh, R. (1981). *The Mathematical Experience*. Boston and New York: Houghton Mifflin.

Funtowicz, S. & Ravetz, J. (1993). Science for the Post-Normal Age. *Futures*, 25(7), 735–755.

Galison, P. (1997). *Image and Logic: A Material Culture of Microphysics*. Chicago: University of Chicago Press.

Galison, P. & Stump, D. (1996). *The Disunity of Science*. Stanford University Press.

Gieryn, T. F. (1983). Boundary-work and the demarcation of science from non-science: strains and interests in professional ideologies of scientists. *American Sociological Review*, 48, 781–795.

Gieryn, T. F. (1999). *Cultural boundaries of science: credibility on the line*. Chicago: University of Chicago Press.

Golinski, J. (1998). *Making Natural Knowledge. Constructivism and the History of Science*. Cambridge: Cambridge University Press.

Gregory, J. & Miller, S. (2001). Caught in the Crossfire? The Public's Role in the Science Wars. In J. A. Labinger & H. Collins (Eds.), *The One Culture? A Conversation about Science* (pp. 61–72). Chicago & London: University of Chicago Press.

Guston, D. H. (1999). Stabilising the Boundary Between U.S. Politics and Science: The Role of the Office of Technology Transfer as a Boundary Organisation. *Social Studies of Science*, 29, 87–111.

Hacking, I. (1999). *The Social Construction of What?* Cambridge: Harvard University Press.

Hess, D. J. (1997). *Science Studies: An advanced introduction.* New York: New York University Press.

Hessen, B. (2009/1931). The Social and Economic Roots of Newton's Principia. In G. Freudenthal & P. McLaughlin (Eds.), *The Social and Economic Roots of the Scientific Revolution: Texts by Boris Hessen and Henryk Grossmann* (pp. 41–102). Springer.

Holton, G. (2003). What Historians of Science and Science Educators Can Do for One Another. *Science & Education*, 12(7), 603–616.

Kitcher, P. (1998). A Plea for Science Studies. In N. Koertge (Ed.), *A House Built on Sand. Exposing Postmodernist Myths about Science* (pp. 32–56). Oxford: Oxford University Press.

Knorr-Cetina, K. (1981). *The Manufacture of Knowledge. An Essay on the Constructivist and Contextual Nature of Science.* Oxford: Pergamon.

Knorr-Cetina, K. (1993). Strong Constructivism—From a Sociologist's Point of View: A Personal Addendum to Sismondo's Paper. *Social Studies of Science*, 23(3), 555–563.

Knorr-Cetina, K. (1999). *Epistemic Cultures. How the Sciences Make Knowledge.* Cambridge: Harvard University Press.

Kuhn, T. S. (1970). *The Structure of Scientific Revolutions (2nd edition).* Chicago: The University of Chicago Press.

Kuhn, T. S. (2000a). The Road since *Structure*. In J. Conant & J. Haugeland (Eds.), *The Road since Structure* (pp. 90–104). Chicago: Chicago University Press.

Kuhn, T. S. (2000b). Commensurability, Comparability, Communicability. In J. Conant & J. Haugeland (Eds.), *The Road since Structure* (pp. 33–57). Chicago: Chicago University Press.

Lakatos, I. (1970). Falsification and the Methodology of Scientific Research Programmes. In I. Lakatos & A. Musgrave (Eds.), *Criticism and the Growth of Knowledge* (pp. 91–196). Cambridge: Cambridge University Press.

Latour, B. (1987). *Science in Action: How to Follow Scientists and Engineers through Society.* Cambridge, Harvard University Press.

Latour, B. (1993). *We Have Never Been Modern.* Cambridge: Harvard University Press.

Latour, B. (1999). For David Bloor... and Beyond: A Reply to David Bloor's 'Anti-Latour'. *Studies in the History and Philosophy of Science*, 30(1), 113-129.

Latour, B. & Woolgar, S. (1979). *Laboratory Life: The Social Construction of Scientific Facts.* Beverly Hills: Sage.

Lotka, A. J. (1926). The frequency distribution of scientific productivity. *Journal of the Washington Academy of Sciences*, 16, 317.

Machamer, P. (1998). Philosophy of Science: An Overview for Educators. *Science & Education*, 7(1), 1–11.

Mannheim, K. (1936). *Ideology and Utopia.* London: Routledge & Kegan Paul.

Masterman, M. (1970). The Nature of a Paradigm. In I. Lakatos & A. Musgrave (Eds.), *Criticism and the Growth of Knowledge* (pp. 59–89). Cambridge: Cambridge University Press.

Matthews, M. R. (Ed.) (1997). Philosophy and Constructivism in Science Education: A Special Issue. *Science & Education*, 6(1–2).

Matthews, M. R. (Ed.) (2000). Thomas Kuhn and Science Education: A Special Issue. *Science & Education*, 9(1–2).

Matthews, M. R. (2004). Thomas Kuhn and Science Education: What Lessons can be Learnt? *Science Education*, 88(1), 90–118.

McComas, W. F. (2000). *The nature of science in science education: Rationales and strategies.* Dordrecht: Kluwer.

Merton, R. K. (1938a). Science and the social order. *Philosophy of Science*, 5, 321–337.

Merton, R. K. (1938b). *Science, Technology and Society in Seventeenth-Century England.* Bruges, Belgium: Saint Catherine Press.

Merton, R. K. (1942). Science and technology in a democratic order. *Journal of Legal and Political Sociology*, 1, 115–126.

Merton, R. K. (1968). The Matthew Effect in Science. *Science*, 159, 56–63.

Morris, C. (1960). On the history of the International Encyclopedia of Unified Science. *Synthese*, 12(4), 517–521.

Novotny, H., Scott. P. & Gibbons, M. (2001). *Re-Thinking Science: Knowledge and the Public in an Age of Uncertainty*. Cambridge: Polity Press.

Pickering, A. (1984). *Constructing Quarks*. Chicago: University of Chicago Press.

Pickering, A. (1995). *The Mangle of Practice: Time, Agency, and Science*. Chicago: University Of Chicago Press.

Pinch, T. (1986). *Confronting Nature: The Sociology of Solar-Neutrino Detection*. Dordrecht: Kluwer.

Price, D. J. de S. (1986). *Little Science, Big Science... and Beyond*. New York: Columbia University Press.

Shapin, S. (2001). How to Be Antiscientific. In J. A. Labinger & H. Collins (Eds.), *The One Culture? A Conversation about Science* (pp. 99–115). Chicago & London: University of Chicago Press.

Shapin, S. & Schaffer. S. (1985). *Leviathan and the Air-Pump*. Princeton: Princeton University Press.

Slezak, P. (1994a). Sociology of Scientific Knowledge and Science Education: Part I. *Science & Education*, 3(3), 265–294.

Slezak, P. (1994b). Sociology of Scientific Knowledge and Science Education. Part II: Laboratory Life Under the Microscope. *Science & Education*, 3(4), 329–356.

Sokal, A. D. (1996). Transgressing the Boundaries: Towards a Transformative Hermeneutics of Quantum Gravity. *Social Text*, 46/47, 217–252.

Sokal, A. D. & Bricmont, J. (1998). *Intellectual Impostures: Postmodern Philosophers' Abuse of Science*. London: Profile Books. – In America: *Fashionable Nonsense: Postmodern Intellectuals' Abuse of Science*. New York: Picador USA.

Star, S. L & Griesemer, J. R. (1989). Institutional Ecology, 'Translations' and Boundary Objects: Amateurs and Professionals in Berkeley's Museum of Vertebrate Zoology, 1907–39. *Social Studies of Science*, 19(4), 387–420.

Ziman, J. M. (2000). Postacademic Science: Constructing Knowledge with Networks and Norms. In U. Segerstråle (Ed.), *Beyond the Science Wars. The Missing Discourse about Science and Society* (pp. 135–154). Albany: State University of New York Press.

Zuckermann, H. & Merton, R. K. (1971). Patterns of evaluation in science: institutionalization, structure and functions of the referee system. *Minerva*, 9(1), 66–100.

Gábor Kutrovátz Assistant Professor at Eötvös Loránd University of Budapest (ELTE), Department of History and Philosophy of Science. With a background in physics, astronomy, and philosophy at graduate level, he received his Ph.D. in HPS from the Budapest University of Technology and Economics. He taught over 20 different courses on logic, argumentation, the history of sciences, and science studies. He has studied the impact of Hungarian historians of science (Imre Lakatos, Árpád Szabó). His recent field of interest is framed by argumentation studies and the social studies of expertise. He was a Bolyai Postdoctoral Fellow in Budapest and a Lakatos Fellow at LSE London. He is on the editorial board of the journal *Argumentation in Context*.

Gábor Áron Zemplén Associate Professor at the Budapest University of Technology and Economics (BME). After his Ph.D. (BME) and various teaching positions (in England, Switzerland, and Bavaria), he received postdoctoral fellowships at the Max

Planck Institute for the History of Science, Berlin, and the Collegium Budapest, Hungary. He works on the history of optics and theories of color (sixteenth to nineteenth century), on models of argumentation, and on specific aspects of NOS education. He is on the editorial board of the journal Argumentation and is editor of the book series *History and Philosophy of Science* published by L'Harmattan, Hungary.

Chapter 35
Generative Modelling in Physics and in Physics Education: From Aspects of Research Practices to Suggestions for Education

Ismo T. Koponen and Suvi Tala

35.1 Introduction

The past decade in science education research has seen extensive discussion of the model-based view (MBV) of science education. Much of the inspiration of the model-based view derives from the notion that models are central knowledge structures in science and vehicles for developing, representing and communicating ideas. Hopefully, focusing attention on models as core components of knowledge will make it possible to produce a more 'authentic' picture of science than that currently offered in school science. Consequently, educational researchers expect the model-based approach to deeply affect future curricula,[1] instructional methods and teaching and learning in general, as well as teachers' conceptions of the nature of science.[2] The views put forward in favour of the model-based view have generated discussion of the epistemological goals of science education as well as produced new approaches to science education. However, question remains: In what areas has the MBV brought us closer to an authentic picture of science and in what areas is there still work to be done?

Many researchers advocating the MBV have sought support from the philosophy of science for the focused and coherent use of a philosophical framework for purposes of research in learning,[3] for practical teaching[4] and for designing didactical approaches in general (Izquierdo-Aymerich and Adúriz-Bravo 2003). However, not only are views of the philosophy of science of interest but also the philosophical

---

[1] See, e.g. Gobert and Buckley (2000), Justi and Gilbert (2000), and Izquierdo-Aymerich and Adúriz-Bravo (2003).

[2] See, e.g. Justi and Gilbert (2002), Oh and Oh (2011), and Van Driel and Verloop (2002).

[3] See, e.g. Adúriz-Bravo and Izquierdo-Aymerich (2005), Develaki (2007), and Nola (2004).

[4] See, e.g. Crawford and Cullin (2004), Halloun (2007), Hestenes (1987, 1992), and Sensevy et al. (2008).

I.T. Koponen (✉) • S. Tala
Department of Physics, University of Helsinki, Helsinki, Finland
e-mail: ismo.koponen@helsinki.fi

M.R. Matthews (ed.), *International Handbook of Research in History, Philosophy and Science Teaching*, DOI 10.1007/978-94-007-7654-8_35,
© Springer Science+Business Media Dordrecht 2014

underpinnings of the recent science studies on scientific practices still need attention and reconsideration. Such reconsiderations are needed to define the essential epistemic aspects of models and modelling in building scientific knowledge and how education could highlight such views on modelling practices. For example, what the MBV has not yet taken into account is that the value of models in scientific practice largely depends on how they can serve as somewhat autonomous, freely developing tools of creative thinking and for exploring theoretical ideas. Indeed, from the viewpoint of the goals of science education to promote higher learning and creative thinking, models as tools of thinking deserve attention. Consequently, this article focuses on the role of models as tools of thinking and as vehicles for creative thought, which we refer to here as 'generative modelling'. Such generative modelling shares many aspects with the 'constructive modelling' and 'generic modelling' discussed by Nersessian (1995, 2008). Following Nersessian, we also draw support from recent science studies of ways of thinking and of practising scientists' use of models. The viewpoints of practising scientists are often flexible and dynamic (as we will discuss later in this article), allowing the employment and development of different kinds of models in the different states of knowledge-building practices. In generative and constructive modelling, the 'running of a model' – either mentally in the form of simulative reasoning or more methodically by using computing algorithms and computers – unfolds the system's dynamical behaviour. The dynamic unfolding of the model in simulations is important because it plays a role both in developing the model and in understanding the processes behind various phenomena through modelling. However, although the purpose of such modelling is to understand phenomena, the models may not be always realistic representations of the systems of the processes. Nevertheless, such models can often play an important role in creative knowledge building because they act as tools for exploring theoretical ideas.

In Sect. 35.2 of this article, we first briefly describe contemporary views of models and modelling in the philosophy of science while emphasising the autonomy of models and their use as tools of thinking and reasoning and in exploring conceptual ideas. A brief summary of current model-based views in science education appears in Sect. 35.3. Because both areas are too broad for a thorough review here, we discuss them insofar as to show that the practices of modellers support the idea that in modelling, (1) models serve in semi-autonomous ways, (2) models mediate between experiments and theory and (3) models are instrumentally reliable rather than 'true'. Sect. 35.4 develops these theses in more detail on the basis of the background outlined in Sects. 35.2 and 35.3. Ultimately, Sect. 35.5 embodies the claims in Sect. 35.4 by providing empirical data from interviews with modellers in the field of materials physics and nanophysics. Lastly, Sect. 35.6 discusses the implications for science education.

35.2 Views of Models and Reality

The philosophical underpinnings of models and modelling directly touch upon philosophical issues concerning the relationship of theory to the world as experienced or how we access reality through experiments and how we express this understanding in abstract theoretical form. In this process, models play a central role. Current science education discusses these relationships and the role of models in it under various names, such as the 'new view on theories' (Grandy 2003), 'new history and philosophy of science' (Izquierdo-Aymerich and Adúriz-Bravo 2003) and 'cognitive theory of science' (Giere 1988). All these views are more or less related to the semantic view of theories (SVT) that originates from the work of Suppes (1962), Suppe (1977), van Fraassen (1980) and Giere (1988). In the SVT, the task of theory is to present a description of phenomena within its 'intended scope' so that one can answer questions about the phenomena and their underlying mechanisms (Suppe 1977). In the SVT, the phenomena (as isolated physical systems) are addressed in terms of models, and theory is identified with a set of models (van Fraassen 1980, Giere 1988, 1999). The best-known positions within the SVT are probably the constructive realism of Giere (1988) and the constructive empiricism of van Fraassen (1980), which offer somewhat different answers to the question of how models represent and what their relationship is to theory.

Within the SVT, researchers in science education have recognised the realistic position of Giere (1988, 1999) as a viewpoint which may more closely link science and science education.[5] One reason for this is that Giere's philosophy of science focuses attention on the cognitive and pragmatic factors involved in *doing science*. Another reason is that Giere outlines the relationship between models and reality through realism as embodied in the *notion of similarity*. The model presumably represents, in some way, the behaviour and structure of a real system; the structural and process aspects of the model are similar to what it models. According to Giere, one can describe the theory in general as a cluster of models or 'as a population of models consisting of related families of models' (Giere 1988, p. 82). In Giere's model-based view, the question of theory is therefore not central, because there is already 'enough conceptual machinery to say anything about theories that needs saying' (Giere 1988, p. 83). It seems that such a semantic view, with its accompanying realism, is the favoured position of current views within science education that seek an authentic image of science.[6]

Currently, the SVT and versions of it within philosophical realism provide a robust background to use models and modelling for purposes of science education in the context of predicting and in explaining. However, the current positions must be reconsidered in order for it to correspond to scientific knowledge building. For example, the most favoured contemporary views say little about the methodological aspects of producing a relationship between models and the world, as one accesses it through experiments. However, this question is of central importance for both building scientific knowledge and learning science and thus deserves further consideration. A suitable departure point for this is to examine scientific practices in modelling.

The SVT and versions of it within philosophical realism are thus promising candidates for a robust background philosophy for science education, but they require further development. At least in physics and physics education, Giere's conception

---

[5] See, e.g. Adúriz-Bravo and Izquierdo-Aymerich (2005), Crawford and Cullin (2004), and Izquierdo-Aymerich and Adúriz-Bravo (2003).

[6] See, e.g. Gilbert et al. (2000), Matthews (1994/2014, 1997), and Nola (1997).

of models (as well as the SVT itself) requires some revision for the following reasons. The SVT does not consider the bidirectional relationship between the model and the experimentally accessible phenomenon. It is helpful to compare the bi-directionality with Giere's views of how models are matched with experiments (Giere et al. 2006). In that picture, the similarity or fit of models with real systems is evaluated on the basis of agreement between experimental data and model predictions, but these schemes entail no feedback between the construction of models and the design of experiments or the process of isolating phenomena by altering the experimental setup (Giere et al. 2006, pp. 29–33). To properly and explicitly take into account the bi-directionality, one must realise that when models serve to explain phenomena, the phenomena are not only modelled but are also fitted to the models (Cartwright 1999). Indeed, the concept of similarity (Giere 1988) must be clarified in order to understand the process of matching a model to real systems – and what such matching really achieves. Taking similarity in as strictly realistic a way as possible (as a similarity of representation) is questionable in physics education, because physics offers no compelling reason to include this requirement among the attributes of good models. In what follows, we discuss in greater detail what such empirical reliability means and how it is achieved, as well as the extent to which these are pragmatic, methodological and epistemological questions that can be approached by examining scientific practice.

A second revision – or rather extension – of the standard SVT involves the semi-autonomy of models (from both theory and experimentation) noted in science studies focusing on the usage and development of models as tools in scientific practice (Morrison and Morgan 1999; Morrison 1999). The autonomy of models is closely related to the question of the relationship between models and theory. In the description provided by Morrison and Morgan (1999), models are largely independent of theory. As Morrison argues, the autonomy of models hinges on two notions:

> (1) the fact that models function in a way that is partially independent of theory and (2) in many cases they are constructed with a minimal reliance on high level theory. (Morrison 1999, p. 43)

However, this does not mean that theory plays a negligible or small role in model building and design. Morrison also emphasised this point noting that models 'are not strictly "theoretical" in the sense of being derived from a coherent theory, some make use of a variety of theoretical and empirical assumptions', which means that models 'obviously incorporate a significant amount of theoretical structure' (Morrison 1999, p. 45). Physics offers several examples of this kind of model-theory relationship. For example, models within complex systems – in particular the archetypical cases of the Ising model and spin-glass models – cannot be 'derived' from any theory, nor it is necessary to define exact 'similarity' between the spin elements of the model and their counterparts in the reality they are intended to model. Instead, the elements and the interactions of the model are meant to stand in some relevant structural relationship to each other, corresponding to the real system, so that the model can reproduce the generic features of the real system (e.g. collective behaviour in some physical disordered or magnetised system but equally well in a sociological or ecological system). Although such a model is not derived from theory, the construction of such models entails many theoretically guided steps.

The theory-to-model relationship has been discussed also by Winsberg (2001, 2003, 2006), who remarks that even though models are not derived from theory, theory nevertheless guides model development. The examples Winsberg provides are typically instances in which a rather well-established theory, such as hydrodynamics or continuum mechanics, already exists, and which allows derivation of a set of dynamic equations in the form of differential or difference equations, but which are not necessarily solvable without further modelling. Consequently, Winsberg argues that such models and modelling are rather semi-autonomous than autonomous. In addition, Winsberg seems to hold position that modelling aims to represent real systems, which, on the other hand, he takes as the most essential epistemological dimension of the use of models and modelling. There is no doubt that Winsberg's position is good departure point in cases that involve established theory as a starting point for modelling a known target, such as hydrodynamic flows, nuclear explosions or the behaviour of plasma, where one can derive a governing dynamic equation for the collective behaviour of a system or its parts. Upon closer inspection, however, Winsberg's examples differ substantially from cases encountered, for example, in complex system behaviour, or even more traditional field of nanosystems, where neither an established theory nor well-known targets exist in the sense discussed by Winsberg. In such fields of research, modelling is more a tool for thinking than a tool for realistic describing and predicting. Consequently, the purpose of modelling is to provide a means for simulative reasoning (cf. Nersessian 1995, 2008) and extended means for performing thought experiments. To perform in the desired manner, such models and modelling must be sufficiently autonomous. However, as Winsberg (2001, 2003, 2006) noted, this does not mean complete independence from theory; theory plays the role of providing guidance, but at the same time, much freedom to use non-theory-derived elements is necessary.

In what follows, we focus on the example of nanophysics, which provides an interesting border case; we have every reason to expect a clear theory-to-model relationship (the research field is, after all, rather conventional branch of materials science), a preference to strive for realistic descriptions and attempts to establish similarity. Somewhat contrary to these expectations, however, simulation practices in nanophysics reveal several extra-theoretical elements, clear instrumental rather than realistic positions are recognised, and instead of similarity with real systems, practical values for reasoning and a certain type of mimetic similarity (though not in the sense of direct visual similarity) emerge as a part of it. Since these findings are from the field of materials science, similar attitudes and stances may also be typical of many other branches of the physical sciences. After all, using models as tools of thinking rather than as tools for realistic representations, or striving for instrumental and practical values instead of realism, may in practice prove far more common than that envisioned by philosophers holding to realism and especially to representational realism.

Nersessian (1995, 2008), who viewed models as means (and tools) to represent ideas as well as reality, emphasised the cognitive aspect of using models with no overarching emphasis on philosophical realism. For this, the SVT is a suitable framework, as it suits views inclined towards philosophical empiricism, realism and pragmatism equally well. The SVT can be seen as a structure or net of models related to each other in different ways and in which mutual relationships are under development. Viewing the SVT in this way leaves room for the generative, mediating models. Such an interpretation of the SVT permits greater independence of models from theory as well as a loose relationship within the set of models, yet without requiring a hierarchical relationship of models, as Giere describes (1988, 1999). Such interpretation provides sufficient, if not complete, freedom from theory and empiry to warrant the phrase 'autonomous models'. In what follows, we show that the modelling practices encountered in physics quite well support the argument that models are semi-autonomous but that at the same time, their design is theoretically guided (without being derived or deduced from theory). Such semi-autonomous models are sufficiently free from theory to be useful for developing of theory and making them interesting enough as objects of research in their own right.

From this background, we propose here a revision which, for the most part, still fits within the SVT but relaxes some of its restrictions. The decision to concentrate on physics stems from the notion that in physics, the role of models and modelling is epistemologically and methodologically essential. Consequently, the process whereby theoretical predictions are linked to the outcome of measurements plays an important role (Koponen 2007). One goal of the present study is to encourage science education to reflect as much as possible the epistemological aspects of doing science as well as aspects with which practising scientist can also agree. This goal can be called providing an 'authentic picture of science'. The current study attempts to provide just such an authentic picture of models and modelling in physics. From this viewpoint, there is no compelling reason to limit views only to those within philosophical realism, which seems to be the preferred way for the current science education literature to view models and modelling.

This article discusses models and modelling from the viewpoint of physics, which then serves as a philosophical background for developing physics education. We first scrutinise recent views of models and modelling in science education in order to outline their philosophical underpinnings and to support the argument that this philosophical basis requires revision. Next, we discuss the question of making a match between theory and experiment, and the role of models and modelling therein, from the viewpoint of recent science studies. We then explore these ideas in the light of an empirical study of modelling scientists' views of their knowledge-building practices. Against this background, we argue that an authentic image of models and modelling in physics requires a certain bi-directionality; models, which respect central, established theoretical ideas, are developed to match phenomena as they take place in laboratory experiments. Indeed, what matching produces is both a functional model, a particular experimental system which can then be modelled, and new knowledge about the phenomenon under consideration. Finally, we suggest that, at least in physics education, the requirements of empirical reliability and empirical success in advancing experimentation are attributes that better correspond to an authentic image of scientific modelling than those of realism and truth, as understood in the philosophy of science.

35.3 Models and Modelling in Science Education

The role of models in providing explanations and predictions is perhaps the most common area in which epistemological questions are explicitly discussed, and in this role, models and modelling are often seen from the point of view of realism.[7] The moderate realistic viewpoint is clearly reasonable from the viewpoint of learning. However, the preferred role of realism in supporting a model-based view is often based on the notion that realism is also the preferred view of the scientists who employ models and modelling and that therefore this kind of approach ensures an 'authentic picture of scientific modelling' (see, e.g. Schwarz et al. 2009; Pluta et al. 2011). Such claims are mostly not justified by argumentation based on the philosophy of science nor are there evidence based on analyses of science practice.[8] Of course, claims of authenticity then rest on weak standing. Before we suggest how to expand viewpoints on models and modelling, we briefly discuss how the relationships between theory, model and experiments are conceived in certain practical solutions for school science.

One of the first suggestions for how to use models and modelling in physics education is Hestenes's (1987, 1992) approach, which underscores the relationship of models to theory and experiment. According to Hestenes, model construction follows comprehensible rules (the rules of the game); models are then validated by matching them with experiments. Hestenes draws insights from Bunge's conception of models and parallels Bunge (1983) when he emphasises the mathematical structure of models and their subordination to theory. Hestenes's ideas reflect a clear predominance of the verificative justification of knowledge, and the truth value of models is judged according to the success of such theory-based predictions. This reflects scientific realism, which adopts not only ontological but also epistemological and methodological realism. When known and accepted theory serves as a basis for making predictions, the aspects Hestenes stresses in modelling clearly constitute an authentic way of modelling in physics. Such theory-based modelling of physical situations also occurs in empirical research in mathematics education that aims to combine mathematical and physical modelling, most often in the field of classical physics (e.g. Hestenes 2010; Neves 2010). This view is justified in the context of making predictions and providing explanations, which are surely the most important aspects of school science, which typically employs models and modelling and presents them as a means to illustrate scientific content (Hestenes 2010). In addition, relying on Hestenes's modelling rules makes it relatively clear how to develop teaching solutions and design teaching activities. Following the modelling methods recommended by Hestenes, traditional physics teaching where theory comes first evidently leads to improvement (Wells et al. 1995; Halloun 2007). However, such an approach loses its teeth in the context of scientific knowledge building,

---

[7] See, e.g. Gilbert et al. (2000), Hestenes (1992), Justi and Gilbert (2000, 2002), and Nola (2004).

[8] Analyses of science practice in fact provide evidence pointing to the opposite conclusions (see, e.g. Koponen (2007) and Tala (2011) and references therein).

where the main interest is in the construction of theory or the acquisition of knowledge not already captured by existing theory.

A related approach from theory to models, but somewhat less theory subordinated, is found in the work of Crawford and Cullin (2004), where the main epistemological role of models is in explaining and developing understanding of the phenomena of nature. According to them, the scientific process can be depicted as a sequence of making observations, identifying patterns in data and then developing and testing explanations of these patterns; in their words 'such explanations are called scientific models' (Crawford and Cullin 2004). In the modelling activity described by Crawford and Cullin, students observe, indentify variables and conceive ideas about the relationship based on their own ideas, which they then test. These authors do not explain their epistemological position with regard to models, but the realistic stance and role of models in providing explanations is apparent on the basis how they refer to the philosophical works of Giere (1988, 1999), Hesse (1963), and Black (1962) in what they discuss about model-to-reality correspondence or similarity. In all cases, a question is the realistic one: How do models represent real systems and how can one improve this correspondence? An adherence to realism is also evident in their general purpose of 'investigating real-world phenomena; then designing, building, and testing computer models related to the real-world investigation' (Crawford and Cullin 2004, p. 1386). They do not explain how these models are produced or how they relate to theory, but taking into account the philosophical underpinnings to which they refer, something close to Giere's concept of similarity between the models and the real world seems to be involved.

The picture of models and modelling proposed by Halloun (2007) is closely related to Hestenes's views in that it sees the role and use of models in making predictions and explanations in a realistic sense. In addition, Halloun focuses to the role of models in investigating and controlling physical systems and phenomena, as well as in influencing the development of a scientific theory. These aspects are important for modelling, which is not theory constrained, but is instead theory generative (cf. Koponen 2007). In agreement with this, Halloun notes that

> For meaningful learning of science, students need to systematically engage in identifying and modeling physical patterns and explicitly structure any scientific theory around a well-chosen set of models. (Halloun 2007, p. 655)

To achieve this, he proposes a modelling scheme, which begins with (1) physical realities and then (2) identifies a pattern for which (3) a model is constructed. This continues with (4) analysing a model and (5) inferring pertinent laws on which one makes predictions (and perhaps also inferences) of the physical realities. Finally, on this basis, (6) the model is refined and integrated into the existing theory (theory may then change) (Halloun 2007, p. 663). In fact, these three steps can be seen as basic steps in creative modelling, which are capable of generating new knowledge, with the ultimate goal not only of predicting or explaining but also of intervening. The success in manipulating and intervening in the phenomenon then provides a basis for developing our understanding of the phenomenon in question. These aspects of modelling are seldom discussed in the context of the model-based view of science education, although such aspects are arguably the most central in advanced levels of scientific modelling (cf. Koponen 2007; Tala 2011).

One can hardly discuss models and modelling in physics without acknowledging the role of mathematics and mathematical thinking in knowledge in physics. To present modelling of a real system, the recent literature on mathematics education, often uses steps quite similar to those Halloun has suggested for modelling in physics. The steps constitute the modelling cycle of the iterative mathematical modelling process for a phenomenon as perceived in the real world. Mathematical modelling cycles are presented as including the identification and simplification of the problem or purification of the variables (cf. Halloun's step 1), formulation of the mathematical model of the real situation (cf. Halloun's steps 2 and 3), solving or studying the mathematical model (cf. Halloun's step 4) and then interpreting the mathematical results in the real world, which is then followed by an evaluation and the development of the model in a repeated modelling cycle.[9] Such mathematical modelling highlights the role of mathematising the phenomena and role of creativity involved in mathematical reasoning and thinking. Empirical research on such mathematical modelling education has proved to be an effective means for preparing students to deal with unfamiliar situations by thinking flexibly and creatively when solving concrete real-world problems by using applied mathematics (Haines and Crouch 2010; Uhden et al. 2012). Theoretical understanding requires focusing on the mathematical framework of modelling, which guides the analysis and simplification of the physical situation and interpretation of the results. Therefore, the viewpoint provided by mathematical modelling education can focus attention on the important role of mathematics in the physical modelling process and the development of mathematical or computational models in a modelling process.

Finally, discussing the creative process of modelling requires that one consider cognitive viewpoints. Nersessian (2008) has studied how scientists use models in discovery processes and to create theories and notes that in 'scientific discovery, models and modeling come first, with further analysis leading to formal expression in the laws and axioms and theories' (Nersessian 2008, p. 205). Scientists use models as cognitive tools, especially when creating, thinking and reasoning their way to novel conceptual representations by means of models.

For practising scientists, an important role of models functions as tools for intervention in and the manipulation of phenomena. Cartwright (1983, 1999) and Hacking (1983) and more recently many other authors (see Morrison and Morgan 1999) discuss this practical role of models. In addition some studies in science education (Izquierdo-Aymerich and Adúriz-Bravo 2003; Crawford and Cullin 2004) now also recognise this practical role of models as tools of thinking. For example, Izquierdo-Aymerich and Adúriz-Bravo (2003) discuss the role of scientific activity and scientific research as an attempt to transform nature and interact with it, rather than as an activity for arriving at truths about the world. They note that models (and theory) which fail to achieve these goals have little value in science education for students and teachers. Unsurprisingly, they follow Hacking (1983) in emphasising

---

[9] See, e.g. Blum and Ferri (2009), Haines and Crouch (2010), and Uhden et al. (2012).

that models serve to make sense of the world, with the ultimate objective of actively transforming the ways to see nature, where 'facts of the world are heavily reconstructed in the framework of theoretical models' (Izquierdo-Aymerich and Adúriz-Bravo 2003). These new aspects concerning the use of models and modelling agree with the physicist's conception of acquiring and justifying knowledge.[10] However, understanding these aspects of transforming, manipulating and intervening in phenomena requires that models serve as autonomous or semi-autonomous agents, much like research instruments.

The viewpoint which sees models as research tools for constructing, understanding and generating knowledge is interesting for education, because it assigns models an active role in the generative and creative process of knowledge construction. Sensevy et al. (2008) have discussed such use of models and modelling in school science from a viewpoint of how the modelling process enables transitions between the abstract and the concrete. In that, they make use of the views put forward by Cartwright (1983, 1999), and Hacking (1983) in a manner similar to that proposed by Koponen (2007), who also endorses Cartwright's and Hacking's views among those put forward by van Fraassen (1980). Sensevy et al. called this emerging view of models and modelling the New Empiricism, which emphasises the integral role of empirical knowledge in model building. This New Empiricism captures quite well the bidirectional roles of empirical and theoretical knowledge, not only in model building and interpretation but also how theory guides the empirical explorations as well as framing of phenomena (cf. Koponen 2007). For this type of picture, a weak realistic position, which emphasises the pragmatic aspects of knowledge, fits very well. Indeed, in this way, education allows room for students to construct their knowledge in an equally flexible way as do the more radical constructivist positions.

The generative and creative use of models is challenging for practical solutions of education, because such use of models requires the means to explore the behaviour of the model and to unfold the dynamic evolution of the model system. The behaviour of the model is then compared to the behaviour of the real system in order to understand which aspects of the system model capture and how the model describes. This requires simulations. Insofar as simulations make the dynamic evolution of the model system explicit, science education literature seldom addresses them. Usually, computer 'simulations' in science education are either ready-made programs, controllable animations of theoretical situations or programs being based on object-oriented programming rather than real simulations[11] (Crawford and Cullin 2004). What is missing from these applications is the possibility to transform the original mental constructions first to more formal model structures (mathematical structures) and then these structures to a form which enables one to 'run' (i.e. simulate) the models. Nevertheless, when models serve as vehicles generating new ideas and as tools in the creative exploration of ideas, simulations are indispensable.

---

[10] See, e.g. Chang (2004), Heidelberger (1998), and Riordan (2003).

[11] See, e.g. Bozkurt and Ilik (2010), Finkelstein et al. (2005), Perkins et al. (2006), and White (1993).

Simulations in this sense, however, must not be conceived as ready-made computer programs or applets, which are rather simple animations with control parameters, but as computational or algorithmic tools which unfold or generate the dynamics the model relations contain, yet remain largely unseen in the model itself.

35.4 Generative Modelling and Simulations

Models in the role of constructing new knowledge, generating theory and as tools for discovery have seen relatively less discussion than have model-to-theory relationships, semantic conceptions of models or model-to-experiment relationships. According to Nersessian (1995, 2008), the role of models in constructing knowledge is most important for creative scientists but seldom addresses in teaching science. Of course, creative development and the construction of models are more difficult to discuss than, for example, theory-subordinated modelling, but on the other hand, having access to such modelling in teaching and learning science adds an important creative dimension to science education.

In her account of constructive modelling, Nersessian (1995, 2008) draws on historical analysis and its cognitive interpretation. According to Nersessian, constructive modelling is a kind of reasoning process that aims to integrate different types of knowledge structures, such as analogies, visual modelling and thought experiments, in order to provide cognitive means to explore and to reason about phenomena or target problems for which no direct model is available. The way Nersessian describes the content and purpose of such constructive modelling also applies into cases not only where direct analogies are missing but also where the ways in which to apply the theory remain obscure or attempts to do so fail (unlike in situations described by Winsberg, where seeing how the theory supports and guides model development is relatively straightforward). An important part of constructive modelling is so-called generic modelling, where 'the generic model represents what is common among the members of specific classes of physical systems' (Nersessian 1995, p. 212), which allows one to transform known models as a basis for exploring new unknown situations. Thus, generic modelling guides scientists' problem solving or helps them recognise similarities between physical systems for new regularities. Constructive modelling, on the other hand, involves

> constructing analogous cases until the constraints fit the target problem. The models thus constructed are proposed interpretations of the target problem. Further, the ability to construct and reason with generic models is a significant dimension of the constructive modeling process. (Nersessian 1995, p. 209)

However, according to Nersessian learning such constructive modelling is challenging in teaching and learning, because it requires sound background knowledge and deep engagement to modelling and possibly also to simulations, knowledge and skills which cannot be expected in introductory levels of learning. These requirements shed some doubt on success of teaching solutions based on very simple ways to use models, very straightforward modelling tasks or on use of ready-made simulations, where the learners' own role in construction of models is moderate.

In what follows, we propose a view of modelling sharing many characteristics with Nersessian's constructive modelling but, here, in more restricted scope as they appear in higher education and in advanced education. In this type of modelling, simulations play a central role but now in the form of algorithmic simulations and computer simulations. Models and modelling, when used as tools to generate understanding and especially in the role of creating new knowledge, often require running models with a computer, namely, simulations. The purpose of such simulations is very practical: to acquire new knowledge of already familiar systems and to probe the behaviour of the simulations (used as tool of investigation) in new situations. This process also develops and transforms the simulation model itself. In simulations, models and modelling play a central role: Simulations are the process of 'running the models' in the virtual world in order to reveal the dynamic behaviour of a system of models. Here, such a combination of modelling and simulations is called *generative modelling*. The practical character of generative modelling makes it a promising case for learning about the epistemological purposes of modelling, to be implemented also as a part of modelling in physics education.

The epistemological value of generative modelling in physics derives from its ability to bridge theoretical ideas – conceptual reality – and real systems. To understand this role, we must consider the relationship of generative modelling to experiments and theory in a balanced manner. Interestingly, this relationship is autonomous; although theory is deeply involved in constructing the models behind the simulations, the models only seldom derive from some underlying theory. This means that a model and the virtual environment provided by simulation respect the central theoretical ideas, but are not directly derived or deduced from theory. In simulations of nanostructure growth, for example, the principles of statistical physics, such as general conditions for equilibrium states and detailed balance, are respected, but the form of probabilities determining the dynamics of the system comprises often purely phenomenological models. Similarly, the relationship of models and simulations to experiments is intriguing; although simulation results are often compared to experimental results in order to verify or develop the simulation model or the simulation environment, this comparison often does more in that it may even affect experimental design and strongly influence which aspects of experiments will be considered. This suggests that the epistemological importance of generative modelling is related to its semi-autonomous role in theory and experiments and that it is this semi-autonomous position which enables generative modelling to 'mediate between the theory and experiment' (Morrison 1999). The way this mediating presents itself involves not only constructing the models to fit the phenomena but also finding and isolating phenomena to fit the models used in the simulations (cf. Cartwright 1999). Consequently, the models need not be faithful to all experimental results; there is freedom to ignore experimental aspects considered irrelevant without strict criteria for how to assess such relevance. On the other hand, the models remain free from theory in the sense that they need not be derived from theory (they must be in accordance with most important theoretical principles, although the criteria for 'most important' change from one situation to another). It is this semi-autonomy and flexibility of models which makes them so useful and practical in fitting together the conceptual (theoretical) and material (experimental) control of the phenomena under study and, through that control, providing new knowledge and advance theory construction.

Recent philosophical analyses of simulations and modelling provide several insights into how to frame the idea of generative modelling. First, generative modelling aims to be representative, but does not aim towards a realistic representation of physical systems and their behaviour. Instead, modelling involved in simulations purposely distorts some aspects of the systems to be represented in order to achieve its goal of more effectively representing the features of reality under scrutiny. Such a position can be described as selective realism (Humphreys 2004) or moderate realism (Koponen 2007). Second, generative modelling uses nested structures of different types of models (some of which are theoretically guided, and others, purely phenomenological) but in such a way that they constitute co-ordinated parts of the model. This resembles Hughes's notion of the nested structure of successive modelling steps with nested representations in an ascending level of abstraction (Hughes 1997, 1999). Also, running simulations in order to unfold the process aspects of the phenomena is central to Hughes's scheme. Third, unfolding the processual aspects of model systems in simulations is closely related to the notion of mimetic capabilities and mimetic similarity (Humphreys 2004). On this basis, notions which must be taken into account when discussing the purposes and capabilities of simulative modelling are as follows:

1. *Mimetic similarity.* Simulative modelling aims to establish partial mimetic similarity between the processual evolution of systems in simulations and corresponding (though not exactly similar) systems in experiments.
2. *Instrumentally reliable models.* Simulative modelling serves to construct and validate *instrumentally reliable models for the processes* behind experimentally accessible phenomena. These models embody the knowledge achieved by producing and developing simulation models.
3. *Generative modelling mediates* between high-level generic models (or theory) and experimentally accessible phenomena by constructing instrumentally reliable models and fitting them to phenomena. In that process, laboratory phenomena are also designed to fit the models better.
4. *Generative modelling is an instrument* of investigation relating to the world of concepts and theories (and experimentation) in parallel fashion to the way in which the measurement instruments relate to real systems; both are probes in their own worlds, about which they deliver information.

The first notion takes mimetic capabilities as central to knowledge production. Mimetic similarity between processes in simulations and those observed in experiments is at the core of the model justification process. It should be noted that mimetic similarity as understood here, however, is not simply visual mimetic similarity. As Winsberg (2003) has noted, mere visual similarity is of little epistemic value and deserves little attention. Similarity in generic behaviour or in the succession of unfolding events or processes between model systems and real systems, however, is of epistemic value, because such similarity between connected events carries information about causal knowledge or the determination of events. Hughes, for example, refers to such mimetic similarity, achieved through simulations, when he notes:

> The dynamic has an epistemic function: it enables us to draw conclusions about the behaviour of the model, and hence about the behaviour of its subject. (Hughes 1999, p. 130)

Epistemologically, however, this is possible only if models have *representative capabilities*. But 'to represent' now means something other than simply picturing, mirroring or mimicking physical systems, or being 'similar' to a real system. Instead, as Morrison and Morgan outline,

> a representation is seen as a kind of rendering – a partial representation that either abstracts from, or translates into another form, the real nature of the system or theory, or one that is capable of embodying only a portion of a system. (Morrison and Morgan 1999, p. 27)

In their words, such a 'model functions as a "representative" rather than a "representation" of a physical system' (Morrison and Morgan 1999, p. 33). In nanophysics, for example, such mimetic similarity is achieved when a spontaneous selection of order or sizes of nanostructures are generically similar in both real measurable systems and a simulated generic model of self-organisation. Successfully acquiring this similarity in simulations leads to greater insight and understanding, thereby producing knowledge. Similar type of mimetic similarity and its importance for knowledge generation without visual similarity is discussed also by Humphreys (2004) and Hughes (1997, 1999).

The second notion claims that instrumentally reliable models are the goal of modelling. It calls attention to the notion contained in selective or moderate realism that models must be empirically reliable descriptions of only certain aspects of processes and that only some of these selected aspects must be empirically validated. Such instrumentally reliable models are often the products of simulative modelling and also enable one to match theoretical ideas with experimental data, thus bridging the conceptual and the real. Instrumental reliability is important to the credibility of models and their epistemic credentials. On such instrumental reliability, Suárez (1999) notes that the degree of confidence rather than degree of confirmation is essential. The degree of confidence in the model, on the other hand, increases through its successful applications. This is a pragmatic virtue, achieved through use of models. In addition, methodological questions about how the production and design of models relate to pre-existing models and modelling methods are also important (Winsberg 2006). This is a kind of methodological continuity which increases the models' reliability. For scientists, such reliability, achieved through pragmatic and methodological credentials, is often sufficient, and questions of 'truth' in the philosophical sense do not play a major role (see, e.g. Riordan 2003). As Winsberg has noted: 'The success of these models can thus provide a model of success in general: reliability without truth' (Winsberg 2006, p. 16). Together these positions mean that in terms of the reliability of and confidence in models, the 'truth' of the models is less important than the realistic position assumes.

The third notion details the well-known claim that 'models mediate' and suggests how to clarify the vague notion of 'mediating'. Namely, modern experiments provide opportunities for precise control, thus allowing us to 'build our circumstances to fit our models' (Cartwright 1999) and thus narrow the gap between models and real systems. As Cartwright notes:

...we tailor our systems as much as possible to fit our theories, which is what we do when we want to get the best predictions possible. (Cartwright 1999, p. 9)

She argues further that it is this very aspect of experimentation that makes theories (and models) successful:

We build it [the system] to fit the models we know work. Indeed, that is how we manage to get so much into the domain of the laws we know. (Cartwright 1999, p. 28)

Accepting these views – evident in the practice of physics – means understanding models as matchmaking tools as well as for manipulating and isolating phenomena.

However, between the real world of entities and phenomena and theory (with its concepts), no direct connection or correspondence exists. Neither are the entities of the real world or its phenomena directly accessible through observation and experimentation. Only through laboratory experiments and measurable quantities do the regularities contained in phenomena or the entities behind them become accessible, observable (or detectable) and discernible. It is such an experimental law – a kind of 'model of data' – that the theoretical models constructed in physics are meant to be matched. The form of models we are interested in here mediates between high-level theory and experimental laws in the above sense. Several philosophers[12] have recently discussed the role of models as mediators between theory and experiment. These views reminded us that models are seldom constructed or derived from theory, nor do experimental data necessitate the models; rather, the models are built using knowledge from many independent sources, sometimes even contradicting the theory.[13] Nevertheless, models carry a substantial amount of well-articulated theoretical knowledge through the theoretical principles involved in their construction; otherwise, they would fail to perform their task of mediating between theory and experiment (Morrison and Morgan 1999; Morrison 1999). To focus on the semi-autonomous role of models, relaxing only the models' close dependence on theory contained within the SVT seems sufficient.

If experimental laws are also taken as models representing the data in suitable form, the emerging picture begins to resemble Suppes's (1962) view, where a hierarchy of models mediates between theory and measurements. This is a process of mutual matching, which entails sequentially adjusting and transforming both kinds of models and involves different levels of models. An essential feature of this bidirectional process is that models can fulfil their task of connecting experimental results to theory 'only because the model and the measurement had already been structured into a mutually compatible form' (Morrison and Morgan 1999, p. 22).

---

[12] See, e.g. Cartwright (1999), Hughes (1997), Morrison (1999), and Morrison and Morgan (1999).

[13] See, e.g. Cartwright (1999), Morrison (1999), and Morrison and Morgan (1999).

Such a process of sequential matchmaking is inherently connected to the use of measuring instruments and the theoretical interpretation of their functioning, an aspect Duhem (1914/1954) already emphasised.

For Suppes, comparison between theory and experiment consists of a sequence of comparisons between logically different types of models. Darling (2002) has described Suppes's view by using the scheme of a 'data path' and 'theory path', which converge at a point where comparisons of data and theoretical predictions are possible. On the theory path, one begins by extracting from a physical theory principles or conditions relevant to the class of experiments under question. These data sets (or models of theory) are the theoretical predictions which, in the end, are compared to the results of the measurements. On the data path, one begins with the actual experimental setup. The essence of this process is that the measurement data (generated by the experiment) cannot be directly compared to the theoretical predictions. To make the comparison, the data must be transformed so that they form a 'model of the data' (Darling 2002).

For Duhem (1914/1954), the experimental results, the measurement itself and the instrumentation used in the measurements are all of central importance. Consequently, Duhem begins with experiments and introduces a sequence of 'translations' which transform experimental results into a form that theory can ultimately annex. The essence of Duhem's viewpoint is that the theoretical interpretation of the use of instruments and how they function is indispensable in every step of the translation sequence; the whole process of interpretation requires a number of theoretical propositions (Darling 2002). Yet, both employ a sequence of modelling steps needed to narrow the gap between actual measurements and the theoretical predictions; there is a mutual fitting of theoretical models to empirical results, as well as models of empirical results to theoretical models. Moreover, in the latter, not only are results idealised, but the experiments themselves are often altered, as is the manner in which the phenomena are produced.

The fourth notion summarises the role of simulative modelling by reiterating the notion of models as instruments (Morrison 1999; Morrison and Morgan 1999) and clarifying what it means to be a 'computational instrument of investigation'. These models emerge in the same cyclical process, where they serve as creative tools: Models bridge the experimental and theoretical world, because they embody both empirical and theoretical arguments in the modelling process. Simulations then serve as a tool for inferring relevant knowledge of the system and its emergent properties. Similarly, in Nersessian's (2002, 2008) view, the most valuable aspect of constructive modelling is that it provides cognitive means for simulative reasoning, mentally exploring or simulating the consequences of the model assumptions. Here, this capability of mental reasoning and simulating takes on a more formal appearance, as algorithmic simulation with the intent to reason and unfold dynamically the consequences of the model's assumptions.

The above four notions clarify the usage and development of models, what models are and how they relate to observable aspects of reality and theory. They also help to characterise the concept of 'model', which should be sufficient for practical purposes, but without attempting to define them. Several reasons compel one to resist the temptation to 'define' models; attempts to do so have turned out to be too restrictive in describing actual modelling practices. Rather than clear-cut definable objects in the philosophy of science, models are cognitive and pragmatic tools for exploring, understanding and manipulating the world with the purpose of constructing knowledge; they are undefinable, but understandable through the ways in which they serve these purposes.

However, an important question remains: How are models developed in the form of simulation models, which can be run in the virtual world of computers? In this we meet the technological limits and especially the idea of the computability, which Humphreys (2004) discusses extensively and in depth. When a model is implemented in a computer, it is simplified and rewritten in a form the computer can handle. Humphreys makes these ideas more transparent at the methodological level by suggesting that in simulations, one should distinguish between the *computational model*, *computational template* and *correction set* to the template. In Humphreys's scheme, the computational model is the model run in the computer, which is constructed on the basis of computational template. These computational templates are mathematical structures, such as formulas, which can serve fruitfully in different contexts; they are well known, and familiar moulds void enough content that they carry over to new unfamiliar fields of inquiry, but with the assurance of how they will work. In many cases, templates as such are insufficient and they must be augmented with correction sets. The correction set serves to match the simulation to the experimental process, thus allowing one to better pursue selectively realistic representations.

On the one hand, in this picture, the models are considered reliable, accurate and correct, but on the other hand, the model may contain sub-models in the form of local parametric models and parametric relationships which must – in a strictly realistic sense – have a direct counterpart in real systems. This, however, does not compromise the reliability of the models, nor does it mean that the models do not represent at all, because the purpose was to represent the real system only to a predetermined degree of realism. Already from the beginning, several possibilities to make the model more realistic are always known, but the increased degree of realism is nearly always sacrificed for the sake of the transparency of the pertinent phenomena, for the greater ease of its mathematical handling and, finally, for better computational tractability. Such a position is well described as selective realist (Humphreys 2004) or moderate realist (Koponen 2007).

The question of which follows is the one of how such selectively realistic simulations provide access to reality. To access real systems in practice, it is enough that they be controllable and manipulable and that new similar systems can be designed with desired features. This goal is realised through simulative modelling, which serves as an inference tool for producing instrumentally reliable models. First, by producing instrumentally reliable models, simulations help to isolate suitable real systems and phenomena to which these models apply. Second, through the bi-directionality of modelling and experimentation, simulative modelling provides a means to manipulate these systems through model predictions. Third, through these steps, simulations provide access to the real, as a means of manipulating and intervening, and for all practical purposes, it is here that the conceptual meets the real.

Thus far, we have discussed generative modelling on the basis of theoretical studies to understand what these ideas mean in the practice of knowledge building. It is now time to consider some examples drawn from practitioners' use of models and modelling.

35.5 Authenticity of Generative Modelling

The picture of the generative modelling painted here differs substantially not only from theory-subordinated verificative modelling but also from the strong realist position, where models' similarity to real systems is extremely important. Finally, this picture challenges the claim that the realistic position grants the authenticity. Of course, we must now respond to this challenge of authenticity and discuss the empirical evidence to support the views promoted here. We did this by exploring the modelling practices and practitioners working in the rapidly developing field of condensed matter physics. We interviewed (altogether ten) PhDs (apprentices, A) and more senior researchers (experts, E) about their knowledge-building practices through modelling (Tala 2011). This in-depth study will reveal clearly a stance of selective realism. The notion that the modelling topics studied are for systems in material sciences strengthens the weight of the argument put forward.

The interviewees' modelling activity focuses on understanding the nanoscale phenomenon through 'hands-on' manipulation of the various models by means of simulations. As one of the informants mentioned:

N1: our simulations are quite down-to-earth, but at the same time [they] make it possible to study general phenomena. (E)[14]

Such bi-directionality is at the core of the generative modelling; models can serve a twofold function: to improve both our understanding of a phenomenon in particular experimental settings and our understanding of the phenomenon on a more general level. Achieving such understanding also plays a role in the generic mental modelling that guides experts' problem solving in new situations (cf. Nersessian 1995):

N2: when beginning to simulate something new, we have often noted with happiness that we have already learnt this [kind of phenomenon] with metals. (E)

Selective realism is evident in many of model builders' comments; in comparisons of experimentation and modelling, only part of the models' features is selected. One interviewed expert explained what makes the models he develops valuable:

N3: In a certain way, in some cases the model is workable/usable for estimating certain things… What makes it interesting is that we can take into account certain important aspects – or the simplicity. (E)

---

[14] Nine interviews were in Finnish, and one in English; the authors translated the excerpts.

35 Generative Modelling in Physics and in Physics Education... 1161

In following excerpts, a couple of younger researchers explain further:

N4: The whole of molecular dynamics, would not function if the frequently repeated calculations were not made as simple as possible and quick for the computer to calculate. (A)
N5: If it [the model] becomes too complicated, it is no longer intuitively clear. I am looking for an intuitively clear model which includes the essential [features of] processes and little else. (A)
N6: The models used to represent different viewpoints of the same phenomenon need not be consistent with each other. Nor do even the parallel models need be commensurable. (A)

Modelling thus captures the partial similarity between the dynamic of the model and the respective part of the laboratory phenomenon under consideration. These remarks suggest that the mimetic similarity of models and real systems is reached through a selective realistic attitude towards models and modelling; only part of the systems' behaviour was of interest in one modelling process.

Modellers' views also contain information about how they see models as autonomous instruments of thinking and exploration, sufficiently free from theory to relax some of its constraints, yet sufficiently close to theory to adhere to the modelling assumptions based on sound physical reasoning and a theoretically motivated basis; all aspects need derive from theory, but they do need to be acceptable and make sense from the viewpoint of the theory. As tools of thinking, models are semi-autonomous constructs that function in the conceptual or virtual world and allow one to explore theoretical possibilities. Note that this would be impossible if models were deduced and derived from theory; then, nothing new would come out of modelling. Moreover, if the connection to theory is thin, it would be impossible to say which aspect of the theory is explored and how the model helps to increase knowledge. The interviewees described several examples of how playing in the virtual world provides them opportunity to construct and study systems with properties which may not be (or at least are not yet known to be) real, but could be real within the given theory. Sometimes such systems extremely reduced or idealised, and when needed, even contradict some theoretical principles. Modellers also highlighted this autonomous role at the general level:

N7: A model lives its own life. (E)

Such remarks are typical among physicists. Indeed, they are unmistaken signs of the semi-autonomous use of models as tools of thinking and knowledge construction.

The modellers' conceptions of the ways in which the models mediate between theory and experiment also typically fall in selective realism and instrumental use of models. As three experts put it:

N8: Basically, what my experimental colleagues give me — as an example — Could you please explain why it [an experimental result] is like that. Not simply to explain, but sometimes [to support] a proof of why this [result] can be used as an effective rule for the construction of some materials and not others.... Not simply to explain, but also to understand. (E)

N9: It's very common that a theory creates reality in the sense that modellers propose some sort of explanation for phenomena, not necessarily only for this one, but a general one. (E)

N10: It is a model which explains a particular physical phenomenon. And then everyone follows that model. The model is probably fine and correct, and sometimes it is not. But experimenters start thinking in terms proposed by this'theory'. (E)

Thus, the experimenters need a model to plan the experiment and to interpret the data obtained from it. Some responses also mentioned that, at best:

N11: the simulation predicts something that will be found out through experimentation in the future. (E, A)

The modellers also described how they:

N12: develop new ideas and try them out [in the virtual world] to see which ones are worth testing in experiments. (A)

N13: discuss with the experimenter what can be done and what should be done. (E)

Such developments and discussions are conducted before launching a new research project. Both particular models and particular experiments can thus be more or less purpose-built to each other already from the beginning.

The interviewees also described how the computer model and experiment then become more closely fitted in the cyclic, iterative processes in which running the simulation plays a central role in understanding and developing both the model and the experimental process. Such a modelling process employs many sources of knowledge. In addition to available experimental data available or theoretical calculations, the interviewees compared simulations with other simulations and simulation results and employs practical knowledge:

N14: Everything available; even plain hand waiving in situations where we do not know what exactly is taking place. (E)

Naturally, these 'hand-waving explanations' are educated guesses based on scientists' experience in modelling practices and their understanding of the principles underlining physics. Referring to the origin of such constructive knowledge building, one expert noted:

N15: We should not forget that we are studying mental projections; we study mental pictures which are foundationally mental. It is what we see. (E)

In the interviews, instrumental reliability is most often about computational or algorithmic reliability or functionality. The capability of models to mediate, on the other hand, rests heavily on such instrumental reliability.

N16: A model doesn't care about the actual conditions or claims to explain them, since the only important property of a model is its functionality. (A)

In fact, the ideas expressed by the modellers are very close to the views what Humphreys has discussed under the rubric 'the template', meaning an established

35 Generative Modelling in Physics and in Physics Education...

and shared algorithmic or computational structure. As the following response puts it,

N17: Simplified mathematical gizmos are the elements shared by the different models. (A)

N18: The harmonic oscillator is the most generally used model in physics; it is used nearly everywhere.... After all, nearly all interactions in these simulations are modelled by harmonic potential. (A)

N19: When a physical template is fitted to a computer, it becomes a kind of new theory... Owing to the digital nature of computers, the discretised template – which is the physical model fitted in the computer – is never the same as the original physical template which provided the starting point. (E)

The choice of 'simplified mathematical gizmos' (see N17) is justified by its tractability and ease of use:

N20: This mathematical model is used because it is computationally very undemanding and effective, ... these models are naturally quite simple, and thus do not represent a system very well. (A)

Here, again selective realism is the unmistaken underpinning of the attitude towards modelling; the practical limitations of simulative modelling override even the desire for realism. And finally, what is achieved is an instrumentally reliable and functional model as well as knowledge about its functions. Moreover, the interviewed scientists frequently refer to practical values; a new model is good if it operates as intended, namely, if it produces the events observed in experiments or predicted by more general theoretical models.

Models and their simulation serve as valuable tools in the creative work of scientific knowledge building. They embody both theoretical and empirical ideas developed in the flexible virtual world. It is quite natural that only semi-autonomous models would be able to serve as an instrument. Few remarks refer to these aspects, but remarks N7 and N14–N17 do. Indeed, we noted that experienced modellers perceive the semi-autonomic role of models as a more covering aspect than novices do; understanding this autonomy seems to be characteristic of the expert-like approach.

The semi-autonomy of models is linked to the fact that scientists favour the functionality and computability of the models and selective realism over strict realism. Consequently, although the interviewees wield the models with a substantial, well-articulated theoretical knowledge, they are not mere deductions from the theoretical structures, nor are they deduced or induced from experimental results; rather, they are rich physical constructs which mediate between theory and experiment. Indeed, the bi-directionality of the model with the adjustments based on empirical evidence, coupled with the usage of the model as a tool for investigation – in the spirit of selective realism or weak realism – seems to be an obvious perspective for modelling practitioners working in many fields of physics. The picture that emerges from the practitioners' views lends little support to a strong realistic stance on modelling, or modelling seen as an essentially theory-driven endeavour. Rather, the picture contains many elements discussed here under the rubric of generative modelling, with a selective realistic position emphasising the instrumental use of models, and sees models as semi-autonomous, mediating tools of reasoning and thinking.

35.6 Implications for Physics Education

The generative modelling in physics could promote a more authentic view of models and modelling than the more traditional picture of modelling based on a strong realistic position with its emphasis on prediction and explanation. In short, we suggest that the models and modelling should not be introduced only as tools for explaining scientific content, but also as instruments for creative thinking with the purpose of creating scientific knowledge. In education that takes this notion into account, the points to be highlighted are (1) mimetic similarity, (2) instrumentally reliable models that are sufficient with very moderate realism, (3) generative modelling that mediates between theory and experimentation as an independent approach and (4) generative modelling that is an instrument of creative investigation. In implementing this generative perspective of modelling in science teaching, it is natural to use the research-based resources for science learning and teaching that we already have. For example, cognitive demands must be taken into account in planning, which means that such an approach is not easily adapted to preliminary levels of science learning. It becomes accessible perhaps in upper secondary school level, desirable in the first year at the university level, necessary at the end of advanced studies and, finally, indispensable at the expert level, where the role of generative modelling is easiest to see in new fields of research. Modelling in contemporary physics is still quite a challenging theme, even for teachers, and even more so in the current situation where implementing theory-derived and more straightforward modelling approaches into education is still relatively new.

One very clear requirement of generative modelling activity is that it should be a dynamic and generative activity that produces (processual) mimetic similarity with phenomena or with processes occurring in real world. At the practical level, this means that model relationships are couched in terms of changes and differences, technically as difference equations or differential equations with constraints. At a more untraditional level, the dependencies and interactions in model elements and their relationships come in the form of updating rules for states of the model, as in cellular automata or in agent-based models. Such models have no direct one-to-one relationship with the system that could be studied with the model and where one could assess the similarity of the system and model. In high school-level teaching and in undergraduate teaching, generative and simulative modelling means that more attention should focus on how certain kinds of dependencies of the model level generate certain dynamic behaviours in the simulation run. For example, what kind of motion is related to linear restoring forces, to inverse power law forces or to exponentially decreasing forces, how do these types of forces act in combinations, and what features of generic behaviour one can detect in real phenomena? Such a model-based approach in practice is quite similar to approaches suggested by Halloun (2007) and Nersessian (1995). In Halloun's and Nersessian's suggestions for perceiving modelling, the goal is to understand the generic mechanism behind the most important features of the phenomena. In this sense, simulative modelling through its capability to provide mimetic similarity provides new tools and new ways of thinking about how to see the world and how to make sense of its regular features. There is also considerable room for one's own invention and construction of models. Such objectives, implicated by the generative nature of scientific modelling, are unattainable with modelling that is too simplified. Indeed, the success of practical solutions requires sufficient domain knowledge (Nersessian 1995). Thus the difficulty of the suggested content of such examples of generative modelling makes implementing these ideas challenging at the lower levels of education.

Moreover, using the models as tools for creative thinking and the construction of knowledge can be challenging in the tradition of science education that emphasises realistic views of models and modelling and where the visual similarity of a model to the target system or the quantitative agreement of model predictions with a measured property of the system is of greater importance. The integration of mathematical or IT modelling lessons with modelling in physics could provide a natural place to introduce new perspectives. In mathematics and IT lessons, students may engage more easily in studying the dynamics of models and modelling in the virtual or mathematical world without striving for a direct one-to-one relationship with the physical world, thus enjoying more freedom to explore theoretical ideas. Practising scientists enjoy such freedom, so why not permit the same freedom and joy of invention in teaching and schooling. The role of mathematics in physics is then perceived as essential means to create and develop physical ideas, where mathematical structures themselves can provide new ideas, rather than of seeing mathematics only as a technical tool for making calculations (or, as sometimes seems to be the case, a nuisance which prevents the capture of the true 'conceptual understanding'). Furthermore, emphasising generative modelling may encourage the effective and efficient reorganisation and employment of mathematics in physics lessons, not as a rival to empirical activities, but as a natural counterpart on a par with experimentation – as it is in doing science.

Teachers' views strongly affect the ways in which models and modelling are used in practice in education, which then affects students' views of physics and learning physics. Therefore, generative modelling should be discussed not only in the education of experts in modelling but also in science teacher education. One obvious way to address generative modelling is to use examples drawn from practice (as in Sect 35.5 here) to show that models and modelling can serve not only as a means to demonstrate achieved and agreed consensus on scientific ideas but also as thinking tools in the construction of scientific knowledge, as well as justification strategies in discourse. Even if students have insufficient background knowledge in physics and mathematics to engage in generative modelling, in many cases they can follow the ideas behind a generative modelling process when it is reconstructed and discussed. Reconstructing such examples of a modelling process and how it relates to empirical evidence can serve in science teaching as well as, for example, reconstructed scientific historical examples. Students can also participate in the shared construction and development of a model by creating analogies and forming idealisation and generic abstractions. Indeed, in creative activities, teachers should be able to understand different ways of thinking (cf. Blum and Ferri 2009) when supporting students in constructing and developing ideas in modelling.

In summary, if we want school science to reflect useful and fruitful aspects of modelling in physics, we should focus much more to new types of creative, generative and simulative modelling. Instead of trying to show how models are produced and refined by relying on established theory, we should be able to show how to produce interesting and suggestive new models and how they can guide generation of new theoretical insights and guide us in seeking new empirical regularities in phenomena. Of course, because the more traditional modelling entails many similar steps, what has been said of this type of modelling remains valid and is not overturned.

35.7 Conclusion

The notions of simulative modelling discussed here call attention to several aspects of modelling one must take into account in producing modelling activities for teaching purposes. First, the essential part of modelling is its close relationship with experiments and experimentation: Models are mediators between the conceptual and the real and serve as means of intervention and inference. This means that in teaching solutions, we should carry out parallel and mutually supportive activities of modelling and experimentation. Second, in developing models, theory is not the only starting point of modelling. Rather, model development requires a variety of sources: theoretical, empirical and computational. The practice and purpose of modelling guides how these sources can be used and employed. This leaves much room for the creativity of modelling and also emphasises the constructive and cognitive aspects of modelling, fitting well constructively oriented teaching that supports students' own knowledge construction. Transferring the mathematical models between different fields of physics in modelling also provides room for understanding the role of mathematical thinking and reasoning in physics. Third, and finally, simulative modelling enhances the view that much of science involves not so much of finding the fundamental truths of nature, but rather constructing reliable and functional knowledge which can help us to cope with nature.

Acknowledgements This work has been supported by grant 1136582 from the Academy of Finland.

References

Adúriz-Bravo, A. & Izquierdo-Aymerich, M. (2005). Utilising the '3P-model' to Characterise the Discipline of Didactics of Science, *Science & Education* 14, 29–41.

Black, M. (1962). *Models and Metaphors*, Ithaca, NY: Cornell University Press.

Blum, W. & Ferri, R. B. (2009). Mathematical Modelling: Can It Be Taught And Learnt?, *Journal of Mathematical Modelling and Application* 1, 45–58.

Bunge, M. (1983). *Epistemology & Methodology II: Understanding the World*. Treatise on Basic Philosophy 6, Dordrecht: Reidel.

Bozkurt E. & Ilika A. (2010). The Effect of Computer Simulations over Students' Beliefs on Physics and Physics Success. *Procedia Social and Behavioural Sciences* 2, 4587–4591.

Cartwright, N. (1983). *How the Laws of Physics Lie*, Oxford, New York: Clarendon Press.

Cartwright, N. (1999). *The Dappled World*, Cambridge: Cambridge University Press.

Chang, H. (2004). *Inventing Temperature: Measurement and Scientific Progress*, Oxford: Oxford University Press.

Crawford, B. & Cullin, M. (2004). Supporting Prospective Teachers' Conceptions of Modeling in Science, *International Journal of Science Education* 26, 1379–1401.

Darling, K. M. (2002). The Complete Duhemian Underdetermination Argument: Scientific Language and Practice. *Studies in History and Philosophy of Science* 33, 511–533.

Develaki, M. (2007). The Model-Based View of Scientific Theories and the Structuring of School Science Programmes, *Science & Education* 16, 725–749.

Duhem, P. (1914/1954). *The Aim and Structure of Physical Theory* (translation of La Théorie Physique: Son Objet, Sa Structure, 2nd ed., 1914 Paris). Princeton: Princeton University Press.

Finkelstein, N. D., Adams, W. K., Keller, C. J., Kohl, P. B., Perkins, K. K., Podolefsky, N. S. & LeMaster, R. S. (2005). When learning about the Real world is Better Done Virtually: A Study of Substituting Computer Simulations for Laboratory Equipment, *Physical Review* ST – PER 1, 010103.

Giere, R. N. (1988). *Explaining Science: A Cognitive Approach*, Chicago: University of Chicago Press.

Giere, R. N. (1999). *Science without Laws*, Chicago: University of Chicago Press.

Giere, R. N., Bickle, J. & Mauldin, R. F. (2006). *Understanding Scientific Reasoning*, 5th ed., Belmont, CA: Thomson Wadsworth,

Gilbert, J. K., Pietrocola, M., Zylbersztajn, A. & Franco, C. (2000). Science and Education: Notions of Reality, Theory and Model, in J.K. Gilbert & C.J. Boulter (eds.), *Developing Models in Science Education*, Dordrecht: Kluwer Academic Publisher.

Gobert, J. D. & Buckley, B. C. (2000). Introduction to Model-based Teaching and Learning in Science Education, *International Journal of Science Education* 22, 891–894.

Grandy, R. E. (2003). What Are Models and Why Do We Need Them?, *Science & Education* 12, 773–777.

Hacking, I. (1983). *Representing and Inventing: Introductory Topics in the Philosophy of Natural Science*, Cambridge: Cambridge University Press.

Haines, C. R. & Crouch R. (2010). Remarks on a Modelling Cycle and Interpreting Behaviours, In R. Lesh, P. L. Galbraith, C. R. Haines, and A. Hurford (eds), *Modeling Students' Mathematical Modeling Competencies* (ICTMA 13), Heidelberg: Springer. pp. 123–128.

Halloun, I. A. (2007). Mediated Modeling in Science Education, *Science & Education* 16, 653–697.

Heidelberger, M. (1998). From Helmholtz's Philosophy of Science to Hertz' Picture Theory. In D. Baird, R.I.G. Hughes and A. Nordmann (eds.), *Heinrich Hertz: Classical Physicist, Modern Philosopher*, Dordrecht: Kluwer Academic Publisher.

Hesse, M. B. (1963). *Models and Analogies in Science*, London: Seed and Ward.

Hestenes, D. (1987). Toward A Modeling Theory of Physics Instruction, *American Journal of Physics* 55, 440–454.

Hestenes, D. (1992). Modeling Games in the Newtonian World, *American Journal of Physics* 60, 732–748.

Hestenes, D. (2010). Modeling Theory for Math and Science Education. In R. Lesh, C. R. Haines, P. L. Galbraith and A. Hurford (eds.), *Modeling Students' Mathematical Modeling Competencies* (ICTMA 13), Heidelberg: Springer, pp. 13–41.

Hughes, R. I. G. (1997). Models and Representations, *Philosophy of Science* 67, S325-S336.

Hughes, R. I. G. (1999). The Ising Model, Computer Simulation, and Universal Physics. In M. S. Morgan and M. Morrison (eds.), *Models as Mediators*, Cambridge: Cambridge University Press.

Humphreys, P. (2004). *Extending Ourselves*, Oxford: Oxford University Press.
Izquierdo-Aymerich, M. & Adúriz-Bravo, A. (2003). Epistemological foundations of school science, *Science & Education* 12, 27 – 43.
Justi, R.S. & Gilbert, J.K. (2000). History and Philosophy of Science through Models: Some Challenges in the Case of "the atom", *International Journal of Science Education* 22, 993–1009.
Justi, R.S. & Gilbert, J.K. (2002). Modelling, Teachers' Views on the Nature of Modelling and Implications for the Education of Modellers, *International Journal of Science Education* 24, 369–387.
Koponen, I. T. (2007). Models and Modelling in Physics Education: A Critical Re-analysis of Philosophical Underpinnings and Suggestions for Revisions, *Science & Education* 16, 751–773.
Matthews, M. R. (1994/2014). *Science Teaching: The Role of History and Philosophy of Science*, New York: Routledge.
Matthews, M.R. (1997). Introductory Comments on Philosophy and Constructivism in Science Education, *Science & Education* 6, 5–14.
Morrison, M. (1999). Models as autonomous agents. In M. S. Morgan and M. Morrison M. (eds.), *Models as Mediators*, Cambridge: Cambridge University Press.
Morrison, M. S. & Morgan, M. (1999). Models as Mediating instruments. In M. S. Morgan and M. Morrison (eds.), *Models as Mediators*, Cambridge: Cambridge University Press.
Nersessian, N. (1995). Should Physicists Preach What They Practice? Constructive Modeling in Doing and Learning Physics, *Science & Education* 4, 203–226.
Nersessian, N. (2008). *Creating Scientific Concepts*. Cambridge, MA: MIT Press.
Neves, R. G. (2010). Enhancing Science and Mathematics Education with Computational Modelling, *Journal of Mathematical Modelling and Application* 1, 2–15
Nola, R. (1997). Constructivism in Science and Science Education: A Philosophical Critique, *Science & Education* 6, 55–83.
Nola, R. (2004). Pendula, Models, Constructivism and Reality, *Science & Education* 13, 349–377.
Oh, P. S. & Oh, S. J. (2011). What Teachers of Science Need to Know about Models: An Overview, *International Journal of Science Education* 33, 1109–1130.
Perkins K., Adams W., Dubson M., Finkelstein N., Reid S., Wieman C. & LeMaster R. (2006). PhET: Interactive Simulations for Teaching and Learning Physics. *The Physics Teacher* 44, 18–22.
Pluta, W. J., Chinn. C. A. & Duncan, R. G. (2011) Learners' Epistemic Criteria for Good Scientific Models, *Journal of Research in Science Teaching* 48, 486–511.
Riordan, M. (2003). Science Fashions and Scientific Fact', *Physics Today*, August 2003, 50.
Schwarz C. V., Reiser B. J., Davis E. A., Kenyon L., Acher A., Fortus D., Shwartz Y., Hug B. & Krajcik, J. (2009). Developing a Learning Progression for Scientific Modeling: Making Scientific Modeling Accessible and Meaningful for Learners, *Journal of Research in Science Teaching* 46, 632–654.
Sensevy, G., Tiberghien, A., Santini, J., Laube, S. & Griggs, P. (2008). An Epistemological Approach to Modeling: Case Studies and Implications for Science Teaching, *Science Education* 92, 424–446.
Suárez, M. (1999). The Role of Models in Application of Scientific Theories: Epistemological Implications. In M. S. Morgan and M. Morrison (eds.), *Models as Mediators*, Cambridge: Cambridge University Press.
Suppe, F. (1977). *The Structure of Scientific Theories*, 2nd ed., Urbana: University of Illinois Press.
Suppes, P. (1962). Models of Data. In E. Nagel, P. Suppes and A. Tarski (eds.), *Logic, Methodology and Philosophy of Science: Proceedings of the International Congress*, Stanford, CA: Stanford University Press.
Tala, S. (2011). Enculturation into Technoscience: Analysis of the Views of Novices and Experts on Modelling and Learning in Nanophysics, *Science & Education* 20, 733–760.
Uhden, O., Karam R., Pietrocola M. & Pospiech G. (2012). Modelling Mathematical Reasoning in Physics Education, *Science & Education* 21, 485–506.

Van Driel, J. & Verloop, N. (2002). Experienced Teachers' Knowledge of Teaching and Learning of Models and Modelling in Science Education, *International Journal of Science Education* 24, 1255–1272.

Van Fraassen, B. C. (1980). *Scientific Image*, Oxford: Clarendon Press.

Wells, M., Hestenes, D. & Swackhamer, G. (1995). A modeling method for high school physics instruction, *American Journal of Physics* 63, 606–619.

White, B. Y. (1993). ThinkerTools: Causal Models, and Science Education Conceptual Change, *Cognition and Instruction* 10, 1–100.

Winsberg, E. (2001). Simulations, Models and Theories: Complex Physical Systems and their Representations, *Philosophy of Science* 68, 442–454.

Winsberg, E. (2003) Simulated Experiments: Methodology for a Virtual World, *Philosophy of Science* 70, 105–125.

Winsberg, E. (2006). Models of Success versus the Success of Models: Reliability without Truth, *Synthese* 152, 1–19.

Ismo T. Koponen is university lecturer in physics and physics teacher education at the Department of Physics, University of Helsinki, Finland. He received his Ph.D. in physics and worked during 1990–2000 as a researcher in materials physics and statistical physics, in modelling and in simulations. From 2000 onwards his teaching and research interests have been related to physics education. In the Department of Physics, he is leading a research group concentrating on problems in physics education, in particular on problems related to concept learning and conceptual change. The focus of his research work is mainly on the university-level teaching.

Suvi Tala is researcher and teacher educator at the Department of Physics in the University of Helsinki. She has an M.Sc. degree in physics and expertise in teaching physics, chemistry, mathematics, philosophy and ethics. Her postgraduate research focuses on the role of technology in science and science education, including the nature of science and technology. Her research interest includes also learning in research groups. In addition, she is participating on a national project aiming to improve the public's awareness about science and technology.

Chapter 36
Models in Science and in Learning Science: Focusing Scientific Practice on Sense-making

Cynthia Passmore*, Julia Svoboda Gouvea*, and Ronald Giere

36.1 Introduction

Over the last few decades, there has been a "practice turn" in the philosophy of science and, more recently, in science education. That is, there has emerged in both fields an effort to understand and apply ideas about how science is actually practiced to issues in philosophy and education.[1]

What this practice turn has meant is that philosophers, along with other scholars in science studies, have turned from seeking an account of science as a singular, logical system for knowledge generation and evaluation and instead have begun to focus more carefully on an examination of the nuances and context dependencies of what scientists actually do to further their aims of making sense of the world. This turn has been described as naturalistic or pragmatic, because it abandons some of the assumptions and constraints of a more traditional philosophical approach in favor of a more empirically based one and, in this way, offers more authentic descriptions of the scientific endeavor.

Similarly, in science education, there has been much debate and discussion about the distorted and decontextualized version of science that has come to be known as "school science."[2] There is an emerging consensus that the overarching emphasis on a singular "scientific method" combined with a focus on memorization and test

---

*Note: The first two authors contributed equally to the creation of this manuscript.

[1] See, for example, Giere (1988), Nersessian (1992, 1999, 2002), Morrison and Morgan (1999) from philosophy of science and Duschl (2008), Gilbert (2004), Matthews (1992), Osborne et al. (2003), and Hodson (1992) from science education.

[2] See Duschl (2008), Hodson (1996, 2008), Rudolph (2005), and Windschitl et al. (2008b).

C. Passmore (✉) • J.S. Gouvea
School of Education, University of California, Davis, USA
e-mail: cpassmore@ucdavis.edu

R. Giere
Department of Philosophy (Emeritus), University of Minnesota, Minneapolis, USA

M.R. Matthews (ed.), *International Handbook of Research in History,*
Philosophy and Science Teaching, DOI 10.1007/978-94-007-7654-8_36,
© Springer Science+Business Media Dordrecht 2014

preparation has contributed to a crisis in science education. Many able students are turning away from science, and more worrisome, there is an alarming lack of scientific literacy among the general public (e.g., Bauer 1992). The practice turn in science education, like the turn in the philosophy of science, has manifested itself in calls for science learning environments to become more authentic to science as actually practiced.

There are two primary reasons to situate science education reform in a consideration of scientific practice. The first is that by engaging students—in explicit ways—with a version of school science that is authentic, students may emerge with a more accurate view of the scientific enterprise. Having such a view will be useful to students as they consider new advances in science and in making decisions about their own futures. Second, because there is a great deal of alignment between how scientists think and reason and the powerful learning mechanisms that all humans use to navigate the world, engaging students in contexts that are authentic should produce a deeper understanding of and insight into the content of science. Indeed, there is good reason to believe that these two aims can be achieved with science education that has been carefully crafted to be authentic.[3]

Meeting these goals requires developing learning environments that engage students in intellectual work that mirrors the work that scientists actually do. And, this means having a clear and coherent picture of the scientific enterprise, one that is developed by looking closely at how science is actually practiced with an eye to how that work can be productively *translated and enacted* in an educational context.

Thus, the practice turns in both science studies and in science education have parallel aims, to attend to the authentic nature of science as it is actually practiced, as well as a parallel challenge, to make sense of the messy intellectual endeavor so that some coherent understanding of the scientific enterprise can be described. Scholars in science studies have examined the complexity of scientific practice from a range of perspectives. Authors have focused on the social structure of science, the cultural and epistemological norms, the day-to-day practices and routines of scientists, the role of tools and material forms, and the reasoning and problem-solving strategies used in scientific practice. Given the nuances and complexities of scientific practice, it can be difficult to conceive of how it can inform the design of science learning environments without leaving us with an account of science that is so diffuse and unstructured to be of practical use.

One way to address the complexity problem is to emphasize the cognitive endeavor of science by focusing on the practice of science and how it supports making sense of how the world works. Such a view does not ignore other perspectives; as Giere and Nersessian have argued, scientific cognition is necessarily embedded in sociocultural contexts where it is shaped and supported by complex interactions with other humans and with material forms. However, a focus on cognition can provide one clear avenue for translating scientific practice into science classrooms.

---

[3] See, for example, Stewart et al. (2005), Duschl (2008), Engle and Conant (2002), Ford (2008), Duschl and Grandy (2008), Lehrer and Schauble (2004), and Roth and Roychoudhury (1993).

This account makes supporting sensemaking primary and asks how can the social, cultural, and material aspects of science classrooms can be structured so that scientific reasoning is supported. It is from this starting point—that a central aim of science education is to engage students in scientific sense-making—that we move forward with for this chapter.

Increasingly, scholars in the history and philosophy of science have turned to examining the pivotal role that models and modeling play in organizing the cognitive activities of practicing scientists. A recognition of the importance of models and modeling in science education has been on the rise as well. There are important connections to be made between the science studies efforts and those in science education around the centrality of models and modeling in the practice of science. A careful consideration of the nature and use of models in science can provide one way to organize our understanding of scientific practice and frame the way we translate this practice into science classrooms to support meaningful sense-making.

36.1.1 Driving Question and Overview

In this chapter we will examine how historians, philosophers, and psychologists have viewed the role of models in science. In particular we are interested in how models function as reasoning tools that allow one to bound, explore, organize, and investigate phenomena and to develop explanations, generalizations, abstractions, and causal claims about those phenomena. We hope to draw out the nature and function of models as context-dependent tools that productively organize a range of sense-making work that scientists undertake in their practice. Sections 36.1 and 36.2 of this chapter are intended to answer a particular driving question:

How do models function to structure and organize scientific practice around sense-making?

To address this question, we first present a rationale for focusing on models; then address ontological and epistemological questions about the nature, form, and development of models; and finally examine how models are actually used in scientific practice.

Ultimately, our goal with this chapter is to unify the views from the science studies literature on how models operate in scientific work and to explore the implications of this view for science education. In order for learning environments to reflect authentic science, they need to be designed to mirror the cognitive activities of scientists. In Sect. 36.3, we address the question:

How can a model-based view of scientific practice be leveraged to organize and focus classroom activity in support of sense-making?

To address this question, we propose a framework for organizing ideas about model functions in science education and apply that framework to a collection of studies in the science education literature. The chapter ends with a consideration of a number of practical and theoretical implications and recommendations.

36.1.2 Rationale for Model Focus: "Why Models?"

To begin, we briefly examine work in history and philosophy of science that motivates our examination of models in science and science education. Much of this scholarship draws on Giere's seminal work in this area, *Explaining Science*. In it he made a deliberate turn away from the "general program" in philosophy of science to a more naturalistic one that examined how scientists actually go about their work on a cognitive level. He began with the assumption that "the representations that scientists construct cannot be radically different in nature from those employed by humans in general" (Giere 1988, p. 62); that is, the sense-making apparatus common to all humans is at work in science as well. The layperson's mental model is not different in kind from the widely accepted scientific model; rather, scientific models are more carefully constructed and systematically evaluated extensions of a more basic cognitive strategy.

This turn toward understanding the meaning making that scientists, as humans, do, rather than characterizing the products of their work on a structural level, emerged from what became "the cognitive study of science," in which mental models play a central role. There was a historically parallel, but independent, move in the philosophy of science from a "syntactic" to a "semantic" view of scientific theories. The latter moves beyond the abstract structure of theories to include issues of the meaning of scientific terms and the truth of scientific statements. In the semantic view of theories, models, still understood as logical rather than mental constructs, are central. Melding these two traditions has been a complicated process (see Downes 1992; Knuuttila 2005). Neither tracking these historical developments nor analyzing their various commitments is our intent here as this has been skillfully done in a number of recent papers for the science education audience.[4] For our purposes, the significance of these developments is that they gave prominence to the idea that models play a central role in scientific sense-making. Fully unpacking why a focus on models in science is useful and how models operate in scientific sense-making requires moving beyond purely philosophical concerns and bringing together a more integrated science studies approach that combines cognitive-historical, psychological, and ethnographic methods.

For example, over the past 15 years, Nancy Nersessian and her colleagues have undertaken a psychological approach to studies of actual scientific practice using both cognitive-historical and contemporary ethnographic techniques. They have spent years observing, documenting, and talking to scientists as they do their day-to-day work (e.g., Osbeck et al. 2010). From these studies has emerged a clear sense that models are at the center of the day-to-day work of science; they are the functional units of scientific thought. As Nersessian explains, mental modeling is the underlying cognitive machinery that makes model-based reasoning so fundamental

---

[4] See Adúriz-Bravo (2012), Bottcher (2010), (Develaki 2007), and Koponen (2007 and this volume). Please also see a special issue of *Science & Education* (Matthews 2007) for a careful treatment of models and modeling for the education audience.

to human sense-making. It is the general machinery that underlies our ability to engage in the more formalized scientific strategies of generating representations, using analogies and thought experimentation.

Building on the foundational works of Giere and Nersessian, there has been a proliferation of scholarship in science studies related to uncovering the role of models in science. For example, Morgan and Morrison (1999), in their edited volume, *Models as Mediators*, pulled together a range of articles that explored the ways in which models function in a variety of disciplinary contexts. And, numerous other scholars situated in biology, chemistry, physics, and economics have undertaken both historical and contemporary descriptions of model use in science.[5] Similarly, in the science education community, there is an emerging movement that acknowledges models and modeling as important aspects of scientific practice (e.g., NRC 2011; Windschitl et al. 2008b). Taken together, these studies provide a rationale for organizing science instruction around models and the primary motivation for this chapter: Models are central to scientific sense-making. They provide a way to organize our understanding of scientific practices and a way to understand the purpose of scientific activity.

Alternative organizing frameworks that centralize other aspects of scientific practice are certainly possible. To focus on models is a choice that, as we explore in this chapter, has particular affordances. Specifically, an explicit focus on modeling helps organize scientific practices such as representation, experimentation, and argumentation around the purpose of making sense of phenomena rather than as discrete activities. Although, this unification is seamless in actual scientific practice, in science education, unification can be more challenging. In the next section we draw on science studies to support the claim that models are central to the sense-making practices of scientists and examine the implications for science education.

36.2 Models in Scientific Practice and Science Learning

This section builds toward an understanding of models as context-dependent tools for making sense of phenomena by drawing on the work of philosophers and historians who emphasize the functional role of models in science. First, we address ontological questions about the nature and form of models, we then address epistemological concerns related to model construction and evaluation, and finally, we turn to a functional account of how models are used in scientific practice. What emerges from this account is a definition of models that emphasizes their utility as tools for sense-making as well as a description of the specific ways in which models

---

[5] In biology, see Cooper (2003), Lloyd (1997), and Odenbaugh (2005, 2009); in chemistry see Suckling et al. (1980); in physics see Cartwright (1997, 1999), Hughes (1999), and Nersessian (1999, 2002); in economics see Boumans (1999) and Morrison (1999). See also Auyang (1998) for comparison across biology, physics, and economics.

serve this function in science. This view of models, embedded in scientific practice, can provide a productive framework for organizing science education environments which is the focus of Sect. 36.3.

36.2.1 What Are Models? Ontological Concerns

> Few terms are used in popular and scientific discourse more promiscuously than "model." (Nelson Goodman 1976, p. 171, as cited in Odenbaugh 2009)

It can be challenging to define models in a concise way. Nevertheless, drawing on pragmatist philosophers, we identify several key attributes of scientific models:

1. *Models are defined by the context of their use.*
2. *Models are partial renderings of phenomena.*
3. *Models are distinct from the representational forms they take.*

In this section we discuss each of these features and discuss the implications of this definition of scientific models for education.

36.2.1.1 Models Are Defined by the Context of Their Use

Nersessian (2002) refers to the cognitive processes involved in deciding how to construct a model as mental modeling. Note that she makes a distinction between mental models, often described as knowledge structures stored in the long-term memory, and mental model*ing* as a *process* of human sense-making. We take up the latter view of model-based cognition as a flexible, context-dependent process whereby humans interpret and reason about situations by selecting and drawing together cognitive resources. Conceived of in this way, models are dynamic entities that are constructed and used as needed.

Although the word model is used to describe a wide range of entities in the sciences, one cannot actually provide a clear definition of what is and is not a model in an abstract sense. As Teller states:

> The point is that when people demand a general account of models, an account which will tell us when something is a model, their demand can be heard as a demand for those intrinsic features of an object which make it a model. But there are no such features. WE make something into a model by determining to use it to represent. (Teller 2001, p. 397, emphasis in original)

For these reasons we fall back on a very basic framework for a model, that models are sets of ideas about how some aspect of the world works. Models are entities that represent some aspects of a phenomenon to some degree. But which of those aspects and to what degree will not be uniform across contexts. Thus, while some philosophers have attempted to specify this relationship as an isomorphism between some source (the world) and some target (the model), we use the more relaxed

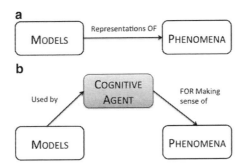

Fig. 36.1 From a dyadic to a triadic understanding of the relationship between models and phenomena. (**a**) The dyadic model focuses on defining the relationship between the model and some phenomenon. (**b**) The triadic model reframes the problem by shifting the focus to the cognitive agent who will ultimately be responsible for determining the nature of the relationship between the model and the phenomenon in a way that is useful given a particular aim

criterion of *similarity* (Giere 1988) to describe the relationship between models and phenomena. A slightly more precise version of the similarity criterion is proposed by Teller who describes similarity in terms of properties: Systems have properties, and some of these (depending on the objective of the modeler) will be part of the model (2001, p. 399).

Perhaps more important than defining what a model is defining what it is not. Appealing to similarity does not imply that any object that is similar to a natural phenomenon is a model. A globe, for example, is not a model of the Earth by default. It *becomes* a model when it is used to make sense of some puzzling pattern or answer some question. The same object can both be a model and not be a model depending on how it is being used.

Pragmatist philosophers, beginning with Giere (1988), have argued that the relationship between models and the world only makes sense in the context of their intended use *by some cognitive agent*. In defining models we cannot simply consider how the model relates to the phenomenon it represents; we must explicitly consider the role played by the cognitive agent (see Fig. 36.1, Giere 2004; Knuuttila 2005). By explicitly drawing attention to the cognitive agent in the system, we end up with a definition of models that foregrounds their function in reasoning. It is the cognitive agent, the modeler, who will decide how to bound, filter, simplify, and represent the phenomenon to generate a model. Which features need be shared and to what degree will depend on the way in which the model user wants to understand that phenomenon.

36.2.1.2 Models Are Partial Renderings of Phenomena

Understanding models in terms of their use can also help clarify the relationship between models and real-world phenomena. It is common to see models referred to as abstractions, simplifications, idealizations, or simply representations of phenomena. Specifying the exact nature of the relationship between models and the world has been a central point of debate in the philosophy of science literature (Downes 1992; Knuttila 2005). As Downes describes, the motivation for this debate has been to say something philosophically robust about models as knowledge structures—to answer the question what can models really tell us about the world? Many proponents of the semantic view have attempted to define the relationship between models and the world as isomorphic. However, for Downes and other pragmatists, the attempt to define a singular relationship between models and real-world phenomena misses the mark. Neither an account of models that focuses exclusively on isomorphism nor an account of models as purely analogical nor any other general account will apply in a universal sense to the variety of different kinds of scientific models that have been historically and continue to be used by practicing scientists (Downes 1992).

Cartwright (1999) similarly challenges the notion that phenomena can be mapped to general theories. Instead, she acknowledges that different phenomena may require models, which vary in the degree to which they make different simplifying assumptions. Cartwright's account is a rejection of the universalism of laws and an acknowledgement of the diversity of phenomena themselves, each of which require its own model formulation. Cartwright describes how a coin dropped from a height can perhaps reasonably be modeled using a simple Newtonian model. But the same model cannot help account for the motion of a dollar bill. The result is that reality is covered by a "patchwork" of models (see Fig. 36.2). This metaphor begins to complicate the possibility that there is a single way to characterize the relationship between models and the world.

Rather than attempt to map a one-to-one relationship between models and phenomena, Morrison and Morgan (1999) describe models as "partial renderings" that can differ widely in the extent to which they accurately represent real systems. They describe how a model of a pendulum can be simple and abstract when used as a means of making sense of simple harmonic motion but can be refined with a series of corrections that increase the complexity of the model as well as its success in making accurate predictions. There is no singular model of a pendulum; rather there are a group of overlapping models of a pendulum, each of which can be used to reason about a pendulum in a different way.

When the importance of a cognitive agent is recognized, a better metaphor for the relationship between models and phenomena is a geometric one proposed by Auyang (1998). As Auyang describes, attempts to a define a one-to-one correspondence between the world and our theoretical understanding of it necessitate a *finite geometry* in which our understanding maps onto the world like "a single global coordinate system covers an entire manifold" (1998, p. 74). As an alternative, Auyang proposes we think of science in terms of a *differential geometry,* in which the manifold is covered by overlapping local coordinate systems (see Fig. 36.2). Thus, Cartwright's patchwork is best understood not as a regular quilt with patches stitched together to create a complete understanding. Instead, the patchwork is much more irregular, with patches of different sizes and shapes overlapping with

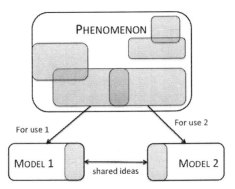

Fig. 36.2 Models related to phenomenon by an overlapping patchwork each of which is used for a slightly different purpose and makes different features of the phenomenon salient to the modeler (After Auyang 1998). *Grey boxes* represent different aspects of some phenomenon that a modeler has chosen to focus on for some particular use

one another, creating areas with many layers of coverage and possibly other areas with large gaps. It is the model user who decides which patch to apply to the world, depending on her aim.[6]

36.2.2 Models Are Distinct from the Representational Forms They Take

Just as there are potentially multiple models that can be used to make sense of a phenomenon, there are also multiple representational modes that any given model can take. Knuuttila (2005) describes modeling as involving two levels of representation. The first level involves choosing the attributes of the system that are relevant to include in a model. Such choices are dependent on the aim of the modeler as noted above. The second level involves making a choice about *how* to represent the relevant attributes of the system in some material form. Often these two levels cannot be separated—choices about what features are important will suggest a particular form just as the choice of a material form will afford or constrain what is attended to. Nevertheless it is worth pointing out that essentially the same ideas can be conveyed in a variety of representational modes including diagrams, equations, physical models, or written text. These different forms should not be conceived as different models per se; rather one should focus on the differences in representational mode.

---

[6] There are two recent books that develop the "patchwork" idea in quite rich directions for those readers who might want an even more sophisticated version of these ideas: Mark Wilson, *Wandering Significance*, Oxford Univ Press, 2006, and William Wimsatt, *Re-engineering philosophy for limited beings*, Harvard Univ Press, 2007.

Too great a focus on the material form of a model can be problematic because it tends to collapse the triadic relationship (between model, cognitive agent, and phenomenon) back into a dyadic one (model and phenomenon only) (see Fig. 36.1). A diagram of the cell is referred to as a "model *of* the cell," but this diagram by itself is merely a depiction of a physical object because it does not suggest what such a diagram is good *for*. Thus, despite the widespread usage in education circles of the word model as applied exclusively to physical objects (like Watson and Crick's tin and cardboard DNA molecule), it is not the material aspect or embodiment of the object that makes it into a model.

In the DNA example, the physical object was deployed to figure out something in the world; it was the physical manifestation of the key features of DNA that were relevant to understanding the mechanisms of inheritance. The material aspect of Watson and Crick's model was absolutely critical, but it was not the *materiality* that made it a model; rather it was how the abstract ideas about the structure and function of DNA along with Franklin's x-ray data were embodied in the material object and how it could then become a tool *for* reasoning about how the molecule functioned. The physical mode of representation was important because it allowed Watson and Crick to visualize precisely how the bond angles among atoms fit together and allowed them to feel confident that their structural understanding was correct. But, as the final line of their manuscript makes clear, that understanding was crucial for making sense of how DNA could possibly function in transmitting genetic information (Watson and Crick 1953).

To return to the of/for distinction made earlier, the three-dimensional structure of DNA is best understood not simply as a model *of* DNA but as a model that included the relevant structural elements of the molecule, thus allowing them to use it *for* developing a deeper understanding of its function. Educators tend to refer to the representational forms themselves as models (i.e., this is a 3-D model of DNA) rather than referring to representational systems as a whole (i.e., this is a physical representation of a DNA molecule that allows one to reason about how this molecule's structure relates to its function in biological inheritance). Unfortunately, this shorthand can promote confusion by foregrounding the form and backgrounding the intention. Being unclear about this issue can lead teachers to conclude that they are doing the scientific practice of modeling in their classrooms whenever they have their students working with physical objects. In fact, it is the cognitive activity— the sense-making—that should provide the primary criterion for determining if students are engaged in modeling. This kind of sense-making goes well beyond merely labeling parts and memorizing functions.

36.2.2.1 What Is a Model? Implications for Science Education

We can use this expanded understanding of models to gain some additional traction in defining models in science education: Models are not simply *of* phenomena, they are tools to be used *for* some reasoning about that phenomenon (Fig. 36.1). This distinction has important implications for decisions about what kinds of models to include in science curricula and for how we assess students' abilities as modelers.

A number of recent articles in the science education literature attempt to provide a typology of models in science toward a goal of finding a list that is inclusive and useful educators for making decisions about what kinds of models to bring into science classrooms.[7] These typologies emphasize the importance of carefully considering the context when making decisions about the model form and content in instructional settings. However, one challenge of these lists is that they can be difficult to interpret on a philosophical level and use in science education because they do not distinguish models from their representational forms. For example, Harrison and Treagust (2000) present a typology that orders models from concrete scale models, through abstract theoretical models, to complex dynamic systems simulations. This ordering is meant to reflect conceptual demand; concrete scale models (e.g., a scale boat) are less challenging than process models (e.g., a chemical equation) and are therefore positioned lower in the typology (p. 219). Harrison and Treagust suggest a learning progression that first introduces students to concrete models and moves toward introducing more abstract models.

The general point that the models used in educational contexts must be chosen in accordance with the abilities of students is an important one but is made more clear when models and their representational forms are kept distinct. A scale model of a boat can be a less sophisticated reasoning tool than a chemical equation, but this has less to do with the representational form than it has to do with the ways in which the models are used. If the scale model of the boat is used to merely represent surface features of a boat, this is not very sophisticated, and indeed the cognitive demand may be relatively low. However, if the scale model of the boat is used to highlight how boat shape relates to buoyancy, then more intellectual challenge is introduced. A scalar representation *of* a boat is not a scientific model at all, whereas a scalar representation of a boat that can be used *for* reasoning about the phenomenon of floating by illustrating scientific ideas about displacement could be one way of representing a model for buoyancy. Thus, the cognitive demand of a particular model has less to do with the form it takes and more to do with the function it serves.

By defining models in the context of their use, the focus shifts to choosing a model that can be used to make sense of the target phenomenon in a way that is appropriate for the cognitive agent. This might mean that in some classroom contexts, a smaller set of constructs are introduced and simpler relationships are highlighted than are present in the model versions used in the scientific community. This point is made by Gilbert and colleagues in their discussion of curricular versus scientific models. Curricular models are simplified versions of scientific models that are specifically adapted for classroom use. Gilbert (2004) suggests that teachers must choose these curricular versions with an understanding of "the scope and limitations of each of these models: the purposes to which they can be put and the quality of the explanations to which they can give rise" (p. 126). That is, teachers and curriculum designers need to carefully select or construct versions of scientific models with which students can productively think.

---

[7] See, for example, Boulter and Buckley (2000), Coll and Lajium (2011), Gilbert (2004), and Harrison and Treagust (2000).

Introducing a model into a classroom also includes, as Harrison and Treagust (2000) suggest, the need to consider the representational form. However, because of the dual nature of the representational role of models, two sets of questions about what kinds of models to introduce to students must be asked: First, with what set of ideas do we want students to engage, and second, what representational mode or modes can support interaction with those ideas?

It is important to separate the notion of the model from the particular representational form it takes. One reason for this has been to keep students focused on the success of the model as a reasoning tool as opposed to particular features of the representational form that can sometimes distract them from the salient conceptual elements of the model. For example, when first graders were asked to design a physical object that "works like an elbow," they tended to focus on the surface features of the representation, adding details to the models that had only to do with physical resemblance between their replicas and real elbows (Penner et al. 1997). Part of the reason they did so seems to be because the task was purely a representational task—make the elbow—and students responded by making physical replicas *of* elbows.

Imagine if the task had been rephrased so that it foregrounded a sense-making aim and backgrounded the representation, e.g., how is it that an elbow allows you to pick up something? This could have been done by introducing some flexibility in the choice of representational form instead of requiring students to build a physical model or by reframing the task around explanation as opposed to design. In such a scenario, the task would have been framed such that the purpose—to understand something about how elbows work—would have been highlighted. A follow-up study by Penner and colleagues demonstrates a shift in this direction where the students began to explore ideas about the elbow as a fulcrum, introducing ideas about torque and distance. In this second study, the model was no longer seen as one *of* the elbow, it was *for* understanding how lever systems like the elbow actually work (Penner et al. 1998).

The reason the *of/for* distinction is so powerful for education is that, again, it situates the model in the context of its use. It highlights function over form. In addition, it helps to keep the focus on reasoning and making sense *with* the model rather than reducing models to just another thing to be learned by rote in the science classroom. Models should be deployed in science classrooms as dynamic entities that help organize and focus a class of cognitive activities toward a clear sense-making goal.

The second major implication of defining models in terms of their intended use is that there is no single model of any particular object or system but rather many possible models, each of which has different affordances and constraints for reasoning about that phenomenon. In the science education literature, mental models are sometimes used to refer to static representations of students' ideas stored in long-term memory. Students are often described as having models "of" particular phenomena. For example, Gilbert (2004) states that "[a]ll students of chemistry must have a mental model, of some kind, of an 'atom,' all those of biology of a 'virus,' all those of physics of a 'current of electricity'" (p. 117). Often students are asked to externalize their internal mental models as drawings (e.g., Coll and Treagust 2003; Gobert 2005). We have seen this work interpreted to mean that students have a singular view of a particular entity or process with the implication that if this view is incorrect, it must be replaced with the correct consensus model.

Defining models in terms of their purpose allows for the possibility that students, like scientists, have multiple sets of ideas about scientific phenomena. The ideas students have about an atom are likely to vary depending on how they are being asked to think about atoms. For this reason it is important when asking students to generate models to be clear about the purpose of the activity. Asking students to depict a generic model (e.g., "Draw me your model of an atom.") is an underspecified task because it does not help them, as cognitive agents, make informed decisions about which features and relationships are important to represent. Consider the difference in the salient ideas that the student would need to draw on if she was trying to reason about bonding versus nuclear radiation.

If instead models are defined as sets of ideas that are activated in the working memory in response to a particular aim, we shift the focus from whether or not students possess a correct mental model to helping draw out the productive ideas that students have for making sense of particular phenomena. This can help orient educators to drawing out and building on students' ideas rather than attempting to replace misconceptions (c.f. Hammer 1996). Modeling in the science classroom has the potential to draw upon the powerful learning and reasoning resources that all students bring to the classroom and to create an environment in which the students are active learners. This approach could result in students who develop their capacity to reason about the complex and interesting world and could go a long way toward addressing the rote approach seen in so many contemporary classrooms.

36.2.3 What Makes a Good Model? Epistemological Concerns

Given that models are defined only in the context of their use, it follows that there is no context-independent way to evaluate a model. Models are built with an understanding of the epistemological criteria that are relevant to the question at hand, and they are evaluated with an understanding of their intended use. This leads us to consider two epistemological concerns:

1. *A focus on models means merging the contexts of discovery and justification.*
2. *Models must balance trade-offs in epistemological criteria.*

36.2.3.1 Merging the Contexts of Discovery and Justification

Practice-based philosophers acknowledge that there is no meaningful distinction *in practice* between the model development and evaluation. In his 1999 chapter in *Models as Mediators*, Marcel Boumans explores the relationship between model building and model justification. He argues that "models integrate a broader range of ingredients" that include theory and empirical data but also metaphors, analogies, mathematical concepts, and techniques. The central claim of his chapter is that this integration is satisfactory when the resulting model can be (1) used as a solution to a theoretical problem, (2) an explanation of an empirical phenomenon, (3) an indication of some possibilities, and (4) a way to mathematically conceptualize a problem. This is an account of modeling that is situated in the context of function, aims, and cognitive payoffs.

Boumans' central thesis is that despite the way in which stories about model development get told, in practice, the "context of discovery" and the "context of justification" are completely intertwined. It is by simultaneously attending to both the theoretical/empirical world and the more pragmatic aims of the modeler that progress on model development is made and justified. The steps are not distinct. One does not build something and then check to see if it does what it is meant to do. Rather one builds something with ongoing and critical attention to the purposes it is supposed to serve. In this way, Boumans explains that "justification is built-in."

Similarly, Nersessian (1992, 2002), by focusing on the cognitive activities of scientists, combines the contexts of discovery and justification into the context of "development" where ideas are articulated and evaluated in a process that is fundamentally creative. New ideas arise in this context not completely de novo but in conversation with existing ideas. She describes how, for example, Maxwell's revolutionary ideas about electromagnetism were borne out of analogies with existing models in mechanics. Further, in the context of development, emerging ideas are not simply held up against a set of rigid standards of justification, but they can interact with those standards to change the rules of the game. Einstein's new framework of relativity fundamentally shifted the criteria against which models would be judged.

36.2.3.2 Models Must Balance Trade-Offs in Epistemological Criteria

Both the Boumans and Nersessian accounts point to the contingent and contextual nature of scientific reasoning and suggest that models will be subject to different epistemological criteria depending on how one intends to use them. They also suggest, as ecologist Richard Levins argued, that "[t]here is no single, best all-purpose model" (1966, p. 7). Levins argued that for both cognitive and methodological reasons, modelers must often choose among the desirable, but often conflicting, epistemic aims of realism, precision, and generality. For example, a fisheries biologist interested in population projections of a species of interest might choose to sacrifice generality in order to construct a model that can generate accurate predictions of population fluctuations, while an ecologist, like Levins himself, might forgo predictive precision in the interest of general explanatory power. The main point to take away from Levins' argument is that modelers will and should build different models depending on their particular aims. The implication is that there is not a single type of model or modeling that can address all biological problems equally well; depending on the question at hand, a biologist will want to choose the model that is the best tool for the particular job.

36.2.3.3 Implication: The Need to Contextualize Meta-modeling Knowledge

When engaging students in the context of model development (i.e., model construction and evaluation) it is important to make them aware of criteria used to judge models but also to help them develop the expertise to recognize which of these criteria are relevant for their purposes at a specific point. A recent study by Pluta et al. (2011) highlights the importance of helping students develop ideas about what makes models "good" in ways that make the context explicit. Without significant instruction, middle school students were able to generate a variety of epistemic criteria for evaluating scientific models. When prompted to list the features of a "good" model, students responded with criteria such as communication, explanatory power, and fit to data. However, the most common responses had to do with the amount of detailed information presented in the models, suggesting that students were thinking of models primarily as useful for conveying information, much as a textbook diagram would.

Looking closer at the nature of the task in this study, students were asked to evaluate a variety of static representations of models including diagrams, pictures, and text similar to what they might see in textbooks (Pluta et al. 2011, p. 500). The task was framed without reference to a particular problem, question, or aim. Given that in the context of this task students were interacting with static, final form models, it is not surprising that many students described models as tools that help communicate ideas, rather than objects to support scientific inquiry. Nevertheless, this study suggests that students do have some resources for thinking about using models in a variety of ways and it supports the argument that a goal of instruction should be to reinforce and refine these ideas with reference to particular scientific aims. We caution against teaching epistemic criteria to students as a normative list of characteristics of "good models" in an abstract and universal sense but instead advocate for instruction that helps students develop and attend to such criteria in the course of developing and using models in context.

36.2.4 How Are Models Used? Functional Concerns

The primary utility of the practice turn is that it has begun to specify, in more detail, the ways in which models function in scientific practice. Once one takes up the "models *for*" orientation, then a crucial next step is to consider what the cognitive agent is doing in more detail. In what follows we build on what has so far been a general argument that models are context-dependent tools for reasoning and now turn to a more specific account of the ways in which these tools can support sense-making in science.

Here the focus is on three scholars from the science studies literature who have taken up the challenge of elaborating a functional analysis of models in science. Jay Odenbaugh (2005) presents an argument from contemporary philosophy of biology for the legitimacy of modeling in biological practice and the range of uses they are put to in that discipline. Nancy Nersessian approaches the problem from the perspective of cognitive science, using cognitive-historical case studies of physicists and contemporary ethnographic methods to unpack the affordances of model-based reasoning. Stella Vosniadou (2001) considers how models function in scientific sense-making by examining similarities between the reasoning of young children/lay adults and scientists.

36.2.4.1 Odenbaugh: Cognitive Benefits of Modeling in Biology

Philosopher of biology Jay Odenbaugh (2005) presents an argument emphasizing the functional utility of modeling in ecology. He states, "model building is first and foremost a strategy for coping with an extraordinarily complex world" (p. 232). He unpacks the strategies of modeling in ecology and the associated cognitive benefits of engaging in these strategies. While, his analysis draws on work in biology, we find it useful for exploring the role of models in science more generally.[8]

Drawing on the work of Levins (1966), Odenbaugh explores five major pragmatic uses for models in biology and their associated benefits: (1) simple, unrealistic models help scientists explore complex systems, (2), models can be used to explore unknown possibilities (3) models can lead to the development of conceptual frameworks, (4) models can make accurate predictions, and (5) models can generate causal explanations. The focus of his argument is that the first three roles of models have been underemphasized in comparison to the latter two.

In his exploration of the first point, Odenbaugh describes how simplification is a purposeful strategy that scientists use in a number of different ways. For example, Odenbaugh describes how simplistic optimality models, which assume that natural selection is the only mechanism shaping natural systems, are used as a baseline from which to consider and explore deviations. That is, simple models can help scientists by allowing them to begin to unpack the reasons why a false model is wrong (see also Wimsatt 1987).

A simple model can be compared to successively more complex models as a systematic strategy for locating error. Odenbaugh illustrates this point with an account of how understanding the deficiencies in the simplest version of the Lotka-Volterra predator–prey model, which is empirically unrealistic, has led to a productive elaboration of increasingly detailed models. Importantly, these models not only make more sense empirically, they also include assumptions that are much more plausible given what is known about natural populations.

In posing a second role for models, Odenbaugh examines how models afford opportunities for exploring possibilities. Rather than representing what is, models can help scientists think about what might be. For example, Odenbaugh describes how ecologist Robert May explored the possible patterns that would emerge from a simple logistic model of population growth. His analysis revealed that increasing

---

[8] See Svoboda and Passmore (2011) for a much more thorough treatment of Odenbaugh's framework.

the per capita rate of increase (R) could yield chaotic dynamics. The significance of this finding was that it oriented ecologists to the possibility that even relatively simple ecological systems could exhibit complex chaotic patterns for certain parameter values.

The third role for models is in leading to the development of new concepts. Odenbaugh (2005) describes how biologist Robert May chose to represent the overall number and degree of interactions in an ecological community in terms of a "connectance" parameter C, which he defined as the proportion of all pairwise species interactions that were not equal to zero. May's analysis suggested that C played a key role in the stability of the community over time. While May's model was later criticized, Odenbaugh describes how his attempt to operationalize and interpret the role of C opened up a discussion in the ecological community surrounding the appropriate ways to conceptualize community complexity and stability. This analysis marked the beginning of a proliferation of ideas in the ecological community as well as a marked increase in experimental work in community ecology that extended well beyond the original model.

In sum, the essence of Odenbaugh's argument is that matching reality is not the only role for models in biology. Making predictions and explanations are important roles for models, but there are others as well that do just as much to support sense-making. Odenbaugh wants to ensure that the utility of the *exploratory* role of models in generating new ideas and new ways of thinking is recognized as well.

36.2.4.2 Nersessian: Model-Based Reasoning in Physics

Nancy Nersessian and her colleagues have investigated the role of models in science both through cognitive-historical case study analyses (e.g., 1992, 1999, 2002) and more recently in laboratory settings (Osbeck et al. 2010). One of her primary aims has been to explore how models help scientists reason about phenomena by attending to the cognitive processes of scientists and how these processes are situated in scientific practice. In her work Nersessian has identified three types of modeling practices that commonly co-occur in case studies of scientific problem solving: (a) visual reasoning, (b) analogical reasoning, and (c) thought experimentation (simulative reasoning).

In a case analysis of the development of electric-field theory, Nersessian describes the strategies used by Faraday and Maxwell (for a detailed account see Nersessian 1992). Faraday and Maxwell were motivated by a desire to make sense out of a puzzling phenomenon: apparent attractions between objects at a distance. This phenomenon is easily observable by, for example, rubbing a balloon against some fabric and noting that it can now "stick" to the wall. Both scientists made extensive use of diagrams to organize and visualize their emerging understanding of how electric phenomena might work. These representations served to highlight the important structures and relationships between them and served as external objects that could be actively reasoned with. As Nersessian (2002) explains, visual representations do more than hold the ideas in a model—they help focus the reasoner on salient features. They also support simulative reasoning by helping create a visual image that can be animated in the mind. Finally, visual representations provide a way to share ideas with the community. Preparing for this sharing event can force the modeler to make ideas clear and the act of sharing such representations is a productive means of extending the reasoning process out to the larger scientific community. It is the interaction between these externalizations and the underlying model ideas that can lead to breakthroughs for the reasoner.

Analogical reasoning is a form of reasoning that is common in many of these analyses. In forming his understanding of the concept of electricity, Maxwell leveraged analogies from classical Newtonian mechanics. He reasoned that electricity could be analogous to other continuous-action phenomena such as heat, fluid flow, and elasticity (see 1992). For example, using a fluid flow model to map out a similar model of electricity was a crucial part of early work in developing understanding of electric forces.

It is also evident from Nersessian's analysis of his writings that Maxwell relied on simulative thought experimentation to reason through the consequences of his model. This strategy of imagining how a phenomenon might change if certain conditions are changed was also famously used by Galileo. As Nersessian describes:

> According to Aristotelian theory, heavier bodies fall faster than lighter ones. This belief rests on a purely qualitative analysis of the concepts of 'heaviness' and 'lightness'. Galileo argued against this belief and constructed a new, quantifiable representation through sustained analysis using several thought experiments and limiting case analyses.....He calls on us to imagine we drop a heavy body and a light one, made of the same material, at the same time. We could customarily say that the heavy body falls faster and the light body more slowly. Now suppose we tie the two bodies together with a very thin—almost immaterial—string. The combined body should both fall faster and more slowly. It should fall faster because a combined body should be heavier than two separate bodies and should fall more slowly because the slower body should retard the motion of the faster one. Clearly something has gone amiss in our understanding of 'heavier' and 'lighter.' (Nersessian 1992, p. 28)

Galileo used this strategy to explore the meaning of the concepts of heavy and light and ultimately reveal the flaws in the Aristotelean model.

Nersessian's focus on model-based reasoning in science draws out some specific cognitive strategies that scientists have at their disposal for making sense of the world. Her analyses make clear that models and the suite of cognitive strategies that they support have helped scientists organize and extend their ideas in ways that have been extremely productive. Further, her account suggests that these same strategies can be productive for students of science (Nersessian 1989, 1995).

36.2.4.3 Vosniadou: Models and Learning Science

Stella Vosniadou has explicitly applied ideas about the cognitive utility of models to science learners. In her 2001 paper, she explores the analogous ways that children and scientists reason with models. Like others, she puts the mental models of children and lay adults on the same dimension as the models of practicing scientists. In her analysis she explores the functions that models play in children's reasoning.

She summarizes her findings with three related functions of models in reasoning: "(a) as aids in the construction of explanations, (b) as mediators in the interpretation and acquisition of new information, and (c) as tools to allow experimentation and theory revision" (p. 359).

In the first sense, Vosniadou notes that models serve a generative function in that they allow the cognitive agent to reason about situations or phenomena that are beyond his or her experience. In her studies of children and lay adults and their views of the shape of the earth, she found that the model served as the "vehicle through which implicit physical knowledge enters the conceptual system" (p. 361). Once this knowledge was articulated in the form of a model, most of her subjects answered questions using a consistent form of this model for the remainder of the study session and were observed to use the abstract ideas to answer specific questions about the earth and objects on it.

Just as the cognitive agent uses the model to generate explanations, so, too, does the model provide a strong filter through which new experiences or information is interpreted. In a study of children's ideas about the day/night cycle, Vosniadou found that the models children had clearly influenced how they interpreted the questions asked of them, just as scientists' models influence how new data are interpreted.

And finally, Vosniadou explores how existing models serve to inform and constrain new ideas and models. In her studies she found that children used their existing ideas to formulate new ones so that a clear connection could be drawn from initial ideas to how those ideas changed over time in the face of new information.

The importance of Vosniadou's contribution is to point out that reasoning with models "is a basic characteristic of the human cognitive system and the use of models by children is the foundation of the more elaborate and intentional use of models by scientists" (p. 367). The fundamental role of models in interpreting and generating new knowledge is central to science and from this premise more specific accounts of the function of models (as delineated above) are possible. Vosniadou's connections between the cognitive work of scientists and children imply that using a modeling framework in education is not only viable but desirable.

From these three scholars, there are a number of ways of describing how models function in science and how this functionality might extend into learning environments. In this final section, we synthesize across these ideas to propose a framework that demonstrates how models, modeling, and model-based reasoning can serve to organize classroom science and focus students' scientific reasoning on making sense of the natural world.

36.3 A Framework for Models in Science Learning

What follows from Sect. 36.2 is that what makes something a model and how a model is developed over time is inextricably linked to the ends it is put to. That is, at a fundamental level, the focus should not be on a model *of* something as an end in itself; rather models are *for* particular sense-making aims. Making sense of a

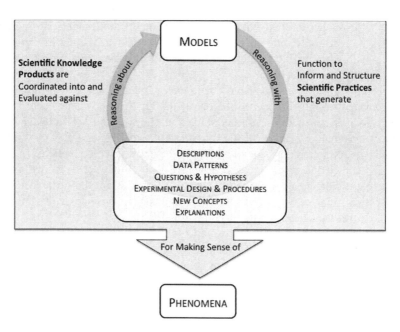

Fig. 36.3 Reasoning with and about models to make sense of phenomena

phenomenon does not typically happen all at once. Rather, one makes sense of a phenomenon by taking the problem in pieces. As these individual pieces come together to inform and constrain the model, both the model and the knowledge products that result from model use become more productive toward the ultimate aim of understanding how something in the world works. Thus, broadly conceived, we delineate two major classes of model-based reasoning. In Fig. 36.3, we have depicted the act of reasoning *with* the model to produce knowledge products and the act of reasoning *about* the model—developing and refining it—using these knowledge products, always toward the ultimate end of sense-making. Thus, the model is a means to an end, not an end in itself.

If models (whether explicit or not) serve as the basis for most, if not all, sense-making endeavors, then getting specific about the myriad ways that model-based reasoning plays out can productively inform science educators as they endeavor to craft more authentic learning experiences for students. Figure 36.3 above is one way to parse reasoning and the products of that reasoning. Of central importance in this figure is the idea that it is through *iterating* between reasoning *with* the model and reasoning *about* the model that one makes progress on coming to understand the phenomenon/phenomena of interest.[9]

---

[9] A major caution about the business of categorizing: We do this for the purpose of discussion and because we believe that a consideration of these different cognitive aims is potentially fruitful in the context of education. However, whenever something is presented in a list of categories, a

The challenge with depicting something as an iterative cycle is that it is difficult to know where to begin in discussing it. We do begin with an assumption from cognitive science that in general reasoning is "theory laden" or model based (Vosniadou 2001). Humans are incapable of interacting with ideas or other stimuli without simultaneously doing some level of integrating of those inputs into the cognitive architecture that is already present. In this sense, the cognitive agent does always begin with a model, although it is often implicit.

This does not imply, however, that instruction must always begin with an explicit model. Quite often instruction may begin with a puzzling phenomenon or an interesting data pattern. It may even begin with an explanation or prediction. However, at some point in the instructional sequence, the model must be drawn out in explicit ways so that its full potential as a reasoning tool can be realized.

36.3.1 Reasoning with Models

A major class of reasoning with models is to use them as filters on the world that serve to constrain and bound the problem space. This role of the model is in line with Odenbaugh's exploratory vision of model use (2005). What it is exactly that one finds intriguing about a particular phenomenon is wholly dependent on the model that the phenomenon is examined through. For example, when examining certain physical attributes of an organism, one may be interested in and attend to the genetic basis of a particular trait if the phenomenon is viewed through the lens of genetics models. Alternatively, one might be interested in the selective advantage of that trait in a particular environment if the phenomenon is viewed through the lens of evolutionary models.

An outcome of this kind of reasoning can be a particular description of the phenomenon. In one sense this is a representational task, but prior to any representation, the observer must bound and filter what it is that is worth noticing. This is done with regard to the model and is often the first stage in the sense-making process: defining what it is exactly that is of interest. In educational contexts, this stage is often done for the students prior to instruction in that the articulated learning goals imply a focus on specific models. To return to the example of organism traits, the way in which those traits are defined or described will vary depending on whether the students are supposed to be learning genetics or evolutionary models.

In the process of making sense of what is observed, one typically interacts with the world in particular ways. In figuring out how to manipulate and generate information from the world, one is guided by a current understanding of how the system functions. Interventions are based on models and an explicit attention to them can clarify what it is one wants to know more about and why that information may be

---

common interpretation is that that format implies an order. This is not our intention. The point here is that models organize a broad array of cognitive aims beyond representing and explaining which seem to be the two most commonly associated with models (Odenbaugh 2005; Knuuttila 2005).

useful in attaining a higher-order goal of sense-making. Reasoning with a model often points to areas that need further exploration and allows scientists to ask meaningful questions. From these questions, then, a scientist can derive and carry out investigative plans whether those involve data collection or simulations.

Two studies by Metz suggest that the degree to which there is an explicit focus on an underlying model may alter the degree to which students can productively engage in generating meaningful questions and investigations. Metz (2004) reports on a classroom study of elementary school children (second and fifth graders) designed to support the children in articulating and designing their own inquiries. Students had the opportunity to observe crickets and were able to generate many questions about them, but, according to Metz, few of them were scientific (e.g., What different crickets like different foods? What color do crickets seem to prefer? Where would crickets go on the playground?) (Metz 2004, p. 240). Metz does not interpret lack of scientific sophistication of these questions as attributed solely to students' age but rather to a lack of curricular support around question generation. In fact, some students did come up with potentially meaningful scientific questions (e.g., Where do crickets spend their time, in the shade or in the sunlight?) suggesting that students did have some scientifically interesting ideas about crickets but that they may have lacked the support needed to make those questions meaningful, reflecting "the failure of this initial version of the animal behavior curriculum to systematically scaffold theory-building" (Metz 2004, p. 267).

In contrast, Metz (2008) describes a classroom vignette in which students engage in thought experiments to explain the phenomenon that ants walk along the same line to get back to their nest. The focus of this activity was on generating plausible explanations for this phenomenon and then using those tentative explanations to develop a research question that could be tested. Students' tentative explanations were simple models with a relatively small number of constructs and relationships (e.g., ants are detecting some smell in the environment to lead them back to the nest). With these ideas explicitly articulated, the students could focus on attending to how they would collect evidence to support or refute various possibilities. For example, when one set of students devised a broom test that would sweep away the trail, other students could critically evaluate that suggestion in light of the emerging smell model by considering whether or not smell was something that could be removed with a broom. In these two studies, the difference between a focus on sensemaking and question generation in absence of a larger sense-making aim illustrates the potential importance of reasoning *with* a model and in making that explicit in the classroom.

When scientists reason with the model to carry out thought experimentation, they consider "how possibly" something might work. As Odenbaugh points out, this may involve adding particular conceptual elements to the broad framework which then find their way into the model itself. Scientists may then use their current models to develop explanations for how a system might work or to make predictions about the future behavior of that system.

In a study by Berland and Reiser (2011), students use simulation output of an ecological system containing organisms at three trophic levels (fox, rabbit, and grass) to determine the trophic level of a fourth unknown invading organism. In this report on the curriculum, it seems that the model that governed the simulation was largely implicit. This is an example where explicit attention to the underlying ecological models may have altered the student discourse during this activity. The students were engaged in sensemaking, but because the model was not made explicit, an important tool for that sensemaking aim was invisible in the classroom discourse. The primary resource the students had for defending claims was the data representation itself.

Berland and Reiser report on how students were engaged in trying to present arguments in support of their preferred claims. For example, one student countered another student's claim that the invader eats rabbits by explicitly referring to the graph: "you claim that the invader eats rabbits, right? Well, at the end of the graph when the rabbits are dead, how do the invader keep going up?" (Berland and Reiser 2011, p. 202). Underneath the first student's claim that the invader eats rabbits is a model in which the invasive organism shares a trophic level with the fox. Making sense of what this implies could have been supported by first developing a robust understanding of the three organism system and then reasoning through what could possibly happen if an invader was added to the system. This kind of thought experimentation with an explicit focus on the model at work behind the scenes of the simulation may have allowed the students to interpret the invader data more clearly. Such a model would explicitly link population numbers in the fox to the abundance of its food source, rabbit, and would make the prediction that if the rabbit population were to decrease, we would expect a decrease in the fox population as well. Being able to imagine how the system might change in response to changes in variables could help students make theoretically justified claims that they could then hold up against the simulation output.

The operations and outcomes of reasoning with models depicted in Fig. 36.3, collectively, are at the core of sense-making. That is, if a phenomenon is thoroughly described, investigated, and explained, then it has been reduced to some kind of order, or it has been made sense of in the broadest sense. Reasoning *with* a model is core to the scientific practices of observing, describing, asking questions, designing investigations, explaining, and predicting.

36.3.2 Reasoning About Models

Reasoning *about* models refers to the integrated practices of developing, evaluating, and revising models. In Fig. 36.3, reasoning *about* models means making decisions about how to synthesize ideas from a number of different sources in the service of more clearly articulating a model. Boumans (1999) makes the important claim that this work is done iteratively and concurrently with model use and so, again, recalls that these different cognitive strategies are teased apart in the diagram for the purposes of discussion.

Reasoning about a model involves making and justifying a number of decisions. One major class of model-based reasoning involves making representational decisions (Nersessian 2002; Knuuttila 2005). In order to share a model within a community of practice, the model must be externalized in some way. The types of things a scientist attends to in creating these externalizations are often central to the formulation of the model itself. For example, if one chooses to represent a model in mathematical form, there are a range of decisions one must make about how different aspects of the model are laid out and the precise ways that each model idea, as represented in a mathematical expression, relates to others. Then, once a mathematical model is analyzed or a computational model is run, the scientist must interpret the results using the initial framework, checking to see if the model output makes sense and is useful in figuring out something about the system under study.

In an undergraduate context, Svoboda and Passmore describe how the question and the model coevolved over time (2011). Specifically, students in this context were attempting to model vaccination-disease dynamics. The students never had a firm hypothesis that they were attempting to prove or disprove. Rather, the group was engaged in a creative and dynamic process of trying to make sense of a complex phenomenon. Throughout the course of their months-long inquiry, they spent a lot of time reasoning *about* the model with close attention to their initial intentions for modeling the system. They had been inspired to undertake this particular project after reading an article that described how media attention about a possible link between autism and the measles, mumps, and rubella (MMR) vaccine led to reduced vaccination and recent disease outbreaks in countries that have voluntary vaccination strategies.

As they went about their work, they continually returned to their initial goal of making sense of the phenomenon of disease-vaccine dynamics. This attention allowed them to make a series of decisions about what to include in their model, how to represent various aspects of the system, and how to interpret results of their modeling activities.

In the process of crafting a specific articulation of a model and communicating it to others, it is not uncommon to come across aspects of the model that need further expansion. Odenbaugh (2005) delineates this as a process of conceptual development. Deciding on, describing, and defining the working pieces of the model are all involved at this point.

Another class of reasoning *about* the model comes when one considers the relationship between particular phenomena and models. Often, a model is developed in the context of examining a very specific phenomenon. For example, one might model the relationships between a set of organisms in a particular environment. Doing so, by necessity, involves attending to the specific details of those organisms in that environment. The model may be deemed useful for explaining one very specialized situation, but from there one might wonder if the model could be applied to other similar phenomena. So, another way in which reasoning about the model occurs is to consider the generalizability of a particular model. One aspect of this may be to make representational decisions about how to broaden the model focus beyond a particular phenomenon to make it useful to explain a larger class.

To come back to the intertwined issues of "discovery and justification" (Boumans 1999), it is in the process of thinking explicitly *about* the model that model development and evaluation come together. Iterating between model use and meta-level processing about the extent to which the model is achieving one's aims is at the core of evaluating a model. The scientist reasons with a model to develop explanations and/or predictions and from there considers whether those knowledge products are useful or not. If not, it may be necessary to consider if the issue is related to the model itself, thus suggesting a need for revision, or if it is an issue of translating the model into an explanation that is at play.

In all cases, issues around reasoning about the model must be inextricably tied to the intention of the modeler. There is no context-free way to make reasonable decisions about the attributes and form of a model. These decisions, in practice, must always be made with regard to the purpose the model is put to in the context of its use.

A study by Hmelo-Silver and Pfeffer illustrates this point. They describe how students had trouble constructing models of aquaria that extended beyond superficial structural features, while experts tended to include deeper functional relationships (Hmelo-Silver and Pfeffer 2004). One way to understand this difference is to ask how each of these groups interpreted the purpose of this model development task. It is clear that experts brought a particular set of aims and purposes to the task. For example, expert aquarium hobbyists were concerned with the aim of maintaining healthy fish and tended to include variables and the relationships among variables in aquaria that are related to fish health. In contrast, ecologists constructed models that could be used to explain ecosystem stability over time.

That students attended to surface features suggests that they viewed the task as primarily descriptive (i.e., they were supposed to be building models *of* aquaria). Rather than propose that students are not as good as experts at attending to deeper features, this as an example where the students needed additional scaffolding around the purpose of model development.

While this work is important because it gave students the opportunity to take responsibility for model development, without an explicit aim, students had no way to productively bound what they were modeling or make decisions about what to include and what to leave out. If instructors do not make the purpose of a modeling activity clear to students, they will bring their own frame to it, and that framing, without explicit attention, may be idiosyncratic rather than shared across the group. If model development and justification cannot be decoupled, then there is simply no robust way to engage in model development in the classroom in the absence of a clear aim that can guide the evaluation/justification of the model.

In contrast, Smith and colleagues (1997) describe how groups of ecology students constructed models relevant to an aphid-wasp-fungus system. Crucially, prior to constructing models, the students were first asked to develop questions that they were interested in investigating. The students were then able to make decisions about the degree of detail and complexity to add into their models. Different groups of students, depending on the question they proposed and the aims they prioritized, then constructed very different models.

A growing number of scholars have written about the importance and utility of allowing students to construct and critique models.[10] What the science studies lens brings to such environments is the importance of coupling model development and evaluation to a clear sense-making aim. In order for students to make the appropriate decisions about how to bound and describe a system, they need to have a clear sense of what the model will ultimately used to do.

What these examples are meant to draw out is how, when contextualized in this way, the practices of constructing, representing, evaluating, and revising models can support productive sensemaking work. These examples highlight how reasoning about models is inextricably linked to considerations of what those models can then be used to do. Thus, reasoning *about* the model is done with critical attention to the output of reasoning *with* the model to achieve a particular aim. When these practices are not linked in instructional settings, then the outcome may not realize the full potential of model-based reasoning in supporting learning.

36.4 Major Implications and Recommendations

36.4.1 Implications

To consider a view of the various aims that a modeler can have and pay explicit attention to the two major classes of model-based reasoning identified above has implications for the way in which science educators approach science instruction and the degree of student ownership and autonomy. Moreover, a focus on models provides a framework in which the various practices of science can be organized and put to productive use in the classroom.

Much has been made over the past few decades about teaching for conceptual change. As part of this approach, many science educators have been involved in determining a canon of scientific ideas. What if the criteria for what counts as the canon were shifted? Obviously scientists do not undertake their cutting-edge work aiming at an established canon of models. To present the task in science classrooms as one in which the students are trying to uncover or guess the canonical model seems disingenuous at best. Maybe there is, however, a canon of classes of phenomena that a scientifically literate student should be able to explain, and those explanations are based on a developmentally appropriate set of models. In this way instruction could actually be crafted authentically, and the intellectual environment of the classroom could reflect the particular sensemaking aims of the community of learners at any given time while concurrently fostering deep understanding of important science concepts.

---

[10] See, for example, Baek et al. (2011), Clement (1989, 2000), Gilbert et al. (1998a, b), Hogan and Thomas (2001), Passmore and Stewart (2002), Schwarz et al. (2009), Svoboda and Passmore (2011), and White (1993).

As students engage in the kind of reasoning described here, they should develop a greater sense of ownership over ideas as they develop them iteratively to address specific reasoning aims that their classroom community has identified. Many of the studies cited above bear witness to the fact that students will engage in complex reasoning when tasks are designed that require it. By organizing these tasks in explicit ways with regard to reasoning with and about models, students may develop a sense of autonomy and begin to demand learning environments that are fundamentally about figuring things out about how the world works.

The framework presented here unites and organizes a number of practices that science educators have been focused on for the past several years (NRC 2007, 2011). The centrality of models to the practices of asking questions, designing investigations, developing explanations, and arguing from evidence becomes clear for both students and teachers. Sadly, higher-level goals of science do not seem to drive science education which tends to focus on the practices in ways that are potentially isolated from higher-order sensemaking goals of explanation (Metz 2006). Placing a model-based view of science at the center of the practices has the potential to counter the treatment of each practice as isolated and unite them in service of sense-making.

36.4.2 Research Recommendations

From the view of models in science presented here, there are a number of research avenues for investigating how models operate in learning science. In Sect. 36.3 we explored the real or imagined role of an explicit focus on models in learning environments that have been described in the science education literature. However, more detailed work is needed in this area to understand exactly how a focus on models and modeling interacts with learning both the conceptual content of particular scientific disciplines and how it may influence students' epistemological views. A recent article by Eve Manz (2012) is one example of this type of scholarship.

If the field can agree on a set of guiding principles for what modeling entails in science learning, then a series of deep investigations into learning environments will be possible that simultaneously attend to the particulars of each context and provide insights into a broader framework. As it stands now, there is a wide range of conceptualizations around modeling and its relationship to other scientific practices, and thus, it is sometimes difficult to understand how different studies speak to one another.

Further, there will need to be additional research on teacher conceptualizations and enactments around model-based inquiry. This work is underway,[11] but it is clear that teachers (both preservice and in-service) have had very little experience with a view of science as a model-based enterprise and thus may be challenged to enact

---

[11] See, for example, Danusso et al. (2010), Nelson and Davis (2012), and Schwarz and Gwekwerere (2006).

model-based curricula (Windschitl et al. 2008a). As researchers uncover and delineate some of the challenges teachers face in supporting students in model-based inquiry, additional resources can be developed to support teachers, including comprehensive curriculum.

36.5 Conclusion

This chapter was crafted to answer questions about the function of models in science and how that view could be translated for educators. The practice turn in science studies has been fruitful in focusing the scholarly community on the importance of models and in identifying the particular ways in which models function in the intellectual lives of scientists. The science studies' work has informed science educators and the time has come to more fully incorporate the findings from philosophical, historical, and cognitive studies of science into science education.

So, in the end we are hoping to make what may seem a subtle shift but one we find incredibly important and powerful in thinking about science education. Instead of listing the kinds of models there are and arguing about what the canonical set of target models for instruction might include, we suggest that the dialogue becomes one that is centrally about the context-dependent roles that models are playing in the students' reasoning/sense-making about phenomena. In making instructional and curricular decisions based on a "models *for*" orientation, we expect a more productive and hopefully authentic version of school science to emerge.

If it is relatively uncontested that models form the basis of most reasoning in science, then it seems obvious that they should form the basis of reasoning in science classrooms. And, although this is often the case when we examine productive classroom activity, the models are rarely made explicit. There may be much to be gained from changing this state of affairs, but ultimately that is an empirical question. The presence of and attention to models as used by cognitive agents *for* specific purposes both focuses and organizes the cognitive activity that is primarily aimed at sense-making. As science educators take seriously the "practice turn" and call for authenticity, it will be a central focus not only on models but also on what they are being used *for* in the sense-making process that will provide a way forward.

References

Adúriz-Bravo, A. (2012). A "Semantic" View of Scientific Models for Science Education. *Science & Education*. doi:10.1007/s11191-011-9431-7

Auyang, S. Y. (1998). *Foundations of complex-system theories: in economics, evolutionary biology, and statistical physics*. Cambridge: Cambridge University Press.

Baek, H., Schwarz, C. V., Chen, J., Hokayem, H., & Zhan, L. (2011). Engaging elementary students in scientific modeling: The MoDeLS fifth-grade approach and findings. In M. S. Khine & I. M. Saleh (Eds.), *Models and Modeling in Science Education: Cognitive Tools for Scientific Enquiry* (pp. 195–218). Dordrecht: Springer Netherlands. doi:10.1007/978-94-007-0449-7

Bauer, H. H. (1992). The so-called scientific method. *Scientific Literacy and the Myth of the Scientific Method* (pp. 19–41).

Berland, L. K., & Reiser, B. J. (2011). Classroom communities' adaptations of the practice of scientific argumentation. *Science Education, 95*(2), 191–216. doi:10.1002/sce.20420

Boulter, C.J., & Buckley, B.C. (2000). Constructing a typology of models for science education. In J.K. Gilbert & C.J. Boulter (Eds.) *Developing models in science education.* (pp. 41–57). Dordrecht: Kluwer Academic Publishers.

Bottcher, F. & Meisert, A. (2010). Argumentation in science education: A model-based framework. *Science & Education 20*(2) 103–140.

Boumans, M. (1999). Built-in justification. In *Models as mediators: Perspectives on natural and social science* (Vol. 52). Morgan, M. S., & Morrison, M. (Eds.). Cambridge University Press.

Cartwright, N. (1997). Models: The blueprints for laws. *Philosophy of Science, 64*(4), S292–S303. University of Chicago Press.

Cartwright, N. (1999) *The dappled world: A study of the boundaries of science.* Cambridge: Cambridge University Press.

Clement, J. J. (1989). Learning via model construction and criticism. In G. Glover, R. Ronning, & C. Reynolds (Eds.), (pp. 341–381). New York: Plenum Publishers.

Clement, J. J. (2000). Model Based Learning as a Key Research Area for Science Education. *International Journal of Science Education, 22*(9), 1041–1053.

Coll, R. K., & Lajium, D. (2011). Modeling and the Future of Science Learning. In M. S. Khine & I. M. Saleh (Eds.), *Models and Modeling in Science Education: Cognitive Tools for Scientific Enquiry* (pp. 3–21). Dordrecht: Springer Netherlands. doi:10.1007/978-94-007-0449-7

Coll, R. K., & Treagust, D. F. (2003). Learners' mental models of metallic bonding: A cross-age study. *Science Education, 87*(5), 685–707. doi:10.1002/sce.10059

Cooper, G. J. (2003). *The science of the struggle for existence: On the foundations of ecology.* Cambridge: Cambridge University Press.

Danusso, L., Testa, I., & Vicentini, M. (2010). Improving prospective teachers' knowledge about scientific models and modelling: Design and evaluation of a teacher education intervention. *International Journal of Science Education, 32*(7), 871–905.

Develaki, M. (2007). The Model-Based View of Scientific Theories and the Structuring of School Science Programmes. *Science & Education, 16*(7), 725–749. doi:10.1007/s11191-006-9058-2

Downes, S. (1992). The Importance of models in theorizing: a deflationary semantic view. *PSA: Proceedings of the Biennial Meeting of the Philosophy of Science Association* (Vol. 1992, pp. 142–153). Chicago: The University of Chicago Press.

Duschl, R. A. (2008). Science Education in Three-Part Harmony: Balancing Conceptual, Epistemic, and Social Learning Goals. *Review of Research in Education, 32*(1), 268–291. doi: 10.3102/0091732X07309371

Duschl, R. A., & Grandy, R. E. (2008). *Teaching scientific inquiry: Recommendations for research and implementation.* Sense Publishers.

Engle, R. A., & Conant, F. R. (2002). Guiding Principles for Fostering Productive Disciplinary Engagement: Explaining an Emergent Argument in a Community of Learners Classroom. *Cognition and Instruction, 20*(4), 399–483. Lawrence Erlbaum Associates (Taylor & Francis Group). doi:10.2307/3233901

Ford, M. (2008). Disciplinary authority and accountability in scientific practice and learning. *Science Education, 92*(3), 404–423.

Giere, R. N. (1988). *Explaining Science: A Cognitive Approach.* University of Chicago Press.

Giere, R. N. (2004). How models are used to represent reality. *Philosophy of Science, 71,* 742–752.

Gilbert, J. K. (2004). Models and Modelling: Routes to More Authentic Science Education. *International Journal of Science and Mathematics Education, 2*(2), 115–130. doi:10.1007/s10763-004-3186-4

Gilbert, J. K., Boulter, C., & Rutherford, M. (1998a). Models in explanations, Part 2: Whose voice? Whose ears? *International Journal of Science Education, 20*(2), 187–203. doi:10.1080/0950069980200205

Gilbert, J. K., Boulter, C., & Rutherford, M. (1998b). Models in Explanations, Part 1: Horses for Courses? *Science Education, 20*(1), 83–97. [Part I should be (a), no?]

Gobert, J. D. (2005). The Effects of Different learning Tasks on Model-building in Plate Tectonics: Diagramming Versus Explaining. *Journal of Geoscience Education, 53*(4), 444–455.

Hammer, D. (1996). Misconceptions or P-Prims: How May Alternative Perspectives of Cognitive Structure Influence Instructional Perceptions and Intentions. *Journal of the Learning Sciences, 5*(2), 97–127. doi:10.1207/s15327809jls0502_1

Harrison, A. G., & Treagust, D. F. (2000). A typology of school science models. *International Journal of Science Education, 22*(9), 1011–1026.

Hmelo-Silver, C. E., & Pfeffer, M. G. (2004). Comparing expert and novice understanding of a complex system from the perspective of structures, behaviors, and functions. *Cognitive Science, 28*, 127–138.

Hodson, D. (1992). In search of a meaningful relationship: an exploration of some issues relating to integration in science and science education. *International Journal of Science Education, 14*, 541–562. doi:10.1080/0950069920140506

Hodson, D. (1996). Laboratory work as scientific method: three decades of confusion and distortion. *Journal of Curriculum Studies, 28*(2), 115–135.

Hodson, D. (2008). *Towards scientific literacy*. Rotterdam, The Netherlands: Sense Publishers.

Hogan, K., & Thomas, D. (2001). Cognitive comparisons of students' systems modeling in ecology. *Journal of Science Education and Technology, 10*(4).

Hughes, R. I. G. (1999). The Ising Model, Computer Simulation, and Universal Physics. In M. Morrison & M. S. Morgan (Eds.), *Models as Mediators: Perspectives on Natural and Social Science* (pp. 97–145). Cambridge: Cambridge University Press.

Knuuttila, T. (2005). *Models as epistemic artefacts: Toward a non-representationalist account of scientific representation. Philosophical Studies*. Helsingin yliopisto.

Koponen, I. (2007). Models and Modelling in Physics Education: A Critical Re-analysis of Philosophical Underpinnings and Suggestions for Revisions. *Science & Education,16*, 751–773.

Lehrer, R., & Schauble, L. (2004). Modeling Natural Variation Through Distribution. *American Educational Research Journal, 41*(3), 635–679. doi:10.3102/00028312041003635

Levins, R. (1966). The strategy of model building in population biology. *American Scientist, 54*(4), 421–431.

Lloyd, E. A. (1997). *The structure and confirmation of evolutionary theory*. Princeton University Press.

Matthews, M. R. (1992). History, philosophy, and science teaching: The present rapprochement. *Science & Education, 1*(1), 11–47. doi:10.1007/BF00430208

Matthews, M. R. (Ed.). (2007). Models in Science and Science Education. [Special Issue] *Science & Education, 16*(7–8).

Manz, E. (2012). Understanding the codevelopment of modeling practice and ecological knowledge. *Science Education, 96*(6) 1071–1105. doi: 10.1002/sce.21030

Metz, K. E. (2004). Children's Understanding of Scientific Inquiry: Their Conceptualization of Uncertainty in Investigations of Their Own Design. *Cognition and Instruction, 22*(2), 219–290. doi:10.1207/s1532690xci2202

Metz, K. E. (2006). The knowledge building enterprises in science and elementary school science classrooms. In L. B. Flick & N. G. Lederman (Eds.), *Scientific inquiry and nature of science* (pp. 105–130). Springer.

Metz, K. E. (2008). Narrowing the gulf between the practices of science and the elementary school science classroom. *The Elementary School Journal, 109*(2), 138–161.

Morgan, M. S., & Morrison, M. (Eds.). (1999). *Models as mediators: Perspectives on natural and social science* (Vol. 52). Cambridge University Press.

National Research Council (2007). *Taking science to school: Learning and teaching science in grades K-8*. Washington, DC: The National Academies Press.

National Research Council; Committee on Conceptual Framework for the New K-12 Science Education Standards. (2011). *A Framework for K-12 Science Education : Practices, Crosscutting*

Concepts, and Core Ideas. Washington D.C.: The National Academies Press. Retrieved from http://www.nap.edu/catalog.php?record_id=13165

Nelson, M. M., & Davis, E. A. (2012). Preservice Elementary Teachers' Evaluations of Elementary Students' Scientific Models: An aspect of pedagogical content knowledge for scientific modeling. *International Journal of Science Education, 34*(12), 1931–1959.

Nersessian, N. J. (1989). Conceptual change in science and in science education. *Synthese, 80*(1), 163–183. doi:10.1007/BF00869953

Nersessian, N. J. (1992). How do scientists think? Capturing the dynamics of conceptual change in science. In R.N. Giere (Ed) *Cognitive models of science: Minnesota studies in the philosophy of science,* Vol XV. Minneapolis: University of Minnesota Press.

Nersessian, N. J. (1995). Should Physicists Preach What They Practice? Constructive Modeling in Doing and Learning Physics. *Science & Education,* 4, 203–226.

Nersessian, N. J. (1999). Model-based Reasoning in Conceptual Change. In L. Magnani, N. Nersessian, & P. Thagard (Eds.), (pp. 5–22). New York: Kluwer Academic/Plenum Publishers.

Nersessian, N. J. (2002). The cognitive basis of model-based reasoning. *The cognitive basis of science* (pp. 133–153). Cambridge University Press.

Odenbaugh, J. (2005). Idealized, Inaccurate but Successful: A Pragmatic Approach to Evaluating Models in Theoretical Ecology. *Biology & Philosophy, 20*(2–3), 231–255. doi:10.1007/s10539-004-0478-6

Odenbaugh, J. (2009). Models in biology. In E. Craig (Ed.), *Routledge Encyclopedia of Philosophy.* London: Routledge.

Osbeck, L., Nersessian, N. J., Malone, K. R., & Newstetter, W. (2010). *Science as Psychology: Sense-making and identity in science practice.* New York: Cambridge University Press.

Osborne, J., Collins, S., Ratcliffe, M., Millar, R., & Duschl, R. A. (2003). What "ideas-about-science" should be taught in school science? A Delphi study of the expert community. *Journal of Research in Science Teaching, 40*(7), 692–720. doi:10.1002/tea.10105

Passmore, C., & Stewart, J. (2002). A modeling approach to teaching evolutionary biology in high schools. *Journal of Research in Science Teaching, 39*(3), 185–204. doi:10.1002/tea.10020

Passmore, C., & Svoboda, J. (2011). Exploring Opportunities for Argumentation in Modelling Classrooms. *International Journal of Science Education,* (October), 1–20. doi:10.1080/09500693.2011.577842

Penner, D. E., Giles, N. D., Lehrer, R., & Schauble, L. (1997). Building functional models: Designing an elbow. *Journal of Research in Science Teaching, 34*(2), 125–143.

Penner, D. E., Lehrer, R. & Schauble, L. (1998) From physical models to biomechanics: A design modeling approach. *The Journal of the Learning Sciences, 7* (3 & 4), 429–449.

Pluta, W. J., Chinn, C. A., & Duncan, R. G. (2011). Learners' epistemic criteria for good scientific models. *Journal of Research in Science Teaching, 48*(5), 486–511. doi:10.1002/tea.20415

Roth, W.-M., & Roychoudhury, A. (1993). The development of science process skills in authentic contexts. *Journal of Research in Science Teaching, 30,* 127–152. doi:10.1002/tea.3660300203

Rudolph, J. L. (2005). Epistemology for the Masses: The Origins of "The Scientific Method" in American Schools. *History of Education Quarterly, 45*(3), 341–376. doi:10.1111/j.1748-5959.2005.tb00039.x

Schwarz, C. V., & Gwekwerere, Y. N. (2006). Using a guided inquiry and modeling instructional framework (EIMA) to support preservice K-8 science teaching. *Science Education, 91*(1), 158–186.

Schwarz, C. V., Reiser, B. J., Davis, E. A., Kenyon, L. O., Achér, A., Fortus, D., Shwartz, Y., et al. (2009). Developing a learning progression for scientific modeling: Making scientific modeling accessible and meaningful for learners. *Journal of Research in Science Teaching,* 46(6), 632–654. doi:10.1002/tea.20311

Smith, E., Haarer, S., & Confrey, J. (1997). Seeking diversity in mathematics education: Mathematical modeling in the practice of biologists and mathematicians. *Science & Education,* 6, 441–472.

Stewart, J., Cartier, J., & Passmore, C. (2005). Developing Understanding Through Model-based Inquiry. In M. S. Donovan & J. Bransford (Eds.), *How Students Learn: History, Mathematics, and Science in the Classroom* (pp. 515–565). Washington D.C.: National Research Council.

Suckling, C.J., Suckling, K.E. & Suckling, C.W. (1980). *Chemistry through models*. Cambridge: Cambridge University Press.

Svoboda, J., & Passmore, C. (2011). The strategies of modeling in biology education. *Science & Education*. doi:10.1007/s11191-011-9425-5

Teller, P. (2001). Twilight of the perfect model model. *Erkenntnis*, *55*, 393–415.

Vosniadou, S. (2001). Models in conceptual development. In L. Magnangi & N. Nersessian (Eds.) *Model-based reasoning: Science, technology, values*. New York: Kluwer Academic.

Watson, J. D. and Crick, F. H. C. (1953). A structure for Deoxyribose Nucleic Acid. Nature, 171, 737–738.

White, B. Y. (1993). Thinker Tools: Causal models, conceptual change, and science education. *Cognition and instruction*, *10*(1), 1–100. Routledge.

Wimsatt, W. C. (1987). False models as means to truer theories. In M. Nitecki and A. Hoffman (Eds.), *Neutral Models in Biology* (pp. 23–55). New York: Oxford University Press.

Windschitl, M., Thompson, J., & Braaten, M. (2008a). How Novice Science Teachers Appropriate Epistemic Discourses Around Model-Based Inquiry for Use in Classrooms. *Cognition and Instruction*, 26(3), 310–378. Routledge. doi:10.1080/07370000802177193

Windschitl, M., Thompson, J., & Braaten, M. (2008b). Beyond the Scientific Method : Model-Based Inquiry as a New Paradigm of Preference for School Science Investigations. Science *Education*, 1–27. doi:10.1002/sce

Cynthia Passmore is currently an Associate Professor specializing in science education in the University of California, Davis School of Education. She did her doctoral work at the University of Wisconsin, Madison and prior to that she was a high school science teacher. Her research focuses on the role of models and modeling in student learning, curriculum design and teacher professional development. She investigates model-based reasoning in a range of contexts and is particularly interested in understanding how the design of learning environments interacts with students' reasoning practices. She has been the principal investigator of several large grants and has coauthored several papers on modeling in science education that have been published in journals such as *Science & Education, The International Journal of Science Education* and *School Science and Mathematics*.

Julia Svoboda Gouvea received a Ph.D. in science education from the University of California, Davis. She also holds an M.A. in population biology from the same university and a B.A. in ecology, evolution, and behavior from Princeton University. Her dissertation work explored students' reasoning with mathematical models in the context of biology and how experiences with modeling influenced students' developing epistemologies of science. Her current research continues to explore the role that models and modeling play in supporting student reasoning and curriculum development in K-12-, undergraduate-, and graduate-level science education. Gouvea is currently an assistant project scientist at UC Davis and holds a concurrent position as a postdoctoral researcher at the University of Maryland, College Park.

Ronald Giere is a professor emeritus of philosophy and a member and former director of the Center for Philosophy of Science at the University of Minnesota. In addition to many papers in the philosophy of science, he is the author of *Understanding Scientific Reasoning* (5th ed 2006), *Explaining Science: A Cognitive Approach* (1988), *Science Without Laws* (1999), and *Scientific Perspectivism* (2006). His current research focuses on agent-centered accounts of models and scientific representation.

Chapter 37
Laws and Explanations in Biology and Chemistry: Philosophical Perspectives and Educational Implications

Zoubeida R. Dagher and Sibel Erduran

37.1 Introduction

The teaching of history and philosophy of science (HPS) in science education has been advocated for several decades.[1] In recent years, however, there has been increasing interest in the philosophical examination of biology and chemistry as distinct branches of science that differ epistemically from physics in significant ways. Philosophers of biology (Hull 1973; Mayr 2004; Ruse 1988) and philosophers of chemistry (Bhushan and Rosenfeld 2000; van Brakel 2000; Scerri and McIntyre 1997) have offered insights into the epistemologies of biology and chemistry. However, these insights have not been integrated sufficiently into biology and chemistry education research, curriculum materials and classroom practice. Research on the nature of science in science education could benefit from such insights in order to improve understanding of not only the disciplinary knowledge but also the meta-level characterisations of scientific knowledge at large.

As science educators we are concerned with the question of how philosophical insights into scientific knowledge can inform science teaching and learning. The goal is not to contribute to the debates in the philosophy of biology and chemistry,

---

An earlier version of this paper was presented at the 2011 IHPST conference and published in F. Seroglou, V. Koulountzos and A. Siatras (Eds), *Science & culture: Promise, challenge and demand*. Proceedings for the 11th International IHPST and 6th Greek History, Philosophy and Science Teaching Joint Conference. 1–5 July 2011, Thessaloniki, Greece: Epikentro

[1] See for example Duschl (1990), Hodson (1988), Matthews (1994/2014) and Schwab (1958, 1978).

Z.R. Dagher (✉)
School of Education, University of Delaware, Newark, DE, USA
e-mail: zoubeida@udel.edu

S. Erduran
University of Limerick, Department of Education and Professional Studies, Ireland

Bogazici University, Department of Primary Education, Turkey
e-mail: Sibel.Erduran@ul.ie

M.R. Matthews (ed.), *International Handbook of Research in History, Philosophy and Science Teaching*, DOI 10.1007/978-94-007-7654-8_37,
© Springer Science+Business Media Dordrecht 2014

but rather to draw out some aspects of these debates that are relevant for education in light of evidence from empirical studies in science education (e.g. Dagher and Cossman 1992; Erduran and Jimenez-Aleixandre 2008; Sandoval and Millwood 2005). In doing so, we problematise the current state of under utilisation of the epistemological aspects of disciplinary knowledge in science education and illustrate with examples how it can be practically addressed. It is hoped that the discussion will assist other science education researchers in exploring the philosophical literature for clarifying and justifying educational goals that relate to scientific knowledge claims.

According to Irzik and Nola (Chap. 30), science can be perceived as a cognitive-epistemic system and as a social system. Scientific knowledge, which constitutes one component of the cognitive-epistemic system, is the culmination of scientific inquiry and includes laws, theories and models. Focusing on these structural elements in the context of any single discipline would be necessary to understand the nature of that discipline. Among these elements, explanations and particularly laws have been understudied from an epistemological perspective in science education research. For instance, while there is a substantial body of literature focused on models (e.g. Justi 2000), the study of the particular epistemological aspects of models has been scarce (e.g. Adúriz-Bravo 2013; Adúriz-Bravo and Galagovsky 2001; Erduran and Duschl 2004). Similarly, despite the importance of laws and explanations in the science disciplines, relevance of their epistemic nature to educational practice is seldom explored (e.g. McComas 2003; Sandoval and Reiser 2004).

One often-cited misunderstanding of the nature of science (NOS) concerns scientific laws. Classified as the number one NOS myth by McComas (1998), many individuals tend to believe 'that with increased evidence there is a developmental sequence through which scientific ideas pass on their way to final acceptance as mature laws' (p. 54). Involved in this belief is the thought that science starts out with facts, progresses to hypotheses, then theories, then, when confirmed, to laws. Another myth pertains to the idea that scientific laws are absolute (McComas 1998). These beliefs represent only two of many other misunderstandings about the nature of scientific knowledge and pose challenges regarding the best approach to deconstruct them. Several approaches have been proposed for countering these and other nature of science misconceptions (Clough 1994; Khishfe and Abd-El-Khalick 2002; Schwartz et al. 2004), but it remains unclear whether efforts to enhance student understandings of the nature of science have resulted in significant or lasting improvements (Lederman 2007).

The context of laws provides a crucial and relevant nexus for promoting the epistemological aspects of biology and chemistry in the classroom. Focusing on the nature of laws in biology education, for example, not only serves to clear existing misconceptions (as the ones mentioned earlier) but offers insight into basic metaphysical and ontological aspects of the discipline which can enhance student understanding of the subject. The inclusion of 'laws' in chemistry education not only elaborates on this important philosophical thesis but also offers some insight into how students' interest in philosophical aspects of chemistry might be stimulated.

Scientific explanations on the other hand refer often to how and why something happens (Chinn and Brown 2000). Typically scientists explain phenomena by determining how and why they occur along with the conditions surrounding the observed events (Nagel 1961). Explanations are important components of scientific theories. They are the backbone of scientific claims and are consequently a central target for epistemological disputes. It is through the refutation or support of components of scientific explanations that the fabric of theories is woven. In science education, considerable emphasis is placed on developing students' ability to substantiate their explanations using reasons and evidence.

Despite the separation of laws and explanations for contrast in biology and chemistry in this chapter for educational purposes, the distinction of these concepts in the history of philosophy of science is not straightforward. For example, the covering laws in Hempel's positivistic framework function not only as core explanatory components (*explanans*) but also as the targets of explanation (*explananda*). In more recent work, lawlike regularities among properties are considered to be a kind of explanation in their own right. For instance, Bird (1998) calls them 'nomic explanations'. He argues that inferring a law from observation is a form of *inference to the best explanation* (IBE), a common form of scientific reasoning.

The task in this paper is not to articulate the distinctions between laws and explanations from a philosophical perspective. Indeed, as educators, it is beyond the scope of our engagement in philosophy of science to contribute to or resolve existing debates or to generate new knowledge in the field. This task is left to the professional philosophers. Rather, the purpose of this analysis is to draw out some themes around laws and explanations, discussed in philosophy of biology and philosophy of chemistry, in ways that are relevant for science education. For example, Mendel's Laws and the Periodic Law are chosen as examples because of their prominence in science curricula at the secondary school level, which is our primary area of interest. At times, the discussion will refer to some contentious characterisations of laws and explanations. Again, here the discussion is reflecting ongoing debates to inform the science education community of the sorts of issues that are of concern to philosophers of science. The implications for science education could include problematising the nature of laws, explanations or indeed the contrast itself. However, given the typically separate reference to laws and explanations in the science curricula, the goal in this chapter is to interrogate the existing literatures for particular and explicit references to either laws or explanations thus informing subsequent analysis of how they are depicted in science education.

Furthermore, while discussions of NOS in the science education literature typically focus on the relationship between laws and theories (specifically on how they are different), they tend to neglect the conceptual disciplinary-based features that pertain to them. Shifting the discussion in this paper from laws and theories to laws and explanations underscores the following key ideas/assumptions: (1) Explanation is a key purpose of science. Theories are developed not as ends in themselves but as powerful explanatory and predictive tools. (2) Laws express regularities that can serve predictive and/or explanatory functions. (3) Explanations are building blocks of scientific theories that can be explored pedagogically at multiple organisational levels. (4) explanations are pragmatic and contextual (de Regt 2011).

Focusing on explanations rather than theories in this paper allows for a nuanced and contextual discussion of their characteristics across disciplines and subdisciplines from philosophical and educational perspectives. The significance of explanations in science curriculum and instruction is recognised by science educators in a variety of ways. In some cases, concern is expressed about linguistic and epistemic aspects, as with Horwood's (1988) illustration of the lack of consistency between the terms 'explain' and 'describe' in teaching materials and Jungwirth's (1979) findings that high school students tend to equate anthropomorphic and teleological explanations with causal explanations. At the level of instruction, teachers are reported to use a wide range of explanatory types some of which are scientific and some are not, calling for further examination of the appropriateness of these explanations (Dagher and Cossman 1992). More recent work presents evidence for the difficulties experienced by students in generating and justifying scientific explanations (Sandoval and Millwood 2005). In addition, there have been ongoing efforts focused on designing instructional models for supporting student development of scientific explanations (Land and Zembal-Saul 2003; McNeill et al. 2006; McNeill and Krajcik 2012).

In summary, the purpose of this chapter is to discuss characterisations of laws and explanations in biology and chemistry and extract some implications for teaching, learning and curriculum. The goal is to demonstrate how some of the ongoing debates about the nature of laws and explanations in biology and chemistry can have useful contributions to teaching these disciplines without necessarily resolving these debates. Exploring the arguments in these debates allows the articulation of how laws and explanations as products of scientific knowledge might be addressed more meaningfully in educational settings by discussing current coverage of laws and explanations in typical biology and chemistry textbooks. The chapter concludes with recommendations for revising textbooks and instruction in ways that restore the grounding of subject matter knowledge in its epistemological context.

37.2 The Nature of Laws and Explanations in Science

Volumes have been written about the nature of laws and explanations in science, mostly using physics as a basis for analysis. Views about the purpose and nature of these entities have changed over time and some aspects of them continue to undergo some debate. Attempting to summarise this vast literature or represent the diversity of views in few paragraphs is impossible without doing grave injustice to the field. It is necessary, however, to highlight key ideas before discussing the characteristics of laws and explanations in biology and chemistry with the understanding that this brief overview is not exhaustive or representative of extant viewpoints.

What distinguishes a law of nature from any other regularity? Traditional definitions of a scientific law typically refer to 'a true, absolute and unchanging relationship among interacting elements' (Dhar and Giuliani 2010, p. 7). This traditional view has been challenged on several bases. Lange (2005) argues that the condition of truth alone does not help make this distinction since other regularities are true also. He proposes the following four criteria to aid in distinguishing laws of nature from other regularities: necessity, counterfactuals, explanatory power and inductive confirmations. Mahner and Bunge (1997) have argued that laws are said to be 'spatially and temporally boundless', where other laws may be 'bounded in space and time'. Cartwright's critique of 'the limited scope of applicability of physical laws' (Ruphy 2003) problematises the 'truth' aspect of laws. Giere (1999) on the other hand holds the view that what has come to be known as 'laws of nature' is in fact historical fossils, holdovers from conceptualisations first proposed in the Enlightenment. He proposes the consideration of models, which he argues are more reflective of how science is actually *practised.*

The significance of the debates about the nature of scientific laws becomes most relevant when discussing the role they play in supporting explanations in the specific sciences. Attempting a balanced description of scientific laws is a complex undertaking considering debates among philosophers about criteria invoked to distinguish between various types of laws such as strict versus *ceteris paribus* laws and empirical versus a priori laws. Such criteria include mathematical models, necessity, and explanatory and predictive potential. Some of these debates will be revisited in the context of the specific sciences later in this chapter.

In an insightful paper written more than five decades ago, Bunge (1961) classified lawlike statements from various philosophical standpoints into more than seven-dozen kinds. He concluded his detailed analysis with calling for less stringent philosophical restrictions regarding what could be classified as a law:

> There are as many classifications of law statements as viewpoints can be profitably adopted in their regard, and there seems to be no reason—save certain philosophical traditions—why most law statements should be regarded as nonlaw statements merely because they fail to comply with either certainty, or strict universality, or causality, or simplicity, or any other requisite found necessary in the past, where science seemed to concern itself exclusively…. That lawlike (*a posteriori* and general in some respect) statements be required corroboration and systematicity in order to be ranked as law statements, seems to fit contemporary usage in the sciences. (Bunge 1961, p. 281)

Bunge's pragmatic view regarding what constitutes scientific laws is a profound one. Continued debates about what counts as a scientific law, argued with core propositions of particular science disciplines, seem to be fundamentally grounded in normative or pragmatic standpoints and from this perspective cannot be said to have been fully resolved. Perhaps the most valuable context for such debates has been relative to the role of laws in generating or supporting scientific explanations (Press 2009).

Explanation is often hailed as one of the main goals of the scientific enterprise. Nagel (1961) articulates the central role of explanations in science when he states: 'It is the desire for explanations which are at once systematic and controllable by factual evidence that generates science; and it is the classification and organisation of knowledge on the basis of explanatory principles that is the distinctive goal of the sciences' (p. 4). While this stance towards explanation may seem obvious, it does not represent a united or a longstanding view. Logical positivists for instance, led by Ernest Mach, held that the aim of science is not to explain but rather to describe and predict phenomena (de Regt 2011). Discussions of the components that distinguish scientific explanation from other forms of explanation have spurred significant philosophical debate and led to a variety of accounts that have expanded understanding of their diversity.

The following discussion focuses on describing three main families or models of explanation. These are nomological explanation, causal explanation and functional explanation. According to de Regt (2011), these models are not mutually exclusive but can be used to explain the same phenomenon or explain phenomena in different disciplines.

The deductive-nomological (D-N) or covering law model of explanation proposed by Hempel and Oppenheim (1948) frames explanation as a logical argument in which the conclusion, or *explanandum,* follows from a set of premises, or *explanans.* The premises that constitute the *explanans* have to include at least one general law and other relevant preconditions. The general law in Hempel's account is a key component of the explanation process and has to be a 'true universal generalisation', allowing for the explanation and/or prediction of various events. One of the unresolved issues in the D-N model of explanation, pointed out by de Regt (2011), is that science is usually concerned with the explanation of laws, which necessitates the use of other more general laws. This proves to be problematic without 'giving an adequate criterion for the generality of laws' (p. 159). One derivative of nomological explanation is the inductive-statistical explanation (I-S), in which the law used in the *explanans* contains high probability that subsequently gives rise to an inductive (as opposed to deductive) support to the *explanandum.*

The Causal Mechanical (CM) model of explanation moves away from the conception of explanation as an argument (Salmon 1984). In generating this model, Salmon abandoned the attempt to characterise explanation or causal relationships in purely statistical terms. The CM model employs several central ideas. A causal process is a physical process, like the movement of a ball through space, that is characterised by the ability to transmit a mark in a continuous way. A mark is some local modification to the structure of a process. A process is capable of transmitting a mark if, once the mark is introduced at one spatiotemporal location, it will persist to other spatiotemporal locations even in the absence of any further interaction.

Causal processes contrast with pseudo-processes that lack the ability to transmit marks. An example is the shadow of a moving physical object. The other major element in Salmon's model is the notion of a *causal interaction.* A casual interaction involves a spatiotemporal intersection between two causal processes which modifies the structure of both—each process comes to have features it would not have had in the absence of the interaction. A collision between two cars that dents both is a paradigmatic causal interaction. According to the CM model, an explanation of some event E will trace the causal processes and interactions leading up to E (Salmon calls this the *etiological* aspect of the explanation), or at least some portion of these, as well as describing the processes and interactions that make up the event itself (the *constitutive* aspect of explanation). In this way, the explanation shows how E 'fit[s] into a causal nexus' (Salmon 1984, p. 9).

Functional explanations typically account 'for the role or presence of a component item by citing its function in the system' (de Regt 2011, p. 164). This type of explanation is commonly employed in the life and social sciences because these domains typically deal with 'complex organized systems, the components of which contribute to the working of the system (organisms, human minds, societies and so forth)' (de Regt 2011, p. 164). Achinstein (1983) presents three categories of functional explanations: the good-consequence doctrine in which the function confers some good on something or someone; the goal doctrine in which the function contributes to a goal that something, its designer or its user has; and the explanation doctrine in which the function includes causes or reasons or consequences. These categories probably make it easier to differentiate between functional explanations with teleological goal-oriented tendencies (the second doctrine) and other functional explanations. Achinstein further distinguishes between three types of functions: design functions, use functions, and service functions, allowing for a more nuanced and contextual differentiation between different functional explanations.

Philosophers of science have discussed a plethora of explanation models.[2] Additional contributions have came from philosophers of biology (e.g. Rosenberg and McShea 2008; Schaffner 1993; Sober 2008) and philosophers of chemistry (e.g. Goodwin 2008; Scerri and McIntyre 1997; van Brakel 2000) presenting and defending explanatory models that communicate the uniqueness of their disciplines. The following section describes the characteristics of laws and explanations in biology and chemistry focusing on aspects that have direct implications for science education.

37.3 The Nature of Laws and Explanations in Biology and Chemistry

37.3.1 Laws in Biology

There has been considerable discussion among philosophers regarding the appropriateness or meaningfulness of the concept of law in biology. Mayr takes the stance that 'laws play a rather small role in theory construction in biology'. He attributes this 'to the greater role played in biological systems by chance and randomness. Other reasons for the small role of laws in biology are the uniqueness of a high percentage of phenomena in living systems as well as the historical nature of events' (Mayr 2004, p. 28). The fact that biological systems are governed by 'dual causation' imposed by natural laws and by 'genetic programmes' makes the

---

[2] For example, see Giere (1988), Harré (1988), Hesse (1970), Pitt (1988), Salmon (1987), and Scriven (1970).

theories that explain them distinct from those pertaining to physical systems (Mayr 2004, p. 30). For Mayr, the matter is not one of nomenclature but one of substance. He concedes that even though some of the important concepts in biology 'can be phrased as laws, they are something entirely different from the Newtonian natural laws' (Mayr 2004, p. 30).

Garvey (2007) takes a position similar to Mayr's when he states that 'Biology does not have strict mathematical laws of its own. There are, as in any science, generalisations. But these generalisations have a habit of proving to be: (i) not distinctive to biology; (ii) not strict, exceptionless, mathematical laws; or (iii) not laws at all. Put in more positive terms, the generalisations found in biology are: (i) laws that belong to other sciences, (ii) *ceteris paribus* laws[3]; or (iii) true by definition' (p. 157–158). Others such as Uzman (2006) maintain that some biological observations tend to be presented as theories at a time when they should be considered laws of nature. He identifies four laws:

First law: "All phenomena of life are consistent with the laws of chemistry and physics."
Second law: "The cell is the fundamental unit of life."
Third law: "Life is continuous across generations."
Fourth law: "Life evolves – populations of organisms change genetically and irreversibly through time."

Örstan (2007) attributes to E. O. Wilson the claim that biology has 2 main fundamental laws: '1. All of the phenomena of biology are ultimately obedient to the laws of physics and chemistry, and 2. All of the phenomena of biology have arisen by evolution thru[sic] natural selection'. It can be argued, using Garvey's criteria, that even these 'laws' constitute generalisations that are nonmathematical and/or true by definition.

Reasons advanced in support or opposition to the concept of laws in biology can be found in various subdisciplines (evolutionary biology, systems biology, molecular biology, ecology). While the complexity of biological systems is widely acknowledged, constant efforts are being undertaken to establish fundamental biological organising principles that exhibit lawlikeness. For example, Dhar and Guiliani (2010) present an approach for uncovering fundamental organising principles in systems biology. Dodds (2009) has identified 36 laws in ecology to minimise the perceived complexity in interpreting ecological systems. McShea and Brandon (2010) recently proposed a detailed account for a 'Zero Force' evolutionary law that they believe to be to biology what Newton's first law is to physics. These efforts demonstrate that the complexity and contingencies inherent in biological systems have not discouraged biologists and philosophers of biology from trying different approaches to generating 'fundamental organising principles'.

To attain the ideal status of a universal law, Dhar and Guiliani (2010) believe that what biologists need to do but is difficult to attain is to construct generalisations

---

[3] The Latin *ceteris paribus* stands for 'all things being equal': *ceteris paribus* laws are laws that have exceptions, often contrasted with strict or 'real' laws (Garvey 2007).

that connect all levels from atoms to ecosystems. From their perspective, Mendel's Laws provide a reasonable framework at the phenotypic level, but an equivalent framework is absent at the cell-cell level. They believe, however, that just because this framework is currently absent does not mean it is not attainable in principle. Thus, there is optimism about the possibility of identifying powerful generalisations at the different levels of organisation (Dhar and Giuliani 2010) or developing empirical biological laws (Elgin 2006). In Elgin's (2006) view, a 'distinctively biological law' is one in which 'two conditions must be met: (1) all non-biological concepts in it must be mathematical, (2) It must contain at least some biological concepts and these concepts must be essential to its truth' (Elgin 2006, p. 130).

Other philosophers of biology take a pragmatic approach that 'replaces a definitional norm with an account of the *use* of scientific laws' (Mitchell 2000, p. 259). From Mitchell's perspective, 'the requirements for lawfulness fail to reflect the reality of scientific practice. As a consequence, the traditional understanding of laws is incomplete and fails to account for how humans have knowledge of the complexity of the world' (Mitchell 2009, p. 53). Mitchell characterises the reasons used to deny the existence of laws in biology as rooted in a normative orientation that regard laws along the Popperian tradition: 'bold, universal, exceptionless' (Mitchell 1997, p. S473). Arguing against the privileging of a very special type of generalisation that meets very stringent conditions and occurs very rarely even in physics, she affirms that 'the contingency of generalizations in biology or other sciences does not preclude their functioning as 'laws' – generalizations that ground and inform expectations in a variety of contexts' (Mitchell 1997, p. S478).

Mendel's Laws of segregation and independent assortment are popular topics for debating the nature of biological laws. Briefly stated,

> Mendel's first law, the Law of Segregation, states that while an organism may contain a pair of contrasting alleles, e.g. Tt, these will segregate (separate) during the formation of gametes, so that only one will be present in a single gamete, i.e. T or t (but not both or neither). Mendel's second law, the Law of Independent Assortment, states that the segregation of alleles for one character is completely random with respect to the segregation of alleles for other characters. (Dictionary of Botany 2003)

However, neither their highly contingent nature (Mitchell 2009) nor the historical ambiguity surrounding their ascendance from principles to laws (see Footnote 2 and Marks 2008) is adequate for demoting them to accidental generalisations. Rather than deny the existence of biological laws, it is more useful in the context of the variation inherent in biological systems to provide "a better understanding of contingency so that we can state the many ways in which laws are not always 'universal and exceptionless" (Mitchell 2009, p. 63).

37.3.1.1 The Case of Mendel's Laws in Biology Textbooks

Mendel's Laws of segregation and independent assortment are often described in biology textbooks to various levels of detail with some reference to Mendel's profile and pea-plant experiments. Inheritance is a classic topic in middle and high school biology curriculum materials. A typical chapter in one high school biology book (BSCS 2003) begins with getting students to consider similarities of features between members of different generations in families. Starting with discussions about familiar experiences, the chapter invites students to read the tragic case of haemophilia in the family of the last Czars then leads them to work on different scenarios in order to predict inheritance patterns. Next, students are engaged in simulations, using beans, to help them understand the inheritance of one and two traits. The chapter explores inherited patterns, defines gametes, describes meiosis, tracks genes and chromosomes through meiosis by guiding students to construct a physical representation that tracks the genotype of the newly divided 'cells' and addresses the role of sample size in leading to more accurate predictions. Next the book introduces Gregor Mendel and a video segment that provides data for students to use to make predictions then compare their predictions with actual results provided by the teacher. Additional exercises pertaining to linked and sex-linked traits are offered to deepen and elaborate the concepts before the chapter ends with a discussion of the genetic basis for human variation.

Of particular interest to this paper is the text book's reference to Mendel's second law. The following excerpt represents one of two occasions in which it is mentioned in one chapter: 'When you follow the inheritance of 2 traits (a di-hybrid cross), more complex patterns result. In garden peas, the genes for the traits of pod color and pod shape are on different chromosomes. As a result, their inheritance conforms to Mendel's Law of independent assortment' (BSCS 2003, p. 438). The only other significant reference to this law appears later in the book in an essay at the end of the unit. The essay discusses the concepts of phenotype and genotype and concludes with an example that demonstrates the random inheritance of traits demonstrating how a baby rabbit may inherit different genes for particular traits from the father and mother like fur colour and eye colour independently of each other. The essay concludes with the following historical narrative:

> The principle of independent assortment was discovered more than 150 years ago in a small European monastery garden. A scholarly monk named Gregor Mendel used pea plants to study patterns of inheritance. Mendel experimented with many generations of pea plants. His insights later became the cornerstone for explaining basic patterns of inheritance. (BSCS 2003, p. 499)

As seen in these excerpts, the same concept is referred to as a law in one section of the book and as principle in another. It is not clear whether the shift in language is accidental or whether it refers to the historical evolution of this idea from 'principle' at the time of its inception to 'law' in later references to this idea.[4] While inconsistency in how textbook authors use categorisations of scientific knowledge has been already

---

[4] In his review of early textbooks, Marks (2008) notes that, initially, Mendel's Law was often presented in the singular in contrast to Galton's Law of Ancestral Heredity as evident in Punnett's 1905 textbook and most other genetics textbooks of the first generation. In his 1909 book, Bateson contrasted Galton's Law against the Mendelian ''scheme', 'principles', 'phenomena', 'methods', 'analysis', 'facts''. (Marks 2008, p. 250). First references to Mendel's Law of Segregation and Law of Independent Assortment appeared in Morgan's second book in 1916 and were further detailed in his 1919 book *The Physical Basis of Heredity*.

documented (e.g. McComas 2003), noting this inconsistency in this paper underscores what appears to be confusion or lack of clarity about the purpose that this law/principle serves, as demonstrated in both excerpts. The textbook provides no further details about what the 'law' or 'principle' of independent assortment entails or the role it plays in explaining phenomena. It is stated casually as a claim that explains some observations, no different than another claim in terms of its generalisability (or lack thereof) or ability to explain or predict. It is not made explicitly clear that this generalisation can be used to explain or predict phenotypes and genotypes of new generations of siblings that go beyond the specifics of the examples discussed in the chapter.

The main issues in the examples quoted earlier are (1) the striking lack of clarification about what a law entails, (2) the unexplained switch from 'law' in the chapter to 'principle' in the historical anecdote and (3) lack of explicit reference to the relative strong explanatory power (Woodward 2001) expressed that is qualitatively and quantitatively different from other concepts presented in the textbook.

In another high school biology book (SEPUP 2011), the discussion of Mendel's work is devoid of references to laws or principles. Some 60 pages after describing the historical work of Mendel, and only in the context of discussing genes and chromosomes, there is a brief reference to one of Mendel's Laws:

> When the chromosomes line up before division, the paternal and maternal chromosomes in the pair line up randomly and separate independently of each other. This is called independent segregation of the chromosomes. Independent segregation of chromosomes explains the behavior of genes that follow Mendel's law of independent assortment. It also accounts for the fact that genes that are linked on the same chromosome don't follow the law of independent assortment.... (SEPUP 2011, p. 356)

This passage illustrates how Mendel's Law of independent assortment exerts an explanatory function. But it is the only place other than the 'Glossary' section at the end of the book where this law is identified. The apparent retreat from explicit emphasis on Mendel's Laws in both textbooks either reflects a general trend of accidental nature or an outcome of authors' awareness of the philosophical controversy about laws in biology. In both cases, avoidance of explication of the significance of laws or principles in biology in the context of a specific content, such as the one explored here, reduces the likelihood that students will understand the usefulness of this aspect of scientific knowledge relative to its explanatory function. One way to rectify this matter is by using language consistently in textbooks, clarifying what the referents mean and educating readers about the distinctive nature of laws in biology, noting their relevance to explaining observations and predicting new ones. Alternatively, teachers can problematise terms like 'laws' and 'principles' and guide the students into a discussion that addresses their meaning and significance.

37.3.2   Laws in Chemistry

Until fairly recently, the status of laws in chemistry has received little attention within philosophy of science (e.g. Cartwright 1983). With the upsurge of philosophy of chemistry in the 1990s (Erduran & Mugaloglu, this handbook), there has been more focus on what might (or not) make laws distinctly chemical in nature. Some philosophers of chemistry (e.g. Christie and Christie 2000) as well as chemical educators (e.g. Erduran 2007) have argued that there are particular aspects of laws in chemistry that differentiate them from laws in other branches of science with implications for teaching and learning in the science classroom. A topic of particular centrality and relevance for chemical education is the notion of 'Periodic Law' which is typically uncharacterised as such:

> Too often, at least in the English speaking countries, Mendeleev's work is presented in terms of the Periodic Table, and little or no mention is made of the periodic law. This leads too easily to the view (a false view, we would submit), that the Periodic Table is a sort of taxonomic scheme: a scheme that was very useful for nineteenth century chemists, but had no theoretical grounding until quantum mechanics, and notions of electronic structure came along. (Christie and Christie 2003, p. 170)

A 'law' is typically defined as 'a regularity that holds throughout the universe at all places and at all times' (Salmon et al. 1992). Some laws in chemistry like the Avogadro's Law (i.e. equal volumes of gases under identical temperature and pressure conditions will contain equal numbers of particles) are quantitative in nature while others are not. For example, laws of stoichiometry are quantitative in nature and count as laws in a strong sense. Others rely more on approximations and are difficult to specify in an algebraic fashion. As a key contributor to philosophy of chemistry, Eric Scerri (2000a) takes the position that some laws of chemistry are fundamentally different from laws in physics (Scerri 2000a). While the emphasis in physics is on mathematisation, some chemistry laws take on an approximate nature:

> The periodic law of the elements, for example, differs from typical laws in physics in that the recurrence of elements after certain intervals is only approximate. In addition, the repeat period varies as one progresses through the periodic system. These features do not render the periodic law any less lawlike, but they do suggest that the nature of laws may differ from one area of science to another. (Scerri 2000a, p. 523)

Viewed from the perspective of physics, the status of the periodic system may appear to be far from lawlike (Scerri and McIntyre 1997). Significantly, the periodic law seems not to be exact in the same sense as are laws of physics, for instance, Newton's laws of motion. Loosely expressed, the Periodic Law states that there exists a periodicity in the properties of the elements governed by certain intervals within their sequence arranged according to their atomic numbers. The crucial feature which distinguishes this form of 'law' from those found in physics is that chemical periodicity is approximate. For example, the elements sodium and potassium represent a repetition of the element lithium, which lies at the head of group I of the periodic table, but these three elements are not identical. Indeed, a vast amount of chemical knowledge is gathered by studying patterns of variation that occur within vertical columns or groups in the periodic table. Predictions which are made from the so-called periodic law do not follow deductively from a theory in the same way in which idealised predictions flow almost inevitably from physical laws, together with the assumption of certain initial conditions.

Scerri further contrasts the nature of laws in physics such as Newton's Laws of Gravitation. Even though both the Periodic Law and Newton's Laws of Gravitation have had success in terms of their predictive power, the Periodic Law is not axiomatised in mathematical terms in the way that Newton's Laws are. Part of the difference has to do with what concerns chemists versus physicists. Chemists are interested in documenting some of the trends in the chemical properties of elements in the periodic system that cannot be predicted even from accounts that are available through contributions of quantum mechanics to chemistry. Christie and Christie (2000), on the other hand, argue that the laws of chemistry are fundamentally different from the laws of physics because they describe fundamentally different kinds of physical systems. For instance, Newton's Laws described above are strict statements about the world, which are universally true. However, the Periodic Law consists of many exceptions in terms of the regularities demonstrated in the properties and behaviours of elements. Yet, for the chemist there is a certain idealisation about how, for the most part, elements will behave under particular conditions. In contrast to Scerri (2000a) and Christie and Christie (2000), Vihalemm (2003) argues that all laws need to be treated homogeneously because all laws are idealisations regardless of whether or not they can be axiomatised. van Brakel further questions the assumptions about the criteria for establishing 'laws':

> If one applies "strict" criteria, there are no chemical laws. That much is obvious. The standard assumption has been that there are strict laws in physics, but that assumption is possibly mistaken . . . Perhaps chemistry may yet provide a more realistic illustration of an empirical science than physics has hitherto done. (van Brakel 1999, p. 141)

Christie and Christie (2000) indicate that taking physics as a paradigmatic science, philosophers have established a set of criteria for a 'law statement', which 'had to be a proposition that (1) was universally quantified, (2) was true, (3) was contingent, and (4) contained only non-local empirical predicates' (p. 35). These authors further argue that such a physics-based account is too narrow and applies only to simple systems. More complex empirical sciences do not necessarily conform to such accounts of laws:

> The peculiar character of chemical laws and theories is not specific to chemistry. Interesting parallels may be found with laws and theories in other branches of science that deal with complex systems and that stand in similar relations to physics as does chemistry. Materials science, geophysics, and meteorology are examples of such fields. (Christie and Christie 2000, p. 36)

The debates around the nature of laws in chemistry are ongoing and it is beyond the scope of this paper to contribute to this debate. However, it is important to problematise the complexity in the ways that philosophers of chemistry dispute the nature of chemical knowledge at large and the nature of laws in particular.

In summary, the suggestion offered by Christie (1994) is considered useful:

> Ultimately the best policy is to define 'laws of nature' in such a way as to include most or all of the very diverse dicta that scientists have chosen to regard as laws of their various branches of science. If this is done, we will find that there is not a particular character that one can associate with a law of nature. (Christie 1994, p. 629)

37.3.2.1 The Case of Periodic Law in Chemistry Textbooks

This section describes a case study of how a typical textbook covers the Periodic Table and how the discussion on the nature of the Periodic Law from a philosophical perspective could inform textbook revision. The purpose of this example is to illustrate how the philosophical dimensions of chemistry can be better captured in textbooks so as to ensure understanding of the epistemological aspects of chemistry. The coverage of the Periodic Table in chemistry textbooks has been highlighted to be problematic from a range of perspectives. For instance, Brito and colleagues (2005) argue that the important distinction of accommodation versus prediction in the context of Periodic Table is not covered in chemistry textbooks. In *A Natural Approach to Chemistry*, a textbook that is in current use in secondary schools in the USA, Hsu and associates (2010) dedicate a whole 31-page chapter to 'Elements and the Periodic Table'. The chapter begins with a section on the origin of elements in the universe. There are numerous occasions where the discussion on elements is linked to everyday contexts including the nature of metals on the hull of a boat, the human body and nutrition. A significant portion of the chapter is dedicated to the discussion on electronegativity, ionisation energy, the groups and series in the Periodic Table and an explanation of why compounds form using the Lewis dot notation. The chapter concludes with a set of open-ended and multiple-choice questions.

The coverage of the Periodic Table does mention the notion of patterns but not laws. In the section describing the development of the Periodic Table and the contributions of Dimitri Mendeleev, the authors state that he *'was trying to figure out if there was any kind of organisation to the elements, some kind of pattern he could use to help organize them in a logical way'* (p. 171). This reference is in contrast to the earlier discussion about the approximate nature of the Periodic Law. Indeed, predictions which are made from the so-called Periodic Law do not follow deductively from a theory in the same way in which idealised predictions flow almost inevitably from physical laws. In this respect, the reference to a 'logical way' in the textbook can be misleading in communicating the approximate nature of periodicity. The characterisation of the word 'periodic' is equally devoid of any specification of the approximate nature of the pattern: *'The Periodic Table is named for this because the rows are organised by repeated patterns found in both the atomic structure and the properties of the elements'* (p. 171). The explanation of the Periodic Table in terms of the atomic theory further stresses the logical ordering that the authors are emphasising throughout the text. In the discussion on the Modern Periodic Table, the authors state the following:

> At the time of Mendeleev, nothing was known about the internal structure of atoms. Protons were not yet discovered so the more logical ordering by atomic number was not possible. Today's table includes many more elements, and is ordered not by atomic mass, but by atomic number. However two things are still true of the periodic table, each column represents a group of elements with similar chemical properties, and each row (or period) marks the beginning of some repeated pattern of physical and chemical properties. While elements can be broadly categorized into metals, non-metals, and metalloids, an understanding that

37 Laws and Explanations in Biology and Chemistry...

each column has similar chemical properties had lead to names for some of these element groups. (Hsu et al. 2010, p. 175)

What follows is the quantum mechanical models and the use of orbitals in explaining the organisation of the groups of elements. Considerable discussion is dedicated to establishing the role of valency in explaining periodicity including the introduction of the Lewis dot diagrams. The coverage of this textbook in terms of the viewing of the Periodic Table as a taxonomical tool and a scheme without any explicit emphasis on the character of periodicity as a lawlike feature of chemistry is consistent with observations of Christie and Christie (2003) mentioned earlier.

In his critique of Atkins' chemistry textbook coverage of quantum mechanical explanations, Scerri (1999) highlights a tendency among chemistry textbook writers to ignore the irregularities in the patterns in the Periodic Table:

> One is tempted to protest that in fact the proffered explanation does indeed require a new principle, namely the strange notion whereby the d- and f-subshells do not need to be complete for the shell itself to be classified as complete./.../Surely it would not have detracted from the triumph of science to admit at this point, or anywhere in the book, that the assignment of electrons to particular orbitals is an approximation. In fact, Atkins could have made his story of the Periodic Kingdom all the more interesting if he had stated that even though his discussion was based around an approximate concept, we are still able to use it to remarkable effect to explain so many macroscopic and microscopic features connected with trends in the periodic table. (Scerri 1999, p. 302)

When the textbooks do cover the peculiarities, they are left undiscussed, as exemplified in the textbook mentioned earlier. Consider, for instance, the following excerpt:

> The transition metals illustrate a peculiar fact: the 3d orbitals have higher energy than the 4 s orbitals!/../Energy is the real, physical quantity that determines how the electrons act in atoms. The real energy levels correspond to the rows of the periodic table. The quantum number is an important mathematical construction, but is not the same as the energy level. (Hsu et al. 2010, p. 179)

Here there is a missed opportunity to raise some philosophical insights into the role of empirical evidence in model building in contrast to the mathematical and theoretical grounds for quantum mechanical models in chemistry. Scerri highlights this issue by inviting textbook writers to consider the grounds on which orbital models are related to periodicity:

> In addition, the failure to provide an adequate explanation of the 4 s/3d question or a deductive explanation of the precise places where the elements appear to 'recur' should give us and Atkins grounds for suspecting that this model is not even all that empirically adequate. (Scerri 1999, p. 303)

So what would a revised chemistry textbook look like in light of this discussion so far? There are at least two issues that this coverage of the nature of the Periodic Law raises for consideration in textbook writing. First, the textbooks should elicit the approximate nature of the Periodic Law and specify the reference to the patterns in periodicity as an instance of law while highlighting the difference of interpretations of law in different branches of science. Second, the juxtaposition of the empirical versus theoretical dimensions of the orbital models should be teased out to clarify the different epistemological status of the Periodic Table in light of its historical and empirical foundation versus the incorporation of theoretical and mathematical characterisations since the advance of quantum mechanical models. Erduran (2007) has proposed that an argumentation framework could offer a useful pedagogical strategy for eliciting different characterisations of laws and suggested a potential activity could be structured as follows:

Claim 1: The Periodic Law and the law of gravitation are similar in nature. The term 'law' can be used with the same meaning for both of them.
Claim 2: The Periodic Law and the law of gravitation are different in nature. The term 'law' cannot be used with the same meaning for both of them.

These claims could be presented with evidence that would support either claim, both or neither. For example, the statement 'a law is a generalisation' could support both claims while 'the periodic law cannot be expressed as an algebraic formula while the law of gravitation can be' could support the second claim. The task for the students would be to argue for either claim and justify their reasoning. Further statements can be developed that would act as evidence for either, both or neither claim.

The inclusion of a framework that simulates the philosophical debate on the nature of laws in a comparative context between physics and chemistry will carry into the classroom the ways in which philosophers have conceptualised the nature of this particular aspect of scientific knowledge in these domains. Without a sense of a debate, textbooks reinforce the 'received view' of science that projects a perception of a consensus when there is none. In summary, the inclusion of meta-perspectives offered by philosophical accounts of laws can provide insights into textbook accounts of laws whereby the particular nuances of chemical knowledge are better framed in terms of consistency with epistemological accounts on chemical knowledge.

37.4 The Nature of Explanation in Biology and Chemistry

37.4.1 Explanation in Biology

Explanation in biology differs from explanation in physics in that it does not aim to provide the typical 'necessary and sufficient conditions'. Instead biological explanations aim to 'gain partial, but ever increasing insights into the causal workings of various *life processes*' (Brigandt 2011, p. 262). Mayr's (1961) distinction between proximate explanation and ultimate explanation provides a basic dichotomy between at least two ways of explaining biological systems. In asking about how a phenomenon happens, the proximate explanation would address physiological or other processes that underlie the cause, while the ultimate explanation would address the phenomenon based on the organism's evolutionary history. The explanations do not contradict but rather complement each other by adding a different dimension: one causal and another historical.

Further expansion of explanatory breadth can be found in Tinbergen's (1952, cited in Sterelny and Griffiths 1999) four explanatory projects in biology, according to which it is possible to address questions about any behaviour by proposing 4 different explanations: proximal, developmental, adaptive and evolutionary:

> Tinbergen distinguished four questions we could have in mind in asking why a bittern stands still with its bill pointed directly at the sky. (1) We could be asking for a *proximal* explanation: an explanation of the hormonal and neural mechanisms involved in triggering and controlling this behavior. (2) We could be asking for a *developmental* explanation: an explanation of how this behavior pattern emerges in a young bittern. (3) We could be asking for an *adaptive* explanation: an account, that is, of the role this behavior currently plays in the bittern's life. (4) Finally, we could be asking for an explanation of how and why this behavior evolved in this bittern and in its ancestors. (Sterelny and Griffiths 1999, p. 50)

Press (2009) suggests that one of the ways in which philosophers contrast physics and biology is by appealing to differences between their respective explanations as they relate to the covering law model. He describes divergent views among philosophers of biology, as represented by Sober, Kitcher and Rosenberg regarding the applicability of the covering-law model in the context of biological explanations. After analysing the different positions, Press (2009) concludes that there is a good fit between Hempel's covering law model and biological explanations, stating that 'the differences between biological and physical explanations are merely a matter of degree. …. biologists, who deal with extremely complex systems, will need to rely relatively heavily on various sorts of approximation if they are to explain anything at all' (Press 2009, p. 374).

Branching off of these distinctions, philosophers of biology have detailed a number of explanatory types that support the aim of gaining insights about the 'causal workings' of biological systems without limiting their discussion to causal explanations. Wouters (1995), for example, outlines five different types of explanation: Physiological, Capacity, Developmental, Viability and Historical/Evolutionary. These different types of explanation approach the same phenomena from different perspectives. To explain the circulatory system of a given organism, for example, Wouters argues that physiological explanations focus on the types of events in the individual organism's life history, whereas a capacity explanation focuses on underlying causal explanations having to do with the structure of the heart and valves. A developmental explanation would focus on the development of the system (heart and vessels), while a viability explanation would focus on why structural differences between systems occur in different organisms. Finally, an evolutionary explanation would focus on differences in systems between organisms in the same lineage.

More recently, Wouters (2007) has proposed a sixth type, design explanation, in which a system in a real organism might be compared to a hypothetical one. Calcott (2009) makes the case for an additional type of explanation that he names lineage explanation. This type of explanation aims to make plausible a series of incremental changes that lead to evolutionary change, focusing on a sequence of mechanisms that lead to the successive changes. Lineage explanations 'show how small changes between ancestral and derived mechanisms could have produced different behavior, physiology and morphology' (Calcott 2009, p. 74). Consequently, they provide an additional 'explanatory pattern' to account for evolutionary change.

Rose (2004) offers a fable that supports the discussion of how biological systems can sustain a variety of explanations. In this fable, five biologists are having a picnic when they noticed a frog jump into a nearby pond. Posing the question of what caused it to do so led to five different answers. The physiologist reasoned that impulses travelled from its retina to the brain and then to the leg muscles. The biochemist pointed out the properties of the proteins, actin and myosin, whose fibrous nature enable them to move in a predictable way. The developmental biologist attributed it to the ontogenetic processes that occurred during early stages of cell division. The animal behaviourist attributed the cause to the snake that was lurking by, whereas the evolutionist discussed the role of natural selection in favouring those frogs that escaped their prey due to their ability to detect them quickly and move fast in response, allowing them to survive and reproduce.

Of course, the question of legitimacy of teleological explanation in biology is important because of historical and pedagogical reasons. This is because attributing purpose to non-purposeful things or events or attributing human qualities to nonhumans can lead to questioning the credibility of the proposed explanation. The human tendency to assign purpose to everything seems to be nourished by the 'sheer efficiency of biological structures [which] reinforces the illusion of purpose' (Hanke 2004, p. 145). Use by some biologists in a metaphorical sense makes these explanations likely to be misunderstood by non-experts, especially in educational settings. Some philosophers of biology differ in their degree of opposition to the use of these explanations—perhaps because they are well aware of their semantic affordances and limitations. Few philosophers strongly object to their use as expressed in Hanke's (2004) viewpoint that teleology is 'bad not so much because it's lazy and wrong (which it is) but because it is a straightjacket for the mind, restricting truly creative scientific thinking' (p. 155).

The philosophical debates around these ideas have implications for educational settings but empirical findings can assist in making informed judgments regarding their use in educational contexts. Some science educators have cautioned against the use of anthropomorphic and teleological explanations in biology teaching out of concern for engendering misconceptions that can interfere with learning (Jungwirth 1979). However, a recent study called for the 'removal of the taboo' regarding teleological and anthropomorphic explanations, arguing that results of an empirical cognitive study has shown that high school students' use of anthropomorphic or teleological explanations is not indicative of teleological reasoning but seems to serve a heuristic value for learning as gleaned from the students' perspective (Zohar and Ginossar 1998).

The range of explanations described by Wouters (1995, 2007), Calcott (2009) and Rose (2004) illustrates the significance of invoking a diverse set of explanations for providing more comprehensive understanding of biological systems. Perhaps one of the overarching attributes of biological explanations is the notion of consilience in which different explanations need not be subsumed under one another and need not contradict with one another. The notion of consilience attributed to Wilson by Rose (2004) can perhaps be viewed as a pragmatic adaptation of the notion of 'consilience of inductions' developed by Whewell in his *Novum Organon Renovatum* (Morrison 2000). The diversity of explanatory types in biology is perhaps reflective of the 'epistemological pluralism' (Rose 2004, p. 129) that is characteristic of the study of biological systems. This diversity is often obscured in biology education, not because it is difficult to communicate, but mostly due to the way the school biology curriculum is chopped up and structured in ways that limit reference at a given point in time to one or two explanatory emphases. This in turn limits teachers' and students' ability to experience the value of epistemological pluralism as a powerful vehicle for explaining and understanding phenomena in the life sciences.

37.4.1.1 Explanation in Biology Textbooks

Topics typically covered in high school biology textbooks in the United States include evolution, genetics, cell biology and ecology. The approach to explicating these topics and the order in which they are presented varies significantly from one publisher to another.[5] For example, in the SEPUP (2011) book, the ecology unit, typically presented in other books as the last unit, is second to a first unit on sustainability, a topic rarely addressed so explicitly in biology textbooks. However, explanations in the three textbooks consulted (see footnote) are similar to one another in that they are not differentiated from the rest of the text, but are blended in the narrative, becoming rather 'invisible'. The explanations follow the topical narrative, but there is no discernible attempt to provide a broader synthesis, weave cross-topical themes, or illustrate the notion of explanatory consilience.

37.4.2 Explanation in Chemistry

In the *Stanford Encyclopedia of Philosophy*, Weisberg and colleagues (2011) review the recent developments in the formulation of chemical explanations. These authors state that from the nineteenth century onwards, chemistry was commonly taught and studied with physical models of molecular structure. Beginning in the twentieth century, mathematical models based on classical and quantum mechanics were applied to chemical systems. The use of molecular models has helped chemists to understand the significance of molecular shape (Brock 2000) and aided visual representation of structure and function of matter. One of the key scientific achievements of the twentieth century, the discovery of the double helical structure of DNA,

---

[5] The three textbooks reviewed in this section are BSCS (2003), Campbell et al. (2009), and SEPUP (2011).

was possible because of the use of physical models as explanatory tools (Watson 1968). The focus of chemical explanations entered a new phase with the advent of quantum mechanical theories and their applications in chemistry. The notion of 'explanation' in chemistry has centred in key debates involving not only models but also philosophical themes such as supervenience and reduction (Earley 2003) which will be referenced briefly later in the paper.

According to Weisberg and colleagues (2011), while exact solutions to the quantum mechanical descriptions of chemical phenomena have not been achieved, advances in theoretical physics, applied mathematics and computation have made it possible to calculate the chemical properties of many molecules very accurately and with few idealisations. This perspective is in contrast to those chemists who argue for employing simple, more highly idealised models in chemistry, which stem from the explanatory traditions of chemistry. In developing this point, Hoffmann illustrates two modes of explanation that can be directed at chemical systems: horizontal and vertical (Hoffman 1998). Vertical explanations are what philosophers of science call 'deductive-nomological explanations'. These explain a chemical phenomenon by deriving its occurrence from quantum mechanics. Calculations in quantum chemistry are often used to make predictions, but insofar as they are taken to explain chemical phenomena, they follow this pattern. By showing that a molecular structure is stable, the quantum chemist is reasoning that this structure was to be expected given the underlying physics.

In contrast to the vertical mode, the horizontal mode of explanation attempts to explain chemical phenomena with chemical concepts. For example, Weisberg and colleagues (2011) use the example of SN_2 reactions as an example of horizontal explanations. The first year organic chemistry curricula include the relative reaction rates of different substrates undergoing SN_2 reactions. They state that an organic chemist might ask 'Why does methyl bromide undergo the SN_2 reaction faster than methyl chloride?' One answer is that 'the leaving group Br^- is a weaker base than Cl^-, and all things being equal, weaker bases are better leaving groups'. This explains a chemical reaction by appealing to a chemical property, in this case, the weakness of bases. Vertical explanations demonstrate that chemical phenomena can be derived from quantum mechanics. They show that, given the (approximate) truth of quantum mechanics, the phenomenon observed had to have happened. Horizontal explanations are especially good for comparing and contrasting explanations, which allow the explanation of trends. In the above example, by appealing to the weakness of Br^- as a base, the chemist invokes a chemical property. This allows the chemist to explain methyl bromide's reactivity as compared to methyl chloride, and also methyl fluoride, methyl iodide and so on. Insofar as chemists want to explain trends, they make contrastive explanations using chemical concepts.

Apart from Hoffmann, earlier chemists argued that the nature of chemical explanations need not be overshadowed by quantum mechanical and reductive approaches. Consider, for instance, the perspective taken by Coulson:

> The role of quantum chemistry is to understand these concepts and show what are the essential features in chemical behavior. [Chemists] are anxious to be told … why the H–F

bond is so strong, when the F–F bond is so weak. They are content to let spectroscopists or physical chemists make the measurements; they expect from the quantum mechanician that he will explain why the difference exists. … So the explanation must not be that the computer shows that [the bonds are of different length], since this is not an explanation at all, but merely a confirmation of experiment. (Coulson 1960)

Although both Coulson (1960) and Hoffmann (1998) defend the use of simple, idealised models to generate horizontal explanations, it is not clear that quantum calculations can never generate contrastive explanations (Weisberg et al. 2011). Although single vertical explanations are not contrastive, a theorist can conduct multiple calculations and, in so doing, generate the information needed to make contrastive explanations. However, the status of quantum mechanical explanations in chemistry is likely to be challenged for some time yet to come given the history of chemists' position on this issue. For example, Brown (2003) has drawn from cognitive sciences to illustrate that chemical explanations are metaphorical in nature and have a character that is distinguishable from representations employed in other fields of science: '…data are not explanatory in themselves. For the chemist to make effective use of powerful computational resources there must still be an underlying metaphorical model of what is happening in the conventional (chemical) sense' (p. 216). Even though chemical explanations involve the use of models and modelling (Erduran 2001), the meaning of the term 'model' or its function is not so straightforward particularly in relation to its import in chemical education. The presence of models in different disciplines related to chemical education, such as cognitive psychology and philosophy of science, makes it even more difficult to come up with a single definition for the term 'model' (Erduran and Duschl 2004).

A particular approach to chemical explanations includes the reference to 'structural explanations' (Harré 2003). Goodwin (2008) explains that in organic chemistry, the phenomena are explained by using diagrams instead of mathematical equations and laws. In that respect, the field of organic chemistry poses a difference in terms of the content of explanations from those in other physical sciences. Goodwin investigates both the nature of diagrams employed in organic chemistry and how these diagrams are used in the explanations. The diagrams particularly mentioned are structural formulas and potential energy diagrams. Structural formulas are two-dimensional arrangements of a fixed alphabet of signs. This alphabet includes letters, dots and lines of various sorts. Letters are used as atomic symbols, dots are used as individual electrons, and lines are used as signs for chemical bonds.

Structural formulas in organic chemistry are mainly used as descriptive names for the chemical kinds. Thus, a structural formula has a descriptive content consisting of a specification of composition, connectivity and some aspects of three-dimensional arrangement. Structural formulas are also used as models in organic chemistry. For example, a ball and stick model is used in explanations. After characterising some features of structural formulas, Goodwin presents a framework of explanations in organic chemistry and describes how both structural formulas and potential energy diagrams contribute to these explanations. Then he gives the examples of 'strain' and 'hyperconjugation' to support his idea about the role of diagrams in capturing the nuances of explanations through structures in organic chemistry.

Debates on reduction—i.e. 'reduction of axioms or laws of one science to the axioms and laws of a deeper putative science' (Scerri 2000b, p. 407)—have taken chemical explanations to its core in understanding what makes an explanation 'chemical'. Similar debates on reductionism have centred on issues related to philosophy of mind, particularly in the context of Multiple Realisability (e.g. Fodor 1974). Educational applications of such debates have been promoted though not yet realised at the level of schooling (e.g. Erduran 2005). One aspect of this debate has concerned the notion of supervenience. Two macroscopic systems that have been constructed from identical microscopic components are assumed to show identical macroscopic properties, whereas the observation of identical macroscopic properties in any two systems need not necessarily imply identity at the microscopic level. Chemical explanations have often been regarded as including microscopic, macroscopic and symbolic dimensions (e.g. Jacob 2001). The main position promoted in this debate is that the asymmetry in the way that properties and kinds of chemical entities are conceptualised suggests that chemical explanations cannot necessarily be reduced to explanations of physics— a realm of epistemology—even if ontologically chemistry might be reliant on physical principles.

37.4.2.1 Explanation in Chemistry Textbooks

Kaya and Erduran (2013) believe that structural explanations as discussed by Goodwin (2008) have relevance for chemistry textbooks. In their study of secondary chemistry textbooks across grade levels, they noted, for example, that for the 9th grade textbooks, topics such as 'development of chemistry', 'compounds', 'chemical changes', 'mixtures' and 'chemistry in our life' all included structural explanations. Similar ways of coverage are noted in the textbook by Hsu and colleagues (2010). In the chapter on organic chemistry, for example, there are sections that illustrate and define 'structural isomers'. The Appendix depicts the textbook reference to structural isomers of 2-pentane and isopentane. The description is 'When a molecule has the same number and type of atoms, but a different bonding pattern, it is a structural isomer' (p. 541). The rest of the text is similar in terms of providing definitions for the characterisation of isomers. There are two types of representations that are both two-dimensional but one represents the C and H atom balls, while the other does not. In this sense, there is potential for confusion for what counts as 'structure' when different levels of representations are superimposed.

37.4.3 Summary

This paper focused on the context, the definitions and the types of laws and explanations in biology and chemistry and described some emphases and patterns that illustrate a number of similarities and differences between biological and chemical laws and explanations. For example, when the types of explanation in chemistry and biology are contrasted, the result is a diversity of types that are distinctive to the science in question. While biological explanations include viability and developmental explanations that draw closely from the nature of biological content, chemical explanations focus on the structural and representational explanations that are based on either quantum mechanical or simple chemical models. The context of debates around the nature of biological and chemical laws and explanations are also rather particular. Whereas reference to principles is common in biologists' discussions of laws, the chemists are preoccupied with questions regarding axiomatisation and approximation.

37.5 Implications for Biology and Chemistry Education

This section provides some suggestions for how biology and chemistry education can be informed by investigations into the nature of laws and explanations. It illustrates the implications of the preceding discussion for teaching, curriculum and learning in biology and chemistry education. Design of instructional activities can exemplify more explicitly the role that variation and chance play in biological systems and enable students to explore the contribution of this uniqueness to shaping the formulation of biology's 'laws' or principles. Awareness of the function of generalisations and principles in biology allows students to appreciate their role in the construction of biological knowledge and enables them to realise that the scarcity of 'laws' does not diminish the 'power' of the generalisations/principles they study or reduce the status of biology to a 'soft science'.

In terms of chemistry teaching, the goals of teaching could include the broader aims of promoting students' understanding of how some chemical laws like the Periodic Law are generated and how they differ from laws in chemistry or other sciences. Lesson activities could acknowledge the observation that, for instance, the Periodic Law will be manifest in the classroom via comparative discussions about the trends in the chemical and physical properties of elements. Furthermore, engaging students in the process of the derivation of some of these trends is likely to give them a sense of how laws are generated and refined in chemistry. How can such discussions of laws, then, be manifested in the classroom? Earlier work has identified strategies such as questioning and discussion in chemistry teaching (Erduran 2007) that can be extended to biology teaching due to their broad pedagogical scope. For example, students could be presented with alternative accounts of scientific laws—those derived deductively and those that are derived with approximation and induction in mind—and asked to question, compare, evaluate and discuss them in relation to other products of scientific knowledge.

This review also has implications for the design of curricula for the inclusion of biological and chemical content knowledge. With respect to biology, curriculum materials should attempt to communicate more explicitly elements of 'epistemological pluralism' and how biologists search for consilience among a proposed family of explanations. Including these ideas in the curriculum should not be limited to presenting isolated narratives about how biologists work but should be reflected in developing more integrated and coherent content frameworks. This is necessary for promoting a more holistic and contextual understanding of structures and processes in biotic systems. Even though the chemistry curriculum typically covers structural explanations as described by Goodwin (2008) across various levels of schooling, the meta-perspectives on the nuances of these explanations are not typically part of either curriculum materials or textbooks (Kaya and Erduran 2013).

The discussion about the power and limits of biological and chemical laws can be initiated in curriculum resources by focusing more explicitly on what Mendel's Laws or the Periodic Law do and fail to do. Curricula and textbooks tend to cover laws in quite an ambiguous and limited manner (i.e. McComas 2003) and often present laws in different science fields on equal footing. That is, when certain generalisations are labelled as laws, textbook authors do not contextualise or explore what that label means. From the point of view of a teacher or student, a law in physics (e.g. Newton's) carries the same epistemic and ontological significance as a law in biology (e.g. Mendel's) or chemistry (e.g. Periodic). In some cases, neither Mendel's Law nor the Periodic law is introduced as laws, and consequently an opportunity to discuss the implications of these terms is lost. A study of Turkish chemistry curricula and textbooks, for example, revealed that there is little or no differentiation of the meta-perspectives on the nature of knowledge (Kaya and Erduran 2013). Other studies on textbooks (e.g. Niaz and Rodríguez 2005) point to the lack of attention to NOS features in general, let alone the nuanced distinctions addressed here. Understanding the relationship of laws and explanations to theories in biology and chemistry demands a deliberate undertaking from historical and contemporary perspectives.

Furthermore, chemistry curricula often contain conceptual mistakes and thus demand closer examination. For example, the notion 'one molecule, one shape' is widespread and results in students' construction of a misconception where molecules are static, not oscillating and taking on different shapes (Kaya and Erduran 2013). The dynamic nature of molecular shape is an inherent aspect of chemical explanations. Coverage of structural explanations with meta-level perspectives is likely to minimise misconceptions about the dynamic nature of molecules. Chemistry curricula also need to scaffold students' understanding of how chemical explanations can rest on structural models and how these differ from, for instance, historical or evolutionary explanations in biology. Design of instructional activities would, then, need to acknowledge the observation that explanations will be manifest in the classroom via discussions of the signs and symbols that make up the alphabet of structures represented in chemistry. Engaging students in the generation, evaluation and application of structural explanations in chemistry is likely to improve their understanding of how chemical language and explanation relate to each other.

There are important reasons for why biology and chemistry learning should be informed by the issues raised in the paper. Familiarising students with different *types* of explanation in biology may mitigate against straying into teleological sidetracks, favouring the capacity/causal type, or privileging some types of explanation over others (those dealing with the how over those dealing with the why). The tendency of students to favour experimental over historical explanations, for example, has been documented in the context of evolutionary theory (see, for example, Dagher and BouJaoude 2005). Thus, biology learning could focus on constructing and utilising a broad range of biological explanations for a given phenomenon and applying this kind of reasoning to multiple contexts/phenomena. In support of this kind of learning, there needs to be a restructuring of the content/curriculum, so that explanations addressing different aspects of the phenomenon under study are not isolated from each other as is typically the case (e.g. evolutionary and ecological concepts are rarely discussed in relation to each other or to physiological concepts in school science). With respect to chemistry learning, the articulation of structural explanations with meta-level perspectives is likely to assist in understanding the dynamic nature of molecules. As discussed earlier, a common problem in chemical education concerns the interpretation of molecular models and straying onto static notions of molecular structures as sidetrack in learning outcomes. Given the centrality of molecular structure and modelling in chemistry, improvement in the learning of the structural explanations is likely to have positive impact on understanding other related areas of chemistry.

37.6 Conclusion

In summary, the aspects of laws and explanations in biology and chemistry emphasised in this paper are not exhaustive but are representative of the types of issues that concern us as science educators interested in improving students' understanding of science. This understanding will be enriched if students are provided multiple opportunities to develop meta-level understanding of how particular domains of science engage with some of the key aspects of scientific knowledge such as laws and explanations. There has been a long-standing criticism of science education in failing to enable students to understand the nature of science, scientific knowledge and scientific knowledge development. While science educators have acknowledged that perspectives from history and philosophy of science can promote a deeper understanding of the nature of science, the role of the nature of disciplinary knowledge has been under-investigated within the science education research community. The aim of this chapter was to articulate the nature of laws and explanations in biology and chemistry so as to extend and enrich the previous agendas for teaching the nature of science using domain-specific epistemologies to describe key debates and features related to disciplinary knowledge. Further research in this area is needed to further clarify, refine, challenge and expand some of the claims presented in this paper.

Acknowledgements The authors wish to thank the five anonymous referees who provided valuable feedback on earlier versions of this chapter.

Appendix

Source of potential confusion about structural explanations in a high school chemistry textbook (Reproduced from Hsu et al. 2010, p. 541).

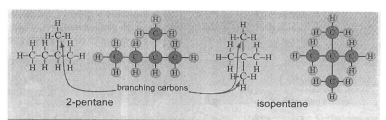

References

Achinstein, P. (1983). *The nature of explanation*. New York, NY: Oxford University Press.
Adúriz-Bravo, A. (2013). A 'semantic' view of scientific models for science education. *Science & Education*, 22(7), 1593–1611.
Adúriz-Bravo, A. & Galagovsky L. (2001) Modelos y analogías en la enseñanza de las ciencias naturales. El concepto de modelo didáctico analógico, *Enseñanza de las ciencias*, 19, 231–242.
Bird, A. (1998). *Philosophy of science*. London & New York: Routledge.
Bhushan, N. & Rosenfeld, S. (2000). *Of minds and molecules: New philosophical perspectives on chemistry*. Oxford, UK: Oxford University Press.
Brigandt, I. (2011). Philosophy of biology. In S. French & J. Saasti (Eds.), *The continuum companion to the philosophy of science* (pp. 246–267). London, UK: Continuum Press.
Brito, A., Rodriguez, M.A. & Niaz, M. (2005). A reconstruction of development of the periodic table based on history and philosophy of science and its implications for general chemistry textbooks. *Journal of Research in Science Teaching*, 42(1), 84–111.
Brock, W.H. (2000). *The chemical tree: A history of chemistry*. New York, NY: W. H. Norton.
Brown, T.L. (2003). The metaphorical foundations of chemical explanation. *Annals of the New York Academy of Science*, 998, 209–216.
Bunge, M. (1961). Kinds and criteria of scientific laws. *Philosophy of Science*, 28 (3), 260–281.
BSCS. (2003). *Biology: A human approach* (Second Edition). Dubuque, IA: Kendall Hunt.

Calcott, B. (2009). Lineage explanations: Explaining how biological mechanisms change. *British Journal of Philosophy of Science*, 60, 51–78.

Campbell, N., Reece, J., Taylor, M., Simon, E. & Dickey, J. (2009). *Biology: Concepts and connections*. San Francisco, CA: Pearson.

Cartwright, N. (1983). *How the laws of physics lie*. Oxford, UK: Clarendon Press.

Chinn, C. & Brown, D.E. (2000). Learning in science: A comparison of deep and surface approaches. *Journal of Research in Science Teaching*, 37(2), 109–138.

Christie, M. & Christie, J. (2003). Chemical laws and theories: a response to Vihalemm. *Foundations of Chemistry*, 5, 165–174.

Christie, M. & Christie, J. (2000). "Laws" and "Theories" in Chemistry do not obey the rules. In N. Bhushan and S. Rosenfeld (Eds.), *Of minds and molecules: New philosophical perspectives on chemistry* (pp. 34–50). New York: Oxford University Press.

Christie, M. (1994). Chemists versus philosophers regarding laws of nature. *Studies in History and Philosophy of Science*, 25, 613–629.

Clough, M. (1994). Diminish students' resistance to biological evolution. *The American Biology Teacher*, 56(7), 409–415.

Coulson, C.A. (1960). Present state of molecular structure calculations. *Reviews of Modern Physics*, 32, 170–177.

Dagher, Z. & BouJaoude, S. (2005). Students' perceptions of the nature of evolutionary theory. *Science Education,* 89, 378–391.

Dagher, Z. & Cossman, G. (1992). Verbal explanations given by science teachers. *Journal of Research in Science Teaching*, 29 (4), 361–374.

De Regt, H. (2011). Explanation. In S. French & J. Saasti (Eds.), *The continuum companion to the philosophy of science* (pp. 157–178). London: Continuum Press.

Dhar, P.K. & Giuliani, A. (2010). Laws of biology: Why so few? *Systems Synthetic Biology*, 4, 7–13.

Dictionary of Botany. (2003). Mendel's laws. Retrieved on September 10, 2012 from http://botanydictionary.org/mendels-laws.html

Dodds, W. (2009). *Laws, theories, and patterns in ecology*. Berkeley, CA: University of California Press.

Duschl, R. (1990). Restructuring science education: the importance of theories and their development. New York, NY: Teachers College Press.

Earley, J. E. (Ed.) (2003). *Chemical explanation: Characteristics, development, autonomy*. New York: Annals of the New York Academy of Sciences.

Elgin, M. (2006). There may be strict empirical laws in biology, after all. *Biology and Philosophy,* 21, 119–134.

Erduran, S. (2001). Philosophy of chemistry: An emerging field with implications for chemistry education. *Science & Education*, 10(6), 581–593.

Erduran, S. (2005). Applying the philosophical concept of reduction to the chemistry of water: implications for chemical education. *Science & Education*, 14(2), 161–171

Erduran, S. (2007). Breaking the law: Promoting domain-specificity in science education in the context of arguing about the Periodic Law in chemistry. *Foundations of Chemistry*, 9(3), 247–263.

Erduran, S. & Duschl, R. (2004). Interdisciplinary characterizations of models and the nature of chemical knowledge in the classroom. *Studies in Science Education*, 40, 111–144.

Erduran, S. & Jimenez-Aleixandre, M. P. (Eds.). (2008). *Argumentation in science education: Perspectives from classroom-based research*. Dordrecht: Springer.

Fodor, J. (1974). Special sciences and the disunity of science as a working hypothesis. *Synthese*, 28, 77–115.

Garvey, B. (2007). *Philosophy of biology*. Montreal, Canada: McGill-Queens University Press.

Giere, R.N. (1988). *Explaining science: A cognitive approach*. Chicago and London: The University of Chicago Press.

Giere, R.N. (1999). *Science without laws*. Chicago: University of Chicago Press.

Goodwin, W. M. (2008). Structural formulas and explanation in organic chemistry. *Foundations of Chemistry,* 10, 117–127.

Hanke, D. (2004). Teleology: The explanation that bedevils biology. In J. Cornwell (Ed.), *Explanations: Styles of explanation in science* (pp. 143–144). Oxford, UK: Oxford University Press.

Harré, R. (1988). Modes of explanation. In D. J. Hilton (Ed.), *Contemporary science and natural selection: Common sense conceptions of causality* (pp. 129–144). New York, NY: New York University Press.

Harré, R. (2003). Structural explanation in chemistry and its evolving forms. *Annals of the New York Academy of Sciences,* 988(1), 59–70.

Hempel, C. & Oppenheim, P. (1948). Studies in the logic of explanation. *Philosophy of Science,* 15, 135–175.

Hesse, M.B. (1970). *Models and analogies in science.* Milwaukee, WI: University of Notre Dame Press.

Hodson, D. (1988). Towards a philosophically more valid science curriculum. *Science Education,* 72, 19–40.

Hoffman, R. (1998). Qualitative thinking in the age of modern computational chemistry–or what Lionel Salem knows. *Journal of Molecular Structure,* 424, 1–6.

Horwood, R.H. (1988). Explanation and description in science teaching. *Science Education,* 72, 41–49.

Hsu, T., Chaniotakis, M., Carlisle, D. & Damelin, D. (2010). *A natural approach to chemistry.* Ronkonkoma, NY: Lab-Aids, Inc.

Hull, D. (1973). *Darwin and his critics.* Cambridge, MA: Harvard University Press.

Jacob, C. (2001). Interdependent operations in chemical language and practice. *HYLE--International Journal for Philosophy of Chemistry,* 7(1), 31–50.

Jungwirth, E. (1979). Do students accept anthropomorphic and teleological formulations as scientific explanations? *Journal of College Science Teaching,* 8, 152–155.

Justi, R. (2000). Teaching with historical models. In J. K. Gilbert & C. J. Boutler (Eds.), *Developing models in science education* (pp. 209–226). Dordrecht: Kluwer.

Kaya, E. & Erduran, S. (2013). Integrating epistemological perspectives on chemistry in chemical education: The cases of concept duality, chemical language and structural explanations. *Science & Education,* 22(7), 1741–1755.

Khishfe, R. & Abd-El-Khalick, F. (2002). The influence of explicit reflective versus implicit inquiry-oriented instruction on sixth graders' views of nature of science. *Journal of Research in Science Teaching,* 39(7), 551–578.

Land, S.M. & Zembal-Saul, C. (2003). Scaffolding reflection and articulation of scientific explanations in a data-rich, project-based learning environment: An investigation of Progress Portfolio. *Educational Technology Research & Development,* 51(4), 65–84.

Lange, M. (2005). Ecological laws: what would they be and why would they matter? *Oikos,* 110(2), 394–403.

Lederman, N. (2007). Nature of science: Past, present, future. In S. Abell & N. Lederman (Eds.), *Handbook of research on science education* (pp. 831–879). Mahwah, NJ: Lawrence Erlbaum.

Mahner, M. & Bunge, M. (1997). *Foundations of biophilosophy.* Berlin, Germany: Springer.

Marks, J. (2008). The construction of Mendel's laws. *Evolutionary Anthropology,* 17, 250–253.

Matthews, M. R. (1994/2014). Science teaching. The role of history and philosophy of science. New York, NY: Routledge.

Mayr, E. (1961). Cause and effect in biology. *Science,* 134, 1501–1505.

Mayr, E. (2004). *What makes biology unique?* Cambridge, UK: Cambridge University Press.

McComas, W. (1998). The principal elements of the nature of science: Dispelling the myths. In W. McComas (Ed.), *The nature of science in science education: Rationales and strategies* (pp. 53–70). Dordrecht, The Netherlands: Kluwer Academic Publishers.

McComas, W. (2003). A textbook case of the nature of science: Laws and theories in the science of biology. *International Journal of Science and Mathematics Education,* 1(2), 141–155.

McNeill, K. & Krajcik, J. (2012). *Supporting Grade 5–8 students in constructing explanations in science: The claim, evidence, and reasoning framework for talk and writing.* Boston, MA: Pearson.

McNeill, K. L., Lizotte, D. J, Krajcik, J. & Marx, R. W. (2006). Supporting students' construction of scientific explanations by fading scaffolds in instructional materials. *The Journal of the Learning Sciences,* 15(2), 153–191.

McShea, D. & Brandon, R. (2010). *Biology's first law.* Chicago, IL: University of Chicago Press.

Mitchell, S. (1997). Pragmatic laws. *Philosophy of Science,* 64, S468-S479.

Mitchell, S. (2000). Dimensions of scientific law. *Philosophy of Science,* 67(2), 242–265.

Mitchell, S. (2009). *Unsimple truths.* Chicago, IL: University of Chicago Press.

Morrison, M. (2000). *Unifying scientific theories: Physical concepts and mathematical structures.* Cambridge, UK: Cambridge University Press.

Nagel, E. (1961). *The structure of science: Problems in the logic of scientific explanation.* New York, NY: Harcourt, Brace and World.

Niaz, M. & Rodríguez, M.A. (2005). The oil drop experiment: Do physical chemistry textbooks refer to its controversial nature? *Science & Education, 14,* 43–57.

Örstan, A. (2007). Wilson's fundamental laws of biology. Retrieved on March 25, 2011 from http://snailstales.blogspot.com/2007/01/wilsons-fundamental-laws-of-biology.html

Pitt, J. (Ed.) (1988). Theories of explanation. New York, NY: Oxford University Press.

Press, J. (2009). Physical explanations and biological explanations, empirical laws and a priori laws. *Biology & Philosophy,* 24, 359–374.

Rose, S. (2004). The biology of the future and the future of biology. In J. Cornwell (Ed.), *Explanations: Styles of explanation in science* (pp. 125–142). Oxford, UK: Oxford University Press.

Rosenberg, A. & McShea, D. (2008). *Philosophy of biology: A contemporary introduction.* New York, NY: Routledge.

Ruphy, S. (2003). Is the world really "dappled"? A response to Cartwright's charge against "cross-wise reduction". *Philosophy of Science,* 70, 57–67.

Ruse, M. (1988). *Philosophy of biology today.* Albany, NY: State University of New York Press.

Sandoval, W. A. & Millwood, K. A. (2005). The quality of students' use of evidence in written scientific explanations. *Cognition and Instruction,* 23(1), 23–55.

Sandoval, W. A. & Reiser, B. J. (2004). Explanation-driven inquiry: Integrating conceptual and epistemic scaffolds for scientific inquiry. *Science Education,* 88, 345–372.

Salmon, M.H., Earman, J., Glymour, C., Lennox, J.G., Machamer, P., McGuire, J.E., Norton, J.D., Salmon, W.C. & Schaffner, K.F. (1992). *Introduction to the philosophy of science.* Englewood Cliffs, NJ: Prentice Hall.

Salmon, W. C. (1984). *Scientific explanation and the causal structure of the world.* Princeton, NJ: Princeton University Press.

Salmon, W. C. (1987). Why ask, "Why?" An inquiry concerning scientific explanation. In A. J. Kourany (Ed.), *Scientific knowledge: Basic issues in the philosophy of science* (pp. 88–104). Englewood Cliffs, NJ: Prentice Hall.

Scerri, E. (2000a). Philosophy of chemistry: A new interdisciplinary field? *Journal of Chemical Education,* 77, 522–526.

Scerri, E. (2000b). The failure of reduction and how to resist disunity in the sciences in the context of chemical education. *Science & Education,* 9, 405–425.

Scerri, E.R. (1999). A critique of Atkins' Periodic Kingdom and some writings on electronic structure. *Foundations of Chemistry,* 1, 297–305.

Scerri, E. & McIntyre, L. (1997). The case for the philosophy of chemistry. *Synthese,* 111(3), 213–232.

Schaffner, K.F. (1993). *Discovery and explanation in biology and medicine.* Chicago: University of Chicago Press.

Schwab, J. J. (1958). The teaching of science as inquiry. Bulletin of Atomic Scientists, 14, 374–379.

Schwab, J. J. (1978). Education and the structure of the disciplines. In J. Westbury & N. Wilk of (Eds.), Science, curriculum and liberal education: Selected essays (pp. 229–272). Chicago: University of Chicago Press.

Schwartz, R., Lederman, N. & Crawford, B. (2004). Developing views of the nature of science in an authentic context: An explicit approach to bridging the gap between nature of science and scientific inquiry. *Science Education*, 88(4), 610–645.

Scriven, M. (1970). Explanations, predictions, and laws. In B. A. Brody (Ed.), *Readings in the philosophy of science* (pp. 88–104). Englewood Cliffs, NJ: Prentice Hall.

SEPUP. (2011). *Biology: Science and global issues*. Berkley, CA: Lawrence Hall of Science.

Sober, E. (2008). *Evidence and evolution: The logic behind the science*. Cambridge, MA: Cambridge University press.

Sterelny, K. & Griffiths, P. (1999). *Sex and death: An introduction to the philosophy of biology*. Chicago, IL: University of Chicago Press.

Uzman, A. (2006). Four laws of biology. Retrieved on March 25, 2011 from http://hunblog.typepad.com/hunblog/2006/09/four_laws_of_bi.html

van Brakel, J. (1999). On the neglect of the philosophy of chemistry. *Foundations of Chemistry, 1*, 111–174.

van Brakel, J. (2000). *Philosophy of chemistry: Between the manifest and the scientific image*. Louvain: University of Louvain Press.

Vihalemm, R. (May 2003). Natural kinds, explanation and essentialism in chemistry. *Annals of the New York Academy of Sciences*, 988(1), 59–70.

Watson, J.D. (1968). *The double helix: A personal account of the discovery of the structure of DNA*. New York, NY: New American Library.

Weisberg, M., Needham, P. & Hendry, R.F. (2011). Philosophy of chemistry. *Stanford Encyclopedia of Philosophy*. http://plato.stanford.edu/entries/chemistry/

Woodward, J. (2001). Law and explanation in biology: Invariance is the kind of stability that matters. *Philosophy of Science*, 68, 1–20.

Wouters, A. (1995). Viability explanation. *Biology & Philosophy,* 10, 435–457.

Wouters, A. (2007). Design explanation. *Erkenntnis, 67,* 65–80.

Zohar, A. & Ginossar, S. (1998). Teleology and anthropomorphism in biology education—heretical suggestions. *Science Education*, 82, 679–697.

Zoubeida R. Dagher is Professor at the School of Education, and a Faculty Fellow at the Center for Science, Ethics & Public Policy at the University of Delaware. She serves currently as President Elect of the International History and Philosophy of Science Teaching (IHPST) Group. She has been a Visiting Scholar at Curtin University of Technology, Perth, Australia, and the Lebanese University and the American University of Beirut in Lebanon. She also served as Deputy Dean in 2007–2008 at the College of Education at Qatar University, Qatar. Dr. Dagher has served as an elected member to the Board of Directors of the National Association for Research in Science Teaching (NARST) and as member of the Advisory Council of IHPST. She has also served as member of editorial review boards in lead science education journals. Dr. Dagher received her PhD in science education from the University of Iowa, USA, and her Masters and Bachelor degrees from the American University of Beirut and the Lebanese American University, respectively. She has taught science to middle and high school students in Lebanon. Her research interests include the nature of school science inquiry and representation of scientific epistemology in science curricula. Dr. Dagher has co-edited a book (BouJaoude and Dagher 2009, Sense) on science education in Arab states. She is currently co-authoring a book to be published by Springer on the nature of science in science education with Sibel Erduran.

Sibel Erduran is Chair of STEM Education at University of Limerick, Ireland. She has had Visiting Professorships at Kristianstad University, Sweden, and Bogazici University, Turkey. She has also worked at University of Pittsburgh and King's College, University of London and University of Bristol. She is an Editor for *International Journal of Science Education*, Section Editor for *Science Education* and serves as Director on the IHPST council. Her higher education was completed in the USA at Vanderbilt (PhD Science Education & Philosophy), Cornell (MSc Food chemistry) and Northwestern (Biochemistry) Universities. She has worked as a chemistry teacher in a high school in northern Cyprus. Her research interests focus on the applications in science education of epistemic perspectives on science in general and in chemistry in particular. She has co-edited a book (Erduran, S., and Jimenez-Aleixandre 2008, Springer) on argumentation in science education, an area of research for which she has received an award from NARST. In 2013, she guest edited *Science & Education* (Volume 22 Number 7) consisting of 17 articles with the editorial entitled 'Philosophy, Chemistry and Education: An Introduction'. She is currently co-authoring a book to be published by Springer on the nature of science in science education with Zoubeida Dagher.

Chapter 38
Thought Experiments in Science and in Science Education

Mervi A. Asikainen and Pekka E. Hirvonen

38.1 Introduction

When Albert Einstein was 16, he considered the following thought experiment. He imagined chasing after a beam of light with the velocity of c. He would then catch the light wave and be moving with it and the light wave would seem to be frozen. Einstein noted both his experiences and Maxwell's electromagnetic theory, which suggests that such a stationary wave does not exist. In addition, he noted that if an observer were to see him riding on a light wave with a velocity of c, Einstein himself would not be able to observe the velocity.

This example illustrates the essence of a thought experiment.[1] The thought experiment describes an imaginary, hypothetical situation. The thought experiment cannot always be performed as a physical experiment, in this particular case, because it is impossible for such a massive object (Einstein) to have the velocity of c. In several respects, however, the thought experiment resembles a physical experiment. Its premise of c as the velocity of light is an empirically measured theoretical result using Maxwell's theory of electromagnetism as a starting point. It rests on the hypothesis of riding a light wave, which inevitably fails as a result of the empirical observations and impossibilities contained in theories associated with physics.

The purpose of what follows in this chapter is to examine the role of thought experiments in science and science education. First, different definitions of the concept of thought experiment will be discussed. Second, it will be argued that TEs form an essential part of scientific methodology, a special case of scientific experimentation. Third, attention will be paid to the role played by TEs in the

---

[1] A scientific experiment can be either a thought experiment performed in thought or a physical experiment performed in the laboratory.

M.A. Asikainen (✉) • P.E. Hirvonen
Department of Physics and Mathematics, University of Eastern Finland, Joensuu, Finland
e-mail: mervi.asikainen@uef.fi

M.R. Matthews (ed.), *International Handbook of Research in History, Philosophy and Science Teaching*, DOI 10.1007/978-94-007-7654-8_38,
© Springer Science+Business Media Dordrecht 2014

development of scientific theories. Subsequently, attention will also be paid to the pedagogical benefits of the use of thought experiments in science learning and the reported studies on the use of thought experiments in science teaching. It will be argued that, as a result of the various benefits of TEs, their use should be increased in science teaching. Finally, our discussion will focus on the challenges posed by TEs in the teaching and learning of science.

38.2 Descriptions of the Concept of the Thought Experiment

Thought experiments have a long history, starting from the time of pre-Socratics (Rescher 1991; Brown and Fehige 2010). It has been argued that during the Middle Ages, thought experimenting was one of the main methods used in science (King 1991). In the seventeenth century, Galileo Galilei and Isaac Newton used thought experiments (TEs) as part of their scientific methodology.[2] The rise of relativity and quantum physics would not have been possible without thought experiments, and famous thought experimenters include Niels Bohr, Erwin Schrödinger, and Albert Einstein (Brown 1986; Matthews 1988; Zeilinger 1999).

Modern science philosophers and scientists have attempted to frame a general description of the concept of thought experiment (TE). Roy A. Sorensen (1992) sees TEs in a broad light: the only difference between an actual experiment and a thought experiment is that a thought experiment attempts to achieve its aim without the benefit of actual implementation. However, as Galili (2009) has criticised, Sorensen's definition of an experiment goes beyond the realm of science. According to Sorensen, a scientific experiment is "a procedure for answering or raising questions about the relationship between variables by varying one (or more) of them and tracking any response by the other or others". Galili states that physical experiments are based on certain theoretical assertions and this is how TEs are also used in science.

Sören Häggqvist (1996) claims that philosophers and scientists often see TEs as something different from "genuine", "actual", or "real" experiments but rather as or a species of experiments similar to "laboratory experiments" or "cyclotron experiments". He characterises a TE loosely as an experiment that aspires to test some hypothesis or theory: it is performed in thought, but paper or pencil, encyclopedias, or computers may also be used (Häggqvist 1996, p. 15).

According to Irvine (1991, pp. 158–159), TEs have to possess at least several, if not all, of the characteristics of a scientific experiment. This means that not all varieties of hypothetical reasoning concerning the natural world can be considered to be TEs. A TE has to bear a special relationship both to previous empirical observations and also to the background theory of TE. A thought experiment cannot ever

---

[2] See, e.g. Newton (1728), Newton (1863), Gendler (1998), Palmieri (2003), Palmier (2005), Hall (2000), and Norton (1991).

replace observations or a physical experiment because a thought experiment rests on auxiliary presuppositions considered to be true but whose failure changes the result of the thought experiment per se.

Nancy Nersessian's view of TEs differs slightly from the views presented above in that she sees a TE as a mental model that enables the dissemination of a possible physical event that is often unrealisable in one's imagination (Nersessian 1989, p. 175). Nersessian claims that Galilei, for instance, acted in this way in the case of his TE concerned with falling bodies. According to Nersessian (1992, p. 27), mental simulation is needed for a thought experiment to be both thought and experimental. The original scientific thought experiment is executed by a scientist who imagines a sequence of events and constructs a mental model. Then she/he constructs it in a narrative form in order to describe the thought experiment to others.[3]

Nersessian's view is fascinating because it makes a connection between thought experiments and mental models. On the other hand, the connection makes her definition somewhat complicated because there is no consensus about how individuals possess their knowledge. Is it in form of models (Nersessian 1989, 1992; Nersessian and Patton 2009), "in pieces" (di Sessa et al. 2004; diSessa and Sherin 1998; diSessa 2002), or as coherent and organised naive theories related to particular topics (Vosniadou 1994; Vosniadou et al. 2008)? Some researchers even think that a mental model is an individual's inner, private model that cannot be expressed exactly; when an individual presents his/her model to an audience, the model is not a mental model but *an expressed model* (Gilbert et al. 1998).

Having analysed several definitions of thought experiment, Igal Galili proposes the following definition: "Thought experiment is a set of hypothetico-deductive considerations regarding phenomena in the world of real objects, drawing on a certain theory (principle or view) that is used as a reference of validity" (Galili 2009, p. 12).

Galili's (2009) definition is not concerned with the reality existing outside scientific theories, and it also excludes a pure, formal analysis manipulating with theoretical entities without addressing the real objects. The definition includes scientific TEs that are part of the scientific process and excludes philosophical TEs. Even if we mostly agree with Galili's definition, we do not think that phenomena should be restricted to include only the world of real objects because that would exclude from the definition Maxwell's demon or hypothetical entities, for instance, whose existence cannot be verified. The definition of a thought experiment should also include a more explicit statement about its mental implementation. In sum, TEs are an essential part of scientific methodology, a special form of scientific experiments. Like other scientific experiments, they are based on a particular background theory. The main difference between physical experiments and TEs is that TEs are performed in thought. In addition, TEs can also be devised with hypothetical entities that have not yet been verified or cannot even be verified at all.

---

[3] Racher Cooper (2005) and Tamara Szabó Gendler (1998, 2004) are also supporters of this kind of mental model account.

38.3 The Role of Thought Experiments in Science

Ernst Mach and Pierre Duhem were the first to consider the value of thought experiments in science. According to Mach, thought experiments are needed because they precede physical experiments by preparing for their actual implementation (Mach 1976). Duhem's view is the opposite: he considered thought experiments useless because they cannot be presented in symbolic form and hence they cannot replace scientific experiments (Duhem 1990).

Mach (1976) stated that the possibility of a thought experiment rests on our mental images, which are more or less copies of facts. When reminiscing, we may even find new properties of the physical facts that we had not noticed previously. Our mental images are easier and faster to use than the physical facts. Thus, it can be said that thought experiments precede physical experiments and prepare for them. This means that every experimenter has to be aware of the details of the experiment before its actual implementation.

Mach thought that if thought experiments are reported sincerely, they will be true even if two thought experimenters report different sequences. In addition, errors can only occur when the results of thought experiments are compared with the physical reality. Sorensen (1992) argues that Mach's account overemphasises the subjectivity of a thought experiment. Sorensen considers that thought experiments can also be fallacious in their reporting phase.

Sorensen also argues that if a TE always precedes a physical experiment, the concept of the thought experiment has to be so wide that it covers "any kind of forethought about an experiment". He thinks that it is not normal in science to perform the full experiment in thought, but the mental processing is more like the planning stage of a physical experiment. We agree with Sorensen with regard to both of his arguments. Thought experiments conducted by two different thought experimenters can lead to different results in the first stage of TEs, and as often as not they cannot both be true. In addition, experimental scientists undoubtedly plan their physical experiments mentally, but these thoughts are frequently more like schemes of action rather than being full thought experiments.

Mach argues that in some cases, the result of a thought experiment can appear so sure and final that its implementation as a physical experiment may even seem unnecessary (Mach 1976). Duhem sees the role of thought experiments differently than Mach. He considered that thought experiments could not replace physical experiments and that they should be forbidden in science and in science teaching (Duhem 1990). Duhem alleged that only mathematical argument was precise and unambiguous, while the language of concrete observation is not: "The facts of experience, in all their native brutality, cannot be used in mathematical reasoning. To feed such reasoning they must be transformed and put into symbolic form". Duhem's view is sometimes termed sceptical objection (see Brown and Fehige 2010).

This kind of sceptical view of thought experiment is not common amongst science philosophers, but critical views of thought experiments can be found (see, e.g. Brown and Fehige 2010). Hull (1998, 2001) argues, for instance, that TEs are nothing but simple illustration and they end up persuading people; but TEs often contain deficiencies, such as incoherence and missing details. Norton (2004b) agrees with Hull over the fallibility of TEs. In addition, he claims that TEs are simply arguments, and hence they cannot offer any kind of special information that could not also be uncovered by conventional argumentation.

Cooper (2005) argues that thought experiments are needed for several reasons. Some thought experiments are practically possible to implement as physical experiments, but there may be sound reasons (ethical reasons, e.g. or the monetary expense of physical experiment) for performing them only in the mind. Other thought experiments may be impossible to implement as real experiments because they involve idealisations. Cooper (2005, p. 344) argues that Galileo's TE demonstrating that bodies continue moving with constant velocity in the absence of a force – a ball rolling in a frictionless U-bend – includes an idealisation. In science, the idealisations are often similar to the limiting case imposed by extrapolation of the results of the physical experiment. Other thought experiments may be impossible to implement as real experiments because they involve the violation of a physical law. According to Cooper, these TEs resemble the computer simulations that scientists run to discover how phenomena may behave if the laws of nature are slightly different. Cooper states that simulations and thought experiments that cannot be implemented as physical experiments can nevertheless be used as potential sources of knowledge in science.

Buzzoni (2009) claims that thought experiments and real-world experiments form a dialectical unity: without thought experiments there would be no real experiments because we would not know how to ask about their nature, and without real experiments we would not find answers to these questions. According to Galili (2009), thought experiments play a heuristic role. They are free from the constraints imposed by reality (heat, friction, etc.), and the thought experimenter can also forget the technical restrictions (equipment, costs, availability, etc.). In a sense, a person conducting a thought experiment mentally models theoretical physics (Peierls 1980).

From the above it follows that a thought experiment is a special case of scientific experiment that can precede a physical experiment and help the experimenter to conduct it. In some cases, a physical experiment may not be possible and TE may then be the only way to experiment. TEs can also be used to idealise complex physical situations and remove constraints imposed by reality. The physical experiment may either confirm the results of a TE or show that the TE was fallible; both types of TEs are important in constructing an understanding of scientific knowledge. This view of thought experiments can then be used as a starting point in science teaching.

38.4 The Epistemological Role of Thought Experiments in Science

If we approve of TEs as one special part of scientific methodology, we need also to discuss whether TEs play a special epistemic role in the knowledge construction processes of science. The theoretical framework of TEs is of great importance because it determines the image of the nature of science that TEs convey when used in science teaching.

There are two different views of the epistemological status of TEs. The argument-based view states that knowledge comes only via sensory experiences, while the intuition-based (Platonic) view argues that TEs provide information beyond our senses. The argument-based views rely on the idea that TEs can be reconstructed as arguments or that they function via their connection to arguments, so they are unable to provide more information than argumentation in general. On the other hand, the intuition-based view argues that a special group of TEs, Platonic thought experiments, go beyond our senses to acquire a priori information about nature.

38.4.1 Argument-Based Views

Argument-based views rely on the idea that TEs are unable to provide more information than argumentation in general. The supporters of the argument-based view do not mean that TEs are meaningless in science; rather, they are meaningless in an epistemological sense.

John Norton is probably the best-known supporter of the view that thought experiments are basically arguments. He thinks that TEs are not epistemic wonders, but they do tell us about our world using our normal epistemic resources[4] (Norton 1996, 2004a). Norton (1996, p. 339) has formulated his claim in a more precise form, referred to as a *Reconstruction Thesis*, as follows:

> *Reconstruction Thesis:* All thought experiments can be reconstructed as arguments based on tacit or explicit assumptions. Belief in the outcome-conclusion of the thought experiment is justified only insofar as the reconstructed argument can justify the conclusion.

TEs draw on hypothetical and counterfactual situations that essentially separate them from physical experiments (Norton 1991, 1996). These unnecessary particulars are needed for the experimental nature of thought experiments; without them, TEs would not be experimental. These particulars can be psychologically useful, but they are unnecessary for the thought experiment itself.

Norton claims that the epistemological potential of TEs is the same as that of argumentation, since every TE can be reconstructed as an argument (Norton 2004a, b). Because TEs do not involve new empirical data, they can only reorganise or generalise the old data (Norton 1991, p. 335). This prior knowledge, based on our previous experiences, can enter into thought experiments as assumptions. Hence, thought experiments are devices that reorganise or generalise these assumptions to achieve the result of the thought experiment. Norton regards these "devices" as arguments.

If the TE simply reorganises, it is a *deductive* argument or a *reductio ad absurdum* argument, where the particular conclusion follows deductively from the premises.

---

[4] Epistemic recourses are processes and tools that we use to decide that we know something or to create knowledge (Redish 2004, p. 31).

For instance, thermodynamics includes some powerful TEs because the first, second, and third laws of thermodynamics can be formulated as "assertions of impossibilities" (Norton 1991, p. 131). The first law can be expressed as an assertion as follows: "It is impossible to design a perpetual motion machine of the first kind, that is, a machine whose sole effect is to produce more energy than it consumes". Norton explains that consequences can be derived from the assertions included in a *reductio* argument, which then almost automatically becomes a thought experiment.

If the TE generalises on a wider scale, it is an *inductive* argument. This kind of TE includes an inductive step that frees the conclusion of its particulars. Norton (1991) suggests that Einstein's magnet and conductor and Einstein's elevator thought experiments belong to this class. According to Norton (1991, p. 137), Einstein's elevator can be constructed as arguments as follows:

1. In an opaque chest, an observer will see free bodies move identically in the case where the box is uniformly accelerated in a gravitation-free space and where the box is at rest in a homogenous gravitational field.
2. Inductive step: (a) the case is typical and will hold for all observable phenomena and (b) the presence of the chest and observer are inessential to the equivalence.
3. A uniformly accelerating frame in gravitation-free space and a frame at rest in a homogenous gravitational field are observationally identical but theoretically distinguished, which is self-contradictory.
4. The verifiability heuristic for theory construction (version 2[5]).
5. A uniformly accelerating frame in a gravitation-free space and a frame at rest in a homogenous gravitational field are the same thing (which becomes a postulate of a new theory).

According to Norton, the inductive step (2), which proceeds from a finite number of specific facts to a general conclusion, is quite problematic but "masked by the thought experiment format". He continues: "The extension from the motion of bodies in free fall to arbitrary processes is quite a leap, especially in view of the bizarre consequences that follow". Based on this example, it seems that constructing a thought experiment as arguments may also contain challenging phases that may not be unambiguous.

Brown and Fehige (2010) present three objections to Norton's claims. First, they consider Norton's view too vague. Second, they argue that Norton reaches far ahead of established facts: every real-world experiment can be represented as a thought experiment but nobody claims that thought experiments are unnecessary. Furthermore, Norton's view does not tell where the arguments come from. Brown and Fehige (2010) admit that a thought experiment can be an essential phase in the building of Nortonian reconstruction, but a thought experiment expressed as an argument loses its power. Arthur (1999) also disagrees with Norton by arguing that

---

[5] States of affairs that are not observationally distinct should not be distinguished by the theory (Norton 1991, p. 135).

if TEs are constructed as arguments, there will be an epistemic loss: the original thought experiment is not epistemically similar to the constructed arguments.

Nersessian (1992, p. 27) argues that a Nortonian reconstruction cannot be performed before the actual thought experiment has been executed. This means that TEs really have experimental power. By claiming that a TE contains particulars irrelevant to the conclusions, Norton fails also to see the constructive function of the narrative form in which thought experiments are presented.

Häggqvist (1996) claims that thought experiments are not arguments because something that is a process, an event, or a procedure cannot, by its nature, be an argument; TEs function, however, via their connection with arguments. He argues that thought experiments work in the same way as experiments in general, by affording premises for their associated arguments. For a successful experiment, the premises are true. Only arguments as truth-valued, linguistic entities matter when the truth-value of a scientific or philosophical theory or hypothesis is evaluated.

The argument-based view of TE as presented by Norton (1991, 1996, 2004a, b) seems to be problematic with regard to its potential use in science teaching. The reconstruction process, in particular, would be rather demanding for students because, in practice, students would already need to understand the original TE quite well in order to be able to perform the reconstruction. This does not mean that we do not appreciate the basic skills of scientific argumentation that constitute important learning goals in science education. The argument view may, however, be useful for science educators and science teachers in regarding the nature of the counterpoint of the argument-based view, i.e. Brown's destructive and constructive TEs, which will be examined next.

38.4.2   Brown's Destructive and Constructive TEs

James Robert Brown (1991) classified TEs according to their role in building scientific theories as destructive and constructive TEs. *A destructive TE* is an argument against a theory; it destroys or at least indicates serious problems in the particular theory. According to Brown, Einstein's chasing the light beam, presented in the introduction to this chapter, and Schrödinger's cat are examples of this kind of TE. Erwin Schrödinger presented a cat paradox where a cat in a box exists in a superposition of two states: dead and alive (Schrödinger 1935). His aim was to question the limitations and conceptual difficulties of quantum mechanics.

In contrast, *constructive* TEs break down into three further types: *direct, conjectural*, and *meditative* TEs. *A meditative TE* helps in the drawing of a conclusion from a specified, well-articulated theory. It may illustrate some counter-intuitive aspects of the theory, making it seem more satisfying, or it may act like a diagram in a geometrical proof that helps to support understanding, or even in the discovery of, the formal proof. Brown uses Maxwell's demon as an example of meditative TE.

The demon sits between the chambers of a gas vessel, which are filled with gas. The demon opens a trapdoor between the chambers by allowing the faster molecules to move to one side and the slower molecules to the other side. The TE shows that if this kind of demon existed, it would decrease the entropy of the gas system and cause a violation of the second law of thermodynamics. James Clerk Maxwell used this thought experiment to discuss the second law of thermodynamics at molecular level and to show that it possessed only statistical certainty (Schlesinger 1996; Radhakrishnamurty 2010). According to Schlesinger, Maxwell's intention was to use the demon to dramatise his claim concerning the statistical nature of thermodynamics.

A *conjectural TE* establishes some phenomenon and hypothesises a theory to explain the theory thereafter. The events of conjectural TEs have a presumed explanation. A *direct TE* begins with an unproblematic phenomenon and ends with a well-articulated theory. Brown considers Newton's bucket to be a prime example of a conjectural TE. Newton suggested that the existence of absolute space could be substantiated by hanging a bucket of water from a rope and spinning it. The concave shape of the water's surfaces caused Newton to assume that it was spinning with respect to something. Furthermore, according to Brown (1991), Stevin's inclined plane[6] and Einstein's elevator[7] belong to this class of TEs.

A small group of TEs are both destructive and constructive at the same time. These thought experiments are termed Platonic TEs (Brown 1991, p. 34). According to Brown (1991), in a few special cases we may go well beyond existing data to obtain a priori information about nature. Brown and Fehige (2010) explain that this information is a priori information about nature since, because the thought experiment does not contain new information, the conclusion does not draw on old data and it is not some sort of logical truth. This view of thought experiments can be further developed by combining a priori epistemology to recent views about the laws of nature, according to which laws consist of objectively existing relations between abstract entities. This view is, therefore, Platonic.

According to Brown, Galileo's free fall and the EPR (Einstein, Podolsky, Rosen) paradox may be regarded as examples of Platonic TEs. Brown argues that Galileo's free fall extinguished Aristotle's view and generated a new view, while EPR seriously challenged the Copenhagen interpretation and established the incompleteness of quantum mechanics. Brown and Fehige (2010) have characterised Brown's view as an intuition-based view.[8]

Galileo's free fall TE indeed revealed an inconsistency in the Aristotelian view, but it could not say anything about the actual descent of objects, which indeed fall at different rates of acceleration relative to the ground. According to McAllister (2004), Galileo's TE merely verified that if the rate of fall of simple and compound bodies was simply a function of their mass, then the rate of fall of bodies would be

---

[6] Stevin's TE discusses the forces that are needed to keep a weight on an inclined plane (see, e.g. Gilbert and Reiner 2000).

[7] If a man is in a windowless elevator, he cannot tell whether the sensation of weight is due to gravity or acceleration.

[8] Intuition can be defined as a capacity for attaining direct knowledge or understanding without the apparent intrusion of rational thought or logical inference (Sadler-Smith and Shefy 2004).

independent of mass. This is an important point that needs to be grasped in physics education (Lehavi and Galili 2009). Hence, Galileo's free fall TE is not actually a Platonic TE. Furthermore, Albert Einstein, Boris Podolsky, and Nathan Rosen attempted to show that quantum mechanics is incomplete, but, instead, a definition of nonlocality was found (Einstein 1918). Quantum mechanics is, however, generally regarded as complete. Bokulich (2001) has discussed both the essence and also further modifications of EPR.

Our view is aligned with that of Arthur (1999), who does not agree with the epistemological power of Platonic TEs but thinks that TEs can go beyond arguments by offering an effortlessly understandable imaginative reconstruction of the phenomenon. According to Arthur (1999, p. 27), there are no pre-existing concepts but rather some sort of presentiment or intuition of them. This does not mean that such ideas would really exist and that we could not yet understand them. Rather, we have not succeeded in formulating them.

Norton (2004a) has questioned the reliability of the use of those TEs that are supposed to be "the glimpsing a Platonic world". Brown's counterargument is that even ordinary vision can be mistaken (1991, p. 65–66). Norton sees this differently: the TE that fails is simply an argument that contains an erroneous assumption. Brown's Platonism has also been criticised for not presenting criteria for good and poor thought experiments (Brown and Fehige 2010). Brown and Fehige argue that this objection will be weak if the intuitions do the work in thought experiments, since rationalists and empiricists do not have a theory of the validity of intuitions.

Brown's (1991) categorisation of TEs as constructive and destructive has already been used in the analysis of thought experiments in physics textbooks and popular physics books by Velentzas, Halkia, and Skordoulis (2007). When they analysed 25 books to discover how the 11 most essential thought experiments in the domains of relativity and quantum mechanics are presented, they found all of the thought experiments contained in the books to be constructive.

The use of Brown's categorisation shows that it has potential in science education. We believe that it could also be used in science teaching as a theoretical framework of thought experiments for understanding how scientific knowledge is constructed. In the following section, thought experiments are discussed from the perspective of science education.

38.5 Thought Experiments in Science Education

In the course of the past 10 years, there has been a slight increase in research activities related to thought experiments in science education, and thought experiments have received more attention in scientific discussions. Here we argue why and how TEs might be used in science teaching in supporting student learning and offering an authentic image of science. In addition, the possible challenges involved in the teaching and learning of TE will form part of the discussion.

38.5.1 Pedagogical Benefits of Thought Experiments

Ernst Mach was the first to realise that thought experiments might have a high didactical value (Mach 1976). He emphasised in particular the role played by students in thought experimenting (Matthews 1988, 1990). By using thought experiments as a teaching method, a teacher can keep students guessing. In addition, this method provides a significant support to the teacher in coming to know his/her students better. Some students are able to guess the next phase immediately, while some will present extraordinary guesses. Through thought experiments, students will learn to distinguish solvable from unsolvable concerns.

The use of TEs introduces an authentic image of the culture of science (Galili 2009; Reiner et al. 1995; Reiner and Gilbert 2008). TEs can be used to address the essential characteristics of physical theories (Galili 2009). They often employ representative models that eliminate technical details, errors, and impeding factors such as heat or friction. By introducing TEs before real experiments, students may develop an ability to appreciate real experiments and perceive the focus of the experiments, which is otherwise frequently difficult to see because of the sheer quantity of details. Naive observers' difficulties in differentiating between non-relevant and relevant details may impede them from finding out the aimed observations, results, and conclusions (see, e.g. Kozma and Russell 1997; McDermott 1993). Klassen (2006) believes that by devising their own thought experiments, students are mentally engaged in the concepts to be learned, and this, in turn, may help them to construct a deeper understanding of science. Nersessian (1992) claims that "the historical processes provide a model for learning activity itself" and may assist students in constructing representations of scientific theories. Social discussions of TEs may lead students to conceptual refinement and construction of reliable knowledge, as would be the case in science itself (Reiner and Gilbert 2008).

Reiner and Burko (2003) claim that both the TEs devised by physicists and also those formulated independently by students are important in the learning of physics. Scientifically correct TEs constructed by famous physicists enable students to familiarise themselves with the potential of TEs and to see them as a special mode of argumentation. In contrast, incorrect TEs prepare them for the existence of logical and conceptual stumbling stones, the temporary state of knowledge in physics, and the meaning of self- and peer criticism in the construction of physical knowledge. By working on thought experiments independently, students also work through the processes that underlie erroneous reasoning and learn to negotiate over the processes and conclusions with their peers in a relevant form of social interaction (Reiner et al. 1995). Procedures such as these all contribute to the clarification of concepts.

It has also been claimed that the use of TE in teaching stimulates students' interest (Lattery 2001; Velentzas et al. 2007; Velentzas and Halkia 2011) and helps their imaginations to develop (Galili 2009). By introducing situations that are impossible to reproduce despite the sophistication of the available equipment, TEs also become an irreplaceable tool of teaching. According to Galili (2009), this applies especially in the teaching of relativity and quantum physics, where real experiments are not widely used in the classroom, and the use of the multimedia often fails to promote enhanced understanding. Encouragement is also given to the use of thought experiments in teaching if the aim of the teaching is to activate students' cognitive processes with situations that would otherwise be beyond their everyday experiences (Velentzas and Halkia 2010).

38.5.2 The Use of Thought Experiments in Science Textbooks

It has been noted that in some domains of physics such as relativity and quantum mechanics, thought experiments are the main method of presenting the concepts in physics textbooks and popular physics books (Velentzas et al. 2007). Because science teachers often base their teaching on textbooks (Levitt 2002; Yore 1991), textbook studies are an important method for understanding the premises of science teachers' use of thought experiments. In addition, it would appear that studies concerned with teachers' use of TEs are still absent from in the literature.

The extent to which thought experiments are used in science textbooks and the ways in which they have been exploited have been studied by Gilbert and Reiner (2000) and Velentzas, Halkia, and Skordoulis (2007). Gilbert and Reiner's study focused on popular physics textbooks[9] while Velentzas and colleagues looked at both popular science books and physics textbooks.[10]

Gilbert and Reiner (2000) discovered that textbooks often miss opportunities to develop thought experiments suitable for teaching even though there were numerous suitable opportunities to do so. Thought experiments in textbooks frequently turn into thought simulations that lack two essential elements of thought experiments: recognition and approval of the imposed problem and conclusions based on the results. Instead of drawing on the six elements of TEs,[11] the textbook thought simulations typically consisted of the following parts:

i. Statement of the conclusion reached
ii. Creation of the imagined world
iii. Conflation of the design and running elements
iv. Statement of the results obtained, often with an optional restatement of the conclusions reached (Gilbert and Reiner 2000, p. 279)

---

[9] The books analysed were Breithaupt's *Understanding Physics for Advanced Level* and Ohanion's *Physics* and *Conceptual Physics* by Hewitt.

[10] The books were either written in Greek or translated into Greek from English. The study aimed at finding out how the books represented the 11 most essential thought experiments in the domains of relativity and quantum mechanics. A total of 25 books were included in the study.

[11] The six elements of a TE: (1) posing a question or a hypothesis, (2) creating an imaginary world, (3) designing the TE, (4) performing the TE mentally, (5) producing an outcome of the TE, and (6) drawing a conclusion.

According to Gilbert and Reiner (2000), this may be the result of the textbook writers not understanding the actual potential of using thought experiments. Indeed, thought experiments can be a successful way to enhance students' cognitive engagement, which is the key to developmental success. Thought experiments offer opportunities for creating new ontological entities, developing reasoning skills, and adopting epistemological engagements. These skills are claimed to be essential for gaining an understanding of physics (Driver et al. 1994). It might also be asked whether this kind of one-sided deductive approach to thought experiments is pedagogically valid.

Velentzas, Halkia, and Skordoulis (2007) observed that all of the thought experiments that they had found in the physics textbooks and popular physics books in their study were constructive. In addition, the authors had modernised numerous thought experiments: for example, Einstein's chest TE was examined in the form of a spaceship thought experiment. The authors had also invented thought experiments independently. The mathematical level of thought experiment was low and the terminology, language, and abstraction level were all modified to match their readers' perceived skills. The use of narratives was typical of the popular textbooks, whereas the other textbooks tended to avoid narratives by using scientific language and terminology.

38.5.3 Studies on the Use of Thought Experiments in Teaching

Thought experiments have been used in science teaching in different ways, and some of the possibilities have been reported. In the following we describe a few of these: using written tasks to help students to understand well-known TEs,[12] constructing historical physics experiments as thought experiments in narrative form (Klassen 2006), and students' own TEs in the context of experimental work (Reiner 1998).

Velentzas, Halkia, and Skordoulis (2007) used the famous TE known as Einstein's elevator thought experiment to introduce the concepts of the equivalence principle to 9th grade students. A group of six students studied the thought experiment as it was presented in a selected popular physics book[13] and replied to related questions, first individually and then as a group. The results indicate that the pupils achieved a reasonable understanding of the concepts. They were also surprisingly enthusiastic about performing the given task. The researchers supposed that this reaction may have been a consequence of the nonmathematical, narrative representation of the task. It seems, then, that popularised thought experiments can be used to inspire pupils in the case of concepts and principles that are discussed in greater depth later in the teaching process.

---

[12] See, e.g. Velentzas et al. (2007), Lattery (2001), Velentzas and Halkia (2011), and Velentzas and Halkia (2012).

[13] Stannard, R. (1991). *Black Hole and Uncle Albert*. London: Faber and Faber Ltd.

Velentzas and Halkia (2011, 2012) have also successfully used thought experiments as a teaching tool in physics teaching for upper secondary students. They studied the ways in which the uncertainty principle and the basic concepts of the theory of relativity could be taught to upper secondary school students. The uncertainty principle was introduced via Heisenberg's microscope thought experiment (Velentzas and Halkia 2011), while the theory of relativity was approached via Einstein's train and Einstein's elevator thought experiments (Velentzas and Halkia 2012). In the case of the uncertainty principle, the students were able to derive the uncertainty principle, and by the end of the teaching, they understood it as a general principle in nature (Velentzas and Halkia 2011). Furthermore, Einstein's TEs concerning relativity enabled students to realise situations related to the world beyond their everyday experiences and to gain a basic understanding of the theory of relativity (Velentzas and Halkia 2012).

Lattery (2001) used Galileo's TE Law of Chords (rates of descent along certain curves) as a basis for a student project at the university level. A group of three students discussed the TE and made predictions, following which they tested the predictions experimentally. Subsequently, they wrote a paper, prepared a poster, and made an oral presentation for their peers and the faculty concerned with the project. Lattery concludes that it offered a positive learning experience for the students themselves, for their peers, and for faculty in general.

Klassen (2006) argued that thought experiments could be expressed as stories. To test his hypothesis, he wrote a story about Benjamin Franklin's life and experiments in a form that invited students to render Franklin's experiments as thought experiments. He believed that this kind of narrative construction would help students to become mentally engaged in the concepts to be learned and that this, in turn, would them help to construct a more profound understanding of science. Even if a method of this kind seems to be rather interesting, its effectiveness should still be assessed scientifically by examining students' learning processes before further conclusions.

Reiner (1998) studied grade 11 students' self-devised thought experiments. A total of 12 students were given the following task. Using a computer-based simulator and hands-on equipment, they were required to design a periscope with a wide visual field. To solve the task, the students worked in groups of three. Analysis of the processes produced by one group showed that the students' thought experiments developed because of a collaborative problem-solving process in which the students used the computer system to validate potential events and results. The system helped the students to make their intentions visible to their peers and also to test hypothetical events. Furthermore, the four different student groups displayed a considerable variety of thought experiments, e.g. the logic and contexts that the students used and the conclusions that they drew varied considerably. It was also typical of the four groups that the students' thought experiments were partial and incomplete; they did not contain all three parts of the typical thought experiment: hypothesis, results, and conclusions. Reiner claims, however, that the results show that the thought experiments, which consisted of episodes, were general rather than random, even if they missed out one or two of the three parts. According to Reiner, a collaborative environment helps students to construct thought experiments as a shared construction that is based on individual students' contributions.

These examples of the implementation of thought experiments in teaching are illustrative; but in actual classroom teaching, some limitations may occur. Teachers need to take into account the fact that students' cognitive processes may lead to erroneous conclusions (Velentzas and Halkia 2011). In analysing some of the famous TEs of physics, Reiner and Burko (2003) have discovered cognitive processes that also lead to erroneous conclusions. At least three of this kind were found: strong *intuition* of a kind that induces the abandonment of theory-based reasoning, *incompleteness* of the basic assumption of the thought experiment, and *irrelevance* of the system's properties in the thought experiment.

Reiner and Burko (2003) claim that the processes that are characteristic of physical thinking are likely to be found in physics learning as well. The use of intuition instead of logical, theory-based reasoning is even stronger in the case of naive physics learners than amongst famous physicists in the history of physics. In addition, research has shown that students more often apply concrete, experiential knowledge rather than using logical reasoning (e.g. DiSessa 1993; Gilbert and Reiner 2000). The incompleteness of the students' TEs relates to the narrowness of the learners' physical world. Their readiness to conclude is insufficient because assumptions integrated into knowledge structures are partial instead of being comprehensive; the learners may not have sufficient knowledge of the physical world to make sense of the TE. Reiner and Burko (2003) argue that the use of TEs in physics learning is important, because it allows students to experience the destructive and constructive role of physical intuitions, incompleteness, and the importance of relevancy.

We agree with Reiner and Burko and Velentzas and Halkia (2011), who recommend the use of TEs in cases where the performance of a physical experiment is impossible, harmful, and dangerous or has nothing to offer in the end for the result. They also suggest the use of TEs in situations that require students to mentally surpass their everyday experiences.

In sum, thought experiments can be used in science teaching to help students to develop their conceptual understanding of science.[14] Thought experiments may increase students' interest in learning science[15] and to activate and support their thinking processes.[16] In addition, the construction of students' own thought experiments can be supported by creating a collaborative environment that enables students to construct thought experiment together with their peers (Reiner 1998). Students' erroneous conclusions should, however, be taken into account in teaching; they can be used as a basis for discussion about the destructive and constructive intuitions in thought experimenting (Reiner and Burko 2003).

---

[14] See Galili (2009), Velentzas et al. (2007), and Velentzas and Halkia (2011, 2012).

[15] See, e.g. Gilbert and Reiner (2000), Velentzas et al. (2007), and Lattery (2001).

[16] See, e.g. Reiner and Burko (2003), Reiner and Gilbert (2008), and Velentzas and Halkia (2011, 2012).

38.6 Conclusion

This article has examined the role played by thought experiments in science and science education. It has been argued that TEs are a natural part of scientific methodology, a special type of scientific experimentation that may play either a constructive or a destructive role in the construction of scientific theories. The important role played by TEs in science should also be discussed in science teaching. In addition, TEs have been used in science education in various ways to foster the development of students' reasoning, mental modelling, and conceptual understanding; to teach them about the nature and processes of science; and to stimulate their interest in science. Thought experiments also provide opportunities for focusing on the epistemology and ontology of science in the teaching of science.

TEs are a special variety of scientific experiment that can, at its best, precede a physical experiment and help the experimenter in conducting it. In some cases, physical experimentation may not yet be possible and the TE can be the only way to experiment; TEs are free from the constraints imposed by the learning environment and by technical restrictions (Cooper 2005; Galili 2009). In addition, a physical experiment may be considered useless if it is unlikely to substantially improve understanding gained from a TE (Sorensen 1992). These statements also hold true in science education: TEs can be used as an effective tool for teaching. By performing a TE before the physical experiment per se, students may develop their ability to see the focus of the physical experiment (Galili 2009). At times, the experiment can only be made mentally as a TE for practical reasons: the school may not have certain equipment or the experiment is too laborious to be conducted during a lesson. In some cases, thought experimenting is the only way to experiment because the situation cannot be performed as a physical experiment, regardless of the sophistication of the equipment available (Galili 2009). TEs also frequently involve idealisations such as technical details, errors, and impeding factors such as heat or friction; these factors can be eliminated by using TEs (Cooper 2005; Galili 2009). This particular use of TEs in school teaching may already be more common than might be expected.

TEs in science can be fallible, but the mistakes can also teach important lessons that help scientists to develop scientific theories. For instance, erroneous conclusions in famous TEs can be explained in terms of three different cognitive processes: strong *intuition*, which induces the abandonment of theory-based reasoning; the *incompleteness* of the basic assumptions of thought experiment; and the *irrelevance* of the properties of the system in the thought experiment (Reiner and Burko 2003). This kind of erroneous reasoning is also likely in the case of students; teachers should also be prepared to take it into account in their teaching (Velentzas and Halkia 2011). Teachers should also be prepared to encourage students to experiment mentally. As Ozdemir's (2009) results have shown, even physics graduates may tend to think that mental simulations cannot be used correctly to explain the phenomena of physics. Hence, teachers should be ready to help their students to become more open-minded and to be undaunted by errors in their reasoning. Teachers need to help their students to gain an insight into the value of thought experiments in scientific reasoning since they may otherwise remain unaware of it (Reiner 2006).

Thought experiments can be used in science teaching to allow students to see that scientific intuitions can play both destructive and constructive roles. It has, however, been observed that authors of science textbooks and popular science books may be in the habit of using only constructive TEs (Velentzas et al. 2007). This rather one-sided use of TEs may bias the image of science that the books attempt to convey. If the authors of textbooks aim at conveying an image of the processes of science, then the use of TEs in textbooks should be carefully designed to include both destructive and constructive TEs.

It must also be emphasised that, when conceptually demanding thought experiments have been simplified for teaching a particular student group, it has been noted that thought experiments stimulate the students' interest (Velentzas et al. 2007). Our own approach tends to agree with that of other researchers who acknowledge that this use of TEs works well if the concepts are taught in greater detail at a later stage. Reconstruction of historical physical experiments as thought experiments has also been reported to enhance students' interest (Klassen 2006).

The role played by a skilful teacher is pivotal in the use of thought experiments in science teaching. Students' own thought experimenting needs to be supported by the teacher by means of the selection of suitable resources, the structuring of the learning activities, and guidance of the students' experimentation (Hennessy et al. 2007). A skilful teacher is able to observe instances of erroneous reasoning and knows how to guide students' learning processes in the right direction. To be able to evaluate thought experiments in science textbooks and also thought experiments implemented by students, a teacher should present or formulate the theoretical background and criteria for the elements of a TE. Gilbert and Reiner (2000, p. 268) provide a system of categorisation for thought experiments that appears to be promising for understanding the use of TEs in science teaching. The categorisation is briefly as follows. An *expressed thought experiment* is a TE that has been placed in the public domain by an individual or a group of researchers. A *consensus thought experiment* is a TE accepted by at least some of the scientific community and one that has been scientifically justified, that is, published in a scientific journal.

In addition, a *historical thought experiment* is a TE that has already been replaced in science but may still be used to explain particular phenomena economically. A *teaching thought experiment* contains "the criterion by the teacher (or, indeed, the taught) of the TE based on the situations familiar to or imaginable by the students, through which to develop an understanding of a given consensus TE". Gilbert and Reiner emphasise that all of the different types of TEs include the six elements of TEs described by Reiner (1998).

As Gilbert and Reiner (2000) point out, although different types of thought experiment exist in science, they all have a certain structure. Hence, thought experiments devised and conducted by students should also include these common elements in order to qualify as genuine thought experiments; if some of the elements are missing, then the exercise should be termed a thought simulation rather than a thought experiment (Gilbert and Reiner 2000). According to some studies, historical TEs have sometimes been modernised in textbooks to be more readily understandable (see, e.g. Velentzas et al. 2007). This reconstruction may, however, lead to another problem: textbooks do not always include all of the necessary elements of thought experiments, with the result that TEs that have been reduced as thought simulations will lead to loss of the necessary cognitive engagement (Gilbert and Reiner 2000). Such thought simulations may nevertheless be used to some extent in science teaching if the primary goal of the teaching is not the actual subject matter or to foster students' understanding of the processes of science but rather to stimulate the students' interest in the science per se. Naturally, it would be unreasonable to assume that, for instance, secondary students would be able to perform thought experiments as effectively as, say, university students. It is perhaps self-evident that the science teacher should have the freedom to decide just how accurate students' mentally performed experiments need to be for them to fulfil the criteria of a thought experiment.

Undoubtedly, TEs need to be considered carefully in the context of science teacher education, and in-service education would need to be organised for practicing teachers. Both pedagogical and subject-matter departments could introduce TEs to students as part of the history and philosophy of science teaching. In addition, many subject-matter courses, such as mechanics, thermodynamics, and quantum physics, offer good opportunities for the use of TEs in the teaching of subject matter. In this way, TEs could become better integrated into the knowledge structures of future science teachers, who could then use thought experiments flexibly in their own science teaching. As Matthews (1992, p. 28) suggests, "A historically and philosophically literate science teacher can assist students to grasp just how science captures, and does not capture, the real, subjective, lived world".

Systematic research into the use of TEs in science teaching is, however, definitely needed so that we can acquire further research-based, valid information on their effective use at various educational levels. In particular, the notion of a *teaching thought experiment* is interesting from the perspective of science teaching as conducted in schools. It would be interesting to discover the kind of TEs that teachers use and how they use them, and whether teachers use thought simulations (TSs) rather than TEs. It is likely that consensus and historical TEs are not widely used in teaching at secondary school level, but teaching TEs may nevertheless prove to be more common than is thought. Thus far, the groups participating in the studies have been small and they have varied from lower secondary school pupils to university students. In consequence, the results cannot be readily compared; and hence our recommendations for the use of TEs in teaching are inevitably still rather loosely based. Nevertheless, analysis of students' thought experiments has interesting possibilities that may help us to understand better the challenges posed by science learning. There is undoubtedly a need for further studies of how science teachers actually use TEs in their teaching. This gap in the literature deserves to be filled.

Acknowledgements We wish to thank Mick Nott, Tarja Kallio, John A. Stotesbury, and also the anonymous reviewers for their helpful critical comments.

References

Arthur, R. (1999). On thought experiments as a priori science. *International Studies in the Philosophy of Science, 13*(3), 215–229.

Bokulich, A. (2001). Rethinking thought experiments. *Perspectives on Science, 9*(3), 285–207.

Brown, J. R. (1986). Thought experiments since the scientific revolution. *International Studies in the Philosophy of Science, 1*(1), 1–15.

Brown, J.R. (1991). Thought experiments: A Platonic account. In T. Horowitz and G.J. Massey (Eds.), *Thought experiments in science and philosophy* (pp. 119–128). Unspecified. http://philsci-archive.pitt.edu/id/eprint/3190. Accessed June 7th 2012.

Brown, J.R., & Fehige, Y. (2010). Thought experiments. In E.N. Zalta (Ed.), *The Stanford Encyclopedia of Philosophy* (Winter 2010 Edition). http://plato.stanford.edu/archives/win2010/entries/thought-experiment/. Accessed November 1st 2011.

Buzzoni, M. (2009). Empirical thought experiments: A trascendental-operational view. *In Thought experiments: A workshop* (Toronto, May 22–23, 2009).

Cooper, R. (2005). Thought experiments. *Metaphilosophy, 36*(3), 328–347.

diSessa, A.A. (1993). Towards an epistemology of physics. *Cognition and Instruction, 10*(2 & 3), 105–225.

diSessa, A.A. (2002). Why "Conceptual Ecology" is a good idea. In M. Limón & L. Mason (Eds.), *Reconsidering conceptual change: Issues in theory and practice* (pp. 28–60).

diSessa, A.A., Gillespie, N.M, & Esterly, J.B. (2004). Coherence versus fragmentation in the development of the concept of force, *Cognitive Science, 28*(6), 843–900.

diSessa, A.A. & Sherin, B.L. (1998). What changes in conceptual change? *International Journal of Science Education, 20*(10), 1155–1191.

Driver, R., Leach, J., Scott, P., & Wood-Robinson, C. (1994). Young people's understanding of science concepts: implications of cross-age studies for curriculum planning. *Studies in Science Education, 24*(1), 75–100.

Duhem, P. (1990). Logical examinations of physical theory. *Synthese, 83*, 183–188.

Einstein, A. (1918) "Dialog über Einwände gegen die Relativitätstheorie", *Die Naturwissenschaften, 48*, 697–702. English translation: Dialog about objections against the theory of relativity. http://en.wikisource.org/wiki/Dialog_about_objections_against_the_theory_of_relativity. Accessed June 1st 2012.

Galili, I. (2009). Thought experiments: Determining their meaning. *Science & Education, 18*, 1–23.

Gendler, T. S. (1998). Galileo and the indispensability of scientific thought experiment. *British Journal of Philosophy of Science, 49*, 397–424.

Gendler, T. S. (2004). Thought experiments rethought – and reperceived. *Philosophy of Science, 71*(5), 1152–1163.

Gilbert, J. K., Boulter, C. & Rutherford, M. (1998). Models in explanations, Part 1: Horses for courses? *International Journal of Science Education, 20*(1), 83–97.

Gilbert, J. K. & Reiner. M. (2000). Thought experiments in science education: potential and current realization. *International Journal of Science Education, 22*(3), 265–283.

Hall, A.R. (2000). *Isaac Newton, adventurer in thought.* Cambridge: Cambridge University Press.

Hennessy, S., Wishart, J., Whitelock, D., Deaney, R., Brawn, R., la Velle, L., McFarlane, A.m Ruthven, & K., Winterbottom, M. (2007). Pedagogical approaches for technology-integrated science teaching, *Computers & Education, 48*(1), 137–152.

Hull, D. L. (1998). *Science as a process: an evolutionary account of the social and conceptual development of science.* Chicago: University of Chicago Press.

Hull, D. L. (2001), *Science and Selection. Essays on Biological Evolution and the Philosophy of Science.* Cambridge: Cambridge University Press.

Häggqvist, S. (1996). *Thought experiments in philosophy.* Stockholm: Almqvist & Wiksell International.

Irvine, A. D. (1991). Thought experiments in scientific reasoning. In T. Horowitz and G. Massey (Ed.), *Thought experiments in science and philosophy* (pp. 149–165).

King, P. (1991). Medieval thought-experiments: the metamethodology of medieval science. In T. Horowitz and G. Massey (Eds.), *Thought experiments in science and philosophy* (pp. 43–64). Unspecified. http://philsci-archive.pitt.edu/id/eprint/3190. Accessed June 7th 2012.

Klassen, S. (2006). *The science thought experiment: How might it be used profitably in the classroom? Interchange, 37*(1–2), 77–96.

Kozma, R.B.T & Russell, J. (1997). Multimedia and understanding: Expert and novice responses to different representations of chemical phenomena. *Journal of Research in Science Teaching, 34*(9), 949–968.

Lattery, M.J. (2001). Thought experiments in physics education: A simple and practical example. *Science & Education, 10*, 485–492.

Lehavi, Y. & Galili, I. (2009). The status of Galileo's law of free-fall and its implications for physics education. *American Journal of Physics, 77*(5), pp. 417–423.

Levitt, K.E. (2002). An analysis of elementary teachers' beliefs regarding the teaching and learning of science. *Science Education, 86*(1), 1–22.

Mach, E. (1976). On thought experiments. In W.O. Price and W. Krimsky (translated and adopted), *Knowledge and Error* (pp. 449–457).

Matthews, M.R. (1988). Ernst Mach and thought experiments in science. *Research in Science Education, 18*, 251–257.

Matthews, M.R. (1990). Ernst Mach and contemporary science education reforms. *International Journal of Science Education, 12*(3), 317–325.

Matthews, M.R. (1992). History, philosophy, and science teaching: The present rapprochement, *Science & Education, 1*(1), 11–47.

McAllister, J.W. (2004). Thought experiments and the belief in phenomena. *Philosophy of Science, 71*, 1164–1175.

McDermott, L.C. (1993). How we teach and how students learn. *Annals of the New York Academy of Sciences, 701*, 9–20.

Nersessian, N.J. (1989). Conceptual change in science and in science education. *Synthese, 80*, 163–183.

Nersessian, N.J. (1992). How do scientists think? Capturing the dynamics of conceptual change in science. In R. Giere (Ed.), *Cognitive Models of Science* (pp. 3–44). Minneapolis: University of Minnesota Press.

Nersessian, N.J. & Patton, C. (2009). Model-based reasoning in interdisciplinary engineering, in A. Meijers (Ed.) *Handbook of the Philosophy of Technology and Engineering Sciences* (pp. 687–718). Amsterdam: Elsevier.

Newton, I. (1728). *A treatise of the system of the world.* Printed for F. Fayram.

Newton, I. (1863). *Newton's principia. Sections I. II. III.* Cambridge and London: McMillan.

Norton, J. D. (1991). Thought experiments in Einstein's work. In T. Horowitz, & G. Massey (Eds.), *Thought experiments in science and philosophy* (pp. 129–148). Unspecified. http://philsci-archive.pitt.edu/id/eprint/3190. Accessed June 7th 2012.

Norton, J. D. (1996). Are thought experiments just what you thought? *Canadian Journal of Philosophy, 26*(3), 333–366.

Norton, J. D. (2004a). On thought experiments: Is there more to the argument? *Philosophy of Science, 71*, 1139–1151.

Norton, J. D. (2004b). Why thought experiments do not transcend empiricism. In C. Hitchcock (Ed.), *Contemporary Debates in the Philosophy of Science.* Bodmin: Blackwell.

Ozdemir, O. F. (2009). Avoidance from thought experiments: Fear of misconception, *International Journal of Science Education, 31*(8), 1049–1068.

Palmieri, P. (2003). Mental models in Galileo's early mathematization of nature. *Studies in History and Philosophy of Science, 34*, 229–264.

Palmieri, P. (2005). Spuntar lo scoglio piu` duro: did Galileo ever think the most beautiful thought experiment in the history of science? *Studies in History and Philosophy of Science, 36*, 223–240.

Peierls, R. (1980). Model-making in physics. *Contemporary Physics*, 21, 3–17.

Radhakrishnamurty, P. (2010). Maxwell's demon and the second law of thermodynamics. *Resonance*, June, 548–560.

Redish, E. F. (2004). A theoretical framework for physics education research: Modeling student thinking. In E.F. Redish & M. Vicentini (Eds.), *Proceedings of the International School of Physics "Enrico Fermi", Course CLVI, Research on Physics Education, volume 156* (pp. 1–63). Bologna: Societa Italiana di Fisica/IOS Press.

Reiner, M. (1998). Thought experiments and collaborative learning in physics, *International Journal of Science Education*, 20(9), 1043–1058.

Reiner, M. (2006). The context of thought experiments in physics learning. *Interchange*, 37(1), 97–113.

Reiner, M., & Burko, L. M. (2003). On the limitations of thought experiments in physics and the consequences for physics education. *Science & Education*, 12, 365–385.

Reiner, M. & Gilbert, J.K. (2008). When an image turns into knowledge: The role of visualization in thought experimentation. In Gilbert, J.K., Reiner, M. & Nakhleh, M. (Eds.), *Visualization: Theory and Practice in Science Education* (pp. 295–309). Surrey: Springer.

Reiner, M., Pea, R.D., & Shulman, D.J. (1995). Impact of simulator-based instruction on diagramming in geometrical optics by introductory physics students. *Journal of Science Education and Technology*, 4(3), 199–226.

Rescher, N. (1991). Thought experiments in presocratic philosophy. In Horowitz and Massey (Eds.), *Thought Experiments in Science and Philosophy* (pp. 31–42).

Sadler-Smith, E., & Shefy, E. (2004). The intuitive executive: Understanding and applying 'gut feel' in decision making. *Academy of Management Executive*, 18, 76–91.

Schlesinger, G.N. (1996). The power of thought experiments. *Foundations of Physics*, 26(4), 467–482.

Schrödinger, E. (1935). Die gegenwärtige Situation in der Quantenmechanik. *Die Naturwissenschaften*, 23, 823–828.

Sorensen, R.A. (1992). *Though experiments*. New York: Oxford University Press.

Stannard, R. (1991). *Black holes and uncle Albert*. London: Faber and Faber.

Velentzas, A. & Halkia, K. (2010). The use of thought and hands-on experiments in teaching physics. In M. Kalogiannakis, D. Stavrou & P. Michaelidis (Eds.) *Proceedings of the 7th International Conference on Hands-on Science*. 25–31 July 2010, Rethymno-Crete, pp. 284–289.

Velentzas, A., & Halkia, K. (2011). The 'Heisenberg's Microscope' as an example of using thought experiments in teaching physics theories to students of the upper secondary school. *Research in Science Education*, 41, 525–539.

Velentzas, A., & Halkia, K. (2012). The use of thought experiments in teaching physics to upper secondary-level students: Two examples from the theory of relativity, *International Journal of Science Education*. DOI:10.1080/09500693.2012.682182

Velentzas, A., Halkia, K., & Skordoulis, C. (2007). Thought experiments in the theory of relativity and in quantum mechanics: Their presence in textbooks and in popular science books. *Science & Education*, 16(3–5), 353–370.

Vosniadou, S. (1994). Capturing and modeling the process of conceptual change, *Learning and Instruction*, 4(1), 45–69.

Vosniadou, S., Vamvakoussi, X., & Skopeliti, I. (2008). The framework theory approach to the problem of conceptual change. In S. Vosniadou (Ed.), *International Handbook of Research on Conceptual Change* (pp. 3–34). New York: Routledge.

Yore, L.D. (1991). Secondary science teachers' attitudes toward and beliefs about science reading and science textbooks. *Journal of Research in Science Teaching*, 28(1), 55–72.

Zeilinger, A. (1999). Experiment and the foundations of quantum physics. *Reviews of Modern Physics*, 71(2), S288–297.

Mervi A. Asikainen is a Senior Lecturer at the Department of Physics and Mathematics, University of Eastern Finland. She is a key person in the Physics Education Research Group of the University of Eastern Finland (PERG-UEF). She is responsible for the methodological issues of physics and mathematics education research in the group. She has M.Sc. in physics and a competence to teach physics, mathematics, and chemistry in secondary schools. Her Ph.D. thesis in physics focused on the learning of quantum phenomena and objects in physics teacher education. She has taught physics in various in-service teacher education courses for both primary and secondary school teachers and worked as a school teacher. At the moment, she teaches courses Basic Physics II, Basic Physics IV, and Quantum and Atomic Physics. Her main interests are the methodology of science education research, teacher knowledge in physics and mathematics, and teaching and learning of university physics. She has published in *European Journal of Physics, American Journal of Physics, Research in Science Education*, and *Journal of Science Teacher Education*.

Pekka E. Hirvonen Associate Professor Hirvonen is the leader of the Physics and Mathematics Education Research Group of the University of Eastern Finland and the education unit of the Department of Physics and Mathematics. He received his Master's degree in 1996 with the pedagogical studies of the subject teacher, Licentiate degree in 1999, Ph.D. in physics in 2003, and the title of docent in 2010. He has published in journals such as *Science & Education, European Journal of Physics, American Journal of Physics, Research in Science Education*, and *Journal of Science Teacher Education*. The research concentrates on two main themes: teaching and learning different topics of physics and mathematics and research-based development work of physics and mathematics teacher education.

Part X
Theoretical Studies: Teaching, Learning and Understanding Science

Chapter 39
Philosophy of Education and Science Education: A Vital but Underdeveloped Relationship

Roland M. Schulz

> *It was through the feeling of wonder that men now and at first began to philosophize. ... but he who asks and wonders expresses his ignorance ... thus in order to gain knowledge they turned to philosophy.*
>
> —Aristotle *(Metaphysics)*

39.1 Introduction

This chapter examines the relationship between the two fields of science education and philosophy of education to inquire about how philosophy could better contribute to improving science curriculum, teaching, and learning, above all teacher education. The value of philosophy *for* science education in general remains underappreciated at both pedagogical levels, whether the research field or classroom practice. While it can be admitted that philosophy has been an area of limited and scattered interest for researchers for some time, it can be considered a truism that modern science teacher education has tended overall to bypass philosophy and philosophy of education for studies in psychology and cognitive science, especially their theories of learning and development (which continue to dominate the research field; Lee et al. 2009). A major turn encompassing philosophy would thus represent an *alternative approach* (Roberts and Russell 1975).

Science education is known to have borrowed ideas from pedagogues and philosophers in the past (e.g., from Rousseau, Pestalozzi, Herbart, and Dewey; DeBoer 1991); however, the subfield of *philosophy of education* has been little canvassed and remains on the whole an underdeveloped area. At first glance such a state of affairs may not seem all too surprising since science education is mainly concerned

R.M. Schulz (✉)
Imaginative Education Research Group (IERG), Faculty of Education,
Simon Fraser University, Vancouver, BC, Canada
e-mail: rmschulz@shaw.ca

M.R. Matthews (ed.), *International Handbook of Research in History,*
Philosophy and Science Teaching, DOI 10.1007/978-94-007-7654-8_39,
© Springer Science+Business Media Dordrecht 2014

with educating students about particular science subjects or disciplines. But this necessarily implies a tight link between content and education. Hence, if education is to mean more than mere instructional techniques with associated texts to encompass broader aims including ideals about what constitutes an educated citizen (i.e., defining "scientific literacy") or foundational questions about the nature of education, learning, knowledge, or science, then philosophy *must* come into view (Nola and Irzik 2005). As is known, an *education in* science can be, and has been, associated with narrow technical training, or with wider liberal education, or with social relevance (STSE), or lately with "science for engineers" (US STEM reforms), an updated version of the older vocational interest.[1] Yet all these diverse curricular directions imply or assume a particular educational philosophy which is rarely clearly articulated (Matthews 1994a/2014; Roberts 1988; Schulz 2009a).

At second glance then, and viewing science education in a broader light, being principally at home in education unavoidably implies an excursion into philosophy of education. In fact, it avoids this subfield of philosophy at its own peril, as argued elsewhere (Matthews 2002; Schulz 2009a, b). Equally, there are lessons to be learned from its own past, yet most science teachers and too many researchers seem little aware, or even concerned to know, about the rich educational philo-historical background of science education as it has developed to the present, whether in North America, Europe, or elsewhere (some examples are Mach, Dewey, Westaway, and Schwab; DeBoer 1991; Gilead 2011; Matthews 1990b). In fact recent critical reviews insist educators must acknowledge and respond to how past historical developments have molded science education while continuing to adversely shape the current institutionalization of school science (Jenkins 2007; Rudolph 2002). A central concern of this chapter is to emphasize the value of philosophy in general and philosophy of education in particular. It will be claimed an awareness of the worth of these fields can have positive results for further defining the *identity* of both the science teacher as professional (Van Driel and Abell 2010; Clough et al. 2009) and science education as a research field (Fensham 2004). The perspective to be taken on board is that to teach science is to have a philosophical frame of mind—about the subject, about education, and about one's identity.

39.2 Philosophy of Science Education Framework

To be clear from the start, there is no attempt made here at formulating a particular philosophical position thought appropriate for science education, in contrast to such discussions having taken place in mathematics education for some time. In that field

---

[1] The prominent US *National Science Teachers Association* (NSTA) has made STEM a central reform emphasis: www.nsta.org/stem. References for the other more common science classroom curricular emphases are Aikenhead (1997, 2002, 2007), Carson (1998), DeBoer (1991), Donnelly (2001, 2004, 2006), Pedretti and Nazir (2011), Roberts (1982), Schwab (1978), Witz (2000), and Yager (1996).

several educators have articulated and debated the notion of a "philosophy of mathematics education," for example, Platonism and foundationalism versus social constructivism and fallibilism (Ernest 1991; Rowlands et al. 2011). On the other hand, it will be stressed that the development of a "philosophy *of* science education," that is, an "in-house philosophy" for the field, could be significant for reforming science education. It can be acknowledged that math educators have been in the forefront of attempting to establish a "philosophy of" for their educational discipline, while science educators in the main have not yet come to consider or value such an overt evolution in their field. Such an endeavor urges exploration of new intellectual territory.

The sign of the times seems ripe for such an investigation ever since the science educational field became staked out by opposing, even irreconcilable positions "from positivism to postmodernism" (Loving 1997).

In the past constructivism was once seen by many educators as a kind of "philosophy" (though not expressed as such) which was to serve the role as a "new paradigm" of science education. Today, however, this view is considered mistaken, although the topic is divisive (Matthews 2002; Phillips 1997, 2000; Suchting 1992).[2] This judgment has come about largely because many supporters at the time did not reflect seriously enough about the philosophical underpinning of its various forms—cognitive, metaphysical, and epistemological.[3] Constructivism remains a dominant and controversial topic in education, but one lesson to be had from the heated debate of the past three decades is that absence of philosophical training among science educators became apparent (Matthews 2009b; Nola and Irzik 2005). Another lesson learned is the absence of any explicit discussion regarding educational philosophy, even though constructivism in some corners was brashly substituted for one. In hindsight it surprises that constructivism—which after all still finds its principal value as learning theory (and perhaps teaching method)—could be mistaken as a dominant kind of "philosophy of" science education at the neglect of broader aims and concerns relevant to educational philosophy, as to what it *means* to educate someone in the sciences. And science education once again showed unawareness of its own history, since Dewey (1916, 1938, 1945) and Schwab (1978) had previously addressed such concerns. At minimum the case of constructivism had illustrated—although not widely recognized—how interwoven, if not dependent, science education in the academy had become with certain psychological ideas and philosophy of science (notably its Kuhnian version; Matthews 2003a).

In light of this background, it will be of some interest to teachers and researchers to raise anew the question of developing a "philosophy of" science education (PSE),

---

[2] "Regrettably, much of the constructivist literature relating to education has lacked precision in the use of language and thereby too readily confused theories of knowledge with ideas about how students learn and should be taught" (Jenkins 2009, p. 75).

[3] The literature on constructivism is vast. Critiques are found in Davson-Galle (1999), Phillips (2000), Grandy (2009), Kelly (1997), Matthews (1998b, 2000), and Scerri (2003). Also see chapter 31 in Handbook.

by asking here what that could *mean* and could *offer* the discipline. The intent is to address these concerns and help sketch out contours. With this project in mind, one can draw attention to two useful aspects pertaining to philosophy in general which can come to our aid and contribute to improving science education and developing such a philosophical perspective: the ability of philosophy to provide a synthesis of ideas taken from associated disciplines with their major educational implications and providing what can be called "philosophies of." In this way it will be shown how philosophical thought can be brought to bear directly on educational ideas and practice.

39.2.1 The Synoptic Framework

The role and value that philosophy itself and its two important subdisciplines of *philosophy of science* (PS) and *philosophy of education* (PE) can have is illustrated by the representation below. Note that "philosophy of science education" (PSE) can then be understood as the *intersection* or *synthesis* of (at least) three academic fields. For each respective field of study, some individual points are stressed which comprise core topics of interest to science education pertinent to each, but is meant to be illustrative not exhaustive:

The framework in itself assumes neither prior philosophical positions (e.g., metaphysical realism or epistemological relativism) nor pedagogical approaches (e.g., constructivism, multiculturalism, sociopolitical activism). As a graphic organizer it does provide science teachers and researchers a holistic framework to undertake analysis of individual topics and perhaps help clarify their own thinking, bias, and positioning with respect to different approaches and ideas. The main point is to show that any particular PSE as it develops for the teacher or researcher should take into consideration, and deliberate upon, the discourses pertinent to the three other academic fields when they impinge upon key topics in science education. At minimum it should contribute to helping develop a philosophic mind-set.

In sum (as Fig. 39.1 shows), any philosophy of science education (PSE) is foremost a *philosophy* ("P") and as such receives its merit from whatever value is assigned to philosophy as a discipline of critical inquiry. (This value may not appear at all obvious to science educators.) Furthermore, such a philosophy would need to consider issues and developments in the philosophy, history, and sociology of science ("PS")[4] and analyze them for their appropriateness for improving learning *of* and *about* science. Finally, such a philosophy would need to consider issues and developments in the philosophy of education and curriculum theory ("PE") and analyze them for their appropriateness for education in science, as to what that can *mean* and how it could be conceived and best achieved. A fully developed or

---

[4] This component is meant to include the associated disciplines and not just the philosophy discipline itself.

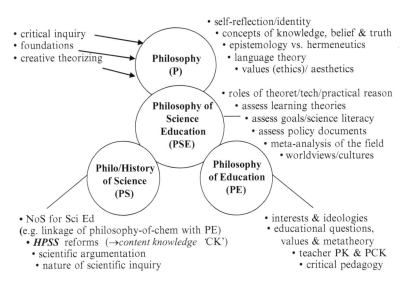

Fig. 39.1 Philosophy of science education (PSE)

"mature" PSE can be understood as an integration of all three fields. It ultimately aims at improving science education as a research field as well as assisting teachers in broadening their theoretical frameworks and enhancing their practice.

39.2.2 Providing "Philosophies of"

Philosophy today has evolved into several specialized subdisciplines. These include philosophy of science, of education, of mathematics, of technology, of history, of religion, and others, which can collectively be called "philosophies of." It is especially the first two that are of immediate concern to us when developing one for ourselves, as Fig. 39.1 illustrates. And yet this conceptualization is not as new as it may appear. Over 40 years ago, the philosopher Israel Scheffler summarized the value of these "philosophies of" for science educators:

> I have outlined four main efforts through which philosophies-of might contribute to education: (1) the analytical description of forms of thought represented by teaching subjects; (2) the evaluation and criticism of such forms of thought, (3) the analysis of specific materials so as to systematize and exhibit them as exemplifications of forms of thought; and (4) the interpretation of particular exemplifications in terms accessible to the novice. (Scheffler 1970, p. 392)

He understood these "philosophies of" would provide invaluable components to a science teacher's identity and preparation, in addition to the common three of (i) subject matter competence, (ii) practice in teaching, and (iii) educational methodology. Especially the inclusion of philosophy of science (PS) topics he considered vital to allow teachers to be "challenged to reflect deeply on the foundations" of their subjects and "to relate their reflections to the task of teaching" (p.388).

Matthews (1994a, b, 1997) is known to have helped popularize Scheffler's earlier vision, whose call for inclusion of PS has been broadly acknowledged today though unfortunately little implemented worldwide in teacher education programs.[5] He has expanded upon Scheffler's line of reasoning to include additional pedagogical and professional arguments. An improved pedagogy, for example, should include several aspects: wisely evaluating constructivism and the educational aims of curricular documents, integrating HPSS topics, developing critical thinking, allowing science courses to show a "human face," and at minimum making science more interesting and understandable. Enhancing professionalism requires teachers to develop a wider perspective of their subject and its role in education, including becoming versed with topics and questions associated with science and society concerns. These would include religion and science, "multicultural science," feminism, techno-science, environmental ethics, animal rights, and others.

In short, philosophical questions concerning both education and science are at the heart of the science education profession, many of which have kept, and continue to keep, teachers, researchers, and curriculum developers engaged. Broadly speaking, they encompass essential concerns immediately identifiable with the two fields of philosophy of education (PE) and philosophy of science (PS):

> ... educational ones about the place of science in the curriculum, and how learning science contributes to the ideal of an educated citizen and the promotion of a modern and mature society. The questions also cover the subject matter of science itself. What is the nature of science? What is the status of its knowledge claims? Does it presuppose any particular worldview? The first category of questions constitutes standard philosophy of education (PE); the second category constitutes philosophy of science (PS) or history and philosophy of science (HPS). (Matthews 2002, p. 342)

The teacher's professional role today has in some cases also come to include cocreating, advising, and assessing so-called national science "standards" documents. Since the 1990s several countries around the world have sought to define curriculum "standards," which harbor considerable agreement on nature-of-science policy statements (McComas and Olson 1998). "Clearly all these curricular exhortations depend on teachers having philosophical acumen and knowledge in order to understand, appraise, and enact them. This requires a mixture of philosophy of science (to understand the substantial claims), and philosophy of education (to interpret and embrace the objectives of the curricula)" (Matthews 2002, p. 343). The same arguments and considerations apply to mathematics education where philosophy *of* mathematics is integral to what, why, and how mathematics is taught and assessed and how teachers understand their professional role and responsibilities.

In the sections below, the intention is to further elaborate on the worth of philosophy as a subject in general, but especially philosophy of education since

---

[5] Whether or not science students themselves should be presented with PS ideas and controversies is still being debated among researchers (Hodson 2009). One philosopher of education has reversed his earlier standpoint (Davson-Galle 1994, 2004, 2008a).

this topic is usually overlooked. Philosophy of science for educators will only be glossed (above all its newer subspecialties) as this topic has been an active area of research.

39.3 The Value of Philosophy

Philosophy is an academic discipline that seeks to establish a systematic reflection on reality however it may be construed. Its analytic function, often termed rational inquiry, involves critical appraisal of different topics, beliefs, and schools of thought.[6] Because of the complexity of the world around us (both natural and artificial), philosophy has been traditionally divided into separate major fields of study (first accredited to Aristotle) such as metaphysics, epistemology, logic, ethics, aesthetics, and politics. These fields individually have either major or lesser bearing on science education directly. The *first two* have played a significant historical role pertaining to our understanding of the nature of reality, of knowledge, and of science:[7]

- *Ontology*: the branch of philosophy (metaphysics) that concerns itself with the most general questions of the nature or structure of reality: what "is" or "what is *being*?" and existence. It examines natural and supernatural claims and asks about the feature of essences (e.g., are natural kinds, like species, universal or nominal?). Questions regarding *scientific* ontology are concerned with ascertaining the status (or validity) of the products of human creativity or discovery; included are scientific models and theoretical entities (e.g., gene, field, black hole, tectonic plates), evaluated as to their truth (realism) or merely useful (fictive) construct to solve problems and "fit" experimental data (empirical adequacy).
- *Epistemology*: the branch of philosophy that studies the nature of knowledge, its scope, foundations, and validity; it deals with theories of knowledge, distinctions between believing and knowing, and justification. *Scientific* epistemology is concerned with describing and ascertaining the nature of both the body of known scientific facts and theories (degree of certainty) *and* the production of new knowledge (i.e., scientific inquiry). *Personal* epistemologies are commonly taken to include individual beliefs, views, and attitudes about a particular subject; hence, they can be considered a "personal knowledge framework"

---

[6] It has also been historically associated with particular schools of thought (e.g., idealism, rationalism, empiricism, existentialism); hence, particular *philosophies* which themselves are often associated with individual philosophers (e.g., Plato, Kant, Marx, Nietzsche).

[7] This is not meant to discount the next three. Logic has made a renewed appearance in science education under the guise of critical thinking and scientific argumentation; those in ethics intersect with discussions of values and socio-ethical issues (Allchin 2001; Corrigan et al. 2007; Witz 1996; Zeidler and Sadler 2008); even aesthetics has been considered for the field (Girod 2007).

(i.e., "what do you know about 'X', and how do you know (it)?"). Two competing views of epistemic justification are *foundationalism* and *coherentism*.[8]

As mentioned, the significance of *philosophy of science* for science education is generally recognized today—though moreso among researchers than science teachers themselves (Duschl 1994, 1988; Hodson 2008; Matthews 1994a)—while philosophy per se is accorded much lesser importance, notwithstanding the limited forays by some researchers into its subfields, which can be acknowledged (e.g., language studies, post-structuralism, hermeneutics, scientific argumentation, "critical theory"). Why this situation has arisen and persists is an open question and would require its own socio-empirical research, and hence is not of immediate concern of this review. But it remains an important question that should be pursued as it could reveal much about our community, about how science education is perceived and undertaken. In other words, it aims at the core of the self-understanding of science education as profession and identity (Fensham 2004).

A familiar question posed by preservice and science teachers alike is: "What does philosophy have to do with science?" or more succinctly and less pejoratively "how can any sort of 'philosophy' contribute to helping my students better understand difficult *scientific* concepts?" Such questions implicitly assume of course a deep divide between science and philosophy, certainly between science education and philosophy.[9] While science teachers need not be openly hostile to philosophy, they certainly appear indifferent. Much responsibility can be laid at the door of the academy, its structure, culture, and teacher training. Their attitudes and preparation effectively expose much about how teacher identity is formed,[10] about preconceptions of knowledge, but also about the nature of university science education and scientific specialization, including the nature and influence of science textbooks.[11]

Classroom teachers tend to be more concerned with valuable but mundane matters of decision-making regarding immediate instruction, learning, and assessment. For them as pertains their professional duties and identity, these concerns have little if anything to do with philosophy—or so it would seem. A consequence of this disregard makes providing educational rationales of their thinking and practice a

---

[8] See the chapters in Bonjour and Sosa (2003) for a concise overview; Sect. 39.5.2.2 targets the former.

[9] That one must inevitably justify the value of philosophy for teachers and many researchers suggests a cultural predicament already exists concerning what constitutes "education" in our present age.

[10] Which includes essentially their "orientations" towards teaching, identified in science teacher education research as formative dispositions attached to identity (Van Driel and Abell 2010; Witz and Lee 2009)

[11] Probably the ongoing reality of the academic divide between the "two cultures" maintained as two solitudes in universities to this day (as described by C.P. Snow; Shamos 1995; Stinner 1989) contributes to the hostility or indifference since science teachers are not generally required to endure Arts faculty courses. All this in combination with the common negative *image* that academic philosophy is preoccupied with obtuse speculation, arcane technical jargon, and unresolved disputes are remote from everyday matters. Certainly quite different, encouraging evaluations can be had (Matthews 1994a; Nola and Irzik 2005).

challenge: "When planning lessons, teachers often struggle when asked to express how they decide what science content within a discipline is worth teaching. Rationales are post-hoc and rarely reflect deep thinking about the structure of the discipline, or how students learn ..." (Clough et al. 2009, p. 833).[12] Their struggles become quite apparent when further asked to give an explicit account of their "philosophy of teaching" or "philosophy of learning." And this counts not just for content teaching and conceptions of learning but equally for providing truly *educational* objectives for either their individual courses or overall science education.[13] Seldom are the contextual aspects of teaching the subject matter made explicit even though *seven* competing "curriculum emphases" have been identified in science educational history (Roberts 1988). In effect, particular curricular emphases bear witness to buried educational philosophies. The teaching profession itself is mired in a scenario of what Roberts (2007) has astutely identified as two substantial conflicting "visions" of science education.[14] These facts alone warrant developing philosophical acumen for teachers.

If this picture as sketched is indicative of teacher training and science education culture, then emphasizing the significance of philosophy, especially philosophy of education, would require a "paradigm shift" in thinking. Exactly this sort of thing had been recommended by Jenkins (2000) for effective reform of that culture, although the present proposal would encompass a wider scope than was initially suggested.[15]

---

[12] They continue: "... Too often the selected textbook defines course scope, sequence, and depth implying that a textbook's inclusion of information, in part, legitimizes teaching that content. Textbooks also exert a significant influence on *how* content is taught..." (ibid).

[13] Many teachers would probably declare "science for all" or "scientific literacy" though seldom with awareness these slogans are replete with ambiguities—the latter goal even suffering inherent incompatibilities due to serious shifts in connotation, and this despite its ultimate prominence in worldwide "standards" documents (Jenkins 2009; Schulz 2009b; Shamos 1995). The science for all theme arguably partially appropriate for junior science nonetheless vanishes when specialty upper secondary or tertiary courses are reached, for here the status quo is maintained as "technical pre-professional training" (Aikenhead 1997, 2002, 2007). In this case an extreme narrowing of the "literacy" notion is found, HPSS aspects are distorted or abused, while the concealment of existent curriculum ideologies remains unrecognized in absence of educational philosophy (e.g., scientism, academic rationalism, "curriculum as technology" or social utility; Eisner 1992).

[14] In his comprehensive review, the categories "vision I" and "vision II" were postulated to account for two major competing images of science literacy behind many curricular reforms. The former designates those conceptions which are "internally oriented," that is, towards science as a knowledge- and inquiry-based discipline and including the image of science education as heavily influenced by the identity, demands, and conceptions of the profession. The latter vision, alternatively, is "outward looking," towards the application, limitation, and critical appraisal of science in society—the image influenced instead by the needs of society and the majority of students not headed for professional science-based careers. Here the question of the "social relevance" of the curriculum is paramount. He claims that while the second vision can encompass the first, the opposite is not true.

[15] For linked views, see Anderson (1992), Fensham (2004), Matthews (1994a/2014, 2002), and Schulz (2009a).

Philosophy in truth cannot be avoided, and not just for analyzing national "standards" documents, providing coherent rationales or detecting curricular ideologies. Science teachers inadvertently find themselves in its territory when confronted by diverse events, such as (i) explaining common scientific *terms* (like "law," "theory," "proof," "explanation," "observation"), or (ii) student-driven *quandaries* ("how do we know X?"; "do models reflect reality?"; "why are we studying this?", etc.), or (iii) when teacher and pupil together come across science-related public *controversies* (e.g., climate change, nuclear weapons, evolution versus intelligent design)—never mind popular beliefs and media reports (e.g., astrology or alien abductions). Such occurrences usually illustrate that "philosophy is not far below the surface" in any classroom (Matthews 1994a, p. 87). Moreover, the scientific tradition (as an integral part of Enlightenment culture) based on rationality, objectivity, and skepticism, which teachers have inherited, is equally challenged by strands of pseudoscience, irrationality, and credulity of the times (Hodson 2009; Slezak and Good 2011). How can teachers illustrate these differences, especially the distinction between valid and reliable knowledge claims from invalid ones (or natural from supernatural claims), without philosophical preparation? Yet it is not just the classroom, contemporary media discourse, or pop culture that is infused with questions, beliefs, claims, and counterclaims of philosophical significance, but likewise the evolution of science itself.

When science is seen historically, its development has always been interwoven with philosophical interests and debates, whether concerning epistemology, logic, metaphysics, or ethics (the major subfields of philosophy proper). A quick survey makes this evident: from debates on the nature of matter or motion in Ancient Greece to questions of logic, method, and truth with Galileo and Kepler during the Copernican revolution (or Descartes and Newton in the Enlightenment), also Lyell and Darwin concerning the age of the Earth or origin of species, respectively, in the nineteenth century (which saw the realist controversy about atoms in chemistry revived). Right down to our present age, philosophical controversies exist whether concerning the onto-epistemological debates in quantum mechanics or reduction in chemistry.[16]

The history of science, furthermore, is not simply a survey of fantastic discoveries, ideas, and theories as too many textbooks would imply, but is equally littered with discarded concepts and discredited theories (e.g., ether, epicycles, phlogiston, phrenology, caloric, Lamarck, Lysenkoism). Can teachers distinguish between quasi-histories and pseudo-histories or unmask how subject matter is organized to reflect the typical linear, non-controversial, and progressive accumulation of scientific knowledge, imitating the myth of "convergent realism?" (Kuhn 1970, 2000; Laudan 1981). The textbook's and one's personal view of scientific knowledge and

---

[16] It should not be forgotten that the seventeenth-century scientific revolution introduced "science" as a field of research and study under the academic umbrella of *Natural Philosophy* to distinguish it from the reigning scholasticism of the universities, hermeticism, and Neoplatonism. Our modern conception of the term and the severance of philosophy from science are of relatively recent origin. The division emerged historically as a development in intellectual thought and specialization, which evolved within European industrial society in the mid-nineteenth century.

its development both presume prior philosophical commitments (e.g., positivism? empiricism? naïve realism? critical realism? social constructivism?) (Duschl 1988; Säther 2003; Selley 1989; Smolicz and Nunan 1975).

Regarding ethics, one should not forget that Socrates was condemned on moral and religious grounds—as were Bruno, Galileo, and Darwin (though not exclusively). Eugenics, once the scientific "hard core" of the social Darwinism movement, was considered a legitimate topic of scientific research less than a century ago. Even modern physics cannot escape this subject, ever since Oppenheimer made the self-incriminating remark that physicists "had known sin" by developing the atomic bomb. The American philosopher C.S. Pierce had stated: "Find a scientific man who proposes to get along without any metaphysics . . . and you have found one whose doctrines are thoroughly vitiated by the crude and uncriticised metaphysics with which they are packed."[17] Studies in history, philosophy, and sociology of the sciences (HPSS) have made this claim abundantly apparent. These fields cannot be either ignored or glossed during science teacher education, but require time and attention for the emergence of an adequate PSE.

We have already noted the worth of philosophy (along with key aspects mentioned above) to lie in providing teachers with both (i) the perspective for synthesis of their educational enterprise by developing a PSE framework and (ii) making available to them in-depth studies termed "philosophies of." Linked to the latter, coming again to philosophy of science (appearing as the "PS" corner of the Fig. 39.1 triangle), teachers need to be made aware that in the past 20 years, new avenues of scholarship have been developing *within* the subfield itself to help them expand their foundational understanding of their specialty (e.g., philosophy of chemistry, philosophy of biology).[18] Here questions concerning major issues in subject matter content that bear directly on senior courses are being discussed. For example, there is dissention whether laws and explanations in biology and chemistry are of the same order and function as those in physics—normally taken for granted in PS literature.[19] Such "cutting-edge" philosophical research has acute ramifications for secondary and postsecondary education, expressly *subject epistemology*, including nature-of-science discourse (Irzik and Nola 2011; Jenkins 2009; Matthews 1998a).[20]

---

[17] Quoted in Matthews (1994a), p. 84.

[18] Scientists and philosophers alike have found it necessary to launch important new *subdisciplines* to address foundational questions and concerns arising from their scientific areas of expertise—notwithstanding those scientists who disparage the study of PS overall (e.g., Weinberg 1992). Philosophy of physics (Cushing 1998; Lange 2002), philosophy of chemistry (McIntyre 2007; Scerri 2001), and philosophy of biology (Ayala and Arp 2009) are becoming established research fields, including philosophy of technology (Scharff 2002), likewise lauded for teachers today (De Vries 2005).

[19] Refer to the respective chapter in this Handbook.

[20] Unfortunately it appears that science education worldwide and many science teachers themselves have tended not to keep abreast of these advances and what they possibly offer for curriculum design, instruction, and reform efforts. One might hope these subdisciplines offer, minimally, deeper and improved insights about subject content but, moreso, a better understanding of the essence of the discipline, the core of which teachers are required to inspire and impart to their

In addition to the above mentioned reasons, the worth of philosophy plainly lies in self-reflection. This means nothing less than to reassess one's own practice, educational ideas and aims; even going so far as to reevaluate one's own constructed sociocultural science teacher *identity*. Along with suggesting "philosophies of," Scheffler also argued that science teachers require philosophy as a "second-order" reflective capacity into the nature of their work, their understanding of science, and their educational endeavors. He considered this capacity analogous to the role philosophy of science plays when examining science:

> The teacher requires ... a general conceptual grasp of science and a capacity to formulate and explain its workings to the outsider ... No matter what additional resources the teacher may draw on, he needs at least to assume the standpoint of philosophy in performing his work ... Unlike the researcher [or the academic] he cannot isolate himself within the protective walls of some scientific specialty; he functions willy-nilly as a philosopher in critical aspects of his role. (Scheffler 1970, p. 389)[21]

These proposals of Scheffler can equally be associated today with requirements to enhance teachers' "pedagogical content knowledge" (PCK: Abell 2007; Van Driel et al. 1998; Shulman 1987), which not only means developing *their epistemology* of science (Matthews 1994a/2014, 1997) but in addition their familiarity with philosophy of education topics (Matthews 2002; Schulz 2009a; Waks 2008). Again, Fig. 39.1 displaying the philosophy of science education (PSE) framework identifies these important aspects and illustrates how they are related to, and embedded within, the three corresponding dimensions of P, PS, and PE.

Philosophy in a nutshell then corresponds to the ancient Socratic dictum to examine oneself, and that "the unexamined life is not worth living." Transposing this motto, "the unexamined pedagogy is not worth doing"; in fact it is unsuccessful (as conceptual change research has uncovered)—if not harmful (i.e., indoctrination into scientism[22]).

Such an examination aligns with Kant's famous definition of Enlightenment as the emergence from one's self-imposed immaturity (due to reliance upon

---

students. Certainly these are less well known to science teachers and not canvassed by science education researchers to the extent of interest shown in the post-structuralist and "science studies" literature. See Allchin (2004), Collins (2007), Hodson (2008), Holton (2003), Kelly et al. (1993), Nola and Irzik (2005), Ogborn (1995), Roth and McGinn (1998), and Slezak (1994a, b).

[21] With such a faculty, teachers could better function in their role as *mediator* between the scientific establishment and their pupils, also between public discourse about science with pupils or adults not conversant either how science evolves or the nature of modern techno-science (see also Hodson 2009).

[22] The term "scientism" can be interpreted in different ways; most construe it negatively (Bauer 1992; Haack 2003; Habermas 1968; Matthews 1994a/2014). Nadeau and Desautels (1984) attribute five components. Irzik and Nola (2009) are careful to distinguish legitimate scientific worldviews from illegitimate *scientistic* ones: "A scientific worldview need not be scientistic. Scientism, as we understand it, is an exclusionary and hegemonic worldview that claims that every worldview question can be best answered exclusively by the methods of science... that claims to be in no need of resources other than science. By contrast, a scientific worldview may appeal to philosophy, art, literature and so on, in addition to science. For example, scientific naturalism can go along with a version of humanism in order to answer worldview questions about the meaning of life."

outside authority), the ability to freely make use of one's own faculty of reason, to "have courage to use your own understanding!"(Kant 1784). This ambition is inherent of course to the *liberal education tradition* (Anderson 1980; Carson 1998; Matthews 1994a; Stinner 1989), the objective sought after when teachers desire students "think for themselves"—easily an identifiable historical goal of science education (DeBoer 1991, 2000; Schwab 1978). This is inclusive of the newer critical thinking movement (Bailin 2002; Siegel 1988, 1989; Smith and Siegel 2004). The primary focus here, however, is upon the further development of teachers' critical thinking and competence and their own capacity to judge not only curricular and policy documents, but above all their pedagogy, epistemological assumptions, and educational beliefs (whether implied by their textbooks, e.g., naïve realism, inductivism, pseudo-history, or proposed by science educational literature, e.g., STEM, STSE, constructivism, postmodernism, science for social action). The topic of *critical thinking* is well-trodden ground in philosophy of education, although researchers seldom avail themselves of this literature (Bailin and Siegel 2003; Siegel 2003).[23]

Finally, as Wittgenstein (1953) stated, philosophy can even be *therapeutic*. Implied for our theme, this means it can alert science teachers to implicit *images* of science and philosophies of education they may hold unaware. Perhaps they have internalized these through practice or originally picked up through teacher training from university professors promoting their own pet educational ideas and theories. Indeed, the teacher may have developed strong opinions about HPSS or "social justice" topics, "but the point of education is to develop students' minds, which means giving students the knowledge and wherewithal to develop informed opinions" (Matthews 1997, p. 171). In any case, translating Pierce's statement above with science educators in mind, one can write: "Find a science educator who proposes to get along without any philosophy-of-education … and you have found one whose goals, perceptions and methods are thoroughly vitiated by a crude and uncriticised one with which they are packed."[24] While the textbook epistemology is often concealed, a teacher's epistemology and educational theory is usually pieced together during their career and rarely made explicit.

In summary, philosophy cannot be gone around, for as a discipline of critical inquiry, it allows analysis into curriculum, textbooks, learning, best practice, and identity. Relooking at our previous PSE triangle (Fig. 39.1), this includes (i) offering conceptual clarity; (ii) unmasking ideologies (social, political, educational); (iii) sorting out foundational aims, values, and teacher identities; (iv) providing perspectives and theoretical frameworks, as well as synoptic and integrative approaches; and (v) possibly even utilizing *creative* theorizing as solutions to pressing problems (discussed below on educational theory).

---

[23] Refer to the chapter contribution in this Handbook.

[24] What is being suggested here can be taken to correspond with a key objective of critical pedagogy, popularized by the Marxist teacher educator Paulo Freire (1970), their advance to "critical consciousness."

39.4 Philosophy of Education and Science Education

Philosophy of education, as mentioned, is a branch of philosophy. It seeks to address questions relating to the aims, nature, and problems of education. As a discipline it is "…Janus-faced, looking both inward to the parent discipline of philosophy and outward to educational practice … This dual focus requires it to work on both sides of the traditional divide between theory and practice, taking as its subject matter both basic philosophical issues (e.g., the nature of knowledge) and more specific issues arising from educational practice (e.g., the desirability of standardized testing)" (Siegel 2007). Thoughtful consideration of educational practice and assessing science curriculum is normally considered part of a teacher's professional competence; hence, some sort of philosophical thinking can be justifiably attributed to educators and researchers. What is of issue is the view that science educators can be encouraged to philosophize on a broader and systematic scale, and they can profit from philosophy of education (PE) studies (using their in-depth deliberations on theory and practice).

39.4.1 The Neglect of Philosophy of Education

If as Aristotle (1998) had intimated (by the opening quote) philosophy begins when one is filled with wonder—a state of being which can arise when confronted with some dilemma (hence one's *lack* of knowledge)—then the neglect to articulate a *systematic philosophy of* (PSE) for one's own science pedagogy (let alone the research field) causes one to ponder why so little effort and time have been invested into the subject. The consequences have not been a minor matter—confusion over educational *aims* including the "science literacy" debate, its meaning and competing "visions"[25]; science education's dependence on socio-utilitarian ideologies and competing group interests; science teachers' confusion about their identity and purpose, including the divide between belief and practice; etc.[26]

Jenkins (2001) has rightfully complained the research field is too narrowly construed and suffers from "an over-technical and over-instrumental approach" at the expense of other perspectives, such as neglecting historical studies. Although some recent research work can be taken as mitigating this charge (Gilead 2011; Jenkins 2007; Olesko 2006), even his perceptive critique had failed to mention the worth of

---

[25] Science education to this day has been unable to resolve the principal dilemma concerning the conflict of the two competing "visions" of its purpose (hence competing conceptions of "scientific literacy"). Roberts (2007, p. 741) admits the community must "somehow resolve the problems associated with educating two very different student groups (at least two)."

[26] Refer to Aikenhead (1997, 2007), Bybee and DeBoer (1994), Donnelly (2004), Donnelly and Jenkins (2001), Pedretti et al. (2008), Schulz (2009a), Shamos (1995), Witz and Lee (2009), and Yager (1996).

philosophical studies. The inertia of traditionalism[27] at the upper levels had prompted Jenkins surprising call for a "paradigm shift," as mentioned—but this is serious talk, nothing less than a plea for somber philosophical contemplation and reorientation. Even at the postsecondary level, the need to reform introductory science classes has received increased attention especially with some new findings in Physics Education Research (PER) indicating that the dominant textbook- and lecture-based instruction in large classrooms is unwittingly producing an antiscientific mind.[28] The appearance in time of three identified public "crises" regarding school science education (1957, early 1980s, late 1990s; Schulz 2009a) and the apparent inability of different "reform waves" to provide for major, long-lasting changes could in turn suggest that a shift towards a more concentrated educational-philosophical examination of the problems lies at hand. It can be argued that the general lack of consideration of educational philosophy and theory, that is, a *philo-educational failure*, could help account for why curricular reforms are particularly vulnerable to the political whims (or "ideologies") of various stakeholder groups, an enduring situation several researchers have taken notice of.[29] It could, for example, better inform policy deliberations when diverse stakeholders are at odds over what should "count" as science education (Fensham 2002; Roberts 1988).[30]

Fensham (2004) argues in his important book *Defining an Identity* that science education is still searching for ways to characterize its own "identity" as a

---

[27] Grade 11 and 12 specialist science courses continue to serve primarily a gatekeeping function for college and university entrance, and their purpose, structure, and content usually replicate first-year tertiary courses—their chief rationale is exclusively with "science for scientists," and not concerned with the large majority who will not specialize. In other words, as induction into pure academic science but at the neglect (if not deliberate omission) of discussing (never mind integrating), the epistemologies, social practices, and proper history of the sciences—otherwise termed *nature-of-science* perspectives (Hodson 2008; Irzik and Nola 2011; Lederman 2007; Matthews 1998a). Reform movements like *Science-Technology-Society* (STS), *Science-Societal Issues* (SSI), and (lately) scientific argumentation studies have been attempting to counter this dominant school paradigm for some time but continue to make only small inroads.

[28] Yet despite these disturbing findings, researchers in these newer fields of study (also Chemical Education Research) still struggle uphill for respect and acceptance in their academic departments, where educational studies and research continue to be afforded a low priority (Gilbert et al. 2004; Hestenes 1998).

[29] See Aikenhead (2006), Bencze (2001), Donnelly and Jenkins (2001), Fensham (2002, 2004), Roberts (1988), and Roberts and Oestman (1998). Laugksch (2000) draws attention to different social group interests in defining "science literacy." Ernest (1991) also identifies several interest groups as determinants of mathematics education.

[30] Fensham's (2002) paper "Time to changing drivers for scientific literacy" (movement away from the academic driver to "social" and industry-based drivers) provoked a lively response from researchers about the "educo-politics" of curriculum development, especially about what role academic scientists should play, if any (Aikenhead 2002; Gaskell 2002); such a suggestion though would reorientate science education back towards the recurrent (and contentious) "social relevancy" goal and the progressivism of Deweyan-type philosophy (Darling and Nordenbo 2003; DeBoer 1991)—whose educational theory is often concealed. It may even involve a Faustian bargain with industrial- and vocational-driven interests. Gaskell believes the risk is worth it. But given the complexity of techno-science and the great diversity of vocations and business interests today leaves one wondering if any sort of meaningful consensus on curriculum is achievable, even locally.

discipline. (His comprehensive survey canvasses the views and backgrounds of 76 prominent researchers in 16 countries, active from the 1960s to the present.) One would like to suppose that helping to define such an identity would include philosophy, especially a *philosophy of science education* (PSE). And it is not only the identity of the *discipline* that is of issue here, but as referred to in the previous section, that of the classroom professional as well. Hence, it might appear the time has come for science education to return to some philosophical ground work, to come to value philosophy of education (PE), and, in turn, for the research field to inaugurate and develop a new *fourth area* of inquiry—philosophic-historical. This one added next to the common three of quantitative, qualitative, and emancipatory, in support of arguments made previously by others for its development as a "mature discipline" (Good et al. 1985; Kyle et al. 1992).

But Fensham's book, with the sole entry of PE on one page alone (where the significance of Dewey is also cited), bears ample evidence of the disregard of this subject topic for researchers and science teachers alike.[31] One can infer from the evidence to date that the worth of any sort of meta-analysis of their discipline and pedagogy seems to hold little value for the majority, thereto the need to bring systematic educational-philosophical reflection to bear on research, curriculum, and teaching.

This claim is further evidenced by a simple perusal of research *Handbooks* published thus far, where the subject of philosophy of education (including topics "philosophy," "educational theory," "curriculum theory") is missing entirely (Fraser et al. 2012; Abell and Lederman 2007; Fraser and Tobin 1998; Gabel 1994). This absence is likewise attested by recent publications of European Handbooks of research in the field (Boersma et al. 2005; Psillos et al. 2003). Crossing over the other way, most handbooks or "guides" of philosophy of education (PE) exhibit the same paucity by avoiding science education, though art education, moral education, knowledge, feminism, postmodernism, critical thinking, and critical pedagogy as subjects remain prevalent.[32] Two exceptions exist: Curren (2003) and Siegel (2009). Comparing both fields, the claim is reinforced by an inspection of the respective leading research journals in both philosophy of education and science education for the past 30 years, which exhibit an almost complete disregard of the opposing field (barring exceptions). What one finds is that only a handful of philosophers write for the science education journals, and even fewer science educators publish in philosophy of education.[33]

---

[31] Fensham in fact suggests that it is the "dominance of psychological thinking in the area" which attests to why Dewey is *not* cited more frequently among respondents in the USA (still the most prominent philosopher of education linked with science education in North America).

[32] Important works are Bailey et al. (2010), Blake et al. (2003), Chambliss (1996a), and Winch and Gingell (1999).

[33] Authors in alphabetical order include Bailin, Burbules, Davson-Galle, Garrison, Grandy, Hodson, Matthews, McCarthy, Norris, Phillips, Scheffler, Schulz, Siegel, and Zembylas (see respective references).

If an examination of the preparation of science education researchers is any indication of the kind of academic preparation science teachers themselves receive (before they become researchers), then another look at Fensham's *Identity* book as commented on by Matthews (2009b, p. 23) is revealing. He notes that "the interviews reveal that the overwhelming educational pattern for current researchers is: first an undergraduate science degree, followed by school teaching, then a doctoral degree in science education" (citing Fensham 2004, p. 164). As Matthews observes, unfortunately "most have no rigorous undergraduate training in psychology, sociology, history or philosophy." Fensham himself comments that at best, "as part of their preparation for the development tasks, these teachers had opportunities to read and reflect on materials for science teaching in schools and education systems that were different from their own limited experience of science teaching."[34] Matthews concludes that Fensham's survey reveals an overall "uncritical adoption of idealist and relativist positions" among researchers and that poor academic preparation is a reason why "shallow philosophy is so evident in the field."[35] It certainly appears as if the inadequate science teacher preparation in philosophy of education is mirrored by the widely recognized fact of the inadequate preparation with respect to philosophy, history, and sociology of science.

39.4.2 Historical Background of Philosophy of Education and Science Education

With an eye fixed solely on the mutual historical developments of both fields, this neglect is rather difficult to explain especially because science education is after all about *education*, with natural focus on the science specialty. But philosophy and education have roots that are intertwined in history long past, convincingly traceable back to Plato (*Meno*; *Republic*). Every major philosopher in the Western tradition from Plato (in Ancient Greece) to Kant (European Enlightenment) to Dewey (modern industrial America) has proposed educational projects of some kind (Rorty 1998; Frankena 1965; Whitehead 1929). As Amelie Rorty correctly points out (1998, p. 1): "Philosophers have always intended to transform the way we think and see, act and interact; they have always taken themselves to be the ultimate educators of mankind." Understood in this way, Dewey was on the mark when he famously phrased the view that the *definition* of philosophy is "the theory of

---

[34] Matthews comments this may be the significant reason why the science education research literature "is dominated by psychological, largely learning theory, concerns" (ibid). Others have also cited the domination of psychology and conceptual change research (Gunstone and White 2000; Lee et al. 2009).

[35] The typical tendency is to adopt philosophical or ideological views from well-known authors outside the field but often not accompanied by critical appraisal of such authors: "… the work of Kuhn, von Glasersfeld, Latour, Bruner, Lave, Harding, Giroux and others is appropriated but the critiques of their work go unread: it is rare that science education researchers keep up with psychological and philosophical literature" (ibid, p. 35).

education in its most general phases" (1916, p. 331)—although most professional philosophers today would probably not construe it as such.

It was the Enlightenment's "project of modernity" (Habermas 1987)—first begun in the seventeenth century—that was expressly formulated as an *educational project* and which saw in the new science of the day an instrument for personal and sociopolitical liberation (Gay 1969; Matthews 1989). It is of course in full awareness of this intellectual and cultural heritage that postmodernists like Lyotard (1984) would outright dismiss the "grand narrative" of this project with its associated role and *image* of science as an emancipatory and positive force, including those science educators convinced by his critique (Loving 1997; Nola and Irzik 2005; Rorty 1984; Schulz 2007).[36] In fact the popularity of strands of post-structuralist Foucault, 1972/1989, 1980; Nola, and postmodernist thinking among some researchers bears witness to the recent discovery of the value of philosophy for the field (Zembylas 2000, 2006).

Looking much further back in time (again at the *Metaphysics*), Aristotle identifies the man of knowledge—one who has attained expertise either via *techné* or *theoria* (instrumental or theoretical reason)—as the one who is plainly able to teach what he has learned and as such draws one distinguishing feature of the philosopher. To be a philosopher was to be a teacher. Conversely, to be a teacher implies one must do philosophy (of one form or other). Science educators seen in this light are inescapably located within a venerable philosophical tradition *along with* the newer scientific one which they usually and exclusively tend to associate themselves with—though, here too, not fully aware of the latter's cultural roots and significance.

The first mention of philosophy of education as a distinct field of study was in Paul Monroe's *Cyclopedia of Education*, published 1911–1913 (Chambliss 1996b). Philosophy of education, depending upon the given nation and its educational traditions, can be viewed as a relatively new discipline or not. As Hirst (2003, p. xv) points out, "philosophical inquiry into educational questions" was more established in the USA, Germany, and Scandinavia, whereas in the UK philosophy of education as a discipline first came into its own in the 1960s. It was dominated by analytic philosophy and accounts of schooling, although in ethics Kantianism was the major influence. In the USA, the *American Philosophy of Education Society* had already been founded earlier in 1941, along with the Deweyan journal *Educational Theory* in 1951. It was the pragmatist philosopher and educationalist John Dewey in his influential work *Democracy and Education* (1916/44) who had conceived of PE to be a study worked out on an experiential basis—in other words, that educational ideas were to be applied and tested in practice. He also considered that theory and

---

[36] Related to this topic is the question of what worldview(s) science assumes or requires in order to be sustained, hence which one(s) educators need to be supportive or cognizant of (Matthews 2009a). This further raises the question of the *universalism* of "Western science," whether or not its knowledge and truth claims are necessarily culturally confined, or merely *evolved*. Disputes over the interpretations of "multicultural science" will not be addressed here, but again science educators require philosophical training in order to adequately tackle these controversial topics. Philosophical treatment of this subject can be found in Hodson (2009), Matthews (1994a), Nola and Irzik (2005), and chapters in this Handbook.

39 Philosophy of Education and Science Education...

practice were interdependent in a kind of feedback loop mutually learning from and reinforcing each other. This stood in contrast to the earlier views of the Englishman Herbert Spencer who instead conceived of education as an inductive science and where PE would serve as a kind of scientific method.

Alternatively, on the continent in Northern Europe, very different views about education had been developing. The ideas of Kant, Schiller, Herder, Herbart, and others had contributed to create the influential *Bildung* paradigm in the nineteenth century.[37] It has become established as the *Bildung/Didaktik* tradition whose conception of education dominates the German-speaking world and the Nordic countries.[38] Today this paradigm is not without its detractors, for by the 1960s this tradition had itself begun to clash with the "critical theory" of the Frankfurt school (Blake et al. 2003; Blake and Maschelein 2003; Smeyers 1994). It continues to engender much debate among educational thinkers and philosophers alike, both in Europe and English-speaking countries. Thereto, advocates of both traditions—Anglo-American "curriculum" and *Bildung/Didaktik*—came together in the 1990s to open dialogue comparing the relative benefits of each (Gundem and Hopmann 1998; Jung 2012; Vásquez-Levy 2002).

The *Bildung* paradigm itself actually represents an *educational metatheory* (Aldridge et al. 1992), a type of "grand theory" in education of which very few have been constructed in modern times (inclusive of Dewey and Egan; Polito 2005). It immediately raises the question of the worth and relation of educational theory to practice, whose merits are currently being contested in philosophy of education (Carr 2010).

The direct link between *Bildung* and science education[39] has been drawn only recently, notably in Fensham's *Identity* Book (2004) and by Witz (2000).[40] Fensham provides a highly informative discussion, explaining the concept and significance of *Bildung* when contrasting the Norse/German tradition with the content knowledge-driven Anglo-American tradition. He contends that a serious shortcoming of the so-called "curriculum tradition" of the English-speaking world is its consistent disregard of metatheory (discussed further below).[41] *He advises science education*

---

[37] The literature on *Bildung* and *Didaktik* is extensive. Some references to its historical development are Barnard (2003), Beiser (1998), Gadamer (1960/1975), and Schiller (1795/1993).

[38] "On the one hand, the concept *Bildung* describes how the strengths and talents of the person emerge, a development of the individual; on the other, *Bildung* also characterizes how the individual's society uses his or her manifest strengths and talents, a "social" enveloping of the "individual" (Vásquez-Levy 2002, p. 118). Given this interpretation, one could in fairness associate the values and aims of the *Bildung* tradition with two prevalent "curriculum ideologies" identified by Eisner (1992) as "rational humanism" and the "personal" stream within progressivism.

[39] Science education and *Bildung* in Germany have been examined by Benner (1990) and Litt (1963).

[40] One Canadian study involving science teachers had sought to fuse the *Bildung* ideal with the STS paradigm and cross-curricular thinking (Hansen and Olson 1996).

[41] "In the one, the maturing young person is the purpose of the curriculum. In the other, the teaching of subjects is the purpose. In the one case, disciplines of knowledge are to be mined to achieve its purpose; in the other, disciplines of knowledge are the purposes" (2004, p. 150).

should acquire one. The same arguments have long been raised in Germany by Walter Jung (2012).

Another interesting aspect about the *Bildung* paradigm can be noted: it exercised an indirect influence via Herbart's ideas on the philosopher-scientist Ernst Mach. While Mach's impact on Einstein's thinking is generally recognized, his educational ideas are hardly known in the English-speaking world. Already back in the late nineteenth century, he had been politically active for educational reforms, including improving teacher education, and is credited with founding and coediting the very first science education journal in 1887 *Journal of Instruction in Physics and Chemistry* (Matthews 1990b, 1994a). Siemsen and Siemsen (2009) argue his rediscovery at present could provide significant contributions to current European reform efforts.

On a final note, for the English-speaking nations, the USA was in the forefront of the establishment of both disciplines (science education and philosophy of education) that have developed in tandem—simultaneously but separately in the early twentieth century. One would think that because of this pedigree, and in some cases of clearly overlapping interests (as exhibited in the important case of Dewey), science education would be more cognizant, and science teacher training more reflective, of their common roots. Unfortunately, on this matter science education seems to suffer amnesia on both counts, for if it can be admitted that "philosophy of education is sometimes, and justly, accused of proceeding as if it had little or no past" (Blake et al. 2003, p. 1), then this certainly rings true of science education.[42]

The call for a philosophy *of* science education (PSE) is not only to raise awareness of this forgotten earlier period, but *to identify the need to create a subdiscipline within educational studies* that, although new, nonetheless has substantial historical roots going back into the science-educational but especially the philosophical-educational past.

Why science educators do not associate themselves just as intimately with philosophy of education is a fascinating question, one that cannot be pursued here. It almost certainly has a lot to do with several factors (such as the prestige of science in society, how disciplinary knowledge is structured, how their own university science education proceeded, and, not least, how they were trained as educational professionals).[43] What is called "foundations in education" courses, which usually include studies in the history and philosophy of education, are often optional for preservice science teachers, depending upon the prerequisites of their attending institutions.[44]

---

[42] Jenkins (2009) notes the same problem with reform movements and policy documents. This complaint (although dated but still relevant) was earlier attested by DeBoer in his Preface to his insightful *History of Ideas in Science Education* (1991).

[43] Roberts (1988, p. 48) draws attention to where teacher *loyalties* commonly lie: "The influence of the subject community is an especially potent force in science education. In general, the 'hero image' … of the science teacher tends to be the scientist rather than the educator [or philosopher]."

[44] Hirst (2008b) has recently complained that in some countries such as England, there are now moves afoot to delist such courses for teacher training altogether. It would not be a stretch to conclude that

39.4.3 Philosophy of Education Today

Coming at last to the present historical culmination, philosophy of education has today progressed to become a respectable, established subdiscipline in philosophy. It comprises evolving research fields, a sizeable literature, professorial chairs, professional associations (e.g. PES), and several leading journals.[45]

There now exists two *Handbooks* (Bailey et al. 2010; Siegel 2009) but also a *Guide* (Blake et al. 2003), *Companion* (Curren 2003), and *Dictionary* of key concepts (Winch and Gingell 1999). An *Encyclopedia* of PE is also on hand (Chambliss 1996a). These can be sought out by science educators to familiarize themselves with the current discussion, inclusive of disputes regarding different topics of individual interest to them. Several newer and older *Introduction* texts are also available (e.g., Barrow and Woods 2006/1975; Tibble 1966), including Carr (2003) and Noddings (2011). For educators seeking immediate information, several encyclopedia articles exist providing succinct, comprehensive overviews of PE (accessible online: Phillips 2008; Siegel 2007).

39.4.4 The Value of Philosophy of Education

Philosophical questions bearing on the different facets of science curriculum, teaching, and learning must be addressed and inspected by the thoughtful educator:[46] questions pertaining to (i) chief educational goals, content selection, and course objectives, or (ii) assessing learning theories, or (iii) bearing on nature-of-science- and techno-science-related issues—thereto, the character of scientific research, knowledge, and societal applications as related to curriculum or policy reforms. Hence, questions also pertaining to who enacts and benefits from such reforms with respect to interests and ideologies. And all this often in spite of, not because of, state-mandated and prepackaged "content knowledge" curricula:

> What are the aims and purposes of science education? What should be the content and focus of science curricula? How do we balance the competing demands of professional training versus everyday scientific and technological competences versus the past and present interactions of science with society, culture, religion and worldviews? What is the structure of science as a discipline and what is the status of its knowledge claims? What are the ethical constraints on scientific research and what are the cognitive virtues or intellectual dispositions

---

such a downgrade in the general value of philosophy-of-education cannot fail to negatively impact science teacher professional development.

[45] The leading journals of the English-speaking world are *Studies in Philosophy and Education, Educational Theory, Educational Philosophy and Theory,* and *Journal of the Philosophy of Education.*

[46] Some classroom case examples are Hadzigeorgiou et al. (2011), Kalman (2010), and Ruse (1990). Bailin and Battersby (2010), Giere (1991), and Kalman (2002) offer science teacher educators rich material for enhancing science subject-related critical thinking:

required for the conduct of science? What is the meaning of key scientific concepts such as theory, law, explanation, and cause? (Matthews 2002, p. 342)

If it is indeed true, for example, that precollege and first-year college level science courses are primarily about "technical preprofessional training," then vital questions need to be asked about what differences should exist between training and education in science. It raises cultural, epistemological, and political questions about the nature of school science: whether, for instance, it is truly reflective of the nature of science (in some form) or more reflective instead about courses performing a "gatekeeping" function by limiting access to higher education (a sociopolitical role)—this in turn reflecting norms of school culture and assimilation (as critical pedagogy perspectives contend).[47] Does a hidden cultural bias exist (as "cultural studies" perspectives contend)? Should the worth of school physics and chemistry education, say, be mainly determined by "political/instrumental value" (prerequisites to college entrance courses; Aikenhead 2006)? If so, this would raise more disturbing questions about the nature of, or links between, socialization, training, and perhaps indoctrination (into scientism). There can be little doubt that in such cases a given "vision" of what constitutes "science education" is in place (with hidden "companion meanings"; Roberts and Oestman 1998).

At minimum it should raise questions about subject epistemology or the preeminent *value* placed upon a certain kind (Gaskell 2002). Such topics, though, have been a staple of PE disputes for quite some time—inclusive of deliberating the difference between hidden aims and genuine educational aims of curriculum and schooling (Apple 1990; Posner 1998), or the differences between education and indoctrination (Snook 1972). Not to forget, previous science education reforms have too often been associated with several past "crises" (as cited) which were themselves linked with wider socioeconomic problems in society: were these just pseudocrises manipulated by science education stakeholders and their interest groups? What educational values/views inform such groups and their policies?[48] Again, similar questions are addressed in PE.

39.4.4.1 Philosophy of Education and the Nature-of-Science Debate

Just focusing on one fundamental topic, the *nature-of-science* (NoS) debate, and zeroing in only on one aspect of this debate, the key question is: "who defines science for science educators?" The scientific experts within isolated academic disciplines (as is common)? Philosophers of science? Historians? Sociologists? Or those within cultural and women's studies? Postmodernist-type thinkers and critics?

---

[47] "Domination, resistance, oppression, liberation, transformation, voice, and empowerment are the conceptual lenses through which critical theorists view schooling and pedagogy" (Atwater 1996, p. 823).

[48] Different kinds of answers are provided by Aikenhead (2006, 2007), Apple (1992), Bencze (2001), Donnelly and Jenkins (2001), Gaskell (2002), Gibbs and Fox (1999), Klopfer and Champagne (1990), Roberts and Oestman (1998), Schulz (2009a), and Zembylas (2006).

Or possibly students and teachers themselves, according to some versions of social constructivist theory?

The NoS topic alone has been recognized as one chief aim of science education for over 50 years, yet to this day, there exists a poor record of achievement world-wide (Lederman 2007). This fact is due to several interrelated causes, not least of which is the entrenchment of traditionalism (conventional discipline-based paradigm)—but moreso the reality that NoS is itself a contested field in HPSS studies. The "science wars" (initially launched by the Sokal hoax 1996a, b) and their aftermath have made the issue public, and science teachers are inadvertently involved in a contest that is being fought in the academy.[49] Researchers can certainly be found on either side, running the gauntlet from "positivism to postmodernism" (Loving 1997; Turner and Sullenger 1999).[50]

These polarized camps have made the business of science education a messy and complicated affair—it has become increasingly difficult to navigate a peda-gogical course between competing views "from diehard realism to radical con-structivism" (Rudolph 2000, p. 404). At best consensus can be found that several common classroom *myths* must be exposed, including talk of "scientific method" (Bauer 1992; Feyerabend 1975; Hodson 1998; Jenkins 2007). Teachers clearly require substantial philosophical background to familiarize themselves with the issues, but even *if* consensus could be achieved (which seems unlikely), the ques-tion cannot be solely confined and determined on HPSS grounds. This decision would leave entirely untouched the related *pedagogical question* how that (would be) conception of science plays a role in the education of the student, as to what educational *aim(s)* school science is ultimately expected to achieve.[51] In other

---

[49] For examples of teachers caught in the debate, see Sullenger et al. (2000) and Witz and Lee (2009). For different perspectives on the debate in the academy, see Brown (2001), Giere (1999), Gross et al. (1995), Laudan (1990), Nola (1994), C. Norris (1997), Siegel (1987a, b), and Sokal and Bricmont (1998).

[50] Science educators continue to quarrel whether basic NoS statements *can* or *should* be defined, even where a measure of recognized consensus is said to exist—inclusive of those now written into global policy documents. The dispute centers on how to determine "consensus" (among which experts?), or questions regarding disciplinary distinctions, or about NoS cultural dependence on "Western" science and Enlightenment traditions, among others (Hodson 2008; Irzik and Nola 2011; Matthews 1998a; Rudolph 2000, 2002). Good and Shymansky (2001) make the case NoS statements found in "standards" documents like NSES and *Benchmarks* could be read from oppos-ing positivist- or postmodernist-type perspectives.

[51] This viewpoint aligns to an extent with Hodson's view (2009, p. 20) except for the fact he ignores relating his desired outcomes to educational philosophy and theory: "In my view, we should select NOS items for the curriculum in relation to other educational goals ... paying close attention to cognitive goals and emotional demands of specific learning contexts, creating opportunities for students to experience *doing* science for themselves, enabling students to address complex socio-scientific issues with critical understanding...." On what philo-educational grounds the selection is to be undertaken, we are not told though he considers students' "needs and interests" (overlap with progressivism?), views of experts ("good" HPSS—the Platonic knowledge aim?), and "wider goals" of "authentic representation" of science and "politicization of students." His lofty ambition for science education (thus his notion of "literacy"), however, includes too many all-encompassing and over-reaching objectives. These must clash and become prioritized (or so it seems) once his

words, for the educational setting, the question "what counts as science?" must be allied with "what counts as science education?"[52] The historian may have something to say (e.g., correcting pseudo-history in textbooks), at other times the philosopher of science (e.g., correcting misleading epistemology inherent to textbooks), other times the sociologist, etc., each depending upon the context of instruction and in coordination with desired educational objectives and policy deliberations of stakeholders.

The issue is precisely that subject content (inclusive of disciplinary structure) must be "problematized" during curriculum decision-making, and for *two* reasons:

(i) It must be broadened to function as a more authentic and appropriate knowledge base.
(ii) It must be transposed into a form that considers the culture and age developmental stage of learners along with desired educational aims (Englund 1998; Schulz 2011).

That the curriculum needs to be made problematic implies that a *philosophical* (and not just instructional) problem initially lies at hand which requires resolution. This problem lands us squarely in philosophy of education (PE) territory. It requires a close linkage of questions found in PS with those found in PE (the base of the triangle in Fig. 39.1). The philo-pedagogical problem concerns the appropriate or *best didactic transposition* of epistemic content knowledge (CK) into an appropriate form accessible to the learner in accordance with educational aims and theory.[53] There are some educational thinkers who argue this cannot be suitably achieved without educational *metatheory* (Carr 2010; Dewey 1916/1944; Egan 1997).

As an example, while a teacher's content knowledge (CK) in chemistry may need to be better informed by research in the philosophy of chemistry (one crucial component of PSE would involve stressing this factor), nonetheless a PSE is more concerned with how such CK can be made to fit with the requirements of an educational metatheory and its concern with the cognitive-emotive *developmental stage* of the learner, with respect to this subject matter. In other words, a teacher's CK and the curriculum are not at the forefront for learning science (although they are invaluable dimensions), as is commonly done. Rather, they are evaluated in light of philosophy of education and the learner's age developmental mind-frame as befits

---

three stated criteria for subordinating goals force them under his socio-techno-activist umbrella of politicizing students—the ghost of Dewey beacons.

[52] The focus here is on the normative nature of the question (i.e., what do policy documents, researchers, or theorists stipulate?), as opposed to the empirical (i.e., what is going on in classrooms now?).

[53] This important topic is too often overlooked in curriculum theory or in the science education literature. See Fensham (2004), Geddis (1993), Klafki (1995), Lijnse (2000), Schulz (2011), Vásquez-Levy (2002), and Witz (2000); Dewey, Mach, and Schwab all in their day also identified the issue that the logic of the discipline does not conform with the psychologic of learning the subject matter of the discipline. Thereto, Aikenhead (1996) has argued that learning science involves a culturally rooted "border-crossing" on the part of the student, to negotiate the transition from the personal "lifeworld" to the "school-science world."

what it means to *educate* a person in the sciences. This emphasis necessarily shifts the focus to the substance of a teacher's pedagogical content knowledge (PCK) and educational philosophy.

If, say, NoS knowledge is taken to be an *end* (an aim in itself), then an implicit "philosophy" would be "academic rationalism" (Eisner 1992)—whose objective could be associated with "knowledge-for-knowledge sake," building "mind" (possibly even critical thinking), and likewise similar-sounding ideals coupled to a typical knowledge-driven educational metatheory (Egan identifies it with Plato's historic project).[54] This *can* equally be squared with science teaching within the conventional academic paradigm, though providing subject content with *context* (Roberts "vision I"); on the other hand, NoS combined with "critical thinking" as *means* to create critical-minded citizenry to strengthen democracy in society would couple NoS teaching with Deweyan-type educational metatheory (Egan identifies this educational tendency with a form of socialization; Roberts "vision II"). There are tensions here which may not be reconcilable[55]—tensions also inherent to liberal education (e.g., aims for the individual and society can clash considerably); they are certainly topics of concerned debate in PE. Not to be forgotten, there are those who wish to teach NoS because it stands alone—the *intrinsic* worth to learn about authentic science (or science as a cultural force); others however see it subservient to other ends—for advancing critical thinking (itself), or chiefly addressing science-societal issues (Zeidler et al. 2005), or yet again, for emancipation (critical pedagogy) and sociopolitical action (Hadzigeorgiou 2008; Hodson 2009; Jenkins 1994).[56]

What is really of issue here, though hardly recognized, is how (and which) *epistemic aims* of science education (e.g., knowledge, truth, justification)[57] can or should be met, either apart from, or linked with, or perhaps subordinated to, other identified *moral* and *political* aims of education (e.g., autonomy, human flourishing, citizenship, social justice).[58] A common and depressing feature of several reform programs (e.g., STS, SSI, sociopolitical activism) is the notable confused state of their several suggested educational aims. Moreover, it can be asserted that such avowed and increasingly popular projects for science education as identified presuppose educational metatheory of some kind, whose existence is either assumed or overlooked.[59] Engagement with philosophy of education debates about,

---

[54] See discussion on the topic of epistemic aims by Adler (2002), Hirst (1974), and Robertson (2009).

[55] See discussion in Egan (1997) and Pring (2010). Smeyers (1994) discusses the European account.

[56] Driver et al. (1996, pp. 16–23) offer five rationales for teaching NoS in classrooms, yet they either assume or overlook their dependence upon different, prior educational theories.

[57] See Nola and Irzik (2005), Robertson (2009), and Siegel (2010) for discussion of these subjects.

[58] See Brighouse (2009) and Pring (2010) for discussion of these subjects. Donnelly (2006) only scratches the surface of the problem with his defined dual clash between "liberal" and "instrumental" educational aims behind community reforms.

[59] This remark also targets research concerning situated cognition models, where it has often been asserted; practice was either *prior* to theorizing or *without* theory. See critiques of Roth by Sherman (2004, 2005).

and analyses of, *indoctrination* can be an antidote to such political-activism programs simply replacing unthinking science lessons with uncritical acceptance of whatever causes teachers or researchers might be energized about. As Erickson has stated (2007, p. 33), the science education community "needs to develop pedagogical models that make explicit the normative premises about aims" in its discourse on scientific literacy.[60] Whenever the topic of educational aims arises, the neglect and need of philosophy of education become only too evident.[61] The time has come for the community to strive for clarity and prioritization concerning which fundamental aims the field can and should achieve (Bybee and DeBoer 1994).

In any event, NoS raises foundational *philo-educational* questions: "What is the ultimate aim of science education?" (or, e.g., of physics education?). "What does it *mean* to be educated in science?" "How is such an education related to human flourishing?" These should ideally be addressed before the subsidiary question "what do we educate people in science *for*?"—often the common starting point of curriculum thinking and policy decision-making, which begins first with the prior value, with its linked presumption of the overall importance, of *social utility*. (The difference so stated is one of choosing between deontologic or teleologic rationales.) The former should not be approached as "mere academic questions" during teacher preparation, for they aim at the heart of what the profession and teacher identity is all about. Yet it should be clear that they cannot be answered without reference to educational philosophy and theory—while the utility rationale, alternatively, presupposes a particular one. In other words, it requires of the science educator a *philosophical valuation* of subject content and aims and an awareness of the broader educational purpose of the science educational field, including some personal positioning among available educational/curriculum theories (Scott 2008).

39.4.5 Overview of Philosophy of Education Subjects and Questions

It is the view of the present author that teachers as well as researchers when becoming more conversant with the ideas and disputes as argued by philosophers of education will help them (at minimum) gain insight and perhaps (at maximum) resolve

---

[60] He continues: "Too often we try to simply derive pedagogical practices from theoretical positions on learning, or diversity, or language, or the latest research on the functioning of the brain, etc." (ibid).

[61] An example of the confusion which results in science education research when PE is ignored is the paper by Duschl (2008). Here empirical research from the learning sciences and science studies is confused with educational goals, which must be chosen on a normative basis. Such research may very tell us *how* students (and scientists) learn but expressly not *why and what* goals they *should* learn. And to argue for a "cultural imperative" is to *make* a normative claim extrapolated from such research—one is dabbling in PE without its recognition. Moreover, whether the avowed economic, democratic, epistemic, "social-learning" goals, etc. (as they have been historically articulated for the field) can be "balanced" as Duschl simply assumes is by no means obvious—PE debates show quite the opposite (Egan 1997; Levinson 2010; Schulz 2009a).

problems related to issues of *common interest* (the nature and kinds of aims; the nature of language and learning, knowledge and truth, educational theory; feminism, multiculturalism; education for citizenship; critical thinking; ideology, interests, and curriculum; indoctrination, etc.). The field of philosophy of education is a veritable mine of ideas, posed problems, and suggested solutions. This holds true whether the *approach* to PE is simply to:

(i) Study prominent philosophers and their views on education
   (e.g., Plato, Aquinas, Rousseau, Kant, Whitehead, Scheffler, Foucault)[62]
(ii) Study educational thinkers and their philosophical positions
   (e.g., Schiller, Herbart, Dewey, Peters, Freire, Hirst, Egan, Noddings)
(iii) Study sub-branches of philosophy and their relevance to education
   (e.g., philosophy of science, moral and political philosophy, or aesthetics)
(iv) Study "schools of thought" in education
   (e.g., idealism, realism, Thomism, Marxism, existentialism, critical theory, postmodernism)[63]
(v) Study the philosophical questions of ultimate concern (e.g., the nature of being, of knowledge and cognition, the ideal of an educated person, autonomy)

There is intellectual insight and pedagogical profit to be had in any of these approaches (Barrow 2010). For the more practical-minded science educator though, the approach to PE could imply instead a focus on specific, contemporary educational questions. Here Amélie Rorty's (1998, pp. 1–2) list of essential PE questions serves to illustrate the "down-to-earth" PE approach, when *transposed* onto science education:

> What are the directions and limits of public [science] education in a liberal pluralist society? … Should the quality of [science] education be supervised by national standards and tests? Should public [science education] undertake moral education?[64]… What are the proper aims of [science education]? (Preserving the harmony of civic life? Individual salvation? Artistic creativity? Scientific progress? Empowering individuals to choose wisely? Preparing citizens to enter a productive labor force?) Who should bear the primary responsibility for formulating [science] educational policy? (Philosophers, …, rulers, a scientific elite, psychologists, parents, or local councils?).[65] Who should be educated [in science]?[66] How does the structure of [scientific] knowledge affect the structure and sequence of learning? … What interests should guide the choice of [science] curriculum?

---

[62] To name just some in the Western tradition; Eastern and other traditions have of course their own major philosophers who have concerned themselves with education.

[63] A classic source of material for this orientation are the essays in Henry (1955).

[64] As those in the *SocioScientific Issues (SSI)*, reform movement today insists (Zeidler and Sadler 2008).

[65] See DeBoer (2000), Fensham (2002), Gaskell (2002), Jenkins (1994), and Roberts and Oestman (1998), for responses to such questions.

[66] Recall the ongoing past disputes between "science for scientists" and "science for all" perspectives on curriculum, goals, and policy (ByBee and DeBoer 1994; DeBoer 1991). The most recent STEM reform movement in the USA can be justifiably accused of redefining science education as "science for engineers."

It is quite clear that common questions and concerns exist and one would have expected more cross-disciplinary discourse than has heretofore existed.

On the other hand, it is not here being suggested that a consensus is to be found among philosophers of education on such questions. In fact there are important disagreements and even diversity of interest and approaches to the solutions, as different PE "schools of thought" display (analytic, existential, phenomenological, postmodern, critical theory, etc.). Indeed, philosophy more often "divides" than it unites, and as one contemporary education philosopher admits: "missing in the present world of diversity of interests is the classic sense of a quest for philosophic unity" (Chambliss 1996b). As Scheffler stressed, "philosophies of" are not forged by some harmony of agreed-upon, sealed discourses. Instead they

> do not provide the educator with firmly established views ... on the contrary, they present him with an array of controversial positions. But this array, although it does not fix his direction, liberates him from the dogmatisms of ignorance, gives him realistic apprehension of alternatives, and outlines relevant considerations that have been elaborated in the history of the problem. (Scheffler 1970, p. 391)

The point is not that some sort of philosophical unity should be either expected or had among philosophers or science educators, although of course consensus on common fundamental issues is to be desired. Rather, the nature of the discourse and sophistication of the debate can help illuminate those problems and issues which science educators are confronted by and continue to struggle with or have misconstrued, have overlooked, or for too long avoided.

39.5 Some Major Philosophy of Education Perspectives and Science Education

39.5.1 Educational Theory and Science Education

To talk of "educational theory" is first of all to recognize that it has undergone shifts in meaning ever since Western philosophy began contemplating educational matters in Ancient Greece. For the sake of brevity (and hazarding oversimplification), one charts a course from there to the current age by noting how its worth and purpose have undergone several changes, not only when specifying what *aims* to target, but *who* should carry the prime duty, namely, either philosophers, educationalists, or empirical scientists (Carr 2010; Phillips 2009).

The priority in Antiquity (Plato, Aristotle, Cicero) was to establish the grounds for knowledge to improve moral virtue (the "Good") but conceived more along a priori philosophical lines—hence the emphasis on reason and rationality. This tendency took "an empirical turn" with Rousseau, progressivism, and the rise of the scientific Enlightenment. This science-inspired propensity has continued right down to the primacy of developmental psychology in our age, "the view that the study of human cognition, emotional and social growth and learning ought to be scientifically grounded" (Carr 2010, p. 38). Largely lost sight of along the way was the previous prominence of moral virtue required to remodel society—reclaimed later in different guises by Deweyan theory (of social adaptation or reconstruction), critical theory/pedagogy, and *Bildung*. The postwar positivistic, language-based "analytic revolution" in philosophy (or "linguistic turn" as Rorty opined) which arose in the US and England facilitated the "new" philosophy of education in the 1960s (e.g., Scheffler and R.S. Peters, respectively).

The "analytic school" in education had sought to improve teacher professionalism by augmenting the usual study of the "doctrines of the great educators" with added philosophical analytical skills to help sort out educational language and thinking (which they had diagnosed as incredibly confused). They also sought to combine their reform effort with guidance sought from research in the social sciences. It allowed for neat separation between the roles of philosopher and scientist, a dualism between theory and practice, and essentially pictured *educational theory as applied science* (a view Piaget held into the 1970s). Needless to say, the "post-analytic revolt" which came afterwards challenged and rejected many of the previous guiding views and assumptions, including its dualism, its epistemological objectivism and deficient language theory, and its philosophy of science (the so-called received view).

In its wake diverse, contemporary "schools of thought" (Barrow 2010) have championed various anti-theory, anti-foundationalist and assorted postmodernist, constructivist, and sociopolitical views. These in turn certainly suffer problems of their own (not to be appraised here), suffice to note others have recently come to reprieve the status of theory.[67] Its proponents not only take issue with anti-theory and postmodern-type arguments but also equally with previous analytic inspired views and dismiss the secondary reliance of educational theory on the social sciences, or worse, its reduction to a mere branch of the field (Carr 2010; Egan 1983, 2002, 2005).[68] They have reasserted the worth of philosophy to deliberate upon educational theory independent from constraints they see placed upon it, especially from scientific psychology.[69] They advocate in spirit that philosophy of education should once again claim its own unique, rightful place, neither accepting subordi-

---

[67] So that it may "engage in explorations of what [science] education might be or might become: a task which grows more compelling as the 'politics of the obvious' grow more oppressive. This is the kind of thing that Plato, Rousseau and Dewey are engaged in on a grand scale" (Blake et al. 2003, p. 15).

[68] Carr holds that educational theory might be better suited to ethics (moral reasoning) than with any sort of empirical science, which is not to dismiss the worth of some empirical work: "On closer scrutiny, it seems that many modern social scientific theories of some educational influence are often little more than normative or moral accounts in thin empirical disguise" (2010, pp. 51–2). This deduction leaves unanswered the important question as to what the proper role and value of empirical research for educational theorizing is to be. The topic is controversial and engenders debate in PE. See Egan (2002) and Hyslop-Margison and Naseem (2007) for a negative assessment and Phillips (2005, 2007, 2009) for a positive view.

[69] "We have suffered from tenuous inferences drawn from insecure psychological theories for generations now, without obvious benefit" (Egan 2002, pp. 100–101).

nate status nor intending to displace the social sciences, rather seeking complimentary standing.

On a related issue, because "theory" is often ill-defined in education (Thomas 1997) and usually strictly identified with learning theory in science education (e.g., Norris and Kvernbekk 1997), one needs to distinguish this term from "grand theory" or *metatheory*—the sort of thing Plato, Rousseau, and Dewey were concerned with (Schulz 2009a).[70] The original emphasis on the requirement for a metatheory in education had been discussed by Aldridge and associates (1992) following the proposal first put forth by Egan in the early 1980s encompassing his critique of "scientific psychology" and the demand educational studies stake out independent territory (Egan 1983). Such a theory could very well insist on the difference between psychological and educational development. *The essential merit of metatheory lies in creating curricular coherence, properly transposing subject content knowledge for the learner, and steering educational aims.*

Any educational metatheory must need be a normative one, for it seeks to *prescribe* an educational process to ultimately yield a certain outcome or *aim* (Hirst 1966). This is usually a kind of person or the ideal of what an educated individual should aspire to become given the values and dispositions to be cultivated and methods employed in the specified program (Frankena 1965). Further, it is in the worth of that final aim that the pedagogical methods of the educational project are justified, which traditionally have themselves been framed within the values and aspirations a society has deemed of ultimate importance: "The *value* of this end-product *justifies* the stages that lead towards its realization. Becoming a Spartan warrior justifies training in physical hardship. Becoming a Christian gentleman justifies exercise in patience and humility" (Egan 1983, p. 9; original italics).

In Western civilization a succession of diverse aims or ideals have historically followed since the time of Ancient Greece, and some of the greatest Western minds have been preoccupied with formulating various philosophies of education to define their respective ideal and suggest ways to realize it (Lucas 1972): Plato, the (philosopher-king) man of knowledge; Aristotle, the "good" or "happy" active citizen; Augustine and Aquinas, the Christian saint; Locke, the successful Christian mercantile gentleman; Rousseau and romanticism, the natural development of self-actualization; Kant, the autonomous individual, self-ruled by moral "good will"; and Dewey, personal and social "growth" through ever-changing experience, as the basis for democratic living.[71]

---

[70] Phillips (2002, p. 233) terms these "classic theories of teaching and learning."

[71] It should be noted that Dewey's aim is among the least predetermined of the others, although it could reasonably be argued that Kant's ideal is also dynamic insofar as he allows for education's dual aim, the "perfecting" of man *qua* man plus the improvement of society and "the human race." In addition, Frankena (1965, p. 156) also notes that such a dual aim in Dewey could considerably conflict—that the expected growth of the individual and society may clash—in anticipation of Egan's critique, which claims the clash is inevitable insofar as modern schooling is molded according to progressivist precepts. Alternatively, for Dewey, but also for Aristotle and Kant, such a possible conflict was thought to be reconcilable in principle.

Frankena (1965) insists any philosophy of education must ask itself three basic questions: *what* dispositions (or "excellences") to cultivate, *how* to cultivate them, and *why*?[72] When examining the position of the educational theorist Kieran Egan (1983), he seems to have these same in mind but reformulates and generalizes them with a slight shift in accent. Instead of using terms like "dispositions to be cultivated" and "ideal," he talks in terms of "end product" and "aims" while explicitly raising the important fourth component of *development*—it is of the essence of an educational *metatheory*, he writes, that it answers four key questions: what to teach (curriculum), how to teach (instruction), when to teach it (stages of learner development), and most importantly, why to teach it (specification of the end product, aim, or ideal). That said, the similarity in questions and intent is obvious.

Egan (1997) has further argued that *three* long-standing yet venerable and operative *ideas* in education (themselves inexorably embedded within science education) are undermining each other.[73] Schools in the West as educational projects are ineffectual primarily because they are caught between three chief objectives (or rationales) which successfully serve to check or undercut each others' intended aims: whether to teach science for (1) intellectual development (knowledge), or (2) for individual fulfillment (character), or (3) for socioeconomic benefit. (The first can be associated with the original knowledge-based educational project of Plato, the second with Rousseau, and the last is a cross-cultural and timeless expectation of most societies.)[74]

39.5.1.1 Educational Metatheory and Scientific Literacy

When science educational goals are examined historically, these three are ubiquitous; they persistently present themselves albeit in different guises, and they certainly can be identified throughout science educational reform history (Bybee and

---

[72] Such questions are actually the purview of what is demanded of an educational *theory*. Philosophy of education properly understood is a much broader field of inquiry that encompasses an analysis of such theories and questions (Peters 1966), which today usually overlaps with curriculum studies. Frankena seems to have been working with a constricted conception solely at the level of theory.

[73] Smeyers (1994) identifies the same quandary for Western European education.

[74] In brief, socialization conflicts with the "Platonic" (knowledge-focused) project because the former seeks the conformity to values and beliefs of society while the latter encourages the questioning of these; socialization also conflicts with the "Rousseauian project" since the latter argues that personal growth must conflict with social norms and needs. It sees growth and hence education in *intrinsic* terms instead of as utility for other socially defined ends. (Here exists the principal tension between the *Bildung* tradition and the dominating utility view of education and science literacy of the English-speaking world.) The Platonic and Rousseauian projects conflict because the former assumes an epistemological model of learning and development and the latter a psychological one. In the former "mind" is created and the aim is *knowledge*; in the latter it develops naturally, requiring only proper guidance, and the aim is *self-actualization*.

DeBoer 1994).[75] Considering the current controversies about prioritizing goals in science education, one may be surprised to learn that even educational debates have a long history. Once again, PE can offer insight into long-standing science educational dilemmas. Aristotle records:

> But we must not forget the question of what that education is to be, and how one ought to be educated. For in modern times there are opposing views about the task to be set, for there are no generally accepted assumptions about what the young should learn, either for virtue or for the best life; nor yet is it clear whether their education ought to be conducted with more concern for the intellect than for the character of the soul. (Aristotle, *Politics*, VIII ii: 1337a33; 1962/1981, p. 453)

It is remarkable to contemplate how his discussion mirrors the debate of values and aims that has steered science education since its inception in the nineteenth century. Consider if you will the conflicting meanings (post-WW2) of "science literacy,"[76] still identified as the overall objective of science education as discipline and practice: whether it is to be primarily understood as personal self-fulfillment (i.e., "virtue" as its own intrinsic worth) or for "critical citizenship" in a democracy (i.e., as instrumental worth; "the best life": STS), or rather solely for development of "mind" per se, as mastery of subject-based formal knowledge and as a tool for developing inductive (later redefined as "critical") reasoning (i.e., "intellect" development; science "processes": traditionalism; "scientific argumentation"). Lastly, whether it should encompass foremost moral development when arguing "socioscientific issues" (SSI) or "science education as/for sociopolitical activism" (i.e., "character of the soul"—always seen by Aristotle in terms of sociopolitical *activity*).

Note as well that the three fundamental goals underlying education (as elaborated above) can be identified here and mapped onto the corresponding conceptions of literacy and onto existing school science educational paradigms.[77] Some critical

---

[75] No one normally holds exclusively to one or the other, although usually one or the other is emphasized over the other two at a given time (depending upon the defined "crisis" at hand and under influence of respective social group interests), and the modern school and indeed many "standards" documents aim at a sort of *balance* between them. Roberts (1988), too, holds that "balance" is both desirable and achievable during public policy curriculum deliberations. Egan though insists that the attempts to achieve "balance" are illusory and must undermine the strengths of any one at the cost of the others.

[76] The term itself first came into use in the late 1950s. Initially broadly framed in terms of science, culture, and society relationships, it soon came however to mean learning technical, subject-specific knowledge: "This emphasis on disciplinary knowledge, separated from its everyday applications and intended to meet a perceived national need, marked a significant shift in science education in the post-war years. The broad study of science as a cultural force in preparation for informed and intelligent participation in a democratic society lost ground in the 1950s and 1960s to more sharply stated and more immediate practical aims" (DeBoer 2000, p. 588). By the 1980s the phrase had become commonplace: "Yet despite the problems of definition, by the 1980s scientific literacy had become the catchword of the science education community and the centerpiece of virtually all commission reports deploring the supposed sad state of science education" (Shamos 1995, p. 85).

[77] As can the seven "curriculum emphases" behind science curricula, identified by Roberts (1988)

observers had thus come to the conclusion that already by the late 1980s, the usefulness of the literacy concept had exhausted itself.[78] *We have a situation here where a discipline cannot agree on the most fundamental purpose and goal of its educational endeavor.*

One can therefore conclude, given this consistent mode of discourse about "science literacy," that the community is placed before one of *three* choices:

(i) *Exclusivist* option: one chooses either an already given or hoped for curricular paradigm; this could be the knowledge-based, specialist "vision I" literacy conception (the given: traditionalism) or, at the other end of the spectrum, opting for an "extreme" form of "vision II" (as Roberts (2007, p. 769) remarks), by redefining literacy as "collective praxis"—such as the (hoped for) image held by Roth and Barton (2004).

(ii) *Inclusivist* option: one agrees instead to hold fast to as many conflicting meanings as possible (e.g., Hodson 2009). Along with DeBoer (2000), one simply accepts the term stands for "a broad and functional understanding of science for general education purposes" (p. 594), and "because its parameters are so broad, there is no way to say when it has been achieved. There can be no test of scientific literacy because there is no body of knowledge that can legitimately define it. To create one is to create an illusion" (p. 597). Rather, only specific goals can be achieved in a piecemeal fashion, where his historically identified *nine* different conceptions are chosen as in a smorgasbord, attentive to the context of school culture and society wishes, and where "schools and teachers need to set their priorities" (ibid.). With this option, divergence is chosen. It is then assumed that "consensus about one definition throughout the worldwide science education community is a goal not worth chasing" (Roberts 2007, p. 736).[79]

(iii) *Abandonment* option: one chooses to reject the term as both useless and meaningless for educational purposes, along with Shamos (1995) and Solomon (1999).

In any case, if an educational metatheory is to be of service to science education, it must also acknowledge and address these options in the deadlock.[80] It may also

---

[78] Shamos has insightfully argued that its common conception tied to citizenship is fundamentally flawed, that the community is chasing a utopia, that it continues to refuse to accept the grounds why it has failed in achieving it, and finally that many rationales typically put forth to justify it are a *myth.*

[79] Option two although seemingly attractive on the surface does not seem viable, and one can imagine numerous problems associated with it. Just mentioning one, it assumes a degree of autonomy for schools and teachers which they generally lack and which in the climate of "accountability" and standardized testing and under the influence of powerful outside social groups would seem to check their ability to make the kind of choices DeBoer would like. A reversion to option one would in all likelihood result, namely, the default traditionalist position.

[80] A series of papers presented at a recent conference attempting to articulate "a more expansive notion of scientific literacy" illustrate the problems associated with this deadlock once more and why the sought-after solutions remain so elusive; discussions including educational theory and philosophy are conspicuously absent (Linder et al. 2007).

39.5.1.2 Educational Metatheory and Advance of Science Education as a Research Field

Fensham's *Identity* book (2004), interestingly enough, also offers an important look at the role of theory (Ch.7) within the science education research community. He admits that the development of theory is a significant indicator of a discipline's advance as a research field:

> If the existence of theory and its development is a hallmark of a mature research field there is some evidence that the research in which the respondents have been engaged in science education has reached this point. On the other hand, the role that theory plays in the respondent's remarks was so variable that it is not possible to attach this hallmark in a simple way to much of their research. (Fensham 2004, p. 101)

With that admission he acknowledges that the usage of theory is restricted and there was little interest on the part of researchers to develop their theory of choice further. What is significant though is the range of *borrowed* theories from outside research fields that the researchers have heavily relied upon.[81] The spectrum stretches from social anthropology, ethnology, and cultural theory to psychology, cognitive science (e.g., information processing; schema restructuring), and philosophy of science (e.g., conceptual change theory).[82] He notes those researchers employing a "political framework" to curriculum, or concepts of power and ideology, shift the common focus of science education onto entirely different factors that influence science teaching and learning. Essential PE-type questions like "what counts as science education?" or "how are ideological meanings reproduced in science education?" are raised, but surprisingly not addressed with that perspective or discipline in mind. One observes rather that in all cases educational theory and philosophy of education nowhere make an appearance.

To the point of the subject at hand, Fensham does mention the topic of "grand theory" (p. 107). He writes that only *one* respondent had admitted to theorizing on this scale, namely, the biologist and educator Joseph Novak, who had earlier published *The Theory of Education* (1977).[83] Novak has today continued to hold, as

---

[81] "This borrowing can have the healthy effect of bringing new insights to bear on the problems of science education, but it can also lead to superficial descriptions that do not seem to be pushing for deeper understanding" (2004, p. 101). He fails to mention a *third* possibility that outside theories can do outright damage to education, as Egan (1983, 2002) argues for the cases of behaviorism, Piaget, and progressivism. The presumed relevance of cognitive science has lately come into question as well (Slezak 2007).

[82] Reliance upon psychology is clearly predominant, primarily Bruner, Gagne, and Piaget in the 1960s and 1970s and the significant role they played marking the revolt against behaviorism.

[83] This book, however, as is familiar today, is based on the psychologist Ausubel's quasi-neural theory of meaningful learning in combination with Toulmin's philosophy of science and principally restricted to learning theory.

Fensham comments, to the value of this theory and the belief that "theories in science education would be developed that have predictive and explanatory power, just as theories in the natural sciences have" (p. 106). This belief closely aligns educational theory with empirical theories in the natural or social sciences,[84] an arrangement both Hirst (1966) and Egan (1983, 2002) explicitly reject.

39.5.1.3 Educational Philosophy and Science Education as "Sociopolitical Activism"

One contemporary reform movement (spearheaded by some international researchers and popular with some policy advocates), namely, "science education as/for sociopolitical action," has been articulated with intentional philosophical perspectives. It could reasonably be interpreted as a rudimentary sort of *"philosophy of"* science education (PSE) as here elucidated (though granted, not formulated in this fashion). The position that science education *should* be oriented (if not exclusively so) to perform sociopolitical action is a normative claim argued on philosophical grounds, justified because of the apparent promise/claim of enhancing critical-minded citizenship and forwarding democracy. It patently stipulates categorical answers to the key questions: "What counts as scientific literacy?" "What counts as science education?" "What is it for?" Whether or not such a muscular and singularly focused PSE can do justice to the other historically identified aims associated as central to science education (including the *aesthetic* component of science; DeBoer 2000; Girod 2007), and therefore the best option for policy deliberations and reform, is a matter for some dispute—although a considered debate especially one involving philosophy of education (PE) is surprisingly lacking to date.[85]

That this sort of politicized PSE represents a "radical program" to challenge common school science education is understood (Jenkins 2009; Levinson 2010). Here our focus is to ask: is such a "program" an adequate PSE?[86] Science education, for example, could plausibly "do" sociopolitical action at times while rejecting "as" and "for." In any event, does politico-social activism as put forth substitute ideology

---

[84] It is admitted that Novak's writings offered an important counter-theory in support of the growing dissatisfaction with the dominance of Piagetian theory arising in the late 1970s (although some science educators continue to hold neo-Piagetian views). With the growth of conceptual change and constructivist research in the 1980s and the influence of Kuhnian philosophy-of-science, this dominance was gradually displaced in the research community. On the other hand, Erickson (2000) cautions there is much common ground between Piaget and the newer constructivist theories. Egan's cultural-linguistic metatheory (1997) is inclusive of learning theory but goes beyond it and outright rejects Piaget (Schulz 2009b).

[85] Leaving aside questions if its individual educational claims are either warranted or empirically validated. Strong advocates for this kind of politico-social activist PSE (just naming some researchers) are Hodson (2009) and Roth and Desautels (2002). Criticisms leveled against it are provided by Hadzigeorgiou (2008) and Levinson (2010).

[86] Does it fully take into consideration the three dimensions of the synoptic framework shown in Fig. 39.1?

for philosophy?[87] Does it presuppose educational metatheory? The present author would argue it must (although this feature is seldom articulated; i.e., social reconstruction). Stepping back, must *any* methodical PSE presuppose metatheory (of some kind)—or can it be gone around for, say, a list of rationales, principles, and exhortations? That debate has not yet begun, but would be welcomed.[88]

One of the responsibilities of a philosophy of science education (PSE) at the research level would be to expose educational theories (especially metatheories), as well as better clarify the relationships between such theories in PE and theories in other (empirical) disciplines (as to their nature, value, and limits), whether one of independence or interdependence.[89] In other words, a philosophical appraisal of several domains, such as conceptual clarification and the validity of borrowed ideas; scrutiny of epistemic and/or moral and political aims—their character and prioritization; analysis of the theory-practice dilemma; also the character, quality, and significance of kinds of assessments or tests employed (range of usefulness), etc.; and hence the question of boundaries, applicability, and relevance.

39.5.2 Epistemology, Knowledge, Understanding, and Hermeneutics

39.5.2.1 Epistemology, Belief, and Epistemic Aims

That science instructors and their technical textbooks are so concerned with accurate and exhaustive transmission of canonical scientific knowledge clearly reveals the central significance of epistemology to science education.[90] One can identify this preoccupation of academic sciences courses (a chief aim of school and college science) with the constricted and popular rendition of the customary *knowledge aim*.

---

[87] Roberts (1988, p. 50) had earlier cautioned the research community about the "*individual ideological preference* of professors of science education" which can "indoctrinate science teachers into believing that what counts as science education is the ideology of a single curriculum emphasis (or perhaps a few emphases)" (original italics).

[88] It seeks as well to address the common blurring of lines between "descriptive" and "normative" research work, the expectation *that* classroom research *should* change classroom teaching and learning, as Sherman (2005) points out, but strictly in accordance with a specified (ideological) program. This academic conflation may indeed be due to our culturally inherited situation, i.e., "if we can't be objective, we'll be openly ideological" (p. 205), but regrettably real "openness" is rare. The argument here in a nutshell is that science education avoids (c)overt ideology for candid philosophy.

[89] Such a conversation can be considered an extension of one already discussing the difference between epistemology and psychology (Duschl et al. 1990; Matthews 2000; Southerland et al. 2001) or critiquing the assumed validity of cognitive science theories for science education (Slezak 2007).

[90] For some time a major portion of science education research has in fact been focused on analyzing and critiquing the strengths and weaknesses of school science epistemology, whether of subject content, or of the student, or of the teacher.

Here is another area where PE discourse can provide relevance, for the knowledge aim or truth aim has been fundamental in the traditional view of education, including its *liberal* construal—notwithstanding significant attacks on that objective from different educational perspectives (e.g., progressivism, post-analytic, postmodernist).

> Although transmitting knowledge is not the only aim of education, it is surprisingly substantial in its ramifications. Because we can compare various educational practices to determine which ones better advance students' knowledge, the knowledge-aim offers educational guidance, justifies central educational practices, and exposes complexities in the educational policies it supports. (Adler 2002, p. 285)

Science teachers plainly assume their courses or textbooks provide (technical) knowledge, indeed substantially *true* knowledge—and for the most part, they would be correct (e.g., propositional knowledge of final form science; Duschl 1990).[91] Yet being philosophically inclined means giving pause to reflect on what *basis* this can be claimed (expertise of the authors? Authority of the scientific community?). HPSS-based reforms do insist, of course, that *content knowledge* (CK, of teacher or curriculum) requires expansion and corrections (e.g., historical and epistemological *context* to be properly understood and learned).[92] But stepping back and asking about justifying CK, or "what is knowledge?"[93], is to venture into both philosophy (P) and PE territory (the right segment of the triangle in Fig. 39.1). The kinds of answers to these questions have vital educational ramifications. How, for example, can one justify teaching evolutionary theory if its stake in knowledge and truth cannot be established against intelligent design claims? Or taking the "culture wars" into view, is cultural indigenous knowledge of nature *true* scientific knowledge? Are there other kinds? If so, how are they legitimated? How to best distinguish them from science?[94]

---

[91] This has also been referred to in the research literature as the "disciplinary view of knowledge" in contrast to "personal learner epistemology" and "social practice views of epistemology" (Kelly et al. 2012). The latter defers to science studies research and how knowledge is attained and justified through discourse practices within epistemic cultures (Knorr Cetina 1999). What is significant is that "within this perspective, knowledge is seen as competent action in a situation rather than as a correct, static representation of the world" (p. 286). What is not being acknowledged is that the two stated perspectives are themselves beholden to two different epistemological philosophies, namely, pragmatism and objectivism. While science education has traditionally been in the thrall of the second and is now expected to shift to the first, it could better take advantage of the respective benefits of each.

[92] Even when basic science "subject matter" is taught, it is always accompanied by some context that may operate covertly (e.g., preparatory, socio-utility, etc.). Such contexts have been called "meta-lessons" (Schwab), "curriculum emphases" (Roberts 1988), and "companion meanings" (Roberts and Oestman 1998).

[93] Also, what kind of science knowledge is of *most* worth (a key question of prioritizing subject content)?

[94] A very informative discussion on such questions, including examining beliefs, learning, knowledge, and critical inquiry pertaining to the aims of science education, can be found in Nola and Irzik (2005). The comments which follow can be considered supplemental to their work.

Students, when not just assuming the authority of the textbook or teacher, occasionally wish to have explained to them the grounds for knowing, grounds that can only partially be established when "doing science" (i.e., scientific inquiry). Four possible harmful *dispositions* to knowledge students can develop from science classrooms are cynicism, dogmatism, skepticism, and relativism, and Norris (1984) rightfully asks "can all these be avoided?" Teachers require philosophical intelligence not just for telling these apart, but awareness when they crop up during instruction and for strategies to overcome them.[95] Thankfully there already exists a tradition in PE that can assist them, which has sought to demonstrate the relevance of epistemology for education (Adler 2002; Carr 1998, 2009; Siegel 1988).

The standard account of knowledge is "justified true belief" (JTB), which stipulates three conditions in order for someone to say they "know X." For instance, science educators would not be satisfied if a student stated they "know" the Earth orbits the Sun but could not provide any evidence for this proposition. In this case the student has a *true belief* (two conditions met), but without justification could not be said to have attained knowledge. Even if philosophers have brought forth serious challenges to JTB[96], this doctrine of traditional epistemology still retains its value in assisting science teachers' thinking about the differences between knowledge, belief, and justifying conditions in the classroom as they arise (Southerland et al. 2001). It highlights the drawbacks of traditional instruction which can overstress the value of rote learning, algorithmic problem solving, and decontextualized subject content, especially if tied to a policy of exaggerated standardized testing (Hofer and Pintrich 1997; Mercan 2012).

JTB can equally shed light on other cases which can occur where knowledge and belief appear conflated, such as when a student has learned content but refuses to believe it (e.g., "I understand evolution, but I don't believe it"; "I can explain the Bohr model but don't believe atoms exist"). Southerland and associates (2001) have provided an overview of the differing conceptions and occasional clashing views concerning how "knowledge" and "belief" are employed as terms in the separate research fields of philosophy, educational psychology, and science education. They also raise the important pedagogical question whether science education should limit its aim to providing knowledge (or understanding) and not demand changing student beliefs (as required by conceptual change research). An interesting exchange of views between Smith and Siegel (2004) and Cobern (2000, 2004) on this topic illuminates that science teachers need to sort out not just their own presuppositions about knowledge and beliefs but require sensitivity to historical and cultural dimensions of these concepts while attending to philosophical arguments.

Within the field of science education research, Norris (1995, 1997) has analyzed how the JTB view of knowledge finds expression in the aim of *intellectual independence*, one key content-transcendent goal articulated since Dewey and progressivism.

---

[95] Certainly the relatively recent research studies to enhance *scientific argumentation* in the classroom also aim towards resolution of the issues and questions raised here, but are not of present concern.

[96] These will not be discussed here; instead see Siegel (2010) and Norris (1997).

He identifies several serious shortcomings of past and recent formulations of this goal (e.g., as found in constructivism and notions of scientific literacy). Norris notes especially the philosophical controversy surrounding the question to what extent, if any, non-experts can reason independently of experts' knowledge and community—hence, to what extent they can be justified to trust in authority and yield to scientists' judgments (and by association, their textbooks). The outcome of the dispute remains contested, but it appears some reliance is indeed unavoidable.

The degree to which intellectual independence is attainable (or not) has major ramifications for the character and educational aims of science education reform movements (like STS, SSI, HPS, social action). It could impose severe limitations, depending upon the stipulated objectives and overall ambition they desire to advance for the discipline, notably which independence-based goals they mistakenly assume school students can rightfully achieve.[97]

Returning to a previous point, Smith and Siegel (2004) in their paper had also named *understanding*—along with knowledge, and *not* (changing) belief—as primary goals for science instruction. The focus here though is not to address their position nor the dispute with Cobern (but noting its significance) rather to point out that "understanding" as both concept and goal has been largely overlooked in the research literature. Its merit with respect to epistemology and the traditional preoccupation with "knowledge" yields a checkered history, too (Toulmin 1972).[98] Yet its prominence does come to the fore in *philosophical hermeneutics* (Gadamer 1976, 1960) as well as Egan's educational metatheory. A systematic investigation of "understanding," its contrast to knowledge, and its merit for science education has yet to be presented.[99]

One fertile perspective on "understanding" has been provided by the late physicist and philosopher Martin Eger (1992, 1993a, b). He had insightfully shown the relevance of Gadamer's "philosophy of the humanities" for science education with regard to the *interpretation* of nature but especially of science *texts*. Hermeneutics, an age-old scholarly discipline, ties understanding to the ability to achieve personal meaning when interpreting text (utilizing the "hermeneutical circle" method). The significance of his ideas lies in offering an alternative approach to viewing science learning and knowing, drawing science education away from psychological and cognitive science perspectives and towards philosophy and the humanities (Bontekoe 1996; Donnelly 2001; Gallagher 1992). Today his ideas are finding useful expression in some research work (Borda 2007; Kalman 2011). He explicitly shifts the emphasis away from epistemology towards *ontology*, away from "knowing" in the

---

[97] Kuhn (1970) was skeptical about what science education could achieve in terms of developing independent thought and argued instead the conservative view of reinforcing the conventional paradigm—in part because this furthered "progress" and in part because students had no competence to do otherwise. Schwab held a different view and thought students could be educated to become "fluid inquirers" within and about a discipline. Siegel (1978) has admirably contrasted the two opposing positions.

[98] Mason's (2003) "*Understanding* understanding" is one of the few to explore the contrast.

[99] Some researchers have ventured into this territory; see, for example, Wallace and Louden (2003).

objectivist sense to interpreting, meaning, and being. This shift, or "interpretative turn" (Hiley et al. 1991), has not been entirely endorsed as regards questions surrounding the nature of language, ontology, and the relationship between epistemology and hermeneutics. The next section provides science educators with an unconventional but updated outlook regarding these major topics.

39.5.2.2 Epistemology and/Versus Hermeneutics

Any discussion involving philosophical hermeneutics recognizes two current state of affairs, namely, the ongoing unresolved dispute over the self-conception of philosophy and the so-called interpretative turn from epistemology *to* hermeneutics.

To the first, one identifies that the modern Anglo-analytic philosophical tradition has fractured into two differing schools of thought as to what the nature and role of modern philosophy *is* and can accomplish (represented by the opposing views of Dummett and Rorty; Bernstein 1983). This opposition is reflected as well in contrasting perspectives on language theory—which Charles Taylor has characterized as the *designative* and *expressive* traditions (Medina 2005, p. 39). That said, authors like Bernstein, Rorty, and Taylor nonetheless all comment on the convergence of thinking in both the Anglo-American and Continental traditions which reject *foundationalism* or the former project of grounding philosophy, knowledge, and language ("objectivism"), as Descartes, Kant, Russell, and the early Wittgenstein sought but failed to do.

With the current preoccupation of repudiating this formerly eminent epistemological tradition[100], the task of "overcoming epistemology" has come to mean different things to different thinkers (Baynes et al. 1987). Dewey and Bentley (1949), for instance, sought to overcome subject/object dualism with his pragmatic focus on "transaction," the active/practical behavior taking place between the knower and known. Taylor (1987) correctly views both Quine and Rorty as abandoning foundationalism (with the former attempting to "naturalize" epistemology), while he solely targets overcoming the conception of knowledge as *representation* that lies behind the ambition of the foundationalist project since Descartes:[101] "If I had to sum up this understanding in a single formula, it would be that knowledge is to be seen as correct representation of an independent reality.

---

[100] "Current attitudes toward foundationalism, as they have been since Descartes, are sharply divided. The minoritarian conviction (Chisholm, Apel, Habermas, Haack, Swinburne, and others) that some version of foundationalism is or is at least potentially viable is outweighed by the majoritarian belief that in the debate since Descartes, foundationalism has died a natural death and cannot be revived" (Rockmore 2004, p. 56).

[101] Rorty, of course, also surfaces representation, but he explicitly ties it to philosophy as a profession whose role as a foundational discipline (with its "theory of knowledge" being essentially a "general theory of representation") was to adjudicate all cultural knowledge claims, eventually including scientific ones. His view is comparable to Taylor's "To know is to represent accurately what is outside the mind; so to understand the possibility and the nature of knowledge is to understand the way in which the mind is able to construct such representation" (1979, p. 3).

In its original form it saw knowledge as the inner depiction of an outer reality" (p. 466).[102] One notes representation plays a significant role in science and science education, and Giere (1999) argues, in contrast, for its continued importance in science independent of foundationalism. Indeed, some philosophers and science educators have argued for a "fallibilist epistemology" as a viable alternative to opposing foundationalist and radical constructivist views of knowledge and belief (Siegel 2001, 2010; Southerland et al. 2001). The collection of papers in Carr (1998) intends to help guide curriculum policy beyond "rational foundationalism" and "promiscuous postmodernism." The discussions in these works can contribute to advancing teachers' epistemological conceptions and deliberations, whether concerning science, curriculum, or student learning.

The second aspect, as mentioned, acknowledges an "interpretative turn" to have taken place not only in philosophy (due initially to Heidegger 1977) but in the natural and social sciences as well (inclusive of language theory)—though granted, still subject to much dispute—that also seeks to move "beyond objectivism and relativism" (according to Bernstein 1983).[103] Such a move can be considered a shift in the philosophical emphasis entirely "from epistemology to hermeneutics," as both Rorty and Gadamer have claimed[104]; certainly it can be admitted the relation between the two modes of inquiry is contentious and differing conceptions of language inform both.

Furthermore, although there are many similarities in Rorty's and Gadamer's positions, there exist important differences as well as to the nature and task of epistemology and hermeneutics, which is instructive. For example, while Rorty would agree that Anglo-analytic philosophy of language has slowly come to abandon the notion of language as correct "picture of the world"[105], he would disagree with Gadamer's universalist perspective of philosophical hermeneutics (with its inherent view of language as the *medium* of all understanding). Both agree that hermeneutics is not to be considered a successor to epistemology, rather that it involves an entirely different approach to comprehend the world—indeed Rorty construes it as a kind of "paradigm shift" (one that is holistic, historicist, and pragmatic). While Rorty makes

---

[102] Taylor links the success of "knowledge as correct representation" standpoint with two factors: its link with the rise of mechanistic science in the seventeenth century, whose mechanized worldview overthrew the Aristotelean one with its notion of "knowledge as participation" ("being informed by the same *eidos*, the mind participated in the being of the known object, rather than simply depicting it," p. 467); secondly, the influence of Cartesian philosophy that insisted a new reliable "method" was required that could guarantee certainty of the representation. Yet this method entailed, unlike in philosophical antiquity, the reflective and critical cast of individual *mind* performing a subjectivist inward turn. Rorty's view is similar (1979, p. 248).

[103] He cites such authors as Rorty and Taylor (in philosophy), Gadamer (in language theory), and Kuhn and Hesse (in philosophy of science). Other philosophers of science endorsing hermeneutics are Heelan (1991) and Ihde (1998).

[104] See especially Rorty (1979, Chap. 7) and Gadamer (1989, p. 235).

[105] "Putnam now agrees with Goodman and Wittgenstein: to think of language as a picture of the world—a set of representations which philosophy needs to exhibit as standing in some sort of nonintentional relation to what they represent—is *not* useful in explaining how language is learned or understood" (1979, p. 295; original italics).

a sharp distinction between the two but sees them as complementary and mutually supportive (epistemology for "normal discourse" and hermeneutics for "abnormal"), Gadamer views them rather as antagonists: hermeneutics as the universal condition of understanding (and hence of *being; Dasein*)[106] but epistemology as a failed *epistéme*-based, historico-philosophical venture whose time has come and gone. The project has died and should be buried. Rorty correctly stresses that Gadamer had emphasized *Bildung* as historical enculturation (hence the crucial role of education) as a proper goal of hermeneutics—construed as an open project of how understanding takes place through interpretation and dialogue, a form of *intersubjectivity*. This is seen in contrast to "knowledge" possession and obsession of isolated, individual cognition (the foundationalist project), but he would not consent that such "understanding" entails knowledge. Rorty is clear that "knowledge" is fallible and constrained to the "normal discourse" of a particular (historical) sociocultural paradigm (explicitly referencing Kuhn's ideas).[107] But taking such a position on a *standard* of knowledge one can argue, alternatively, must implicate Rorty's outlook as committed to the epistemic assumptions of Cartesian foundationalism.[108]

There is certainly more that can be surveyed here in the debate about the shift "from epistemology to hermeneutics." Siegel (2010), for instance, takes issue with Taylor's arguments for "overcoming" epistemology, while Suchting (1995) criticizes many of the "lessons" supposedly drawn from hermeneutics. Several very important questions exist that still need addressing, such as if the common division between *explanation* and *understanding* is abandoned—which has long been accepted as *the* major difference between the natural and social sciences (Mason 2003)—and "interpretation" comes to characterize all human inquiry, does or should a "contrast class" exist in opposition to it? Thereto, how can or should one demarcate the lines between the humanities and the different sciences? Moreover, how does one adjudicate between better and worse interpretations? Is hermeneutics[109] really an alternative

---

[106] This hermeneutic perspective on learning and understanding corresponds with the newer epistemological perspectives of the field: "… increasingly, science education researchers are viewing meaning as public, interpreted by participants (and analysts) through interaction of people via discourse including signs, symbols, models, and ways of being" (Kelly et al. 2012, p. 288).

[107] Hence his complaint that one can distinguish between "systematizers" (those engrossed in normal discourse) and "edifying" philosophers (anti-foundationalists like Dewey and hermeneutic thinkers like Heidegger, Gadamer, who disrupt it) within the tradition—the latter whose status as "true" philosophers is often questioned by academic professionals.

[108] Rockmore (2004, p. 57) writes that Rorty maintains "a strict but wholly arbitrary distinction between epistemology and hermeneutics in order to equate the failure of foundationalism with a form of skepticism that cannot be alleviated through a hermeneutical turn." He accuses Rorty of still clinging to a standard of knowledge that he admits cannot be met. Rorty freely concludes that one can no longer hope to bring the mind in contact with the real and that *interpretation* must be the alternative, but just denies this will lead to knowledge in the conventional sense. Alternatively, Rockmore argues that "the main strategy for knowledge is, and always has been, interpretation" (ibid), not to be taken as tantamount to skepticism.

[109] This is not meant to imply this field of study is monolithic, and commentators commonly distinguish between "right-wing" (Gadamer) and "left-wing" (Derrida) factions. Yet such a categoriza-

paradigm to epistemology (as Gadamer and Rorty insist) or another albeit extraordinary version of epistemology itself, just not of the classical foundationalist sort (as Rockmore (2004) and Westphal (1999) contend)?[110]

There are fundamental issues and concerns identified here that a philosophy of science education (PSE) would equally need to consider and evaluate, which have necessarily arisen in the dispute between the advocates of epistemology, hermeneutics, and their different perspectives on language, knowing, and understanding.

39.6 Conclusion

Philosophy and philosophy of education continue to remain outside the mainstream of thinking in science education. The chief purpose of this chapter has been to bring them closer into the fold. Philosophy is, on the one hand, underappreciated and ignored by science teachers, on the other, occasionally raided, used, and abused by science education researchers. Philosophy of education by contrast (and when compared to philosophy of science) has the dubious distinction of being disregarded by both groups.

Philosophy as a discipline of critical inquiry enables teachers to develop a thoughtful, critical capacity to reflect upon curricular, epistemological, and popular media issues as they arise, whether during classroom discourse or professional policy deliberations. Philosophy is not far below the surface in any classroom, and in truth cannot be avoided. This holds especially when discussing common terms like "law," "theory," and "proof," or justifying content knowledge, or analyzing national "standards" documents, or providing coherent educational rationales for their courses, or for detecting curricular ideologies and conveyed textbook myths (e.g., academic rationalism, indoctrination into scientism, epistemological positivism, historically defined convergent realism, evolution versus intelligent design arguments, ambiguities and hazards of modern techno-science, cultural and personal bias). Philosophy of education as a subdiscipline prepares a forum of informed analysis and discussion on a range of topics and issues that bear directly on science education as an educational project, which has deep roots in the historico-philosophical past.

---

tion is equally overly simplified. Those in educational studies—see Gallagher (1992)—distinguish four separate schools: conservative (Dilthey; Hirsch), moderate (Gadamer; Ricoeur), radical (Derrida; Foucault), and critical (Habermas; Apel).

[110] Rockmore maintains that the shift leads to a *redefinition* of epistemology, from "knowing the way the mind-independent world is" to "the interpretation of experience" which is justified by the standards in use in a given cognitive domain. In this reformulation "then epistemology as hermeneutics presents itself as a viable successor to the traditional view of epistemology—indeed as the most likely approach at the start of the new century" (p. 11). Westphal criticizes Rorty for failing to distinguish between classical epistemology and hermeneutics seen as a generic epistemological task, hence, to differentiate the replacement of only one type (foundationalism): "*hermeneutics is epistemology,* generically construed … it belongs to the same genus precisely because like them it is a meta-theory about how we should understand the cognitive claims of common sense, of natural and social sciences, and even metaphysics and theology" (p. 416; original italics).

A philosophy of science education (PSE) can be understood as a *synthesis* of (at least) three academic fields of philosophy (P), philosophy of science (PS), and philosophy of education (PE), each of which have distinctive contributions to make in its development. It can be interpreted as a "second-order" reflective capacity on the part of the teacher, as an extension of their pedagogical content knowledge. The research field requires this capacity to think deeper and more systematically about the unique educational dimensions of teaching and learning of science as philosophy, as profession, and as practice. It should be inaugurated as a new *fourth* area of research inquiry.

PSE is ultimately concerned with the explicit *problematizing* of school science and its epistemology for two substantive reasons: (i) to recognize the current inadequate portrayal as inauthentic science and so to improve the content knowledge (CK) of both the curriculum and teacher through HPSS studies and integration and (ii) to allow for the effective didactic transposition of subject content for the culturally rooted, age-appropriate learner in accordance with educational aims, philosophy, and theory.

Pertaining to performing useful functions, its value is taken as being *threefold:* it serves to, first of all, provide a platform for both researchers and practitioners (in their separate ways) to perform meta-analysis (critical function); secondly, to reconceptualize, remake, and reform curriculum and instruction (creative function); and lastly, to implement, as an example, effective critical thinking for teacher and student, appropriate to subject content and age level (pragmatic function). In the process it is understood such a philosophy when developed would be articulating in essence its meaning of "scientific literacy" and thus specifying and prioritizing essential objectives for science education. Whether or not it could successfully perform these functions without an explicit educational metatheory at hand is open to challenge and debate. In any event, it would ultimately aim at improving science education by broadening the research field and opening new territory for exploration, as well as assisting teachers in broadening their theoretical frameworks, sharpening their critical acumen, and enhancing their pedagogical content knowledge.

References

Abell, S.K. and Lederman, N.G. (eds.) (2007). *Handbook of research on science education.* Mahwah, NJ: Lawrence Erlbaum Associates.

Abell, S.K, (2007). Research on science teacher knowledge. In: S.K. Abell and N.G. Lederman (eds), *Handbook of research on science education* (pp. 1105–45). Mahwah, NJ: Lawrence Erlbaum Associates.

Adler, J.E. (2002). Knowledge, truth and learning. In: R.R. Curren (ed.), *A companion to the philosophy of education* (pp. 285–303). Oxford: Blackwell Publishers.

Aikenhead, G. (2007). Humanistic perspectives in the science curriculum. In S.K. Abell & N.G. Lederman (Eds.), *Handbook of research on science education* (pp. 880–930). Mahwah, N.J.: Lawrence Erlbaum Associates.

Aikenhead, G. (2006). *Science education for everyday life. Evidence-based practice.* New York and London: Teachers College Press.

Aikenhead, G. (2002). The educo-politics of curriculum development: A response to Fensham's 'time to change drivers for scientific literacy.' *Canadian Journal of Science, Mathematics and Technology Education, 2(1)*, 49–57.

Aikenhead, G. (1997). STL and STS: Common ground or divergent scenarios? In E. Jenkins (ed.), *Innovations in scientific and technological education* (pp. 77–93). *(Vol. VI)*, Paris: UNESCO Publishing.

Aikenhead, G. (1996). Science education: Border crossing into the sub-culture of science. *Studies in Science Education, 27*, 1–52.

Aldridge, J., Kuby, P. & Strevy, D. (1992). Developing a metatheory of education. *Psychological Reports, 70*, 683–687.

Allchin, D. (2004). Should the sociology of science be rated X? *Science Education, 88*, 934–946.

Allchin, D. (2001). Values in science: An educational perspective. In: F. Bevilacqua, E. Giannetto and M. Matthews (eds.), *Science education and culture: The contribution of history and philosophy* (pp. 185–196). Dordrecht: Kluwer.

Anderson, R.N. (1992). Perspectives on complexity: an essay on curricular reform. *Journal of Research in Science Teaching, 29(8)*: 861–876.

Anderson, J. (1980). *Education and inquiry* (edited by D.Z. Phillips). Oxford: Blackwell.

Apple, M. (1992). Educational reform and educational crisis. *Journal of Research in Science Teaching, 29(8)*: 779–789.

Apple, M. (1990/1975). *Ideology and curriculum. 2^{nd}ed.* New York: Routledge.

Aristotle. (1998). *The metaphysics.* (Translated and introduction by Hugh Lawson-Tancred). London: Penguin books.

Aristotle. (1962/1981). *The politics.* (Translated by T.A. Sinclair. Revised and re-presented by Trevor J. Saunders). London: Penguin Books.

Atwater, M.M. (1996). Social constructivism: Infusion into the multicultural science education research agenda. *Journal of Research in Science Teaching, 33(8)*, 821–837.

Audi, R. (2009). Science education, religious toleration, and liberal neutrality toward the good. In: H. Siegel (ed.), *The Oxford handbook of philosophy of education* (pp. 334–357). Oxford and New York: Oxford University Press.

Ayala, F.J. and Arp, R. (eds.) (2009). *Contemporary debates in philosophy of biology.* Malden, MA: John Wiley & Sons.

Bailey, R., Barrow, R., Carr, D. and McCarthy, C. (eds.) (2010). *The SAGE handbook of the philosophy of education.* London: SAGE.

Bailin, S. and Battersby, M. (2010). *Reason in the balance: An inquiry approach to critical thinking.* Toronto: McGraw-Hill.

Bailin, S. and Siegel, H. (2003). Critical thinking. In: N. Blake, P. Smeyers, R. Smith and P. Standish (eds), *The Blackwell guide to the philosophy of education* (pp. 181–193). UK: Blackwell.

Bailin, S. (2002). Critical thinking and science education. *Science & Education, 11*, 361–375.

Barnard, F.M. (2003). *Herder on nationality, humanity and history.* Montreal & Kingston: McGill-Queen's University Press.

Barrow, R. (2010). Schools of thought in philosophy of education. In: R. Bailey, R. Barrow, D. Carr and C. McCarthy (eds.) *The SAGE handbook of the philosophy of education* (pp. 21–36). London: SAGE Publications Limited.

Barrow, R. & Woods, R. (1975/2006). *An introduction to philosophy of education, 4th Ed.*, London and New York: Routledge.

Bauer, H.H. (1992). *Scientific literacy and the myth of the scientific method.* Urbana and Chicago: University of Illinois Press.

Baynes, K., Bohman, J., & McCarthy, T. (eds.) (1987). *After philosophy. End or transformation?* Cambridge, Massachusetts: MIT Press.

Beiser, F. (1998). A romantic education. The concept of *Bildung* in early German romanticism. In: Amelie O. Rorty (ed.), *Philosophers on education. Historical perspectives* (pp. 284–299). London and New York: Routledge.

Bencze, L. (2001). Subverting corporatism in school science. *Canadian Journal of Science, Mathematics and Technology Education, 1(3)*: 349–355.

Benner, D. (1990). Wissenschaft und Bildung. Überlegungen zu einem problematischen Verhältnis und zur Aufgabe einer bildenden Interpretation neuzeitlicher Wissenschaft. (Science and education. Reflections on a problematic relationship and on the task of an educative interpretation of modern science). *Zeitschrift für Pädagogik, 36*(4), 597–620.

Bernstein, R.J. (1983). *Beyond objectivism and relativism. Science, hermeneutics and praxis.* Philadelphia: University of Pennsylvania Press.

Beyer, L.E. and Apple, M.W. (1998) (Eds.). *The curriculum: Problems, politics and possibilities.* Albany, NY: Suny. Press.

Blake, N., Smeyers, P., Smith, R. & Standish, P. (eds.) (2003). *The Blackwell guide to the philosophy of education.* UK: Blackwell.

Blake, N. & Masschelein, J. (2003). Critical theory and critical pedagogy. In: *The Blackwell guide to the philosophy of education* (pp 38–56). UK: Blackwell.

Boersma, K, Goedhart, M, De Jong, O, Eijkelhof, H. (eds.) (2005). *Research and the quality of science education.* Dordrecht: Springer.

Bonjour, L. and Sosa, E. (2003). *Epistemic justification. Internalism vs. externalism, foundations vs. values.* Oxford: Blackwell Publishing.

Bontekoe, R. (1996). *Dimensions of the hermeneutic circle.* New Jersey: Humanities Press.

Borda, E.J. (2007). Applying Gadamer's concept of dispositions to science and science education. *Science & Education, 16*(9–10), 1027–1041.

Brighouse, H. (2009). Moral and political aims of education. In: H. Siegel (ed.), *The Oxford handbook of philosophy of education* (pp. 35–51). Oxford & New York: Oxford University Press.

Brown, J.R. (2001) *Who rules in science? An opinionated guide to the wars.* Cambridge, MA: Harvard University Press.

Burbules, N.C. & Linn, M.C. (1991). Science education and philosophy of science: Congruence or contradiction? *International Journal of Science Education, 13,* 227–241.

Burns, D. P. and Norris, S.P. (2009). Open-minded environmental education in the science classroom. *Paideusis: Journal of the Canadian Philosophy of Education Society, 18,* 35–42.

Bybee, R. & DeBoer, G.E. (1994). Research on goals for the science curriculum. In: D. Gabel (Ed.), *Handbook of research on science teaching and learning* (pp. 357–387). New York: Macmillan.

Carr, D. (2010). The philosophy of education and educational theory. In: R. Bailey, R. Barrow, D. Carr and C. McCarthy (eds.) *The SAGE handbook of the philosophy of education* (pp. 37–54). London: SAGE Publications Limited.

Carr, D. (2009). Curriculum and the value of knowledge. In: H. Siegel (ed.), *The Oxford handbook of philosophy of education* (pp. 281–299). Oxford & New York: Oxford University Press.

Carr, D. (2003). *Making sense of education.* London: Routledge Farmer.

Carr, D. (ed.) (1998). *Education, knowledge and truth: Beyond the postmodern impasse.* London: Routledge.

Carson, R.N. (1998). Science and the ideals of liberal education. In: B.J. Fraser & K.G. Tobin (eds.), *International Handbook of Science Education. Part II* (pp. 1001–1014). Dordrecht: Kluwer Academic Publishers.

Chambliss, J.J. (ed.) (1996a). *Philosophy of education: An encyclopedia.* New York: Garland Publishing Company.

Chambliss, J.J. (1996b). Philosophy of education, history of. In: J.J. Chambliss (ed.), *Philosophy of education: An encyclopedia.* New York: Garland Publishing.

Clough, M., Berg, C. & Olson, N. (2009). Promoting effective science teacher education and science teaching: A framework for teacher decision-making. *International Journal of Science and Mathematics Education, 7,* 821–847.

Cobern, W.W. (2004). Apples and oranges. A rejoinder to Smith and Siegel. *Science & Education, 13,* 583–589.

Cobern, W.W. (2000). The nature of science and the role of knowledge and belief. *Science & Education, 9*(3), 219–246.

Collins, H. (2007). The uses of sociology of science for scientists and educators. *Science & Education, 16,* 217–230.

Corrigan, D., Dillion, J. & Gunstone, D. (Eds.). (2007). *The re-emergence of values in the science curriculum*. Rotterdam, Holland: Sense Publishers.

Curren, R.R. (ed.) (2003). *A companion to the philosophy of education*. Oxford: Blackwell Publishers.

Cushing, J.T. (1998). *Philosophical concepts in physics. The historical relation between philosophy and scientific theories*. Cambridge: Cambridge University Press.

Darling, J. & Nordenbo, S.E. (2003). Progressivism. In: N. Blake, P. Smeyers, R. Smith and P. Standish (eds), *The Blackwell guide to the philosophy of education* (pp. 288–308). UK: Blackwell.

Davson-Galle, P. (2008a). Why compulsory science education should *not* include philosophy of science. *Science & Education, 17*, 667–716.

Davson-Galle, P. (2008b). Against science education: The aims of science education and their connection to school science curricula. In: T. Bertrand & L. Roux (eds.), *Education Research Trends* (pp. 1–30), Nova Publishers.

Davson-Galle, P. (2004). Philosophy of science, critical thinking and science education. *Science & Education, 13*(6), 503–517.

Davson-Galle, P. (1999). Constructivism: A curate's egg. *Educational Philosophy and Theory, 31*(2), 205–219.

Davson-Galle, P. (1994). Philosophy of science and school science. *Educational Philosophy and Theory, 26*(1): 34–53.

DeBoer, G.E. (2000). Scientific literacy: Another look at its historical and contemporary meanings and its relationship to science education reform. *Journal of Research in Science Teaching, 37(6):* 582–601

DeBoer, G.E. (1991). *History of ideas in science education. Implications for practice*. New York and London: Teachers College Press.

deVries, M.J. (2005). *Teaching about technology. An introduction to the philosophy of technology for non-philosophers*. Dordrecht: Springer.

Dewey, J. & Bentley, A.F. (1949). *Knowing and the known*. Boston: The Beacon Press.

Dewey, J. (1945). Method in science teaching. *Science Education, 29*, 119–123.

Dewey, J. (1938/1997). *Experience and education*. New York: Touchstone.

Dewey, J. (1916/1944). *Democracy and education. An introduction to the philosophy of education*. New York: Free Press.

Donnelly, J. (2006). The intellectual positioning of science in the curriculum, and its relationship to reform. *Journal of Curriculum Studies, 38(6)*, 623–640.

Donnelly, J. (2004). Humanizing science education. *Science Education, 88*, 762–784.

Donnelly, J. (2001). Instrumentality, hermeneutics and the place of science in the school curriculum. In: F. Bevilacqua, E. Giannetto, and M. Matthews (eds.), *Science education and culture: the contribution of history and philosophy* (pp. 109–127). Dordrecht: Kluwer Academic Publishers.

Donnelly, J. & Jenkins, E. (2001). *Science education. Policy, professionalism and change,* London: Paul Chapman Publishing, Ltd.

Driver, R., Leach, J., Millar, R. & Scott, P. (1996). *Young people's images of science*. Buckingham, UK: Open University Press.

Duschl, R. & Hamilton, R.J. (eds.) (1992). *Philosophy of science, cognitive psychology, and educational theory and practice*. Albany, N.Y.: State University of New York Press.

Duschl, R., Hamilton, R. & Grady, R.E. (1990). Psychology and epistemology: match or mismatch when applied to science education? *International Journal of Science Education, 12*(3), 230–243.

Duschl, R. (2008). Science education in three-part harmony: Balancing conceptual, epistemic, and social learning goals. *Review of Research in Education, 32*, 268–91

Duschl, R. (1994). Research on the history and philosophy of science. In: D. Gabel (ed.), *Handbook of research on science teaching and learning* (pp. 443–465). New York: Macmillan.

Duschl, R. (1990). *Restructuring science education: the role of theories and their importance*. Columbia University, New York: Teachers College Press.

Duschl, R. (1988). Abandoning the scientistic legacy of science education. *Science Education, 72(1)*, 51–62.

Egan, K. (2005). Students' development in theory and practice: the doubtful role of research. *Harvard Educational Review, 75*(1): 25–42

Egan, K. (2002). *Getting it wrong from the beginning. Our progressivist inheritance from Herbert Spencer, John Dewey, and Jean Piaget*. New Haven and London: Yale University Press.

Egan, K. (1997). *The educated mind. How cognitive tools shape our understanding*. Chicago: University of Chicago Press.

Egan, K. (1983). *Education and psychology. Plato, Piaget, and scientific psychology*. New York and London. Teachers College Press.

Eger, M. (1993a). Hermeneutics as an approach to science: Part I. *Science & Education, 2*, 1–29.

Eger, M. (1993b). Hermeneutics as an approach to science: Part II. *Science & Education, 2*, 303–328.

Eger, M. (1992). Hermeneutics and science education: An introduction. *Science & Education 1*, 337–348.

Eisner, E. (1992). Curriculum ideologies. In: P. W. Jackson (ed.), *Handbook of research on curriculum* (pp. 302–326). New York: Macmillan.

Englund, T. (1998). Problematizing school subject content. In: D.A Roberts and L. Oestman (eds.), *Problems of meaning in science curriculum* (pp. 13–24). New York and London: Teachers College Press.

Erickson, G. (2007). In the path of Linnaeus: Scientific literacy re-visioned with some thoughts on persistent problems and new directions for science education. In: Linder et al. (eds.), *Promoting scientific literacy: Science education research in transaction. Proceedings of the Linnaeus Tercentenary Symposium, Uppsala, Sweden* (May 28–29; pp. 18–41). Uppsala: Geotryckeriet.

Erickson, G. (2000). Research programmes and the student science learning literature. In: R. Millar, J. Leach and J. Osborne (eds.), *Improving science education. The contribution of research* (pp. 271–292). Buckingham: Open University Press.

Ernest, P.A. (1991). *Philosophy of mathematics education*. London: Routledge-Falmer.

Fensham, P. (2004). *Defining an identity. The evolution of science education as a field of research*. Dordrecht: Kluwer Academic Publishers.

Fensham, P. (2002). Time to change drivers for scientific literacy. *Canadian Journal of Science, Mathematics and Technology Education, 2*(1): 9–24.

Feyerabend, P. (1975/1988). *Against method*. (Revised edn). London: Verso.

Foucault, M. (1972/1989). *The archaeology of knowledge*. London: Routledge.

Foucault, M. (1980). *Power/knowledge. Selected interviews and other writings 1972-77*. New York: Pantheon Books.

Frankena, W.K. (1970). A model for analyzing a philosophy of education. In: J.R. Martin, *Readings in the philosophy of education: A study of curriculum* (pp. 15–22). Boston: Allyn and Bacon.

Frankena, W.K. (1965). *Three historical philosophies of education. Aristotle, Kant, Dewey*. Chicago: Scott, Foresman and Company.

Fraser, B.J., Tobin, K.G. and McRobbie, C.J. (eds.) (2012). *2nd International handbook of science education*. Springer international handbooks of education, 24, Springer.

Fraser, B.J. and Tobin, K.G. (eds.). (1998). *International handbook of science education. Two volumes*. Dordrecht: Kluwer Academic Publishers.

Friere, P. (1970). *Pedagogy of the oppressed*. New York: Continuum.

Gabel, D. (ed). (1994). *Handbook of research on science teaching and learning*. New York: Macmillan.

Gadamer, H. (1976). *Philosophical hermeneutics*. (Translated and edited by David Linge). Berkeley: University of California Press.

Gadamer, H. (1960/1975/1989). *Truth and method. Second revised edition*. (Translation revised by J. Weinsheimer and D.G. Marshall). New York & London: Continuum.

Gallagher, S. (1992). *Hermeneutics and education*. New York: State University Press.

Garrison, J.W. and Bentley, M. L. (1990). Science education, conceptual change and breaking with everyday experience. *Studies in Philosophy and Education, 10*(1), 19–36.

Gaskell, J. (2002). Of cabbages and kings: Opening the hard shell of science curriculum policy. *Canadian Journal of Science, Mathematics and Technology Education, 2(1)*, 59–66.

Gay, P. (1969/1996). *The enlightenment: An interpretation. The science of freedom.* New York: Norton & Company.

Geddis, A.N. (1993). Transforming subject-matter knowledge: The role of pedagogical content knowledge in learning to reflect on teaching. *International Journal of Science Education, 15,* 673–683.

Gibbs, W.W. & Fox, D. (1999). The false crisis in science education. *Scientific American, 281*(4), 87–92.

Giere, R.N. (1999). *Science without laws.* Chicago: University of Chicago Press.

Giere, R.N. (1991). *Understanding scientific reasoning.* 3rded. Orlando, FL: Harcourt Brace Jovanovich.

Gilbert, J.K., Justi, R., Van Driel, J., De Jong, O. & Treagust, D.F. (2004). Securing a future for chemical education. *Chemistry Education: Research and Practice, 5(1),* 5–14.

Gilead, T. (2011). The role of education redefined: 18th century British and French educational thought and the rise of the Baconian conception of the study of nature. *Educational Philosophy and Theory, 43*(10), 1020–1034.

Girod, M. (2007). A conceptual overview of the role of beauty and aesthetics in science and science education. *Studies in Science Education, 43,* 38–61.

Good, R. & Shymansky, J. (2001). Nature-of-science literacy in *Benchmarks* and *Standards:* post-modern/relativist or modern/realist? *Science & Education, 10:* 173–185.

Good, R., Herron, J., Lawson, A. & Renner, J. (1985). The domain of science education. *Science Education, 69,* 139–141.

Grandy, R. (2009). Constructivisms, scientific methods, and reflective judgment in science education. In: H. Siegel (ed.), *The Oxford handbook of philosophy of education* (pp. 358–380). Oxford and New York: Oxford University Press.

Gross, P.R., Levin, N., & Lewis, M.W. (Eds.). (1995). *The flight from science and reason.* Baltimore and London: Johns Hopkins University Press.

Gundem, B.B. & Hopmann, S. (eds.) (1998). *Didaktik and/or curriculum. An international dialogue.* Peter Lang: New York.

Gunstone, R. & White, R. (2000). Goals, methods and achievements of research in science education. In: R. Millar, J. Leach, J. Osborne (eds.), *Improving science education: The contribution of research* (pp. 293–307). Buckingham, UK: Open University Press.

Haack, S. (2003). *Defending science—within reason. Between scientism and cynicism.* Amherst, N.Y.: Prometheus Books.

Habermas, J. (1987). *The philosophical discourse of modernity.* (F. G. Lawrence, trans.) Cambridge, MA: MIT Press.

Habermas, J. (1968). *Knowledge and human interests.* Oxford: Polity Press.

Hadzigeorgiou, Y. (2008). Rethinking science education as socio-political action. In: M. Tomase (Ed.), *Science Education in Focus.* New York: Nova Pubs.

Hadzigeorgiou, Y., Klassen, S., and Klassen, C. (2011). Encouraging a "romantic understanding" of science: The effect of the Nikola Tesla story. *Science & Education, 21*(8), 1111–1138.

Hansen, K.-H. and Olson J. (1996). How teachers construe curriculum integration: the Science, Technology, Society (STS) movement as *Bildung. Journal of Curriculum Studies, 28(6),* 669–682.

Heidegger, M. (1977). *Basic writings.* (Edited and general introduction by David Krell). San Francisco: Harper Collins.

Henry, N.B. (Ed.) (1955). *Modern philosophies and education: The Fifty-fourth yearbook of the national society for the study of education.* Chicago: University of Chicago Press.

Hestenes, D. (1998). Who needs physics education research? *American Journal of Physics, 66,* 465–467.

Hirst, P. (2008a). In pursuit of reason. In: L.J. Waks (ed.), *Leaders in philosophy of education. Intellectual self portraits* (pp. 113–124). Rotterdam: Sense Publishers.

Hirst, P. (2008b). Philosophy of education in the UK In: L.J. Waks (ed.), *Leaders in philosophy of education. Intellectual self portraits* (Appendix B, pp. 305–310). Rotterdam: Sense Publishers.

Hirst, P. (2003). Foreword. In: Blake, N., Smeyers, P., Smith, R. & Standish, P. (eds.) *The Blackwell guide to the philosophy of education* (pp. xv–xvi). UK: Blackwell.

Hirst, P. (1974). *Knowledge and the curriculum*. London: Routledge.

Hirst, P. (1966). Educational theory. In: J.W. Tibble (ed.), *The study of education* (pp. 29–58). New York: Routledge and Kegan Paul Ltd.

Heelan, P.A. (1991). Hermeneutical phenomenology and the philosophy of science. In: H.J. Silverman (ed.), *Continental philosophy IV. Gadamer and Hermeneutics. Science, culture, literature* (pp. 213–228). New York and London: Routledge.

Hesse, M. (1980). *Revolutions and reconstructions in the philosophy of science*. Brighton, Sussex: Harvester, Press.

Hiley, D.R., Bohman, J.F. & Shusterman, R. (eds.) (1991). *The interpretative turn. Philosophy, science, culture*. Ithaca and London: Cornell University Press.

Hodson, D. (2009). *Teaching and learning about science. Language, theories, methods, history, traditions and values*. Rotterdam: Sense Publishers.

Hodson, D. (2008). *Towards scientific literacy. A teacher's guide to the history, philosophy and sociology of science*. Rotterdam: Sense Publishers.

Hodson, D. (1998). Science fiction: The continuing misrepresentation of science in the school curriculum. *Curriculum Studies, 6(2)*, 191 – 216.

Hodson, D. (1994). Seeking directions for change: The personalisation and politicisation of science education. *Curriculum Studies, 2(1)*, 71 – 98.

Hodson, D. (1988). Experiments in science and science teaching. *Educational Philosophy & Theory, 20*, 53–66.

Hofer, B.K., & Pintrich, P.R. (1997). The development of epistemological theories: Beliefs about knowledge and knowing and their relation to learning. *Review of Educational Research, 67(1)*, 88–140.

Holton, G. (2003). What historians of science and science educators can do for one another. *Science & Education, 12*, 603–616.

Hyslop-Margison, E.J. & Naseem M.A. (2007). Philosophy of education and the contested nature of empirical research: A rejoinder to D.C. Phillips. *Philosophy of Education*, 310–318.

Ihde, D. (1998). *Expanding hermeneutics. Visualism in science*. Evanston, Ill.: Northwestern University Press.

Irzik, G. and Nola, R. (2011). A family resemblance approach to the nature of science for science education. *Science & Education, 20*, 591–607.

Irzik, G. and Nola, R. (2009). Worldviews and their relation to science. In: M.R. Matthews (Ed.), *Science, worldviews and education*. Dordrecht: Springer.

Jenkins, E. (2009). Reforming school science education: A commentary on selected reports and policy documents. *Studies in Science Education, 45*(1), 65–92.

Jenkins, E. (2007). School science: A questionable construct? *Journal of Curriculum Studies, 39(3)*, 265–282.

Jenkins, E. (2001). Science education as a field of research. *Canadian Journal of Science, Mathematics and Technology Education, 1*(1): 9–21.

Jenkins, E. (2000). "Science for all": time for a paradigm shift? In: R. Millar, J. Leach, J. Osborne (eds), *Improving science education. The contribution of research* (pp 207–226). Buckingham, UK: Open University Press.

Jenkins, E. (1994). Public understanding of science and science education for action. *Journal of Curriculum Studies, 26*(6): 601– 611.

Jung, W. (2012). Philosophy of science and education. *Science & Education, 21*(8).

Kalman, C. (2011). Enhancing student's conceptual understanding by engaging science text with reflective writing as a hermeneutical circle. *Science & Education, 20*(2), 159–172.

Kalman, C. (2010). Enabling students to develop a scientific mindset. *Science & Education, 19*(2), 147–164.

Kalman, C. (2002). Developing critical thinking in undergraduate courses: A philosophical approach. *Science & Education, 11*, 83–94.

Kant, I. (1784). "Beantwortung der Frage: Was ist Aufklärung?" In: E. Bahr (ed.) (1974), *Was ist Aufklärung? Thesen und Definitionen*. Stuttgart: Reklam.

Kelly, G.J., McDonald, S., & Wickman, P. O. (2012). Science learning and epistemology. In: K. Tobin, B. Fraser, & C. McRobbie (Eds.), *Second international handbook of science education* (pp. 281–291). Dordrecht: Springer.

Kelly, G., Carlsen, W. & Cunningham, C. (1993). Science education in sociocultural context: Perspectives from the sociology of science. *Science Education, 77*: 207–20.

Kelly, G.J. (1997). Research traditions in comparative context: A philosophical challenge to radical constructivism. *Science Education, 81*, 355–375.

Klafki, W. (1995). Didactic analysis as the core of the preparation for instruction (Didaktische Analyse als Kern der Unterrichtsvorbereitung). *Journal of Curriculum Studies, 27* (1), 13–30.

Klopfer, L.E. and Champagne, A.B. (1990). Ghosts of crisis past. *Science Education, 74*(2), 133–153.

KnorrCetina, K. (1999). *Epistemic cultures. How the sciences make knowledge*. Cambridge, Massachusetts: Harvard University Press.

Kuhn, T. (2000). *The road since structure: Philosophical essays, 1970–1993 with an autobiographical interview*. (Edited by J. Conant and J. Haugeland). Chicago: University of Chicago Press.

Kuhn, T. (1977). *The essential tension. Selected studies in scientific tradition and change*. Chicago and London: University of Chicago Press.

Kuhn, T. (1970). *The structure of scientific revolutions*. 2nd ed. Chicago: University of Chicago Press.

Kyle, W.C. Jr., Abell, S.K, Roth, W-M., and Gallagher, J.J. (1992). Toward a mature discipline of science education. *Journal of Research of Science Teaching, 29*: 1015–1018.

Lange, M. (2002). An introduction to the philosophy of physics. Cornwall: Blackwell.

Laudan, L. (1990). *Science and relativism. Some key controversies in the philosophy of science*. Chicago: University of Chicago Press.

Laudan, L. (1981). A confutation of convergent realism. In: M. Curd and J. A. Cover (Eds.) (1998), *Philosophy of science. The central issues* (pp. 1114–1135). New York: W.W. Norton & Company.

Laugksch, R. (2000). Scientific literacy: A conceptual overview. *Science Education, 84*, 71–94.

Lederman, N.G. (2007). Nature of science: past, present and future. In: S.K. Abell and N.G. Lederman (eds.), *Handbook of research on science education* (pp. 831–879). Mahwah, NJ: Lawrence Erlbaum Associates.

Lee, M., Wu, Y, & Tsai, C. (2009). Research trends in science education from 2003 to 2007. A content analysis of publications in selected journals. *International Journal of Science Education, 31*(15), 1999–2020.

Levinson, R. (2010). Science education and democratic participation: An uneasy congruence? *Studies in Science Education, 46*(1), 69–119.

Lijnse, P. (2000). Didactics of science: The forgotten dimension in science education research? In: R. Millar, J. Leach & J. Osborne (eds.), *Improving science education: The contribution of research*. Buckingham: Open University Press.

Linder, C., Oestman, L. and Wickman P. (Eds.) (2007). *Promoting scientific literacy: Science education research in transaction. Proceedings of the Linnaeus Tercentenary Symposium, Uppsala, Sweden* (May 28–29). Uppsala: Geotryckeriet.

Litt, T. (1963). *Naturwissenschaft und Menschenbildung*, 3rd ed. (Science and education) Heidelberg: Quelle und Meyer.

Locke, J. (1693). *Some thoughts concerning education*. Reprinted in *John Locke on education*, edited by Peter Gay (1964). New York: Teachers College Press.

Loving, C.C. (1997). From the summit of truth to its slippery slopes: Science education's journey through positivist-postmodern territory. *American Educational Research Journal*, 34(3), 421–452.

Lucas, C. (1972). *Our western educational heritage*. New York: Macmillan.

1310   R.M. Schulz

Lyotard, J-F. (1979/1984). *The postmodern condition. A report on knowledge*. (G. Bennington & B. Massumi, trans.). Minneapolis: University of Minnesota Press.

Mason, R. (2003). *Understanding* understanding. New York: Suny Press.

Matthews, M. R. (ed.) (2009a). *Science, worldviews and education*. Dordrecht: Springer.

Matthews, M. R. (2009b). Book review of Fensham's (2004) *Defining an identity*. Newsletter of the IHPST group, May 2009, pp. 21–39. http://ihpst.net/

Matthews, M.R. (2009c).The philosophy of education society of Australasia (PESA) and my intellectual growing-up. *Educational Philosophy and Theory, 41*(7), 777–781.

Matthews, M.R. (2005). What can be rescued from the positivist bathwater? In: C. Higgins (ed.), *Philosophy of education 2004* (pp. 223–232). Champaign IL: Philosophy of Education (POE) Society.

Matthews, M.R. (2004). Reappraising positivism and education: The arguments of Philipp Frank and Herbert Feigl, *Science & Education 13*(1–2), 7–39.

Matthews, M. R. (2003a). Thomas Kuhn's impact on science education: What can be learned? *Science Education, 88(2),* 90–118.

Matthews, M. R. (2003b). Data, phenomena, and theory: How clarifying the concepts can illuminate the nature of science. *Philosophy of Education,* 283–292.

Matthews, M.R. (2002). Teaching science. In: R.R. Curren (ed.), *A companion to the philosophy of education* (pp. 342–353). Oxford: Blackwell Publishers.

Matthews, M.R. (2001). Learning about scientific methodology and the "big picture" of science: The contribution of pendulum motion studies. In: S. Rice (ed.) *Philosophy of Education 2001* (pp. 204–213). Champaign, IL.: Philosophy of Education Society.

Matthews, M. R. (2000). Appraising constructivism in science and mathematics education. In: D.C. Phillips (ed.), *Constructivism in education. Opinions and second opinions on controversial issues* (161–192). NSSE: University of Chicago Press.

Matthews, M. R. (1998a). The nature of science and science teaching. In: B.J. Fraser and K.G. Tobin (Eds.), *International handbook of science education. Part II* (pp. 981–999). Dordrecht: Kluwer.

Matthews, M. R. (Ed.). (1998b). *Constructivism in science education: A philosophical examination*. Dordrecht: Kluwer.

Matthews, M.R. (1997). Scheffler revisited on the role of history and philosophy of science in science teacher education. In: H. Siegel (ed.), *Reason and education: Essays in honor of Israel Scheffler* (pp. 159–173). Dordrecht: Kluwer Publishers.

Matthews, M.R. (1996). Charles Darwin. In: J.J. Chambliss (ed.), *Philosophy of education: An encyclopedia*, (pp. 135–137). New York: Garland Publishing.

Matthews, M. R. (1994a/2014). *Science teaching: The role of history and philosophy of science*. New York: Routledge.

Matthews, M.R. (1994b). Philosophy of science and science education. In: T. Husen & T.N. Postlethwaite (eds.) *The international encyclopedia of education*, Second Edition (pp. 4461–4464). London: Pergamon Press.

Matthews, M. R. (1990a). History, philosophy and science teaching. What can be done in an undergraduate course? *Studies in Philosophy and Education, 10,* 93–97.

Matthews, M. R. (1990b). Ernst Mach and contemporary science education reforms. *International Journal of Science Education, 12*(3), 317–325.

Matthews, M. R. (1988). A role for history and philosophy in science teaching. *Educational Philosophy and Theory, 20*(2), 67–81.

Matthews, M.R. (1987). Galileo's pendulum and the objects of science. In: B. & D. Arnstine (eds.), *Philosophy of education* (pp. 309–319). Normal, IL: POE Society.

Matthews, M. R. (1980). Knowledge, action and power. In: R. Mackie (ed.), *Literacy and revolution: The pedagogy of Paulo Freire* (pp. 82–92). London: Pluto Press.

Matthews, M.R. (ed.). (1989). *The scientific background to modern philosophy*. Indianapolis, IN: Hackett Publishing Company.

McCarthy, C.L. (2007). Meaning, mind, and knowledge. A pragmatic view. In: C. Higgins (ed.), *Philosophy of education 2007* (pp. 421–433). Champaign IL: Philosophy of Education (POE) Society.

39 Philosophy of Education and Science Education... 1311

McComas, W.F. & Olson, J.K. (1998). The nature of science in international science education standards documents. In: W.F. McComas (Ed.), *The nature of science in science education: Rationales and strategies* (pp. 41–52). Dordrecht: Kluwer.

McIntyre, L. (2007). The philosophy of chemistry: ten years later. *Synthese, 155*, 291–292.

Medina, J. (2005). *Language. Key concepts in philosophy.* New York: Continuum.

Mercan, F.C. (2012). Epistemic beliefs about justification employed by physics students and faculty in two different problem contexts. *International Journal of Science Education, 34*, 8, 1411–1441.

Nadeau, R. and Désautels, J. (1984). *Epistemology and the teaching of science.* Ottawa: Science Council of Canada.

Nejadmehr, R. (2009). *Education, science and truth.* New York: Routledge.

Noddings, N. (2011). *Philosophy of education.* 3rd ed. Westview Press.

Nola, R. and Irzik, G. (2005). *Philosophy, science, education and culture.* Dordrecht: Springer.

Nola, R. (1994). Post-modernism, a French cultural Chernobyl: Foucault on power/knowledge. *Inquiry, 37*: 3–43.

Norris, C. (1997). *Against relativism. Philosophy of science, deconstruction and critical theory.* Oxford: Blackwell.

Norris, S.P. and Burnes, D.P. (2009). (See Burnes and Norris).

Norris, S.P., Falk, H., Federico-Agrasso, M., Jiménez-Aleixandre, M.P., Phillips, L.M., and Yarden, A. (2009). Reading science texts—epistemology, inquiry, authenticity—a rejoinder to Jonathan Osborne. *Research in Science Education, 39*, 405–410.

Norris, S.P., Phillips, L.M., & Osborne, J.F. (2008). Scientific inquiry: The place of interpretation and argumentation. In: J. Luft, R.L. Bell, & J. Gess-Newsome (Eds.), *Science as inquiry in the secondary setting* (pp. 87–98). Arlington, VA: NSTAPress.

Norris, S.P. and Korpan, C.A. (2002). Philosophy or science: A response to Matthews. *Philosophy of Education 2001, 56*, 214–216.

Norris, S.P. & Korpan, C.A. (2000). Science, views about science, and pluralistic science education. In R. Millar, J. Leach, & J. Osborne (Eds.), *Improving science education: The contribution of research* (pp. 227–244). Buckingham, UK: Open University Press.

Norris, S.P. and Kvernbekk, T. (1997). The application of science education theories. *Journal of Research in Science Teaching, 34*, 977–1005.

Norris, S.P. (1997). Intellectual independence for non-scientists and other content-transcendent goals of science education. *Science Education, 81*, 239–258.

Norris, S.P. (1995). Learning to live with scientific expertise: Towards a theory of intellectual communalism for guiding science teaching. *Science Education, 79*, 201–217.

Norris, S.P. (1992). Practical reasoning in the production of scientific knowledge. In: R. Duschl and R. Hamilton (Eds.), *Philosophy of science, cognitive psychology, and educational theory and practice* (pp. 195–225). Albany, NY: NYU State Press.

Norris, S.P. (1988). How the laws of physics lie and other value issues. *Philosophy of Education, 43*, 321–325.

Norris, S.P. (1985). The philosophical basis of observation in science and science education. *Journal of Research in Science Teaching, 22*, 817-833.

Norris, S.P. (1984a). Cynicism, dogmatism, relativism, and scepticism: Can all these be avoided? *School Science and Mathematics, 84*, 484-495.

Norris, S.P. (1984b). Defining observational competence. *Science Education, 68*, 129-142.

Norris, S.P. (1982). A concept of observation statements. *Philosophy of Education, 37*, 132-142.

Novak, J. (1977). *A theory of education.* Ithaca, New York: Cornell University Press.

Ogborn, J. (1995). Recovering reality. *Studies in Science Education, 25*, 3–38.

Olesko, K.M. (2006). Science pedagogy as a category of historical analysis: past, present and future. *Science & Education, 15*, 863–880.

Pedretti, E.G. and Nazir, J. (2011). Currents in STSE education: Mapping a complex field, 40 years on. *Science Education, 95*, 601–626.

Pedretti, E.G., Bencze, L., Hewitt, J., Romkey, L., & Jivraj, A. (2008). Promoting issues-based STSE perspectives in science teacher education: problems of identity and ideology. *Science & Education, 17*, 941–960.

Peters, R. S. (1966). The philosophy of education. In: Tibble, J.W. (ed), *The study of education* (pp 59–89). New York: Routledge and Kegan Paul Ltd.

Phillips, D.C. (2010). What is philosophy of education? In: R. Bailey, R. Barrow, D. Carr and C. McCarthy (eds.) *The SAGE handbook of the philosophy of education* (pp. 3–19). London: SAGE.

Phillips, D.C. (2009). Empirical educational research: Charting philosophical disagreements in an undisciplined field. In: H. Siegel (ed.), *The Oxford handbook of philosophy of education* (pp. 381–408). Oxford & New York: Oxford University Press.

Phillips, D.C. (2008). Philosophy of education. In: *Stanford encyclopedia of philosophy.* http://plato.stanford.edu/entries/education-philosophy/

Phillips, D.C. (2007). Getting it wrong from the beginning, but maybe (just maybe) it's a start. *Philosophy of Education,* 319–322.

Phillips, D.C. (2005). The contested nature of empirical research. *Journal of Philosophy of Education, 39* (4): 577–97.

Phillips, D.C. (2002). Theories of teaching and learning. In: R.R. Curren (ed.), *A companion to the philosophy of education* (pp. 233–237). Oxford: Blackwell Publishers.

Phillips, D.C. (ed.), (2000). *Constructivism in education. Opinions and second opinions on controversial issues.* NSSE: University of Chicago Press.

Phillips, D.C. (1997). Coming to terms with radical social constructivisms. *Science & Education, 6*(1/2), 85–104.

Plato. (1975). *Protagoras and Meno.* (Translated by W.K.C. Guthrie). Penguin Classics.

Plato. (1974). *The Republic.* 2nd Ed. (Translated by Desmond Lee). Penguin Classics.

Plato. (1970). *The Laws.* (Translated by Trevor Saunders). Penguin Classics.

Polito, T. (2005). Educational theory as theory of culture: A Vichian perspective on the educational theories of John Dewey and Kieran Egan. *Educational Philosophy and Theory, 37(4),* 475–494.

Posner, G.J. (1998). Models of curriculum planning. In: L.E. Beyer & M.W. Apple (Eds.*), The curriculum: Problems, politics and possibilities.* (2nd ed., pp. 79–100). Albany, NY: Suny. Press.

Pring, R. (2010). The philosophy of education and educational practice. In: R. Bailey, R. Barrow, D. Carr and C. McCarthy (eds.) *The SAGE handbook of the philosophy of education* (pp. 56–66). London: SAGE.

Psillos, D., Kariotoglou, P., Tselfes, V., Hatzikraniotis, E., Fassoulopoulos, G. & Kallery, M. (eds.) (2003). *Science education research in the knowledge-based society,* Dordrecht: Kluwer Academic Publishers.

Roberts, D.A. & Oestman, L. (eds.) (1998). *Problems of meaning in science curriculum.* New York and London: Teachers College Press.

Roberts, D.A. (2007). Scientific literacy/science literacy. In: S.K. Abell and N.G. Lederman (eds.), *Handbook of research on science education* (pp 729–780). Mahwah, NJ: Lawrence Erlbaum Associates.

Roberts, D.A. (1988). What counts as science education? In: P. Fensham (ed.), *Development and dilemmas in science education* (pp. 27–54). Philadelphia: Falmer

Roberts, D.A. (1982). Developing the concept of "curriculum emphases" in science education. *Science Education, 66,* 243–260.

Roberts, D.A. and Russell, T.L. (1975). An alternative approach to science education: drawing from philosophical analysis to examine practice. *Curriculum Theory Network, 5*(2): 107–125.

Robertson, E. (2009). The epistemic aims of education. In: H. Siegel (ed.), *The Oxford handbook of philosophy of education* (pp. 11–34). Oxford & New York: Oxford University Press.

Rockmore, T. (2004). *On foundationalism. A strategy for metaphysical realism.* Oxford: Rowman & Littlefield, Publishers, Inc.

Rorty, A.O. (ed.) (1998). *Philosophers on education. Historical perspectives.* London: Routledge.

Rorty, R. (1979). *Philosophy and the mirror of nature.* Princeton: University Press.

Rorty, R. (1984). Habermas and Lyotard on Postmodernity. In Hoesterey, I. (Ed.) (1991). *Zeitgeist in babel. The postmodernist controversies* (pp. 84–97). Indiana Press.

Roth, W-M. & Barton, A. (2004). *Rethinking scientific literacy.* NY: Routledge-Falmer.

Roth, W-M. and Desautels, J. (Eds.) (2002). *Science education as/for sociopolitical action.* New York: Peter Lang.

Roth, W-M. & McGinn, M. (1998). Knowing, researching and reporting science education: Lessons from science and technology studies. *Journal of Research in Science Teaching, 35*(2), 213–235.

Rousseau, Jean-Jacque. (1762/1979). *Emile.* (Translated and an introduction by Allan Bloom). Basic Books.

Rowlands, S., Graham, T. & Berry, J. (2011). Problems with fallibilism as a philosophy of mathematics education. *Science & Education, 20 (7–8),* 625–686.

Rudolph, J.L. (2002). Portraying epistemology: school science in historical context. *Science Education, 87,* 64–79.

Rudolph, J.L. (2000). Reconsidering the 'nature of science' as a curriculum component. *Journal of Curriculum Studies, 32(3),* 403–419.

Ruse, M. (1990). Making use of creationism. A case study for the philosophy of science classroom. *Studies in Philosophy of Education, 10*(1), 81–92.

Säther, J. (2003). The concept of ideology in analysis of fundamental questions in science education. *Science & Education, 12,* 237–260.

Scerri, E. (2003). Philosophical confusion in chemical education research. *Journal of Chemical Education, 80*(5), 468–194.

Scerri, E. (2001). The new philosophy of chemistry and its relevance to chemical education. *Chemistry Education: Research and Practice in Europe, 2(2),* 165–170

Scharff, R.C. (ed.) (2002). *Philosophy of technology: the technological tradition. An anthology.* London: Blackwell.

Scheffler, I. (1973/1989). *Reason and teaching.* Indianapolis: Hackett.

Scheffler, I. (1970/1973). Philosophy and the curriculum. In: *Reason and teaching.* London: Routledge, 1973 (pp. 31–44). Reprinted in *Science & Education* 1992, *1*(4), 384–394.

Schiller, F. (1795/1993). Letters on the aesthetic education of man. In: F. Schiller *Essays.* (Edited by W. Hinderer and D. O. Dahlstrom; pp. 86–178). New York: Continuum.

Schulz, R.M. (2011). Developing a philosophy of science education. In: F. Seroglou, V. Koulountzos, & A. Siatras (Eds.), *Science and Culture: Promise, Challenge and Demand –* Proceedings of the 11[th]IHPST & 6[th]Greek History, Philosophy & Science Teaching Joint Conference, July 2011, Thessaloniki, Greece (pp. 672–677). Thessaloniki: Epikentro Publications.

Schulz, R.M. (2009a). Reforming science education: part I. The search for a *philosophy* of science education. *Science & Education, 18,* 225–249.

Schulz, R.M. (2009b). Reforming science education: part II. Utilizing Kieran Egan's educational metatheory. *Science & Education, 18,* 251–273.

Schulz, R.M. (2007). Lyotard, postmodernism and science education. A rejoinder to Zembylas. *Educational Philosophy and Theory, 39(6),* 633–656.

Schwab, J.J. (1978). *Science, culture and liberal education. Selected essays.* (Edited by In Westbury and Neil J. Wilkof). Chicago: Chicago University Press.

Scott, D. (2008). *Critical essays on major curriculum theorists.* New York: Routledge.

Selley, N.J. (1989). The philosophy of school science. *Interchange, 20 (2),* 24–32.

Shamos, M.H. (1995). *The myth of scientific literacy.* New Brunswick, NJ: Rutgers University Press.

Sherman, W. (2005). A reply to Roth.*Canadian Journal of Science, Mathematics and Technology Education, 5(2),* 199–207.

Sherman, W. (2004). Science studies, situatedness, and instructional design in science education. A summery and critique of the promise. *Canadian Journal of Science, Mathematics and Technology Education, 4,* 443–465.

Shulman, L.S. (1987). Knowledge and teaching: Foundations of the new reform. *Harvard Educational Review, 57* (1), 1–22.

Siegel, H. (2010). Knowledge and truth. In: R. Bailey, R. Barrow, D. Carr and C. McCarthy (eds.) *The SAGE handbook of the philosophy of education* (pp. 283–295). London: SAGE.

Siegel, H. (ed.) (2009). *The Oxford handbook of philosophy of education.* Oxford and New York: Oxford University Press.

Siegel, H. (2007). Philosophy of education. In: *Encyclopædia Britannica.* Retrieved from http://www.britannica.com/EBchecked/topic/179491/philosophy-of-education

Siegel, H. (2004). The bearing of philosophy of science on science education, and *vice-versa*: the case of constructivism. *Studies in History and Philosophy of Science, 35*: 185–198.

Siegel, H. (2003). Cultivating reason. In: R.R. Curren (ed.), *A companion to the philosophy of education* (pp. 305–319). Oxford: Blackwell Publishers.

Siegel, H. (2002). Multiculturalism, universalism, and science education: In search of common ground. *Science Education, 86(6), 803–820.*

Siegel, H. (2001). Incommensurability, rationality, and relativism: in science, culture and science education. In: P. Hoyningen-Huene & H. Sankey (eds.), *Incommensurability and related matters* (pp. 207–24). Dordrecht: Kluwer.

Siegel, H. (ed.) (1997). *Reason and education: Essays in honor of Israel Scheffler.* Dordrecht: Kluwer Academic Publishers.

Siegel, H. (1992). Two perspectives on reason as an educational aim: the rationality of reasonableness. *Philosophy of Education*, 225–233.

Siegel, H. (1989). The rationality of science, critical thinking, and science education. *Synthese, 80*(1): 9–32.

Siegel, H. (1988). *Educating reason: Rationality, critical thinking and education.* London: Routledge.

Siegel, H. (1987a). Farewell to Feyerabend. *Inquiry, 32*, 343–369.

Siegel, H. (1987b). *Relativism refuted. A critique of contemporary epistemological relativism.* Dordrecht: Kluwer.

Siegel, H. (1978). Kuhn and Schwab on science texts and the goals of science education. *Educational Theory, 28(4),* 302–309.

Siemsen, H. and Siemsen, K.H. (2009). Resettling the thoughts of Ernst Mach and the Vienna Circle in Europe: The cases of Finland and Germany. *Science & Education, 18,* 299–323.

Slezak, P. and Good, R. (eds.) (2011). Thematic special issue: Pseudoscience in society and schools. *Science & Education, 10* (5/6).

Slezak, P. (2007). Is cognitive science relevant to teaching? *Journal of Cognitive Science, 8,* 171–208.

Slezak, P. (1994a). Sociology of scientific knowledge and scientific education: Part I. *Science & Education, 3*(3): 265–294.

Slezak, P. (1994b). Sociology of scientific knowledge and science education. Part II: Laboratory life under the microscope. *Science & Education, 3*(4): 329–355.

Smeyers, P. (1994). Philosophy of education: Western European perspectives. In: T. Husen & T. Postlethwaite (eds.), *The international encyclopedia of education,* 2nded., Vol. 8 (pp. 4456–4461). London: Pergamon Press.

Smith, M. & Siegel, H. (2004). Knowing, believing and understanding. *Science and Education, 13*(6), 553–582.

Smolicz, J.J. and Nunan, E.E. (1975). The philosophical and sociological foundations of science education: The demythologizing of school science. *Studies in Science Education, 2,* 101–143.

Snook, I.A. (Ed.). (1972). *Concepts of indoctrination.* London: Routledge & Kegan Paul.

Sokal, A. and Bricmont, J. (1998). *Fashionable nonsense. Postmodern intellectuals' abuse of science.* New York: Picador.

Sokal, A. (1996a). Transgressing the boundaries: Toward a transformative hermeneutics of quantum gravity. *Social Text, 46/7.* (14.1–2): 217–52.

Sokal, A. (1996b). A physicist experiments with cultural studies. *Lingua Franca.* (July/August): 62–64.

Solomon, J. (1999). Meta-scientific criticisms, curriculum innovation and the propagation of scientific culture. *Journal of Curriculum Studies, 31*(1), 1–15.

Southerland, S.A., Sinatra, G.M., and Matthews, M.R. (2001). Belief, knowledge, and science education. *Educational Psychology Review, 13(4),* 325–351.

Stinner, A. (1989). Science, humanities and society—the Snow-Leavis controversy. *Interchange, 20(2),* 16–23.

Suchting, W. (1992). Constructivism deconstructed. *Science & Education, 1*(3), 223–254.

Suchting, W.A. (1995). Much ado about nothing: Science and hermeneutics. *Science & Education, 4*(2), 161–171.

Sullenger, K., Turner, S., Caplan, H., Crummey, J., Cuming, R., Charron, C., and Corey, B. (2000). Culture wars in the classroom: Prospective teachers question science. *Journal of Research in Science Teaching, 37*(9), 895–915.

Taylor, C. (1991). *The malaise of modernity.* Toronto: Anansi Press.

Taylor, C. (1987). Overcoming epistemology. In: K. Baynes, J. Bohman, & T. McCarthy (Eds.), *After philosophy. End or transformation?* (pp. 464–485). London: MIT Press.

Thomas, G. (1997). What's the use of theory? *Harvard Educational Review, 67*, 75–104.

Tibble, J.W. (ed.), (1966) *The study of education.* New York: Routledge and Kegan Paul Ltd.

Toulmin, S.E. (1972). *Human understanding.* Princeton, N.J.: Princeton University Press.

Turner, S. & Sullenger, K. (1999). Kuhn in the classroom, Latour in the lab: science educators confront the nature-of-science debate. *Science, Technology and Human Values, 24* (1), 5–30.

Van Driel, J.H. and Abell, S. (2010). Science teacher education. In: Peterson, D., Baker, E. and McGraw, B. (Eds.), *International encyclopedia of education* (pp. 712–718). Elsevier.

Van Driel, J.H., Verloop, N. & de Vos, W. (1998). Developing science teachers' pedagogical content knowledge. *Journal of Research in Science Teaching, 35* (6), 673–695.

Vásquez-Levy, D. (2002). Essay review. *Bildung*-centred Didaktik: a framework for examining the educational potential of subject matter. *Journal of Curriculum Studies, 34* (1), 117–128.

Waks, L.J. (ed.) (2008). *Leaders in philosophy of education. Intellectual self-portraits.* Rotterdam: Sense Publishers.

Walker, D.F. (2003). *Fundamentals of curriculum. Passion and professionalism.* 2nd ed. Mahwah, New Jersey: Lawrence Erlbaum and Associates.

Wallace, J. and Louden, W. (2003). What we don't understand about teaching for understanding: Questions from science education. *Journal of Curriculum Studies, 35(5)*, 545–566.

Weinberg, S. (1992). *Dreams of a final theory.* New York: Pantheon Books.

Westaway, F.W. (1929). *Science Teaching.* Blackie and Son: London.

Westphal, M. (1999). Hermeneutics as epistemology. In: J. Greco and E. Sosa (eds.), *The Blackwell guide to epistemology* (pp. 415–435). Oxford, UK: Blackwell.

Whitehead, A.N. (1929/1957). *The aims of education and other essays.* New York: Free Press.

Winch, C. and Gingell, J. (1999). *Key concepts in the philosophy of education.* London and New York: Routledge.

Wittgenstein, L. (1953/1958). *Philosophical investigations,* 3rded. (Translated by G.E.M. Anscombe). Englewood Cliffs, NJ: Prentice Hall.

Witz, K. and Lee, H. (2009). Science as an ideal: Teacher's orientations to science and science education reform. *Journal of Curriculum Studies, 41*(3), 409–431.

Witz, K. (2000). OP-ED The 'academic problem'. *Journal of Curriculum Studies, 32*(1), 9–23.

Witz, K. (1996). Science with values and values for science education.*Journal of Curriculum Studies, 28(5),* 597–612.

Yager, R.E. (Ed.) (1996). *Science/technology/society as reform in science education.* New York: New York State University Press.

Zeidler, D.L., & Sadler T.D. (2008). Social and ethical issues in science education: A prelude to action. *Science & Education, 77*(8–9), 799–803.

Zeidler, D.L, Sadler, T.D., Simmons, M.L. & Howes, E.V. (2005). Beyond STS: A research-based framework for socioscientific issues education. *Science Education, 89,* 357–377.

Zembylas, M. (ed.) (2006). Special issue: Philosophy of science education. *Educational Philosophy and Theory, 38*(5).

Zembylas, M. (2002). Constructing genealogies of teachers' emotions in science teaching. *Journal of Research in Science Teaching, 39,* 79–103.

Zembylas, M. (2000). Something 'paralogical' under the sun: Lyotard's postmodern condition and science education. *Educational Philosophy and Theory, 32,* 159–184.

Roland M. Schulz is a researcher with the *Imaginative Education Research Centre (IERG)* at Simon Fraser University in Vancouver and a sessional instructor in science teacher education. He holds a Ph.D. in Curriculum Theory focusing on philosophy of education, philosophy of science, and science education. His undergraduate degrees are in physics and in physical science education. He has been active as a secondary science teacher, having taught physics and chemistry for many years in Canada and physics in Istanbul, Turkey. His research interests include science education reform, philosophy of education, teacher education, hermeneutics and language theory, and the use of narrative and models to incorporate the history and philosophy of science for improving curriculum and instruction. Previous publications have appeared in the international journals *Science & Education* and *Educational Philosophy and Theory.*

Chapter 40
Conceptions of Scientific Literacy: Identifying and Evaluating Their Programmatic Elements

Stephen P. Norris*, Linda M. Phillips, and David P. Burns

'Scientific literacy is a programmatic concept' (Norris and Phillips 2009, p. 271). Programmatic concepts have elements that point in a valued direction or name a desired goal. In the case of scientific literacy, it points to goals that educators, scientists, and politicians want for citizens and society. It should not be surprising, then, that scientific literacy is contested. Not everyone possesses the same sense of valued directions and desired goals, so different individuals and groups urge their views on others. The question raised in this chapter is, 'By what means can the programmatic elements of conceptions of scientific literacy be identified and evaluated?'

First, we provide a detailed analysis of the nature of programmatic concepts and provide examples of the programmatic elements found in conceptions of scientific literacy. Given that definitions of scientific literacy bear upon what is taught in science education, a lens is needed through which to identify and judge these programmatic elements. Specifically, what values underlie these elements and what theories of value might be brought to bear in assessing them? The answer to this latter question will compose the second major section of the paper in which we present an analysis of approximately 70 conceptions of scientific literacy found in the literature since the year 2000. We identify the goals that each of these conceptions of scientific literacy implies and uncover the programmatic elements that are used to justify these goals. Our purpose here is not to be exhaustive in presenting

---

*Sadly Stephen Norris passed away prior to publication of the handbook

S.P. Norris
Centre for Research in Youth, Science Teaching and Learning, University of Alberta, Edmonton, AB T6G 2G5, Canada

L.M. Phillips (✉)
Canadian Centre for Research on Literacy, University of Alberta, Edmonton, AB T6G 2G5, Canada
e-mail: linda.phillips@ualberta.ca

D.P. Burns
Department of Educational Studies, Kwantlen Polytechnic University, Surrey, BC V3W 2M8, Canada

M.R. Matthews (ed.), *International Handbook of Research in History, Philosophy and Science Teaching*, DOI 10.1007/978-94-007-7654-8_40, © Springer Science+Business Media Dordrecht 2014

conceptions of scientific literacy but to present a sufficiently wide range of views to have a good representation of goals and programmatic elements. Third, we point to a number of pitfalls in any attempt to make preferential selections among the programmatic elements of conceptions of scientific literacy.

40.1 Programmatic Concepts

Programmatic concepts are of notable importance in education because of education's practical orientation toward 'social practices and habits of mind' (Scheffler 1960, p. 19). Programmatic concepts 'are not recognizable as such by their linguistic form alone: reference to the context needs to be made' (Scheffler 1960, p. 19). A study of the context in which a concept is used can reveal whether the concept has the effect of implying practical consequences or whether it does not serve such a purpose. Thus, programmatic concepts require 'independent, practical evaluation' (Scheffler 1960, p. 21) because they can raise serious moral and practical issues.

Like any concept in education, we expect programmatic concepts to meet certain linguistic and logical standards such as consistency, suitableness, and non-arbitrariness. However, programmatic concepts carry the extra burden of expressing value choices that embody programmes of action. We thus understand that to evaluate a programmatic concept, such as scientific literacy, it is necessary to consider both its linguistic features and value implications. Thus, an overriding question that we shall consider is, 'What is promised as a result of adopting a particular conception of scientific literacy?' We shall argue that any adoption should follow, rather than precede, the evaluation of what is promised in terms of sound theories of value.

Robert Ennis (1969) claimed that a programmatic concept takes the form of a definition, but 'it is more than this' (p. 178). Ennis's is a key point of focus, because it reminds us that programmatic concepts are not neutral descriptions of usage, compared let us say to dictionary definitions, which are intended objectively to describe linguistic practice. Programmatic concepts are not neutral, because they both imply a programme of what should or ought to be done and support that programme with explicit or implicit terms embedded within the meaning of the concept. According to Ennis, a programmatic concept effectively is 'a proposal (that is, a request, or command, or entreaty, etc.) for adoption of a program or point of view' (p. 179). For example, part of the conception of scientific literacy offered by Millar and Osborne (1998) is that a scientifically literate person has 'sufficient scientific knowledge and understanding to … read simple newspaper articles about science…' (p. 9). In this concept, we can see both of Ennis's elements at work. First, the concept reads like a definition. Thus, we can raise the questions of fact: Is the definition accurate? Is this what people mean by 'scientifically literate person'? Second, the concept reads like a proposal: We should take scientifically literate people to be able to read simple articles about science; we ought not to settle for less. Thus, we can raise the value questions: Is this how we ought to think of scientifically literate people? Should we consider reading simple newspaper articles to be important to scientific literacy? Is being able to read simple newspaper articles about science sufficient? The answers to these questions are not straightforward, but what is important is to acknowledge that these two elements are entailed: the questions of fact and the questions of value.

Consider a concept of our own to see again these two elements working. We theorized a fundamental sense of scientific literacy that 'means comprehending, interpreting, analyzing, and critiquing [scientific] texts' (Norris and Phillips 2003, p. 229). So, we can ask, 'Is this what scientific literacy means?' – the factual question about usage. More important, however, is that we were making a proposal that the science education field ought to include this way of thinking about scientific literacy in setting curricular goals. So we can ask whether this is a sensible, worthwhile, and productive way to think of scientific literacy – the value questions about adoption.

One might wonder whether all concepts are programmatic to some extent. If they are not, what are some concepts that are not programmatic? Consider the chemical concept of an element – that is, a substance consisting of atoms, all of which have the same number of protons. To argue that the concept of element is programmatic requires showing how it expresses value choices that embody programs of action. It might be argued, for example, that the concept arose from the desire for simplicity in theories that explain the natural world. However, unlike the concept of scientific literacy, over which we can exercise almost limitless discretion in adjusting its meaning, the concept of element captures the very real state of affairs that the substance of the world comes in discrete types – such as gold, which has exactly and always 79 protons in its atoms. Consider the concept of π, that is, the ratio of the circumference of circles to their diameters. No doubt, the concept arose from thinking driven by a set of values, including perhaps curiosity. Yet, it is difficult to imagine alternative conceptions to π. Alternatives to programmatic concepts are not only easy to imagine but seem invited by programmatic concepts such as scientific literacy. The concept of π refers to an unalterable fact about a precise geometric shape – its meaning, as the unique ratio of circumference to diameter, is beyond debate (although its exact value is actually beyond knowing). Thus, although most, or perhaps all, of our concepts contain in their history some elements of judgement and choice based upon values, that does not make all of our concepts programmatic. Only those concepts whose value choices embody programs of action about which we can expect significant and indefinitely extended debate fall into the programmatic category.

To summarize: to advance a conception of education, of any sort, is a programmatic task. That is, we promote certain educational ideas for consideration because we assume that certain paths and goals of development and growth are more valuable than others. These assumptions are underpinned by a wide range of premises about what counts as epistemically, morally, or politically valuable. As a result of this reality in educational discourse and thought, responsible practice requires an awareness of, and justification for, these values that underlie our assumptions. Science education is no exception to bearing this burden of responsibility. Discussion of scientific literacy, in particular, would benefit from precisely the sort of analysis we are recommending.

40.2 Programmatic Elements in Conceptions of Scientific Literacy

Using Google Scholar™, we searched for articles published since 2000 using the search phrase 'scientific literacy'. We identified 74 articles in which we could isolate a definition of scientific literacy. Within these 74 articles, we identified a subset of 62 for which we could classify either the scientific literacy objectives espoused, the justification for espousing those objectives, or both of these. That is, there are two main categories of value discussion at play in contemporary conceptions of scientific literacy. The first regards the value *classification* of the desired outcomes of scientific literacy. The second regards the value *justification* provided (or implied) for such desired outcomes. These two categories, it will be shown, serve to draw attention to the important value analysis required of various conceptions of scientific literacy.

40.2.1 Value Classifications of Desired Outcomes

A focused reading of the 62 sources yielded the observation that many in the field would expect: the outcomes of scientific literacy do not belong uniformly to one value category. Rather, we saw repeated use of a relatively small set of key concepts: knowledge, ability, understanding, independence, participation, appreciation, among several others. One can discern several groupings in this list. Knowledge and understanding are (loosely speaking) epistemological concepts dealing with a mental state someone might possess. One might possess, for example, an understanding of particular scientific propositions or have knowledge of how to calculate the propagation of errors of measurement. Ability likely is interpreted best as a capacity to engage in particular actions. Independence and appreciation, in contrast to the preceding two categories, are states of personal character. We describe persons as more or less independent, and we say that persons are more or less appreciative, with reference to descriptions of whom those persons are and of what they are said to do. One might have knowledge or ability without the concomitant disposition to do anything in particular to use them. Independence and appreciation are, on the other hand, dispositional ways of being (Riveros et al. 2012). It is for this reason that we find it valuable to categorize the proposed outcomes of scientific literacy into three categories of values: values regarding the states of *knowing* one might obtain, values regarding the *capacities* one might refine, and values regarding the personal *traits* one might develop. We call them all values because they refer to ends of science education that are judged desirable.

This schema, or closely related ones, has been identified by others. For example, Laugksch (2000) produced a similar division. After examining the historical development of conceptions of scientific literacy, he concluded that three categories of what it means to be scientifically literate are applicable: being learned, competent, and 'able to function minimally as consumers and citizens' (p. 82). His schema, similar to the one proposed by us above, denotes a move from possessing particular knowledge and understanding, to the possession of certain capacities, to the fulfilling of certain roles. This third category, it should be noted, is not explicitly dispositional (as our third category is). Laugksch maintained ability as the key element of the third category, rather than explicitly separating capacity from disposition to act. Nevertheless, it is difficult to imagine that he desires people who are *able* to function as citizens while being indifferent to whether they do function in that way. Thus, we believe that Laugksch likely had dispositions in mind when formulating his schema, even if he did not explicitly use dispositional language.

The division between knowledge, capacities, and traits of character is pedagogically significant because it draws attention to the unique status of these three forms of valuable outcomes. To say that students should know certain things is clearly pedagogically distinct from saying that they should be able to do certain things. To say that students should embody certain traits, and hence be certain sorts of persons, is strikingly different from both knowledge and capacity.

Table 40.1 indicates which categories of objectives, if any, were evident in each of the sources. We have indicated the presence of objectives with bullets. In those cases marked with a bullet, we are confident there is positive evidence that the referenced objective is included. In those cases without indicator bullets, we did not uncover positive evidence, but another read, or another reader, might find implied evidence present. The knowledge and capacity objectives were widely present, occurring in 59 and 60 of the sources, respectively. The traits objective was also represented well by 38 of the sources.

40.2.1.1 Knowledge

Even though nearly every source said that knowledge is a goal of scientific literacy, it was frequently difficult to find statements of exactly what knowledge was desired. Take, for example, the following statement from Foster and Shiel-Rolle (2011, p. 85): 'At its simplest, the concept of "scientific literacy" refers to the fundamental knowledge that the general public needs to understand about science so that individuals can use that information to make informed decisions regarding personal, civic, and economic matters...'. Such a statement provides a very general idea of the function that the scientific knowledge is intended to serve, but it provides no indication of what that knowledge actually is. Is the knowledge substantive scientific knowledge? Is it, as the words invite one to think, knowledge about science, rather than knowledge of substantive science itself? Later in the same article, however, when describing a summer science camp designed to 'enhance scientific literacy in rural communities' (p. 87), the authors are very specific about the sorts of knowledge required: for example, 'identify the four species of mangroves found within the Bahamas; understand how mangroves serve as nurseries for juvenile fish and stingrays ... a basic geological understanding of cave formation in the Bahamas' (p. 90). One might then wonder whether this very specific knowledge is the sort that

1322

Table 40.1 Presence of scientific literacy objectives, political justifications, and moral justifications within cited works

Citations	Objectives			Political justification		Moral justification		
	K	Ca	T	L	Cm	V	P	U
Baker et al. (2009)	●	●			●			
Baram-Tsabari and Yarden (2005)	●	●	●	●				
Bhathal (2011)	●	●	●		●		●	
Bonney et al. (2009)	●	●	●					
Britsch (2009)	●	●						
Brossard and Shanahan (2006)	●	●		●				
Brown et al. (2005)	●	●	●	●	●	●	●	●
Bybee (2008)	●	●	●	●		●	●	●
Bybee (2009)	●	●	●	●			●	●
Bybee and McCrae (2011)	●	●	●	●				
Bybee et al. (2009a)	●	●	●	●				●
Bybee et al. (2009b)	●	●	●	●				●
Cajas (2001)	●	●		●				
Cavagnetto (2010)	●	●		●	●			
Chen et al. (2009)		●		●	●			●
Colucci-Gray et al. (2006)	●	●	●	●	●			●
Cook et al. (2011)	●	●						●
Correia et al. (2010)	●	●	●	●			●	●
DeBoer et al. (2000)	●	●	●	●	●		●	●
Derry and Zalles (2011, April)	●	●		●	●			
Dillon (2009)	●	●	●	●	●	●		●
Dos Santos (2009)	●	●	●	●	●	●	●	●
Eijick and Roth (2007)	●	●	●					
Evans and Rennie (2009)	●	●	●					
Fang (2005)	●	●	●		●			
Feinstein (2011)	●	●		●	●		●	●
Foster and Shiel-Rolle (2011)	●	●	●	●				●
Fuselier and Nelson (2011)	–	–	–	–	–	–	–	–
Gawalt and Adams (2011)	–	–	–	–	–	–	–	–
George and Brenner (2010)	●	●	●	●				●
Greenleaf et al. (2011)	●	●	●			●	●	●
Hobson (2008)	●	●		●				●
Holbrook and Rannikmae (2009)	●	●	●	●		●	●	●
Hondou et al. (2011)	●							
Howes et al. (2009)	–	–	–	–	–	–	–	–
Knain (2006)		●			●			
Krajcik and Sutherland (2010)	●	●		●				●
Lau (2009)	●	●	●	●				●
Laugksch (2000)	●	●	●	●	●	●	●	●
Lee (2004)	–	–	–	–	–	–	–	–
Lee et al. (2005)	–	–	–	–	–	–	–	–
Lee and Roth (2003)	●	●	●	●	●	●	●	

(continued)

40 Conceptions of Scientific Literacy...

Table 40.1 (continued)

	Objectives			Political justification		Moral justification		
Citations	K	Ca	T	L	Cm	V	P	U
Lima et al. (2010)	●	●	●	●	●			●
Marks and Eilks (2009)	–	–	–	–	–	–	–	–
Mbajiorgu and Ali (2003)		●		●				●
McConney et al. (2011)	●	●	●	●	●		●	●
Millar (2006)	●	●	●	●	●			
Millar (2011)	–	–	–	–	–	–	–	–
Murcia (2009)	●	●	●	●	●	●		●
O'Neill and Polman (2004)	●	●	●	●			●	
Patrick et al. (2009)	–	–	–	–	–	–	–	–
Pearson et al. (2010)	●	●						●
Porter et al. (2010)	●	●						
Rannikmae et al. (2010)	●							
Rennie and Williams (2002)	●	●		●				
Reveles and Brown (2008)	●	●			●			
Reveles et al. (2004)	●	●	●		●			
Rheinlander and Wallace (2011)	–	–	–	–	–	–	–	–
Ritchie et al. (2011)	●	●	●	●				●
Roth and Lee (2002)	●	●			●			
Rughinis (2011)	●	●	●	●				
Sadler (2011)	–	–	–	–	–	–	–	–
Sadler and Zeidler (2009)	●	●	●	●	●			
Schroeder et al. (2009)	●	●	●	●	●			
Shwartz et al. (2006)	●	●	●	●	●			
Soobard and Rannikmäe (2011)	●	●	●	●				
Sullivan (2008)	●	●						
Thomson and De Bortoli (2008)	–	–	–	–	–	–	–	–
Turner (2008)	●	●	●	●	●		●	●
Wallace (2004)	●	●			●			
Wang et al. (2011)	●	●						
Webb (2010)	●	●						
Yarden (2009)	–	–	–	–	–	–	–	–
Yore et al. (2007)	●	●	●	●	●			●
Totals	59	60	38	41	28	9	15	29

Objectives- *K* knowledge, *Ca* capacities, *T* traits
Justifications- *L* liberal, *Cm* communitarian, *V* virtue, *P* principled, *U* utilitarian

is needed 'to make informed decisions regarding personal, civic, and economic matters' as the authors want scientific literacy to provide.

Lau (2009) recognizes that the expression 'scientific knowledge' can be used pejoratively, perhaps to mean content learned by rote: 'Local science educators found that the junior secondary science curriculum was dominated by scientific knowledge, leaving many important scientific processes and understanding of the nature of science untouched' (p. 1062). At times, Lau appears to prefer the term 'understanding' to pick out what is valuable for scientific literacy, but he especially presses the distinction between knowledge of science and knowledge about science. The former he sees as referring to the substantive content of science; the latter as referring to meta-scientific knowledge, that is, to knowledge about scientific knowledge.

40.2.1.2 Capacities

Gräber et al. (2001) suggest 'a competency based model of scientific literacy' (p. 209). Including neither knowledge nor traits on the face of it, but instead focusing on capacities, this could be one of the most radical positions in the field. When examined more closely, however, a three-way distinction similar to our own underlies their model. They suggest that scientific literacy is needed 'for the individual to cope with our complex world' (p. 209), and to shape the sort of scientific literacy needed for this task, three questions must be answered: What do people know? What can people do? What do people value? They answer each question with sets of competencies, even the question about knowledge, for which they cite the need for subject competence and epistemological competence. The subject and epistemological competence fit under knowledge as we define it. The competencies they include under their second question are what we mean by capacities: social competence, procedural competence, and communicative competence.

Capacities often shade into knowledge, depending upon how the nature of the capacity is formulated. For example, Bybee (2009) says the following: 'the student with less developed scientific literacy might be able to recall simple scientific factual knowledge about a physical system and to use common science terms in stating a conclusion' (p. 2). On the one hand, Bybee is speaking of capacities (being able to recall something, and using something); on the other hand, he is thinking about the factual and terminological knowledge a student has. We see no deep theoretical point in this discrepancy that arises from vagaries of English usage nor any important issue that hangs on settling it.

40.2.1.3 Traits

The traits we identified fell into two major groups: intellectual and moral traits, the second of which are often termed 'moral virtues'. Intellectual traits are characteristics of people that promote intellectual flourishing. These traits might include inquisitiveness, open-mindedness, and carefulness. Virtues are characteristics that promote moral flourishing. These virtues might include honesty, generosity, and courage. Clearly, there are connections between the categories because a virtue such as honesty is central to intellectual flourishing, and a trait such as open-mindedness can be seen as morally superior to its opposite. Evans and Rennie (2009) associate scientific literacy with traits when they say that it is the capacity 'to be interested in, and understand, the world around them ... to be sceptical and questioning of claims made by others about scientific matters ...' (pp. 25–26).

40.2.2 Value Justifications for Desired Outcomes

This is the point at which our second categorization becomes relevant. If it is the case that there are three broad sorts of valued outcomes being considered in discussions about scientific literacy, then these differing values might very well demand and attract differing forms of justification. Table 40.1 indicates that we identified two broad categories of justification for scientific literacy objectives: moral and political. 'Morality' can be used in a descriptive or normative sense. In the descriptive sense, 'morality' refers to a code of conduct put forward by or followed by a society. You might imagine an anthropologist observing and interacting with a society and from the data collected inferring the code of conduct adopted in that society. If the descriptive sense is taken to exhaust the meaning of what is moral, then what is moral simply refers to the code of conduct that any group or person adopts. If the descriptive sense is taken in this way, then it conflicts with the normative sense of 'morality'. In the normative sense, what is moral is taken to apply universally, that is, to all those 'who can understand it and govern their behavior by it' (Gert 2011, p. 1). In the normative sense, it is assumed that actual codes of behaviour do not necessarily capture what it means to be moral. Actual codes of behaviour can be analysed and critiqued for falling short on morality, where morality here is thought of as an ideal that exists outside human practices and to which those practices can be held accountable. In the normative sense, moral justifications can never be overridden by other non-moral considerations.

Moral justifications of the objectives of scientific literacy can thus refer either to codes of conduct that are actually adopted or to codes of conduct that are not adopted but which, upon reflection, ought to be adopted. In the first type of case, the moral justification would be cast in terms of whether the objective led to behaviour that was in accord with an accepted code. In the latter type of case, the justification would be cast in terms of whether the objective led to behaviour that was considered moral, regardless of its conformity to an accepted code. If the behaviour was in conflict with an accepted code, the justification would in effect be saying that the code was morally deficient.

On one account, political justifications reduce to moral ones. That is, a behaviour is politically justified if, and only if, it is morally justified. On another account, moral justification is a necessary but insufficient condition for political justification: a behaviour is justified politically, only if it is morally justified. Consider an example outside of science education. If it is accepted that it is morally wrong to kill other human beings outside of situations of immediate threat to one's own life, then no reference to possible beneficial consequences can ever provide a political justification sufficient to override the moral condemnation of the act.

Politics can be defined roughly as the method by which groups of people make collective decisions. In advanced societies, politics is largely about how governments function, the business of collective decision making having been assigned to governments for many of the societies' resources. In this light, a political justification can be seen as one that defends a particular distribution of resources or the means of that distribution. So conceived, political justification requires more than mere consent from those affected by the decisions taken. The justification requires the offering of publicly accessible reasons that can be understood and challenged. It is through the public display and vetting of reasons that the political decisions acquire legitimacy; legitimacy cannot be bestowed by acquiescence.

A tight connection between political and moral justification is difficult to avoid, because the distribution of limited resources needs to conform to a sense of justice. So, although the distribution of resources is a practical matter and the form that distribution takes can be defended in part on practical grounds, it is impossible to avoid questions of fairness in the distribution. Education is one of the limited resources that our society has available for distribution. Therefore, political justifications of decisions of how to distribute that resource must ultimately conform to the demands of justice and be seen as legitimate on the basis of the reasons used to defend them.

40.2.2.1 Types of Moral Justification

The most prominent form of moral justification we uncovered in this analysis was *utilitarian*, found in roughly one-half the cases. Classical utilitarianism holds that actions are good insofar as they produce the most overall happiness (McLachlan 2010) or the greatest good for the greatest number. Debates rage over what constitutes happiness and over whether happiness is the greatest good that is to be sought. Whatever the particulars, utilitarianism is a form of consequentialism, which is a family of views that holds that the goodness or morality of an action is to be judged by its effects. In the case of justifying the objectives of scientific literacy, utilitarianism is manifested primarily on economic grounds, via an analysis of the sort of literacy that contributes to a healthy economy and personal competitiveness in the job market. Laugksch (2000) notes, for example, that advanced economies require technologically skilled professionals and only a scientifically literate populace can produce such professionals. Scientific literacy, on this understanding, is good because it strengthens economies and personal economic competitiveness. These consequences are, presumably, thought to be conducive to overall good or to human happiness, bringing us back to the touchstone of utilitarianism. Thus, the moral justification for scientific literacy might be outlined as follows: a society is economically stronger in the long term when its citizens are educated to a certain level of scientific literacy. An economically strong society is better for the persons encompassed by it. Therefore, it is justified to pursue scientific literacy for citizens.

Foster and Shiel-Rolle (2011) provide a utilitarian argument for scientific literacy that is almost textbook in its adoption of the utilitarian view described above:

> The importance of developing a scientific literate society is multifaceted. First, increasing scientific literacy has been considered to be a critical strategy for maintaining a country's technological and economic standing … Second, [i]t is increasingly important to have a scientifically literate society to make informed decisions regarding policy development and

its implementation. Lastly, … [s]cientifically literate international communities can … potentially use the scientific insight to improve their local agricultural and marine practices, economies and educational systems. (p. 86)

So, although they see the reasons for promoting scientific literacy as multifaceted, all of their reasons fall into the same category: utilitarianism.

Bybee (2008) provides the following utilitarian moral justification for scientific literacy in the context of environmental issues and PISA 2006:

Scientific literacy is essential to an individual's full participation in society. The understandings and abilities associated with scientific literacy empower citizens to make personal decisions and appropriately participate in the formulation of public policies that impact their lives. Assertions such as these provide a rationale of scientific literacy as the central purpose of science education. Too often, however, the rationale lacks connections that answer questions such as "personal decisions—" *concerning what?*" "fully participate—*in what?*" or "formulate policies—*relative to what?*" One could answer these questions using contexts that citizens daily confront; for example, personal health, natural hazards, and information at the frontiers of science and technology. Two other domains stand out— national resources and environmental quality. (pp. 567–568)

Readers would be correct to object that this evidence is insufficient to establish that Bybee relies upon a utilitarian justification for scientific literacy. Thus, it is necessary to examine his text in greater detail. Further examination of his document leaves little doubt about the nature of his justification. For example, he frames four policies that are supported by fostering scientific literacy as he conceives of it. The first policy is the fulfilment of basic human 'physiological needs such as clean air and water and sufficient food' (p. 578). The second and third policies come down to the same justification. The second policy deals with 'maintaining and improving the physical environment' (p. 578), which is 'the common heritage of humankind, and they are essential to fulfilling basic needs' (p. 579). The third policy focuses on the wise use of natural resources, which 'is closely related to … fulfillment of both the physical environment and to fulfillment of basic needs' (p. 579). Finally, the fourth policy aims 'toward establishing a greater sense of community' (p. 579). As such, the fourth policy might appear to find its grounding in communitarianism. In addition, the policy refers to the reduction of 'prejudice, such as racism, sexism, ethnocentricism and nationalism' (p. 579). This aim leans toward some sort of principled justification of avoiding forbidden behaviour. However, it is clear that both the push for community and the reliance on principles are themselves grounded on the same urge to fulfil basic human needs: 'If fulfillment of human needs and improvement of the environment … are to become realities, we must increase community involvement… one of the first steps … is the elimination of prejudicial barriers to community' (p. 579). Thus, the second to fourth policies reduce to the first, which clearly is motivated on utilitarian grounds. Feinstein (2011) makes a very deliberate effort to argue that the utilitarian justification for scientific literacy is the most robust:

This essay examines the idea that science education is useful in daily life . . . I focus on usefulness for two reasons. First, claims about the usefulness of science education are more testable than claims about its cultural, aesthetic, or moral value. In other words, when someone says science education is useful in a particular way, we should be able to find evidence for or against that claim, at least in theory. Second, the idea that science education is useful exerts a powerful political influence: People, particularly people with money and resources, seem to believe in it . . . It is important to specify what I mean by "useful in daily life," because that phrase has several possible interpretations. I am referring to the very specific notion that science education can help people solve personally meaningful problems in their lives, directly affect their material and social circumstances, shape their behavior, and inform their most significant practical and political decisions. (p. 169)

Feinstein's argument is very interesting. He claims to prefer utilitarian justifications because they are more testable than moral justifications. However, as we have shown, utilitarianism is one form of morality, namely, the form that focuses on the consequences of actions to decide on their morality, so Feinstein's attempt to separate the good from the useful fails to recognize that usefulness is just one way to interpret goodness. As Feinstein claims, consequences might also prove testable, supposedly making the judgement that science education is useful more clear-cut. Yet, the calculation is not so easy as Feinstein envisages. He sees usefulness as measured by helping people solve personally meaningful problems and by informing their most significant decisions. Deciding whether a problem is personally meaningful or a decision is of greatest significance also involves moral judgement that cannot be reduced to utility calculation.

The next most frequently found type of moral justification was based upon *principles*. Principled justification (deontology, in technical terms) concerns itself with what is morally forbidden, required, and permitted. It stands in contrast to varieties of consequentialism in that it holds that conformity to moral norms, such as particular duties or principles, makes actions morally praiseworthy (Alexander and Moore 2008). In its Kantian form, for instance, one seeks to ascertain the principle (or *maxim*) underlying a particular proposed action. This principle is then tested by asking if it could be applied universally, including to oneself. Praiseworthy actions, in Kantianism's simplified form, are those that satisfy these two tests. It is because of these reasons that deontologists hold that actions cannot be judged by their effects alone: according to this view, some actions are forbidden no matter how good their consequences might turn out.

Laugksch (2000) notes, for example, that it is often argued that citizens ought to be scientifically literate because they are affected by science, because they help to fund scientific research through tax dollars, and because they can participate in public deliberation when they are informed about scientific issues. In each of these cases, the fundamental justification is not that greater good is produced through scientific literacy (as in utilitarian justifications), but rather that certain principles are best satisfied when citizens are helped to become scientifically literate: such as, we should be informed about those matters that affect our lives, we should understand what we help to support through tax dollars, or that all citizens ought to participate in public decision making. These principles generally refer to conceptions of ethical governance. Since the criteria in this reasoning are most related to the satisfaction of certain ethical principles (such as democratic sovereignty), this sort of reasoning is best categorized as principled or deontological. Thus, a person holding that participation in public deliberation is a right might not be deterred in holding this view just because some people are not scientifically literate, are unable to participate in deliberations about scientific issues, but nonetheless show no ill effects from their lack of participation. Nor need the person be deterred by public deliberation that reached poorer choices than might be reached by a much narrower blue-ribbon panel. Once more, abiding by the principle is much more important than the consequences.

Bhathal (2011) makes the argument that Indigenous persons are underrepresented in science classes and in scientific careers, the implication being that this situation in principle is wrong. Although not explicitly argued by Bhathal, the underrepresentation seems to be thought of as a violation of the moral right to non-discrimination and equality. Bhathal describes a study in which 15 secondary school Aboriginal students were brought to a university six times during a semester to conduct projects in astronomy. The projects drew upon both Aboriginal astronomy and modern scientific astronomy and aimed to improve the students' scientific literacy. It is mildly curious to us that in the outcomes section at the end of the article, several utilitarian ends were cited in an apparent justification of the intervention that had taken place: students developed more positive attitudes toward science, gained knowledge of Aboriginal astronomy, were interested in the projects, and more were disposed at the end of the projects than at the beginning of them to continue with their high school education. Of course, all of these outcomes are positive and important. Our mild curiosity stems from the fact that none of these findings refers to the original justification for undertaking the projects, which was principle-based rather than based on utilitarian outcomes. Presumably, the justification runs as follows: the utilitarian ends achieved are indicators that teaching scientific literacy through astronomy addresses the underrepresentation of Aboriginal students in science, which is deplorable on principle. However, none of this argumentation is made explicit, so it is possible that we are putting words into the authors' mouths that they would not accept.

The same sort of principle-based justification is also present with respect to gender representation in George and Brenner (2010). As stated by the authors, 'The goal of the project was to create opportunities in college curricula that urge women to take science seriously whether or not they are science majors, to learn how science orders and explores the world, and to question it along the way' (p. 28). Although it is never put so explicitly, it seems the principle justifying this goal is that there ought not be a major difference in the representation of males and females in science. At the end of their report, the authors say: '... we cannot conclude that introducing feminist science studies improved the students' learning compared with a more traditional science course. Still, by introducing this course into the women's studies and science for the liberal arts curricula, we have built at least one two-way street across what has been an intimidating intellectual divide' (p. 34). Improving learning was a utilitarian goal of this project, but it was secondary by the authors' own admission. The primary goal was the increased representation of women in science. Thus, the authors seem to hold fast to their original principled stance in the previously quoted extract. Although the authors do not explicitly say, 'Even though we cannot show improved learning ...', it appears that this is what they have in mind. That is, they can nevertheless claim success despite no evidence of improved learning because they have found evidence that their approach can successfully address lower female participation rates in science, their principled objection to which is their primary justification.

In addition to utilitarian and principled justifications, one also finds conceptions of scientific literacy supported by *virtue theoretical* arguments, though these are the least frequently appearing type of justification. Virtue theoretical arguments are about human character and excellence (Hursthouse 2010). As such, they stand in contrast to arguments about doing what one is obliged to do or acting so as to produce desirable consequences. Also as such, the virtuous person is not simply one who practises virtuous acts, such as truth telling. Rather, truth telling is practised because it is valued for its own sake. We also need to distinguish two broad categories of virtue: namely, moral virtue and intellectual virtue. Among the first category we might find benevolence, compassion, empathy, gentleness, and selflessness. Among the second category we might find detachment, determination, flexibility, open-mindedness, perseverance, and reliability. We do not mean these lists to be comprehensive or mutually exclusive. For example, open-mindedness is intellectually virtuous in the conduct of science because it can prompt the consideration of alternative explanations that might be more powerful than the one under consideration. Likewise, open-mindedness is morally virtuous in everyday dealings with others because it can lead to the respect of moral agents who nevertheless hold different views and practise different customs than oneself.

Laugksch (2000), for example, discusses 'the intellectual, aesthetic, and moral benefits of scientific literacy to individuals' (p. 86). He describes these mostly in terms of intellectual virtues such as 'cultivated mind' and 'educated person', and perhaps a moral virtue, 'not merely wiser but better' (p. 86). Notions of cultivation and betterment are, at their root, notions of human excellence. Scientific literacy, in this case, is valuable insofar as it contributes to making students better persons. This emphasis places his reasoning within the category of virtue theory and places the focus on individuals' characters as opposed to, say, their knowledge and capacities.

It is striking that Dillon (2009) noted the concern of members of the 2007 Linné Scientific Literacy Symposium to the effect that there was a lack of emphasis on virtue in school science: 'There is little flavour in school science of the importance that creativity, ingenuity, intuition or persistence have played in the scientific enterprise' (Members of the Linné Scientific Literacy Symposium 2007, p. 7). In the same discussion, Dillon notes a particular emphasis on virtue in the reformed science education curriculum of Turkey and remarks that this emphasis would be 'unusual to a Western European eye' (2009, p. 208). He exemplifies the unusual emphasis with the following desired outcomes among others: 'Self-disciplined (Self-controlled, prompt, self-evaluating, sincere, consistent)' (Taşar and Atasoy 2006, p. 9). All of these outcomes belong among those on lists of intellectual and moral virtues.

Murcia (2009) says that scientific literacy is 'about a way of thinking and acting' (p. 219). This statement suggests she might be providing a virtue justification for scientific literacy. Although in Table 40.1 we have marked this work as providing virtue justification for scientific literacy objectives, we are not completely confident that it does. Murcia develops a framework of scientific literacy and says: 'The aim of this framework was to clarify the type of knowledge, roles and abilities required to act scientifically in a contemporary context' (p. 218). Among her categories, one of the 'roles' is the one most closely related to the nature of character. She refers to the work of Ford and Forman (2006), who have named the roles of Constructor and Critiquer for scientifically literate people. The latter of these, in particular, is easily interpreted in terms of the intellectual virtue of criticalness, the former can perhaps be interpreted in terms of creativeness. We are tentative in these recommended interpretations of Murcia's work because we are not sure that is what she meant. Nevertheless, interpretation in terms of intellectual virtue does make sense of this aspect of her work.

Although Holbrook and Rannikmäe (2009) call them 'skills', we take their 'personal skills related to creativity, initiative, safe working' (p. 283) to be personal qualities much the same as virtues. Similarly, although Dos Santos (2009) never uses the word 'virtue', his entire article is about a particular sort of virtue, namely, commitment: 'The conclusions of those works [Freire's] reveal a political commitment to struggle for liberation' (p. 369); 'students need to take part directly in SSI discussions, so they can interact with the world, discuss their living conditions, and become committed to social change' (p. 374). The clear implication of these lines is to point to the desirability of a particular sort of character trait, that is, one of commitment to social change.

One article (Greenleaf et al. 2011) presented all three types of moral justification.

Regarding principled moral justification, there is a repeated emphasis on inequity and underrepresentation as a motivating force. Specifically, they note underrepresentation of certain cultural groups in science and inequity of resources and training for teachers. The core argument is that illiteracy and lack of scientific literacy are objectionable on the grounds that they foster inequity (a deontological concern when framed in this way). Here is an example excerpt:

> Withdrawing adolescents from instruction in science to remediate reading difficulties threatens to exacerbate historic inequities in achievement for populations of students traditionally underrepresented in the sciences … There is therefore increasing urgency to investigate how the integration of reading instruction into science learning at the high school level might advance the reading and science achievement of underachieving youth. (p. 649)

Utilitarian justifications are found straightforwardly. The article opens with the standard argument about how the health of the nation (the United States in this case) requires this sort of education for scientific literacy on grounds of both economic and democratic utility: 'Our democracy and future economic well-being depend on a literate populace, capable of fully participating in the demands of the twenty-first century … Yet National Assessment of Educational Progress (NAEP) results indicate that most American youth lack the skills to successfully engage in the higher-level literacy, reasoning, and inquiry needed for an information-generating and information-transforming economy' (p. 648).

Although they are the most ambiguous of the three sorts of justification found in this article, we believe a case can be made that Greenleaf et al. offer virtue justifications for scientific literacy. They refer on several occasions to dispositional outcomes, suggesting that they are concerned with individuals' characters, but they seem to go further and to frame some dispositions in virtue terms. For example, they explicitly say that they want 'resilient' learners and students with 'stamina' (pp. 657–658) and construct the issue as one of personal identity construction. When we use terms of personal excellence, and we note that we seek to have people develop into certain sorts of people, we have moved from discussion of ordinary cognitive and attitudinal traits to discussion of virtues, either prudential or moral.

40.2.2.2 Types of Political Justification

Approximately two-thirds of the articles appeal to political liberalism to justify their scientific literacy objectives. Liberal theory places a fundamental emphasis on the exercise of personal freedom. One ought to be free to determine for oneself important elements of personal belief and lifestyle (such as religious belief). Rawls (1993), for example, famously argued that a political community ought to be arranged so that free persons could participate in rational public debate and dialogue in spite of their diverse beliefs. The educational corollary to this position is that persons must be nurtured into the knowledge, capacities, and traits required to participate in such critical deliberation. When public questions regard scientific practice or regulation or depend on scientific knowledge and method, it follows that scientific literacy specifically is required. One might argue, therefore, that the requirement to support public democratic deliberation is not just a principled moral imperative (as we argued prior to this section) but is also a liberal political imperative.

Correia et al. (2010) provide a textbook case of a liberal justification for scientific literacy:

> Scientific literacy (SL) is necessary in post-industrial society to nurture an autonomous citizenry . . . We are negotiating a new contract between society and science, and all citizens must have the right and the ability to make their own judgments about the ethical aspects of scientific and technological issues. (p. 680)

> Scientific literacy is a novel requirement for producing informed and autonomous citizens in post-industrial societies. Moreover, it is necessary that a student achieve scientific literacy during his or her career in higher education to be able to achieve the education for sustainability. Universities striving to teach sustainability must graft a holistic perspective onto the traditional specialized undergraduate curriculum. This new integrative, inter/trans-disciplinary epistemological approach is necessary to allow autonomous citizenship, that is, the possibility that each citizen understand and participate in discussions about the complex issues posed by our contemporary post-industrial society. (p. 685)

Bybee (2009) expresses a clear emphasis on the capacity of individuals to take a position on personal and political decisions and to formulate arguments in support of it:

> . . . as people are presented with more, and sometimes conflicting, information about phenomena, such as climate change, they need to be able to access collective scientific knowledge and understand, for example, the scientific basis for evaluations . . . versus the

basis for perspectives by individuals representing oil, gas, or coal companies. Finally, citizens should be able to use the results of scientific reports and recommendations about issues such as health, prescription drugs, and safety to formulate arguments supporting their decisions about scientific issues of personal, social, and global consequence. (pp. 3–4)

These examples from Correia et al. and Bybee are clear instances of liberal political justification because they explicitly emphasize individual decision-making capacity, not collective meaning or collective agency. It is worth noting that, even though most of the articles contained in our review have a liberal justification at play for the objectives of scientific literacy, these justifications were nearly always vague references to individual decision-making capacity rather than explicit invocations of liberal theory. We think it is fair to say that the scientific literacy literature assumes liberal politics in most cases, in contrast to defending liberalism.

Also, there is not always a clear divide between liberal justifications and moral ones. There seems to be a moral implication in O'Neill and Polman (2004) when they discuss why people need scientific literacy to understand their personal choices:

We suggest that on a societal scale, schools would function more effectively if they covered less content, in ways that would allow students to build a deeper understanding of how scientific knowledge claims and theories are constructed. This would be of use to all students in their decision making outside of school, and beneficial to those pursuing postsecondary studies in science as well. (p. 237)

The primary justification here for teaching about the nature of science seems to be the assistance this knowledge would bring to decision making outside of school. This justification is phrased in fairly standard liberal civic terms, meaning something like 'all agents should be able to access and understand information relevant to their personal choices'. Such a justification can be interpreted as well in Kantian moral terms, which is not surprising given that liberal political theory has a strong deontological heritage. This often displayed union of political liberalism and Kantian ethics in education points to a linkage between the discussion in this section on types of political justification with the foregoing section on types of moral justification.

The most explicit uses of political theory in these articles were instances in which communitarians wanted to distance themselves from the assumption of liberal politics. About one-half of the articles appealed to communitarian justifications for scientific literacy. Liberalism in educational theory often is contrasted with communitarianism (e.g., Strike 2000). Liberalism holds primary the fair distribution of liberties and resources to enable the individual selection of forms of life to lead. Fairness according to communitarianism must be judged within traditions and thus can vary from society to society and from time to time. In communitarian political thought, the emphasis is placed primarily on the collective determination of the community, not on the autonomous choice of the individual. Sectarian schools are, for example, a prominent manifestation of communitarian religious thought. Rather than emphasizing individual choice and deliberation across communities, communitarian politics seeks to nurture a single community. Of course, given the presumption of communitarianism that standards of value are community-specific, communitarians in plural states must be prepared to accept many communities. Such political reasoning can be applied also to scientific literacy. When science educators speak of fostering scientific values, for instance, they are at least partially speaking about creating and nurturing a particular kind of shared community – in this instance, one based upon science. At its root, this is a communitarian stance.

There are two broad kinds of communitarianism reflected among the articles. The first is represented in articles like Dos Santos (2009). This article uses a Freirean approach to emphasize community meaning and knowledge. The point of the article is to advance the ways in which a particular community constructs its problems and knowledge. Dos Santos draws attention to the existence of landfills in many Brazilian cities to illustrate how Freirean pedagogy could turn this situation into an important science lesson. 'Teachers could take their students to visit landfills, to interview people that work there, and later discuss in the classroom how that community could change the situation' (p. 373). The assumption of the collective determination of the community in setting its future is quite explicit in this quotation. Consider another situation mentioned by Dos Santos in which a school is situated in a location without a sewer system. 'The search for solutions for this problem will inevitably point out the need of mobilizing the school, and local community for political actions aimed at providing that community with sewage' (p. 374). Again, we see the emphasis on community decision making and action as opposed, let us say, to individual decision making and action, such as individual home owners installing septic systems or composting toilets. Note also that the political theory at play in the Dos Santos' document is much more explicit than in nearly all of the examples that drew upon liberal theory.

The second kind of communitarianism found in the articles places an emphasis on drawing students into scientific culture. In articles such as these, scientific literacy is justified on the grounds that it draws people into science as a cultural act. Such a justification seems centrally communitarian and quite distinct from saying that scientific literacy is important because it fosters individual decision making. Cavagnetto (2010), for example, argues that skills in scientific argumentation and understanding scientific processes and principles are insufficient bases in scientific literacy. Such skill and knowledge do not initiate one into the culture of science, which can only be entered by engaging in scientific practices. So, the justification for teaching argumentation is not that it teaches argumentative skill, but that it introduces students into a fundamental practice of science: 'Therefore, the goal of argument instruction in the context of scientific literacy is not the transfer of argument skills but rather the transfer of an understanding of scientific practice' (p. 352).

40.3   Choosing Among Conceptions of Scientific Literacy

We have shown thus far that the programmatic elements of various conceptions of science literacy contain at least three categories of valued objectives: knowledge, capacities, and traits. These three categories of objectives are justified either

Table 40.2 Frequencies of political and moral justifications offered for each scientific literacy objective

Objectives	Political justifications		Moral justifications		
	L	Cm	V	P	U
Knowledge	40	26	9	15	27
Capacities	41	28	9	15	29
Traits	31	19	9	14	22

Key. L liberal, *Cm* communitarian, *V* virtue, *P* principled, *U* utilitarian

morally or politically. On the moral side, we documented utilitarian, principled, and virtue theoretical justifications. On the political side, we documented liberal and communitarian justifications. There is no necessary link between any particular category of valued objective and any particular form of justification, and we found no strong correlation in the data, as can be seen by checking the frequencies of types of justifications for each objective in Table 40.2. Nor is it the case that these justifications contradict one another, although they might in certain renditions. Thus, it is possible at least sometimes to use without contradiction more than one type of justification to support the same objective. One could support the development of valuable traits of personality not only on the grounds that they contribute to personal excellence (a virtue theoretical justification) but also because they enable persons to contribute to the overall social good (a utilitarian justification) or also because persons have a moral right to be so supported (a principled justification). Yet, we note that among moral justifications, fewer utilitarian justifications were provided for the trait objective than for other objectives, which makes sense if you consider moral traits to be favoured more for their desirability than their usefulness. These traits might enhance personal participation in democratic deliberation (a liberal justification), or they might allow one to appreciate scientific culture (a communitarian justification). However, we note that between the political justifications, there were fewer communitarian justifications for trait objectives than for other objectives, which makes sense if traits are viewed as individual accomplishments.

Our schema is not totally exhaustive of all the objectives and justifications that do or could exist, so it is possible to propose analytic divisions beyond what we have offered. Returning to the opening of the chapter, the point is that scientific literacy is highly programmatic. This programmatic character is manifested in different kinds of valued objectives and justifications for them. To understand what is at stake in debates about scientific literacy and to decide one's own position, one needs to identify the valued objectives and the justifications for them, to understand how each type of justification functions, and to critique each. The canvass we have done of several examples should provide some useful starting points for such identification, understanding, and critique.

Understanding how the goals of scientific literacy and their justifications are linked allows us better to understand scientific literacy itself, and it allows scholars of science education to ask more fruitful and illuminating questions about this educational ideal. As we have demonstrated, each form of justification focuses on certain aspects of value and not others. The choice of whether to accept a particular justification is educationally significant. The choice is not always easy, and the alternatives among various justifications are not always clearly distinguishable in terms of quality, although we hope that we have provided guidance for making such distinctions.

So how might choices be made among alternative versions of scientific literacy objectives and their possible justifications? Among all the possibilities of contrasting positions defined by crossing the three types of scientific literacy objectives with the five types of justifications, we will consider three contrasts for illustration. The first contrast is between political and moral justifications, the second within political justifications between liberal and communitarian justifications, and the third within moral justifications between virtue, principled, and utilitarian justifications. The entire domain is essentially contested, so that there are not always clearly right or even better positions, but, rather, a panoply of good positions with some clearly poor or not so good ones. This is the nature of the scientific literacy domain; there is no way to make the ambiguity and uncertainty disappear. The point is to identify which definition of scientific literacy comparatively speaking is most educationally significant.

40.3.1 Contrast Between Political and Moral Justifications

At a very general level of consideration, ignoring any specifics for the moment and assuming all other things are equal, moral justifications in the normative sense that we prefer, based as they are upon what is right, should trump political justifications, based as they are upon what is possible. Thus, a moral justification for a science curriculum directed toward a universal scientific literacy that respected gender and racial equality might be seen, all other things being equal, as stronger than a political justification for a curriculum that promoted scientific literacy for social activism. Similarly, aiming to foster individual intellectual and moral virtues such as curiosity, open-mindedness, and valuing fair tests might be seen, *ceteris paribus*, as a more important reason to foster scientific literacy than the creation of critical and informed citizens for the promotion of the democratic state. The *ceteris paribus* clause covers a host of nuance and qualifications. For moral justifications to trump political ones, they have to be sound moral justifications. Thus, for example, if the justification offered is based on providing a gender and racially equal scientific literacy, then the curriculum that is envisaged to do this must be able to result in that desired outcome, or the justification for using that curriculum to reach the desired objectives breaks down. Also, it is one matter to respect gender and racial equality in providing opportunity for scientific literacy development and quite a different matter to demand equal outcomes. Perhaps equal scientific literacy outcomes across gender and race is not the most morally desirable outcome. An argument is needed to make the case one way or the other and that argument needs to be assessed.

Perhaps the best position for science educators is to consider moral justifications offered for scientific literacy objectives before considering the political ones.

There is no calculus, unfortunately – a theme we will repeat – for choosing between strong moral and strong political justifications if they result in the recommendation of different objectives and curricula for obtaining them. The situation is even more complicated, as shown in the following section.

40.3.2 Contrast Within Political Justifications

The distinctions between moral and political justifications are not always so easy to make. The political justifications based on liberalism and communitarianism themselves draw upon moral theory. Liberalism celebrates justice for all in the service of individual autonomy. Communitarianism sees collective decision making as a higher good than individual choice. Theoretical arguments do not exist that can help us choose definitively between these positions. Making the situation more complicated is that proponents on either side often acknowledge the values advocated by the other.

A central feature of liberalism is its neutrality concerning conceptions of what is good. Individuals should be free to follow their own preferences so long as they do not impede others' freedom to follow their preferences, but liberalism makes no judgement on the quality of those preferences. On this conception, scientific literacy has to be seen as a tool to help individuals to choose and pursue their preferred life courses. Communitarians believe that goods must be rank ordered and that there are goods in common. On this view, scientific literacy is seen as one of the common goods that helps build a society.

Does liberal justice trump communitarian benevolence? Does scientific literacy for individual self-determination trump scientific literacy for social cohesion and progress? It is difficult for educators to choose between these positions, because we are used to understanding education itself as good for both individuals and for society. The situation is such that many educators have comfortably chosen both justifications as being sound. We know of no strict contradiction between the positions. Furthermore, we find it difficult ourselves as educators to conceive of a justification for scientific literacy objectives that did not take into account the benefits for both the individual and for the society. This position is not one that we can defend here, but one we are confident would serve well science education. Our comfort rests on several considerations. First, there is no scholarly consensus on which version of political and moral life among those we have discussed is the preferable. Second, although we believe that education cannot be neutral, we also believe that it is not the role of education to select only one version of political and moral life to espouse, if there is more than one viable alternative, which there is. Third, we do believe it is the role of education to introduce students to the various forms of political and moral life that find widespread justification within our society. After all, it is the generation that is now in school that will be the one that has to grapple with these difficult and age-old issues, and for this reason is the one in need of broad exposure to more than one formulation of democratic politics and ethical living.

40.3.3   Contrast Within Moral Justifications

Under certain moral justifications scientific literacy might be argued to be *intrinsically* good, while under others it might not be an ultimate good itself but be an *instrument* for the good. If scientific literacy is part of a flourishing human life (a virtue theoretical argument), then one might conceive of such literacy as a good thing on its own (intrinsically good). If scientific literacy is valuable only insofar as it helps the economy, then it is only an instrument to the good (in this case, wealth or employment and the happiness that flows from those benefits). Such a utilitarian justification leads to an entirely different vision of science education than the preceding virtue theoretical justification. Why, for instance, should we help students to become broadly informed about science when specialization is often more highly rewarded in the job market? The answer to this question, and many like it, requires one to make the kind of distinctions we have begun to introduce here.

This contrast is very similar to the one described in the previous section. Some moral theorists argue that happiness is the greatest good and that the best society is the one in which the total utility is maximized and that utility is spread as evenly as possible across all people. Other theorists argue that individual excellence, in the form of virtues, is the greatest good, and the best society is the one in which people are the most virtuous. We argue for reasons similar to those in the previous section that education cannot choose between virtue and utility. Education can choose moral over immoral or amoral behaviour, but it is not its role to favour one version of morality over another, if all those versions find strong justification within the scholarly community. Therefore, a version of scientific literacy that sought only utilitarian goals could be critiqued on the grounds that it ignored individual excellence; a version focussed entirely on fostering virtue could be faulted for ignoring other important goals, such as the contribution to social utility resulting from making an effort to become scientifically literate.

40.4   Conclusion

We find ourselves in a position of not being able to make, all things being equal, definitive choices among justifications for the objectives of scientific literacy. Education in a way is like liberalism – it cannot (or, perhaps better, should not) choose between comprehensive conceptions of the good. MacLeod (1997) provides an excellent account of the meaning of comprehensive conceptions of the good:

> Conceptions of the good are views about the nature and constitutive elements of a valuable life. Conceptions of the good may be comprehensive or partial. A comprehensive conception of the good attempts to delineate a complete account of the sources and nature of a good life. A partial conception of the good merely identifies particular activities or projects that contribute to the realization of human excellence. Commitment to a religion can constitute a comprehensive conception of the good since adoption of a faith is sometimes viewed as grounding the meaning of a person's entire life. The idea that the appreciation of

fine music is a valuable human activity is likely to constitute only a partial conception of the good. Aesthetic appreciation is a possible component in a good life but it does not constitute a full account of the good life. (p. 529, fn 2)

Therefore, ideally, a true education would introduce students to as many comprehensive conceptions of the good as are available. Thus, regarding moral justifications for the objectives of scientific literacy, science education should not privilege either a virtue, principled, or utilitarian justification. Likewise, regarding political justifications, science education cannot favour either political liberalism or communitarianism. Rather, science education must recognize each of these versions of moral and political life as viable and must structure the goals of scientific literacy so as to promote each alternative or at least to make the alternatives available to students so as to keep their futures as open as possible. However, science education rightly can take a negative stand against justifications for scientific literacy that are based on partial conceptions of the good. Thus, for example, a justification for the objectives of scientific literacy based only upon their role in securing employment for the individual or economic prosperity for society does not appeal to a comprehensive conception of the good unless it situates employment and economic prosperity within a more thoroughgoing conception of the good life. For example, the employment and the prosperity might be seen to lead to happier, more fulfilling, and more self-directed lives than otherwise would be possible. Alternatively, they might be seen to promote the possibility for fuller communitarian living.

It may seem ironic that science education must be designed upon a liberal footing if it is to support more than a liberal political agenda and to make available for consideration by students more than one version of moral life. This is so because between the versions of political and moral life under consideration, only liberalism takes the view that it cannot uphold one comprehensive version of the good over another. Even communitarian-motivated education, such as sectarian education, if it is to count as education and not indoctrination, must acknowledge and make its students aware of alternative versions of the good. Nothing can count as a true education that deliberately attempts to circumscribe students' futures by trying to keep morally and politically defensible options hidden or, even worse, by trying to denigrate them. This programmatic stance is the proper one for any version of science education vying for our support.

References

Alexander, L. & Moore, M. (2008). Deontological ethics. In E.N. Zalta (Ed.), *The Stanford Encyclopedia of Philosophy (Fall 2008 Edition)*. Retrieved November 27, 2011, from http://plato.stanford.edu/archives/fall2008/entries/ethics-deontological/

Baker, D. R., Lewis, E. B., Purzer, S., Watts, N. B., Perkins, G., Uysal, S., et al. (2009). The Communication in Science Inquiry Project (CISIP): A project to enhance scientific literacy through the creation of science classroom discourse communities. *International Journal of Environmental and Science Education, 4*(3), 259–274.

Baram-Tsabari, A., & Yarden, A. (2005). Text genre as a factor in the formation of scientific literacy. *Journal of Research in Science Teaching, 42*(4), 403–428. doi:10.1002/tea.20063

Bhathal, R. (2011). Improving the scientific literacy of Aboriginal students through astronomy. *The Role of Astronomy in Society and Culture, Proceedings of the International Astronomical Union Symposium, 260*, 679–684. doi:10.1017/S1743921311003012

Bonney, R., Cooper, C. B., Dickinson, J., Kelling, S., Phillips, T., Rosenberg, K. V., et al. (2009). Citizen science: A developing tool for expanding science knowledge and scientific literacy. *BioScience, 59*(11), 977–984. doi:10.1525/bio.2009.59.11.9

Britsch, S. (2009). Differential discourses: The contribution of visual analysis to defining scientific literacy in the early years classroom. *Visual Communication, 8*(2), 207–228. doi:10.1177/1470357209102114

Brossard, D., & Shanahan, J. (2006). Do they know what they read? Building a scientific literacy measurement instrument based on science media coverage. *Science Communication, 28*(1), 47–63. doi:10.1177/1075547006291345

Brown, B. A., Reveles, J. M., & Kelly, G. J. (2005). Scientific literacy and discursive identity: A theoretical framework for understanding science learning. *Science Education, 89*(5), 779–802. doi:10.1002/sce.20069

Bybee, R. W. (2008). Scientific literacy, environmental issues, and PISA 2006: The 2008 Paul F-Brandwein Lecture. *Journal of Science Education and Technology, 17*(6), 566–585. doi:10.1007/s10956-008-9124-4

Bybee, R. W. (2009). Program for International Student Assessment (PISA) 2006 and scientific literacy: A perspective for science education leaders. *Science Educator, 18*(2), 1–13.

Bybee, R., Fensham, P., & Laurie, R. (2009). Scientific literacy and contexts in PISA 2006 science. *Journal of Research in Science Teaching, 46*(8), 862–864. doi:10.1002/tea.20332

Bybee, R., & McCrae, B. (2011). Scientific literacy and student attitudes: Perspectives from PISA 2006 science. *International Journal of Science Education, 33*(1), 7–26. doi:10.1080/09500693.2010.518644

Bybee, R., McCrae, B., & Laurie, R. (2009). PISA 2006: An assessment of scientific literacy. *Journal of Research in Science Teaching, 46*(8), 865–883. doi:10.1002/tea.20333

Cajas, F. (2001). The science/technology interaction: Implications for science literacy. *Journal of Research in Science Teaching, 38*(7), 715–729. doi:10.1002/tea.1028

Cavagnetto, A. R. (2010). Argument to foster scientific literacy: A review of argument interventions in K-12 science contexts. *Review of Educational Research, 80*(3), 336–371. doi:10.3102/0034654310376953

Chen, F., Shi, Y., & Zu, F. (2009). An analysis of the Public Scientific Literacy study in China. *Public Understanding of Science, 18*(5), 607–616. doi: 10.1177/0963662508093089

Colucci-Gray, L., Camino, E., Barbiero, G., & Gray, D. (2006). From scientific literacy to sustainability literacy: An ecological framework for education. *Science Education, 90*(2), 227–252. doi:10.1002/sce.20109

Cook, S. B., Druger, M., & Ploutz-Snyder, L. L. (2011). Scientific literacy and attitudes towards American space exploration among college undergraduates. *Space Policy, 27*(1), 48–52. doi:10.1016/j.spacepol.2010.12.001

Correia, R. R. M., do Valle, B. X., Dazzani, M., & Infante-Malachias, M. E. (2010). The importance of scientific literacy in fostering education for sustainability: Theoretical considerations and preliminary findings from a Brazilian experience. *Journal of Cleaner Production, 18*(7), 678–685. doi:10.1016/j.jclepro.2009.09.011

DeBoer, G. E. (2000). Scientific literacy: Another look at its historical and contemporary meanings and its relationship to science education reform. *Journal of Research in Science Teaching, 37*(6), 582–601. doi:10.1002/1098-2736(200008)37:6<582::AID-TEA5>3.0.CO;2-L

Derry, S., & Zalles, D. (2011, April). *Scientific literacy in the context of civic reasoning: An educational design problem*. Presidential address at the Annual Meeting of the American Educational Research Association, New Orleans, LA.

Dillon, J. (2009). On scientific literacy and curriculum reform. *International Journal of Environmental and Science Education, 4*(3), 201–213.

Dos Santos, W. L. P. (2009). Scientific literacy: A Freirean perspective as a radical view of humanistic science education. *Science Education, 93*(2), 361–382. doi:10.1002/sce.20301

Eijick, M., & Roth, W.-M. (2007). Rethinking the role of information technology-based research tools in students' development of scientific literacy. *Journal of Science Education and Technology, 16*(3), 225–238. doi: 10.1007/s10956-007-9045-7

Ennis, R.H. (1969). *Logic in teaching*. Englewood Cliffs, NJ: Prentice-Hall, Inc.

Evans, R. S., & Rennie, L. J. (2009). Promoting understanding of, and teaching about, scientific literacy in primary schools. *Teaching Science, 55*(2), 25–30.

Fang, Z. (2005). Scientific literacy: A systemic functional linguistics perspective. *Science Education, 89*(2), 335–347. doi:10.1002/sce.20050

Feinstein, N. (2011). Salvaging science literacy. *Science Education, 95*(1), 168–185. doi:10.1002/sce.20414

Ford, M., & Forman, E. (2006). Re-defining disciplinary learning in classroom contexts. In J. Green & A. Luke (Eds.). *Review of Research in Education, 30*, (pp. 1–32). Washington, DC: American Educational Research Association.

Foster, J. S., & Shiel-Rolle, N. (2011). Building scientific literacy through summer science camps: A strategy for design, implementation and assessment. *Science Education International, 22*(2), 85–98.

Fuselier, L., & Nelson, B. (2011). A test of the efficacy of an information literacy lesson in an introductory biology laboratory course with a strong science-writing component. *Science and Technology Libraries, 20*(1), 58–75. doi:10.1080/0194262X.2011.547101

Gawalt, E. S., & Adams, B. (2011). A chemical information literacy program for first-year students. *Journal of Chemical Education, 88*(4), 402–407. doi:10.1021/ed100625n

George, L.A., & Brenner, J. (2010). Increasing scientific literacy about global climate change through a laboratory-based feminist science course. *Journal of College Science Teaching, 39*(4), 28–34.

Gert, B. (2011). The definition of morality. In E.N. Zalta (Ed.), *The Stanford Encyclopedia of Philosophy (Summer 2011 Edition)*. Retrieved February 16, 2012, from http://plato.stanford.edu/archives/sum2011/entries/morality-definition.

Gräber, W, Erdmann, T., & Schlieker, V. (2001). *ParCIS: Aiming for scientific literacy through self-regulated learning with the internet*. University of Kiel, Institute for Science Education: Kiel, Germany [ERIC Reproduction No. ED466362].

Greenleaf, C. L., Litman, C., Hanson, T. L., Rosen, R., Boascardin, C. K., Herman, J., et al. (2011). Integrating literacy and science in biology: Teaching and learning impacts of reading apprenticeship professional development. *American Educational Research Journal, 48*(3), 647–717. doi:10.3102/0002831210384839

Hobson, A. (2008). The surprising effectiveness of college scientific literacy courses. *The Physics Teacher, 46*(7), 404–406. doi:10.1119/1.2981285

Holbrook, J., & Rannikmäe, M. (2009). The meaning of scientific literacy. *International Journal of Environment and Science Education, 4*(3), 275–288.

Hondou, T., Sekine, T., & Suto, S. (2011). What are the limits of validity in science? New lab class to improve scientific literacy of humanities students. *Latin-American Journal of Physics Education, 5*(2), 348–351.

Howes, E. V., Lim, M., & Campos, J. (2009). Journeys into inquiry-based elementary science: Literacy practices, questioning, and empirical study. *Science Education, 93*(2), 189–217. doi:10.1002/sce.20297

Hursthouse, R. (2010). Virtue Ethics. In E. N. Zalta (Ed.), *The Stanford Encyclopedia of Philosophy (Winter 2010 Edition)*. Retrieved November 27, 2011, from http://plato.stanford.edu/archives/win2010/entries/ethics-virtue

Knain, E. (2006). Achieving scientific literacy through transformation of multimodal textual resources. *Science Education, 90*(4), 656–659. doi: 10.1002/sce.20142

Krajcik, J. S., & Sutherland, L. M. (2010). Supporting students in developing literacy in science. *Science, 328*(5977), 456–459. doi:10.1126/science.1182593

Lau, K. (2009). A critical examination of PISA's assessment on scientific literacy. *International Journal of Science and Mathematics Education, 7*(6), 1061–1088. doi:10.1007/s10763-009-9154-2

Laugksch, R. (2000). Scientific literacy: A conceptual overview. *Science Education, 84*(1), 71–94. doi:10.1002/(SICI)1098-237X(200001)84:1<71::AID-SCE6>3.0.CO;2-C

Lee, O. (2004). Teacher change in beliefs and practices in science and literacy instruction with English language learners. *Journal of Research in Science Teaching, 41*(1), 65–93. doi:10.1002/tea.10125

Lee, O., Deaktor, R. A., Hart, J. E., Cuevas, P., & Enders, C. (2005). An instructional intervention's impact on the science and literacy achievement of culturally and linguistically diverse elementary students. *Journal of Research in Science Teaching, 42*(8), 857–887. doi:10.1002/tea.20071

Lee, S., & Roth, W.-M. (2003). Science and the "good citizen": Community-based scientific literacy. *Science Technology Human Values, 28*(3), 403–424. doi:10.1177/0162243903028003003

Lima, A., Vasconcelos, C., Felix, N., Barros, J., Mendonca, A. (2010). Field trip activity in an ancient gold mine: Scientific literacy in informal education. *Public Understanding of Science, 19*, 322–334. doi:10.1177/0963662509104725

Marks, R., & Eilks, I. (2009). Promoting scientific literacy using a sociocritical and problem-oriented approach to chemistry teaching: Concept, examples, experiences. *International Journal of Environmental and Science Education, 4*(3), 231–245.

Mbajiorgu, N. M., & Ali, A. (2003). Relationship between STS approach, scientific literacy, and achievement in biology. *Science Education, 87*(1), 31–39. doi:10.1002/sce.10012

McConney, A., Olive, M., Woods-McConney, A., & Schibeci, R. (2011). Bridging the gap? A comparative, retrospective analysis of science literacy and interest in science for Indigenous and Non-Indigenous Australian students. *International Journal of Science Education, 33*(14), 2017–2035. doi:10.1080/09500693.2010.529477

McLachlan, J. A. (2010). *Ethics in action: Making ethical decisions in your daily life*. Toronto: Pearson Canada.

MacLeod, C.M. (1997). Liberal neutrality or liberal tolerance? Law and Philosophy, 16, 529–559.

Members of the Linné Scientific Literacy Symposium. (2007). Statement of Concern. In C. Linder, L. Östman, & P.-O. Wickman (Eds.), *Promoting scientific literacy: Science education research in transaction* (pp. 7–8). Uppsala, SE: Uppsala University.

Millar, R. (2006). Twenty first century science: Insights from the design and implementation of a scientific literacy approach in school science. *International Journal of Science Education, 28*(13), 1499–1521. doi:10.1080/09500690600718344

Millar, R. (2011). Reviewing the National Curriculum for science: Opportunities and challenges. *Curriculum Journal, 22*(2), 167–185. doi:10.1080/09585176.2011.574907

Millar, R., & Osborne, J. (Eds.) (1998). *Beyond 2000: Science education for the future: The report of a seminar series funded by the Nuffield Foundation*. London: King's College London. Retrieved February 16, 2012 from, http://www.nuffieldfoundation.org/sites/default/files/Beyond%202000.pdf

Murcia, K. (2009). Re-thinking the development of scientific literacy through a rope metaphor. *Research in Science Education, 39*(2), 215–229. doi:10.1007/s11165-008-9081-1

Norris, S. P., & Phillips, L. M. (2003). How scientific literacy in its fundamental sense is central to scientific literacy. *Science Education, 87*(2), 224–240. doi:10.1002/sce.10066

Norris, S.P. & Phillips, L.M. (2009). Scientific literacy. In D.R. Olson & N. Torrance (Eds.). *Handbook of research on literacy* (pp. 271–285). Cambridge: Cambridge University Press.

O'Neill, D. K., & Polman, J. L. (2004). Why educate "little scientists"? Examining the potential of practice-based scientific literacy. *Journal of Research in Science Teaching, 41*(3), 234–266. doi:10.1002/tea.20001

Patrick, H., Matzicopoulos, P., & Samarapungavan, A. (2009). Motivation for learning science in kindergarten: Is there a gender gap and does integrated inquiry and literacy instruction make a difference? *Journal of Research in Science Teaching, 46*(2), 166–191. doi:10.1002/tea.20276

Pearson, P. D., Moje, E., & Greenleaf, C. (2010). Literacy and science: Each in service of the other. *Science, 328*(5977), 459–463. doi:10.1126/science.1182595

Porter, J. A., Wolbach, K. C., Purzycki, C. B., Bowman, L. A., Agbada, E., & Mostrom, A. M. (2010). Integration of information and scientific literacy: Promoting literacy in undergraduates. *CBE-Life Sciences Education, 9*, 536–542. doi:10.1187/cbe.10-01-0006

Rannikmäe, M., Moonika, T., & Holbrook, J. (2010). Popularity and relevance of science education literacy: Using a context-based approach. *Science Education International, 21*(2), 116–125.

Rawls, J. (1993). *Political liberalism*. New York: Columbia University Press.

Rennie, L. J., & Williams, G. F. (2002). Science centers and scientific literacy: Promoting a relationship with science. *Science Education, 86*(5), 706–726. doi:10.1002/sce.10030

Reveles, J. M., & Brown, B. A. (2008). Contextual shifting: Teachers emphasizing students' academic identity to promote scientific literacy. *Science Education, 92*(6), 1015–1041. doi:10.1002/sce.20283

Reveles, J. M., Cordova, R., & Kelly, G. J. (2004). Science literacy and academic identity formulation. *Journal of Research in Science Teaching, 41*(10), 1111–1144. doi:10.1002/tea.20041

Rheinlander, K., & Wallace, D. (2011). Calculus, biology and medicine: A case study in quantitative literacy for science students. *The Electronic Journal of the National Numeracy Network, 4*(1), Art. 3. doi:10.5038/1936-4660.4.1.3

Ritchie, S. M., Tomas, L., & Tones, M. (2011). Writing stories to enhance scientific literacy. *International Journal of Science Education, 22*(5), 685–707. doi:10.1080/09500691003728039

Riveros, A., Norris, S.P., Hayward, D.V., & Phillips, L.M. (2012). Dispositions and the quality of learning. In J.R. Kirby & M.J. Lawson (Eds.), *The quality of learning* (pp. 32–50). Cambridge: Cambridge University Press.

Roth, W.-M., & Lee, S. (2002). Scientific literacy as collective praxis. *Public Understanding of Science, 11*(1), 33–56. doi:10.1088/0963-6625/11/1/302

Rughinis, C. (2011). A lucky answer to a fair question: Conceptual, methodological, and moral implications of including items on human evolution in scientific literacy surveys. *Science Communication*. (Published on-line, not yet in-print). doi:10.1177/1075547011408927

Sadler, T. D. (2011). Socio-scientific issues-based education: What we know about science education in the context of SSI. *Contemporary Trends and Issues in Science Education, 39*, 355–369. doi:10.1007/978-94-007-1159-4_20

Sadler, T. D., & Zeidler, D. L. (2009). Scientific literacy, PISA, and socioscientific discourse: Assessment for progressive aims of science education. *Journal of Research in Science Teaching, 46*(8), 909–921. doi:10.1002/tea.20327

Scheffler, I. (1960). *The language of education*. Springfield, IL: Charles C. Thomas Publisher.

Schroeder, M., McKeough, A., Graham, S., Stock, H., & Bisanz, G. (2009). The contribution of trade books to early science literacy: In and out of school. *Research in Science Education, 39*(2), 231–250. doi:10.1007/s11165-008-9082-0

Shwartz, Y., Ben-Zvi, R., & Hofstein, A. (2006). The use of scientific literacy taxonomy for assessing the development of chemical literacy among high-school students. *Chemistry Education Research and Practice, 7*(4), 203–225.

Soobard, R., & Rannikmäe, M. (2011). Assessing student's level of scientific literacy using interdisciplinary scenarios. *Science Education international, 22*(2), 133–144.

Strike, K. (2000). Liberalism, communitarianism, and the space in between. *Journal of Moral Education, 29*(2), 133–231. doi:10.1080/713679340

Sullivan, F. R. (2008). Robotics and science literacy: Thinking skills, science process skills and systems understanding. *Journal of Research in Science Teaching, 45*(3), 373–394. doi:10.1002/tea.20238

Taşar, M.F., & Atasoy, B. (2006, November). *Turkish educational system and the recent reform efforts: The example of the new science and technology curriculum for grade 4–8*. Paper presented at the meeting of Asia Pacific Educational Research Association, Tai Po, New Territories, Hong Kong.

Thomson, S., & De Bortoli, L. (2008). Exploring scientific literacy: How Australia measures up. The PISA 2006 survey of students' scientific, reading and mathematical literacy skills. *OECD Programme for International Student Assessment (PISA Australia)*.

Turner, S. (2008). School science and its controversies; or, whatever happened to scientific literacy? *Public Understanding of Science, 17*, 55–72. doi:10.1177/0963662507075649

Wallace, C. S. (2004). Framing new research in science literacy and language use: Authenticity, multiple discourses, and the "Third Space". *Science Education, 88*(6), 901–914. doi:10.1002/sce.20024

Wang, J., Chen, S., T., R., Chou, C., Lin, S., & Kao, H. (2011). Development of an instrument for assessing elementary school students' written expression in science. *The Asia-Pacific Educational Researcher, 20*(2), 276–290.

Webb, P. (2010). Science education and literacy: Imperatives for the developed and developing world. *Science, 328*(5977), 448–450. doi:10.1126/science.1182596

Yarden, A. (2009). Reading scientific texts: Adapting primary literature for promoting scientific literacy. *Research in Science Education, 39*(3), 307–311. doi:10.1007/s11165009-9124-2

Yore, L. D., Pimm, D., & Tuan, H. (2007). The literacy component of mathematical and scientific literacy. *International Journal of Science and Mathematics Education, 5*(4), 559–589. doi:10.1007/s10763-007-9089-4

Stephen P. Norris is Professor and Canada Research Chair in Scientific Literacy at the University of Alberta, Canada. He has undergraduate degrees in physics and in science education, a master's degree in science education, and a Ph.D. in the philosophy of education. He has published extensively in philosophy of education and science education for over three decades. During the last 20 years he has published extensively on interpreting scientific text. His most recent books are: a co-authored volume with Linda Phillips and John Macnab, *Visualization in mathematics, reading and science education* published by Springer; an edited volume, *Reading for evidence and interpreting visualizations in mathematics and science education* published by Sense; and a forthcoming volume, *Adapted primary literature: The use of authentic scientific texts in secondary schools*, co-authored with AnatYarden and Linda Phillips and to be published by Springer.

(Sadly Stephen Norris died in February 2014.)

Linda M. Phillips is Centennial Professor and Director of the Canadian Centre for Research on Literacy at the University of Alberta. She holds two undergraduate degrees and a master's from Memorial University of Newfoundland and a PhD in cognition and reading from the University of Alberta. She is the recipient of national and international awards for outstanding research, teaching, and service. She has served on several editorial boards including Reading Research Quarterly; and was the senior editor for the Handbook of Language and Literacy Development: A Roadmap from 0 to 60 Months. Linda has published books, chapters and articles in top tier language and literacy venues. Her current research includes: the development of a Test of Early Language and Literacy (TELL) for children ages 3–8 years; the study of children's reasoning when reading in conventional and dynamic assessment contexts; the exploration of scientific literacy (reading when the content is science); the study of emergent and family literacy; and the use of fMRI studies in understanding reading development.

David P. Burns is a Faculty Member in the Department of Educational Studies, Kwantlen Polytechnic University, Surrey, British Columbia, Canada. He began his career teaching high school social studies in Edmonton, Alberta, Canada. He received his M.Ed. and Ph.D. in Educational Policy Studies from the University of Alberta for his research on moral education. His research interests include philosophy of education, pedagogical ethics, moral education, educational law, environmental education, science education, and teacher education.

Chapter 41
Conceptual Change: Analogies Great and Small and the Quest for Coherence

Brian Dunst and Alex Levine

41.1 Introduction

Science enjoys a privileged and unique position in Western culture. One basis for this privilege is the widely held perception that science continually produces progressively more reliable knowledge about the way our world works. The success of any new tool inspires imitators, and so the epistemic success of science has inspired analogies to other processes of epistemic and conceptual growth and change, including those that presumably take place in the developing mind of the human child. Owing to a thread running throughout Thomas S. Kuhn's *Structure of Scientific Revolutions*, psychologists, as well as historians and philosophers of science, have proposed an assortment of theories for modeling conceptual change in both science and childhood over the last half-century. Many of these models assume a robust bidirectional analogy between conceptual change in childhood development and theory change in the history of science. This analogy suggests that the relevant kinds of changes that a child experiences while acquiring and revising her conceptual makeup are sufficiently similar to the relevant kinds of changes that a scientific community undergoes over the course of its historical progression.

Enabled by this analogy, philosophers and historians of science have helped themselves to the work of developmental psychologists, and likewise developmental psychologists have helped themselves to the work of philosophers and historians of science. The fact that, in any given discussion, the analogy is usually drawn in only one direction works to conceal a potentially vicious circularity: if the analogy lends significantly to insights and inferences in one domain (say, in drawing on an

---

This paper has profited greatly from comments, suggestions, and corrections from Anna-Mari Rusanen and three anonymous referees. Any remaining errors are our own.

B. Dunst (✉) • A. Levine
Department of Philosophy, University of South Florida,
4202 E. Fowler Ave., FAO 226, Tampa, USA
e-mail: brian.dunst@gmail.com; alevine@cas.usf.edu

M.R. Matthews (ed.), *International Handbook of Research in History,
Philosophy and Science Teaching*, DOI 10.1007/978-94-007-7654-8_41,
© Springer Science+Business Media Dordrecht 2014

analogy from research on conceptual change in childhood in devising philosophical accounts of historical theory change in science), while the codomain from which it draws its insights and inferences relies upon an inverse analogical mapping (from the philosophical explanation of historical theory change in science to research on conceptual change in childhood), we may worry whether, after all, all this theorizing amounts to more than mere theoretical circularity.

If we are to ground either of these domains in methodologically acceptable terms, we must examine and justify the hidden assumptions that underlie this apparent reciprocal analogy between theory change in science and conceptual change in development. For instance, we can investigate whether the availability of a particular learning strategy in childhood (or science) suggests its availability in science (or childhood).[1] If no such guarantee can be maintained, reliance on such analogies is surely excessive and mistaken. In this chapter, we will show that there is perhaps a more interesting and less problematic relation between two distinct and specific ways that researchers invoke the concept of analogy: *argument by way of analogy* and reference to *analogical reasoning*. These, we argue, are distinct objects of research, but there may be an inference allowing results from the domain of analogical reasoning to inform and adjudicate the legitimacy of the arguments by analogy to which philosophers, historians of science, and developmental psychologists have helped themselves. If such an inference is possible, it means that the circularity that threatens researchers in both camps may not be a *vicious* circularity after all. In this chapter, we develop the following preliminary hypothesis: both the analogical reasoning that takes place during conceptual change in childhood and the successful arguments from analogy deployed by scientists express a quest shared by many if not all epistemic subjects: the search for coherence.

In our first section, we consider some of the more influential treatments of the relationship between conceptual change in childhood and science in the conceptual change literature, drawing attention to the extent to which both historians of science and cognitive developmental psychologists, though articulating their approaches in different technical vocabulary, have focused on conceptual change as involving not merely a change in *concepts*, but in *conceptions*. The following section turns to models of analogical reasoning of the sort that have struck participants in the conceptual change literature as promising ways of accounting for conceptual change. In our final section, we turn to the *motives* of the epistemic subject (whether child or adult scientist) embarking on the project of conceptual change.

---

[1] A fairly straightforward example – one to which we will return later – is whether mechanisms for the adoption of novel conceptions that maximize conceptual coherence and internal consistency in childhood development are necessary or sufficient for the process of scientific theory change. Similarly, we may ask whether the kinds of theoretical "paradigm shifts" that have historically occurred in science constrain the types of conceptual mechanisms that can allow for successful conceptual change in development. Stella Vosniadou and her colleagues have done extensive empirical research on these issues (cf., e.g., Vosniadou 2007; Christou and Vosniadou 2005; Vosniadou et al. 2004).

41.2 Conceptual Change as Change in Conception

As one of us has argued elsewhere (Levine 2000), Kuhn considered his early work on theory change in science to be deeply resonant with developmental psychologist Jean Piaget's genetic epistemology. For his part, Piaget saw his stage theory of genetic epistemology as attempting "to explain knowledge, and in particular scientific knowledge, on the basis of its history, its sociogenesis, and especially the psychological origins of the notions and operations upon which it is based" (Piaget 1970, p. 1). Thus, Piaget too understood his own hypotheses on childhood development as deriving at least partially from an analogy to the ways in which scientific knowledge changes historically. One can then wonder to what extent the analogy to childhood development that Kuhn meant as a support for his wider position on scientific revolutions – and through which he reasoned in formulating this position – is ultimately rooted in Piaget's intuitions about scientific revolutions. The worry is that Kuhn's appeal to the developmental analogy might be nothing but an appeal to intuition, which would weaken his arguments substantially.

Kuhn's arguments may be salvaged if the analogy runs unidirectionally (viz., from childhood development to scientific theory change, but not vice versa). As Kitcher (1988) observes, the conceptual tools forged during childhood development are available for adult scientists in their theoretical (re)conceptualizations during scientific revolutions (as well as normal science); but it is less plausible that highly nuanced and complex scientific theories are available conceptual resources for developing children. Intuitively, this places developmental psychology as *epistemically* prior to the study of historical scientific change. Unfortunately for Kuhn, while many developmental psychologists may agree with this order of epistemic priority, Piaget did not – and Kuhn himself appears to have realized this. In a telling footnote in *Structure of Scientific Revolutions*, Kuhn approvingly cites Piaget's historical sensitivity: "Because they displayed concepts and processes that also emerge directly from the history of science, two sets of Piaget's investigations proved particularly important" (Kuhn 1996, p. vii). Presumably, Kuhn believed that the close relationship that Piaget saw between childhood development and theory change throughout the history of science was evidence in support of his own thesis. The two approaches likely appeared to Kuhn as mutually reinforcing and suggestive of a more broadly *coherent* theoretical picture – which itself would have lent itself to the plausibility of Kuhn's philosophical project.

For our purposes, a paradigm case of an account that addresses the relationship between Kuhnian scientific concept change and (psychological) conceptual change in individual agents may be found in Posner, Strike, Hewson, and Gertzog's 1982 paper, "Accommodation of a Scientific Conception: Toward a Theory of Conceptual Change." In it the authors construct their account to show how individual proto-scientists go about making radical changes in their central or "core" concepts (e.g., going from thinking of physics in terms of Newtonian mechanics to thinking in terms of Einsteinian relativity), but without revising more mundane or peripheral concepts and beliefs (such as the belief that it is in fact not raining outside right now). Thus, their argument points toward the now familiar differences between (mere) concepts and (full) conceptions. Briefly, a *concept* is one node within a conceptual framework, and a *conception* is the relatively modular constellation or web of relations that define a particular knowledge domain. "Force" is a concept within a "Newtonian Mechanical" conception of physics. Posner and colleagues rely heavily on the legitimacy of an analogy drawn *from* Kuhnian scientific change literature *to* the psychological processes involved in scientific learning.

Their use of this analogy is unidirectional. They wish to show that the processes involved in Kuhnian paradigm shifts from normal science, through a scientific revolution, and into a new paradigm *are of the same kind as* those taking place within novice students of science. Initially they draw the analogy between the kinds of process involved both in historical episodes of concept change and individual episodes of scientific learning – both are processes requiring changes to be made to "core" concepts. A "core" concept is one whose character is in some causal way determinant of an entire conception. Change a "core" concept and the result is that the entire conception is qualitatively affected.

Here Posner and colleagues appropriate Imre Lakatos's distinction between the *assimilation* of "recalcitrant" (anomalous) data by making changes to the "protective belt" of auxiliary hypotheses, methodologies, and beliefs and *accommodation* of the "research program" (including making changes to its "hard core") to the recalcitrant data. In the former (assimilation), current theoretical commitments, methodologies, and scientific practices provide the "background" upon which new concepts are to be understood. If the new concepts can be reconciled in a way harmonious with these background commitments, the new concept can become accepted into the general theoretical framework of the research program without any radical revision. In the latter (accommodation), the new concept cannot be reconciled with the background commitments harmoniously, and the new concept either poses a threat to the very identity or "hard core" of the research program or is dismissed as observational error. Posner and collaborators make analogical use of "assimilation" and "accommodation" in their account of scientific learning. Learners "assimilate" by using their current conceptual commitments as a basis or background in assessing new phenomena or concepts. When the new phenomena or concepts resist cohering with background conceptual commitments, they must instead be "accommodated."

Piaget (1953) also prominently features the concepts of assimilation and accommodation in his theory of genetic epistemology. For Piaget, the processes of assimilation and accommodation form a dialectically coupled adaptive system. An attempt at assimilation is an attempt to fit one's occurrent state of affairs to one's occurrent belief system. Piaget likened successful assimilation to successful biological adaptation. When assimilation fails as an adaptive mechanism or strategy, learners must attempt accommodation – changing their occurrent belief structure to fit with the occurrent state of affairs. Importantly, Piaget saw *both* processes of assimilation and accommodation as necessarily altering one's conceptual structure. Thus, any assimilation necessarily requires *some* accommodation, and any accommodation is accompanied by assimilation. This would seem to contrast with the Posner group's usage of "accommodation," unless we understand their emphasis as a matter of degree: the sense in which they use the term accommodation is said to occur only under conditions of radical revision in conceptual structure.

Borrowing from Stephen Toulmin (1972), Posner and colleagues refer to the causal interdependence between one's concepts and conceptions as one's "conceptual ecology." The issue central in their discussion is identifying the conditions and features of conceptual ecologies under which scientific learners come to *accommodate* new core concepts. Toward this end, they outline four primary conditions severally necessary and jointly sufficient for accommodation to occur:

1. There must be dissatisfaction with existing conceptions.
2. A new conception must be intelligible.
3. A new conception must appear initially plausible.
4. A new concept should suggest the possibility of a fruitful research program. (Posner et al. 1982, p. 214)

It should be clear that these conditions are meant to align to similar Kuhnian conditions for adopting a new scientific paradigm. These conditions would look something like the following:

1. There must be dissatisfaction with the existing theoretical scientific framework(s).
2. A new scientific paradigm must be intelligible.
3. A new scientific paradigm must appear initially plausible
4. A new scientific paradigm should suggest the possibility of a fruitful research program.

Each of these four conditions needs to be met in order for a new theoretical paradigm to be taken up by a scientific community. Kuhn famously belabored how difficult it is for a single person to bridge the divide between two paradigms. However, if Posner and colleagues' analogy is to hold, this is precisely what occurs in the process of conceptual accommodation. Drawing directly from Kuhn's arguments, Strike and Posner later argue that – for similar reasons – changes to one's scientific conceptions, or to core concepts, are likewise difficult: "If one assumes that misconceptions are similar to paradigms, these views provide obvious reasons why misconceptions will be resistant to change, even given contrary instruction" (Strike and Posner 1992, p. 153). This is one reason, they say, that explains why established misconceptions are so robust and persistent in the context of science learning.

Posner and colleagues identify other constraints that keep learners from assimilating (and thus push toward accommodation). These constraints are features of the learner's conceptual ecology that are particularly influential in deciding which concepts come across to the learner as the most plausible. Thus, their model of conceptual succession is recursive: the character of the current conceptual ecology casually influences or determines the successor conceptual ecology. In the case of core concept change, this means that there are conceptual ecological factors that constitutively factor into shaping the character of the new (successor) core concept(s).

There are two assumptions worth mentioning about the kind of learners that Posner and colleagues are talking about here: first, they are rational. The decision procedure involved in their account of core concept change depends on the learner's being capable of rationally weighing the evidence and choosing the best alternative on well-reasoned grounds. Thus, the authors assume with Allison Gopnik that science learners are already "little scientists." The second assumption is that conceptual knowledge is essentially representational. Elsewhere (Levine and Schwarz 1993; Brooks 1991), both of these assumptions have been problematized, but it is worth noting that the view Posner's group put forth is fundamentally tied to them.

With this in mind, the remaining conceptual ecology constraints are the character of the anomalies affecting current concepts, analogies, and metaphors used to make novel concepts intelligible; epistemological commitments such as current explanatory ideals and views about the character of knowledge; the metaphysical beliefs and concepts that make sense within the current conceptual ecology, whether knowledge in other fields coheres with new concepts; and the character of concepts in competition for accommodation. Given these myriad conflicting constraints, concept change is difficult. Whenever possible, learners will attempt to assimilate rather than accommodate new concepts. Only when assimilation doesn't work does accommodation become a practical possibility – and only when a novel concept can be seen as intelligible does accommodation become a plausible action.

Just as Kuhn later qualified his claims about incommensurability, Posner and colleagues qualify their claims about the processes involved with accommodation. Specifically, they preserve the possibility of partial accommodation, in which the process of accommodation is not a binary, all-or-none affair. This means that for any given conceptual core, there may be unresolved inconsistencies or incomplete explanations. It follows that accommodation needn't be abrupt. It may take some time to make all the revisions necessary to move from a change in core *concept* to a change in overall *conception*. Given the number of causal variables and interactions in a conceptual ecology, the process of accommodation will also most likely be nonlinear and involve a process of trial and error and revision in order to cohere the core and peripheral concepts of a conception.

All of these qualifications fit nicely with Kuhn's later (1970) amendments to his account of theory change in scientific communities wherein he makes room for external beliefs and values in playing a (nonlinear) causal role in the processes of theory adoption. As is now well known, Kuhn's later views downplayed radical incommensurability, which in turn opened the door for closer analogies between scientific practice and developmental psychology (cf., Hoyningen-Huene 1993; Levine 2000). There is no question that Posner and colleagues' account of core concept change as the causal inter-workings of a learner's conceptual ecology draws a strong analogy from the Kuhnian literature. Strike and Posner later write:

> We have been substantially influenced by those theories of rationality that have been developed by authors such as Kuhn, Toulmin, and Lakatos…the substantive conceptions (of Kuhn's paradigms and Lakatos' research programmes) suggest what are to count as problems and what is to count as relevant evidence. Indeed, they provide the perceptual categories by which the world is perceived…Such accounts of rationality can be easily turned into accounts of rational learning suitable for pedagogical purposes. (Strike and Posner 1992, pp. 151–2)

While Strike's and Posner's optimism is commendable, and their utilization of the general Kuhnian framework is interesting and fruitful as a research program, what remains to be seen is whether such a strong analogy is warranted. Are the kinds of learning experiences science learners undergo sufficiently like historical changes in scientific communities? In order to assess this question, we would do well to understand exactly what features are meant to be analogous and which aren't. And to do this, we must first get clear on the kind of relationship an analogy conveys.

41.3 Analogy

For the purposes of this chapter, we will be utilizing a variant of the Structure-Mapping approach to analogical reasoning originally developed by Dedre Gentner (1983). This approach employs a basic distinction between what it calls "source" and "target" domains. In discussing conceptual change in childhood or science, the domains are typically supposed to be conceptions or theories, together with modes of conceptual or theoretical change.[2] For ease of exposition, we will be employing the term "theory" to represent both scientific theories, as well as conceptions. The presumed analogy between theories operates as a functional mapping from the source domain to the target domain. Typically, the source domain is taken to constitute a more reliable object of knowledge than the target domain. These analogical mappings can accommodate different degrees of structural stringency. The strongest mapping, and upper bound, for stringency in an analogy is identity between source and target domains. If the domains are *identical*, the mapping is not appropriately called an analogy (rather, it is an endomorphism – a mapping from a domain back to itself). Short of identity, the next most stringent class of mappings is *isomorphisms*, mappings that maintain source domain relations among elements of the target domain.

Less stringent are *homomorphic mappings* for which the mapping function defines a transformation wherein some, but not all, of the relational structure is maintained. There are at least two interesting kinds of homomorphic structural mappings: those whose domains *differ in kind* and those whose domains *differ in degree*. In a homomorphic mapping between different kinds, the source and target domains

---

[2] This presupposition is typical of, though not exclusive to, "theory-theory" approaches to the developmental attainment of conceptual tools for dealing with human behavior, where the theory most children acquire is variously called "folk psychology" or "theory of mind." In such contexts, the debt owed to the history and philosophy of science is often explicitly acknowledged, as in the title of Piaget student Annette Karmiloff-Smith's (1988) essay, "The Child is a Theoretician, not an Inductivist." In another example, Allison Gopnik observed that the tendency of developmental psychologists to refer to a child's conceptual knowledge base as a "theory" is a reflection of the extent to which developmental psychology must be previously informed by the history of science. She writes (1996): "cognitive and developmental psychologists have *invoked the analogy of science itself*. They talk about our everyday conceptions of the world as implicit and intuitive theories, and about changes in those conceptions as theory changes" (Gopnik 1996, p. 485, emphasis added).

are taken to be independent knowledge stores. There are very few if any recognizable entities common to both source and target domains. Homomorphic mappings between domains that differ in degree are taken to draw from one conceptual knowledge store but in ways that preclude identity (allowing us to consider the source and target domains as effectively distinct).

A good example of an analogy whose source and target domains differ only in degree is Darwin's argument by analogy in *The Origin of Species*. There are two analogies that comprise the backbone of Darwin's argument. First is his analogy for the mechanism of evolution. He draws from the source domain of "artificial" selection (animal husbandry) whose effects were well known and documented to the target domain of "natural" selection which was to work via the very same functional mechanisms (selective retention of favorable characteristics) as artificial selection. The identity of the underlying mechanism of change in both domains (selection) suggests the difference between them is merely one of degree. What differs is the amount of time required for the phenotypic effects of either process to noticeably accrue and the physical means by which selective retention operates (human breeders selecting breeding pairs in artificial selection and random mutation and the "struggle for existence" in natural selection). Darwin applies a second analogy between the conventional taxonomic units *species* and *variety*. He argues by way of analogy that different varieties differ from each other just as different species differ from each other – that is, the differences (all around) are *of a* kind (not *in* kind). The only *difference* between variety-difference and species-difference is in the degree of phenotypic divergence. As Darwin puts it, "a well-marked variety may be justly called an incipient species" (Darwin 1859, p. 52).

Even less stringent than homomorphic mappings are *congruence relations*, which do not provide well-defined transformative functions directly between source and target domains, but for which an intermediate domain can be constructed. A pair of homomorphisms can then be given defining transformations from source to intermediate and from intermediate to target domains. Important in congruence relations is that transitivity strictly cannot be maintained through both transformations (viz., from source, through the intermediate, to the target domain). We will call the intermediate domain the *transfer domain*. The limiting case of analogical mapping is *disanalogy*, wherein there is no readily apparent congruence relation.

For present purposes, this taxonomy of analogies will serve two distinct but related functions. First, it will help us to get clear on the nature of the often invoked if seldom scrutinized analogy between conceptual change in science and conceptual change in childhood. Different sorts of analogy support different sorts of inference. If we suppose that the two domains of conceptual change are isomorphic, the analogy between cognitive development and the history of science ought in principle to support inferences in both directions with the risks outlined above. If, on the other hand, they are connected only by congruence relations, such risks are mitigated – but then arguments that invoke such an analogy carry significantly less weight than often assumed.

But our discussion of analogy serves a second purpose. Recognition of the prevalence and significance in the conceptual change literature, of an analogy between conceptual change in childhood and conceptual change in science, led us to consider the role of *argument* by analogy in general. Confusing matters is that within that same literature, processes such as *argument from analogy* and *analogical reasoning* have also been touted as playing central roles in the actual cognitive processes of conceptual change. Let us assume, for the moment, that whatever the precise nature analogy between childhood development and scientific change, it is close enough to allow us to treat argument from analogy *as a species of* analogical reasoning (Gentner 1983). In that case, the precise entailments of this analogy may well be illuminated by an account of the role of analogical reasoning in conceptual change. That is, by getting clearer on the ways in which analogical reasoning functions, we will be in a position to better understand whether arguments by analogy relating conceptual change in science and childhood development are as problematic as they seem.

Analogical reasoning is often used in mapping congruently related conceptual domains and, for the purposes of analysis, often requires carefully constructing a transfer domain. But how are transfer domains constructed? Or, to address our specific concerns more directly, how can a novel theory be discovered, developed, or changed? Many commentators (notably Susan Carey) think it is reasonable to suppose that this task necessitates some sort of "bootstrapping" process imaginatively drawing upon available cognitive resources. As we have seen, Posner and colleagues require that successor core concepts causally depend on the current state of one's conceptual ecology. There are many accounts of the possibilities and constraints constitutive of such available resources, though the proper exposition of even the most plausible of these theories is far too ambitious for the scope of this discussion (see instead, e.g., Carey 2009 and Strike and Posner 1992). Additionally, we may wonder whether the necessary bootstrapping process is systematic enough to give rise to what might plausibly be called a theory. It would be enough, for starters, to articulate a theory of analogical reasoning consistent with the constraints operating in any one of the domains of conceptual change, if not all of them at once.

As an example of the construction of a transfer domain, we consider an idealized analogy between the solar system and the atom, as Ernest Rutherford would come to understand it. For our source domain, we select the solar system as described by Newtonian mechanics. Presumably, we are constructing an analogy between the solar system (source domain) and the atom (target domain) in order to explain or understand some aspect of the atom that has previously escaped our grasp (viz., the experimental evidence that it is composed mostly of empty space). In such cases, an analogical transfer domain can be used as a cognitive resource to model the relevant aspects of the target domain. It should be noted that the reason we use the solar system–atom analogy as an example of a congruence mapping (rather than the stronger homomorphic mapping) is because much of the relational structure is not maintained between the two models. Electrons possess substantially different properties and behave in drastically different ways from planets in their orbits about the Sun. Since the analogical relation is more metaphorical than literal, we are put in a position of needing to "spell out" more precisely how the analogy is to work. To do this, we draw an intermediate analogy to some other domain that helps clarify the analogical relation in both domains. In general, the need for a transfer mapping is indicative of a congruence analogy for the precise reason that the difficulty or impossibility of an adequate direct mapping precludes a homomorphic approach.

Through a process of abstraction, we can compile a set of comparable attributes between the source and target domains. This set will comprise the transfer domain. First, let us state the source and target domains:

(S): The solar system contains less massive planets that centripetally orbit a more massive Sun, with most of the volume of the solar system occupied by empty space.

(A): The atom is comprised of constituents similar in various ways to positively and negatively charged particles (α and β particles) recently observed. The whole is largely transparent to energetic α particles (see Geiger and Marsden 1909; Rutherford 1911) and must thus be mostly empty.

We have purposefully worded these two descriptions so as to easily lend themselves to common abstraction. In vivo, the process of honing the language used for comparison would likely be a more complex process involving multiple steps, each utilizing analogical reasoning. One can imagine a child (or scientific community) working through a series of partial, transient, provisional, and defeasible analogical reasoning processes in order to formulate the *source* domain of a more general transfer mapping. Finding an effective transfer domain is likely a hard-won victory. Additionally, there may be many alternative abstractive partitionings of specific contexts or problem situations for each of (S) and (A). Our task is to create one, (T), for which we can simultaneously define homomorphic mapping functions from (S) to (T) and from (T) to (A). There needn't be one and only one mapping of this kind. It stands to reason that with a greater number of potential mappings comes increased likelihood of the successful construction of a transfer domain – as an increase in potential mappings corresponds to an increase in available cognitive resources (such as concepts, conceptions, and conceptual relations, among others). For the sake of this example, we now provide one such transfer domain:

(T): The model contains entities that orbit a central object, bound by an attractive force exerted from the central object on the orbiting entities.

We are now able to construct a pair of homomorphic mappings (analogies): one abstraction from (S) to (T) and one instantiation from (T) to (A). So long as we succeed at constructing such mappings, we are guaranteed to meet certain relevancy criteria – in fact, the transfer model itself functions as a minimal relevancy constraint for the analogical task.

Now that each of the source and target model can be expressed in terms of the abstract transfer model, we have grounds for inductive hypothesis testing to help determine whether the relevancy constraint established by the transfer domain survives further observation. By discovering relevant similarities and dissimilarities through this process, we have assembled a procedure by which we can come to understand the target domain better. The point is just that such analogical comparison establishes a mechanism for generating testable hypotheses. More generally, such a process of mapping from a source conception to a transfer conception and then again from the transfer conception to a target conception allows for a learner to make the requisite transition from one conceptual ecology to a new one – or to shift from a misconception to a correct conception. A larger worry remains with the question as to whether analogical transfer mappings are *actually employed* by children or scientific communities in vivo. Nancy Nersessian has argued that in at least some contexts of scientific modeling, this does occur (Nersessian 2008, pp. 206–207). However, it remains less obvious that childhood conceptual development employs such a mechanism.

Gentner and Markman have devoted significant attention to the structures and relations that impose psychological constraints on analogical reasoning. In particular, they understand analogy as dependent on our psychological abilities to "align" or to bring into attunement the basic structures (or gestalts) of the domains involved in the attempted analogy (Gentner and Markman 1997). They identify three specific psychological constraints on analogical reasoning:

1. *Structural consistency*: It must observe parallel connectivity and one-to-one correspondence. Parallel connectivity requires that matching relations must have matching arguments. One-to-one correspondence limits any element in one representation to at most one matching element in the other representation.
2. *Relational focus*: Analogies must involve common relations but need not involve common object descriptions.
3. *Systematicity*: Analogies tend to match connected systems of relations. The systematicity principle captures a tacit preference for coherence and causal predictive power in analogical processing. (Gentner and Markman 1997)

Important for Gentner and Markman are not only the *similarities* in structures but the *differences* as well – both similarities and differences are alignable. They understand analogical reasoning to involve an iterable process of connecting one domain to another. On the taxonomy we have developed here, what is interesting in Gentner's and Markman's "structural alignment framework" is that it implies that through the processes of analogical reasoning, we can *massively* and *in parallel* employ multiple *transfer mappings* in an iterable fashion – one transfer mapping after another, connected in serial. Presumably, the only stop to such a process occurs when some psychological state of relative equilibrium is reached. The question that remains is what the conditions for such equilibrium could be (i.e., *in what* is there equilibrium? External coherence? Internal consistency? Practical applicability?)

Susan Carey, for her part, writes extensively on the kinds of mechanisms required to satisfy what she believes are the constraints on childhood learning. For her, an innate conceptual core in conjunction with a "Quinean bootstrapping" process that incorporates available external resources is necessary for the possibility of coming to understand the kinds of complex scientific theories that members of our scientific communities clearly require. Carey is concerned with the conditions and constraints operative in cognitive development that allow (or in some cases prohibit) an agent to come to understand, for example, continuous rather than discrete conceptions of number. Such "theoretical" understandings underwrite the more complex scientific theories that utilize and employ them. So for Carey, theory development and construction always recursively operate on current cognitive resources, which include an agent's prior theoretical attainments, a mode of communication capable of translating from a standing theory (source) to a novel one (target), as well as physically and socially available resources (e.g., external tools for learning, teachers). In many cases, the "novel" theory is not truly novel in that it is included in some of the available resources; for example, teachers already possess a solid understanding of the target theoretical domain (cf., Carey 2009, pp. 413–445).

Crucial to Carey's and others' formulations of theory change is that incommensurability is always local. Largely in response to criticisms of *structure*, the later Kuhn came to think of theory incommensurability as analogous to localized problems of translatability between languages, in contradistinction to his early claims of global incommensurability (roughly analogous to Quinean radical indeterminacy of translation). Carey's "Quinean bootstrapping" process precludes global incommensurability and untranslatability in that more basic concepts, theories, language use, and practices remain relatively unaffected by specific theoretical alterations. For example, in the midst of a conceptual theory change from discrete to continuous number theories, a child's beliefs and theoretical understanding about colors may be unaffected at first. It is possible that a theoretical conceptual change causes a cognitive agent's understanding to undergo global change – but not all at once. Eventually, the child may come to understand difference in hue as a continuous difference, but perhaps not, and perhaps not at first. Carey is at pains to emphasize that the process of Quinean bootstrapping (and the associated conceptual change it enacts) is difficult, often incomplete, and takes a lot of time.

An alternative approach toward the dynamics and constraints involved in conceptual change is the "Multiple-Interaction" approach discussed by Nira Granott (1993). She suggests that highly important to factors formative in a learner's conceptual development are the kinds of social contexts and interactions fostered and engaged in by both novices and experts. Instead of referring to individual and isolated psychological processes, Granott proposes that analysis should reflect the *social* etiology of cognitive processes. Her model differentiates (as a matter of degree) first between collaborative and disruptive interactions. Second, she tracks the degrees and types of expertise involved in interactions and the degrees of involvement of each participant. Some of the categories involved in her multidimensional analysis include *mutual collaboration* between equal peers; *symmetric counterpoint* relationships between interactants who take varying approaches to solving the same problem; *parallel activity*, in which individuals attempt to solve a problem in a relatively isolated (but coordinated) way; *asymmetric collaborations* between novices and experts; and *asymmetric counterpoint* relationships between novices and experts. Granott acknowledges that such relationships may also turn disruptive and that there are myriad other social factors that affect conceptual development (such as expertise in irrelevant domains, social roles, gender, race, personal histories of previous interactions, dynamic personality traits). But the general point is that perhaps Quinean bootstrapping is not the only factor required in conceptual development or at least that it shouldn't be focused on to the exclusion of other social factors.

41.4 Coherence

Our discussion thus far has pointed to the significance of conceptual change not as a change in concepts but as a change in *conceptions*. Such change, we have argued, involves the construction of mappings across conceptual domains. We have yet to consider the epistemic subject's *motives* for constructing such mappings. With this third piece in place, we will finally be in a position to articulate the hypothesis we advertised at the beginning of this chapter concerning the relationship between conceptual change in childhood and science: both express the epistemic subject's quest for coherence.

In a well-known discussion, Paul Thagard has argued that in conceptual change, cognitive agents are often motivated by a need to cohere all theoretical beliefs. He defines, and for the purposes of this chapter, we follow him in defining, the notion of conceptual coherence as follows:

1. Conceptual coherence is a symmetric relation between the pairs of concepts.
2. A concept coheres with another concept if they are positively associated, i.e., if there are objects to which they both apply.
3. The applicability of a concept to an object may be given perceptually or by some other reliable source.
4. A concept incoheres with another concept, if they are negatively associated, i.e., if an object falling under one concept tends not to fall under the other concept.
5. The applicability of a concept to an object depends on the applicability of other concepts (Thagard 1992; Thagard et al. 2002; Thagard and Verbeurgt 1998).

Reconsidering the example of the transformation of color conceptions discussed previously, Thagard's notion of conceptual coherence (and incoherence) suggests that the developing agent's naïve theory of color was incommensurable (or "incoheres") with a new understanding of continuity in number as well as with phenomenal and empirical evidence (perhaps from recently seeing a rainbow and its continuous blending of hues). Kuhnian historians of science, as well as more recent contributors to cognitive developmental psychology, might see such motives at work in participants in the "crises" that accompany scientific revolutions. To Thagard, what makes a crisis is that incoherence appears as problematic for us both as individual subjects and as members of scientific communities – conceptual incoherence is something that begs for remedy. In this, way we may finally see an anchor for an appropriate transfer domain in a useful analogy between concept change in childhood development and theory change in scientific communities: in both cases, the primary motivation is to cohere our understanding about the world. We might understand this drive toward coherence as underwriting scientific methodology as well as childhood curiosity or even as the primary underlying thread in analogical reasoning – especially in congruence mappings necessitating the utilization of a transfer domain. The institutionalization of scientific practice is nothing over and above the value that each of its communities' members places on coherence. This valuation of coherence by each member of a scientific community is no different *in kind* from the importance attached to coherence by a curious child. If there is a difference between these two, it is of *degree* or in the mode of implementation; and the analogy between them remains strong. If there is no difference, the analogy collapses into an identity – the claim that children are engaged in precisely the same coherence-resolving activities as adult members of scientific communities.

Considered from the vantage point of research in conceptual change, learning scientific concepts is often understood as replacing one's state of confusion, misconception, or overly simplistic or "common sense" understanding of a knowledge domain with more complex or theoretically sophisticated newer models. Strike and Posner (1992), for example, regard the replacement process as necessarily involving cognitive dissonance between actively conflicting conceptions. Generally, such conceptual change is understood as the comprehensive replacement of entire conceptions within a particular knowledge domain, rather than the piecemeal replacement of individual concepts within an overall conceptual or theoretical structure whose structure is preserved (cf., e.g., Vosniadou and Brewer 1992; Carey 1985; Chi 1992; diSessa 1988; Posner et al. 1982). Such a conceptual structure is thought to undergo a process of holistic reorganization until a stable or consistent set of interrelated and more scientific conceptions is attained. Again, Strike and Posner (1992) see this process as a decision (of sorts) between competing conceptions – one of which provides a more intelligible, coherent, plausible, and fecund outlook for the knowledge domain than the alternatives.

More recently, there has been substantial debate about the extent to which conceptual change in childhood succeeds at fulfilling the conditions of intelligibility, coherence, plausibility, and fecundity. As recently canvassed by Rusanen and colleagues (2008), "There is [now] a large body of experimentally based literature where it has been argued that the difference between the consistency or coherence of the belief systems of novices and experts is one of degree, not of kind" (Rusanen et al. 2008, p. 65; cf., e.g., Vosniadou 2007; Christou and Vosniadou 2005; Vosniadou et al. 2004). Some researchers, for instance, have proposed that novice "explanatory frameworks" are themselves already internally coherent, consistent, and interrelated sets of beliefs in a sense evocative of Kuhnian "Paradigms" (Samarapungavan and Wiers 1997). Others, such as Chi and her colleagues, have argued that novice belief systems exhibit "ontological" conceptual coherence (Chi 1992; Chi and Slotta 1993; Slotta and Joram 1995). Alternatively, there are also those who disagree with these characterizations of novices' belief systems. Andrea diSessa (1993) and Smith et al. (1994), for instance, describe novice learning as relying not on coherent, systematic theories but rather as more or less unorganized, context-sensitive elements belief-like primitives that are systemically inconsistent and thus do not cohere. In diSessa's "knowledge-in-pieces" account, naïve physical knowledge is organized into phenomenological primitives or "p-prims" – a novice's simple explanations abstracted from experiences and uncritically accepted (diSessa 1983).[3]

---

[3] We are grateful to Anna-Mari Rusanen for pointing us toward many of the works cited in this paragraph.

There is much to be said about whether and to what extent the value of coherence is socially inculcated (perhaps as a resource for Quinean bootstrapping of concept change) during childhood development. But our discussion has perhaps more clearly shown that as a matter of both causal influence and epistemic priority, the relation between childhood development and historical theory change in science is something of a chicken-and-egg problem. One cannot be resolved without necessary reference to the other. If there is something, then, that underwrites the analogy between the two, it is that both developing children and scientific communities share the (perhaps often implicit) underlying valuing of coherence, which acts to regulate the quest for knowledge. So, while the particular material and conceptual resources available to developing children on the one hand (e.g., social and cultural educational tools, practices, and resources) and adult members of a scientific community (e.g., the institutions and the tools and practices deployed scientific research) on the other are clearly distinct and different from each other, the underlying coherence-seeking processes by which each domain operates contribute to genetically informing and reinforcing the same process in the other. Thus, this relationship between Kuhnian theory change in the history of science and developmental conceptual changes in children should perhaps not be seen or understood as an *argument by analogy* as much as a complex causal interrelationship in which both domains materially depend on *analogical reasoning*. Along with hosts of other cognitive and methodological resources, these analogical reasoning methods causally factor into and influence the relations that genetically inform and reinforce particular epistemic developments in both developing children and adult scientists.

References

Brooks, R.: 1991, 'Intelligence without Representation', *Artificial Intelligence* 47: 139–157.
Carey, S. (1985). Conceptual change in childhood. Cambridge, Mass: MIT Press.
Carey, S.: 2009, *The Origin of Concepts*, Oxford University Press, Oxford.
Chi, M. T. H. (1992). Conceptual change within and across ontological categories: Examples from learning and discovery in science. In R. Giere (Ed.), Cognitive Models of Science: Minnesota Studies in the Philosophy of Science, (pp. 129–186). University of Minnesota Press: Minneapolis, MN.
Chi, M. & Slotta, J.: 1993. The Ontological Coherence of Intuitive Physics. *Cognition & Instruction* 10 (2 & 3): 249–260.
Christou, K. P., & Vosniadou, S.: 2005. 'How Students Interpret Literal Symbols in Algebra: A Conceptual Change Approach'. B. G. Bara, L. Barsalou, & M. Bucciarelli (Eds.), *Proceedings of the XXVII Annual Conference of the Cognitive Science Society*, Italy.: 453–458.
Darwin, C.: 1859, *On the Origin of Species*, John Murray, London.
diSessa A. A. (1983) Phenomenology and the Evolution of Intuition in Gentner, D., & Stevens, A. L. (1983). Mental models. Hillsdale, N.J: Erlbaum.
diSessa, A. A. (1988). Knowledge In Pieces. In G. Forman & P. Pufall (Eds.), Constructivism in the Computer Age (pp. 49–70). Hillsdale, NJ: Erlbaum.
diSessa, A. A. (1993). Toward an Epistemology of Physics. Cognition and Instruction, 10, (2 & 3), 105–225.
Geiger, H. and Marsden, E.: 1909, 'On a Diffuse Reflection of the α-Particles,' *Proceedings of the Royal Society, Series A*, 82: 495–500.

Gentner, D.: 1983 'Structure-mapping: A theoretical framework for analogy', *Cognitive Science*. 7(2): 155–170.

Gentner, D, Markman, A.B.: 1997. 'Structure Mapping in Analogy and Similarity', *American Psychologist* 52(1): 45–56.

Gopnik, A.: 1996, 'The Scientist as Child,' *Philosophy of Science* 63(4).

Granott, N.: 1993, 'Patterns of Interaction in the Co-Construction of Knowledge: Separate Minds, Joint Effort, and Weird Creatures,' in Wozniak, R.H., & Fischer, K.W. (Ed.s) *Development in Context: Acting and Thinking in Specific Environments*. Hillsdale, NJ. Earlbaum: 183–207.

Hoyningen-Huene, P.: 1993, *Reconstructing Scientific Revolutions*, University of Chicago Press, Chicago

Karmiloff-Smith, A.: 1988, 'The Child is a Theoretician, Not and Inductivist," *Mind and Language* 3(3).

Kitcher, P.: 1988, 'The Child as the Parent of the Scientist', *Mind and Language* 3(3).

Kuhn, T. S. (1970). The structure of scientific revolutions. Chicago: University of Chicago Press.

Kuhn, T.S.: 1996., *The Structure of Scientific Revolutions*, 3rd edn., University of Chicago Press, Chicago.

Levine, A.: 2000, 'Which Way is Up? Thomas S. Kuhn's Analogy to Conceptual Development in Childhood,' *Science & Education* 9: 107–122

Levine, A. and Schwarz, G.: 1993, 'Three Inferential Temptations', *Behavioral and Brain Sciences* 16 (1).

Nersessian, N.: 2008, *Creating Scientific Concepts*, MIT Press, Cambridge, MA.

Piaget, J.: 1953. *The origin of intelligence in the child*. New Fetter Lane, New York: Routledge & Kegan Paul.

Piaget, J. (1970). Genetic epistemology. New York: Columbia University Press.

Posner, G. J., Strike, K. A., Hewson, P. W., & Gertzog, W. A. (April 01, 1982). Accommodation of a scientific conception: Toward a theory of conceptual change. Science Education, 66, 2, 211–227

Rusanen, A.-M., Lappi, O., Honkela, T., & Nederström, M.: 2008, 'Conceptual Coherence in Philosophy Education—Visualizing Initial Conceptions of Philosophy Students with Self-Organizing Maps,' in Love, B.C. et al. (Eds.), *Proceedings of the 30th Annual Conference of the Cognitive Science Society*, Cognitive Science Society, Austin, TX.

Rutherford, E.: 1911, 'The Scattering of α and β Particles by Matter and the Structure of the Atom,' *Philosophical Magazine* 21: 669–688.

Samarapungavan, A., & Weirs, R. W. (1997). Children's thoughts on the origin of species: A study of explanatory coherence. Cognitive Science, 21, 147–177

Slotta, J. D., & Joram, E. (January 01, 1995). Assessing Students' Misclassifications of Physics Concepts: An Ontological Basis for Conceptual Change. Cognition and Instruction, 13, 3, 373–400.

Smith, J. P., diSessa, A. A., & Roschelle, J.: 1994. Misconceptions reconceived: A constructivist analysis of knowledge in transition. *Journal of the Learning Sciences*, 3(2): 115–163.

Strike, K.A., and Posner, G.J.: 1992, 'A Revisionist Theory of Conceptual Change,' in R. Duschi and R. Hamilton (Eds.), *Philosophy of Science, Cognitive Psychology, and Educational Theory and Practice*, SUNY Press, Albany

Thagard, P.: 1992, *Conceptual Revolutions*, Princeton University Press, Princeton, NJ.

Thagard, P., Eliasmith, C., Rusnock, P., & Shelley, C.: 2002. 'Knowledge and Coherence'. R. Elio (Ed.), *Common Sense, Reasoning and Rationality*, New York: Oxford University Press: 104–131.

Thagard, P., & Verbeurgt, K.: 1998. 'Coherence as Constraint Satisfaction'. *Cognitive Science*, 22 (1): 1–24.

Toulmin, S. (1972). Human understanding. Princeton, N.J: Princeton University Press.

Vosniadou, S., & Brewer, W. F. (1992). Mental models of the earth: A study of conceptual change in childhood. Cognitive Psychology, 24, 535–585.

Vosniadou, S.:2007. 'The cognitive-situative divide and the problem of conceptual change', *Educational Psychologist*, 42(1): 55–66.

Vosniadou, S., Skopeliti, I. & Ikospentaki K.: 2004. 'Modes of Knowing and Ways of Reasoning in Elementary Astronomy.' *Cognitive Development*, 19: 203–222.

Brian Dunst is a Visiting Assistant Professor of Philosophy at Jefferson College in Hillsboro, Missouri. His research interests include the interactive relationships between epistemic agents and their environments. He has recently received his doctorate in philosophy from the University of South Florida.

Alex Levine is Professor of Philosophy at the University of South Florida, Tampa. He earned his M.A. and Ph.D. in philosophy at the University of California, San Diego, and his B.A. in philosophy and mathematics at Reed College. His scholarship has contributed to the history and philosophy of science as well as to the philosophy of mind and includes several papers on the connections between the two fields pertaining to accounts of conceptual change. He has recently published two coauthored books on the reception of Darwinisim in Latin America, *From Man to Ape* (Chicago, 2010) and *Darwinistas* (Brill, 2012). He has also been active as a scholarly translator.

Chapter 42
Inquiry Teaching and Learning: Philosophical Considerations

Gregory J. Kelly

Indeed, the very word 'cognition' acquires meaning only in connection with a thought collective.

Ludwik Fleck 1935

42.1 Inquiry in Science Education Reform

Debates regarding science education go through various stages of reform, perceived change, and more reform (DeBoer 1991). These changes have centered on the extent to which students' interests, autonomy, and knowledge are balanced against the cultural knowledge of the legitimizing institutions. Dewey (1938a), Schwab (1960), Rutherford (1964), and more recently the (USA) National Research Council [NRC] (1996, 2011) have, in various ways, called for engaging students in the scientific practices of professional scientists. These calls for reform conceptualize inquiry differently, and each can be viewed as making a set of assumptions about knowledge, science, students, and learning – thus suggesting the need for examining epistemological issues in science teaching and learning. In this chapter I consider some of the opportunities afforded by an inquiry-oriented science education but also the constraints to successful implementation of inquiry in schooling.

Inquiry in science entails conducting an investigation into the natural or designed world, or even into the applications of scientific knowledge to societal issues. Such investigations typically concern a domain for which at least some of the participating inquirers do not know the results prior to the investigation. Dewey (1929, 1938a) characterized inquiry as dialectical processes emerging from problematic situations

---

G.J. Kelly (✉)
College of Education, Pennsylvania State University, University Park, PA, USA
e-mail: gkelly@psu.edu

M.R. Matthews (ed.), *International Handbook of Research in History, Philosophy and Science Teaching*, DOI 10.1007/978-94-007-7654-8_42,
© Springer Science+Business Media Dordrecht 2014

aimed at reaching some resolution.[1] Inquiry has been characterized as engaging learners in scientifically oriented questions, formulating and evaluating evidence and explanations, and communicating results (National Research Council 1996). As such, inquiry is derived from views of knowledge, is underwritten by interpretations of knowledge, and instantiates perspectives on knowledge. Furthermore, the referent for what counts as inquiry activity need not be limited to the work of professional scientists, as other members of society can be viewed as engaging in scientific practices. Thus, inquiry science poses epistemological questions, and with a focus on science education, these questions can be addressed from a philosophy of science point of view.

Interesting questions arise as to whether inquiry science teaching is directed at learning knowledge and practices of science or at aspects of the nature of science or both. We can speak of learning science through inquiry, where inquiry is the means to learn knowledge and practice. Or we can view the pedagogy as inquiry about science where the intent is to communicate lessons about the nature of science. Often these perspectives on inquiry purposefully brought together, so that learning knowledge and practices through inquiry serves to inform students about science by engaging in the practices constituting scientific activity. I will refer to the dual purpose approach as teaching science as inquiry. As each of these views of inquiry presupposes views of scientific knowledge and thus manifests an epistemological orientation, we would expect to find implications of the philosophy of science for teaching science in this manner. Nevertheless, the relationship of inquiry teaching and philosophy of science is not straightforward.

42.2 Educational Challenges of Teaching Science as Inquiry

There are a number of important challenges to teaching science as inquiry. First, through many years of research and across different learning theories, it is clear that students need concepts to learn concepts. Students learn concepts in bunches, and these cannot be typically investigated one at time through (even careful) classroom-based practice activities or empirical investigations. Educators should not assume that students are able to induce sophisticated scientific concepts from empirical phenomena. While few educational programs explicitly assert that students construct knowledge in the absence of more knowing others, a number of perspectives suffer from this assumption, often under various banners such as hands-on learning, discovery, or radical constructivism (Kelly 1997). As some knowledge is required to learn, then inquiry approaches that situate the student at the center of investigation need to recognize that only with sufficient, relevant background knowledge can answerable questions be posed by students. Thus, inquiry approaches to science

---

[1] Dewey's (1938a) definition is as follows: "Inquiry is the controlled or directed transformation of an indeterminate situation into one that is so determinate in its constituent distinctions and relations as to convert the elements of the original situation into a unified whole" (pp. 104–105).

learning need to consider the importance of learning through engaging in activities and discourse of science with more knowing others.

A second challenge for inquiry instruction is that learning science entails more than learning the final-form knowledge of scientific communities (Schwab 1960; Duschl 1990). While propositional knowledge (*knowing that*) is important, knowing how to engage in scientific practices and how to make epistemic judgments ought not be neglected. Therefore, science learning should include conceptual, epistemic, and social goals (Duschl 2008; Kelly 2008). While much of inquiry has focused on students' engagement in practical or laboratory activities, pedagogies focused on socioscientific issues and science in social contexts pose important opportunities to learn through investigations in unknown domains (Sadler and Fowler 2006). Inquiry can arguably include evaluation of expertise, certainty, and reliability of scientific claims of others.

A third challenge to learning science as inquiry concerns the nature of the intended propositional or procedural knowledge (*knowing how*) in the curriculum. Science topics and community practices may be more or less appropriate for an inquiry approach. Some knowledge and practices may be attainable through student-centered approaches, while others require the direction of more knowing others. Clearly, at least some scientific practices can be learned only through intensive effort, which may require extensive participation in a community of learners. Other topics might be suited for other forms of instruction. Furthermore, methods of assessment, either formative or summative, need to be carefully chosen to match the learning goals appropriate to the knowledge sought.

Fourth, learning the conceptual knowledge, epistemic criteria, and social practices over time in science domains may require coordination of scope vertically (across grades over time) and horizontally (across subject matter areas at a given grade) across the curriculum. While academics find ways to separate disciplines, and there may be interesting epistemological distinctions, students experience schooling as a whole. Science may not be separate from views and knowledge of history, mathematics, reading, writing, and so forth. Thus, the challenge for teaching science as inquiry includes understanding how such approaches can be supported or undermined by other curricular decisions and pedagogies, both from within and from outside science programs.

Despite these challenges, inquiry teaching and learning have been advocated in different forms many times across generations (most recently, see NRC 2011). The potential for learning knowledge and practices of disciplines through engagement in purposeful activity has been recognized both as a means to learn science and as a way to develop student interest. The linguistic turn in philosophy and the continual rediscovery of the importance of learning through participation in discourse practices of epistemic communities have led educators to examine ways that inquiry can be enacted in various settings. This potential of engaging in discourse practices as inquiry has not always been realized, and there is still considerable debate about the nature of inquiry and its overall merits (Blanchard et al. 2010; Kirschner et al. 2006; Kuhn 2007; Minner et al. 2010). Much of the debate fails to recognize the relationship and disagreement among the learning goals, limited measures of assessment, and the purposes of education – that is, rhetorically, the interlocutors argue past each other. Much of this debate regarding differences in traditional and experiential education was identified in Dewey's (1938b) *Education and Experience*. Has the field advanced since? How can philosophy of science help? To address these issues, I consider some challenges for using philosophy of science in science education.

42.3 Challenges for Using Philosophy of Science to Inform Inquiry Science Teaching

Just as inquiry poses challenges because of the realities of teaching and learning science, drawing from the philosophy of science to inform science education poses challenges because of the nature of philosophy. Educators have called for developing philosophically informed science curricula (Hodson 2009). While this is a welcome perspective, in this section I examine the assumptions of the application of philosophy of science to science education and note that some of the difficulty lies not with educators' misunderstandings about philosophy but rather with the nature of philosophy as a discipline. I identify four dimensions of this difficulty.

First, the philosophy of science treats a number of technical issues that may not directly inform educational practices. Throughout the history of the philosophy of science, issues such as inference, perception, and abductive reasoning form the basis for a number of technical arguments conducted by specialists. These arguments are important for the development of the field of philosophy of science and may advance understanding about the nature of science, but do not necessarily lend themselves readily to educational applications. For example, one debate concerns arguments for an instrumental versus realist view of scientific theories (van Fraassen 1980; Boyd 1991): Do theories serve as predicting devices or rather do they refer to real objects in the natural world independent of our theory-dependent views of such objects? While there is something at stake in philosophy, and indeed plausibly for education, regarding instrumentalism and realism, the technical arguments do not necessarily lead to specific implications for education. For example, scientific realism and constructive empiricism recognize the strong theory dependence of scientific methods. Procedures and inferences about actions in the course of an investigation are dependent on the extant theoretical knowledge of the inquirers. This level of consensus may be enough to develop science curricula that propose reasonably informed experiences for students, without a final answer to the instrumentalist-realist debates. While the particulars of the debate may not have easy answers for education, there are useful tools and ways of thinking in philosophy of science that have merit for education.

Second, philosophy of science includes different perspectives and knowledge that change over time. As philosophy of science changes, educators need to work to understand those changes and update their own of philosophies of science. Furthermore, this effort will be complicated by the number of philosophical positions. For example, Laudan (1990) broadly identifies four major research traditions: positivist, realist, relativist, and pragmatist. Within any one of these perspectives, there is considerable variation. For example, Dewey's (1938a) pragmatism refers to science as an approach to reasoning; Toulmin's (1972) pragmatic point of view provides historical evidence from the history of science to examine conceptual change over time; Rorty's (1991) pragmatism seeks to change the nature of the conversation from technical philosophical debate to thinking about the usefulness of knowledge, be it science or other. Thus, the nature of philosophy of science is itself variable and like science fields experiences changes through research.

Third, philosophy of science has historically been normative and relatively apolitical (with a few exceptions, see Matthews 2009; Rouse 1996). Some of the central goals of philosophy of science concern questions about how science should be practiced, rather than the actual practices occurring in real settings. While some motivation for the study of scientific reasoning emerged from the realization of scientific knowledge as remarkably (and perhaps uniquely) reliable, the focus of philosophy of science has historically been on studying structure and change of scientific theories (Suppe 1977). Machamer (1998) characterized philosophy of science as concerned with the nature and character of scientific theories, the history and nature of inquiry, the value systems of scientists, and the effects and influences of science in society. While such a view expands beyond a focus on theory, the focus of the discipline has traditionally been normative – thinking about ways that reasoning should occur to lead to reliable results. This poses challenges to educators. Developing an inquiry orientation around socioscientific issues requires some consideration of the messy, ill-formed reasoning and ambiguity that surrounds science in society. Additionally, even in highly controlled settings, the reasoning patterns of students are likely to vary from the logical rigor demonstrated in philosophy. Therefore, models of conceptual change from science disciplines can at best be viewed as analogies for promoting thinking about student learning.

Fourth, the complexity of philosophy of science, and science studies more generally, particularly the empirical study of scientific practices (such as that found in the sociology and anthropology of science), poses challenges about how to characterize the nature of scientific knowledge and practices for students (Kelly et al. 1993). The rich debates within philosophy of science require specialized knowledge and an understanding of the history of ideas in this domain. Furthermore, the nature of science within philosophy changes. The complexities of science suggest that there is no one nature of science, but rather natures of the sciences (Kelly 2008), and that learning about the knowledge and practices of scientific disciplines requires engaging with such practices in particular domains (Rudolph 2000; Schwab 1960). Philosophy of science offers insight into knowledge in the various disciplines, but is not readily applicable to inquiry science teaching.

42.4 Philosophy of Science and Inquiry

I have argued that teaching science as inquiry poses a number of serious challenges. I have subsequently argued that drawing implications from the philosophy of science similarly for inquiry is problematic. But surely a field dedicated to understanding the bases of scientific knowledge should have something important to say to those seeking to teach science. Issues such as observation, experimentation, inference, and explanation seem relevant to learning about the workings of science. Yet, such practices pose challenges for novice learners who may not have the conceptual and epistemic bases to engage in such scientific practices in inquiry settings.

What can the philosophy of science offer? I argue that despite potential problems of implementation, philosophy of science contributes much, including methods for posing questions about science, models for serious thinking about science, understandings about aspects of scientific inquiry, and a skeptical orientation regarding ways that science is characterized in curriculum materials and instruction.

42.4.1 An Inquiry Stance Toward the Nature of Inquiry

Philosophy of science provides methods for posing questions about science, scientific activity, and values entailed in such inquiry. Philosophy of science steps back from the details of specific scientific investigations, debates, and controversies and seeks to examine the rational basis for theory choice. Over time, the characterization of theory change as depicted in philosophy of science has changed, and the debates continue. For example, certain versions of early understandings of logical empiricism sought to understand the logic of theory choice. This perspective attempted to view theories as predicting devices and focused on the cognitive content (often viewed as the empirical consequences) of particular theories. Alternatives of various sorts to this depiction emerged after Kuhn's (1962/1996) influential view of theories as connected to overarching paradigms that influence the nature of observation. Recognizing the importance of theories, beyond their empirical consequences, led to a number of developments in empiricism and scientific realism, along with various social constructionist views of science. Across the perspectives, philosophy of science continues to engage in inquiry into the inquiry processes of science.

Modeling inquiry into inquiry has two implications for science teaching and learning. First, question posing serves as a model for school science pedagogy and research into learning science as inquiry. For pedagogy, inquiry requires finding ways to pose questions and problems. Indeed, recognizing what is a good question to ask is often a key feature of inquiry. For research into inquiry, posing questions about the inquiry process and examining ways that inquiry changes over time can advance educational thinking about science education. Second, inquiry into inquiry in philosophy of science demonstrates the importance of thinking about epistemic practices within a community and the value of shared repertoires for investigations and argumentation.

42.4.2 Development of Understandings About Aspects of Scientific Inquiry

Philosophy of science may identify educational perspectives on science that are not readily available through causal observation, or even participation. Careful analysis of theory change, induction, and explanation in the field of philosophy of science can lead to understandings about the nature of science. Furthermore, increasingly philosophy of science is being influenced by the empirical study of scientific practice (Fuller 1988). These studies are informing philosophy of science in ways that bring further relevance to the consideration of inquiry approach in education. Four examples illustrate this case.

First, across perspectives in the philosophy of science, there is wide agreement about the theory dependence of scientific methods. Hypotheses are not tested one by one, but rather a set of auxiliary hypotheses are held constant for a given domain of knowledge for each investigation. Disagreements about results, say for a tested hypothesis, include evaluations of plausibility of the auxiliary hypotheses, as much as the meaning of empirical results for the tested hypothesis. Part of what is at stake in advancing knowledge is understanding how theory, methods, and specific results map onto the plausibility of background theoretical knowledge. Furthermore, such investigations are the product of persuasive arguments and knowledge emerging out of (often) strenuous debates. Thus, theory dependence advances in knowledge situated within a relevant epistemic community.

Second, scientists engage in social practices for years before learning to recognize phenomena from the point of view of the discipline (that is to "see as") (Goodwin 1994; Kuhn 1962//1996; Wittgenstein 1953/1958). Such socialization provides stability in the field and provides the basis for inquiry. Becoming a relevant observer or speaker or member generally requires a significant apprenticeship, as a new member of a community learns the practices and applied knowledge of the research area in question. This view builds on the work of Wittgenstein (1958) and has been shown from historical (Hanson 1958; Kuhn 1962/1996) and sociological (Collins 1985) perspectives. Importantly, engaging in social practices entails learning the discourse processes and nuanced meanings of a field. This has led to careful examination of the ways that discourse processes make visible events for observers (Lynch 1993).

Third, the use of models has become recognized as important for scientific inquiry (Giere 1999). Models in science are viewed as holding an internal structure that represent aspects of some phenomenon or mechanism (Machamer 1998). These models come in different sorts (e.g., analogous physical conditions, mathematical representations, idealized cognitive models) and serve different roles at various stages of knowledge construction (Schwarz et al. 2009). Modeling in science education draws from philosophy of science and cognitive theory. For example, Windschitl et al. (2008) proposed a view of science that focuses student discourse on learning scientific concepts. They identified several epistemic characteristics of scientific knowledge represented in models. Such models are "testable, revisable, explanatory, conjectural, and generative" (p. 943). Windschitl and colleagues propose a model-based inquiry approach that uses a set of conversations to organize knowledge, generate testable research questions, seek evidence, and construct an argument. This model-based approach to inquiry offers the possibility of moving students beyond learning only theoretical knowledge by situating them in a community that considers the epistemic criteria for scientific models (Pluta et al. 2011). Such a view is consistent with the dialogical perspectives in social epistemology.

Finally, the complexities and variety of activities that might count as science have made characterizing these activities as a whole increasingly problematic. While at one time physics may have served as a model of science, emerging views of science recognize important disciplinary differences. Furthermore, the disunity of science and the range of the many fields that can properly be called science require that understandings, such as the nature of science, and disciplinary inquiry, such as the philosophy of science, look at specific ways the actual work of science is accomplished. This issue has been brought to science education in reviews of the nature of science (Kelly et al. 1998) and in specific applications to disciplinary knowledge within fields of inquiry such as biology education (Rudolph and Stewart 1998), chemistry education (Erduran 2001), and geology education (Ault 1998).

42.4.3 Values of Scientific Communities

Philosophy of science identifies values undergirding scientific inquiry. Such values are relevant to inquiry in science education. As an illustrative example, I consider the identification of values in science and the importance of establishing discourse ethics for fair debate in science fields. Longino's (1990, 2002) social epistemology articulates ways that productive discourse can be accomplished in scientific communities. In her work Longino (1990) examined both constitutive values internal to scientific communities and contextual values that influence assumptions in science. Her work considered how values for discourse could be established to promote reason and objectivity given the deeply value-laden work of science. Her solution was to propose a set of four social norms for social knowledge (Longino 1990, 2002): The *venue* refers to the need for publicly recognized forums for the criticism of evidence, methods, assumptions, and reasoning. Everyday venues may include research meetings, conference presentations, and publications. *Uptake* refers to the extent to which a community tolerates dissent and subjects its beliefs and theories to modification over time in response to critical discourse. This value is somewhat contested, as in some areas dissent can be interpreted as not adhering to the best available explanation. *Publicly recognized standards* are needed as a basis for criticism of the prevailing theories, hypotheses, and observational practices. These standards contribute to framing debates regarding how criticism is made relevant to the goals of the inquiring community. One would expect public standards to evolve over time as research groups, communities, and disciplines develop new knowledge and practices. Finally, Longino (2002) argued for communities characterized by *equality of intellectual authority*. This equality needs to be tempered, so differing levels of expertise and knowledge are appropriately considered. While these are values identified as prescriptive for public discourse in science, such values may be applicable to inquiry in science education (Kelly 2008).

42 Inquiry Teaching and Learning: Philosophical Considerations

42.4.4 Developing Skepticism Toward Portrayals of Science in Curriculum Materials and Instruction

Philosophy of science can help educators promote a healthy skepticism regarding how science is characterized in curriculum materials and instruction. Inquiry in science education is often seen as a means to realizing understandings about the nature of science – importantly this often entails opportunities to raise issues about science (Crawford et al. 2000). Machamer (1998) characterizes the philosophy of science as "the discipline that studies the history and structure of inquiry" (p. 2). The study of inquiry, thus, should evince aspects of the ways that disciplinary knowledge is constructed, assessed, used, and communicated. These issues have been taken up in science education, relying on the philosophy of science and science studies more generally. A fundamental question is whether there can be a consensus view characterizing the nature of science as a set of declarative statements, or if inquiry can serve as a means for engaging in aspects of disciplinary practice where epistemological issues arise. For example, Rudolph (2000) cautions about assuming a generalized view of science or a standard set of assumptions about the nature of science, given the disciplinary differences and the heterogeneous practices across the workings of science in its many forms and disciplines. Irzik and Nola (2011) make similar arguments against a consensus view of the nature of science. Their perspective takes a family resemblance view to account for the many ways science differs across disciplinary perspectives. Importantly, these authors note that while actual inquiry practices vary, engaging in "data collecting, classifying, analyzing, experimenting, and making inferences" (p. 593) is central to developing understandings of science. Considerations of the criteria for which such practices are relevant to a given situation, and under what conditions, can lead to productive conversations about how to characterize science for the various educational purposes of different science education programs. For example, Van Dijk (2011) proposed that a family resemblance view of the nature of science offers the flexibility for the fields of science communication where promoting scientific literacy is a key goal. This perspective recognizes the disunity of science and argues against viewing science as a set of declarative statements, suggesting that such a perspective offers ways of communicating the nuances in the variation across images of science.

Allchin (2011) suggests that achieving a robust view of science requires abilities to make sense and assess the validity of scientific claims. As suggested in the preceding section on inquiry into inquiry, philosophy of science can model the reasoning needed to understand the complexities of science while supporting skepticism toward generalized statements about science. Allchin proposes methods for evaluating students' understanding through engaging students in case studies of assessment of scientific claims, thus showing how the substantive knowledge and explanatory ideals of a given discipline is related to the inquiry methods (Ault and Dodick 2010; Kelly et al. 2000). This view of inquiry entails engagement with knowledge of the natural, designed, or socioscientific worlds, for a given task, and thus takes the expanded view of inquiry (beyond just hands-on science) described in the introduction of this chapter.

42.5 Toward a Sociocultural Philosophy of Science for Education

42.5.1 Shift in Epistemic Subject from the Individual to a Collective

Philosophy of science has shifted the epistemic subject from the individual learner to the relevant social group (Fuller 1988; Longino 2002). Such a shift provides the basis for a thoroughly social view of knowledge and practice in science (Lynch 1993) and science education (Kelly and Chen 1999). There are clear curricular implications for a social epistemology. These include creating practical experiences that take into account the extant knowledge of the students, designing investigations that acknowledge the interpretative flexibility of empirical evidence, and situating decisions about experimental results and socioscientific issues in a dialogical process (Kelly 2008). The social basis of scientific knowledge has a long history. From Fleck's (1935/1979) thought collective, Wittgenstein's (1958) language games, Kuhn's (1962/1996) paradigms, to Toulmin's (1972) constellation of explanatory procedures, to Longino's (1990) shared values, a continuous thread runs through twentieth-century philosophy of science: the sociocultural basis for scientific progress.

There are many examples that illustrate the importance of the sociocultural basis of scientific progress. Three examples highlight some of the relationships with inquiry: the *sociohistorical contexts of scientific discovery*, the *acculturation of new members to a community*, and the *relevance of epistemic criteria and evaluation of knowledge claims*. Before reviewing their implications, it is important to recognize the distinction between the aims of scientific groups, which are orientated toward producing new knowledge, and the aims of education, which include acculturating novices into ways of understanding the natural world. Scientific and educational institutions have different purposes, and failing to recognize the differences confounds aspects of inquiry with discovery, learning, and so forth. Inquiry in science activity may lead to new knowledge. Inquiry in education serves to instruct members how to engage in relevant specific processes of investigation, use concepts in context, and develop means for understanding community practices. Under some circumstances, inquiry in educational settings generates new knowledge within the local community, thus showing some similarity with scientific communities.

Advances in science emerge from *sociohistorical contexts* where relevant groups of inquirers draw from extant knowledge, design and execute ways of collecting evidence, and propose solutions and evaluate solutions to outstanding, communally recognized problems. Fleck's (1935/1979) analysis of the science of syphilology provides a telling case. A variety of notions of the origins and causes of syphilis emerged from various social constituents. Religious, astrological, and medical communities proposed ways of understanding the origins and nature of the disease. The eventual development of the idea of syphilis as an infectious disease occurred through agonistic debates in which both the nature of the causal entity and the relevance of certain preconditions were simultaneously examined. For any experimental result to be taken as evidence, a whole set of preconditions and assumptions of the thought collective need to be taken into consideration. The eventual success of the identification of the infectious agent was the result of the collective effort of a community of health officials, whose contributions and work "cannot easily be dissected for individual attribution" (p. 41). The debate had to be won around the epistemic criteria for evidence – not just around the nature of the evidence from the different perspectives.

A second example of the epistemic shift relevant to inquiry for education is the manner that newcomers are acculturated into particular ways of seeing, communicating, and being. This realization about the substantive and important socialization into the ways of being in science counters forms of positivism (Ayer 1952) that based scientific progress on logic and objective experimental facts (although see Carnap 1950). These ways of being are dependent on the social practices of a relevant community (Mody and Kaiser 2008). Much of the work of apprenticeship for the ways of seeing, communicating, and being entails active participation in the practices of a relevant community. Learning to participate and become a member involves collective action. Understanding the ways that the language of a group operates, the nuances in meaning, and the path to modification in such meaning involves use of discourse in contexts. Furthermore, the completion of such an apprenticeship may be critical to being taken seriously by peers (Collins 1985).

A third example of social processes involved in scientific progress concerns the epistemic criteria for the evaluation of knowledge claims. Rather than viewing reasoning in science as a logical process of hypothesis testing, contemporary philosophy of science recognizes the dialectical processes of persuasion, debate, and critique. Indeed, scientific knowledge is social knowledge to the extent that knowledge claims are judged in relevant disciplinary communities. Longino (2002) and Habermas (1990) each have proposed norms for productive conversations in communities that respect alternatives but focus clearly on the strength of marshalling evidence. This leads to implications for inquiry centered on the social basis for decisions and the importance of using evidence in science. A dialectic approach to the construction of knowledge claims has plausible relevance to education. Nevertheless, such an approach needs to consider the local context and participants. Interesting questions about inquiry can be raised about students' developmental ages and abilities and variations regarding the science topic at hand.

42.5.2 Philosophy of Science and Learning

The relationship of philosophy of science and learning has been a central part of numerous developments in science education. One intersection occurred during a focus on constructivist learning in science education. Constructivism entered science education through a focus on students' ideas and understandings, building initially on Piaget (for review, see Kelly 1997). These learning theories and their close cousins, such as conceptual change theory, brought a welcomed focus on students' conceptions. Through careful attention to how students made sense of science phenomena, researchers were able to examine learning from the learners' point of view. This had a significant impact on science education and brought in philosophy of science. For example, the development of the alternative conceptions movement and conceptual change theory both used the work of Kuhn (1962/1996) and others to consider how students' constellation of conceptions served as framework for sensemaking. These foci led to pedagogy attending to students' sensemaking and provided opportunities for students to be actively involved in knowledge construction.

Despite the many positive contributions of constructivism to science education, there were two central philosophical problems. First, many forms of constructivism, particularly radical constructivism, set their epistemological commitments on the mind of the individual learner. This view conceptualized the problem of knowledge and learning as a cognizing subject making sense through exploration. This epistemological orientation ignored the important contributions from philosophy of language and other social views. Thus, by committing to a Cartesian subject, the constructivist orientation was ill equipped to integrate discourse and consider the value of social practice (Kelly 1997). Rather than viewing learning as socialization into a community, constructivists tended to view learning as changes in the cognitive structure of an individual mind. Second, some forms of constructivism confounded the construction of knowledge with ontological questions about reality and world making. Radical constructivism in particular was clear about its commitment to an idealist ontology and failed to understand the nuanced ways other ontological commitments could adhere to similarly reasonable pedagogies (see contributions in Matthews 1998).

A serious competitor to constructivist theories of learning emerged in the form of sociocultural theory. This view of learning conceptualizes the problem of learning as one of participation and appropriation of knowledge and practices of some relevant group. Central to this view is the important role of discourse processes through which everyday events are constructed (Kelly and Green 1998). By viewing learning as acculturation, the role of social processes and cultural practices are emphasized. From this point of view, as groups affiliate over time, they form particular ways of speaking, acting, and being that are defined by the group membership and evolve as the group changes (Gee and Green 1998; Kelly 2008; Kelly et al. 1998). Discourse practices established by the group become cultural tools for members to construct knowledge. These cultural tools, signs, and symbols mediate social interaction, which forms the basis for learning (Vygotsky 1978). Learning does not occur only for individuals because the cultural tools themselves serve as resources for members and evolve as members internalize the common practices and transform them through externalization (Engestrom 1999). Thus, this view of learning entails more than changes in the internalized cognitive structure of individual minds; instead, participants learn to be members of a group with common knowledge, identity, and affiliation through shared cultural practices that constitute membership in a community.

Sociocultural psychology and philosophy of science share some important central tenets and premises about science, knowledge, and inquiry. Both represent a shift in the epistemic subject from the individual learner or scientist to the relevant epistemic *community*, the relevance of agency within the potential created by a social language, and the value of dialectical processes for proposing, evaluating, and testing knowledge claims. Perspectives from Vygotsky (1978) and neo-Vygotskians (Cole and ·Engestrom 1993) evince the importance of considering how interpsychological processes can be internalized by individual learners. Thus, much like the social epistemology in the philosophy of science (Fuller 1988; Longino 2002; Toulmin 1972), the individual has agency and plays a key role in the development of knowledge but does so within the social languages of a relevant community. This suggests that instructional design for inquiry should consider how social practices are established and used to communicate ways of inquiring into the natural world. Such communication occurs across events leading to the development of knowledge, including the problem-posing phase of inquiry, the sensemaking talk around investigations, deliberation around meaning of results, and evaluation of the epistemic criteria for assessing proposed ideas, models, and theories.

42.6 Conclusion: Philosophical Considerations for Inquiry Teaching and Learning

Science education has considered inquiry as a goal for reform a number of times across decades – for examples, see Dewey (1929), Schwab (1960), Rutherford (1964), and NRC (1996). Whether or not inquiry was in the foreground, we have seen proposed educational change in the form of goals, standards, and frameworks. Reforms come and go and sometimes come back (Cuban 1990), yet careful consideration of aims should always be present in the conversation about education. This chapter examined philosophical considerations of inquiry, yet science education reform in any form or name can be informed by philosophy of science. Reform in education should not be aimed to reach final resolution of the issues around curriculum, instruction, and assessment once and for all. Rather, reform is a process that can include participants as part of a vibrant democracy where agency and identity are formed through active engagement in educational decision-making (Strike 1998).

This chapter argued for a view of philosophy informed by the empirical study of everyday practice (Fuller 1988; Kelly and Chen 1999; Lynch 1993). I conclude by first considering ways that this view of philosophy can inform science education. I then offer some research directions for the field of history, philosophy, and sociology of science and science teaching.

Philosophy has the potential to inform educational practice and ways of thinking about reform in educational policy. First, philosophy offers ways of posing questions. Posing questions and examining implications represent a contribution of such philosophical considerations. A number of central questions continue to be posed: What counts as understanding? What does it mean to learn? What is knowledge?

How can disciplinary knowledge and practice be assessed? Posing questions and examining in detail any proposed reform offer a contribution to the overall debate in educational reform. Second, philosophy can contribute through conceptual sorting. Through philosophical analysis of the conceptual content of educational texts (policy, curriculum, frameworks, standards) and of education events (research, teaching), philosophy can bring clarity or identify areas of ambiguity. Developing understandings about the nature of knowing, inquiry, and meaning is central to reform that progresses and advances thinking about education. While such meanings can be informed by empirical study, understanding the meaning of inquiry requires careful thought and analysis. The study of everyday practice in science and education settings (Kelly et al. 2012; Lynch 1993) can inform our views about the nature of science, inquiry, and meaning; nevertheless, there is considerable theoretical work needed to render empirical results informative. Thus, normative decisions about directions for science education cannot be answered by empirical study alone, or even more empirical studies – a balance must be struck between careful, descriptive studies and philosophical considerations of meaning. Third, philosophy of science can inform our field by scrutinizing the nature of education research, including the important work of understanding ways to develop productive conversations across theoretical traditions (Kelly 2006). Science and education are human endeavors that require ideas to be generated and assessed through dialectic processes. The field of educational research should consider ways to enhance discourse around educational practice.

With these philosophical considerations in mind, I now consider some plausible research directions for science education regarding inquiry. Inquiry in science education has taken many forms and served different goals (Abd-El-Khalick et al. 2004). In this chapter I identified a number of problematic aspects to thinking about learning science as inquiry. By drawing from a social epistemology in the philosophy of science, I have examined reasons why inquiry as an instructional approach has both potential and drawbacks. The efficacy of this approach depends crucially on how it is implemented, for whom, under what conditions, and for what purposes. I propose four areas for research regarding inquiry in science education.

First, we learned much as a field from the detailed, analytical work of the anthropology and sociology of science (e.g., Knorr-Cetina 1999). The study of everyday practice makes clear the social processes by which *what counts* as science is discussed, debated, and determined. Inquiry contexts, such as the model-based inquiry approach of Windschitl and colleagues (2008), provide a context to examine empirically the value of such approaches for science education. While science studies have their drawbacks, they offer insights into the inner workings of the various sciences. The methodological orientation to examine inquiry as it is interactionally accomplished in everyday life suggests that a similar approach in science education can be fruitful. Close, careful studies of the discourse events around inquiry can illustrate how inquiry is enacted. Contexts such as design challenges, investigations, and studies of socioscientific issues provide potentially inventive pedagogies that can be investigated empirically.

Second, there is a persistent lack of interest among students in pursuing science (Sjøberg and Schreiner 2010). Inquiry models for science instruction have been proposed as a means to address such concerns, beginning with Schwab (1960) and continuing thereafter. Yet, it is not clear that engaging students in inquiry, either into the natural world through investigations or into the socioscientific world through debate, will necessarily increase student interest in science. Research derived from philosophy of science may make science more real, authentic, or consistent with professional practice, but this may not take into account students' views and interests. Furthermore, studies examining the referent for science beyond that of professional science may point to directions that are better at engaging students – for example, ways that citizens use science to address everyday environmental concerns. Such studies would pose a new set of questions about what counts as science for the field.

Third, striving to meet the conceptual, epistemic, and social goals of science education (Duschl 2008) requires a critical analysis and discussions about the nature of inquiry. Such research would need to be reflexive about inquiry into inquiry. Work in science studies and the philosophy of education may be helpful for understanding how inquiry can be conceptualized in science education. I have argued both for the descriptive, empirical studies of science and science education and for the importance of the normative or moral arguments for reason, science, and education. The field of science education can be informed by both perspectives.

Fourth, inquiry most broadly construed entails learning and self-actualization. The educational goal of inquiry should not only be to meet specific standards, concepts, or procedures, but rather to develop the capacity for further learning. Through engagement in the sociocultural resources of other people and through interaction with the natural, designed, or social world, learners can develop an enhanced capacity to learn and develop new ideas. Education from inquiry should develop the ability to engage in more inquiry.

Acknowledgments I would like to thank William Carlsen, Christine Cunningham, Richard Duschl, and Beth Hufnagel for their helpful comments on an earlier draft of this chapter.

References

Abd-El-Khalick, F., BouJaoude, S., Duschl, R., Lederman, N.G., Mamlok-Naaman, R., Hofstein, A., Niaz, M., Treagust, D., and Tuan, H. (2004). Inquiry in science education: International perspectives. *Science Education, 88*, 397–419.

Allchin, D. (2011), Evaluating knowledge of the nature of (whole) science. *Science Education, 95*, 518–542.

Ault, C.R. (1998). Criteria of excellence for geological inquiry: The necessity of ambiguity. *Journal of Research in Science Teaching, 35*, 189–212.

Ault, C. R. and Dodick, J. (2010), Tracking the Footprints Puzzle: The problematic persistence of science-as-process in teaching the nature and culture of science. *Science Education, 94*, 1092–1122.

Ayer, A. J. (1952). *Language, Truth, and logic*. New York Dover.

Blanchard, M. R., Southerland, S. A., Osborne, J. W., Sampson, V. D., Annetta, L. A. and Granger, E. M. (2010), Is inquiry possible in light of accountability?: A quantitative comparison of the relative effectiveness of guided inquiry and verification laboratory instruction. *Science Education, 94*, 577–616.

Boyd, R. (1991). Confirmation, semantics, and the interpretation of scientific theories. In R. Boyd, P. Gasper. and J. D. Trout (Eds.), *The philosophy of science* (pp. 3–35). Cambridge, MA: MIT Press.

Carnap, R. (1950). Empiricism, semantics, and ontology. *Revue Internationale de Philosophie, 4*, 20–40.

Cole, M. & Engestrom, Y. (1993). A cultural-historical approach to distributed cognition. In G. Salomon (Ed.) *Distributed cognitions: Psychological and educational considerations* (pp. 1–46). Cambridge, UK: Cambridge University Press.

Collins, H. M. (1985). *Changing order: Replication and induction in scientific practice*. London: Sage.

Crawford, T., Kelly, G. J., & Brown, C. (2000). Ways of knowing beyond facts and laws of science: An ethnographic investigation of student engagement in scientific practices. *Journal of Research in Science Teaching, 37*, 237–258.

Cuban, L. (1990). Reforming again, again, and again. *Educational Researcher, 19*, 3–13.

DeBoer, G. E. (1991). *A history of ideas in science education*. New York: Teachers College Press.

Dewey, J. (1929). *The Quest for Certainty: A Study of the Relation of Knowledge and Action*. New York: Minton, Balch.

Dewey, J. (1938a). *Logic: The Theory of Inquiry*. New York: Holt, Rinehart, & Winston.

Dewey, J. (1938b). *Experience and education*. New York: MacMillan.

Duschl, R. A. (1990). *Restructuring science education: The importance of theories and their development*. New York: Teacher's College Press.

Duschl, R. A. (2008). Science education in three-part harmony: Balancing conceptual, epistemic, and social learning goals. *Review of Research in Education, 32*, 268–291.

Engestrom, Y. (1999). Activity theory and individual and social transformation. In Y. Engestrom, R. Miettinen, & R.-L. Punamaki (Eds.) *Perspectives on activity theory* (pp. 19–38). Cambridge, UK: Cambridge University Press.

Erduran, S. (2001). Philosophy of Chemistry: An Emerging Field with Implications for Chemistry Education. *Science & Education 10*, 581–593.

Fleck, L. (1935/1979). *Genesis and development of a scientific fact*. (F. Bradley & T. J. Trenn, Trans.). Chicago: University of Chicago Press.

Fuller, S. (1988). *Social epistemology*. Bloomington: Indiana University Press.

Gee, J. P. & Green, J. L. (1998). Discourse analysis, learning, and social practice: A methodological study. *Review of Research in Education, 23*, 119–169.

Giere, R. (1999). *Science without laws*. Chicago: University of Chicago Press.

Goodwin. (1994). Professional vision. *American Anthropologist, 96*(3), 606–663.

Habermas, J. (1990). *Moral consciousness and communicative action*. (translated by C. Lenhardt & S. W. Nicholsen). Cambridge, MA: MIT press.

Hanson, N. R. (1958). *Patterns of discovery*. Cambridge: Cambridge University Press.

Hodson, D. (2009). *Teaching and learning about science*. Rotterdam: Sense.

Irzik, G., & Nola, R. (2011). A family resemblance approach to the nature of science for science education. *Science & Education, 20*, 591–607.

Kelly, G. J. (1997). Research traditions in comparative context: A philosophical challenge to radical constructivism. *Science Education, 81*, 355–375.

Kelly, G. J. (2006) Epistemology and educational research. In J. Green, G. Camilli, & P. Elmore, (Eds.), *Handbook of Complementary Methods in Education Research* (pp. 33–55). Mahwah, NJ: Lawrence Erlbaum Associates.

Kelly, G. J. (2008). Inquiry, Activity, and Epistemic Practice. In R. Duschl & R. Grandy (Eds.) *Teaching Scientific Inquiry: Recommendations for Research and Implementation* (pp. 99–117; 288–291). Rotterdam: Sense Publishers.

Kelly, G. J., Carlsen, W. S., & Cunningham, C. M. (1993). Science education in sociocultural context: Perspectives from the sociology of science. *Science Education, 77*, 207–220.

Kelly, G. J., & Chen, C. (1999). The sound of music: Constructing science as sociocultural practices through oral and written discourse. *Journal of Research in Science Teaching, 36*, 883–915.

Kelly, G. J., Chen, C., & Crawford, T. (1998). Methodological considerations for studying science-in-the-making in educational settings. *Research in Science Education, 28*(1), 23–49.

Kelly, G. J., Chen, C., & Prothero, W. (2000). The epistemological framing of a discipline: Writing science in university oceanography. *Journal of Research in Science Teaching, 37*, 691–718.

Kelly, G. J., & Green, J. (1998). The social nature of knowing: Toward a sociocultural perspective on conceptual change and knowledge construction. In B. Guzzetti & C. Hynd (Eds.), *Perspectives on conceptual change: Multiple ways to understand knowing and learning in a complex world.* (pp. 145–181). Mahwah, NJ: Lawrence Erlbaum Associates.

Kelly, G.J., McDonald, S., & Wickman, P. O., (2012). Science learning and epistemology. In K. Tobin, B. Fraser, & C. McRobbie, (Eds.) *Second International Handbook of Science Education* (pp. 281–291). Dordrecht: Springer.

Kirschner, P. A., Sweller, J., & Clark, R. E. (2006). Why minimal guidance during instruction does not work: An analysis of the failure of constructivist, discovery, problem-based, experiential, and inquiry-based teaching. *Educational Psychologist, 41*, 75–86.

Knorr-Cetina, K. (1999). *Epistemic cultures: How the sciences make knowledge.* Cambridge, MA: Harvard University Press.

Kuhn, D. (2007), Reasoning about multiple variables: Control of variables is not the only challenge. *Science Education, 91*, 710–726.

Kuhn, T. S. (1962/1996). *The structure of scientific revolutions* (3rd ed.). Chicago: University of Chicago Press.

Laudan, L. (1990). *Science and relativism: Some key controversies in the philosophy of science.* Chicago: University of Chicago Press.

Longino, H. E. (1990). *Science as social knowledge: Values and objectivity in science inquiry.* Princeton: Princeton University Press.

Longino, H. E. (2002). *The fate of knowledge.* Princeton: Princeton University Press.

Lynch, M. (1993). *Scientific practice as ordinary action: Ethnomethodology and the social studies of science.* Cambridge: Cambridge University Press.

Machamer, P. (1998). Philosophy of science: An overview for educators. *Science & Education, 7*, 1–11.

Matthews, M.R. (Ed.): 1998, *Constructivism and science education: A philosophical examination.* Kluwer Academic Publishers, Dordrecht.

Matthews, M. R. (Ed.) (2009). Politics and philosophy of science [Special issue]. *Science & Education, 18*(2).

Minner, D. D., Levy, A. J. & Century, J. (2010). Inquiry-based science instruction -- What is it and does it matter? Results from a research synthesis years 1984 to 2002. *Journal of Research in Science Teaching, 47*, 474–496

Mody, C. C. M., & Kaiser, D. Scientific training and the creation of scientific knowledge. In E. J. Hackett, O. Amsterdamska, M. Lynch, & J. Wajcman (Eds.) (2008). *Handbook of science and technology studies (3rd ed).* Cambridge, MA: MIT press.

National Research Council (1996). *National science education standards.* Washington DC: National Academy Press.

National Research Council. (2011). *A Framework for K-12 Science Education: Practices, Crosscutting Concepts, and Core Ideas.* Committee on a Conceptual Framework for New K-12 Science Education Standards. Board on Science Education, Division of Behavioral and Social Sciences and Education. Washington, DC: The National Academies Press.

Pluta, W. J., Chinn, C. A. and Duncan, R. G. (2011), Learners' epistemic criteria for good scientific models. *Journal of Research in Science Teaching, 48*, 486–511.

Rorty, R. (1991). *Objectivity, relativism, and truth.* New York: Cambridge University Press.

Rouse, J. (1996). Beyond epistemic sovereignty. In P. Galison & D. J. Stump (eds.), *The disunity of science* (pp. 398–416). Stanford, CA: Stanford University Press.

Rudolph, J. (2000). Reconsidering the 'nature of science' as a curriculum component. *Journal of Curriculum Studies, 32*, 403–419.

Rudolph, J. L. and Stewart, J. (1998), Evolution and the nature of science: On the historical discord and its implications for education. *Journal of Research in Science Teaching, 35*, 1069–1089.

Rutherford, F. J. (1964). The role of inquiry in science teaching. *Journal of Research in Science Teaching, 2*, 80–84.

Sadler, T. D. and Fowler, S. R. (2006), A threshold model of content knowledge transfer for socioscientific argumentation. *Science Education, 90*, 986–1004.

Schwab, J. (1960). The teaching of science as enquiry. In J Schwab and P. Brandwein (eds.) *The teaching of science* (pp. 3–103). Cambridge, MA: Harvard University Press.

Schwarz, C. V., Reiser, B. J., Davis, E. A., Kenyon, L., Achér, A., Fortus, D., Shwartz, Y., Hug, B. and Krajcik, J. (2009). Developing a learning progression for scientific modeling: Making scientific modeling accessible and meaningful for learners. *Journal of Research in Science Teaching, 46*, 632–654.

Sjøberg, S. & Schreiner, C. *The ROSE project An overview and key findings*. Downloaded from the internet,1.11.11,at:http://roseproject.no/network/countries/norway/eng/nor-Sjoberg-Schreiner-overview-2010.pdf.

Strike, K. A. (1998). Centralized goal formation, citizenship, and educational pluralism: Accountability in liberal democratic societies. *Educational Policy, 12*, 203–215.

Suppe. F. (1977). *The structure of scientific theories* (2nd ed.). Urbana, IL: University of Illinois.

Toulmin, S. (1972). *Human understanding, Vol. 1: The collective use and evolution of concepts*. Princeton: Princeton University Press.

Van Dijk, E. M. (2011). Portraying real science in science communication. *Science Education, 95*, 1086–1100.

Van Fraassen, B. C. (1980). *The scientific image*. Oxford: Clarendon Press.

Vygotsky, L. (1978). *Mind in society: The development of higher psychological processes*. Cambridge, MA: Harvard.

Windschitl, M., Thompson, J. and Braaten, M. (2008). Beyond the scientific method: Model-based inquiry as a new paradigm of preference for school science investigations. *Science Education, 92*, 941–967.

Wittgenstein, L. (1953/8). *Philosophical investigations* (3rd ed.). (G. E. M. Anscombe, Trans.). New York: Macmillan Publishing.

Gregory J. Kelly is the Associate Dean for Research, Outreach, and Technology in the College of Education at the Pennsylvania State University. He completed his BS degree from the State University of New York in Albany in physics and his Ph.D. in science education from Cornell University. His research and teaching explore issues of knowledge and discourse in science education settings, teaching and learning science in secondary schools, and uses of history, philosophy, sociology of science in science education. Greg previously served as editor for the journal *Science Education*.

Chapter 43
Research on Student Learning in Science: A Wittgensteinian Perspective

Wendy Sherman Heckler

[W]e cannot avoid thinking about [Wittgenstein's] methods and their rationale if we are to find his philosophy intelligible.

(Minar 1995, p. 416)

43.1 Introduction

In the science education research literature, there is an overwhelming tendency among those who reference Wittgenstein's work to do so in ordinary ways. In other words, Wittgenstein is cited as if he has offered up a corrective theory that we can and should apply to our studies of the human condition; "meaning-in-context," "family resemblance," or "language games" come first to mind. But this chapter will suggest that such a tendency misses the central point of Wittgenstein's work: to model an alternative orientation for the philosophical project.

Wittgenstein warned against our becoming "bewitched" by natural language. He felt that many of the time-honored problems of philosophy were not problems at all, but only puzzles that result from a lack of clarity about how our language works. For example, we tend to see an expression such as "My back hurts!" solely as a report of a private experience. This is due to our picture of language as serving primarily or even exclusively to name things and, in particular, to name states of mind. Because of this representational notion of language, we tend to think of such expressions as reports of prior introspections or private experiences. As with "pain," so too with perceptions and thoughts: each person has access to an "inner world"

W.S. Heckler (✉)
Office of Academic Affairs, Otterbein University, Westerville, OH, USA
e-mail: wshermanheckler@otterbein.edu

M.R. Matthews (ed.), *International Handbook of Research in History, Philosophy and Science Teaching*, DOI 10.1007/978-94-007-7654-8_43, © Springer Science+Business Media Dordrecht 2014

1381

of phenomena that no one else can breach. This raises difficult philosophical questions about how one's inner world coordinates with an outer world of external experiences; how we communicate with others, for instance, or how we "internalize" what we learn about the world. It underwrites venerable programs of skepticism, asking what we can know of other lives and worlds.

Wittgenstein thought that talk of private inner worlds was misguided and that philosophical misconceptions like this one arise from the grammar of our common expressions in tandem with this wrong-headed view of language. Pronouncements of being in pain are not reports of a check conducted over an internal state of affairs. Rather, Wittgenstein encourages us to think of such pronouncements as they are typically used: as a type of pain behavior, simply a way of acting when one is in pain. Describing one's pain is an alternative to crying out, not the result of an inward examination and processing of phenomena known only to oneself (cf., Hacker 1999).

Wittgenstein described the proper role of philosophy as akin to therapy. He thought the philosopher's task was to clear up conceptual confusions at play in our language or to "show the fly the way out of the bottle." Rather than affirming one side or another of well-known philosophical debates, Wittgenstein questioned the assumptions on which the debates took place and attempted to dissolve puzzles by revealing them as artifacts of unexamined grammars.

It is understandably tempting for science education researchers to dismiss Wittgenstein's advice. After all, research in science education isn't doing philosophy; rather, we attempt to solve actual, empirical problems by conducting studies about teaching and learning and developing theoretical explanations of the results. But what if the assumptions underlying our empirical work suffer from conceptual confusion? For instance, if private worlds are suspect, what can be said of studies of how learners construct personal meaning of scientific concepts? Or, to suggest a more appropriately Wittgensteinian manner of stating the issue: what if the question of "how learners construct personal meaning of scientific concepts" is examined not as an empirical question at all but for its merits as a sensible conceptual one? This chapter takes interest in questions such as these in an effort to encourage better understanding of Wittgenstein's unique contributions to philosophy and by extension, to the social sciences.

43.2 Wittgenstein's Life and Philosophical Contribution

Ludwig Wittgenstein's (1889–1951) life was extraordinary in its range of experiences; in contrast, his philosophical work remains notable for its single-minded focus on the implications of his alternative view of the project of philosophy (Monk 1990, 2011). Born in Vienna to an aristocratic family, Wittgenstein regularly interacted with musicians, artists, and intellectuals. While he is known for his career as a Professor of Philosophy at the University of Cambridge, Wittgenstein also sought fulfillment in various other vocations throughout his lifetime. He served in the

Austrian Army during World War I, including some time on the front lines in Russia and Italy and nearly a year spent as a prisoner of war. After the war, Wittgenstein took up elementary school teaching for a time in rural, largely impoverished Austrian villages in an attempt to restore his humanity and improve his soul. During World War II, and although it meant abandoning his professorship at Cambridge, he took a job at Guy's Hospital (in London) where he prepared pharmaceutical ointments and delivered medicine to patients. Wittgenstein was naturally inclined toward technical-mechanical expertise, having originally aimed to study aeronautical engineering as a young university student. While working for a time as an architect, he designed elements such as windows, doors, and radiators for his sister Gretl's home in Vienna. Wittgenstein's emotional life mirrored this vacillating search for vocational contentment: he was sensitive, temperamental, and by turns suicidal. He was also driven to perfection in each task and relationship he undertook and therefore routinely profoundly disappointed with the inability to achieve it (Monk 1990).

Wittgenstein's introduction to philosophical study resulted from his focus on mathematics. After beginning an education in engineering at the University of Manchester, he took interest in mathematics and its philosophical foundations. This interest led to a correspondence with Gottlob Frege, and most biographers credit Frege with encouraging Wittgenstein to work with Bertrand Russell at Cambridge (Monk 1990; Richter 2004; alternatively see Anscombe 1995). His studies eventually led to the ideas published in the *Tractatus Logico-Philosophicus,* a work Anscombe locates "halfway between Frege and Russell – at least in some ways" (1995, p. 396). In the *Tractatus* Wittgenstein is credited with advancing a "picture theory of meaning," in which the sense of a proposition is founded on its accurate portrayal of empirical matters (Richter 2004). According to common interpretation, Wittgenstein's *Tractatus* argues that logic can be shown through language but at the same time, logical propositions themselves are nonsensical (since logical propositions do not reflect actual worldly objects). The *Tractatus* was highly influential on members of the Vienna Circle, so that these logical empiricists were inspired to dismiss the philosophical ambition to find metaphysical truths and to equate philosophy with understanding "the logic of scientific language" (Hacker 2007a). Further, based on ideas from the *Tractatus* the Vienna Circle "proposed the principle of verification as the key to the notion of linguistic meaning and invoked verifiability as a criterion of meaningfulness" (Hacker 2007a, p. 2).

After the publication of the *Tractatus*, Wittgenstein gave up philosophy for a time but was eventually lured back to Cambridge and to revisiting his ideas based on conversations he had with members of the Vienna Circle and with Frank Ramsey (Monk 1990). As he lectured and worked with colleagues at Cambridge from the 1930s on, his thinking began to depart significantly from his earlier work.[1] In particular,

---

[1] But see Crary and Read (2000) for an exploration of the continuity in Wittgenstein's early and later philosophy.

Wittgenstein developed a wide-ranging and forceful rebuttal of the representationalist view of language. His conception of the nature of philosophy itself also matured, leading him to argue that the discipline is not characterized by foundational problems that need universal solutions, but rather consists in various and persistent missteps in the use of language which muddle our perspective. Although he prepared numerous manuscript versions of these ideas and came close to publishing them on several occasions, he instead left instructions with a cadre of trusted students for posthumous publication of his later writings (Monk 1990). The *Philosophical Investigations* and other subsequent publications drawn from his *Nachlass* express Wittgenstein's later philosophy, which argued essentially that, properly conceived, philosophy was a methodology for clearing conceptual confusions rather than a search for theoretical or generalized truths. Through philosophical analysis, we seek to understand how language contributes to – in fact, engenders – supposed philosophical puzzles such as the "mysteries" of consciousness, or introspective, private access to inner states of being. Philosophy's task was ideally to dissolve puzzles and not to write alternate theories that might account for them (Monk 2011).

Wittgenstein's later works[2] are often cited for their insights into a wide variety of philosophical topics, including meaning and understanding, rule following, the "inner" and "outer" realms of human activity, and the grounds of "certainty" (see Kenny 2006). In contrast to the representationalist view of language expressed in his earlier work, these writings identify the meaning of an expression as its rule-governed use in language, inextricably tied to its use in our lives (Diamond 1989). As Hacker summarizes,

> The meaning of an expression, with marginal qualifications, is its use. It is also what is understood by anyone who understands or knows what an expression means. And it is what is given by an explanation of meaning. An explanation of meaning, even a humdrum explanation given by means of a series of examples, is a rule – a standard of correctness – for the use of the explanandum. (2007b, p. 4)

Language does work other than representing; most simply, language expresses (Hacker 1999). This focus on the nature of language becomes the basis for Wittgenstein's therapeutic philosophical method: under careful scrutiny of their roles in language, ordinary concepts "dressed up" as theoretical ones lose their beguiling nature.

This chapter will focus primarily on Wittgenstein's later writings, for how they have obtained relevance in the science education research literature, and in particular as related to students' learning of science. Central to the treatment – and the idea will be revisited – is Wittgenstein's aversion to the "craving for generality" that marks normative academic discourses. The concepts emanating from his philosophy should not be read as pronouncements or contributions to a general theory. Rather, Wittgenstein's writings serve as exemplars in their own contexts and as lessons on how to "do philosophy" as a conceptual interrogation on any next relevant occasion (Hacker 1999, 2007b; Monk 1990).

---

[2] See, e.g., Wittgenstein (1958, 1960, 1967, 1969, 1980).

43.3 The Place of Wittgenstein's Writings in Science Education

Wittgenstein remains a relatively marginal figure in science education research. However, his work has been invoked with increasing regularity, especially since what might be termed the "social turn" in research on learning in (science) education (see Gergen 1985 for a formulation of this movement in psychology more generally). As Vygotsky's theories of learning became more prominent in educational studies, researchers sought to understand the social aspects of human development, especially through discursive interaction. Wittgenstein's writings on "private language" and the character of language more generally are a natural resource for consultation in this line of inquiry. Another source of Wittgenstein's influence has been through the social studies of science movement. Wittgenstein's corpus has been a central resource to sociological studies of science for the last two decades.[3] In turn, these studies have influenced science educators and the study of classroom science.

43.3.1 Wittgenstein as a Corrective to Conceptual Change Theory's Appeal to Rationality

Early mention of Wittgenstein in science education was in the context of the debate over the newly posited theory of conceptual change. "Conceptual change" theory was introduced as a disciplinary-specific alternative to Piaget's developmental theory of learning, where accommodation (or how perturbations could leverage significant revision of learners' conceptual knowledge schema) was hypothesized to mirror the process of theory change in professional science (Posner et al. 1982). According to Posner and colleagues' foundational formulation, scientific conceptions can only be integrated by a learner when there is dissatisfaction with existing (prescientific) conceptions and when the scientific concept in question is intelligible, plausible, and carries the potential for further productive use.

Critics of the conceptual change theory of learning invoked Wittgenstein in different ways in order to question Posner, Strike, Hewson, and Gertzog's characterization of learning as "a rational activity" that involves "a kind of inquiry" resulting in a learner "accept[ing] ideas because they are seen as intelligible and rational" (1982, p. 212). West and Pines (1983) argued for considering nonrational elements of students' ideas and learning and specifically suggested that learning involves "feelings of power, simplicity in complexity, aesthetics, and personal integrity" (p. 38). They cited the collection of Wittgenstein's writings published as *On Certainty* in their claim that even such rational-sounding notions as "doubt" and "certainty" are imbued with both rational and nonrational elements.

Garrison and Bentley (1990) provided a similar critique of the place of rationality in conceptual change, noting that Wittgenstein's discussion of language learning

---

[3] See (Bloor 1983, 1992; Lynch 1992a, b, 1993).

allows for the existence of "preconceptual" learning. Preconceptual learning includes children's use of primitive forms of language (taught by imitation and training) when learning to talk. They argued that, in contrast, *rational* learning should be considered "learning-by-explanation" and that it would come later, with conceptual understanding. The distinction is important, according to Garrison and Bentley, because they contend that leaning science is similar to a child's first learning a language. In their view, the concepts of science are so unlike the concepts of everyday life that the "language games" taking place in each context are vastly different, even incommensurable. For this reason, Garrison and Bentley speculated that learning science may consist of initial-language-type, preconceptual learning relying on processes *other than* rational choice between competing conceptions. Prior to conceptual learning, individuals learn to participate in language games via imitation and drill, and they acquire a "world picture" (or rule-like system of enabling beliefs) through exemplars and persuasion.[4]

Cobern (1995) adopted Garrison and Bentley's starting point that science and everyday life are premised on very different world pictures, or as Cobern saw it, different cultural worldviews. According to Cobern, "science…is fundamentally an issue of culture" that "requires learning to see the world in a new and very different way" (1995, p. 292). Along with Garrison and Bentley, Cobern accepted Wittgenstein's assessment that "the acquisition of a fundamental viewpoint is an issue of persuasion and internalization, rather than instruction and learning" and that this was "due to the lack of sufficient common ground" (Cobern 1995, p. 293).

As a sociocultural perspective on learning gained prominence in science education, more researchers began turning from a view of rationality as the evidence of a universal logic of science to a view of rationality as an expression of various local, collective rationalities – including those of groups of scientists and ultimately individual classrooms. The Wittgensteinian-instructed philosopher Stephen Toulmin was quite influential in this development, a point that was outlined clearly in the work of Kelly and his colleagues:

> Analysis of Toulmin also showed that his view of conceptual change is based on a theory of rationality of science in which science is viewed, not as a universal set of inference rules or commitments to central theories, but as a collective set of commonly held concepts, practices, and actions of members of a group called "scientists." Thus, conceptual change can be viewed as a theory of rationality in that it makes visible what counts as reasons for changes in knowledge within a group. (Kelly and Green 1998, pp. 148–149)

Furthermore, by the late 1990s, the focus on rationality in conceptual change underscored some trends seen in the science education research literature on student learning more generally:

> Rationality does not mean that scientists nor students follow stepwise inferences from data to ontological truths. History, philosophy, and sociology of science all suggest that scientific theory choice is problematic, contextual, ill-structured (Barnes and Edge 1982; Collins and Pinch 1993; Kuhn 1970). We should not assume that the actions of individual

---

[4] See Wagner (1983, pp. 610–611) for an earlier but similar argument based on Kuhn's philosophy.

scientists are necessarily rational. Likewise, the possibility that rationality is local is accorded; that is, choices governing what counts as rational action are subject to the conventionalized practices of particular social groups (Rorty 1989; Strike 1984; Wittgenstein 1958). Nevertheless, incorporating rationality into educational theory allows for a way of talking about learning that appeals to the social nature of knowledge and the normative ideals for theory choice. (Kelly 1997, p. 365)

As will be discussed below, the impact of the field of science studies on views of students' science learning came into play more extensively during this time. In addition, science educators shifted from being concerned about learners as individual rational actors to focusing on rationality as a social practice and therefore on social aspects of learning science.

43.3.2 Wittgenstein's Philosophy as Supporting Views of Learning as a Sociolinguistic Process

In science education, Wittgenstein's work is often used to lend support to the notion that "learning is social" or some similar formulation that draws a contrast between the picture of learning as an individual comprehending and "internalizing" knowledge of the world and a picture of learning as something more like "the ability to participate in group life in a meaningful way." Wittgenstein's writings about the nature of language have been particularly influential in this regard, since talk can be indicative of either successful or marginal participation in group life. Notably too, Wittgenstein argued against the possibility of a private language known only to an introspecting subject (see, e.g., Hacker 1999). Initial interest in Wittgenstein's notion of "language games" was a precursor to these developments.

Using Wittgenstein's corpus, Stenhouse (1986) also critiqued the conceptual change literature in science education. He was concerned first with the nature of concepts, and second, with the nature of conceptual change. Citing Wittgenstein, Stenhouse introduced science educators to the idea that the acquisition of concepts is not to be thought of as the establishment of mental representations in the minds of individual learners. Rather, concept mastery is performative and linguistic: to understand a concept is to be able to use it correctly in context. In other words, "using a concept correctly" implies that one is able to follow the rules for the language game in play, e.g., science. For instance, a student who spoke earnestly of "the force on an object resulting from its acceleration" rather than the other way around would be judged to not yet understand the concept of "force." Stenhouse argued that this sense of concept attainment should already be familiar to teachers, who use schoolwork and test performance to assess students' conceptual understanding; teachers don't look at brains to assess understanding. He noted, "insofar as they are inter-personal and 'public,' all educational transactions must take place though language-games" (p. 417).

The language games Stenhouse (1986) envisioned for science and everyday life were varied and yet continuous with one another. He argued that what counts as using a concept correctly can be different in different language games, so that, for example, a child would not talk about "water" in the same way as a chemist would. In fact, Stenhouse suggested that language games were developmental, in the sense that the "language-games of 'ordinary life' are needed not only for initiating students into specialist language-games, but also in the setting up of these language games de novo by those who invented them" (p. 417). His recommendation for teachers was to take great care in understanding a child's language game, so as to initiate conceptual change from within it.[5]

The notion of conceptual attainment as successful participation in language games has developed in the research literature since Stenhouse's initial arguments. Kelly and Chen (1999) studied high school science classroom work from the perspective that "the meaning of a word, symbol, or construct is situationally defined by its use in a particular discourse practice (language game)" (p. 909). For high school students learning to construct evidence-based arguments in science, Kelly and Chen argued "that practice using terms in multiple contexts is central to understanding" (1999, p. 909). Relatedly, in advocating that classroom discourse structures change from the familiar initiation-response-evaluation (or I-R-E; see Mehan 1979) format to one of "transformative communication," Polman and Pea (2001) characterized both types of discourse as Wittgensteinian language games. In the first case, students and teachers orient to IRE structures as the familiar – and allegedly problematic – language game of modern institutionalized schooling. On the other hand, using the transformative communication practices espoused by the authors is said to allow teachers' work to instead focus on "transforming students' actions into more successful 'moves' in the 'language game' of science" (Polman and Pea 2001, p. 227). Thus, some language games are scientific, and some not, and in this way the formulation of "language games" preserves our normative dispositions.

Although their focus was on the "complex of representations, tools, and activities of a discipline" that they label "disciplinary discourse," Airey and Linder (2009) used the concept of language game to develop their treatment of how students learn (or attain fluency in) this disciplinary discourse. They argued that a "mutually accepted system" like a language game "can only occur if both student and lecturer have experienced the ways of knowing of some part of the discipline" and that "such ways of knowing may perhaps only be holistically experienced through certain types of disciplinary discourse" (p. 40). They differentiated between students' imitation of discourse and their holistic understanding of it, saying that successful teachers notice when students are not playing a language game correctly and seek to remedy this in order for students to achieve fluent disciplinary understanding. Thus, the game formulation is tied to levels or stages of learning.

Gyllenpalm and Wickman (2011) employed a sociocultural perspective of analysis to study what they name the "inquiry emphasis conflation" in science education: Teachers in classrooms tend to mistakenly equate "experiment" with any pedagogical

---

[5] In contrast, as noted earlier, Garrison and Bentley (1990) used Wittgenstein's notion of language games to argue that learning science was a fundamentally different process than learning in everyday life. They promoted the necessity of a "break" with everyday experience for learning science.

"laboratory task" and "hypothesis" with "prediction about the outcome of a laboratory task." From their analytic perspective, a focus on language games highlights the ways in which individual actions are tied to cultural institutions; by looking at "pivot terms" that tie to "central aspect[s] of two or more cultural institutions and their associated activities" (p. 910), they can better understand the ways in which different meanings may impact educational goals.

While for some researchers the notion of "language game" in science education is used largely to underscore the idea of conceptual understanding as linguistic performance, for others, this participation in language necessarily shifts our focus on learning itself from an "individual process" to a "social" one. In characterizing the philosophical commitments of conceptual change theory in science education, Kelly (1997) argued that "learning science is an acculturation, an initiation into a set of language games" (p. 367). Through Toulmin's philosophy, Kelly also emphasized the public nature of language games and (per Wittgenstein) rejected the notion of individualized, "private" language games. He contrasted the radical constructivist attempt to map "social" processes onto presumed private language games with conceptual change theory's commitment to public, shared concepts "because of the socialization that learners experience in the process of learning science" (1997, p. 369). At the same time, students "may need to *internalize* these [acculturated, scientific] conceptions and language games" (p. 370, emphasis added), a notion which he differentiates from the radical constructivist claim that students must originally create or construct this knowledge on their own.[6]

Kelly and Crawford (1997) directly addressed the "socio-cognitive discontinuity" in educational theories about cognition and indeed argued for "resisting any association of cognition as separate from social" (p. 555). Ivarsson et al. (2002) continued in this direction. They integrated some of Wittgenstein's views with Vygotsky's learning theories in order to illustrate how some conceptual change research may be advancing misguided conclusions, due to a focus on individual cognition. In particular, Ivarsson and colleagues looked at students' understanding of the earth's shape and gravity when they were given maps to use as part of an interview protocol. They argued for rejecting the notion that inner processes underlie external (discursive and nondiscursive) action:

> There is no sense, following [Vygotskian and Wittgensteinian] perspectives, in assuming that there is a level of thinking that is "pure" and that underlies reasoning in human practices. We cannot separate thought processes, say in the context of doing geometry or playing chess, from the conceptual tools that are applicable to such activities. Thinking is the use of tools. Or, as Wittgenstein so suggestively put it in the context of the use of language: "When I think in language, there aren't 'meanings' going through my mind in addition to the verbal expressions: the language is itself the vehicle of thought" (Wittgenstein 1953, §329). (Ivarsson et al. 2002, p. 78)

---

[6] However, even before Kelly's attempts at sorting out philosophical differences between these views, some radical constructivists hinted at a willingness to concede the sociality of language; Wheatley approvingly cited Bloor's pronouncement in his "commentary on Wittgenstein, 'The real source of 'life' in a word is provided not by the individual but by the society'" (Wheatley 1991, p. 11).

Students who were shown maps in Ivarsson, Schoultz, and Säljö's study had no difficulty identifying the shape of the earth as round, even though previous research indicated that a significant number of students have difficulty thinking of the earth in this way. In drawing attention to the interactional nature of their interviews, these researchers maintained that an attempt to study "mental processes" apart from cultural tools used in discourse is worthless: "There is nothing to be gained by positing such a level of inquiry as the one implied by a notion of pure cognition underpinning our thinking" (Ivarsson et al. 2002, p. 97).

In sum, as views about the social nature of cognitive processes became more commonplace in science education research, references to Wittgenstein were used to support those general arguments. The study of social aspects of learning science necessitated examination of real-world interactions between students and teachers. For many researchers, participation in "language games" became a way to conceive of learning that provided a welcome alternative to a view of learning science as acquiring suitable mental representations of scientific concepts. "Language games" have thus been used to reference the discourse structure in classrooms as well as to emphasize the location of conceptual understanding. Many studies of science learning that focus on classroom discourse theorize that learning involves a combination of social and individual elements; e.g., meaning is constructed through discourse and interaction but must be internalized by the individual. However, a few researchers have used Wittgenstein to support an argument for abandoning the individual-social dichotomy altogether.

43.3.3 Wittgenstein's Alignment with Science Studies and Its Influence in Science Education

In the late twentieth century, sociology of science gained prominence with science educators as a model of how to conceptualize and study science and thus study science learning in the classroom. This seemed a natural extension of the project that began with the conceptual change theory of learning; after all, students' attempts at scientific meaning making were seen to parallel those of scientists. Furthermore, as noted above, changing views about rationality – from one, universally correct version to many local kinds – were seen to necessitate close examination of knowledge production and use in an assortment of contexts, including science classrooms (cf., Kelly 1997).

Two sociologists studying scientific practice and noted in the science education literature are especially relevant in assessing Wittgenstein's influence on this research genre. Perhaps the most obvious connection between science studies and Wittgenstein occurs through the writings of David Bloor on the Strong Programme in the Sociology of Scientific Knowledge (SSK). Bloor relied on Wittgenstein to justify his SSK project because Wittgenstein's later writings discuss and acknowledge the rule-governed nature of human forms of life (specifically, of language use and mathematics). Bloor (1983) interpreted Wittgenstein as positing that social consensus determines the correct interpretation of a rule; if this is so, then sociology would naturally be in a position to explain the genesis of this consensus and thus to explain human knowledge (including knowledge of logic, mathematics, and science). Another frequently cited, Wittgensteinian-inspired sociologist who studied scientists at work is Michael Lynch. Interestingly, Lynch critiqued Bloor's interpretation of Wittgenstein's writings on rule following and their implications for science study. Lynch argued for understanding Wittgenstein not as having "made science and mathematics safe for sociology" but instead as having "made things entirely unsafe for the analytic social sciences" (1992a, p. 232). According to Lynch, Wittgenstein does not license sociology to be the final arbiter of explanations of knowledge construction. On this account, Wittgenstein's relevance for science education may be not only about our conceptualizations but our conceptualizations of social scientific analysis.

With his colleagues, Kelly integrated Wittgenstein's perspectives into sociocultural views of learning in studies conducted over the late 1990s and early 2000s. In addition to arguing for a social conception of learning, many of these studies drew a direct parallel between studies of classroom science learning and studies of science as conducted in professional laboratory settings. Kelly and Crawford (1997) described an approach to studying science classrooms that emulated the success of sociological studies of professional science:

> A common thread among at least two of these research traditions, the empirical program of relativism (Collins, 1981), and the Strong program (Bloor, 1976), is that they are empirical; that is, they explore the scientific enterprise as it unfolds through investigations of the actors and actions (for review, see Kelly, Carlsen, & Cunningham, 1993). We take a similar research methodological stance in this study: We explore school science through an ethnographic approach, studying the actions and actors that comprise the culture of a conceptual physics classroom. Through detailed, over-time analysis of the social and discursive practices of the members of one high school physics class, we seek to document the actual practice of school science as it is constructed, signaled, and acknowledged. (Kelly and Crawford 1997, p. 534)

Kelly and Crawford cited the Garfinkel et al. (1981) study of Cocke and Disney's discovery of the optical pulsar as exemplary of an approach for describing how a cultural object of science took shape from a series of inscriptions, over the course of data runs. Their own study of data runs in a conceptual physics classroom described students' ways of determining "what counted as the acceleration of carts rolling down an inclined ramp" (p. 545). Kelly, Brown and Crawford (2000) used a study of university chemistry laboratory work by Lynch et al. (1983) for inspiration in analyzing the ways in which "science" was "situationally defined" in a third grade classroom during an experiment on algae growth. In both the university chemistry lab and the elementary school algae experiment, "Written instructions and the suggested experimental design underspecified the actions required to make sense of and accomplish the task of a science experiment," so that "something more" was needed to complete the practical inquiry (p. 650). Kelly, Chen and Prothero (2000) noted that their perspective was informed by ethnomethodology (Lynch's analytic orientation) and that their empirical study of writing in a university oceanography course "provide[d] a specification of the epistemic activities" (p. 702) at play in the setting, a reference to Lynch's (1992a) discussion of the ways in which ethnomethodological studies of work provide an empirical extension of Wittgenstein's philosophy. Lynch (1992a) himself argued that this extension is "not a move into empirical sociology so much as an attempt to rediscover the sense of epistemology's central concepts and themes" (pp. 257–258) through their description as practical activities.

Lynch's (1992a) discussion of ethnomethodology as an extension of Wittgenstein's work was developed into his conception of "epistopics" (Lynch 1993). His intent in suggesting investigations of "observation, description, replication, testing, measurements, explanation, proof, and so on" was to "divorce [these terms] from a 'metatheoretical' aura and to attend to the manifest fact that they are *words*" (1993, p. 280, emphasis in original). In science education, Tapper (1999) used Lynch's epistopics as an organizing theme in his analysis of students' talk during laboratory practical work. Tapper noted that Wittgenstein's notions of language game and family resemblance are foundational to the concept of "epistopics," which Lynch talks about as practices that look different across different settings but share a family resemblance across a set of similar language games (Tapper 1999, p. 449).

Studies in the sociology of science used Wittgenstein's writings to authorize their project of careful examination of scientific practice, albeit in different ways. Science studies have served as exemplars and sources of analytic insight for researchers studying the "social construction of science" in classrooms. Although science educators have acknowledged the link between science studies and Wittgenstein's philosophy, it is not clear that the differences in interpretations of Wittgenstein's program by sociologists of science have been fully appreciated.[7]

43.3.4 Wittgenstein's Philosophy as the Basis for Analytic Approaches to the Study of Learning

A final theme to be examined is the use of Wittgenstein's writing in developing specific analytic programs or tools in science education. Aside from offering concepts that change our theoretical views of what is happening during science learning, Wittgenstein's philosophy has been taken as offering up imperatives for how to study the world around us. For example, Kelly (2005) focused on Wittgenstein's advice to "don't think, look!" as confirming the importance of empirical study of "*what* people come to know" and "*how* people come to know in various settings" (p. 86, emphasis added). Through empirical exemplars, he argued that descriptive studies of learning in science classrooms serve three purposes: to "illustrate how situated practices define meanings" (p. 88), to "make visible the practices involved in constructing and learning scientific knowledge" (p. 91) in order for science educators to direct interest in them, and to "focus on the everyday social practices of actual people" (p. 93), perhaps ultimately in order to "develo[p]

---

[7] Papayannakos (2008) is an exception.

ways of understanding and empathizing, and thus improving human conditions" (p. 99). Notably, however, Kelly's (2005) focus on "don't think, look!" presented description as an alternative to nonempirical studies but did not specifically point to the divergence of a program of analytic *description* from the "craving for generality" (often via theoretical *explanations*) that typically marks all scholarly investigation, whether empirical or not (see *Zettel*, §314, and Wittgenstein 1960, p. 17).

On the other hand, Macbeth's (2000) study of an interview between a student and conceptual change researcher (a scene from the professional development materials developed as part of the Harvard-Smithsonian Private Universe Project) is descriptive and, although not explicitly Wittgensteinian, shows an attempt to "bring into view, and even clarify, the conceptualizations of a research literature" (p. 236). The study described a scene in which a researcher constructed a "completely dark room" in order to challenge a student's conception that she will be able (eventually) to see there. Macbeth invoked Wittgenstein throughout the study, not to identify particular constructs as objects of analysis, but rather to compare the scene at hand with various exemplars given by Wittgenstein in his later writings.[8] Notably, and in a manner quite different from most studies of students' conceptions, the diagnostic interview examined in the study is not described as being about the meanings of terms, e.g., "light" and "dark" and "vision." Rather, the interview discourse is considered for what work it is doing: the work of a diagnostic interview and simultaneously the work of producing a professional development exhibit for teachers. Descriptions are given of how questions are asked and heard differently by the student subject and teacher audience, but the analysis is not used to promote a new theoretical understanding of meaning or discourse. In taking science education research itself as the topic of interest, and showing how researchers attach "beliefs" and "understandings" to students through diagnostic interviews, Macbeth attempted to provide clarity about our professional analytic concepts.

The most sustained use of Wittgenstein's work as basis for an analytic program in science education originated with Wickman and Östman (e.g., 2002a, b), who introduced Wittgenstein's writings to science educators as the foundation for a "tools taken from Wittgenstein to analyze discourse change" (2002b, p. 603). Their acknowledged frustration was with a sociocultural view of learning, which addressed the connection between people's actions and the sociohistorical contexts of which they are a part, but did not specifically posit a process for changes in learning to occur. Since they accepted that meaning is given through discourse processes, Wickman and Östman proposed looking closely at discourse in order to pinpoint the circumstances of discourse change. They examined transcripts of (classroom-based) interaction in order to determine how meanings change for students in the course of a classroom activity.

---

[8] For example, Macbeth describes how the student hears her interviewer's question "Have you ever been in a completely dark room?" as a reasonable query rather than as a professional science educator's attempt to ascertain whether she harbors a misconception about light and vision. He references Wittgenstein's comment "on how asking an Englishman if he has ever been to Budapest is quite different than asking if he has ever been to the moon" and notes that the student "hears the question about a "completely dark room" as of the former kind" (p. 242).

The methodology developed by Wickman and Östman involves analyzing student talk in order to determine which meanings appear to be "standing fast" for students and conversely where students apparently perceive gaps in understanding. They suggested that these gaps are filled when "participants establish new relations in terms of similarities and differences to what is standing fast" (2002b, p. 605). Finally, the analysis considers and sorts the gaps which are filled from those which continue to linger after the educational encounter. The authors cite their significant debt to Wittgenstein as follows:

> Wittgenstein explored the idea that it is relations that make the immediate intelligible. He noticed that "What stands fast does so, not because it is intrinsically obvious or convincing; it is rather held fast by what lies around it" (Wittgenstein 1969, p. 144). He elaborates this idea by referring to "family resemblances" when explaining the meaning of language-games (Wittgenstein 1953/1967, pp. 66–67): There is no single similarity that is common to all "games." Instead there is "a complicated network of similarities overlapping and criss-crossing: sometimes overall similarities, sometimes similarities of detail." These similarities connect everything we call games in the same way as "the various resemblances between members of a family." However, relations are not just similarities. Wittgenstein also refers to differences in his description of language-games.
>
> Hence, both similarities and differences beget "what is standing fast." However, Wittgenstein shows that an exhaustive description of all these relations for all possible contexts is not possible. Neither is there a single definition that is meaningful in all contexts. An explanation can only make sense if it relates to similarities and differences to what is already standing fast in a specific context. Meaning is tied to the context of a language-game. *Knowledge* could thus be understood as *relations of similarities and differences in what is immediately intelligible* and *learning* as *constructing new relations to what is immediately intelligible*. (Wickman and Östman 2002b, p. 605, emphasis added)

In other words, "language game" was conceived as a theoretical object employing the Wittgensteinian constructs of "standing fast" and "family resemblance" in order to provide alternative formulations of knowing and learning in science. But importantly, these constructs were used to determine the procedure for analysis of transcribed discourse occurring during science learning scenarios. Lines of transcript were scrutinized for terms that appeared to be understood by students (i.e., those that stand fast) and terms that appeared to need clarification (i.e., demonstrating a gap in understanding). Documentation of the ways in which students closed the gaps (by constructing relations to what is known) provided the authors with ways to describe science learning that could focus on issues related to specific science content lessons (e.g., chemical reactions or biological structures) or that could be used to answer larger questions about science education (e.g., the nature of student-teacher interaction or in situ misconceptions).

Originally, this method was used to study the ways in which students produce generalizations during laboratory work (Wickman and Östman 2002a) and to determine what students were able to learn about insect morphology in a university-level lab practical (Wickman and Östman 2002b). Over the next decade, other studies borrowed and refined the methodological approach. Wickman (2004) reframed the methodology as "practical epistemology analysis" and with it studied university students' learning during a chemistry laboratory activity in which unknowns were to be identified through (qualitative) analysis of reaction properties. Lidar et al. (2006) employed practical epistemology analysis to study the ways in which teachers' actions and students' apparent views of school science knowledge interacted in a seventh grade class engaged in chemistry activities.

Jakobson and Wickman (2007) studied children's spontaneous use of metaphors in learning science and determined that metaphors were used to construe relations between familiar and unfamiliar terms, to make aesthetic descriptions and evaluate classroom norms. In a related study, Jakobson and Wickman (2008) used practical epistemology analysis to further develop the connection between aesthetic judgment and normative participation in elementary school science. Hamza and Wickman (2008) asked whether the assumption in science education that misconceptions impact learning is true during real-time practical lab work and found through practical epistemology analysis that "misconceptions in electrochemistry did not constrain the ways students established relations to fill gaps noticed during the practical" (p. 160). Finally, Lidar et al. (2010) also used practical epistemology analysis to revisit the misconceptions literature in science education. Specifically they sought to illuminate a way in which individual experiences in learning might be synthesized with social and institutional ones, especially though examining the role of context and the use of artifacts in learning processes (see p. 693). Lidar and colleagues (2010) focused on children's conceptions of the earth and gravity and identified points at which different student groups 'take the learning path in different directions' perhaps "due to previous experiences or interaction with the teacher, peers, books, artifacts, or nature" (p. 706).

Combining practical epistemology analysis with analysis of teachers' epistemological moves (cf., Lidar et al. 2006) and with the analysis of companion meanings (or what might be otherwise known as the "hidden curriculum"), Lundqvist et al. (2009) developed an analytic program they labeled "communication analyses of companion meanings, or CACM" (p. 860). Again, Wittgenstein was cited for the insight that "people act without hesitation or doubt" such that most meanings can be seen to "stand fast" in talk (p. 863). As assumed in other practical epistemology analysis studies, Lundqvist and colleagues argued that "To learn something new, people have to create *relations* between the new and what already stands fast for them" (p. 864, emphasis in original). However, for students, teachers have a role in the learning process, and Lundqvist, Almqvist, and Östman noted that teachers' turns in discourse may serve to direct students and result in "changes in the students' practical epistemologies" in what can be seen as "confirming, instructional, reorienting, reconstructing, and generating moves" (p. 864, cf. Lidar et al. 2006). However, Lundqvist and colleagues emphasized the importance of analyzing epistemological moves in situ, since per Wittgenstein, the meaning of such moves is contextual:

> In a different situation, the same kind of epistemological move might have a completely different meaning. This design accords with Wittgenstein's methodological advice, which is to *look* at the circumstances in which words and sentences are used (e.g., Wittgenstein, 1953/2001: §66, 1969: §501). (Lundqvist et al. 2009, p. 864, emphasis in original)

With an understanding of students' practical epistemologies and teachers' epistemological moves in hand, CACM was used to uncover the local norms at play in the science classroom and then "to identify and problematize norms as resources in the practice studied, as well as constituting specific worldviews and being potential consequences of students' socialization" (Lundqvist et al. p. 865). Through actual analysis of empirical materials, Lundqvist and colleagues showed how naïve empiricist and naïve rationalist views of science may be at play in classroom science lessons, a result that "could attract considerable criticism" for the way in which it contradicts "many countries' policy documents with regard to how to portray science and its activities" (p. 870).

In these many ways, Wittgenstein's corpus has thus served as an inspiration to science educators seeking new ways of researching learning in science, not just in the theoretical assumptions made about learning, but in actual analytic practices. Wittgenstein has been read as telling us to focus our efforts on describing language use in practice. He has also been interpreted as authorizing an analysis of meanings that stand fast versus those which must be resolved through building relations. In this way, educators have used Wittgenstein's philosophy to first describe classroom life on a micro-interactional scale and in some cases to use these descriptions to answer larger questions about science education or learning.

43.3.5 Summary

The discussion above is not exhaustive with respect to Wittgenstein's influence in science education research; others have considered the implications of the *Tractatus* in science education (e.g., Besson 2010; Rowlands et al. 2007) or the import of Wittgenstein's concepts for NOS pedagogy (e.g., Irzik and Nola 2011). But this chapter's focus is on the ways in which Wittgenstein's later writings have informed our conceptions of and research on students' science learning. The impact of Wittgenstein's philosophy in science education, while certainly not ubiquitous, can be seen in substantial strands of argument across the literature of the last 20 years. First, Wittgenstein was influential in a shift among science educators away from a singular definition of rationality as it relates to knowledge, learning, and particularly scientific rationality. Second, Wittgenstein is often invoked by researchers who aim to study learning from a sociocultural and discursive perspective. His notion of "language games" helped researchers transition from thinking of conceptual under-standing as "the acquisition of mental schema" to conceiving of it as "the ability to use language correctly in various contexts." Wittgenstein was tangentially influential in the movement that sought to build studies of science classroom learning in the model of sociological studies of professional laboratory science, since his philosophy was central to the science studies movement itself. Finally, researchers have used Wittgenstein's philosophy to argue for a different model of analysis, for example, studies that focus on description and studies that focus on the ways in which words do or do not "stand fast" in actual instances of talk.

43.4   A Critical Assessment of Wittgenstein's Role (So Far) in Science Education

The discussion so far suggests that many of the arguments or lines of inquiry introduced above tend to make use of Wittgenstein in "ordinary" ways, that is, as a source of analytic conceptual innovation for science education research investigations. Most of these projects use Wittgenstein in ways that are familiar in our pursuit of new topics or distinctions. But when we read Wittgenstein as yet another resource or authority for speaking differently, we may risk missing what he has to say, as a program aimed at rewriting the very terms of doing philosophy. In this way, straying from Wittgenstein's disciplined form of analysis is not difficult to do. Indeed, whether one has or not is a central contest in the commentaries on his corpus.[9] In this light, it is a virtual certainty that inconsistencies can be found in treatments of Wittgenstein's work in science education. Often we see them within paragraphs, or even sentences, of suggestions for following Wittgenstein's insights as questionable formulations of the implications of these insights. This is not necessarily because specific passages or citations are misunderstood. To the contrary, Wittgenstein *does* talk about the meaning of a word being its use in a language game or the impossibility of a private language, for example.

However, a major source of incongruity is the difference between Wittgenstein's project – his method of doing philosophy "as therapy" – and the project that drives most work in science education. For Wittgenstein, the goal of philosophy is to untangle conceptual confusions. His writings critique the very enterprise of philosophy as we know it, a reading that is very difficult to find in science education. In education research, the identifying task is to build a scientific account of various aspects of the educational experience, e.g., a theory of learning that can explain and predict human action or at least the acquisition of scientific understanding. Yet it is just this kind of theory building that was a target of Wittgenstein's efforts at conceptual clarification; it was precisely these "totalizing" accounts and ambitions that he was dissolving. What Wittgenstein pointed to over and over again in his work was our tendency to theorize the world and to inflate otherwise ordinary words so as to appear extraordinary and foundational. It was the pursuit of such foundations that he was inveighing against; he was writing a critique of the very possibility:

> When philosophers use a word – "knowledge," "being," "object," "I," "proposition," "name" – and try to grasp the *essence* of the thing, one must always ask oneself: is the word ever actually used in this way in the language-game which is its original home? –
>
> What *we* do is to bring words back from their metaphysical to their everyday use. (Wittgenstein 1958, §116, emphasis in original)[10]

---

[9] For instance, see Malcolm (1989) and Baker and Hacker (1990); or see Sharrock and Button (1999) on the Lynch-Bloor exchange cited earlier.

[10] Note also the use of "language game" in this passage: not as a theoretical imperative but something mundane and ordinarily recognizable.

Wittgenstein's recommendation for science education researchers, then, would undoubtedly have been to step back and "battle against the bewitchment of our intelligence by means of language" (Wittgenstein 1958, §109). The "bewitchment" will not be broken by the introduction of new topics, but rather by an examination of old habits, and especially those deeply taken for granted in grammatical confusions.

43.4.1 *"Meaning" as a Source of Conceptual Confusion*

Although there are many candidate terms in this battle, "meaning" seems to cast a particularly charming spell on contemporary theories of learning. This handbook chapter has focused on the ways in which Wittgenstein's writings have been used in research on learning in science; all of it are in some way connected to accounts of meaning. Meaning is subject to local forms of rationality, which must be investigated empirically, uncovered, and articulated. The concept of "language game" serves as an alternative to "mental images," as a vehicle through which meaning is constructed or conveyed. "Meaning-in-use" usually refers to the performative and public nature of meaning; however, the contextual and temporary nature of meaning is also often linked to this phrase as well. Meanings that "stand fast," according to the use of this expression in science education, are those that are not questioned during talk.

In science education research, each of the above concepts or phrases associated with Wittgenstein's philosophy has been pressed into service on behalf of a driving interest in students' meaning making. "Meaning making" typically refers to a notion that the meanings of words or expressions are constructed or created via a process involving an interaction between prior understanding and engagement with new phenomena, whether those phenomena are physical or discursive. The picture is something like this: meanings result from engaging in practices; or, people assign meanings (typically, to words or perhaps gestures in discourse) as an outcome of interacting with others and/or the physical world. Alternatively, one might say that for science education researchers, meaning making involves a process of verifying and then attaching or assigning an ostensive definition to a "concept," whether this concept is said to be located in discourse or in mental schema.

To wit, Kelly and Green (1998) write, "Members of a group ascribe meaning to the processes, artifacts, practices, and signs and symbols that they construct in and through everyday activity" (p. 147). And Wickman (2004) notes of his "practical epistemology analysis" methodology that "The focus here is…to use a formal theory of meaning-making in illuminating the *connection* between *how* people produce meaning and *what* meaning is produced in a specific practice" (p. 327, emphasis in original). Lidar et al. (2010) outline the basic analytic practice:

> An encounter is where meaning making happens…The meaning-making process is thus analyzed by studying encounters where gaps could occur, and how people to fill the gaps, establish relations between the new in an encounter, and what stands fast. Biesta and Burbules (2003, p. 36) describe meaning making as "the way in which the organism responds to the environment." In practical epistemology terminology, this could be expressed: Meaning making is to create relations between the new in an encounter and what is standing fast. (p. 694–695)

In this way, Wittgenstein's philosophy has been tied to a generally "constructivist" logic of meaning making in science education, where uncertainty is continuously resolved by individuals linking their objects of inquiry to known meanings and/or objects in the environment. And the study of meaning, on this view, is rightfully an empirical investigation of "how" people make meaning and "what" meanings they make.[11]

However, from a Wittgensteinian viewpoint, this picture of meaning making in science education is entirely misleading. Again, recall that Wittgenstein championed seeking conceptual clarity about the terms which scholars were routinely tempted to use in a theoretical way. Thus, as an alternative to taking an unquestioned empirical interest in "meaning," we may first want to grasp – conceptually – the way in which we might think of the term. Recall Hacker's summary of Wittgenstein's philosophy in this regard:

> The meaning of an expression, with marginal qualifications, is its use. It is also what is understood by anyone who understands or knows what an expression means. And it is what is given by an explanation of meaning. An explanation of meaning, even a humdrum explanation given by means of a series of examples, is a rule – a standard of correctness – for the use of the explanandum. (Hacker 2007b, p. 4)

Wittgenstein's philosophy implies that "meaning" is something we should think of as *conventional* and in that sense *stable*, not arbitrarily "made" (or constructed) on the spot, as a result of some kind of process. What is meant by "meaning" in our ordinary, everyday discourse is not mysterious or in need of scientific explanation. When we are uncertain of the meaning of an expression, we find recourse in exemplars that illustrate its use; in other words, the meaning of an expression can be sought in imagining its typical use in language.

Now, of course, an expression can be misunderstood by or unfamiliar to persons or groups of persons. The meaning of an expression can change, or new expressions with new meanings can be created for specific purposes. These can, over time, become convention. And certainly, one can use double entendres, puns, or other wordplay to joke about meanings – but again, these too are orthodox. "Being ironic" on the one hand, or "getting the joke" on the other, in fact depends on the conventional nature of meaning. But these examples license neither the picture of a general, pervasive process of "meaning making" nor a science which seeks to explain individual (or community) engagement in a constant skeptical inquiry

---

[11] Does conceptualizing meaning making as being "social" change its nature for the purposes of empirical investigation? Kelly and Crawford (1997) state: "We view meaning as of a group, not an individual, and therefore view the substance of cognition as social (Wittgenstein 1958)" (p. 536). Social constructivism is often based on the view – said to be taken from Wittgenstein – that groups "make meaning" by creating rules for using expressions (and this is the process to be empirically investigated; cf. Bloor 1983). On this point, Sharrock and Button's (1999) and Francis's (2005) critiques of Bloor's program are instructive. Briefly, constructivism treats consensus as a precondition of rule following (and thus in need of explanation) rather than seeing the regular agreement about meaning as part of the use of language in our lives (see also Richter 2004; similarly Diamond 1989; Lynch 1992a). In this way, social constructivism shares with "individual" constructivism the same wrong-headed view of meaning as resulting from a process that requires theoretical explanation.

with the world, pursuing a means of anchoring words with meanings in order to proceed with learning (or even with basic communication).[12]

In other words, a first distinction to be drawn here is between "meaning" as an object of empirical analysis and "meaning" as we might *ordinarily* think of it: We ask the meaning of an unfamiliar word or expression we encounter in everyday life, and it is explained to us via examples or definition. Ryle similarly deflates the meaning of "meaning": "To know what an expression means involves knowing what can (logically) be said with it and what cannot (logically) be said with it. It involves knowing a set of bans, fiats and obligations, or, in a word, it is to know the rules of the employment of that expression" (1971, p. 363).

And there is a second sense in which to think of "meaning" as knowing how to use an expression. It is helpful to understand the context of Wittgenstein's words and that they took place in a larger discussion about other notions of what "meaning" might be. In part, what Wittgenstein was doing in emphasizing meaning-in-use was contrasting this with a notion that the meaning of a word or expression is identical with the object denoted by it (cf., Ryle 1971). Richter (2004) notes that "the main rival views" to be contrasted with Wittgenstein's notion of meaning-in-use were "that the meaning of a word is some object that it names" or "that the meaning of a word is some psychological feeling" (no page number). In other words, Wittgenstein was drawing a distinction between meaning as *doing* and meaning as *representing*. He was critiquing representational notions of language. Yet analytic orientations in science education that theorize meanings as being "assigned to processes, artifacts, practices, and signs and symbols" or meaning making as consisting in construing relations between concepts return us to a picture of "meanings" as representational "links" – between the world and talk and between known and new. Of course, explaining the meaning of a word *can* consist in giving an ostensive definition. But Hacker reminds us of the proper way to think about an ostensive definition, which is "*not [as] a link* between word and object, or language and reality, but a rule for the use of a word" (Hacker 2007b, p. 4, emphasis added). Again, Ryle is useful here: "the meanings or significations of many kinds of expressions are matters not of *naming* things but of *saying* things" (1971, p. 362, emphasis in original).

---

[12] Here it is instructive to consider how the science studies literature has been looked to as exemplary of studying new meaning creation. Specifically, Garfinkel et al. (1981) study of the discovery of an optical pulsar has been recommended as a model for the study of "the processes by which scientific objects are discursively created" (Kelly and Crawford 1997, p. 535) and thus for how the "school science knowledge creation" of classroom cohorts may be similarly examined. This tendency to link science studies and research in science education perhaps stems from our history of connecting our understanding of how scientists create knowledge to how people create meaning (e.g., a la conceptual change and various constructivist theories of science learning). But Garfinkel, Lynch, and Livingston's examination of talk recorded during the optical pulsar discovery occurred *after* the event had been taken up and affirmed within the relevant science community. With hindsight the "night's work" can get flagged as the "origin" (or part of the origin) of a new conventional understanding about pulsars, but we shouldn't then imagine that new meanings are created this way in *every* lab encounter. Neither should we focus on students' ordinary science classroom interaction as an incessant creation of new meanings. See Greiffenhagen and Sherman (2008) on the problem with conceiving of students' knowledge of the natural world as having "the form of a systematic, causal (proto-scientific) theory" (p. 16).

The upshot of these considerations is that science educators' efforts to explain "meaning making" appear misguided, especially among those who find instruction in Wittgenstein's later works. The nature of "meaning" is not mysterious when we consider its ordinary use. Meanings are not connections, associations, names, or links between language and the world. In contrast to the prevailing *theoretical* orientation, when we *ordinarily* focus on meanings, we focus on their conventional nature; we focus on the typical rules for using an expression, and not on how meanings might be idiosyncratically "made" by individuals or groups. In short, there is no empirical imperative for the study of meaning making because there is no general, overarching *process* at play in need of explanation.

43.4.2 Meanings "Standing Fast"

Yet, the impulse to explain in matters of meaning persists, perhaps most vividly among the advocates of practical epistemology analysis. Practical epistemology analysis theorizes just such a process and turns on Wittgenstein's notion of "standing fast" to anchor it. Wickman (2004) proposes that when "a word is 'clearly understood' in a specific language-game [it] can be operationalized as 'standing fast,' which means that it can be seen to be used without hesitation or questioning" (p. 328). Words that do stand fast act as anchors to those that do not; and by establishing these different standings, we can examine how new meanings can be created. In other words, "standing fast" is the marker of certain knowledge, and sequences in which students speak without running into apparent difficulty are taken as evidence for words – and therefore meanings – "standing fast."

But, in what context was Wittgenstein using the phrase "standing fast"? The passage from *On Certainty* cited by Wickman and Östman (2002b) reads:

> The child learns to believe a host of things. I.e. it learns to act according to these beliefs. Bit by bit there forms a system of what is believed, and in that system some things stand unshakeably fast and some are more or less liable to shift. What stands fast does so, not because it is intrinsically obvious or convincing; it is rather held fast by what lies around it. (Wittgenstein 1969, §144)

On the surface, it seems reasonable that these remarks could leverage the kind of methodology developed as "practical epistemology analysis": find out what students seem to know and not know as evidenced in their ways of speaking, and investigate what lies around it to make this knowledge certain (or not).

But importantly, this passage is actually part of a larger argument that Wittgenstein is making about the nature of certainty and doubt. He was drawing a contrast between words that *stand fast* and ones that are *known* or *learned*. Quoting at length from *On Certainty* illustrates more of this context:

> How does someone judge which is his right and which is his left hand? How do I know that my judgment will agree with someone else's? How do I know that this color is blue? If I don't trust myself here, why should I trust anyone else's judgment? Is there a why? Must I not begin to trust somewhere? That is to say: somewhere I must begin with not-doubting; and that is not, so to speak, hasty but excusable: it is part of judging.

I should like to say: Moore does not *know* what he asserts he knows, but it stands fast for him, as also for me; regarding it as absolutely solid is part of our *method* of doubt and enquiry.

I do not explicitly learn the propositions that stand fast for me. I can discover them subsequently like the axis around which a body rotates. The axis is not fixed in the sense that anything holds it fast, but the movement around it determines its immobility.

No one ever taught me that my hands don't disappear when I am not paying attention to them. Nor can I be said to presuppose the truth of this proposition in my assertions etc., (as if they rested on it) while it only gets sense from the rest of our procedure of asserting. (1969, §150–153, emphasis in original)

Wittgenstein is calling attention to the skepticist's quest for a formal grounding of our knowledge. How do we know we have a body? How do we know that my "blue" is your "blue"? and so on. Wittgenstein's Cambridge colleague G. E. Moore addressed philosophical skepticism by famously declaring that his hands exist because he knows they exist. But in the passage above (and throughout *On Certainty*), Wittgenstein questions the grammar of this assertion. He argues that "knowing" is sensible only when "not knowing" is possible; "Doubting and non-doubting behavior. There is the first only if there is the second" (Wittgenstein 1969, §354).

In the specific case of knowing one's hands exist, "If Moore were to pronounce the opposite of those propositions which he declares certain, we should not just not share his opinion: we should regard him as demented" (1969, §155). This is not to claim that it is impossible to imagine a context in which such a statement is coherent (e.g., as a joke or, more seriously, as being said by someone just awakening from a procedure where extremities had been amputated). Wittgenstein's point is that one does not generally consider possession of body parts to be in question. So, "standing fast" for Wittgenstein was an alternative way to describe what Moore said that he *knows* and an attempt to dispel the skepticist orientation altogether by illustrating that such a statement is not in need of any kind of grounding. Moore was looking for assurances that no actual language game needs.

With this very specific purpose for the use of the phrase "standing fast" in mind, it is difficult to support the interpretation of it being advanced in science education research. Since at least the work of Ausubel, educational theorists have been arguing that learning is a process of linking new knowledge to existing knowledge. "Practical epistemology analysis" represents a contemporary interpretation of this theory, more invested in language and social interaction than its predecessors, but still focused on the idea that knowledge must be secured in order for learning, or even communication itself, to take place. Wittgenstein on the other hand explicitly rejected a foundational view of knowledge in general:

But since a language-game is something that consists in the recurrent procedures of the game in time, it seems impossible to say in any individual case that such-and-such must be beyond doubt if there is to be a language-game—*though it is right enough to say that as a rule some empirical judgment or other must be beyond doubt*. (1969, §519, emphasis added)

A doubt without an end is not even a doubt. (1969, §625)

For why should the language game rest on some kind of knowledge? (1969, §477)

Of course, there are often times in conversation when meanings are unclear, and it is sensible to suggest that students who encounter unfamiliar words or usages during a classroom task may infer meanings from more familiar terms. What Wittgenstein

43.4.3 Meaning, Interpreting, and Understanding

The discussion above suggests that any program that aims to formally explain students' meaning making is ill advised. This seems to be Wittgenstein's counsel, though it may be a decidedly unsatisfactory conclusion for science education researchers. It is tempting to think that the matter has been oversimplified in this discussion by focusing on the ordinary and conventional notion of "meaning." After all, Wittgenstein also emphasized meaning-in-context, presumably pointing to the sometimes unexpected and unique use of words. Furthermore, science educators' interests in this regard are understandable, as students in the science classroom do not yet know the conventions of science or even school science. It could be argued that what we are really after in science education is how students *interpret* meanings in various contexts, and particularly in the classroom setting, e.g., how they interpret "gene" or "heredity." Certainly, we "interpret" the meaning of talk in interaction. This may be particularly true in instances when one encounters an unfamiliar expression, a likely occurrence in the science classroom. If our interest is not in meanings, do we instead need a series of empirical investigations and ultimately a theory to explain *interpretations*? In other words, perhaps the question is what *interpretations* are made by science students, and how are they made?

The Lidar et al. (2010) study of students' discussions about the earth's shape and gravity provides delightful examples of what could be construed as novices' interpretive work. When asked "If you were to walk for many days in a straight line, where would you end up?" some second, fourth, and fifth grade students were able to produce the sought-after answer; one would end up back where he or she started. But other students mention running into a house or the ocean. And one group of students finds a "straight" road outward from their town on a map and "walks" along it with their fingers for "20 steps" to arrive at an answer to the question. The latter group's answer to the question is unique and unexpected – but is being able to explain and predict an occurrence like this the goal of our empirical investigation of science learning? Is explaining and predicting such interpretations even possible?

Again, we can examine Wittgenstein's writings for some guidance. In the case of learning a new meaning, he discusses how ostensive teaching can be variously interpreted:

> Let us then explain the word 'tove' by pointing to a pencil and saying 'this is tove'…Now the ostensive definition 'this is tove' can be interpreted all sorts of ways…The definition can then be interpreted to mean:
>
> "This is a pencil",
> "This is round",
> "This is wood",
> "This is one",
> "This is hard," etc., etc. (Wittgenstein 1960, p. 2)

In other words, it is entirely possible that even in the most basic of interactions – pointing and naming – interpretations of what is meant can be diverse. Nevertheless, none of these interpretations is unreasonable, and all point to a sensible conclusion reached from witnessing the pointing and talking described. The meanings of various interpretations of "tove" are *still* conventional. Likewise, the interpretations of some students in Lidar and colleagues' (2010) study may be unexpected, and while it is instructive for teachers to be aware of a range of possible student interpretations of questions asked in a classroom activity, it is difficult to know what about them is in need of explanation or what explanations set them apart.

The temptation of Wittgenstein's vignette about "tove," or a focus on it as a play on interpretation – and it is a dangerously compelling one – is to assume that *every* interaction in language consists in making an interpretation. Words or expressions *can* be interpreted, *but they don't have to be.* Here again is where we want to be reminded of Wittgenstein's philosophical project to "bring words back from their metaphysical to their everyday use" (Wittgenstein 1958, §116). Wittgenstein cautions against the tendency to assume that an interpretation always lies behind sense and meaning. He uses the example of the command "fetch me a red flower from the meadow" and asks, must one conjure up an image of a red patch and compare it to the flowers before him in order to comply with the order? In other words: must we interpret, e.g., "red" in order to obey the command? His clever response: "consider the order '*imagine* a red patch'" (Wittgenstein 1960, p. 3). What process is to be imagined here? Wittgenstein continues, "Now you might ask: do we interpret the words before we obey the order? And in some cases you will find that you do something which might be called interpreting before obeying, in some cases not" (1960, p. 3).

As with "meaning," it would appear that there is nothing foundational or extraordinary about "interpretation" that calls for empirical investigation and theory building. The astute reader will surmise that any next candidate term, once considered in the context of its typical use in the language, will be "deflated" in a similar way, e.g., understanding[13]:

> 'What happens when a man suddenly understands?' – The question is badly framed. If it is a question about the meaning of the expression 'sudden understanding,' *the answer is not to point to a process that we give this name to.* – The question might mean: what are the tokens of sudden understanding; what are its psychical accompaniments?…The question what the expression means is not answered by such a description; and this misleads us into concluding that understanding is a specific indefinable experience. (Wittgenstein 1958, §321–322, emphasis added)

> Trying to understand may involve (polymorphous) processes, but actually understanding is neither an act nor a process of any kind. We distinguish between thinking that one has understood and actually having understood: thus, understanding carries *no subjective sovereignty* for its claimant. Since understanding in these cases (we exempt the notion of 'empathy' here) does not designate *any* process, the notion that understanding speech is a matter of 'processing' it (the concept favored by cognitivists infatuated with computational jargon) cannot pass muster. (Coulter 2008, p. 28, emphasis in original)

---

[13] Or, "learning" (Macbeth et al. 2011).

In light of these discussions, a conceptual analysis of "meaning" would seem to recommend that we not think of meanings as individually or situationally "constructed" as the product of some internal or social process or any combination of the two. The meaning of an expression can be interpreted, and people can understand or misunderstand an utterance. A particular expression can have different uses and thus different meanings. But meanings themselves are not "negotiated" or "assigned" in interaction, in the sense of having been created from an internal or external process that must then be explained. What we *can* do in conversation is to ascertain whether or not someone understands how to use an expression and whether someone (even ourselves) misunderstood what was expressed. But in ordinary life we don't make these assessments by looking "behind" or "underneath" talk for the meanings assigned to an expression, as if these were separate from the talk itself. Per Wittgenstein, "When I think in language, there aren't 'meanings' going through my mind in addition to the verbal expressions: the language is itself the vehicle of thought" (1958, §329). Or, as Coulter reminds us, "it is the scenic performance (or its possibility) which comprises the criterion for having understood something, not any phenomena postulated as 'internal' to the person" (2008, p. 28). In other words, we judge understanding or misunderstanding by linguistic performance, by whether or not someone has followed the correct rule for using an expression.

43.5 Conclusion

Wittgenstein's later philosophy has served to inspire new interest in language and thinking across the social sciences, and science education is no exception. Specifically in science education, Wittgenstein's writings have been used to argue for an alternate conception of rationality, to support theories examining the discursive and social nature of learning, to advocate for investigations of science classrooms that parallel ethnographic and sociological investigations of professional science labs, and to develop alternative research methodologies. And yet, this chapter has argued that efforts to merge concepts from Wittgenstein's philosophy into the typical project of science education research often can be misguided. Rather than understand Wittgenstein's writings as a penetrating critique of our familiar ways of doing "analysis," science education researchers have routinely attempted to recruit selected arguments and formulations for use in familiar programs of research. But this habit likely results in perpetuating conceptual confusions, if not introducing new candidates. Minar argues that "we cannot avoid thinking about [Wittgenstein's] methods and their rationale if we are to find his philosophy intelligible" (1995, p. 416). In the observation is a recommendation for how the science education literature might proceed differently with Wittgenstein's insights.

The "ordinary ways of researching" in science education are tied to a historical sense of the project of education research more broadly, which is to "scientize" teaching: education researchers essentially do what teachers do, but with the power of "theory" behind their judgments (cf. Sherman 2005). This urge to theorize was an anathema to Wittgenstein. In an extensive passage early in *The Blue and Brown Books*, he laments what he calls the "craving for generality" in academic endeavors[14]:

> Now what makes it difficult for us to take this line of investigation [i.e., Wittgenstein's philosophical method] is our craving for generality.
>
> This craving for generality is the resultant of a number of tendencies connected with particular philosophical confusions. There is—
>
> (a) The tendency to look for something in common to all the entities which we commonly subsume under a general term....The idea of a general concept being a common property of its particular instances connects up with other primitive, too simple, ideas of the structure of language. It is comparable to the idea that *properties* are *ingredients* of the things which have the properties; e.g., that beauty is an ingredient of all beautiful things as alcohol is of beer and wine, and that we therefore could have pure beauty, unadulterated by anything that is beautiful.
>
> (b) There is a tendency rooted in our usual forms of expression, to think that the man who has learnt to understand a general term, say, the term "leaf", has thereby come to possess a kind of general picture of a leaf, as opposed to pictures of particular leaves.... We say that he sees what is common to all these leaves; and this is true if we mean that he can on being asked tell us certain features or properties which they have in common. But we are inclined to think that the general idea of a leaf is something like a visual image, but one which only contains what is common to all leaves....This again is connected with the idea that the meaning of a word is an image, or a thing correlated to the word....
>
> (c) Again, the idea we have of what happens when we get hold of the general idea 'leaf,' 'plant,' etc. etc., is connected with the confusion between a mental state, meaning a state of a hypothetical mental mechanism, and a mental state meaning a state of consciousness (toothache, etc.).
>
> (d) Our craving for generality has another main source: our preoccupation with the method of science....Philosophers constantly see the method of science before their eyes, and are irresistibly tempted to ask and answer questions in the way science does....I want to say here that it can never be our job to reduce anything to anything, or to explain anything. Philosophy really *is* 'purely descriptive'....
>
> Instead of 'craving for generality' I could also have said "the contemptuous attitude toward the particular case." (1960, pp. 17–18)

Taking up Wittgenstein's writings means seeing them in light of *his* project: to "bring words back from their metaphysical to their everyday use" (Wittgenstein 1958, §116). In science education, we see the tendencies toward generality described above in studies which aim to formally explain students' "meaning making," even in studies which purport to be informed by Wittgenstein's writings. An alternative program of analysis resting on "therapeutic" descriptions of the ordinary use of concepts has been suggested in the critiques advanced in this chapter.

Citing Norman Malcolm, Richter (2004) identifies four primary ways in which Wittgenstein's therapeutic method of philosophy could be employed: "describing circumstances in which a seemingly problematic expression might actually be used in everyday life, comparing our use of words with imaginary language games, imagining fictitious natural history, and explaining psychologically the temptation

---

[14] Reproduced by kind permission of the copyright holder, HarperCollins Publishers.

to use a certain expression inappropriately" (no page number). In the discussions above, an attempt has been made to contrast the use of "meaning" and related terms in science education research with their ordinary, mundane use in order to reconsider not only the topics of our research but the research program itself. Wittgenstein advocated a very different kind of descriptive enterprise.

It is important to clarify that in taking this position, Wittgenstein should not be seen as "antiscience" or "'anti-philosophy."[15] Scientific practice is completely appropriate for making sense of the natural world. What Wittgenstein is arguing is that the assumptions we make about the natural world in studying it the way we do are not assumptions we can make about the way language works; thus, the study of people is necessarily different from the study of the natural world (cf. Hutchinson et al. 2008; Winch 1958). Rather than aim for producing general theoretical explanations of human language-based action, we should aim for careful, patient examination and description of the use of concepts in our lives (cf., Diamond 1989).

> Here we come up against a remarkable and characteristic phenomenon in philosophical investigation: the difficulty—I might say—is not of finding the solution but rather that of recognizing as the solution something that looks as if it were only a preliminary to it....This is connected, I believe, with our wrongly expecting an explanation, whereas the solution to our difficulty is a description, if we give it the right place in our considerations. If we dwell upon it, and do not try to get beyond it.
> The difficulty here is: to stop. (Wittgenstein 1967, §314)

References

Airey, J. & Linder, C. (2009). A disciplinary discourse perspective on university science learning: Achieving fluency in a critical constellation of modes. Journal of Research in Science Teaching, 46(1), 27–49.

Anscombe, G. E. M. (1995). Cambridge philosophers II: Ludwig Wittgenstein. *Philosophy, 70* (273), 395–407.

Baker, G. P., & Hacker, P. M. S. (1990). Malcolm on language and rules. *Philosophy, 65*(252), 167–179.

Besson, U. (2010). Calculating and understanding: Formal models and causal explanations in science, common reasoning, and science teaching. *Science & Education, 19*, 225–257.

Bloor, D. (1992). Left and right Wittgensteinians. In A. Pickering (Ed.), *Science as practice and culture* (pp. 266–282). Chicago, IL: University of Chicago Press.

Bloor, D. (1983). *Wittgenstein: A social theory of knowledge.* New York, NY: Columbia University Press.

Cobern, W. W. (1995). Science education as an exercise in foreign affairs. *Science & Education, 4*, 287–302.

Coulter, J. (2008). Twenty-five theses against cognitivism. *Theory, Culture and Society, 25*(2), 19–32.

Crary, A. & Read, R., Eds. (2000). *The new Wittgenstein.* London, UK: Routledge.

Diamond, C. (1989). Rules: Looking in the right place. In D. Z. Phillips and P. Winch (Eds.), *Wittgenstein: Attention to Particulars. Essays in honour of Rush Rhees (1905–1989)*, pp. 12–34. London, UK: Macmillan.

---

[15] Of course, allowing that philosophy be understood as therapeutic conceptual analysis.

Francis, D. (2005). Using Wittgenstein to respecify constructivism. *Human Studies, 28*, 251–290.

Garfinkel, H., Lynch, M. & Livingston, E. (1981). The work of a discovering science construed with materials from the optically discovered pulsar. *Philosophy of the Social Sciences, 11*, 131–158.

Garrison, J. W. & Bentley, M. L. (1990). Science education, conceptual change, and breaking with everyday experience. *Studies in Philosophy and Education, 10*, 19–35.

Gergen, K. J. (1985). The social constructionist movement in modern psychology. *American Psychologist, 40*(3), 266–275.

Greiffenhagen, C. & Sherman, W. (2008). Kuhn and conceptual change: On the analogy between conceptual changes in science and children. *Science & Education, 17*, 1–26.

Gyllenpalm, J. & Wickman, P.-O. (2011). "Experiments" and the inquiry emphasis conflation in science teacher education. *Science Education, 95*(5), 908–926.

Hacker, P. M. S. (2007a). Analytic philosophy: Beyond the linguistic turn and back again. In M. Beaney (Ed.), *The Analytic Turn: Analysis in Early Analytic Philosophy and Phenomenology*. London, UK: Routledge.

Hacker, P. M. S. (2007b). The relevance of Wittgenstein's philosophy of psychology to the psychological sciences. *Proceedings of the Leipzig Conference on Wittgenstein and Science*. [online resource]. http://info.sjc.ox.ac.uk/scr/hacker/docs/Relevance%20of%20W%27s%20phil.%20 of%20 psychol.%20to%20science.pdf

Hacker, P. M. S. (1999). *Wittgenstein: On human nature*. New York, NY: Routledge.

Hamza, K. M. & Wickman, P.-O. (2008). Describing and analyzing learning in action: An empirical study of the importance of misconceptions in learning science. *Science Education, 92*, 141–164.

Hutchinson, P., Read, R., & Sharrock, W. (2008). *There is no such thing as a social science: In defense of Peter Winch*. Burlington, VT: Ashgate.

Ivarsson, J. Shoultz, J., & Säljö, R. (2002). Map reading versus mind reading. In M. Limon and L. Mason (Eds.), *Reconsidering conceptual change: Issues in theory and practice*, pp. 77–99. Dordrecht, The Netherlands: Kluwer Academic Publishers.

Irzik, G. & Nola, R. (2011). A family resemblance approach to the nature of science for science education. *Science & Education, 20*, 591–607.

Jakobson, B. & Wickman, P.-O. (2008). The roles of aesthetic experience in elementary school science. *Research in Science Education, 38*(1), 45–65.

Jakobson, B. & Wickman, P.-O. (2007). Transformation through language use: Children's spontaneous metaphors in elementary school science. *Science & Education, 16*, 267–289.

Kelly, G. J. (2005). Discourse, description and science education. In R. K. Yerrick & W.-M. Roth, *Establishing scientific classroom discourse communities: Multiple voices of teaching and learning research*, pp. 79–104. Mahwah, NJ: Lawrence Erlbaum.

Kelly, G. J. (1997). Research traditions in comparative context: A philosophical challenge to radical constructivism. *Science Education, 81*(3), 355–375.

Kelly, G. J., Brown, C. & Crawford, T. (2000). Experiments, contingencies, and curriculum: Providing opportunities for learning through improvisation in science teaching. *Science Education, 84*(5), 624–657.

Kelly, G. J. & Chen, C. (1999). The sound of music: Constructing science as sociocultural practices through oral and written discourse. *Journal of Research in Science Teaching, 36*(8), 883–915.

Kelly, G. J., Chen, C. & Prothero, W. (2000). The epistemological framing of a discipline: Writing science in university oceanography. *Journal of Research in Science Teaching, 37*(7), 691–718.

Kelly, G. J. & Crawford, T. (1997). An ethnographic investigation of the discourse processes of school science. *Science Education, 81*(5), 533–559.

Kelly, G. J. & Green, J. (1998). The social nature of knowing: Toward a sociocultural perspective on conceptual change and knowledge construction. In B. Guzzetti and C. Hynd (Eds.), *Perspectives on conceptual change: Multiple ways to understand knowing and learning in a complex world*, pp. 145–181. Mahwah, NJ: Lawrence Erlbaum Associates.

Kenny, A. (2006). *The Wittgenstein reader (second edition)*. Malden, MA: Blackwell Publishing.

Lidar, M., Almqvist, J. & Östman, L. (2010). A pragmatist approach to meaning making in children's discussions about gravity and the shape of the earth. *Science Education*, doi: 10.1007/s11191-010-9298-z

Lidar, M., Lundqvist, E. & Östman, L. (2006). Teaching and learning in the science classroom: The interplay between teachers' epistemological moves and students' practical epistemology. *Science Education, 90*(1), 148–163.

Lundqvist, E., Almqvist, J. & Östman, L. (2009). Epistemological norms and companion meanings in science classroom communication. *Science Education, 93*(5), 859–874.

Lynch, M., (1993). *Scientific practice and ordinary action: Ethnomethodology and social studies of science*. New York, NY: Cambridge University Press.

Lynch, M. (1992a). Extending Wittgenstein: The pivotal move from epistemology to the sociology of science. In A. Pickering (Ed.), *Science as practice and culture* (pp. 215–265). Chicago, IL: University of Chicago Press.

Lynch, M. (1992b). From the "will to theory" to the discursive collage: A reply to Bloor's "Left and Right Wittgensteinians." In A. Pickering (Ed.), *Science as practice and culture* (pp. 283–300). Chicago, IL: University of Chicago Press.

Lynch, M., Livingston, E., & Garfinkel, H. (1983). Temporal order in laboratory work. In K. Knorr-Cetina & M. Mulkay (Eds.), *Science observed: Perspectives on the social study of science* (pp. 205–238). Beverly Hills, CA: Sage Publications.

Macbeth, D. (2000). On an actual apparatus for conceptual change. *Science Education, 84*(2), 228–264.

Macbeth, D., Lee, Y.-A., Heckler, W. S., Shon, M., Bjelic, D., & Yahsi, Z. (2011, July). *Against learning: Some notes and exhibits on behalf of instructed action*. Paper presented at the International Institute for Ethnomethodology and Conversation Analysis (IIEMCA), Fribourg, Switzerland.

Malcolm, N. (1989). Wittgenstein on language and rules. *Philosophy, 64*(247), 5–28.

Mehan, H. (1979). *Learning lessons: Social organization in the classroom*. Cambridge, MA: Harvard University Press.

Minar, E. (1995). Feeling at home in language (What makes reading Philosophical Investigations possible)? *Synthese, 102*, 413–52.

Monk, R. (2011). Ludwig Wittgenstein. In *Encyclopædia Britannica* [online resource]. http://www.britannica.com/EBchecked/topic/646252/Ludwig-Wittgenstein

Monk, R. (1990). *Ludwig Wittgenstein: The duty of genius*. New York, NY: Penguin.

Papayannakos, D. P. (2008). Philosophical skepticism not relativism is the problem with the Strong Programme in Science Studies and with educational constructivism. *Science & Education, 17*, 573–611.

Polman, J. L. & Pea, R. D. (2001). Transformative communication as a cultural tool for guiding inquiry science. *Science Education, 85*(3), 223–238.

Posner, G. J., Strike, K. A., Hewson, P. W. & Gertzog, W. A. (1982). Accommodation of a scientific conception: Toward a theory of conceptual change. Science Education, 66(2), 211–227.

Richter, D. J. (2004). Ludwig Wittgenstein (1889–1951). In *Internet Encyclopedia of Philosophy (IEP)* [online resource]. http://www.iep.utm.edu/wittgens/

Rowlands, S., Graham, T., Berry, J. & McWilliam, P. (2007). Conceptual change through the lens of Newtonian mechanics. *Science & Education, 16*, 21–42.

Ryle, G. (1971/1957). The theory of meaning. In G. Ryle, *Collected papers, Volume II*, pp. 350–372. London, UK: Hutchinson.

Sharrock, W. & Button, G. (1999). Do the right thing! Rule finitism, rule skepticism and rule following. *Human Studies, 22*, 193–210.

Sherman, W. (2005). A reply to Roth. *Canadian Journal of Science, Mathematics & Technology Education, 5*(2), 199–206.

Stenhouse, D. (1986). Conceptual change in science education: Paradigms and language games. *Science Education, 70*(4), 413–425.

Tapper, J. (1999). Topics and manner of talk in undergraduate practical laboratories. *International Journal of Science Education, 21*(4), 447–464.

Wagner, P. A. (1983). The nature of paradigmatic shifts and the goals of science education. *Science Education, 67*(5), 605–613.

West, L. H. T. & Pines, A. L. (1983). How "rational" is rationality? *Science Education, 67*(1), 37–39.

Wheatley, G. H. (1991). Constructivist perspectives on science and mathematics learning. *Science Education, 75*(1), 9–21.

Wickman, P.-O. (2004). The practical epistemologies of the classroom: A study of laboratory work. *Science Education, 88*(3), 325–344.

Wickman, P.-O. & Östman, L. (2002a). Instruction as an empirical problem: How students generalize during practical work. *International Journal of Science Education, 24*(5), 465–486.

Wickman, P.-O. & Östman, L. (2002b). Learning as discourse change: A sociocultural mechanism. *Science Education, 86*, 601–623.

Winch, P. (1958/1990). *The idea of a social science and its relation to philosophy* (2nded.). London: Routledge.

Wittgenstein, L. (1980). *Remarks on the philosophy of psychology (volume II)*. Chicago, IL: University of Chicago Press.

Wittgenstein, L. (1969). *On certainty*. G. E. M. Anscombe & G. H. von Wright, Ed., D. Paul & G. E. M. Amscombe, Trans. New York, NY: Harper & Row.

Wittgenstein, L. (1967). *Zettel*. G. E. M. Anscombe & G. H. von Wright, Ed., G. E. M. Anscombe, Trans. Berkeley, CA: University of California Press.

Wittgenstein, L. (1960). *The blue and brown books: Preliminary studies for the 'Philosophical Investigations'* (2nd ed.). New York, NY: Harper & Row.

Wittgenstein, L. (1958). *Philosophical investigations*. G. E. M. Anscombe, Trans. Engelwood Cliffs, NJ: Prentice Hall.

Wendy Sherman Heckler is Associate Vice President for Academic Affairs and Dean of University Programs at Otterbein University in Westerville, Ohio, USA. Her M.A. and Ph.D. in science education are from the Ohio State University and her classroom teaching experience is in physics, chemistry, and mathematics. Her research explores issues of conceptual clarity in education research design and empirical claims, through an integration of the perspectives of analytic philosophy and ethnomethodological sociology. At Otterbein, she has taught courses in research methods and design and the philosophy of research, as well as courses in science education pedagogy, physics, and integrative studies.

Chapter 44
Science Textbooks: The Role of History and Philosophy of Science

Mansoor Niaz

44.1 Introduction

Gerald Holton's (1952) *Introduction to Concepts and Theories in Physical Science* provided a glimpse for students and teachers as to how science evolves through the interactions of theories, experiments, and the work of actual scientists within a history and philosophy of science (HPS) perspective. This textbook has been a source of inspiration for many students of HPS. Looking back after almost 50 years, Holton (2003) considered that the textbook facilitated understanding of science as a coherent story based on the thoughts and work of living scientists. More recently, a new edition of this textbook has presented science as a human adventure, from Copernicus to Einstein and beyond (Holton and Brush 2001).

Writing in a special issue dedicated to *Students' Models and Epistemologies of Science,* Linn et al. (1991) pointed out: "Gerald Holton infused Harvard Project Physics with marvelous historical examples of scientific investigation. These were heralded as ground breaking, but were rarely imitated. Instead, if they discuss process at all, *science textbooks describe an outmoded and incorrect view of scientific knowledge acquisition*" (p. 729, italics added). The importance of history and philosophy of science was generally recognized by science educators in the 1960s (Klopfer 1969; Robinson 1969), and in the last 20 years, there has been a worldwide sustained effort to provide a rationale for its inclusion in the science curriculum (Matthews 1990, 1994/2014; Scheffler 1992).

Many researchers and teachers have endorsed the inclusion of history and philosophy of science (HPS) in the science curriculum, textbooks, and classroom

---

M. Niaz (✉)
Epistemology of Science Group, Department of Chemistry,
Universidad de Oriente, 6101 Cumaná, Estado Sucre, Venezuela
e-mail: niazma@gmail.com

M.R. Matthews (ed.), *International Handbook of Research in History,*
Philosophy and Science Teaching, DOI 10.1007/978-94-007-7654-8_44,
© Springer Science+Business Media Dordrecht 2014

practice (Hodson 1985, 1988, 2008, 2009; Matthews 1994/2014, 1998; Niaz 2008, 2009a, b, 2011). Interestingly, however, Bevilacqua and Bordoni (1998) have stated: "We are not interested in adding the history of physics to teaching physics, as an optional subject: the history of physics is 'inside' physics" (p. 451). Guisasola et al. (2005) have endorsed a similar thesis. This is an important argument, which has generally been ignored in the science education literature and especially that related to textbooks. In other words, we do not have to justify or request for the inclusion of HPS in the science curriculum and textbooks. Teaching science without the historical context in which ideas and theories develop comes quite close to what Schwab (1962, 1974) has referred to as "rhetoric of conclusions." Matthews (1998) has argued that philosophy is not far below the surface in any science classroom, as most textbooks and classroom discussions deal among others, with concepts, such as law, theory, model, explanation, cause, hypothesis, confirmation, observation, evidence, and idealization (p. 168). Similarly, Niaz and Rodríguez (2001) based on a historical framework have shown that HPS is already "inside" chemistry, and we do not need separate courses for its introduction.

At this stage it is important to clarify that the approach (HPS being inside the science curriculum) suggested here does not rule out the possibility of designing separate courses dealing with various aspects of the history of science both at the undergraduate and graduate level. It is worthwhile to provide here some examples of such an approach. Abd-El-Khalick (2005) explored the views of preservice science teachers by including a discussion of historical episodes such as dinosaur extinction controversy, Copernican revolution, Michelson-Morley experiment, cold fusion, and verification of the relativity theory. Pocoví (2007) studied the effect of history-based instructional material (Faraday's writings) to facilitate freshman students' understanding of field lines. Hosson and Kaminski (2007) developed a teaching strategy based on the history of the optical mechanism of vision (Alhazen's writings) for facilitating high schools students' reasoning about vision. Niaz (2009b) developed a course for facilitating in-service chemistry teachers' understanding of nature of science by including historical controversies related to atomic models and the oil drop experiment. This clearly shows that the two approaches can be used to facilitate an understanding based on HPS.

Abd-El-Khalick (2005) has expressed the relationship between nature of science (NOS) and history and philosophy of science (HPS), in cogent terms:

> … if we want teachers to address NOS instructionally, our efforts to help them develop the necessary understandings need to go beyond a few hours of NOS-related instruction in a science- methods course. Naturally, history and philosophy of science, which are the 'stuff' of NOS, are primary candidates for enriching the development of science teachers in the area of NOS. (p. 38)

The role of history of science in science education and especially the textbooks has been the source of considerable discussion in the literature.[1] At this stage it is

---

[1] For example, see Brush (2000), Bensaude-Vincent (2006), Chiappetta and colleagues (2006), Gooday and colleagues (2008), Kindi (2005), Niaz (2010a), Siegel (1978), and Zemplén (2007).

important to differentiate between two types of studies related to science textbooks that can be classified as:

(a) Domain specific: These studies are based on a historical reconstruction of a given topic of the science curriculum. The following are some examples of such studies presented in this chapter: quantum hypothesis (Brush 2000), photoelectric effect (Niaz et al. 2010a), periodic table (Brito et al. 2005), and atomic structure (Niaz 1998, 2000a; Justi and Gilbert 2000; Padilla and Furio-Mas 2008).
(b) Domain general: These studies are based on a series of nature of science (NOS) dimensions, which are in turn derived from the history and philosophy of science. Such dimensions consider NOS to be empirical, tentative, inferential, creative, theory driven, social, and culturally embedded. The following examples of this research are presented in this chapter: Abd-El-Khalick et al. (2008) and Leite (2002).

Although there is an overlap between these two types of studies, the distinction between the domain specific and domain general is important. Domain-general dimensions can be used to evaluate various parts of the textbooks, irrespective of the science content selected. In contrast, domain-specific studies are based on a historical reconstruction of a topic, and, for example, criteria developed for atomic structure cannot be used for evaluating the photoelectric effect.

Purpose of this chapter is to review research based on analyses of science textbooks that explicitly use a history and philosophy of science framework. This review has focused on studies published in the 15-year period (1996–2010) and has drawn on the following major science education journals: *International Journal of Science Education, Journal of Research in Science Teaching, Science Education,* and *Science & Education.* In order to select a study for review, I used the following list of criteria: title of the article, abstract, keywords, theoretical rationale, method, conclusion, and references. Furthermore, it was not just one criterion or certain terms but rather the overall contribution of the study towards facilitating an understanding of the science topic, within an HPS perspective that determined its selection. It is important to note that there are some issues such as science literacy, science-technology-society (STS), conceptual change, pedagogical content knowledge, analogies, and constructivism which are important within an HPS perspective. However, articles related to such issues were included only if they had an explicit bearing on some aspect of HPS and thus facilitated greater understanding. Based on these criteria 52 articles were selected for review, and Table 44.1 provides a distribution of the articles based on the year of publication and the journal. Table 44.2 provides a distribution of the selected articles according to the following subjects: biology, chemistry, physics, and school science. It can be observed that over half the studies(28 out of 52) were published in *Science & Education,* which clearly shows the importance of HPS for this journal. In the following sections discussion of each of these studies is presented. Table 44.2 shows that 19 studies dealt with school science (primary, secondary, and high school). However, 15 of these studies explicitly dealt with biology, chemistry, and physics textbooks at the high school level and hence were discussed along with other studies in these subjects.

1414 M. Niaz

Table 44.1 Distribution of articles based on textbooks published in major science education journals*, 1996–2010 (n=52)

| Year | Number of articles | | | |
	IJSE	JRST	SE	S&E
1996	–	1	1	1
1997	–	1	–	–
1998	–	1	1	–
1999	1	–	2	1
2000	4	1	–	2
2001	3	–	–	3
2002	–	–	–	3
2003	–	–	–	3
2004	–	–	–	–
2005	1	1	–	5
2006	–	–	–	1
2007	1	–	–	3
2008	1	1	–	3
2009	1	–	1	1
2010	–	–	1	2
Total	12	6	6	28

*IJSE international journal of science education, *JRST* journal of research in science teaching, *SE* science education, *S&E* science & education

Table 44.2 Distribution of articles according to subject published in major science education journals, 1996–2010 (n=52)

Subject	n	Articles
Biology	2	Flodin (2009), Hofmann and Weber (2003)
Chemistry	14	Brito et al. (2005), De Berg (2006), De Berg (2008a), De Berg (2008b), Furió-Más et al. (2005), Niaz (1998), Niaz (2000a), Niaz (2001a), Niaz (2001b), Niaz and Fernández (2008), Niaz and Rodríguez (2005), Padilla and Furió-Más (2008), Rodríguez and Niaz (2002), Schwartz (1999)
Physics	17	Arriassecq and Greca (2007), Assis and Zylbersztajn (2001), Brush (2000), Coelho (2010), Cotignola et al. (2002), Galili (2001), Galili and Tzeitlin (2003), Gilbert and Reiner (2000), Guisasola et al. (2005), Niaz et al. (2010a), Niss (2009), Pocoví and Finley (2003), Tampakis and Skordoulis (2007), Tarsitani and Vicentini (1996), Treagust and Harrison (2000), Velentzas et al. (2007)
School science	19	Abd-El-Khalick et al. (2008), Barberá et al. (1999), Chiappetta and Fillman (2007), Dagher and Ford (2005), De Posada (1999), Furió et al. (2000), Gardner (1999), Gericke and Hagberg (2010), Irez (2009), Justi and Gilbert (2000), Knain (2001), Koliopoulos and Constantinou (2005), Leite (2002), German et al. (1996), Milne (1998), Moody (1996), Shiland (1997), Skoog (2005), Van Berkel et al. (2000)

Notes
1. Textbooks included in the subjects of biology, chemistry, and physics are based on university freshman or higher levels
2. Textbooks included in the subject of school science are based on primary, secondary, or high school textbooks in biology, chemistry, and physics

44.2 University and High School Biology Textbooks

According to Chiappetta and Fillman (2007): "For over half a century, high school biology textbooks have played a critical role in science education because most students enroll in this course and use the adopted textbook that is a central component of the curriculum" (p. 1848). Similarly, Bybee (1989) has emphasized the role of textbooks in reforming biology education.

Chiappetta and Fillman (2007) analyzed five best-selling high school biology textbooks published in the USA (2002–2004), in order to study the inclusion of the following four NOS aspects: (a) science as a body of knowledge, (b) science as a way of investigating, (c) science as a way of thinking, and (d) science and its interactions with technology and society. These authors discuss the issue of inter-coder agreements and analyzed a much larger sample (than the 5 % in previous studies) of the textbooks from the following major topics: methods of science, ecology, cells, heredity, DNA, and evolution. It is concluded that these textbooks have a better balance of presenting the four NOS aspects as recommended by the reform documents (AAAS 1989; NRC 1996) than a previous study conducted 15 years ago, especially with regard to devoting more text to engaging students in finding out answers and learning how scientists do science.

Nature of science as depicted in five Turkish secondary school biology textbooks was studied by Irez (2009). These textbooks were widely used and published in the period 2006–2007. The following 11 themes regarding NOS were identified: description of science, characteristics of scientists, scientific method, empirical NOS, tentative NOS, nature of scientific theories and laws, inference and theoretical entities in science, subjective and theory-laden NOS, social and cultural embeddedness of science, and imagination and creativity in science. Based on these themes, the author also generated cognitive maps regarding NOS that provided an overall picture of how science was described in each textbook. Results obtained revealed that discussions regarding NOS represented a very small part of the textbooks and science was generally portrayed as a collection of facts and not as a dynamic process of generating and testing alternative explanations about nature. Of the 11 NOS themes studied, the author considered the following to be particularly misrepresented: scientific method and the tentative nature of scientific knowledge.

Coverage of human evolution in high school biology textbooks published in the USA (1900–2000) has been studied by Skoog (2005). Despite the sequencing of the human genome (Roberts 2001), human evolution continues to be a controversial topic in most parts of the world. Mayr (1997) has expressed this dilemma within a historical perspective: "No Darwinian idea was less acceptable to the Victorians than the derivation of man from a primate ancestor" (p. 25). Taking a longitudinal approach, Skoog has analyzed 113 high school textbooks in the following time periods: (a) 1900–1919 (n = 8): Only two textbooks stressed the validity of evolution. However, none of the textbooks contained material on human evolution or human fossil record. (b) 1920–1929 (n = 14): Five of the textbooks included some material concerned with human evolution and the human fossil record. The uniqueness of humans was noted in 11 of the 14 textbooks. This period was also characterized by widespread attempts to ban the teaching of evolution or its inclusion in textbooks. (c) 1930–1939 (n = 15): Four textbooks provided brief passages regarding human evolution. Discussions of the human fossil record were found in six textbooks. (d) 1940–1949 (n = 15): As compared to the previous time periods, these textbooks emphasized evolution to a greater extent. However, five textbooks failed to include any material on human evolution. (e) 1950–1959 (n = 15): Overall, eight textbooks lacked a discussion on both human evolution and the fossil record. According to Skoog (2005) this lack could be attributed to the following: "The 1950s were characterized by growing social unrest and insecurity as the Cold War and anti-communist fervor were building. Communism was associated with godlessness as was evolution by some" (p. 404). (f) 1960–1969 (n = 17): This period saw the publication of three textbooks and their revised editions, funded by the National Science Foundation under the auspices of the *Biological Science Curriculum Study* (BSCS). These textbooks gave unprecedented emphasis to human evolution, and this led the competing publishers to follow suit. (g) 1970–1989 (n = 17): Due to legislative attempts in various states to gain equal time for creationism, the unprecedented emphasis given to evolution did not continue in this period. Even the new versions of BSCS textbooks were characterized by less directness and certainty with respect to human evolution. (h) 1990–2000 (n = 12): Overall, the textbooks representing this period emphasized human evolution in a very comprehensive manner. Following N. Bohr, who considered science to be "the gradual removal of prejudice," Skoog (2005) concluded: "It is imperative that policy-makers, administrators, teachers, authors, and publishers work together to provide biology textbooks … [an opportunity] … to pursue the 'removal of prejudices' in the future citizens of America" (p. 419).

The study by Hofmann and Weber (2003) was included as it critiques Wells (2000), which is often cited by creationists who object to evolution in the science curriculum. Furthermore, Wells evaluated 10 science textbooks with respect to the notion of "universal common ancestry," and all textbooks that considered common descent as a "fact" were poorly graded. Hofmann and Weber (2003) have argued that fossil and molecular evidence is more than sufficient to warrant science educators to consider common descent as a well-established scientific fact.

According to Moody (1996) exposition of evolutionary theory in secondary school biology textbooks published in the USA has been adversely affected due to the influence of some nonscientific beliefs in the general public. Reviewing previous research on this topic, the author points out that it has been primarily concerned with measuring the quantity of space devoted to evolution in such textbooks during several decades. In contrast, the present study devised procedures to assess the overall role of evolution in the structure of the textbook. Results obtained revealed that the role of evolution in textbooks published in the 1990s increased considerably. This coincides with the results of Skoog (2005) who also reported a similar comprehensive increase in the coverage of evolution. Skoog, however, does not cite the work of Moody (1996).

German et al. (1996) analyzed nine high school biology laboratory manuals, published in the USA, during the period 1986–1993, to assess the promotion of scientific inquiry. It was hypothesized that open-ended laboratory activities (including pre-laboratory) which stress process skills reflect the nature of science more accurately. Inventory used for evaluating laboratory manuals was based on the following topics of the biology curriculum: cell structure, diffusion/osmosis, fermentation/cell respiration, leaves, tropisms, circulation, respiration, learning/behavior, hormones, and field studies. Authors used inter-rater agreements to establish the reliability of the inventory. Results of this study indicated that in general, high school biology laboratory manuals are highly structured in that they provide step-by-step detailed instructions. These manuals seldom provided opportunities for students to pose a question to be investigated, formulate a hypothesis to be tested, predict experimental results, work according to their design, or formulate a new question based on their own investigation.

Barberá and colleagues (1999) have traced the developments in the Spanish biology curriculum in the twentieth century (period of 100 years) based on the official publications of the nine national curricula and main textbooks used in this period. The main objective of the study was to focus on the relationship between socially controversial biological issues and the decision making procedures used by the different Spanish governments. Special attention was given to one of the most sensitive issue in biology education, namely, teaching of evolution. Authors found the political, social, and religious beliefs held by powerful and influential social groups to be particularly important in the elaboration of curriculum guidelines for socially controversial issues. It was concluded that such studies provide the necessary background for understanding biology education and its future development.

It seems that evolution is one of the most difficult and controversial topic of biology textbooks. Interestingly, its coverage in textbooks varies according to the prevailing sociopolitical environment. For example, during the cold war (1950s), the coverage of evolution decreased in textbooks published in the USA. Similarly, in Spain the coverage of evolution seems to depend on the political, social, and religious beliefs of influential social groups. A recent review has endorsed similar difficulties in teaching evolution (Smith 2010).

Flodin (2009) has studied different meanings of the gene concept within different subdisciplines in one biology textbook (Campbell and Reece 2005). This textbook is considered to be the most widely used English language science textbook in the world. According to the publisher it has reached two-thirds of all biology students in the USA. The author found the following interpretations of the gene concept in the textbook, depending on the subdiscipline: a trait (transmission genetics), an information structure (molecular biology), an actor (genomics), a regulator (developmental biology), and a marker (population genetics). These different functions of the gene concept are intermingled in the textbook, and the differences are not dealt with explicitly. The author concluded that such presentations in biology textbooks can become an obstacle to understanding for students and teachers. Findings of this study clearly show the importance of alternative interpretations of the gene concept and the need for a better explanation. Furthermore, it appears that textbook analyses are important and can improve even "the most widely used English language science textbook in the world."

Conceptual variation in the depiction of gene function in upper secondary school biology and chemistry textbooks was studied by Gericke and Hagberg (2010). The study is based on 20 textbooks published in different countries (Sweden = 13, Australia = 2, Canada = 2, the UK = 2, the USA = 1). Of the 13 textbooks from Sweden, five were chemistry textbooks, as genetics forms part of the chemistry syllabus. The phenomenon of gene function can be described with multiple models, such as Mendelian, classical, biochemical classical, neoclassical, and the modern. Each of these models can be characterized by various epistemological features. Authors used content analysis (including inter-coder agreements) to determine the degree to which the epistemological features are represented in the subject matter and the five historical models. Results obtained revealed that most textbooks adopted a holistic approach that integrated various scientific frameworks, while ignoring conceptual variation and incommensurability between multiple models. It is concluded that such presentations can lead to cognitive conflicts for both the students and teachers, especially if they lack adequate knowledge of history and philosophy of science.

44.3 University and High School Chemistry Textbooks

Van Berkel and colleagues (2000) have explored the *hidden structure* of school chemistry based on the following research questions: (a) Why school chemistry textbooks from different countries look so remarkably similar? (b) What does the school chemistry curriculum look like? (c) Why is school chemistry so resistant to reforms? On the basis of content analysis of school chemistry textbooks (published in the Netherlands and UK) and syllabi, these authors have identified school chemistry as a form of normal science education (NSE), which is in turn based on Kuhn's "normal science." NSE is considered to be "dangerous" in that it isolates the learner from the history and philosophy of science and, as such, is narrow and rigid and tends to instill a dogmatic attitude towards science. The role of Kuhn's normal science has been the subject of considerable research in science education and the analyses of science textbooks (Niaz 2011; Siegel 1978). Interestingly, Kuhn himself was somewhat ambiguous with respect to the relationship between NSE and science textbooks. Nevertheless the following statement from Kuhn (1977) is quite illustrative of the dilemma involved:

> The objective of a textbook is to provide the reader, in the most economical and easily assimilable form, with a statement of what the contemporary scientific community believes it knows and of the principal uses to which that knowledge can be put. *Information about how that knowledge was acquired (discovery) and about why it was accepted by the profession (confirmation) would at best be excess baggage. Though including that information would almost certainly increase the 'humanistic' values of the text and might conceivably breed more flexible and creative scientists*, it would inevitably detract from the ease of learning the contemporary scientific language. (p. 186, italics added)

44 Science Textbooks: The Role of History and Philosophy of Science   1419

This clearly shows that despite Kuhn's extraordinary contribution to HPS, he was perhaps not in tune with respect to how inclusion of HPS in science textbooks could invigorate students' and teachers' understanding of science.

Abd-El-Khalick and colleagues (2008) have drawn attention to the importance of including nature of science (NOS) in high school chemistry textbooks. These authors analyzed 14 textbooks (published in the USA, 1966–2005) including five "series" spanning one to four decades, with respect to the following NOS aspects: empirical, tentative, inferential, creative, theory driven, myth of the scientific method, nature of scientific theories and laws, and the social and cultural embeddedness of science. Based on the scoring rubric designed for this study, all three authors analyzed all textbooks independently and attained an inter-rater agreement of 86 %. Results from this study revealed that high school chemistry textbooks fared poorly in their representation of NOS, which led the authors to conclude: "These trends are incommensurate with the discourse in national and international science education reform documents (AAAS 1989; NRC 1996) ..." (p. 835). Authors considered the following finding to be the most disturbing: All textbooks (except Toon et al. 1968) espoused the diehard myth of the "scientific method" (p. 848). Interestingly, Niaz and Maza (2011) in a study designed for evaluating nature of science found that Toon and Ellis (1978) had the highest score in a sample of 75 general chemistry textbooks. Abd-El-Khalick and colleagues (2008) refer to this as an "author effect" as compared to a "publisher effect." In other words science educators could approach textbook authors with well-formulated and documented arguments so as to facilitate the inclusion of such facets of NOS and HPS in their textbooks.

Justi and Gilbert (2000) analyzed high school chemistry textbooks (nine from Brazil and three from the UK, published 1993–1997) to study the presentation of atomic models. These authors report the use of hybrid models in textbooks based on various historical developments, such as Ancient Greek, Dalton, Thomson, Rutherford, Bohr, and quantum mechanics (Schrödinger's equation). Hybrid models do not provide students an opportunity to understand the dynamical nature of science, in which different approaches to understand phenomena are contrasted and critiqued. The authors concluded: "Hybrid models, by their nature as composites drawn from several distinct historical models, do not allow the history and philosophy of science to make a full contribution to science education" (p. 993).

Based on a historical reconstruction of the atomic models of Thomson, Rutherford, and Bohr, Niaz (1998) has analyzed 23 general chemistry textbooks published in the USA (1971–1992). All textbooks were evaluated on eight criteria which were validated by inter-rater agreements. Results obtained revealed that most textbooks emphasize experimental details based on observations, leading to the presentation of scientific progress as a *rhetoric of conclusions* (Schwab 1962), based on irrevocable truths. Such presentations in textbooks lack the conceptualizations of *heuristic principles* that led the scientists to design and interpret their experiments. For example, one of the criteria dealt with the Thomson-Rutherford controversy with respect to the single/compound scattering of alpha particles. Both Rutherford (1911) and Thomson performed similar experiments on the scattering of alpha particles, but their interpretations were entirely different. Thomson propounded the hypothesis of *compound scattering*, according to which a large angle deflection of an alpha particle resulted from successive collisions between the alpha particles and the positive charges distributed throughout the atom. Rutherford, in contrast, propounded the hypothesis of *single scattering*, according to which a large angle deflection resulted from a single collision between the alpha particle and the massive positive charge in the nucleus. This rivalry led to a bitter dispute between the two proponents (Wilson 1983). Rutherford's dilemma was that, on the one hand, he was entirely convinced and optimistic that his model of the atom provided a better explanation of experimental findings, and yet it seems that the prestige, authority, and even perhaps some reverence for his teacher (Thomson) made him waver in his conviction. A science student may wonder why Thomson and Rutherford did not meet over dinner (they were well know to each other) and decide in favor of one or the other model. These issues, if discussed in class and textbooks, could make the presentation of science much more human and motivating. Interestingly, none of the general chemistry textbooks (Niaz 1998) presented this historical episode, and two general physics textbooks (Rodríguez and Niaz 2004a) made a satisfactory presentation. One of these textbooks was by Cooper (1970), a Nobel Laureate in physics, who has endorsed a history and philosophy of science perspective in science textbooks (cf. Niaz et al. 2010b, for Cooper's perspective). In our efforts to study textbooks published in different cultures, next we analyzed 21 general chemistry textbooks published in Turkey and found that none of the textbooks referred to the Thomson-Rutherford controversy (cf. Niaz and Coştu 2009).

In a subsequent study, Rodríguez and Niaz (2002) reported that the importance of history of chemistry was recognized in the literature as early as the 1920s. O. Reinmuth (1932), editor of the *Journal of Chemical Education*, recognized the importance of the historical approach to teaching chemistry:

> It is much more important that he be shown how conclusions are reached on the basis of experimental evidence than that he memorize the conclusions. Too many students acquire the idea that scientific laws, theories and hypotheses spring full-armed from the brains of geniuses as the result of some occult phenomenon which the average man never experiences. (p. 1140)

Interestingly, this comes quite close and antecedes by about 30 years Schwab's (1962) advice that science cannot be taught as "rhetoric of conclusions." Indeed, Brush (1989) has cautioned that the historical approach does not consist in merely "assertion of the conclusions" that scientists have reached in the past, but rather "... to show how they were reached and what alternatives were plausibly advocated ...." In order to pursue further the results reported by Niaz (1998) with respect to atomic models in general chemistry textbooks, Rodríguez and Niaz (2002) evaluated 30 textbooks published in the USA, in the period 1929–1967. Once again, results obtained revealed that most of the textbooks published in this period ignored the history and philosophy of science and lacked an understanding of the fact that students do not need to memorize experimental details but rather understand what the scientist was trying to do. It seems that despite the rhetoric with respect to the importance of history of chemistry, general chemistry textbooks have not changed much with respect to atomic models over 60 years (1929–1992).

A historical reconstruction of the determination of the elementary electrical charge and the ensuing controversy between R. Millikan and F. Ehrenhaft has been reported by Niaz (2000a). Both Millikan and Ehrenhaft had very similar experimental data, and still Millikan postulated the existence of a universal charged particle (electron) and Ehrenhaft postulated fractional charges (sub-electrons). Holton (1978) has provided the following insight on the impasse:

> It appeared that the same observational record could be used to demonstrate the *plausibility of two diametrically opposite theories*, held with great conviction by two well-equipped proponents and their respective collaborators. Initially, there was not even the convincing testimony of independent researchers. (pp. 199–200, emphasis added)

Niaz (2000a) has reported that of the 31 general chemistry textbooks analyzed (published in the USA, 1968–1999), none mentioned the Millikan-Ehrenhaft controversy. Similarly, none of the 43 general physics textbooks (published in the USA, 1970–2001) analyzed by Rodríguez and Niaz (2004b) mentioned the controversy. At this stage it could be argued that only advanced textbooks can be expected to include controversial aspects of scientific progress. To follow up on this, Niaz and Rodríguez (2005) analyzed 28 physical chemistry textbooks (published 1951–2002) and once again found that none mentioned the controversy. Interestingly, all these textbooks not only ignore the controversy but also consider the experiment to be characterized by its simplicity and precise results. For students, who still perform this experiment in their labs, such textbook presentations are quite perplexing (cf. Klassen 2009). Some textbooks even state that Millikan found no fractional charges and that the oil drop experiment was characterized by its simplicity and precise results. This in our opinion comes quite close to "distortion" of the historical events. Indeed, according to Kragh (1992) textbooks, "… not only distort historical reality, but they do so in a systematic way" (p. 359). Holton (2000) has endorsed the inclusion of such controversies in introductory science courses: "… introduction of the history and methodology of physics into the physics classroom, not least in terms of important controversies – is completely congenial to me … I agree one should teach research methodology in introductory physics and Millikan's case is certainly a well documented case that would lend itself to this purpose" (p. 1).

Padilla and Furio-Mas (2008) based on a historical reconstruction have traced the origin of concepts such as "chemical equivalent," "mole," and the "amount of substance." During the nineteenth century there was a widespread tendency (led by W. Ostwald) to replace the theoretical concept of "atom" with measurable notions of "volume" and "chemical equivalents," which led to considerable controversy. The progressive acceptance of the atomic-molecular theory was the origin of the magnitude, "amount of substance," and its unit, the "mole." These authors analyzed 30 general chemistry textbooks (published in the USA, 1980–2004) and found that a majority of the textbooks present the following: (a) amount of substance and mole within an ahistoric and unproblematic perspective and (b) misconceptions with respect to the mole concept, which is confused with number of elementary entities. It is concluded that lack of a historical perspective leads to distorted views of science, which does not facilitate meaningful learning. In an earlier study, Furió et al. (2000) analyzed 87 high school chemistry textbooks (published in Spain, 1976–1996), in order to determine the difficulties in teaching the concepts of "amount of substance" and "mole." It was observed that a majority of the textbooks do not introduce these concepts in a meaningful way within a historical perspective.

The degree to which high school and university level general chemistry textbooks distort the image of science, while presenting acid–base reactions, is the subject of a study by Furió-Más et al. (2005). In order to understand acids and bases, these authors identified three historical models: macroscopic, Arrhenius, and Brönsted-Lowry. Authors have argued cogently with respect to understanding the macroscopic model based on early empirical knowledge of electrical conductivity of ionic solutions, before the later models. The study is based on 19 high school textbooks published (1975–2001) in Spain and 18 general chemistry textbooks published (1968–2000) in the USA. Results based on inter-rater agreements revealed the following: (a) Arrhenius and Brönsted-Lowry theories are described in all textbooks; however only 60 % justify the change from one theory to another; (b) 66 % of the textbooks ignored that the concept of hydrolysis helps to understand the conjugate acid–base pair in the Brönsted-Lowry theory; and (c) most textbooks presented a socially "neutral" description based on inductive generalizations that lead to a linear accumulative perspective of progress in science. It is plausible to suggest that the inclusion of the concept of acids and bases by G. N. Lewis would have provided a better historical perspective.

In the late nineteenth century, there was a dispute as to whether a mathematical description of osmotic pressure could contribute any chemical information with respect to what was happening at the molecular level during osmosis. This was particularly significant as the van't Hoff osmotic pressure law, $\pi V = nRT$, was mathematically analogous to the ideal gas law, $PV = nRT$. By the middle of the twentieth century, based on thermodynamic concepts of osmosis, mathematical models provided greater chemical understanding of osmosis. However, by the end of the twentieth century, some chemists preferred a kinetic-molecular approach to osmosis rather than the thermodynamic approach. In order to understand these historical disputes, de Berg (2006) has analyzed 11 physical chemistry textbooks published in the period 1963–2002 (the USA=7, UK=4). Results obtained revealed the following: (a) A majority of the textbooks referred to and discussed the analogy between the van't Hoff law and the ideal gas law; (b) except one, all textbooks provided a mathematical derivation of the van't Hoff law; and (c) none of the textbooks referred to kinetic-molecular ideas. The author concluded that the kinetic-molecular approach is more accessible to students than the mathematical thermodynamic approach and thus deserves more attention in chemistry education, in order to facilitate greater understanding. Furthermore, thermodynamics should not just be a matter of manipulating symbols, but rather it is important to provide students with a qualitative sense of the problem.

Heat (an extensive property) and temperature (an intensive property) are an important part of the science curriculum at both the high school and university freshman level, and most students have considerable difficulty in differentiating between the two. An underlying issue that makes understanding difficult is the conceptual rivalry in the history of science between the caloric (heat as a substance) and the kinetic theories. According to Brush (1976), "The kinetic theory could not flourish until heat as a substance had been replaced by heat as atomic motion" (p. 8). De Berg (2008a) has analyzed the concepts of heat and temperature in 10 general chemistry textbooks published (1993–2006) in the USA. Results obtained revealed that heat is considered to be as follows: (a) *energy that flows* from an object at the higher temperature to one at the lower temperature; (b) *energy that is transferred* from an object at the higher temperature to one at a lower temperature; (c) process of energy transfer from a higher temperature object to a lower temperature object; (d) only three textbooks illustrated the difference between heat and temperature; and (e) no uniform picture of "heat" is presented and almost no exposure is given to the caloric theory. The author concluded that terms such as *heat flow* or *energy flow* are remnants of the caloric theory of heat and the language used: "…even in current textbooks is reminiscent of the old caloric theory…" (p. 86).

Chemistry students are generally not exposed to different ways of understanding a chemical reaction. De Berg (2008b) has analyzed eight chemistry textbooks published between 1758 and 1891 in order to understand the chemistry of the oxides of tin. This period is significant in chemical history as it covers the two chemical revolutions, associated with A. Lavoisier (1770–1790) and J. Dalton (1855–1875). Selected textbooks were from the UK, European continent, and the USA. Unlike the textbooks of today, these fulfilled multiple functions such as the teaching of chemistry to secondary, college, medicine, and pharmacy students. Results obtained provided insight into the foundation of a number of chemical ideas such as nomenclature and composition used in modern chemistry. Furthermore, four major preparation techniques for the production of tin oxides emerged: the heating of tin in air, the addition of nitric acid to tin, the alkaline hydrolysis of tin (II) and tin (IV) salts, and the hydrolysis of alkaline stannate salts. Early textbooks of the period give lengthy descriptions and explanations for these reaction schemes. It would be interesting to study this and similar topics in current textbooks.

In the early nineteenth century, Dalton's research program in order to be operationalized required the following items as part of the positive heuristic (cf. Lakatos 1970): (a) chemical formulae, (b) atomic weights (masses), and (c) composition by weight of the compound. Early in Dalton's career, only the third item was known and hence the inductivist approach concluded that only combining "equivalents" or "measures based on volumes" were important. The law of definite proportions in chemistry is basically an elaboration of the third item of Dalton's positive heuristic. In contrast, Gay-Lussac's law of combining volumes provided the antiatomists a rationale for accepting the laws of definite and multiple proportions without the "superfluous" atomic theory of Dalton. Niaz (2001a) analyzed 27 general chemistry textbooks (published in the USA, 1969–1999) to determine if they followed one of the following interpretations: (a) *Inductivist*: Gay-Lussac's law of combining volumes provided a rationale for accepting the laws of definite and multiple proportions. (b) *Lakatosian*: Dalton's atomic theory predicted and explained Gay-Lussac's law of combining volumes. Results obtained (based on inter-rater agreements) revealed that only two textbooks followed the inductivist interpretation and the remaining 25 textbooks simply ignored the historical perspective.

The periodic table of chemical elements is considered to be a conceptual tool that helps to organize a great deal of information, leading to a better understanding of chemistry. Despite this overall positive picture, most chemistry teachers and textbooks give the impression that for almost 100 years (1820–1920), scientists had no idea or never asked the question as to whether there could be an underlying pattern to explain periodic properties of the elements. Brito et al. (2005) have presented a historical reconstruction of the development of the periodic table. The basic idea behind this study was the hypothesis that even before the electronic structure of the atom (Thomson, Lewis, Bohr, Moseley, and others) was discovered, different explanations were offered for periodicity. It is important to note that when D. Mendeleev (and others) started working on the periodic table, he had the following sources of information: Dalton's atomic theory, law of multiple proportions, Cannizaro's Karlsruhe lecture, fairly reliable atomic weights, atomicity (valence), and various physical and chemical properties of chemical elements. In his famous Faraday lecture, Mendeleev (1889) explained his hypothesis cogently: "... the veil which conceals the true conception of mass, it nevertheless indicated that the explanation of that conception must be searched for in the masses of atoms; the more so, *as all masses are nothing but aggregations, or additions, of chemical atoms ...*" (p. 640, emphasis added). This clearly shows that among other factors, Mendeleev considered the atomic theory to be an important cause of periodicity of chemical elements. Based on inter-rater agreements, Brito and colleagues (2005) analyzed 57 general chemistry textbooks published (1968–2002) in the USA and found that a majority of the textbooks ignored the role played by the atomic theory in the development of the periodic table. It is concluded that it is more fruitful to present a more balanced picture to the students by highlighting how Mendeleev solved the dilemma by looking for underlying patterns to explain and understand periodicity.

The role of authors and publishers as agents of change in Spanish science pedagogical reform and textbooks has been explored by De Posada (1999). Fifty-eight high school chemistry textbooks published (1974–1998) in Spain were analyzed to understand the topic of metallic bonding. Results obtained revealed the following: (a) Almost half the textbooks simply define the metallic bond and thus obscure the relationship between models and experimental data; (b) theoretical models employed by textbooks are metaphorical in nature (similar to analogies) and are thus open to misinterpretations; (c) drawings used in textbooks need to be more explicit with respect to nature of metallic bonding; (d) the topic of metallic bonding needs to be integrated with other topics, in order to provide students meaningful learning; and (e) based on the General Act for the Educational System of 1990 (Spain), some textbooks had improved by including constructivist guidelines. This study could be extended by including the presentation of metallic bonding in university general chemistry textbooks.

In the early twentieth century, all chemical bonds were considered to be ionic (transfer of electrons), and even bonds in compounds such as methane and hydrogen were believed to be polar, despite their lack of polar properties. Lewis's (1916) theory of sharing electrons (covalent bond) when first proposed was completely out of tune with established belief. In order to understand the sharing of electrons,

44 Science Textbooks: The Role of History and Philosophy of Science

Lewis postulated the cubic atom (a theoretical device) that provided the rationale for the octet rule. For some chemists the idea of two sharing negative electrons was simply absurd and bizarre. Based on a historical reconstruction of the origin of the covalent bond, Niaz (2001b) has analyzed 27 general chemistry textbooks published (1968–1999) in the USA. Results obtained revealed (based on inter-rater agreements) that most textbooks did not deal adequately with the following aspects: (a) Postulation of the covalent bond by Lewis in 1916 posed considerable conceptual difficulties; (b) Lewis used the cubical atom in order to understand the sharing of electrons (octet rule); (c) sharing of electrons had to compete with the transfer of electrons (ionic bond), considered to be the dominant paradigm until about 1920; and (d) Pauli's exclusion principle (1925 and after) provides a theoretical explanation of the sharing of electrons, just as the cubical atom did previously. It is concluded that the transition from Lewis's cubic atom → Pauli's exclusion principle → what next provides an illustration of how scientific knowledge is tentative.

Shiland (1997) has analyzed eight high school chemistry textbooks published (1964–1994) in the USA, to evaluate the degree to which introduction of the quantum mechanical model of the atom complies with a conceptual change model (Posner et al. 1982). Four aspects of the model were operationalized by the following guidelines: (a) *Dissatisfaction* was measured by examining how the Bohr model was shown to be unsatisfactory and what specific evidence was provided; (b) *intelligibility*, which is the learner's ability to represent an idea, was measured by determining the number of pages required to present the theory; (c) *plausibility* was measured by examining whether the inadequacies of the Bohr model were addressed by the quantum model; and (d) *fruitfulness* was measured by listing problems or questions which required quantum theory to explain or predict an observable phenomena. Results obtained revealed that most textbooks did not fulfill the conditions required for conceptual change. It is concluded that high school chemistry teachers cannot rely on their textbooks to create the conditions necessary for conceptual change while introducing quantum mechanics.

Most students face considerable difficulty in understanding quantum mechanics and as a consequence use quantum numbers and electron configurations as algorithms that provide little insight in understanding progress in science. Cushing (1991) has referred to this as the level of empirical adequacy, namely, having a formula or algorithm that is capable of reproducing experimental data. Similarly, Posner and colleagues (1982) have outlined a series of criteria for promoting conceptual change that were used by Shiland (1997) to understand quantum mechanics in high school chemistry textbooks. Based on this conceptual framework, Niaz and Fernández (2008) have analyzed 55 general chemistry textbooks published (1968–2002) in the USA. Criteria based on the following aspects were elaborated to evaluate the textbooks: (a) origin of the quantum hypothesis, (b) alternative interpretations of quantum mechanics, (c) differentiation between an orbital and electron density, (d) differentiation and comparison between classical and quantum mechanics, and (e) introduction of quantum numbers based on electron density. Results obtained (based on inter-rater agreements) showed the following: (i) Most textbooks provide low dissatisfaction, intelligibility, plausibility, and fruitfulness on criteria a, b, and c;

(ii) on criterion d, some textbooks were partially intelligible (providing analogies) and partially plausible (providing thought-provoking ideas); and (iii) on criterion e, some textbooks were partially plausible (providing experimental data of ionization energies), and only one textbook fulfilled the conditions for fruitfulness by showing how heights of the peaks from photoelectron spectroscopy correspond to the number of electrons.

Teaching of chemistry at the college and university level in most parts of the world is oriented towards the preparation of chemists and other science-based professionals. *Chemistry in Context: Applying Chemistry to* Society, a textbook for non-science majors, has been developed by Schwartz and colleagues (1997) under the sponsorship of the *American Chemical Society*. Schwartz (1999) summarizes the origin, development, content, pedagogy, evaluation, and influence of this textbook and considers its potential implications for other disciplines and the instruction of science majors. The book introduces the principles of chemistry within the context of socially significant issues, such as global warming, ozone depletion, alternate energy sources, nutrition, and genetic engineering. The chemistry is included as needed to inform an understanding of chemical principles, and an additional feature is the inclusion of student-centered activities designed to promote critical thinking.

44.4 University and High School Physics Textbooks

According to Gardner (1999), although as science teachers we are generally not trained in history and philosophy of science, still "our implicit theories about the nature of science and technology influence the stories we tell our students" (p. 330). This clearly shows the need to take into consideration teachers' prior epistemological beliefs. Indeed, this is an important guideline for both textbook authors and curriculum developers. This study discusses the meanings attached to the terms "science" and "technology" and outlines four views of the nature of their relationship: (a) *idealist view* (technology as applied science), (b) *demarcationist view* (separate fields), (c) *materialist view* (technology as a necessary precursor to science), and (d) *interactionist view* (scientists and technologists working together). Five high school physics textbooks published in Canada (1986–1992) were analyzed within this perspective. Results obtained revealed various positive features, such as recognition of the human face of science and technology, frequent reference to careers in which knowledge of physics might be useful, and many illustrations of interesting technological artifacts. However, some of the textbooks are dominated by an idealist storyline which represents a limited view.

Newtonian mechanics has been the subject of critical appraisal by E. Mach (1960/1883). Mach's work was deeply rooted in his philosophy which assumed that only sensations can be known and are real, and he was particularly critical of Newtonian concepts of absolute time and space and inertial mass. Mach's views with respect to matter and motion were later known as Mach's principle, which was of heuristic value for Einstein's development of his general theory of relativity.

Assis and Zylbersztajn (2001) have analyzed five general physics textbooks published (1963–1995) in the USA, in order to trace Mach's influence in the teaching of classical mechanics. Results obtained revealed that all five textbooks are influenced by Mach and especially with respect to the dynamical operational definition of inertial mass. Interestingly, however, none of the textbooks recognized that Mach was the originator of this definition of inertial mass.

Starting from Newton to Poincaré, the concept of force in physics has been the source of considerable discussion. At the International Congress for Philosophy in Paris in 1900, Poincaré asked if the fundamental equation of dynamics, $F = ma$, is verifiable experimentally. Later he himself responded by pointing that the problem was difficult as we do not even know what force and mass are. In order to scrutinize this difficulty, Coelho (2010) has analyzed about a hundred textbooks published in the twentieth and twenty-first centuries in different parts of the world. Knowing the difficulties involved some textbooks do not provide a definition of force. Most of the textbooks, however, define force to be the cause of acceleration. It is argued that as acceleration is observable, its cause must be something real. Thus, force is real. This interpretation has been questioned by some physicists and philosophers of science. According to Coelho (2010), "… the mere fact that these two kinds of theses coexist, shows the difficulty in 'seeing' force in phenomena" (p. 91).

Galili and Tseitlin (2003) have traced the historical origin of Newton's first law (NFL), which is the law of inertia, by consulting new translations of Newton's work in the twentieth century. This literature shows the richness of NFL in understanding the meaning of inertia. These authors analyzed 40 introductory physics textbook (high school, college, and university) published in Israel and the USA. Results obtained revealed that most textbooks (even those widely used) did not refer to NFL as it was considered to be a special case of Newton's second law. Finally, the authors concluded: "Newton's First Law is far from just a trivial special case of Newton's Second Law. As such, NFL can, and should, be carefully preserved and studied in the corpus of physics knowledge transmitted through the generations" (p. 68).

The presentation of the weight concept as a gravitational force has been questioned by Galili (2001). Even popular general physics textbooks (e.g., Sears & Zemansky) through many editions used worldwide have endorsed the following definition of weight as a gravitational force: "The weight of a body is the total gravitational force exerted on the body by all other bodies in the universe." According to Galili (2001), "This obscure definition, never introduced by Newton, can be neither empirically employed nor theoretically validated" (p. 1081). As an alternative based on a historical reconstruction, Galili has suggested an operational definition that distinguishes between weight and gravitational force in the following terms: "Weight of the body is the force, which acts downwards and causes spontaneous falling. Numerically, weight is given by the product mg^*, with g^* – the acceleration of a free fall, as it is measured in a particular frame of reference" (pp. 1082–1083). The study revealed that very few textbooks (e.g., Keller, Lerner) present this definition. Furthermore, given the complexity of the issues involved, some textbooks have simply excluded the definition of the weight concept.

The concept of electric field has been the subject of a study by Pocovi and Finley (2003), based on a historical reconstruction of the ideas of M. Faraday, J. C. Maxwell, and A. Einstein. Maxwell was able to give more impetus to Faraday's ideas of lines of force by expressing them mathematically. At this stage, for the action-at-a-distance theorists, forces were transmitted at a distance, whereas for the field theorists, the transmission took place through a medium. Einstein gave the field the status of a fundamental entity, so that any electromagnetic problem was completely described by the field equations. These authors analyzed two well-known and widely used general physics textbooks published (1978, 1991) in the USA. Results obtained revealed that both textbooks did not adequately explain the field concept in the description of electromagnetic phenomena. Furthermore, the equivalence of the action-at-a-distance and field views within electrostatics was generally ignored. According to the authors such presentations are problematic, as textbooks are assumed to be accurate, complete, coherent, and the primary source of information for students.

The development of the concept and theories of magnetic field within a historical perspective has been explored by Guisasola et al. (2005). These authors analyzed 30 introductory physics and electromagnetism textbooks published (1972–1999) in the USA. Results obtained (based on inter-rater agreements) revealed that a majority of the textbooks (a) present the introduction to the theory of magnetic field in a nonproblematic, nonhistorical, and linear accumulative manner. Development of the theory of the magnetic field had to overcome many difficulties, and various theories were proposed; (b) ignore the problems which occurred in identifying the sources of stationary magnetic field and the equivalence between charges in movement and magnets; (c) do not relate the theory of magnetic field to theories from other areas of physics, such as mechanics or optics; (d) do not discuss Maxwell's laws as an attempt to unify characteristics common to electricity and magnetism; and (e) possible limitations of the theory are not discussed.

Vaquero and Santos (2001) have analyzed 38 high school and general physics textbooks published (1844–1900) in Spain in order to explore the role played by heat and kinetic theory. Authors used the following questions for evaluating the textbooks: (1) use of imponderable fluids (18 textbooks were classified as affirmative), (2) use of the term caloric to refer to heat (14 textbooks were classified as affirmative), (3) use of the concept of energy in a general form (14 textbooks were classified as affirmative), (4) use of the mechanical theory of heat (21 textbooks were classified as affirmative), and (5) use of the kinetic theory of gases (5 textbooks were classified as affirmative). According to the authors affirmative responses to the first two questions would indicate traditionalism and to questions 3 and 4, modernity. If the kinetic theory of gases appeared, then the textbooks were considered as "quite complete." This state of the textbooks is attributed to the curricular plans of Spanish universities and the overall political climate in which physics was relegated to a minor faculty. It is interesting to note that even current chemistry textbooks published in the USA (as reported by De Berg 2008a, see previous section) and physics textbooks (cf. Cotignola et al. 2002) have difficulties in the presentation of heat. Similarly, the presentation of kinetic theory in current textbooks (published in the USA, cf. Niaz 2000b) lacks the historical context in which the kinetic theory developed. Further analyses of textbooks published in Spain after 1900 could provide evidence as to how kinetic theory was introduced progressively.

An analysis of thermodynamics in general physics textbooks has led Tarsitani and Vicentini (1996) to suggest that "... several mental representations of the same subject do exist, even in the case of a 'mature' science, and in fact are often found in scientific literature ... different textbooks may expound the same subject agreeing on many phenomenological and theoretical aspects, yet disagreeing not only on the logical structure and definition of fundamental concepts, but also on the general view of the scope and object of the subject matter" (p. 51). Based on a historical reconstruction, these authors have found two approaches to presenting thermodynamics in textbooks: (a) statistical approach (following among sources, J. C. Maxwell's *theory of heat*) and (b) phenomenological approach (following primarily M. Planck's *Treatise on Thermodynamics*). Two examples of textbooks (published in the UK & USA) that follow the statistical approach are discussed. Three examples of textbooks (two published in the USA and one in Switzerland) that follow the phenomenological approach are presented. A comparison of the two types of textbooks shows that in the phenomenological approach, physical theory is nearest to experience and the kinetic theory does not play a central role. On the other hand, the statistical approach has its own problems, as it works with abstract models, which makes the explanation of the second law and entropy problematic.

Niss (2009) has emphasized the importance of metamodeling in statistical mechanics textbooks, as it facilitates a better understanding of the purpose of modeling and hence the nature of science. Textbooks were selected according to the following criteria: inclusion of a chapter on phase transitions, articulate modeling issues, represent a class of textbooks, and not too idiosyncratic in approach. Five textbooks were selected, considered to be classics in the field, and published (UK = 1; USA = 4), over a period of 70 years (1936–2001). Results obtained revealed the following: (a) Different messages are sent with respect to what constitutes a good model; (b) models are used to elaborate phenomenological theories of phase transitions; however these theories differ widely; (c) present different notions of what it means to understand a physical phenomenon; and (d) recent textbooks pay more attention to universality that is capturing different physical systems under the same umbrella. It was concluded that the five textbooks provide different metamodeling knowledge, which is not presented explicitly.

Cotignola et al. (2002) have analyzed thermodynamic concepts in seven general physics textbooks published in the USA (1991–1998). According to the authors, "Most books now use a definition of heat closer to the presently accepted one: a process of energy transfer associated with a temperature difference between the system under study and its surroundings. In spite of a correct initial definition, many authors (Resnick, Tipler, Giancoli, Serway and Hewitt) finally succumb to 'heat is a form of energy'" (p. 285). Joule's experiments referred to the different equivalent processes capable of producing the same increase in the system's temperature. Given the central role heat had in the beginnings of thermodynamics, five of the textbooks while describing Joule's experiments do not correctly differentiate between heat and energy. Only three of the textbooks correctly explain internal energy, namely, energy associated with the internal structure of the system under study.

It is well known that Thomas Kuhn directed the project "Sources for History of Quantum Physics," a valuable archive now available at various institutions around the world. Based on this experience, Kuhn raised a provocative question: Who first proposed the quantum hypothesis? Kuhn (1978) stated categorically:

> ... the arguments in Planck's first quantum papers [Planck 1990] did not, as I now read them, seem to place any restrictions on the energy of the hypothetical resonators that their author had introduced to equilibrate the distribution of energy in the black-body radiation field. Planck's resonators, I concluded, absorbed and emitted energy continuously at a rate governed precisely by Maxwell's equations. His theory was still classical .... (p. viii)

Kuhn concluded that it was P. Ehrenfest and A. Einstein who first recognized that the blackbody law could not be derived without restricting resonator energy to integral multiples of $h\nu$. In other words, Planck in 1900 simply introduced an approximate mathematical quantization in doing the calculations. Kuhn's thesis has been endorsed by some historians (e.g., Brush 2000; Kragh 1999). However, physicists have in general resented the attempt to deprive Planck of credit for the quantum hypothesis. In order to evaluate the support for Kuhn's thesis, Brush (2000) has analyzed 28 general physics textbooks published in the USA (1990–1997). Results obtained showed that only six textbooks supported Kuhn's hypothesis with respect to the origin of the quantum hypothesis. In comparison, Niaz and Fernández (2008) found that of the 55 general chemistry textbooks published in the USA, only one reluctantly supported Kuhn's hypothesis.

Tampakis and Skordoulis (2007) have examined the reception of quantum mechanics in the Greek scientific community through nine physics textbooks published in Greece (1913–1963). Authors found that the quantum theory appeared first in a textbook in 1925, with a brief mention of Planck's hypothesis and the photoelectric effect. It was not until 1962 that quantum mechanics was finally established in Greek textbooks. Authors attribute this delayed appearance to three factors: scientific, social, and ideological. It is concluded that the debate between the political left and right and the church organizations led to an extremely idealistic misinterpretation of the theory, before a more technical interpretation appeared in the textbooks.

Thought experiments (TEs) have played an important role in the history of science and have also been recognized in science education (Gilbert and Reiner 2000). Velentzas et al. (2007) have investigated the role of thought experiments in the theory of relativity and quantum mechanics in general physics textbooks (ten published in the USA, 1961–1997, and one in Greece, 1999). In general, these textbooks considered TEs to be an important tool when presenting these topics, and the following were used quite frequently: Einstein's train and elevator and Heisenberg's microscope. Despite the use of TEs, some textbooks preferred real experiments, especially the Michelson-Morley (MM) experiment in the context of the theory of relativity. Indeed, the MM experiment shows how in the history of science experiments are difficult to understand and interpret. According to Lakatos (1970, p. 162), starting from 1905, it took almost 25 years for the MM to be understood and considered as the "greatest negative experiment in the history of science." Brush (2000) has analyzed 26 general physics textbooks published in the USA (1990–1997), with respect to the "genetic" relationship between MM and Einstein's special theory of relativity (STR). Results obtained revealed that nine textbooks still attributed Einstein's theory to the negative result of the MM. Nevertheless, it is a cause of concern that these nine textbooks included those widely used all over the world (e.g., Serway, Sears, Zemansky). More recently, Arriassecq and Greca (2007) have analyzed the MM experiment in high school physics textbooks published in Argentina (n = 9, 1980–2000) and general physics textbooks published in the USA (n = 6, 1971–2001). Most textbooks in their study suggested that the starting point for Einstein's STR was the MM experiment, which contributes to generate a distorted view of the dynamics of scientific research. Even textbooks written by famous physicists (e.g., *Feynman's Lectures on Physics*) contribute to this empiricist perspective of science. The genetic relationship between the Michelson-Morley experiment and Einstein's STR in physics textbooks published in three different countries (Argentina, Greece, USA) is a good example of what Holton (1969) has referred to as the myth of experimenticism. According to this myth, progress in science is presented as the inexorable result of logically sound conclusions derived from experimentally indubitable premises.

Treagust and Harrison (2000) have analyzed Feynman's (1994) Lecture 1 "atoms in motion" in his *Six Easy Pieces* and consider it to be "a classic example of expert pedagogical content knowing" (p. 1162). The *Lecture* was found to contain effective explanations based on science content, educational context, and teacher- and student-related factors. Furthermore, these authors differentiate between scientists' explanations and science teaching explanations. Science explanations are characterized as strictly theory and evidence driven. In contrast, the *Lecture* includes rich and creative metaphors, analogies, and models based on anthropomorphisms and teleological expressions.

Millikan's (1916) determination of Planck's constant h (photoelectric effect) has been the subject of a study by Niaz et al. (2010a). Of the 103 general physics textbooks (published in the USA, between 1950s and 2000s) analyzed, only five made a brief mention of Millikan's presupposition and belief in the classical wave theory of light. A historical reconstruction shows that Millikan recognized the validity of Einstein's photoelectric equation and at the same time questioned the underlying hypothesis of light quanta. Very few textbooks mentioned that Millikan's experimental data provided support for Einstein equation but not his theory. Again, none of the textbooks mentioned that scientific theories are underdetermined by experimental evidence. Only one of the textbooks came close to referring to the dilemma faced with respect to the lack of acceptance of Einstein's quantum hypothesis in the scientific community, precisely because of the rivalry with the classical wave theory of light.

44.5 School Science Textbooks

Leite (2002) has analyzed five high school physics textbooks published (1996–1998) in Portugal based on criteria, such as historical experiments, analyses of data from historical experiments, integration of historical references within the text, use of original historical sources, evolution of science, and sociopolitical context in scientific research, among others. Results obtained (based on inter-rater agreement) led the author to conclude that the historical content included in these textbooks hardly provided students with an adequate image of science and the work of the scientists.

Milne (1998) has emphasized the role of stories in science textbooks as presentation of science cannot be reduced to just including facts. Stories can take various forms: (a) Heroic: focus on a hero who single-handedly contributed to the development of science. (b) Discovery: scientific knowledge is presented as having occurred as the result of an accident. (c) Declarative: processes or scientific concepts as objects that are open to observation by anyone. (d) Politically correct: a critical examination of the interaction between science and society. Examples from high school textbooks are provided to illustrate heroic stories, such as Galileo as an individual who had the courage to stand against the dark forces of the inquisition. Milne (1998) concluded: "… if we wish to involve students more in thinking about the enterprise that we call science, we would do well to tell stories that emphasize the human aspects of the development of scientific knowledge" (p. 186). Milne also presented her own version of the life and work of Galileo. This version has been critiqued by Whitaker (1999), as it lacked some historical details and the necessary differentiation between "observations" and "experimentation." The role of experimentation in Galileo's work is the subject of considerable debate. Milne (1999) responded by pointing out that her story did not represent the truth and that there are various ways to elaborate stories in science.

The role of ideologies in Norwegian grade 8 science textbooks (n = 4, published in 1997) has been explored by Knain (2001). Ideologies are considered to be grounded in worldviews (Cobern 1996) which are culturally influenced and shared by people through social interaction. For each textbook, 30 pages were selected from the following topics: nature of science (introductory chapter), nature (astronomy), and society (fighting diseases). For example, portraying scientific development as dependent on the work of individual scientists doing crucial experiments is evident from the following historical note in one of the textbooks: "An Englishman, Alexander Fleming, discovered that a certain mould fungus with the name *Penicillin* precipitates a substance that kills bacteria" (p. 325). For most textbooks in this study, experiments and observations are what make science different from other ways of knowing. Furthermore, controversies constitute an important part of science as experimental results are difficult to interpret and their relevance in the sociopolitical context is not obvious.

Importance of biographies of scientists in science education has been explored by Dagher and Ford (2005). Authors analyzed the images of science and scientists in 12 biographies of historic and contemporary scientists written for primary and middle school children, published in the USA (1987–2003), and addressed the following research questions: (1) How is the scientist described in the biography? (2) What is the nature and process of scientific knowledge? (3) How are social processes related to science portrayed? Results obtained revealed a marked difference in how different authors portrayed their subject depending on the age of the target audience. Biographies for primary school (e.g., M. Curie & A. Einstein) emphasize personal characteristics and childhood of the scientists. Biographies for secondary school provide more details of nature of scientific work and processes of science. Authors concluded that biographies can be used: "… to provide useful springboards for arousing student curiosity and interest in exploring the historical record" (p. 391).

As a topic of the science curriculum, teaching of the pendulum incorporates conceptual, methodological, and cultural aspects (Matthews et al. 2005). Based on these aspects Koliopoulos and Constantino (2005) have analyzed school science textbooks (primary and secondary) published in Greece and Cyprus. Results obtained revealed that in both countries the pendulum is confined mainly to the study of the simple pendulum, which is incidental and limited and never introduced as a comprehensive unit. At the gymnasium level the textbooks espouse an empirical approach in which the dependence of the period on the length of the string of the simple pendulum and the acceleration of gravity emerge from a simple observation of the pendulum motion. At the lyceum level apart from the experimental activity, textbooks include the derivation of mathematical relations. Given the pendulum's historical connection with clock making, time measurement, the longitude problem, and navigation, it readily constitutes "…a window on the scientific revolution" (Matthews 2000, p. 293). Despite this rich cultural context, we found only one study dealing with the presentation of the pendulum in textbooks.

44.6 Educational Implications

Various topics of the science curriculum provide an opportunity to illustrate the tentative nature of scientific knowledge, and still very few textbooks referred to this important aspect of nature of science. American Association for the Advancement of Science has expressed this in cogent terms: "The notion that scientific knowledge is always subject to modification can be difficult for students to grasp. It seems to oppose the certainty and truth popularly accorded to science, and runs counter to the yearning for certainty that is characteristic of most cultures, perhaps especially so among youth" (AAAS 1993, p. 5). Indeed, divergent opinions that often lead to controversies are one of the most important aspects of scientific progress (Niaz 2009a). Modern philosophers of science have referred to this facet of nature of science in explicit terms:

> Many major steps in science, probably all dramatic changes, and most of the fundamental achievements of what we now take as the advancement or progress of scientific knowledge have been controversial and have involved some dispute or another. Scientific controversies are found throughout the history of science. This is so well known that it is trivial. What is not so obvious and deserves attention is a sort of paradoxical dissociation between science as actually practiced and science as perceived or depicted by both scientists and philosophers. While nobody would deny that science in the making has been replete with controversies, the same people often depict its essence or end product as free from disputes, as the uncontroversial rational human endeavor par excellence. (Machamer et al. 2000, p. 3)

Similarly, Dybowski (2001), a practicing teacher, has designed an innovative course in the history of physical chemistry that facilitates an appreciation of how scientific inquiry actually happens in research laboratories and recounted his experience: "… we sometimes find that we come away with two (or more) differing views of events that cannot be reconciled. At first, this failure to bring closure is disconcerting to some students, but one underlying theme of the course is the appreciation of the *possibility of divergent opinions on certain issues, even in a science like chemistry*" (p. 1623, emphasis added). Different approaches to teaching thermodynamics (cf. Tarsitani and Vicentini 1996, physics section) are illustrative of this dilemma.

Most textbooks in this review presented the experimental details, without the conceptualization that progress in science is based on competing frameworks of understanding that clash in the face of evidence. Writing in a special issue dedicated to *Nature of Science in Science Education*, Lederman et al. (1998) asked a very pertinent question, *"what seems to be the problem?"* and then responded:

> Most texts briefly, and inadequately, discuss the nature of science in the opening chapter, and then portray science in a distorted, positivistic, and 'final form' fashion throughout the rest of the book. (p. 507)

Despite some improvement, it seems that problems with respect to most textbooks seem to be the same. In this context, Kubli (2005) has emphasized the role of "real" experiments: "The element of uncertainty – the sure sign of a *real experiment* – is drastically reduced if we repeat an experiment whose result can be deduced from the textbook. But even in the classroom, we can regenerate something of the pioneering spirit if we show how difficult it originally was to arrive at a result, and if we can convey the fascination of breaking new ground" (pp. 515–516, italics added). Based on wave-particle duality, Niaz and Marcano (2012) have provided further evidence on this aspect.

Almost half a century ago, Polanyi (1964) had drawn our attention to an important facet of textbooks, by emphasizing the degree to which established knowledge in textbooks departs from the events associated with the original discovery:

> Yet as we pursue scientific discoveries through their consecutive publication on their way to the textbooks, which eventually assures their reception as part of established knowledge by successive generations of students, and through these by the general public, we observe that the intellectual passions aroused by them appear gradually toned down to a faint echo of their discoverer's first excitement at the moment of Illumination … A transition takes place here from a heuristic act to the routine teaching and learning of its results, and eventually to the mere holding of these as known and true, in the course of which the personal participation of the knower is altogether transformed. (pp. 171–172)

Drawing on the "events associated with the original discovery," "excitement at the moment of illumination," and teaching science as practiced by scientists can indeed be an important guideline for future science textbooks (Niaz 2010b).

44.7 Conclusion

This chapter analyzed 52 studies published over a period of 15 years (1969–2010) in four major science education journals, and the following are some of the important findings:

44 Science Textbooks: The Role of History and Philosophy of Science

(a) Most biology, chemistry, physics, and school science textbooks lack a historical perspective required to facilitate a better understanding of the dynamics of scientific progress.
(b) Most of the textbooks analyzed were published in the USA and to a much lesser extent from the following countries: Argentina, Australia, Brazil, Canada, Cyprus, Greece, Israel, the Netherlands, Norway, Portugal, Spain, Sweden, Switzerland, Turkey, and the UK.
(c) Few studies provided details of the procedure and reliability of the application of the criteria/rubric for analyzing textbooks. One study (Chiappetta and Fillman 2007) used Cohen's *kappa* statistic, and some studies provided details of inter-rater agreements.
(d) Studies analyzed in this chapter refer to a wide range of 21 subjects that form part of the science curriculum, such as the following: nature of science = 9, atomic structure = 8, Newtonian mechanics = 5, quantum mechanics = 4, special theory of relativity (Michelson-Morley experiment) = 4, evolution = 4, gene concept = 1, oil drop experiment = 2, heat and temperature = 2, chemical bonding = 2, electromagnetism = 2, thermodynamics = 2, science stories = 2, laboratory manuals (biology) = 1, acids and bases = 1, osmotic pressure = 1, periodic table = 1, kinetic theory = 1, statistical mechanics = 1, photoelectric effect = 1, and pendulum = 1 (note this gives a total of 56 instead of 52, as some studies dealt with more than one subject). In all these studies, a majority of the textbooks lacked a history and philosophy of science (HPS) perspective. However, it is important to note that a small number of textbooks did provide material based on HPS that can further students' understanding of science. This shows that HPS is already "inside" the science curriculum, provided textbook authors and teachers make an effort to scrutinize the historical record.
(e) Review of the literature in this chapter revealed that if a topic/concept was difficult/controversial, textbooks try to avoid presenting it, and the following are some of the examples: (i) concept of force (Coelho 2010), (ii) concept of weight (Galili 2001), (iii) Newton's first law (Galili and Tseitlin 2003), (iv) inclusion of evolution which can facilitate "removal of prejudices" (Barberá et al. 1999; Skoog 2005), (v) difference between heat and temperature (Cotignola et al. 2002; de Berg 2008a), (vi) different approaches to thermodynamics (Tarsitani and Vicentini 1996), (vii) role of atomic theory in the development of the periodic table (Brito et al. 2005), (viii) postulation of the sharing of electrons, covalent bond (Niaz 2001b), (ix) origin of the quantum hypothesis (Brush 2000; Niaz and Fernández 2008), (x) special theory of relativity and the Michelson-Morley experiment (Brush 2000; Arriassecq and Greca 2007), (xi) interpretation of alpha particle experiments by Thomson and Rutherford (Niaz 1998), (xii) interpretation of oil drop experiment by Millikan and Ehrenhaft (Holton 1978; Niaz 2000a), and (xiii) photoelectric effect. Millikan's data supported Einstein's equation but not his theory (Niaz et al. 2010a).

References

Abd-El-Khalick, F. (2005). Developing deeper understandings of nature of science: The impact of a philosophy of science course on preservice science teachers' views and instructional planning. *International Journal of Science Education, 27*, 15–42.

Abd-El-Khalick, F., Waters, M., & Le, A. (2008). Representation of nature of science in high school chemistry textbooks over the past four decades. *Journal of Research in Science Teaching, 45*(7), 835–855.

American Association for the Advancement of Science, AAAS (1989). *Project 2061: Science for all Americans*. Washington, DC: AAAS.

American Association for the Advancement of Science, AAAS (1993). *Benchmarks for science literacy: Project 2061*. New York: Oxford University Press.

Arriassecq, I., & Greca, I.M. (2007). Approaches to the teaching of special relativity theory in high school and university textbooks of Argentina. *Science & Education, 16*(1), 65–86.

Assis, A.K.T., & Zylbersztajn, A. (2001). The influence of Ernst Mach in the teaching of mechanics. *Science & Education, 10*(1–2), 137–144.

Barberá, O., Zanón, B., & Pérez-Plá, J.F. (1999). Biology curriculum in twentieth century Spain. *Science Education, 83*(1), 97–111.

Bensaude-Vincent, B. (2006). *Textbooks* on the map of science studies. *Science & Education, 15*, 667–670.

Bevilacqua, F., & Bordoni, S. (1998). New contents for new media: Pavia project physics. *Science & Education, 7*, 451–469.

Brito, A., Rodríguez, M.A., & Niaz, M. (2005). A reconstruction of development of the periodic table based on history and philosophy of science and its implications for general chemistry textbooks. *Journal of Research in Science Teaching, 42*(1), 84–111.

Brush, S.G. (1976). *The kind of motion we call heat: A history of the kinetic theory of gases in the 19th century*. New York: North-Holland.

Brush, S.G. (1989). History of science and science education. *Interchange, 9*, 39–58.

Brush, S.G. (2000). Thomas Kuhn as a historian of science. *Science & Education, 9*(1–2), 39–58.

Bybee, R. W. (1989). Teaching high school biology: Materials and strategies, In W. G. Rosen (Ed.), *High school biology today and tomorrow* (pp. 165–177). Washington, DC: National Academy Press.

Campbell, N.A., & Reece, J.B. (2005). *Biology* (7th ed.). San Francisco: Pearson.

Chiappetta, E.L., Ganesh, T.G., Lee, Y.H., & Phillips, M.C. (2006, April). Examination of science textbook analysis research conducted on textbooks published over the past 100 years in the United States. Paper presented at the Annual Conference of the National Association for Research in Science Teaching (NARST), San Francisco.

Chiappetta, E.L., & Fillman, D.A. (2007). Analysis of five high school biology textbooks used in the United States for inclusion of the nature of science, *International Journal of Science Education, 29*(15), 1847–1868.

Cobern, W.W. (1996). Worldview theory and conceptual change in science education. *Science Education, 80*(5), 579–610.

Coelho, R.L. (2010). On the concept of force: How understanding its history can improve physics teaching. *Science & Education, 19*(1), 91–113.

Cooper, L.N. (1970). *An introduction to the meaning and structure of physics* (short edition). New York: Harper & Row.

Cotignola, M.I., Bordogna, C., Punte, G., & Cappannini, O.M. (2002). Difficulties in learning thermodynamic concepts: Are they linked to the historical development of this field? *Science & Education, 11*(3), 279–291.

Cushing, J.T. (1991). Quantum theory and explanatory discourse: Endgame for understanding. *Philosophy of Science, 58*, 337–358.

Dagher, Z.R., & Ford, D.J. (2005). How are scientists portrayed in children's science biographies? *Science & Education, 14*(3–5), 377–393.

De Berg, K.C. (2006). The kinetic-molecular and thermodynamic approaches to osmotic pressure: A study of dispute in physical chemistry and the implications for chemistry education. *Science & Education, 15*(5), 495–519.

De Berg, K.C. (2008a). The concepts of heat and temperature: The problem of determining the content for the construction of an historical case study which is sensitive to nature of science issues and teaching-learning issues. *Science & Education, 17*(1), 75–114.

De Berg, K.C. (2008b). Tin oxide chemistry from Macquer (1758) to Mendeleeff (1891) as revealed in the textbooks and other literature of the era. *Science & Education, 17*(2–3), 265–287.

De Posada, J.M. (1999). The presentation of metallic bonding in high school science textbooks during three decades: Science educational reforms and substantive changes of tendencies. *Science Education, 83*, 423–447.

Dybowski, C.R. (2001). A course in the history of physical chemistry with an emphasis on writing. *Journal of Chemical Education, 78*(12), 1623–1625.

Feynman, R.P. (1994). *Six easy pieces*. Reading, MA: Helix Books.

Flodin, V.S. (2009). The necessity of making visible concepts with multiple meanings in science education: The use of the gene concept in a biology textbook. *Science & Education, 18*(1), 73–94.

Furió, C., Azcona, R., Guisasola, J., & Ratcliffe, M. (2000). Difficulties in teaching the concepts of 'amount of substance' and 'mole'. *International Journal of Science Education, 22*(12), 1285–1304.

Furió-Más, C., Calatayud, M.L., Guisasola, J., & Furió-Gómez, C. (2005). How are the concepts and theories of acid–base reactions presented? Chemistry in textbooks and as presented by teachers. *International Journal of Science Education, 27*(11), 1337–1358.

Galili, I. (2001). Weight versus gravitational force: Historical and educational perspectives. *International Journal of Science Education, 23*(10), 1073–1093.

Galili, I., & Tseitlin, M. (2003). Newton's first law: Text, translations, interpretations and physics education. *Science & Education, 12*(1), 45–73.

Gardner, P.L. (1999). The representation of science-technology relationships in Canadian physics textbooks. *International Journal of Science Education, 21*(3), 329–347.

Gericke, N.M., & Hagberg, M. (2010). Conceptual variation in the depiction of gene function in upper secondary school textbooks. *Science & Education, 19*(10), 963–994.

German, P.J., Haskins, S., & Auls, S. (1996). Analysis of nine high school biology laboratory manuals: Promoting scientific inquiry. *Journal of Research in Science Teaching, 33*(5), 475–499.

Gilbert, J.K., & Reiner, M. (2000). Thought experiments in science education: Potential and current realization. *International Journal of Science Education, 22*(3), 265–283.

Gooday, G., Lynch, J.M., Wilson, K.G., & Barsky, C.K. (2008). Does science education need the history of science? *Isis, 99*, 322–330.

Guisasola, J., Almudía, J.M., & Furió, C. (2005). The nature of science and its implications for physics textbooks: The case of classical magnetic field theory. *Science & Education, 14*(3–5), 321–338.

Hodson, D. (1985). Philosophy of science, science, and science teaching. *Studies in Science Education, 12*, 25–57.

Hodson, D. (1988). Toward a philosophically more valid science curriculum. *Science Education, 72*, 19–40.

Hodson, D. (2008). *Towards Scientific Literacy: A Teachers Guide to the History, Philosophy and Sociology of Science*. Rotterdam: Sense Publishers.

Hodson, D. (2009). *Teaching and learning about science: Language, theories, methods, history, traditions and values*. Rotterdam: Sense Publishers.

Hofmann, J.R., & Weber, B.H. (2003). The fact of evolution: Implications for science education. *Science & Education, 12*(8), 729–760.

Holton, G. (1952). *Introduction to concepts and theories in physical science*. New York: Addison-Wesley.

Holton, G. (1969). Einstein and 'crucial' experiments. *American Journal of Physics, 37*(10), 968–982.

Holton, G. (1978). Subelectrons, presuppositions, and the Millikan-Ehrenhaft dispute. *Historical Studies in the Physical Sciences, 9*, 161–224.

Holton, G. (2000). Personal communication, September.

Holton, G. (2003). The project physics course, then and now. *Science & Education, 12*, 779–786.

Holton, G., & Brush, S.G. (2001). *Physics, the human adventure: From Copernicus to Einstein and beyond* (3rd ed.) New Brunswick, NJ: Rutgers University Press.

Hosson, C., & Kaminski, W. (2007). Historical controversy as an educational tool: Evaluating elements of a teaching-learning sequence conducted with the text "Dialogues on the ways that vision operates." *International Journal of Science Education, 29*, 617–642.

Irez, S. (2009). Nature of science as depicted in Turkish biology textbooks. *Science Education, 93*, 422–447.

Justi, R.S., & Gilbert, J.K. (2000). History and philosophy of science through models: Some challenges in the case of 'the atom'. *International Journal of Science Education, 22*, 993–1009.

Kindi, V. (2005). Should science teaching involve the history of science? An assessment of Kuhn's view. *Science & Education, 14*, 721–731.

Klassen, S. (2009). Identifying and addressing student difficulties with the Millikan oil drop experiment. *Science & Education, 18*, 593–607.

Klopfer, L.E. (1969). The teaching of science and the history of science. *Journal of Research in Science Teaching, 6*, 87–95.

Knain, E. (2001). Ideologies in school science textbooks. *International Journal of Science Education, 23*(3), 319–329.

Koliopoulos, D., & Constantinou, C. (2005). The pendulum as presented in school science textbooks of Greece and Cyprus. *Science & Education, 14*(1), 59–73.

Kragh, H.A. (1992). Sense of history: History of science and the teaching of introductory quantum theory. *Science & Education, 1*, 349–363.

Kragh, H. (1999). *Quantum generations: A history of physics in the twentieth century*. Princeton, NJ: Princeton University Press.

Kubli, F. (2005). Science teaching as a dialogue --- Bakhtin, Vygotsky and some applications in the classroom. *Science & Education, 14*, 501–534.

Kuhn, T.S. (1977). The function of measurement in modern physical research (first published in 1961). In T.S. Kuhn (Ed.), *The essential tension* (pp. 178–224). Chicago: University of Chicago Press.

Kuhn, T.S. (1978). *Black-body theory and the quantum discontinuity: 1894–1912*. New York: Oxford University Press.

Lakatos, I. (1970). Falsification and the methodology of scientific research programmes. In I. Lakatos & A. Musgrave (Eds), *Criticism and the growth of knowledge* (pp. 91–195). Cambridge, UK: Cambridge University Press.

Lederman, N.G., McComas, W.F., & Matthews, M.R. (1998). Editorial. *Science & Education, 7*(6), 507–509.

Leite, L. (2002). History of science in science education: Development and validation of a checklist for analyzing the historical content of science textbooks. *Science & Education, 11*(4), 333–359.

Lewis, G.N. (1916). The atom and the molecule. *Journal of American Chemical Society, 38*, 762–785.

Linn, M.C., Songer, N.B., & Lewis, E.L. (1991). Overview: Students' models and epistemologies of science. *Journal of Research in Science Teaching, 28*, 729–732.

Mach, E. (1960). *The science of mechanics --- A critical and historical account of its development*. La Salle, IL: Open Court.

Machamer, P., Pera, M., & Baltas, A. (2000). Scientific controversies: An introduction. In P. Machamer, M. Pera & A. Baltas (Eds.), *Scientific controversies: Philosophical and historical perspectives* (pp. 3–17). New York: Oxford University Press.

Matthews, M.R. (1990). History, philosophy, and science teaching: A rapprochement. *Studies in Science Education, 18*, 25–51.

Matthews, M.R. (1994/2014). *Science teaching: The contribution of history and philosophy of science.*New York: Routledge.

Matthews, M.R. (1998). In defense of modest goals when teaching about the nature of science. *Journal of Research in Science Teaching, 35*, 161–174.

Matthews, M.R. (2000). *Time for science education.* New York: Kluwer/Plenum.

Matthews, M.R., Gauld, C.F., & Stinner, A. (2005). Eds., *The pendulum: Scientific, historical, philosophical and educational perspective.* Dordrecht, The Netherlands: Springer.

Mayr, E. (1997). *This is biology.* Cambridge, MA: Harvard University Press (Belnap).

Mendeleev, D. (1889). The periodic law of the chemical elements (Faraday lecture, delivered on 4 June, 1889). *Journal of the Chemical Society, 55*, 634–656.

Millikan, R.A. (1916). A direct photoelectric determination of Planck's *h. Physical Review, 7*, 355–388.

Milne, C. (1998). Philosophically correct science stories? Examining the implications of heroic science stories for school science. *Journal of Research in Science Teaching, 35*(2), 175–187.

Milne, C. (1999). "Only some facts matter for my given pattern": The fact of stories in school science. A response to Whitaker. *Journal of Research in Science Teaching, 36*(10), 1155–1157.

Moody, D.E. (1996). Evolution and the textbook structure of biology. *Science Education, 80*(4), 395–418.

National Research Council, NRC (1996). *National science education standards.* Washington, DC: National Academy Press.

Niaz, M. (1998). From cathode rays to alpha particles to quantum of action: A rational reconstruction of structure of the atom and its implications for chemistry textbooks. *Science Education, 82*, 527–552.

Niaz, M. (2000a). The oil-drop experiment: A rational reconstruction of the Millikan-Ehrenhaft controversy and its implications for chemistry textbooks. *Journal of Research in Science Teaching, 37*, 480–508.

Niaz, M. (2000b). A rational reconstruction of the kinetic molecular theory of gases based on history and philosophy of science and its implications for chemistry textbooks. *Instructional Science, 28*(1), 23–50.

Niaz, M. (2001a). How important are the laws of definite and multiple proportions in chemistry and teaching chemistry? A history and philosophy of science perspective. *Science & Education, 10*, 243–266.

Niaz, M. (2001b). A rational reconstruction of the origin of the covalent bond and its implications for general chemistry textbooks. *International Journal of Science Education, 23*(6), 623–641.

Niaz, M. (2008). *Teaching general chemistry: A history and philosophy of science approach.* New York: Nova Science Publishers.

Niaz, M. (2009a). *Critical appraisal of physical science as a human enterprise: Dynamics of scientific progress.* Dordrecht, The Netherlands: Springer.

Niaz, M. (2009b). Progressive transitions in chemistry teachers' understanding of nature of science based on historical controversies. *Science & Education, 18*, 43–65.

Niaz, M. (2010a). Science curriculum and teacher education: The role of presuppositions, contradictions, controversies and speculations vs Kuhn's 'normal science.' *Teaching and Teacher Education, 26*, 891–899.

Niaz, M. (2010b). Are we teaching science as practiced by scientists? *American Journal of Physics, 78*(1), 5–6.

Niaz, M. (2011). *Innovating science teacher education: A history and philosophy of science perspective.* New York: Routledge.

Niaz, M., & Coştu, B. (2009). Presentation of atomic structure in Turkish general chemistry textbooks. *Chemistry Education Research and Practice, 10*, 233–240.

Niaz, M., & Fernández, R. (2008). Understanding quantum numbers in general chemistry textbooks. *International Journal of Science Education, 30*(7), 869–901.

Niaz, M., Klassen, S., McMillan, B., & Metz, D. (2010a). Reconstruction of the history of the photoelectric effect and its implications for general physics textbooks. *Science Education, 94*, 903–931.

Niaz, M., Klassen, S., McMillan, B., & Metz, D. (2010b). Leon Cooper's perspective on teaching science: An interview study. *Science & Education, 19*(1), 39–54.

Niaz, M., & Marcano, C. (2012). *Reconstruction of wave-particle duality and its implications for general chemistry textbooks*. Dordrecht, The Netherlands: Springer Briefs in Education.

Niaz, M., & Maza, A. (2011). *Nature of science in general chemistry textbooks*. Dordrecht, The Netherlands: Springer Briefs in Education.

Niaz, M., & Rodríguez, M.A. (2001). Do we have to introduce history and philosophy of science or is it already 'inside' chemistry? *Chemistry Education: Research and Practice in Europe, 2*, 159–164.

Niaz, M., & Rodríguez, M.A. (2005). The oil drop experiment: Do physical chemistry textbooks refer to its controversial nature? *Science & Education, 14*, 43–57.

Niss, M. (2009). Metamodelling messages conveyed in five statistical mechanical textbooks from 1936 to 2001. *International Journal of Science Education, 31*(5), 697–719.

Padilla, K., & Furio-Mas, C. (2008). The importance of history and philosophy of science in correcting distorted views of 'amount of substance' and 'mole' concepts in chemistry teaching. *Science & Education, 17*(4), 403–424.

Planck, M. (1900). Zur theorie des gesetzes der energieverteilung im normalspectrum. *Verhandlungen der Deutschen Physikalische Gesellschaft, 2*, 237–245.

Pocoví, M.C. (2007). The effects of a history-based instructional material on the students' understanding of field lines. *Journal of Research in Science Teaching, 44*, 107–132.

Pocoví, M.C., & Finley, F.N. (2003). Historical evolution of the field view and textbook accounts. *Science & Education, 12*(4), 387–396.

Polanyi, M. (1964). *Personal knowledge*. Chicago: University of Chicago Press (first published 1958).

Posner, G.J., Strike, K.A., Hewson, P.W., & Gertzog, W.A. (1982). Accommodation of a scientific conception: Toward a theory of conceptual change. *Science Education, 66*, 211–227.

Reinmuth, O. (1932). Editor's outlook. *Journal of Chemical Education, 9*, 1139–1140.

Roberts, L. (2001). The human genome. *Science, 2991*(5507), 1177–1188.

Robinson, J.T. (1969). Philosophy of science: Implications for teacher education. *Journal of Research in Science Teaching, 6*, 99–104.

Rodríguez, M.A., & Niaz, M. (2002). How in spite of the rhetoric, history of chemistry has been ignored in presenting atomic structure in textbooks. *Science & Education, 11*(5), 423–441.

Rodríguez, M.A., & Niaz, M. (2004a). A reconstruction of structure of the atom and its implications for general physics textbooks. *Journal of Science Education and Technology, 13*, 409–424.

Rodríguez, M.A., & Niaz, M. (2004b). The oil drop experiment: An illustration of scientific research methodology and its implications for physics textbooks. *Instructional Science, 32*, 357–386.

Rutherford, E. (1911). The scattering of alpha and beta particles by matter and the structure of the atom. *Philosophical Magazine, 21*, 669–688.

Scheffler, I. (1992). Philosophy and the curriculum. *Science and Education, 1*(4), 385–394.

Schwab, J.J. (1962). *The teaching of science as enquiry*. Cambridge, MA: Harvard University Press.

Schwab, J.J. (1974). The concept of the structure of a discipline. In E.W. Eisner & E. Vallance (Eds.) *Conflicting conceptions of curriculum* (pp. 162–175). Berkeley, CA: McCutchan Publishing Corp.

Schwartz, A.T. (1999). Creating a context for chemistry. *Science & Education, 8*(6), 605–618.

Schwartz, A.T., Bunce, D.M., Silberman, R.G., Stanitski, C.L., Straton, W.J., & Zipp, A.P. (1997). *Chemistry in context: Applying chemistry to society* (2nd ed.). Dubuque, IA: Wm. C. Brown.

Shiland, T.W. (1997). Quantum mechanics and conceptual change in high school chemistry textbooks. *Journal of Research in Science Teaching, 34*(5), 535–545.

Siegel, H. (1978). Kuhn and Schwab on science texts and the goals of science education. *Educational Theory, 28*, 302–309.

Skoog, G. (2005). The coverage of human evolution in high school biology textbooks in the 20th century and in current state science standards. *Science & Education, 14*(3–5), 395–422.

Smith, M.U. (2010). Current status of research in teaching and learning evolution: I. Philosophical/epistemological issues. *Science & Education, 19*, 523–538.

Tampakis, C., & Skordoulis, C. (2007). The history of teaching quantum mechanics in Greece. *Science & Education, 16*(3–5), 371–391.

Tarsitani, C., & Vicentini, M. (1996). Scientific mental representations of thermodynamcis. *Science & Education, 5*(1), 51–68.

Toon, E.R., & Ellis, G.L. (1978). *Foundations of chemistry.* Toronto: Holt, Rinehart & Winston.

Toon, E.R., Ellis, G.L., & Brodkin, J. (1968). *Foundations of chemistry.* New York: Holt, Rinehart & Winston.

Treagust, D.F., & Harrison, A.G. (2000). In search of explanatory frameworks: An analysis of Richard Feynman's lecture 'atoms in motion'. *International Journal of Science Education, 22*(11), 1157–1170.

Van Berkel, B., De Vos, W., & Verdonk, A.H., & Pilot, A. (2000). Normal science education and its dangers: The case of school chemistry. *Science & Education, 9*(1–2), 123–159.

Vaquero, J.M., & Santos, A. (2001). Heat and kinetic theory in 19th-century physics textbooks: The case of Spain. *Science & Education, 10*(3), 307–319.

Velentzas, A., Halkia, K., & Skordoulis, C. (2007). Thought experiments in the theory of relativity and in quantum mechanics: Their presence in textbooks and in popular science textbooks. *Science & Education, 16*(3–5), 353–370.

Wells, J. (2000). *Icons of evolution, science or myth? Why much of what we teach about evolution is wrong.* Washington, DC: Regnery.

Whitaker, R.J. (1999). Reflections on Catherine Milne's "Philosophically correct stories? Examining the implications of heroic science stories for school science". *Journal of Research in Science Teaching, 36*(10), 1148–1154.

Wilson, D. (1983). *Rutherford: Simple genius.* Cambridge, MA: MIT Press.

Zemplén, G.A. (2007). Conflicting agendas: Critical thinking versus science education in the international baccalaureate *theory of knowledge* course. *Science & Education, 16*, 167–196.

Mansoor Niaz received his B.Sc. (1966) and M.Sc. (1968) from Peshawar University, Pakistan, and then took graduate courses (1969–1971) in physical chemistry at Yale University. He worked as a Senior Fulbright Research Scholar at Purdue University (1988–1989). At present he is a Professor of Science Education at Universidad de Oriente, Venezuela. His research interests include the application of history, philosophy of science, and cognitive psychology to science education. He has served/been serving on the editorial boards of the following journals: *Science & Education, Journal of Research in Science Teaching, International Journal of Science Education, Journal of Science Education and Technology, Interchange, Chemistry Education Research and Practice, International Journal of Science and Mathematics Education, Educación Química, and Revista de Educación en Ciencias.* Niaz has published over 150 articles in international refereed journals. Three recent publications: (a) Niaz (2009). *Critical appraisal of physical science as a human enterprise: Dynamics of scientific progress.* Springer, Dordrecht; (b) Niaz (2011). *Innovating science teacher education: A history and philosophy of science perspective.* Routledge, New York; and (c) Niaz (2012). *From 'science in the making' to understanding the nature of science: An overview for science educators.* Routledge, New York.

Chapter 45
Revisiting School Scientific Argumentation from the Perspective of the History and Philosophy of Science

Agustín Adúriz-Bravo

45.1 Argumentation in Science and in Science Education

The purpose of this chapter is to examine the notion of scientific argumentation as it is applied in the realm of science education nowadays, but this examination is done – in accordance with the thematic thread of this handbook – shifting from the extensively used discursive perspective to one centred on *metatheoretical* issues. In order to set an initial consensus for the discussion that follows, it might be convenient to advance here a broad definition of argumentation, which will be eventually revisited to incorporate more theoretical elements. Using the phrasing on the back cover of Myint Swe Khine's (2012, n/p) compilation, scientific argumentation could be loosely identified with 'arriving at conclusions on a topic through a process of logical reasoning that includes debate and persuasion'. This definition points out that an argument typically involves (a) supporting an assertion on other elements, (b) a range of options when choosing such elements and (c) strategies to convince the argument's recipients that the favoured option is appropriate.

Literature reviews around argumentative practices in the science classroom rapidly conduct to acknowledging that argumentation is a central issue or focus – or more properly a 'line of research' (Jiménez-Aleixandre and Erduran 2008) or a 'strand' (Nielsen 2011) – within current didactics of science (i.e. science education as an academic discipline). However, such reviews show, at the same time, that 'argumentation in the field of science education has constituted itself into a multi-disciplinary topic, most profoundly approached from language sciences' (Archila 2012, p. 363;

A. Adúriz-Bravo (✉)
GEHyD-Grupo de Epistemología, Historia y Didáctica de las Ciencias Naturales,
CeFIEC-Instituto de Investigaciones Centro de Formación e Investigación en Enseñanza de las Ciencias, Universidad de Buenos Aires, Ciudad Autónoma de Buenos Aires, Argentina
e-mail: aadurizbravo@cefiec.fcen.uba.ar

M.R. Matthews (ed.), *International Handbook of Research in History, Philosophy and Science Teaching*, DOI 10.1007/978-94-007-7654-8_45,
© Springer Science+Business Media Dordrecht 2014

my translation). Hence, the interest of this chapter to recover an *epistemic* focus, which could be broadly defined, borrowing Greg Kelly and Charles Bazerman's words, as the recognition

> that writing and argument play important roles in scientists' and technologists' thinking and forming knowledge communities [...]. The forms of expression, invention, and knowledge are responsive to the particular argumentative fields of the professions and disciplines. The epistemic activity of researchers is shaped by rhetorical concerns of who is to be convinced of what, how others respond to novel work, what the organization of their communicative activity is, and what the goals of community cooperation are [...]. The representation and role of evidence in relation to generalizations and claims has been a particularly crucial matter in the development of scientific argument. (Kelly and Bazerman 2003, pp. 28–29)

Indeed, argumentation has been recognised by some traditions, authors and texts in the philosophy of science as a key epistemic feature of the scientific enterprise,[1] i.e. a feature constitutive of its very nature, which serves to *demarcate* science from other human activities. It could arguably be stated that

> the majority of philosophical conceptions on the structure of a scientific theory, as well as some of the most important models of [scientific] explanation, incorporate argumentation (understood as justifying inferences) as a central piece in the scientific machinery. (Asti Vera and Ambrosini 2010, p. 6; my translation)

This argumentation-based perspective on the nature of science is apparent in Stephen Toulmin's (1958) famous book, *The uses of argument*, especially in essay IV, where he examines 'substantial arguments' in the experimental sciences. But it should be noted that although argumentation-like processes have been consistently considered in the metatheoretical discussion of scientific processes and products by philosophers (e.g. Giere et al. 2005), the use of the expression 'scientific argumentation' is not as extended as it could be expected within the philosophy of science – at least until very recently. This may be partly due to the concealment of the more elaborate communicative aspects of science in the rather formalist, syntactic view 'received' from the Vienna Circle. In the philosophy of science, the idea of scientific argumentation has been very usually rephrased in terms of explanation, justification, debate, controversy, judgement, persuasion, rhetoric, etc.

Many portrayals of science-in-the-making have pointed to the existence of an extremely elaborate, social, use of *evidences* to give support to our complex, articulated understandings of the natural world (i.e. *scientific explanations*) and, at the same time, to *convince* other people that such understandings are plausible

---

[1] See the following 'focussed' philosophy of science textbooks for more or less extensive discussions around philosophers that inspect the centrality of argumentation in science: Asti Vera and Ambrosini (2010), Føllesdal and Walløe (1986), and Salmon (1995). Also of particular interest for this chapter are the portrayals of the 'combined' scientific practice of argumentation-explanation that revolve around the notion of *abductive* reasoning (cf., Adúriz-Bravo 2005; Aliseda 2006; Bex and Walton 2012; Giere 1988; Giere et al. 2005; Lawson 2009; Samaja 1999).

and fruitful.[2] Such accounts of the nature of science share four main characteristics:

1. They consider explanation – in argumentative contexts – as one of the core epistemic practices of science (cf., Bricker and Bell 2008; Jiménez-Aleixandre and Erduran 2008; Khine 2012, who all cite the *philosophical* origins of this idea that has been imported into didactics of science).
2. They revolve around the notion of evidence (or data, proof, reasons, supporting assertions, warrant and a host of other phrasings) as a key to understand scientific semiosis (i.e. meaning production).
3. They highlight the constituent intentions of the 'acts of speech' (*à la* John Searle) or 'language games' (*à la* Ludwig Wittgenstein)[3] included in the very fabric of the scientific activity (cf., Asti Vera and Ambrosini 2010).
4. They acknowledge the social and situated character of the aforementioned processes, which are developed at the interior of specific knowledge communities with their rules and values.

In accordance with this pre-eminent role given to argumentation in science, it has been repeatedly suggested from didactics of science that argumentation should be incorporated as a major component in a high-quality science education for all (cf., Erduran and Jiménez-Aleixandre 2008; Jiménez-Aleixandre 2010; Osborne 2005). The consideration of argumentation as a central process of 'scientists' science' has permitted didacticians of science (i.e. science educators as researchers) to advance at least three main reasons for the inclusion of argumentation in 'school science'[4] (cf., von Aufschnaiter et al. 2008, p. 102):

1. Meaningful and critical science learning requires argumentation. In this sense, 'learning to argue is seen as a core process [...] in learning to think and to construct new understandings [, since] comprehending why ideas are wrong matters as much as understanding why other ideas might be right' (Osborne 2010, p. 464). Thus, mastering the argumentative aspects of science and examining actual pieces of scientific argumentation would help distinguish claims and statements that are supported from those that are not, and also to assess the

---

[2] Leema Kuhn Berland and Brian Reiser (2009) also present a three-element characterisation of argumentation, which is very similar to the one proposed here. They talk about: '(1) using evidence and general science concepts to *make sense of the specific phenomena being studied*; (2) *articulating these understandings*; and (3) *persuading others of these explanations* by using the ideas of science to explicitly connect the evidence to the knowledge claims' (p. 29; emphasis in the original).

[3] These two theoretical constructs refer to the communicative activity as a whole, with all its pragmatic constraints, where different types of texts – among them, arguments – are produced.

[4] The distinction here between 'scientists' science' and 'school science' (cf., Izquierdo-Aymerich and Adúriz-Bravo 2003) is based on the French tradition in *didactique des sciences*. In the theory of didactical transposition (Chevallard 1991), there is a 'savoir savant' constructed within the disciplines and a 'savoir enseigné', taught at school, which emerges from transposing (i.e. performing adaptive operations on) the former. Thus, science as done at school resembles in some aspects, and differs in some others, from science as performed by scientists.

quality and pertinence of the supports provided. It could be safely stated that this first reason is very general, goes beyond scientific argumentation and its epistemology and values arguing in all its cognitive, metacognitive and communicative dimensions,[5] linked to 'fostering the development of students' rationality' (Siegel 1995, p. 159).

2. Since scientists produce and evaluate arguments all the time in order to do science, a school science that is structured around argumentation would convey important messages about the nature of science, hence the need to inform argumentation-based instruction with findings from the philosophy and history of science. In coherence with this second reason, in science classes, a non-negligible part of students' activity would be to construct arguments around their understandings of the natural world, and to share, defend and criticise such arguments as it is done in actual scientific practice (cf., Driver et al. 2000, for school science, and Giere 1988, for scientists' science). Here we could use the distinction proposed by Marilar Jiménez-Aleixandre and colleagues (2000) between doing authentic school science and 'doing the lesson', the first one being characterised by 'the generation and justification of knowledge claims, beliefs, and actions taken to understand nature' (p. 758). It should be noted, of course, that the resulting nature of science that would circulate in the classroom *would heavily depend on the notion of argumentation that is being implemented*, be it more 'rationalist' or more 'constructivist' (see the 'tensions' defined in Sect. 45.2).

3. When considering science education as a tool for scientific literacy and citizen education, it is suggested that students need to engage in argumentation in order to tackle decision-making and to participate in socioscientific debates similar to those that they will encounter in their adult lives. As Jiménez-Aleixandre and colleagues (2000) point out: one of the most currently valued educational goals is 'equip[ping] students with capacities for reasoning about problems and issues, be they practical, pragmatic, moral and/or theoretical' (p. 757); it has been repeatedly proposed that argumentation would foster such capacities. Those capacities would involve evaluating different pieces of scientific evidence and judging their relative importance in making decisions around key issues of personal and social importance. Along this line, and closely following the French linguist Christian Plantin (2005, 2011), Pablo Archila states that

argumentation has been positioning itself as a social imperative, if it is considered as a way to treat differences, eliminating them, or moving them forward towards collective welfare [...]; [education for citizenship] can resort to argumentation to justify, on the basis of shared values, the existence of positions on debated issues that are socially sensitive, such as racism, abortion, the defence of the environment, war, women and children, animal rights, among others. (Archila 2012, p. 364; my translation)

Thus, there is strong consensus that 'student participation in argument develops communication skills, metacognitive awareness, critical thinking [reason 1 above],

---

[5] An anonymous reviewer of this chapter suggested the inclusion of this remark. Emphasis on this central 'learning to learn' aspect of argumentation is probably a cause for the blurring of its more specific epistemic aspects, linked to the nature of science.

45 Revisiting School Scientific Argumentation from the Perspective of the History... 1447

an understanding of the culture and practice of science [reason 2], and scientific literacy [reason 3]' (Cavagnetto 2010, p. 336).

Due to this interest in the diverse contributions of argumentation to science education, in the last decade a vast and rapidly expanding corpus of literature has accumulated in didactics of science.[6] Several possible approaches to the study of argumentation in school science have been put forward, related to the theoretical conceptualisations utilised and to the practical aims sought.[7] In this sense, '[a]ccording to different conceptualizations in this domain [of argumentation studies] instructional accounts to promote argumentative abilities of students also differ considerably' (Böttcher and Meisert 2011, p. 104). It could be added that, in consistency with those different conceptualisations, the 'natures' of science propounded for instruction also differ.

Underneath the variety of approaches, different intellectual threads can be recognised. A number of disciplines, fields of study or theoretical frameworks have converged to help didacticians of science in the task of defining, fostering and assessing argumentation in science education.[8] Nevertheless, the epistemic perspective, where an HPS[9] background would be of use, has been somewhat obscured by active discussion from linguistic, cognitive, ethnographic or pedagogical perspectives. Indeed, as stated above, most research around the place of argumentation in science education has been developed within the area of 'research with a focus on classroom discourse during the teaching and learning of science' (von Aufschnaiter et al. 2008, p. 103), with some studies also focussing on written argumentative products (cf., Adúriz-Bravo et al. 2005; Bell and Linn 2000; Erduran et al. 2004). Thus, the interest has been mainly put in the strictly *linguistic* aspects.

---

[6] In Archila (2012), Buty and Plantin (2008a), Erduran and Jiménez-Aleixandre (2008), Jiménez-Aleixandre (2010), Jiménez-Aleixandre and Díaz de Bustamante (2003), Khine (2012), Nielsen (2011), Sampson and Clark (2006, 2008), and Sanmartí (2003), there are rather comprehensive literature reviews on the subject, with more than three hundred references in English, French and Spanish.

[7] For example: Abell and colleagues (2000), Adúriz-Bravo and colleagues (2005), Bell and Linn (2000), Driver and colleagues (2000), Duschl (1990), Duschl and Osborne (2002), Fagúndez Zambrano and Castells Llavanera (2009), García Romano and Valeiras (2010), Henao and Stipcich (2008), Islas and colleagues (2009), Konstantinidou and colleagues (2010), Lawson (2003), Linhares Queiroz and Passos Sá (2009), Newton and colleagues (1999), Osborne and colleagues (2001), Revel Chion and colleagues (2005), Ruiz and colleagues (2011), Sanmartí (2003), Sasseron and Carvalho (2011), and Schwarz and colleagues (2003).

[8] See, for instance, Cadermártori and Parra (2000), Candela (1999), Kuhn (1992), Martins (2009), Mason and Scirica (2006), and Pontecorvo and Girardet (1993), among a host of others, for theoretical foundations ranging from psychologist James F. Voss to semiotician Mikhail Bakhtin, going through argumentation theorist Frans van Eemeren and social anthropologist Jean Lave.

[9] I will here use the acronym HPS (history and philosophy of science in/for science education) to denote the area of research within didactics of science that strives to incorporate a metatheoretical perspective in science education (cf., Matthews 1994/2014, 2000). This area would mainly draw from the meta-sciences (philosophy, history and sociology of science), but it would also include elements from the science studies and from other 'less disciplined' metatheoretical endeavours, such as science-technology-society (STS), feminist epistemologies or public understanding of science.

In order to transcend this discursive approach, and to recover substantive links between scientific argumentation and metatheoretical reflection on the nature of science, the aim of this chapter is threefold:

1. Identifying and characterising a subset of literature on argumentation in science education where connections to HPS are apparent or can be unproblematically proposed.
2. Spotting there some of the 'bridges' that are explicitly announced or can be implicitly recognised between mainstream HPS and argumentation in the science classroom, such as evidence-based science education, inquiry, nature of science and scientific explanation and justification.
3. On the basis of the two previous points, 'revisiting' some defining aspects of school scientific argumentation with an epistemic perspective, using categories from HPS that may help in the re-conduction of this issue towards convergence with the area of research of this handbook.

As stated above, the current state of development of the emerging line of research around argumentation within didactics of science is impressive, with several hundreds of papers accumulated (cf., Osborne et al. 2012). Consequently, this chapter does not purport to be a comprehensive literature review in all aspects of argumentation,[10] but rather an account of some productions on school scientific argumentation selected due to their possibility to be 'tuned' to the discussions in our own field, HPS. At the same time, the chapter makes an effort to incorporate into the English-speaking discussion in science education some less visible contributions from the continental, 'Didaktik' tradition (cf., Westbury et al. 2000), to a great extent shared by Germanic, Scandinavian, Latin, Greek and Slavic countries.

45.2 The Notion of School Scientific Argumentation

In this chapter, I call 'school scientific argumentation' (cf., Adúriz-Bravo 2011) the argumentative *processes* (i.e. discursive practices) and *products* (i.e. texts in any semiotic register) that occur in the science classrooms of all educational levels – from Kindergarten to University. In this sense, 'argumentation' here refers both to argu*ment* and argu*ing*, i.e. 'the product, statement or piece or reasoned discourse [...] and [...] the social process or activity' (Jiménez-Aleixandre and Erduran 2008, p. 12).

---

[10] The tables of content of the three available *handbooks* on school scientific argumentation (i.e. Buty and Plantin 2008a; Erduran and Jiménez-Aleixandre 2008; Khine 2012) can give readers an idea of the current lines of research within the strand. These lines would be, once chunked and retitled, argumentation, learning and concept formation; argumentation, learning environments and communities of practice; argumentation, discourse and language games; argumentation, social interactions and meaning negotiation; argumentation and scientific reasoning; argumentation and socioscientific and moral issues; argumentation and science teacher education; argumentation-based instruction; argumentation quality and assessment; and argumentation and epistemic criteria and practices.

From now on, the chapter will be restricted to the argumentation intentionally generated so that students understand and use scientific theories and models for problem-solving within the boundaries of science. What we can call 'socioscientific argumentation' will thus be purposefully excluded, since such kind of argumentation has epistemological traits that cannot be totally captured with the elements discussed in this chapter.[11] Among those special traits of socioscientific argumentation, the following could be mentioned: (a) it is heavily context dependent; (b) it usually results from a co-construction by different utterers; (c) it draws upon moral reasoning; and (d) it does not have as main reference 'the scholarly societies acknowledged to create and validate scientific knowledge' (Tiberghien 2008, p. xi), but rather social representations and knowledge from different disciplined and undisciplined sources.

The installation of school scientific argumentation as a central issue of science education can be attributed to what may be seen as an 'argumentative turn'. That is to say, in the last four decades or so, social sciences, and social interests and debates more generally, seem to be moving in the direction of recognising argument and arguing as key features of our post-modern culture in general and of science in particular. Within the argumentative turn, at least three fields that are important for the endeavours of our community of didacticians of science are shifting towards the consideration of the nature of science as strongly argumentative (cf., Adúriz-Bravo 2010):

1. Firstly, new school science curricula point at scientific argumentation as one of the central competencies to be achieved during compulsory education (cf., Buty and Plantin 2008b; Jiménez-Aleixandre and Federico-Agraso 2009). True citizenship is now being characterised by the ability to engage in (socio-)scientific argumentation and to make informed decisions in fields such as environment, climate, energy, sustainability, public and individual health, food and pollution. It could be argued that these curricula express the current social expectations (i.e. the 'social imperative' of which Archila [2012] talks) on the education of critical citizens.

2. Secondly, meta-sciences (philosophy, history and sociology of science) and other metatheoretical perspectives have turned towards the study of the scientific language and have directly challenged the received view that considers it an *ex post facto* labelling system that operates after clear and distinct ideas and concepts have been construed. The language of science is now 'problematised'; it is seen as a rich and complex set of cultural tools that enable semiosis: giving meaning to the natural world and making sense to the users (cf., Sutton 1996, who speaks about language as an 'interpretive system'). Within this context, where a 'linguistics of science' is emerging, argumentation is considered a paradigmatic genre in science.

---

[11] For authoritative works on argumentation in connection with socioscientific issues, see Zeidler (2003, especially Chaps. 3, 4, 5, and 7) and Sadler (2011, especially Chaps. 11 and 12).

3. And thirdly, with direct bearings to the corpus of knowledge examined in this chapter, didactics of science and other educational studies (learning psychology, classroom ethnography, etc.) have been paying increasing attention, at least in the last 15 years, to the so-called cognitive-linguistic ability (cf., Sanmartí 2003) of scientific argumentation, analysing 'argumentation discourse in science learning contexts' (Jiménez-Aleixandre and Erduran 2008, p. 4). The science classroom is now depicted as a cultural system where language has a structuring function and thus 'talking science' (cf., Lemke 1990) should be turned into content to be explicitly and specifically taught.

It could be contended that the first of these three fields – new curricula that express new social mandates – has installed argumentation as a central issue for science education; the second field – metatheoretical studies on the language of science – has enriched our image of the nature of science by acknowledging the existence of argumentative games; and the third field – educational studies on argumentation – has equipped didactics of science with theories and methods, and it has at the same time promoted the over-emphasis on the discursive aspects.

Consistent with this prior analysis, it is the contention here that the notion of school scientific argumentation can be broadly characterised through resorting to the idea of evidence; it can then be more concretely defined using a distinct linguistic stance, and, afterwards, it can be inspected from a metatheoretical perspective, ascertaining its participation in the construction of science.

For a broad definition, this chapter resorts to Jiménez-Aleixandre and Díaz de Bustamante (2003), who see scientific argumentation as 'the ability to relate data and conclusions, to evaluate theoretical propositions in the light of empirical data or data from other sources' (p. 361, my translation).

The term 'evidence' will be used here to designate not only empirical data arising from observation and experimentation but also theoretical reasons, authoritative claims, elements from worldviews, ethical considerations, stakeholders' interests and other kinds of 'supporting assertions'.[12] Thus, evidence collectively denotes the *grounds* provided to justify the assertion or claim that is being argued for:

> Evidences are the observations, facts, experiments, signs, samples, or reasons with which we intend to show that a statement is true or false. (Jiménez-Aleixandre 2010, p. 20; my translation)

This initial, general, characterisation identifies scientific argumentation as one of the basic processes of knowledge construction, a process that

> recasts the role of evidence and data in scientific classrooms: rather than being used to demonstrate the scientific canon or even to guide students to construct correct scientific principles, it is the grounds on which claims – generated by students in the process of argumentation – are warranted. (Atkins 2008, p. 63)

---

[12] A conception of evidence that is broader than 'experimental data' on the one hand better captures the history of scientific activity and on the other hand is essential in order to account for argumentation in socioscientific contexts.

This approach to argumentation represents a sophistication of the definition presented in Sect. 45.1, at least in the line of its first highlighted element – 'arriving at conclusions [...] through a process of logical reasoning' – as it underlines the *functional* role played by evidence in the derivation of such conclusions.

For a more specific definition, it is useful to adhere to the one presented by the research group LIEC (*Lectura i Ensenyament de les Ciències*, 'Reading and Science Teaching') from the Universitat Autònoma de Barcelona in Spain:

> Argumentation is a social, intellectual, and verbal activity that allows justifying or rebutting a claim; it consists of making statements taking into account the recipient and the aim with which they are transmitted. In order to argue, one must choose between different options or explanations and reason the criteria that permit evaluating the chosen option as the most adequate. (Sanmartí 2003, p. 123; my translation)

According to this strongly linguistic approach, arguing would then be elaborating a text (be it oral, written or multi-semiotic) with the aim of changing the epistemic value of the ideas sustained by an audience (or a single recipient) on an issue or matter. Such a change is sought through providing meaningful reasons so that the audience or recipient see that a new set of ideas is 'justified' by evidence in its most general sense, introduced above. The weight attributed here to justifying and convincing to some extent mirrors the other two highlighted elements of the definition in Sect. 45.1: 'a process [...] that includes debate and persuasion'.

This theoretical conceptualisation on scientific argumentation, and a host of others to which didactics of science has resorted, stem from 'a range of relevant disciplines' (Bricker and Bell 2008, p. 474). According to Bricker and Bell's (2008) classic article, the most relevant of such disciplines are formal logic, argumentation theory, science studies (and here the philosophy of science would be included) and the 'learning sciences'. The next paragraphs draw on the contributions of the first three, which are more pertinent for an HPS approach.

In order to characterise scientific argumentation from a didactical point of view, some 'tensions' (cf., Adúriz-Bravo 2010) that underlie the notion of argumentation – within and outside the science classroom – need to be discussed; such tensions are unveiled when analytical tools from the aforementioned disciplines are employed. It could be safely said that these tensions have many times been dismissed or underrepresented in the literature of didactics of science, partly perhaps as a result of the hegemony of the so-called Toulmin's argumentation pattern (or 'TAP') as the preferred theoretical and methodological framework (see Sect. 45.2.1). The generalised use of TAP has fixed the discussion around semiformal reconstructions of arguments akin to those propounded by the theory of argumentation of mid-twentieth century or, rather, around a highly stylised didactical version of such reconstructions.

The four tensions that are developed in the following subsections are:

1. The opposition between two intellectual traditions to study argumentation, namely, the *Anglo-Saxon* (e.g. Stephen Toulmin, Henry W. Johnstone Jr., Ralph H. Johnson, Douglas Walton, G. Thomas Goodnight) and the *continental* (e.g. Arne Naess, Chaïm Perelman, Oswald Ducrot, Frans van Eemeren & Rob

Grootendorst, Christian Plantin).[13] These two traditions would represent complementary ways of going beyond the classical, neo-Aristotelian, approach to the study of arguments: in the first case, by 'softening' the requirements of syllogistic logic, and in the second, by opening the floor to pragmatic and rhetorical constraints.

2. *Logic* versus *dialogic* argumentation. The opposition between two extreme forms of argumentation – argumentation as explanation and argumentation as debate – is traditionally presented as the existence of 'analytical' and 'dialectical' arguments.[14] Such opposition is usually conflated with the distinction between the use of formal and informal logic in order to analyse such arguments, revised in the fourth tension.

3. Arguing as *explaining* versus arguing as *justifying*, partially connected to the former, and pointing at Jiménez-Aleixandre and Erduran's (2008, p. 9) distinction between producing scientific knowledge about the world and giving 'rhetorical significance' to that knowledge. The 'explanatory' part of argumentation, in this context, would entail making sense of a phenomenon on the basis of data, while the 'justification' part would mean supporting the claim that the data are consistent with the proposed explanation and therefore convincing an audience of its validity (cf., Osborne and Patterson 2011, p. 629, who use similar phrasings, but sharply separate these two operations).

4. Arguments as texts of *'harder'* versus *'softer' syntax*. This refers to the clash between the existence of sanctioned patterns with an a priori rationality dictated by formal logic, leading to heavily 'idealised notions of arguments' (Jiménez-Aleixandre and Erduran 2008, p. 15), and the pragmatic use of what we can call *para-logical* (i.e. ampliative) techniques to capture argumentation 'as it is practiced in the natural languages' (Jiménez-Aleixandre and Erduran 2008, p. 14). Among these 'real' argumentative practices, scientists', teachers' and students' discourse would be included.

45.2.1 Anglo-Saxon Versus Continental Approach to Argumentation

Since the three traditions that follow this first one can be said to hinge to some extent on an *ab initio* divergence between theoretical approaches to argumentation, this subsection is longer and more detailed than the rest; in those, cross-references to the ideas exposed here are made.

---

[13] For other authors not mentioned in this list, see Reygadas and Haidar (2001), Santibáñez (2012).

[14] This opposition is in turn based on Aristotle's division of 'perspectives' on argumentation that has been thoroughly used in continental studies and retrieved by the Anglo-Saxon tradition: logical, dialectical and rhetorical (cf., Harpine 1985; van Eemeren and Houtlosser 2003). The chapter concentrates only in the first two classes of arguments.

45 Revisiting School Scientific Argumentation from the Perspective of the History... 1453

The Anglo-Saxon tradition in argumentation studies was long based on the assumption that arguments are more or less 'syllogistic' (i.e. deductive-like) in nature (this restrictive requirement of 'deductivity' is still retained in the general definition of argumentation presented in Sect. 45.1). Arguments were usually portrayed as a tight structure in which a key assertion is logically inferred from a set of supporting assertions (Asti Vera and Ambrosini 2010). As Stephen Toulmin critically remarks,

> [T]he assumption [...] made by most Anglo-American academic philosophers [was] that any significant argument can be put in formal terms: not just as a syllogism, since for Aristotle himself any inference can be called a 'syllogism' or 'linking of statements', but a rigidly *demonstrative deduction* of the kind to be found in Euclidean geometry. Thus was created the Platonic tradition that, some two millennia later, was revived by René Descartes. (Toulmin 2003, p. vii; my emphasis)

Accordingly, classical argumentation theory among Anglo-Saxon authors more or less overlapped in scope and methods with the discipline of logic – the main aim being to ascertain the *validity* of arguments using formal techniques.

In the Anglo-Saxon tradition, the main connecting threads would be the attention paid to the *syntactic* aspects of the language used to argue and the aim of analysing individual propositions and their structural relations in order to justify and assess theoretical arguments, dialogic exchanges and informed judgements set against the backdrop of their social contexts. The evolution of this tradition could be seen as an expansion of the traditional apparatus to study argumentation – which strictly resorted to formal logic – towards the use of 'para-logical' tools, moving then onto 'informal logic'. The focus is thus to capture 'natural' arguments, to formulate

> [the] statements [referred to in those arguments] in a 'normal' (philosophical, universal) language in some canonical form [, since a]fter 2,300 years of formal logic, [argumentation theory is] still infinitely remote from having a clear idea of what such a language should look like. (Bar-Hillel 1970, p. 204)

This Anglo-Saxon approach to argumentation will be here characterised through rapidly examining the work of the British-born philosopher of science Stephen Toulmin, with a peripheral mention to the Canadian argumentation theorist Douglas Walton and the American educational psychologist Deanna Kuhn.

Toulmin's (1958) framework hinges upon a naturalistic approach to the rationality of practical arguments (which he calls 'substantial' arguments). Substantial arguments are opposed to 'theoretical' arguments, which are analytic and necessary. This means that, in the latter, the argued assertions are the conclusions of *sensu stricto* inferences; such assertions are deductively connected to a set of premises providing the evidence for it (hard data or other grounds, but always satisfying the relationship of logical necessity with the conclusion). Thus, what is being sustained is already 'contained' in what we know.

Substantial arguments, on the contrary, seek to offer 'justification' for an assertion that is deemed to be of interest, in a specified and recognisable context. Thus, Toulmin suggests going beyond formal logic when modelling arguments and proposes an 'argumentation pattern' with tightly interrelated components:

the *claim* (which is the statement in need of justification), *data* to support such claim and a *warrant* that allows the 'legitimate' transition from data to claim. Even more 'real' arguments in the natural language are heavily modalised and include qualifiers, rebuttals and backing to the warrant.

It could be stated that, in Toulmin's framework, the claim – 'conclusion' *sensu lato* – has more content than that of the evidences provided, and thus it is only partially sustained by them. Accordingly, it is convenient to portray the 'movement' from the premises containing the evidence to the conclusion as an ampliative inference, which should be captured with inductive, analogical, abductive, etc. reasoning patterns (cf., Stadler 2004; Diéguez Lucena 2005).

In turn, the goal of Walton's (1996) framework is more related to understanding persuasive arguments, for example, in legal contexts. Walton is thus more interested in *dialogic*, conversational argumentation (see next subsection), where 'actors exchange replies and counter-replies' (Asti Vera and Ambrosini 2010, p. 133; my translation). Walton's *schemes* for 'presumptive reasoning' refer to strategies used in hypothetic, non-demonstrative, argumentation. To capture those schemes, he enumerates a variety of categories; for instance, he talks about 'arguments based on experts' opinions', which might be instrumental both for scientists' science and school science. *Pertinence* of the utterances – and of the reasons given therein – is a key theoretical element of his framework.

As a complement to the general Anglo-Saxon perspective, D. Kuhn (1993, 2010), moving markedly away from philosophical and linguistic considerations, proposes a conceptualisation of science and of science education as argumentative endeavours that resorts to psychological and cognitive foundations. In this sense, she is a good example of contributions to argumentation from the 'learning sciences'.

Opposing the Anglo-Saxon tradition, we can talk of a 're-emergence' of a continental approach to argumentation studies, which occurs after World War II and is of course favoured by external, socio-cultural, factors (cf., Jiménez-Aleixandre and Erduran 2008). Chaïm Perelman's life story – he was a Polish Jew who immigrated to Brussels – is a good example of this. The continental tradition will here be represented in the works of the expert in rhetoric Perelman, the Dutch scholars in 'speech communication' Frans van Eemeren and Rob Grootendorst and Christian Plantin. The connecting threads of this tradition would be the introduction of the audience as a key element and the attention to pragmatic and rhetorical aspects.

Perelman publishes, together with Lucie Olbrechts-Tyteca, his *Traité de l'argumentation* in 1958 (the same year of Toulmin's *The uses of argument*). In this book, the authors propose a 'new rhetoric', understood as an art of persuading and convincing; with this, they also intend to abandon formal logic in the evaluation of argument validity. But, differing from the Anglo-Saxon perspective, persuasion is highlighted; in order to characterise arguments, Perelman constructs new concepts around this idea, such as argumentative force and relevance or the 'intensity of adherence of an audience'. The introduction of the audience as 'a genuine actor in the argumentative phenomenon' (Asti Vera and Ambrosini 2010, p. 110; my translation) is generally considered to be Perelman's main contribution.

Van Eemeren and Grootendorst, at the Universiteit van Amsterdam, develop what they call a *pragma-dialectical theory* of argumentation; like Perelman, they seek to analyse and assess argumentation as a natural practice of language. Pragma-dialectics takes into account the fact that arguments are usually presented within interactive, dialogic discussion. These authors also confront the use of syllogistic structures to study argumentation, since formal logic would be opaque to the subtleties of the social practice of arguing. Scientific argumentation would also need this approach, since scientists direct their arguments to convince peers (or other audiences) so that they accept the point of view that is being offered. Carlos Asti Vera and Cristina Ambrosini (2010) recognise a very 'fecund' starting point in pragma-dialectics, since 'it proposes not abstracting arguments of any of their dimensions, in order to analyse and evaluate them as they are presented in the social theatre, in their empirical, dialogic and contextual determinations' (p. 133).

Plantin is also interested in a rhetorical study of dialogic argumentation (he calls it 'dialogale' in French: cf., Plantin 2011) and again focuses on persuasion as one of its central characteristics. He interprets argumentation as a way of producing speech in situations where doubt, debate and confrontation predominate. It is interesting to remark that Plantin wants to redeem rhetoric from its reputation as a 'sorceress' (Buty and Plantin 2008b, p. 21); according to him, rhetoric has been stereotypically discredited, being repeatedly associated with manipulation, void words and politicians' clichés (for these he uses the very graphic French expression of 'langue de bois').

45.2.2 Logic Versus Dialogic Argumentation

What I call 'logic argumentation' – where arguments are practically confounded with explanations or inferences – can be described, using Richard Duschl's terminological choices (cf., Duschl et al. 1999; Duschl 2008), as the production of analytical arguments. These arguments are grounded in (formal) logic, and they constitute a movement from a set of premises to a conclusion (cf., Asti Vera and Ambrosini 2010). What I call 'dialogic argumentation' – where arguing is seen as exchange of ideas or confrontation – fits with the idea of dialectical arguments, which are 'those that occur during discussion or debate and involve reasoning with premises that are not evidently true' (Duschl 2008, p. 163). It could arguably been said that it was in order to understand this latter kind of arguments that the field of (new) argumentation theory emerged in the 1950s, somewhat vanishing its boundaries with informal logic.

This broad distinction made under this tension can be related to the two major scholarly approaches to argumentation in Sect. 45.2.1 as follows: the stereotypical Anglo-Saxon approach was almost restricted to analytical arguments and logic argumentation (as is apparent in Toulmin's critique), while the best-known continental frameworks over-emphasised dialectical arguments and dialogic argumentation. This simplified, one-to-one relationship tends to relax in more recent texts.

For didactical purposes, it seems convenient to blur this watertight distinction and consider that school scientific argumentation combines in itself the long-standing Greco-Latin traditions of arguing as producing 'any piece of reasoned discourse' (Jiménez-Aleixandre and Erduran 2008, p. 12) and arguing as 'dispute or debate between people opposing each other with contrasting sides to an issue' (Jiménez-Aleixandre and Erduran 2008, p. 12). Thus, on the logic side, argumentation evokes the etymological meaning of the Latin verb 'arguere': 'make clear through discourse'; such meaning stems from the Indo-European root 'arg-', meaning 'brilliant' (conserved in modern terms such as the Italian 'argento', 'silver' or the French 'argille', 'clay'). On the dialogic side, argumentation points at one of the standard meanings of the English verb 'argue': 'discuss', 'dispute' and 'disagree'. But these two aims of clarifying and debating coexist – and are virtually impossible to divorce from each other – in the language game of argumentation in science.

45.2.3 *Arguing as Explaining and Arguing as Justifying*

When argumentation is seen as a vehicle for scientific explanation, the emphasis is put on the sharing of theoretical elements that permit us to understand the world. Arguments are seen as 'solid', i.e. with a claim well supported by foundations and backings (cf., Asti Vera and Ambrosini 2010), and such a view purports to be context and audience independent.[15] In this first perspective, Toulmin's idea of warrant is paramount: warrants serve as the explanatory elements; their aim is to give testimony of the legitimacy of the transition from data to claim. Warrants provide general, abstract and *uniform* transitions, which are relatively autonomous of (i.e. not referring directly to) particular sets of data.

When argumentation is seen as an act of speech where justification is demanded and offered (cf., Tindale 1999, who examines this idea based on Michael Billig and Chaïm Perelman), the focus is moved to the recipient's or audience's adherence to the claim presented. In this second perspective, more akin to continental studies, 'argumentation is a feature of social relations and shares in the complexity of those relations' (Tindale 1999, p. 75).

In science education, the distinction between argumentation as explanation and argumentation as justification can be partially aligned with what Nussbaum and colleagues (2012) call the 'two faces of [school] scientific argumentation'. According to these authors, argumentation is on the one hand *explanatory*, when it presents and debates scientists' theories about reality. On the other hand, argumentation is *prescriptive*, when it informs scientific (and socioscientific) debates, where decision-making is often required. These authors distinguish between 'theoretical discourse, pertaining to what theories of the world best fit the data and practical, deliberative discourse, regarding how to apply those theories to reach practical goals' (Nussbaum

---

[15] This is what Constanza Padilla (2012) calls 'demonstrative dimension' of argumentation.

et al. 2012, p. 17). Accordingly, students and teachers together would use scientific arguments in the science classroom to explain theoretically *and* to circulate and share understandings and applications.

45.2.4 Hard and Soft Arguments

This last tension, as advanced above, has to do with the capacity attributed to formal, abstract structures to capture real discourse. The classical, positivistic approach of categorical rationalism 'supposes enthroning formal logic as the *exclusive* model of rationality' (Asti Vera and Ambrosini 2010, p. 110; my translation, emphasis in the original). Through the lens of formal logic, only what we might call 'hard arguments' survive: those that are 'fully explicit [and] neatly packaged into premises and conclusions' (Smith 2003, p. 34).

If one adheres to this restriction, real argumentation practices are almost always subsumed into the realm of material (or informal) fallacies. There is an *ab initio* 'half-empty glass' metaphor operating here, since – from the point of view of hard rationality – most arguments are considered to be logically non-pertinent, only psychologically persuasive, and often intended to deceive (cf. Asti Vera and Ambrosini 2010). Even in the case of (empirical) science, most relevant arguments do not measure up to the extremely restrictive standards of demonstrative argumentation, since they contain in their fabric elements that are not bound by the relationship of necessity, and therefore cannot be completely formalised without consideration of their empirical content.

Two options arise to oppose this 'hard' approach: in the first place, rationality can be resigned altogether, slipping down the irrational slopes of contextualism, relativism or constructivism. A 'third way', which seems more productive for science education, would be to broaden the scope of arguments that can be considered well supported. This third way would imply a 'temperate', non-aprioristic, rationality, which resorts to the use of 'para-logical' techniques, i.e. non-demonstrative patterns of inference such as induction or abduction. Softening the syntax admitted for arguments is, in all cases, allowing a richer study of argumentation as it occurs in the real world. This would constitute a *naturalisation* of argumentation theory.

For this last tension, the link to the Anglo-Saxon-continental dispute is not straightforward. One might be tempted to assume that the Anglo-Saxon approach closes up the number and variety of patterns of argumentation that are admissible and is therefore more identifiable with the idea of 'harder syntax'. This might be the case for the classical studies, those that fall under Toulmin's critique, but it is certainly not applicable to post-Toulminian accounts of scientific argumentation among English-speaking scholars. On the other hand, a pairing of what I have proposed to call 'softer syntax' to continental accounts would be too hasty, since the examination of the structure and components of an argument is seldom a concern among authors who zoom out to rhetorico-pragmatic considerations.

45.3 The Epistemics of School Scientific Argumentation

This section is devoted to dissecting some of the epistemic aspects of school scientific argumentation, aspects that can be theorised through the lens of HPS.[16] The section discusses different constituting elements of the *epistemics* (i.e. epistemology) of argumentation, identified on the basis of a review of the literature in didactics of science that is heavily theory driven. That is to say, the review is guided by an attention to metatheoretical perspectives and especially to the philosophy of science. As it was advanced in the introduction to the chapter, in order to organise such review, possible 'bridges' between argumentation and HPS are defined.

Under the five bridges enumerated here, studies on school scientific argumentation with an interest in one or more particular epistemic aspects are grouped. The studies may or may not present an explicit HPS background, and this will be indicated for each case. The five resulting groups are:

1. *Argumentation as an epistemic practice.* In this first approach, undoubtedly the most exploited one, the bridge consists in identifying argumentation as a paradigmatic example of epistemic practice, i.e. a practice of knowledge construction that gives its character to the scientific activity. Richard Duschl (1998, 2008), Marilar Jiménez-Aleixandre (Jiménez-Aleixandre and Federico-Agraso 2009; Bravo-Torija and Jiménez-Aleixandre 2011), Gregory Kelly (Kelly and Chen 1999; Kelly and Takao 2002), Victor Sampson and Douglas Clark (2006, 2008), and William Sandoval (Sandoval 2003; Sandoval and Reiser 2004; Sandoval and Millwood 2005, 2008), among many others, have advocated for a conceptualisation of argumentation along this line.

2. *Argumentation as a feature of the nature of science.* In this second, more encompassing approach, the bridge consists in describing the 'non-natural' nature of science,[17] at least partially, through inspecting the role that argumentation (both in the senses of explaining and of justifying) plays in doing, thinking and talking about the natural world. Authors who can be located within this perspective[18] identify science not with the 'discovered' facts of the world, but rather with an extremely elaborate inferential and discursive construction regarding the ways in which scientists appropriate and transform those facts.

3. *Argumentation in scientific inquiry.* In this third approach, school science is designed as an inquiry-based endeavour aiming at genuine scientific literacy (see public policy documents such as AAAS 1993; NRC 1995). The bridge here

---

[16] The name of this section is a paraphrasis of an expression by Sandoval and Millwood (2008, p. 72).

[17] Both Lewis Wolpert (1992) and Lydia Galagovsky (2008) refer to this 'non-naturality' of science in the titles of their books. Nevertheless, the meanings of the expressions that they use are quite distinct from each other. Wolpert's thesis, positivistic in its foundations, is that science is a way of thinking far away from common sense. Galagovsky's compilation of chapters aims at showing how science is a very elaborate human construction and not a mere expression of the way the world is.

[18] For example, Allchin (2011), Duschl (1990, 1998), Hodson (2009), Lawson (2003, 2005, 2009), and McDonald (2010)

is the attention to the inclusion of argumentative skills in such an endeavour. A grasp of the nature of science in science education

involves understanding *how knowledge is generated, justified, and evaluated* by scientists and *how to use such knowledge to engage in inquiry* in ways that reflect the practices of the scientific community. (Clark et al. 2010, p. 1; emphasis in the original)

The two elements of the nature of science italicised in this quote could be somehow referred to the two poles of tension 3: on the one hand, students need to comprehend the epistemic practice of knowledge generation (explanation); on the other hand, students need to apply that knowledge in school scientific inquiry (justification). Proposals along this line[19] strive to meaningfully connect argumentation and inquiry through the introduction of evidence- and argument-based practices in the science classroom.

4. *Model-based argumentation.* In this fourth approach,

the general model-based perspective in [...] the philosophy of science [is used in order to] understand arguments as reasons for the appropriateness of a theoretical model which explains a certain phenomenon. (Böttcher and Meisert 2011, p. 103)

The bridge here is that argumentation is regarded as a tool to assess and apply the models that constitute the content of school science. Authors who use this perspective (Adúriz-Bravo (2011), Böttcher and Meisert (2011) and much less directly Lehrer and Schauble (2006), who talk about 'model-based reasoning' and Windschitl et al. (2008), who talk about 'model-based inquiry') conceptualise models using *semantic* tools from the philosophy of science of the last three decades.

5. *Argument-based school science.* This fifth approach is rather unspecific; it suggests that argumentation should be a substantive part of the (social) activity in the science classroom (and in science teacher education). Authors adhering to this perspective talk about 'argumentation-based' teaching or instruction.[20] The bridge here are the reasons provided in favour of this position, drawn mainly from the sociology of science (with references to Helen Longino or Bruno Latour, for instance) and to a lesser extent from other metatheoretical perspectives.

A proviso should be made here: in the very biased selection of literature in which the bridges between argumentation and HPS have been identified, papers that use HPS elements for the design of instructional units and materials, but then fail to use those elements to characterise or justify the presence of argumentation in those units and materials, were purposefully excluded. For instance, Bell and Linn (2000), Monk and Osborne (1997) and Revel Chion and colleagues (2009) use the history and philosophy of science to lay the foundations for the teaching of different

---

[19] For example, Clark and colleagues (2010), Duschl and Grandy (2008), Sampson and Clark (2007), Sandoval and Reiser (2004), and Windschitl and colleagues (2008).

[20] Cf., Driver and colleagues (2000), Izquierdo-Aymerich (2005), Newton and colleagues (1999), Ogunniyi (2007), and Ogunniyi and Hewson (2008).

scientific topics (Darwin's ideas, light, the bubonic plague, etc.), and then – more or less independently of those foundations – they propose to implement argumentation as a teaching strategy.

In the subsections that follow, the five aforementioned bridges are explicated through one or two epitomic examples of each of them.

45.3.1 Argumentation as an Epistemic Practice

Richard Duschl's work locates explanation at the vertex of the pyramid of the activities in science (cf., Duschl 1990), identifying it as a privileged aim of the scientific enterprise. In his framework, and following Gregory Kelly and Deanna Kuhn, argumentation would constitute one of the most favoured epistemic (i.e. knowledge-producing) practices. Consistent with this conceptualisation of scientists' science, Duschl proposes, for school science,

> [s]hifting the dominant focus of teaching from what we know (e.g., terms and concepts) to a foc[us] that emphasizes how we know what we know and why we believe what we know (e.g., using criteria to evaluate claims). (Duschl 2008, p. 159)

School science would then require 'epistemic apprenticeship' (Jiménez-Aleixandre and Erduran 2008, p. 9): students should appropriate criteria to evaluate arguments in the light of evidence. Accordingly, science in the classroom could be structured as a set of 'epistemological and social processes in which knowledge claims can be shaped, modified, restructured and, at times, abandoned' (Duschl 2008, p. 159). Duschl talks about 'knowledge-building rules' that represent or embody the epistemic practices of the community formed by students and teacher(s).

Thus, the core of this conceptualisation of argumentation as an exemplar of educationally valuable epistemic practice would be captured in questions such as

> What counts as a claim? What counts as evidence? How do you decide what sort of evidence supports, or refutes, a particular claim? How are individual claims organized to produce a coherent argument? What kinds of coordination of claims and evidence make an argument persuasive? (Sandoval and Millwood 2008, p. 72)

One of the most favoured strategies in the studies allocated in this first group has been to recognise epistemic *statuses*, *criteria* or *levels* in students' argumentative practice, with the aim of 'assessing the nature or quality of arguments in the context of science education' (Sampson and Clark 2008, p. 449). Such assessment is done, for instance, in terms of their complexity, robustness, validity, etc.

For this first bridge, explicit recurrence to authors from the area of HPS has been somewhat low. In Sandoval and Millwood (2008), for instance, of almost 30 cited references, only three are to authors with a meta-scientific perspective: Philip Kitcher, Bruno Latour and Stephen Toulmin. In Duschl (2008), of around 45 cited references, again only three are to texts in the realm of HPS (Derek Hodson, Nicholas Rescher and Toulmin). In Sampson and Clark (2008), among circa 65 references, only two 'meta-scientists' feature: Latour and Thomas Kuhn. The relationship between favouring argumentative practices in science education and metatheoretically characterising those as epistemic practices is therefore *indirect*: most authors that develop this first bridge refer to some seminal texts in didactics of science (e.g. Driver et al. 2000; Duschl and Osborne 2002; Kelly and Takao 2002) that have acknowledged the philosophical foundations of that relationship, but then do not go on developing such foundations.

45.3.2 Argumentation as a Feature of the Nature of Science

There is a substantive connection between this second approach and the first one, since a widespread hypothesis in science education considers that 'students' epistemological beliefs [i.e. their conceptions on the nature of science] are developed through their own epistemic practices of making and evaluating knowledge claims' (Sandoval and Millwood 2008, p. 85). Epistemic practices in general, and argumentation in particular, would then be, at the same time, a specific feature of the nature of science (cf., Hodson 2009, Chap. 8) and a powerful means to access to a coherent and robust conceptualisation of such nature.

Both Jonathan Osborne and Sibel Erduran, in many of their papers (cf., Erduran et al. 2004; Osborne et al. 2001), have enumerated different links between the nature of science and argumentation. Osborne and colleagues (2001), for instance, subordinate those links to the need to teach the nature of science *explicitly*,[21] since 'contact with school science is insufficient to generate an understanding of how science functions' (p. 69). For such teaching, argumentation becomes a privileged tool, insofar as it permits presenting students with opportunities to examine and discuss epistemological issues such as evidence, prediction, analytical thinking, controversy, reasoning, evaluation and critical thinking.

From a more focussed point of view, Anton Lawson points out that nature-of-science instruction should teach to science students 'that the best [scientific] argument considers all of the alternatives and explicitly includes the relevant evidence and reasoning supporting and/or contradicting each' (Lawson 2009, p. 337). He suggests introducing, in science education, what he calls an 'if/then/therefore' argumentative pattern. His theoretical framework, which he deems valid both for scientists' science and for school science,

> distinguishes among an argument's declarative elements (i.e., puzzling observations, causal questions, hypotheses, planned tests, predictions, conducted tests, results, and conclusions) and its procedural elements (i.e., abduction, retroduction, deduction, and induction). (Lawson 2009, p. 358)

---

[21] As one of the anonymous reviewers of this chapter pointed out, considering the nature of science or argumentation important goals of science education does not imply deciding to teach these issues explicitly. The contention that school scientific skills are not developed by 'exposure' and deserve 'direct instruction' is still debated; nevertheless, such contention seems to be finding some support coming from recent empirical studies (e.g. Kirschner et al. (2006), at a general level, and McDonald (2010), for the case of nature of science and argumentation).

It should be noted that Lawson provides extensive HPS backing to his framework, using the history of science in order to construct case studies of scientific reasoning, argumentation and discovery and – to a lesser extent – the philosophy of science to understand those three processes.

In my own work, I portray scientific argumentation as the textual counterpart of the epistemic operation of scientific explanation (Adúriz-Bravo 2005, 2010, 2011). I define argumentation as the *subsumption* of some phenomenon of the natural world under a theoretical model (in the sense of the semanticist family), which is seen as a good candidate to 'explaining' it (and hence there is direct connection with bridge 4). Similarly to Lawson, my argument is that some discoveries and inventions, as reported by scientists through history, can be reconstructed as cases of abductive and analogical thinking; these kinds of inferences would then be the mechanism to subsume the 'phenomenon-case' under a 'model-rule'. I distinguish between abduction *sensu lato*, as any ampliative, non-monotonic, inference producing or evoking hypotheses and abduction *sensu stricto*, as a 'reverse' deductive schema *à la* Peirce (cf., Adúriz-Bravo 2005; Aliseda 2006; Samaja 1999).

45.3.3 Argumentation in Scientific Inquiry

School scientific inquiry can be broadly conceptualised as a 'knowledge building process in which explanations are developed to make sense of data and then presented to a community of peers so they can be critiqued, debated and revised' (Clark et al. 2010, p. 1). In this sense, inquiry would function as a reconciliation of the two poles of the second (logic-dialogic) and third (explain-justify) tensions. Within this framework of ideas, argumentation nicely fits when understood as

> the ability to examine and then either accept or reject the relationships or connections between and among the evidence and the theoretical ideas invoked in an explanation or the ability to make connections between and among evidence and theory [...]. (Clark et al. 2010, p. 1)

From this perspective, argumentation is seen as an artefact to develop and evaluate explanations (cf., Kuhn Berland and Reiser 2009; Osborne and Patterson 2011; Windschitl et al. 2008). In other words, in this third approach the practices of explanation and argumentation would be *complementary*:

> First, explanations of scientific phenomena can provide a product around which the argumentation can occur, as proponents of an explanation attempt to persuade their peers of their understandings. Second, argumentation creates a context in which robust explanations – those with which the community (the students) can agree – are valued. (Kuhn Berland and Reiser 2009, p. 28)

For this third bridge, it should be noted that Kuhn Berland and Reiser's (2009) paper has an extensive and developed HPS background. These authors show how several philosophers of science, in the last six decades, extended

> [t]he everyday sense of argumentation[, which] typically suggests a competitive interaction in which participants present claims, defend their own claims, and rebut the claims of their

opponents until one participant (or side) "wins" and the other "loses". [Instead, i]ndividuals compare conflicting explanations with the support for those explanations and work to identify/construct an explanation that best fits the available evidence and logic. (Kuhn Berland and Reiser 2009, pp. 27–28)

45.3.4 Model-Based Argumentation

In model-based argumentation, scientific arguments are understood as the 'reasons for the appropriateness of a theoretical model which explains a certain phenomenon' (Böttcher and Meisert 2011, p. 103), and argumentation 'is considered to be the process of the critical evaluation of such a model if necessary in relation to alternative models' (Böttcher and Meisert 2011, p. 103). Here, the second and fourth tensions are apparent: on the one hand, models that explain are judged in terms of the reasons for their justification; on the other hand, critical evaluation of the appropriateness of those models would require the use of some analytical tools arising from classical or modern logic.

Central to this approach to school scientific argumentation is the thesis that

[t]he model-based theory represents a suitable theoretical framework for describing arguments and argumentation referring to the similarity between models and empirical data as the central reference for model evaluation. (Böttcher and Meisert 2011, p. 137)

Derek Hodson (2009) provides a detailed description of the role attributed to argumentation in a model-based depiction of the nature of science. Closely following Ronald Giere (Giere 1988; Giere et al. 2005), he states that

[r]eaching consensus about the most acceptable model involves a cluster of interacting, overlapping and recursive steps: (i) collection of data via observation and/or experiment, (ii) reasoning, conjecture and *argument*, (iii) calculation and prediction, and (iv) critical scrutiny of all these matters by the community of practitioners. Language plays a role in all these steps [...]. As an integral part of these activities, *arguments* are constructed and evaluated at a number of different levels. (Hodson 2009, p. 259; my emphasis)

Such description, explicitly based on HPS, justifies the use of argumentative strategies within the framework of model-based science education.

From a slightly different perspective, but also stressing the role of models in scientific argumentation, Jiménez-Aleixandre (2010) focuses on 'arguments on explanatory models', stating that such arguments intend to identify cause-effect relations in the explanations and interpretations on natural phenomena.

45.3.5 Argumentation in School Science

School scientific argumentation is brought to the centre of the arena of teaching practices ('pedagogy') when the pre-eminently *empirical* conception of students' activity in the science classroom is abandoned in favour of a more theory-laden, social and discursive depiction of school science. Rosalind Driver and her colleagues (2000) accurately explain this shift in the following quote:

> Our contention is that, to provide adequate science education for young people, it is necessary to reconceptualize the practices of science teaching so as to portray scientific knowledge as socially constructed. This change in perspective has major implications for pedagogy, requiring discursive activities, especially argument, to be given a greater prominence. Traditionally, in the UK (and other Anglo-Saxon countries), there has been considerable emphasis on practical, empirical work in science classes. Reconceptualizing the teaching of science in the light of a social constructivist perspective requires, among other matters, the reconsideration of the place of students' experiments and investigations. Rather than portraying empirical work as constituting the basic procedural steps of scientific practice (the "scientific method"), it should be valued for the role it plays in providing evidence for knowledge claims. (Driver et al. 2000, p. 289)

In Mercè Izquierdo-Aymerich's work,[22] argumentation is incorporated as a central feature of her general theoretical framework for didactics of science (developed with colleagues at the Universitat Autònoma de Barcelona). She labels such framework, following Ronald Giere (1988), the 'cognitive model of school science'; this and other authors from the so-called semantic view of scientific theories in contemporary philosophy of science provide her with the conceptualisation of theoretical models that she deems to be most fruitful for science education (and hence the intersection with bridge 4).

Within this framework, school scientific arguments are cognitive and discursive tools that permit making meaningful connections between the realm of facts in the world and the models that can give meaning to those facts:

> Students reason according to their initial models, which generally have an *iconic* relationship with phenomena; a simple image may function as a model for students. Experimentation and its written reconstruction bring students to a new epistemic level, in which non-iconic (i.e., *symbolic*) signs are much more relevant. Symbols can only connect correctly with their referents if the first, more concrete step is done [...]. In order to give momentum to this process, it is necessary that students learn how to use argumentation in their discourse [...]. (Izquierdo-Aymerich and Adúriz-Bravo 2003, p. 38; emphasis in the original)

45.4 Conclusion: Towards Convergence of Argumentation with HPS

The purpose of this short conclusive section is to revisit six characterisations of school scientific argumentation with the ideas provided by an HPS-informed approach, which were discussed throughout this chapter. For each of the excerpts revisited, connections with the five bridges are made, and some HPS references (mainly from the philosophy of science) are suggested that could help in furthering the discussion only sketchily initiated here.

---

[22] See Izquierdo-Aymerich (2005), Izquierdo-Aymerich and Adúriz-Bravo (2003), and Izquierdo-Aymerich and colleagues (1999).

Sampson and Clark (2008) propose to use 'the term «argument» to describe the artefacts students create to articulate and justify claims or explanations and the term «argumentation» to describe the complex process of generating these artefacts' (p. 448). This first terminological clarification reminds us of the fact that in order to fully understand school scientific argumentation, we should consider it as *a product that arises from a highly elaborate process and is therefore shaped by the very nature of that process.* Here the connection with bridge 1 is direct: an epistemic characterisation of the argumentation process is required, be it more 'internalist', focussing on inferences (e.g. Charles Sanders Peirce, Stephen Toulmin or Nancy Nersessian) or more 'externalist', looking at social interactions within the scientific communities (e.g. Thomas Kuhn, Bruno Latour or Helen Longino).

Marilar Jiménez-Aleixandre (2010) starts her book on key ideas about argumentation with a working definition of the notion; she considers it the 'ability to relate explanations and evidences' (p. 11, my translation). In this kind of phrasing, the evidence-based character of the scientific enterprise is highlighted: *evidence (in its broadest sense) becomes a key epistemic factor, one of the cornerstones of scientists' activity.* This emphasis can lead, in science education, to fruitful discussion around the notion of rationality, with questions such as what counts as 'valid' support for scientific claims, and how is this support obtained and shared? To answer such questions, related mainly to bridges 2 and 3, a postpositivistic notion of rationality can be introduced. For this kind of discussion, ideas from Stephen Toulmin,[23] William H. Newton-Smith or Ronald Giere seem appropriate.

Rosalind Driver, in one of her posthumous papers (Driver et al. 2000), advocates for a 'situated perspective', where 'argument can be seen to take place as an individual activity, through thinking and writing, or as a social activity taking place within a group – a negotiated social act within a specific community' (pp. 290–291). When arguing, *scientists give meaning to the world and communicate such meaning to peers and other audiences*; this should be a guiding idea of the nature of science discussed in the science classroom. Again, this double cognitive and social perspective can be inspected with tools from the philosophy of science and from science studies, anchoring the discussion in selected episodes from the history of science.

Anton Lawson, distinguishing himself from Toulmin's ideas on argumentation, so hegemonic in didactics of science, prefers to see

> the primary role of argumentation, not as one of convincing others of one's point of view (although that is certainly part of the story) but rather as one of discovering which of several possible explanations for a particular puzzling observation should be accepted and which should be rejected. (Lawson 2009, p. 337).

In such preference, the explanatory and theoretical aspects of argumentation are highlighted, and this might constitute a possible connection with bridge 4. Arguments *propose a way of 'seeing' the world that is structured around theoretical views.* Here, the so-called semanticist family (Giere, Frederick Suppe, Bas van

---

[23] Here I refer to Toulmin (2001).

Fraassen), with their various conceptualisations of scientific theories, might prove a powerful background.

Izquierdo-Aymerich and myself accept a 'relaxation' of the requirements for an argument to be considered scientific, in tune with the naturalistic approach introduced in the fourth tension:

> An argumentation is formed by a set of reasons that convey a statement and reach a conclusion. Scientific arguments are hardly ever strictly formal (logical or mathematical); they are generally analogical, causal, hypothetico-deductive, probabilistic, abductive, inductive... One of their functions is to make a theoretical model plausible, convincingly connecting it to a growing number of phenomena. (Izquierdo-Aymerich and Adúriz-Bravo 2003, p. 38)

This approach reminds us that *there is variety and richness in the language games that have been used in science through history*. Studies around the linguistics of science, especially those following Wittgenstein's ideas, may be of use to reflect on the issues posed here.

In the last characterisation of argumentation that is reviewed for this chapter, Kuhn Berland and Reiser (2011) recover the centrality of the aim of persuasion when arguing:

> The process of attempting to persuade the scientific community of an idea reveals faults in the argument (i.e., evidence that is unexplained by the idea or misapplication of accepted scientific principles), and identifying these faults creates opportunities for the community to improve upon the ideas being discussed. (Kuhn Berland and Reiser 2011, p. 212)

It can be argued that scientific disciplines are such inasmuch as they have disciples: therefore, *it is constitutive of their very nature the will to communicate, convince, persuade and teach*. This last input for science education can find support in texts from the science studies, especially in those situated in pragmatic and rhetorical perspectives.

References

AAAS [American Association for the Advancement of Science] (1993). *Project 2061: Benchmarks for science literacy.* Washington, DC: American Association for the Advancement of Science.

Abell, S.K., Anderson, G. & Chezem, J. (2000). Science as argument and explanation: Exploring concepts of sound in third grade. In Minstrell, J. & Van Zee, E.H. (Eds.). *Inquiry into inquiry learning and teaching in science* (pp. 100–119). Washington, D.C.: American Association for the Advancement of Science.

Adúriz-Bravo, A. (2005). *Una introducción a la naturaleza de la ciencia: La epistemología en la enseñanza de las ciencias naturales.* Buenos Aires: Fondo de Cultura Económica.

Adúriz-Bravo, A. (2010). Argumentación científica escolar: Herramientas para su análisis y su enseñanza. Plenary lecture presented at the *Seminario Internacional sobre Enseñanza de las Ciencias*, Cali, Colombia, June.

Adúriz-Bravo, A. (2011). Fostering model-based school scientific argumentation among prospective science teachers. *US-China Education Review*, 8(5), 718–723.

Adúriz-Bravo, A., Bonan, L., González Galli, L., Revel Chion, A. & Meinardi, E. (2005). Scientific argumentation in pre-service biology teacher education. *Eurasia Journal of Mathematics, Science and Technology Education*, 1(1), 76–83.

45 Revisiting School Scientific Argumentation from the Perspective of the History... 1467

Aliseda, A. (2006). *Abductive reasoning: Logical investigations into discovery and explanation.* Dordrecht: Springer.

Allchin, D. (2011). Evaluating knowledge of the nature of (whole) science. *Science Education, 95*(3), 518–542.

Archila, P.A. (2012). La investigación en argumentación y sus implicaciones en la formación inicial de profesores de ciencias. *Revista Eureka sobre Enseñanza y Divulgación de las Ciencias, 9*(3), 361–375.

Asti Vera, C. & Ambrosini, C. (2010). *Argumentos y teorías: Aproximación a la epistemología.* Buenos Aires: CCC Educando.

Atkins, L.J. (2008). The roles of evidence in scientific argument. In *AIP Conference Proceedings: 2008 Physics Education Research Conference*, Volume 1064 (pp. 63–66). Edmonton: American Institute of Physics.

Bar-Hillel, Y. (1970). *Aspects of language: Essays and lectures on philosophy of language, linguistic philosophy and methodology of linguistics.* Jerusalem: The Magnes Press.

Bell, P. & Linn, M.C. (2000). Scientific arguments as learning artifacts: Designing for learning from the web with KIE. *International Journal of Science Education, 22*(8), 797–817.

Bex, F.J. & Walton, D.N. (2012). Burdens and standards of proof for inference to the best explanation: Three case studies. *Law, Probability & Risk, 11*(2–3), 113–133.

Böttcher, F. & Meisert, A. (2011). Argumentation in science education: A model-based framework. *Science & Education, 20*(2), 103–140.

Bravo-Torija, B. & Jiménez-Aleixandre, M.P. (2011). A learning progression for using evidence in argumentation: An initial framework. Paper presented at the *9th ESERA Conference*, Lyon, France, September.

Bricker, L.A. & Bell, P. (2008). Conceptualizations of argumentation from science studies and the learning sciences and their implications for the practices of science education. *Science Education, 92*(3), 473–498.

Buty, C. & Plantin, C. (Eds.) (2008a). *Argumenter en classe de sciences: Du débat à l'apprentissage.* Paris: Institut National de Recherche Pédagogique.

Buty, C. & Plantin, C. (2008b). Introduction: L'argumentation à l'épreuve dans l'enseignement des sciences et vice-versa. In Buty, C. & Plantin, C. (Eds.). *Argumenter en classe de sciences: Du débat à l'apprentissage* (pp. 17–41). Paris: Institut National de Recherche Pédagogique.

Cademártori, Y. & Parra, D. (2000). Reforma educativa y teoría de la argumentación. *Revista Signos, 33*(48), 69–85.

Candela, A. (1999). *Ciencia en el aula: Los alumnos entre la argumentación y el consenso.* Mexico: Paidós.

Cavagnetto, A.R. (2010). Argument to foster scientific literacy: A review of argument interventions in K-12 science contexts. *Review of Education Research, 80*(3), 336–371.

Chevallard, Y. (1991). *La transposition didactique: Du savoir savant au savoir enseigné.* Grenoble: La Pensée Sauvage Éditions.

Clark, D.B., Sampson, V.D., Stegmann, K., Marttunen, M., Kollar, I., Janssen, J., Weinberger, A., Menekse, M., Erkens, G. & Laurinen, L. (2010). Scaffolding scientific argumentation between multiple students in online learning environments to support the development of 21st century skills. In Ertl, B. (Ed.). *E-collaborative knowledge construction: Learning from computer-supported and virtual environments* (pp. 1–39). New York: IGI Global.

Diéguez Lucena, A. (2005). *Filosofía de la ciencia.* Madrid: Biblioteca Nueva.

Driver, R.A., Newton, P. & Osborne, J.F. (2000). Establishing the norms of scientific argument in classrooms. *Science Education, 84*(3), 287–312.

Duschl, R.A. (1990). *Restructuring science education: The importance of theories and their development.* New York: Teachers College Press.

Duschl, R.A. (1998). La valoración de argumentaciones y explicaciones: Promover estrategias de retroalimentación. *Enseñanza de las Ciencias, 16*(1), 3–20.

Duschl, R.A. (2008). Quality argumentation and epistemic criteria. In Erduran, S. & Jiménez-Aleixandre, M.P. (Eds.). *Argumentation in science education: Perspectives from classroom-based research* (pp. 159–175). Dordrecht: Springer.

1468   A. Adúriz-Bravo

Duschl, R.A., Ellenbogen, K. & Erduran, S. (1999). Understanding dialogic argumentation among middle school science students. Paper presented at the *American Educational Research Association Annual Conference*, Montreal, Canada, April.

Duschl, R.A & Grandy, R. (Eds.) (2008). *Teaching scientific inquiry: Recommendations for research and implementation.* Rotterdam: Sense Publishers.

Duschl, R.A. & Osborne, J.F. (2002). Supporting and promoting argumentation discourse. *Studies in Science Education, 38*(1), 39–72.

Erduran, S. & Jiménez-Aleixandre, M.P. (Eds.) (2008). *Argumentation in science education: Perspectives from classroom-based research.* Dordrecht: Springer.

Erduran, S., Simon, S. & Osborne, J.F. (2004). TAPping into argumentation: Developments in the application of Toulmin's argument pattern for studying science discourse. *Science Education, 88*(6), 915–933.

Fagúndez Zambrano, T.J. & Castells Llavanera, M. (2009). La enseñanza universitaria de la física: Los objetos materiales y la construcción de significados científicos. *Actualidades Investigativas en Educación, 9*(2), 1–27.

Føllesdal, D. & Walløe, L. (1986). *Rationale Argumentation: Ein Grundkurs in Argumentations- und Wissenschaftstheorie.* Berlin: Walter de Gruyter. (Norwegian original from 1977.)

Galagovsky, L. (Ed.) (2008). *¿Qué tienen de "naturales" las ciencias naturales?* Buenos Aires: Biblos.

García Romano, L. & Valeiras, N. (2010). Lectura y escritura en el aula de ciencias: Una propuesta para reflexionar sobre la argumentación. *Alambique, 63*, 57–64.

Giere, R.N. (1988). *Explaining science: A cognitive approach.* Chicago: University of Chicago Press.

Giere, R.N., Bickle, J. & Mauldin, R.F. (2005). *Understanding scientific reasoning* (5th edition). Belmont: Wadsworth Publishing Company.

Harpine, W.D. (1985). Can rhetoric and dialectic serve the purposes of logic? *Philosophy and Rhetoric, 18*(2), 96–112.

Henao, B.L. & Stipcich, M.S. (2008). Educación en ciencias y argumentación: La perspectiva de Toulmin como posible respuesta a las demandas y desafíos contemporáneos para la enseñanza de las ciencias experimentales. *Revista Electrónica de Enseñanza de las Ciencias, 7*(1), 47–62.

Hodson, D. (2009). *Teaching and learning about science: Language, theories, methods, history, traditions and values.* Rotterdam: Sense Publishers.

Islas, S.M., Sgro, M.R. & Pesa, M.A. (2009). La argumentación en la comunidad científica y en la formación de profesores de física. *Ciência & Educação, 15*(2), 291–304.

Izquierdo-Aymerich, M. (2005). Hacia una teoría de los contenidos escolares. *Enseñanza de las Ciencias, 23*(1), 111–122.

Izquierdo-Aymerich, M. & Adúriz-Bravo, A. (2003). Epistemological foundations of school science. *Science & Education, 12*(1), 27–43.

Izquierdo-Aymerich, M., Sanmartí, N., Espinet, M., García, M.P. & Pujol, R.M. (1999). Caracterización y fundamentación de la ciencia escolar. *Enseñanza de las Ciencias*, extra issue, 79–92.

Jiménez-Aleixandre, M.P. (2010). *10 ideas clave: Competencias en argumentación y uso de pruebas.* Barcelona: Graó.

Jiménez-Aleixandre, M.P., Bugallo Rodríguez, A. & Duschl, R.A. (2000). "Doing the lesson" or "doing science": Arguments in high school genetics. *Science Education, 84*, 757–792.

Jiménez-Aleixandre, M.P. & Díaz de Bustamante, J. (2003). Discurso de aula y argumentación en la clase de ciencias: Cuestiones teóricas y metodológicas. *Enseñanza de las Ciencias, 21*(3), 359–370.

Jiménez-Aleixandre, M.P. & Erduran, S. (2008). Argumentation in science education: An overview. In Erduran, S. & Jiménez-Aleixandre, M.P. (Eds.) *Argumentation in science education: Perspectives from classroom-based research* (pp. 3–27). Dordrecht: Springer.

Jiménez-Aleixandre, M.P. & Federico-Agraso, M. (2009). Justification and persuasion about cloning: Arguments in Hwang's paper and journalistic reported versions. *Research in Science Education, 39*(3), 331–347.

45 Revisiting School Scientific Argumentation from the Perspective of the History... 1469

Kelly, G.J. & Bazerman, C. (2003). How students argue scientific claims: A rhetorical-semantic analysis. *Applied Linguistics*, *24*(1), 28–55.

Kelly, G.J. & Chen, C. (1999). The sound of music: Constructing science as sociocultural practices through oral and written discourse. *Journal of Research in Science Teaching*, *36*(8), 883–915.

Kelly, G.J. & Takao, A. (2002). Epistemic levels in argument: An analysis of university oceanography students' use of evidence in writing. *Science Education*, *86*(3), 314–342.

Khine, M.S. (Ed.) (2012). *Perspectives on scientific argumentation: Theory, practice and research.* Dordrecht: Springer.

Kirschner, P.A., Sweller, J. & Clark, R.E. (2006). Why minimal guidance during instruction does not work: An analysis of the failure of constructivist discovery, problem-based, experiential, and inquiry-based teaching. *Educational Psychologist*, *41*(2), 75–86.

Konstantinidou, A., Cerveró, J.M. & Castells, M. (2010). Argumentation and scientific reasoning: The "double hierarchy" argument. In Taşar, M.F. & Çakmakci, G. (Eds.). *Contemporary science education research: Scientific literacy and social aspects of science* (pp. 61–70). Ankara: Pegem Akademi.

Kuhn, D. (1992). Thinking as argument. *Harvard Educational Review*, *62*(2), 155–179.

Kuhn, D (1993). Science as argument. *Science Education*, *77*(3), 319–337.

Kuhn, D. (2010). Teaching and learning science as argument. *Science Education*, *94*(5), 810–824.

Kuhn Berland, L. & Reiser, B. (2009). Making sense of argumentation and explanation. *Science Education*, *93*(1), 26–55.

Kuhn Berland, L. & Reiser, B. (2011). Classroom communities' adaptation of the practice of scientific argumentation. *Science Education*, *95*(2), 191–216.

Lawson, A.E. (2003). The nature and development of hypothetico-predictive argumentation with implications for science teaching. *International Journal of Science Education*, *25*(11), 1387–1408.

Lawson, A.E. (2005). What is the role of induction and deduction in reasoning and scientific inquiry? *Journal of Research in Science Teaching*, *42*(6), 716–740.

Lawson, A.E. (2009). Basic inferences of scientific reasoning, argumentation, and discovery. *Science Education*, *94*(2), 336–364.

Lehrer, R. & Schauble, L. (2006). Cultivating model-based reasoning in science education. In Sawyer, R.K. (Ed.). *Cambridge handbook of the learning sciences* (pp. 371–387). Cambridge: Cambridge University Press.

Lemke, J. (1990). *Talking science: Language, learning, and values.* Norwood: Ablex.

Linhares Queiroz, S. & Passos Sá, L. (2009). O espaço para a argumentação no ensino superior de química. *Educación Química*, *20*(2), 104–110.

Martins, I. (2009). Argumentation in texts from a teacher education journal: An exercise of analysis based upon the Bakhtinian concepts of genre and social language. *Educación Química*, *20*(2), 26–36.

Mason, L. & Scirica, F. (2006). Prediction of students' argumentation skills about controversial topics by epistemological understanding. *Learning and Instruction*, *16*, 492–509.

Matthews, M. (1994/2014). *Science teaching: The role of history and philosophy of science*, New York: Routledge.

Matthews, M. (2000). *Time for science education: How teaching the history and philosophy of pendulum motion can contribute to science literacy.* New York: Plenum Publishers.

McDonald, C. (2010). The influence of explicit nature of science and argumentation instruction on preservice primary teachers' views on nature of science. *Journal of Research in Science Teaching*, *47*(9), 1137–1164.

Monk, M. & Osborne, J.F. (1997). Placing the history and philosophy of science on the curriculum: A model for the development of pedagogy. *Science Education*, *81*(4), 405–424.

Newton, P., Driver, R. & Osborne, J.F. (1999). The place of argumentation in the pedagogy of school science. *International Journal of Science Education*, *21*(5), 553–576.

Nielsen, J.A. (2011). Dialectical features of students' argumentation: A critical review of argumentation studies in science education. *Research in Science Education*, on-line first.

NRC [National Research Council] (1995). *National science education standards.* Washington, DC: National Academy Press.

Nussbaum, E.M., Sinatra, G.M. & Owens, M.C. (2012). The two faces of scientific argumentation: Applications to global climate change. In Khine, M.S. (Ed.) *Perspectives on scientific argumentation: Theory, practice and research* (pp. 17–37). Dordrecht: Springer.

Ogunniyi, M.B. (2007). Teachers' stances and practical arguments regarding a science-indigenous knowledge curriculum: Part 1. *International Journal of Science Education, 29*(8), 963–986.

Ogunniyi, M.B. & Hewson, M.G. (2008). Effect of an argumentation-based course on teachers' disposition towards a science-indigenous knowledge curriculum. *International Journal of Environmental and Science Education, 3*(4), 159–177.

Osborne, J. (2005). The role of argument in science education. In Boersma, K. Goedhart, M., de Jong, O. & Eijkelhof, H. (Eds.). *Research and the quality of science education* (pp. 367–380). Dordrecht: Springer.

Osborne, J.F. (2010). Arguing to learn in science: The role of collaborative, critical discourse. *Science, 328*, 463–466.

Osborne, J.F., Erduran, S., Simon, S. & Monk, M. (2001). Enhancing the quality of argument in school science. *School Science Review, 82*(301), 63–70.

Osborne, J.F., MacPherson, A., Patterson, A. & Szu, E. (2012). Introduction. In Khine, M.S. (Ed.). *Perspectives on scientific argumentation: Theory, practice and research* (pp. 3–16). Dordrecht: Springer

Osborne, J.F. & Patterson, A. (2011). Scientific argument and explanation: A necessary distinction? *Science Education, 95*(4), 627–638.

Padilla, C. (2012). Escritura y argumentación académica: Trayectorias estudiantiles, factores docentes y contextuales. *Magis, 5*(10), 31–57.

Plantin, C. (2005). *L'argumentation: Histoire, théories et perspectives.* Paris: PUF.

Plantin, C. (2011). "No se trata de convencer sino de convivir": L'ère post-persuasion. *Rétor, 1*(1), 59–83.

Pontecorvo, C. & Girardet, H. (1993). Arguing and reasoning in understanding historical topics. *Cognition and Instruction, 11*(3 & 4), 365–395.

Revel Chion, A., Adúriz-Bravo, A. & Meinardi, E. (2009). Análisis histórico-epistemológico de las concepciones de salud desde una perspectiva didáctica: Narrando la "historia" de la peste negra medieval. *Enseñanza de las Ciencias*, extra issue, 168–172.

Revel Chion, A., Couló, A., Erduran, S., Furman, M., Iglesia, P. & Adúriz-Bravo, A. (2005). Estudios sobre la enseñanza de la argumentación científica escolar. *Enseñanza de las Ciencias*, extra issue *VII Congreso Internacional sobre Investigación en la Didáctica de las Ciencias*, Oral presentations, Section 4.1., n/pp.

Reygadas, P. & Haidar, J. (2001). Hacia una teoría integrada de la argumentación. *Estudios sobre las Culturas Contemporáneas, VII*(13), 107–139.

Ruiz, F.J., Márquez, C. & Tamayo, O.E. (2011). Teachers' change of conceptions on argumentation and its teaching. In *E-book ESERA 2011*, Strand 6, pp. 86–92. doi: http://lsg.ucy.ac.cy/esera/e_book/base/ebook/strand6/ebook-esera2011_RUIZ-06.pdf

Sadler, T.D. (Ed.) (2011). *Socioscientific issues in the classroom: Teaching, learning and research.* Dordrecht: Springer.

Salmon, M.H. (1995). *Introduction to logic and critical thinking.* Fort Worth: Harcourt Brace.

Samaja, J. (1999). *Epistemología y metodología: Elementos para una teoría de la investigación científica* (3rd edition). Buenos Aires: EUDEBA.

Sampson, V.D, & Clark, D.B. (2006). Assessment of argument in science education: A critical review of the literature. In Barab, A., Hay, K.E. & Hickey, D.T. (Eds.). *Proceedings of the Seventh International Conference of the Learning of Science: Making a difference* (pp. 655–661). Mahwah: Lawrence Erlbaum.

Sampson, V.D. & Clark, D.B. (2007). Incorporating scientific argumentation into inquiry-based activities with online personally-seeded discussions. *The Science Scope, 30*(6), 43–47.

45 Revisiting School Scientific Argumentation from the Perspective of the History... 1471

Sampson, V.D. & Clark, D.B. (2008). Assessment of the ways students generate arguments in science education: Current perspectives and recommendations for future directions. *Science Education*, 92(3), 447–472.

Sandoval, W.A. (2003). Conceptual and epistemic aspects of students' scientific explanations. *Journal of the Learning Sciences*, 12(1), 5–51.

Sandoval, W.A. & Millwood, K.A. (2005). The quality of students' use of evidence in written scientific explanations. *Cognition and Instruction*, 23(1), 23–55.

Sandoval, W.A. & Millwood, K.A. (2008). What can argumentation tell us about epistemology. In Erduran, S. & Jiménez-Aleixandre, M.P. (Eds.). *Argumentation in science education: Perspectives from classroom-based research* (pp. 71–88). Dordrecht: Springer.

Sandoval, W.A. & Reiser, B. (2004). Explanation-driven inquiry: Integrating conceptual and epistemic scaffolds for scientific inquiry. *Science Education*, 88(3), 345–372.

Sanmartí, N. (Ed.) (2003). *Aprendre ciències tot aprenent a escriure ciència*. Barcelona: Edicions 62.

Santibáñez, C. (2012). Teoría de la argumentación como epistemología aplicada. *Cinta de Moebio*, 43, 24–39.

Sasseron, L.H. & Carvalho, A.M.P. (2011). Construindo argumentação na sala de aula: A presença do ciclo argumentativo, os indicadores de alfabetização científica e o padrão de Toulmin. *Ciência & Educação*, 17(1), 97–114.

Schwarz, B.B., Neuman, Y., Gil, J. & Ilya, M. (2003). Construction of collective and individual knowledge in argumentative activity: An empirical study. *The Journal of the Learning Sciences*, 12(2), 221–258.

Siegel, H. (1995). Why should educators care about argumentation. *Informal Logic*, 17(2), 159–176.

Smith, P. (2003). *An introduction to formal logic*. Cambridge: Cambridge University Press.

Stadler, F. (Ed.) (2004). *Induction and deduction in the sciences*. Dordrecht: Kluwer.

Sutton, C. (1996). Beliefs about science and beliefs about language. *International Journal of Science Education*, 18(1), 1–18.

Tiberghien, A. (2008). Preface. In Erduran, S. & Jiménez-Aleixandre, M.P. (Eds.). *Argumentation in science education: Perspectives from classroom-based research* (pp. ix-xv). Dordrecht: Springer.

Tindale, C.W. (1999). *Acts of arguing: A rhetorical model of argument*. Albany: State University of New York Press.

Toulmin, S.E. (1958). *The uses of argument*. Cambridge: Cambridge University Press.

Toulmin, S.E. (2001). *Return to reason*. Cambridge: Harvard University Press.

Toulmin, S.E. (2003). *The uses of argument* (updated edition). Cambridge: Cambridge University Press.

van Eemeren, F.H. & Houtlosser, P. (2003). The development of the pragma-dialectical approach to argumentation. *Argumentation*, 17, 387–403.

von Aufschnaiter, C., Erduran, S., Osborne, J.F. & Simon, S. (2008). Arguing to learn and learning to argue: Case studies of how students' argumentation relates to their scientific knowledge. *Journal of Research in Science Teaching*, 45(1), 101–131.

Walton, D.N. (1996). *Argumentation schemes for presumptive reasoning*. Mahwah: Lawrence Erlbaum Associates.

Westbury, I., Hopmann, S. & Riquarts, K. (Eds.) (2000). *Teaching as a reflective practice: The German Didaktik tradition*. Mahwah: Lawrence Erlbaum Associates.

Windschitl, M., Thompson, J. & Braaten, M. (2008). Beyond the scientific method: Model-based inquiry as a new paradigm of preference for school science investigations. *Science Education*, 92(5), 941–967.

Wolpert, L. (1992). *The unnatural nature of science: Why science does not make common sense*. London: Faber and Faber.

Zeidler, D.L. (Ed.) (2003). *The role of moral reasoning on socioscientific issues and discourse in science educationn* Dordrecht: Kluwer.

Agustín Adúriz-Bravo is Professor of Didactics of Science at the Research Institute CeFIEC of the Universidad de Buenos Aires, Argentina. He received an M.Sc in Physics Teaching from the Universidad de Buenos Aires and a Ph.D in Didactics of Science from the Universitat Autònoma de Barcelona, Spain. He is a regular Visiting Professor at several Universities in the Americas and Europe. His general research focus is on the contributions of the philosophy of science to science teacher education. He has extensively published in Spanish and English.

Chapter 46
Historical-Investigative Approaches in Science Teaching

Peter Heering and Dietmar Höttecke

46.1 Introduction

The notion of the historical-investigative approach to science teaching was first introduced by Kipnis (1996). According to his theory, teaching and learning science within a context of history and philosophy of science (HPS) should be mastered through students' practical investigations. He suggests a middle ground between highly structured verification laboratories, which aim at generating "true" results, and open-ended experiments, where students are not guided at all. The basic idea of students doing practical work while learning science, along with its history, had been developed earlier in similar ways (e.g., Pedzisz and Wilke 1993; Rieß and Schulz 1994; Teichmann 1979, 1999).

We use the notion of the historical-investigative approach here as a broad idea that characterizes a variety of perspectives for teaching and learning science with HPS. Two central aspects are tightly joined together, the first being that science is embedded within a historical context. A central objective of this idea is to broaden students' understanding of scientific concepts and theories, to promote their interest in science, and to foster their general historical awareness. These are all ideas with a long tradition (Conant 1957; Mach 1912; Ramsauer 1953). The second aspect is concerned with the development of procedural knowledge and process skills, which are often highlighted in standard documents (Barth 2010).

P. Heering (✉)
Institute of Mathematic, Scientific, and Technical Literacy,
University of Flensburg, Flensburg 24943, Germany
e-mail: peter.heering@uni-flensburg.de

D. Höttecke
Department of Education, University of Hamburg, Hamburg 20146, Germany
e-mail: dietmar.hoettecke@uni-hamburg.de

M.R. Matthews (ed.), *International Handbook of Research in History, Philosophy and Science Teaching*, DOI 10.1007/978-94-007-7654-8_46,
© Springer Science+Business Media Dordrecht 2014

There is also a long-standing tradition in science education which calls for practical work to be done: students plan investigations, perform or even design experiments, collect and analyze data, draw conclusions, and discuss and negotiate their results. In particular, the idea of inquiry-based teaching and learning science is based on the analogy of students' activities to those of scientists. Inquiry refers to a diverse set of activities in which scientists study (e.g., Anderson 2007), explore, explain, or even manufacture (Knorr-Cetina 1981) the natural world. Scientists propose explanations based on both evidence and inference.

Several curricular standard documents, such as the NSES (National Research Council 2000), have stressed that learners should be engaged through scientifically oriented questions. They should prioritize evidence while proposing explanations which have to be critically evaluated, communicated, and justified. Inquiry learning and learning science with HPS share this special focus of student centeredness in science education. At the same time, inquiry learning and learning with and about HPS have both been shown to be useful means of learning about the nature of science (NOS). Research has indicated that NOS has to be addressed explicitly and reflectively in order to enhance the efficiency of teaching (e.g., Abd-El-Khalick and Lederman 2000; Seker and Welsh 2005).[1] Science teachers should confront students with their deeply held beliefs about science and how science works. During the last three or four decades, an extensive body of literature about NOS has been written.[2] Our analysis reveals that there are strong relationships of teaching and learning NOS and HPS on a curricular level (e.g., National Research Council 1996), with research and development in science education.[3]

Our analysis is based on two major threads of discourse: the discourse of research in HPS and the discourse of research in science education. Both are relevant to the HPSST community and for the development of ideas of how to teach science within a historical-investigative framework. In the second section of this chapter, a brief analysis of the recent discourse in HPS is presented, showing that experiments became an important focal point of historical and philosophical analysis in the early 1980s. Since then, from an epistemological point of view, experiments have no longer been reduced to a means for testing mere theoretical knowledge. The material and instrumental procedures in science are currently regarded as a central feature for explaining what science is all about, how science proceeds, and in which respect science is a cultural and human endeavor. The presentation of this

---

[1] For the efficacy of inquiry-based techniques for learning about NOS, see, for example, Akerson et al. (2007), Khishfe and Abd-El-Khalick (2002), and Schwarz, Lederman, and Crawford (2004). Evidence for the efficacy of HPS for learning about NOS has been presented, for example, by Allchin (1999), Clough (2011), Galili and Hazan (2001), Howe and Rudge (2005), Irwin (2000), Kruse and Wilcox (2011), Lin and Chen (2002), and Rudge and Howe (2009).

[2] For a recent and extended overview of research on students' views on NOS, see Hodson (2009). A critical review of research methods with regard to different orientations of NOS is given by Deng et al. (2011). See also Lederman (2007), Lederman and Lederman (2012), and McComas and Almazroa (1998).

[3] See, for example, Hodson (2009), Höttecke et al. (2012), Irwin (2000), and Rudge and Howe (2009).

46 Historical-Investigative Approaches in Science Teaching

"experimental turn" in HPS is followed by an overview of recent research in science education regarding practical work and learning demonstrating that practical work in science education has a long-standing tradition. Central problems which have been identified by research throughout recent decades will be discussed as far as they relate to the historical-investigative approach.

Based on the analysis of these two threads of discourse, a structured discussion of several approaches in science education is presented, approaches which all have one common feature: merging the idea of teaching science with HPS with practical and investigative learning activities.

46.2 The Experimental Turn in HPS

46.2.1 Investigating Science as Practice and Culture

Understanding science as a cultural activity gained increasing importance during the second half of the twentieth century. Such a view breaks with positivistic accounts, which describe the development of science and scientific knowledge as progressing towards the truth, and allows a description of science as something that is shaped or even influenced by factors which might have been regarded as nonscientific. Such a cultural perspective is closely related to a constructivist understanding of science. A number of studies aim at describing and understanding science as a sociocultural practice. Among them are laboratory studies (Knorr-Cetina 1981; Latour and Woolgar 1979), as well as social science studies (e.g., Shapin 1994). The analysis of science under a sociocultural perspective has shown, in particular, different ways in which a scientific consensus might be reached and the role that social factors, such as authority and reputation, might play in the establishment of scientific knowledge (e.g., Collins 1985; Galison 1987).

Besides social and cultural aspects of science, the role of material objects, such as scientific instruments and material procedures, received scant attention by those studying the history and philosophy of science until about 1980. Since then, they have become increasingly important for the development of a comprehensive understanding of science as a process. For decades, instruments and their use, development, and history had been an issue only for a small community of historians of scientific instruments. The recognition of material and instrumental procedures in science has been a more recent development in HPS. Accordingly, laboratory notebooks have become an important source for detailed accounts of scientific practice at the workbench.[4] The consideration and analysis of science as a visual practice is

---

[4] For discussions of laboratory notes, see, in particular, Holmes et al. (2003). See also Steinle's studies on Faraday (Steinle 1996) and Dufay (Steinle 2006). For further case studies based on Faraday's laboratory diaries, see Gooding (1990), Tweney (1985), and Höttecke (2001).

another recent tendency in HPS. Images and their role in shaping and communicating scientific ideas and results have, therefore, been analyzed in detail.[5]

Even though instruments and instrument collections have been a topic of research for quite some time, there are some recent developments in this field which deserve a closer look. Traditionally, research on scientific instruments had two major directions of impact: either there was a scope on the construction of particular instruments, since they were regarded as highly relevant to the general development of science (e.g., development of theories, confirmation of laws, instrumental procedures), or instrument collections were analyzed in order to explore and characterize distinguishing features they had in common (e.g., collections of microscopes). Accounts on the history of scientific instruments that followed one of these two research traditions remained, by and large, descriptive. The material and procedural aspects were scarcely considered to be an important issue.

Besides these two approaches, there were a few early studies which focused on experimental practices in science, probably the first one being Settle's experimental analysis of Galileo's inclined plane experiment (Settle 1961). The methodological approach that he used was particularly innovative: he reenacted the experiment with a reconstructed device. Several similar studies followed which also analyzed experiments described by Galileo (Drake 1970; MacLachlan 1973). The experimental approach, while being limited to experiments that had been described by Galileo, can be understood when the aim and context of the experimental analysis are taken into consideration. During the 1960s and 1970s, Galileo's experimental accounts were regarded as fictitious. Historians and philosophers of science held the notion that Galileo, being predominantly a theoretician, used accounts of experiments merely as devices to strengthen and clarify his theoretical arguments. It was doubted that Galileo had ever carried out the experiments he described in his publications. Consequently, the reenactments were intended to demonstrate that Galileo had, in fact, been able to carry out these experiments according to his description.[6]

Since the early 1980s, experimentation in science has received more attention than ever before from historians and philosophers of science.[7] Initially, history of science was concerned foremost with the development of theories. According to this view, the only role experiments could ever play in science was limited to confirming theories or even just providing the possibility of falsification (Popper 1934). During the early 1980s, some studies, however, began to focus on experimentation as an independent aspect of the formation of scientific knowledge, alongside the development of theoretical knowledge. In this respect, Ian Hacking's

---

[5] For a discussion on the role of images in the sciences, see, in particular, Heßler (2006) and Busch (2008). For discussions of images and their role in nanotechnology, see Bigg and Hennig (2009).

[6] There are, of course, some earlier examples which analyze materials, as well as practices, through reenactment. The field in which this method was likely first applied is archeology, the most prominent example being the demonstration by the Norwegian Thor Heyerdahl, in the late 1940s, of the possibility of travelling from the American Pacific coast to Polynesia with a simple raft.

[7] See, for example, Collins (1985), Galison (1987), Gooding (1990), Gooding and James (1985), Gooding et al. (1989a), Hacking (1983), Radder (2003), Schaffer (1983), and Shapin and Schaffer (1985).

dictum that "experimentation has a life of its own" (1983, p. 150) became well known. Experiments were regarded as a central element in the process of knowledge production. Manipulations of the natural world have multiple facets and cannot be used merely to test assumptions.

Such a wide view of experimentation also contributed to a new awareness of instruments in the history of science. Instruments were, at the time, considered to be artifacts which were designed and constructed according to explicit objectives, their function and shape possibly even having changed over the course of their use. They could not be taken for granted as objects anymore, as they were believed to have been a crucial part of experimental and manipulative practices of the past. Their meanings were not regarded as anything established or given anymore but as developed in laboratories and elsewhere as a result of material, social, and cultural practices. Hence, their materiality and contingent use were seen as relevant objects of historical science research.

The general experimental or even material turn in HPS described above also led to new perspectives for science museums. Historians of science criticized the restoration of a historical instrument as a mere transformation into a mint condition. During the process of restoring an old instrument, traces of its former use should be kept rather than displaying it as a cleanly polished instrument. Historical analysis focused on such traces because they were regarded as signs of the historic progression of an instrumental procedure. Instrumental procedures were not regarded as unproblematic, stable, or even fixed but as probably changing over the course of the development of an experiment.

46.2.2 Investigating Material Procedures

The experimental turn has not resulted in a diversified research focus. Various different directions and research currents can instead be identified. Research focusing on the historical analysis of scientific experimentation either emphasized the material or the methodological character of experiments. The latter idea can be "found in studies that understood science in terms of activity rather than contemplation. Science was to be understood not as a body of knowledge but as a network of embodied practices" (Morus 2010, p. 775). Such a historical perspective is related to a broader understanding of the role of experiments in knowledge production: "The aim of the experimenter is to transform … inscriptions to a stage where they seem capable of but one reading and become powerful weapons in argument" (Gooding et al. 1989b, p. 5).

A central approach can be seen in various attempts to analyze historical experiments by the method of replication. This approach follows a tradition founded by the studies of Settle, Drake, and others regarding Galileo's work, already mentioned above. The general focus of the historical investigations of experimental practice, however, shifted from a mere reproduction of experimental results presented in original papers towards detailed accounts of experimentation as both process and practice. Material manipulations, instrumental procedures, and their general relationship to the development of theories or models were under scrutiny. Yet, it has to be noted that even in his first paper, Settle had already made another aspect explicit: "To get a better appreciation for some of the problems he [Galileo, P.H.] faced I have tried to reproduce the experiment essentially as Galileo described it" (Settle 1961, p. 19). Settle's point is remarkable for his time. He identifies a central aspect of this kind of research: the idea is not to question the experimental findings but to develop an understanding of the crucial details of the experiment and the difficulties that Galileo might have had to face due, in part, to choosing adequate materials and instrumental design and to developing the respective skills that are required for a successful performance.

One of the first researchers to apply this methodology more systematically was David Gooding, who analyzed experiments described by Michael Faraday. While doing this, Gooding developed a particular focus:

> [S]ince skills cannot be recovered from the familiar literary and material forms of evidence - manuscripts, publications and instruments - historians of science should, if possible, venture beyond these, to study the activity that produced them. … Empirical results never are entirely independent of the practices that let to their production. Facts are practice-laden as well as theory-laden. … If much of what experimentalists do cannot be recorded, it cannot be recovered by reading texts or even by studying apparatus (Gooding 1989, pp. 63 f).

Besides Gooding, it was the Oldenburg group, in particular, established by Falk Rieß, which systematically developed the methodology of reenacting experiments of the past. Their so-called replication method[8] was based on the reconstruction of an apparatus as close as possible to the available historical evidence, which might be presented in written sources or even surviving instruments. Such reconstructions were used for authentically reconstructing material procedures and instrumental manipulations. The major objective was to write case studies about reconstructed and analyzed scientific practices. This replication method has been established during the last two and a half decades. Meanwhile, other researchers apply this (or an adapted) methodology, which is particularly prominent in the physical sciences.[9] There are also examples of it being used in chemistry[10] and some in biology.[11]

---

[8] For systematic discussions of the method, see, in particular, Breidbach et al. (2010), Heering (1998), Rieß (1998), and Sichau (2002). For a slightly different methodological interpretation, see Frercks (2001). Case studies are collected in Breidbach et al. (2010) or Heering et al. (2000). Moreover, see for single cases Engels (2006), Heering (1992, 1994, 2002, 2005, 2006, 2007, 2008, 2010), Heering and Osewold (2005), Hennig (2003), Höttecke (2000, 2001), Kärn (2002), Müller (2004), Sibum (1995, 1998), Sichau (2000a), Staubermann (2007), Voskuhl (1997), and Wittje (1996).

[9] See, for example, Cavicchi (2006), Fiorentini (2005), Lacki and Karim (2005), Martínez (2006), Palmieri (2008, 2009), and Staubermann (2011). The Jena group, led by Breidbach, focuses more strongly on the reconstruction of the instrument than on the practice with the instruments. For a detailed account, see Breidbach et al. (2010), as well as Frercks and Weber (2006).

[10] See, for example, Fatet (2005), Principe (2000), Tweney (2005), Usselman et al. (2005), Chang (2011), and Eggen et al. (2012).

[11] See, for example, Maienschein (1999) and Maienschein et al. (2008).

Moreover, the approach is no longer limited to the history of science; it has also gained importance in other professional fields, most notably in technology[12] and cultural studies.[13]

Manual procedures are usually hard to communicate, and complex manual procedures cannot be explicitly communicated at all because of their tacit nature (Collins 1985; Polanyi 1966). One may think, for instance, on how hard or even impossible it is to explain exactly how to ride a bike without tumbling. Detailed accounts on laboratory practice, including their tacit dimensions, require new methods. The replication method, however, allows for an analysis of the complex interactions of materials, instruments, rooms and spaces, people, their bodies, and the associated theoretical ideas. Additionally, the analysis comprises an account of the social and cultural meanings to which an experiment or instrument may be related, such as norms and values in science and beyond.

In summary, history and philosophy of science have embraced scientific experiments and instruments, as well as practical manipulations of scientists, in a new way during the last three decades. As a result, our current understanding of their role in the production of scientific knowledge and the establishment of scientific practices has increased. As we will show in Sect. 46.4, the role of experiments, instruments, and practical manipulations corresponds to recent developments in HPS-informed science education. The next step is to recapitulate briefly what we know about practical work in science education in general.

46.3 Research About Practical Work in Science Education: A Background for Historical-Investigative Approaches

Experiments and practical work are of great importance in science education. Scholars have long argued in favor of its predominant role. For natural studies, practical fieldwork became important in the USA towards the end of the nineteenth century (Kohlstedt 2010). In physics, practical work became an accepted part of high-school education in the USA by 1910 (Rosen 1954).[14] A commission of the German Society of Natural and Medical scientists released a document with recommendations (GDNÄ 1905 according to Willer 1990) in the early twentieth century. More than a 100 years ago, scholars had already emphasized the importance of practical work in science education for the enhancement of process skills, as well as for general attitudes of accuracy and exact observation. The commission emphasized that science should be taught as an exemplar of how knowledge is generally

---

[12] One of the best-known examples in this field is Wright's analysis of the Antikythera mechanism (Wright 2007); for other examples, see the contributions in Staubermann (2011).

[13] Most prominent is the discussion on the role that the camera obscura may have played in painting. For a summary of the discussion, see Lefèvre (2007).

[14] For the process of transition from textbook to practical work, see, for example, Hoffmann (2011), Kremer (2011), and Turner (2011).

acquired in the empirical sciences. While the first idea is still accepted today, the idea of a clear-cut, single epistemic methodology appears to be obsolete.[15] This does not mean that science is not driven by a limited number of methodological rules, like, for instance, the use of controlled experiments or avoiding ad hoc revisions to theories (Irzik and Nola 2011).

Even today, there is wide agreement among science educators that practical work is of general and great importance for teaching and learning science. A European Delphi study, for instance (Welzel et al. 1998), gathered empirical data on the main teaching objectives for laboratory work in science education, as recognized by science teachers at the upper-secondary and first-year university level. According to this study, the great value of experiments and laboratory courses is seen as important for reaching several educational objectives. Through them, students can learn how to connect theory and practice in science and how to test knowledge. There are opportunities for the enhancement of process and social skills, scientific thinking skills, motivation, and personality development. Besides the general appreciation of practical work in science education, the teachers in this study valued structured and guided instructions of students' laboratory activities, which were seen to be of high importance for the development of experimental skills and insights into the relationship between theory and practice.

Recent research in the history of science that focuses on experimentation has strongly stressed the idea that scientific experimentation is a multifaceted activity with many possible relationships between experimentation and observation, on the one hand, and inference and theory development, on the other (e.g., Hacking 1983; Heidelberger and Steinle 1998). The Kantian idea of an experiment where a scientist directs questions to nature like a judge to a witness can no longer be supported, since nature cannot be regarded as an unaffected and independent agency (Kutschmann 1994). According to our current understanding, a scientific experiment is, instead, an act of intervention, where questions, interests, public and private perspectives, background knowledge and skills, an experimenter's body, instruments, rooms and spaces, material and theoretical entities, and procedures interact to develop science within a cultural and societal context (see Sect. 46.2 of this paper).

The actual role and use of experiments in science education presents a somewhat problematic situation. An extended video-based study focusing on physics education in Germany has shown that experiments usually play a major role in physics teaching (Tesch and Duit 2004); however, teachers appreciate the role of experiments in science teaching only in a rather general way. Jonas-Ahrend (2004), reporting her findings from an interview study with physics teachers, asserts that the educational purpose and relationship between students' experimentation and their learning are hardly considered by teachers. Strong guidance of lab activities in science education has often been criticized as cookbook-style[16] or even as a verification laboratory (Kang and Wallace 2005; Metz et al. 2007) because it fosters a

---

[15] See, for example, Feyerabend (1972), Hentschel (1998), Pickering (1995), and Ziman (2000).

[16] See, for example, Clough (2006), Hofstein and Kind (2012), Hofstein and Lunetta (2004), and Metz and Stinner (2006).

portrayal of science as a rhetoric of conclusions or even indoctrinates students in correct procedures (Nott and Wellington 1996). Inquiry or "discovery" learning may pose the danger of misrepresenting science as primarily an inductive endeavor. According to Harris and Taylor (1983), a pure inductivist idea of science may bear several pedagogical pitfalls. They consider the problem that in following an inductive perspective on science, a certain chain of inferences from observations to conclusions rules out alternative explanations of phenomena. They criticize the lack of a coherent philosophy of science implicit in curricular material. Science, if outlined as inductive, might even justify an alleged "progressive" view on education, according to which the child is misleadingly regarded as autonomous and experiences practical work as authentic.

It often happens that students do not really know what the purpose of an experimental procedure they follow or the meaning of the data they collect might be.[17] According to Gallagher and Tobin (1987), high-school teachers rarely consider whether their students really understand what they are doing and what their experimental results might indicate. Science teachers hardly exhibit behavior that encourages students to think about the nature of scientific inquiry in a reflective manner. Thus, it is hardly surprising that research has indicated that students have a limited understanding of the nature and purposes of experimentation in science.[18] Concerning the current situation in science teaching, Hofstein and Kind (2012, p. 192) have concluded that "practical work meant manipulating equipment and materials, but not ideas."

Whenever students perform experiments, neither their ideas nor their performances resemble what scientists in their laboratories are actually doing. Chinn and Malhotra (2002) have compared students performing so-called "simple" experiments with scientists doing authentic inquiry. They conclude that the cognitive operations of both are rather different. While scientists generate their own questions, questions are often directed to students. Students usually investigate only one given variable, while scientists have to select or even invent variables to investigate. Students are usually told what and how to measure; scientists, however, incorporate multiple measures of independent, intermediate, and dependent variables. Whereas students usually draw conclusions from a single experiment, scientists coordinate results from multiple studies. Such differences reveal the deep gap between regular inquiry activities at school and the ways scientists perform their experiments.

This gap is also mirrored in teacher students' understanding of experiments. In a study based on focus group interviews, Gyllenpalm and Wickmam (2011) found out that Swedish university teacher students understand the notion experiment rather as a method of teaching than a method of scientific inquiry. According to this study an "'experiment' was never explicitly associated with a particular methodology for producing new knowledge about causal relationships" (ibd., p. 920) and rather used

---

[17] See, for example, Flick (2000), Hart et al. (2000), Hofstein and Kind (2012), Hofstein and Lunetta (2004), Lunetta et al. (2007), and Schauble et al. (1995).

[18] See, for example, Carey et al. (1989), Lubben and Millar (1996), Meyer and Carlisle (1996), Milne and Taylor (1995), and Solomon et al. (1996).

in an everyday sense of the word. As long as science education lays claim to the idea that the practical work of students and experiments in science should have anything in common, there is a bridge to be built between different views of experiments in science and in science education.

Scholars have stressed the role of open-ended activities within an inquiry framework, which has been shown to be superior to strongly guided practical work (e.g., Berg et al. 2003). Trumper (2003) demands that from a constructivist perspective of teaching and learning, the way that we teach in the laboratory should be rethought. Anderson (2007) summarizes the features of a new student orientation towards a becoming more self-directed learner. Under such an orientation, students process information, as well as interpret and explain data. They design their own activities, form interpretations of data, and share authority for answers. Still, inquiry learning should not be unguided since research from the field of educational psychology (Kirschner et al. 2006) warns us that the advantage of guidance begins to recede only if learners already possess sufficient prior knowledge to cope with a certain problem. Researchers have generally emphasized that science teachers should focus on a stronger process perspective, instead of the restricted perspective of scientific content (Flick 2000). Practical work in science education should aim towards an understanding of scientific evidence (Gott and Duggan 1996). The general role of metacognitive activities (Hofstein and Lunetta 2004) should be more greatly appreciated. The same holds for the teacher's supportive role in cognitive scaffolding – the interactive instructional strategy where teachers provide tailored instructions based on a diagnosis of the abilities and problems of their students.[19]

Several educational consequences follow from the view of science as a social endeavor. Scientific inquiry appears to be an activity where one makes sense of material and empirical and theoretical entities which have to be presented to a community of experts. Problems of validity cannot be solved by a single scientist in his or her laboratory. Instead, communities of peers have to criticize, negotiate, debate, and even revise the meaning and prevalence of any entity in science. Such a sociocultural view of science accompanies a Vygotskian perspective (Vygotsky 1978) of teaching and learning, one that stresses the social context of cognitive development. Students interact with each other and build communities of practice in order to promote learning. Accordingly, Duschl (2000) calls for instructional sequences in science teaching and learning to be more strongly oriented to the epistemic practices of science. Discussions, debates, and arguments about what counts as evidence deserve a more prominent role in teaching and learning.

Several approaches in science education, and especially in the field of history and philosophy of science in science teaching (HPSST), have been developed throughout recent decades. They continue the general appreciation of student-centered laboratory courses, open-ended inquiry activities, and manipulative investigations of material objects within a context of HPS. In the following section, the character and role of such approaches are analyzed and discussed.

---

[19] See, for example, Flick (2000), Valk and Jong (2009), and van de Pol et al. (2010). Tao (2003) explicitly calls for actively scaffolding students' understanding while using science stories about NOS.

46.4 Historical-Investigative Approaches in Science Education

There are several approaches to teaching and learning science which are closely related to the historical-investigative approach. Each of them stresses different aspects of teaching, learning, students' activities, experimentation, scientific instruments, or HPS in general. For each of these approaches, we will discuss the role played by HPS, on the one hand, and the role of students' investigative activities, on the other. To illustrate both extremes, some approaches strongly focus on investigative activities or inquiry-orientated learning, which is inspired only by scientific experiments of the past. Other approaches stress the role of historical context in teaching and learning science, while students' own investigations are clearly instructed by historical patterns.

In brief, historical-investigative approaches in science teaching are characterized by balancing the following aspects or a combination of them in a particular way:

- Contextualizing science with its history and philosophy
- Stressing material, social and/or cultural aspects of science
- Enabling teaching and learning about NOS
- Allowing for students' more or less guided own practical explorations of natural or technical phenomena
- Basing students' investigations on research activities related to past science
- Enabling students' critical reflections of their own actions and learning, as well as fostering their reasoning skills
- Using aspects of HPS to allow students to deduce their own meaning of their experiences with material entities and their manipulations and vice versa

In the following section, several approaches will be discussed which essentially fit into the general category of historical-investigative teaching and learning. We are aware that the authors of the work presented below might consciously have avoided the notion "historical-investigative" as an accurate account of their own work. Nevertheless, since we want to discuss the breadth of connate approaches in different educational fields without neglecting fundamental differences, we have decided on subsuming all these approaches under this common topic.

Besides the conceptual differences which will arise in the discussion below, one has to remember that science teachers favoring or rejecting a certain approach usually depend on the availability of resources and not necessarily on conceptual considerations.

46.4.1 Historical Investigations Within a Narrative Approach

Recently, stories have received more attention from educators favoring a historical approach. While most of the respective case studies clearly distinguish between a historical narration and an (independent) inquiry with modern materials, there are also a few exceptions that can be seen as a historical inquiry approach.

An example of such an approach has been developed by Metz and Stinner (2006), who have adapted the replication method. The general structure of the activities is retained, but while the replication method emphasizes a reconstruction of instruments, materials, and procedures as close to the available sources as possible, Metz and Stinner argue for a method called "historical representation." Historical representations are specific forms of case studies which recommend the reproduction of historical experiments with modern materials. A central idea is that students interact with a narrative about the history of science through the experiments they are performing. The activities they design and perform alternate between their own ideas and prior knowledge and those presented by the narrative. This means that students formulate hypotheses on their own, design tests, and, finally, compare and contrast their own ideas, experiences, and measurements with the original work. Explorations performed independently from the original are, therefore, welcomed. The use of alternative materials and innovative adaptations of the original instruments, materials, or procedures is encouraged. The experiments the students perform are not intended as verification labs but as a means to address the nature of science explicitly. The historically based investigation comprises four parts or phases: introduction, experimental design, experimental results, and analysis and interpretation of data and explanation.

Narrative approaches are, however, not without any danger. Metz and his colleagues (2007) mention that "we cannot expect students on their own to develop a critical stance towards narrative; thus mediation to guide students through a process of critical analysis should be an essential component of the narrative process" (Metz et al. 2007, p. 320). Accordingly, Tao (2003), in a study with 150 secondary-school students, found that science stories are useful contexts for students to argue for their preexisting views on NOS. As a consequence, the teacher should actively scaffold students' understandings in order to support students' cognitive development.[20]

The strategy of redoing scientific experiments of the past is also employed by Chang (2011). Here, history of science serves as a source for the exploration of peculiar and supposedly well-known natural phenomenon, which concerns accounts about how knowledge in science not only is generated but also forgotten. He demonstrates that a seemingly clear scientific "fact," such as the boiling point of water, becomes questionable or even obscured but may be demystified in the light of evidence of science from the past. Chang recovers past scientific knowledge, which has been forgotten instead of being integrated into the body of accepted knowledge we call current science. He shows that the boiling temperature of water notably depends on the form of the vessel in which the water is heated. He considers that "[g]etting involved in historical experiments will almost invariably teach students (and teachers) that things are more complicated that they had been led to be believe" (Chang 2011, p. 322). As a result, the experimental difficulties of a historical situation can serve as a basis for inquiry-based activities that focus on the peculiarities of the boiling temperature of water, yet, even without necessarily referring explicitly to the historical context.

---

[20] See footnote 19.

A thoughtful consideration of past science helps students to realize that modern science deals with a restricted range of objects, methods, and materials. Likewise, Vera and his colleagues (2011) describe a "simple experiment with a long history," the combustion of a candle in an inverted vessel that is partly immersed in water. This experiment was initially described by Lavoisier, though it can be traced back to antiquity. Similar to Chang's examples, they demonstrate the historical development of a classical experiment and expose potential difficulties that may have played a role in the historical discourse, as well as in generating a misconception about the explanation that can be found even in fairly recent textbooks. Yet, the methodological approach of Vera et al. is somewhat different in that it also involves computer simulations in communicating their experiments.

Experimental approaches of elementary-level scientific content immersed in a historical context appear to be a straightforward and unproblematic variation of a historical-investigative approach. Yet, there are some problems with respect to the historical context that, on first sight, appears to be clear and evident. A good illustration in this respect is Kipnis' discussion of Oersted's experiment on electromagnetic interaction (Kipnis 2005). He argues that Oersted's discovery of electromagnetism is an example of the role chance might play in the development of scientific knowledge. Even though Oersted's experiment is frequently described as being based on chance, this perception might be caused by a lack of an adequate consideration for the Romantic movement in physics and its role in theory development. Martins (1999), for example, argues strongly that Oersted's work has been based on theoretical considerations influenced by Romantic natural philosophy.

46.4.2 Historical Investigations Starting from Laboratory Diaries

Original manuscripts and, in particular, laboratory notes can provide an authentic basis for structuring a conceptual and procedural understanding of science. The collection of Michael Faraday (1932–1936) is a prominent example of such manuscripts due to the extensive detail and the availability of his notes and even more so because his laboratory diary was published in the 1930s. Crawford (1993), as well as Barth (2000), uses different sections of Faraday's notebooks in their classroom as a starting point for students' investigative activities. According to Crawford, there are at least three aspects that turned out to be beneficial: "In fact, some of them [Faraday's problems] were problems the pupils shared, but said that they had previously felt stupid about declaring their puzzlement" (Crawford 1993, p. 205). When following Faraday's problems, however, Crawford "had shaken their faith that someone somewhere, at least in teaching/learning situations, would eventually produce an answer. They had to think for themselves" (Crawford 1993, p. 205). The students did not only develop a conceptual understanding from these lessons but "they also learnt and recognized for themselves many things about science itself" (Crawford 1993, p. 205). Faraday's experimental accounts were used as a kind of "starter" (Crawford 1993, p. 206) of students' own inquiry-based activities.

Likewise, Barth (2000) used Faraday's laboratory diary to enhance students' understanding of electromagnetic induction. Suitable chapters from Faraday's notebook were presented to the students as a basis for their own inquiry activities on electromagnetic induction. Following Faraday's notes, the students made their own discoveries about electromagnetic induction using modern equipment while guided by Faraday's original experimental ideas. Moving back and forth between the diary and the apparatuses – reconstructions of an experimental apparatus and a modern version – the students were able to develop their own understanding. Moreover, they experienced the difficulties of stabilizing and amplifying an effect adequately, difficulties comparable to those that Faraday and his contemporaries faced. It is remarkable that Barth even managed to guide his students to the point where they developed insight into an initial mistake that Faraday had made. Faraday, in a letter to his friend, Philipps, discussed the directions of an electric current induced by a primary current in a parallel wire but then predicted the wrong directions of electromagnetically induced currents. While Faraday, himself, corrected his mistake a short time later in his official paper to the Royal Society, his initial incorrect prediction of an electric current has only been recognized by a few science historians (Romo and Doncel 1993).

46.4.3 Historical Investigations and Instruments from Past Science

Experimenting with historical or reconstructed instruments opens up new ways of understanding science as an experimental practice. Reconstructed historical devices, as well as original apparatuses, can be used in formal and informal science education. A leading approach in this respect has been recognized by Devons and Hartmann, who developed a "laboratory devoted to repeating crucial experiments in the history of physics with apparatus reconstructed according to the original descriptions" (Hartmann Hoddeson 1971, p. 924). One of the criteria they use is similar to the ones claimed in historical studies based on the replication method. According to them, "methods and materials used in these experiments are essentially those used originally" (Devons and Hartmann 1970, p. 44). Their work[21] serves as a starting point for several followers who reenacted historical experiments for educational purposes. Among the major objectives are both an enhanced understanding of experimental practice with a special focus on reconstructions of material and performative aspects of science. The approach is particularly motivated by the general objective that learners should thoroughly understand how knowledge in science is generated or even manufactured (see, e.g., Heering 2000, 2007; Höttecke 2000, 2001; Kipnis 1993; Rieß et al. 2006). Experiments do not appear to be simple devices for answering questions in a yes-or-no manner but allow for detailed accounts on how instrumental and material manipulations in science interact with theoretical and cultural entities.

---

[21] See also Devons and Hartmann Hoddeson (1970) and Hartmann Hoddeson (1971).

Reenacting historical experiments is an approach that has been systematically used by the Oldenburg group, led by F. Rieß for research in HPS (see the Sect. 46.2) and for educational purposes. The method was mainly used in physics teacher-training courses[22] although there are also some instances where this approach has been used at a secondary-school level (Heering 2000; Höttecke et al. 2012). This approach is based on replicas which were largely developed in the process of analyzing historical experiments with the replication method. Replicas of historical instruments are often characterized by material cultures of a certain time and space. Materials favored by natural philosophers of the eighteenth century, for instance, were amber, natural resin, shellac, or glass, but not PVC. Replicas, therefore, allow for new experiences with materials and their possibly peculiar qualities, with which students are often unfamiliar. Thus, this approach aims to explore the material culture of a particular time. Moreover, replicas are usually not designed for teaching purposes but originate from authentic laboratory practices of the past. Replicas still represent theoretical ideas which are embodied in instruments, materials, and the ways that they were used at the workbench. Hence, they are a rich resource for investigating the distinctive interaction between material and theoretical entities, as well as between instrumental and social practices in science. An important feature of this approach generally is that knowledge, data, and the experiences of the students working with the replicas are not isolated. The Oldenburg group used them as a rich resource for contextualizing experiments, materials, and instruments from specific cultural and societal perspectives.

During the past three decades, a rich resource of replicas, mainly from the history of physics, has been constructed. The STeT project (Science Teacher e-Training) has been looking for new ways to provide teachers with materials connected to the history of instruments and experiments (Kokkotas and Bevilacqua 2009). Among the materials are short films, which demonstrate how the replicas work in action. The project HIPST (History and Philosophy in Science Teaching) uses the idea of developing case studies. Some of the cases developed in HIPST enable teaching and learning about science and NOS with an explicit focus on students' historical investigations with replicas. A wide range of student-centered activities and materials have been developed for teaching and learning, including role-play, films, and methods for explicitly reflecting the NOS (Höttecke et al. 2012).

46.4.4 *Historical Investigations with Modern Materials*

Several approaches use the working principle of historical experiments for educational purposes without using replicas, historic materials, or even original instruments. For these approaches, historical context still becomes the focal point around

---

[22] See, for example, Rieß (2000), also Sichau (2000b) on thermodynamics and Höttecke (2000) on electricity.

which teaching and learning must be organized. Such approaches[23] can, therefore, be seen as modifications of those discussed in the former section.

Tsagliotis (2010) has developed a case study where primary-school students construct microscopes with modern materials, such as PVC tubes and plastic lenses. They make observations and relate them to simplified chapters of Robert Hook's *Micrographia*. Maiseyenka and her colleagues (2010) show how the historical context of cooling and ice production may be enhanced by students' own investigation of physical principles, such as the effect of the dilution of salt on evaporation or on freezing mixture. The case study is addressed with students in grades five to seven where the topic of producing and enjoying ice cream is of interest.

According to Kipnis' (1996, 2002) idea of historical-investigative teaching, science students would rather imitate scientists than use artificial or contrived experiments found in regular science education. He suggests that several modifications should be allowed to be made to the original experiments. Thus, using such an approach means that students begin their own investigations from a historical context rather than reenacting an experiment. Kipnis stresses that through this approach, students have the chance to become discoverers. Students should realize that they are capable of repeating certain important steps of famous scientists on their own. This approach enables students' decision-making regarding the proper result of historic scientific disputes. Building their self-confidence is another important expected outcome. Furthermore, students should appreciate the great discoveries of the past and learn about the strategic elements of experimenting. On the other hand, Kipnis suggests quite a strict structure for integrating experiments into historic contexts. According to his strategy, during an investigation, students formulate a problem based on a specific historical background. Then, they must identify and select relevant variables. All variables should be examined independently in order to test a hypothesis. Finally, the students must draw general conclusions based on their analysis of the variables and their effect on a certain phenomenon. While the students learn how to control variables in a clear-cut way, many aspects of scientific experimentation discussed in Sect. 46.2 are in danger of not being adequately considered. The step-by-step method aimed at "producing true results" (1996, p. 281), combined with the act of restricting experimentation to the examination of variables and test of hypotheses, hardly matches the constructive role of practical manipulations, material procedures, or explorative strategies of experimentation in science. This is not surprising since Kipnis stresses science as a "drama of ideas" (1996, p. 288).

Allchin (1999) describes an introductory science lab course for nonscience majors using history of science as a curricular guide. Within the year-long interdisciplinary course, outside the regular curricular teaching structure, instructors enjoyed significant freedom. The designed "historically inspired labs" focused, among others, on topics like basic astronomy, early medicine, density of matter, Galileo's ideas on pendulum motion, production of paint, electrophysiology, titration of vinegar, and the ballistic pendulum. Most instruments were adapted from existing introductory laboratory course materials. The instruments, even though

---

[23] See, for example, Achilles (1996), Kipnis (1993, 1996, 2002), Teichmann (1979), Teichmann et al. (1990), and Wilke (1988).

modern, were "set more explicitly in their historical context and with added emphasis on reflecting the investigatory process" (Allchin 1999, p. 620). The main objective of the course was "a coupled understanding of the content and process of science" (Allchin 1999, p. 621). Therefore, the investigatory activities were related to real science in a wider sense; for example, during a lecture, an instructor demonstrated the hydrostatic paradox, which means that the hydrostatic pressure of a water column only depends on the depth of the column and not on the shape or the volume of a vessel. After the demonstration, the students started to suggest a variety of different tubes in order to explore their effect on pressure.

While all approaches discussed in this section start with a historical context from the very start of teaching, Allchin (1999) mentions an alternative option. According to him, history of science has often been introduced retrospectively as a way of reviewing the students' investigations, their experiences, and results. Thus, history of science was not directly related to students' investigations from the outset but used as a means for reflecting experiences with instruments, experiments, and phenomena retroactively. Nevertheless, as the students indicated, the labs were integral for understanding how science works.

Lin and colleagues (2002) present evidence of the effectiveness of a teaching approach based on practical activities of students in an HPS context. They conducted a 1-year study, with grade eight students, about the efficacy of promoting students' problem-solving ability through history of science teaching. A treatment group was compared to a control group. The students were randomly assigned to one of the groups. The treatment group was taught with an emphasis on the development of scientific content. Students learned details about how previous scientists discussed, debated, and hypothesized and how they conducted experiments. Either teacher demonstrations or students' hands-on activities were analogous to the experiments and ideas of previous scientists. The students simulated historic experiments and were required to predict their results. They formulated hypotheses, explained their own reasoning, and reflected on analogies and differences between their own hands-on activities and those of previous scientists. The control group was taught in a somewhat traditional manner. There, students worked according to cookbook-style labs, aiming at getting correct answers. They had to follow predetermined procedures, listen to lectures, and solve problems presented in textbooks. Statistical analysis of pre- and posttest data indicated significant effects of the treatment regarding the problem-solving abilities of the students. Unfortunately, there is still a lack of empirical evidence about the efficacy of the practical work of students framed by the history of science. The results of this study are, nevertheless, encouraging.

46.4.5 Historical Investigations in Science Museums and Instrument Collections

Original historical instruments are rarely used in science education. A reason for this is probably their status of having historical and heritage significance. Therefore, they have to be preserved and saved from any damage. There are, however, a few examples where instruments have been used in reenacting historical experiments in science education, most notably an experiment on the decomposition of water, performed in a history of science course at the university level (Eggen et al. 2012). Here, part of the apparatus for decomposing water was borrowed from the historical teaching collection, while the voltaic pile was reconstructed according to descriptions from 1800. The authors concluded that the exercise enabled valuable insight, both into the nature of the devices they had used and the experiment as a whole. Likewise, original microscopes and telescopes are used in science museums for educational purposes. There, they illustrate the optical quality, as well as the magnification, of these devices.

Cavicchi (2008) describes the visit of a group of students with a disparate background in science education to the MIT Museum's collection of historical telephones. The students explored several historic devices, such as telephones and a nineteenth-century Voltaic pile. They even "had to innovate distinctive actions such as unscrewing a lid, releasing a crank, picking a lock" (Cavicchi 2008, p. 726) in order to surmise the workings of the telephones in relation to their construction and materiality. Lissajous figures were produced by reflecting light off orthogonal nineteenth-century tuning forks. Students' investigations were enriched through the use of historic texts. The main focus of the approach is to enable intensive experiences with physical phenomena while exploring the historical context of their emergence. Cavicchi calls her approach "critical exploration," according to earlier work from cognitive psychology. As the undergraduate students engage with complex phenomena and materials, they experience, firsthand, the relationship among the critical components: the materials, their own actions, and history. The students' actions put them in the shoes of past historical investigators who had also developed an understanding of science through experimentation with materials.

Reconstructions of past experimental instruments have also entered museums. In those cases, they might be placed close to the original items in an exhibition. Visitors are allowed to use them instead of the original devices. In this way, a similar experience as would be had with the original device can be achieved without threatening the museum curator's responsibility to preserve the heritage of the device (Heering and Müller 2002). An alternative might be to offer visitors the opportunity to reconstruct instruments on their own in a simple manner. Such a procedure is based on historical artifacts (or their reconstructions) and thus enables new insight into the material culture of scientific experimentation (Heering and Sauer 2012).

Barbacci and her colleagues (2010) present a case study called "Discovery of Dynamic Electricity and the Transformation of Distance Communications," which is designed as an informal educational activity. It is addressed to high-school students and teachers and aims to bring out interrelations between history of science and general social history. The case study has been developed and tested in the Fondazione Scienza e Tecnica's physics laboratory collection in Florence, Italy. During an interactive workshop at the Fondazione, it is possible to explore several of the central topics in electrodynamics from a historical perspective, like the invention of the galvanic battery, the use of electromagnetic coils, and the early application of the galvanoscope, galvanometer, and electromagnet. The lesson moves on to the invention of the telegraph, which is placed into a context of the development of railways, the problem of establishing standard time, and submarine telegraphy, and the sociocultural implications of these phenomena. The lesson presented at the museum follows a narrative-experimental approach: historical background information is presented together with certain scientific discoveries and their technical applications. Experiments and instruments matter on two levels, as a hands-on activity and as a presentation of original devices from the scientific instrument collection. Hands-on activities were designed to help solve several practical problems presented to the students. They built a type of Volta pile and tried to verify its operation; they repeated Oersted's experiment with a current-carrying wire in order to deflect a magnetic needle in a particular way; and they studied optical telegraphy and built a simple model of a Morse telegraph. Having been presented with the original scientific instruments, students were able to compare their own material manipulations with them and understand them as witnesses of past developments in science, their application, and their sociocultural significance.

Additionally, there have been recent attempts to teach through the analysis of historical instruments. In a fairly simple manner, this can be done by examining devices in their showcase in a museum. There have also been attempts to teach material culture through the analysis and contextualization of historical artifacts that may even be (or at least have been) part of everyday life (Anderson et al. 2011; Cavicchi 2007, 2011).

46.5 Conclusion

The analysis of several historical-investigative approaches has shown that the ideas of science as both practice and culture are often considered in science education. Over the past three decades, research into history and philosophy of science has strongly emphasized the constitutive role of experiments, instruments, and material procedures in science and their close relationship to the development of theoretical ideas. During that time, experiments in science have not been restricted to a mere means of testing ideas or hypotheses. In the aftermath of the practical and material turn in HPS, metaphors like the "mangle of practice," as Andy Pickering (1995) puts it, guide our understanding of science as a complex interaction of theoretical, material, and human agencies. Such complex views on how science works hardly match any cookbook-style idea of a single scientific method.

Straightforward epistemologies do not match what recent research in HPS has discovered about the idiosyncratic roles of experiments, instruments, and material procedures in science; hence, a single "big story" about science as a linear inductive, deductive, or hypothetic-deductive endeavor can hardly be told anymore. In this respect, the idea of a case study or a narrative about a particular event, scientist, experiment, or any other occurrence in science might actually generalize ideas about science, rather than define what science is or should be.

Even though the educational benefits of practical work in science education have been stressed for a long time, the current status of experiments in science teaching, in general, as a means for enhancing students' understanding of NOS is still weak. On the other hand, "verification labs" have been widely refuted by historical-investigative approaches. Practical work, according to historical-investigative approaches, is not restricted to the manipulation of equipment and materials or to a mere inductive endeavor. Manipulations of materials and instruments are, instead, a means for the development of the scientific ideas and skills of students who reflect simultaneously on their own investigations and their multiple relations to historical contexts.

In the practical and material turn in HPS and science studies, scholars focused more on detailed accounts of processes and practices in science. Laboratory diaries, notebooks, materials, and instruments, alongside the more traditional sources of letters and publications of scientists, become regarded as a rich resource for detailed accounts on how science works. This development is mirrored by recent attempts to make use of such resources for educational purposes. Through the analysis of laboratory notebooks and practical work with replicas of historical instruments, scientific investigations transform into a detailed process.

Alongside the practical and material turn in HPS, case studies about how science works gained increasingly more attention in science education. The Harvard case studies (Conant 1957) often have been quoted on the subject. Recently, case studies have had a stronger focus on learning about NOS in an explicit reflective manner. They offer several opportunities for students to develop their own investigations in close relation to the history of science (e.g., Allchin 2012; Clough 2011; Höttecke et al. 2012). Other approaches, like "critical exploration," (Cavicchi 2006, 2007) emphasize the role of reflected investigations of phenomena or technical devices, framed by historical contexts.

Regarding the use of instruments in science education for investigating material, as well as historical entities, a wide array of options have been offered, beginning with the use of authentic materials and instruments up to the use of modern ones. Each option has its specific advantages and suffers from particular problems. While the use of modern materials and instruments is hardly useful for reenacting and exploring the material culture of science, the use of replicas or even original instruments from the past suffer from a lack of availability or even usability for several teaching purposes. Hence, the relevance of each approach depends highly on where learning will take place (e.g., formal learning at school or informal learning in a science museum) and which teaching objectives will be targeted (e.g., NOS, conceptual knowledge, process skills).

In science museums, historical-investigative approaches appear to be quite innovative. Since the traditional goal of science museums was usually to balance the objective of protecting a heritage with that of enabling science learning, new ways for exploring the past with hands-on activities have been established which encompass both objectives. Here, approaches have been developed which allow visitors' hands-on activities, together with a careful display, or even use, of instruments of the past.

46 Historical-Investigative Approaches in Science Teaching

Whether or to what degree historical-investigative approaches will be disseminated and implemented widely is a matter of future research, as are the concomitant limiting factors or problems of such implementation. Scholars (Barab and Luehmann 2003; Höttecke and Silva 2011) have pointed out that teachers are the gatekeepers for any curricular innovation. Whether a wide and successful implementation will ever happen depends, at least in school science teaching, not only on the future design of curricular and standard documents but also on whether historical-investigative approaches actually will meet the needs and desires of science teachers. Therefore, materials for teaching and learning should be designed in a way that allows for a flexible and open use (Valk and Jong 2009).

Some of the above-discussed approaches explicitly refer to inquiry-based learning (Allchin 2012; Höttecke et al. 2012). From research regarding teachers' perspectives on inquiry-based learning, we already know that their general beliefs about successful science learning are linked to their beliefs about laboratory work and inquiry (Wallace and Kang 2004). Open-ended activities, which comprise discussions about scientific controversies or uncertain scientific evidence, will possibly alienate science teachers (Newman et al. 2004). Teachers often make detailed plans for instructional units which result in inflexibility in their reactions to students' ideas and products (Schwartz and Crawford 2004). This makes it a challenge to present science as inquiry or an open-ended activity (Roehrig and Luft 2004). The science teachers' expectation of having ready-made answers and safe content-knowledge (Höttecke and Silva 2011) may discline them to accept historical-investigative approaches in science teaching. Science teachers need to know how to deal with new, emerging perspectives in the context of NOS. They should be able to organize open-ended investigations and moderate student-centered discussions. In general, they have to develop specific professional repertoires for teaching NOS successfully within any of the historical-investigative frameworks. Whether or not they succeed will depend on their readiness and willingness to leave established trails of teaching science behind and on their future professional development for teaching HPS in science education. The design of teaching materials following any of the historical-investigative approaches has to consider these conditions when planning for and establishing their successful implementation.

References

Abd-El-Khalick, F., & Lederman, N. G. (2000). Improving Science Teachers' Conceptions of Nature of Science: A Critical Review of the Literature. *International Journal of Science Education, 22* (7), 665–701.

Achilles, M. (1996). *Historische Versuche der Physik: Funktionsfähig nachgebaut.* Frankfurt/Main: Wötzel

Akerson, V. L., Hanson, D. L., & Cullen, T. A. (2007). The Influence of Guided Inquiry and Explicit Instruction on K–6 Teachers' Views of Nature of Science. *Journal of Science Teacher Education, 18,* 751–772.

Allchin, D. (1999). History of Science - With Labs. *Science & Education, 8,* 619–632.

Allchin, D. (2012). The Minnesota Case Study Collection: New Historical Inquiry Case Studies for Nature of Science Education. *Science & Education*, 21(9), 1263–1281.

Anderson, K., Frappier, M., Neswald, E., & Trim, H. (2011). Reading Instruments: Objects, Texts and Museums. *Science & Education, DOI* 10.1007/s11191-011-9391-y, 1–23.

Anderson, R. D. (2007). Inquiry as an Organizing Theme for Science Curricula. In S. K. Abell & N. G. Lederman (Eds.), *Handbook of Research in Science Education* (pp. 807–830). New York, London: Routledge.

Barab, S.A., & Luehmann, A.L. (2003). Building Sustainable Science Curriculum: Acknowledging and Accommodating Local Adaptation. *Science Education*, 87(4), 454–467.

Barbacci, S., Bugini, A., Brenni, P., & Giatti, A. (2010). The Discovery of Dynamic Electricity and the Transformation of Distance Communications. Case study of the HIPST project (www.hipst.eu), retrieved from: http://hipstwiki.wetpaint.com/page/Case+Study+1 (2012-07-17).

Barth, M. (2000). Electromagnetic Induction Rediscovered Using Original Texts. *Science & Education*, 9(4), 375–387.

Barth, M. (2010). *Process Skills: Taking up the cudgels for historical approaches.* Paper presented at the "History and Philosophy in Science Teaching Conference", University of Kaiserslautern/ Germany, March 11–14, 2010, retrieved from http://www.hipst.uni-hamburg.de/archive/Barth.pdf (23.08.2012).

Berg, C., Anders R., Bergendahl, C. B., Lundberg, B. K. S., & Tibell, L. A. E. (2003). Benefiting from an Open-ended Experiment? A Comparison of Attitudes to, and Outcomes of, an Expository versus an Open-Inquiry Version of the same Experiment. *International Journal of Science Education*, 25 (3), 351–372.

Bigg, C., & Hennig, J. (Eds.) (2009). *Ikonografie des Atoms in Wissenschaft und Öffentlichkeit des 20. Jahrhunderts.* Göttingen: Wallstein-Verlag.

Breidbach, O., Heering, P., Müller, M., & Weber, H. (Eds.). (2010). *Experimentelle Wissenschaftsgeschichte.* München: Wilhelm Fink Verlag.

Busch, W. (Ed.). (2008). *Verfeinertes Sehen: Optik und Farbe im 18. und frühen 19. Jahrhundert.* München: R. Oldenbourg.

Carey, S., Evans, R., Honda, M., Jay, E., & Unger, C. (1989). 'An experiment is when you try it and see if it works': a study of grade 7 student's understanding of the construction of scientific knowledge. *International Journal of Science Education*, 11 (special issue): 514–529.

Cavicchi, E. M. (2006). Nineteenth-Century Developments in Coiled Instruments and Experiences with Electromagnetic Induction. *Annals of Science*, 63, 319–361.

Cavicchi, E. M. (2007). Mirrors, swinging weights, light bulbs…: Simple experiments and history help a class become a scientific community. In P. Heering & D. Osewold (Eds.), *Constructing scientific understanding through contextual teaching* (pp. 47–63). Berlin: Frank & Timme.

Cavicchi, E. M. (2008). Historical experiments in students' hands: Unfragmenting science through action and history. *Science & Education*, 17, 717–749.

Cavicchi, E. M. (2011). Classroom Explorations: Pendulums, Mirrors, and Galileos Drama. *Interchange,* 42(1), 21–50.

Chang, H. (2011). How Historical Experiments Can Improve Scientific Knowledge and Science Education: The Cases of Boiling Water and Electrochemistry. *Science & Education*, 20, 317–341.

Chinn, C. A., & Malhotra, B. A. (2002). Epistemologically Authentic Inquiry in Schools: A Theoretical Framework for Evaluating Inquiry Tasks. *Science Education,* 86(2), 175–218.

Clough, M. P. (2006). Learners' Responses to the Demands of Conceptual Change: Considerations for Effective Nature of Science Instruction. *Science Education*, 15, 463–494.

Clough, M. P. (2011). The Story Behind the Science: Bringing Science and Scientists to Life in Post-Secondary Science Education. *Science & Education*, 20(7–8), 701–717.

Collins, H. M. (1985). *Changing order: replication and induction in scientific practice.* London; Beverly Hills: Sage Publications.

Conant, J. B. E. (1957). *Harvard Case Histories in Experimental Science.* Cambridge (Mass.): Harvard University Press.

Crawford, E. (1993). A critique of curriculum reform: Using history to develop thinking. *Physics Education*, 28, 204–208.

Deng, F., Chen, D.-T., Tsai, C.-C., & Chai, C. S. (2011). Students' Views of the Nature of Science: A Critical Review of Research. *Science Education*, 95(6), 961–999.

Devons, S., & Hartmann, L. (1970). A history-of-physics laboratory. A laboratory in which students can reproduce historically significant physics experiments provides them with a useful change of viewpoint. *Physics Today*, 23(2), 44–49.

Drake, S. (1970). Renaissance Music and Experimental Science. *Journal of the History of Ideas Band 31, No. 4, S.* 483–500.

Duschl, R. A. (2000). Making the Nature of Science explicit. In R. Millar, J. Leach, J. Osborne, *Improving Science Education. The Contribution of Research* (pp. 187–206). Buckingham, Philadelphia: Open University Press.

Eggen, P.-O., Kvittingen, L., Lykknes, A., & Wittje, R. (2012). Reconstructing Iconic Experiments in Electrochemistry: Experiences from a History of Science Course. *Science & Education*, 21(2), 179–189.

Engels, W. (2006). Die Nebelkammeraufnahme - das automatisch generierte Laborbuch? In M. Heßler (Ed.): *Konstruierte Sichtbarkeiten: Wissenschafts- und Technikbilder seit der Frühen Neuzeit* (pp. 57–74). München: Wilhelm Fink.

Faraday, M. (1932–1936). *Faraday's Diary. Being the Various Philosophical Notes of Experimental Investigation made by Michael Faraday*, ed. By T. Martin, 7 Vol., London: G. Bell and Sons, LTD.

Fatet, J. (2005). Recreating Edmond Becquerel's electrochemical actinometer. *Archives des sciences*, 58(2), 149–158.

Feyerabend, P. K. (1972). Von der beschränkten Gültigkeit methodologischer Regeln. Neue Hefte für Philosophie. Dialog als Methode, H2/3, pp. 124–171.

Fiorentini, E. (2005). *Instrument des Urteils zeichnen mit der Camera Lucida als Komposit*. Berlin: Max-Planck-Inst. für Wissenschaftsgeschichte.

Flick, L.B. (2000). Cognitive Scaffolding that Fosters Scientific Inquiry in Middle Level Science. *Journal of Science Teacher Education*, 11 (2), 109–129.

Frercks, J. (2001). *Die Forschungspraxis Hippolyte Fizeaus: Eine Charakterisierung ausgehend von der Replikation seines Ätherwindexperiments von 1852*. Berlin: Wissenschaft und Technik.

Frercks, J., & Weber, H. (2006). Replication of Replicability. Schmidt's Electrical Machine. *Bulletin of the Scientific Instrument Society*, 89, 3–8.

Galili, I., & Hazan, A. (2001). The Effect of a History-Based Course in Optics on Students' Views about Science. *Science & Education*, 10, 7–32.

Galison, P. (1987). *How experiments end*. Chicago and London: Chicago University Press.

Gallagher, J. J., & Tobin, K. (1987). Teacher management and student engagement in high school science. *Science Education*, 71, 535–555.

Gooding, D. (1989). History in the laboratory: Can we tell what really went on? In F. James (Ed.): *The development of the laboratory. Essay on the Place of Experiment in Industrial Civilization* (pp. 63–82). Houndmills: Macmillan.

Gooding, D. (1990). *Experiment and the Making of Meaning: Human Agency in Scientific Observation and Experiment*. Dordrecht/Boston/London: Kluwer Academic Publishers.

Gooding, D. & James, F. (1995). *Faraday Rediscovered. Essays on the Life and Work of Michael Faraday 1791–1867*. New York: Stockton Press.

Gooding, D., Pinch, T., & Schaffer, S. (Eds.). (1989a). *The Uses of Experiment*. Cambridge: Cambridge University Press.

Gooding, D., Pinch, T., & Schaffer, S. (1989b) Introduction: some uses of experiment. In D. Gooding, D., T. Pinch, & S. Schaffer (Eds.), *The Uses of Experiment* (pp. 1–27). Cambridge: Cambridge University Press.

Gott, R., & Duggan, S. (1996). Practical work: its role in the understanding of evidence in science. *International Journal of Science Education*, 18(7), 791–806.

Gyllenpalm, J. & Wickmam, P.-O. (2011). "Experiments" and the Inquiry Emphasis Conflation in Science Teacher Education. *Science Education*, 95, 908–926.

Hacking, I. (1993 [1983]). *Representing and Intervening: Introductory Topics in the Philosophy of Natural Science*. Cambridge, New York, Port Chester, Melbourne, Sydney: Cambridge University Press.

Harris, D., & Taylor, M. (1983). Discovery Learning in School Science: The Myth and the Reality. *Journal of Curriculum Studies*, 15(3), 277–289.

Hart, C., Mulhall, P., Berry, A., Loughran, J., & Gunstone, R. (2000). What is the Purpose of this Experiment? Or Can Students Learn Something from Doing Experiments? *Journal of Research in Science Teaching*, 37 (7), 655–675.

Hartmann Hoddeson, L. (1971). Pilot experience of teaching a history of physics laboratory. *American Journal of Physics*, 39, 924–929.

Heering, P. (1992). On Coulomb's inverse square law. *American Journal of Physics 60*, 988–994.

Heering, P. (1994). The replication of the torsion balance experiment: The inverse square Law and its refutation by early 19th-century German physicists. In C. Blondel & M. Dörries (Eds.), *Restaging Coulomb: Usages, Controverses et réplications autour de la balance de torsion.* Firenze: Leo S. Olschki, 47–66.

Heering, P. (1998). *Das Grundgesetz der Elektrostatik: Experimentelle Replikation und wissenschaftshistorische Analyse.* Wiesbaden: DUV.

Heering, P. (2000). Getting shocks: Teaching Secondary School Physics through History. *Science & Education*, 9(4), 363–373.

Heering, P. (2002). Analysing Experiments with Two Non-canonical Devices: Jean Paul Marat's Helioscope and Perméomètre. *Bulletin of the Scientific Instrument Society 74*, 8–15.

Heering, P. (2005). Weighing the Heat: The Replication of the Experiments with the Ice-calorimeter of Lavoisier and Laplace. In M. Beretta (Ed.), *Lavoisier in Perspective* (pp. 27–41). München: Deutsches Museum.

Heering, P. (2006). Regular Twists: Replicating Coulomb's Wire-Torsion Experiments. *Physics in Perspective*, 8, 52–63.

Heering, P. (2007). Public experiments and their analysis with the replication method. *Science & Education*, 16, 637–645.

Heering, P. (2008). The enlightened microscope: re-enactment and analysis of projections with eighteenth-century solar microscopes. *British Journal for the History of Science,* 41(150), 345–368.

Heering, P. (2010). An Experimenter's Gotta Do What an Experimenter's Gotta Do - But How? *ISIS,* 101(4), 794–805.

Heering, P., & Müller, F. (2002). Cultures of Experimental Practice - An Approach in a Museum. *Science & Education*, 11, 203–214.

Heering, P., & Osewold, D. (2005). Ein Problem, zwei Wissenschaftler, drei Instrumente. *Centaurus*, 47, 115–139.

Heering, P., & Sauer, F. (2012). Das Projekt Galilei: Konzeption und Umsetzung. In S. Bernholt (Ed.), *Konzepte fachdidaktischer Strukturierung für den Unterricht* (pp. 119–121). Münster: LIT-Verlag, 2012.

Heering, P., Riess, F., & Sichau, C. (Eds.). (2000). *Im Labor der Physikgeschichte: zur Untersuchung historischer Experimentalpraxis.* Oldenburg: Bis, Bibliotheks- und Informations system der Universität Oldenburg.

Heidelberger, M., & Steinle, F. (Eds.) (1998). *Experimental Essays – Versuche zum Experiment.* Baden-Baden, Nomos.

Hennig, J. (2003). *Der Spektralapparat Kirchhoffs und Bunsens.* München: Deutsches Museum.

Hentschel, K. (1998). Feinstruktur und Dynamik von Experimental systemen. In M. Heidelberger & F. Steinle (Hg.), *Experimental Essays* (pp. 325–354). Baden-Baden: Nomos

Heßler, M. (Ed.). (2006). *Konstruierte Sichtbarkeiten: Wissenschafts- und Technikbilder seit der Frühen Neuzeit.* München: Wilhelm Fink.

Hodson, D. (2009). *Teaching and Learning about Science. Language, Theories, Methods, History, Traditions and Values.* Rotterdam: Sense Publishes.

Hoffmann, M. (2011). Learning in the Laboratory: The Introduction of "Practical" Science Teaching in Ontario's High Schools in the 1880s. In P. Heering and R. Wittje (Eds.), *Experiments and Instruments in the History of Science Teaching* (pp. 177–205). Stuttgart, Franz Steiner Verlag.

46 Historical-Investigative Approaches in Science Teaching

Hofstein, A., & Kind, P. M. (2012). Learning In and From Science Laboratories. In B. J. Fraser, K. G. Tobin & C. J. McRobbie (Eds.), *Second International Handbook of Science Education*, Vol. I (pp. 189–207). Dordrecht: Springer.

Hofstein, A., & Lunetta, V. N. (2004). The Laboratory in Science Education: Foundations for the Twenty-First Century. *Science Education*, 88, 28–54.

Holmes, F. L., Renn, J., & Rheinberger, H.-J. (Eds.) (2003). *Reworking the Bench: Research Notebooks in the History of Science*. Dordrecht: Kluwer.

Höttecke, D. (2000). How and what can we learn from replicating historical experiments? A case study. *Science & Education*, 9(4), 343–362.

Höttecke, D. (2001). *Die Natur der Naturwissenschaften historisch verstehen. Fachdidaktische und wissenschaftshistorische Untersuchungen* (Understanding the nature of science historically. Didactical and historical investigations). Berlin: Logos-Verlag, Theses.

Höttecke, D., & Silva, C. C. (2011). Why Implementing History and Philosophy in School Science Education is a Challenge - An Analysis of Obstacles. *Science & Education*, 20(3-4), 293–316.

Höttecke, D., Henke, A., & Rieß, F. (2012). Implementing History and Philosophy in Science Teaching - Strategies, Methods, Results and Experiences from the European Project HIPST. *Science & Education*, 21(9), 1233–61.

Howe, E. M., & Rudge, D. W. (2005). Recapitulating the History of Sickle-Cell Anemia Research. *Science & Education*, 14(3-5), 423–441.

Irwin, A. R. (2000). Historical Case Studies: Teaching the Nature of Science in Context. *Science Education*, 84(1), 5–26.

Irzik, G., & Nola, R. (2011). A Family Resemblance Approach to the Nature of Science for Science Education. *Science & Education*, 20, 591–607.

Jonas-Ahrend, G. (2004). *Physiklehrervorstellungen zum Experiment im Physikunterricht* (Physics teachers' ideas of experiments in physics teaching). Berlin: Logos-Verlag.

Kang, N.-H., & Wallace, C. S. (2005). Secondary Science Teachers' Use of Laboratory Activities: Linking Epistemological Beliefs, Goals, and Practices. *Science Education*, 89(1), 141–165.

Kärn, M. (2002). Das erdmagnetische Observatorium in der Scheune: Messungen mit dem originalgetreuen Nachbau eines Magnetometers von Gauß und Weber. *Mitteilungen der Gauss-Gesellschaft Göttingen 39*, 23–52.

Khishfe, R., & Abd-El-Khalick, F. (2002). Influence of Explicit and Reflective versus Implicit Inquiry-Oriented Instruction on Sixth Graders' Views of Nature of Science. *Journal of Research in Science Teaching*, 39(7), 551–578.

Kipnis, N. (1993). *Rediscovering Optics*. Minneapolis: Bena Press.

Kipnis, N. (1996). The ‚Historical-Investigative' Approach to Teaching Science. *Science & Education*, 5(3), 277–292.

Kipnis, N. (2002). A History of Science Approach to the Nature of Science: Learning Science by Rediscovering it. In W. F. McComas (Ed.), *The Nature of Science in Science Education* (pp. 177–196). Dordrecht: Kluwer.

Kipnis, N. (2005). Chance in Science: The Discovery of Electromagnetism by H.C. Oersted. *Science & Education*, 14, 1–28.

Kirschner, P. A., Sweller, J., & Clark, R. E. (2006). Why minimal guidance during instruction does not work: An analysis of the failure of constructivist, discovery, problem-based, experiential, and inquiry-based teaching. *Educational Psychologist*, 41(2), 75–86.

Knorr-Cetina, K. (1981). *The manufacture of knowledge. an essay on the constructivist and contextual nature of science*. Oxford: Pergamon Press.

Kohlstedt, S. G. (2010). *Teaching children science: hands-on nature study in North America, 1890-1930*. Chicago; London: University of Chicago Press.

Kokkotas, P., & Bevilacqua, F. (2009). *Professional Development of Science Teachers: Teaching Science Using Case Studies From the History of Science*. CreateSpace

Kremer, R. L. (2011). Reforming American Physics Pedagogy in the 1880s: Introducing 'Learning by Doing' via Students Laboratory Exercises. In: P. Heering and R. Wittje (Eds.), *Experiments and Instruments in the History of Science Teaching* (pp. 243–280). Stuttgart, Franz Steiner Verlag.

Kruse, W. J., & Wilcox, J. (2011). *Using Historical Science Stories to Illuminate Nature of Science Ideas and Reduce Stereotypical Views in a Sixth Grade Classroom.* Paper presented at the 2011 Association for Science Teacher Educators International Meeting, Minneapolis, MN, January 19-22, retrieved from https://www.box.com/shared/q1zyn2hfxu (201208-28).

Kutschmann, W. (1994). Erfinder, Entdecker oder Richter. Wandlungen des Subjektverständnisses in den Naturwissenschaften. In: W. Misgeld, K.P. Ohly, H. Rühaak & H. Wiemann (Eds.), *Historisch-genetisches Lernen in den Naturwissenschaften* (pp. 287–308), Weinheim: Deutscher Studienverlag.

Lacki, J., & Karim, Y. (2005). Replication of Guye and Lavanchy's experiment on the velocity dependency of inertia. *Archives des sciences, 58*(2), 159–169.

Latour, B., & Woolgar, S. (1979). *Laboratory Life: The Construction of Scientific Facts.* Princeton: Princeton University Press.

Lederman, N. G. (2007). Nature of Science: Past, Present, and Future. In S. K. Abell & N. G. Lederman (Eds.), *Handbook of research on science education* (pp. 831–879). Mahwah, NJ: Erlbaum.

Lederman, N. G., & Lederman, J. S. (2012). Risk, Uncertainty and Complexity in Science Education. In B. J. Fraser, K. G. Tobin & C. J. McRobbie (Eds.), *Second International Handbook of Science Education*, Vol. I (pp. 355–359). Dordrecht: Springer.

Lefèvre, W. (Ed.). (2007). *Inside the camera obscura : optics and art under the spell of the projected image.* Berlin: Max-Planck-Inst. für Wissenschaftsgeschichte.

Lin, H.-S., & Chen, C.-C. (2002). Promoting Preservice Chemistry Teachers' Understanding about the Nature of Science through History. *Journal of Research in Science Teaching, 39* (9), 773–792.

Lin, H.-S.; Hung, J.-Y., & Hung, S.-C. (2002). Using the history of science to promote students' problem-solving ability. *International Journal of Science Education, 24*(5), 453–464.

Lubben, F., & Millar, R. (1996). Children's ideas about the reliability of experimental data. *International Journal of Science Education, 18*(8), 955–968.

Lunetta, V.N., Hofstein, A., & Clough, M.P. (2007). Learning and teaching in the school science laboratory: an analysis of research, theory, and practice. In S.K. Abell & N.G. Lederman (Eds.), *Handbook of research on science education* (pp. 393–441). Mahwah, NJ: Lawrence Erlbaum.

Mach, E. (1912). *Die Mechanik in ihrer Entwicklung. Historisch-kritisch dargestellt. 7th ed.* Leipzig: Brockhaus.

MacLachlan, J. (1973). A Test of an "Imaginary" Experiment of Galileo's. *ISIS 64*, 374–379.

Maienschein, J. (1999). The Value of practicing practical history. *Endeavour, 23*(1), 3–4.

Maienschein, J., Laubichler, M., & Loettgers, A. (2008). How can history of science matter to scientists? *ISIS, 99*(2), 341–349.

Maiseyenka, V., Henke, A., Launus, A., Riess, F., & Höttecke, D. (2010). *History of Cooling Technology, Ice Production.* Case study developed within the European project HIPST (History and Philosophy in Science Teaching), hipst.eled.auth.gr/hipst_docs/cooling_tech_engl.pdf (23.08.2012).

Martínez, A. A. (2006). Replication of Coulomb's Torsion Balance Experiment. *Archive for History of Exact Sciences, 60*(6), 517–563.

Martins, R. d. A. (1999). Resistance to the Discovery of Electromagnetism: Ørsted and the Symmetry of the Magnetic Field. In F. G. Bevilacqua, Enrico (Ed.), *Volta and the History of Electricity* (pp. 245–265). Milan: Ulrich Hoepli.

McComas, W. F., & Almazroa, H. (1998). The Nature of Science in Science Education: An Introduction. *Science & Education, 7*, 511–532.

Metz, D., & Stinner, A. (2006). A role for historical experiments: capturing the spirit of the itinerant lecturers of the 18th century. *Science & Education, 16*, 613–624.

Metz, D., Klassen, S., McMillan, B., Clough, M., & Olson, J. (2007). Building a Foundation for the Use of Historical Narratives. *Science & Education, 16*, 313–334.

Meyer, K., & Carlisle, R. (1996). Children as experimenters. *International Journal of Science Education, 18*(2), 231–248.

Milne, C., & Taylor, P. C. (1995). Practical Activities Don't Talk to Students: Deconstructing a Mythology of School Science. In F. Finley & D. Allchin. (Eds.), *Proceedings of the Third*

International History, Philosophy, and Science Teaching Conference (pp. 788-801). Minneapolis/Minnesota.

Morus, I. R. (2010). Placing Performance. *Isis, 101*(4), 775–778.

Müller, F. (2004). *Gasentladungsforschung im 19. Jahrhundert*. Berlin: Verlag für Geschichte der Naturwissenschaften und der Technik.

National Research Council (1996). *National Science Education Standards*. Washington D.C.: National Academic Press, retrieved from: http://www.nap.edu/openbook.php?record_id= 4962&page=23 (2010-02-02).

National Research Council (2000). *Inquiry and the National Science Education Standards. A Guide for Teaching and Learning*. Washington: National Academy Press, retrieved from: http:// books.nap.edu/openbook.php?record_id=9596&page=R1 (2012-05-31).

Newman, W. J. Jr., Abell, S. K., Hubbard, P. D., McDonald, J., Otaala, J., & Martini, M. (2004). Dilemmas of Teaching Inquiry in Elementary Science Methods. *Journal of Science Teacher Education*, 15(4), 257–279.

Nott, M. & Wellington, J. (1996). When the black box springs open: practical work in schools and the nature of science. *International Journal of Science Education*, 18(7), 807–818.

Palmieri, P. (2008). *Reenacting Galileo's experiments: rediscovering the techniques of seventeenth-century science*. Lewiston: Edwin Mellen Press.

Palmieri, P. (2009). Experimental history: swinging pendulums and melting shellac. *Endeavour, 33*(3), 88–92.

Pedzisz, B., & Wilke, H.-J. (1993). Historische Freihandexperimente. Beschreibung einer Auswahl von historischen Freihandexperimenten aus der Elektrizitätslehre. *Physik in der Schule* 31(1), 20–27.

Pickering, A. (1995). *The Mangle of Practice. Time, Agency, and Science*. Chicago, London: University of Chicago Press.

Polanyi, M. (1966). *The tacit dimension*. New York.

Popper, K. R. (1934), *The logic of scientific discovery*, London

Principe, L. M. (2000). Apparatus and Reproducibility in Alchemy. In F. L. L. Holmes, Trevor Harvey (Ed.), *Instruments and Experimentation in the History of Chemistry* (pp. 55–74). Cambridge (Mass.) & London: MIT Press.

Radder, H. (Ed.) (2003). *The Philosophy of Scientific Experimentation*. Pittsburgh: University of Pittsburgh Press.

Ramsauer, C. (1953). *Grundversuche der Physik in historischer Darstellung*. Berlin Göttingen Heidelberg: Springer.

Rieß, F. (1998). Erkenntnis durch Wiederholung: eine Methode zur Geschichtsschreibung des Experiments. In M. S. Heidelberger, Friedrich: (Ed.), *Experimental Essays - Versuche zum Experiment* (1. Aufl. ed., pp. 157–172). Baden-Baden: Nomos.

Rieß, F. (2000). History of Physics in Science Teacher Training in Oldenburg. *Science & Education*, 9, 399–402.

Rieß, F., & Schulz, R. (1994). Naturwissenschaftslernen als Textverstehen und Geräteverstehen - Naturwissenschaftsdidaktik in hermeneutischer Absicht und die Rekonstruktion historischer Experiemetierpraxis. In W. Misgeld, K. P. Ohly, H. Rühaak & H. Wiemann (Eds.), *Historisch-genetisches Lernen in den Naturwissenschaften* (pp. 185–204). Weinheim: Studienverlag.

Rieß, F., Heering, P., & Nawrath, D. (2006). *Reconstructing Galileo's Inclined Plane Experiments for Teaching Purposes*. Paper presented at the 8th International History and Philosophy of Science and Science Teaching Conference in Leeds, 2005, retrieved from http://www. ihpst2005.leeds.ac.uk/papers/Riess_Heering_Nawrath.pdf (2012-08-31).

Roehrig, G. H., & Luft, U. A. (2004). Constraints experienced by beginning secondary science teachers in implementing scientific inquiry lessons. *International Journal of Science Education*, 26, 3–24.

Romo, J., & Doncel, M. G. (1993). Faraday's initial mistake concerning the direction of induced currents, and the manuscript of Series I of his Researches. *Archive for History of Exact Sciences*, 47(4), 291–385.

Rosen, S. (1954). A History of the Physics Laboratory in the American Public High School (to 1910). *American Journal of Physics*, 22(194), 194–204.

Rudge, D. W., & Howe, E. M. (2009). An explicit and reflective approach to the use of history to promote understanding of the nature of science. *Science & Education*, 18, 561–580.

Schaffer, S. (1983). Natural Philosophy and Public Spectacle in the Eighteenth Century. *History of Science*, 21, 1–43.

Schauble, L., Glaser, R., Duschl, R., Schulz, S., & Johnson, J. (1995). Students' Understanding of the Objectives and Procedures of Experimentation in the Science Classroom. *Journal of the Learning Sciences*, 4, 131–166.

Schwartz, R. S., & Crawford, B. A. (2004). Authentic Scientific Inquiry as Context for Teaching Nature of Science. In L.B. Flick & N. G. Lederman (Eds.), *Scientific Inquiry and Nature of Science. Implications for Teaching, Learning, and Teacher Education* (pp. 331–355). Dordrecht: Kluwer Academic Publishers.

Schwartz, R. S., Lederman, N. G., & Crawford, B. A. (2004). Developing Views of Nature of Science in an Authentic Context: An Explicit Approach to Bridging the Gap Between Nature of Science and Scientific Inquiry. *Science Education*, 88(4), 610–645.

Seker, H., & Welsh, L. C. (2005). The Comparison of Explicit and Implicit Ways of Using History of Science for Students Understanding of the Nature of Science. Paper prepared for the Eighth International History, Philosophy, Sociology & Science Teaching Conference (IHPST), Leeds, UK 2005, July 15 - 18, 2005, retrieved from http://www.ihpst2005.leeds.ac.uk/papers/Seker_Welsh.pdf (2012-08-29).

Settle, T. B. (1961). An Experiment in the History of Science. *Science 133*(1), 19–23.

Shapin, S. (1994). *The social history of truth*. Chicago: University of Chicago Press.

Shapin, S., & Schaffer, S. (1985). *Leviathan and the Air-Pump*. Princeton: University Press.

Sibum, H. O. (1995). Reworking the Mechanical Value of Heat: Instruments of Precision and Gestures of Accuracy in Early Victorian England. *Studies in the History and Philosophy of Science*, 26, 73–106.

Sibum, H. O. (1998). Les Gestes de la Mesure: Joule, les pratiques de la brasserie et la science. *Annales HSS*, 745–774.

Sichau, C. (2000a). Industry and Industrial Relations within the Laboratory: The Material Conditions of Joule-Thomson Experiments, In M. Lette & M. Oris (Eds.), *Proceedings of the XXth International Congress of History of Science. Vol. 7: Technology and Engineering* (pp. 49–59). Turnhout (Belgien): Brepols.

Sichau, C. (2000b). Practicing Helps: Thermodynamics, History, and Experiment. *Science & Education* 9, 389–398.

Sichau, C. (2002). *Die Viskositätsexperimente von J.C. Maxwell und O.E. Meyer: Eine wissenschaftshistorische Studie über die Entstehung, Messung und Verwendung einer physikalischen Größe*. Berlin: Logos.

Solomon, J., Scot, L., & Duveen, J. (1996). Large-Scale Exploration of Pupils' Understanding of the Nature of Science. *Science Education*, 80 (5), 493–508.

Staubermann, K. (2007). *Astronomers at work: a study of the replicability of 19th century astronomical practice*. Frankfurt am Main: Deutsch.

Staubermann, K. B. (Ed.). (2011). *Reconstructions : recreating science and technology of the past*. Edinburgh: NMS Enterprises.

Steinle, F. (1996). Work, Finish, Publish: The Formation of the Second Series of Faraday's Experimental Researches in Electricity. *Physis*, 33, 141–220.

Steinle, F. (2006). Concept Formation and the Limits of Justification: "Discovering" the Two Electricities. In J. Schickore & F. Steinle (Eds.), *Revisiting Discovery and Justification* (pp. 183–195). Dordrecht: Springer.

Tao, P.-K. (2003). Eliciting and developing junior secondary students' understanding of the nature of science through a peer collaboration instruction in science stories. *International Journal of Science Education*, 25(2), 147–171.

Teichmann, J. (1979). Die Rekonstruktion historischer Modelle und Experimente für den Unterricht - drei Beispiele. *Physik und Didaktik*, 4, 267–282.

Teichmann, J. (1999). Studying Galileo at Secondary School: A Reconstruction of His ‚Jumping Hill' Experiment and the Process of Discovery. *Science & Education*, 8(2), 121–136.

Teichmann, J., Ball, E., & Wagmüller, J. (1990). *Einfache physikalische Versuche aus Geschichte und Gegenwart* (5 ed.). München: Deutsches Museum.

Tesch, M., & Duit, R. (2004). Experimentieren im Physikunterricht - Ergebnisse einer Videostudie. *Zeitschrift für Didaktik der Naturwissenschaften*, 10, 51–69.

Trumper, R. (2003). The Physics Laboratory – A Historical Overview and Future Perspectives. *Science & Education*, 12, 645–670.

Tsagliotis, N. (2010). *From Hooke's Micrographia towards the construction of a simple microscope for the teaching and learning of primary science.* Case study developed within the European project HIPST (History and Philosophy in Science Teaching), retrieved from http:// hipstwiki.wetpaint.com/page/auth+case+2 (2012-08-23).

Turner, S. (2011). Changing Images of the Inclined Plane, 1880-1920: A Case Study of a Revolution in American Science Education. In P. Heering and R. Wittje (Eds.), *Experiments and Instruments in the History of Science Teaching* (pp. 207–242). Stuttgart, Franz Steiner Verlag

Tweney, R. D. (1985). Faraday's Discovery of Induction: A Cognitive Approach. In D. Gooding & F. James, *Faraday Rediscovered. Essays on the Life and Work of Michael Faraday 1791-1867* (pp. 189–209). New York: Stockton Press.

Tweney, R. D. (2005). On replicating Faraday: experiencing historical procedures in science. *Archives des sciences, 58*(2), 137–147.

Usselman, M. C., Renhart, C., & Foulser, K. (2005). Restaging Liebig: A Study in the Replication of Experiments. *Annals of Science*, 62, 1–55.

Valk, T.v.d., & Jong, O. d. (2009). Scaffolding Science Teachers in Open-inquiry Teaching. *International Journal of Science Education*, 31(6), 829–850.

van de Pol, J., Volman, M., & Beishuizen, J. (2010). Scaffolding in Teacher-Student Interaction: A Decade of Research. *Educational Psychology Review, 22*, 271–296.

Vera, F., Rivera, R., & Nâuänez, C. (2011). Burning a Candle in a Vessel, a Simple Experiment with a Long History. *Science & Education, 20*(9), 881–893.

Voskuhl, A. (1997). Recreating Herschel's actinometry: An essay in the historiography of experimental practice. *British Journal for the History of Science*, 30, 337–355.

Vygotsky, L. S. (1978). *Mind in society: The development of higher psychological processes* (ed. by M. Cole et al.). Cambridge, MA: Harvard University Press.

Wallace, C.S. & Kang, N.-H. (2004). An Investigation of Experienced Secondary Science Teachers' Beliefs about Inquiry: An Examination of Competing Belief Sets. *Journal of Research in Science Teaching,* 41 (9), 936–960.

Welzel, M., Haller, K., Bandiera, M., Hammelev, D., Koumaras, P., Niedderer, H., Paulsen, A., Robinault, K. & Aufschnaiter, S.v. (1998). Ziele, die Lehrende mit experimentellem Arbeiten in der naturwissenschaftlichen Ausbildung verbinden. - Ergebnisse einer europäischen Umfrage. *Zeitschrift für Didaktik der Naturwissenschaften*, 4(1), 29–44.

Wilke, H.-J. (Ed.). (1988). *Physikalische Schulexperimente: Historische Experimente*. Berlin: Volk u. Wissen.

Willer, J. (1990). *Physik und menschliche Bildung. Eine Geschichte der Physik und ihres Unterrichts*. Darmstadt: Wiss. Buchgesellschaft.

Wittje, R. (1996). *Die frühen Experimente von Heinrich Hertz zur Ausbreitung der "Elektrischen Kraft": Entstehung, Entwicklung und Replikation eines Experiments.* Unpublished diploma thesis, University of Oldenburg/Germany.

Wright, M. T. (2007). The Antikythera Mechanism Reconsidered. *Interdisciplinary Science Reviews, 32*(1), 27–44.

Ziman, J. (2000). *Real Science. What it is, and What it means.* Cambridge: Cambridge University Press.

Peter Heering is professor of physics and physics didactics at the University of Flensburg, Germany. He completed his habilitation in history of science at the University of Hamburg in 2006, earned his Ph.D. in physics education and history of physics from the University of Oldenburg in 1995, and was initially trained as a teacher for physics and chemistry. His research interests include the history of scientific experimentation, the relation between research and educational experiments, and the implementation of history of science in science education. Among his recent publications are Heering, P., and Wittje, R. (Eds.) (2011), *Learning by doing: experiments and instruments in the history of science teaching* Stuttgart: Steiner, and Heering, P. (2010), An Experimenter's Gotta Do What an Experimenter's Gotta Do – But How? *ISIS*, 101(4), 794–805.

Dietmar Höttecke is a professor for physics education at the University of Hamburg, Germany. He earned his Ph.D. in physics education and history of physics from the University of Oldenburg in 2001 and is trained as a teacher for physics and German language and literature. His research interests include applications of history and philosophy of science in science teaching, nature of science in science education, judgment and decision-making in socioscientific issues, education for sustainable development, and science teachers' professional development. Among his recent publications are two papers about problems and solutions of implementing history and philosophy of science in science education: Höttecke, D. and Silva, C.C. (2011), Why Implementing History and Philosophy in School Science Education is a Challenge – An Analysis of Obstacles. *Science & Education*, 20(3–4), 293–316 and Höttecke, D., Henke, A., and Rieß, F. (2012), Implementing History and Philosophy in Science Teaching – Strategies, Methods, Results and Experiences from the European Project HIPST. *Science & Education*, 21(9), 1233–1261.

Chapter 47
Science Teaching with Historically Based Stories: Theoretical and Practical Perspectives

Stephen Klassen and Cathrine Froese Klassen

47.1 Introduction

In recent decades, a trend has evolved in educational literature that emphasizes the potential of narratives, especially the story, to improve teaching and learning.[1] These studies approach the subject from both a methodological and theoretical perspective. Their authors purport that science stories illustrating the related abstract concepts engage and motivate the learner emotionally and intellectually. Further to that, this chapter examines the nature and structure of the science story and its capacity to provoke certain kinds of student responses that promote the learning of science. Historically based stories, in particular, promote the desired student interest and motivation by presenting humanistic episodes that explicitly include scientific content.[2]

The reasons for incorporating stories in teaching are summarized by Noddings and Witherall, who assert that

> we learn from stories. More important, we come to understand—ourselves, others, and even the subjects we teach and learn. Stories engage us. … Stories can help us to understand by making the abstract concrete and accessible. What is only dimly perceived at the level of principle may become vivid and powerful in the concrete. Further, stories motivate us. Even that which we understand at the abstract level may not move us to action, whereas a story often does. (Noddings and Witherell 1991, pp. 279–280)

Well-constructed and effective stories stimulate students' imagination and, thereby, produce affective engagement during learning episodes. The motivating effect and the vivid and powerful perceptions of stories take place largely at the

---

[1] For example, see Egan (1986, 1989b), Kenealy (1989), Klassen (2009a), Kubli (1999), Metz et al. (2007), Solomon (2002), and Stinner (1992).

[2] See Hadzigeorgiou et al. (2012), Klassen (2009a), Kubli (2006), and Solomon (2002).

S. Klassen (✉) • C. Froese Klassen
Faculty of Education, The University of Winnipeg, Winnipeg, MB, Canada
e-mail: s.klassen@uwinnipeg.ca

M.R. Matthews (ed.), *International Handbook of Research in History, Philosophy and Science Teaching*, DOI 10.1007/978-94-007-7654-8_47, © Springer Science+Business Media Dordrecht 2014

1503

emotional level, which is an insight in education that has only recently been explored. Yet, it has, likely, always been apparent to good teachers that stories make learning experiences memorable. The story approach originated within a general educational perspective and has, subsequently, been adapted for use in science education.

This chapter is about stories in science teaching and learning that are, at least initially, in written form and may be communicated to students in that form or narrated by a teacher or student. As readers will appreciate, the task of writing an effective story is a highly technical and challenging creative undertaking which is not the forte of every science teacher or educator. Before such writing can be carried out, the ambiguities that exist in the use of the terms "narrative" and "story" as applied in the science education context must be examined and removed. The terms "story" and "narrative" in this context are used in the literary sense and not in the broad, generic, and undefined sense that prevails in common usage.

There is significant evidence supporting the effectiveness of science stories to improve learning, and appropriate methods exist for integrating them with instruction systematically. The model of the story as the reenactment of a type of learning process and the role of the story in generating a romantic understanding of science are two pedagogical bases supporting the story approach. While stories serve, among other functions, to improve student motivation, the elaboration of theories and research on student engagement, motivation, or interest are beyond the scope of this chapter. The story form that is elaborated contains science concepts and, at the same time, emphasizes the humanistic dimension of science—an indispensible element of such stories and a critical feature of the only two empirical studies undertaken in this area to date. The humanistic aspect requires the incorporation of the historical accomplishments of notable scientists, necessitating the use of history of science to form the basis of the science stories. Stories need to be reasonably faithful to nature of science criteria, which is a reason for using history-of-science cases as the background for stories.

This chapter deals with stories that are historically based, as opposed to strictly fictional stories, in accordance with most of the literature in the field. The discovery of a good historically based story is like the discovery of a hidden treasure. Perhaps, the fascination for students arises from the romance of far-removed events with participants who had the same kinds of hopes, dreams, and struggles as we, and yet, in a very different environment. Good teachers often employ such stories in their teaching. They have found, like Swiss science educator Fritz Kubli, that "bare bones do not make an appetizing meal" for students (Kubli 2005, p. 520). Historically based stories (sometimes called "cases") are a valuable, some would say essential, part of providing a rich and diversely connected context for student learning.

Early theoretical work on stories has, to a large extent, centered on explaining why stories are expected to make learning experiences memorable. Some studies have focused on the idea that stories stimulate an emotional response in the listener or reader (Egan 1986, 1989a, b; Miall and Kuiken 1994). Others have argued that the story is a device that imposes coherence on a set of events (Kenealy 1989), which makes it a suitable vehicle for the integration of history of science in science instruction. The inclusion of historically based stories in science teaching has been advocated and attempted by a growing number of science educators.[3] Recently, a theoretical foundation for the use of narratives in the science classroom has been developed.[4] Yet, the observation of Bruner, in (1986), that "in contrast to our vast knowledge of how science and logical reasoning proceeds, we know precious little in any formal sense about how to make good stories" (p. 14) still rings disturbingly true today. This work is, in some sense, a response to the 25-year-old lament of Bruner.

47.2 An Overview of the Literature

A review of the scholarly literature on the theoretical basis, desirability, and use of stories (and narratives in general) in science classrooms over the past 15 years reveals that surprisingly little work has been done in this area (see Table 47.1). The list includes all relevant, peer-reviewed papers published in scholarly journals that deal with what are purported to be science stories, as opposed to other narrative forms. Not included in the list is the paper by Tao (2003), who used comic strips based on various historical "stories."

Bruner's early observation concerning how little was known about creating good stories was followed with his recommendation a decade later that we "convert our efforts at scientific understanding into the form of narratives" (1996, p. 125). In the literary field, "making good stories" is a matter of course, but this is not so in science education. The literature search yielded only 18 scholarly papers—which we categorized as advocacy, theoretical, descriptive, exemplar, or empirical—that have been published since Bruner's 1996 recommendation. Table 47.1 reveals that the literature on the use of stories in science teaching has been sparse up until 2005. Thereafter, the publication rate has increased, although not to a dramatic extent. Much of the literature is theoretical in nature, attempting to establish the features of stories that will likely contribute to their effectiveness in teaching science content and the nature of science. It describes instances in which stories and narratives have been used in classrooms but without a rigorous analysis of their success. In some cases, the literature is more explicit in describing the use of stories and includes the actual story used in the classroom, serving as an exemplar of the approach. The study of Negrete and Lartigue (2010) presents data of students' performance after listening to what the researchers call "stories" in comparison to factual summaries of the science content. The data, however, serve only to establish a rating scheme for the effectiveness of stories, the "stories" themselves being excerpts of much longer popular science writings that were not written to portray science in the sense being discussed here.

---

[3] For example, see Kenealy (1989), Klassen (2009a), Kubli (1999), Martin and Brouwer (1991), Stinner (1992), and Wandersee (1990).

[4] For example, see Avraamidou and Osborne (2009), Klassen (2010), Kubli (2001), Metz et al. (2007), and Norris et al. (2005).

Table 47.1 A listing and summary of scholarly publications on the use of stories in science teaching

Author(s)	Year	Type	Summary
Hadzigeorgiou, Klassen, and Froese Klassen	2012	Empirical	Presents the romantic understanding framework for science stories and a quasi-experimental study demonstrating significant advantages for student learning about ac electricity through the story of Nikola Tesla
Clough	2011	Descriptive	Describes the process of developing 30 historical short stories
Klassen	2011	Exemplar, descriptive	Presents the design of a historical story surrounding the photoelectric effect and describes its use in the classroom
Klassen	2010	Theoretical	Presents a theoretical description of the linkage of the science story to a type of conceptual change
Kokkotas, Rizaki, and Malamitsa	2010	Descriptive	Presents the theoretical basis and describes the classroom implementation of a historical story on electromagnetism that yielded positive results described through examples of student journal entries
Negrete and Lartigue	2010	Theoretical, descriptive	Presents an evaluation scheme for using science stories with a brief description of a classroom study using stories
Avraamidou and Osborne	2009	Theoretical	Presents a case for the use of narrative in science education as a way of making science meaningful, relevant, and accessible to the public
Klassen	2009a	Theoretical, exemplar, empirical	Develops a basis for writing historical science stories along with an example, its use in a classroom, and the results in terms of student-generated questions
Isabelle	2007	Descriptive, exemplar, advocacy	Describes a classroom study teaching through a story and recommends the strategy to teachers, although not discussing demonstrated benefits of the case at hand
Klassen	2007	Exemplar, descriptive	Describes the design and classroom use of a historical story surrounding the first Atlantic cable
Metz et al.	2007	Theoretical	Presents a basis for constructing and using stories in science teaching
Hadzigeorgiou	2006	Descriptive, exemplar	Discusses the potential of stories in teaching and learning physics and presents a historical planning framework for teaching current electricity
Kubli	2006	Theoretical	Presents a basis for effective use of storytelling in the science classroom
Norris et al.	2005	Theoretical	Presents a discussion of the problems and potential of using science stories with explanatory content and the theoretical basis of such stories
Solomon	2002	Theoretical, advocacy	Illustrates types of science stories that might be used and their purposes
Kubli	2001	Theoretical	Presents a theoretical basis which science teachers can employ to become better storytellers
Milne	1998	Theoretical	Discusses the assumptions behind certain perspectives contained in stories and the desirability of these perspectives
Knox and Croft	1997	Descriptive	Describes the use of story in a meteorology classroom with positive responses from students but did not demonstrate learning benefits

All of the literature listed in Table 47.1 utilizes or advocates historically based stories, with the exception of Avraamidou and Osborne (2009), who advocate for stories that are exclusively fictional. All of the exemplars or empirical studies are historically based.

While some of the papers overlap between or among the designated categories, as outlined in Table 47.1, overall, ten of the papers are theoretical, eight descriptive (with or without exemplars), and only two empirical. Five papers include exemplars. With the exception of Hadzigeorgiou, Klassen, and Kubli, researchers have not pursued their initial study with further research in the area. The table reflects the movement towards empirical approaches, which is vital in the development of this research area. The two empirical studies are summarized in greater detail below.

47.2.1 Empirical Studies Utilizing Stories in Science Teaching

The 2009 study of Klassen (2009a) presents a literary framework for the construction of stories that is based on Norris and coauthors (2005) and Kubli (2001). A story about Louis Slotin and the beginnings of the science of radiation protection is presented as an exemplar, and a classroom study utilizing the story as a door opener for an experimental investigation is described. Student responses to the story were gathered in the form of student-generated questions at the end of the story. Klassen sees a major purpose for using door-opening science stories as raising questions in students' minds. Not only is the raising of good questions important from a constructivist, pedagogical point of view, but also there is reason to believe that questions are implicitly involved in theory formation. Therefore, evidence for the generation of good questions as a result of listening to the story would serve as a major indicator that stories tend to enhance learning. The data presented in the study support the conclusion that good questions were, indeed, generated as a response to the story. The largest number of questions were of a type that suggested higher-level thinking, rather than simple, factual ones typically introduced by the interrogatives "when," "where," and "who." There was, however, evidence that the students were inexperienced with generating well-framed questions, as indicated by a number of questions which appeared, on the surface, to be polar questions (requiring "yes" or "no" responses).

Hadzigeorgiou, Klassen, and Froese Klassen (2012) provide details of an empirical study utilizing a story, based on Nikola Tesla's life and work, that was written to embody the elements of romantic understanding (see Sect. 47.5.2 for elaboration of romantic understanding) in order to teach the concepts of alternating current electricity. The study design was quasi-experimental, with the experimental group of students ($n=95$) only listening to the Tesla story and the control group ($n=102$) receiving conventional instruction with the mastery-teaching technique. Both groups were given the same open-ended, paper-and-pencil concept test on alternating current electricity. All students were encouraged to make journal entries.

The authors determined that the students who were encouraged to understand the concept of alternating current romantically—that is, through the story—became more engaged with the science content compared to students who were taught explicitly by conventional instruction. A quantitative analysis of students' journal entries revealed that of the students in the experimental group, 96 % made relevant journal entries, whereas only 55 % in the control group did so. Some of these students also demonstrated imagination and curiosity in their journal entries, which was not observed in the control group. Additionally, they undertook relevant independent research on their own initiative, which was indicative of the transformative effect of the storytelling instruction. This was a notable difference from the students in the control group who did not undertake any related reading outside of class, according to the analyses of the teachers' observations. The results also affirmed the effectiveness of teaching alternating current electricity through the Tesla story in the statistically superior concept-test results of the experimental group over the control group. Data on the question requiring an explanation of the concept of alternating current, for example, showed that 72 % of the experimental group versus 41 % in the control group gave an acceptable response. When the test was repeated 8 weeks later, 67 % of the experimental group versus 31 % in the control group gave an acceptable response, suggesting not only superior understanding but also greater long-term retention. A content analysis of the student journals of the experimental group revealed that at least two characteristics of romantic understanding were identified in all journals and that two thirds of the students in the experimental group made associations with the science content in their demonstration of the characteristics of romantic understanding.

The authors conclude that

> [t]he implication of this study for science education is that a particular type of a science story, that is, a romantic story, has significant potential for improving learning in the discipline. Such a story should be based upon human qualities, heroic or otherwise, that evoke wonder and give students the opportunity, through the plot, to associate science content with such qualities and simultaneously experience a sense of wonder. (Hadzigeorgiou et al. 2012, p. 1134)

47.3 Explication of Key Concepts: Narrative, Story, and Science Story

Narrative is a humanistic mode of expression that has as its core purpose the recounting of related events involving characters. In the most elementary sense, a narrative tells of someone having done something. Traditionally, narrative has been distinguished from exposition, description, and persuasion, which are other modes of communication (Connors 1981). Various scholars have used the terms "narrative" (Kubli 1998; Martin and Brouwer 1991), "story" (Egan 1989b; Kenealy 1989; Kubli 1999; Stinner 1995), "thematic" (Holbrow et al. 1995), or "storyline" (Arons 1988; Coleman and Griffith 1997; Stinner 1995) to describe their approach in using this mode in science teaching. The narrative approach has a spectrum of possible adaptations, ranging from the smallest stand-alone story element, such as the vignette

(Wandersee 1992) or anecdote (Shrigley and Koballa 1989), to the largest story-like structure, such as a curriculum unit unified by a theme (Gorman and Robinson 1998; Holbrow et al. 1995) or storyline (Coleman and Griffiths 1997; Stinner and Williams 1998). These various proposed approaches, other than the story itself, have not been put to the test in formal research studies and will not be elaborated in this chapter. There is no recent science education literature describing the use of other forms of narrative. Some time ago, Shrigley and Koballa (1989) described the anecdote and Wandersee (1990) described the vignette and its use in science teaching. Neither of these two forms was developed any further in the literature nor studied through research.

The notion of narrative is broad, and the term does not have a categorical definition in the literature, in contrast to the literary notion of story. Moreover, in the narratological literature, it is not distinguished clearly from the act of storytelling, which is intricately connected to voice, narrator, and other oratorical devices. For reasons such as these, the discussion in this chapter, of necessity, has certain delimitations. The first of these is the definition of "narrative" itself. The literature on narratives is complicated by theorists using varying terminology for similar concepts, making the task of definition challenging. Specifically, the terms "narrative" and "story" are frequently used interchangeably and then either not defined or defined in a way that is not adequate for application in a particular context. To aid in the sorting out of related and, possibly, conflated terminology, the core constituents of narrative, namely, its "raw material," will be distinguished from the delivery of narrative material, the reception and personal reconstruction of narratives, and the specific ways and contexts of using narrative material. Defining narrative in terms of core constituents has the added benefit of providing a "litmus test" for educational material that claims to be "narrative." The definition of narrative constructed here incorporates the insights of Altman (2008), who presents a particularly succinct history of narrative definitions and divides the definition of "narrative material" (p. 10) into categories.

Whereas the concept of "narrative" is broad and generic, the concept of "story" has specifically defined attributes. In concurrence with Altman (2008), this chapter makes a distinction between "some" narrative and "a" narrative (p. 17), in that "some" narrative is a narrative excerpt and "a" narrative is a complete narrative episode with a beginning, middle, and ending—one of the special requirements of a story. In this chapter, the term "story" is not used in the generic and loose sense, as in popular and even in some scholarly literature, but in its defined sense that is identified with a specific literary form. Such a story also requires a central role for the main character, who is involved in some type of conflict that compels him or her to make a critical decision, which, ultimately, determines the outcome of the story.

The definitions of narrative and story presented below additionally make use of the insights of Klassen (2009a, 2010), Kubli (2001), and Norris and coauthors (2005) but utilize only elements and characteristics that are essential for the construction of narrative on the part of the teller or writer, as opposed to its reception or its expression in particular settings or for particular purposes. In this respect, the definitions are, of necessity, based on structuralist approaches to narrative—an insight which is crucial for the operationalization of narrative theory in the science education context and which has heretofore not been recognized by researchers attempting to formulate a definition of narrative or story.

47.3.1 The Concept of Narrative

In view of the considerations presented above, narratives have five fundamental elements, as outlined below, in some instances with illustrations. Since stories constitute a class within the narrative genre, the aspects of narratives apply equally well to stories. The purpose of presenting a definition of narrative is not only for clarification but for subjecting writing that purports to be narrative, or even story, to a criterion-based test that can ascertain whether it can, justifiably, be characterized as a narrative. For writing that consists, for example, of a mixture of narrative and exposition, it would be relatively easy and uncontroversial to analyze the text sentence by sentence for its degree of narrativity. While a narrative does not necessarily contain all of the elements of a specific class of narratives, such as the story, it must, by definition, include the features of characters, actions, situations, consequential coherence, and past time.

47.3.1.1 Characters

The most basic element of any narrative is that it involves at least one character who is the representation of a person or persons. In historically based science stories, the characters are real people (usually scientists), and their story is a form of historical interpretation. Altman (2008) points out that while some definitions of narrative have omitted the role of characters and concentrate, instead, on the events which give rise to plot, characters are essential and central since they are the agents that produce the action.

47.3.1.2 Actions

The actions of the characters, sometimes called "events," are the "raw material" of the story, but, by themselves, they produce only chronologies that lack interest; they are not narratives. In historically based narratives, the actions of the characters are obtained from the historical records. Agency, of necessity, creates a personal dimension; therefore, it becomes necessary to attribute thoughts or spoken words to the protagonist, especially. The preferable method of obtaining such expressions is to adapt them from writings of that person or other relevant historical records. If thoughts or spoken words not obtained from the historical record are added for effect, these must not contradict the history.

Example: Eighteen-year-old Alessandro Volta was passionate about electricity, indeed so passionate that he announced to his somewhat startled family, "I'm not going to university! I would rather spend my time on investigating electrical phenomena."[5] Alessandro's family was used to his surprising turns.[6] As a child, Alessandro had not learned to speak until he was age four, which had led his alarmed parents to think that he might be slow-witted, but then little Alessandro suddenly began to develop at a furious pace, out-performing all his school mates.[7] In his father's words, "We had a jewel in the house and did not know it!" And so, at age eighteen, the gifted young Volta launched into a scientific career by beginning to write to the leading scientists of the day about his ideas, and, surprisingly, they replied. (Klassen 2009b)

In the above example, all of the details are historically accurate or plausible. The words attributed to Volta are plausible based on the record. The words attributed to Volta's father are a part of the historical record explicitly.

47.3.1.3 Situations

Narrative has situations or states insofar as the character responds to them or helps to create them. The situational aspects of narratives that represent a state of affairs or a state of being are often not separable from the actions from which they result. An example extracted from the story "The Soul of Solar Energy: Augustin Mouchot," written by the authors, is given here. The first segment ends with a state of being, followed by an action, followed by a changed state of affairs in the last segment. The situational sentences are italicized.

As Augustin arose in the chill of the dawn, his thoughts drifted to what he had just been reading about the energy of the sun. Physicist Claude Pouillet had written that every square meter of the Earth's surface receives about 10 Calories of energy every minute. Augustin chuckled, "Not a very useful fact on a cloudy day like today!" Then, a flash of inspiration crossed his mind: "It's not cloudy every day. Wouldn't it be possible to heat enough water with the sun's light and spare the fire that is only meant to heat the house?" *While he made the last preparations to teach his geometry class, he could not get his mind off the energy issue.*

...

Over the next few months, Augustin immersed himself in his new project of building a solar energy collector despite having to teach his regular classes.

...

Soon, Augustin had completed the construction of his first solar water heater, which was capable of holding three litres of water. Lucky for him, it happened to be a cloudless day! Excitedly, he placed the boiler and mirror in the direct sunlight. To his amazement, the water, which he had initially measured to be 15 degrees, boiled in just an hour and a half. From then on—on sunny days—Augustin saved himself the bother and the expense of coal-heated water when he bathed. (Klassen and Froese Klassen 2012, italics not in original)

---

[5] The discourse is imaginary but plausible.

[6] A reasonable assumption, based on the historical record.

[7] Part of the historical record (from here to the end of the segment).

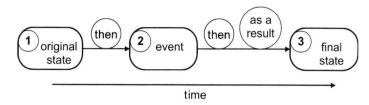

Fig. 47.1 The causal structure of a story (From Klassen 2010)

47.3.1.4 Consequential Coherence

As mentioned above, by themselves, the actions of the characters produce only chronologies, which lack interest. They are neither stories nor narratives. Narratives and stories require that the events be causally linked (see Fig. 47.1 for a representation of the causal structure of a story). The story is written in such a way that it is clear that the actions of the characters produce the changes in the state of affairs. Besides the fact that the events of the story follow one another in time, it is especially the causative linkage of events that produces the perception of the flow of time. Of course, the entire story need not be structured in a strict chronological sequence, as the story might contain flashbacks and flash-forwards; however, it must have a consequential sequence.

> *Example of a chronology without consequential coherence:*
> Ohm took up a teaching position in Cologne;
> he performed an experiment to determine resistance;
> and he published his experiment.
>
> *Example of a passage with consequential coherence:*
> Ohm understood the mathematical theory of heat;
> he applied the theory to electricity by analogy;
> and as a result, Ohm understood the mathematical theory of electrical resistance.

47.3.1.5 Past Time

The events of a story take place in the past and are recounted. The science-based story is historical, and it would be a contradiction in terms to write it in the present tense. Use of the present tense is reserved for other modes of communication, for example, exposition.

47.3.2 Science as Historical Narrative

The scientific work done and the results achieved by investigating scientists have to be communicated to other scientists. Their results cannot readily be separated from the experiments that were undertaken. The carrying out of experiments is a human endeavor, replete with false starts, errors, and frustrations. This phenomenon significantly increases the potential for historical narratives to supplement the published results in the scientific literature. R. G. Collingwood (1945) has carried this line of reasoning further by claiming that history is, in fact, a more fundamental form of thought than the science about which it reports. He writes that

> the scientist who wishes to know that … an event has taken place in the world of nature can know this only by consulting the record left by the observer and interpreting it, subject to certain rules, in such a way as to satisfy himself that the man whose work it records really did observe what he professes to have observed. This consultation and interpretation of records is the characteristic feature of historical work. … I conclude that natural science as a form of thought exists and always has existed in a context of history, and depends on historical thought for its existence. (Collingwood 1945, pp. 176–177)

What Collingwood describes about scientific work appears very much like the process of historiography. Yet, current-day scientific publications usually exclude the human dimension of the work that brought about certain scientific conclusions or theories in science. In that sense, scientific publications are a specialized, rational reconstruction of history. Perhaps, it would not be an exaggeration to say that science has become a form of dehumanized and decontextualized history. An attractive method of interesting the disinterested young student in science is to portray it differently—more realistically, from the human and contextual perspective as a specialized narrative or story.

It is important to emphasize that historical chronologies, by themselves, are not narratives, as they have no causal structure. In the absence of clearly defined criteria, such as causality, many of the terms in the scholarly literature, such as thematic approach, storyline, vignette, anecdote, or story-like structure, must be subjected to closer scrutiny. The existence of this definitional vagueness emphasizes the importance of establishing a "litmus test" for both narrative and story, in order to produce a standardized vehicle for the delivery of narratives and stories and for the purposes of research. The definitions that have been given above may be taken as such a litmus test.

47.3.3 The Concept of the Science Story

The essential elements of a story include those of narratives that have been described. Two further components are necessary to create the classification of story within the broader narrative form, namely, a defined plot structure and agency—the critical choice made by the protagonist. The science story, a classification within the class of story, includes an additional element, namely, science and NOS content. Aspects of narratives or stories that are specific to their method of delivery or to their reception by the listener or reader, while potentially important for the effectiveness of the story, are not considered as core constituents in its written form.

47.3.3.1 Plot Structure

The plot structure of a story consists of a beginning-middle-end structure that frames the story and creates a stand-alone unit. In general, the story structure typically includes an introduction, rising action that includes some sort of conflict, a climax, and a resolution or conclusion. In the conceptualization and construction of the story, planning the plot is crucial and involves a creative interpretation of the history that identifies fascinating elements in the record and combines them with the relevant science. Planning a plot requires selecting historical details to form a coherent story. Specifically, a typical plot (a) begins by "setting the scene" in some manner, (b) followed by the presentation of a problematic situation (c) which reaches a crisis, (d) necessitating a critical decision made by the main character (e) that results in a climactic moment, and (f) concludes in a resolution of the situation, which can be either positive or negative.

47.3.3.2 Agency: Critical Choice Made by the Main Character

In a story, the role of the main character is crucial and will affect the outcome of the story through a consequential, critical choice that he or she has made. Crucial choices are integral to the typical plot structure as outlined above.

> *Example:* Grabbing the hemispherical beryllium shell by the thumb-hole on the top, Lou carefully lowered the top half onto the bottom half covering a hemispherical plutonium shell, which, in turn, covered the polonium initiator, holding them apart with a screwdriver. As he rotated the screwdriver slightly this way and that, the shell moved up and down. From across the room the familiar crunching sound of the Geiger counters swelled and ebbed. Then it happened. No one knows what broke Lou's concentration, but something did. The screwdriver slipped and clattered to the floor and a blue flash filled the room as the top shell touched the bottom, releasing a torrent of neutrons and gamma–rays. Time seemed to come to a screeching halt. Almost instinctively, Lou, using both hands, grabbed the lethal assembly and flipped the bomb-shell off the table and onto the floor with what seemed a deafening crash. "Well, that does it—I'm dead!" Lou heard himself say. "Tell me this is a nightmare," he thought. *But it wasn't.* (Klassen 2009a, p. 421)

47.3.3.3 Science and NOS Content

The scientific aspects of the historical episode that are contained in the story should be embedded at the appropriate points in the story in such a way that they flow naturally with the story. In some cases, it will be possible to include the scientific concepts developed sufficiently to allow students to comprehend them fully. In other cases, the scientific questions and issues will be included as questions, issues, and problems which need to be developed more fully in activities that follow the story. Such content should be included from the perspective of the characters of the story and be interwoven in a narrative manner as much as possible.

Similar considerations reflect the NOS issues that are contained in the story. Sometimes these issues may be stated explicitly in the flow of the story, and at other times, these will be raised more indirectly and will need to be followed with a student activity that will focus on the issue and clarify it. The following example is taken from the story used for the experimental intervention in Hadzigeorgiou and colleagues (2012) (the reader is referred back to Sect. 47.2.1 for a description of the study):

> *Example of embedding science content:* In an attempt to demonstrate the dangers of AC power, Edison sponsored an electrical engineer to travel the country electrocuting animals with both DC and AC. Because the frequency of AC confuses the heart, animals that are electrocuted by AC die, whereas animals that are electrocuted by DC are stunned but survive. Edison used these so-called "experiments" to contrast the danger of AC with the relative safety of DC. We know that the effect of any type of electric current on a human being is very difficult to predict, as it depends on a number of factors (for example, the condition of the skin, amount of fluid in the body, and the point of contact). Tesla, however, had been experimenting with very high frequency currents, which, as he, showed, did no harm. With his theatrical flair, Tesla could draw sparks to his own fingers and even walk through sparks without being hurt. He had realized that the high frequency of the current kept it on his skin. It was this strange effect, known as the skin effect, which made Tesla famous. He even sent sparks to the audience, making people realize that AC current, at least as used by him, was safe. (Hadzigeorgiou, Froese Klassen, and Klassen 2011)

47.4 The Non-scripted and Scripted Approaches to Using Stories

The story approach in teaching science, as described in this chapter, is based on the use of science stories that are constructed according to the concept outlined above. What is not included in this approach is that of the non-scripted story, told spontaneously, randomly, and, perhaps, frequently. In the story-permeated approach—the spontaneous injection of non-scripted narratives—the narratives may range from a recounting of the teacher's experiences to anecdotes that may or may not pertain to curricular content or concepts. Teachers who use stories in this fashion must have expertise not only in the telling of stories, but they must have a wide repertoire of good story material at their disposal. Such an approach is largely informal and more difficult to fit into a definitive framework than the scripted-story approach. Because of the number of uncontrolled variables, this approach is virtually beyond the scope of research studies.

The scripted-story approach—the formal, planned integration of scripted stories—pertains specifically to the curricular concepts taught. Particularly teachers inexperienced in storytelling and lacking a repertoire of science-story material can enhance their instruction with the scripted-story approach. Scripts allow for a measure of structure, which is not necessarily the case for spontaneously injected narratives. The distinction between the two vehicles is fundamental to each approach for integrating narratives and stories in science instruction. A "narrative," as already defined, consists of the recounting of actions produced by characters, while a story is more rigidly defined by its structure—that the events that happen, revolving around the protagonist, must be interrelated and consequential (i.e., have a consequence for the protagonist and, possibly, others).

Both of these approaches also concern the act of storytelling itself. This chapter deals primarily with story *crafting*, as opposed to story *telling*. The latter consists of the performance aspect of storytelling where the narrator becomes the performer and requires proficiency in the relevant performance art. While it is indisputable that the manner in which the story is told plays a vital role and is certain to influence the students' level of enthusiasm and possible engagement with the story, it is not within the scope of this discussion to deal with the oral presentation of stories as it is presented in the literary or theatrical tradition (learning how to become an effective storyteller, both in terms of retelling existing stories and spontaneously creating new stories). For the storyteller of the spontaneously injected narrative, the challenge of telling the story from a first-person participant perspective is enormous, especially when coupled with the fact that this method, when dealing with historical information, already presupposes expert knowledge of the history of science on the teacher's part. Conversely, the planned injection of scripted stories demands that the teacher must become aware of the existence of available science stories that pertain to particular science concepts and be able to integrate them meaningfully and strategically. This can be performed by a novice teacher.

Both the story-permeated and scripted-story approaches share a facet of the narrative act that is largely beyond the control of the teacher, other than that he or she may provide the stimuli for it, and that element is the inner narrative taking place in the student's mind in response to any new learning from the narrative itself or the embedded concepts. This inner dialogue is affected by the manner in which the story is told or read and may influence the level of student engagement and learning. The inner dialogue, which may focus on the storytelling act alongside the story, is complex, spontaneous, and beyond the storyteller's control.

47.5 Theoretical and Pedagogical Reasons for Using Stories

There is good evidence that in order to engender meaningful learning, it is essential that teaching and learning methods be imbedded in appropriate contexts (Kenealy 1989; Martin and Brouwer 1991; Roth and Roychoudhury 1993). Historical contexts address the "why" and "how" aspects of the development of science in a way that includes the scientists as living persons who are concerned with personal, ethical, sociological, and political issues. It is generally accepted that this form of presentation is likely to engender increased motivation in students. Such historical materials must not consist of mere chronologies but rather expose the settings in which discoveries were made in the form of stories (Stinner et al. 2003; Metz et al. 2007). The use of stories to teach science has both theoretical and evidential support apart from the contextual argument.[8] It is the literary story form, in particular,

---

[8] See Egan (1986, 1989a), Hellstrand and Ott (1995), Kubli (2005), Miall and Kuiken (1994), and Norris et al. (2005).

that is known to produce consistent affective engagement (Miall and Kuiken 1994). This engagement also has a physiological basis. According to Miall and Kuiken, narrative techniques in the literary story "accentuate... activity in cortical areas specialized for affect" (1994, p. 392). The literary story adheres to the normal considerations in literature while making use of scientific and historical materials for its construction. Teachers who use such stories hope to capitalize on affective arousal in the form of increased student motivation.

The important role of the emotions in learning has been recognized only recently. Relevant, in this regard, is the research of neurobiologist Antonio Damasio on emotion and rationality. Damasio studied human subjects who had lost the ability to communicate information about emotions from one part of the brain to the other. He was able to support his hypothesis concerning the relationship between emotion and reason, which states that the emotions act as an arbitrator in rational decision-making and that without access to one's emotions, it is impossible to plan and make rational decisions (Damasio 1994). Educator Douglas Barnes demonstrates, in a research study of student group learning, that "unless pupils are willing to take the risk of some emotional commitment they are unlikely to learn" (1992, p. 87). Cognitive psychologist Pierce Howard, in a popular review of current neurobiological research, further explains the role of emotions in learning this way: "Experience arouses emotion, which fixes attention and leads to understanding and insight, which results in memory" (Howard 2000, p. 549).

At issue is the means by which emotion could be aroused in an appropriate manner in the teaching and learning situation. It is well established that stories have the ability to engage the emotions. Educator Kieran Egan (1989a, b) has long advocated the story form as the principle method of engaging students' emotions. Egan argues that the presentation of curriculum content through stories stimulates the imagination and evokes emotional response, thereby producing learning that more easily assimilates with long-term memory than learning produced by drilling and memorizing. The listener or reader engages with the story because she or he is encouraged to participate vicariously in the experiences of the protagonist. The kind of motivation produced by story is intrinsic, as opposed to the extrinsic motivation produced by a prescriptive teaching and learning episode (Mott et al. 1999). The story also provides an organizing structure for related knowledge and experiences (Mandler 1984).

47.5.1 Story as Reenactment of the Learning Process

It is noteworthy that the causal relationship of elements in a story (Fig. 47.1), which was described in Sect. 47.3.1.4, is structurally analogous to that of a type of conceptual change in the learning process (Fig. 47.2) (Klassen 2010). The schematic representation of conceptual change of Fig. 47.2 elaborates a temporal learning episode. The evolution of an event such as a scientific phenomenon (the learning episode) begins with the observation of the properties of the object (or entity) in question—the original state. The learner then observes a change in the state or properties of

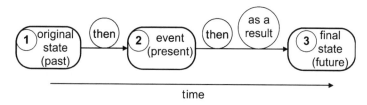

Fig. 47.2 A temporal framework for conceptual change (From Klassen 2010)

the object. The enigma in the learning process is that the learner does not know ahead of time what the final state will be. There are unseen properties of the situation that resulted in or contributed to the change in state. Curiosity or the need to uncover and explain such hidden properties initiates the learning process. Provided that there is sufficient motivation, the learner will attempt to construct explanations for such unanswered questions, usually through a process of reenactment of the event on the basis of some explanatory mental model. The process can, for example, be applied to the experiment of Ohm and his subsequent attempt to explain it:

> The deflection of the magnetometer on Ohm's apparatus indicated a particular resistance to the flow of electricity;
> then a different length of wire was inserted into Ohm's apparatus;
> as a result, the deflection of the magnetometer on Ohm's apparatus indicated a different flow of electricity.

The process here consists of two experimental observations (states) with the experimenter's intervention between them. Experimental intervention often proceeds by an adjustment of one of the parameters. In order to construct a theoretical expression to explain his observations, Ohm wrote an equation that he already had good reason to believe explained the observations. Ohm replayed his series of experiments on paper by first assuming that the current flow through the wires was analogous to the flow of heat, which he understood. As the historical record shows in Ohm's own words (Magie 1965), Ohm was able to reproduce his experimental results by recalculating his model with the parameters set at suitable values, showing that his explanation was a good one. In this way, Ohm justified his theory. Ohm's method can be viewed as the process of reenactment of his theoretical expression using suitable parameters that resulted in a reproduction of the experimental observations. Clearly, the minimal story sequence of Fig. 47.1 is followed; however, the usual requirement that a story involve the state of a human agent is not met. What changes in the case at hand is an experimental observation. Of course, when the subject of explanation is an abstract concept, then the involvement of human agents occurs at a secondary level.

Klassen (2010) shows that just as the story form consists of a temporal sequence, some of the learning involves a similar temporal sequence. The story form will be most useful for learning if it becomes a part of the sensemaking process and contributes to the formation of long-term memory structures. In Ohlsson's model of sensemaking (1999), the learning process includes the reenactment of remembered events in light of new knowledge. The story form can similarly be viewed as a reenactment of the cognitive process that would normally be involved in order to learn the story content. The story consists of the recounting of a chronology of events that includes causative links between succeeding events. Remembering, retelling, hearing, or reading a story will then serve as a form of mental reenactment.

The structure of a story is just one part of the design of a good story. Other features of stories are, for example, the effect of the untold, suspense, irony, and lifelikeness, which might be designated as the story's literary qualities. These features tend to stimulate the emotions of the reader or listener. An artistically crafted story arouses emotions that, in turn, contribute to the integration of the story details with long-term memory (Klassen 2010). A well-crafted story should, therefore, contribute to better learning of a variety of content knowledge items. This is what anthropologists have maintained all along. In the distant past, stories were used as the most effective device for passing on the culture to succeeding generations (Levi-Strauss 1966). In the context of learning science, there is good reason to utilize the story as a productive application of this enduring method.

47.5.2 Story as a Stimulus for Romantic Understanding in Science

As established in Sect. 47.2, relatively little has been written or researched regarding the construction of stories in regard to their relationship to effective learning. A notable exception is the learning theory of Kieran Egan that advances the concept of romantic understanding. Romantic understanding, which by definition takes place in a humanistic context, may be defined as the ability to grasp the meaning of the features of subject matter in a manner that tends to be idealistic in expectation and glamorously imaginary, possibly even exotic and involving the potential for heroic achievement. During the period of their development when they exhibit romantic understanding (approximately between ages 8 and 15), children are attracted, for example, to literary characters who do heroic, but possible, things.

Egan identifies five vital means by which children at this stage make sense of the world and of experience and by which they mediate between the world and the mind. These means, which he calls "cognitive tools," become the specific characteristics of romantic understanding. They are (1) the humanization of meaning, arising from the realization of the humanistic dimension of all knowledge; (2) an association (even identification) with heroes and heroic qualities; (3) a focus on and confrontation of the extremes and limits of reality and experience; (4) a sense of wonder; and (5) the contesting of conventions and conventional ideas.

Hadzigeorgiou and colleagues (2012), in order to reflect these tools and the instructional strategy in the concept itself, conceive of romantic understanding as "the motivating insight that emerges through the combined engagement of the emotions and the intellect in response to a specialized text" (p. 1114). The instructional strategy of the story is the "specialized text," which specifically incorporates the various cognitive tools.

The humanization of meaning in the story is a natural outgrowth of the fact that it is human endeavor which generates scientific knowledge and that human emotions are an integral component of its creation. The human element in the science content is incorporated through the scientist's life and work. In Tesla's case, for instance, it was his personal ambitions, his humanitarian ideals, his uncommon ingenuity, and his frustration with the establishment that together contributed to his ultimate achievements (Hadzigeorgiou et al. 2012). The resulting captivation with Tesla is rooted in his admirable character traits and abilities. It is this aspect with which students identify and which inspires them to emulate notable scientists because they come to understand that they, too, can develop such qualities as they conceive of new human possibilities through the story.

Students are further attracted to the stories because they exemplify the extremes of physical reality, whether they manifest in nature, the lowest temperature or the smallest electric charge, or in the human experience, the fastest athlete or the longest flight. These extremes, along with the many fascinating and mysterious phenomena and astonishing ideas that abound in science, evoke a sense of wonder. Through this kind of experience, students become aware of their incomplete or mistaken knowledge, helping them see the science lesson in a new light and emotionally charge the new information to be learned. Another source of fascination with the story arises from the fact that many scientists had to struggle against firmly held beliefs and ideas that stood in direct opposition to their new-found knowledge. When learning about the obstacles that scientists, such as Galileo, Boyle, Volta, Tesla, Einstein, and Marconi, faced and that they stood firm in their convictions, students can realize the strength of human convictions which bring the human element into the science content.

47.6 Constructing and Utilizing Stories

47.6.1 Utilizing History as the Raw Material for Science Stories

The writing of a story that is meant to utilize history of science cannot proceed without considering what interpretation of history is to guide the selection and adaptation of historical materials. History of science is subject to a broad spectrum of possible interpretations. One end of the spectrum is what Herbert Butterfield (1931/1959) called the Whig approach to history in which history of science is viewed in light of current knowledge. Implicit in this approach is the assumption that current knowledge is superior to the knowledge of past scientists. Critics of the Whig approach object to applying current days' standards to history because historical figures operated in a different environment from that of today, with different assumptions and standards. There are also internal histories of science written primarily by scientists, some of whom participated in the events about which they wrote many years later. The purposes of such histories are to legitimize the science, to aid in the socialization of novices, and to provide exemplars that will be used as

Table 47.2 An evaluative list of features for the construction of a science story

Character(s) taken from history of science
Actions that are consistent with the historical record
Situations or states
Consequential coherence of the characters' actions
Past time
Plot structure with rising action and climax
Critical choice made by the main character
Appropriate science and NOS content

models for problem-solving (Kragh 1987). Internal history often provides an official version of the roots of the discipline that tends to romanticize the events and portray science as an inevitable consequence of the force of progress. Exposing students only to this version of history encourages a distorted view of the nature of science, not to mention of the history, itself. The other end of the spectrum of approaches is the localized view in which history is interpreted only in light of the knowledge and context of the time and place in question. This approach, referred to as horizontal history by Mayr and diachronical history by Kragh, has been criticized on the grounds that history cannot be interpreted when comparisons to the larger context cannot be made (Kragh 1987; Mayr 1990). Furthermore, it has been claimed that purely diachronical history, consisting of a chronology of events restricted to the local context (Kragh 1987; Mayr 1990), is uninteresting to the nonspecialist.

For the purposes of writing a story to serve as an introduction to, or framework for, the teaching of a topic in science, aspects of all of the historical interpretations mentioned may be present to a certain degree. Certainly, the overriding consideration will be to portray the history accurately by using the best original and secondary sources. Any account must also be sensitive to the practices, beliefs, and social mores of the time, albeit, taking into account the limited identification with these on the part of current-day students. Any story arising from the history must be sensitive to such possible areas of misunderstanding by students, all the while not implying current superiority. Of course, the history of an event in the discipline that has been written for students of the discipline cannot help but provide some degree of legitimization and socialization. The goal is to portray scientists as human beings, "warts and all" (Winchester 1989, p. v), in order to give students the opportunity to become affectively involved in the story of science. Usually, the listeners to such stories will have a substantial degree of empathy for the protagonists of the stories.

47.6.2 Guidelines for the Construction of a Story

Story construction is a creative process and should not be formulaic or constrained by a fixed format; however, it must contain essential elements that meet the criteria for a story. A framework for the writing of a science story as described in this chapter should be guided by its characteristic features. The elements that were elaborated in Sect. 47.3 comprise an evaluative list and are summarized in Table 47.2. These can be taken as guidelines against which to judge any science story.

47.6.3 A Framework for Incorporating Stories in Science Teaching

In the case of science teaching, the type of instruction being described here corresponds to *contextual science teaching* (Klassen 2006). The contextual method is described, in detail, elsewhere (Klassen 2006) and is called the story-driven contextual approach (SDCA). A brief overview of the method and its framework will be provided here. Klassen identifies five important contexts that are a part of effective science lessons. These are the (1) practical, (2) theoretical, (3) social, (4) historical, and (5) affective contexts. Each of these contexts should relate, in some way, to the scientific concept being taught and contribute to the overall meaningfulness of the learning episode and to the degree of engagement of the students.

47.6.3.1 The Practical Context

The practical context provides hands-on, laboratory-style investigative activities for the students. Ultimately, the practical context is meant to replace traditional "labs" in the "normal science" curriculum. Students who are potential scientists benefit from practicing as novice scientists, since they are only one step removed from being apprentices. Even students who have no intention of becoming scientists benefit from participating in an authentic activity that includes creativity and some intellectual challenge beyond guessing what the lab manual wants.

47.6.3.2 The Theoretical Context

Each lesson should, if possible, also contain a theoretical element. Research or paper-and-pencil type questions that can be formulated in a brainstorming session after the presentation of a science story tend to be meaningful and memorable for students. Here students are challenged to answer conceptual "why" and "how" questions about the issues raised by the story. In a typical class, students should perform both practical and theoretical investigations.

The manipulation of ideas in the theoretical context is meant to replace paradigm exemplars of the type that students learn in normal science education. Normal science education relies heavily on end-of-chapter questions to provide student learning experiences. These problems are usually contrived and remote from students' life experiences. In contrast, problems in the theoretical context emerge as a natural necessity in the course of investigations. Ideas or concepts take on meaning as they are naturally generated by the context. The theoretical context, however, is dependent on the practical context to provide a well-rounded learning opportunity. As in the practical context, students are cast into the role of being novice researchers.

47.6.3.3 The Social Context

The benefit of cooperation in assisting learning is a relatively uncontroversial fact, and the ability to work together productively in small groups is a skill recognized by science curriculum developers. The benefits are likely to accrue not only in the form of improved learning of academic content but also in the learning of scientific and life skills related to social organization and leadership. In our context, making use of the benefit of cooperative learning requires careful attention to the structuring, organizing, and evaluating of group activities that are a part of the larger context. Working productively in groups is a developmental process which can significantly enhance the individual student's level of learning. It is possible for the social context to be applied to assessment by having group oral presentations to the entire class after they have completed their practical and theoretical investigations.

47.6.3.4 The Historical Context

The historical context is the basis of the science story, as discussed, in depth, above. It will portray science in a more realistic and humanistic light and make learning science a more attractive endeavor for most students.

In view of the preceding discussion, history of science that is to be used for pedagogical purposes must tread a fine line through the pitfalls of extremes that could conceivably arise in interpreting history. Obviously, the origins of ideas must relate to the current understanding, which is the point from which history must, of necessity, be approached in education. A merely logical reconstruction of past events that produces pseudo-history must be avoided. History must be placed in its original context, while relating it to our current views, in a manner that respects the originators and portrays them in a fair and balanced way. The objective of accuracy or faithfulness to the historical record must, in turn, be balanced against the demands of a curriculum that limits the depth to which the history can be probed. It should be realized that the place of history is not only to make a conceptual point but also to introduce the humanistic element into the process of learning science. Portraying scientists as human beings and giving students the opportunity to become affectively involved in the story of science are worthy goals in themselves.

47.6.3.5 The Affective Context

The affective context is provided in numerous ways, initially by the story itself. Good stories are known to engage the emotions and enhance memory. The ability of the story to rouse emotions has already been discussed earlier in this chapter. While the affective context must be recognized, it is not a discrete context, as the social and practical contexts, for example, through aspects such as group work and hands-on activities, can play a notable role in generating student interest and contribute to

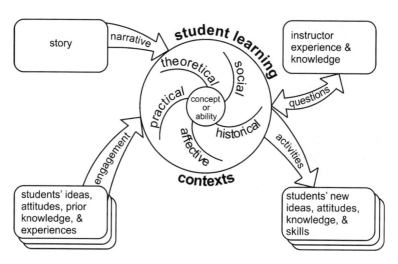

Fig. 47.3 A schema for incorporating stories in science teaching: the SDCA

increasing their motivation and level of participation. It is influenced further by the teacher, not only in her or his posing of questions, which may engender curiosity and wonder, but also in the telling of the story and the design and execution of the instructional activities.

47.6.3.6 Operationalizing the SDCA Framework

The story may be used as a starting point for a lesson, or it may be used in segments to motivate a series of student activities. Figure 47.3 (adapted from Klassen 2006) presents the process of integrating the science story into a lesson. The teacher supplies the story as a motivating, organizing, and contextualizing structure for the lesson. Students naturally bring their prior ideas, attitudes, knowledge, and experiences to the learning episode, which may be relevant to the concepts or abilities addressed. The presence of the various contexts will assure a heightened degree of engagement with the material of the lesson. The teacher contributes supervision in the role of a research leader, establishing the direction of the activities and acting as a resource for required information. After any successful lesson or series of lessons, students will have acquired various new ideas, attitudes, knowledge, and skills relating to the concepts or abilities that are being promoted by the lesson, which will influence the next lesson or series of lessons in an ongoing cycle.

Stories when used as door openers (Kubli 2005) are not used for the primary purpose of explanation but for the purposes of making the science concept being taught more memorable, reducing the distance between teacher and students and illuminating a particular point being made (Kubli 2005; Metz et al. 2007). Door-opening science stories provide "reasons for needing to know." Another, perhaps more significant, purpose behind such stories is to raise questions or leave the student with unresolved problems or issues which form a significant part of the science material being taught. These questions arise not only from the story itself but also from the scientific issues and concepts that the science story contains. According to Gil-Pérez and coauthors (2002), questions play a central role in constructivist pedagogy. In their words, "[f]rom a scientific point of view it is essential to associate knowledge construction with problems: as Bachelard (1938) stresses, 'all knowledge is the answer to a question'" (p. 14). One would then expect that well-told stories would provide an incentive for students to raise a number of questions that they consider both interesting and important.

47.6.3.7 The Science Story as a Stand-Alone Implementation of the SDCA

The five contexts as identified in the SDCA are inherent in science instruction that incorporates well-designed stories in a well-conceived manner. The historical and theoretical contexts rest in the story itself. Since the science is elaborated within the story text, the practical context can be generated through the students' imagination, as shown in the Hadzigeorgiou and colleagues' (2012) research study. It has also already been demonstrated that the affective context emerges in the response to the story in the form of engagement, motivation, and long-term retention. Given that the told story raises questions in the students' minds and produces ongoing group discussion, the social context is developed within the learning environment, as well. The degree to which this is developed is largely in the hands of the teacher who determines in what kind of cooperative learning experiences his or her students will engage.

47.7 Conclusion

Theories of narrative are diverse and do not necessarily precipitate a set of common features of narrative. Because of the complexity and different understandings of narrative and story, a distillation of the common elements among various expositions of narrative theory does not result in a definition that can be applied practically in science education. The analysis of the literature and recent research reveals that the writing of good stories must focus on the structural elements of the story and include the relevant narrative features. The construction of such a list derives particularly useful insights from the narrative theory of Altman (2008) who emphasizes character and distinguishes elements relevant to the writing of a story. The story centers on the character who, in our case, is a real scientist, and the details are provided predominantly by the historical record. The consequential choices made by the scientist have the effect of raising human interest that, for example, stimulate the romantic understanding of students, as demonstrated in the study of Hadzigeorgiou and colleagues (2012). Such stories should also include an interwoven elaboration of the related science concepts. Provided that the story is well written, students can gain, from the told story alone, all the conceptual knowledge that they would have gleaned from conventional classroom instruction and demonstrate improved learning and better long-term retention of knowledge (Hadzigeorgiou et al. 2012). A story containing all the characteristics required to produce an effective contextual learning environment is also likely to rest in or produce the contexts essential to ensure sound science instruction. While the historical and affective contexts are always supplied by the story, teachers can opt to supply the theoretical, practical, and social elements separately from the story when they use it in the role of door opener to the instructional episode in conjunction with the questions that are naturally raised by the story.

Although the past decade has seen significant development in the definition and practical implementation of stories in the science classroom, the definitional structure of story that has been presented here requires wider dissemination and more diverse application. In order for research among various researchers to be comparable, there needs to be a standardized vehicle or framework of story structure and delivery, and the current chapter provides one such implementation method. Without a certain degree of standardization, it will not be possible to judge the effectiveness of the use of stories in science instruction nor to assess the anticipated progress in this research area.

Recent research on the effectiveness of stories in science instruction has already raised topics for further research; for example, the relative effectiveness of the story approach in different curricular contexts, in different national and cultural settings, and among different age groups of students needs to be investigated. The implementation process, of itself, raises additional research imperatives. Among these are the assessment of existing so-called stories, the writing of new science stories that correlate to specific curricula, and the collection and dissemination of such science stories that meet the criteria established in this chapter.

Acknowledgment The writing of this chapter has been supported by a research grant from the University of Winnipeg.

References

Altman, R. (2008). *A theory of narrative*. New York: Columbia University Press.
Arons, A. B. (1988). Historical and philosophical perspectives attainable in introductory physics courses. *Educational Philosophy and Theory, 20*(2), 13–23.
Avraamidou, L., & Osborne, J. (2009). The Role of Narrative in Communicating Science. *International Journal of Science Education, 31(12),* 1683–1707
Bachelard, G. (1938). *La Formation de l'esprit scientifique*. Paris: Vrin.
Barnes, D. (1992). *From Communication to Curriculum, 2nd ed*. Portsmouth, NH: Boynton/Cook Publishers.
Bruner, J. (1986). *Actual minds, possible worlds*. Cambridge, MA: Cambridge University Press.
Bruner, J. (1996). *The Culture of Education*. Cambridge, MA: Harvard University Press.
Butterfield, H. (1931/1959). *The whig interpretation of history*. London: Bell and sons.
Clough, M. P. (2011). The Story Behind the Science: Bringing Science and Scientists to Life in Post-Secondary Science Education. *Science & Education, 20(7–8),* 701–717.

Coleman, L. & Griffith, D. (1997). Physics in context: An IUPP course. In E. F. Redish and J. S. Rigden (Eds.), *The changing role of physics departments in modern universities: Proceedings of ICUPE*. The American Institute of Physics.

Collingwood, R. G. (1945). *The Idea of Nature*. Oxford: Clarendon Press.

Connors, R. J. (1981). The rise and fall of the modes of discourse. *College Composition and Communication, 32*(4), 444–455.

Damasio, A. R. (1994). *Descartes' error: emotion, reason, and the human brain*. New York: G.P. Putnam.

Egan, K. (1986). *Teaching as story telling*. London, Ontario: Althouse Press.

Egan, K. (1989a). Memory, imagination, and learning: Connected by the story. *Phi Delta Kappan, 70*(6), 455–459.

Egan, K. (1989b). The Shape of the Science Text: A Function of Stories. In S. de Castell, A. Luke, & C. Luke, (Eds.), *Language, Authority and Criticism: Readings on the School Textbook* (pp. 96–108). New York: The Falmer Press.

Gil-Pérez, D., Guisásola, J., Moreno, A., Cachapuz, A., Pessoa de Carvalho, A.M., Martínez Torregrosa, et al. (2002). Defending Constructivism in Science Education. *Science & Education, 11*, 557–571.

Gorman, M. & Robinson, J. K. (1998). Using history to teach invention and design: The case of the telephone. *Science & Education, 7*(2), 173–201.

Hadzigeorgiou, Y. (2006). Humanizing the teaching of physics through storytelling: the case of current electricity. *Physics Education, 41(1),* 42–46.

Hadzigeorgiou, Y., Froese Klassen, C., & Klassen, S. (2011). The Genius who Lit and Transformed our World. Unpublished manuscript.

Hadzigeorgiou, Y., Klassen, S., & Froese Klassen, C. (2012). Encouraging a "Romantic Understanding" of Science: The Effect of the Nikola Tesla Story. *Science & Education, 21*(8), 1111–1138.

Hellstrand, A. & Ott, A. (1995). The utilization of fiction when teaching the theory of relativity. *Physics Education, 30* (5), 284–286.

Holbrow, C. H., Amato, J. C., Galvez, E. J., & Lloyd, J. N. (1995). Modernizing introductory physics. *American Journal of Physics, 63*(12), 1078–1090.

Howard, P. J. (2000). *The owner's manual for the brain: Everyday applications from mind-brain research (2nd ed.)*. Austin: Bard Press.

Isabelle, A. D. (2007). Teaching Science Using Stories: The Storyline Approach. *Science Scope, 31(2),* 16–25.

Kenealy, P. (1989). Telling a coherent "story": A role for the history and philosophy of science in a physical science course. In D. E. Herget (Ed.), *HPSST, Proceedings of the First International Conference* (pp. 209–220).

Klassen, S. (2006). A theoretical framework for contextual science teaching. *Interchange, 37*(1–2), 31–62.

Klassen, S. (2007). The Application of Historical Narrative in Science Learning: The Atlantic Cable Story. *Science & Education, 16*(3–5), 335–352.

Klassen, S. (2009a). The construction and analysis of a science story: A proposed methodology. *Science & Education, 18*(3–4), 401–423.

Klassen, S. (2009b). Alessandro Volta: Electricity Without Limits. Unpublished manuscript.

Klassen, S. (2010). The relation of story structure to a model of conceptual change in science learning. *Science & Education, 19*(3), 305–317.

Klassen, S. (2011). The Photoelectric Effect: Reconstructing the story for the physics classroom, *Science & Education, 20*(7–8), 719–731.

Klassen, S., & Froese Klassen, C. (2012). Augustin Mouchot: The Soul of Solar Energy. Unpublished manuscript.

Knox, J., & Croft, P. (1997). Storytelling in the meteorology classroom. *Bulletin of the American Meteorological Society, 78*, 897–906.

Kokkotas, P., Rizaki, A., & Malamitsa, K. (2010). Storytelling as a strategy for understanding concepts of electricity and electromagnetism. *Interchange: A Quarterly Review of Education, 41(4),* 379–405.

Kragh, H. (1987). *An introduction to the historiography of science*. Cambridge: Cambridge University Press.

Kubli, F. (1998). *Plädoyer für Erzählungen im Physikunterricht*. Köln: Aulis.

Kubli, F. (1999). Historical aspects in physics teaching: Using Galileo's work in a new Swiss project, *Science & Education, 8*(2), 137–150.

Kubli, F. (2001). Can the theory of narratives help science teachers be better storytellers? *Science & Education, 10*, 595–599.

Kubli, F. (2005). Science Teaching as a Dialogue – Bakhtin, Vygotsky and some Applications in the Classroom. *Science & Education, 14*(6), 501–534.

Kubli, F. (2006). Teachers should not only inform but also entertain. *Science & Education*, published online June 14, 2006, DOI 10.1007/s11191-006-9012-3.

Levi–Strauss, C. (1966). *The savage mind*. Chicago: University of Chicago Press.

Magie, W. F. (1965). *A source book in physics*. Cambridge, MA: Harvard University Press.

Mandler, J. (1984). *Stories, scripts, and scenes: Aspects of schema theory*. Hillsdale, NJ: Erlbaum.

Martin, B. E. & Brouwer, W. (1991). The Sharing of Personal Science and the Narrative Element in Science Education, *Science Education, 75*(6), 707–722.

Mayr, E. (1990). When is historiography whiggish? *Journal of the History of Ideas, 51*(2), 301–309.

Metz, D., Klassen, S., McMillan, B., Clough, M., & Olson, J. (2007). Building a Foundation for the Use of Historical Narratives. *Science & Education, 16*(3–5), 313–334.

Miall, D. S., & Kuiken, D. (1994). Foregrounding, defamiliarization, and affect response to literary stories. *Poetics, 22*, 389–407.

Milne, C. (1998) Philosophically Correct Science Stories? Examining the Implications of Heroic Science Stories for School Science. *Journal of Research in Science Teaching, 35(2)*, 175–187.

Mott, B. W., et al. (1999, November). Towards Narrative-Centered Learning Environments. *Proceedings of the AAAI Fall Symposium on Narrative Intelligence, November, 1999.*

Negrete, A., & Lartigue, C. (2010). The science of telling stories: Evaluating science communication via narratives (RIRC method). *Journal of Media and Communication Studies, 2*(4), 98–110.

Noddings, N., & Witherell, C. (1991). Epilogue: Themes remembered and foreseen. In C. Witherell & N. Noddings (Eds.), *Stories lives tell* (pp. 279–280). New York: Teachers College Press.

Norris, S. P., Guilbert, S. M., Smith, M. L., Hakimelahi, S., & Phillips, L. M. (2005). A theoretical framework for narrative explanation in science, *Science Education, 89*(4), 535–554.

Ohlsson, S. (1999). Theoretical commitment and implicit knowledge: Why anomalies do not trigger learning. *Science & Education, 8*(5), 559–574.

Roth, W. & Roychoudhury, A. (1993). The development of science process skills in authentic contexts. *Journal of Research in Science Teaching, 30*(2), 127–152.

Shrigley, R. L., & Koballa, T. R. (1989). Anecdotes: What research suggests about their use in the science classroom. *School Science and Mathematics, 89*(4), 293–298.

Solomon, J. (2002). Science stories and science texts: What can they do for our students? *Studies in Science Education, 37*(1), 85–106.

Stinner, A. & Williams, H. (1998). *History and philosophy of science in the science curriculum.* In K. G. Tobin & B. J. Fraser (Eds.), International Handbook of Science Education (pp. 1027–1045). Boston, MA: Kluwer.

Stinner, A. (1992). Contextual teaching in physics: From science stories to large-context problems. *Alberta Science Education, 26* (1), 20–29.

Stinner, A. (1995). Contextual settings, science stories, and large context problems: Toward a more humanistic science education. *Science Education, 79*(5), 555–581.

Stinner, A., McMillan, B., Metz, D., Jilek, J., & Klassen, S. (2003). The renewal of case studies in science education. *Science & Education, 12*(7), 617–643.

Tao, P-K. (2003). Eliciting and developing junior secondary students' understanding of the nature of science through a peer collaboration instruction in science stories. *International Journal of Science Education, 25(2)*, 147–171.

Wandersee, J. H. (1990). *On the value and use of the history of science in teaching today's science: Constructing historical vignettes.* In D. E. Herget (Ed.). More history and philosophy of science in science teaching (pp. 278–283). Tallahassee, FL: Florida State University.

Wandersee, J. H. (1992). The historicality of cognition: Implications for science education research. *Journal of Research in Science Teaching, 29*(4), 423–434.

Winchester, I. (1989). Editorial—History, science, and science teaching. *Interchange, 20*(2), i–vi.

Cathrine Froese Klassen (B.A. Hons., B.Ed., M.A.) is an Instructor in the Faculty of Education at the University of Winnipeg, with a teaching assignment in Gifted Education. She began her career as a high-school teacher. Besides holding key leadership roles in provincial curriculum committees and professional associations in the Gifted Education, German, and English areas, she was English Department Head, Advanced Placement Coordinator, and Coordinator for Gifted and Talented Programming and Critical Thinking in her school division in Winnipeg. From 2007 to 2011, she held the position of Executive Administrator of The World Council for Gifted and Talented Children at its Headquarters at The University of Winnipeg. She has taught courses in Gifted Education, German Language, and German Curriculum, Instruction, and Assessment at the University of Manitoba and the University of Winnipeg. Her publications include two volumes of conference proceedings, as editor, and numerous articles in the field of gifted education, as well as in science education, particularly on the use of stories in teaching science.

Stephen Klassen is Associate Professor at The University of Winnipeg where he teaches physics and holds an appointment in the Faculty of Education. His Ph.D. (University of Manitoba) is in Science Education, and his background is in experimental physics. Dr. Klassen's current research is on the writing, analysis, and use of stories (from the history of science) in science teaching and the analysis of physics textbooks for their portrayal of the history of science. His work, in part, is published in *Science & Education* (11 papers), *Science Education*, *Interchange*, and the *Canadian Journal of Science, Mathematics, and Technology Education.* Since 1997, he has presented papers at the International Conference on History of Science in Science Education (ICHSSE) and has contributed significantly to its organization, especially in cochairing the Planning Committee. As of 2001, he has been actively involved in the International History, Philosophy, and Science Teaching Group (IHPST) by having presented at each of its conferences, having served as Program Chair in 2003, and joining its governing Council in 2010 and the Editorial Committee of *Science & Education* the following year.

Chapter 48
Philosophical Inquiry and Critical Thinking in Primary and Secondary Science Education

Tim Sprod

48.1 Introduction

This chapter considers two related, yet nonidentical, issues in science education: the role that can be played by philosophical inquiry and the place of critical (or indeed other types of) thinking.

The relationship between the two issues is clear, if you accept the claim by Lipman (1991) that philosophy is the discipline whose central concern is thinking. On this basis, he argues that philosophy should, therefore, be taught in all schools as a means of (among other things) improving the thinking of school students. Broadly accepting Lipman's claim, this chapter assesses its strength in relation to science teaching in schools.

This chapter starts with consideration of educational research into the nature of thinking, exploring the extent to which it is context bound, the problem of transfer across contexts and the nature of excellent thinking. Drawing on this work, I then turn attention to thinking within the context of science education. We shall see that the evidence suggests that present methods of science education, by and large, are not as good as they could be in developing the scientific thinking of students.

Next, I survey three programmes which have shown considerable promise in not only improving the scientific thinking of school students but also in having a positive impact on other desirable outcomes of science education: the Cognitive Acceleration through Science Education (CASE) programme, the Conceptual Challenge in Primary Science project and the Philosophy for Children (P4C) suite of approaches. Behind the success of these approaches seem to lie some common features: they provoke puzzlement or cognitive conflict; they depend heavily on a certain style of classroom discussion which is recognizably philosophical; they make metacognition explicit; and they have ways of encouraging transfer to other contexts.

T. Sprod (✉)
School of Philosophy, University of Tasmania, Hobart, Australia
e-mail: timsprod3@gmail.com

M.R. Matthews (ed.), *International Handbook of Research in History,*
Philosophy and Science Teaching, DOI 10.1007/978-94-007-7654-8_48,
© Springer Science+Business Media Dordrecht 2014

Finally, this chapter draws together the matters discussed, looks at some possible objections and considers the implications in four areas: for future research on effective programmes for scientific thinking and the other outcomes mentioned above, for investigating the impact of attributes of teacher on such programmes, for the provision of training for teachers in the delivery of effective programmes and on the need for the development of more materials to support teachers.

48.2 Thinking

48.2.1 Is Thinking General or Context Bound?

The first issue that arises is the generality of the good thinking at which we aim. Ever since Spearman (1927) introduced the notion of general intelligence g, there has been disagreement among theorists as to whether good thinking is a generic capacity (see, e.g. Ennis 1989, 1990), able to be turned to any discipline or whether good thinking is embedded in particular contexts or situations (McPeck 1990) so that a good thinker in science may not be able to think well in other disciplines. On the answer to this puzzle rests a more practical issue: should schools make space in the curriculum for separate 'thinking' classes, or should thinking be taught within each discipline? It should be noted that this is not an 'either-or' situation: it is possible that both are needed.

More recently, the discussion about whether thinking is domain specific or not has become more nuanced. No longer do theorists and experimental psychologists line up to defend either side. Rather, there seems to be more agreement that there is truth in both accounts. In a recent survey, Adey and colleagues (2007, p. 78) state that '[w]e can find no support in the current psychological literature for either of these extreme views... What emerges from the vast literature that exists on the structure of general abilities [is] that both general and specialised processes are interwoven in the human mind'. Despite this convergence, there still exists considerable difference as to the nature of these two and of the ways in which they interact. Perkins and Saloman conclude:

> The approach that now seems warranted calls for the intimate intermingling of generality and context-specificity in instruction. ... We forecast that wider-scale efforts to join subject-matter instruction and the teaching of thinking will be one of the exciting stories of the next decade of research and educational innovation. (Perkins and Saloman 1989, p. 24)

To claim that there is such a thing as general thinking ability (or general cognitive ability, or intelligence) is not to deny that all thinking is contextual, in the sense that it must take place in some context, using some content. Rather, it is to claim that there is a core set of capacities and dispositions which have the potential to be applied across a large variety of contexts. These general abilities are then augmented by capacities that are more specialised to particular contexts and applications.

Moreover, there is growing agreement that complex thinking in a domain requires a reasonable grasp of content (Bailin 1998; Paul and Elder 1999; Willingham 2007).

In relation to education in science, two questions arise that will be addressed later in this chapter. Firstly, there is the question of transfer. If students develop the capacity to use a certain type of thinking well in one context, how easily can they transfer that capacity to different contexts? If such transfer does not happen spontaneously, then what are the conditions and classroom activities that will enhance such transfer? Secondly, there is the issue of plasticity. Do individuals have a genetically determined capacity for good thinking, or is their ability susceptible to environmental influences, including educational ones? Does the development of general cognitive capacities run according to a set timetable, or can educational interventions speed up their acquisition?

Before we turn our attention to these questions, we need to survey the ways in which good thinking has been characterised.

48.2.2 *What Is Thinking, and What Makes It Good?*

If we are aiming to improve the scientific thinking of school students, then we had better have a clear idea of the target. This is, however, less easy than it might seem on the surface, since there is considerable disagreement about what to call such thinking, let alone of what it consists. Lipman (1991), for example, in a chapter interestingly titled 'The cornucopia of cognitive performance', talks variously of higher-order thinking, good thinking and excellent thinking.

The title to this chapter has used what is probably the most popular name: critical thinking. The popularity of this name probably stems from John Passmore's highly influential article 'On teaching to be critical' (1967). Yet, within what might be called the critical thinking movement, there exists quite a bit of internal disagreement about how exactly 'critical thinking' should be characterised, particularly at the level of specific characteristics a critical thinker has.[1] Despite this diversity, different theorists often mention the same or similar elements, differing more on their arrangement or interrelations. A common thread that runs through many of these accounts is a distinction between capacities (or skills, or capabilities)[2] and dispositions (or habits, or traits). The former refer to cognitive moves that can be made, and the latter to the desire to make them. It is commonly claimed that a good thinker needs both.

The label critical thinking, however, has been questioned for an overly rational emphasis (de Bono 1986). Some of the critical thinking theorists – see, for example,

---

[1] See, for example, Ennis (1987), Paul and Elder (2007), Resnick (1987), and Siegel (1990).

[2] Some theorists object to the term 'thinking skills' (e.g. Hart (1993); see also Lipman (1991, pp. 78–80),where he discusses Hart), as reducing a complex, interwoven human activity to a series of atomistic technical skills. I will not enter into this discussion: in what follows, I will use 'skills', 'capacities', 'capabilities' and so on interchangeably.

Siegel (1990) – claim that critical thinking is a very wide concept, including all cognition and quite a bit beyond, such as the affective. Others prefer to confine the adjective 'critical' to the rational, convergent type of thinking and to refer to the more imaginative, divergent type as creative thinking (Swartz et al. 1998). In these accounts, it can seem as though there are two distinct types of thinking. Lipman (1991), whose favoured term is higher-order thinking, prefers to talk about the critical and creative *aspects* of higher-order thinking, saying 'there is no creative thinking that is not shot through with critical judgments, just as there is no critical judgment that is not shot through with creative judgments' (p. 193). He also added a further aspect – caring thinking (Lipman 1995) – to acknowledge the important role of the emotions in higher-order thinking. In my own work, I argue for the term 'reasonableness' and identify five aspects: critical, creative, committed (roughly equivalent to Lipman's 'caring'), contextual and embodied thinking (Sprod 2001).

Lipman (1991, pp. 50–51) makes the point that good thinking is contextual and nonhierarchical. By this, he means that skills which may seem to be low level in one context may be crucial in another. Further, a good thinker not only has the capacity to use skilful thinking and the disposition to do so but also must have the judgement and discrimination to recognise the appropriate moment to call upon each skill. Using the analogy of the carpenter, Lipman (1991) comments that it is not enough to know how to wield a hammer and a plane skilfully: the judgement to know which situations require a hammer, and which a plane, is also essential. Ritchhart and Perkins (2000, p. 30) analyse the dispositions needed to think well as having three aspects – the inclination to think well, the ability to do so and the sensitivity to know where and when to use those abilities – making what appears to be the same point in somewhat different language.

Further, Paul and Elder (2007) point out that thinking which is good by technical standards (pragmatically successful) can be 'intellectually flawed' if not grounded in 'fair-mindedness and intellectual integrity'. The former may merely involve 'the technical perfecting of isolated cognitive skills' (Lipman 1991, p. 51).

Characterising accurately the nature of thinking is well beyond the scope of this chapter, but the foregoing discussion has highlighted that, if we have the aim of teaching good thinking in science, we need to be aware that the sort of thinking we are seeking to encourage and inculcate is quite a complex set of capabilities and dispositions, backed by sound judgement. It is now time to turn our attention to a matter that merges the concerns of these last two sections: the nature of scientific thinking.

48.2.3 Scientific Thinking

We have seen above that there are differences of opinion as to whether good scientific thinking is good general thinking applied to science[3] or is a distinctive type of

---

[3] Huxley (1894) is often quoted giving his opinion that science is 'nothing but trained and organised common sense'. See also Royer (1987).

thinking.[4] Once again, many commentators seem to be converging on a view that scientists utilise many general thinking skills and dispositions but also have developed more specific ones that either relate more to scientific contexts or have been developed to a greater and more precise degree by scientists.[5]

Adey and colleagues (2007) make a powerful case that we can consider scientific (and indeed, all) thinking as a combination of central, shared competence (which they call general intelligence g) and a suite of subsidiary competences that are much more subject specific. Exactly how these are related is less certain, and they survey a number of models. We should note in passing here that it is hard to see how a sharp distinction can be drawn between the central and the subsidiary elements of thinking. Are some of them merely more sophisticated versions of general thinking (e.g. ability to discriminate), while others are rather different from any skills of general thinking (e.g. statistical inference)? Is there a difference in kind that separates the drawing of distinctions between characters in literary analysis and distinguishing between possible confounding variables in science?

Research scientists at the cutting edge have clearly developed their scientific thinking to a high degree. It is also clear that the precise mix of capacities used in expert scientific thinking will vary from one field to another within the sciences – the development of these subsidiary competences depends heavily on the requirements of the field.

Much work has been done on the acquisition of scientific thinking by children as they mature and are educated,[6] yet only a small proportion of it extends beyond the middle years of schooling. I am unaware of any study that traces in detail the development of scientific thinking in the transition from senior secondary school student to fully fledged scientist (though there are studies that take single-stage 'snapshots', particularly in early undergraduate years). Indeed, such a study would be a major one, requiring a very long-term longitudinal design. Moreover, much of the research focuses on, as Zimmerman (2007) puts it, 'the three major cognitive components of scientific discovery: Searching for hypotheses, searching for experiments (i.e., data or evidence from experiments or investigations more generally), and evidence evaluation'. Thus, the model of scientific thinking here is quite narrow, being based on a singular conception of The Scientific Method.[7] Nor is it clear that searching for

---

[4] Wolpert (1992) argues that 'science involves a special mode of thought and is unnatural for two main reasons … Firstly, the world just is not constructed on a common-sensical basis. This means that 'natural' thinking – ordinary, day-to-day common sense – will never give an understanding about the nature of science…. Secondly, doing science requires a conscious awareness of the pitfalls of 'natural' thinking. For common sense is prone to error when applied to problems requiring rigorous and quantitative thinking ….'

[5] See, for example, Adey et al. (2007), Adey (2006), Bailin (2002), Ennis (1990), Gazzard (1993), McPeck (1990), Norris (1992), Paul and Elder (2007), Perkins and Saloman (1989), Siegel (1989), and Siegel (1991).

[6] See, for example, Kuhn et al. (1988), Kuhn and Pearsall (2000), Zimmerman (2000), and Zimmerman (2007). The latter is a recent, comprehensive survey of such work.

[7] Many school science texts refer to what I am calling here 'The Scientific Method', which takes experimental work with tightly controlled variables as the model for all science (Lederman and Lederman 2004). See further discussion in Sprod (2011, pp. 4, 66).

possible explanations, seeking data to test them and evaluating the evidence adduced are skills that apply solely to science.

Searching ERIC with the descriptors 'science education' and 'thinking skills' returns 649 matches. Using the above terms with a variety of words derived from 'category', 'type' and 'taxonomy' revealed no empirical studies that tried to categorise the different scientific thinking skills that students develop through their science education. There were, however, quite a few studies[8] that took a particular thinking skill – or, occasionally, disposition – and studied the ways in which some intervention may strengthen it. Hence, researchers and teachers clearly have some typology of scientific thinking on which they draw.

The most comprehensive typology of science thinking skills that I could discover on the Internet (Table 48.1) is found on the Boston Museum of Science (2001) site. The single source that they quote for their list is the Californian State Department of Education (1990) document 'Science Framework for California Public Schools' (see Table 48.2). The members of the team who wrote the document appear to have compiled this list by drawing on their own experience and ideas, judging by the lack of references to any papers which have developed a typology of science thinking skills.

Again, a perusal of the lists in Tables 48.1 and 48.2 leads one to wonder if any or, at the very least, many of these skills are unique to science. Certainly, in their more developed guises, such skills as categorisation may become much more sophisticated and specialised than their everyday counterparts, but it is hard to see how their roots can be anywhere other than in more general thinking skills that can be applied across many domains. There appears to be a need for empirical work to delineate the thinking skills that scientists actually use in their work and to explore how they develop these skills through their education, right up to the level of employment as a scientist.

Hence, we may well feel that children's scientific thinking in the early years of education leans heavily on those shared competencies that are utilised in all the disciplines and that as they grow older and become more educated in science, not only will their general competence increase, but they will also start to develop some of the more context-dependent thinking competencies that apply more specifically to science. Science education, then, ought to have the dual roles both of strengthening the general ability to think well (a role it ought to share with the other disciplines) and of developing those modes of thinking which have more specific application in science.

48.2.4 Developing Scientific Thinking in Schools

In the light of compelling evidence (see, e.g. Lyons (2006), Pell and Jarvis (2001), Tytler (2007)) that students find school science boring, largely requiring the regurgitation of facts, there have been calls, as Tytler puts it, to 'reimagine' science

---

[8] For example, Venville and Dawson (2010) on argumentation and informal reasoning, Settelmaier (2003) on ethical dilemmas, May et al. (2006) on analogical reasoning and Vieira et al. (2011) on conceptual clarification.

48 Philosophical Inquiry and Critical Thinking in Primary and Secondary Science... 1537

Table 48.1 Science thinking skills typology (Adapted from Boston Museum of Science (2001))

Seeing the unseen – observation
Becoming active observers
Using all senses in making observations
Developing observational skills
Awareness of strengths and limitations of observations
Using tools to extend senses
Transferring observational skills to other contexts

Finding the pattern – classification
Observing, comparing and sorting objects and phenomena in meaningful ways
Recognising that systems for organising and classifying objects and phenomena reveal underlying
   meaning
Using organised material collections to answer questions and solve problems
Searching for the 'hidden' meaning in objects and phenomena

Making models – description
Recognising the presence and value of models
Becoming familiar with several types of models: physical models, conceptual models, mathematical
   models and computer simulations
Practising four specific science thinking skills associated with making and using models:
   Recognising the similarity between models and the things they represent
   Assessing the strengths and limitations of models in explaining and predicting the behaviour of
      the objects or phenomena they represent
   Using models to raise questions, communicate ideas and test hypotheses in many different
      contexts
   Creating models to explain things they cannot be observed directly

Testing the theory – experimentation
Formulating and testing ideas about the world
Asking questions and generating ideas
Formulating hypotheses
Gathering and weighing evidence
Using instruments to design experiments
Recording and interpreting data
Drawing conclusions
Being aware of what it means to experiment
Transferring scientific habits of mind (curiosity, respect for evidence, scepticism and open-
   mindedness) to other settings
Participating in ongoing scientific research

Putting it to work – application
[No list of component skills given]

Playing with ideas – imagination
[No list of component skills given]

education. One of the common phrases used in these calls is to make science lessons more 'minds-on' (Sprod 2011). Such a characterisation builds on a broad-brush history of science education which sees the early approach as 'ears-and-eyes-on' (lectures and demonstration experiments), to which, from the mid-twentieth century, was added 'hands-on' (student experimentation). Towards the end of the last

Table 48.2 List of science thinking skills adapted from the California Department of Education's *Science Framework for California Public Schools* (1990, p. 151)

Observing: the scientific thinking process from which fundamental patterns of the world are construed

Communicating: the scientific thinking process that conveys ideas through social interchanges

Comparing: the scientific thinking process that deals with concepts of similarities and differences

Ordering: the scientific thinking process that deals with patterns of sequence and seriation

Categorising: the scientific thinking process that deals with patterns of groups and classes

Relating: the scientific thinking process that deals with principles concerning interactions

Inferring: the scientific thinking process that deals with ideas that are remote in time and space

Applying: the scientific thinking process by which we use knowledge

century, in the light of more research (e.g. Jurd 2004; Abrahams and Millar 2008) that showed much practical work in schools was unaccompanied by little thought about the science or even the purpose of the experiment, the phrase 'minds-on' started to appear more frequently. Indeed, in much of the literature, the two latter phrases are often conjoined into 'hands-on, minds-on' (e.g. in Andre (1997), Jegede and Taylor (1995), and Pedersen and McCurdy (1992)).

So, even granted that we have some reasonably shared idea of what constitutes good scientific thinking, we can ask whether students learn this well by following a traditional science education or whether some new or different ('minds-on') peda-gogical approach needs to be added to science curricula. There is quite a bit of evidence that traditional methods are not enough (Kuhn 1999), though most of this evidence is in the form of studies that show some particular intervention has a positive effect on scientific thinking skills over and above that of traditional methods.[9] The implication, then, is that those traditional methods are failing to make the most of the possibilities for developing scientific thinking.

Many of these studies are quite small scale and carried out over a relatively short time frame. Hence, methodological questions arise about experimenter effects, confounding variables, generalisability to other contexts and so on. Moreover, it is not clear that all researchers are using key terms such as 'direct instruction', 'transfer', 'inquiry method', 'scaffolding' and 'conventional teaching' in the same sense (Zimmerman 2007, p. 215), making comparisons between studies difficult.

However, perhaps the major shortcoming of small-scale studies is that they do not deal with the most important question: even granted that some statistically significant effects can be demonstrated after intervention, are these effects long lasting? In other words, do they transfer to other contexts at much later times? Only if it can be demonstrated that certain changes to teaching and learning make a real difference to children's abilities over significant timescales and, in novel situations, will there be a strong case for making those changes.

---

[9] For example, Balcaen (2008), Cavagnetto et al. (2011), Choi et al. (2010), Dawson (2010), Hand and Choi (2010), Lee and She (2010), Miri et al. (2007), Mitchell (2010), Pithers and Soden (2000), Sadler (2004), She and Liao (2010), Songer et al. (2009), and Sprod (1998). See also note 8.

There are, as mentioned, many suggestions as to what changes to scientific education need to be made. Many of these form the subject of other chapters in this handbook. In the next section of this chapter, I will survey some approaches[10] that can make claims to have demonstrated lasting changes in reasoning. As we shall see, scientific reasoning is not their sole target.

48.3 Philosophical Inquiry and Science Education

The three programmes on which this section will focus are as follows: Cognitive Acceleration through Science Education (Adey et al. 1995) and its related spin-offs; the Conceptual Challenge in Primary Science program; and approaches derived from Philosophy for Children and its methodology, the community of inquiry (Lipman et al. 1980). As stated in the introduction, programmes designed for science classes in schools which involve some form of philosophical inquiry can have many more aims that merely improve the scientific thinking of students. Nevertheless, given Lipman's claim that philosophy is the natural home of thinking, such programmes will have as one of their central aims the expansion and augmentation of students' ability to think scientifically.

48.3.1 Cognitive Acceleration Through Science Education (CASE)

If educational managers require solid empirical evidence before introducing changes to curricula, then the CASE project is compelling. CASE and the cognitive acceleration (CA) projects that grew out of it[11] have been among the most intensively studied innovations in science education,[12] with a most impressive track record.

---

[10] Unfortunately, the survey that follows will, due to my linguistic limitations, be largely limited to work published in English. Certainly, important work has been carried out in other languages – see, for example, Vieira et al. (2010). There are also many projects in science education that include in their aims the improvement of scientific thinking but for which, to my knowledge, no empirical research has been done to test the claims – for example, Aikenhead's *Logical Reasoning in Science and Technology* (Aikenhead 1990).

[11] 'Other members of the growing family include CAME (in mathematics for junior secondary), PCAME (mathematics for Years 5 and 6, ages 9–11 years), *Let's Think!* (science/general reasoning for Year 1, 5–6 year olds), *Let's Think through Science!* (for Years 3 and 4, 7–9 years) – all developed at King's College, London – and CATE (technology), and ARTS (junior secondary music, drama, and visual arts)' developed elsewhere (Adey 2005).

[12] Adey (1997, 2004, 2005), Adey et al. (2002), Adey and Shayer (1990, 1994), and Shayer and Adey (1993)

The original CASE project involved using one lower secondary science class every second week for the first two secondary school years in activities based around Piagetian-based operations such as control and exclusion of variables, ratio and proportion, compensation and equilibrium, correlation, probability, classification and formal models. In other words, the major focus was not on any particular scientific content,[13] but rather on some aspect of scientific thinking. A typical lesson moves from a 'concrete preparation', where new equipment and vocabulary are introduced, through an activity designed to create cognitive conflict in their students' heads, onto the core phase where students strive jointly to construct a satisfactory explanation, paying attention to the thinking that is going on (metacognition) and seeking to apply their insights to other similar situations, all under the guidance of the teacher.

The major claim to arise from the project is that it is possible to improve the thinking skills of students significantly through this sort of programme and that the effects of these improvements can still be measured 3 years after the intervention finishes, by their effect on student performance in external examinations – not only in science but also in mathematics and English GCSEs. This is long-term, far transfer (Adey and Shayer 1994).

There are two aspects which need attention here. The first is the theoretical underpinnings of the project, and the second is the key factors which the researchers have identified as being vital to the programme's effects.

CASE's theoretical base lies in psychology rather than philosophy. The initial theoretical driver of the project was Piagetian stage theory. The materials were designed to accelerate the passage of students who used them through the stages, by provoking cognitive conflict and assisting assimilation. Specifically, it was thought that students could be assisted to move from the concrete operational to the formal operational stage; hence, the project was targeted at students around the age of 12 (the first 2 years of secondary school).

Drawing on the work of Vygotsky (1962) on the social nature of cognitive development and some little recognised work of Piaget and his collaborators on social interaction (Adey et al. 2007), the means for accelerating cognitive development were devised, trialled extensively and assessed in a multistage longitudinal study.

The thoughts of the principal investigators in the suite of CA projects on why the programmes work will be of primary interest to readers of this handbook.

Firstly, we have to address the question, raised in 2.1, of transfer. Does CASE really demonstrate long-term, far transfer of effects on scientific thinking, when much of the literature points to the difficulty of doing so? Adey (2006, p. 2) says 'we can answer a clear 'yes': stimulating higher order thinking in science improves students' general intellectual ability across the board'. Adey et al. argue that while

---

[13] Of course, each lesson does contain content, so that control of variables might be studied through consideration of the effects of the length, width and material of a pipe on the pitch of the note produced by blowing across it, or probability via flipping coins.

general abilities are acquired and/or practised in particular contexts, on specific content, which might be thought to limit their general application, by

mining more deeply into the insights and models of developmental psychology and paying attention to the general intellectual processors of the mind (both 'executive' and 'central') … (Adey et al. 2007, p. 92)

we can create conditions under which transfer is much more likely. Thus,

moving along the scale of "subject matter versus general ability" towards the direction of developing general abilities actually opens up broad opportunities for raising levels of traditional academic achievement. … you have to have both concrete content and reflective abstraction. If you teach the specifics with abstraction in mind, the general is learned, but if you try to teach the general directly, the specifics often are not learned. (Adey et al. 2007, p. 92)

These observations may give us cause to wonder: is using the CA materials necessary to get the gains their research has shown, or are there essential features of the CA approach which can be used to design other programmes or to influence teaching methodologies more generally? If so, what are the key features? In a number of articles, the team has identified these factors, calling them the pillars of CA. Here is one characterisation of the pillars:

1. *Cognitive Conflict.* Piaget suggested that one of the mechanisms by which cognition develops is through a challenge to existing cognitive structures by experiences which make demands somewhat beyond the child's current processing capability. The same idea is encompassed by Vygotsky's zone of proximal development ('the only good learning is that which is advance of development'). CASE activities are designed to provide such challenge, in scientific contexts, on a slope of increasing difficulty such that, at some point, students of different abilities all encounter cognitive conflict.
2. *Social Construction.* Both Piaget and Vygotsky stressed the role of social interaction in cognitive development, although it is Vygotsky's claim that 'ideas appear first in the social space and then become internalised by the individual' that is best remembered. CASE pedagogy emphasises the importance of collaborative learning in the class, with groups of students interacting with one another, positive argument and critical questioning encouraged and every student's contribution valued.
3. *Metacognition.* Another notion central to the Piagetian model of cognitive development, especially for the emergence of formal operations, is 'reflective abstraction', the idea that the individual reaches a higher level of thinking by reflecting on their own thinking. The Vygotskyan notion that language acts as a mediator of learning also suggests that putting thoughts into words (the conscious explication of thought) is a powerful driver of cognitive development. CASE teachers encourage their students to explain what they are thinking, what they find difficult, what they have learned, what mistakes they have made and how they corrected them (Adey 2006, pp 2–3).

To these three pillars, they have elsewhere added 'two subsidiary pillars (*concrete preparation*, introducing the topic, and *bridging* – showing how the same sort of thinking can be used elsewhere)' (Adey 2005).

One might wonder, at this point, whether the domain through which such cognitive acceleration is delivered needs to be science and, if not, whether there are advantages in doing it through science. Again the CA team has addressed such thoughts. Adey says that, while 'in principle there is no reason why such an approach should not be taken through any subject domain' (Adey 2006), 'science in schools is a domain which may be peculiarly well adapted for the development of, at least, a general understanding of problem solving which would be expected to be useful across both school domains and everyday life' (Adey 1997). This is, in part, because 'the schemata of formal operations described by Inhelder and Piaget (1958) (control and exclusion of variables, proportionality, classification systems, equilibrium, and the others) have a very science-y look to them' (Adey 2006).

Indeed, the CA team have identified several other programmes which seem to fit their pillars well[14] and which have a track record of improving cognitive functioning:

> Long ago Vygotsky claimed that all learning in school from the early years onward should be directed as much to children's cognitive development as to their subject learning (Shayer 2002). The transfer evidence from Cognitive Acceleration, Instrumental Enrichment, and Philosophy for Children suggests that the technology is now being developed to make that a practical and realisable aim. Each of these interventions clearly stimulates something much deeper than domain specific systems, and that 'something', we would claim, is general mental ability, or general intelligence. (Adey et al. 2007, p. 92)

As noted previously, one of the great strengths of CASE and related programmes is that there are a number of longitudinal studies demonstrating their efficacy in long-term, far transfer. As Adey et al. (2007) themselves comment, it is a pity that other promising educational programmes have not been the subject of similar research, particularly in terms of attempting to measure whether the effects seen in short-term studies are long lasting. Clearly, large longitudinal studies are not easy to carry out. Issues of funding, continuity of staffing and the like have been discussed in a special issue of *Research in Science Education* (vol. 35, no. 1, 2005).

Longitudinal studies also have the disadvantage, as Tytler and Peterson (2005) put it, that 'there are many points at which we wish we had gathered different or better data at the earlier stages'. For example, Adey (1994, personal communication) began to wish that the CASE team had collected more data on the use and type of discussion in CASE classes, as they realised that this second pillar was one of the more powerful features explaining the effects of the intervention.

Of interest, while considering longitudinal studies that have demonstrated robust long-term effects, is the comparison of CASE with Novak and Musonda's study of science concept learning, based on the theories of Ausubel (Novak and Musonda 1991; Novak 2005). Tytler et al. (2005) comment that '[t]he success of both interventions, situated as they are in quite different theories, gives pause for thought. They perhaps suggest the existence of common and significant principles underlying both Piagetian and Ausubelian theories'. Piaget was concerned

---

[14]There are other programmes, not identified by the CA team, that also share many of these features and have been shown to have positive effects, for example, work by Carol McGuinness and colleagues in Northern Ireland (McGuinness 2006).

with the structures of thought, Ausubel with acquisition of concepts. I would suggest that the link is that both studies also draw heavily on Vygotskyan ideas on how children internalise both the capacities of thought and concepts through social interaction.

48.3.2 The Conceptual Challenge in Primary Science Project

A second project that can be seen to share the Cognitive Acceleration pillars referred to above is the *Conceptual Challenge in Primary Science* project, originating at Oxford Brooks University in the United Kingdom (Mant et al. 2007; Wilson et al. 2004).

The four emphases in this project are:

1. Increased time spent in discussion of scientific ideas
2. An increased emphasis on the encouragement of higher order thinking
3. More practical work and investigations
4. More focused and purposeful recording by pupils, less writing (Mant et al. 2007, p. 1712)

In the first three, we can recognise the pillars social construction, metacognition and the opportunity for cognitive conflict respectively. Two key teachers from each of 16 primary schools participated in ongoing professional development sessions run by the researchers that concentrated on developing 'cognitively challenging, practical science lessons with plenty of space for thinking and discussion' (Mant et al. 2007, p. 1712). One of the major innovations was the introduction of 'Bright Ideas Time' discussion slots based on a discussion prompt (more details of the methods are to be found in Wilson and Mant 2006). The time to include more practical work and discussion sessions came largely from taking a close look at how much writing and recording students were required to do so that 'teachers focused on the *purpose* of recording by the pupils, so that although less was demanded it was of a higher quality' (Mant et al. 2007, p. 1716, emphasis added).

As with the CASE research, the main measure of effectiveness of the method was to use an externally set and marked test. In this project, the test was the UK's national science assessment test, and the relevant measure was the proportion of students achieving level 5 on the test. Specifically, the project found that, when matched against similar schools, the 16 experimental schools recorded a significantly greater average increase in level 5s of 11.8 %, against 2.0 % in the control schools. Semi-structured focus group interviews with project students showed that they rated the lessons as more challenging, active, collaborative and requiring more thinking.

In more recent research, the Oxford Brookes group has polled over 5,000 12-year-old students to identify exemplary science teachers, measured by the degree to which the students reported being engaged and motivated by science lessons. Through questionnaires administered to the students of those teachers (Wilson and Mant 2011a) and through group interviews with those teachers

(Wilson and Mant 2011b), they sought to determine the teacher-mediated factors that make science education effective. The four factors that both students and teachers agreed upon make interesting reading in the context of the concerns of this chapter: they are the use of discussion, an emphasis on thinking, more student practical work and the contextualisation of science in terms of students' lives (Wilson and Mant 2011b, p. 118).

48.3.3 Philosophy for Children (P4C) and the Community of Inquiry (CoI)

Philosophy for Children had its genesis in the work of Lipman et al. (1980). The basic structure of the programme, as Lipman initially conceived it, is a series of novels accompanied by teacher's manuals, delivered in the classroom through a community of inquiry. The initial novel, *Harry Stottlemeier's Discovery* (Lipman 1974), had a 'spine' concerned with formal syllogistic logic, though plenty of other philosophical puzzles were also woven into it. These were 'unpacked' for the teachers through explanations, discussion plans and exercises in the manual (Lipman and Sharp 1975).[15]

The community of inquiry is a specific type of whole-class discussion. Lipman (1991) identifies five stages: (I) reading a portion of the story; (II) gathering questions from the students as a means of constructing an agenda; (III) the students taking responsibility for discussion; (IV) the teacher taking responsibility for prompting the discussion to be rigorous, rich and meaningful; and (V) follow up activities.[16]

It is worth commenting on these stages. In (I), the stories have been written to contain what my son Liam called 'philosoful' hooks (Sprod 1993). However, instead of the teacher identifying them and asking the students questions about them, it is the students who set the agenda for the discussion to follow by asking questions about those incidents in the story that puzzle or interest them. In this way, the community of inquiry starts in a narrative about children which supplies the concrete situation (CA's first subsidiary pillar) and identifies those ideas that have caused cognitive conflict (CA's first major pillar) for the students – especially when they choose from the questions asked one that most students want to try to answer.

Stages (III) and (IV), which interleave, use the whole-class discussion as a means of providing social construction of concepts, thinking skills and dispositions and

---

[15] It is worth noting at this point that the Philosophy for Children field has diversified considerably since Lipman's model was devised – so much so, in fact, that different theorists and practitioners have suggested broader names, such as philosophy with children (Murris 2008), philosophy in schools (Hand and Winstanley 2009) and dialogical philosophy (Stone 2011). Moreover, there has been an explosion in classroom materials that use many different materials instead of Lipman's purpose-written novels, such as specially written short stories (e.g. Cam 1997; Worley 2011), picture books (e.g. Murris 1992; Sprod 1993; Wartenberg 2009) and film (e.g. Wartenberg 2007).

[16] The discussion that follows draws in part on a much fuller discussion of the community of inquiry in Sprod (2001), Chaps. 7, 8, and 9, especially pp. 183–189.

many more elements (CA's second major pillar). The teacher's role is vital[17]: to keep the discussion in the zone of proximal development. If the discussion loses its edge of inquiry, then the students fall below the ZPD (Gardner 1996), yet it could also lose the students by making too great a demand on their thinking so that they cannot follow.

The community of inquiry handles the third of CA's major pillars, and the second minor one, in two ways. At times, the metacognition and bridging can be tacit. For example, the teacher may label a certain student's idea as an assumption, without exploring explicitly with the students what an assumption is. At other times, though, the discussion can turn its focus on the tools of thinking being used (Sprod 2001, pp. 190–191). Using assumptions again as the example, the teacher may ask students to consider what the role of assumptions is or why it matters that we identify assumptions. Similarly, for bridging, the teacher or a student may introduce another situation when the analysis just completed can also be applied, or the teacher can ask students explicitly to think about the analysis and to find other areas in which to apply it.

From this brief outline of the key features of Philosophy for Children and the community of inquiry, we can see that it exhibits the features identified by Adey (2005) for effective cognitive acceleration. It is worth, however, considering the relation of the CA programmes to P4C in a little more detail.

One of the rich and interesting comments by Adey and colleagues, when considering the relation of CA to other programmes, is the following:

> Lipman's [sic] Philosophy for Children has shown good transfer effects with a general population but is essentially a pragmatic, under-theorised approach. Both [it and Feuerstein's Instrumental Enrichment] face the practical problem in school curriculum terms of requiring scheduled time for "thinking lessons".
>
> We suggest that the content-based approach offers more promise for large scale implementation …. (Adey et al. 2007, p. 92)

In the first sentence above, Adey and his colleagues, as psychologists themselves, clearly have psychological theory in mind when they say Philosophy for Children is under-theorised. Anybody who has seriously read Lipman's magnum opus, *Thinking in Education* (1991), as well as a host of other theoretical texts in the field,[18] would be loath to say that P4C is *philosophically* under-theorised. Indeed, I would claim that, while the CASE authors may have theorised the underlying processes of thinking in a more thorough psychological way, P4C writers have a better theoretical grasp on the philosophical bases of rigorous whole-class discussions. Clearly, each camp can learn from the other.

The second claim is that P4C requires separate 'thinking lessons'. If philosophy is not considered a central subject in a good education (philosophy commonly does not appear in curricula in the English-speaking world, but it does have a place in many other countries), then it is true that philosophical communities of inquiry may

---

[17] See Sprod (2001), Chap. 3, for a philosophical treatment of this 'pedagogical action'.

[18] For example, Hand and Winstanley (2009), Lipman (1993), Lipman et al. (1980), Matthews (1982, 1994); McCall (2009), Pritchard (1996), Splitter and Sharp (1995), and Sprod (2001).

be seen as extra add-ons to the curriculum: separate thinking lessons. Even so, they would still be 'content based' in the sense that Adey et al. use – the content being philosophy.

Nevertheless, there is a good point here: do we need more subjects in an already crowded curriculum? Where do we make room for separate thinking – or philosophy – lessons? As the CA programme recognises, it might be better to infuse the subjects already taught with thinking.[19] Given the specific focus of this handbook, then the case for scientific communities of inquiry needs to be explored. I will do so below.

Before doing that, though, let us look at the evidence for the effectiveness of P4C, both in terms of improving thinking and in other areas. Over many years, both theorists and teachers in P4C have sought to demonstrate empirically the effects of the programme. Anecdotal evidence is abundant, but not compelling, especially to outsiders. While many studies have been carried out,[20] most are small scale and often poorly designed. Hence, despite the near unanimity of their findings of improvement, their reliability can be questioned. A meta-analysis of 18 of the most robust of these studies (Garcia-Moriyon et al. 2004) found that there was consistent evidence of improvement in reasoning, with a mean effect size of 0.58 ($p<.001$). However, in no case was the intervention more than 1 year, and no longitudinal data were collected to test for retention of the effect.

Since then, a well-designed, longitudinal study has been carried out in Scotland. Prior to the intervention, a survey of ten previous studies, considered to be rigorous in methodology, found a mean effect size of 0.43, 'indicating a consistent moderate positive effect for P4C on a wide range of outcome measures' (Trickey and Topping 2004). Subsequent reports on the intervention (1 h per week over 16 months) found the groups exposed to Philosophy for Children had 'significant standardized gains in verbal and also in non-verbal and quantitative aspects of reasoning' (Topping and Trickey 2007a) and exhibited 'increased use of open-ended questions by the teacher, increased participation of pupils in classroom dialogue, and improved pupil reasoning in justification of opinions' (Topping and Trickey 2007c), while there was a significant gain in 'self-esteem as a learner, significant reduction in dependency and anxiety and of greater self-confidence' (Trickey and Topping 2006). The control groups showed no significant changes on any of these measures. Following up 2 years later, they found that '[t]he significant pre-post cognitive ability gains in the experimental group in primary school were maintained towards the end of their second year of secondary school' (Topping and Trickey 2007b).

Thus, there is now robust evidence that P4C has a similar positive effect on reasoning and long-term transfer to CASE. There are some differences. Trickey and Topping did not measure effects on external examination results, but CASE did not look at the wider range of outcomes, such as classroom interaction and socio-emotional factors. The CA programmes focus fairly narrowly on cognitive and subject-specific factors,

---

[19] One might also add ethics.

[20] Citations for 74 empirical studies can be found at <http://cehs.montclair.edu/academic/iapc/research.shtml>.

while the P4C approach encompasses a much wider range of factors (as we shall see in the next section). Reasoning is not the only target.

So far, we have been considering the application of the community of inquiry methodology with philosophy as its primary focus. Let us look at the idea of bringing science to the centre.

48.3.4   The Scientific Community of Inquiry

A distinction has been made earlier between the Philosophy for Children programme – whether this be Lipman's original set of eight novels[21] or some modification using other materials,[22] but which retains philosophy as the central content focus – and the teaching methodology employed, the community of inquiry.

As a pedagogical technique, the community of inquiry (CoI) can be used with other subject areas as the core content. So, we can talk about an historical CoI, an artistic CoI, a mathematical CoI, an ethical CoI and so on. However, due to the nature of the questioning and inquiry that goes on in a CoI, many (e.g. Cam 1995, pp. 14–15; Splitter and Sharp 1995, p. 24) have argued that the CoI inevitably leads into an exploration of the philosophical roots of the central discipline, at least at times. So, as we shall see, it will prove with the scientific community of inquiry.

Compared to the wealth of materials developed for communities of philosophical inquiry, there has not, to date, been a great deal of work done on communities of inquiry in the other disciplinary areas, including science.[23] Given that Lipman derives the phrase community of inquiry from Peirce (Lipman 1991, p. 15), and that Peirce used the phrase (in *The Fixation of Belief*) in relation to inquiry within the community of scientists (Peirce 1955, pp. 5–22), this is perhaps a little puzzling.

Gazzard (1993) contends that 'Philosophy is an integral part of every discipline and therefore similarly should be an integral part of its instruction … science more than any other discipline need the complement of philosophy … for it is scientific knowledge more than most which is accepted by the general population as being true' (p. 619). To support her view, she points to the generative and fallibilistic nature of scientific knowledge, contrasting this with the teaching of science as revealed truth. Hence, science teaching should include consideration of the epistemology and methodology of science. Through the use of a scientific

---

[21] The Lipman novels, with the year group for which they are intended in brackets, are the following: Elfie (1), Kio & Gus (2–3), Pixie (3–4), Nous (4–6), Harry Stottlemeier's Discovery (5–6), Lisa (7–8), Suki (9–10) and Mark (11–12) – full details at 'http://cehs.montclair.edu/academic/iapc/docs/Curriculum_Brochure.pdf'. Each has an accompanying manual. Lipman's intention was that they be studied consecutively throughout schooling.

[22] See note 15 for some examples.

[23] However, there are a few articles discussing P4C and science education. See especially Lipman (1988) chapter 7 'Philosophy and Science Education at the Elementary School Level' (pp. 87–99) but also Clark (1994), Liao (1999), Novemsky (2003), Smith (1995), Weinstein (1990a, b, 1992) and the Ed.D. thesis of Ferreira (2004 – to be discussed below).

community of inquiry, she says, 'students [will] ... realize *for themselves* that science does not provide answers in the sense of closure ... and that science itself is perhaps best conceived of as perpetual inquiry' (p. 624, italics in original). This is not to say the CoI should supplant science instruction, but it should complement it (p. 629).

Despite this call, only a few people have gone on to develop materials in the science area. Often, the materials are one-off ideas for discussion starters, or local applications where the materials have not been made more widely available.[24] One of these was intensively researched for a doctoral dissertation (Ferreira 2004; see Ferreira 2012, for a summary).[25] Using chapter 1 of Lipman's *Harry Stottlemeier's Discovery* (1974) as a starting point, Ferreira wrote four additional chapters with a focus on classification, observation and inference in science. She then used the stories, accompanied by multisensory practical activities, in a semester-long Year 5 Brazilian science class, integrated into the school's normal science (biology) programme. Using mainly qualitative methods, Ferreira showed that the use of these stories, within a community of inquiry, facilitated children's learning of the target science process skills as well as other thinking skills, and encouraged reflection on those skills. Moreover, the students identified with the fictional characters, used them as models for their own thinking and also increased their abilities to self-correct and to build on the ideas of other students.

Below, I shall explore four programmes that have taken a more systematic approach to the use of a scientific community of inquiry (a fifth is explored in Hunt and Taylor's article in this handbook). As we shall see, each of these approaches follows the Lipman classroom pattern of stimulus material, question gathering and discussion phase pretty closely, with the major variation between them being the nature of the stimulus.

48.3.4.1 Lipman's *Kio & Gus*

The first is one of Lipman's novels, *Kio & Gus* (1986), and its accompanying manual *Wondering at the World* (Lipman and Sharp 1986). To say this is a science programme is a little misleading, though there is a considerable emphasis on zoology and ecology, as the title of the manual implies. Aimed at children of around ages 7–8, the novel contains a wealth of philosophical and ethical hooks, as well as the scientific ones. In the preface to the manual, Lipman and Sharp

---

[24] For example, the UK-based website p4c.com, which contains a resource area onto which teachers can upload materials they have developed, contains 20 one-off P4C science lessons. Web searches unveil references to other uses of the CoI in science education, e.g. Ling (2007), Cunningham (2011) and Phillipson and Poad (2010), but I have not been able to see the classroom materials used, beyond the description in the papers cited.

[25] Ferreira, now at the Universidade de Brasília, Brasília, Brazil, is overseeing several projects developing further P4C-based science education materials and researching their contributions to science education.

state that 'it is designed to be a complement to these sciences, rather than a substitute for them' (p. 1).

They canvass two ways in which the scientific community of inquiry can help: firstly, by providing 'logically disciplined, reasonable discussions' in which children can test their scientific hypotheses, models and concepts (at times, against the standard scientific ones). Just inviting them to 'exchange their myths for the truths offered them by science' is not enough: they must be 'allowed to think these matters through for themselves' (pp. 1–2). Special attention is paid to scientific concepts. They comment that 'scientific concepts, while generally definable by means of specific criteria and classificatory procedures, tend to appear to the early elementary student as inert rather than dynamic' (p. 3: one might extend this observation to much older students as well). However, scientific concepts need to be assimilated into the child's present understanding, and so 'exercises are provided to compel students to reason about the information essential to the concept under discussion' (p. 8).

Secondly, children need to 'think scientifically'. Their 'cognitive skills need to be cultivated … they cannot be cultivated in isolation from the discipline to which they must subsequently be applied … efforts to strengthen thinking skills in a 'content-free' manner are futile'. Scientific thinking should be addressed within science – it cannot 'be expected to develop naturally, and … be in place when needed' (p. 2).

Having said this, it must be noted that science only sporadically appears in the rest of the Lipman corpus. Lipman was aware of this (pers. comm.) and hoped that others would take up the job of writing such material. Moreover, the science in *Kio & Gus* is restricted to the life sciences, with no physics, chemistry or earth science. Lipman's approach requires dedicated time in the school timetable, as working through the novel *Kio & Gus* would take at least a lesson week for 1–2 years. While this is feasible in the early childhood years as an adjunct to science sessions, it would become less so in upper primary, and unlikely to be possible in secondary school.

48.3.4.2   Nevers and Colleagues' *Philosophizing with Children About Nature*

A long-running project in Germany has been concerned with the use of the community of inquiry in biology teaching.[26] In this case, the leaders of the project have had a dual aim: a pedagogical one of using the community of inquiry to deepen students' understanding of biology and its ethical implications while improving their scientific thinking and a research aim of understanding the ways in which children of different ages think about such matters, with a focus on issues of environmental ethics.

---

[26] Gebhard et al. (1997, 2003), Nevers et al. (1997, 2006), Nevers (1999, 2005, 2009). Their work, in part, builds on the work of Helmut Schreier, who has been philosophising with primary school children about nature (among other issues) for many years (see, e.g. Schreier 1997; Schreier and Michalik 2008). None of his stories have, to my knowledge, been published in English.

The methodology used is similar to the Lipman approach outlined above. Purpose-written stories feature children involved in various different ethical dilemma situations characterised by a conflict of interest between a child or young person and a living organism or system. Hence, the major emphasis is on the interplay of biology and ethics. For research purposes, three sets of stories were used involving a plant, an animal or an ecosystem as the source of conflict, and each basic story was presented to different age groups (6–8, 10–12, 14–16). An example of such a story is in Gebhard et al. (2003, p. 95), and that paper, plus Nevers et al. (2006), provides copious transcripts of children's comments in discussion.

From recording, transcribing and analysing more than 150 group discussions, Nevers and her colleagues have been able to identify basic philosophical positions to which students subscribe, the nature and strength of the reasoning that they bring to bear on these positions, the consistency with which they hold them and their susceptibility to change in the light of discussions (Nevers et al. 1997, 2006).

A key finding is that children rely on anthropomorphism to moralise nature (Gebhard et al. 2003). This is effective when an animal or a tree is the source of conflict, but not with an ecosystem. In the latter case children tend to anthropomorphise individual organisms within the ecosystem and thus indirectly attribute moral status to it. Furthermore, the evidence indicates that anthropomorphism is not as apparent among older students. However, a follow-up study using a dilemma story involving trees and a questionnaire distributed randomly among university students suggests that anthropomorphic thinking is still prevalent among a large number of these students and drawn upon to moralise organisms, even if it is not expressed publicly.

In an individual case study, it was found that appealing to an aesthetically pleasing image from nature may serve to transform the discussion from unproductive circularity towards constructive compromise (Nevers 1999, p. 20; Nevers 2000).

Doctoral work by Hausberg (2012) under Nevers' supervision investigated the potential of philosophising with children for encouraging creative thinking (a sample of such work is to be found in Hausberg and Calvert 2009). The theoretical basis was a multifaceted model of creativity proposed by Urban (2004). Philosophical discussions with fifth and sixth graders were recorded and examined with a computer-supported text analysis system to identify various different categories of creative thinking and action. These included areas of cognitive thinking such as analysis and synthesis, analogy formation and metacognition as well as personal qualities such as humour and persistence. Group dynamic qualities such as the ability to elaborate and coordinate contributions from others were also identified.

A second phase attempted to assess whether the creative qualities expressed and trained by philosophising were applied in a different learning situation. Middle school students who had been philosophising in separate sessions for several years were presented with an open problem of biological nature and asked to discuss the problem in small groups, looking for possible solutions. Afterwards, the solutions were evaluated by the students in a moderated plenary discussion. Although it is not possible to conclusively demonstrate a transfer effect by these means, the evidence indicates that this is highly plausible. Almost all the forms of creative expression used when philosophising were also applied when the students dealt with an open biological problem, and solutions were proposed that went far beyond those usually found in biology classes.

48.3.4.3 University of Ulster's *Science in Society* Projects

At the University of Ulster, the *Science in Society* team has developed several packages based around the use of the community of inquiry, including:

- *Primary Community of Scientific Enquiry* (ages 7–11, for use in the subject The World Around Us, which includes science, geography and history)[27]
- *Forward Thinking* (ages 11–14, for use in science and/or citizenship)[28]
- *Community of Scientific Enquiry: Learning Science Through Dialogue* (ages 15–16, for use in GCSE Biology)[29]

While these resources are aimed at different age groups, all share a common structure: one or more starter activities; a trigger experience for students which consists of scientific information and/or experiments; a method for collecting the students' questions and deciding on the one to start discussing; an inquiry/discussion phase; and some sort of reflection on, or evaluation of, the inquiry. This is very similar to the Lipman model, except for the use of information sheets or experiments as the trigger for inquiry, instead of a story about similar-aged children discussing science.

As the overall title of the project implies, one of the emphases in the materials (particularly the *Forward Thinking* package) is on ethical implications and social impacts of science. Even with this package, though, the experience of the team is that students also want to explore philosophical questions about the nature of science and inquire into the science involved, including the meaning of scientific concepts (Dunlop 2012; Dunlop et al. 2011). The primary materials concentrate more on scientific content and processes – especially physics and chemistry – while the GCSE materials consciously target all these areas.

An integral part of the project involves evaluation of the impact of the scientific community of inquiry, though this is at an early stage. In the only peer-reviewed evaluation study published at the time of writing, Dunlop et al. (2011) present evidence that lower secondary students taking the programme enjoy their science more, engage with each other more, deepen their understanding of the science covered and correct and question each other more. Teachers agreed that better learning takes place and the quality of scientific reasoning improves and found that they had clearer access to student misconceptions. Several commented that the discussions worked best when the students had already been exposed to factual background

---

[27] See http://www.ulster.ac.uk/scienceinsociety/pcose.html, where you may read the teacher support material and student handouts, and also Dunlop et al. (2011).

[28] See http://www.ulster.ac.uk/scienceinsociety/forwardthinking.html, also Dunlop (2012).

[29] Still in development

information, suggesting that the best time for a community of scientific inquiry may be after prior learning or research.

The materials are, at present, used by teachers involved in the project, who have engaged in training in their use and often have the in-class support of the researchers. The potential to publish the materials more widely exists, but it seems to me that they would require considerably more support material before teachers could use them effectively, independently. Such an outcome is desirable but may depend on continued funding.

48.3.4.4 Sprod's *Discussions in Science*

Finally, I have developed a suite of 18 connected short stories,[30] with teacher support material, for use in middle schooling (roughly, ages 10–14) (Sprod 2011). These include material from physics, chemistry, biology and the earth sciences and are designed to be used within a science programme, as they contain links to related experiments and theory. Their use follows the standard Lipman pattern of reading the story around the class, gathering the students' questions and entering a discussion based around the question chosen by the class. The teacher support material includes information about the scientific, philosophical, ethical and social issues that lie behind the stories, discussion guides (lists of questions that teachers may draw upon to unpack the issues) and suggestions for related activities, such as experiments or research projects.

Underpinning this approach, and in particular a number of the stories, are several research studies in which I participated. The first stories were written for an investigation of the impact of the scientific community of inquiry on students' ability to reason scientifically, which demonstrated statistically significant gains for the experimental group over a control group in a pretest/posttest design (effect size: 0.70) (Sprod 1998). Discourse analysis of the use of epistemic episodes (Sprod 1997a) in the discussion provided support for the interpretation that the improvements were due to the internalisation of improved group thinking in the class discussions. Both these papers contain transcripts of discussion excerpts from the experimental Year 7 (lower secondary) class.

The content of three of the stories in the collection are based on an investigation of how children come to understand the way light and vision are coordinated.[31] Characters in the stories represent different conceptions of the nature of light and how objects and the eye interact with light. Further stories likewise draw on a wide range of children's misconceptions, scientific theories, philosophical accounts, ethical controversies on scientific applications and my own experience of teaching science over many years.

The stories are linked through the narrative device of a group of children discussing their experiences of science and the world, trying to make sense of it. Thus, the

---

[30] These stories can be read at www.acer.edu.au/discussions-in-science/.

[31] Collis et al. (1998), Jones et al. (1997), Sprod (1997b), and Sprod and Jones (1997).

stories make use of the power of narrative, though they are different in conception from stories drawn from the history of science as discussed by Klassen and Klassen elsewhere in this handbook. Matters that are embedded in the stories include the teasing out of the meaning of scientific concepts, puzzles about scientific methodologies and the nature of science, ethical concerns about the practice and applications of science, and issues about the links between the science students are learning and their everyday lives. Throughout, both through the encouragement and modelling of rigorous thinking in the discussions and through explicit consideration, the skills and dispositions of scientific thinking are addressed.

In this list, we can see that the community of inquiry holds great promise for dealing with many of the concerns that form the subjects of chapters in this handbook. Further, by harnessing the power of narrative and drawing on the method of allowing the students to set the agenda from what they find interesting in the story, they can do so in a way that lends them ecological validity – they fit into students' expressed interests and connect to their lived experience. To pick only a handful of examples, I can instance conceptual puzzles such as those about energy (Bevilacqua), light (Galili) or ecology (Korfiatis, Lefkaditou & Hovardas); considerations of the methods and nature of science such as controversies in earth science (Dolphin & Dodick), understanding the purpose of practical work (Ford) or getting a feel for scientific inquiry (Kelly); inquiry into ethical issues (Cuoló); exploration of scientific modes of thinking such as the use of models (Kopenen & Tala), thought experiments (Asikainen & Hirvonen), coordination of the macro and micro levels (Guisasola) and the nature of scientific argumentation (Bravo); and finally links to the everyday through developing scientific literacy (Norris, Phillips & Burns) or considering multicultural issues in science (Horsthemke). All of these are raised at some level in one or other of my stories, for possible consideration in the scientific community of inquiry.

48.4 Summary

48.4.1 Survey of Conclusions

This chapter has surveyed briefly some of the research into the improvement of thinking through education and whether and how it might be taught, particularly in the context of science education.

As we have seen, the preponderance of the evidence is that thinking is complex – having at least critical, creative and emotional (caring) aspects – and always takes place within a context. Nevertheless, there is a core of general thinking that can be learned and used within one context and then, provided the conditions are right, transferred to other contexts. Moreover, considerable evidence indicates that we can improve children's thinking by developing their capacities and dispositions for higher-order thinking and strengthening their judgement. Whether we are right in claiming that we are accelerating development that would otherwise take place somewhat more slowly or that we are equipping them in ways that they may not otherwise achieve is not yet entirely clear.

The evidence also supports the view that good general thinking is supplemented by more specialised thinking capacities within different domains, though it seems these capabilities probably develop from more general ones. We might draw the conclusion from this that any general thinking skills programmes ought to be supplemented by attention to improving thinking within each discipline and hence in all subjects in schools. Alternatively, we might conclude that if each subject strives to improve thinking within its own domain, then there will be no need for a general thinking skills programme. Either way, there is a strong argument for including an explicit element designed to improve thinking in science courses at all levels of education. A meta-analysis of critical thinking courses (Abrami et al. 2008) bears this out in its finding that the most effective critical thinking courses combined an explicit focus on thinking capacities and dispositions with application to a particular content area.[32]

Concentrating further on thinking within the domain of science, we have seen that scientific thinking seems to draw on both the general thinking capacities and dispositions that apply across all domains and more specialised scientific thought processes (which are, in any case, not identical in all scientific categories). Just what the categories and attributes of such advanced scientific thinking are does not seem to have been researched in sufficient detail. Moreover, there has been little work done on exactly how young children – who in their science lessons are presumably utilising generic thinking – develop through their education (primary, secondary and tertiary) the more specialised scientific thinking required by professional scientists or, more importantly for most, scientifically literate citizens.

Nevertheless, we do have a good deal of evidence that the teaching of scientific thinking is not done particularly well in schools at present. In large degree, this seems to be because the improvement of scientific thinking needs to have an explicit focus in science lessons, while generally it does not. While there are copious case studies in the literature of individual teachers whose teaching style does encourage better thinking by making it explicitly visible to students, such approaches have not been widely implemented at a system level.

However, we have seen that there are good models, supported by robust research, for such an approach, and have looked in some detail at two: the Cognitive Acceleration through Science Education and the Philosophy for Children programmes. It seems clear that both programmes (together with the Conceptual Challenge in Primary Science project) share some important and powerful features. They 'begin in wonder' (puzzlement, cognitive conflict), as Aristotle (1998, 982b12) put it in Book 1 of the Metaphysics; use rigorous dialogue (social construction); turn attention explicitly on the thinking taking place (metacognition); and address transfer or bridging, all in the context of science. In other words, such programmes make both student and teacher thinking – and the quality of that thinking – the

---

[32] They also found that including explicit thinking outcomes in the aims of the course and providing professional development for teachers in the improvement of critical thinking were important factors.

subject of an inquiry visible to all: it is 'minds-on' science. The obvious conclusion is that other science programmes, provided they incorporate similar features, are likely to be similarly effective.

48.4.2 A Challenge to These Conclusions

Reconsidering his 2004 paper that largely agrees with the conclusions above, Davson-Galle (2008a, b) has called into question, firstly, whether it is legitimate to teach the nature of science and critical thinking within science at school and, secondly, even whether we can justify compulsory science courses at all. There are two major grounds for these challenges. The less radical is based on a cost-benefit analysis: would the time allocated to such aims be better spent in teaching something else – say, more science content? More controversially, he questions the legitimacy of overriding students' autonomy in compelling them to study science. As he acknowledges, this latter challenge can be applied to much, if not all, of education.

Turning to the first challenge, I believe that Davson-Galle has created a false dichotomy: either we teach scientific thinking and the nature of science or we teach science content. Doing one can only be at the expense of the other. Consideration of the research cited above – particularly the longitudinal studies in CASE – seems to show that sacrificing curriculum time to the improvement of scientific thinking does not result in students learning less scientific content, at least as measured by outcomes in later science external exams. Rather, it makes the learning of science content faster and more effective. The evidence for learning about the nature of science is less clear cut, as it has not been as explicitly tested, but if that learning is merged with the improvement of scientific thinking, through a dialogical pedagogy, there are reasons for thinking that similar claims are justified. However, there is probably a need to test this hypothesis more directly.

Davson-Galle's second challenge is somewhat beyond the scope of this review. However, I believe that it is based on a misconception of the notion of autonomy, which assumes that it is a characteristic that students just have. In my view, autonomy is rather a capacity which students develop, particularly through improving their ability to think well – see my analysis of communicative autonomy in Sprod (2001, especially Sects. 2.4, 2.5, and 3.2). Our justification for compulsory education is then that it is an effective way to build autonomy and that the more an educational programme contributes to such building, the more justified it is.

48.4.3 Further Work

Given that such programmes are not as widely included in science courses as perhaps they ought to be, I will finish by considering what further work is needed. It seems to fall into three categories.

Firstly, the research foundation for improving the teaching of scientific thinking could be considerably stronger. Both CASE and P4C (though, for the latter, not in the context of science education) have research to indicate that long-term, far transfer of the improvement of thinking takes place. However, these studies have not been fine grained enough to explore the relative impact of the various factors I have mentioned above, let alone variations that arise from differences in the way they are implemented or the cultural context in which they are applied.[33] Adey et al. (2007) are surely right to call for more research that uses large-scale longitudinal studies, well designed to tease out such factors, across a variety of social settings. We need to refine our understanding of the conditions that are truly effective in developing scientific thinking in all its aspects, and we need to study whether the various programmes have other desirable effects, such as strengthening learning communities, enhancing enjoyment of science, improving ethical judgement in matters scientific, illustrating the nature of science and deepening conceptual understanding.

Moreover, we can question how we ought to interpret 'long-term, far transfer'. Certainly, both the CASE and the Tricky and Topping P4C research show measurable effects several years after intervention finishes. But is this long term enough? Surely we would want to see effects that last well into adulthood, and the research to show this has not been done (indeed, such research faces considerable practical difficulties). Additionally, the CASE studies did not test the long-term maintenance of improvements in thinking directly. Rather, results of external exams or testing were used as a proxy. We are only surmising that the experimental students still think significantly better than their control peers.

Of course, showing effects on more general tests does strengthen the claim for far transfer, especially when the CASE work shows effects outside the science domain. But we can question whether the effects extend even further: into thinking in everyday life, or in making citizenship decisions, to choose two examples.[34] Such claims are often made, but research backing for them is lacking. While it is desirable to improve students' exam results and to foster more and better research scientists, for a far greater proportion of students, the justification for seeking to improve scientific thinking rests rather on creating thoughtful and scientifically literate citizens. We ought to be gathering evidence to allow us to ascertain whether such outcomes occur.

Secondly, one of the factors not addressed in much of the research done so far is the impact on the success of such programmes of the background, attitude and competence of the teachers who deliver the programme. Much of the research surveyed in this chapter investigated the impact of the researcher's own teaching on thinking (especially in the small-scale studies) or that of teachers who had volunteered to

---

[33] Indeed, we should note that, as correlational studies, such research does not show conclusively that improving students' scientific thinking through dialogue causes better science learning and hence exam results. It is possible that some other factor – such as an improved attitude to science – is at play.

[34] Note that this possibility depends on such programmes encouraging generalisation of thinking abilities across contexts – a matter discussed in Sect. 48.2.1 above.

take part, often because they were already experienced in the programme being studied or because they had a long-standing sympathy for such approaches. Clearly, such studies are open to experimenter effects. If such programmes are to be widely implemented, then teachers from all sorts of backgrounds, with a variety of prior training and quite diverse attitudes towards what constitutes good science teaching, are going to be involved. Only large-scale studies that recruit all (or a random sample of) the teachers in the study population to implement the programme are going to be able to tell us about the impacts of such teacher attributes on the success of the programmes. Indeed, the research design of such studies will need to ensure the right data are collected to give us better knowledge of teacher effects.

Such considerations bring us to the third reflection. Successful implementation of effective ways of improving scientific thinking will depend on having capable, well-trained teachers delivering them. Research repeatedly shows that teachers are the most important in-school factor in student achievement: the 'evidence supports the assertion that the effects of teachers far exceed the independent effects of schools' (Manzano 2000, p. 60).

Yet Adey and his colleagues comment that teaching in a way that improves students' scientific thinking:

> is difficult to do. Teaching for cognitive stimulation is far more demanding, and seems far more risky in the classroom than is efficient instruction in content matter. Amongst others, Adey et al. (2004) have described the extent of the conceptual-pedagogical change that teachers must make to move from one form of teaching to another. (Adey et al. 2007, p. 93)

Given this, the training provided to teachers needs to be effective in helping them to make permanent modifications and additions to their practice, whether it is delivered during initial teacher training or to teachers already working in schools. Such effective professional development, of course, requires capable providers and can be resource hungry. Again, research into the efficacy of professional development and the factors that make it more effective could assist us to improve such provision.

Finally, there is a need to develop more materials to support teachers in this modified approach to science education. While the CASE materials do provide a coherent set of activities for the first 2 years of secondary school, as well as the *Let's Think Through Science* books for 7/8-year-olds and 9/10-year-olds,[35] these activities address only parts of the science that students could be learning and are tied to the English National Curriculum. Moreover, they tend to target particular scientific thinking capabilities, rather than content areas. In the P4C field, the materials available are even less systematic. Thus, teachers looking for an activity to engage in 'minds-on' science that specifically fits into a unit they are developing are unlikely to be able to find one. Science teaching has multiple aims, among them ones like covering content and developing practical skills, which are probably not best approached through the discussion-based activities considered in this chapter. A teacher who is trying to put together a unit of work in science that covers all these

---

[35] See http://www.cognitiveacceleration.co.uk/resources/other_subject_resources.html.

aims will be looking for an activity that fits in neatly: at present, the choice is limited. Moreover, there is not an even coverage over all the subdisciplines of science, with perhaps an overemphasis on the life sciences.

In summary, then, this chapter makes the case that it is possible, by making the right sorts of additions to the pedagogical toolboxes of teachers, to deliver science education courses in a way that strengthens students' scientific thinking, as well as other desirable outcomes: a deeper understanding of the concepts of science, the nature of the scientific endeavour, the ethical implications of science and greater links between the science learned and the students' everyday lives – arguably, without sacrificing student mastery of scientific content and skills. At the core of the successful methods lies discussion that is recognisably philosophical.

References

Abrahams, I., & Millar, R. (2008). Does practical work really work? A study of the effectiveness of practical work as a teaching and learning method in school science. *International Journal of Science Education, 30*(14), 1945–1969.

Abrami, P. C., Bernard, R. M., Borokhovski, E., Wade, A., Surkes, M. A., Tamim, R., & Zhang, D. (2008). Instructional interventions affecting critical thinking skills and dispositions: A stage 1 meta-analysis. *Review of Educational Research, 78*(4), 1102.

Adey, P. (1997). It All Depends on the Context, Doesn't It? Searching for general, educable dragons. *Studies in Science Education, 29*, 45–91.

Adey, P. (2004). Evidence for long-term effects: Promises and pitfalls. *Evaluation & Research in Education, 18*(1-2), 83–102.

Adey, P. (2005). Issues arising from the long-term evaluation of cognitive acceleration programs. *Research in Science Education, 35*(1), 3–22.

Adey, P., Csapó, B., Demetriou, A., Hautamäki, J., & Shayer, M. (2007). Can we be intelligent about intelligence?: Why education needs the concept of plastic general ability. *Educational Research Review, 2*(2), 75–97.

Adey, P., Robertson, A., & Venville, G. (2002). Effects of a cognitive acceleration programme on Year I pupils. *British Journal of Educational Psychology, 72*(1), 1–25.

Adey, P., & Shayer, M. (1990). Accelerating the development of formal thinking in middle and high school students. *Journal of Research in Science Teaching, 27*(3), 267–285.

Adey, P., & Shayer, M. (1994). *Really raising standards.* Routledge.

Adey, P., Shayer, M., & Yates, C. (1995). *Thinking science: the curriculum materials of the Cognitive Acceleration through Science Education (CASE) project.* Nelson.

Adey, P. (2006). Thinking in science - thinking in general? *Asia-Pacific Forum on Science Learning and Teaching, 7*(2).

Aikenhead, G. S. (1990). *Logical Reasoning in Science and Technology.* Toronto: John Wiley.

Andre, T. (1997). Minds-on and Hands-on Activity: Improving Instruction in Science for All Students. Presidential Address, 1995. *Mid-Western Educational Researcher, 10*(2), 28-34.

Aristotle. (1998). *Metaphysics* (H. Lawson-Tancred, Trans.). London: Penguin.

Bailin, S. (1998). *Skills, generalizability and critical thinking.* Proceedings from Philosophy of Education Society of Great Britain: Conference Papers 1998.

Bailin, S. (2002). Critical thinking and science education. *Science & Education, 11*(4), 361–375.

Balcaen, P. (2008). Developing Critically Thoughtful, Media-Rich Lessons in Science: Process and Product. *Electronic Journal of e-Learning, 6*(3), 161–170.

Boston Museum of Science (2001). Science Thinking Skills: Providing visitors practice in science thinking skills. Retrieved 9 November, 2011 from http://www.mos.org/exhibitdevelopment/skills/index.html

48 Philosophical Inquiry and Critical Thinking in Primary and Secondary Science... 1559

California State Department of Education (1990). Science Framework for California Public Schools. Retrieved 9 November, 2011 from http://www.eric.ed.gov/PDFS/ED325324.pdf

Cam, P. (1995). *Thinking Together: Philosophical Inquiry for the Classroom.* Alexandria, NSW: Hale & Iremonger.

Cam, P. (1997). *Thinking Stories 3.* Sydney: Hale & Iremonger.

Cavagnetto, A. R., Hand, B., & Norton-Meier, L. (2011). Negotiating the Inquiry Question: A Comparison of Whole Class and Small Group Strategies in Grade Five Science Classrooms. *Research in Science Education, 41*(2), 193–209.

Choi, A., Notebaert, A., Diaz, J., & Hand, B. (2010). Examining arguments generated by year 5, 7, and 10 students in science classrooms. *Research in Science Education, 40*(2), 149–169.

Clark, M. A. (1994). Bat Milk and Other Life Stories: Philosophy for Children Applied to the Teaching of University Science. *Analytic Teaching 15*(1), 23–28.

Collis, K. F., Jones, B. L., Sprod, T., Watson, J. M., & Fraser, S. P. (1998). Mapping Development in Students' understanding of vision using a cognitive structural model. *International journal of science education, 20*(1), 45–66.

Cunningham, R. (2011). *Deliberative Democracy and Sustainable Design: Why should these be central to a school curriculum for the twenty first century?* Proceedings from Education and Citizenship in a Globalising World, London.

Davson-Galle, P. (2004). Philosophy of science, critical thinking and science education. *Science & Education, 13*(6), 503–517.

Davson-Galle, P. (2008a). Against science education: the aims of science education and their connection to school science curricula, in *Education Research Trends*, Bertrand, T. & Roux, L. (ed), Hauppauge, New York: Nova Science Publishers, pp. 1–30.

Davson-Galle, P. (2008b). Why compulsory science education should not include philosophy of science. *Science & Education, 17*(7), 677–716.

Dawson, V. (2010). Measuring the impact of instruction about argumentation and decision-making in high-school genetics. *Genomics Education for Decision-making.*

de Bono, E. (1986). Beyond Critical Thinking. *Curriculum Review, 25*(3), 12–16.

Dunlop, L., Humes, G., Clarke, L., & McKelvey-Martin, V. (2011). Developing communities of enquiry: dealing with social and ethical issues in science at key stage 3. *School Science Review, 93*(342).

Dunlop, L. (2012). P4C in secondary science. In L. Lewis & N. Chandley (Eds.), *Philosophy for Children Through the Secondary Curriculum.* London: Continuum.

Dunlop, L., Clarke, L., & McKelvey-Martin, V. (2011). Using communities of enquiry in science. *Learning & Teaching Update, 49*, 4–6.

Ennis, R. H. (1990). The Extent to Which Critical Thinking is Subject Specific: Further Clarification. *Educational Researcher, 19*(4), 13–16.

Ennis, R. H. (1987). A Taxonomy of Critical Thinking Dispositions and Abilities. In J. B. Baron & R. J. Sternberg (Eds.), *Teaching thinking skills: Theory and practice.* New York: WH Freeman and Company.

Ennis, R. H. (1989). Critical thinking and subject specificity: clarification and needed research. *Educational Researcher, 18*(3), 4–10.

Ferreira, L. B. M. (2004). *The Role of a Science Story, Activities and Dialogue Modeled on Philosophy for Children in Teaching Basic Science Process Skills for Fifth Grade*, dissertation for Ed.D., Montclair State University (unpublished).

Ferreira, L. B. M. (2012). Philosophy for children in the science class: children learning basic science process skills through narrative. *Thinking: The Journal of Philosophy for Children, 20*(1&2), 71–79.

Garcia-Moriyon, F., Rebollo, I., & Colom, R. (2004). Evaluating Philosophy for Children: A meta-analysis. *Thinking: The journal of philosophy for children, 17*(4), 14–22.

Gardner, S. (1996). Inquiry is no mere conversation. *Analytic Teaching, 16*(2), 41–47.

Gazzard, A. (1993). Thinking Skills in Science and Philosophy for Children. In M. Lipman (Ed.), *Thinking Children and Education* (pp. 619–631). Dubuque: Kendall/Hunt.

Gebhard, U., Billmann-Mahecha, E., & Nevers, P. (1997). Naturphilosophische Gespräche mit Kindern. Ein qualitativer Forschungsansatz. In H. Schreier (Ed.), *Mit Kindern über die Natur philosophieren* (pp. 130–153). Heinsberg: Dieck.

Gebhard, U., Nevers, P., & Billmann-Mahecha, E. (2003). Moralizing trees: Anthropomorphism and identity in children's relationships to nature. In S. Clayton & S. Opotow (Eds.), *Identity and the natural environment: The psychological significance of nature* (pp. 91–111). Cambridge, MA: MIT Press.

Hand, B., & Choi, A. (2010). Examining the impact of student use of multiple modal representations in constructing arguments in organic chemistry laboratory classes. *Research in Science Education, 40*(1), 29–44.

Hand, M., & Winstanley, C. (Eds.). (2009). *Philosophy in schools.* London: Continuum.

Hart, W. A. (1993). Against Skills. In M. Lipman (Ed.), *Thinking Children and Education* (pp. 632–644). Dubuque: Kendall-Hunt.

Hausberg, A. (2012). *Fressen Katzen Rotklee? Untersuchung kreativer Ausdrucksformen beim Philosophierenmit Kindern und Jugendlichen und ihr Transfer bei der Lösung einer offenen Aufgabe mit biologischem Bezug.* PhD. University of Hamburg, Hamburg.

Hausberg, A., & Calvert, K. (2009). PhiNa: Aspects of Creative Philosophising with Children About Nature. In W. C. Turgeon (Ed.), *Creativity and the Child: Interdisciplinary perspectives* (pp. 227–236). Oxford: Inter-Disciplinary Press.

Huxley, T. H. (1894). On the educational value of the natural history sciences. Retrieved from <http://aleph0.clarku.edu/huxley/CE3/EdVal.html>

Inhelder, B., & Piaget, J. (1958). *The growth of logical thinking.* London: Routledge Kegan Paul.

Jegede, O. J., & Taylor, P. C. (1995). The Role of Negotiation in a Constructivist-Oriented Hands-On and Minds-On Science Laboratory Classroom. Paper presented at the Annual Meeting of the American Educational Research Association, San Francisco, 17–21 April 1995.

Jones, B. L., Sprod, T., Collis, K. F., & Watson, J. M. (1997). Singaporean and Australian Students' Understanding of Vision. *Asia Pacific Journal of Education, 17*(2), 85–101.

Jurd, E. (2004). Are the children thinking. *Primary Science Review, 82*(3/4), 12–14.

Kuhn, D. (1999). A developmental model of critical thinking. *Educational Researcher, 28*(2), 16.

Kuhn, D., Amsel, E., & O'Loughlin, M. (1988). *The development of scientific reasoning skills.* Orlando, CA: Academic.

Kuhn, D., & Pearsall, S. (2000). Developmental origins of scientific thinking. *Journal of Cognition and Development, 1*(1), 113–129.

Lederman, N. G., & Lederman, J. S. (2004). The nature of science and scientific inquiry. In G. Venville & V. Dawson (Eds.), *The art of teaching science* (pp. 2–17). Sydney: Allen & Unwin.

Lee, C. Q., & She, H. C. (2010). Facilitating Students' Conceptual Change and Scientific Reasoning Involving the Unit of Combustion. *Research in Science Education, 40,* 479–504.

Liao, B. (1999). Stages of Wonder: A Lesson in Physics. *Thinking 14*(4), 49.

Ling, Y. (2007). *Philosophy in Children (P4C) and Pupils' Learning in Primary Science in Singapore.* Proceedings from Redesigning Pedagogy: Culture, Knowledge and Understanding, Singapore.

Lipman, M. (1974). *Harry Stottlemeier's Discovery.* Upper Montclair, NJ: Institute of the Advancement of Philosophy for Children.

Lipman, M. (1986). *Kio & Gus.* Upper Montclair, NJ: First Mountain Foundation.

Lipman, M. (1988) *Philosophy Goes to School* Philadelphia: Temple University Press, 1988.

Lipman, M. (1991). *Thinking in education.* Cambridge: Cambridge University Press.

Lipman, M. (Ed.). (1993). *Thinking children and education.* Dubuque, Iowa: Kendall/Hunt.

Lipman, M. (1995). Caring as thinking. *Inquiry: Critical thinking across the disciplines, 15*(1), 1–13.

Lipman, M., & Sharp, A. (1975). *Philosophical Inquiry: Instruction Manual to Accompany Harry Stottlemeier's Discovery.* Upper Montclair, NJ: Institute of the Advancement of Philosophy for Children,.

Lipman, M., & Sharp, A. M. (1986). *Wondering at the World: Instructional Manual to Accompany Kio and Gus.* Lanham, MD: University Press of America.

Lipman, M., Sharp, A. M., & Oscanyan, F. S. (1980). *Philosophy in the Classroom.* Philadelphia: Temple University Press.

Lyons, T. (2006). Different countries, same science classes: Students' experiences of school science in their own words. *International Journal of Science Education, 28*(6).

Mant, J., Wilson, H., & Coates, D. (2007). The Effect of Increasing Conceptual Challenge in Primary Science Lessons on Pupils' Achievement and Engagement. *International Journal of Science Education, 29*(14), 1707–1719.

Manzano, R. J. (2000). *A New Era of School Reform: Going Where the Research Takes Us.* Aurora, CO: Mid-continent Research for Education and Learning.

Matthews, G. B. (1982). *Philosophy and the young child.* Cambridge, MA: Harvard UP.

Matthews, G. B. (1994). *The philosophy of childhood.* Cambridge, MA: Harvard UP.

May, D. B., Hammer, D., & Roy, P. (2006). Children's analogical reasoning in a third-grade science discussion. *Science Education, 90*(2), 316–330.

McCall, C. C. (2009). *Transforming Thinking: Philosophical Inquiry in the Primary and Secondary Classroom* (1 ed.). London: Routledge.

McGuinness, C. (2006). Building Thinking Skills in Thinking Classrooms. *Teaching and Learning Research Briefing, 18.* Retrieved from www.tlrp.org

McPeck, J. E. (1990). Critical thinking and subject specificity: a reply to Ennis. *Educational Researcher, 19*(4), 10–12.

Miri, B., David, B. C., & Uri, Z. (2007). Purposely teaching for the promotion of higher-order thinking skills: A case of critical thinking. *Research in science education, 37*(4), 353–369.

Mitchell, I. (2010). The Relationship Between Teacher Behaviours and Student Talk in Promoting Quality Learning in Science Classrooms. *Research in Science Education, 40*(2), 171–186.

Murris, K. (1992). *Teaching philosophy with picture books.* London: Infonet.

Murris, K. S. (2008). Philosophy with Children, the Stingray and the Educative Value of Disequilibrium. *Journal of Philosophy of Education, 42*(3–4), 667–685.

Nevers, P. (2005). Wozu ist Philosophieren mit Kindern und Jugendlichen im Biologieunterricht gut? In C. M. Hößle, Kerstin (Ed.), *Philosophieren mit Kindern und Jugendlichen Didaktische und methodische Grundlagen des Philosophierens* (pp. 24–35). Hohengehren: Schneider Verlag.

Nevers, P. (2009). Transcending the Factual in Biology by Philosophizing with Children. In G. Y. M. Iversen, Gordon & G. Pollard (Eds.), *Hovering over the face of the deep: philosophy, theology and children* (pp. 147–160). Münster: Waxmann.

Nevers, P. (1999, September 21). *How Children and Adolescents Relate to Nature.* Proceedings from Center for the Study of Ethics in Society, Kalamazoo MI.

Nevers, P., Billmann-Mahecha, E., & Gebhard, U. (2006). Visions of Nature and Value Orientations among German Children and Adolescents. In R. J. G. van den Born, W. T. de Groot, & R. H. J. Lenders (Eds.), *Visions of Nature: A scientific exploration of people's implicit philosophies regarding nature in Germany, the Netherlands and the United Kingdom* (pp. 109–127). Münster: Lit Verlag.

Nevers, P., Gebhard, U., & Billmann-Mahecha, E. (1997). Patterns of Reasoning Exhibited by Children and Adolescents in Response to Moral Dilemmas Involving Plants, Animals and Ecosystems. *Journal of Moral Education, 26*(2), 169–186.

Nevers, P. (2000). Naturethik und Konfliktbewältigung bei Kindern: Ergebnisse, Fragen und Spekulationen aus einer hermeneutischen Untersuchung. In K. Ott & M. Gorke (Eds.), *Spektrum der Umweltethik* (pp. 191-213). Marburg: Metropolis-Verlag.

Norris, S. (1992). *The Generalisability of Critical Thinking.* New York: Teachers College Press.

Novak, J. D. (2005). Results and implications of a 12-year longitudinal study of science concept learning. *Research in Science Education, 35*(1), 23–40.

Novak, J. D., & Musonda, D. (1991). A twelve-year longitudinal study of science concept learning. *American Educational Research Journal, 28*(1), 117.

Novemsky, L. (2003). Using a Community of Inquiry for Science Learning, or the Story of 'It'. *Thinking 16*(4), 45–49.

Passmore, J. (1967). On teaching to be critical. *The concept of education,* 192–211.

Paul, R., & Elder, L. (1999). Content is Thinking: Thinking is Content. Retrieved 11 November, 2011 from http://www.criticalthinking.org/pages/content-is-thinking-thinking-is-content/958

Paul, R. W., & Elder, L. (2007). Defining Critical Thinking. Retrieved from www.criticalthinking.org/aboutCT/define_critical_thinking.cfm

Pedersen, J. E., & McCurdy, D. W. (1992). The effects of hands-on, minds-on teaching experiences on attitudes of preservice elementary teachers. *Science Education, 76*(2), 141–146.

Peirce, C. S. (1955). *Philosophical writings of Peirce.* New York: Dover.

Pell, T., & Jarvis, T. (2001). Developing attitude to science scales for use with children from five to eleven years. *International Journal of Science Education, 23*(8), 847–862.

Perkins, D. N., & Saloman, G. (1989). Are cognitive skills context bound? *Educational Researcher, 18*(1), 16–25.

Phillipson, N., & Poad, G. (2010). Use of Dramatic Enquiry to explore controversies in science. *School Science Review, 92*(339), 65–74.

Pithers, R. T., & Soden, R. (2000). Critical thinking in education: A review. *Educational Research, 42*(3), 237–249.

Pritchard, M. S. (1996). *Reasonable Children: Moral Education and Moral Learning.* Lawrence, KS: University Press of Kansas.

Resnick, L. B. (1987). *Education and learning to think.* Washington, DC: National Academies Press.

Ritchhart, R., & Perkins, D. N. (2000). Life in the mindful classroom: Nurturing the disposition of mindfulness. *Journal of Social Issues, 56*(1), 27–47.

Royer, R. (1987). Science Begins with Everyday Thinking. *Thinking, 7*(2), 46–49.

Sadler, T. D. (2004). Informal reasoning regarding socioscientific issues: A critical review of research. *Journal of Research in Science Teaching, 41*(5), 513–536.

Schreier, H. (Ed.). (1997). *Mit Kindern über die Natur philosophieren.* Heinsberg: Dieck.

Schreier, H., & Michalik, K. (2008). In Pursuit of Intellectual Honesty with Children: Children's Philosophy in Hamburg's Elementary Schools Encouraged by Dewey's Ideas. In *Pragmatism, education and children: international philosophical perspectives* (pp. 127–141). Amsterdam: Editions Rodopi.

Settelmaier, E. (2003, March 23-26). *Dilemmas with Dilemmas. Exploring the Suitability of Dilemma Stories as a Way of Addressing Ethical Issues in Science Education.* Proceedings from Annual Meeting of the National Association for Research in Science Teaching, Philadelphia.

Shayer, M. (2002). Not just Piaget; not just Vygotsky, and certainly not Vygotsky as alternative to Piaget. In M. Shayer & P. Adey (Eds.), *Learning intelligence: Cognitive acceleration across the curriculum from 5 to 15 years.* Milton Keynes: Open University Press.

Shayer, M., & Adey, P. S. (1993). Accelerating the development of formal thinking in middle and high school students IV: Three years after a two-year intervention. *Journal of Research in Science Teaching, 30*(4), 351–366.

She, H. C., & Liao, Y. W. (2010). Bridging scientific reasoning and conceptual change through adaptive web-based learning. *Journal of Research in Science Teaching, 47*(1), 91–119.

Siegel, H. (1989). The rationality of science, critical thinking, and science education. *Synthese, 80*(1), 9-41.

Siegel, H. (1990). *Educating reason.* Routledge.

Siegel, H. (1991). The Generalizability of Critical Thinking. *Educational Philosophy and Theory, 23*(1), 18–30.

Smith, G. (1995). Critical Thinking, a Philosophical Community of Inquiry and the Science/Maths Teacher. *Analytic Teaching 15*(2), 43–52.

Songer, N. B., Kelcey, B., & Gotwals, A. W. (2009). How and when does complex reasoning occur? Empirically driven development of a learning progression focused on complex reasoning about biodiversity. *Journal of Research in Science Teaching, 46*(6), 610–31.

Spearman, C. (1927). 'General Intelligence', objectively determined and measured. *American Journal of Psychology, 15*, 201–293.

Splitter, L., & Sharp, A. M. (1995). *Teaching for better thinking: The classroom community of inquiry.* Camberwell, VIC: Australian Council for Educational Research.

Sprod, T. (1997a). 'Nobody really knows': the structure and analysis of social constructivist whole class discussions. *International Journal of Science Education, 19*(8), 911–924.

Sprod, T. (1997b). Longitudinal research and development: Selley on children, light and vision. *International Journal of Science Education, 19*(6), 739–740.

Sprod, T. (1998). "I can change your opinion on that": Social constructivist whole class discussions and their effect on scientific reasoning. *Research in Science Education*, 28(4), 463–480.

Sprod, T., & Jones, B. L. (1997). The sun can't bounce off a bird': Young children and their understanding of vision. *Australian Journal of Early Childhood*, 22, 29–33.

Sprod, T. (1993). *Books into Ideas*. Melbourne: Hawker Brownlow.

Sprod, T. (2001). *Philosophical discussion in moral education: the community of ethical inquiry*. London: Routledge.

Sprod, T. (2011). *Discussions in Science*. Camberwell, Victoria: ACER.

Stone, J. (2011). *Questioning Education: A Critique of Philosophy for Children*. MA dissertation, Institute of Education, London.

Swartz, R. J., Fischer, S. D., & Parks, S. (1998). *Infusing the Teaching of Critical and Creative Thinking into Secondary Science: A Lesson Design Handbook*. Pacific Grove, CA: Critical Thinking Books and Software.

Topping, K. J., & Trickey, S. (2007a). Collaborative philosophical enquiry for school children: cognitive effects at 10–12 years. *British Journal of Educational Psychology*, 77(2), 271–288.

Topping, K. J., & Trickey, S. (2007b). Collaborative philosophical inquiry for schoolchildren: Cognitive gains at 2-year follow-up. *British Journal of Educational Psychology*, 77(4), 787–796.

Topping, K. J., & Trickey, S. (2007c). Impact of philosophical enquiry on school students' interactive behaviour. *Thinking Skills and Creativity*, 2(2), 73–84.

Trickey, S., & Topping, K. J. (2004). Philosophy for children: a systematic review. *Research Papers in Education*, 19(3), 365–380.

Trickey, S., & Topping, K. J. (2006). Collaborative Philosophical Enquiry for School Children: Socio-Emotional Effects at 11 to 12 Years. *School Psychology International*, 27(5), 599.

Tytler, R. (2007). Re-imagining science education: Engaging students in science for Australia's future. *Australian Education Review*, 51.

Tytler, R., Arzi, H. J., & White, R. T. (2005). Editorial–Longitudinal Studies on Student Learning in Science. *Research in science education*, 35(1), 1–2.

Tytler, R., & Peterson, S. (2005). A longitudinal study of children's developing knowledge and reasoning in science. *Research in science education*, 35(1), 63–98.

Urban, K. K. (2004). *Kreativität: Herausforderung für Schule, Wissenschaft und Gesellschaft*. Münster: LIT-Verlag.

Venville, G. J., & Dawson, V. M. (2010). The impact of a classroom intervention on grade 10 students' argumentation skills, informal reasoning, and conceptual understanding of science. *Journal of Research in Science Teaching*, 47(8), 952–977.

Vieira, R. M., Tenreiro-Vieira, C., & Martins, I. (2010). Pensamiento crítico y literacia científica [Critical Thinking and Scientific Literacy]. *Revista Alambique - Didáctica de las Ciencias Experimentales*, 65, 96–103.

Vieira, R. M., Tenreiro-Vieira, C., & Martins, I. P. (2011). Critical thinking: Conceptual clarification and its importance in science education. *Science Education International*, 22(1), 43–54.

Vygotsky, L. S. (1962). *Thought and Language*. Cambridge, MA: MIT Press.

Wartenberg, T. E. (2007). *Thinking on screen: film as philosophy*. London: Taylor & Francis.

Wartenberg, T. E. (2009). *Big ideas for little kids: teaching philosophy through children's literature*. R&L Education.

Weinstein, M. (1990a). Critical Thinking and Scientific Method. *Inquiry: Critical Thinking Across the Disciplines,* 5(2).

Weinstein, M. (1990b). Towards an Account of Argumentation in Science. *Argumentation,* 4(3), 269–298.

Weinstein, M. (1992). Critical Thinking and the Goals of Science Education. *Inquiry: Critical Thinking Across the Disciplines,* 9(1), 3.

Willingham, D. T. (2007). Critical thinking: Why is it so hard to teach? *American Educator*, Summer, 8–19.

Wilson, H., & Mant, J. (2006). *Creativity and excitement in science: Lessons from the AstraZeneca Science Teaching Trust project*. Oxford: Oxford Brookes University.

Wilson, H., & Mant, J. (2011a). What makes an exemplary teacher of science? The pupils' perspective. *School Science Review*, 93(342), 121–125.

Wilson, H., & Mant, J. (2011b). What makes an exemplary teacher of science? The teachers' perspective. *School Science Review, 93*(343), 115–119.

Wilson, H., Mant, J., & Coates, D. (2004). There's Nothing More Exciting Than Science: An AstraZeneca Science Teaching Trust Project. *Primary Science Review, 83*, 20–23.

Wolpert, L. (1992). The unnatural nature of science: Why science does not make (common) sense. London: Faber & Faber.

Worley, P. (2011). *The If Machine: Philosophical Enquiry in the Classroom.* London: Continuum.

Zimmerman, C. (2000). The Development of Scientific Reasoning Skills. *Developmental Review, 20*(1), 99–149.

Zimmerman, C. (2007). The development of scientific thinking skills in elementary and middle school. *Developmental Review, 27*(2), 172–223.

Tim Sprod runs workshops internationally for teachers of philosophy in schools (focusing particularly on science and ethics education), and the International Baccalaureate's Theory of Knowledge programme. Previously, he was the inaugural coordinator of the IB Diploma at The Friends' School in Hobart, Tasmania. He holds a B.Sc. in geology/geophysics from the University of Tasmania; a PGCE from the University College of Wales, Aberystwyth; an M.Ed. investigating the efficacy of discussion in science teaching from the University of Oxford; and a Ph.D. exploring the role of philosophical discussion in ethics education from the University of Tasmania.

His science and philosophy teaching experience covers more than 25 years in Australia, United Kingdom, Papua New Guinea and the Bahamas, at all levels from kindergarten to university undergraduates. As a past secretary of the International Council for Philosophical Inquiry with Children and past chair of the Federation of Australasian Philosophy in Schools Associations, he has an international reputation in the Philosophy for Children movement.

Tim's most recent publication is *Discussions in Science: Promoting conceptual understanding in the middle school years*, a practical book for teachers on the use of philosophical discussion in science classrooms (ACER, 2011). Past books include *Books into Ideas* (Hawker Brownlow, 1993), *Places for Thinking* (with Laurance Splitter, ACER, 1999), *Philosophical Discussion in Moral Education* (Routledge, 2001) and *IB Prepared: approach your assessment the IB way – Theory of Knowledge* (with Antonia Melvin, IB, 2010).

Chapter 49
Informal and Non-formal Education: History of Science in Museums

Anastasia Filippoupoliti and Dimitris Koliopoulos

> *I, Clio the Renowned, eldest daughter of Mnemosyne, Muse of History, warden of memory, wish to teach men lest they neglect the past of their knowledge as of their ignorance.*
>
> (Jean Marc Levy-Léblond 2012)

49.1 Introduction

History of science has a long presence in formal science education. During the late 1960s and early 1970s, an educational movement emerged (mainly in the Anglo-Saxon literature) that argued for the benefits of using the history of science in secondary education. Initial references also carry some preliminary perspectives on the advantages and disadvantages of such a partnership (Brush 1969, 1974; Klopfer and Cooley 1963). These perspectives characterise the research field diachronically, but the issues of instructional strategy choices and methodological techniques with which history of science can be effectively linked to science education are still open research questions.

The use of history of science in formal education is related to three trends in educational research:

1. A humanistic approach to science teaching that aims to contribute to the 'broad cultivation' and scientific literacy of pupils as citizens (e.g. Klopfer 1969; Langevin 1964; Matthews 1994/2014)

A. Filippoupoliti (✉)
Democritus University of Thrace, Alexandroupolis, Greece
e-mail: afilipp@yahoo.gr

D. Koliopoulos
University of Patras, Patras, Greece

M.R. Matthews (ed.), *International Handbook of Research in History, Philosophy and Science Teaching*, DOI 10.1007/978-94-007-7654-8_49, © Springer Science+Business Media Dordrecht 2014

2. The development of student understanding of the nature and characteristics of scientific knowledge, mainly via the 'nature of science' educational movement (e.g. Hodson 2008; Lederman 2007)
3. The cognitive development of pupils and the shift of interest from methodological to conceptual dimensions of scientific knowledge (e.g. Monk and Osborne 1997; Nersessian 1992; Strauss 1988)

Despite the increasing influence of the history of science in formal science education during recent decades, one cannot ignore the difficulties and the obstacles that a broader educational use of the history of science faces. Among these, Hottecke and Silva (2010) refer to the negative stance of educators to any proposed change to the traditional teaching culture and the boundaries imposed upon educators by the official science curriculum that either ignores or degrades the role and importance of history of science in teaching.

It is interesting therefore to examine what happens with the kind of dissemination of history of science that originates or relates closely to the modern science museum. The dissemination of history of science is related in this case with informal and non-formal educational approaches.[1] What are the aims of this sort of dissemination, how are they achieved and how are they related to non-formal and informal education? The present review aims to bring forward these issues and open a potential academic discussion. We first discuss the types of museums that have been developed; we then analyse the history of science as an exhibition and communication element; and finally, we approach the subject as an educational element. The review will not address how the science museum is being treated as a research subject itself by historians of science.

49.2 A Definition of a Science Museum and the Types of Science Museums

Museum studies have grown since the late 1960s following a considerable rise in the number and types of museums worldwide. Museum studies literature offers a wealth of definitions and classifications of museums organised mainly according to

---

[1] In the present article, the terms *informal education* and *non-formal education* are considered as distinct terms (Coombs and Ahmed 1973; Escot 1999; Eshach 2006). An *informal* educational process is not an organised and systematic one that occurs in different educational settings (schools, museums etc.). It is a process – quite often unintentional – offered by the personal environment of an individual. The interrelationship between the individual and the exhibition during a museum visit is a typical example of an informal educational process. In contrast, *non-formal* educational environments are related to autonomous cultural institutions that provide scientific knowledge, such as museums, and are environments that offer organised educational activities (as in the case of educational programmes in museums or programmes that are organised between school and museum).

the academic disciplines to which they refer through their collections, exhibitions and public programmes.

The science museum is not a homogenous entity. The nature and characteristics of the science museum can be studied through the variety of categorisations produced by both museum professionals and museum researchers. These categorisations group museums based either on the way in which these institutions confront collecting, displaying and interpretation of objects and the way they conceive exhibition space (Wagensberg 2004) or on the evolution of the science museum (de Clercq 2003; Friedman 2010). The latter are significant not solely because the history of the museum as social institution as demonstrated by the related literature on the history of museums and collections is a vital subject (Arnold 2006; Findlen 1989, 1994; Impey and MacGregor [1985] 2001; Yanni 1999), but also because this literature can be used to interpret the function of modern science museums by either researchers coming from fields of inquiry other than museum studies (i.e. science educators) or by science teachers (Koliopoulos 2003).

A history of the science museum goes back to the Renaissance collections of curiosities and learned cabinets (e.g. the cabinet of Francesco I de Medici in Florence (Findlen 2000; Pearce 1993) and the collections of seventeenth-century philosophical and scientific institutions (e.g. collections held by the Royal Society of London). During the second half of eighteenth century, along with the founding of the first public museums, a number of museums of natural history were established. Unlike the earlier cabinets, these were public institutions allowing a large number of visitors into their exhibition spaces. In addition, the galleries exhibited objects according to a classification system that was closely adapted to distinct academic disciplines.

These institutions praised the collected object (e.g. scientific instruments, natural history specimens and technological artefacts), accumulated natural curiosities and man-made artefacts and favoured the wooden or glass-case presentation. The *Musée des Arts et Métiers* in Paris is an exemplary case reflecting this exhibition philosophy (Ferriot and Jacomy 2000). There the visitor was considered a passive admirer of a glorious scientific past. The act of interpretation was not facilitated by the museum curator, although some interpretation was provided by a few means such as the object's label. In this context, scientific objects were displayed as art objects and admired by the upper class (Bennett 1995). The *Natural History Museum* in London took a similar approach.

University science collections fall into the same category given that most of them have been created to act as repositories of worn and outdated scientific apparatus once used in the teaching of physics and chemistry or collections of objects related to the natural sciences (e.g. stuffed animals). The museum of the King's College London that was founded to host the King George III science collections in mid-nineteenth-century London is an interesting case in point, yet by the end of the century it had become a mere repository (Filippoupoliti 2011).

Between the middle of the nineteenth century and World War II, another type of museum emerged that differed from the traditional museums just described. During this time, museums also embraced an explicit educational mission following the mid-nineteenth-century demand for educating the lay public. Interpretation of the exhibition was performed by presentation of a series of objects that reflected a certain scientific concept or idea, and an attempt was made to form concise units according to certain scientific themes (e.g. energy, power, physics etc.). The *Science Museum* in London (est. 1885) and the *Deutsches Museum* (est. 1903) in Munich are examples of this category, although in recent decades these museums have enhanced the exhibition space with modern design and interactive exhibits. Along with the older galleries, a series of interactive hands-on exhibits are presented to update the established scientific narrative (Durant 2000). This category also includes the *Museum of the History of Science* at Oxford (est. 1925) and the *Whipple Museum of the History of Science* at Cambridge (est. 1944), the former *Istituto e Museo di Storia della Scienza* now the *Museo Galileo* (est. 1927) in Florence and the *Museum Boerhaave* (est. 1928) in Leiden, Holland (de Clercq 2003).

Although science centres differ from science museums, they are usually treated together in the literature. A science centre has a distinct experimental philosophy that moves from the display of the authentic object to create an original/meaningful museum experience through active visitor participation. Beyond object worship, it is the exhibition space that matters more as it assimilates the laboratory, a gallery of research and a place of demonstration. Historically, this type of a science institution can be traced back to the 1930s, when the *Palais de la Découverte* in Paris was founded according to a rationale relevant to the division of academic scientific disciplines, followed by the San Francisco *Exploratorium: The Museum of Science, Art and Human Perception* (est. 1960s), which is regarded as the 'father' of science centres (Hein 1990; Cole 2009). Another example is the *Cité des Sciences et de l'Industrie* in Paris, in which the focus of exhibition activity is the social use of natural sciences and technology (Caro 1997; Zana 2005). This science centre has created a special children's science museum that offers exhibitions and activities designed to address the cognitive and emotional needs of young children (Guichard 1998).

The development of science centres has considerably influenced museological approach and museographical practice of even the most traditional museums. For example, the recently renovated *Museo Galileo* in Florence and the *Museum of the History of Science* at the University of Oxford have improved their approaches to the display of objects. They have modernised the permanent and temporary exhibitions as well as their communications approach to the public (e.g. including new interactive activities as part of an exhibition and providing virtual tours via the museum website). The hybrid form that such museums have become raises the issue of establishing a new educational identity for these institutions (Quin 1993).

We pose the following questions which we will tackle in the following section: How does each of the science museum types implement the history of science in exhibition and educational practice? What sort of interpretation do they offer? Do each of these different interpretation patterns offer the same epistemological status and give a certain communication role to the history of science? Does the history of science constitute one of the seminal elements in the diffusion of scientific knowledge communicated via science museums, or are museums designated solely for the history of science the only appropriate institutions to research, exhibit and diffuse objects, ideas and issues related to the history of science?

49.3 History of Science as an Exhibit and Communication Element

History of science is an exhibited theme found in a variety of museum types. Museums of the history of science distinctly safeguard, interpret and display the material culture of science (Bennett 1997, 2005; Camerota 2011). Museums of the history of science are usually university museums that base their foundation on collections of scientific instruments and apparatuses once used in research and university teaching or on private collections that have been donated to the museum. Two characteristic examples are the *Museum of the History of Science* in Oxford (established 1924) by the gift of the collection of Lewis Evans[2] to the University and the *Whipple Museum of the History of Science* at the University of Cambridge founded in 1956 to house Robert Whipple's[3] collection of scientific instruments and rare books (Bennett 1997; Taub and Willmoth 2006). In these institutions, the history of science is present in many ways, most importantly in the use of elements of the history of science in exhibitions in which a part or the majority of the scientific collections (authentic scientific instruments or biological specimens) are used.

How then does a museum of the history of science differ from a science museum? Bennett (2005) notes that

> museums of the history of science contain old instruments and apparatuses, just like any science museum … If it is not the nature of the collections that is different, it should be the assumptions about what the collections are for, which will inform how they are selected and how they are used. (pp. 606–607)

Because of their privileged relationship with academic history of science, museums of the history of science can certainly provide exhibitions of their collections that gain their meaning from the cognitive, methodological and cultural dimension of the history of science.

Another category of science museum where history of science is present includes those institutions whose historical tradition, collections and particular museological/museographical approaches make possible the presentation of a history of science exhibition narrative even though the history of science is not a distinct part of the institutional mission such as university museums that hold collections of scientific instruments and natural history and biological specimens (Tucci 2002; Lourenço 2005; Subiran et al. 2009). One difficulty that this type of museum confronts in presenting collections to the broader audience is the absence of a unified and coherent theme

---

[2] Lewis Evans (1853–1930) was a collector, brother of the notable archaeologist, Sir Arthur Evans, who excavated the Palace of Knossos, Crete (Greece). See also P. de Clercq, Lewis Evans and the White City Exhibitions, *Sphaere. The online journal of the Museum of the History of Science, University of Oxford,* available at http://www.mhs.ox.ac.uk/sphaera/index.htm?issue11/articl4.

[3] Robert Stewart Whipple (1871–1953) donated more than 1,000 scientific instruments to the University of Cambridge in 1944. See also S. De Renzi (1998). Between the market and the academy: Robert S. Whipple (1872–1953) as a collector of science books. In R. Myers and M. Harris (eds), *Medicine, Mortality and the Book Trade* (pp. 87–108). St. Paul's Bibliographies: Oak Knoll Press.

topic that could become the basis of an institution recognisable by non-experts (Antoine 2010, p. 9). One such theme topic, according to Antoine (2010), is the implementation of the scientific method via elements from the history and philosophy of science.

Non-university museums such as the *Musée des Arts et Métiers* in Paris and the *Science Museum* in London that hold scientific collections are good examples of this category of museum. Although their original aim was not the dissemination of the history of science,[4] today these museums are ideal places for the display of science because of the richness of their collections. Also, institutions such as centres of scientific research and for the popularisation of science (e.g. *Royal Institution of Great Britain*), scientific institutions (e.g. *Royal Observatory*, Greenwich, England) and laboratories or the private premises of eminent men of science that have become house museums (e.g. the *Charles Darwin Down House* in England and the *Maison d'Ampère* in France) are potential places for disseminating the history of science.

The implementation of history of science can differ among museums according to their type. Studying three institutions that display collections of historic astronomical instruments, Maison (2002) suggested three different ways of exhibiting such collections. The *Musée des Arts et Métiers* emphasises the technological dimension of the displayed scientific instruments, and the exhibition is based on historical evidence that presents a holistic view of the technical culture from Renaissance to the present day. In contrast, the *Observatoire de Paris* emphasises the concepts of the physical sciences and how these are intertwined with the function of astronomical instruments. Finally, the *Royal Observatory of Greenwich* displays collections with the aim of presenting the social and economic aspects related to the development of astronomy research over time.

Finally, even though science centres don't hold any permanent collections of authentic/historical objects, occasionally they may host temporary exhibitions that present elements of the history of science. These centres seem to function as contemporary scientific textbooks that, according to Kuhn, can hide the process of how scientific knowledge is obtained. If someone replaces the word 'textbooks' with 'science centres' in the next extract, the meaning would not be twisted:

> Textbooks thus begin by truncating the scientist's sense of his discipline's history and then proceed to supply a substitute for what they have eliminated. Characteristically, textbooks of science contain just a bit of history, either in an introductory chapter or, more often, in scattered references to the great heroes of an earlier age. From such references both students and professionals come to feel like participants in a long-standing historical tradition. Yet the textbook-derived tradition in which scientists come to sense their participation is one that, in fact, never existed. For reasons that are both obvious and highly functional, science textbooks (and too many of the older histories of science) refer only to that part of the work of past scientists that can easily be viewed as contributions to the statement and solution of the texts' paradigm problems. Partly by selection and partly by distortion, the scientists of earlier ages are implicitly represented as having worked upon the same set of fixed problems and in accordance with the same set of fixed canons that the most recent revolution in scientific theory and method has made them seem scientific. (Kuhn 1970, pp. 137–38)

---

[4] Moreover, history of science as an academic discipline emerged later.

What is the mode of history of science as an *exhibition narrative*? Which one of the history of science narratives one occasionally confronts in museum exhibitions? Are historical facts explained and interpreted? Is emphasis being given to historical moments/turning points and the importance of controversies and scientific revolutions? Is it more important to research science as a social action that is formed by the social-historical-cultural context? Or is it more seminal to trace the history of science as a history of ideas or as an exploration of the material culture and non-literary traditions? A first attempt to answer these questions will be presented in the following paragraphs.

Even though history of science as an academic discipline emerged during the first part of the twentieth century, historic scientific instruments were already on display by the second half of the nineteenth century in museums such as the King's College London King George III Museum as well as in international/world exhibitions such as the Special Loan Exhibition in London in 1876. Historian Steven Conn has called the museum exhibition culture of that period an 'object-based epistemology' (Conn 2000). According to that perspective, the exhibited object (e.g. a scientific instrument) is able to confirm and support the 'scientific power' of a phenomenon or an idea and therefore as a historic object can stand as a symbol of scientific progress. For many decades in the early twentieth century, museums preserved the type of museological narrative that they inherited from their nineteenth-century predecessors. For instance, scientific instruments and apparatuses were preferably displayed in a thematic way, and their mode of display reflected an encyclopaedia of natural sciences in which each displayed object stood for a particular scientific phenomenon or process.

During the 1980s, shifts in the museological and museographical approach to science museums (Schiele and Koster 1998) in research trends in the history of science and in the increasing interest of historians of science in science collections led to important changes in the ways museum curators displayed the history of science in exhibitions. At least three epistemological approaches can be identified in these museum exhibitions. The first approach is the traditional one mentioned earlier that treats the history of science as the documentation of objects and facts. The second approach treats the history of science as a history of ideas and is not broadly used to weave a narrative into a science exhibition. In this case, the authenticity of the science collection is of minor importance (i.e. whether objects are historic scientific instruments or reconstructions). Emphasis is being given to how an idea (or ideas) is born, developed and cognitively treated in order to give meaning to objects. The *Grande Galerie de l'Evolution* of the *Muséum National d'Histoire Naturelle* in Paris focuses on the evolution of species (Van Praet 1995). Other examples of such an exhibition approach include the following: The exhibition 'Exploring the World, Constructing Worlds: Experimental Cultures of Physics from the sixteenth to nineteenth Century' in the *Museum of Natural History and Pre-History* in Oldenburg, Germany (Heering and Muller 2002), which addresses issues such as 'astronomical and experimental practice in the sixteenth and seventeenth centuries' and 'the science of precision measurement in the nineteenth century' and the Galilean exhibit of the *Exploratorium* in San Francisco, entitled 'The Gravity-Powered Calculator', which was also reconstructed by Cerretta (2012).

Exhibitions belonging to the above-mentioned two categories aim at disseminating the content, the process and the product of science from an internal point of view, the view of science. In contrast, a third approach considers trends in the history of science literature that view science as an example of culture with particular practices and tools that are affected, developed and transformed according to the cultural and historical context in which they have been developed, including non-scientific factors (Golinski 1998; Galison and Thompson 1999; Daston 2000).

In addition, the emergence of Social Studies of Science since the 1980s has provided researchers with fresh perspectives on understanding the intersection of scientific practice and culture (Latour and Wooglar 1986; Latour 1987). In this context, emphasis is given to how scientific practice is being formulated in the laboratory and in the performance of crucial experiments (Arnold 1996; Chittenden et al. 2004). For instance, the exhibitions hosted at the *Wellcome Collection* of the Wellcome Trust in London and the temporary exhibitions hosted in the Science Museum in London and the Nobel Museum in Stockholm are examples of cultural turns in the reading of the history of science.[5] From the perspective of science education, Pedretti (2002) also refers to the use of the history of science by science museums addressing socioscientific issues.

The above-mentioned modes of introducing the history of science in museums lead to informal education and informal learning. Museum visitors and school groups in particular can gain an interest in science as well as gain a popularised conception of the content and method of science (Stocklmayer et al. 2010). However, this kind of popularisation eliminates the systemic dimension of the meaning of scientific and historic knowledge and consequently sometimes deforms and transforms it to such an extent as to alter totally its meaning and, in still other instances, leads to paradoxical assertions (Jacobi 1999; Jurdant 2009). The risks stemming from the popularisation of scientific and historical knowledge could possibly be reduced if museums place more emphasis on the educational dimension of communication and on their function as institutions for non-formal education (Escot 1999). This issue will be analytically treated in the following section.

49.4 History of Science as an Educational Tool

Science museums are gradually increasing their emphasis on their science education functions (Teichmann 1981; Tran 2007; Stocklmayer et al. 2010). Museums produce a wealth of educational material for all types of visitors, the design of which varies according to type, content and creator. For instance, some materials are composed by

---

[5] See, for example, the Nobel Museum Centennial exhibition *Cultures of Creativity* (Stockholm, Sweden) which examines creativity in science. Available at http://www.nobelmuseum.se/en/exhibitions/cultures-of-creativity

in-house museum professionals linking the programme directly to certain exhibits and perhaps implying that an exhibit can easily be transformed to educational material.

Many science museums design programmes in collaboration with schools and other educational institutions, either because they seek to consider the concerns raised by such institutions or because they seek theoretical and/or practical tools to support exhibit design. University departments that offer postgraduate museum studies courses or science education courses provide essential support towards the design of meaningful educational programmes for museums' visitors. Does the history of science have a specific role in the design of museum educational programmes? Do science museum professionals need formal education about how to give certain meanings to science collections through the aid of history of science, exhibitions and associated narratives? Or, is non-formal/informal education sufficient to act as a means of diffusing scientific knowledge?

Our review of the educational tools used by museums to communicate the history of science elements identified four categories of educational material:

(1) *Guided tours focused on narratives from the history of science.* This is the simplest educational intervention, engaging the history of science in a sequential science museum-guided tour. These tours typically present stories of people, ideas and/or practices from the history of science field and may contribute to raising the interest of visitors for the exhibition or to making meaning from an exhibition.

For instance, Fadel (2011) uses history of science elements in lectures given during the performance of experiments at the *Palais de la Découverte* in Paris. He notes that the history of science can be a very powerful tool for introducing a new concept, idea or theory. Sometimes, stories and anecdotes taken from history are helpful as brief breaks to keep the attention of the audience. In other cases history can help people realise how answers to questions always seem obvious when one already knows the answer but seldom are apparent beforehand (Fadel 2011).

In formal education, the design and narration of stories that introduce elements of the history of science is a common practice (Stinner et al. 2003). Unlike formal education, during a guided tour in the museum, the guide cannot expand the narration to explain a topic in detail. In this context, guided museum tours using narratives from the history of science are the weakest type of educational programme for presenting the history of science.

(2) *Museum educational programmes/workshops.* These activities are designed mostly for students and teachers, not the general public. In many instances, these programmes are developed and performed by specialised museum educators (Tran 2007). The *Deutsches Museum* is one example of a successful implementation of history of science elements in museum educational programmes. Teichmann (1981) points out that

historical objects displayed are to be integrated into the other educational activities of the museum and not simply remain commemorative pieces; i.e. historical collections and modern didactics are to be united according to the following aspects: (a) often modern situations can be clarified by means of historical explanations; (b) the completely different conditions of the past and the then existing specific difficulties in the realization of new knowledge, can offer a valuable lesson in questioning the apparently foregone conclusions of today; (c) the incorporation of modern and historical objects into the framework of human science and cultural development, can exhibit the characteristic position of science and technology (p. 474).

Educational programmes are structured educational environments designed to acquaint students and teachers with scientific and historical knowledge in a systematic way. For example, the context for knowledge could be the experimental history of physical sciences (Sibum 2000), the construction of concepts and methods via the reconstruction of artefacts or historical experiments (Teichmann 1999; Heering and Muller 2002) or the historical development of our understanding of the taxonomy of biological organisms (Faria et al. 2012).

(3) *The collaboration between museums and formal education.* Many researchers have argued that the collaboration between school and museum can promote achieving both cognitive and emotional student outcomes. A number of studies suggest that the museum visit and the children's activities during the visit should be accompanied by school before and after the visit (Griffin and Symington 1997; Anderson and Lucas 1997; Anderson et al. 2000; Guisasola et al. 2005; Guisasola et al. 2009). Other researchers claim that the involvement of teachers in non-formal educational settings such as science museums should be part of teacher training in science and pedagogy (DeWitt and Osborne 2007).

Unfortunately, studies of the development and evaluation of educational programmes in museums that introduce elements of the history of science are few. Anderson and colleagues (2011) describe a museum workshop about the role of artefact analysis/manipulations on research and teaching in the history of science and technology. In this study students from university departments of education also addressed this subject during classroom coursework using Eotvos torsion balance, an instrument used to measure small gravitational variations. Students constructed three narratives related to the science of geodesy and discussed issues related to laboratory practice and the nature of science.

Falomo-Bernarduzzi and colleagues (2012) have developed activities related to Galileo's laboratory that are designed to take place either in the museum or in the school and explain that these

activities do not 'incidentally' interest schools, because they happen to connect with the school curriculum, but they are thought out with each school for the school. These workshops give clues which are the starting points for classroom activities linked to the project but also part of normal school learning. (Falomo-Bernarduzzi et al. 2012)

The researchers describe projects that rely extensively on the history of science in a number of ways using primary and secondary sources, museum exhibitions, multimedia and hands-on reconstructions of historical experiments. More specifically, they present activities that are based on the exhibition 'Laboratorio di Galileo' which includes reproductions of the apparatuses designed and used by Galileo for his experiments in mechanics.

Finally, Paparou (2011) describes lecture-demonstration activities created and performed by teachers in the classroom using collections of scientific instruments from the local *Museum of History and Physics* of the first high school of Chios

Island (Greece). Examples of such lecture-demonstrations include 'The first days of electricity' and 'The history of magnets and compasses'. During these programmes, participants were invited to observe and compare scientific instruments, conduct experiments and evaluate the experimental results, make explanatory hypotheses and explore historical scientific documents (Paparou 2011).

All the educational attempts that were discussed in the previous sections focus on the study of scientific instruments and experiments as tools for educating students and teachers about history of science issues in the context of collaboration between museums and formal education institutions. It is apparent that such a collaboration can play a seminal role in evaluating and transforming scientific collections (original/historical collections, digital collections or collections of reconstructed instruments) from tools of research to tools of education (Heering 2011).

49.5 Conclusion

The variety of reviews that refer to the introduction of elements of history of science in primary and secondary school (Matthews 1994/2014; Duschl 1994; Seroglou and Koumaras 2001; Hottecke and Silva 2010) indicates the systematic and continuous involvement of historians of science and science educators with the issue of introducing elements from the history of science into formal science education. In contrast, as the present review has shown, the study of the role of the history of science in informal and non-formal science education is heterogeneous and fragmentary. It is necessary to raise new research questions and construct new lines of research to investigate the subject in a more systematic way.

We have suggested three lines of research strands below:

(1) *The epistemological research strand.* This strand refers to those research questions primarily of interest to science museum professionals related to the role that history of science can play in the realisation of the communication and education objectives of museums. How and why can the history of science as presented through museum collections contribute to the rescue, preservation and diffusion of scientific heritage and culture at local, national and international levels? Lourenco (2012), for example, suggests that

> the increased interest by the historian of science creates opportunities for a more significant role of history in museums of science, potentially resulting in better documented collections, as well as more meaningful and contextualized exhibitions and educational programmes. However, more history in museums of science requires considerable structural and cultural changes in their traditional missions, roles and practices. (Lourenco 2012)

On the other hand, primary questions that in our opinion should concern science centres that aim at the diffusion and popularisation of modern scientific knowledge are the following: Is it possible, and if so, how could the history of science contribute to reducing the ever-growing gap between the production of scientific knowledge and its understanding by lay people? How could the history of science contribute to restorating the relationship between science and culture that has increasingly soured since the early twentieth century? (Bensaude-Vincent 2001; Lévy-Leblond 2004). Is it possible to incorporate the narrative of the history of scientific ideas into the narrative of the modern world and its relationship to contemporary society, or should they be considered two epistemologically incompatible narratives? These questions are also interrelated to the following research strand.

(2) *The museological/museographical research strand.* This strand is mostly related to the way in which science museums take into account the history of science and translate it into a communication and educational tool to achieve their educational mission. Historians of science, museologists and possibly science educators need to collaborate towards that end. Referring to collections and exhibitions of the *Science Museum* in London, Bud (1997) noted that

> before the Second World War the progressivism of the galleries and the inspiration of its greatest icons mostly matched the views of academics. However, the post-war years, which saw an efflorescence of paper-based historiography of science, saw too a decoupling between the interests of academics interested in intellectual process and of curators focused upon their objects. This decoupling meant that the history of science of which the Museum was the public space, was somewhat distanced from the burgeoning academic discipline. (pp. 50–51)

Bud makes clear that exhibitions of science act as important means of transformation of scientific knowledge, scientific and social practices and authentic objects to content, exhibits and forms of display, so that they could be successfully communicated to broader audiences. The concept of 'mediating transposition' used by Guichard and Martinand (2000) and the 'museographic transposition' used by Simonneaux and Jacobi (1997) constitute a proper context in which exhibitions that introduce elements of history of science used in combination with contemporary communication strategies and museographical techniques could be analysed or designed. In this context, further research questions could be posed in the following broad areas: (a) in relation to the deconstruction and reconstruction of a historical subject in science and the identification of possible related misconceptions often found in exhibitions (i.e. epistemological analysis, see Foss Mortensen 2010) and/or (b) the decoding and recoding of messages, if we regard exhibitions as pedagogical multi-modal texts (i.e. semiotic analysis, see Anyfandi et al. 2010).

(3) *The learning/pedagogical research strand.* In this noteworthy heterogeneous strand, the main issue is the investigation of learning in informal and non-formal settings and more particularly if and how cognitive progress of visitors is achieved during a science museum visit (e.g. Anderson et al. 2003; Martin 2004; Griffin 2004). Can history of science maximise visitors' learning best when designed as a communicational element or as an educational tool? Is it better to use the history of science so that museum visitors can construct understandings of the nature of science and of conceptual elements of science? Studies addressing such questions can inform researchers in the fields of psychology and science education as well as designers of science exhibitions who seek to develop

a museological/museographical approach that maximises visitor learning. An important dimension of this research strand is developmental studies that investigate possible correlations between student learning of the official school programme and the coordinated activities that take place in schools and museums conjointly. In addition, existing didactic models that investigate how the introduction of elements of the history of science into formal education influence students' cognitive progress (e.g. Monk and Osborne 1997; Hottecke et al. 2012) could be altered to include activities in museum settings.

A necessary precondition for the establishment of the above-mentioned research strands is the acceptance of the strong transdisciplinary and interdisciplinary nature of this research and the creation of a collegial environment among the researchers involved. In other words, we need to accept that the intersection of the history of science, scientific museology and science education represents a fruitful set for the consideration of the theoretical background, the methodological approach and the social practices of science learning.

References

Anderson, D. and Lucas, K.B. (1997). The effectiveness of orienting students to the physical features of a science museum prior to visitation. *Research in Science Education*, 27(4), 485–495.

Anderson, D., Lucas, K.B., Ginns, I.S. & Dierking, L.D. (2000). Development of knowledge about electricity and magnetism during a visit to a science museum and related post-visit activities. *Science Education*, 84(5), 658–679.

Anderson, D., Lucas, K.B. & Ginns, I.S. (2003). Theoretical perspectives on learning in an informal setting. *Journal of Research in Science Teaching*, 40(2), 177–199.

Anderson, K., Frappier, M., Neswald, E. & Trim, H. (2011) Reading instruments: Objects, texts and museums. *Science & Education*, doi: 10.1007/s11191-011-9391-y

Antoine, M. (2010). Les universités doivent-elles vraiment exposer leurs collections au grand public? *La Lettre de l'OCIM*, 129, 7–12.

Anyfandi, G., Koulaidis, V. & Dimopoulos, K. (2010). A social-semiotic framework for the analysis of science exhibits. In A. Filippoupoliti (ed.) *Science Exhibitions: Communication and Evaluation* (pp. 102–149). Edinburgh: Museums Etc.

Arnold, K. (1996). Presenting science as product or as process: Museums and the making of science. In S. M. Pearce (ed.) *Exploring Science in Museums* (pp. 57–78). London: Athlone Press.

Arnold, K. (2006). *Cabinets for the Curious. Looking back at Early English Museums.* London: Ashgate.

Bennett, J. (1997). Museums and the establishment of the history of science at Oxford and Cambridge. *British Journal for the History of Science*, 30, 29–46.

Bennett, J. (2005). Museums and the history of science: Practitioner's postscript. *Isis*, 96(4), 602–608.

Bennett, T. (1995). *The Birth of the Museum: History, Theory, Politics*. London & New York: Routledge.

Bensaude-Vincent, B. (2001). A genealogy of the increasing gap between science and the public. *Public Understanding of Science*, 10(1), 99–113.

Brush, S.G. (1969). The role of history in the teaching of physics. *The Physics Teacher*, 7, 271–280.

Brush, S.G. (1974). Should the History of Science be rated X? *Science*, 183 (4130), 1164–1172.

Bud, R. (1997). History of science and the Science Museum. *British Journal of the History of Science*, 30, 47–50.

Camerota, F. (2011). Promoting scientific heritage and disseminating scientific culture: The role of the Museo Galileo in Florence. In F. Seroglou, V. Koulountzos & A. Siatras (eds) *Proceedings of the 11th International IHPST and 6th Greek History, Philosophy and Science Teaching Joint Conference* (pp. 117–119). Thessaloniki: Epikentro.

Caro, P. (1997). Tensions between science and education in museums and elsewhere. In G. Famelo & J. Carding (eds) *Here and Now, Contemporary Science and Technology in Museums and Science Centres* (pp. 219–225). London: The Trustees of the Science.

Cerretta, P. (2012). Gravity-powered calculator, a Galilean exhibit. *Science & Education* special issue 'History of science in museums', doi: 10.1007/s11191-012-9549-2.

Chittenden, D., Farmello, G. & Lewenstein, B. (2004). *Creating Connections, Museums and the Public Understanding of Current Research*. Creek CA: Altamira Press.

Cole, K.C. (2009). *Something Incredibly Wonderful Happens: Frank Oppenheimer and the World He Made up*. Boston: Houghton Mifflin Harcourt.

Conn, S. (2000). *Museums and American Intellectual Life 1876–1926*. Chicago: University of Chicago Press.

Coombs, P.H. & Ahmed, M. (1973). *New Paths to Learning for Rural Children*. New York: International Centre for Educational Development.

Daston, L. (ed.) (2000). *Biographies of Scientific Objects*. Chicago: University of Chicago Press.

De Clercq, J.S. (2003). Museums as a mirror of society: a Darwinian look at the development of museums and collections of science. In P. Tirell (ed.) *Proceedings of the 3rd Conference of the International Committee for University Museums and Collections* (pp. 57–65). Oklahoma: UMAC Publication.

De Clercq, P., Lewis Evans & the White City Exhibitions. *Sphaere. The online journal of the Museum of the History of Science, University of Oxford*. http://www.mhs.ox.ac.uk/sphaera/index.htm?issue11/articl4. Accessed 28 November 2012.

De Renzi, S. (1998). Between the market and the academy: Robert S. Whipple (1872–1953) as a collector of science books. In R. Myers & M. Harris (eds), *Medicine, Mortality and the Book Trade* (pp. 87–108). Newcastle, DE: Oak Knoll Press.

DeWitt, J. & Osborne, J. (2007). Supporting teachers on science-focused school trips: Towards an integrated framework of theory and practice. *International Journal of Science Education*, 29(6), 685–710.

Durant, J. (2000). The Wellcome Wing at the Science Museum London. A breathtaking theatre of contemporary science. *Museologia, An International Journal of Museology*, 1, 43–48.

Duschl, R. (1994). Research on the history and philosophy of science. In D. Gable (Ed.) *Handbook of Research in Science Teaching* (pp. 443–465). New York: Macmillan.

Escot, C. (1999). *La culture scientifique et technologique dans l'éducation non formelle*. Paris: UNESCO Publications.

Eshach, H. (2006). Bridging in-school and out-of-school learning: formal, non-formal and informal. In H. Eshach (ed.) *Science literacy on primary schools and pre-schools* (pp. 115–141). Dordrecht Netherlands: Springer.

Fadel, K. (2011). History of science, education and popular science show: The case of the particle accelerator at the Palais de la Découverte (Paris). Paper presented in the 11th International IHPST and 6th Greek History, Philosophy and Science Teaching Joint Conference, Thessaloniki, 1–5 July 2011.

Falomo-Bernarduzzi, L., Albanesi, G. & Bevilacqua, F. (2012). Museum heroes all: the Pavia approach to school-science museum interactions. *Science & Education* special issue 'History of science in museums', doi: 10.1007/s11191-012-9541-x

Faria, C., Pereira, G. & Chagas, I. (2012). D. Carlos de Bragança, a pioneer of experimental Marine Oceanography: Filling the gap between formal and informal science education. *Science & Education*, 21(6), 813–826.

Ferriot, D. & Jacomy, B. (2000). The Musée des Arts et Métiers. In S. Lindqvist (ed.) *Museums of Modern Science*, (pp. 29–42). New York: Science History Publications.

Filippoupoliti, A. (2011). Premises for exhibition and for use: King's College London Museum, mid to late nineteenth century. *Museum History Journal* special issue 'University museums', 4(1), 11–28.

Findlen, P. (1989). The museum: its classical etymology and Renaissance genealogy, *Journal of the History of Collections*, 1(1), 59–78.

Findlen, P. (1994). *Possessing Nature: Museums, Collecting and Scientific Culture in Early Italy.* Berkeley: University of California Press.

Findlen, P. (2000). The modern muses. Renaissance collecting and the cult of remembrance. In S.A. Crane (ed.) *Museums and Memory* (pp. 161–178). Stanford CA: Stanford University Press.

Foss Mortensen, M. (2010). Museographic transposition: The development of a museum exhibit on animal adaptations to darkness. *Education et Didactique*, 4(1), 115–138.

Friedman, A. (2010). The evolution of the science museum. *Physics Today*, 63(10), 45–51.

Galison, P. & Thompson, E. (eds) (1999). *The Architecture of Science.* Cambridge MA: MIT Press.

Golinski, J. (1998). *Making Natural Knowledge: Constructivism and the History of Science.* Cambridge: Cambridge University Press.

Griffin, J. & Symington, D. (1997). Moving from task-oriented to learning-oriented strategies on school excursions to museums. *Science Education*, 81(6), 763–779.

Griffin, J. (2004). Research on Students and Museums: Looking More Closely at the Students in School Groups. *Science Education*, 88(S1), 59–70.

Guichard, J. (1998). Adapter la muséologie aux enfants. In B. Schiele & E.H. Koster (eds) *La révolution de la muséologie des Sciences: Vers les musées du XXI siècle* (pp. 207–248). Lyon: Presses Universitaires de Lyon.

Guichard J. & Martinand J.-L. (2000). *Médiatique des Sciences.* Presses Universitaires de France.

Guisasola, J., Morentin, M. & Zuza, K. (2005). School visits to science museums and learning sciences: A complex relationship. *Physics Education*, 40(6), 544–549.

Guisasola, J., Solbes, J., Barragues, J.-I., Morentin, M., & Moreno, A. (2009). Students' understanding of the special theory of relativity and design for a guided visit to a Science Museum. *International Journal of Science Education*, 31(15), 2085–2104.

Heering, P. & Muller, F. (2002). Cultures of experimental practice. An approach in a museum. *Science & Education*, 11(2), 203–214.

Heering, P. (2011). Tools for investigation, tools for instruction: Potential transformations of instruments in the transfer from research to teaching. In P. Heering and R. Wittje (eds) *Learning by Doing. Experiments and Instruments in the History of Science Teaching* (pp. 15–30). Stuttgart: Franz Steiner Verlag.

Hein, H. (1990). *The Exploratorium. The Museum as Laboratory.* Washington: Smithsonian Institution.

Hodson, D. (2008). *Towards Scientific Literacy: A Teachers' Guide to the History, Philosophy and Sociology of Science.* Rotterdam: Sense Publishers.

Hottecke, D. & Silva, C. (2010). Why implementing history and philosophy in school science education is a challenge: An analysis of obstacles. *Science & Education*, 20(3–4), 293–316.

Hottecke, D., Henke, A. & Riess, F. (2012). Implementing history and philosophy in science teaching: Strategies, methods, results and experiences from the European HIPST Project. *Science & Education*, 21(9), 1233–1261.

Impey, O. & MacGregor, A. (eds) (1985 [2001]). *The Origins of Museums: The Cabinets of Curiosity in the Sixteenth- and Seventeenth-century Europe.* London: House of Stratus.

Jacobi, D. (1999). *La Communication Scientifique. Discours, Figures, Modèles.* Grenoble: Presses Universitaires de Grenoble.

Jurdant, B. (2009). *Les Problèmes Théoriques de la Vulgarisation Scientifique.* Paris: Editions des Archives Contemporaines.

Klopfer, L. (1969). The teaching of science and the history of science. *Journal of Research in Science Teaching*, 6(1), 87–95.

Klopfer, L. & Cooley, W. (1963). The history of science cases for high schools in the development of student understanding of science and scientists: A report on the HOSG instruction project. *Journal of Research in Science Teaching*, 1(1), 33–47.

Koliopoulos, D. (2003). Blunting the tensions between informal and formal education in science: Reforming the relationship between the school and the science museum in Greece. *Mediterranean Journal of Educational Studies*, 8(1), 81–95.

Kuhn, T. (1970). *The Structure of Scientific Revolutions*. Chicago: University of Chicago Press.

Langevin, P. (1964). La valeur éducative de l'histoire des sciences. In P. Labérenne (ed.) *La Pensée et l'Action* (pp. 193–208). Paris: Les éditeurs Français Réunis.

Latour, B. (1987). *Science in Action: How to Follow Scientists and Engineers through Society*. Milton Keynes: Open University Press.

Latour, B. & Woolgar, S. (1986). *Laboratory Life: The Construction of Scientific Facts*. London: Sage.

Lederman, N. (2007). Nature of science: past, present, and future. In S. Abell & N. Lederman (eds) *Handbook of Research on Science Education* (pp. 831–879). London & New York: Routledge.

Lévy-Leblond, J.M. (2004). *Science in Want of Culture*. Paris: Futuribles.

Lévy-Leblond, J.M. (2012). The muses of science. A utopian oracle. *Science & Education* special issue 'History of science in museums', doi: 10.1007/s11191-012-9522-0

Lourenço, M.C. (2005). *Between Two worlds: The Distinct Nature and Contemporary Significance of University Museums and Collections in Europe*. Unpublished doctoral thesis, Conservatoire National des Arts et Métiers, Paris.

Lourenço, M.C. (2012). Documenting collections: cornerstones for more history of science in museums. *Science & Education* special issue 'History of science in museums', doi: 10.1007/s11191-012-9568-z.

Maison, L. (2002). L'exposition des instruments anciens d'Astronomie: histoire et défis actuels. *La Lettre de l'OCIM*, 84, 39–44.

Martin, L.M.W. (2004). An emerging research framework for studying informal learning and schools. *Science Education*, 88(S1), 71–82.

Matthews, M. (1994/2014). *Science Teaching. The role of History and Philosophy of Science*. London & New York: Routledge.

Monk, M. & Osborne, J. (1997). Placing the history and philosophy of science on the curriculum: a model for the development of pedagogy. *Science Education*, 81(4), 405–424.

Nersessian, N. (1992). How do scientists think? Capturing the dynamics of conceptual change in science. In R. Giere (ed.) *Cognitive Models of Science* (pp. 5–22). University of Minneapolis: Minnesota Press.

Paparou, F. (2011). Shall we stroll to the museum? Educational proposal for the exploration of a historic school scientific instrument collection. In F. Seroglou, V. Koulountzos & A. Siatras (eds) *Proceedings of the 11th International IHPST and 6th Greek History, Philosophy and Science Teaching Joint Conference* (pp. 574–580). Thessaloniki: Epikentro.

Pearce, S.M. (1993). *Museums, Objects and Collections. A Cultural Study*. Leicester: Leicester University Press.

Pedretti, E. (2002). T. Kuhn Meets *T. Rex*: Critical conversations and new directions in science centres and science museums. *Studies in Science Education* 37(1), 1–41.

Quin, M. (1993). Clones, hybrides ou mutants ? L'évolution des grands musées scientifiques européens. *Alliage*, 16–17, 264–272.

Schiele, B. & Koster, E.H. (1998). *La Révolution de la Muséologie des Sciences*. Lyon : Presses Universitaires de Lyon.

Seroglou, F. & Koumaras, P. (2001). The contribution of the history of physics in physics education: a review. *Science & Education*, 10(1–2), 153–172.

Sibum, H.O. (2000). Experimental history of science. In S. Lindqvist (ed.) *Museums of Modern Science* (pp. 77–86). Canton MA: Science History Publications/USA.

Simonneaux, L. & Jacobi, D. (1997). Language constraints in producing prefiguration posters for a scientific exhibition. *Public Understanding of Science,* 6, 383–408.

Stinner, A., MacMillan, B., Metz, D., Jilek, J. & Klassen, S. (2003). The renewal of case studies in Science Education. *Science & Education,* 12(7), 617–643.

Stocklmayer, S., Rennie, L. & Gilbert, J. (2010). The roles of the formal and informal sectors in the provision of effective science education. *Studies in Science Education,* 46(1), 1–44.

Strauss, S. (ed.) (1988). *Ontogeny, Phylogeny and Historical Development.* Norwood, New Jersey: Ablex Publishing.

Subiran, S., Lourenco, M., Wittje, R., Talas. S. & Bremer, T. (2009). Initiatives européennes et patrimoine universitaire. *La Lettre de l'OCIM,* 123, 5–14.

Taub, L. & F. Willmoth (eds) (2006). *The Whipple Museum of the History of Science: Instruments and interpretations, to celebrate the 60th anniversary of R.S. Whipple's gift to the University of Cambridge.* Cambridge: Cambridge University Press.

Teichmann, J. (1981). Deutsches Museum, München. Science, technology and history as an educational challenge. *European Journal of Science Education,* 3(4), 473–478.

Teichmann, J. (1999). Studying Galileo at secondary school: A reconstruction of his 'jumping-hill' experiment and the process of discovery. *Science & Education,* 8(2),121–136.

Tran, L. (2007). Teaching science in museums: The pedagogy and goals of museum educators. *Science Education,* 91(2), 278–297.

Tucci, P. (2002). Role of university museums and collections in disseminating scientific culture. *Museologia,* 2, 17–22.

Van Praet, M. (1995). Les expositions scientifiques miroirs épistémologiques de l'évolution des idées en sciences de la vie. *Bulletin d'Histoire et d'Epistémologie des Sciences de la Vie,* 2, 52–69.

Wagensberg, J. (2004). The basic principles of modern scientific museology. *Museos de Mexico y del Mundo,* 1(1), 14–19.

Yanni, C. (1999). *Nature's Museums. Victorian Science and the Architecture of Display.* London: Athlone Press.

Zana, B. (2005). History of the museums, the mediators and scientific education. *Journal of Science Communication,* 4(4), 1–6.

Anastasia Filippoupoliti Teaches museum education at the Department of Education Sciences in Preschool Age at the Democritus University of Thrace. She collaborates with the Hellenic Open University (Cultural Organisations Management), the University of Athens (Museum Studies) and the Eugenides Foundation. She has obtained a Ph.D. degree in Museum Studies and an M.A. degree in Museum Studies, both from the University of Leicester (UK) and a first degree in History and Philosophy of Science (University of Athens). Among her publications is the two-volume edited book *Science Exhibitions: Communication and Evaluation* and *Science Exhibitions: Curation and Design* (Edinburgh: MuseumsEtc, 2010). Her research interests are interdisciplinary and span the fields of museology, museum education, science communication and cultural history of science.

Dimitris Koliopoulos Teaches Science Education and Scientific Museology at the Department of Educational Sciences and Early Childhood Education at the University of Patras. He collaborates with the Hellenic Open University and the University of Cyprus (Department of Education). He holds a diploma in physics from Aristotle University of Thessaloniki (Greece) and has done postgraduate studies in Science Museum Education at the Jussieu University of Paris (France) and Science Education at the Jussieu University of Paris (France) and University of Patras (Greece), respectively. His research interests concern epistemological and educational aspects of the transformation of scientific knowledge to school science in formal and non-formal educational settings. He has also been involved for many years in the pre-service and in-service training of teachers in the preschool, primary and secondary education.

Part XI
Theoretical Studies: Science, Culture and Society

Chapter 50
Science, Worldviews and Education

Michael R. Matthews

Karl Marx, in the opening of *The Eighteenth Brumaire of Louis Bonaparte,* famously wrote that:

> Men make their own history, but they do not make it just as they please; they do not make it under circumstances chosen by themselves, but under circumstances directly found, given and transmitted from the past. (Marx 1851, p. 595)

Marx's appreciation of the way in which human life – its engagements, politics, culture and economic practices – is shaped by circumstances, and in turn how lives act and transform those circumstances, is a quite general claim that applies also to scientific engagements and practices. Science, broadly speaking, is the effort of people and societies to identify, understand and 'make sense of' the objects and processes in the world around them; to tabulate the properties of natural things and processes; to ascertain what and how causal mechanisms in the world operate; and to achieve some degree of predictive certainty about the course of events and some degree of control over them. Science is conducted by people living in societies in specific historic stages of scientific, philosophical, intellectual (including mathematical), religious, technological, economic and cultural (including ethical and artistic) development. All of these elements bear upon scientists and upon the science they conduct; these elements both limit and put constraints on science and also enhance it. In turn, science bears upon these circumstances: sometimes strengthening, other times modifying, sometimes overthrowing or negating different of these domains. The history of these interactions provides grounds for identifying how some scientific traditions are better than others at achieving their goal of understanding and effectively intervening in the natural world; the history allows some lessons to be learnt about the kinds of social and cultural circumstances that allow

---

The present chapter draws on Matthews (2009a, b, c) and on contributions to Matthews (2009d).

M.R. Matthews (✉)
School of Education, University of New South Wales, Sydney, NSW, Australia
e-mail: m.matthews@unsw.edu.au

M.R. Matthews (ed.), *International Handbook of Research in History,*
Philosophy and Science Teaching, DOI 10.1007/978-94-007-7654-8_50,
© Springer Science+Business Media Dordrecht 2014

science and scientists to flourish and conversely the circumstances that inhibit and curtail scientists and limit the scientific endeavour.

50.1 Science and Culture

Science, formerly 'natural philosophy', has always been a dynamic part of culture; it is affected by culture and has effects on culture; thus, science and worldviews (or *Weltanschauung*)[1] are interrelated, and a good science education should give students some appreciation of this interrelationship.[2] The educational value of such appreciation was recognised by the American Association for the Advancement of Science in its *Project 2061* publication where it said:

> ... Becoming aware of the impact of scientific and technological developments on human beliefs and feelings should be part of everyone's science education. (AAAS 1989, p. 173)

The position was elaborated a year later in its *The Liberal Art of Science*:

> The teaching of science must explore the interplay between science and the intellectual and cultural traditions in which it is firmly embedded. Science has a history that can demonstrate the relationship between science and the wider world of ideas and can illuminate contemporary issues. (AAAS 1990, p. xiv)

These expectations found their way through to the US National Science Education Standards where there was a separate content strand on 'History and Nature of Science Standards' (NRC 1996) that affirmed:

> Students should develop an understanding of what science is, what science is not, what science can and cannot do, and how science contributes to culture. (NRC 1996, p. 2)

And

> The standards for the history and nature of science recommend the use of history in school science programs to clarify different aspects of scientific inquiry, the human aspects of science, and the role that science has played in the development of various cultures. (NRC 1996, p. 107)

Hugh Gauch, an agricultural scientist, wrote the lead essay in a recent volume of *Science & Education* (vol.18, nos.6–7, 2009) dedicated to *Science, Worldviews and Education* where he averred that questions about science's relation to worldviews, either theistic or atheistic ones, are among the most significant of contemporary issues for scientists, science teachers and culture more generally (Gauch 2009). Many people are vitally interested in questions such as whether God exists, whether the world has purpose, whether there are spiritual entities that have causal influence on the world, whether humans have spiritual souls which distinguish them from

---

[1] The German expression for 'world outlook' is more directly connected to feelings, ethics and personal and political action than the more passive Anglo term 'worldview'.

[2] A classic account of the history of these interactions is J. D. Bernal's four volume study *Science in History* (Bernal 1965). See also Crombie (1994), Dewitt (2004), and Randall (1962).

the animal world, whether the world is such that prayers can be answered and natural causal processes interrupted and so on. It is surely important for students and teachers to know if science can give answers, one way or the other, to these questions, or whether science is necessarily mute on the matters. Presumably knowledge of the nature of science should shed some light on whether science can or cannot answer such questions. Gauch surveys opinions of scientists, philosophers and educators and, predictably, finds disagreement within each group on the question of the legitimate purview of science.

Importantly Gauch carefully reports what position papers of the American Association for the Advancement of Science (AAAS) and the US National Research Council (NRC) say about the defining characteristics of science and thus what they say about worldviews and science. He identifies seven 'pillars' of the scientific enterprise that the AAAS and the NRC endorse. These are:

Pillar P1: *Realism.*	The physical world, which science seeks to understand, is real.
Pillar P2: *Presuppositions.*	Science presupposes that the world is orderly and comprehensible.
Pillar P3: *Evidence.*	Science demands evidence for its conclusions.
Pillar P4: *Logic.*	Scientific thinking uses standard and settled logic.
Pillar P5: *Limits.*	Science has limits in its understanding of the world.
Pillar P6: *Universality.*	Science is public, welcoming persons from all cultures.
Pillar P7: *Worldview.*	Science, hopefully, contributes to a meaningful worldview.

Gauch sees these seven pillars as, in part, amounting to the popular view that investigation of the supernatural lies outside of the domain of science; this is the widely held 'nonoverlapping magestria' (NOMA) position put forward by the late Stephen J. Gould (1999). But Gauch also finds an inconsistency with the AAAS position because at the same time the AAAS asserts that 'we live in a directional, although not teleological, universe'. For Gauch this is a denial of the fundamental worldview of the Judaic-Christian-Islamic traditions for which the world is neither purposeless nor ultimately unguided; and it is thus a statement that, contra NOMA, science is not worldview independent. He advances and defends the related thesis that:

> Science is worldview independent as regards its presuppositions and methods, but scientific evidence, or empirical evidence in general, can have worldview import. Methodological considerations reveal this possibility and historical review demonstrates its actuality. (Gauch 2009, p. 679)

The following fundamental questions arise for science teachers and curriculum writers and have been addressed by educators and by historians and philosophers of science:

What constitutes a worldview?
How do worldviews impinge upon and in turn be modified by ontological, epistemological, ethical and religious commitments?
What worldview commitments, if any, are presupposed in the practice of science?
What is the overlap between learning about the nature of science (NOS) and learning about worldviews associated with science?

- # What is the legitimate domain of the scientific method? Should scientific method be applied to historical questions, especially to historical questions concerning scriptures and sacred texts?
- # To what extent should learning about the scientific worldview be a part of science instruction?
- # Should science instruction inform student worldviews or leave them untouched? Should students be just 'border crossers' moving from their own culture with its particular worldviews to the science classroom in order to 'pick up' instrumental or technical knowledge and then back to their 'native' culture without being affected by the worldviews and outlooks of science?
- # What judgement do we make of science education programmes where the scientific view of the world is not affirmed or internalised but only learnt for instrumental or examination purposes, where learning science is akin to an anthropological study where students are not expected to believe or adopt what they are learning but merely be able to manipulate formulae and give correct answers on exams?

50.2 Science, Philosophy and Worldviews: Some Historical Developments

The celebrations in 2009 of the 150th anniversary of the publication of Darwin's *The Origin of Species* generated wide recognition of the interplay of science, culture and worldviews. Internationally – by dint of popular journals, academic symposia, newspaper articles, museum displays, books and television documentaries – the general public came to see what scholars have long recognised, namely, that the *Origin* provided not just a novel account of the origin of species by natural selection but it initiated a transformation of modern worldviews and a new understanding of the place of human beings in the natural world.[3] Versions of Darwin's evolutionary naturalism, reinforced and strengthened by modern genetics,[4] have entered into most modern worldviews, excepting those of many Christian fundamentalists, many Muslims and many indigenous cultures.[5]

Earlier in 2005 with the celebration of the centenary of Einstein's *annus mirabilis*, the public also saw and appreciated the contribution of science to worldviews. People knew, perhaps less clearly and dramatically than with Darwin, that something important began to happen in 1905 with the publication

---

[3] A 2013 Richard Attenborough television documentary series, 'The Galapagos Islands', is promoted as 'The islands that transformed our view of life on earth'.

[4] Learning that that *Homo sapiens* shares 98.4 % of its genes with pygmy chimpanzees can change a person's views of their relationship to the animal world.

[5] Of the voluminous literature on Darwinism and worldviews, see especially Dennett (1995, 2006), Greene (1981), McMullin (1985), and Ruse (1989).

50 Science, Worldviews and Education

of Einstein's three papers, even if few understood the details and could share the opinion of the physicist-philosopher Fritz Rohrlich that:

> The development of quantum mechanics led to the greatest conceptual revolution of our century and probably to the greatest that mankind had ever experienced. It most likely exceeded the great revolutions in our thinking brought about by the Copernican revolution, the Darwinian revolution, and the special as well as the general theory of relativity. Quantum mechanics forced us to reconsider our deepest convictions about the reality of nature. (Rohrlich 1987, p. 136)

Although Darwin and Einstein are the most recent and most widely known cases of science impacting on philosophy and culture, these impacts go right back to the very cradle of Western natural philosophy; there has been a continuous interaction between science, philosophy, metaphysics and ultimately worldviews. The 'science' (natural philosophy) of the classical and Hellenic materialists and atomists – Thales, Anaximander, Leucippus, Democritus, Epicurus, Anaxagoras and others – was in constant struggle with the dualist, teleological philosophy and purposeful worldviews of Platonists and Aristotelians. Karl Popper (Popper 1963, Chap. 5) drew attention to this 'struggle' between the early naturalist and materialist scientific tradition among the pre-Socratics and its dualist, teleological, philosophical opponents, chiefly Plato and Aristotle. The latter pair won, and the former group were for nearly 2,000 years relegated to being just 'pre-Socratics' or the philosophical 'warm-up' or targets for the main Athenian adventure. But to a small extent, their reputation has been recovered, with one representative historian of Greek philosophy writing of the Atomists Leucippus and Democritus that:

> In their atomism, their theory of motion, their distinction between primary and secondary qualities, and most of all, in their insistence that explanation of natural processes shall be mechanical, the atomists anticipated much in the world view of modern science. (Allen 1966, p. 15)

Anaximander's explanation of thunder as noise created not by heavenly Gods or spirits but by the rubbing together of wind particles well represents the division between materialist or naturalist explanatory systems and 'pre-scientific' ones.[6]

For Popper, and many others, the Scientific Revolution was a 'return to the past', a recapturing of materialist ontology and non-teleological causal relations.[7] Wallis Suchting has described these struggles in the cradle of Western science and philosophy as:

> Despite all the differences between Plato and Aristotle the latter carried on the work of the former in essential ways, like that of offering a metaphysical 'foundation' for the sciences and a teleological view of the world. Christianity took up elements of Platonic thought … but, its philosophical high-point, in Thomism, mainly appropriated Aristotle. Atomism carried on a basically marginal existence, … till it was recuperated by Galileo. (Suchting 1994, p. 45)

---

[6] Benjamin Farrington's *Science and Politics in the Ancient World* (1939) is a classic treatment of these themes.

[7] See especially Blumenberg (1987), Mittelstrass (1989), Solmsen (1960), Stove (1991), and Vitzthum (1995, Chap. 2).

The Scientific Revolution of the seventeenth century occurred in a Europe whose cultural, scholarly and religious life was permeated by Aristotelian philosophy, by convictions about ontology, epistemology, ethics and theology that were informed and judged by the texts of Aristotle. Neo-Aristotelian Scholasticism, although not monolithic in its interpretation of Aristotle,[8] dominated medieval and Renaissance universities.[9] Scholastic philosophy was intimately connected with the Catholic Church, but it also held sway in Protestant seminaries and universities (Dillenberger 1961, Chap. 2). As one commentator has observed:

> The Middle Ages mean simply the absolute reign of the Christian religion and of the Church. Scholastic philosophy could not be anything else than the product of thought in the service of the reigning *Credo*, and under the supervision of ecclesiastical authority. (De Wulf 1903/1956, p. 53)[10]

In Aristotelian-informed Scholastic ontology, things were constituted by form and by matter; this was the doctrine or principle of hylomorphism; it was fundamental to the Aristotelian tradition. Frederick Copleston has rightly noted that Aquinas, the greatest of the Scholastics,[11] 'took over the Aristotelian analysis of substance' (Copleston 1955, p. 83) and:

> According to Aquinas, therefore, every material thing or substance is composed of a substantial form and first matter. Neither principle is itself a thing or substance; the two together are the component principles of a substance. And it is only of the substance that we can properly say that it exists. 'Matter cannot be said to be; it is the substance itself which exists'. (Copleston 1955, p. 90)

It was the 'New Science' that led eventually to the unravelling of this settled medieval philosophical-theological worldview. This began with the publication in (1543) of Copernicus's astronomical work *On the Revolution of the Heavenly*

---

[8] The varieties of medieval and renaissance Aristotelianisms arose from efforts to accommodate ever new developments and discoveries in natural philosophy. See Blum (2012) and Schmitt (1983).

[9] A classic work on the doctrines and history of Scholastic philosophy is De Wulf (1903/1956). See also volumes two and three of Frederick Copleston's *History of Philosophy* (Copleston 1950). Etienne Gilson and Jacques Maritain are the best-known twentieth-century exponents of Scholasticism.

[10] Sadly this description, *sans* Church, fitted philosophy departments in most of the former communist states and philosophy departments China where 'Marxist Dialectics' is still a compulsory subject. It is also the situation of philosophy departments in many Islamic states where all philosophical positions in ontology, epistemology and ethics need to conform to state-sanctioned interpretations of the Koran. According to Sheikh Muhammed Salih Al-Munajjid, a known Islamic lecturer and author: 'hence philosophy, as defined by the philosophers, is one of the most dangerous falsehoods and most vicious in fighting faith and religion on the basis of logic, which it is very easy to use to confuse people in the name of reason, interpretation and metaphor that distort the religious texts'. In such Marxist and Islamic regimes, philosophy simply cannot be practised; the regimes are replicating at a state level what the Roman Catholic Church used maintain at a seminary level, namely, rigorous thought control.

[11] On the life and philosophy of Aquinas, see Copleston (1955), Gilson (1929), Kenny (1980), and Weisheipl (1974).

Spheres (Copernicus 1543/1952).[12] But it was almost a century later that the unravelling took dramatic shape with the publication in 1633 of Galileo's *Dialogues Concerning the Two Chief World Systems* followed 50 years later by Newton's *Principia Mathematica*. These two books, separated by a mere 50 years, embodied the intellectual core of the Scientific Revolution; they constituted the Galilean-Newtonian Paradigm, a GNP far more influential than any economic GNP has ever been.

The 'New Science' established the Copernican heliocentric account of the solar system which removed humans from their religiously and culturally privileged place in the centre of the universe; it introduced a mechanical and lawful account of natural processes; it challenged and in many places overthrew the long dominant Aristotelian philosophical system that was, among other things, intimately tied up with Roman Catholic theology and ethics; and famously the GNP caused a reassessment of the role of religious authority in the determination of claims about the world and indeed in any claims.[13]

The new science (natural philosophy) of Galileo, Descartes, Huygens, Boyle and Newton caused a massive change not just in science but in European philosophy that had enduring repercussion for religion, ethics, politics and culture. Early modern philosophers –from Francis Bacon, Thomas Hobbes, John Locke, David Hume, George Berkeley, René Descartes, Gottfried Leibniz up to Immanuel Kant – were all engaged with and reacting to the breakthroughs of early modern science,[14] as of course were the later philosophers of the French, English, German and Scottish Enlightenment; seventeenth-century science was the seed that bore eighteenth-century philosophical and worldview fruit. With the inevitable exceptions and qualifications required when talking of any large-scale transformation or revolution in thought, it can be said that all the major natural philosophers of the time rejected Aristotelianism in their scientific practice, their theorising and in their enunciated philosophy. Overwhelmingly the new philosophy to which they turned was corpuscularian, mechanical and realist – it has rightly been called the 'mechanical worldview'.[15]

In this new worldview, there was simply no place for the entities that Aristotelianism utilised to explain events in the world: Hylomorphism, immaterial substances, unfolding natures, substantial forms, teleological processes and final

---

[12] For the background, context and impact of Copernicus, see Blumenberg (1987), Gingerich (1975, 1993), and Grant (2004).

[13] A classic discussion is Dijksterhuis's *The Mechanization of the World Picture* (1961/1986). On the wider impact of the Galilean-Newtonian method, see Butts and Davis (1970), Cohen (1980), McMullin (1967), and Shank (2008).

[14] Unfortunately these early modern philosophers are frequently studied in isolation from the contemporary science with which they were engaged; early modern philosophy is presented to students as a drawn-out soliloquy, not the dialogue and debate with early modern science that it was. This theme, with texts, is developed in Matthews (1989a).

[15] For historical and philosophical elaboration of the mechanical world view, see Dijksterhuis (1961, 1961/1986), Einstein and Infeld (1938/1966), Hall (1954/1962), Harré (1964), Hatfield (1990), and Westfall (1971).

causes were all banished from the philosophical firmament. René Descartes aptly sums up the new philosophy in the conclusion of his *Principles of Philosophy* (1659) with a clear statement of the new corpuscularian philosophy:

> Nor do I think that anyone who uses his reason will deny that we do much better to judge of what takes place in small bodies which their minuteness alone prevents us from perceiving, by what we see occurring in those that we do perceive [and thus explain all that is in nature, as I have tried to do in this treatise], than, than in order to explain certain given things, to invent all sorts of novelties, that have no relation to those that we perceive [such as are first matter, substantial forms, and all the great array of qualities which many are in the habit of assuming, any of which it is more difficult to understand than all the things which we profess to explain by their means]. (*Principles* Bk.IV, art.101; Haldane and Ross 1931, pp. 297–298)

A foretaste of the coming mechanical worldview can be found in Galileo's distinction between, what will come to be called, the primary and secondary qualities of bodies. Seventy years later, his distinction was repeated by Robert Boyle and was famously articulated by John Locke,[16] and it has had an enduring presence in the subsequent history of Western philosophy. The distinction was at the heart of Galileo's theory of matter, a theory that answers such basic ontological questions as: Of what is matter constituted? And, what are the inherent and necessary properties of matter?[17]

For Aristotle and the Scholastics, matter was ultimately of the one stuff – 'prime matter' – gold, silver, timber, did not differ in their ultimate material; they just differed in how this material was arranged and what forms animated it. For this philosophical tradition, the properties or qualities of bodies were real. Colour, heat and odour belonged to bodies; the quality was a quality of the body. Heated red bodies *were* hot and they were red. These qualities are perceived by the senses, not generated by the senses. Aristotelians were realists, not subjectivists, about qualities.

In contradiction to this, Galileo reached back to pre-Socratic atomistic sources and to more recent medieval nominalist sources, for his account of matter. As a student he had read Democritus, Lucretius and possibly other early atomists such as Leucippus the teacher of Democritus. For them colour and taste were opinions, mere names; what existed in the world was atoms and the void, and atoms had neither colour nor taste. They held a material monist position – all matter was an aggregate of invisible and indivisible 'atoms' each of which was made of the same material and differing among themselves only in size and shape. It was the particular aggregate of atoms that gave bodies their tangible properties; a body's properties were not produced or caused by its form. When new substances are created from different materials, their immutable atoms are just rearranged in different ways; there is no change of form, because there was no form to change. This atomistic ontology was so comprehensively rejected by Aristotle in this *Physics* and his *Metaphysics* that it disappeared from the philosophical firmament for over a

---

[16] See Locke's *Essay Concerning Human Understanding* Book II Chap. 8 (Locke 1689/1924, pp. 64–73).

[17] On ancient, medieval and modern theories of matter, see contributions to McMullin (1963a, b).

thousand years until it was revived by some thinkers on the margins of medieval philosophy such as William of Ockham and Nicholas of Autrecourt.

Galileo's atomism is first and most famously stated in his *The Assayer* (Galileo 1623/1957) where he advances invisible 'atomic' motions as the cause of heat. He says:

> But first I must consider what it is that we call heat, as I suspect that people in general have a concept of this which is very remote from the truth. For they believe that heat is a real phenomenon, or property, or quality, which actually resides in the material by which we feel ourselves warmed. (Galileo 1623/1957, p. 274)

Galileo makes explicit his atomism, or corpuscularianism, when he says:

> Those materials which produce heat in us and make us feel warmth, which are known by the general name of 'fire', would then be a multitude of minute particles having certain shapes and moving with certain velocities. Meeting with our bodies, they penetrate by means of their extreme subtlety, and their touch as felt by us when they pass through our substance is the sensation we call 'heat'. ... I do not believe that in addition to shape, number, motion, penetration, and touch there is any other quality in fire corresponding to 'heat'. (ibid)

Galileo believed that it was the shape, size, motion and collisions of minute, unseen 'atoms' or corpuscles that determined all outward and perceivable states, processes and phenomena. This was his restatement of ancient atomism. Galileo's ontology was simply inconsistent with Scholastic metaphysics and thus with the medieval worldview built upon it. Galileo's distinction between primary and secondary qualities was the beginning of the unravelling of this 'Medieval Synthesis' and its replacement by the 'mechanical' or 'corpuscularian' worldview and ultimately the 'scientific' worldview.

Newton, the greatest of all seventeenth-century scientists, was also a champion of the New Philosophy.[18] Beginning in his student days, Newton embraced Galileo's mathematical methods, his Copernicanism, his experimentalism, his rejection of Aristotle's physics, his rejection of Scholastic philosophy and his embryonic atomism.[19] In the Preface of the *Principia*, Newton identifies himself with the 'moderns, rejecting substantial forms and occult qualities' and endeavours 'to subject the phenomena of nature to the laws of mathematics' (Newton 1729/1934, p. xvii).

In keeping with Boyle's example of experimentally testing and utilising metaphysical positions, Newton in his *Opticks* gave an atomistic account of light and optical phenomena (Newton 1730/1979). After 300-odd pages of optical experiments and investigations, Newton in Query 29 of Book III says:

> Are not Rays of Light very small Bodies emitted from shining Substances? For such Bodies will pass through uniform Mediums in right Lines without bending into the Shadow, which is the Nature of Rays of Light. They will also be capable of several Properties, and be able

---

[18] Numerous works are available on Newton's philosophy and metaphysics, among them are Hughes (1990), McGuire (1995), McMullin (1978), and Stein (2002). Although an atomist, Newton distanced himself from Descartes' interpretation of the theory.

[19] For Newton's early scientific and philosophical formation, see Herivel (1965) and Westfall (1980, Chaps. 3,4,5)

to conserve their Properties unchanged in passing through several Mediums which is another Condition of the Rays of Light. (Newton 1730/1979, p. 370)

Much can be said about atomism, its recovery by philosophical contemporaries of Galileo such as Francis Bacon and Pierre Gassendi and its role in the Scientific Revolution, but for current purposes it suffices to repeat Craig Dilworth's judgement that:

The metaphysics underlying the Scientific Revolution was that of early Greek atomism. … It is with *atomism* that one obtains the notion of a *physical* reality underlying the phenomena, a reality in which *uniform causal* relations obtain. … What made the Scientific Revolution truly distinct, and Galileo … its father, was that for the first time this empirical methodology [of Archimedes] was given an ontological underpinning. (Dilworth 2006, p. 201)

And the role of the New Science in generating the modern worldview is well stated by Herbert Butterfield[20]:

Since that revolution overturned the authority in science not only of the middle ages but of the ancient world – since it ended not only in the eclipse of scholastic philosophy but in the destruction of Aristotelian physics – it outshines everything since the rise of Christianity and reduces the Renaissance and Reformation to the rank of … mere internal displacements … it changed the character of men's habitual mental operations even in the conduct of the non-material sciences, while transforming the whole diagram of the physical universe and the very texture of human life itself. (Butterfield 1957, p. 7)

When Butterfield writes of the New Science changing 'the character of men's habitual mental operations', he is speaking of what the AAAS will later call the 'scientific habit of mind' (Rutherford and Ahlgren 1990, Chap. 12) and what Jawaharlal Nehru and the drafters of the Indian Constitution called for in promoting a 'scientific temper'.[21] Effects on a society's 'mental operations', 'habits of mind' or 'scientific temper' depend on citizens learning about and valuing science, on having a worldview where such ways of thinking can be exercised, and hence ultimately on effective and widespread science education.

50.3 From Science to Heresy: The Catholic Church's Condemnation of Atomism

The worldviews of science and of religion do not always sit easy with each other; accommodations usually need to be made. In recent times, worldview conflicts occasioned by disputes about Creation, Creationism, Teleology, Miracles, the

---

[20] There is a vast literature on the Scientific Revolution, including a debate on whether to capitalise the terms; an informative guide to the different assessments, literature and debates is H. Floris Cohen's *The Scientific Revolution* (Cohen 1994).

[21] The 1948 Indian Constitution makes obligatory the state's 'promotion of scientific temper' among its citizenry, not just scientific knowledge but scientific outlook or habits of mind (Haksar 1981).

existence of souls and spirits and so on have moved from academic corridors to the public domain with bestsellers (Dawkins 2006; Dennett 1995), television programmes, public debates and countless scholarly articles devoted to explicating or defending one side or other of these conflicts.[22]

Many of the major seventeenth-century contributors to the new science – Galileo, Descartes, Boyle and Newton – were believers, although in somewhat tense relations with their respective established churches (Roman Catholic for the first two, Anglican for the second two). Some believers rejected the new science; some wanted the new science but not its associated metaphysics; and some, such as Joseph Priestley, embraced both the new science and its atomistic metaphysics and adjusted their religious ontology accordingly. When the seventeenth-century natural philosophers and the Enlightenment philosophers of the eighteenth century stressed the materialism, mechanism and determinism of the new science, they brought upon themselves the ire of most contemporary religious figures who saw the emerging new worldview as anti-Christian and atheistic.[23] The historian Richard Westfall well summarises the general situation:

> Natural science rested on the concept of natural order, and the line that separated the concepts of natural order and material determinism was not inviolable. The mechanical idea of nature, which accompanied the rise of modern science in the 17th century, contradicted the assertion of miracles and questioned the reality of divine providence. Science, moreover, contained its own criteria of truth, which not only repudiated the primacy of ancient philosophers but also implied doubt as to the Bible's authority and regarded the attitude of faith enjoined by the Christian religion with suspicion. (Westfall 1973, pp. 2–3)

And Westfall proceeds to say:

> every one of the problems could be resolved in a variety of ways to reconcile science with religion. But the mere fact of reconciliation meant some change from the pattern of traditional Christianity (ibid)

These 'grand historic' reconciliations are repeated at the personal level for many science students.

Although Galileo was, in 1615, warned not to hold or teach the Copernican doctrine of a moving Earth, it was only after *The Assayer* and its endorsement of atomism was published in (1623) that he faced serious theological charges. There was a move by opponents from general disquiet to specific repudiation.

Atomism presented particular and grievous problems for Christian belief, but the most basic and important one was the central Roman Catholic, Greek Orthodox and Eastern Uniate teaching on Christ's presence in the Eucharist, the doctrine of Transubstantiation. The Eucharist was the sacramental heart of the Catholic Mass, and the Mass was and is the devotional heart of the Church. Belief in the Real Presence of Christ, brought into being by the priest's consecration of the communion host, underwrites devotional practice and doctrinal authority. Denial of the Real

---

[22] See Michael Ruse (2011) for one informed account of these debates and also the careful review of this book by Peter Slezak (2012).

[23] See, for instance, Brooke (1991, Chap. V), Israel (2001), and Porter (2000).

Presence was a capital offence. It was a litmus test in the Inquisition, where failure to affirm the belief meant a horrible death at the stake.

Scholastic philosophy, with its Aristotelian categories of substance, accidents and qualities, could bring a modicum of intelligibility to this central mystery of faith, as it could also bring a modicum of intelligibility to doctrines such as the Incarnation, the Trinity, and immortality of the soul. Scholasticism held that at consecration the substance of bread changed to the substance of Christ's body, but the accidents remained that of bread. So Christ became truly present, even though no sensible, observable change was apparent.

Thomas Aquinas formulated the orthodox doctrine as:

> All the substance of the bread is transmuted into the body of Christ... therefore, this is not a formal conversion but a substantial one. Nor does it belong to the species of natural mutations; but, with its own definition, it is called transubstantiation. (*Summa Theologica* III, q.75, a.4, in Redondi 1988, p. 212)

This Thomist formulation, along with the Aristotelian philosophical apparatus required for its interpretation, was affirmed as defining Catholic orthodoxy at the Council of Trent in 1551.

The nature of the heresy charges against Galileo, and the degree to which atomism was at odds with established religiosity and theology, can be seen in a condemnation brought anonymously by Father Giovanni de Guevara, a Vatican confidant of Pope Urban VIII. Guevara was a priest of a contemplative order whose very life revolved around adoration of the Eucharistic sacrament. He had minimal philosophical training but enough to see the conflict between Galileo's atomistic position and the orthodox interpretation of the Real Presence – for Guevara they could not both be true (Redondi 1987, pp. 166ff). In his 1624/1625 deposition he charged that:

> [Galileo's position] is in conflict with the entire community of Theologians who teach us that in the Sacrament remain all the sensible accidents of bread, wine, color, smell, and taste, and not mere words, but also, as is known, with the good *judgment* that the quantity of substance does not remain. (Redondi 1987, pp. 333–334)[24]

The charge of atomism against Galileo, with its direct implications of heresy, was made in 1626 by Father Grassi, a prominent Jesuit professor of mathematics and astronomy at the Collegio Roman. Grassi made clear the gravity of the philosophical point by adding that transubstantiation 'constitutes the essential point of faith or contains all other essential points' (Redondi 1987, p. 336).[25] Descartes' matter theory was likewise condemned in 1671 because its categories did not allow an intelligent rendering of the doctrine of Transubstantiation.

John Hedley Brooke, an historian sympathetic to the positive contribution of religion to science, recognised the problem that atomism posed 'especially for the

---

[24] A translation of the deposition, and discussion, is also available in Finocchiaro (1989, pp. 202–204).

[25] This contention echoed through all Catholic teaching and devotional practice; as one Catholic Handbook states the matter: 'The Catholic belief is that the sacrifice of the Mass is the sacrifice of the body and blood of Christ *under the form of bread and wine*' (Lucey 1915, p. 93).

Roman Catholic Church, which took a distinctive view of the presence of Christ at the celebration of the Eucharist' (Brooke 1991, p. 141). He writes:

> With an Aristotelian theory of matter and form, it was possible to understand how the bread and wine could retain their sensible properties while their substance was miraculously turned into the body and blood of Christ. .... But if, as the mechanical philosophers argued, the sensible properties were dependent on an ulterior configuration of particles, then any alteration to that internal structure would have discernible effects. The bread and wine would no longer appear as bread and wine if a real change had occurred. (Brooke 1991, p. 142)

50.4 The Decline of Atomism: Scientific or Philosophic Causes?

No sooner had Newton ceased writing than the philosophy of atomism, and its associated mechanical worldview was augmented and modified. This history is a case study of the relation between science and metaphysics: To what extent did the metaphysics change for philosophical reasons and to what extent did it change for scientific reasons?[26] To the scientific ontology of atoms and the void there was added, after considerable struggle, attractive and repulsive forces. Leibniz famously denounced Newton's attractive forces because he thought they reintroduced Scholastic occult entities to the ontology of natural philosophy (Hall 1980). In the nineteenth century, to this expanded scientific ontology were added magnetic and electric fields. The formulation of electromagnetic field theory by Maxwell, Boltzmann and Hertz fully stretched, and then ruptured, the atomistic ontology; and the energeticist interpretation of thermodynamics had the same result; and at the end of the century Mach, for example, abandoned atoms and denounced the atomic hypothesis as metaphysics.[27] This then provided the full range of scientifically legitimate explanatory and causal entities, at least until the advent of quantum theory.

The expansion of the ontology of science is a case study in the interaction of science and metaphysics. The atomists held on *philosophical* grounds that all legitimate explanation had ultimately to be in terms of the properties of atoms and of their movements and interactions. Their science was constrained by their philosophy. Clearly the addition of forces and fields to the class of existent things was not done on philosophical grounds but on *scientific* grounds; it seemed that only recourse to the latter entities enabled consistent scientific explanation and progress.

This expanded ontology was inconsistent with the metaphysics of *physicalism*: Forces and fields did not have mass; they could not be bumped into; and they had no

---

[26] On the history of atomism and its connections with science on the one hand and with philosophy on the other, see Chalmers (2009), Pullman (1998), and Pyle (1997). An older historical study that concentrates more on the philosophical side of atomism is Melsen (1952).

[27] There are many good accounts of the modification, and eventual breakdown, of the mechanical worldview. See especially Einstein and Infeld (1938, Chap. 2), Harman (1982, Chap. 6), Hesse (1961), and McMullin (1989).

colour. They were not physical objects, the things that physicalism maintained were the only kinds of existing entities. It was also inconsistent with *materialism* in as much as this philosophy maintained that all entities with causal powers were material.[28] But the enriched ontology was consistent with *naturalism*, the view that only those kinds of entities exist that science reliably demonstrates as having consistent, causal and explanatory power. Thus things can be natural while not physical or material. The Mechanical Worldview survived the demise of atomism: There were still pushes and pulls, nature was not unfolding Aristotelian-like from within, but the deterministic pushes and pulls were no longer just those of colliding bodies, but gravitational and electric forces were added (Westfall 1971).

50.5 Philosophy as the Handmaiden of Theology and of Other Systems

Joseph Priestley, one of the luminaries of the British Enlightenment and a lifelong Christian believer, well expressed the ill ease felt about cloaking Christian doctrine in Scholastic clothes. In 1778 he wrote to the Jesuit 'materialist' philosopher Abbé Roger Boscovich saying that:

> the vulgar hypothesis [Aristotelian matter theory], which I combat, has been the foundation of the grossest corruptions of true Christianity; and especially [those] of the church of Rome, of which you are a member; but which I consider as properly *antichristian*, and a system of abominations little better than heathenism. (Schofield 1966, p. 167)

Despite such criticisms, the Catholic Church was guided by the medieval view that 'philosophy was the handmaiden' of theology; philosophy was to be subservient to religious and theological purposes. This was the import of the sixteenth-century Tridentine decrees and curial decisions right through to the twentieth century.[29] Pope Leo XIII promulgated his encyclical *AEterni Patris* that gave the name *philosophia perennis* (perennial philosophy) to Thomism and directed Catholic educational institutions to base their philosophical and theological instruction upon it. In 1914 Pius X issued his *Doctoris Angelici* decree, stating that:

> We desired that all teachers of philosophy and sacred theology should be warned that if they deviate so much as an iota from Aquinas, especially in metaphysics, they exposed themselves to grave risk. (Weisheipl 1968, p. 180)

---

[28] On the history and philosophy of materialism, see Bunge (1981) and Vitzthum (1995).

[29] One hundred years *after* Priestley's complaints to Boscovich, Joseph McCabe, a former Franciscan priest and professor of philosophy who left the Church in the 1890s, well described the state of Roman Catholic theology when he said derisively of his theological training that:

> The various points of dogma which are contained (or supposed to be contained) in Scripture, were first selected by the Fathers, and developed, generally by the aid of the Neo-Platonic philosophy, into formidable structures. The schoolmen completed the synthesis with the aid of Peripatetic philosophy, and elaborated the whole into a vast scheme which they called theology. (McCabe 1912, p. 73)

50 Science, Worldviews and Education

A few years later, the *Code of Canon Law*, promulgated by Pope Benedict XV in 1917, reinforced the position by requiring that all professors of philosophy hold and teach the method, doctrine and principles of St Thomas. The papal endorsement of thirteenth-century philosophy continued through to 1950 when Pope Pius XII in *Humani generis* demanded that future priests be instructed in philosophy 'according to the method, doctrine and principles of the Angelic Doctor' (Weisheipl 1968, p. 183). It was only in the final years of the twentieth century, with Pope John Paul II's 1998 encyclical, *Fides et Ratio*, that the Catholic Church relaxed its attachment to Thomism as official Church philosophy. Thomism was downgraded from Absolute Truth to Highly Probable Truth.[30]

The Thomist tradition had enormous cultural and personal impact in Catholic Europe (especially Ireland, Portugal, Spain, Italy, Poland), Latin America,[31] the Philippines and elsewhere. For centuries Thomism was marshalled to support Church teaching on contraception, abortion, masturbation and homosexuality; where the Church exercised political power and influence, these teachings transferred into national law with the immoral acts becoming illegal and punishable by the State and not just for Catholics but for all citizens. In all cases the reason for condemnation was that the activity was 'unnatural', this whole conceptualisation coming from Aristotle's understanding of objects and actions as having natures which left alone unfolded 'naturally' and when interfered with unfolded 'violently' or 'unnaturally'.[32]

Clearly Thomism and Scholasticism and more generally Aristotelianism survived the Scientific Revolution; belief in the core metaphysical and ethical positions has survived to the present day.[33] Indeed neo-Aristotelianism is perhaps the most substantial and lively current of thought in contemporary ethical theory, with exponents such as Alasdair McIntyre, Elizabeth Anscombe and Martha Nussbaum all contributing substantial books to the Aristotle-sourced project of 'virtue ethics'. But the success of modern science has meant that Thomism in particular and Aristotelianism more generally has had to engage with science. Some have done this while preserving Aristotelianism (van Laer 1953, 1956; 1962; Maritain 1935/1951; Mascall 1956); for others the engagement has led to

---

[30] On John Paul II's encyclical and how it reviewed and revised the status of Thomism, see Ernst (2006).

[31] Concerning early twentieth-century Thomist philosophy in Colombia, Daniel Restrepo wrote: 'To the extent that the Columbian State was governed by theocratic criteria, philosophy, conceived as "servant of theology", played the role of ideological mediator in the political action and principles of those who had held power since 1886' (Restrepo 2003, p. 144). Late into the twentieth century, passing 'Thomism I' was still a requirement for progression in many Latin American universities. Much like passing 'Dialectics I' is such a requirement in present day China.

[32] As an example of this reasoning and mindset, for Aquinas sexual intercourse was 'naturally ordained for procreation' (*Sentences* 4.31.2.2), so even indulging in coitus for reasons of health (a good purpose) nevertheless rendered the act unnatural and thus sinful as it was not done for its primary end. On all of this, see Noonan (1965, Chap. viii).

[33] See, for instance, arguments and literature in Ashley (1991) and Lamont (2009). The philosophy journal *New Scholasticism* was published from 1927 to 1989, *The Thomist* journal has been published continuously since 1939 and *The Modern Schoolman* has been published continuously since 1925. And of course numerous non-Anglo 'scholastic' philosophy journals are still published.

rejection in whole or part of Aristotelian philosophy.[34] This is a substantial example of the impact of science on philosophy and culture and of culture's responses and reactions to such impacts.

The same dynamics have played out in the Muslim world where the medieval view of philosophy as the servant of the Koran still holds. It is simply not possible for a Muslim to entertain or commit to any philosophical system that cannot be reconciled with the assumed ontology, epistemology and ethics of the Koran. The project of 'Islamisation of knowledge' is widely accepted as simply a part of Islam and of being a Muslim. Its purpose is to counter the humanistic and secular foundation of the Western education and culture, which it sees as based on five core principles:

1. The sovereignty of man, as though supreme (humanism)
2. Basing all knowledge on human reasoning and experience (empiricism)
3. Unrestricted freedom of thought and expression (libertarianism)
4. Unwillingness to accept 'spiritual' truths (naturalism)
5. Individualism, relativism and materialism

A representative Islamic appraisal of the Scientific Revolution is Seyyed Nasr's claim that the new science of Galileo and Newton had tragic consequences for the West because it marked:

> The first occasion in human history when a human collectivity completely replaced the religious understanding of the order of nature for one that was not only nonreligious but that also challenged some of the most basic tenets of the religious perspective. (Nasr 1996, p. 130)

Nasr repeats Western religious and Romantic criticisms of the new science when he writes:

> Henceforth as long as only the quantitative face of nature was considered as real, and the new science was seen as the only science of nature, the religious meaning of the order of nature was irrelevant, at best an emotional and poetic response to 'matter in motion'. (Nasr 1996, p. 143)

The engagement of philosophical systems with science has been especially urgent when the systems are tied to political and institutional power, as Thomism has been with the Roman Catholic Church. The same situation has applied with Marxism within the former Soviet state,[35] Maoism and dialectics in contemporary China,[36] Confucianism in Chinese history,[37] National Socialist philosophy in Hitler's

---

[34] The survival of Thomism and the dynamics of its engagement with modern science is discussed in Matthews (2009b, pp. 718–720). In a recent publication, a neo-Aristotelian moves philosophy of science away from philosophising on the content of science to philosophical reflection on the activity of science (Marcos 2012).

[35] For 'official' philosophy in the Soviet Union and its contested relationship to science, see Graham (1973).

[36] 'Introduction to Dialectics of Nature (IDN)', based on Engels' book, was under Mao, a compulsory course for all Chinese graduate students. Under Xiaoping Deng, the course remained compulsory but in 1987 was rebadged 'Philosophy of Science and Technology (PST)' with the same IDN teachers. For the relationship of philosophy, politics and science in China, see Chan (1969) and Gong (1996).

[37] See Kwok (1965).

Third Reich,[38] Hindu philosophy at different times and in different states in modern India[39] and Islamic philosophy in Muslim countries[40] and more loosely when custodians of traditional belief systems control what can be thought and taught in traditional indigenous cultures.

In all of these cases, local science and philosophy was and has been made to answer to the dominant, institutionalised philosophy and worldview; and educational bodies were forced to accept such 'direction from above' as being in the interest of the nation, religion or culture. This cultural-political circumstance poses acute questions for the classroom science teacher: Should they foster independence of thought in their students or become functionaries of whatever the dominant ideological power might be? These are matters requiring a thoughtful and informed philosophy of education, unfortunately something mostly ignored in contemporary science teacher education where not only philosophy of education, but most foundational subjects have been progressively removed and replaced with training in pedagogical technique, classroom management skills and use of new technologies; the 'apprenticeship' model of teacher education allows little opportunity for 'reflection on principles' or for understanding the history and philosophy of the discipline being taught.[41]

50.6 Philosophy and Modern Science

Despite revolutions, paradigm changes, commercialisation and much else, modern science is continuous with the New Science of Galileo and Newton and prompts the same range of philosophical questions: Science and philosophy continue to go hand in hand.[42] Peter Bergmann expressed this point when he said that he learnt from Einstein that 'the theoretical physicist is ... a philosopher in workingman's clothes' (Bergmann 1949, p. v, quoted in Shimony 1983, p. 209).[43] One commentator on the work of Niels Bohr remarked that 'For Bohr, the new theory [quantum theory] was not only a wonderful piece of physics; it was also a philosophical treasure chamber which contained, in a new form, just those thoughts he had dreamed about in his early youth' (Petersen 1985, p. 300). It is no accident that many of the major physicists of the nineteenth and twentieth centuries wrote books on philosophy and

---

[38] See Beyerchen (1977) and Cornwell (2003).

[39] See Nanda (2003).

[40] On the tensions and accommodations between science and Islam, see Edis (2007) and Hoodbhoy (1991).

[41] On philosophy of education in science education, see Schulz (2009) and his contribution to this handbook; on the larger issue of educational foundations, see contributions to Tozer et al. (1990).

[42] Some useful studies on the philosophical dimension of science are Amsterdamski (1975), Buchdahl (1969), Burtt (1932), Dilworth (1996/2006), Smart (1968), Trusted (1991), and Wartofsky (1968).

[43] Paul Arthur Schilpp's anthology on Einstein is titled *Albert Einstein: Philosopher-Scientist* (Schilpp 1951).

the engaging overlaps between science and philosophy.[44] Many less well-known physicists also wrote such books teasing out relations between their scientific work and the ontology, epistemology and ethics that it presupposed and for which it had implications.[45] And not just physicists, many chemists and biologists have made contributions to this genre.[46]

This is not, of course, to say that all these good scientists wrote good philosophy or drew sound conclusions from their scientific work: Some did, others did not. Eighty years ago, Susan Stebbing wrote a classic critique of the hugely influential idealist philosophical conclusions drawn by the renowned British physicists James Jeans and Arthur Eddington (Stebbing 1937/1958). Mario Bunge has developed comparable arguments against the idealist and subjectivist conclusions drawn from quantum mechanics by David Bohm, Niels Bohr and many proponents of the Copenhagen school (Bunge 1967, 2012). The point is not that the major scientists drew common philosophical conclusions, it is rather that they all philosophised; they all reflected on their discipline and their activity, and they saw that such reflection bore upon the big and small questions of philosophy. This fact supports the contention that philosophy is inescapable in good science;[47] it should also suggest that philosophy is inescapable in good science education.

The Oxford philosopher, R. G. Collingwood in his landmark study *The Idea of Nature* wrote on the history of mutual interdependence of science and philosophy and commented that:

> The detailed study of natural fact is commonly called natural science, or for short simply science; the reflection on principles, whether those of natural science or of any other department of thought or action, is commonly called philosophy. ... but the two things are so closely related that natural science cannot go on for long without philosophy beginning; and that philosophy reacts on the science out of which it has grown by giving it in future a new firmness and consistency arising out of the scientist's new consciousness of the principles on which he has been working. (Collingwood 1945, p. 2)

He goes on to write that:

> For this reason it cannot be well that natural science should be assigned exclusively to one class of persons called scientists and philosophy to another class called philosophers. A man

---

[44] See, for instance, Bohm (1980), Bohr (1958), Boltzmann (1905/1974), Born (1968), Duhem (1906/1954), Eddington (1939), Heisenberg (1962), Jeans (1943/1981), Mach (1893/1960), Planck (1932), and von Helmholtz (1995).

[45] See, for instance, Bridgman (1950), Bunge (1998a, b), Campbell (1921/1952), Chandrasekhar (1987), Cushing (1998), Holton (1973), Margenau (1950), Rabi (1967), Rohrlich (1987), Weinberg (2001), and Shimony (1993).

[46] For instance, Bernal (1939), Birch (1990), Haldane (1928), Hull (1988), Mayr (1982), Monod (1971), Polanyi (1958), and Wilson (1998). One recent contribution to the genre is by Francis Collins, the geneticist and leader of the Human Genome Project (Collins 2007).

[47] There are countless books on the worldview of modern physics: see, for example, contributions to Cushing and McMullin (1989), especially Abner Shimony's contribution 'Search for a Worldview Which Can Accommodate Our Knowledge of Microphysics' (Shimony 1989). See also the contributions to the special issue of *Science & Education* dealing with Quantum Theory and Philosophy (vol. 12 nos. 5–6, 2003).

50   Science, Worldviews and Education

who has never reflected on the principles of his work has not achieved a grown-up man's attitude towards it; a scientist who has never philosophized about his science can never be more than a second-hand, imitative, journeyman scientist. (Collingwood 1945, p. 2)

What Collingwood says about the requirement of 'reflecting upon principles' being necessary for the practice of good science can equally be said for the practice of good science teaching. Liberal education promotes just such deeper reflection and the quest to understand the meaning of basic concepts, laws or methodologies for any discipline (mathematics, history, economics, theology) being taught including science.[48]

50.7   Science and the 'Invisible World'

The world's major religions have had an ongoing engagement with science, investigating how their own ontological, epistemological, anthropological and ethical commitments – their worldviews – are to be reconciled with both scientific findings and scientific worldviews. Religion is the most publicly discussed and debated aspect of the science and worldview interaction and the one that most often occupies educators in their writing of national and provincial curricula, in their arguments about multicultural and indigenous science, in their debates about textbook selection and in their classroom teaching and interactions with students and parents. Because modern science emerged out of Christian Europe in the seventeenth century, the arguments and adjustments between Christianity and science – over Creation, Evolution, Providence, Miracles, Revelation, Authority – have been debated longest in this religious tradition, and hence it will be the focus of this chapter.[49] This section of the chapter will deal with just one of the many issues and debates that have arisen in the field: the putative existence and powers of spiritual agencies, spirits, ghosts, poltergeists and angels, inhabitants of what John Wesley, the founder of Methodism, called the 'Invisible World'.

50.7.1   Abrahamic Religions

Belief in a spirit-filled, invisible world is fundamental to the Judaeo-Christian-Islamic tradition. Jewish society simply took over the heavily populated world of demons that the Mesopotamian and Hellenic worlds also recognised with their ontology of beings intermediate between gods and men, these were the *daimones*. The Judaeo-Christian explanation of this realm of troublemakers and evil inducers

---

[48] The Philosophy for Children movement has shown that this reflection and quest can begin in Elementary school (Lipman 1991; Matthews 1982; Sprod 2011).

[49] Among a veritable library of relevant books, see Barbour (1966), Brooke (1991), Haught (1995), Jaki (1978), Mascall (1956), and contributions to Lindberg and Numbers (1986).

was of course the expulsion from heaven of Satan and his fallen angels (Genesis 6:1–4). This was a more than satisfactory explanation of their existence, powers and inclinations. Jinn, or spirits and angels, were an integral part of the Judaic tradition, everyone in pre-Islamic Arabia believed in them; they lived in a world unseen to humans; they eat and drink and procreate; some are righteous while others are evil. Illness, unusual events, misfortunes, catastrophes and so on were attributed to this host of other-worldly ne'er-do-wells.

The New Testament and the early Christian Church which was a sect of Judaism simply carried on belief in the reality and powers of demons, or 'unclean spirits' as they are also called. These demons were responsible for false teaching (1 Timothy 4:1); they performed wonders (Apocalypse 16:14); they rule the kingdom of darkness (Ephesians 1:21, 3:10); and so on.

Of particular account in New Testament demonology is the widespread and frequent occurrence of possession of people by the devil or evil spirits. This continued a Judaic and Mesopotamian belief in diabolical possession, one that routinely attributed psychic illness (as now understood) to such a cause (Mathew 8: 16, 12:27; Mark 1:34; Luke 7:21, 11:19; Acts 19:13–16). The apostles exorcised evil spirits where they could, with the most graphic instance being the exorcism in the Gerasa cemetery where the demons fled the person and possessed the herd of swine that they then drove to their death in the Sea of Galilee. Converts such as Paul also had such powers and exercised them effectively such as when he drove the evil spirit from the girl from Philippi (Acts 16:16). Sometimes they were not successful, as with the boy now seen to be most probably an epileptic (Mathew 17:14–21; Mark 9:14–29; Luke 9:37–43).

John McKenzie, a Catholic commentator (from whom the foregoing textual references are taken), has written: 'The belief in heavenly beings thus runs through the entire Bible and exhibits consistency' (McKenzie 1966, p. 32). And further adds:

> But while the use of popular imagery should be understood to lie behind many details of the New Testament concept of demons, the Church has always taught the existence of personal evil spirits, insisting that they are malicious through their own will and not through their creation. (McKenzie 1966, p. 194)

The Protestant tradition held comparable views. Martin Luther wrote:

> Demons live everywhere, but are especially common in Germany. On a high mountain called Polterberg there is a pool full of them: they are held captive there by Satan. If a stone is thrown in a great storm arises and the whole countryside is overwhelmed. Many deaf persons and cripples were made so by the Devil's malice. Plagues, fevers and all sorts of other evils come from him. As for the demented, I believe it to be certain that all of them were afflicted by him. (In Mencken 1946/1930, p. 244)

John Wesley wrote in his *Journal* in 1768 that: 'The giving up of witchcraft is in effect the giving up the Bible'. He regarded witchcraft as 'one great proof of the invisible world'.

It is hardly surprising that half of all Americans tell pollsters that they believe in the Devil's existence, and 10 % claim to have communicated with him (Sagan 1997, p. 123). The extent of such belief has been more recently documented in the findings of the large-scale 2008 Pew Report on religious belief and practice in the USA.[50] This survey of 35,000 US adults, most of whom would have completed the high school science requirement, found that belief in some form of God was nearly unanimous (92 %) and that this God was not the remote, untouching God of eighteenth-century Deists, but a God who was actively engaged in the affairs of people and of processes in the world. Nearly eight in ten American adults (79 %) agree that miracles still occur today as in ancient times. Similar patterns exist with respect to beliefs about the existence of angels and demons. Nearly seven in ten Americans (68 %) believe that angels and demons are active in the world. Majorities of Jehovah's Witnesses (78 %), members of evangelical (61 %) and historically black (59 %) Protestant churches and Mormons (59 %) are *completely* convinced of the existence of angels and demons.

Belief in such a rich spirit-populated world 'invisible world' is a requirement for the world's 1.5 billion Muslims. Belief in angels is the second of Islam's six Articles of Faith. In Islam, Jinn are spirits made by Allah from smokeless fire; some Muslim scholars say that Jinn populated the earth 2,000 years before the creation of humans out of clay. The Islamic philosopher Seyyed Nasr writes:

> To rediscover the body as the abode of the Spirit…is to re-establish our link with the plants and animals, with the streams, mountains and the stars. It is to experience the Spirit in the physical dimension of our existence. (Nasr 1996, p. 262)

The whole constellation of traditional religious beliefs, especially those affirming an active ongoing engagement of God, angels and spirits with human affairs, requires that the world, including human beings, be constituted in certain ways; that the world has a certain ontology; and that the human beings are so constituted that it can know of and interact with these supernatural agencies. All of this amounts, in part, to a religious worldview, a view about how the world and human beings need to be constituted so as to enable, or ground, religious belief, experience and practice.

Henry Gill, a Catholic priest, philosopher and physics lecturer, gave succinct expression to the kind of worldview held by many of the above-mentioned religious believers:

> It will be useful to recall briefly the Catholic teaching as to the existence of spirits. The Scripture is full of references to both good and bad spirits. There are good and bad angels. Each of us has a Guardian Angel, whose presence, alas, we often forget. Angels, as the Catechism tells us, have been sent as messengers from God to man. Our Lady, at the Incarnation, St. Joseph before the flight into Egypt, both received messages. Our Lord Himself was tempted by Satan …. Finally, it is the certain teaching of the Church that the conditions which depend on whether the human being to whom it belonged has or has not lived according to the dictates of conscience and quitted this life in friendship or at enmity with God. (Gill 1944, pp. 127–128)

This statement implies and presupposes certain ontological, epistemological, anthropological and theological positions which rolled together constitute a statement of the traditional Roman Catholic worldview, a worldview that was 'at home' in

---

[50] The survey was conducted between May and August 2007 and published in June 2008 in the Pew Report at www.pewreport.org.

Thomism and Scholasticism and is professed by a goodly number of the world's 1.2 billion Roman Catholics. The constituent domains of a worldview are meant to cohere. If one's ontology has angels and spirits existing with certain powers, then one's epistemology has to account for the possibility of this knowledge, and further it needs to indicate how the truth or falsity of claims about spirits can be ascertained. Is the epistemological ground for such claims Intuition? Authority? Religious Experience? or Revelation? It is rarely claimed that the ground is Science.

Despite being everywhere and being endowed with amazing powers and being variously credited with causing tsunamis, AIDS, schizophrenia, adultery, paedophilia and much else, such angels and spirits do not show up in laboratories or scientific texts; they have not gained a place in the scientific understanding of the natural, social or personal worlds. This gives rise to a certain disconnect. Such claims are then either discordant with, or orthogonal to, the worldview and conduct of science.

50.7.2 Traditional Societies

In traditional or indigenous cultures, these convictions about the 'invisible world' and interactions between this supernatural world and the everyday world are usually bolstered with animist beliefs where plants and natural objects are endowed with intelligences and spiritual attributes and where natural processes can be swayed by rituals, incantations, charms, potions, magic, sorcery and spells. In most such cultures, spirits are everywhere and have immense powers; they feature in traditional stories, legends and myths and underwrite a wide variety of social and medical practice.

Papua New Guinea is a representative case. In the early months of 2013, there had been a series of horrific sorcery-related gruesome murders committed. In January outside Mt Hagen, the capital of the Western Highlands, a 20-year-old mother was accused of sorcery, doused in petrol and burnt alive atop a pile of rubbish and car tyres. She supposedly had used her powers as a witch to kill a boy who had been admitted to hospital with chest pains. In March a highlands man ate his newborn son in order to bolster his sorcery powers. The same month in the Southern Highlands, six supposed witches were tortured with hot irons and one roasted to death. In April in Bougainville, two elderly women, accused of being witches and causing the death of a school teacher, were tortured for 3 days then beheaded in front of a large mob that included police officers. In just one Highland province, Simbu, there are 150 sorcery-inspired attacks per year.[51]

At the same time, the PNG government released a report on the AIDS epidemic in the country detailing the prevalence, and uselessness, of traditional treatments such as having sufferers sit atop huts inside of which is burnt 'special fires' in expectation that the rising smoke would carry off the evil spirits inhabiting the

---

[51] See accounts and interviews in Elliot (2013).

person and causing the sickness. Such practices are widespread in the country where things are believed to happen not just for physical reasons because there can always be some non-natural trigger or cause for the happening; indeed the latter are so commonplace that to refer to them as 'non-natural' fails to understand the traditional worldview where 'white' and 'black' magic (*Sanguma*) are just part of how things are; *Sanguma* is everywhere and is recognised in the legal code. As one commentator remarked on these practices:

> In a remote world lacking scientific explanation, in which life could be brutish and short, it was natural that people sought not only a way to understand how their world worked, but also to find a way to take a measure of control over it. (Callick 2013, p. 11)

The reality and efficacy of sorcery is recognised in the 1971 *Sorcery Act*. A 1977 PNG Law Commission study on 'Sorcery in PNG' concluded:

> We have written some general ideas about sorcery we know from our own experience as Papua New Guineans. In order to get a balanced view of sorcery we would like to say that sorcery is very much a matter of the innermost belief of the people. Fear of, or the practice of sorcery or various occults is a world-wide phenomenon. Sorcery or black magic exists in Europe, in Asia, in Africa and in North and South America as well as the Pacific.
> Major world religions claim the reality of forces or personalities greater than the human and animal powers. Whether these powers or personalities can be shown to exist is often quite irrelevant to the belief. From these beliefs many practices and procedures follow. (Narokobi 1977, p. 19)

The 2013 revision of the legal code is moving to deny the reality of such powers and make supposed *Sanguma* bashings, torture and killings criminal offences.

A long-time PNG Catholic priest, Philip Gibbs, recognised the incompatibility of Enlightenment-informed scientific worldviews and biblical worldviews when he described PNG culture as having a:

> Pre-Enlightenment, or Biblical, worldview … They don't believe in coincidence or accidents. When something bad happens, they don't ask what did it but who did it. (Elliot 2013, p. 18)

The situation with PNG traditional society is repeated in sometimes more and other times less extreme forms in most traditional societies and other societies, where the spirit world looms large and where centuries, if not thousands of years, of tradition, folklore and superstition are embedded.

Ancient rock art of the Australian aboriginal Worrorra people has recently been re-discovered in the spectacular Kimberley country in north-west Australia. The recurrent image is of Wandjinas, the supreme spirit who created the country during the Dreamtime. The Worrorra belief is:

> The Wandjinas created the animals and the baby spirits that live in the rock pools or sacred Ungud places throughout the Kimberley, and they continue to control everything that happens on the land, sea and sky. (http://wandjinatours.com.au/)

It is routine in most Southeast Asian countries for residential and other buildings to have 'spirit houses' prominently placed so that spirits disturbed in the construction have a new home; a home where food and offerings are left so the annoyed spirits do not do mischief. Some such beliefs and 'the practices and procedures' based on them are benign, while others are dramatically less so.

'Smoking' ceremonies where Australian aboriginal people gather and burn special leaves so the spirit of a deceased can be released and be carried upwards with the smoke to 'heaven' are a benign example.[52] The case of a New Zealand Maori couple who in November 2007 gouged out the eyes of their 14-year-old daughter in order to allow the escape of a bad spirit who supposedly possessed the girl is a less benign example. This 'exorcism' was witnessed by 40 relatives.[53] Of course these beliefs and practices are not just those of 'traditional' societies: Not long after the Maori episode, the Vatican's official exorcist, Father Gabriele Amorth, who has conducted 70,000 exorcisms, claimed that many paedophilia cases were the direct work of devils who possessed or otherwise influenced the offending priests (Amorth 2010).

When politicians, doctors, nurses and educated community members deny the efficacy or existence of 'bad' spirits or devils, they are involved in proto-science. The basic claim is that 'there is no evidence' for such possession and that the evidence (paedophilia or children dying) can be accounted for by other natural causes. This basic claim moves discussion into the field of science and evidence appraisal. But once that move is made, then why not extend the examination to the efficacy or existence of 'good' spirits and angels?

50.7.3 Feng Shui

Non-scientific, and in some cases anti-scientific, worldviews are widespread in advanced economies and cultures where commitment to astrology, parapsychology, levitation, clairvoyance, mediums, extrasensory perception, the paranormal, telepathy, astral travel, Thiaoouba consciousness and so on are common.[54]

One entrenched practice and theory that does not attract the attention it warrants is feng shui (Rossbach 1984). Millions of Chinese believe in its principles, and increasingly it is being adopted outside of the Chinese community. Feng shui advising is a thriving business with thousands of consultants, and law courts determining whether correct or incorrect feng shui advice was given in cases where poor business returns or illness follows occupancy of feng shui-certified commercial and residential buildings.

Feng shui, or Chinese geomancy, derives from an ancient Chinese system of rules, concepts and principles that endeavours to explain the impact on people's lives of the layout and design of their business and home. Its origins lay in the 3,000-year-old writing, *I Ching*, of the ancient sage Fu His who had the inspiration that the diverse fundamental forces of the universe were mirrored in the orderly markings on the shell of the tortoise which when arranged in threes gave eight

---

[52] There is a parallel use of incense in the Roman Catholic burial liturgy.

[53] *Sydney Morning Herald* November 27, 2007

[54] On this phenomena, see Dawkins (1998), Sagan (1996), Shermer (1997) and the classic study by W. E. H. Lecky (1914). Hundreds of thousands of websites are devoted to these 'alternative' science practices and 'theories'.

trigrams corresponding to Heaven, Earth, Fire, Water, Mountain, Lake, Wind and Thunder (Spear 1995, Chap. 5).[55] Unlike the case of chaotic and inconsistent good or bad spirits, feng shui purports to describe regular, lawful natural processes in just the same way that science does. But over and above the mundane energy of science, there is another universal life force called Chi or Mana. One feng shui exponent explains that:

> Chi is the vital force that breathes life into the animals and vegetation, inflates the earth to form mountains, and carries water through the earth's ducts … Without chi, trees will not blossom, rivers will not flow, man will not be. (Birdsall 1995, p. 37)

But chi is a peculiar kind of vital energy as:

> Doors are seen as the entrance of chi for any place. A building, house or room takes in its chi through the doors and, to a lesser extent, the windows. If the doors are too small then not enough chi will get into a place too many doors down a corridor may affect and confuse the flow of chi. (Birdsall 1995, p. 129)

And being more precise:

> One of the classic rules of feng shui…is that if you have a toilet in the wealth corner of your home or business, then you are likely to have financial or abundance troubles. (Birdsall 1995, p. 117)

However there is a correction for such faulty construction:

> Place a crystal in the window of the bathroom to draw in the universal chi. (Birdsall 1995, p. 118)[56]

And so on for 200 pages of this popular book. But the author moves from relatively trivial and harmless nostrums to something more arresting:

> Science can no longer dismiss the concept of our energetic body existing outside our physical bodies, as the former can now be photographed and analysed by Kirlian photographs. Often disease can be seen in the auric body before it shows in the physical body. (Birdsall 1995, p. 38)

But these high-energy photographic effects have everything to do with effects of changes in proximal humidity around bodies and nothing to do with supposed auras; as water vapour is progressively removed from around a hand or head, the aura image disappears. Basing medical diagnosis and treatments on such foolery can do damage, quite apart from wasting people's money.

With its commitment to ontological, epistemological and axiological principles that guide behaviour, feng shui counts as a worldview, one that is held by millions of people. As one advocate writes: 'More than just the practice of geomancy, placement, or spatial arrangement, feng shui is also a philosophy or a way of seeing the world' (Spear 1995, p. 15). And where it is not a self-contained worldview, it is a component that its believers' wider worldviews need to accommodate.

---

[55] There are over 100 English translations of the basic text, *I Ching*.

[56] Another advocate who addresses this problem does admit that 'no feng shui cure can be as powerful as a properly placed, flawless diamond to activate chi in an environment' (Spear 1995, p. 131).

Following feng shui 'principles' often does no harm. It is like the fabled Notre Dame football coach who said that God most answered the team's prayers when 'the forwards were big and the backs were fast'; or knowing that 'with an uncle's blessing and one dollar, you can ride the subway'. So if you build in such a way that your living quarters receive natural sunlight, then it is just a bonus that the room also falls on a good chi line and has auspicious bagua. Ditto if your bedroom has a window that does after all let in fresh air. But feng shui belief can do harm, and it is a distraction: Sometimes illness is caused by an infection, and poor profit results are the outcome of bad business decisions. In such cases rearranging the furniture or putting in an extra door, mirror or even a diamond will make no difference to the malady.

If teachers have some training in history and philosophy of science, the philosophical implications of feng shui can be drawn out. There are a host of questions and tasks that can usefully occupy students:

What are the claims of feng shui and are they scientifically testable?
Do other theories have the same implications and thus are they equally supported by whatever experimental evidence might support feng shui?
Where there are two, or more, such theories consistent with the empirical evidence [people in sunny houses feel well and suffer less colds] can any crucial experiment be devised to evaluate the theories?
Are feng shui predictions and theory elaboration all ad hoc? And what is wrong with ad hoc adjustments in theory?
Is feng shui a progressive research programme making novel claims or a reactive or degenerating one which only embraces claims made on other grounds?
How can the cultural, social and economic pros and cons of feng shui be explicated and appraised?

And this drawing out can be done by social science and science teachers working together.

For science teachers in cultures where feng shui is part of the social fabric, such questions can be used as a way into better understanding of the nature and methods of science, including the question of the function of naturalism in science and in culture. And orthodox religious belief, traditional cultural beliefs, astrology, psychokinesis, aura therapy, psychoanalysis and anything else can be substituted for feng shui in the foregoing questions. Each question draws on routine discussion in philosophy of science, and so each question can be the occasion for elaboration and learning of the latter.

50.7.4 Education and the 'Invisible World'

Thus far no such traditional, or other, spirits have been identified by science as having any causal interaction with the world. Yet they are very much a central part of the worldviews of several billion people. The educational question is what

50 Science, Worldviews and Education

to do about such beliefs? Should nothing be done and the cultural status quo retained unaltered? Should students be encouraged to believe just in good spirits and not in bad ones?[57] And if they are to believe in good spirits, should they believe for ontological reasons (there actually are such things) or for instrumental reasons (such belief is harmless and part of the cultural or religious tradition)? Or not believe in spirits at all?

The last was the choice of Joseph Priestley, the famed eighteenth-century English scientist, historian, philosopher, theologian and Dissenting Church minister[58]:

> The notion of madness being occasioned by evil spirits disordering the minds of men, though it was the belief of heathens, of the Jews in our Savior's time, and of the apostles themselves, is highly improbable; since the facts may be accounted for in a much more natural way. (Rutt 1817–1832/1972, vol. 7, p. 309)

For Priestley, Jesus was simply mistaken when he attributed the cure of madness to driving out evil spirits because subsequent science and philosophy had shown there were no such things to be driven out.[59]

Whether the Abrahamic religious traditions or 'indigenous' traditions can abandon belief in all, most or just some of 'the invisible world' is an engaging theological and cultural question, but the grounds for such discussion can usefully be prepared by discussions in philosophy of science about what constitutes 'hard-core' commitments in a research programme and how these are separated from 'protective belt' commitments – to use the terminology of Imre Lakatos (1970). Philosophers of science have for long dealt with the questions of what are the 'core' commitments of a research programme and what are 'optional' commitments, and familiarity with these discussions and analyses can inform comparable discussion about religion and important cultural beliefs.

50.8 Multiculturalism and Science Education

Examples of spirit-laden cultures and traditions have been given above. It was pointed out that such belief constellations were either discordant with or orthogonal to science, with the latter depending on whether spiritual, or supernatural, agencies had engagement with the world. In these latter cases, the relationship is not orthogonal; once it is claimed that the 'invisible world', or supernatural agencies, connect to the world and have worldly impacts, then they enter the domain or 'magestria' of

---

[57] In March 2013 Indonesia's criminal law statutes were being rewritten so as to make the practice of black magic (where people are harmed by sorcery) illegal but keep white magic legal. This raises not just ethical questions but engaging philosophical ones as well. Does a 'guilty' verdict acknowledge that such powers of mind over matter were exercised?

[58] The basic texts for Priestley's life, writings and achievements are Schofield's (1997, 2004). See also de Berg (2011), Matthews (2009c).

[59] In passing it is worth noting that every account of Priestley's life shows that a deeply 'spiritual' life is possible without any belief in spirits.

science. It was also suggested that students can be encouraged to engage in a number of routine philosophical questions about such belief systems. Such questions and pedagogy raise important matters about the purposes of science education and the distinction between *understanding* science and *believing* science. Some maintain that science education should leave cultural beliefs untouched, that students should simply leave their culture's worldview (ontology, epistemology, metaphysics, authority structure, politics and religion) at the classroom door, then enter inside to learn the instrumentally understood content of science, then go back outside and become again full-believing participants in their culture. This is close to advocating an anthropological approach to learning science. Just as anthropologists can be expected to learn *about* the beliefs and practices of different societies without any expectation that they adopt or come to believe them, some say that students can learn science in the same way, a sort of 'spectator' learning where one learns but does not believe.

Glen Aikenhead, in a much cited paper, has advocated such a strategy calling it 'border crossing' (Aikenhead 1996, 2000). Just as tourists when they cross borders do not lose their cultural identity even though they temporarily adopt foreign customs about driving, eating, dressing and language, so also science students should not lose their cultural identity (as a traditional Catholic, a fundamentalist Christian, an Intelligent Designer, a PNG highlander, a feng shui enthusiast and so on) just because the science laboratory has no place for their own rich beliefs. This 'border-crossing' option is a form of pedagogical NOMA; it is a profoundly anti-Enlightenment view.

Early modern and Enlightenment philosophers thought much would be gained if the method of the New Science might be applied to the seemingly intractable social, political, religious, philosophical and cultural problems of the times. During the period of Galileo's most productive work, the terrible 30 Years' War (1618–1648) raged all over Europe – in German states, France, Italy, Spain, Portugal and the Netherlands – and was also fought out in the Indies and in South America. It is widely accepted that between 15 % and 20 % of the German population, Catholic and Protestant alike, were killed. Along with ferocious religious wars witch crazes also engulfed Europe with the worst excesses occurring in France, Switzerland, Germany and Scotland. In the Swiss canton of Vaud, in the 90 years between 1591 and 1680, 3,371 women were tried for witchcraft and all were executed (Koenigsberger 1987, p. 136). The Salem witch trials took place in Massachusetts in 1692, 5 years after publication of Newton's *Principia*. As late as 1773, nearly 100 years after publication of Newton's *Principia*, the Presbyterian Church of Scotland reaffirmed its belief in witchcraft; but Catholic Spain has the distinction of being the last European country to burn a witch at the stake, this being in the early nineteenth century. And the lamentable practice still goes on in Papua New Guinea, Africa and doubtless many other traditional societies untouched by science and the Enlightenment.

Newton believed that there would be beneficial flow-on effects if the methods of the New Science were applied to other fields. As he stated it: 'If natural philosophy in all its Parts, by pursuing this Method, shall at length be perfected, the Bounds of

Moral Philosophy will be also enlarged' (Newton 1730/1979, p. 405). David Hume echoed this expectation with the subtitle of his famous *Treatise of Human Nature* which reads *Being an Attempt to Introduce the Experimental Method of Reasoning into Moral Subjects* (Hume 1739/1888).

The contrast between the aspirations of the Enlightenment philosophers and contemporary 'border-crossing' science educators is profound and speaks to a major divergence in their appreciation of science. This indeed is the case, with Glen Aikenhead maintaining that 'the social studies of science' reveal science as:

mechanistic, materialist, reductionist, empirical, rational, decontextualized, mathematically idealized, communal, ideological, masculine, elitist, competitive, exploitive, impersonal, and violent. (Aikenhead 1997, p. 220)

This claim is as puzzling as it is disturbing. Is the claim meant to describe the work of Galileo? Newton? Huygens? Darwin? Mendel? Faraday? Mach? Thompson? Lorentz? Maxwell? Rutherford? Planck? Einstein? Bohr? Curie? Does it describe the work of Edward Jenner in developing smallpox vaccine? Or the achievements of Jonas Salk and Albert Sabin in developing polio vaccine? We are not told whose science warrants the description. It is clearly a composite or collage that requires unpicking, but this is not done – the good, the bad and the ugly are all lumped together. Unfortunately there are numerous cases of corrupt science – Nazi and Stalinist science being the best-known examples – but these do not warrant generalisations, they warrant correction. From Aikenhead's undifferentiated description, it is doubtful whether science should even be in the curriculum; it certainly should be rated X, with even border crossing being dangerous for minors. Other prominent and influential science educators share Aikenhead's unfavourable estimation of science. Consider, for instance, the demand that teachers need to:

deprivilege science in education and to free our children from the 'regime of truth' that prevents them from learning to apply the current cornucopia of simultaneous but different forms of human knowledge with the aim to solve the problems they encounter today and tomorrow. (Van Eijck and Roth 2007, p. 944)

Or the claims made in a contribution to a current major science education handbook that:

... one of the first places where critical inquirers might look for oppression is positivist (or modernist) science ... modernist science is committed to expansionism or growth ... modernist science is committed to the production of profit and measurement ...modernist science is committed to the preservation of bureaucratic structures. (Steinberg and Kincheloe 2012, p. 1487)

Science is a force of domination not because of its intrinsic truthfulness, but because of the social authority (power) that it brings with it. (Steinberg and Kincheloe 2012, p. 1488)

Modernism refers to a way of understanding the world produced by Enlightenment thinkers and employing a scientific methodology and the concept of rationality Drawing on dualism, scientists asserted that the laws of physical and social systems could be uncovered objectively; the systems operated apart from human perception, with no connection to the act of perceiving .... This separation of mind and matter had profound and unfortunate consequences. (Steinberg and Kincheloe 2012, pp. 1490, 1491)

Such appraisals demonstrate the need for science educators to be careful and considered in their reading and studies of history and philosophy of science; clearly a little knowledge is a dangerous thing. The above accounts, apart from being confused and contradictory, cannot be sustained. Taking the subject *out* of the picture and relying on measuring instruments (rulers, scales, thermometers, barometers, clocks) instead of subjective appraisals of length, weight, temperature, pressure and duration; utilising mathematics; introducing idealisations and abstractions; valuing objective evidence; and being public, communal, publishing, criticising and debating are all the things that enabled the Scientific Revolution to occur in seventeenth-century Europe and progress to its current international status. And of course Copernicus and Galileo had no social authority enforcing their heliocentricism, on the contrary; while Lysenko had all the oppressive authority of Stalin behind his non-Mendelian genetics and it counted scientifically, and ultimately, for nothing. All of this is missing in the accounts of science given by the above science educators.

50.9 Philosophical Systems and Religious Belief

The juxtaposition of scientific and religious worldviews brings into focus a number of enduring philosophical, religious and cultural issues, among which are at least the following:

1. Do religions make metaphysical claims? Are there preferred philosophical systems for the expression of such claims? Are particular religions tied to such systems?
2. Is there a need for religious claims to be made intelligible and testable? Or is Faith deeply personal, experiential and indefinable? Is Faith apart from and beyond Reason?
3. Should philosophical systems be judged by their theological adequacy or compatibility?
4. Should religious establishments (churches, priests, ministers, imams, ulama) have the authority to proscribe philosophical systems or metaphysics to their co-religionists? Should they be able to proscribe to others outside their religion when their religion exercises or influences state power?

These issues were argued within the Christian churches; they were debated in the Enlightenment and are still debated.[60] For example, Claude Tresmontant in his *Christian Metaphysics* argues that:

> The thesis which I submit to the critical examination of the reader is that there is *one* Christian philosophy and one only. I maintain, in other words, that Christianity calls for

---

[60] For representative literature on this topic of 'Christian Philosophy', see Mascall (1966, 1971), Plantinga (2000, 2011), Tresmontant (1965), and Trethowan (1954). The suitability of Thomism as a vehicle for the interpretation of Christian doctrine is discussed in McInerny (1966). The same debates and literatures can be found in Islamic, Judaic, Hindu and other traditions.

50 Science, Worldviews and Education

a metaphysical structure which is not any structure, that Christianity is an original metaphysic. ... [it is] a body of very precise and very well-defined theses which are properly metaphysical .... (Tresmontant 1965, pp. 19–20)

The Catholic priest, philosopher and historian of philosophy, Fredrick Copleston wrote of his celebrated 1949 debate with A. J. Ayer that:

After all, my defence of metaphysics was largely prompted and certainly strengthened by what I believed to be the religious relevance of metaphysical philosophy. (Copleston 1991, p. 63)

One such common metaphysical position, vitalism, is clearly stated by another Catholic priest and philosopher:

That there is a fundamental difference between living and non-living matter is obvious. Catholic philosophers hold that an organized or living substance is distinguished from inanimate matter in that the former is informed by a 'vital principle' which confers on it the characteristics we associate with life. (Gill 1943, p. 73)

Such a position might be labelled 'privileged' in as much as the metaphysics comes from outside of science, not from within. Privilege for such metaphysical positions is usually derived from Revelation, Theology, Religious Experience, Philosophy, Intuition or Politics. Such privileged metaphysical views can be found enunciated by advocates of Judaic, Islamic, Hindu, Buddhist and a host of other religions, as well as advocates of the maintenance of indigenous belief systems and worldviews. These traditions would formulate the above four fundamental issues in their own terms. For instance, if 'Marxism-Leninism' is substituted for 'religion' and the 'Central Committee' substituted for 'religious authorities', then the above list of issues is applicable to the situation that pertained in the Soviet Union and its satellites and still pertains in China. No authoritarian state, since the Athens that put Socrates to death, has welcomed open and free philosophical study and debate.[61]

50.9.1 Compatibility of Science and Religion

When considering the compatibility of science and religion, we need to distinguish a number of sometimes conflated issues[62]:

First, whether religious claims and understandings have to be adjusted to fit proven scientific facts and theories. There really is no longer any serious debate on this issue; sensible believers and informed theologians acknowledge that religious

---

[61] Anthony Kenny, the British philosopher and former Catholic priest, depressingly relates in his autobiography how as a doctoral student at Rome's Gregorian University he needed his supervisor's permission to read David Hume's *On Religion* and that his degree would not be awarded unless he affirmed an anti-Modernist oath (Kenny 1985, p. 146).

[62] Different taxonomies or ways of classifying science/religion relationships are developed in Barbour (1990), Haught (1995) and Polkinghorne (1986). These, and others, are discussed in Mancy et al. (in press).

claims need to be modified or given a nonliteral interpretation to fit with proven or even highly probable science. Joseph Priestley, the eighteenth-century Enlightened believer, told the story of 'a good old woman, who, on being asked whether she believed the literal truth of *Jonah* being swallowed by the whale, replied, yes, and added, that if the Scriptures had said that *Jonah* swallowed the whale, she would have believed it too'. Priestley thought that such convictions simply indicated that the term 'belief' was being misused in the context: 'How a man can be said to *believe* what is, in the nature of things, *impossible*, on any authority, I cannot conceive (Rutt 1817–1732/1972, vol.6, p. 33). All serious thinkers on the topic, since St Augustine, agree with Priestley.[63]

Second, whether religious believers can be scientists. Again, at one level, there is no debate on this matter. As a simple matter of psychological fact, there have been and are countless believers of all religious stripes who are scientists.[64] But this sense of compatibility is of some but not determinate philosophical interest. The arguments and evidences put forward by these numerous eminent and believing scientists are relevant to the question of rational compatibility but not just the fact that there are such believers. Undoubtedly some scientists are astrologers, others channel spirits, some might think they are Napoleon reincarnated, some are racist and others are sexist and so on for a whole spectrum of beliefs that, as a matter of fact, have been held by scientists. No one doubts that science, as a matter of psychological fact, is compatible with any number of belief systems – recall that the Nobel laureates Philipp Lenard and Johannes Stark were both Nazi ideologues. Scientists are humans, and humans notoriously can believe all sorts of crazy things at the same time; but such psychological compatibility has no bearing on the rationality or reasonableness of their beliefs or the philosophical compatibility between science and belief systems. The latter is a logical or normative matter. The philosophically interesting question is whether a scientist can be a *rational* religious believer (or astrologer, diviner, reincarnationer, racist, sexist, Nazi, etc.).

Third, whether religion is compatible with the metaphysics and worldview of science. Where there is incompatibility between scientific and religious

---

[63] There has been debate about just what degree of proof a factual scientific claim needs to have before it triggers a revision in a competing factual religious claim – Augustine thought revision was needed only in the face of absolutely proven 'scientific' claims. The details of this debate do not bear on the present argument; for the arguments and the debate's literature, see McMullin (2005).

[64] John Polkinghorne, an Anglican priest, could be picked out as an exemplar of a research physicist and believer (Polkinghorne 1986, 1988, 1991, 1996). Many such individuals can be found contributing to journals such as *Zygon: Journal of Religion & Science*. For just one compilation of contemporary Christian scientists, see Mott (1991). There are comparable compilations of Hindu, Islamic, Mormon and Judaic scientists. There may even be compilations of Scientologist scientists and Christian Science scientists. These lists are relevant to the question of the psychological compatibility between scientific and religious beliefs, but not their philosophical or rational compatibility.

metaphysics and worldviews – as in the case of atomism and traditional Roman Catholic doctrine developed above – the options usually taken to reconcile the differences are to claim that:

1. Science has no metaphysics; it deals just with appearances and makes no claims about reality. This is the option made famous by the Catholic positivist Pierre Duhem.[65] It is the claim made by many fundamentalists who say, specifically of evolution, that 'it is just a theory'.
2. The metaphysics of science is false; at least any such purported metaphysics that is inconsistent with religious beliefs. This is the option advocated by the Scholastic tradition discussed above; by Claude Tresmontant and Seyyed Nasr who are quoted above; and by philosophical theologians such as Alvin Plantinga (2011), E. L. Mascall (1956) and numerous others.
3. There can be parallel, equally valid, metaphysics. This is an old option given recent prominence by Stephen Gould in his NOMA formulation (Gould 1999).

Gould's much-repeated claim was that:

> The magisterium of science covers the empirical realm: what the Universe is made of (fact) and why does it work in this way (theory). The magisterium of religion extends over questions of ultimate meaning and moral value. These two magisteria do not overlap, nor do they encompass all inquiry (consider, for example, the magisterium of art and the meaning of beauty). (Gould 1999, p. 6)

The problem for NOMA is that, apart from classical Deists for whom God stays remote in His heaven and has no dealings with His creation, the core conviction of religious traditions is that the two realms overlap: that the supernatural has engagement with the natural; that God engages with His Creation; that certain texts (Torah, Bible, Koran, Book of Mormon, Sikh scriptures) are inspired, if not divinely written; miracles occur; prayers are answered and so on. However as soon as claims are made for supernatural engagement with the natural world and processes, then they come within the magestria of science: The cause might be of another world, but the effect is of this world. Such causal claims need not and should not be ignored by science; science can test claims about miracles, supernatural interventions and even Divine authorship, just as it can test claims about putative paranormal and psychic occurrences.[66]

50.9.2 Scientific Study of Scripture

Because so much hinges upon it, the very possibility of a 'scientific' or scholarly investigation of sacred texts or Scriptures has always been contentious in all

---

[65] See extensive discussion and bibliography in Martin (1991).

[66] The anti-NOMA view that science can test supernatural claims is convincingly argued by many, including Boudry et al. (2012), Fishman (2009), Slezak (2012), and Stenger (1990, 2007).

'scripture-based' religious traditions. One historian of the origins of modern biblical scholarship writes that:

> Thus it is not surprising as may be thought that we can turn firstly to two *scientists* (this shows the importance of the Copernican revolution for the beginning of the historical-critical method) who are not usually thought of in this connection at all: Johannes Kepler and Galileo Galilei. (Andrew 1971, p. 95)

Galileo's 1615 *Letter to the Grand Duchess Christina* was perhaps the first clear statement in the Christian tradition of the view that Revelation should be investigated in the same way that other cultural artefacts and natural objects were investigated, namely, scientifically (Galileo 1615/1957). Galileo, in order to defend Copernicanism against those bringing scriptural objections to a rotating Earth, argued that Scripture had to be interpreted by Reason, and where there was conflict between Scripture and claims established conclusively by Reason, then Scripture had to be reinterpreted accordingly. Galileo was appealing to a well-established interpretative tradition in the Catholic Church going back at least to St Augustine, a tradition that acknowledged that not everything in Scripture was to be read literally as many things simply could not have occurred: Moses authoring the entire Pentateuch including the story of his own death; Jonah surviving in the whale; Methuselah living for 969 years; a flood that covered the whole earth, but of which other nations knew nothing; iron axe heads floating on rivers; numerous people rising from the dead; and so on. These stories had to be read metaphorically or poetically. But the Church was the authority that decreed what text was literal and what was poetic.

Baruch Spinoza's 1670 *Tractatus Theologico-Politicus* (Spinoza 1670/1989) is widely regarded as the first comprehensive attempt to state the exegetical principles of modern, secular, historical study of the Judaeo-Christian scriptures; the study of biblical history removed from theological dogma. He wrote in Chap. 7: 'the method of interpreting Scripture is no different from the method of interpreting Nature, and is in fact in complete accord with it' (*TTP* 3.98). This meant first, attention to the text only and not to tradition; second, adopting the primacy of 'Natural Reason' in interpretation of scriptural text. As a consequence in an ecumenical display, Calvinist ministers and Jewish rabbis combined to drive Spinoza from Amsterdam.[67] Some Enlightenment figures, Edward Gibbon and John Toland, for instance, thought scripture so imperfect that the advance of human reason simply dissolved revelation. Toland claimed that 'there is nothing in the Gospel contrary to reason' and then crucially added 'or above it' (Hyland et al. 2003, p. 60). The postscript ruled out recourse to affirmations of unanchored faith.

The addition was crucial because Kepler, Galileo and others admitted that scripture asserted things that were 'beyond human understanding' and so not contrary to human understanding; in the latter cases, scripture had to be reinterpreted, and in the former, there was no need to do so. This meant a cleavage in the reach of the Galilean method. For Toland and most Enlightenment thinkers, the 'beyond human reason'

---

[67] On Spinoza and the adoption of 'scientific method' in biblical studies, see Bagley (1998) and contributions to Force and Popkin (1994).

category was not countenanced; science was the arbiter separating literal from poetic in biblical texts.

In the mid-nineteenth century at the time of Pius IX, the German schools of Hermeneutics and higher Biblical criticism represented the more radical wing of Enlightenment views about the scholarly study and interpretation of Revelation. The Pope was particularly outraged by Ernest Renan's just published *Vie de Jésus* (Renan 1863/1935)[68] which he said ought to have been suppressed by the French government; it was so suppressed, along with most of the Enlightenment canon, in the Papal States and in other Catholic countries where the Church yielded political and judicial influence.

Salman Rushdie, the Muslim apostate who was condemned to death by the Iranian spiritual leader Ayatollah Khomeini, has recently written that:

> What is needed is a move beyond tradition – nothing less than a reform movement to bring the core concepts of Islam into the modern age. … If, however the Koran were seen as a historical document, then it would be legitimate to reinterpret it to suit the new conditions of successive new ages.

In saying this, Rushdie is claiming no more than what, informed by modern science, European Enlightenment figures of the eighteenth century asserted. One can only believe that the ensuing working out of this claim, to see Scripture as a historical text, will be the same in Islam as it has been in Christianity and in Judaism where the claim has been grappled with for four centuries. For some the Divine content of Revelation will evaporate, for others what will be revealed will be what Reason permits to be revealed, and for others such as those holding to some version of the 'Inerrancy of Scripture' doctrine, there will be some uneasy accommodation between Reason and Revelation.

50.10 Science and Naturalism

The conduct of science presupposes at least methodological naturalism (MN). This is the view that, when doing science, whatever occurs in the world is to be explained by natural mechanisms and entities and that these entities and mechanisms are the ones either revealed by science or in principle discoverable by science. This methodological presupposition does not rule out miracles or Divine interventions or other non-scientific causes; it just means that such processes cannot be appealed to while seeking scientific explanations. There has been historically a transition from more open or mixed methodology to having MN function as a defining principle of scientific investigation. As Robert Pennock states the matter:

> … science has completely abandoned appeal to the supernatural. In large part this is simply the result of consistent failure of a wide array of specific 'supernatural theories' in competition with specific natural alternatives. (Pennock 1999, p. 282)

---

[68] Other contributors to this 'higher criticism' or scientific study of texts were Friedrich Schleiermacher (1763–1834), F. C. Bauer (1792–1860) and David Strauss (1808–1874).

A stricter version of naturalism is ontological naturalism (ON), which sometimes is called metaphysical naturalism. This is the view that there is a scientific explanation for all events, that supernatural explanations (e.g. Divine interventions, miracles) simply do not occur. Many see ON as pure dogmatism, and it can be if it is held in advance as a philosophical principle. But it can be held on less dogmatic two-step grounds:

(i) Thus far no credible evidence has been advanced for the existence of any putative non-natural entity, or entity not within the scientific realm.

Many of course reject (i), and that is a whole separate argument. But some accept (i) and nevertheless say that ON does not follow from it or only follows dogmatically, as no one knows that evidence might turn up. But the nondogmatic holder of ON can add a second step to their argument:

(ii) Do not believe things for which there is no evidence.

If (ii) is granted, then ON does indeed follow. Then the dogmatism claim moves back to belief in (ii) rather than belief in ON. But belief in (ii) need not be dogmatic; it can be the 'default' position and its opposite, namely, the holding of beliefs for which there is no evidence, is dogmatic. This was in essence Bertrand Russell's 'teapot' argument.

> I ought to call myself an agnostic; but, for all practical purposes, I am an atheist. I do not think the existence of the Christian God any more probable than the existence of the Gods of Olympus or Valhalla. To take another illustration: nobody can prove that there is not between the Earth and Mars a china teapot revolving in an elliptical orbit, but nobody thinks this sufficiently likely to be taken into account in practice. I think the Christian God just as unlikely. (Russell 1958)

Both science-informed methodological and ontological naturalists admit the existence of whatever kinds of entities (e.g. atoms, fields, forces, quarks) science reveals as having regular causal relations with the rest of nature. But ontological naturalists do not admit the existence of spiritual or Divine entities or any kind of entity that does not enter into scientifically demonstrated lawful and causal relations with nature.[69] Traditional religious believers must reject ontological naturalism, but of course religious scientists routinely adopt methodological naturalism in the laboratory; to do otherwise would put them outside of the scientific enterprise.[70]

Materialists are a subspecies of ontological naturalists, but they are less relaxed about what can exist. Basic or 'old-fashioned' materialists grant existence only to

---

[69] Although often confused, there is a difference between realism and naturalism (including Materialism). Realism simply asserts that there is a world independent of human thought. Such an independent world might include spirits, minds, universals, mathematical objects, forms or any other independent existent. Realism neither rules in or rules out any particular kind of putatively existing being. A theological realist about angels believes that angels exist, not that the word 'angel' is shorthand for 'makes people behave' or 'strengthens our cultural bonds'. Naturalism is a subspecies of realism, it asserts that the only existing things are the things that science postulates and incorporates into successful and mature theories; materialism in turn is a subspecies of naturalism.

[70] On naturalism see Martin Mahner's contribution to this handbook. See also Fishman and Boudry (2013), De Caro and Macarthur (2008), Devitt (1998), French et al. (1995), Mahner (2012), Nagel (1956), and Wagner and Warner (1993)

material, physical, 'three-dimensional' objects, the kind of things that can be tripped over. They reject the postulation of nonmaterial scientific entities, believing that such postulation is a failure of scientific nerve and it is the slippery slope to idealism.[71] This is clearly as much an a priori metaphysical position as it is a deduction from scientific practice. Emergent materialism is a more sophisticated version where the world is seen as material but stratified. The properties of material aggregations are greater than, and different from, the properties of the building blocks. So cells have different kinds of properties than molecules, brains have different properties than neurons, societies have different properties than individuals and so on. For emergent materialists, the world is changing and evolving, and new properties emerge from more complex material formations.[72]

50.11 Worldviews and Philosophy in Science Classrooms

To state the obvious: It is important for teachers to recognise how science lessons engage with the religious beliefs and worldviews of students. Much has been written on this topic.[73] It needs to be remembered that the more profound philosophical dimensions of science can be approached in classrooms and curricula through small steps; there is no need to go in at the deep end. Any science textbook will contain terms such as 'observation', 'evidence', 'fact', 'controlled experiment', 'scientific method', 'theory', 'hypothesis', 'theory choice', 'explanation', 'law', 'model', 'cause', etc. As soon as students discuss and teachers explicate the meaning of these terms, and related concepts, then philosophy has begun. And the more their meaning, and conditions for correct usage, is investigated then the more sophisticated a student's philosophising becomes. The pupil who asks: 'Miss, if no one has seen atoms, how come we are drawing pictures of them?' has raised just one of the countless philosophical questions to which science gives rise (the relationship of models to reality). Likewise the student who wants to know whether after seeing 20 white swans they can conclude that 'all swans are white' touches upon another enduring philosophical dispute (the problem of induction and evidential support for theory). Similarly the student who, having been told about the force of gravitational attraction that exists between bodies, asks why we cannot see it, touch it, smell it or trip over it is highlighting yet another core philosophical issue (the realist versus instrumentalist debate about theoretical terms).[74]

---

[71] This was Lenin's argument against supposed idealist movements in early twentieth-century science and philosophy (Lenin 1920/1970). On the history and philosophy of materialism, see Vitzthum (1995).

[72] On emergent materialism, see Broad (1925), Bunge (1977, 1981), and Sellars (1932).

[73] See at least Erduran and Jiménez-Aleixandre (2008), Cobern (1991, 1996), Yasri et al. (2013), Preston and Epley (2009), Taber et al. (2011), and Stolberg (2007).

[74] For further discussion of the role of philosophy in science teaching, see Matthews (1994/2014, Chap. 5).

With knowledgeable teachers, all of these questions can be further explored, elaborated and connected to studies in other subjects such as economics, history, theology and so on. And such philosophical preparation or exercises can usefully lead on to the more profound questions concerning science, worldviews and religion; without philosophical and historical preparation, the latter discussion too readily becomes merely the exchange of hot air and the advancement of not much at all.

50.11.1 *From Physics to Metaphysics: The Law of Inertia*

Consider the law of inertia and its related concept of force. The law is the foundation stone of classical physics which is taught in school to every science student. A representative textbook statement is:

> Every body continues in its state of rest or of uniform motion in a straight line except in so far as it is compelled by external impressed force to change that state. (Booth and Nicol 1931/1962, p. 24)

It might be 'demonstrated' by means of sliding a puck on an air table and a puck on an ice sheet or by utilising a version of Galileo's inclined plane demonstration.[75] In a purely technical science education, the law is learnt by heart, and problems worked out using its associated formulae: $F = ma$.

Technical purposes might be satisfied with correct memorisation and mastery of the quantitative skills – 'a force of X newtons acts on a mass of Y kilograms, what acceleration is produced?' – but the goals of liberal education cannot be so easily satisfied.

Just a little philosophical reflection and historical investigation on this routine topic of inertia opens up whole new scientific and educational vistas. The medieval natural philosophers were in the joint grip of Aristotle's physics and of common sense beliefs resulting from their routine everyday experience; indeed Aristotle's physics was more or less just the sophisticated articulation of common sense. Aristotle's empiricism is evident when he says that 'if we cannot believe our eyes what can we believe'. A contemporary Aristotelian says that:

> Aristotle began where everyone should begin – with what he already knew in the light of his ordinary, commonplace experience. .... Aristotle's thinking *began* with common sense, but it did not *end* there. It went much further. It added to and surrounded common sense with insights and understandings that are not common at all. (Adler 1978, pp. xi, xiii)

These understandings resulted in the medieval commitment to the principle of *Omne quod movetur ab alio movetur*: the famous assertion of Aristotle, Aquinas and all the Scholastics which translates as 'Whatever is moved is moved by another (the motor)' and its inverse, if a motor ceases to act, then motion ceases. The principle

---

[75] On Galileo's inclined plane experiments, see Palmieri (2011); on their classroom use, see Turner (2012).

50   Science, Worldviews and Education   1623

grew out of daily experience, common sense and Aristotle's physics. Clagett summarised Aristotle's conviction as:

> for Aristotle motion is a process arising from the continuous action of a source of motion or 'motor' and a 'thing moved'. The source of motion or motor is a force – either internal as in natural motion or external as in unnatural [violent] motion – which during motion must be in contact with the thing moved. (Clagett 1959, p. 425)

Given the fact of motion in the world, then the principle led Aristotle to the postulation of a First Mover. Aquinas and the Scholastics took over this argument and made it an argument for the existence of a prime mover who they identified as God.[76]

Medieval impetus theory was an elaboration of Aristotelian physics: The mover gave something (impetus) to the moved which kept it in motion when the mover was no longer acting (the classic case of a thrown projectile). Some, like da Marchia, thought this impetus naturally decayed, and hence the projectile's motion eventually ceased. Others, like Buridan, thought that the transferred power was only diminished when it performed work, and as pushing aside air was work, then the projectile's motion would also eventually cease. Both theories were consistent with the phenomena: When a stone is thrown from the hand it goes only so far then drops to the ground.[77] Galileo performed a thought experiment by thinking through Buridan's theory to the circumstance of there being no work performed, in which case the projectile once impressed with impetus (force in modern speak) would continue moving forever. But for Galileo it would follow the Earth's contour. He repeated this circumstance with his experiment of a ball rolling down one incline and up another; as the second plane was gradually lowered towards horizontal, the ball moved further and further along it. He supposed that with the smoothest plane and the most polished ball, the ball would just keep moving on the second plane when horizontal; this was the visualisation of his theory of circular inertia.[78]

Galileo had no idea of a body being able to move off the Earth in a straight line away into an infinite void. Like everyone else, Galileo was both physically and conceptually anchored to the Earth. It was only Newton who would make this massive conceptual leap sufficient to have a projectile leave the Earth and move in an infinite void; he moved conceptually from a 'Closed World to the Infinite Universe' (Koyré 1957); by doing so Newton laid the foundation of modern mechanics. The whole 2000-year history of the development of the law of inertia reveals a good deal about the structure and mechanisms of the scientific enterprise, including the process of theory generation and theory choice.[79] Working through this history of argument bears fruit for arguments about worldviews and science.

---

[76] See the elaborate and informative discussion in Buckley (1971).

[77] The classic works on medieval impetus theory are Clagett (1959) and Moody (1975).

[78] The classic treatment is Clavelin (1974).

[79] See Ellis (1965) and Hanson (1965) for excellent discussion of Newton's formulation of inertia. On force, see Ellis (1976), Hesse (1961), Hunt and Suchting (1969), and Jammer (1957).

Apart from interesting and important history, basic matters of philosophy arise in any good classroom treatment of the law of inertia and the concept of force:

Epistemology – we never see force-free behaviour in nature nor can it be experimentally induced, so what is the source and justification of our knowledge of bodies acting without impressed forces?

Ontology – we do not see or experience force apart from its manifestation, so does it have existence? What is mass? What is a measure of mass as distinct from weight?

Cosmology – does such an inertial object go on forever in an infinite void? What happens at the limits of 'infinite' space? Were bodies created with movement?

These are the sorts of considerations that prompted Poincaré to say: 'When we say force is the cause of motion, we are talking metaphysics' (Poincaré 1905/1952, p. 98). And as every physics class talks of force being the cause of acceleration, then there is metaphysics lurking in every classroom, just waiting to be exposed by students who are encouraged to think carefully about what they are being taught and by teachers who know something of the history and philosophy of the subject they teach. Such exposition and engagement of school classes in the fundamental ontological, epistemological and methodological matters of philosophy that are occasioned by teaching and learning the law of inertia can be seen in a number of excellent texts.[80] In a recent publication, Ricardo Lopes Coelho discusses both the historical and pedagogical literature on this topic (Coelho 2007). All of this prepares the ground for a more nuanced and informed discussion of the big issues of science, worldviews and religion.

Thinking carefully and historically about basic principles and concepts is a quite general point about the intelligent and competent mastery of any discipline, be it Mathematics, History, Psychology, Literature, Theology, Economics or anything else. They all have their own, and overlapping, concepts and standards for identifying good and bad practice and judgements; consequently there are philosophical questions (epistemological, ontological, methodological and ethical) about each discipline; there is a philosophy of each discipline. The intelligent learning of any discipline requires some appropriate interest and competence in its philosophy; that is simply what 'learning with understanding' means – an obvious educational point made by Ernst Mach (1886/1986) and more recently by Israel Scheffler (1970).[81] If serious scientists, such as listed earlier in this chapter, feel it important to write books on the philosophy of their subject, then assuredly science teachers and students will benefit from following their example and engaging with the same questions.

---

[80] See especially those of Arnold Arons (Arons 1977, Chaps.14–15; 1990, Chap. 3), Gerald Holton and Stephen Brush (Holton and Brush 2001, Chap. 9) and the Harvard Project Physics texts (Holton et al. 1970).

[81] Mach's argument is discussed in Matthews (1989b), and Scheffler's argument is discussed in Matthews (1997).

The arguments of Mach and Scheffler have belatedly and independently found expression in the wide international calls for students to learn about the 'nature of science' while learning science. One cannot learn about the nature of science without learning philosophy of science, which was precisely Mach and Scheffler's argument.

50.12 Conclusion

Science has contributed immensely to our philosophical and cultural tradition, this is part of the 'flesh' of science; however too often science teaching presents just the 'bare bones' of laws, formulae and problems, the 'final products' of science. This is one reason why, notoriously, advanced 'technical' science is so often associated with religious and ideological fundamentalism and bigotry.[82] The cultural flesh of science should be part of any serious science programme.

Carl Sagan's undergraduate experience at the University of Chicago in the 1950s is no longer available to many students, but it is worth recalling as an ideal and something which might be striven for:

> At the University of Chicago I also was lucky enough to go through a general education program devised by Robert M. Hutchins, where science was presented as an integral part of the gorgeous tapestry of human knowledge. It was considered unthinkable for an aspiring physicist not to know Plato, Aristotle, Bach, Shakespeare, Gibbon, Malinowski, and Freud – among many others. In an introductory science class, Ptolemy's view that the Sun revolved around the Earth was presented so compellingly that some students found themselves re-evaluating their commitment to Copernicus. ... I also witnessed at first hand the joy felt by those whose privilege it is to uncover a little about how the Universe works. (Sagan 1996, pp. xiv-xv)

In a good liberal education, science students, and hopefully other students as well, will learn about the philosophical dimensions of science, beginning with routine matters such as conceptual analysis, epistemology, values and so on. They will also learn about the metaphysical, especially ontological, dimensions of science, some of which have been discussed above. They should also be introduced to and hopefully make decisions about the constitution and applicability of the scientific outlook, habit of mind or the scientific temper. To entertain questions such as: Is a scientific outlook required for the solution of social and ideological problems? By reading about any number of courageous scientists beginning with Galileo and moving through Joseph Priestley and on to Andrei Sakharov (Sakharov 1968), students can be introduced to the issue of the social and cultural requirements for the pursuit of science, the issue that so animated the Enlightenment scientists, philosophers and social reformers.

---

[82] That there is no connection between advanced technology and advanced thinking was sadly demonstrated when numerous spectators to the Papua New Guinea witch burning described above captured the event on their mobile phone cameras and uploaded the burning onto the World Wide Web.

In particular students might think through and re-argue the Enlightenment tradition's claims that on purely epistemological grounds, science, and more generally the pursuit of truth in all human domains, requires legal protection of free speech, freedom of the press and support for diversity, unhindered scholarly publication and freedom of association. To entertain questions such as: Does the promotion and spread of science entail a liberal, secular, democratic, non-authoritarian state?

All of this makes science classes more intellectually engaging, it promotes 'minds-on' science learning, and it enables diverse subjects in a school curriculum (history, mathematics, technology, religion) to be related. The introduction of history and philosophy to science lessons enables students to better understand the science and the scientific methodology they are learning, to better appreciate the role of science in the formation of the modern world and contemporary worldviews and perhaps the knowledge and enthusiasm to support science and the spread of the 'scientific habit of mind'.

Undoubtedly such an education has an impact on, and contributes to, the worldviews of students. So it is worth noting Frederick Copleston's caution:

> It must be recognized, I think, that the creation of world-views is none the less a pretty risky procedure. There is, for example, the risk of making unexamined or uncriticized presuppositions in a desire to get on with the painting of the picture. Again, there are the risks of over-hastily adopting desired conclusions, and also of allowing one's judgements to be determined by personal prejudices or psychological factors. (Copleston 1991, p. 71)

But science teachers are not so much creating worldviews but encouraging students to identify and then to begin to analyse and appraise aspects of worldviews. For educators, it is the student's inquiry and thinking that is important. A good science teacher can agree with Bertrand Russell who famously said in 1916 at the height of the Great War when he criticised the use of schools by both sides for nationalist indoctrination:

> Education would not aim at making them [students] belong to this party or that, but at enabling them to choose intelligently between the parties; it would aim at making them able to think, not at making them think what their teachers think. (Egner and Denonn 1961, pp. 401–402)

References

(AAAS) American Association for the Advancement of Science: 1989, *Project 2061: Science for All Americans*, AAAS, Washington, DC. Also published by Oxford University Press, 1990.

(AAAS) American Association for the Advancement of Science: 1990, *The Liberal Art of Science: Agenda for Action*, AAAS, Washington, DC.

(NRC) National Research Council: 1996, *National Science Education Standards*, National Academy Press, Washington.

Adler, M.J.: 1978, *Aristotle for Everybody*, Macmillan, New York.

Aikenhead, G.:2000, "Renegotiating the culture of school science'. In Robin Millar & Jonathan Osborne (eds.), *Improving Science Education*, Open University Press, Philadelphia, pp. 245–264.

50 Science, Worldviews and Education

Aikenhead, G.S.: 1996, 'Cultural Assimilation in Science Classrooms: Border Crossings and Other Solutions', *Studies in Science Education* **7**, 1–52.

Aikenhead, G.S.: 1997, 'Towards a First Nations Cross-Cultural Science and Technology Curriculum', *Science Education* **81**(2), 217–238.

Allen, R.E. (ed.): 1966, *Greek Philosophy. Thales to Aristotle*, The Free Press, New York.

Amorth, G.: 2010, *The Memoirs of an Exorcist*, Ediciones Urano, Rome.

Amsterdamski, S.: 1975, *Between Experience and Metaphysics: Philosophical Problems in the Evolution of Science*, Reidel Publishing Company, Dordrecht.

Andrew, M.E.: 1971, 'The Historical-Critical Method in the Seventeenth Century and in the Twentieth', *Colloquim: The Australian and New Zealand Theological Review* 4(2), 92–104.

Arons, A.B.: 1977, *The Various Language, An Inquiry Approach to the Physical Sciences*, Oxford University Press, New York.

Arons, A.B.: 1990, *A Guide to Introductory Physics Teaching*, John Wiley, New York.

Ashley, B.M.: 1991, 'The River Forest School and the Philosophy of Nature Today'. In R.J. Long (ed.) *Philosophy and the God of Abraham. Essays in Memory of James A. Weisheipl, OP*, Pontifical Institute of Medieval Studies, Toronto, pp.1-15.

Bagley, P.J.: 1998, 'Spinoza, Biblical Criticism and the Enlightenment'. In J.C. McCarthy (ed.) *Modern Enlightenment and the Rule of Reason*, The Catholic University of America Press, Washington DC, pp. 124–149.

Barbour, I.G.: 1966, *Issues in Science and Religion,* SCM Press, London.

Barbour, I.G.: 1990, *Religion in an Age of Science,* SCM Press, London.

Bergmann, P.: 1949, *Basic Theories of Physics*, Prentice-Hall, New York.

Bernal, J.D.: 1939, *The Social Function of Science*, Routledge & Kegan Paul, London.

Bernal, J.D.: 1965, *Science in History*, 4 vols., (3rd edition), C.A Watts, London.

Beyerchen, A.D.: 1977, *Scientists Under Hitler: Politics and the Physics Community in the Third Reich*, Yale University Press, New Haven.

Birch, L.C.: 1990, *On Purpose*, University of New South Wales Press, Sydney.

Birdsall, G.: 1995, *Feng Shui: The Ancient Art of Placement*, Waterwood Management Proprietary Ltd., Sydney.

Blum, P.R.: 2012, *Studies on Early Modern Aristotelianism,* Brill, Leiden.

Blumenberg, H.: 1987, *The Genesis of the Copernican World*, MIT Press, Cambridge MA.

Bohm, D.: 1980, *Wholeness and the Implicate Order*, Ark Paperbacks, London.

Bohr, N.: 1958, *Atomic Physics and Human Knowledge*, Wiley, New York.

Boltzmann, L.: 1905/1974, *Theoretical Physics and Philosophical Problems*, Reidel, Dordrecht.

Booth, E.H. & Nicol, P.M.: 1931/1962, *Physics: Fundamental Laws and Principles with Problems and Worked Examples*, Australasian Medical Publishing Company, Sydney. 16th edition 1962.

Born, M.: 1968, *My Life & My Views*, Scribners, New York.

Boudry, M., Blancke, S. & Braeckman, J.: 2012, 'Grist to the mill of anti-evolutionism: The failed strategy of ruling the supernatural out of science by philosophical fiat', *Science & Education,* 21, 1151–1165.

Bridgman, P.W.: 1950, *Reflections of a Physicist*, Philosophical Library, New York.

Broad, C.D.: 1925, *The Mind and Its Place in Nature*, Harcourt Brace, New York.

Brooke, J.H.: 1991, *Science and Religion: Some Historical Perspectives*, Cambridge, Cambridge University Press.

Buchdahl, G.: 1969, *Metaphysics and the Philosophy of Science*, Basil Blackwell, Oxford.

Buckley, M.J.: 1971, *Motion and Motion's God*, Princeton University Press, Princeton.

Bunge, M.: 1967, 'Analogy in Quantum Mechanics: From Insight to Nonsense', *The British Journal for Philosophy of Science* 18, 265–286.

Bunge, M.: 1977, 'Emergence and the Mind', *Neuroscience* **2**, 501–509.

Bunge, M.: 1981, *Scientific Materialism*, Reidel, Dordrecht.

Bunge, M.: 1998a, *Philosophy of Science*, Vol. 1, Transaction Publishers, New Brunswick, NJ.

Bunge, M.: 1998b, *Philosophy of Science*, Vol. 2., Transaction Publishers, New Brunswick, NJ.

Bunge, M.: 2012, 'Does Quantum Physics Refute Realism, Materialism and Determinism?', *Science & Education* **21**(10), 1601–1610.

Burtt, E.A.: 1932, *The Metaphysical Foundations of Modern Physical Science* (second edition), Routledge & Kegan Paul, London.

Butterfield, H.: 1957, *The Origins of Modern Science 1300–1800*, G. Bell and Sons, London. (originally 1949)

Butts, R.E. & Davis, J.W. (eds.): 1970, *The Methodological Heritage of Newton*, University of Toronto Press, Toronto.

Callick, R.: 2013, 'Will Horror Force PNG to Protect its Women?', *The Australian*, February 9–11, p. 11.

Campbell, N.R: 1921/1952, *What Is Science?* Dover, New York.

Chalmers, A.F.: 2009, *The Scientist's Atom and the Philosopher's Stone: How Science Succeeded and Philosophy Failed to Gain Knowledge of Atoms*, Springer, Dordrecht.

Chan, W.-T.: 1969, *A Source Book in Chinese Philosophy*, Princeton University Press, Princeton.

Chandrasekhar, S.: 1987, *Truth and Beauty: Aesthetics and Motivations in Science*, University of Chicago Press, Chicago.

Clagett, M.: 1959, *The Science of Mechanics in the Middle Ages*, University of Wisconsin Press, Madison WI.

Clavelin, M.: 1974, *The Natural Philosophy of Galileo. Essay on the Origin and Formation of Classical Mechanics*, MIT Press, Cambridge.

Cobern, W.W.: 1991, *Word View Theory and Science Education Research*, National Association for Research in Science Teaching, Manhattan, KS.

Cobern, W.W.: 1996, 'Worldview Theory and Conceptual Change in Science Education', *Science Education* **80**(5), 579–610.

Coelho, R.L.: 2007, 'The Law of Inertia: How Understanding its History can Improve Physics Teaching', *Science & Education* **16**(9–10), 955–974.

Cohen, I.B.: 1980, *The Newtonian Revolution*, Cambridge University Press, Cambridge.

Cohen, H.F.: 1994, *The Scientific Revolution: A Historiographical Inquiry*, University of Chicago Press, Chicago.

Collingwood, R.G.: 1945, *The Idea of Nature*, Oxford University Press, Oxford.

Collins, F.S.: 2007, *The Language of God: A Scientist Presents Evidence for Belief*, Free Press, New York.

Copernicus, N.: 1543/1952, *On the Revolutions of the Heavenly Spheres*, (trans. C.G. Wallis), Encyclopædia Britannica, Chicago.

Copleston, F.C.: 1950, *A History of Philosophy*, 8 vols., Doubleday & Co., New York.

Copleston, F.C.: 1955, *Aquinas*, Penguin Books, Harmondsworth.

Copleston, F.C.: 1991, 'Ayer and World Views'. In A. Phillips Griffiths (ed.) *A.J. Ayer: Memorial Essays*, Cambridge University Press, Cambridge.

Cornwell, J.: 2003, *Hitler's Scientists: Science, War and the Devil's Pact*, Penguin, London.

Crombie, A.C.: 1994, *Styles of Scientific Thinking in the European Tradition*, three volumes, Duckworth, London.

Cushing, J.T. & McMullin, E. (eds.): 1989, *Philosophical Consequences of Quantum Theory*, University of Notre Dame Press, Notre Dame, IN.

Cushing, J.T.: 1998, *Philosophical Concepts in Physics: The Historical Relation between Philosophy and Scientific Theories*, Cambridge University Press, Cambridge.

Dawkins, R.: 1998, *Unweaving the Rainbow*, Penguin Press, London.

Dawkins, R.: 2006, *The God Delusion*, Bantam Press, London.

De Berg, K.C.: 2011, 'Joseph Priestley Across Theology, Education and Chemistry: An Interdisciplinary Case Study in Epistemology with a Focus on the Science Education Context', *Science & Education* 20(7–8), 805–830.

De Caro, M. & Macarthur, D. (eds.): 2008, *Naturalism in Question*, Harvard University Press, Cambridge MA.

De Wulf, M.: 1903/1956, *An Introduction to Scholastic Philosophy: Medieval and Modern*, (trans. P. Coffey), Dover Publications, New York.

Dennett, D.C.: 1995, *Darwin's Dangerous Idea: Evolution and the Meanings of Life*, Allen Lane, Penguin Press, London.

Dennett, D.C.: 2006, *Breaking the Spell: Religion as a Natural Phenomenon*, Penguin, New York.

Devitt, M.: 1998, 'Naturalism and the A Priori', *Philosophical Studies* 92, 45–65.

Dewitt, R.: 2004, *Worldviews: An Introduction to the History and Philosophy of Science*, Blackwell Publishing, Oxford.

Dijksterhuis, E.J.: 1961, 'The Origins of Classical Mechanics from Aristotle to Newton'. In M. Claggett (ed.) *Critical Problems in the History of Science*, University of Wisconsin Press, Madison.

Dijksterhuis, E.J.: 1961/1986, *The Mechanization of the World Picture*, Princeton University Press, Princeton NJ.

Dillenberger, J.: 1961, *Protestant Thought & Natural Science: A Historical Study*, Collins, London.

Dilworth, C.: 1996/2006, *The Metaphysics of Science. An Account of Modern Science in Terms of Principles, Laws and Theories*, Kluwer Academic Publishers, Dordrecht, (second edition 2006).

Duhem, P.: 1906/1954, *The Aim and Structure of Physical Theory*, trans. P.P. Wiener, Princeton University Press, Princeton.

Eddington, A.: 1939, *The Philosophy of Physical Science*, Cambridge University Press, Cambridge.

Edis, T.: 2007, *An Illusion of Harmony: Science and Religion in Islam*, Prometheus Books, Amherst, NY.

Egner, R.E. & Denonn, L.E. (eds.): 1961, *The Basic Writings of Bertrand Russell*, George Allen & Unwin, London.

Einstein, A. & Infeld, L.: 1938/1966, *The Evolution of Physics*, Simon & Schuster, New York.

Elliot, T.: 2013, 'Witch-Hunt', *Sydney Morning Herald*, Good Weekend, April 20, pp. 16–21.

Ellis, B.D.: 1965, 'The Origin and Nature of Newton's Laws of Motion'. In R.G. Colodny (ed.) *Beyond the Edge of Certainty*, Englewood Cliffs, NJ., pp. 29–68.

Ellis, B.D.: 1976, 'The Existence of Forces', *Studies in History and Philosophy of Science* **7**(2), 171–185.

Erduran, S. & Jiménez-Aleixandre, M.P. (eds.): 2008, *Argumentation in science education: Perspectives from classroom-based research*. Dordrecht: Springer.

Ernst, H.E.: 2006, 'New Horizons in Catholic Philosophical Theology: *Fides et Ratio* and the Changed Status of Thomism', *The Heythrop Journal* **47**(1), 26–37.

Farrington, B.: 1939, *Science and Politics in the Ancient World*, George Allen & Unwin, London.

Finocchiaro, M.A.: 1989, *The Galileo Affair: A Documentary History*, University of California Press, Berkeley.

Fishman, Y.I. & Boudry, M.: (2013) 'Does Science Presuppose Naturalism (or, Indeed, Anything at All)? *Science & Education*

Fishman, Y.I.: 2009, 'Can Science Test Supernatural Worldviews?', *Science & Education* 18(6–7), 813–837.

Force, J.E. & Popkin, R.H. (eds.): 1994, *The Books of Nature and Scripture: Recent Essays on Natural Philosophy, Theology, and Biblical Criticism in the Netherlands of Spinoza's Time and the British Isles of Newton's Time*, Kluwer Academic Publishers, Dordrecht.

French, P.A., Uehling, T.E. & Wettstein, H.K. (eds.): 1995, *Philosophical Naturalism*, University of Notre Dame Press, Notre Dame, IN.

Galileo, G.: 1615/1957, 'Letter to Madame Christina of Lorraine, Grand Duchess of Tuscany Concerning the Use of Biblical Quotations in Matters of Science'. In S. Drake (ed.), *Discoveries and Opinions of Galileo*, Doubleday, New York, pp. 175–216.

Galileo, G.: 1623/1957, *The Assayer*. In S. Drake (ed.), *Discoveries and Opinions of Galileo*, Doubleday, New York, pp. 229–280.

Gauch, Jr., H.G.: 2009, 'Science, Worldviews and Education', *Science & Education* **18**(6–7), 667–695.

Gill, H.V.: 1943, *Fact and Fiction in Modern Science*, M.H. Gill & Son, Dublin.

Gill, H.V.: 1944, *Fact and Fiction in Modern Science*, M.H. Gill & Son, Dublin.

Gilson, E.: 1929, *The Philosophy of St. Thomas Aquinas*, second edition, Dorset Press, New York.

Gingerich, O. (ed.): 1975, *The Nature of Scientific Discovery: A Symposium Commemorating the 500th Anniversary of the Birth of Nicolaus Copernicus*, Smithsonian Institution Press, Washington DC.

Gingerich, O.: 1993, *The Eye of Heaven: Ptolemy, Copernicus, Kepler*, American Institute of Physics, New York.

Gong, Y.: 1996, *Dialectics of Nature in China*, Beijing University Press, Beijing, China.

Gould, S.J.: 1999, *Rock of Ages: Science and Religion in the Fullness of Life*, Ballantine Books, New York.

Graham, L.R.: 1973, *Science and Philosophy in the Soviet Union*, Alfred A. Knopf, New York.

Grant, E.: 2004, *Science and Religion, 400 bc to ad 1550. From Aristotle to Copernicus*, The Johns Hopkins University Press, Baltimore.

Greene, J.C.: 1981, *Science, Ideology and World View: Essays in the History of Evolutionary Ideas*, University of California Press, Berkeley.

Haksar, P.N. and associates: 1981, *A Statement on Scientific Temper,* Nehru Centre, Bombay.

Haldane, E.S. & Ross, G.R.T. (eds.): 1931, *The Philosophical Works of Descartes*, Vol. I, Cambridge University Press, Cambridge

Haldane, J.S.: 1928, *The Sciences and Philosophy*, Hodder & Stoughton, London.

Hall, A.R.: 1954/1962, *The Scientific Revolution: 1500–1800*, 2nd edition, Beacon Press, Boston. (Third updated edition 1983.)

Hall, A.R.: 1980, *Philosophers at War*, Cambridge University Press, Cambridge.

Hanson, N.R.: 1965, 'Newton's First Law: A Philosopher's Door into Natural Philosophy'. In R.G. Colodny (ed.), *Beyond the Edge of Certainty*, Prentice Hall, Englewood-Cliffs, NJ, pp. 6–28.

Harman, P.M.: 1982, *Energy, Force and Matter: The Conceptual Development of Nineteenth-Century Physics*, Cambridge University Press, Cambridge.

Harré, R.: 1964, *Matter and Method*, Macmillan & Co., London.

Hatfield, G.: 1990, 'Metaphysics and the New Science'. In D.C. Lindberg & R.S. Westman (eds.) *Reappraisals of the Scientific Revolution*, Cambridge University Press, Cambridge, pp. 93–166.

Haught, J. F.: 1995, *Science and Religion: from Conflict to Conversation*, Paulist Press, New York.

Heisenberg, W.: 1962, *Physics and Philosophy*, Harper & Row, New York.

Helmholtz, H. von: 1995, *Science and Culture: Popular and Philosophical Essays*, (edited with Introduction by David Cahan; original essays 1853–1892) Chicago University Press, Chicago.

Herivel, J: 1965, *The Background to Newton's 'Principia'*, Clarendon Press, Oxford.

Hesse, M.B.: 1961, *Forces and Fields: The Concept of Action at a Distance in the History of Physics*, Thomas Nelson & Sons, London.

Holton, G. & Brush, S.G.: 2001, *Physics, the Human Adventure. From Copernicus to Einstein and Beyond*, Rutgers University Press, New Brunswick.

Holton, G., Rutherford, F.J. & Watson, F.G. (eds.): 1970, *The Project Physics Course: Text*, Holt, Rinehart, & Winston, New York.

Holton, G.: 1973, *Thematic Origins of Scientific Thought*, Harvard University Press, Cambridge.

Hoodbhoy, P.: 1991, *Islam and Science: Religious Orthodoxy and the Battle for Rationality*, Zed Books, London.

Hughes, R.I.G.: 1990, 'Philosophical Perspectives on Newtonian Science'. In P. Bricker & R.I.G. Hughes (eds.) *Philosophical Perspectives on Newtonian Science*, MIT Press, Cambridge MA, pp. 1–16.

Hull, D.L.: 1988, *Science as a Process: An Evolutionary Account of the Social and Conceptual Development of Science,* University of Chicago Press, Chicago.

Hume, D.: 1739/1888, *A Treatise of Human Nature: Being an Attempt to Introduce the Experimental Method of Reasoning into Moral Subjects*, Clarendon Press, Oxford.

Hunt, I.E. & Suchting, W.A.: 1969, 'Force and "Natural Motion"', *Philosophy of Science* **36**, 233–251.

Hyland, P., Gomez, O. & Greensides, F. (eds.): 2003, *The Enlightenment: A Source Book and Reader*, Routledge, New York.

Israel, J.: 2001, *Radical Enlightenment: Philosophy and the Making of Modernity 1650–1750*, Oxford University Press, Oxford.

Jaki, S.L.: 1978, *The Road of Science and the Ways to God*, University of Chicago Press.

Jammer, M.: 1957, *Concepts of Force: A Study in the Foundations of Dynamics*, Harvard University Press, Cambridge, MA.

Jeans, J.: 1943/1981, *Physics and Philosophy*, Dover Publications, New York.

Kenny, A.: 1980, *Aquinas*, Oxford University Press, Oxford.

Kenny, A.: 1985, *A Path from Rome: An Autobiography*, Sidgwick & Jackson, London.

Koenigsberger, H.G.: 1987, *Early Modern Europe 1500–1789*, Longman, London.

Koyré, A.: 1957, *From the Closed World to the Infinite Universe*, The Johns Hopkins University Press, Baltimore.

Kwok, D.W.Y.: 1965, *Scientism in Chinese Thought: 1900–1950*, Yale University Press, New Haven, CT.

Laer, van P.H.: 1953, *Philosophico-Scientific Problems*, Duquesne University Press, Pittsburgh.

Laer, van P.H.: 1956, *Philosophy of Science Part I Science in General*, Duquesne University Press, Pittsburgh.

Laer, van P.H.: 1962, *Philosophy of Science Part II A Study of the Division and Nature of Various Groups of Sciences*, Duquesne University Press, Pittsburgh.

Lakatos, I.: 1970, 'Falsification and the Methodology of Scientific Research Programmes'. In I. Lakatos & A. Musgrave (eds.) *Criticism and the Growth of Knowledge*, Cambridge University Press, Cambridge, pp. 91–196.

Lamont, J.: 2009 'The Fall and Rise of Aristotelian Metaphysics in the Philosophy of Science', *Science & Education* 18(6–7), 861–884.

Lecky, W.E.H.: 1914, *History of the Rise and Influence of the Spirit of Rationalism in Europe*, two volumes, D. Appleton & Co., New York.

Lenin, V.I.: 1920/1970, *Materialism and Empirio-Criticism: Critical Comments on a Reactionary Philosophy*, 2nd edition, Progress Publishers, Moscow.

Lindberg, D.C. & Numbers, R.L. (eds.): 1986, *God and Nature: Historical Essays on the Encounter between Christianity and Science*, University of California Press, Berkeley.

Lipman, M.: 1991, *Thinking in Education*, Cambridge University Press, Cambridge.

Locke, J.: 1689/1924, *An Essay Concerning Human Understanding,* abridged and edited by A.S. Pringle-Pattison, Clarendon Press, Oxford.

Lucey, J.M.: 1915, 'The Mass. The Proper Form of Christian Worship'. In *Cabinet of Catholic Information*, The Treasury Publishing Company, pp. 84–100. [no editor, no place of publication]

Mach, E.: 1886/1986, 'On Instruction in the Classics and the Sciences'. In his *Popular Scientific Lectures*, Open Court Publishing Company, La Salle, pp. 338–374.

Mach, E.: 1893/1960, *The Science of Mechanics*, (6th edition), Open Court Publishing Company, LaSalle Il.

Mahner, M.: 2012,'The Role of Metaphysical Naturalism in Science', *Science & Education* 21(10), 1437–1459.

Marcos, A.: 2012, *Postmodern Aristotle*, Cambridge Scholars Publishing, Newcastle-upon-Tyne.

Margenau, H.: 1950, *The Nature of Physical Reality: A Philosophy of Modern Physics*, McGraw-Hill, New York.

Maritain, J.: 1935/1951, *The Philosophy of Nature*, New York.

Martin, R.N.D.: 1991, *Pierre Duhem: Philosophy and History in the Work of a Believing Physicist*, Open Court, La Salle, IL.

Marx, K.: 1851/1972, *The Eighteenth Brumaire of Louis Bonaparte*. In R.C. Tucker (ed.) 1972, *The Marx-Engels Reader*, W.W. Norton, New York, pp. 594–617.

Mascall, E.L.: 1956, *Christian Theology and Natural Science: Some Questions in Their Relations*, Longmans, Green & Co., London.

Mascall, E.L.: 1966, *He Who Is. A Study in Traditional Theism*, Darton, Longman & Todd, London.

Mascall, E.L.: 1971, *The Openness of Being: Natural Theology Today*, Darton, Longman & Todd, London.

Matthews, G. B.:1982, *Philosophy and the young child.* Harvard University Press, Cambridge, MA.

Matthews, M.R. (ed.): 1989a, *The Scientific Background to Modern Philosophy*, Hackett Publishing Company, Indianapolis.

Matthews, M.R.: 1989b, `Ernst Mach and Thought Experiments in Science Education', *Research in Science Education* **18**, 251–258.

Matthews, M.R.: 1994/2014, *Science Teaching: The Role of History and Philosophy of Science*, Routledge, New York.

Matthews, M.R.: 1997, 'Israel Scheffler on the Role of History and Philosophy of Science in Science Teacher Education', *Studies in Philosophy and Education* **16**(1–2), 159–173.

Matthews, M.R.: 2009a, 'Science, Worldviews and Education: An Introduction', *Science & Education* **18** (6–7), 641–666.

Matthews, M.R.: 2009b, 'Teaching the Philosophical and Worldview Dimension of Science', *Science & Education* **18** (6–7), 697–728.

Matthews, M.R.: 2009c, 'Science and Worldviews in the Classroom: Joseph Priestley and Photosynthesis', *Science & Education* **18**(6–7), 929–960.

Matthews, M.R. (ed.): 2009d, *Science, Worldviews and Education*, Springer, Dordrecht.

Mayr, E.: 1982, *The Growth of Biological Thought*, Harvard University Press, Cambridge MA.

McCabe, J.: 1912, *Twelve Years in a Monastery*, (3rd edition), Watts & Co., London.

McGuire, J.E.: 1995, *Tradition and Innovation: Newton's Metaphysics of Nature*, Kluwer Academic Publishers, Dordrecht.

McInerny, R.M.: 1966, *Thomism in an Age of Renewal*, University of Notre Dame Press, Notre Dame.

McKenzie, J.L.: 1966, *Dictionary of the Bible*, Geoffrey Chapman, London.

McMullin, E. (ed.): 1963a, *The Concept of Matter in Greek and Medieval Philosophy*, University of Notre Dame Press, Notre Dame.

McMullin, E. (ed.): 1963b, *The Concept of Matter in Modern Philosophy*, University of Notre Dame Press, Notre Dame.

McMullin, E. (ed.): 1967, *Galileo Man of Science*, Basic Books, New York.

McMullin, E.: 1978, *Newton on Matter and Activity*, University of Notre Dame Press, Notre Dame.

McMullin, E.: 1985, 'Introduction: Evolution and Creation'. In E. McMullin (ed.) *Evolution and Creation*, University of Notre Dame Press, Notre Dame, IN., pp. 1–58.

McMullin, E.: 1989, 'The Explanation of Distant Action: Historical Notes'. In J.T. Cushing & E. McMullin (eds.) *Philosophical Consequences of Quantum Theory: Reflections on Bell's Theorem*, University of Notre Dame Press, Notre Dame, pp. 272–302.

McMullin, E.: 2005, 'Galileo's Theological Venture'. In E. McMullin (ed.) *The Church and Galileo*, University of Notre Dame Press, Notre Dame, pp. 88–116.

Melsen, A.G. van: 1952, *From Atomos to Atom*, Duquesne University Press, Pittsburgh.

Mencken, H.L.: 1946/1930, *Treatise on the Gods*, 2nd edition, Alfred A. Knopf, New York.

Mittelstrass, J.: 1989, 'World Pictures: The World of the History and Philosophy of Science'. In J.R. Brown & J. Mittelstrass (eds.) *An Intimate Relation: Studies in the History and Philosophy of Science*, Kluwer Academic Publishers, Dordrecht.

Monod, J.: 1971, *Chance and Necessity: An Essay on the Natural Philosophy of Modern Biology*, Knopf, New York.

Moody, E.A.: 1975, *Studies in Medieval Philosophy, Science and Logic*, University of California Press, Berkeley.

Mott, N. (ed.): 1991, *Can Scientists Believe?* James & James, London.

Nagel, E.: 1956, 'Naturalism Reconsidered'. In his *Logic without Metaphysics*, Freepress, Glencoe, IL, chap.1.

Nanda, M.: 2003, *Prophets Facing Backward. Postmodern Critiques of Science and Hindu Nationalism in India*, Rutgers University Press.

Narokobi, B. (ed.): 1997, 'Occasional Paper No.4, Sorcery', Papua New Guinea Law Commission, Port Moresby.

Nasr, S.H.: 1996, *Religion and the Order of Nature*, Oxford University Press, Oxford.

Newton, I.: 1729/1934, *Mathematical Principles of Mathematical Philosophy*, (translated A. Motte, revised F. Cajori), University of California Press, Berkeley.

Newton, I.: 1730/1979, *Opticks or A Treatise of the Reflections, Refractions, Inflections & Colours of Light*, Dover Publications, New York.

Noonan, J.T.: 1965, *Contraception: A History of Its Treatment by Catholic Theologians and Canonists*, Mentor-Omega Books, New York.

Palmieri, P.: 2011, *A History of Galileo's Inclined Plane Experiment and Its Philosophical Implications*, The Edwin Mellen Press, Lewiston, NY.

Pennock, R.T.: 1999, *Tower of Babel: The Evidence against the new Creationism*, MIT Press, Cambridge MA.

Petersen, A.: 1985, 'The Philosophy of Niels Bohr'. In A.P. French & P.J. Kennedy (eds.) *Niels Bohr: A Centenary Volume*, Harvard University Press, Cambridge, MA. pp. 299–310.

Planck, M.: 1932, *Where is Science Going?*, W.W. Norton, New York.

Plantinga, A.: 2000, *Warranted Christian Belief*, Oxford University Press, Oxford.

Plantinga, A.: 2011, *Where the Conflict Really Lies. Science, Religion and Naturalism,* Oxford University Press, New York.

Poincaré, H.: 1905/1952, *Science and Hypothesis*, Dover Publications, New York.

Polanyi, M.: 1958, *Personal Knowledge*, Routledge and Kegan Paul, London.

Polkinghorne, J.: 1996, *The Faith of a Physicist: Reflections of a Bottom-up Thinker.* Fortress Press, Minneapolis.

Polkinghorne, J.C.: 1986, *One World: The Interaction of Science and Theology*, SPCK, London.

Polkinghorne, J.C.: 1988, *Science and Creation: The Search for Understanding*, SPCK, London.

Polkinghorne, J.C.: 1991, *Reason and Reality: The Relationship between Science and Theology,* SPCK, London.

Popper, K.R.: 1963, *Conjectures and Refutations: The Growth of Scientific Knowledge*, Routledge & Kegan Paul, London.

Porter, R.: 2000, *The Enlightenment: Britain and the Creation of the Modern World*, Penguin Books, London.

Preston, J. & Epley, N.: 2009, Science and God: An Automatic Opposition between Ultimate Explanations. *Journal of experimental Social Psychology, 45*, 238–241.

Pullman, B.: 1998, *The Atom in the History of Human Thought*, Oxford University Press, Oxford.

Pyle, A.: 1997, *Atomism and Its Critics: From Democritus to Newton*, Thoemmes Press, Bristol.

Rabi, I.I.: 1967, *Science the Centre of Culture*, World Publishing Company, New York.

Randall Jr., J.H.: 1962, *The Career of Philosophy*, Columbia University Press, New York.

Redondi, P.: 1987, *Galileo Heretic*, Allen Lane, London.

Redondi, P.: 1988, *Galileo Heretic*, Allen Lane, London.

Renan, E.: 1863/1935, *The Life of Jesus*, Watts & Co., London.

Restrepo, D.H.: 2003, 'Philosophy in Contemporary Colombia'. In G. Fløistad (ed.) *Philosophy of Latin America*, Kluwer Academic Publishers, Dordrecht, pp. 143–154.

Rohrlich, F.: 1987, *From Paradox to Reality: Our Basic Concepts of the Physical World*, Cambridge University Press, Cambridge.

Rossbach, S.: 1984, *Feng Shui*, Rider, London.

Ruse, M.: 1989, *The Darwinian Paradigm: Essays on its History, Philosophy, and Religious Implications*, Routledge, London.

Ruse, M.: 2011, *Science and Spirituality: Making Room for Faith in the Age of Science,* Cambridge University Press, Cambridge

Russell, B.: 1958, 'Letter to Mr Major', in *Dear Bertrand Russell: A Selection of his Correspondence with the General Public, 1950–1968*, Allen & Unwin, London, 1969.

Rutherford, F.J. & Ahlgren, A.: 1990, *Science for All Americans*, Oxford University Press, New York.

Rutt, J.T. (ed.): 1817–32/1972, *The Theological and Miscellaneous Works of Joseph Priestley*. 25 vols., London. (Kraus Reprint, New York, 1972)

Sagan, C.: 1996, *The Demon-Haunted World: Science as a Candle in the Dark*, Headline Book Publishing, London.

Sagan, C.: 1997, *The Demon-Haunted World: Science as a Candle in the Dark*, Headline Book Publishing, London.

Sakharov, A.D.: 1968, *Progress, Coexistence and Intellectual Freedom*, W.W. Norton, New York.

Scheffler, I.: 1970, 'Philosophy and the Curriculum'. In his *Reason and Teaching*, London, Routledge, 1973, pp. 31–44. Reprinted in *Science & Education* 1(4), 385–394.

Schilpp, P.A. (ed.): 1951, *Albert Einstein: Philosopher-Scientist*, second edition, Tudor, New York.

Schmitt, C.B.: 1983, *Aristotle and the Renaissance*, Harvard University Press, Cambridge MA.

Schofield, R.E. (ed.): 1966, *A Scientific Autobiography of Joseph Priestley (1733–1804): Selected Scientific Correspondence*, MIT Press, Cambridge.

Schofield, R.E.: 1997, *The Enlightenment of Joseph Priestley: A Study of His Life and Work from 1733 to 1773*, Penn State Press, University Park, PA.

Schofield, R.E.: 2004, *The Enlightened Joseph Priestley: A Study of His Life and Work from 1773 to 1804*, Penn State Press, University Park, PA.

Schulz, R.M.: 2009, 'Reforming Science Education: Part I. The Search for a Philosophy of Science Education', *Science & Education*, 18 (3–4), 225–249.

Sellars, R.W.: 1932, *The Philosophy of Physical Realism*, Macmillan, New York.

Shank, J.B.: 2008, *The Newton Wars and the Beginning of the French Enlightenment*, University of Chicago Press, Chicago.

Shermer, M.: 1997, *Why People Believe Weird Things: Pseudoscience, Superstition, and other Confusions of Our Time*, W.H. Freemand & Co., New York.

Shimony, A.: 1983, 'Reflections on the Philosophy of Bohr, Heisenberg, and Schrödinger'. In R.S. Cohen & L. Laudan (eds.) *Physics, Philosophy and Psychoanalysis*, Reidel, Dordrecht, pp. 209–221.

Shimony, A.: 1989, 'Search for a Worldview Which Can Accommodate Our Knowledge of Microphysics'. In J.T. Cushing & E. McMullin (eds) *Philosophical Consequences of Quantum Physics*, University of Notre Dame Press, Notre Dame, pp. 25–37

Shimony, A.: 1993, *Search for a Naturalistic World View*, Cambridge University Press, Cambridge.

Slezak, P.: 2012, Review of Michael Ruse *Science and spirituality: Making room for faith in the age of science. Science & Education*, 21, 403–413.

Smart, J.J.C.: 1968, *Between Science and Philosophy: An Introduction to the Philosophy of Science*, Random House, New York.

Solmsen, F.: 1960, *Aristotle's System of the Physical World: A Comparison with His Predecessors*, Cornell University Press, Ithaca.

Spear, W.: 1995, *Feng Shui Made Easy*, HarperCollins, London.

Spinoza, B.: 1670/1989, *Tractatus Theologico-Politicus*, trans. S. Shirley, E.J. Brill, Leiden.

Sprod, T.: 2011, *Discussions in Science: Promoting Conceptual Understanding in the Middle School Years*, ACER Press, Melbourne

Stebbing, L.S.: 1937/1958, *Philosophy and the Physicists*, Dover Publications, New York.

Stein, H.: 2002, 'Newton's Metaphysics'. In I.B. Cohen & G.E. Smith (eds.) *The Cambridge Companion to Newton*, Cambridge University Press, Cambridge, pp. 256–302.

Steinberg, S.R. & Kincheloe, J.: 2012, 'Employing the Bricolage as Critical Research in Science Education'. In B. Fraser, K. Tobin & C. McRobbie (eds), *International Handbook of Science Education*, 2nd Edition, Springer, pp. 1485–1500.

Stenger, V. J.: 2007, *God: The Failed Hypothesis: How Science shows that God does not Exist*, Prometheus Books, Amherst, NY.

Stenger, V.J.: 1990, *Physics and Psychics: The Search for a World Beyond the Senses*, Prometheus Books, Amherst, NY.

Stolberg, T. L.: 2007, 'The Religio-scientific Frameworks of Pre-service Primary Teachers: An Analysis of their Influence on their Teaching of Science. *International Journal of Science Education*, 29(7), 909–930.

Stove, D.C.: 1991, *The Plato Cult and Other Philosophical Follies*, Basil Blackwell, Oxford.

Suchting, W.A.: 1994, 'Notes on the Cultural Significance of the Sciences', *Science & Education* 3(1), 1–56.

Taber, K. S., Billingsley, B., Riga, F., & Newdick, H.: 2011, 'Secondary Students' Responses to Perceptions of the Relationship between Science and Religion: Stances Identified from an Interview Study', *Science Education*, 95(6), 1000–1025.

Tozer, S., Anderson, T.H., & Armbruster, B.B. (eds.): 1990, *Foundational Studies in Teacher Education: A Reexamination*, Teachers College Press, New York.

Tresmontant, C.: 1965, *Christian Metaphysics*, Sheed and Ward, New York.

Trethowan, I.: 1954, *An Essay in Christian Philosophy*, Longmans, Green & Co., London.

Trusted, J.: 1991, *Physics and Metaphysics: Theories of Space and Time*, Routledge, London.

Turner, S.C.: 2012, 'Changing Images of the Inclined Plane: a case study of a revolution in American science education', *Science & Education* 21(2), 245–270.

Van Eijck M. & Roth W.-M.: 2007, 'Keeping the Local Local: Recalibrating the Status of Science and Traditional Ecological Knowledge (TEK) in Education', *Science Education* 91, 926–947

Vitzthum, R.C.: 1995, *Materialism: An Affirmative History and Definition*, Prometheus, Amherst, NY.

Wagner, S. & Warner, R. (eds.): 1993, *Naturalism: A Critical Appraisal*, University of Notre Dame Press, Notre Dame.

Wartofsky, M.W.: 1968, *Conceptual Foundations of Scientific Thought: An Introduction to the Philosophy of Science*, Macmillan, New York.

Weinberg, S.: 2001, *Facing Up. Science and Its Cultural Adversaries*, Harvard University Press, Cambridge MA.

Weisheipl, J.A.: 1968, 'The Revival of Thomism as a Christian Philosophy'. In R.M. McInerny (ed.) *New Themes in Christian Philosophy*, University of Notre Dame Press, South Bend, IN., pp. 164–185.

Weisheipl, J.A.: 1974, *Friar Thomas D'Aquino: His Life, Thought and Works*, Basil Blackwell, Oxford.

Westfall, R.S.: 1971, *The Construction of Modern Science: Mechanisms and Mechanics*, Cambridge University Press, Cambridge.

Westfall, R.S.: 1973, *Science and Religion in Seventeenth-Century England*, University of Michigan Press, Ann Arbor.

Westfall, R.S.: 1980, *Never at Rest: A Biography of Isaac Newton*, Cambridge University Press, Cambridge.

Wilson, E.O.: 1998, *Consilience: The Unity of Knowledge*, Little, Brown & Co., London.

Yasri, P, Arthur, S., Smith, M.U. & Mancy, R.: 2013, 'Relating Science and Religion: An Ontology of Taxonomies and Development of a Research Tool for Identifying Individual Views', *Science & Education* 22(10), 2679–2707.

Michael R. Matthews is an honorary associate professor in the School of Education at the University of New South Wales. He has degrees in Geology, Psychology, Philosophy, History and Philosophy of Science, and Education. His Ph.D. in philosophy of education is from UNSW.

He is Foundation and continuing editor of the journal *Science & Education: Contributions from the History and Philosophy of Science.*

His books include *The Marxist Theory of Schooling: A Study of Epistemology and Education* (Humanities Press 1980); *Science Teaching: The Role of History and Philosophy of Science* (Routledge 1994); *Challenging New Zealand Science Education* (Dunmore Press 1995); and *Time for Science Education: How Teaching the History and Philosophy of Pendulum Motion can Improve Science Literacy* (Plenum Publishers 2000).

His edited books include *The Scientific Background to Modern Philosophy* (Hackett 1989); *History, Philosophy and Science Teaching: Selected Readings* (Teachers College Press 1991); *Constructivism in Science Education: A Philosophical Examination* (Kluwer Academic Publishers 1998); *Science Education and Culture* (Kluwer Academic Publishers 2001, with F. Bevilacqua and E. Giannetto); *The Pendulum: Scientific, Historical, Philosophical and Educational Perspectives* (Springer 2005, with A. Stinner and C. F. Gauld); *Science, Worldviews and Education* (Springer 2009); and the *Springer Handbook of Research in History, Philosophy and Science Teaching* (Springer 2013). In 2010 he was awarded the Joseph H. Hazen Education Prize of the History of Science Society (USA) in recognition of his contributions to the teaching of history of science.

Outside of the academy, he served two terms as an alderman on Sydney City Council (1980–1986).

Chapter 51
What Significance Does Christianity Have for Science Education?

Michael J. Reiss

51.1 Introduction

Worldwide, religion remains of importance to many people, including young people; a survey undertaken in 2011 in 24 countries found that 73 % of respondents under the age of 35 (94 % in primarily Muslim countries and 66 % in Christian majority countries) said that they had a religion/faith and that it was important to their lives (Ipsos MORI 2011).

Furthermore, to the bemusement of many science educators in school and elsewhere, and the delight of some, issues to do with religion seem increasingly to be of importance in school science lessons, science museums and some other educational settings. This chapter begins by examining the nature of religion in general and Christianity in particularly and then examines the nature of science before looking at possible ways in which religion in general and Christianity in particular might relate to science. The chapter then considers whether or not Christianity has implications for science education and, if it does, how teaching might take account of Christian belief.

To many science educators even raising the possibility that religion might be considered within science education raises suspicions that this is an attempt to find a way of getting religion into the science classroom for religious rather than scientific reasons. This is not the intention here. In terms of the nature of science, part of the argument is that considering religion can be, on occasions, useful simply for helping learners better understand why certain things come under the purview of science and others don't.

Another argument for considering religion within science education proceeds much as an argument for considering history in science education might. While

---

M.J. Reiss (✉)
Institute of Education, University of London, 20 Bedford Way, London WC1H 0AL, UK
e-mail: m.reiss@ioe.ac.uk

M.R. Matthews (ed.), *International Handbook of Research in History, Philosophy and Science Teaching*, DOI 10.1007/978-94-007-7654-8_51,
© Springer Science+Business Media Dordrecht 2014

1637

science can be learnt and studied in a historical vacuum, there are a range of arguments for examining science in its historical contexts. For a start, this helps one understand better why certain sorts of science were pursued at certain times. Wars, for instance, have sometimes led to advances in chemistry, physics and information science (e.g. explosives, missile trajectories, code breaking), while certain botanical disciplines, such as systematics and taxonomy, have flourished during periods of colonisation. Much biology is studied in the hope that medical advances will ensue, so studies of anatomy have developed into studies of physiology and, more recently, genetics and molecular biology. Then there is the observation that for many learners understanding science in historical context can aid motivation. Science courses that take contexts and applications into account are now quite widespread (cf. the whole STS movement even if the jury is still out as to the consequences for the understanding of science concepts).

Similarly, while many students enjoy learning about the pure science of genetics and evolution, otherwise are motivated and come to understand the science better if they appreciate something of the diversity of religious beliefs held by such principal protagonists as Charles Darwin, Joseph Hooker, Thomas Huxley and Gregor Mendel and the religious views (including the diversity of religious views) of the cultures in which they lived and worked.

There are a number of places where religion and science interact. Consider, first, the question of 'authority' and the scriptures as a source of authority. To the great majority of religious believers, the scriptures of their religion (the Tanakh, the Christian bible, the Qur'an, the Vedas, including the Upanishads, the Guru Granth Sahib, the various collections in Buddhism) have an especial authority by very virtue of being scripture. This is completely different from the authority of science. Newton's *Principia* and Darwin's *On the Origin of Species* are wonderful books, but they do not have any permanence other than that which derives from their success in explaining observable phenomena of the material world and enabling people to see the material world through Newtonian/Darwinian eyes. Indeed, as is well known, Darwin knew almost nothing of the mechanism of inheritance despite the whole of his argument relying on inheritance, so parts of *The Origin* were completely out of date over 100 years ago.

Then consider the possibility of miracles, where the word is used not in its everyday sense (and the sense in which it is sometimes used in the Christian scriptures), namely, 'remarkable', 'completely unexpected' or 'wonderful' (as in the tabloid heading 'My miracle baby'), but in its narrower meaning of 'contrary to the laws of nature'. Scientists who do not accept the occurrence of miracles can react to this latter notion of miracles in one of the three ways: (i) miracles are impossible (because they are contrary to the laws of nature); (ii) miracles are outside of science (because they are contrary to the laws of nature) and (iii) miracles are very rare events that have not yet been incorporated within the body of science but will be (as rare meteorological events, e.g. eclipses, and mysterious creatures, e.g. farm animals with two heads or seven legs, have been).

This chapter addresses such issues. It focuses on Christianity because other chapters in this volume address others of the world's major religions. At the same time, without wishing to appear triumphalistic or colonial, there are some particular reasons why Christianity deserves consideration in its own right. For a start, and without negating the importance of science to all cultures, and the especial roles in its origins and development played in China and by Islam, Christianity played a major role in the origin of modern science. Historians appear broadly to agree on this although there is continuing debate about other influences and the relative contributions of each (Brooke 1991; Harrison 2001; Hooykaas 1972).

Then there is the fact that throughout the history of science, many scientists have been Christians and have seen their faith as supporting their science. In a sense this is a trivial point given that a large proportion of the population was Christian in the countries where modern science principally arose (i.e. Western Europe). Less trivial is the point that some of these scientists tackled certain scientific issues in ways that connected to their faith, though this is less the case nowadays when there is far more of a clear-cut separation between a scientist's beliefs and their science. A thorough history of science can only be developed if the significance of Christianity for certain scientists is acknowledged. Standard instances include Robert Boyle (e.g. Hunter 2009; MacIntosh 2006), Michael Faraday (e.g. Russell 2000) and Georges Lemaître (Farrell 2005).

The Christian faith represents a significant worldview in many countries, though substantially less so nowadays in Western Europe than in the past. It has helped to shape many modern institutions and provides a framework that exists in contrast to the materialism which is also widespread in many contemporary societies and classrooms. Gauch (2009) argues that 'the presuppositions and reasoning of science can and should be worldview independent, but empirical and public evidence from the sciences and humanities can support conclusions that are worldview distinctive' (p. 27).

Finally, there is the obvious point that there are a number of instances where science and Christianity intersect, whether in the classroom or wider public discourse. One clear instance is the creationism-evolution 'debate'; another is to do with such bioethical issues as the acceptability of genetic engineering, euthanasia, stem cell research and cloning; and another is to do with such philosophical issues as determinism. The importance of Christian theology and practice, noting that the situation is complicated/enriched by the fact that there is rarely a single Christian voice, for debates about determinism, evolution and bioethics are considered below.

51.2 The Nature of Religion

There are many religions, and it is difficult to answer the question 'What is the nature of religion?' in a way that satisfies the members of all religions. Nevertheless, the following, derived from Smart (1989) and Hinnells (1991), are generally characteristic of most religions (Reiss 2008a):

Religions have a *practical and ritual dimension* that encompasses such elements as worship, preaching, prayer, yoga, meditation and other approaches to stilling the self.

The *experiential and emotional dimension* of religions has at one pole the rare visions given to some of the crucial figures in a religion's history, such as that of Arjuna in the *Bhagavad Gita*, the revelation to Moses at the burning bush in *Exodus* and Saul on the road to Damascus in *Acts*. At the other pole are the experiences and emotions of many religious adherents, whether a once-in-a-lifetime discernment of the transcendent or a more frequent feeling of the presence of God either in corporate worship or in the stillness of one's heart.

All religions hand down, whether orally or in writing, vital stories that comprise the *narrative or mythic dimension*, for example the story of the birth, life, death, resurrection and ascension of Jesus in the Christian scriptures. For some religious adherents such stories are believed literally, for others they are understood symbolically.

The *doctrinal and philosophical dimension* arises, in part, from the narrative/mythic dimension as theologians within a religion work to integrate these stories into a more general view of the world. Thus the early Christian church came to its understanding of the doctrine of the Trinity by combining the central claim of the Jewish religion – that there is but one God – with its understanding of the life and teaching of Jesus Christ and the working of the Holy Spirit.

If doctrine attempts to define the beliefs of a community of believers, the *ethical and legal dimension* regulates how believers act. So Sunni Islam has its Five Pillars, while Judaism has the Ten Commandments and other regulations in the Torah and Buddhism its Five Precepts.

The *social and institutional dimension* of a religion relates to its corporate manifestation, for example the Sangha – the order of monks and nuns founded by the Buddha to carry on the teaching of the Dharma – in Buddhism, the umma' – the whole Muslim community – in Islam, and the Church – the communion of believers comprising the body of Christ – in Christianity.

Finally, there is the *material dimension* to each religion, namely the fruits of religious belief as shown by places of worship (e.g. synagogues, temples and churches), religious artefacts (e.g. Eastern Orthodox icons and Hindu statues) and sites of special meaning (e.g. the river Ganges, Mount Fuji and Uluru (Ayers Rock)).

As will be discussed below, the relationship between science and religion has changed over the years (Al-Hayani 2005; Brooke 1991; Szerszynski 2005); indeed, the use of the singular, 'relationship', risks giving the impression that there is only one way in which the two relate. Nevertheless, there are two key issues: one is to do with understandings of reality and the other to do with evidence and authority. Although it is always desperately difficult to generalise (difficult in the sense that one lays oneself open to accusations that one hasn't considered every particular – and yet the alternative is to be submerged in a weight of detail that would surely suffocate all but the most devoted/obsessive of readers), most religions hold that reality consists of more than the objective world, and many religions give weight to personal and/or (depending on the religion) institutional authority in a way that science generally strives not to.

For example, there is a very large religious and theological literature on the world to come, i.e. life after death, (e.g. Hick 1976/1985). However, to labour the point, although some (notably Atkins 2011) have argued that science disproves the existence of life after death, it can be objected that science, strictly speaking, has little or nothing to say about this question because life after death exists or would exist outside of or beyond the realm to which science relates. Furthermore, many religious believers within a particular religion are likely to find the pronouncements on the question of life after death by even the most intelligent and spiritual of their present leaders (let alone reputable scientists) to be of less significant than the few recorded words of their religion's founder(s). In the case of Christianity, while the proportion of believers who take literally the resurrection promises of the New Testament may be less than in previous ages (though high-quality comparative quantitative data are unavailable), it remains the case that literal belief in an afterlife is widespread among believers.

Before moving on to the nature of science, it is worth, in this section on the nature of religion, briefly saying something specifically about the nature of Christianity. While there are many Christian denominations, they all treat the Jewish scriptures and the New Testament as jointly constituting their scriptures, though there are differences as to which books are included in the Christian 'Old Testament'. The core beliefs of Christianity are summed up in the books of the New Testament and subsequent formulations such as the Apostles' and Nicene Creeds. The most widely recited formulation is the Nicene Creed which, in its 1975 ecumenical version, reads:

We believe in one God,
the Father, the Almighty
maker of heaven and earth,
of all that is, seen and unseen.
We believe in one Lord, Jesus Christ,
the only Son of God,
eternally begotten of the Father,
God from God, Light from Light,
true God from true God,
begotten, not made,
of one Being with the Father.
Through him all things were made.
For us and for our salvation
he came down from heaven:
by the power of the Holy Spirit
he became incarnate from the Virgin Mary, and was made man.
For our sake he was crucified under Pontius Pilate;
he suffered death and was buried.
On the third day he rose again
in accordance with the Scriptures;
he ascended into heaven
and is seated at the right hand of the Father.
He will come again in glory to judge the living and the dead,
and his kingdom will have no end.
We believe in the Holy Spirit, the Lord, the giver of Life,
who proceeds from the Father and the Son.
With the Father and the Son he is worshipped and glorified.
He has spoken through the Prophets.
We believe in one holy catholic and apostolic Church.
We acknowledge one baptism for the forgiveness of sins.
We look for the resurrection of the dead,
and the life of the world to come. Amen. (Episcopal Church 1979)

It is evident that some of the issues addressed in the Nicene Creed are to do with science and some are not. However, even those that at first appear not to be (e.g. the Trinity) have been examined from a science and religion perspective (e.g. Polkinghorne 2004). Indeed, if we restrict ourselves to those features of Christianity that are central to the science-religion issue, these include a belief that the triune God who creates, sustains and redeems the world is non-capricious (i.e. there are laws of nature) and deeply concerned with the created order (Poole 1998).

51.3 The Nature of Science

The term 'nature of science' is understood in a number of ways, as discussed elsewhere in this volume, but at its heart is knowledge about how, and to a lesser extent why, science is undertaken. So the nature of science includes issues about the fields of scientific enquiry and the methods used in that enquiry as well as, to a certain extent, something about the purpose of science.

A key point about the fields of scientific enquiry is that these have shifted over time. In large measure this is simply because of developments in instrumentation. We can now study events that happen at very low temperatures, at distances, at speeds and at magnifications that simply were not possible even a few decades ago. What is still unclear is the extent to which certain matters currently outside of mainstream science will one day fall within the compass of science. Take dreams, for example. It may be that these will remain too subjective for science, but it may be that developments in the recording of brain activity will mean that we can obtain a sufficiently objective record of dreams for them to be amenable to rigorous scientific study.

But the scope of science has also shifted for reasons that are more to do with theorisation than with technical advances (Reiss 2013a). Consider beauty. Aesthetics for a long time was not considered a scientific field. But there is now, within psychology and evolutionary biology, growing scientific study of beauty and desire (e.g. Buss 2003). Indeed, a number of the social sciences are being nibbled away at by the natural sciences, and if one believes some scientists, almost the only valid knowledge is scientific knowledge (Atkins 2011).

Despite such movements in the fields of scientific enquiry and in the actual methods employed by scientists, the overarching methods of science (what a social scientist might term its methodology) have shifted far less, certainly for several hundreds of years, arguably for longer than that.

As is well known, Robert Merton characterised science as open-minded, universalist, disinterested and communal (Merton 1973). For Merton, science is a group activity; even though certain scientists work on their own, science, within its various subdisciplines, is largely about bringing together into a single account the contributions of many different scientists to produce an overall coherent model of one aspect of reality. In this sense, science is (or should be) impersonal. Allied to the notion of science being open-minded, disinterested and impersonal is the notion of scientific objectivity. The data collected and perused by scientists must be objective in the sense that they should be independent of those doing the collecting (cf. Daston and Galison 2007) – the idealised 'view from nowhere'. This is the main reason why the data obtained by psychotherapists are (at least at present) not really scientific: they depend too much on the specifics of the relationship between the therapist and the client. The data obtained by cognitive behavioural therapists, on the other hand, are more scientific (cf. Salkovskis 2002).

Karl Popper emphasised the falsifiability of scientific theories (Popper 1934/1972): unless one can imagine collecting data that would allow one to refute a theory, the theory isn't scientific. The same applies to scientific hypotheses. So, iconically, the hypothesis 'All swans are white' is scientific because we can imagine finding a bird that is manifestly a swan (in terms of its anatomy, physiology and behaviour) but is not white. Indeed, this is precisely what happened when early White explorers returned from Australia with tales of black swans.

Popper's ideas easily give rise to a view of science in which knowledge accumulates over time as new theories are proposed and new data collected to distinguish between conflicting theories. Much school experimentation in science is Popperian: we see a rainbow and hypothesise that white light is split up into light of different colours as it is refracted through a transparent medium (water droplets); we test this by attempting to refract white light through a glass prism; we find the same colours of the rainbow are produced, and our hypothesis is confirmed. Until some new evidence causes it to be falsified, we accept it (Reiss 2008a).

Thomas Kuhn made a number of seminal contributions, but he is most remembered nowadays for his argument that while the Popperian account of science holds well during periods of normal science when a single paradigm holds sway, such as the Ptolemaic model of the structure of the solar system (in which the Earth is at the centre) or the Newtonian understanding of motion and gravity, it breaks down when a scientific crisis occurs (Kuhn 1970). At the time of such a crisis, a scientific revolution happens during which a new paradigm, such as the Copernican model of the structure of the solar system or Einstein's theory of relativity, begins to replace (initially to coexist with) the previously accepted paradigm. The central point is that the change of allegiance from scientists believing in one paradigm to their believing in another cannot, Kuhn argues, be fully explained by the Popperian account of falsifiability.

A development of Kuhn's work was provided by Lakatos (1978) who argued that scientists work within research programmes. A research programme consists of a set of core beliefs surrounded by layers of less central beliefs. Scientists are willing to accept changes to these more peripheral beliefs so long as the core beliefs can be defended. So, in biology, we might see in contemporary genetics a core belief in the notion that development proceeds via a set of interactions between the actions of genes and the influences of the environment. At one point, it was thought that the passage from DNA to RNA was unidirectional. Now we know (reverse transcriptase, etc.) that this is not always the case. The core belief (that development proceeds via a set of interactions between the actions of genes and the influences of the environment) remains unchanged, but the less central belief (that the passage from DNA to RNA is unidirectional) is abandoned.

The above account of the nature of science portrays science as what John Ziman (2000) has termed 'academic science'. Ziman argues that such a portrayal was reasonably valid between about 1850 and 1950 in European and American universities but that since then we have entered a phase largely characterised by 'post-academic science'. Post-academic science is increasingly transdisciplinary and utilitarian, with a requirement to produce value for money. It is more influenced by politics, it is more industrialised and it is more bureaucratic. The effect of these changes is to make the boundaries around the domain of science a bit fuzzier. Of course, if one accepts the contributions of the social study of science (e.g. Yearley 2005), one finds that these boundaries become fuzzier still. The argument in this chapter does not *rely* on such a reading of science though someone who is persuaded by the 'Strong Programme' within the sociology of scientific knowledge (i.e. the notion that even valid scientific theories are amenable to sociological investigation of their truth claims) is much more likely to accept the worth of science educators considering the importance of religion as one of many factors that influence the way science is practised and scientific knowledge produced.

51.4 Understandings of Possible Relationships Between Science and Religion

It is clear that there can be a number of axes on which the science-religion issue can be examined. For example, the effects of the practical and ritual dimension are being investigated by scientific studies that examine such things as the efficacy of prayer and the neurological consequences of meditation (e.g. Lee and Newberg 2005); a number of analyses of religious faith, informed by contemporary understandings of evolutionary psychology, behavioural ecology and sociobiology, examine the possibility or conclude that religious faith can be explained by science (e.g. Dennett 2006; Hinde 1999; Reynolds and Tanner 1983); the narrative/mythic dimension of religion clearly connects (in ways that will be examined below) with scientific accounts of such matters as the origins of the cosmos and the evolution of life; the doctrinal and philosophical dimension can lead to understandings that may agree or disagree with standard scientific ones (e.g. about the status of the human embryo); and the ethical and legal dimension can lead to firm views about such matters as land ownership, usury and euthanasia.

Perhaps only the social and institutional and the material dimensions of religion are relatively distinct from the world of science (understand as the natural sciences rather than the social sciences more broadly), in that science has little if anything to say about such manifestations of religion – i.e. in Christianity, the Church and such things as religious artefacts.

There is now a very large literature on the relationship between science and religion (a major overview is provided by Clayton and Simpson 2006). Indeed, the journal *Zygon* specialises in this area, while *Science & Christian Belief* focuses on the relationship between science and Christianity. A frequent criticism by those who write in this area (e.g. Roszak 1994 and regular articles by Andrew Brown and Paul Vallely in the *Church Times*) is of what they see as simplistic analyses of the area by those, often renowned scientists, who write occasionally about it. Indeed, it is frequently argued that the clergy both in the past and nowadays are often far more sympathetic to a standard scientific view on such matters as evolution than might be supposed (e.g. Colburn and Henriques 2006).

A particularly thorough historical study of the relationship between science and religion is provided by John Hedley Brooke (1991). Brooke's aim is 'to reveal something of the complexity of the relationship between science and religion as they have interacted in the past' (p. 321). He concludes:

> Popular generalizations about that relationship, whether couched in terms of war or peace, simply do not stand up to serious investigation. There is no such thing as *the* relationship between science and religion. It is what different individuals and communities have made of it in a plethora of different contexts. Not only has the problematic interface between them shifted over time, but there is also a high degree of artificiality in abstracting the science and the religion of earlier centuries to see how they were related. (Brooke 1991, p. 321)

Perhaps the best known categorisation of the ways in which the relationship between science and religion can be understood was provided by Ian Barbour (1990). Barbour himself updated this book (Barbour 1997), and since 1990 there has been a considerable literature about the ways in which science and religion relate (e.g. Glennan 2009; Haught 1995; Plantinga 2010; Stenmark 2004); indeed, Mark Vernon argues that rather more agnosticism and less dogmatism in the science-religion field would be wise (Vernon 2008). Nevertheless, Barbour's (1990) typology continues to dominate the literature and so is employed here. Barbour, who focuses especially on epistemological assumptions of recent Western authors, identifies four main groupings.

First, there is the relationship of *conflict*; 'first' simply because it is the first in Barbour's list and first, perhaps, also in the minds of many people, whether or not they have a religious faith (cf. McGrath 2005). Barbour doesn't give a reason for the order of his listing, but at least two can be suggested: comprehensibility and familiarity. It is both easy and familiar (given Barbour's declared focus on recent Western authors) to see the relationship between science and religion as one of conflict. However, as one might expect from a professor of science, technology and society giving the Gifford lectures (the result of an 1885 bequest of £80,000 'for the establishment of a series of lectures dealing with the topic of natural religion' (Gifford lectures 2006)), Barbour sees limitations in this way of understanding the science-religion issue. As he memorably puts it:

> In a fight between a boa constrictor and a wart-hog, the victor, whichever it is, swallows the vanquished. In scientific materialism, science swallows religion. In biblical literalism, religion swallows science. The fight can be avoided if they occupy separate territories or if, as I will suggest, they each pursue more appropriate diets. (Barbour 1990, p. 4)

Barbour's second grouping is *independence* (e.g. Gould 1999). Science and religion may be seen as independent for two main reasons: because they use distinctive methods or because they function as different languages. In any event, the result is that each is seen as distinct from the other and as enjoying its own autonomy:

> Each has its own distinctive domain and its characteristic methods that can be justified on its own terms. Proponents of this view say there are two jurisdictions and each party must keep off the other's turf. Each must tend to its own business and not meddle in the affairs of the others. Each mode of inquiry is selective and has its limitations. (Barbour 1990, p. 10)

Barbour's third grouping moves beyond conflict and independence to *dialogue* (cf. Berry 1988; Polkinghorne 2005; Watts 1998; Williams 2001). As an example of dialogue, Barbour points out how our understanding of astronomy has forced us to ask why the initial conditions were present that allowed the universe to evolve. The point is not that the findings of science require a religious faith – that would be for the warthog of religion to swallow the boa constrictor of science. Rather the point is that scientific advances can give rise (no claim is made that they do for all people) to religious questions, so that a dialogue ensues.

Barbour's final grouping is one in which the relationship between science and religion is seen to be one of *integration* (cf. Peacocke 2001; Polkinghorne 1994). For example, in natural theology it is held that the existence of God can be deduced from aspects of nature rather than from revelation or religious experience (e.g. Ray 1691/2005). Natural theology has rather fallen out of favour (but see Polkinghorne 2006). A more modern version is process theology which rejects a view of the world in which purely natural events (characterised by an absence of divine activity) are interspersed with occasional gaps where God acts. Rather, for process theologians, every event is understood 'to be jointly the product of the entity's past, its own action, and the action of God' (Barbour 1990, p. 29). Furthermore, God is not the Unmoved Mover of Thomas Aquinas but instead acts reciprocally with the world.

I think it can be difficult for those who have never had a religious faith, or have only had one rather tenuously, to imagine what a life is like that is lived wholly within a religious ordering. For such a person, the relationship between science and their faith may be described as 'integrated' though this is to give an epistemological framing to the relationship, whereas what may be going on is that the person has little overt interest in the precise nature of the relationship between science and religion other than that there can clearly be no conflict between them.

Anthropologists provide good accounts of what it can be like to live a life where one's religious faith integrates with every aspect of one's life. One of my favourite such accounts is that of du Boulay (2009) who studied life in a Greek Orthodox Village in the late 1960s and early 1970s. Everything that happened in the village needs to be understood by reference to Greek Orthodoxy. To give just one instance, the annual liturgical and agricultural cycles intermeshed, so that after the harvest, the sowing of the seed for next year's harvest was closely related to the Christian calendar:

> The main sowing of the wheat is carried into November, and the Archangel Michael, celebrated on 8 November and seen on his icons with drawn sword, is a formidable figure associated with the darkening November days with the leaves being stripped from the trees and the smoke gusting in ashy draughts down the chimneys; but this is a month named after the preeminent agricultural task – 'The Sower' (Σπορίας). And the Entry of the Mother of

God into the Temple on 21 November, soon after the Christmas fast has begun, is also in the village given the character of the time as the 'Mother of God Half-Way-Through-The-Sowing' (*Παναγια Μισοσπειρτσα*). The task of the sowing of the wheat then continues into the time know as 'Andrew's' (St Andrew, whose day is 30 November, but who has given his name to the following month of December), and can go on up to Christmas – and even beyond, if the weather has not been fit. (du Boulay 2009, p. 106)

Having examined possible relationships between science and religion, and given a flavour of the way in which religion can order a person's understanding of and immersion in the world, I turn now to the issues of determinism and evolution to discuss at a more fine-grained level how Christianity and science can correlate.

51.5   Determinism

The 'science-religion issue' is often examined simply by recourse to certain cause célèbres – Galileo and Copernican heliocentrism, Darwin and evolution and arguments about the sanctity of life, for example. At school level, examinations of such particular instances of the relationship between science and religion, along with a more general consideration about how science and religion can relate, are perfectly appropriate. However, there are certain 'higher-order' questions that teachers and curriculum developers need to consider to decide whether they can be introduced meaningfully at school level. One such central question is about whether nature is deterministic and, if it/she is not, whether that has anything to do with divine action. Theologically speaking, this is part of the more general question as to how (for those who have a religious faith) God acts in nature (Dixon 2008).

The post-Newtonian advent in the early twentieth century of quantum theory and, later in the same century, of chaos theory has led many to wonder whether within either or both of these two frameworks might lie a space for divine action in a way that does not contradict the scientific worldview in the way that miracles seem to. For almost anyone who has not studied quantum physics to at least first-degree level, it is exceptionally difficult to understand what is going on that is relevant to the science-religion issue, but a core concept is that of determinism, which results from the issue of the relationship of measurement to reality (e.g. Bhaskar 1978; Osborn 2005).

As is well known, in 1927 Heisenberg argued that certain key physical variables that had previously been presumed to be independent (e.g. the position and momentum of an object) are linked. Measuring the one to a very high degree of precision necessarily means that the other cannot be so precisely determined. Thus far there is not a great deal that is of interest to the non-physicist – the issue appears to be one of epistemology. However,

Heisenberg himself took a more radical view – he saw this limitation as a property of nature rather than an artefact of experimentalism. This radical interpretation of uncertainty as an ontological principle of indeterminism implies that quantum mechanics is inherently statistical – it deals with probabilities rather than well-defined classical trajectories. Such a view is clearly inimical to classical determinism. (Osborn 2005, p. 132)

Put somewhat loosely, a number of people have tried to find room for divine action in this indeterminacy. No consensus yet exists as to the validity of this search though, on balance, the current views seem to be that such a search is mistaken for reasons both of theology and of physics. A particularly helpful, though demanding, analysis of both the theology and the physics is provided by Saunders (2002). Beginning with the theology, Saunders draws on the widespread distinction between general and special forms of divine action. In the words of Michael Langford, general divine action is 'the government of the universe through the universal laws that control or influence nature, man, and history, without the need for specific or ad hoc acts of divine will' (Langford 1981, p. 11). On the other hand, special divine action is characterised by

> Those actions of God that pertain to a *particular* time and place in creation as distinct from another. This is a broad category and includes the traditional understanding of 'miracles', the notion of particular providence, responses to intercessionary prayer, God's personal actions, and some forms of religious experience. (Saunders 2002, p. 20)

Oversimplifying considerably, all religions are comfortable with the notion of general divine action, but they differ both among and within themselves considerably in their understanding of specific divine action. In particular, many leading theologians (but see Pannenberg 2006) are uncomfortable with the notion of specific divine action so defined for a number of reasons including the particular problems for the occurrence of suffering that it raises (if suffering can sometimes be averted miraculously, why isn't it always or, at least, much more often?) and the apparent shortcomings, including capriciousness, suggested by a divine being who relies on occasional exercises of supernatural activity to keep things moving along (cf. Kenny 1992).

Going onto the physics, Saunders is sceptical of attempts to locate the possibility of specific divine action in quantum or chaos theory. The argument here becomes even more technical and depends, in respect of quantum theory, on whether one accepts the standard (Copenhagen) interpretation of reality (in which Schrödinger's cat is either dead or alive before the box is opened) or the more radical interpretation (in which the cat is both dead and alive). In both cases, though, as well as in the case of chaos theory (sometimes termed 'complexity theory' on the grounds that it deals with systems that are deterministic but unpredictable because of their exquisite sensitivity to small changes in their initial conditions), Saunders rejects attempts to find opportunities for specific divine activity in the science.

51.6 Evolution

51.6.1 The Scientific Consensus Concerning Evolution

As with any large area of science, there are parts of what we might term 'front-line' evolution that are unclear, where scientists still actively work attempting to discern what is going on or has gone on in nature. But much of evolution is not like that. Evolution is a well-established body of knowledge that has built up over 150 years as a result of the activities of many thousands of scientists. The following are examples of statements about evolution that lack scientific controversy:

- All of today's life on Earth is the result of modification by descent from the simplest ancestors over a period of several thousand million years.
- Natural selection is a major driving force behind evolution.
- Evolution relies on those occasional instances of the inheritance of genetic information that help (rather than hinder) its possessor to be more likely to survive and reproduce.
- Most inheritance is vertical (from parents) though some is horizontal (e.g. as a result of viral infection carrying genetic material from one species to another).
- The evolutionary forces that gave rise to humans do not differ in kind from those that gave rise to any other species. (Reiss 2013b)

For those who accept such statements and the theory of evolution, there is much about the theory of evolution that is intellectually attractive. For a start, a single theory provides a way of explaining a tremendous range of observations: for example, why it is that there are no rabbits in the Precambrian, why there are many superficial parallels between marsupial and placental mammals, why monogamy is more common in birds than in fish and why sterility (e.g. in termites, bees, ants, wasps and naked mole rats) is more likely to arise in certain circumstances than in others. Indeed, I have argued elsewhere that evolutionary biology can help with some theological questions, including the problem of suffering (Reiss 2000).

51.6.2 Rejecting Evolution

The theory of evolution is not a single proposition that a person must either wholly accept or wholly reject. At one pole are materialists who, eschewing any sort of critical realist distinction between the empirical, the actual and the real (Bhaskar 1978), maintain that there is no possibility of anything transcendent lying behind what we see of evolution in the results of the historical record (fossils, geographical distributions, comparative anatomy and molecular biology) and today's natural environments and laboratories. At the other pole are the advocates of creationism, inspired by a literal reading of certain scriptures. But in between lie many others (Scott 1999) including those who hold that evolutionary history can be providential as human history is.

In addition, there are a whole set of nonreligious reasons why someone may actively reject aspects of the theory of evolution. After all, it may seem to defy common sense to suppose that life in all its complexity has evolved from non-life. And then there is the tremendous diversity of life we see around us. To many it hardly seems reasonable to presume that giant pandas, birds of paradise, spiders, orchids, flesh-eating bacteria and the editor of this book all share a common ancestor – yet that is what mainstream evolutionary theory holds.

It is, though, for religious reasons that many people reject evolution. Creationism exists in a number of different forms, but between about 10 % of adults in the

Nordic countries and Japan and 50 % of adults in Turkey (40 % in the USA) reject the theory of evolution and believe that the Earth came into existence as described by a literal (i.e. fundamentalist) reading of the early parts of the Bible or the Qur'an and that the most that evolution has done is to change species into closely related species (Lawes 2009; Miller et al. 2006). Christian fundamentalists generally hold that the Earth is nothing like as old as evolutionary biologists and geologists conclude – as young as 10,000 years or so for young Earth creationists. For Muslims, the age of the Earth is much less of an issue.

Allied to creationism is the theory of intelligent design. While many of those who advocate intelligent design have been involved in the creationism movement, to the extent that the US courts have argued that the country's First Amendment separation of religion and the State precludes its teaching in public schools (Moore 2007), intelligent design can claim to be a theory that simply critiques aspects of evolutionary biology rather than advocating or requiring religious faith. Those who promote intelligent design typically come from a conservative faith-based position (though there are atheists who accept intelligent design). However, in their arguments against evolution, they typically make no reference to the scriptures or a deity but argue that the intricacy of what we see in the natural world, including at a subcellular level, provides strong evidence for the existence of an intelligence behind this (e.g. Meyer 2009). An undirected process, such as natural selection, is held to be incapable of explaining all such intricacy.

51.6.3 Evolution in School Science

Few countries have produced explicit guidance as to how schools might deal with the issues of creationism or intelligent design in the science classroom. One country that has is England (Reiss 2011). In the summer of 2007, the then DCSF (Department of Children, Schools and Families) Guidance on Creationism and Intelligent Design received Ministerial approval and was published (DCSF 2007). The Guidance points out that the use of the word 'theory' in science (as in 'the theory of evolution') can mislead those not familiar with science as a subject discipline because it is different from the everyday meaning, when it is used to mean little more than an idea.

The DCSF Guidance goes on to state 'Creationism and intelligent design are sometimes claimed to be scientific theories. This is not the case as they have no underpinning scientific principles, or explanations, and are not accepted by the science community as a whole' (DCSF 2007) and then says:

> Creationism and intelligent design are not part of the science National Curriculum programmes of study and should not be taught as science. However, there is a real difference between teaching 'x' and teaching about 'x'. Any questions about creationism and intelligent design which arise in science lessons, for example as a result of media coverage, could provide the opportunity to explain or explore why they are not considered to be scientific theories and, in the right context, why evolution is considered to be a scientific theory. (DCSF 2007)

This is a key point and one that is independent of country, whether or not a country permits the teaching in schools of religion (as in the UK) or does not (as in France, Turkey and the USA). Many scientists, and some science educators, fear that consideration of creationism or intelligent design in a science classroom legitimises them. For example, the excellent book *Science, Evolution, and Creationism* published by the US National Academy of Sciences and Institute of Medicine asserts 'The ideas offered by intelligent design creationists are not the products of scientific reasoning. Discussing these ideas in science classes would not be appropriate given their lack of scientific support' (National Academy of Sciences and Institute of Medicine 2008, p. 52).

However, just because something lacks scientific support doesn't seem a sufficient reason to omit it from a science lesson. This is a point that holds more widely than with respect to the teaching of evolution; for instance, when teaching about climate change, one might want to examine the argument that sunspot cycles are sufficient to explain all of global warming, even though this is no longer a reputable scientific position. Nancy Brickhouse and Will Letts (1998) have argued that one of the central problems in science education is that science is often taught 'dogmatically'. With particular reference to creationism, they write:

> Should student beliefs about creationism be addressed in the science curriculum? Is the dictum stated in the California's *Science Frameworks* (California Department of Education 1990) that any student who brings up the matter of creationism is to be referred to a family member of member of the clergy a reasonable policy? We think not. Although we do not believe that what people call 'creationist science' is good science (nor do scientists), to place a gag order on teachers about the subject entirely seems counterproductive. Particularly in parts of the country where there are significant numbers of conservative religious people, ignoring students' views about creationism because they do not qualify as good science is insensitive at best. (Brickhouse and Letts 1998, p. 227)

51.6.4   Evolution in Science Museums

Education about evolution does not only take place in schools. It takes place through books, magazines, TV, the Internet, radio and science museums. Science museums have long had exhibits about evolution. Tony Bennett (2004) provides an historical analysis to look at how science museums have presented evolution. He attempts to discern the modes of power that lie behind the manifestations of particular forms of knowledge and concludes that

> In their assembly of objects in newly historicised relations of continuity and difference, evolutionary museums not only made new pasts visible; they also enrolled those pasts by mobilising objects – skulls, skeletons, pots, shards, fossils, stuffed birds and animals – for distinctive social and civic purposes. (Bennett 2004, p. 189)

In one sense this is hardly surprising – museums have to make selections about what to display and how to curate such displays, and these are clearly cultural decisions, whether one is referring to evolution or anything else. However, visitors to science museums can easily presume that they are being presented with objective fact. For example, the classic story about the evolution of the modern horse can be oversimplified to the point that the viewer concludes that evolution is linear and progressive.

Monique Scott too has produced a book about evolution in museums (Scott 2007) though her work, unlike Bennett's, is more to do with the present than with history. Using questionnaires and interviews, Scott gathered the views of nearly 500 visitors at the Natural History Museum in London, the Horniman Museum in London, the National Museum of Kenya in Nairobi and the American Museum of Natural History in New York. Perhaps her key finding is that many of the visitors interpreted the human evolution exhibitions as providing a linear narrative of progress from African prehistory to a European present. As she puts it:

> Despite the distinctive characters of each of the four museums considered here and the specific cultural differences among their audiences, it is clear that museums and their visitors traffic in common anthropological logic – namely the color-coded yardstick of evolutionary progress. In fact, visitors equipped with a weighty set of popular images – imagery derived from such things as *Condé Nast Traveler* magazines, *Planet of the Apes* films, and *National Geographic* images – occupy the nexus between the evolutionary folklore circulating outside the museum and that which has been generated within it. This collection of images often urges Western museum visitors to negotiate between the "people who stayed behind" and their own fully evolved selves (defined often by such culturally coded "evolutionary leaps" as clean-shaven-ness and white skin). (Scott 2007, p. 148)

So how might one hope that science museums would treat religion when putting together exhibitions about evolution? Museums have a number of advantages over classroom teachers; for one thing, they usually have longer to prepare their teaching. So we might hope that a science museum, while not giving the impression that the occurrence of evolution is scientifically controversial today, might convey something of the history of the theory of evolution. This would include the fact that evolution was once scientifically controversial and that religious believers have varied greatly as to how they have reacted to the theory of evolution. On the one hand we have today's creationists; on the other we have Charles Kingsley, the Anglican divine and friend of Charles Darwin, who read a prepublication copy of *On the Origin of Species* and wrote to Darwin:

> I have gradually learnt to see that it is just as noble a conception of Deity, to believe that he created primal forms capable of self development into all forms needful pro tempore & pro loco, as to believe that He required a fresh act of intervention to supply the lacunas w[h]. he himself had made. (Kingsley 1859)

51.7 The Uses to Which Advances in Scientific Knowledge May Be Put

The tremendous growth in scientific knowledge means that we are faced with an ever-increasing number of ethical questions that our predecessors simply did not have to consider. Many of these are in the area of bioethics (e.g. Brierley et al. in press; Mepham 2008). How do we weigh human interests against those of the natural environment and laboratory animals? Is it acceptable to experiment on human embryos? And what role does religion have in answering such ethical questions about our use of scientific knowledge?

In a recent book titled *Dishonest to God: On keeping God out of politics*, Mary Warnock (2010), despite having a certain affection and sympathy for the Church of England, lists many examples where religious arguments have in her view inappropriately been used in parliamentary debates in attempts, some successful, some unsuccessful, to influence national legislation. She concludes:

> The danger of religion, any religion, lies in its claim to absolute immutable moral knowledge which, if justified, would indeed give its adherents a special place in instructing others how to behave, perhaps even a right to do so. (Warnock 2010, p. 165)

Our concern here is not so much with claims to knowledge as with how one makes practical decisions about scientific matters in a world with a multiplicity of values, religious and otherwise. And here religion has a place at the table (Reiss 2012). In just the same way as consequentialists have to learn to accept that many deontologists are not going to accept the consequentialist understanding of ethics as being decisive, and vice versa, so those of no religious persuasion need to accept that significant numbers of people have religious beliefs and hold that these beliefs help shape what is deemed morally right and morally wrong.

In this sense, those of no religious persuasion need to take the same sort of account of religious believers as those who eat meat need to take account of vegetarians. We would deem it unacceptable, nowadays, for the authorities in charge of a prison, a hospital or any other residential establishment to fail to provide vegetarian food on the grounds that vegetarianism is unnecessary, a minority lifestyle choice or a fad. In the same way, a secular society that respects its citizens needs to take account of religious views. Of course, precisely the converse holds too. A theocracy that respects its citizens needs to take account of the views of those who have no religious faith or belong to a minority faith.

I am well aware that to many with a religious faith, this may seem like 'selling out'. To this objection I would respond as follows. First, it is as good as you are going to get nowadays in an increasing number of countries. Secondly, if a religious viewpoint has sufficient validity, it should be capable of holding its own in arguments with those who have no religious faith. For example, while Roman Catholic arguments about the unacceptability of contraception are very difficult to defend to non-Roman Catholics, more broad-based arguments about the sanctity of human life and therefore the unacceptability of euthanasia can receive a more sympathetic hearing among a secular audience so long as 'the sanctity of human life' is not seen as a trump card but is translated into religiously neutral language about respect and the protection of the vulnerable. Thirdly, my own reading of the Christian scriptures is that God's nature is such that there is rarely an easily discerned voice from heaven. Usually, determination of what is morally right and morally wrong, while influenced by the reading of scripture and an understanding of the religious tradition to which one belongs, needs supplanting by broader reflection and study and should be informed, in the case of bioethics, by ongoing advances in the biosciences.

One objection to the line I have been advancing is that it is a relativistic one that depends on the specifics of history and geography. This is a common objection – not just in theology and bioethics but in other disciplines including science and aesthetics – and a standard response is to assert that to deny immutable knowledge is not necessarily to slide inexorably into relativism. One can occupy a middle ground. Indeed, as Parfit (2011) concludes, there are considerable commonalities between the main secular ethical frameworks (Kantian deontology, consequentialism and contractualism) once one gets down to specifics.

There will be some, who may or may not be atheists, who are not convinced that religion has any role to play in bioethics or any other issue to do with the use to which scientific knowledge is put. Religion, it might be maintained, rests on irrational belief in the supernatural and an excessive reliance on tradition, and while notions of respect may require us to tolerate such views, nothing should be done that might allow them to influence public policy. It is fine for people to have freedom of expression (e.g. freedom to attend worship) but that is entirely separate from granting religion a public role. If religion were to enjoy such privileges, we would have to extend them to other odd belief systems, such as those who believe they have been abducted by aliens (Clancy 2005) or those who hold that Elvis Presley is still alive (e.g. Brewer-Giorgio 1988, Elvis Is Alive 2012).

There are several reasons why this line of argument does not work. First, the proportion of the population, even in more secular countries, who have some religious beliefs, is considerably higher than the proportion of the population who believe in alien abductions or Elvis' longevity. Secondly, religious faith has been around for all of human time, whereas conspiracy theories and fads come and go. Thirdly, religious beliefs are often core to a person's being in a way that alien abduction (however upsetting) and Presley mania are but rarely. Fourthly, there is a close connection between many bioethical issues and religious faith which there isn't between bioethical issues and alien abduction or Elvis Presley. Of course, if the state were to set up a publicly funded museum about aliens, then there might well be a case for granting a voice to those who believe they have experienced such abductions (and this would almost certainly be good for business).

51.8 The Approach of Worldviews

Before going on to consider the pedagogical implications of all this, mention should be made of one approach to the science-religion issue that has become prominent within science education and is of considerable pedagogical value – namely, the concept of worldviews. The essence of a worldview, as the word itself implies, is that it is a way of conceiving and understanding the world that one inhabits (cf. Aerts et al. 1994). So, someone with a traditional Christian worldview is likely to believe that the world is fundamentally good but has become corrupted as a result of human sin. However, there is always the hope of redemption, and one of the tasks of Christians is to live their lives so as to help bring about the kingdom of God. On the other hand, someone with an atheistic worldview is likely to believe that the world is morally neutral and that there are no ultimate purposes in life beyond those that we decide for ourselves. Which of these two worldviews one finds the more convincing and conducive says much about oneself.

Creationism can profitably be seen not as a simple misconception that careful science teaching can correct, as careful science teaching might hope to persuade a student that an object continues at uniform velocity unless acted on by a net force, or that most of the mass of a plant comes from air as opposed to the soil. Rather, a student who believes in creationism can be seen as inhabiting a non-scientific worldview, a very different way of seeing the world compared to the scientific perspective. The pedagogical significance of this comes largely from the observation that one very rarely changes one's worldview as a result of one or two lessons, however well taught, whereas one may indeed replace a misconception with its scientifically validated alternative about such a brief teaching sequence (Reiss 2008b).

The probable reason for this difference in the difficulty of replacing worldviews and misconceptions is twofold. First, a student is likely to have far more of personal significance invested in a religious worldview than a scientific misconception. It is clear that the personal implications of abandoning a belief in a literal reading of the chronology of *Genesis*, including the 6 days of creation as 6 periods each of 24 h, are far greater than of discarding a presumption that plants gain most of their mass from the soil. Secondly, many scientific misconceptions are relatively discrete – one can discard one without this affecting much else of one's scientific understanding. Abandoning creationism entails accepting the notion of Deep Time, the relatedness of all life and the realisation that there is no *scala naturae*.

51.9 Pedagogical Implications

The question of the significance of religious issues for science education can be considered at the intended, the implemented and the attained curriculum levels (Robitaille and Dirks 1982). In a school setting there are therefore implications for curriculum developers, for classroom teachers and for learners. In this section I concentrate on teachers (whether in school science classrooms or informal settings) and learners.

Science teaching is demanding for teachers, particularly in a school setting. I have discussed elsewhere whether or not it is realistic to expect science teachers to deal with ethical issues in science lessons (Reiss 1999). Although there are examples of this happening successfully (Jones et al. 2010; Reiss 2008c), this is far from always being the case. It seems even more optimistic to expect science teachers to deal with religious issues, even when these are restricted to religious issues that relate to science. I therefore welcome the current guidelines in England about dealing with creationism in science lessons (DCSF 2007) which do not require but do allow science teachers to deal with the creationism and suggest that this principle be followed when dealing with religious issues in general in science classrooms.

The aim of including religion in science learning is not primarily to teach about religion but to enable richer and more effective ways to enable students to understand certain ideas within science and to help them appreciate better certain topics where science and religion interact. If science teachers, or other communicators of science, do deal with religious issues, or science issues that have religious connotations, I recommend that they be both true to science and respectful of their students and others, irrespective of such people's religious beliefs. Indeed, nothing pedagogically is to be gained by denigrating or ridiculing students.

The principle of respect for students has implications for assessment too. Well-designed examination material should be able to test student knowledge of science and its methods without expecting students to have to convert, or pretend that they have converted, to a materialistic set of beliefs. So, for example, while it is appropriate to ask students to explain how the standard neo-Darwinian theory of evolution attempts to account for today's biodiversity, it is not appropriate to ask students to explain how the geological sciences prove that the Earth is billions of years old.

Perhaps the most important implications of religion in general and Christianity in particular for the teaching of science come when teaching about the nature of science (Black et al. 2007). It can be a useful exercise with some students for science educators to get them to consider whether such topics as astrology, ghosts, paranormal phenomena and miracles fall within the scope of science or not. The aim is not to smuggle such topics into science but to get students more rigorously to think about what science is and how it proceeds. I remember one student of mine who undertook a survey among her peers to see whether, as predicted by astrology, their astrological birth sign was related to their personality, using a validated measure of personality. It wasn't. That student learnt something about science that was of value to her that I suspect she might not have learnt had I told her not to be silly and instead research something within mainstream science.

Skehan and Nelson (2000) point out that science educators generally do not do a good job of providing students with criteria to compare the strength of great scientific ideas. They emphasise the value of enabling students to develop skills of critical thinking when considering controversial topics such as evolution and provide a valuable list of eight criteria for comparing major scientific theories:

1. How many lines of independent evidence support the theory?
2. How many previously unconnected areas of knowledge did a theory tie together?
3. Does the theory make precise predictions?
4. How clear are the causal mechanisms?
5. Does the theory adequately explain the ultimate origin of the systems it describes and explains?
6. Is the theory scientifically controversial, or only publicly or politically controversial?
7. Is the theory fundamental to many practical benefits embraced by our economic system?
8. Is the theory widely understood and accepted by the general public?

This list differs considerably from the criteria discussed earlier (Sect. 51.3). Nevertheless, there would seem to be much of value in encouraging students to consider such questions (or others) when examining the validity of evolution or other major scientific theories.

Stolberg and Teece (2011) write about how to teach the science-religion issue but address their advice to specialist teachers of religion, not science, which provides a useful counterweight to the rest of this section. They point out that religious education teachers often assume that they should be neutral when teaching about controversial issues, yet this can be unrealistic and may not be the most effective way of teaching:

> Teachers may well feel that adopting a neutral stance – focusing on 'the facts', giving a 'balanced' picture – is most likely to be the 'safest' one to adopt. In practice, this is a very difficult strategy to achieve. The choice of facts you present (or withhold), the 'expert' opinions you share with your students and all the other educational judgments – in terms of the resources chosen and time devoted to the issue being explored – makes the effort of teaching religion and science issues in this way unrealistic. As with all controversial issues, however, your students need to be taught to examine critically the information they are given and the attitudes or values that have led to its production. So, rather than seeking to 'not get involved', you should be explicit about the aims and objectives of any exercise so that your students are aware of the circumstances in which they are being asked for their opinions and share the basis for their thinking. (p. 71)

51.9.1 Specific Issues to Do with Creationism

Part of the purpose of school science lessons is to introduce students to the main conclusions of science – and the theory of evolution is one of science's main conclusions. For this reason, school biology and earth science lessons should present students with the scientific consensus about evolution, and parents should not have the right to withdraw their children from such lessons. At the same time, science teachers should be respectful of any students who do not accept the theory of evolution for religious (or any other) reasons.

Science teachers should not to get into theological discussions, for example, about the interpretation of scripture. They should stick to the science, and if they are fortunate enough to have one or more students who are articulate and able to present any of the various creationist arguments against the scientific evidence for evolution (e.g. that the theory of evolution contradicts the second law of thermodynamics, that radioactive dating techniques make unwarranted assumptions about the constancy of decay rates, that evolution from inorganic precursors is impossible in the same way that modern science disproved theories of spontaneous generation), they should use their contributions to get the rest of the group to think rigorously and critically about such arguments and the standard accounts of the evidence for evolution.

My own experience of teaching the theory of evolution for some 30 years to school students, undergraduate biologists, trainee science teachers, members of the general public and others is that people who do not accept the theory of evolution for religious reasons are most unlikely to change their views as a result of one or two lessons on the topic, and others have concluded similarly (e.g. Long 2011). However, that is no reason not to teach the theory of evolution to such people. One can gain a better understanding of something without necessarily accepting it. Furthermore, recent work suggests that careful and respectful teaching about evolution can indeed make students who initially reject evolution considerably more likely to accept at least some aspects of the theory of evolution (Winslow et al. 2011).

For sites of informal education, there are some issues that are the same for schools and some that are different. The principles of respect for students and others, irrespective of their religious beliefs, hold in the same way, but the principle of being true to science manifests itself somewhat differently depending on the nature of the site of informal education. If the site is one that identifies itself as being scientific, for example, a science or natural history museum or centre, then it can validly attempt to convince visitors that the standard scientific position is correct. Other informal education sites may have less of a science agenda. In any event, any museum or other site of informal learning should be able to prepare carefully and access resources in a way that may not be possible for a classroom teacher, so as to ensure that an exhibition, a display, a taught session or an outreach activity does deal with relevant religious issues.

Finally, there are an increasing number of creationist museums (e.g. http://creationmuseum.org/) and zoos (e.g. www.noahsarkzoofarm.co.uk/). Perhaps somewhat optimistically, I would ask those running such creationist places of learning to make one concession to evolution. I do not expect them to promote evolution, but it is reasonable to ask them to make it clear that the scientific consensus is that the theory of evolution and not creationism is the best available explanation for the history and diversity of life. It is perfectly acceptable for those running creationist institutions to critique evolution and to try to persuade those visiting such institutions that the standard evolutionary account is wrong. But just as science teachers with no religious faith should respect students who have creationist views, so creationists should not misrepresent creationism as being in the scientific mainstream. It is not.

References

Aerts, D., Apostel L., De Moor B., Hellemans S., Maex E., Van Belle H. & Van Der Veken J. (1994). *World Views: From fragmentation to integration*. Brussels: VUB Press. Available at http://www.vub.ac.be/CLEA/pub/books/worldviews.pdf. Accessed 19 September 2012.
Al-Hayani, F. A. (2005). Islam and science: contradiction or concordance, *Zygon*, 40, 565–576.
Atkins, P. (2011). *On being: A scientist's exploration of the great questions of existence*. Oxford: Oxford University Press.
Barbour, I. G. (1990). *Religion in an age of science: The Gifford Lectures 1989–1991, volume 1*. London: SCM.
Barbour, I. G. (1997). *Religion and science: Historical and contemporary issues*. London: HarperCollins.
Bennett, T. (2004). *Pasts beyond memory: Evolution, museums, colonialism*. London: Routledge.
Berry, R. J. (1988). *God and evolution*. London: Hodder & Stoughton.
Bhaskar, R. (1978). *A realist theory of science*. Sussex: Harvester Press.
Black, P., Poole, M. & Grace, G. (2007). *Science education and the Christian teacher*. London: Institute of Education, University of London.

du Boulay, J. (2009). *Cosmos, life, and liturgy in a Greek Orthodox village*. Limni, Evia: Denise Harvey.

Brewer-Giorgio, G. (1988). *Is Elvis alive?* New York: Tudor Communications.

Brickhouse, N. W. & Letts IV, W. J. (1998). The problem of dogmatism in science education. In: *Curriculum, religion, and public education: Conversations for an enlarging public square*, Sears, J. T. & Carper, J. C. (Eds). New York: Teachers College, Columbia University, pp. 221–230.

Brierley, J., Linthicum J. & Petros, A. (in press). Should religious beliefs be allowed to stonewall a secular approach to withdrawing and withholding treatment in children? *Journal of Medical Ethics*.

Brooke, J. H. (1991). *Science and religion: Some historical perspectives*. Cambridge: Cambridge University Press.

Buss, D. M. (2003). *The evolution of desire: Strategies of human mating, revised edn*. New York: Basic Books.

Clancy, S. A. (2005). *Abducted: How people come to believe they were kidnapped by aliens*. Cambridge, MA: Harvard University Press.

Clayton, P. & Simpson, Z. (2006). *The Oxford handbook of religion and science*. Oxford: Oxford University Press.

Colburn, A. & Henriques, L. (2006). Clergy views on evolution, creationism, science, and religion. *Journal of Research in Science Teaching*, 43, 419–442.

Daston, L. & Galison, P. (2007). *Objectivity*. New York: Zone Books.

DCSF (2007). Guidance on creationism and intelligent design. http://webarchive.nationalarchives. gov.uk/20071204131026/http://www.teachernet.gov.uk/docbank/index.cfm?id=11890. Accessed 5 March 2012.

Dennett, D. C. (2006). *Breaking the Spell: Religion as a natural phenomenon*. London: Allen Lane.

Dixon, T. (2008). *Science and religion: A very short introduction*. Oxford: Oxford University Press.

Elvis is Alive (2012). Proof. Retrieved from www.elvis-is-alive.com/. Accessed 15 September 2012.

(The) Episcopal Church (1979). *Book of Common Prayer*. Oxford: Oxford University Press.

Farrell, J. (2005). *The day without yesterday: Lemaître, Einstein, and the birth of modern cosmology*. New York: Thunder's Mouth Press.

Gauch, H. G. Jr. (2009). Science, worldviews, and education. *Science & Education*, 18, 667–695.

Gifford Lectures (2006). Overview: history of the Gifford Lectures, http://www.giffordlectures. org/overview.asp. Accessed 5 March 2012.

Glennan, S. (2009). Whose science and whose religion? Reflections on the relations between scientific and religious worldviews. *Science & Education*, 18, 797–812.

Gould, S. J. (1999). *Rocks of ages: Science and religion in the fullness of life*. New York: Ballantine.

Harrison, P. (2001). *The Bible, Protestantism and the rise of natural science*. Cambridge: Cambridge University Press.

Haught, J. F. (1995). *Science and religion: From conflict to conversation*. Mahwah NJ: Paulist Press.

Hick, J. (1976/1985). *Death and eternal life*. Basingstoke: Macmillan.

Hinde, R. A. (1999). *Why gods persist: A scientific approach to religion*. London: Routledge.

Hinnells, J. R. (1991). *A handbook of living religions*. London: Penguin Books.

Hooykaas, R. (1972). *Religion and the rise of modern science*. Edinburgh: Scottish Academic Press.

Hunter, M. (2009). *Boyle: Between God and science*. New Haven CT: Yale University Press.

Ipsos MORI (2011). Religion and globalisation. http://www.fgi-tbff.org/randp/casestudies/religion-globalisation. Accessed 15 September 2012.

Jones, A., McKim, A. & Reiss, M. (Eds) (2010). *Ethics in the science and technology classroom: A new approach to teaching and learning*. Rotterdam: Sense.

Kenny, A. (1992). *What is faith? Essays in the philosophy of religion*. Oxford: Oxford University Press.

Kingsley, C. (1859). Letter to Charles Darwin, 18 November. http://www.darwinproject.ac.uk/entry-2534. Accessed 5 March 2012.

Kuhn, T. S. (1970). *The structure of scientific revolutions, 2nd edn.* Chicago: University of Chicago Press.

Lakatos, I. (1978). *The methodology of scientific research programmes.* Cambridge: Cambridge University Press.

Langford, M. (1981). *Providence.* London: SCM.

Lawes, C. (2009). *Faith and Darwin: Harmony, conflict, or confusion?* London: Theos.

Lee, B. Y. & Newberg, A. B. (2005). Religion and health: a review and critical analysis, *Zygon*, 40, 443–468.

Long, D. E. (2011). *Evolution and religion in American education: An ethnography.* Dordrecht: Springer.

McGrath, A. (2005). Has science eliminated God? Richard Dawkins and the meaning of life. *Science & Christian Belief*, 17, 115–135.

MacIntosh, J. J. (2006). *Boyle on atheism.* Toronto: University of Toronto Press.

Mepham, B. (2008). *Bioethics: An introduction for the biosciences, 2nd edn.* Oxford: Oxford University Press.

Merton, R. K. (1973). *The sociology of science: Theoretical and empirical investigations.* Chicago: University of Chicago Press.

Meyer, S. C. (2009). *Signature in the cell: DNA and the evidence for Intelligent Design.* New York: HarperCollins.

Miller, J. D., Scott, E. C. & Okamoto, S. (2006). Public acceptance of evolution. *Science*, 313, 765–766.

Moore, R. (2007). The history of the creationism/evolution controversy and likely future developments. In: *Teaching about scientific origins: Taking account of creationism*, Jones, L. & Reiss, M. J. (Eds). New York: Peter Lang, pp. 11–29.

National Academy of Sciences and Institute of Medicine (2008). *Science, evolution, and creationism.* Washington, DC: National Academies Press.

Osborn, L. (2005). Theology and the new physics. In: *God, humanity and the cosmos, 2nd edn revised and expanded as A companion to the science-religion debate*, Southgate, C. (Ed.). London: T & T Clark, pp. 119–153.

Pannenberg, W. (2006). Problems between science and theology in the course of their modern history, *Zygon*, 41, 105–112.

Parfit, D. (2011). *On what matters: Volume 1.* Oxford: Oxford University Press.

Peacocke, A. (2001). *Paths from science towards God: The end of all our exploring.* Oxford: Oneworld.

Plantinga, A. (2010). Religion and science. In *The Stanford Encyclopedia of Philosophy*, Zalta, E. N. (Ed.), http://plato.stanford.edu/archives/sum2010/entries/religion-science/. Accessed 9 September 2012.

Polkinghorne, J. (1994). *Science and Christian belief: Theological reflections of a bottom-up thinker. The Gifford Lectures for 1993–4.* London: SPCK.

Polkinghorne, J. (2004). *Science and the Trinity: The Christian encounter with reality.* London: SPCK.

Polkinghorne, J. (2005). The continuing interaction of science and religion, *Zygon*, 40, 43–50.

Polkinghorne, J. (2006). Where is natural theology today? *Science & Christian Belief*, 18, 169–179.

Poole, M. (1998). *Teaching about science and religion: Opportunities within* Science in the National Curriculum. Abingdon: Culham College Institute.

Popper, K. R. (1934/1972). *The logic of scientific discovery.* London: Hutchinson.

Ray, J. (1691/2005). *The wisdom of God manifested in the works of the creation.* London: Ray Society.

Reiss, M. J. (1999). Teaching ethics in science. *Studies in Science Education*, 34, 115–140.

Reiss, M. J. (2000). On suffering and meaning: an evolutionary perspective. *Modern Believing*, 41(2), 39–46.

Reiss, M. J. (2008a). Should science educators deal with the science/religion issue? *Studies in Science Education*, 44, 157–186.

Reiss, M.J. (2008b). Teaching evolution in a creationist environment: an approach based on worldviews, not misconceptions. *School Science Review*, 90(331), 49–56.

Reiss, M. J. (2008c). The use of ethical frameworks by students following a new science course for 16–18 year-olds. *Science & Education*, 17, 889–902.

Reiss, M. J. (2011). How should creationism and intelligent design be dealt with in the classroom? *Journal of Philosophy of Education*, 45, 399–415.

Reiss, M. J. (2012). What should be the role of religion in science education and bioethics? In: *Sacred Science? On science and its interrelations with religious worldviews*, Øyen, S. A., Lund-Olsen, T. & Vaage, N. S. (Eds). Wageningen: Wageningen Academic Publishers, pp. 127–139.

Reiss, M.J. (2013a) Religion in science education. In: *Science Education for Diversity: Theory and Practice*, Mansour, N. & Wegerif, R. (Eds), Springer, Dordrecht, pp. 317–328.

Reiss, M.J. (2013b) Beliefs and the value of evidence. In: *Communication and Engagement with Science and Technology: Issues and Dilemmas*, Gilbert, J.K. & Stocklmayer, S.M. (Eds), Routledge, New York, pp. 148–161.

Reynolds, V. & Tanner, R. E. S. (1983). *The Biology of Religion*. London: Longman.

Robitaille, D. & Dirks, M. (1982). Models for mathematics curriculum. *For the Learning of Mathematics*, 2, 3–19.

Roszak, T. (1994). God and the final frontier. *New Scientist*, 28 March, 40–41.

Russell, C. A. (2000). *Michael Faraday: Physics and faith*. Oxford: Oxford University Press.

Salkovskis, P. M. (2002). Empirically grounded clinical interventions: cognitive-behavioural therapy progresses through a multi-dimensional approach to clinical science. *Behavioural and Cognitive Psychotherapy*, 30, 3–9.

Saunders, N. (2002). *Divine action and modern science*. Cambridge: Cambridge University Press.

Scott, E. (1999). The creation/evolution continuum. *Reports of the National Center for Science Education*, 19, 16–23.

Scott, M. (2007). *Rethinking evolution in the museum: Envisioning African origins*. London: Routledge.

Skehan, J. W. & Nelson, C. E. (2000). *The creation controversy & the science classroom*. Arlington VA: NSTA Press.

Smart, N. (1989). *The world's religions: Old traditions and modern transformations*. Cambridge: Cambridge University Press.

Stenmark, M. (2004). *How to relate science and religion: A multidimensional model*. Grand Rapids, MI: Wm B Eerdmans.

Stolberg, T. & Teece, G. (2011). *Teaching religion and science: Effective pedagogy and practical approaches for RE teachers*. Abingdon: Routledge.

Szerszynski, B. (2005). Rethinking the secular: science, technology, and religion today. *Zygon*, 40, 813–822.

Vernon, M. (2008). *After Atheism: Science, religion and the meaning of life*. Basingstoke: Palgrave Macmillan.

Warnock, M. (2010). *Dishonest to God*. London: Continuum.

Watts, F. (Ed.) (1998). *Science meets faith*. London: SPCK.

Williams, P. A. (2001). *Doing without Adam and Eve: Sociobiology and original sin*. Minneapolis, MN: Fortress Press.

Winslow, M. W., Staver, J. R. & Scharmann, L. C. (2011). Evolution and personal religious belief: Christian university biology-related majors' search for reconciliation. *Journal of Research in Science Teaching*, 48, 1026–1049.

Yearley, S. (2005). *Making sense of science: Understanding the social study of science*. London: SAGE.

Ziman, J. (2000). *Real science: What it is and what it means*. Cambridge: Cambridge University Press.

Michael Reiss is the Pro-Director of Research and Development and Professor of Science Education at the Institute of Education, University of London; Vice President and Honorary Fellow of the British Science Association; Honorary Visiting Professor at the Universities of Leeds and York and the Royal Veterinary College; Honorary Fellow of the College of Teachers; Docent at the University of Helsinki; Director of the Salters-Nuffield Advanced Biology Project; and an Academician of the Academy of Social Sciences. His books include Reiss, M. J. and White, J. (2013) *An Aims-based Curriculum*, IOE Press; Jones, A., McKim, A. and Reiss, M. (Eds) (2010) *Ethics in the Science and Technology Classroom: A New Approach to Teaching and Learning*, Sense; Jones, L. and Reiss, M. J. (Eds) (2007) *Teaching about Scientific Origins: Taking Account of Creationism*, Peter Lang; Braund, M. and Reiss, M. J. (Eds) (2004) *Learning Science Outside the Classroom*, RoutledgeFalmer; Levinson, R. and Reiss, M. J. (Eds) (2003) *Key Issues in Bioethics: A Guide for Teachers*, RoutledgeFalmer; and Reiss, M. J. (2000) *Understanding Science Lessons: Five Years of Science Teaching*, Open University Press. For further information, see www.reiss.tc.